TM 9-2320-289-34P
CUCV
Commercial Utility Cargo Vehicle
Repair Parts and Special Tool Lists
May 1992

The CUCV or Commercial Utility Cargo Vehicle is a US Military vehicle based on readily available commercial trucks. Originally intended to augment the purpose-built, but expensive GAMA Goat 6x6 and older Jeeps. The first generation was based on Dodge / Chrysler trucks.

This book is focused on the M1008 series second generation CUCV which was General Motor's first major light-truck military vehicle production since World War II. They began production in 1984 and ended production in 1996 with most units being produced as 1984 model year units. Later production was focused on replacements for existing CUCV's. The majority of units were built from existing heavy duty light truck commercial parts. The M1009 was an upgraded/up-rated Chevy K5 Blazer with a 3/4 ton capacity. The M1008 series trucks were a 1-1/4 ton or 5/4 ton rated truck. In all 70,000 units were produced with three power trains.

This manual is the repair parts and special tool list for these vehicles. It is published as a convenience to enthusiasts who may wish to have a quality professionally printed copy of the manual.

This publisher has also printed other manuals for this series of vehicles.

Should you have suggestions or feedback on ways to improve this book please send email to Books@OcotilloPress.com

---

Edited 2021 Ocotillo Press
ISBN 978-1-954285-71-2

Cover Photo Credit: Alf Van Beem / Wikipedia, Public Domain Image M1008A1 Chevrolet K30 Pickup with Box

Ocotillo Press
Houston, TX 77017
Books@OcotilloPress.com

Disclaimer: The user of this book is responsible for following safe and lawful practices at all times. The publisher assumes no responsibility for the use of the content of this book. The publisher has made an effort to ensure that the text is complete and properly typeset, however omissions, errors, and other issues may exist that the publisher is unaware of.

TM 9-2320-289-34 P

ARMY TM9-2320-289-34P
AIR FORCE TO 36A12-1A-2084-2
MARINE CORPS TM2320-34P/5

TECHNICAL MANUAL

DIRECT SUPPORT AND GENERAL SUPPORT MAINTENANCE
REPAIR PARTS AND SPECIAL TOOLS LISTS
(INCLUDING DEPOT MAINTENANC E
REPAIR PARTS AND SPECIAL
TOOLS LISTS   )
FO R

TRUCK, CARGO, TACTICAL, 1-1/4 TON, 4X4, M1008
(2320-01-123-6827)
TRUCK, CARGO, TACTICAL, 1-1/4 TON, 4X4, M1008A1
(2320-01-123-2671)
TRUCK, UTILITY, TACTICAL, 3/4 TON, 4X4, M1009
(2320-01-123-2665)
TRUCK, AMBULANCE, TACTICAL, 1-1/4 TON, 4X4, M1010
(2310-01-123-2666)
TRUCK, SHELTER CARRIER, TACTICAL, 1-1/4 TON, 4X4, M1028
(2320-01-127-5077)
TRUCK, SHELTER CARRIER W/PTO, TACTICAL, 1-1/4 TON, 4X4, M1028A1
(2320-01-158-0820)
TRUCK, SHELTER CARRIER W/PTO, TACTICAL, 1-1/4 TON, 4X4, M1028A2
(2320-01-295-0822)
TRUCK, SHELTER CARRIER, TACTICAL, 1-1/4 TON, 4X4, M1028A3
(2320-01-325-1937)
TRUCK, CHASSIS, TACTICAL, 1-1/4 TON, 4X4, M1031
(2320-01-133-5368)

This manual supersedes TM9-2320-289-34P, 9 February 1990, and all changes.

DEPARTMENTS OF THE ARMY, THE AIR FORCE,
AND HEADQUARTERS, MARINE CORPS
01 MAY 1992

TECHNICAL MANUAL

*TM 9-2320-289-34P

DEPARTMENTS OF THE ARMY, THE AIR FORCE,
AND HEADQUARTERS, MARINE CORPS
Washington, D. C., 1 May 1992

DIRECT SUPPORT AND GENERAL SUPPORT MAINTENANCE
REPAIR PARTS AND SPECIAL TOOLS LISTS
(INCLUDING DEPOT MAINTENANCE
REPAIR PARTS AND SPECIAL
TOOLS LISTS)

FOR

TRUCK, CARGO, TACTICAL, 1-1/4 TON, 4X4, M1008
(2320-01-123-6827)
TRUCK, CARGO, TACTICAL, 1-1/4 TON, 4X4, M1008A1
(2320-01-123-2671)
TRUCK, UTILITY, TACTICAL, 3/4 TON, 4X4, M1009
(2320-01-123-2665)
TRUCK, AMBULANCE, TACTICAL, 1-1/4 TON, 4X4, M1010
(2310-01-123-2666)
TRUCK, SHELTER CARRIER, TACTICAL, 1-1/4 TON, 4X4, M1028
(2320-01-127-5077)
TRUCK, SHELTER CARRIER W/PTO, TACTICAL, 1-1/4 TON, 4X4, M1028A1
(2320-01-158-0820)
TRUCK, SHELTER CARRIER W/PTO, TACTICAL, 1-1/4 TON, 4X4, M1028A2
(2320-01-295-0822)
TRUCK, SHELTER CARRIER, TACTICAL, 1-1/4 TON, 4X4, M1028A3
(2320-01-325-1937)
TRUCK, CHASSIS, TACTICAL, 1-1/4 TON, 4X4, M1031
(2320-01-133-5368)

Current as of 7 February 1992

REPORTING ERRORS AND RECOMMENDING IMPROVEMENTS

You can help improve this manual. If you find any mistakes or if you know of a way
to improve the procedures, please let us know, Mail your letter, DA Form 2028
(Recommended Changes to Publications and Blank Forms), or DA Form 2028-2,
located In the back of this manual, direct to: Commander, U.S. Army
Tank-Automotive Command, ATTN: AMSTA-MB, Warren, MI 48397-5000. A reply
will be furnished to you.

Approved for public release; distribution is unlimited.

* This manual supersedes TM 9-2320-289-34P, dated 9 February 1990, and all changes.

TM9-2320-289-34P

TABLE OF CONTENTS

		Page	Illus Fig
SECTION I.	INTRODUCTION	1	
SECTION II.	REPAIR PARTS LIST	1-1	

GROUP 01 ENGINE

0100 - ENGINE ASSEMBLY	1-1	
ENGINE ASSEMBLY MOUNTS, LIFTING BRACKETS, AND MOUNTING HARDWARE	1-1	1
0101 - CRANKCASE, BLOCK, CYLINDER HEAD	2-1	
CYLINDER HEADS, BLOCK PLUGS, AND RELATED PARTS	2-1	2
0102 - CRANKSHAFT	3-1	
ENGINE CRANKSHAFT AND RELATED PARTS	3-1	3
0103 - FLYWHEEL ASSEMBLY	4-1	
FLYWHEEL	4-1	4
0104 - PISTONS AND CONNECTING RODS	5-1	
PISTON AND CONNECTING ROD	5-1	5
0105 - VALVES, CAMSHAFTS, AND TIMING SYSTEM	6-1	
CYLINDER HEAD AND CRANKCASE COVERS	6-1	6
CAMSHAFT VALVES, ROCKER ARMS, AND RELATED PARTS	7-1	7
0106 - ENGINE LUBRICATION SYSTEM	8-1	
ENGINE OIL PUMP ASSEMBLY FILTER, PAN, AND RELATED PARTS	8-1	8
OIL COOLER LINES, SAMPLING VALVE, AND RELATED PARTS	9-1	9
CRANKCASE DEPRESSION VALVE AND RELATED PARTS	10-1	10
0108 - MANIFOLDS	11-1	
INTAKE AND EXHAUST MANIFOLDS	11-1	11

GROUP 03 FUEL SYSTEM

0301 - CARBURETOR, FUEL INJECTOR	12-1	
FUEL INJECTOR AND RELATED PARTS	12-1	12
0302 - FUEL PUMPS	13-1	
SUPPLY PUMP, INJECTOR PUMP, AND RELATED PARTS	13-1	13
VACUUM VALVE AND HOSES	14-1	14
INJECTOR PUMP COMPONENT PARTS	15-1	15
INJECTOR LINES AND RELATED PARTS	16-1	16
0304 - AIR CLEANER	17-1	
AIR CLEANER	17-1	17
0306 - TANKS, LINES, FITTINGS, HEADERS	18-1	
FUEL TANK AND RELATED PARTS (M1009)	18-1	18
FUEL TANK AND RELATED PARTS (ALL EXCEPT M1009)	191	19
REAR FUEL LINES AND RELATED PARTS (M1009)	20-1	20
FRONT FUEL LINES AND RELATED PARTS (M1009)	21-1	21
FUEL LINES AND RELATED PARTS (ALL EXCEPT M1009)	22-1	22
0309 - FUEL FILTERS	23-1	
FUEL FILTER AND RELATED PARTS	23-1	23
0311 - ENGINE STARTING AIDS	24-1	
GLOW PLUGS AND TEMPERATURE SENSOR	24-1	24
0312 - ACCELERATOR, THROTTLE, OR CHOKE CONTROLS	25-1	
ACCELERATOR LINKAGE AND RELATED PARTS	25-1	25

TABLE OF CONTENTS (Con't)

		Page	Illus Fig

GROUP 04   EXHAUST SYSTEM

0401 -MUFFLER AND PIPES 28-1
    EXHAUST MUFFLERS AND PIPES (ALL EXCEPT M1009) 28-1   28
    EXHAUST MUFFLERS AND PIPES (M1009) 27-1   27

GROUP 05   COOLING SYSTEM

0501 - RADIATOR, EVAPORATIVE COOLER, OR HEAT EXCHANGER 28-1
    RADIATOR, COOLANT RECOVERY RESERVOIR, AND RELATED PARTS. . 28-1   28
0502 - COWLING, DEFLECTORS, AIR DUCTS, SHROUDS, ETC. 29-1
    RADIATOR SHROUD 29-1   29
0503- WATER MANIFOLD, HEADERS, THERMOSTATS, AND HOUSING GASKET 30-1
    ENGINE THERMOSTAT AND HOSES 30-1   30
0504 - WATER PUMP 31-1
    WATER PUMP 31-1   31
0505 - FAN ASSEMBLY 32-1
    ENGINE FAN ASSEMBLY 32-1   32

GROUP 06   ELECTRICAL SYSTEM

0801 -GENERATOR, ALTERNATOR 33-1
    ALTERNATOR AND MOUNTING HARDWARE, LEFT SIDE
      (ALL EXCEPT M1010) 33-1   33
    ALTERNATOR AND MOUNTING HARDWARE, RIGHT SIDE
      (ALL EXCEPT M1010) 34-1   34
    ALTERNATOR AND MOUNTING HARDWARE (M1010) 35-1   35
    ALTERNATOR COMPONENT PARTS (ALL EXCEPT M1010) 38-1   36
    ALTERNATOR COMPONENT PARTS (M1010) 37-1   37
0802 -GENERATOR REGULATOR. 38-1
    VOLTAGE REGULATOR AND RELAY (M1010) 38-1   38
0803 - STARTING MOTOR 39-1
    STARTER MOTOR AND RELAY 39-1   39
    STARTER MOTOR COMPONENT PARTS 40-1   40
0807 -INSTRUMENT OR ENGINE CONTROL PANEL 41-1
    INSTRUMENT PANEL WIRING HARNESS 41-1   41
    INSTRUMENT CLUSTER ASSEMBLY 42-1   42
    DOOR AJAR AND GLOW PLUG INDICATOR, VOLTMETER,
      GAS-PARTICULATE FILTER UNIT, AND FLOODLIGHT SWITCH 43-1   43
0808 -MISCELLANEOUS ITEMS 44-1
    ACCESSORY WIRING TERMINAL BOARD AND COMPONENTS 44-1   44
    POWER JUNCTION BOX AND CABLE. 45-1   45
    RADIO FEED HARNESS (M1009) 48-1   48
    BATTERY BOOSTER RESISTOR 47-1   47
    GLOW PLUG RELAY 48-1   48
    HEADLAMP BLACKOUT LAMP AND TAILLAMP SWITCHES AND FUSES. 49-1   49
0809 - LIGHTS 50-1
    BLACKOUT HEADLAMP AND HEADLAMP ILLUMINATING COMPONENTS 50-1   50
    REAR BLACKOUT LAMPS AND TAILLAMPS ILLUMINATING COMPONENTS
      (ALL EXCEPT M1010 AND M1031) 51-1   51
    REAR FENDER SIDE MARKERS M1028A2 AND M1082A3) . 52-1   52
    REAR BLACKOUT LAMPS AND TAILLAMP ILLUMINATING COMPONENTS
      (M1010 AND M1031) 53-1   53

TM9-2320-289-34P

TABLE OF CONTENTS (Con't)

		Page	Illus Fig
0610 - SENDING UNITS AND WARNING SWITCHES		54-1	
	ENGINE OIL PRESSURE AND WATER TEMPERATURE SENDING UNITS	54-1	54
0611 - HORN, SIREN		55-1	
	TRUCK HORN AND RELATED PARTS	55-1	55
0612 - BATTERIES, STORAGE		56-1	
	BATTERIES, BATTERY TRAYS, AND RELATED PARTS	56-1	58
	CABLES AND SLAVE CONNECTOR	57-1	57
0613 - HULL OR CHASSIS WIRING HARNESS		58-1	
	FRONT WIRING HARNESS	58-1	58
	BODY AND FUEL SENDING WIRING HARNESSES (ALL EXCEPT M1009)	59-1	59
	BODY AND FUEL SENDING WIRING HARNESSES (M1009)	60-1	60
	BLACKOUT AND TAILLAMP WIRING HARNESS (ALL EXCEPT M1010 AND M1031) AND TRAILER WIRING HARNESS (ALL EXCEPT M1010)	61-1	61
	REAR FENDER WIRING HARNESS (M1028A2 AND M1028A3)	62-1	82
	BLACKOUT AND TAILLAMP WIRING HARNESSES (M1010 AND M1031)	63-1	63
	INSTRUMENT PANEL EXTENSION WIRING HARNESS (M1010)	64-1	64
	ENGINE WIRING HARNESS (ALL EXCEPT M1010)	65-1	65
	ENGINE WIRING HARNESS (M1010)	66-1	66
	GLOW PLUG WIRING HARNESS	67-1	67
	HEATER CONTROL, DIAGNOSTIC, AND DOOR ALARM WIRING HARNESSES	68-1	68
GROUP 07	TRANSMISSION		
0705 - TRANSMISSION SHIFTING COMPONENTS		69-1	
	SHIFTING LINKAGE	69-1	69
	VACUUM MODULATOR VALVE	70-1	70
0710 - TRANSMISSION ASSEMBLY AND ASSOCIATED PARTS		71-1	
	TRANSMISSION ASSEMBLY AND MOUNTS	71-1	71
	TORQUE CONVERTER, OIL PAN, AND FILLER TUBE	72-1	72
	FORWARD AND DIRECT CLUTCH ASSEMBLIES	73-1	73
	CENTER SUPPORT REACTION CARRIER, AND OUTPUT CARRIER ASSEMBLIES	74-1	74
	TRANSMISSION CASE	75-1	75
	PARKING LOCK	76-1	76
0714 - SERVO UNIT		77-1	
	CONTROL VALVE ASSEMBLY GOVERNOR ASSEMBLY, AND SERVO VALVE UNIT	77-1	77
	CONTROL VALVE COMPONENT PARTS	78-1	78
0721 - COOLERS, PUMPS, MOTORS		79-1	
	OIL PUMP ASSEMBLY AND TRANSMISSION FILTER	79-1	79
	TRANSMISSION OIL COOLER LINES	80-1	80
GROUP 08	TRANSFER, FINAL DRIVE, PLANETARY AND DROP GEARBOX ASSEMBLIES		
0801 - POWER TRANSFER, FINAL DRIVE, PLANETARY OR DROP GEARBOX ASSEMBLIES		81-1	
	TRANSFER CASE ASSEMBLY AND ADAPTER (MODEL 208)	81-1	81
	TRANSFER CASE ASSEMBLY AND ADAPTER (MODEL 205)	82-1	82
	TRANSFER CASE COMPONENT PARTS (MODEL 208)	83-1	83
	TRANSFER CASE COMPONENT PARTS (MODEL 205)	84-1	84
	TRANSFER CASE HOUSING (MODEL 205)	85-1	85
	CASE VENT COMPONENT PARTS	86-1	86
0803 - GEARSHIFT, VACUUM BOOSTER, AND CONTROLS		87-1	
	GEARSHIFT LEVER AND LINKAGE (MODEL 208)	87-1	87
	GEARSHIFT LEVER AND LINKAGE (MODEL 205)	88-1	86

iv

TABLE OF CONTENTS (Con't)

		Page	Illus Fig
GROUP 09	**PROPELLER, PROPELLER SHAFTS, UNIVERSAL JOINTS, COUPLER, AND CLAMP ASSEMBLY**		
	0900 - PROPELLER SHAFTS	89-1	
	FRONT PROPELLER SHAFT ASSEMBLY (ALL EXCEPT M1009, FIRST DESIGN)	89-1	89
	FRONT PROPELLER SHAFT ASSEMBLY (ALL EXCEPT M1009, SECOND DESIGN)	90-1	90
	FRONT PROPELLER SHAFT ASSEMBLY (M1008, FIRST DESIGN)	91-1	91
	FRONT PROPELLER SHAFT ASSEMBLY (M1009, SECOND DESIGN)	92-1	92
	REAR PROPELLER SHAFT ASSEMBLY	93-1	93
GROUP 10	**FRONT AXLE**		
	1000 - FRONT AXLE ASSEMBLY	94-1	
	FRONT AXLE	94-1	94
	FRONT AXLE BREATHER	95-1	95
	FRONT AXLE COMPONENT PARTS AND SPINDLE ASSEMBLIES (ALL EXCEPT M1009)	96-1	96
	FRONT AXLE COMPONENT PARTS AND SPINDLE ASSEMBLIES (M1009)	97-1	97
	1002 - DIFFERENTIAL	96-1	
	DIFFERENTIAL ASSEMBLY RING, PINION, AND RELATED PARTS (ALL EXCEPT M1009)	96-1	98
	DIFFERENTIAL COMPONENTS (M1009)	99-1	99
	RING AND PINION (M1009)	100-1	100
	1004 - STEERING AND LEANING WHEEL MECHANISM	101-1	
	STEERING KNUCKLE ASSEMBLY (ALL EXCEPT M1009)	101-1	101
	STEERING KNUCKLE ASSEMBLY (M1009)	102-1	102
GROUP 11	**REAR AXLE**		
	1100 - REAR AXLE ASSEMBLY	103-1	
	REAR AXLE	103-1	**103**
	REAR AXLE BREATHER (ALL EXCEPT M1009)	104-1	104
	REAR AXLE BREATHER (M1009)	105-1	105
	AXLE SHAFTS (ALL EXCEPT M1009)	106-1	106
	REAR AXLE SHAFTS (M1009)	107-1	107
	1101 - HOUSING, BEAM, HOUSING COVERS, PLUGS, SEALS, ETC.	106-1	
	REAR AXLE HOUSING (ALL EXCEPT M1009)	106-1	106
	REAR AXLE HOUSING (M1009)	109-1	109
	1102 - DIFFERENTIAL	110-1	
	DIFFERENTIAL LOCK ASSEMBLY, RING, PINION, AND RELATED PARTS (ALL EXCEPT M1009, M1028A2, AND M1028A3)	110-1	110
	DIFFERENTIAL LOCK ASSEMBLY RING, PINION, AND RELATED PARTS (M1028A2 AND M1028A3)	111-1	111
	DIFFERENTIAL LOCK ASSEMBLY RING, PINION, AND RELATED PARTS (M1009)	112-1	112
	DIFFERENTIAL LOCK ASSEMBLY COMPONENT PARTS (M1009)	113-1	113
GROUP 12	**BRAKES**		
	1201 - HANDBRAKES	114-1	
	PARKING BRAKE PEDAL AND RELEASE MECHANISM	114-1	114
	1202 - SERVICE BRAKES	115-1	
	REAR BRAKESHOE ASSEMBLY COMPONENTS, FRONT BRAKE PADS, AND RELATED PARTS (ALL EXCEPT M1009)	115-1	115
	REAR BRAKESHOE ASSEMBLY COMPONENTS, FRONT BRAKE PADS, AND RELATED PARTS (M1009)	116-1	116

	Page	Illus Fig
1204 -HYDRAULIC BRAKE SYSTEM.	117-1	
POWER BOOSTER AND MASTER CYLINDER ASSEMBLIES	117-1	117
FRONT BRAKE LINES AND RELATED PARTS	118-1	118
REAR BRAKE LINES, WEIGHT PROPORTIONAL VALVE, AND RELATED PARTS . . .	119-1	119
FRONT BRAKE CALIPER ASSEMBLIES (ALL EXCEPT M1009)	120-1	120
FRONT BRAKE CALIPER ASSEMBLIES (M1009)	121-1	121
REAR BRAKE WHEEL CYLINDER	122-1	122
1208 -MECHANICAL BRAKE SYSTEM	123-1	
BRAKE PEDAL, MOUNTING BRACKETS, AND RELATED PARTS	123-1	123

GROUP 13   WHEELS AND TRACKS

	Page	Illus Fig
1311 - WHEEL ASSEMBLY	124-1	
FRONT HUB, ROTOR, REAR HUB, BRAKEDRUM, AND RELATED PARTS (ALL EXCEPT M1009).	124-1	124
FRONT HUB, ROTOR, BRAKEDRUM, AND RELATED PARTS (M1009)	125-1	125
1313 -TIRES, TUBES, TIRE CHAINS.	126-1	
TIRES	128-1	128

GROUP 14   STEERING

	Page	Illus Fig
1401 - MECHANICAL STEERING GEAR ASSEMBLY	127-1	
STEERING WHEEL AND COLUMN	127-1	127
PITMAN ARM, ABSORBER, AND RELATED PARTS	128-1	128
TIE-ROD ASSEMBLY (ALL EXCEPT M1009)	129-1	129
TIE-ROD ASSEMBLY (M1009)	130-1	130
1407 - POWER STEERING GEAR ASSEMBLY.	131-1	
POWER STEERING GEAR AND COMPONENT PARTS	131-1	131
1410 - HYDRAULIC PUMP OR FLUID MOTOR ASSEMBLY	132-1	
HYDRAULIC PUMP	132-1	132
HYDRAULIC PUMP COMPONENT PARTS	133-1	133
1411 - HOSES, LINES, FITTINGS	134-1	
HYDRAULIC STEERING LINES	134-1	134

GROUP 15   FRAME , TOWING ATTACHMENTS, DRAWBARS, AND ARTICULATION SYSTEMS

	Page	Illus Fig
1501 - FRAME ASSEMBLY	135-1	
FRONT BUMPER AND RELATED PARTS.	135-1	135
REAR BUMPER AND MOUNTING PARTS (ALL EXCEPT M1009 AND M1010)	138-1	138
REAR BUMPER AND MOUNTING PARTS (M1009)	137-1	137
REAR TIE-DOWNS (M1031)	138-1	138
FRAME ASSEMBLY COMPONENTS (ALL EXCEPT M1009)	139-1	139
FRAME ASSEMBLY RELATED PARTS (ALL EXCEPT M1009)	140-1	140
FRAME ASSEMBLY COMPONENTS (M1009)	141-1	141
FRAME ASSEMBLY RELATED PARTS (M1009)	142-1	142
1503 - PINTLES AND TOWING ATTACHMENTS.	143-1	
CLEVISES AND SUPPORTS	143-1	143
PINTLE ASSEMBLY AND CLEVIS (M1008, M1028, AND M1028A1)	144-1	144
PINTLE ASSEMBLY AND CLEVIS (M1009).	145-1	145
1504 - SPARE WHEEL CARRIER AND TIRE LOCK	148-1	
SPARE WHEEL CARRIER (ALL EXCEPT M1009)	148-1	146
SPARE WHEEL CARRIER (M1009)	147-1	147

TABLE OF CONTENTS (Con't)

		Page	Illus Fig
GROUP 16	**SPRINGS AND SHOCK ABSORBERS**		
	1801 - SPRINGS	148-1	
	FRONT SPRINGS AND RELATED PARTS (ALL EXCEPT M1009)	148-1	148
	FRONT SPRINGS AND RELATED PARTS (M1009)	149-1	149
	REAR SPRINGS AND RELATED PARTS (ALL EXCEPT M1009, M1028, M1028A1, M1028A2, AND M1028A3)	150-1	150
	REAR SPRINGS AND RELATED PARTS (M1028,M1028A1, M1028A2, AND M1028A3)	151-1	151
	REAR SPRINGS AND RELATED PARTS (M1009)	152-1	152
	1804 - SHOCK ABSORBER EQUIPMENT	153-1	
	FRONT SHOCK ABSORBERS	153-1	153
	REAR SHOCK ABSORBERS (ALL EXCEPT M1009)	154-1	154
	REAR SHOCK ABSORBERS (M1009)	155-1	155
	1805 - TORCMJE, RADIUS, AND STABILIZER RODS	158-1	
	FRONT STABILIZER BAR	158-1	156
	REAR STABILIZER BAR (M1028A2 AND M1028A3)	157-1	157
GROUP 18	**BODY, CAB, HOOD, AND HULL**		
	1801 -BODY, CAB, HOOD, AND HULL ASSEMBLIES	158-1	
	BRUSH GUARD AND GRILLE	158-1	158
	FRONT MOLDINGS	159-1	159
	FRONT SUPPORT PANELS	160-1	160
	COWL TOP VENTILATOR PANEL	161-1	161
	FRONT HOOD LATCH RELEASE CABLE	162-1	162
	FRONT HOOD, LATCH, AND HINGES	163-1	163
	INSTRUMENT PANEL TRIM	164-1	164
	INSTRUMENT PANEL COMPONENTS	165-1	165
	SIDE DOORS AND COMPONENTS	166-1	166
	INSIDE DOOR TRIM PANEL AND RELATED PARTS	167-1	167
	SIDE DOOR WINDOW REGULATOR	168-1	168
	SIDE DOOR COMPONENTS	169-1	169
	SIDE DOOR LOCK ASSEMBLY	170-1	170
	SIDE DOOR WEATHERSTRIPS	171-1	171
	BODY CAB MOUNTING (M1009)	172-1	172
	CARGO BOX AND COMPONENTS (M1008, M1008A1, M1028, M1028A1, M1028A2, AND M1028A3)	173-1	173
	CARGO BOX MOUNTING (M1008, M1008A1, M1028, M1028A1, M1028A2, AND M1028A3)	174-1	174
	BODY CARGO AND COMPONENTS (M1009)	175-1	175
	CARGO ENDGATE AND COMPONENTS (M1008,M1008A1, M1028, M1028A1, M1028A2, AND M1028A3)	178-1	176
	ENDGATE AND COMPONENTS (M1009)	177-1	177
	ENDGATE WINDOW HANDLE ASSEMBLY (M1009)	178-1	178
	ENDGATE HINGES AND RELATED PARTS (M1009)	179-1	179
	ENDGATE LATCHES AND RELATED PARTS (M1009)	160-1	180
	ENDGATE TORQUE RODS (M1009)	181-1	181
	ENDGATE LATCH HANDLE (M1009)	182-1	182
	ENDGATE WEATHERSTRIPS (M1009)	183-1	183
	TOP ASSEMBLY AND COMPONENTS (M1009)	184-1	184
	FRONT TIE-DOWN BRACKETS (M1028, M1028A1, M1028A2, AND M1028A3)	185-1	185

	Page	Illus Fig
1801 - BODY, CAB, HOOD, AND HULL ASSEMBLIES [Con't]		
REAR TIE-DOWN BRACKETS (M1028,M1028A1, M1028A2, AND M1028A3)	186-1	186
CARGO TIE-DOWNS (M1008 AND M1008A1)	187-1	187
JACK STOWAGE COMPONENTS (ALL EXCEPT M1009)	188-1	188
JACK STOWAGE COMPONENTS (M1009)	189-1	189
1802 - FENDERS, RUNNING BOARDS WITH MOUNTING AND ATTACHING PARTS, OUTRIGGERS, WINDSHIELD, GLASS, ETC.	190-1	
WINDSHIELD AND WINDOW GLASS	190-1	190
FRONT FENDERS, INNER WHEEL PANELS, AND RELATED PARTS	191-1	191
REAR FENDER AND RELATED PARTS (M1028A2 AND M1028A3)	192-1	192
1805- FLOORS, SUBFLOORS, AND RELATED COMPONENTS	193-1	
FLOORMATS, INSULATORS, AND RELATED PARTS	193-1	193
1806 - UPHOLSTERY, SEATS, AND CARPETS	194-1	
SUNVISOR AND INSTRUMENT PANEL PAD	194-1	194
SEATBELTS AND MOUNTING PARTS (ALL EXCEPT M1009 AND M1010)	195-1	195
BENCH SEAT AND COMPONENT'S (ALL EXCEPT M1009 AND M1010)	196-1	198
DRIVER'S SEATBELT AND MOUNTING PARTS (M1009)	197-1	197
PASSENGER'S SEATBELT AND MOUNTING PARTS (M1009)	198-1	198
DRIVER'S SEAT ADJUSTER ASSEMBLY (M1009)	199-1	199
PASSENGER'S SEAT ADJUSTER ASSEMBLY (M1009)	200-1	200
DRIVER'S SEAT ASSEMBLY AND COMPONENTS (M1009)	201-1	201
PASSENGER'S SEAT ASSEMBLY AND COMPONENTS (M1009)	202-1	202
REAR BENCH SEAT ASSEMBLY AND COMPONENTS (M1009)	203-1	203
SEATBELTS AND MOUNTING PARTS (M1010)	204-1	204
SEAT ASSEMBLY, DRIVER'S SEAT, ADJUSTER ASSEMBLY AND RELATED PARTS (M1010)	205-1	205
SEAT MOUNTING PARTS (M1010)	208-1	208
1808 - STOWAGE RACKS, BOXES, STRAPS, CARRYING CASES, CABLE REELS, HOSE REELS, ETC	207-1	
COMMUNICATIONS RACK ASSEMBLY AND RELATED PARTS (M1028A1)	207-1	207
WEAPONS MOUNT (ALL EXCEPT M1010)	208-1	208
BRACKETS AND MOUNTING PARTS	209-1	209
1812 -SPECIAL PURPOSE BODIES	210-1	
BODY ASSEMBLY (M1010)	210-1	210
LEFT SIDE OUTER PANEL COMPONENT PARTS (M1010)	211-1	211
RIGHT SIDE OUTER PANEL COMPONENT PARTS (M1010)	212-1	212
ROOF PANEL COMPONENT PARTS (M1010)	213-1	213
REAR FRAME COMPONENT PARTS (M1010)	214-1	214
LEFT REAR DOOR ASSEMBLY (M1010)	215-1	215
RIGHT REAR DOOR ASSEMBLY (M1010)	216-1	216
REAR BUMPER AND STEP ASSEMBLY (M1010)	217-1	217
LEFT SIDE INNER PANEL RELATED PARTS (M1010)	218-1	218
RIGHT SIDE INNER PANEL RELATED PARTS (M1010)	219-1	219
LEFT SIDE STOWAGE BOX ASSEMBLY AND COMPONENT PARTS (M1010)	220-1	220
RIGHT SIDE STOWAGE BOX ASSEMBLY AND COMPONENT PARTS (M1010)	221-1	221
RELAY PANEL ASSEMBLY COMPONENTS AND RELATED PARTS (M1010)	222-1	222
ATTENDANT'S SEAT ASSEMBLY AND COMPONENT PARTS (M1010)	223-1	223
UPPER LITTER ASSEMBLY AND COMPONENT PARTS (M1010)	224-1	224
FRONT HALF-PARTITION ASSEMBLY (M1010)	225-1	225
INTERIOR FOCUS LIGHT (M1010)	226-1	226
FRONT FRAME ASSEMBLY COMPONENTS (M1010)	227-1	227
INTERIOR SLIDING DOOR ASSEMBLY AND COMPONENT PARTS (M1010)	228-1	228
PULLMAN COLLAR AND RELATED PARTS (M1010)	229-1	229

TABLE OF CONTENTS (Con't)

	Page	Illus Fig

GROUP 22   BODY, CHASSIS, AND HULL ACCESSORY ITEMS

2201 - CANVAS, RUBBER, OR PLASTIC ITEMS 230-1
    CARGO COVER ASSEMBLY (M1008 AND M1008A1) 230-1 230
    COVER FRAME COMPONENTS (M1008 AND M1008A1) 231-1 231
    CARGO COVER MOUNTING PARTS (M1008 AND M1008A1) 232-1 232
2202 - ACCESSORY ITEMS 233-1
    WINDSHIELD WIPER MOTOR AND PUMP ASSEMBLY, WIPER ARM
      LINK ASSEMBLY WIPER BLADE ASSEMBLY AND RELATED PARTS 233-1 233
    WIPER MOTOR COMPONENT PARTS. 234-1 234
    REARVIEW MIRRORS 235-1 235
    ANTENNA MOUNTING (ALL EXCEPT M1009 AND M1010) 236-1 236
    RADIO AND ANTENNA BRACKETS (M1009) 237-1 237
    RADIO, FIRE EXTINGUISHER, AND ANTENNA BRACKETS (M1010) 238-1 238
2207 - WINTERIZATION EQUIPMENT 239-1
    HEATER ASSEMBLY, ... 239-1 239
    HEATER ASSEMBLY COMPONENTS. 240-1 240
    HEATER BLOWER MOTOR ASSEMBLY AND COMPONENT PARTS 241-1 241
    HEATER ASSEMBLY AIR DUCTS 242-1 242
    HEATER CABLE AND CONTROL ASSEMBLY 243-1 243
    HEATER HOSES. 244-1 244
    PERSONNEL HEATER ASSEMBLY (M1010) 245-1 245
    WARM AIR HEATER COMPONENT PARTS (M1010) 246-1 246
    PERSONNEL HEATER CONTROL COMPONENTS (M1010) 247-1 247
    WARM AIR HEATER VALVE ASSEMBLY COMPONENT PARTS (M1010) . . . 248-1 248
    WARM AIR HEATER BLOWER ASSEMBLY COMPONENT PARTS (M1010) 249-1 249
    WARM AIR HEATER BURNER ASSEMBLY COMPONENT PARTS (M1010) .. 250-1 250
    PERSONNEL HEATER FUEL PUMP LINES, AND RELATED PARTS (M1010) 251-1 251
    PERSONNEL HEATER FUEL LINES AND RELATED PARTS (M1010) 252-1 252

GROUP 33   SPECIAL PURPOSE KITS

3301 - REUSABLE SHIPPING CONTAINERS. 253-1
    ENGINE SHIPPING CONTAINER ASSEMBLY 253-1 253
3303 - WINTERIZATION KITS 254-1
    TRANSMISSION AND OIL PAN 254-1 254
    AUXILIARY HEATER FUEL PUMP, LINES, AND RELATED PARTS (M1008) 255-1 255
    WARM AIR HEATER EXHAUST (ALL EXCEPT M1010) 256-1 256
    WARM AIR HEATER EXHAUST (M1010) 257-1 257
    COOLANT HEATER EXHAUST 258-1 258
    COOLANT HEATER REAR EXHAUST.. . . . . 259-1 259
    ENGINE OIL COOLER LINES. 260-1 260
    ENGINE COOLANT CROSSOVER HOUSING, HOSES, AND FITTINGS ., ... 261-1 261
    COOLANT HEATER, FUEL FILTER, FUEL LINE, AND MOUNTING BRACKETS 262-1 262
    ENGINE COOLANT HEATER COMPONENT PARTS 263-1 263
    WARM AIR HEATER WIRING HARNESS. 264-1 264
    COOLANT HEATER WIRING HARNESS 265-1 265
    WARM AIR FUEL PUMP EXTENSION HARNESS (M1010) 266-1 266
    BATTERY BOXES AND CABLES 267-1 267
    PERSONNEL HEATER WIRING HARNESS 266-1 266
    DOMELIGHT LAMP AND WIRING HARNESS (M1008 AND M1008A1) 269-1 269
    HOOD AND RADIATOR INSULATORS 270-1 270

		Page	Illus Fig
3303-	WINTERIZATION KITS (Con't)		
	ROOF COVER, INTAKE DEFLECTOR, WINDOW, VENT AND		
	RELATED PARTS (M1008 AND M1008A1)	271-1	271
	ROOF AND ENDGATE COVERS AND MOUNTING PARTS (M1009)	272-1	272
	REAR DOOR ASSEMBLY (M1008)	273-1	273
	REAR PANEL COMPONENTS (M1008)	274-1	274
	ROOF PANEL COMPONENTS (M1008)	275-1	275
	FLOOR AND SIDEWALL PANELS (M1008)	276-1	276
	FLOOR AND SIDEWALL PANELS (M1009)	277-1	277
	CAB FLOOR INSULATORS	278-1	278
	SEAT AND SPARE TIRE SPACERS (M1009)	279-1	279
	BASE HEATER INLET HOSE, PIPE, AND RELATED PARTS	280-1	280
	BASE HEATER OUTLET HOSE, PIPE, AND RELATED PARTS	281-1	281
	HEATER CONTROL MOUNTING AND CABLE	282-1	282
	HEATER CONTROL ASSEMBLY COMPONENT PARTS	283-1	283
	HEATER AND MOUNTING PARTS	284-1	284
	HEATER BLOWER AND HOSES	285-1	285
	BLOWER COMPONENT PARTS	286-1	286
	INLET HOSE AND EXHAUST PIPE	287-1	287
	PERSONNEL/CARGO HEATER (M1008)	288-1	288
3307 -	SPECIAL PURPOSE KITS	289-1	
	TROOP SEAT ASSEMBLY AND COMPONENT PARTS	289-1	289
GROUP 39	SEARCHLIGHT AND ELECTRICAL ILLUMINATING EQUIPMENT		
3901 -	SEARCHLIGHT OR ILLUMINATING LIGHT ASSEMBLY	290-1	
	SPOTLIGHT ASSEMBLY AND COMPONENT PARTS (M1010)	290-1	290
GROUP 47	GAGES (NONELECTRICAL), WEIGHING AND MEASURING DEVICES		
4701 -	INSTRUMENTS	281-1	
	SPEEDOMETER AND CABLE ASSEMBLY AND RELATED PARTS	291-1	291
GROUP 52	REFRIGERATION, AIR CONDITIONER/HEATER, AND AIR CONDITIONING COMPONENTS		
52011 -	AIR CONDITIONER/HEATER ASSEMBLY AND GAS COMPRESSOR ASSEMBLY	292-1	
	AIR CONDITIONER ASSEMBLY (M1010)	292-1	292
	AIR OUTLET BOX AND CONTROL PANEL ASSEMBLY COMPONENT PARTS (M1010)	293-1	293
	MAIN CASE ASSEMBLY AND COMPONENT PARTS (M1010)	294-1	294
	GAS COMPRESSOR ASSEMBLY AND MOUNTING BRACKETS (M1010)	295-1	285
	GAS COMPRESSOR COMPONENT PARTS (M1010)	296-1	296
5217 -	REFRIGERANT PIPING	297-1	
	GAS COMPRESSOR HOSES (M1010)	297-1	297
5230 -	CONDENSER,	288-1	
	CONDENSER, COVER ASSEMBLY AND COMPONENT PARTS (M1010)	298-1	298
5241 -	EVAPORATOR	299-1	
	EVAPORATOR ASSEMBLY COMPONENT PARTS AND RELATED PARTS (M1010)	299-1	299
5243 -	BLOWER ASSEMBLY	300-1	
	BLOWER ASSEMBLY COMPONENT PARTS (M1010)	300-1	300
GROUP 94	REPAIR KITS		
9401 -	REPAIR KITS	KITS-1	
	REPAIR KITS	KITS-1	KITS

x

TABLE OF CONTENTS (Con't)

		Page	Illus Fig
GROUP 95	GENERAL USE STANDARDIZED PARTS		
	9501- BULK MATERIEL	BULK-1	
	BULK MATERIEL	BULK-1	BULK
SECTION III.	SPECIAL TOOLS		
GROUP 26	TOOLS AND TEST EQUIPMENT		
	2604 - SPECIAL TOOLS	301-1	
	SPECIAL TOOLS	301-1	301
	SPECIAL TOOLS	302-1	302
	SPECIAL TOOLS	303-1	303
	SPECIAL TOOLS.	304-1	304
	SPECIAL TOOLS.	305-1	305
SECTION IV.	CROSS-REFERENCE INDEXES		
	NATIONAL STOCK NUMBER INDEX.	I-1	
	PART NUMBER INDEX	I-54	
	FIGURE AND ITEM NUMBER INDEX.	I-166	

DIRECT SUPPORT AND GENERAL SUPPORT MAINTENANCE REPAIR PARTS AND SPECIAL TOOLS LISTS (INCLUDING DEPOT MAINTENANCE REPAIR PARTS AND SPECIAL TOOLS LISTS)

SECTION I. INTRODUCTION

1. Scope.

This RPSTL lists and authorizes spares and repair parts; special tools; special test, measurement, and diagnostic equipment (TMDE); and other special support equipment required for performance of Unit, Direct Support, and General Support Maintenance of the CUCV Series Truck. It authorizes the requisitioning, issue, and disposition of spares, repair parts and special tools as indicated by the source, maintenance and recoverability (SMR) codes.

2. General.

In addition to Section I. Introduction, this Repair Parts and Special Tools List is divided into the following sections:

a. *Section II. Repair Parts List.* A list of spares and repair parts authorized by this RPSTL for use in the performance of maintenance. The list also includes parts which must be removed for replacement of the authorized parts. Parts lists are composed of functional groups in ascending alphanumeric sequence, with the parts in each group listed in ascending figure and item number sequence. Bulk materials are listed in item name sequence, Repair kits are listed separately in their own functional group within Section Ii. Repair parts for reparable special tools are also listed in the section. Items listed are shown on the associated illustration(s)/figure(s).

b. *Section III. Special Tools List.* A list of special tools, special TMDE, and other special support equipment authorized by this RPSTL (as indicated by Basis of Issue (BOI) information in DESCRIPTION AND USABLE ON CODE column) for the performance of maintenance.

c. *Section IV. Cross-reference Indexes.* A list, in National Item Identification Number (NIIN) sequence, of all National stock numbered items appearing in the listing, followed by a list in alphanumeric sequence of all part numbers appearing in the listings, National stock numbers and part numbers are cross-referenced to each illustration/figure and item number appearance. The figure and item number index lists figure and item numbers in alphanumeric sequence and cross-references NSN, CAGE, and part numbers.

3. Explanation of Columns (Sections II and III).

a. *ITEM NO. (Column (1)).* Indicates the number used to identify items called out in the illustration.

b. *SMR CODE (Column (2)).* The Source, Maintenance, and Recoverability (SMR) code is a 5-position code containing supply/requisitioning information, maintenance category authorization criteria, and disposition instructions, as shown in the following breakout:

Source Code

1st two positions   XXxxx

How you get an item.

Maintenance Code

xxXXx

3d position — Who can install, replace, or use the item.

4th position — Who can do complete repair* on the item.

Recoverability Code

xxxxX   5th position

Who determines disposition action on an unserviceable item.

Complete Repair: Maintenance capacity, capability, and authority to perform all corrective maintenance tasks of the "Repair" function in a use/user environment in order to restore serviceability to a failed item.

(1) *Source Code.* The source code tells you how to get an item needed for maintenance, repair, or overhaul of an end item/equipment. Explanations of source codes follows:

Code	Aplication/Explanation
PA PB PC** PD PE PF PG	Stocked Items; use the applicable NSN to request/requisition items with these source codes. They are authorized to the category indicated by the code entered in the 3d position of the SMR code.

** Items coded PC are subject to deterioration. |
KD KF KB	Items with these codes are not to be requested/requisitioned individually. They are part of a kit which is authorized to the maintenance category indicated in the 3d position of the SMR code. The complete kit must be requisitioned and applied.
MO-(Made at UM/ AVUM Level) MF-(Made at DS/ AVUM Level) MH-(Made at GS Level) ML-(Made at Specialized Repair Activity (SRA)) MD-(Made at Depot)	Items with these codes are not to be requested/requisitioned individually. They must be made from bulk material which is identified by the part number In the DESCRIPTION AND USABLE ON CODE (UOC) column and listed in the Bulk Material group of the repair parts list in this RPSTL. If the item is authorized to you by the 3d position code of the SMR code, but the source code indicates it is made at a higher level, order the item from the higher level of maintenance.
AO-(Assembled by UM/ AVUM Level) AF-(Assembled by DS/AVIM Level) AH-(Assembled by GS Category) AL-(Assembled by SRA) AD-(Assembled by Depot)	Items with these codes are not to be requested/requisitioned individually. The parts that make up the assembled item must be requisitioned or fabricated and assembled at the level of maintenance indicted by the source code. If the 3d position code of the SMR code authorizes you to replace the item, but the source code indicates the item is assembled at a higher level, order the item from the higher level of maintenance.
XA -	Do not requisition an "XA"-coded item. Order its next higher assembly. (Also refer to the NOTE following.)
XB -	If an "XB" item is not available from salvage, order it using the CAGE and part number given.

XC - Installation drawing, diagram, instruction sheet, field service drawing, that is identified by the manufacturer's part number.

XD - Item is not stocked. Order an "XD"-coded item through normal supply channels using the CAGE and part number given, if no NSN is available.

NOTE: Cannibalization or controlled exchange, when authorized, may be used as a source of supply for items with the above source codes, except for those source coded "XA" or those aircraft support items restricted by requirements of AR 700-42.

(2) *Maintenance Code.* Maintenance codes tell you the level(s) of maintenance authorized to USE and REPAIR support items. The maintenance codes are entered in the third and fourth positions of the SMR code as follows:

(a) The maintenance code entered in the third position tells you the lowest maintenance level authorized to remove, replace, and use an item. The maintenance code entered in the third position will indicate authorization to one of the following levels of maintenance.

Code	Application/Explanation
C -	Crew or operator maintenance done within unit maintenance or aviation unit maintenance.
O -	Unit maintenance or aviation unit category can remove, replace, and use the item.
F -	Direct support or aviation intermediate level can remove, replace, and use the item.
H -	General support level can remove, replace, and use the item.
L -	Specialized repair activity can remove, replace, and use the item.
D -	Depot level can remove, replace, and use the item.

(b) The maintenance code entered in the fourth position tells whether or not the item is to be repaired and identifies the lowest maintenance level with the capability to do complete repair (i.e., perform all authorized repair functions). (NOTE: Some limited repair may be done on the item at a lower level of maintenance, if authorized by the Maintenance Allocation Chart (MAC) and SMR codes.) This position will contain one of the following maintenance codes:

Code	Application/Explanation
O -	Unit maintenance or aviation unit is the lowest level that can do complete repair of the item.
F -	Direct support or aviation intermediate is the lowest level than can do complete repair of the item.

H - General support Is the lowest level that can do complete repair of the item.

L - Specialized repair activity is the lowest level that can do complete repair of the item.

D - Depot is the lowest level that can do complete repair of the item.

Z - Nonreparable. No repair is authorized.

B - No repair is authorized, (No parts or special tools are authorized for the maintenance of a "B"-coded item.) However, the item maybe reconditioned by adjusting, lubricating, etc., at the user level.

(3) Recoverability Code. Recoverability codes are assigned to items to indicate the disposition action on unserviceable items. The recoverability code is entered in the fifth position of the SMR code as follows:

Code   Application/Explanation

Z - Nonreparable item. When unserviceable, condemn and dispose of the item at the level of maintenance shown in the 3d position of the SMR code,

O - Reparable item. When uneconomically reparable, condemn and dispose of the item at unit maintenance or aviation unit level.

F - Reparable item. When uneconomically reparable, condemn and dispose of the item at the direct support or aviation intermediate level.

H - Reparable item. When uneconomically reparable, condemn and dispose of the item at the general support level.

D - Reparable item. When beyond lower level repair capability, return to depot. Condemnation and disposal of item not authorized below depot level.

L - Reparable item. Condemnation and disposal of item not authorized below specialized repair activity (SRA).

A - Item requires special handling or condemnation procedures because of specific reasons (e.g., precious metal content, high dollar value, critical material, or hazardous material). Refer to appropriate manuals/directives for specific instructions.

c. *CAGEC (Column (3))*. The Commercial and Government Entity (CAGE) Code (C) is a 5-digit alphanumeric code which is used to identify the manufacturer, distributor, or Government agency, etc., that supplies the item.

d. *PART NUMBER (Column (4))*. Indicates the primary number used by the manufacturer (individual, company, firm, corporation, or Government activity), which controls the design and characteristics of the item by means of its engineering drawings, specifications standards, and inspection requirements to identity an item or range of items.

NOTE: When you use an NSN to requisition an item, the item you receive may have a different part number from the part ordered.

e. *DESCRIPTION AND USABLE ON CODE (UOC) (Column (5)). This* column includes the following information:

(1) The Federal item name and, when required, a minimum description to identify the item.

(2) Physical security classification. Not Applicable.

(3) Items that are included in kits and sets are listed below the name of the kit or set on Figure KIT.

(4) Spare/repair parts that make up an assembled item are listed immediately following the assembled item line entry.

(5) Part numbers for bulk materials are referenced in this column in the line item entry for the item to be manufactured/fabricated.

(6) When the item is not used with all serial numbers of the same model, the effective serial numbers are shown on the last line(s) of the description (before UOC).

(7) The usable on code, when applicable (see paragraph 5, Special Information).

(8) In the Special Tools List section, the basis of issue (BOI) appears as the last line(s) in the entry for each special tool, special TM DE, and other special support equipment. When density of equipments supported exceeds density spread indicated in the basis of issue, the total authorization is increased proportionately.

(9) The statement "END OF FIGURE" appears just below the last item description in Column 5 for a given figure in both Section II and Section III.

f. (QTY *(Column (6))*. The QTY (quantity per figure column) indicates the quantity of the item used in the breakout shown on the illustration/figure, which is prepared for a functional group, subfunctional group, or an assembly. A "V" appearing in this column in lieu of a quantity indicates that the quantity is variable and the quantity may vary from application to application.

4. Explanation of Columns (Section IV).

a. *NATIONAL STOCK NUMBER (NSN) INDEX*.

(1) *STOCK NUMBER column. This* column lists the NSN by National Item Identification Number (NIIN)

sequence. The NIIN consists of the last nine

$$\overset{NSN}{\overline{}}$$

digits of the NSN (i.e., 5305-01-674-14). When

$$\underset{NIIN}{}$$

using this column to locate an item, ignore the first 4 digits of the NSN. However, the complete NSN should be used when ordering items by stock number.

(2) *FIG. column.* This column lists the number of the figure where the item is identified/located. The figures are in numerical order in Section II and Section III.

(3) ITEM *column.* The item number identifies the item associated with the figure listed in the adjacent FIG. column. This item is also identified by the NSN listed on the same line.

b. *PART NUMBER INDEX.* Part numbers in this index are listed by part number in ascending alphanumeric sequence (i.e., vertical arrangement of letter and number combination which places the first letter or digit of each group in order A through Z, followed by the numbers O through 9 and each following letter or digit in like order).

(1) *CAGEC column. The* Commercial and Government Entity (CAGE) Code (C) is a 5-digit alphanumeric code used to identify the manufacturer, distributor, or Government agency, etc., that supplies the item.

(2) *PART NUMBER column.* Indicates the primary number used by the manufacturer (individual, firm, corporation, or Government activity), which controls the design and characteristics of the item by means of its engineering drawings, specifications standards, and inspection requirements to identify an item or range of items.

(3) *STOCK NUMBER column.* This column lists the NSN for the associated part number and manufacturer identified in the PART NUMBER and CAGE columns to the left.

(4) FIG. column. This column lists the number of the figure where the item is identified/located in Sections II and III.

(5) *ITEM column.* The item number is that number assigned to the item as it appears in the figure referenced in the adjacent figure number column.

c. *FIGURE AND ITEM NUMBER INDEX.*

(1) *FIG. column.* This column lists the number of the figure where the item is identified/located in Sections II and III.

(2) *ITEM column.* The item number is that number assigned to the item as it appears in the figure referenced in the adjacent figure number column.

(3) *STOCK NUMBER column.* This column lists the NSN for the item.

(4) *CAGE column.* The Commercial and Government Entity (CAGE) is a 5-digit alphanumeric code used to identify the manufacturer, distributor, or Government agency, etc., that supplies the item.

(5) *PART NUMBER column.* Indicates the primary number used by the manufacturer (individual, firm, corporation, or Government activity), which controls the design and characteristics of the item by means of its engineering drawings, specifications standards and inspection requirements to identify an item or range of items.

5. Special Information.

a. *USABLE ON CODE. The* usable on code appears in the lower left corner of the Description column heading. Usable on codes are shown as "UOC:......" in the Description column (justified left) on the first line following applicable item description/nomenclature. Uncoded items are applicable to all models. Identification of the usable on codes used in the RPSTL are:

Code	Used On
194	M\1008,Type "B"Cargo, Shelter
208	M1008A1,Type "B" Cargo, Troop Seat
209	M1009, Type "A" Utility
210	M1010, Type "C" Ambulance
230	M 1028, Type "E" Shelter Carrier
252	M1028A1,Type"F" Shelter Carrier w/PTO
254	M1028A2, Type "F" Shelter Carrier w/PTO
256	M1028A3, Type "E" Shelter Carrier
231	M1031, Type "D" Chassis

b. *FABRICATION INSTRUCTIONS.* Bulk materials required to manufacture items are listed in the Bulk Material Functional Group of this RPSTL. Part numbers for bulk materials are also referenced in the Description column of the line item entry for the item to be manufactured/fabricated. Detailed fabrication instructions for items source coded to be manufactured or fabricated are found in *TM 9-2320-289-20* or *TM 9-2320-289-34.*

c. *ASSEMBLY INSTRUCTIONS.* Detailed assembly instructions for items source coded to be assembled from component spare/repair parts are found in *TM 9-2320-289-20* or *TM 9-2320-289-34.* Items that makeup the assembly are listed immediately following the assembly item entry or reference is made to an applicable figure.

d. *KITS.* Line item entries for repair parts kits appear in group 9401 in Section II.

e. *INDEX NUMBERS.* Items which have the word BULK in the figure column will have an index number shown in the item number column. This index number is a cross-reference between the National Stock Number/Part Number Index and the bulk material list in Section II.

f. *ASSOCIATED PUBLICATIONS.* The publications listed below pertain to the CUCV Series Truck and its components:

Publication	Short Title
LO 9-2320-289-72	CUCV Series Truck
TM 9-2320-289-10	CUCV Series Truck
TM 9-2320-289-20	CUCV Series Truck
TM 9-2320-289-20P	CUCV Series Truck
TM 9-2320-289-34	CUCV Series Truck

6. How to Locate Repair Parts.

a. *When National Stock Number or Part Number is Not Known:*

(1) *First.* Using the table of contents, determine the assembly group or subassembly group to which the item belongs. This is necessary since figures are pre-pared for assembly groups and subassembly groups, and listings are divided into the same groups.

(2) *Second.* Find the figure covering the assembly group or subassembly group to which the item belongs.

(3) *Third.* Identify the item on the figure and use the Figure and Item Number Index to find the NSN.

b. *When National Stock Number or Part Number is Known:*

(1) *First.* Using the National Stock Number or Part Number Index, find the pertinent National Stock Number or Part Number. The NSN index is in National Item Identification Number (NIIN) sequence (see paragraph 4.a.(1)). The part numbers in the Part Number index are listed in ascending alphanumeric sequence (see paragraph 4. b). Both indexes cross-reference you to the illustration/figure and item number of the item you are looking for.

(2) Second. Turn to the figure and item number, verify that the item is the one you're looking for, then locate the item number in the repair parts list for the figure.

7. Abbreviations.

For standard abbreviations see MIL-STD-12D, *Military Standard Abbreviations for Use on Drawings, Specifications, Standards, and in Technical Documents.*

Abbreviations	Explanation
NIIN	National Item Identification Number (consists of the last 9 digits of the NSN)
RPSTL	Repair Parts and Special Tools List

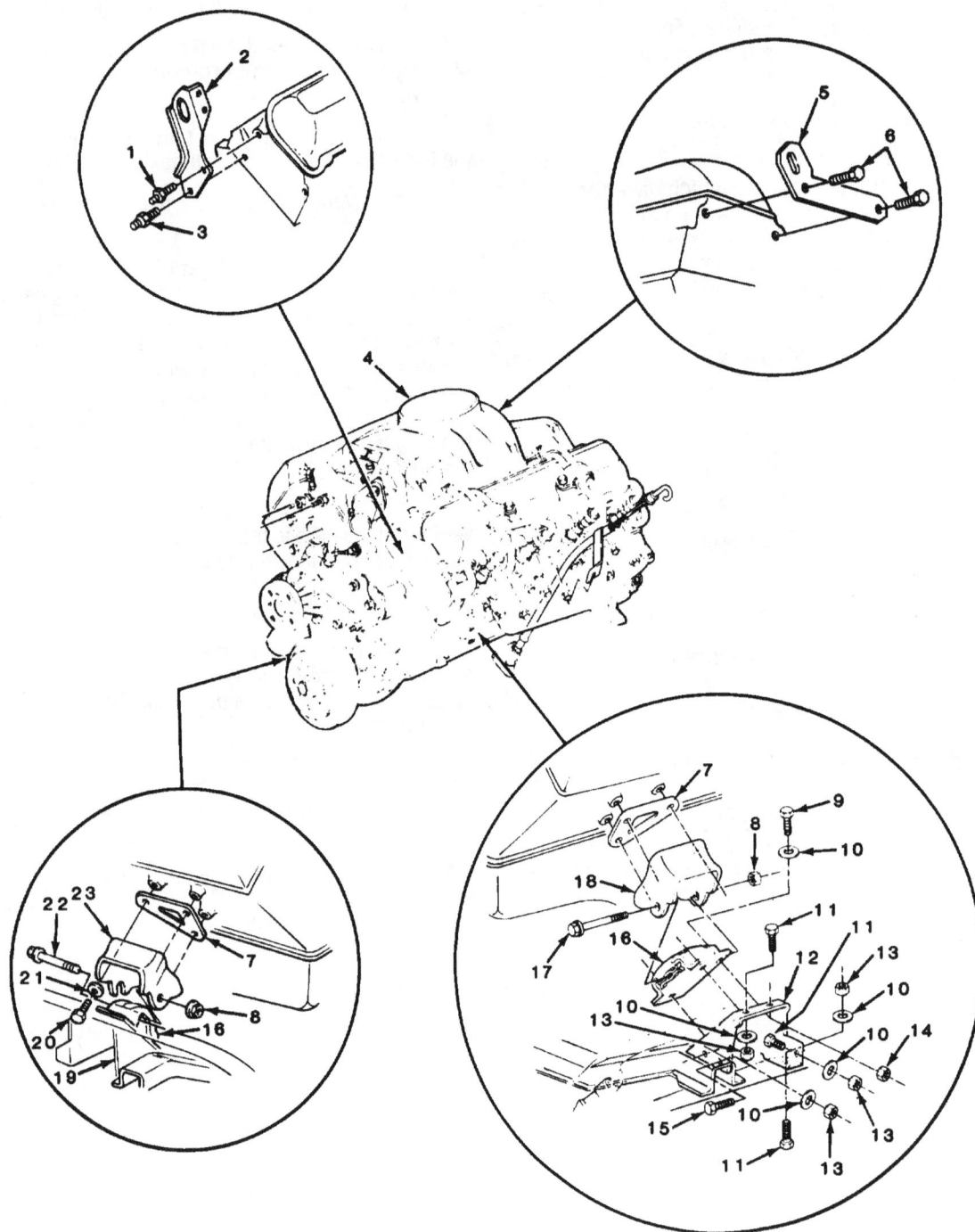

FIGURE 1. ENGINE ASSEMBLY, MOUNTS, LIFTING BRACKETS,
AND MOUNTING HARDWARE.

TA510808

(1)	(2)	(3)	(4)	(5)	(6)
ITEM NO	SMR CODE	CAGEC	PART NUMBER	DESCRIPTION AND USABLE ON CODES (UOC)	QTY

GROUP 01 ENGINE

GROUP 0100 ENGINE ASSEMBLY

FIG. 1 ENGINE ASSEMBLY, MOUNTS,
LIFTING BRACKETS, AND MOUNTING
HARDWARE

Item	SMR	CAGEC	Part Number	Description	Qty
1	PAOZZ	11862	1403394-6	STUD,CONTINUOUS THR	1
2	PAOZZ	11862	14033945	BRACKET,ENGINE LIFT FRONT	1
3	PAOZZ	11862	22535073	STUD,CONTINUOUS THR	1
4	PAFHD	11862	14067734	ENGINE,DIESEL FOR COMPONENT PARTS SEE FIG'S 2 THRU 11	1
5	PAOZZ	11862	14033947	BRACKET,ANGLE REAR	1
6	PAOZZ	11862	11504595	SCREW,CAP,HEXAGON H	2
7	PAFZZ	11862	14037861	SPACER,PLATE	2
8	PAFZZ	11862	6262212	NUT,SELF-LOCKING,EX	2
9	PAFZZ	96906	MS90728-60	SCREW,CAP,HEXAGON H	6
10	PAFZZ	96906	MS27183-16	WASHER,FLAT	18
11	PAFZZ	11862	9440344	BOLT,MACHINE	8
12	PAFZZ	11862	14055501	BRACKET,ENGINE MOUN LEFT	1
13	PAFZZ	11862	9422299	NUT,SELF-LOCKING,HE	12
14	PAFZZ	80205	NAS1408A6	NUT,SELF-LOCKING,HE	6
15	PAFZZ	11862	9440356	BOLT,MACHINE	4
16	PAFZZ	11862	459021	MOUNT,RESILIENT	2
17	PAFZZ	11862	460308	BOLT,MACHINE	1
18	PAFZZ	11862	14071967	BRACKET,ENGINE MOUN LEFT	1
19	PAFZZ	11862	14055502	BRACKET,ENGINE MOUN RIGHT	1
20	PAFZZ	96906	MS90728-61	SCREW,CAP,HEXAGON H	6
21	PAFZZ	96906	MS27183-13	WASHER,FLAT	6
22	PAFZZ	11862	6262211	SCREW,CAP,HEXAGON H	1
23	PAFZZ	11862	14037878	BRACKET,ENGINE MOUN RIGHT	1

END OF FIGURE

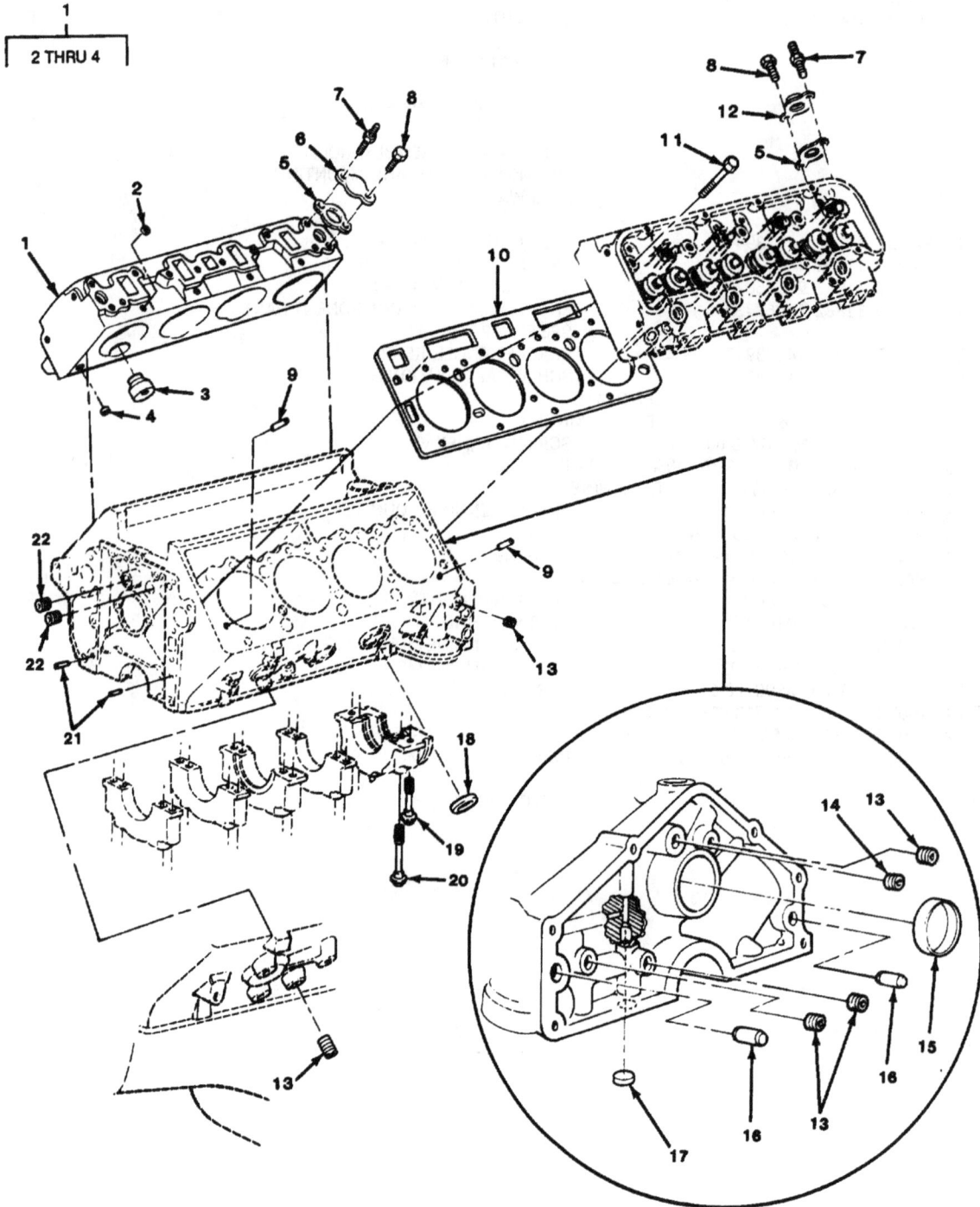

FIGURE 2. CYLINDER HEADS, BLOCK PLUGS, AND RELATED PARTS.

TA510809

(1) ITEM NO	(2) SMR CODE	(3) CAGEC	(4) PART NUMBER	(5) DESCRIPTION AND USABLE ON CODES (UOC)	(6) QTY
				GROUP 0101 CRANKCASE, BLOCK, CYLINDER HEAD	
				FIG. 2 CYLINDER HEADS, BLOCK PLUGS, AND RELATED PARTS	
1	PAFHH	11862	14079304	CYLINDER HEAD,DIESE	2
1	PAFHH	7X677	14071072	CYLINDER HEAD,DIESE	2
2	PAHZZ	11862	14025517	.PLUG,EXPANSION	8
3	PAHZZ	11862	14025518	.CHAMBER,PRE	8
4	PAHZZ	11862	14077163	.PLUG,EXPANSION	2
5	PAFZZ	11862	23500846	GASKET	2
6	PAFZZ	11862	14045274	COVER,ACCESS	1
7	PAFZZ	11862	22535073	STUD,CONTINUOUS THR	2
8	PAFZZ	11862	11504595	SCREW,CAP,HEXAGON H	2
9	PAHZZ	11862	3894327	SPACER,RING	4
10	PAFZZ	11862	1049600	GASKET PART OF KIT P/N 14067590	2
11	PAFZZ	11862	14077193	BOLT,EXTERNALLY REL	34
12	PAFZZ	11862	14028949	FLANGE,PIPE	1
13	PAHZZ	81348	WW-P-471ACBBCB PLUG,PIPE		6
14	PAHZZ	11862	14090911	PLUG,MACHINE THREAD	1
15	PAHZZ	11862	3999200	PLUG,EXPANSION	1
16	PAHZZ	11862	1453658	PIN,STRAIGHT,HEADLE	2
17	PAHZZ	24617	9421745	PLUG,PROTECTIVE,DUS	1
18	PAHZZ	11862	10000462	CAP,PROTECTIVE,DUST	6
19	PAFZZ	11862	14077192	BOLT,SHOULDER	10
20	PAFZZ	11862	14077195	BOLT,EXTERNALLY REL	10
21	PAHZZ	11862	14022670	PIN,STRAIGHT,HEADLE	2
22	PAHZZ	96906	MS9176-13	PLUG,EXPANSION	2

END OF FIGURE

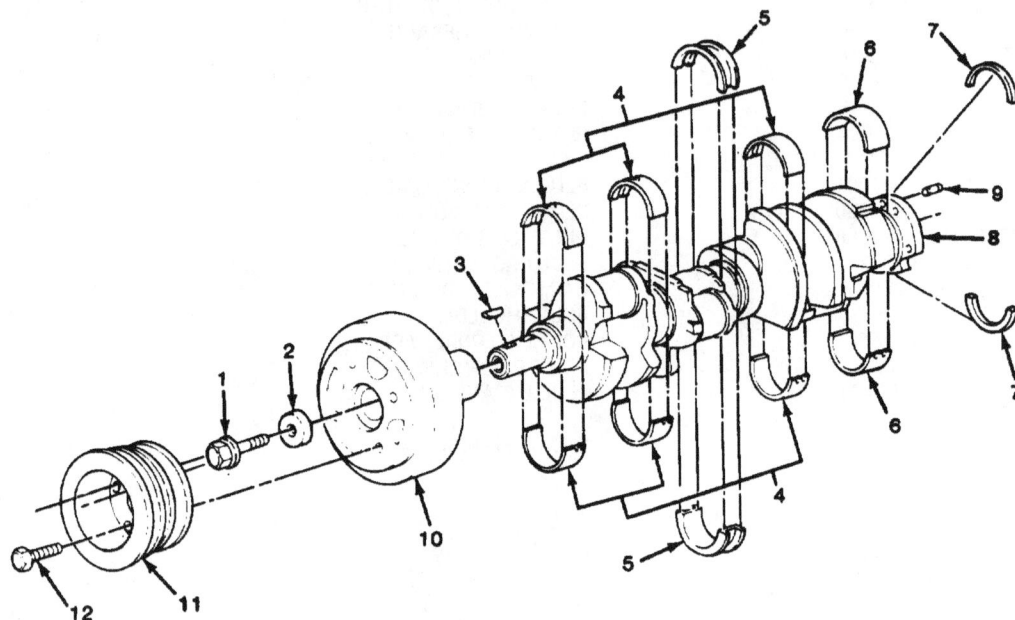

TA510810

FIGURE 3. ENGINE CRANKSHAFT AND RELATED PARTS.

(1) (2) (3) (4) ITEM SMR PART NO CODE CAGEC NUMBER	(5) DESCRIPTION AND USABLE ON CODES (UOC)	(6) QTY
	GROUP 0102 CRANKSHAFT	
	FIG. 3 ENGINE CRANKSHAFT AND RELATED PARTS	
1 PAFZZ 11862 14022672	SCREW,CAP,HEXAGON H	1
2 PAFZZ 11862 14022673 WASHER,FLAT		1
3 PAFZZ 24617 106751	KEY,SEA WATER PUMP	1
4 PAHZZ 11862 18009096 BEARING,SLEEVE		3
4 PAHZZ 11862 18009097	BEARING,SLEEVE (.013 MM US)	1
4 PAHZZ 11862 18009098	BEARING,SLEEVE (.026 MM US)	1
5 PAHZZ 34623 5740201 BEARING,SLEEVE		1
5 PAHZZ 11862 23500111	BEARING,SLEEVE (.026 MM US)	1
5 PAHZZ 11862 14053400	BEARING,SLEEVE (.013 MM US)	1
6 PAHZZ 11862 14055002 BEARING,SLEEVE		1
6 PAHZZ 11862 14055003	BEARING,SLEEVE (.013 MM US)	1
6 PAHZZ 11862 14055004	BEARING,SLEEVE (.026 MM US)	1
7 PAFZZ 11862 23500139	PACKING,PREFORMED PART OF KIT P/N 15633467	2
8 PAHHH 11862 14024271 CRANKSHAFT,ENGINE		1
9 PAHZZ 11862 3701679 .PIN,STRAIGHT,HEADLE		1
10 PAFZZ 11862 14022671	DAMPER ASSEMBLY,TOR	1
11 PAOZZ 11862 14067703 PULLEY,CONE		1
	UOC:194,208,209,230,231,252,254,256	
11 PAOZZ 11862 14067702 PULLEY,CONE		1
	UOC:210,230	
12 PAOZZ 11862 11504595	SCREW,CAP,HEXAGON H	4

END OF FIGURE

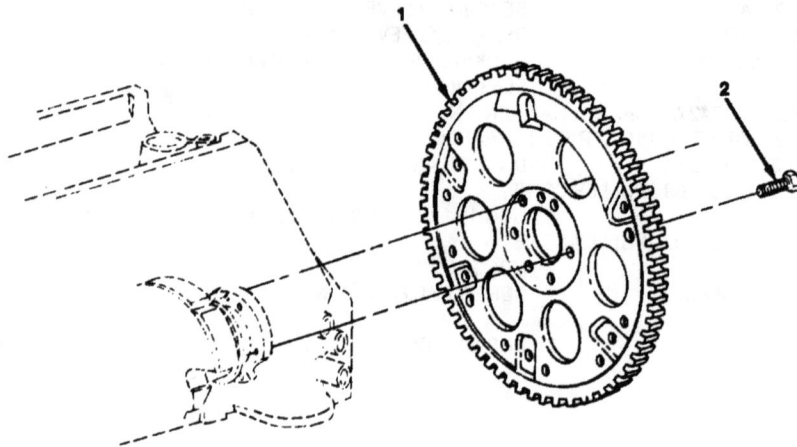

TA510811

FIGURE 4. FLYWHEEL.

(1) ITEM NO	(2) SMR CODE	(3) CAGEC	(4) PART NUMBER	(5) DESCRIPTION AND USABLE ON CODES (UOC)	(6) QTY
				GROUP 0103 FLYWHEEL ASSEMBLY	
				FIG. 4 FLYWHEEL	
1	PAFZZ	11862	14077157	FLYWHEEL,ENGINE	1
2	PAFZZ	11862	3727207	BOLT,MACHINE	6

END OF FIGURE

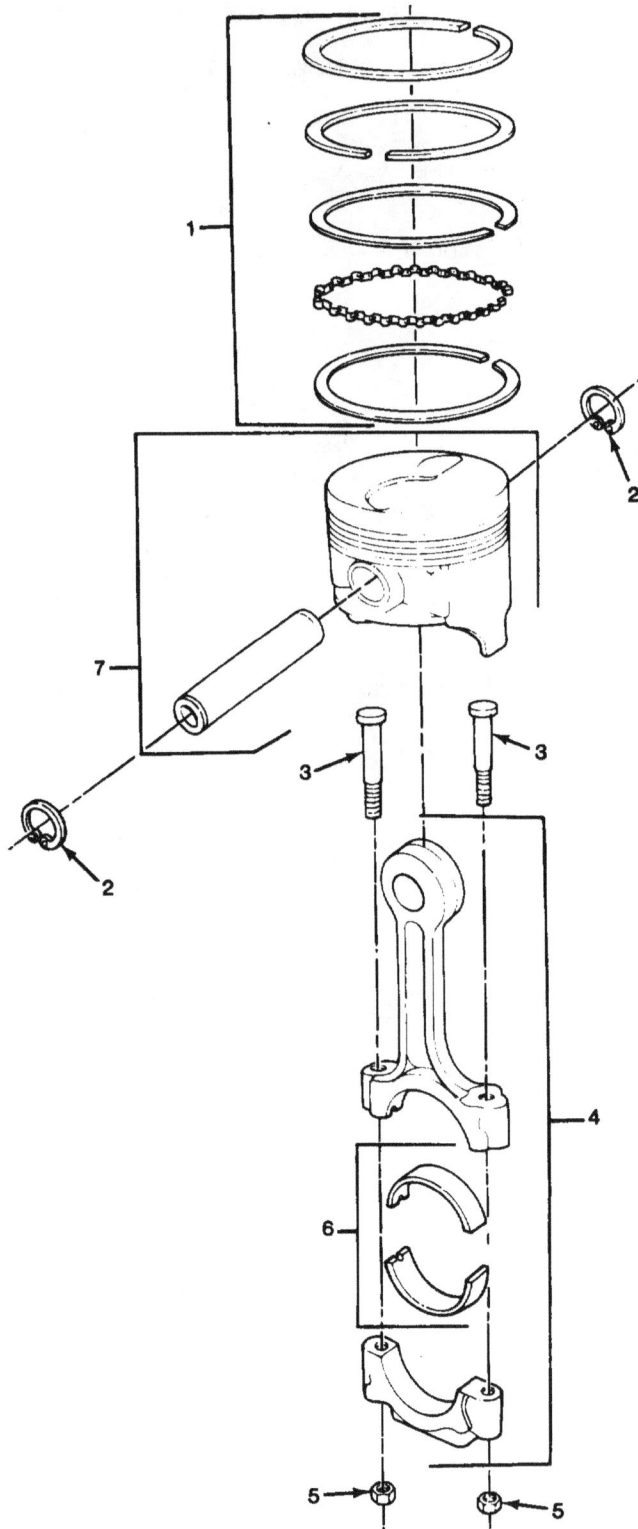

FIGURE 5. PISTON AND CONNECTING ROD.

TA510812

TM9-2320-289-34P

(1) (2) (3) (4) ITEM SMR NO CODE CAGEC	PART NUMBER	(5) DESCRIPTION AND USABLE ON CODES (UOC)	(6) QTY
		GROUP 0104 PISTONS AND CONNECTING RODS	
		FIG. 5 PISTON AND CONNECTING ROD	
1 PAHZZ 11862	15537018	RING SET,PISTON	8
1 PAHZZ 11862	15537020	RING SET,PISTON (0.75 OS)	1
2 PAHZZ 11862	23500298	RING,RETAINING	16
3 PAHZZ 11862	14025526	BOLT,MACHINE	16
4 PAHZZ 11862	14025523	CONNECTING ROD,PIST	8
5 PAHZZ 11862	14025527	NUT,PLAIN,HEXAGON	16
6 PAHZZ 11862	18009094	BEARING HALF,SLEEVE	8
6 PAHZZ 11862	18009095	PARTS KIT,PISTON AS (.026 MM US)	1
7 PAHZZ 11862	23500391	PISTON,INTERNAL COM	8
7 PAHZZ 11862	23500392	PISTON,INTERNAL COM (.13 OS)	1
7 PAHZZ 11862	23500393	PISTON,INTERNAL COM (.75 MM OS)	1

END OF FIGURE

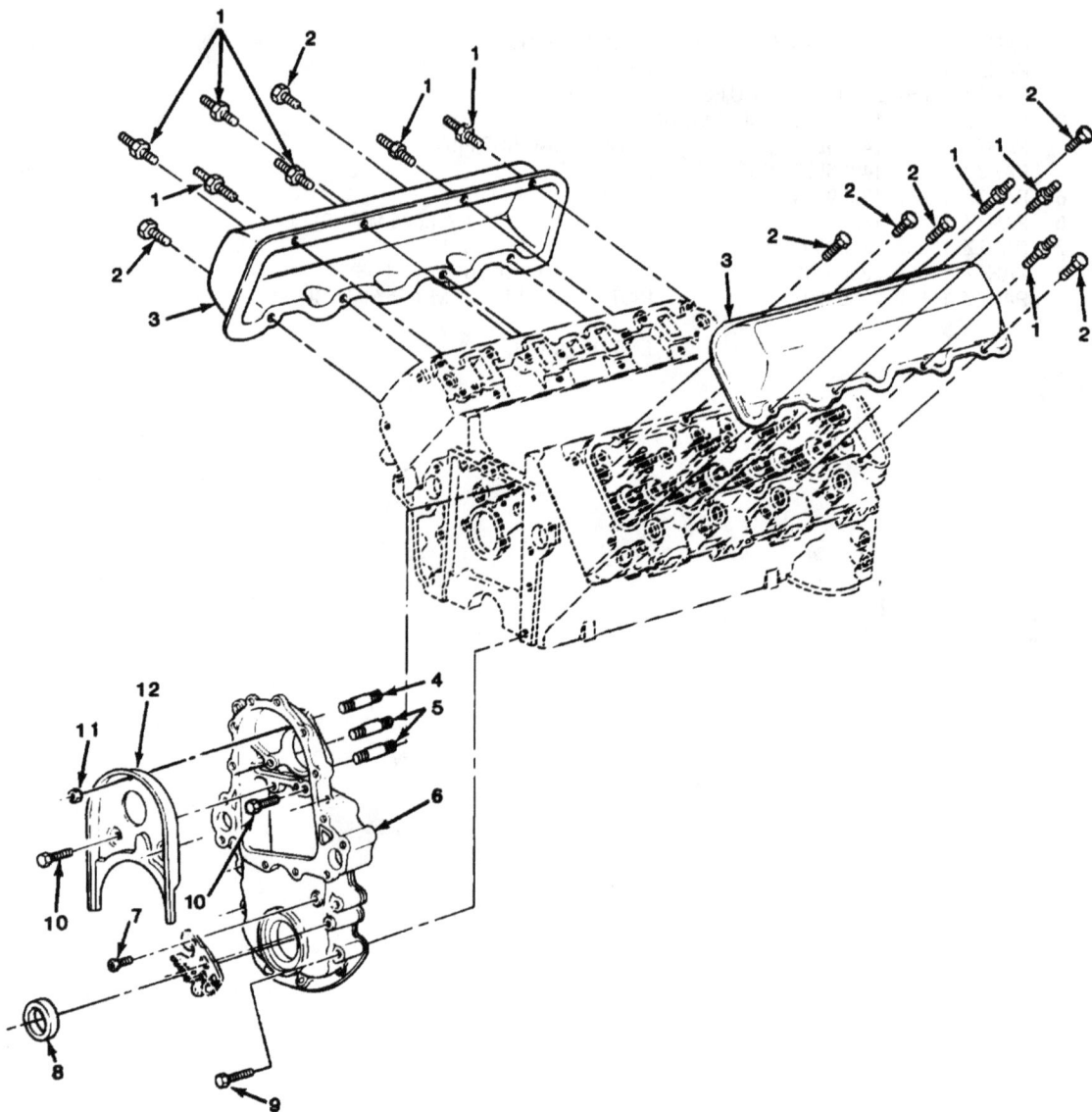

FIGURE 6. CYLINDER HEAD AND CRANKCASE COVERS.

TA510813

(1) (2) (3) (4) ITEM SMR NO CODE CAGEC	PART NUMBER	(5) DESCRIPTION AND USABLE ON CODES (UOC)	(6) QTY
		GROUP 0105 VALVES, CAMSHAFTS, AND TIMING SYSTEM	
		FIG. 6 CYLINDER HEAD AND CRANKCASE COVERS	
1 PAFZZ 11862	14033818	STUD,SHOULDERED	9
2 PAFZZ 24617	11504270	BOLT,MACHINE	7
3 PAFZZ 11862	23501992	COVER,ENGINE POPPET	2
4 PAFZZ 11862	10024396	STUD,PLAIN	1
5 PAFZZ 11862	14024203	STUD,PLAIN	2
6 PAFFF 11862	14044971	COVER,TIMING GEAR,I	1
7 PAFZZ 11862	14003948	.SCREW,MACHINE	2
8 PAFZZ 11862	3860095	.SEAL,PLAIN ENCASED PART OF KIT P/N 15663467	1
9 PAFZZ 11862	11502804	BOLT,CLOSE TOLERANC	4
10 PAFZZ 11862	1635490	BOLT,SELF-LOCKING	3
11 PAFZZ 11862	14005953	SHIELD,EXPANSION	1
12 PAFZZ 11862	14033820	BAFFLE,CRANKCASE,VE	1

END OF FIGURE

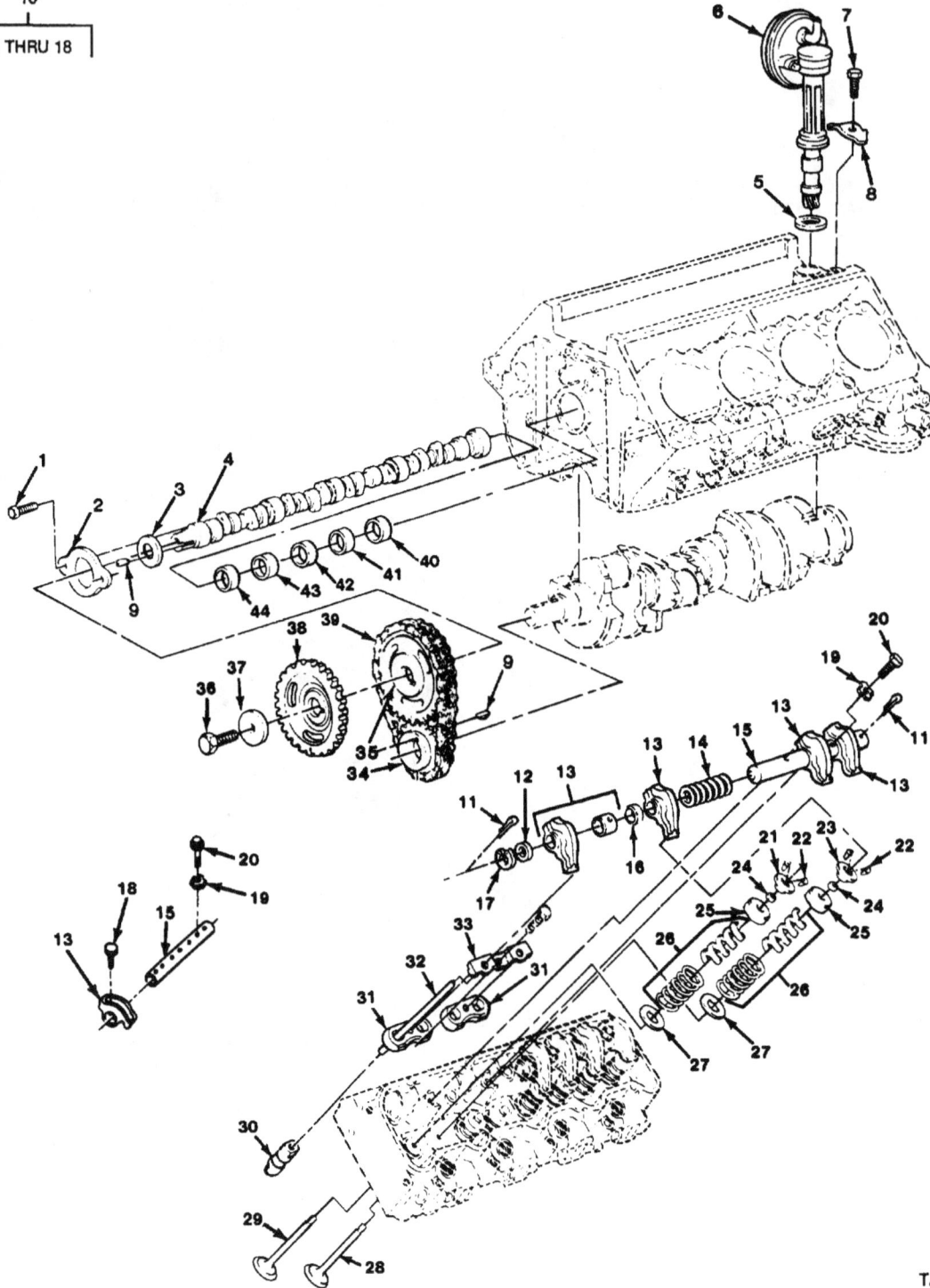

FIGURE 7. CAMSHAFT, VALVES, ROCKER ARMS, AND RELATED PARTS.

TA510814

(1) ITEM NO	(2) SMR CODE	(3) CAGEC	(4) PART NUMBER	(5) DESCRIPTION AND USABLE ON CODES (UOC)	(6) QTY
				GROUP 0105 VALVES, CAMSHAFTS, AND TIMING SYSTEM	
				FIG. 7 CAMSHAFT, VALVES, ROCKER ARMS, AND RELATED PARTS	
1	PAFZZ	24617	11500815	SCREW,CAP,HEXAGON H	2
2	PAHZZ	11862	14022644	BEARING,WASHER,THRU STD	1
3	PAHZZ	11862	14022643	SPACER,RING	1
4	PAHZZ	11862	14066308	CAMSHAFT,ENGINE	1
5	PAOZZ	11862	14022649	GASKET PART OF KIT P/N 15633467	1
6	PAOZZ	11862	7849302	PUMP,AIR,EMISSION O	1
7	PAOZZ	11862	1635490	BOLT,SELF-LOCKING	1
8	PAOZZ	11862	14022650	CLAMP	1
9	PAFZZ	24617	106751	KEY,SEA WATER PUMP	2
10	AFFFF	11862	23500073	ROCKER ARM ASM PART OF ENGINE ASM WITH SERIAL #OJA	4
10	AFFFF	11862	14061505	ROCKER ARM ASM PART OF ENGINE ASM WITH SERIAL #FJA	4
11	PAFZZ	96906	MS24665-353	.PIN,COTTER PART OF AMS P/N 14061505 ONLY	2
12	PAFZZ	11862	14028982	.WASHER,SPRING TENSI PART OF ASM P/N 14061505 ONLY	2
13	PAFZZ	11862	14022634	.ROCKER ARM,ENGINE P PART OF ASM P/N 14061505 ONLY	4
13	PAFZZ	11862	23500074	.ROCKER ARM,ENGINE P PART OF ASM P/N 23500073 ONLY	4
14	PAFZZ	11862	14028981	.SPRING,HELICAL,COMP PART OF ASM P/N 14061505 ONLY	1
15	PAFZZ	11862	14033822	.SHAFT,STRAIGHT PART OF ASM P/N 14061505 ONLY	1
15	PAFZZ	11862	23500075	.SHAFT,STRAIGHT PART OF ASM P/N 23500073 ONLY	1
16	PAFZZ	11862	14057297	.SPACER,SLEEVE PART OF ASM P/N 14061505 ONLY	2
17	PAFZZ	11862	14028983	.WASHER,FLAT PART OF ASM P/N 14061505 ONLY	2
18	PAFZZ	11862	23500076	.RETAINER,ROCKER ARM PART OF ASM P/N 23500073 ONLY	4
19	PAFZZ	11862	14057296	RETAINER,ROCKER SHA	8
20	PAFZZ	24617	11503603	SCREW,CAP,HEXAGON H	8
21	PAFZZ	23862	14003974	CAP,ROTOR	8
22	PAFZZ	11862	3947770	LOCK,VALVE SPRINK R	32
23	PAFZZ	11862	14042575	ROTOR,ENGINE POPPET	8
24	PAFZZ	11862	3835333	SEAL INLET PART OF KIT P/N 14067590	16
25	PAFZZ	11862	14025515	SPACER,SLEEVE	16
26	PAFZZ	11862	14025512	SPRING	16
27	PAFZZ	11862	14003968	WASHER,FLAT	16
28	PAFZZ	11862	10137454	VALVE,POPPET,ENGINE	8
28	PAFZZ	11862	14050661	VALVE,POPPET,ENGINE (.089 MM OS)	1
28	PAFZZ	11862	14050662	VALVE,POPPET,ENGINE (.349 MM OS)	1
29	PAFZZ	11862	14033927	VALVE,POPPET,ENGINE	8

(1) (2) (3) (4)				(5)	(6)
ITEM	SMR		PART		
NO	CODE	CAGEC	NUMBER	DESCRIPTION AND USABLE ON CODES (UOC)	QTY
29	PAFZZ	11862	14050658	VALVE,POPPET,ENGINE (.089 MM OS)	1
29	PAFZZ	11862	14050659	VALVE,POPPET,ENGINE (.394 MM OS)	1
30	PAFZZ	11862	5234530	TAPPET,ENGINE POPPE	16
31	PAFZZ	11862	14077148	PLATE,VALVE LIFTER	8
32	PAFZZ	11862	14057232	PUSH ROD,ENGINE POP	16
33	PAFZZ	11862	14022640	CLAMP,VALVE LIFTER	4
34	PAFZZ	11862	14022646	SPROCKET WHEEL	1
35	PAFZZ	11862	14022645	SPROCKET WHEEL	1
36	PAFZZ	24617	11503316	SCREW,CAP,HEXAGON H	1
37	PAFZZ	11862	14022648 WASHER,FLAT		1
38	PAFZZ	11862	14022653 GEAR,HELICAL		1
39	PAFZZ	11862	TC181 CHAIN,ROLLER		1
40	PAHZZ	11862	14048457	BEARING,SLEEVE NO. 5	1
41	PAHZZ	11862	14048456	BEARING,SLEEVE NO. 4	1
42	PAHZZ	11862	14048455	BEARING,SLEEVE NO. 3	1
43	PAHZZ	11862	14048454	BEARING,SLEEVE NO. 2	1
44	PAHZZ	11862	SH1366S	BEARING,SLEEVE NO. 1	1

END OF FIGURE

22   25

| 23 THRU 31 | 26 THRU 29 |

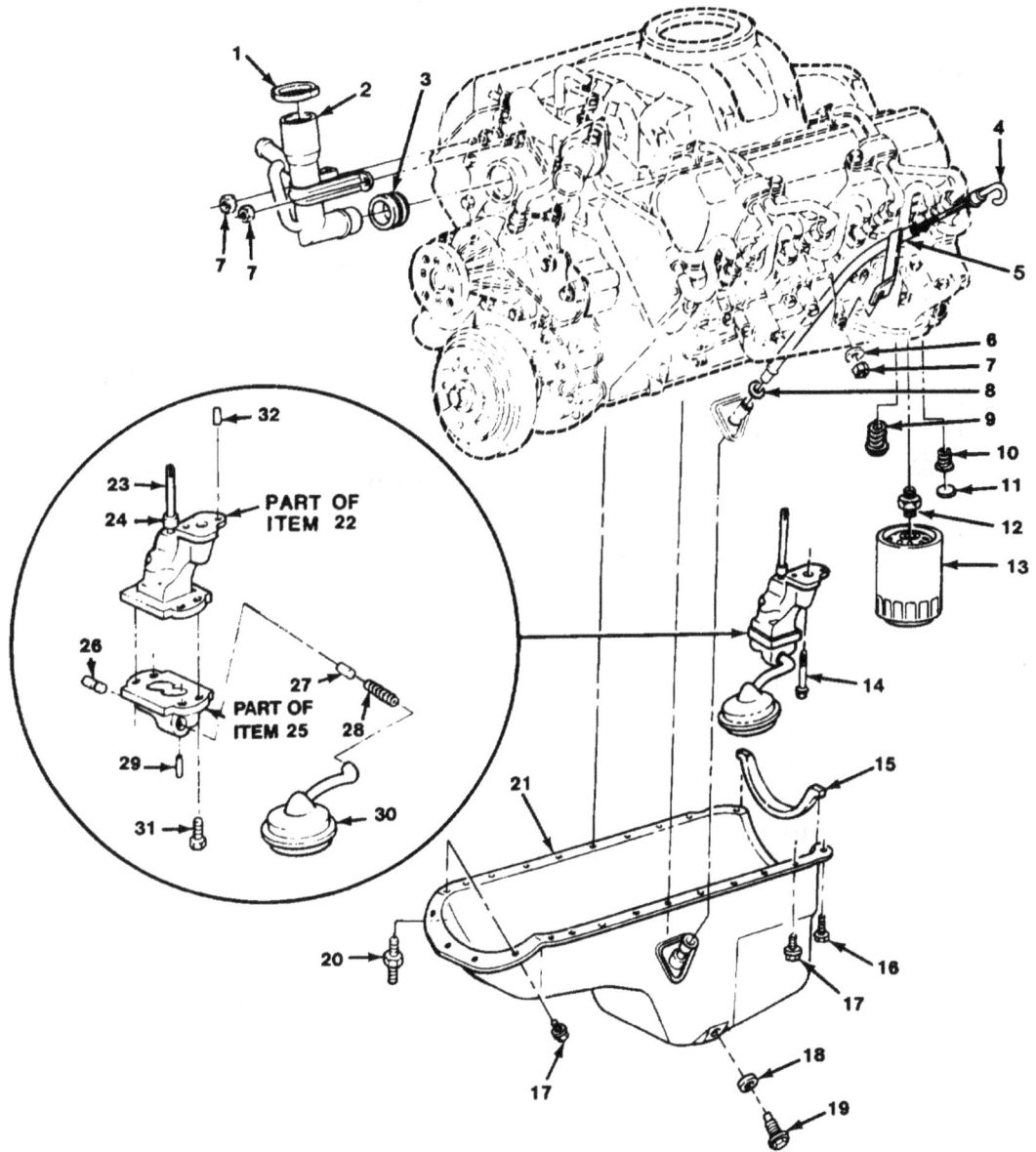

TA510815

FIGURE 8. ENGINE OIL PUMP ASSEMBLY, FILTER, PAN, AND RELATED PARTS.

(1) (2) (3) (4)				(5)	(6)
ITEM SMR			PART	DESCRIPTION AND USABLE ON CODES (UOC)	QTY
NO	CODE	CAGEC	NUMBER		

GROUP 0106 ENGINE LUBRICATION
SYSTEM

FIG. 8 ENGINE OIL PUMP ASSEMBLY,
FILTER,PAN,AND RELATED PARTS

1	PAOZZ 70040 FC106	CAP,FILLER OPENING	1
2	PAOZZ 11862 14071059	FILLER NECK	1
3	PAOZZ 11862 14028942	GROMMET,NONMETALLIC	1
4	PAOZZ 11862 14050523	GAGE ROD-CAP,LIQUID	1
5	PAOZZ 11862 14045268	TUBE,BENT,METALLIC	1
6	PAOZZ 11862 14036369	WASHER,FLAT	1
7	PAOZZ 11862 22521550	NUT,PLAIN,HEXAGON	3
8	PAOZZ 24617 274244	PACKING,PREFORMED	1
9	PAOZZ 11862 25011208	VALVE,REGULATING,OI	1
10	PAOZZ 11862 25011206	VALVE,SAFETY RELIEF	1
11	PAOZZ 11862 14066310	PLUG,PISTON PIN	1
12	PAOZZ 11862 14022700	REDUCER,BOSS	1
13	PAOZZ 70040 PF-35	FILTER,FLUID	1
14	PAFZZ 11862 14028914	BOLT,MACHINE	1
15	PAOZZ 11862 14022683	SEAL,NONMETALLIC SP	1
16	PAOZZ 11862 11508534	BOLT,MACHINE	2
17	PAOZZ 24617 11507029	SCREW,CAP,HEXAGON H	21
18	PAOZZ 11862 14079550	GASKET	1
19	PAOZZ 11862 337185	PLUG,MACHINE THREAD	1
20	PAOZZ 11862 14066307	STUD,SHOULDERED	1
21	PAOZZ 11862 14061649	OIL PAN	1
22	PAFFF 11862 14079426	PUMP AND SCREEN ASS	1
23	PAFZZ 11862 14022699	.SHAFT,STRAIGHT	1
24	PAFZZ 11862 477249	.RETAINER,PACKING	1
25	PAFFF 11862 14028976	.COVER ASSEMBLY,OIL	1
26	PAFZZ 11862 3704871	.DISK,SOLID,PLAIN	1
27	PAFZZ 11862 3702366	.VALVE,RELIEF,PRESSU	1
28	PAFZZ 11862 360582	.SPRING,HELICAL,COMP	1
29	PAFZZ 11862 838839	.PIN,GROOVED,HEADLES	1
30	PAFZZ 11862 14077182	.SCREEN ASSEMBLY,OIL	1
31	PAFZZ 11862 11508600	.BOLT,MACHINE	4
32	PAFZZ 24617 141195	PIN,STRAIGHT,HEADLE	2

END OF FIGURE

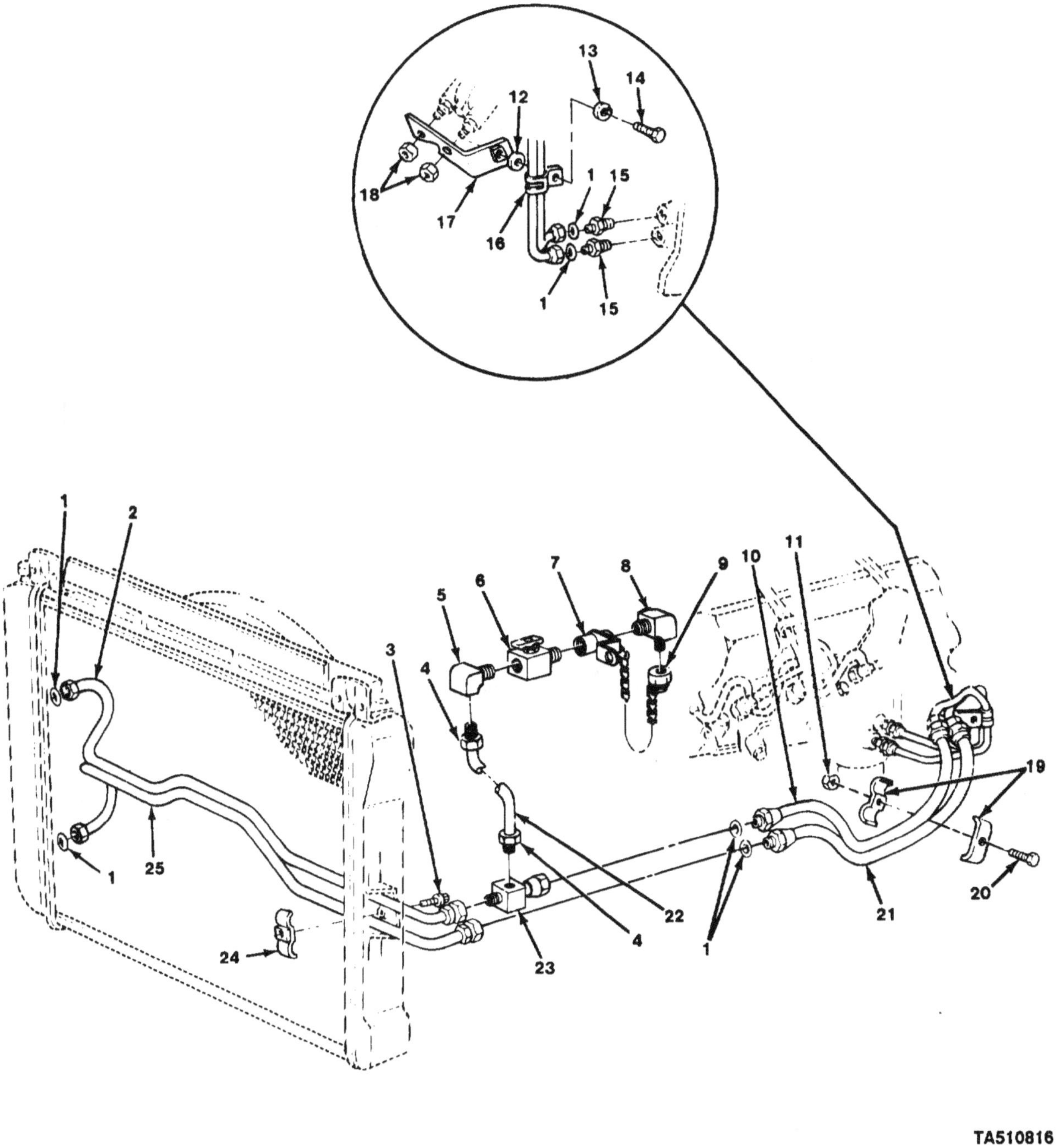

TA510816

FIGURE 9. OIL COOLER LINES, SAMPLING VALVE, AND RELATED PARTS.

(1) ITEM NO	(2) SMR CODE	(3) CAGEC	(4) PART NUMBER	(5) DESCRIPTION AND USABLE ON CODES (UOC)	(6) QTY
				GROUP 0106 ENGINE LUBRICATION SYSTEM	
				FIG. 9 OIL COOLER LINES, SAMPLING VALVE, AND RELATED PARTS	
1	PAOZZ	11862	14055585	SEAL RING,METAL	6
2	PAOZZ	11862	14061344	TUBE ASSEMBLY,METAL	1
3	PAOZZ	96906	MS90728-8	SCREW,CAP,HEXAGON H	1
4	PAOZZ	11862	137396	INVERTED NUT,TUBE C	2
5	PAOZZ	72582	224425	ELBOW,PIPE TO TUBE	1
6	PAOZZ	79470	6820	COCK,SHUTOFF,SCREW	1
7	PAOZZ	11862	15599988	BRACKET,DOUBLE ANGL	1
8	PAOZZ	72582	118754	ELBOW,PIPE TO TUBE	1
9	PAOZZ	70411	SP2489-FM	CAP,FILLER OPENING	1
10	PAOZZ	11862	15548901	HOSE ASSEMBLY,NONME UOC:194,208,209,230,231,252,254,256	1
10	PAOZZ	11862	14063336	HOSE ASSEMBLY,NONME UOC:210,230	1
11	PAOZZ	96906	MS35691-406	NUT,PLAIN,HEXAGON	1
12	PAOZZ	11862	14061350	DISK,OIL COOLER	1
13	PAOZZ	96906	MS35338-44	WASHER,LOCK	1
14	PAOZZ	96906	MS90728-6	SCREW,CAP,HEXAGON H	1
15	PAOZZ	11862	14055586	ADAPTER,STRAIGHT,PI	2
16	PAOZZ	11862	14047899	CLAMP,LOOP	1
17	PAOZZ	11862	14061348	BRACKET,ANGLE	1
18	PAOZZ	11862	22521550	NUT,PLAIN,HEXAGON	2
19	PAOZZ	11862	14061352	STRAP,RETAINING	2
20	PAOZZ	96906	MS90728-10	SCREW,CAP,HEXAGON H	1
21	PAOZZ	11862	15548902	HOSE ASSEMBLY,NONME UOC:194,208,209,230,231,252,254,256	1
21	PAOZZ	11862	14063337	HOSE ASSEMBLY,NONME UOC:210,230	1
22	MFOZZ	11862	15599986	PIPE ASM (9.68"LG) MAKE FROM PIPE, P/N 3696822	1
23	PAOZZ	11862	15599987	TEE,TUBE	1
24	PAOZZ	11862	14036784	STRAP,RETAINING	1
25	PAOZZ	11862	14061345	TUBE ASSEMBLY,METAL	1

END OF FIGURE

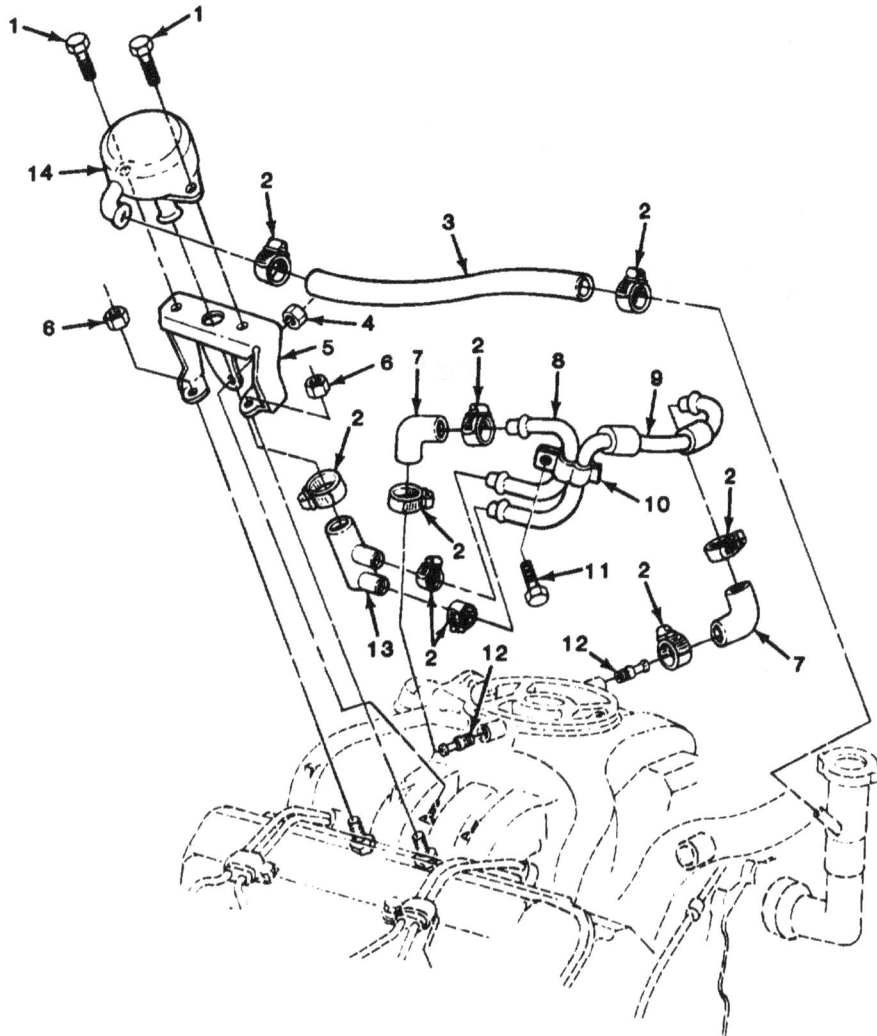

TA510817

FIGURE 10. CRANKCASE DEPRESSION VALVE AND RELATED PARTS.

TM9-2320-289-34P

(1)	(2)	(3)	(4)	(5)	(6)
ITEM NO	SMR CODE	CAGEC	PART NUMBER	DESCRIPTION AND USABLE ON CODES (UOC)	QTY

GROUP 0106 ENGINE LUBRICATION
SYSTEM

FIG. 10 CRANKCASE DEPRESSION VALVE
AND RELATED PARTS

1	PAOZZ	11862	11508353	BOLT,MACHINE	2
2	PAOZZ	96906	MS35842-11	CLAMP,HOSE	9
3	PAOZZ	11862	14067732	HOSE,PREFORMEC	1
4	PAOZZ	24617	11502488	NUT,PLAIN,HEXAGON	1
5	PAOZZ	11862	14067733	BRACKET ASSEMBLY,CR	1
6	PAOZZ	11862	22521550	NUT,PLAIN,HEXAGON	2
7	PAOZZ	11862	14050445	HOSE,PREFORMED	2
8	PAOZZ	11862	14050442	TUBE,BENT,METALLIC	1
9	PAOZZ	11862	14050443	TUBE,BENT,METALLIC	1
10	PAOZZ	11862	14050441	CLAMP,LOOP	1
11	PAOZZ	11862	11509135	SCREW,ASSEMBLED WAS	1
12	PAOZZ	11862	14050444	ADAPTER,STRAIGHT,TU	2
13	PAOZZ	11862	14050446	CONNECTOR,CRANKCASE	1
14	PAOZZ	11862	25042462	SENSOR,AIR CHARGED	1

END OF FIGURE

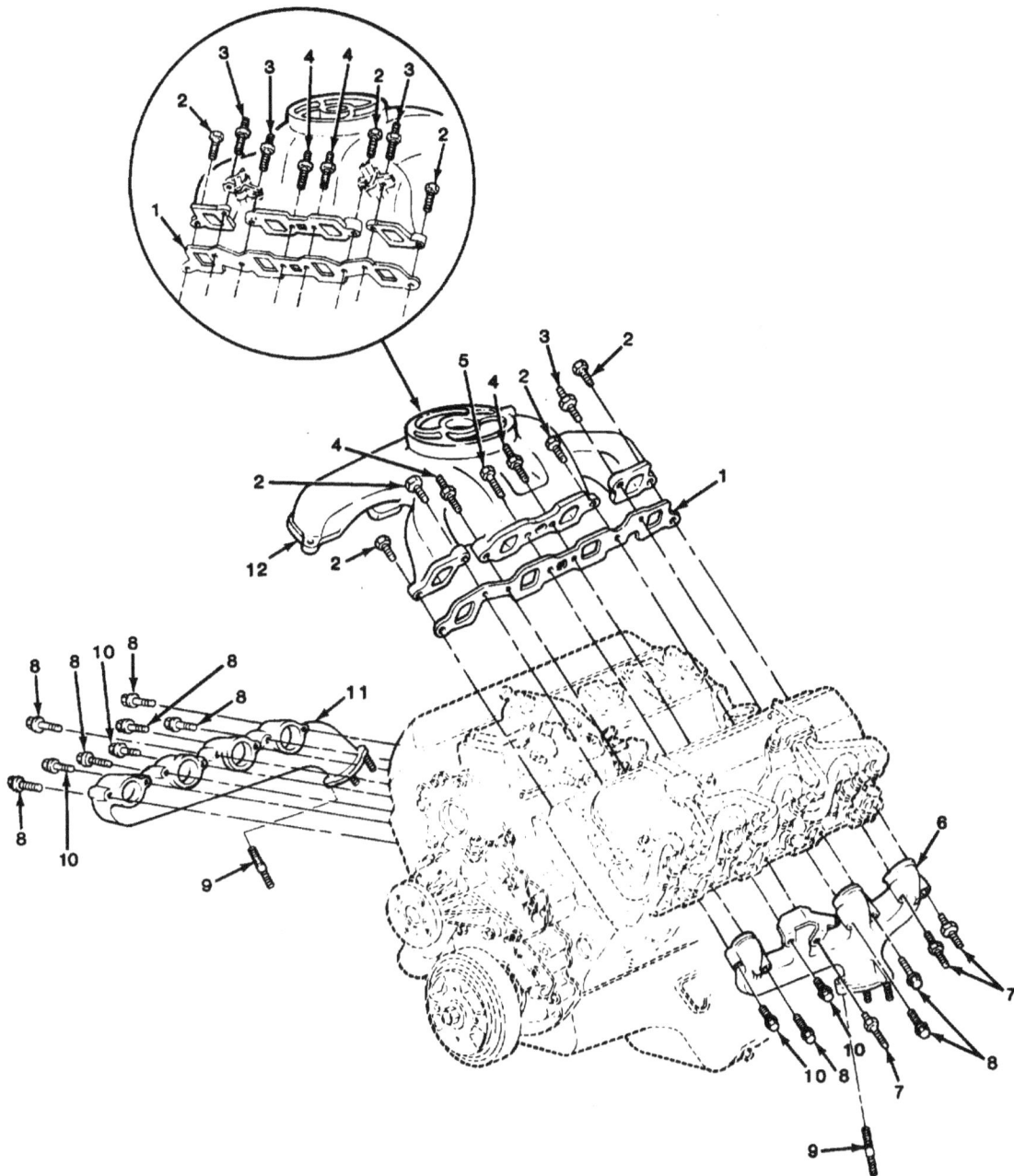

TA510818

FIGURE 11. INTAKE AND EXHAUST MANIFOLDS.

(1) ITEM NO	(2) SMR CODE	(3) CAGEC	(4) PART NUMBER	(5) DESCRIPTION AND USABLE ON CODES (UOC)	(6) QTY

GROUP 0108 MANIFOLDS

FIG. 11 INTAKE AND EXHAUST MANIFOLDS

(1)	(2)	(3)	(4)	(5)	(6)
1	PAFZZ	7X677	15633464	GASKET SET PART OF KIT P/N 14067590	2
2	PAFZZ	11862	1635490	BOLT,SELF-LOCKING	7
3	PAFZZ	11862	22535073	STUD,CONTINUOUS THR	4
4	PAFZZ	89554	23500832	STUD,SHOULDERED	4
5	PAFZZ	11862	11505068	SCREW,CAP,HEXAGON H	1
6	PAOZZ	11862	14022657	MANIFOLD,EXHAUST LEFT	1
7	PAOZZ	11862	14028924	STUD,SHOULDERED AND	3
8	PAOZZ	11862	14028922	BOLT,MACHINE	9
9	PAOZZ	11862	14022654	STUD,PLAIN	6
10	PAOZZ	11862	14028923	SCREW,CAP,HEXAGON H	4
11	PAOZZ	11862	14025568	MANIFOLD,EXHAUST RIGHT	1
12	PAFZZ	11862	14071068	MANIFOLD,INTAKE	1

END OF FIGURE

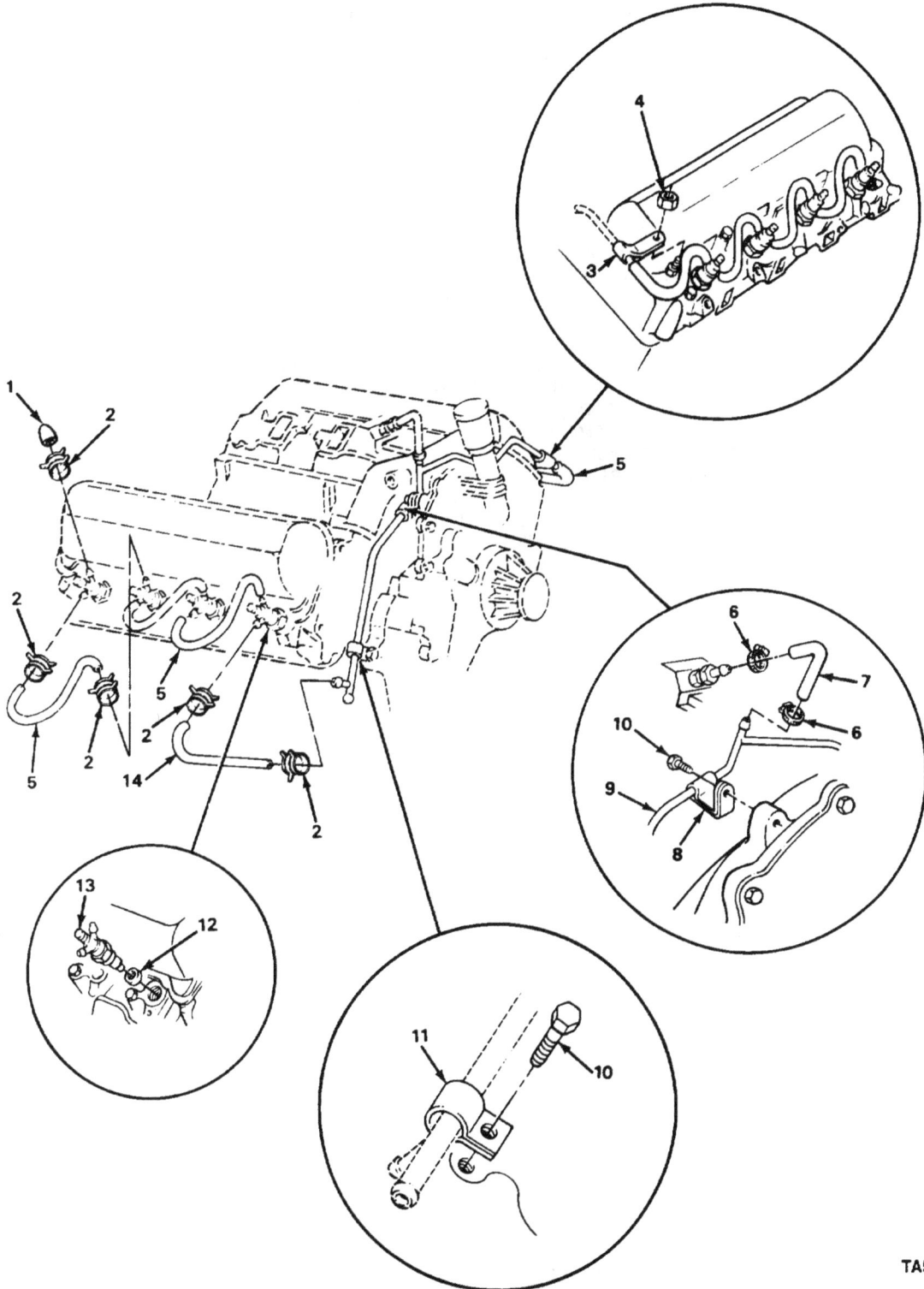

FIGURE 12. FUEL INJECTOR AND RELATED PARTS.

TA510819

(1) ITEM NO	(2) SMR CODE	(3) CAGEC	(4) PART NUMBER	(5) DESCRIPTION AND USABLE ON CODES (UOC)	(6) QTY
				GROUP 03 FUEL SYSTEM	
				GROUP 0301 CARBURETOR, FUEL INJECTOR	
				FIG. 12 FUEL INJECTOR AND RELATED PARTS	
1	PAOZZ	11862	14066301	CAP	2
2	PAOZZ	11862	11663000	CLAMP	18
3	PAOZZ	11862	14033893	CLAMP,LOOP	1
4	PAOZZ	11862	22521550	NUT,PLAIN,HEXAGON	1
5	MOOZZ	11862	14066306	HOSE,NONMETALLIC (7" LG) MAKE FROM P/N 14066305	7
6	PAOZZ	11862	25518880	CLAMP,HOSE	2
7	MOOZZ	11862	9439363	HOSE, NON MET (3" LG) MAKE FROM HOSE,P/N 9439402	1
8	PAOZZ	11862	14033895	CLAMP,LOOP	1
9	PAOZZ	11862	14061569	TUBE,BENT,METALLIC	1
10	PAOZZ	11862	11504512	BOLT,MACHINE	2
11	PAOZZ	96906	MS21333-45	CLAMP,LOOP	1
12	PAFZZ	11862	14025557	GASKET PART OF KIT P/N 14067590	8
13	PAFZZ	63632	6704001	NOZZLE,FUEL INJECTI	8
14	PAOZZ	11862	14066305	TUBING,NONMETALLIC (7.01" LG) CUT TO SIZE	1

END OF FIGURE

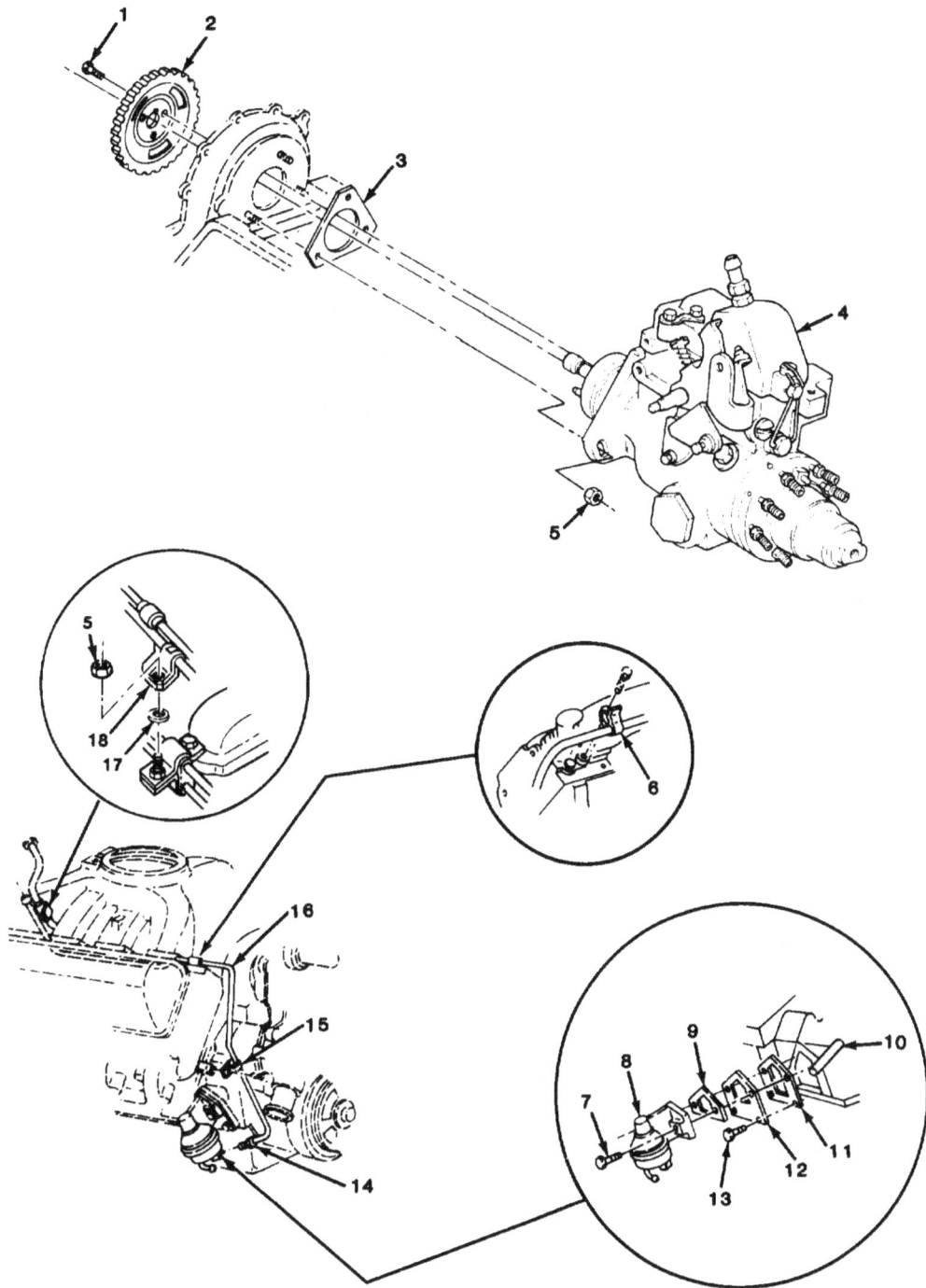

TA510820

FIGURE 13. SUPPLY PUMP, INJECTOR PUMIP,AND RELATED PARTS.

(1) ITEM NO	(2) SMR CODE	(3) CAGEC	(4) PART NUMBER	(5) DESCRIPTION AND USABLE ON CODES (UOC)	(6) QTY
				GROUP 0302 FUEL PUMPS	
				FIG. 13 SUPPLY PUMP, INJECTOR PUMP, AND RELATED PARTS	
1	PAFZZ	11862	11508017	SCREW,ASSEMBLED WAS	3
2	PAFZZ	11862	14022652	GEAR,HELICAL	1
3	PAFZZ	11862	14022651	GASKET PART OF KIT P/N 15633467	1
4	PAFHH	84760	DB2829-4521	PUMP,FUEL,METERING FOR COMPONENT PARTS SEE FIG 15	1
5	PAFZZ	11862	11506101	NUT	4
6	PAFZZ	11862	14033896	CLIP,SPRING TENSION	1
7	PAOZZ	96906	MS90728-64	SCREW,CAP,HEXAGON H	2
8	PAOZZ	11862	6471831	PUMP,FUEL,CAM ACTUA	1
9	PAOZZ	11862	9776705	GASKET PART OF KIT P/N 15633467	1
10	PAOZZ	11862	14050425	PIN,STRAIGHT,HEADLE	1
11	PAOZZ	11862	3705044	GASKET PART OF KIT P/N 15633467	1
12	PAOZZ	11862	3719599	PLATE,FUEL PUMP	1
13	PAOZZ	11862	11509669	BOLT,ASSEMBLED WASH	2
14	PAFZZ	93061	411FS-6	INVERTED NUT,TUBE C	1
15	PAFZZ	11862	22511422	CLAMP,LOOP	1
16	MFFZZ	11862	14061503	TUBE ASSEMBLY,METAL (35.52"LG) MAKE FROM TUBE, P/N 3750950	1
17	PAFZZ	11862	3792381	SPACER,SLEEVE	1
18	PAFZZ	11862	14063340	CLAMP,LOOP	1

END OF FIGURE

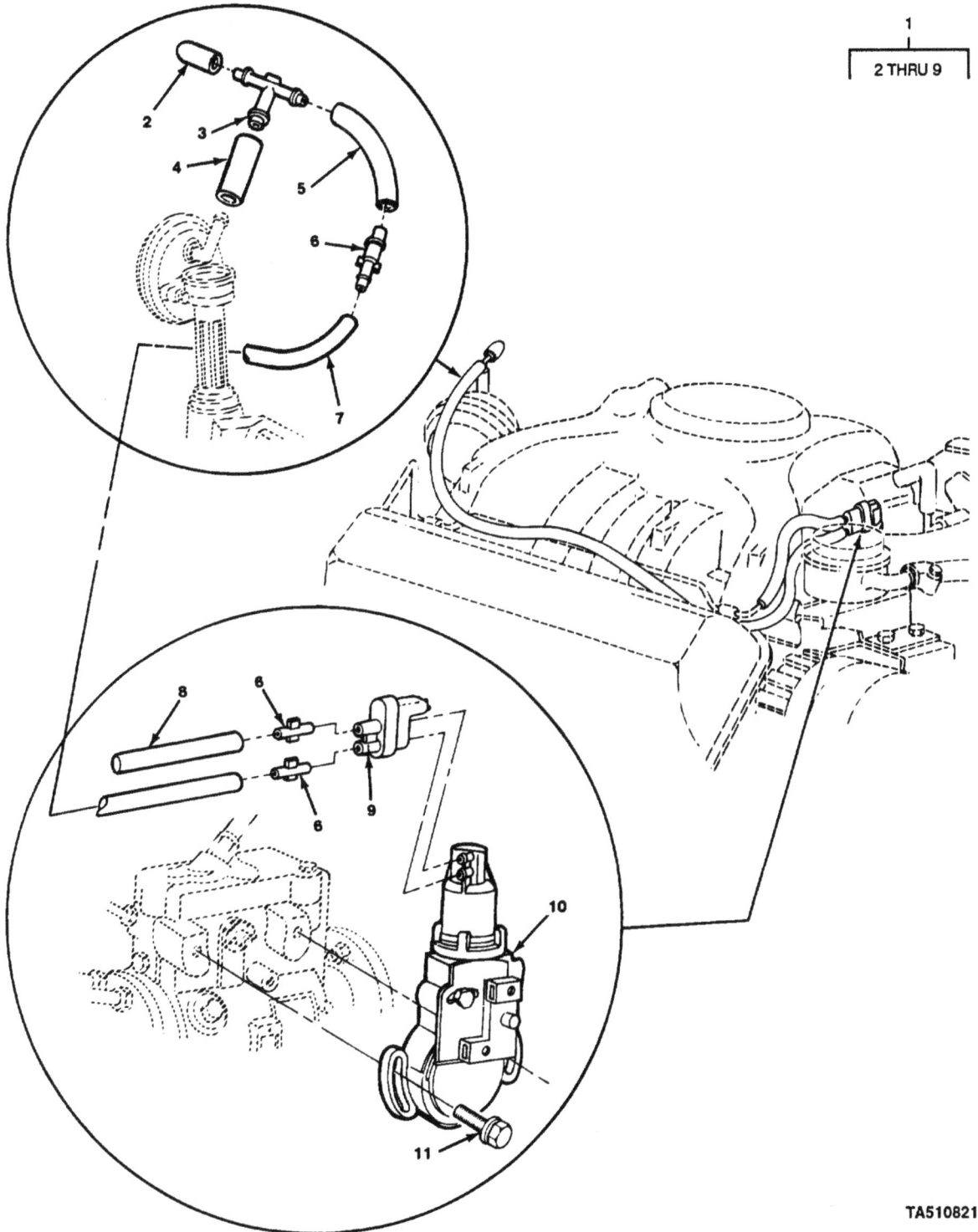

FIGURE 14. VACUUM VALVE AND HOSES.

TA510821

(1) (2) (3) (4) ITEM SMR PART NO CODE CAGEC NUMBER	(5) DESCRIPTION AND USABLE ON CODES (UOC)	(6) QTY

GROUP 0302 FUEL PUMPS

FIG. 14 VACUUM VALVE AND HOSES

Item	SMR	CAGEC	Part Number	Description	Qty
1	XDOOO	11862	14045233	DUPLEX HOSE,RUBBER	1
2	PAOZZ	11862	343350	.CAP,PROTECTIVE,DUST	1
3	PAOZZ	11862	3970076	.TEE	1
4	MOOZZ	11862	M51	.HOSE,RUBBER (2" LG) MAKE FROM 1 HOSE,P/N 9438124	
5	MOOZZ	11862	M140	.HOSE,RUBBER (5.51" LG) MAKE FROM HOSE,P/N 3987364	1
6	PAOZZ	11862	560625	.CONNECTOR,HOSE	3
7	MOOZZ	11862	M495	.HOSE,RUBBER (19.49" LG) MAKE FROM HOSE,P/N 9438381	1
8	MOOZZ	11862	M127	.HOSE,RUBBER (5" LG) MAKE FROM HOSE,P/N 9438381	1
9	PAOZZ	11862	22506637	.ADAPTER	1
10	PAOZZ	11862	14057219	VALVE,VACUUM REGULA	1
11	PAOZZ	96906	MS51869-28	SCREW,TAPPING,THREA	2

END OF FIGURE

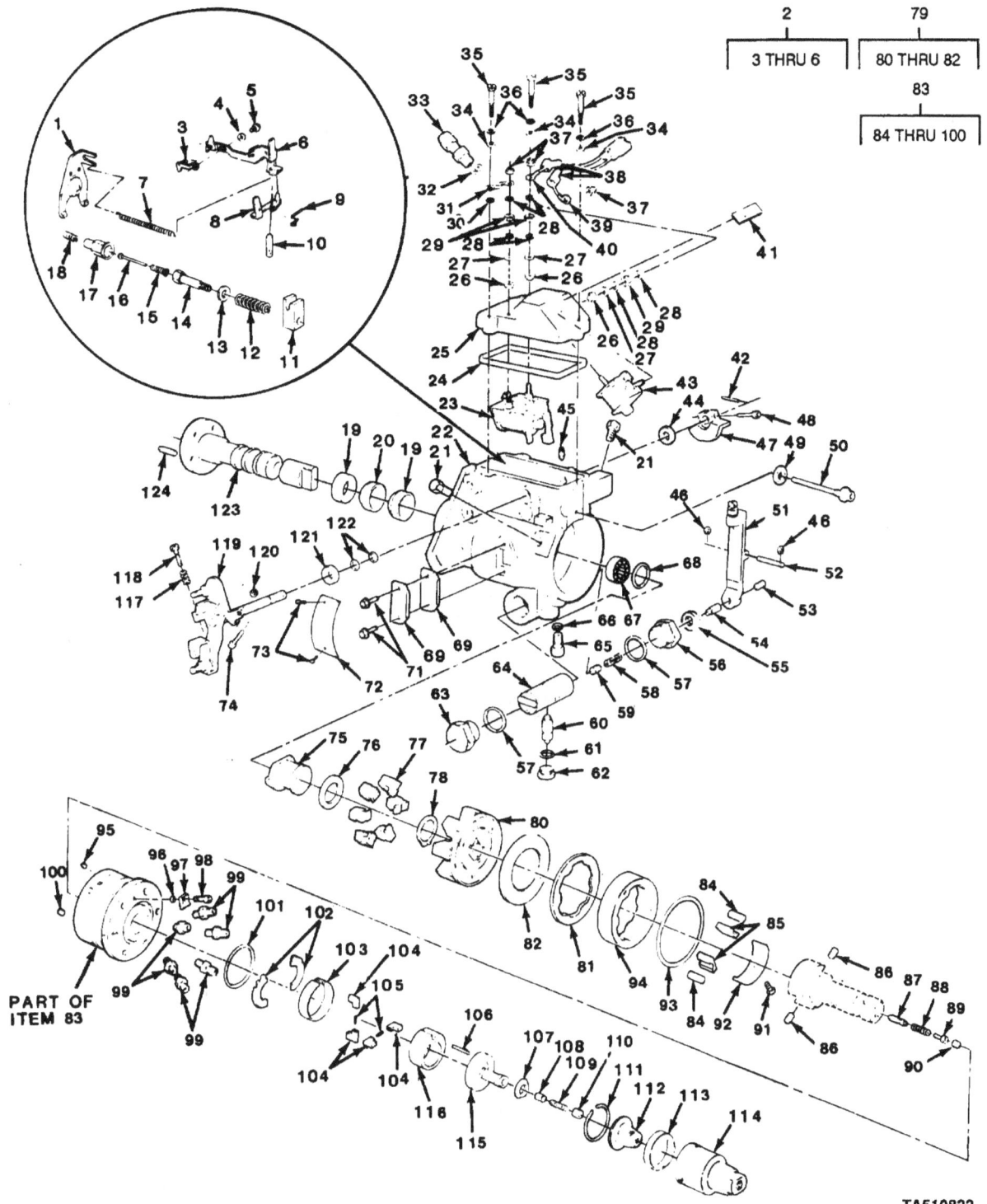

FIGURE 15. INJECTOR PUMP COMPONENT PARTS.

TA510822

(1) (2) (3) (4) ITEM SMR PART NO CODE CAGEC NUMBER	(5) DESCRIPTION AND USABLE ON CODES (UOC)	(6) QTY
	GROUP 0302 FUEL PUMPS	
	FIG. 15 INJECTOR PUMP COMPONENT PARTS	
1 PAHZZ 84760 23461	SHIFTER FORK	1
2 PAHHH 84760 18021	HOOK ASSEMBLY,GOVER	1
3 PAHZZ 84760 18020	.CONNECTING LINK,RIG	1
4 PAHZZ 84760 12362	.WASHER,SPRING TENSI	1
5 PAHZZ 84760 12360	.SCREW,MACHINE	1
6 PAHZZ 84760 12358	.LINK FUEL INJECTION	1
7 PAHZZ 84760 23643	SPRING,HELICAL,EXTE	1
8 PAHZZ 84760 21895	ARM,METERING VALVE	1
9 PAHZZ 84760 21917	SPRING	1
10 PAHZZ 84760 20849	VALVE,METERING OPTIONAL WITH P/N 22256 (STD. USE AS REQUIRE	1
10 PAHZZ 84760 22256	PIN,METERING OPTIONAL WITH P/N 20849 (O.S. USE AS REQUIRE	1
11 PAHZZ 84760 21763	BLOCK,PUMP,INJECTIO	1
12 PAHZZ 84760 22813	SPRING,HELICAL,COMP	1
13 PAHZZ 84760 22325	WASHER,FLAT	1
14 PAHZZ 84760 22326	ROD,GOVERNOR	1
15 PAHZZ 84760 10541	SPRING,HELICAL,COMP	1
16 PAHZZ 84760 22064	PIN,SHOULDER,HEADED	1
17 PAHZZ 84760 22327	SLEEVE,ADJUSTING,GO	1
18 PAHZZ 84760 22125	SPRING,HELICAL,COMP (BLUE)	1
19 PAHZZ 84760 10453	PACKING,PREFORMED	2
20 PAHZZ 84760 21860	PACKING,PREFORMED	1
21 PAHZZ 84760 11331	BOLT,MACHINE	2
22 XAHZZ 84760 27002	HOUSING,INJECTION P	1
22 XDHZZ 84760 27015	HOUSING,FUEL PUMP (.002" O.S.) PISTON BORE	1
23 PAFZZ 84760 21323	SOLENOID (OPTIONAL WITH P/N 21584) PART OF KIT P/N 26214	1
23 PAFZZ 84760 21323	SOLENOID (OPTIONAL WITH P/N 21323) PART OF KIT P/N 26214	1
24 PAFZZ 84760 27244	GASKET	1
25 PAFZZ 84760 22851	COVER,GOVERNOR CONT	1
26 PAHZZ 84760 12500	WASHER,SHOULDERED PART OF KIT P/N 26214	3
27 PAHZZ 22787 10-9858	WASHER,FLAT PART OF KIT P/N 26214	3
28 PAHZZ 96906 MS35338-38	WASHER,LOCK PART OF KIT P/N 26214	6
29 PAHZZ 96906 MS35649-282	NUT,PLAIN,HEXAGON PART OF KIT P/N 26214	3
30 PAHZZ 84760 18493	WASHER,LOCK	1
31 PAHZZ 84760 20951	LEAD,IGNITION,ENGIN PART OF KIT P/N 26214	1
32 PAFZZ 84760 27607	PACKING,PREFORMED	1
33 PAHZZ 84760 23183	VALVE,CHECK	1
34 PAFZZ 84760 13521	WASHER,FLAT	3
35 PAFZZ 84760 22351	SCREW,MACHINE	3
36 PAFZZ 45152 2239H	WASHER,LOCK	3

(1) ITEM NO	(2) SMR CODE	(3) CAGEC	(4) PART NUMBER	(5) DESCRIPTION AND USABLE ON CODES (UOC)	(6) QTY
37	PAHZZ	03350	22FT832	NUT,SELF-LOCKING,HE PART OF KIT P/N 26214	3
38	PAHZZ	84760	21618	INSULATOR,ANGLE BRA	2
39	PAHZZ	84760	22985	TERMINAL,QUICK DISC PART OF KIT P/N 26214	1
40	PAHZZ	84760	24901	TERMINAL,QUICK DISC	1
41	XDHZZ	84760	24680	LABEL	1
42	PAHZZ	84760	21284	PIN,GROOVED,HEADLES	1
43	PAFZZ	84760	22840	SOLENOID,ELECTRICAL (OPTIONAL WITH P/N 23628)	1
43	PAFZZ	84760	23861	SOLENOID,ELECTRICAL (OPTIONAL WITH P/N 22840)	1
44	PAHZZ	84760	14408	SHIM	1
45	PAHZZ	34623	5740572	SCREW ASSY.,VENT (#1 USE AS REQUIRED TO ACHIEVE PROPER FUEL F	1
45	PAHZZ	84760	21661	SCREW,ASSY. VENT (#2 USE AS REQUIRED TO ACHIEVE PROPER FUEL F	1
45	PAHZZ	84760	21662	SCREW ASSY.,VENT (#3 USE AS REQUIRED TO ACHIEVE PROPER FUEL F	1
45	PAHZZ	84760	21663	SCREW ASSY. VENT (#4 USE AS REQUIRED TO ACHIEVE PROPER FUEL F	1
45	PAHZZ	84760	21664	SCREW ASSY. VENT (#5 USE AS REQUIRED TO ACHIEVE PROPER FUEL F	1
45	PAHZZ	84760	21665	SCREW ASSY. VENT (#6 USE AS REQUIRED TO ACHIEVE PROPER FUEL F	1
45	PAHZZ	84760	22733	SCREW,ASSEMBLY PANE (#7 USE AS REQUIRED TO ACHIEVE PROPER FUEL F	1
45	PAHZZ	84760	22734	SCREW (#8 USE AS REQUIRED TO ACHIEVE PROPER FUEL F	1
46	PAHZZ	84760	23566	RING,RETAINING	2
47	PAHZZ	84760	23428	CAM FACE,PUMP	1
48	PAHZZ	84760	22398	SCREW,MACHINE	1
49	XDHZZ	84760	27600	WASHER	1
50	PAHZZ	84760	23352	STUD	1
51	PAHZZ	84760	22721	LEVER,REMOTE CONTRO	1
52	PAHZZ	84760	22397	PIN,GROOVED,HEADLES	1
53	PAHZZ	84760	22642	SETSCREW	1
54	PAHZZ	84760	22917	PLUNGER,DETENT	1
55	PAHZZ	84760	27163	SEAL	1
56	PAHZZ	84760	22693	PLUG,MACHINE THREAD	1
57	PAHZZ	84760	27609	SEAL,DRAIN PLUG	2
58	PAHZZ	84760	23171	SPRING,HELICAL,COMP	1
59	PAHZZ	84760	22367	VALVE,SERVO ADVANCE	1
60	PAHZZ	84760	23925	PIN,CAM ADVANCE	1
61	XDHZZ	84760	27610	SEAL	1
62	PAHZZ	84760	23056	PLUG,MACHINE THREAD	1
63	PAFZZ	84760	23426	PLUG,MACHINE THREAD	1
64	PAHZZ	84760	24433	PISTON ASSEMBLY STD	1
64	PAHZZ	84760	24434	PISTON ASSEMBLY (.002" O.S.)	1
65	PAHZZ	84760	24566	SCREW	1
66	XDHZZ	84760	27602	SEAL	1
67	PAHZZ	84760	23101	BEARING,ROLLER,NEED	1

(1)	(2)	(3)	(4)	(5)	(6)
ITEM NO	SMR CODE	CAGEC	PART NUMBER	DESCRIPTION AND USABLE ON CODES (UOC)	QTY
68	PAHZZ	84760	22937	RING,RETAINING	1
69	XDFZZ	84760	27603	GASKET	1
70	PAFZZ	84760	23107	COVER,ACCESS	1
71	PAFZZ	84760	21194	SCREW,MACHINE	2
72	XDFZZ	84760	10394	PLATE,IDENTIFICATIO	1
73	PAFZZ	84760	24419	SCREW,TAPPING	2
74	PAFZZ	84760	21712	SCREW,MACHINE	1
75	PAHZZ	84760	21312	SLEEVE,GOVERNOR THR	1
76	PAHZZ	84760	20222	BEARING,WASHER,THRU UOC:194,208,209,210,230	1
77	PAHZZ	84760	21201	WEIGHT,COUNTERBALAN	6
78	PAHZZ	96906	MS16624-1093	RING,RETAINING	1
79	PAHHH	84760	27984	RETAINER ASSEMBLY,G	1
80	PAHZZ	84760	23265	.RETAINER,GOVERNOR W	1
81	PAHZZ	84760	22935	.RING,FLEXIBLE,RETAI	1
82	PAHZZ	84760	18986	.HUB ASSEMBLY,GOVERN	1
83	PAHHH	84760	28396	PUMP,ROTARY	1
84	PAHZZ	84760	11141	.PIN,STRAIGHT,HEADLE	2
85	PAHZZ	84760	24569	.SHOE,CAM ROLLER	2
86	PAHZZ	84760	11056	.PLUNGER,ROTOR (USE AS REQUIRED AS INDICATED ON ROTOR HEAD P/N 23122)	2
86	PAHZZ	84760	11057	.PLUNGER,ROTOR,FUEL (USE AS REQUIRED AS INDICATED ON ROTOR HEAD P/N 23122)	1
86	PAHZZ	84760	11058	.PLUNGER,ROTOR (USE AS REQUIRED AS INDICATED ON ROTOR HEAD P/N 23122)	1
86	PAHZZ	84760	11059	.PLUNGER,ROTOR (USE AS REQUIRED AS INDICATED ON ROTOR HEAD P/N 23122)	1
86	PAHZZ	84760	11060	.PLUNGER,ROTOR (USE AS REQUIRED AS INDICATED ON ROTOR HEAD P/N 23122)	1
86	PAHZZ	84760	11062	.PLUNGER,ROTOR (USE AS REQUIRED AS INDICATED ON ROTOR HEAD P/N 23122)	1
86	PAHZZ	84760	11063	.PLUNGER,ROTOR (USE AS REQUIRED AS INDICATED ON ROTOR HEAD P/N 23122)	1
86	PAHZZ	84760	11064	.PLUNGER,ROTOR (USE AS REQUIRED AS INDICATED ON ROTOR HEAD P/N 23122)	1
86	PAHZZ	84760	11065	.PLUNGER,ROTOR (USE AS REQUIRED AS INDICATED ON ROTOR HEAD P/N 23122)	1
87	PAHZZ	84760	26833	.DISK,VALVE (USE AS REQUIRED AS INDICATED ON ROTOR HEAD P/N 23122)	1
87	PAHZZ	84760	26834	.DISK,VALVE (O.S. USE AS REQUIRED AS INDICATED ON ROTOR HEAD ASSY P/N 23122)	1

(1)	(2)	(3)	(4)	(5)	(6)
ITEM NO	SMR CODE	CAGEC	PART NUMBER	DESCRIPTION AND USABLE ON CODES (UOC)	QTY
88	PAHZZ	84760	26071	.SPRING,HELICAL,COMP	1
89	PAHZZ	84760	26070	.STOP,VALVE	1
90	PAHZZ	84760	23124	.SCREW	1
91	PAHZZ	84760	11175	.SCREW,SELF-LOCKING	1
92	PAHZZ	84760	23238	.SPRING,FLAT	1
93	XDHZZ	84760	27245	.SEAL	1
94	PAHZZ	84760	27833	.CAM RING,FUEL INJEC STANDARD	1
94	PAHZZ	84760	23120	.CAM,CONTROL (.006" C.S.)	1
95	PAHZZ	84760	11438	.SETSCREW	1
96	PAHZZ	84760	27601	.PACKING,PREFORMED	1
97	PAHZZ	84760	21283	.PLATE,LOCKING	1
98	PAHZZ	84760	21287	.SCREW,MACHINE	1
99	PAFZZ	84760	20727	.CONNECTOR,FUEL LINE	8
100	PAHZZ	84760	12216	.SETSCREW	1
101	XDHZZ	84760	27608	SEAL	1
102	PAHZZ	84760	20530	RETAINER,ROTOR,TRAN	2
103	PAHZZ	84760	22988	LINER,TRANSFER PUMP UOC:194,208,209,210,230	1
104	PAHZZ	84760	20803	VANE,INJECTION PUMP STD. (USE AS REQUIRED) UOC:194,208,209,210,230	4
104	PAHZZ	84760	20512	BLADE,TRANSFER PUMP	1
105	PAHZZ	84760	15699	SPRING,HELICAL,COMP	2
106	PAHZZ	84760	19837	PIN	1
107	PAHZZ	84760	19832	SEAL RING,METAL	1
108	PAHZZ	84760	19895	PISTON,FUEL REGULAT	1
109	PAHZZ	84760	21198	SPRING,HELICAL,COMP	1
110	PAHZZ	84760	15228	PLUG ASSEMBLY,END P	1
111	PAHZZ	84760	20529	RING,RETAINING	1
112	PAHZZ	84760	20527	FILTER ELEMENT,FLUI	1
113	PAHZZ	84760	20523	PLATE,PUMP PRESSURE	1
114	PAHZZ	84760	21296	CAP,TRANSFER PUMP	1
115	PAHZZ	84760	21200	REGULATOR ASSEMBLY	1
116	PAHZZ	84760	20528	RING,RETAINING	1
117	PAFZZ	84760	21358	SPRING,HELICAL,COMP	1
118	PAFZZ	84760	21646	SCREW,CAP,HEXAGON H	1
119	PAHZZ	84760	24345	SHAFT ASSEMBLY,THRO	1
120	PAFZZ	96906	MS35650-302	NUT,PLAIN,HEXAGON	1
121	PAHZZ	84760	22900	SPACER,SLEEVE	1
122	XDHZZ	84760	24585	PACKING,PREFORMED (USE WITH SHAFT P/N 24625)	2
123	PAHZZ	84760	62538	SHAFT,SHOULDERED	1
123	PAHZZ	84760	24623	SHAFT,DRIVE (USE WITH PIN P/N 23685)	1
124	PAHZZ	84760	23685	PIN,STRAIGHT,HEADLE (USE WITH O.S. SHAFT P/N 24625)	1
124	PAHZZ	84760	23100	PIN,STRAIGHT,HEADLE (USE WITH SHAFT P/N 24623)	1

END OF FIGURE

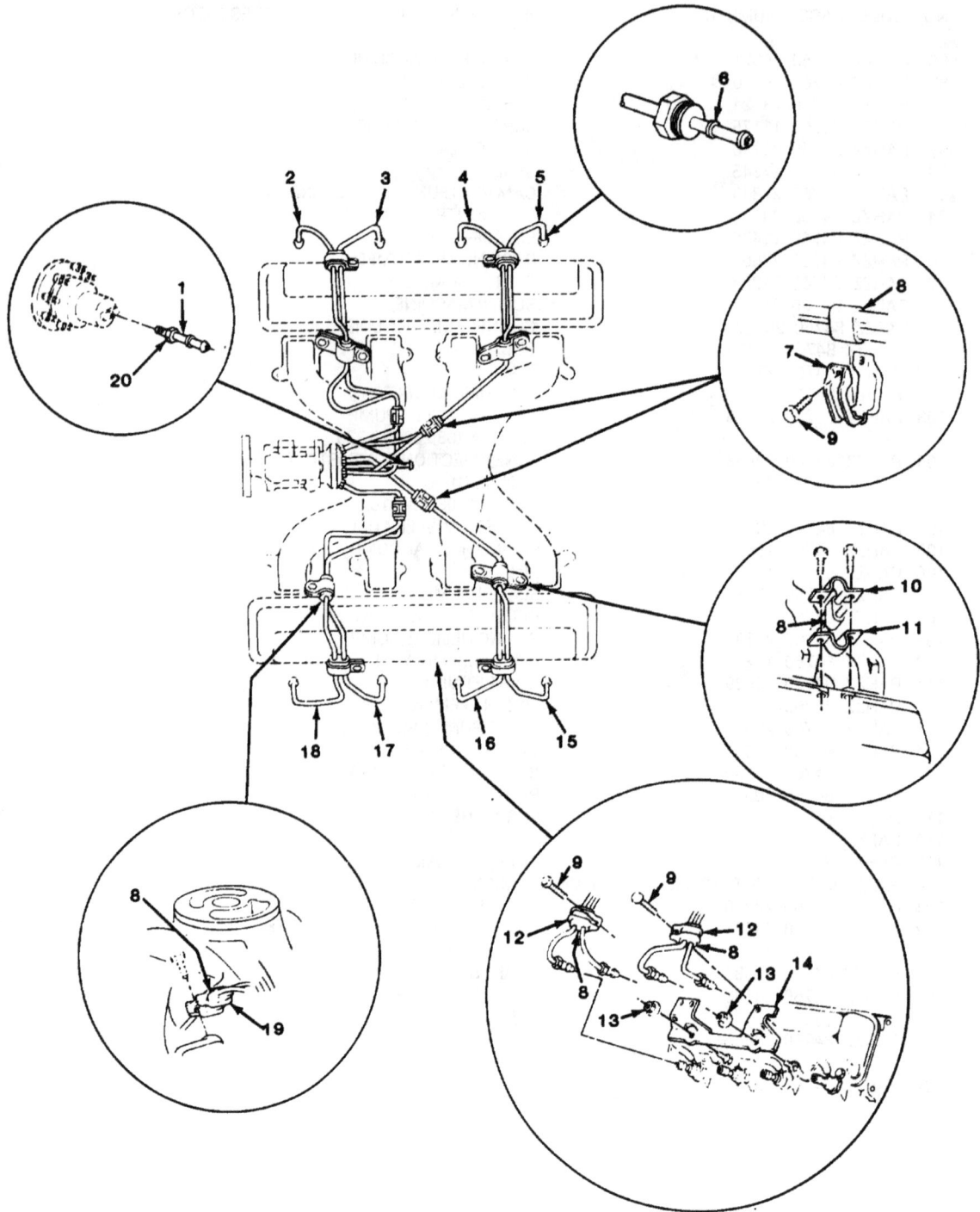

FIGURE 16. INJECTOR LINES AND RELATED PARTS.

TA510823

(1)	(2)	(3)	(4)	(5)	(6)
ITEM	SMR		PART		
NO	CODE	CAGEC	NUMBER	DESCRIPTION AND USABLE ON CODES (UOC)	QTY

GROUP 0302 FUEL PUMPS

FIG. 16 INJECTOR LINES AND RELATED PARTS

1	MFFZZ	11862	14063339	TUBE ASSEMBLY,METAL (2.76" LG.)	1
				MAKE FROM TUBE, P/N 603827	
2	PAFZZ	11862	14033912	TUBE ASSEMBLY,METAL	1
3	PAFZZ	11862	14033914	TUBE ASSEMBLY,METAL	1
4	PAFZZ	11862	14033916	TUBE ASSEMBLY,METAL	1
5	PAFZZ	11862	14033918	TUBE ASSEMBLY,METAL	1
6	PAFZZ	11862	14033920	SPACER,SLEEVE	8
7	PAOZZ	11862	14033955	CLAMP,LOOP	4
8	PAOZZ	11862	560614	GROMMET,NONMETALLIC	12
9	PAOZZ	11862	11503617	SCREW,TAPPING,THREA	8
10	PAOZZ	11862	14033921	STRAP,RETAINING	3
11	PAOZZ	11862	14033922	STRAP,RETAINING	3
12	PAOZZ	11862	560613	CLAMP,LOOP	4
13	PAOZZ	11862	22521550	NUT,PLAIN,HEXAGON	4
14	PAFZZ	11862	14033824	BRACKET,DOUBLE ANGL	2
15	PAFZZ	11862	14033917	TUBE ASSEMBLY,METAL	1
16	PAFZZ	11862	14033915	TUBE ASSEMBLY,METAL	1
17	PAFZZ	11862	14033913	TUBE ASSEMBLY,METAL	1
18	PAFZZ	11862	14033911	TUBE ASSEMBLY,METAL	1
19	PAOZZ	11862	14033953	CLAMP,LOOP	1
20	PAFZZ	24617	137397	INVERTED NUT,TUBE C	1

END OF FIGURE

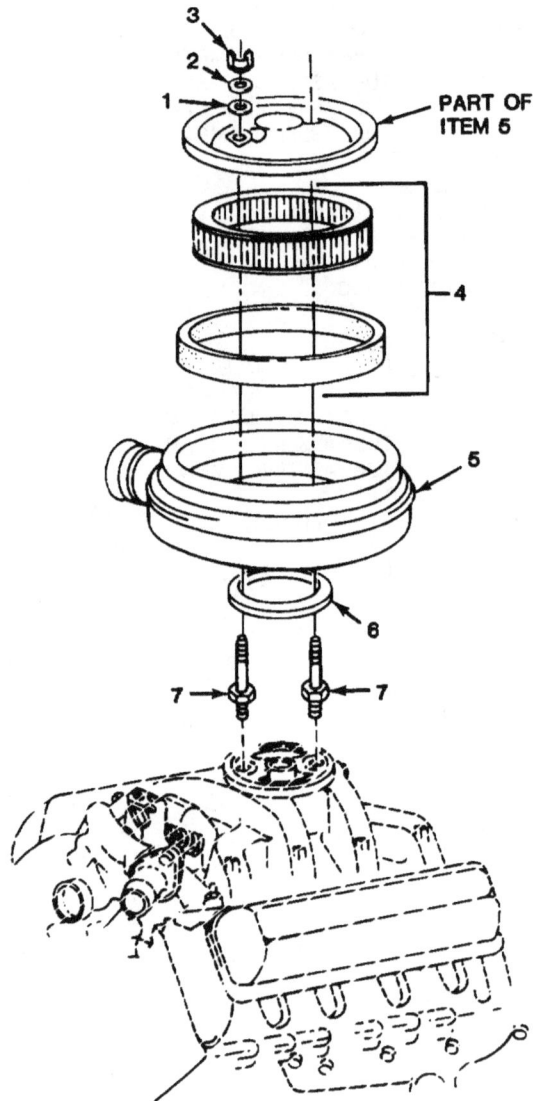

TA510824

FIGURE 17. AIR CLEANER.

(1) ITEM NO	(2) SMR CODE	(3) CAGEC	(4) PART NUMBER	(5) DESCRIPTION AND USABLE ON CODES (UOC)	(6) QTY
				GROUP 0304 AIR CLEANER	
				FIG. 17 AIR CLEANER	
1	PAOZZ	11862	3827499	WASHER,FLAT	2
2	PAOZZ	11862	3790768	WASHER,FLAT	2
3	PAOZZ	11862	14001197	NUT,PLAIN,WING	2
4	PAOZZ	70040	A644C	FILTER ELEMENT,INTA	1
5	PAOZZ	11862	25041910	AIR CLEANER,INTAKE INCLUDES ITEM #4	1
6	PAOZZ	11862	15530620	GROMMET,NONMETALLIC	1
7	PAOZZ	11862	14033948	STUD,SHOULDERED AND	2

END OF FIGURE

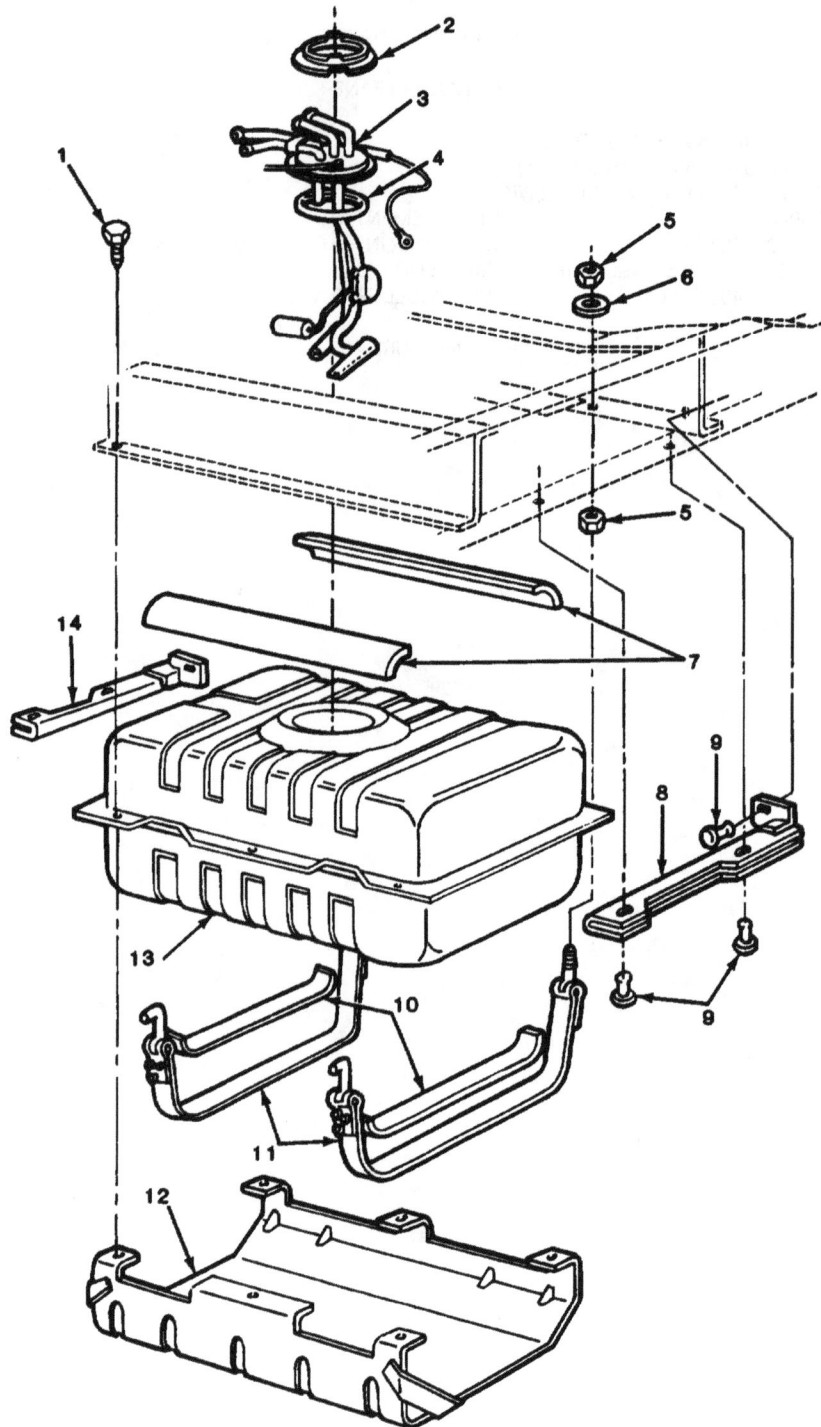

TA510825

FIGURE 18. FUEL TANK AND RELATED PARTS (M1009).

(1) (2) (3) (4) ITEM SMR NO CODE CAGEC	PART NUMBER	(5) DESCRIPTION AND USABLE ON CODES (UOC)	(6) QTY
		GROUP 0306 TANKS, LINES, FITTINGS, HEADERS	
		FIG. 18 FUEL TANK AND RELATED PARTS (M1009)	
1 PAOZZ 96906	MS90728-59	SCREW,CAP,HEXAGON H UOC:209	6
2 PAOZZ 11862	22516548	CAM,FUEL SENDER UOC:209	1
3 PAOZZ 11862	25004140	GAGE,CAS FLOW INCLUDES ITEM #4 UOC:209	1
4 PAOZZ 11862	22515965	GASKET UOC:209	12
5 PAOZZ 96906	MS51967-12	NUT,PLAIN,HEXAGON UOC:209	4
6 PAOZZ 96906	MS27183-15	WASHER,FLAT UOC:209	2
7 PAOZZ 11862	6263877	FELT,MECHANICAL,PRE UOC:209	2
8 PAOZZ 11862	14020491	GUARD ASSEMBLY,FUEL LEFT UOC:209	1
9 PAOZZ 11862	480534	RETAINER,PANEL HOLE UOC:209	6
10 PAOZZ 11862	6260631	ANTISQUEAK,FUEL TAN UOC:209	2
11 PAOZZ 11862	334675	STRAP,RETAINING UOC:209	2
12 PAOZZ 11862	368752	SHIELD ASSEMBLY,FUE UOC:209	1
13 PAOZZ 11862	14050685	TANK,FUEL,ENGINE UOC:209	1
14 PAOZZ 11862	14020492	BRACKET,ENGINE ACCE RIGHT UOC:209	1

END OF FIGURE

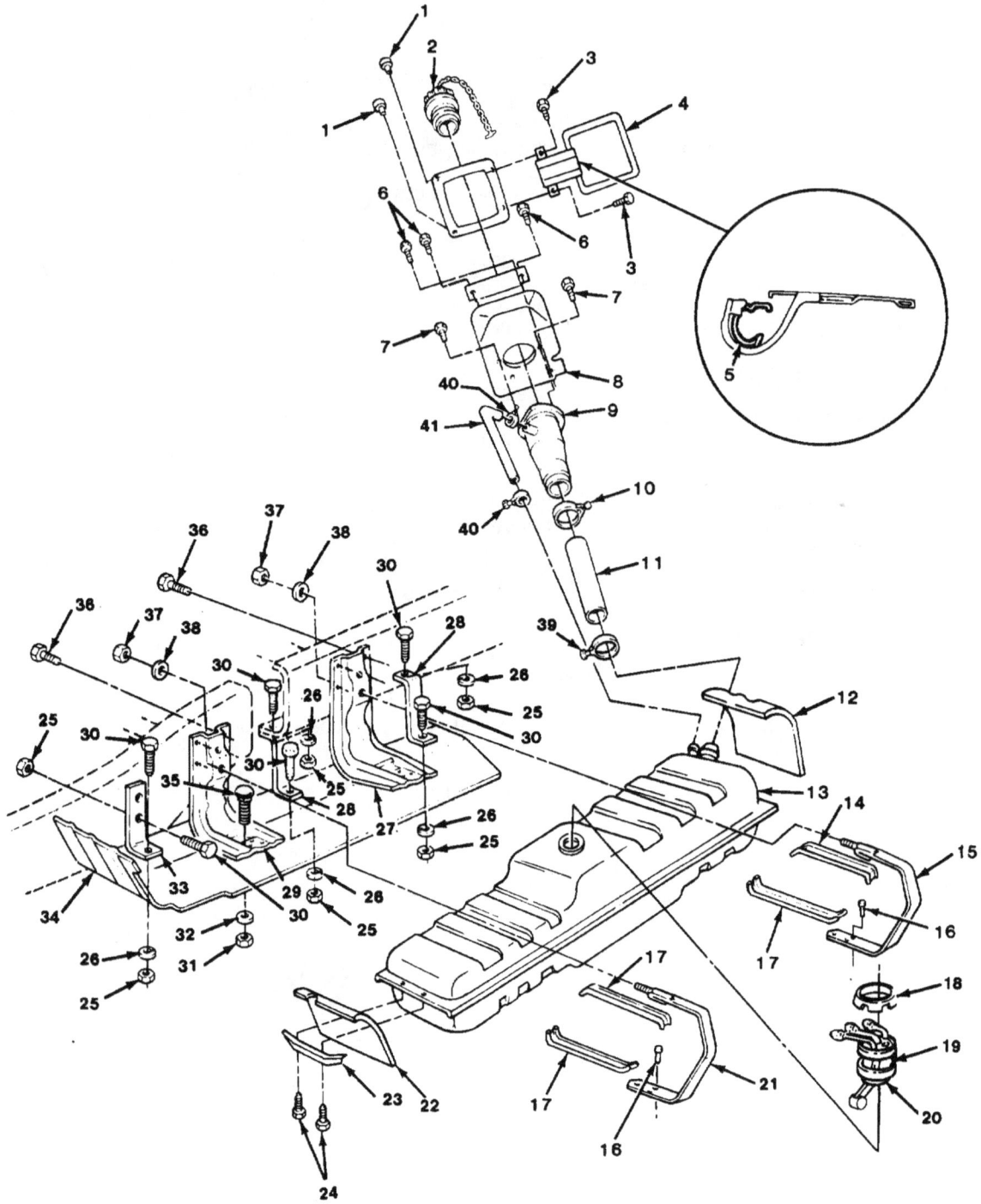

FIGURE 19. FUEL TANK AND RELATED PARTS (ALL EXCEPT M1009).

TA510826

(1) ITEM NO	(2) SMR CODE	(3) CAGEC	(4) PART NUMBER	(5) DESCRIPTION AND USABLE ON CODES (UOC)	(6) QTY
				GROUP 0306 TANKS,LINES,FITTINGS, HEADERS	
				FIG. 19 FUEL TANK AND RELATED PARTS (ALL EXCEPT M1009)	
1	PAOZZ	11862	14010707	BUMPER,NONMETALLIC UOC:194,208,230,252,254,256	2
2	PAOZZ	11862	14063326	CAP,FILLER OPENING UOC:194,208,210,230,231,252,254,256	1
3	PAOZZ	96906	MS51869-24	SCREW,CAP,SOCKET HE UOC:194,208,230,252	2
4	PAOZZ	11862	14052026	DOOR,ACCESS UOC:194,208,230,252,254,256	1
5	PAOZZ	11862	4813235	SPRING,FLAT UOC:194,208,230,252	1
6	PAOZZ	96906	MS51869-24	SCREW,CAP,SOCKET HE UOC:194,208,230,252,254,256	3
7	PAOZZ	11862	14026247	SCREW,ASSEMBLED WAS UOC:194,208,230,252,254,256	3
8	PAOZZ	11862	14063363	HOUSING,FILLER NECK UOC:194,208,230,252,254,256	1
9	PAOZZ	11862	14063333	FILLER NECK UOC:210,231	1
9	PAOZZ	11862	14063327	FILLER NECK UOC:194,208,230,252,254,256	1
10	PAOZZ	96906	MS35842-14	CLAMP,HOSE UOC:194,208,210,230,231,252,254,256	1
11	PAOZZ	11862	14063334	HOSE,PREFORMED UOC:210,231	1
11	PAOZZ	11862	14063328	HOSE,PREFORMED UOC:194,208,230,252,254,256	1
12	PAOZZ	11862	334523	PAD,CUSHIONING UOC:194,208,210,230,231,252,254,256	1
13	PAOZZ	11862	14071984	TANK,FUEL,ENGINE UOC:194,208,210,230,231,252,254,256	1
14	MOOZZ	11862	6263870	INSULATOR F/TNK ST (3.00"XX19.25") MAKE FROM FELT, P/N 6263877 UOC:194,208,210,230,231,252,254,256	1
15	PAOZZ	11862	6262755	STRAP ASSEMBLY,FUEL UOC:194,208,210,230,231,252,254,256	1
16	XDOZZ	24617	9414411	RIVET,SOLID UOC:194,208,210,230,231,252,254,256	6
17	MOOZZ	11862	6263871	INSULATOR F/TNK ST (3.00"X15.75") MAKE FROM FLET, P/N 6263877 UOC:194,208,210,230,231,252,254,256	3
18	PAOZZ	11862	22516548	CAM,FUEL SENDER UOC:194,208,210,230,231,252,254,256	1
19	PAOZZ	11862	22515965	GASKET UOC:194,208,210,230,231,252,254,256	1
20	PAOZZ	11862	25004137	TRANSMITTER,LIQUID INCLUDES ITEM #33	1

(1) (2) (3) (4) ITEM SMR PART NO CODE CAGEC NUMBER	(5) DESCRIPTION AND USABLE ON CODES (UOC)	(6) QTY
	UOC:194,208,210,230,231,252,254,256	
21 PAOZZ 11862 359847	BAND,RETAINING	1
	UOC:194,208,210,230,231,252,254,256	
22 PAOZZ 11862 334521	SHIELD,FUEL TANK,FR	1
	UOC:194,208,210,230,231,252,254,256	
23 PAOZZ 11862 334522	BRACKET,ANGLE	1
	UOC:194,208,210,230,231,252,254,256	
24 PAOZZ 96906 MS51850-86	SCREW,TAPPING,THREA	2
	UOC:194,208,210,230,231,252,254,256	
25 PAOZZ 96906 MS51967-6	NUT,PLAIN,HEXAGON	7
	UOC:194,208,210,230,231,252,254,256	
26 PAOZZ 96906 MS35338-45	WASHER,LOCK	5
	UOC:194,208,210,230,231,252,254,256	
27 PAOZZ 11862 341287	SUPPORT ASSEMBLY,FU	1
	UOC:194,208,210,230,231,252,254,256	
28 PAOZZ 11862 344714	BRACKET,DOUBLE ANGL	2
	UOC:194,208,210,230,231,252,254,256	
29 PAOZZ 11862 467525	SUPPORT ASSEMBLY,FU	1
	UOC:194,208,210,230,231,252,254,256	
30 PAOZZ 96906 MS90728-33	BOLT,MACHINE	7
	UOC:194,208,210,230,231,252,254,256	
31 PAOZZ 24617 9422295	NUT,SELF-LOCKING,CO	2
	UOC:194,208,210,230,231,252,254,256	
32 PAOZZ 96906 MS27183-12	WASHER,FLAT	2
	UOC:194,208,210,230,231,252,254,256	
33 PAOZZ 11862 15599221	BRACKET,ANGLE	1
	UOC:194,208,210,230,231,252,254,256	
34 PAOZZ 11862 474955	SHIELD,FUEL TANK	1
	UOC:194,208,210,230,231,252,254,256	
35 PAOZZ 11862 14072666	BOLT,RIBBED NECK	2
	UOC:194,208,210,230,231,252,254,256	
36 PAOZZ 96906 MS90728-59	SCREW,CAP,HEXAGON H	8
	UOC:194,208,210,230,231,252,254,256	
37 PAOZZ 96906 MS51967-12	NUT,PLAIN,HEXAGON	2
	UOC:194,208,210,230,231,252,254,256	
38 PAOZZ 96906 MS27183-15	WASHER,FLAT	2
	UOC:194,208,210,230,231,252,254,256	
39 PAOZZ 96906 MS35842-13	CLAMP,HOSE	1
	UOC:194,208,210,230,231,252,254,256	
40 PAOZZ 96906 MS35842-11	CLAMP,HOSE	2
	UOC:194,208,210,230,231,252,254,256	
41 PAOZZ 11862 14063335	HOSE,PREFORMED	1
	UOC:210,231	
41 PAOZZ 11862 14036751	HOSE,PREFORMED	1
	UOC:194,208,230,252,254,256	

END OF FIGURE

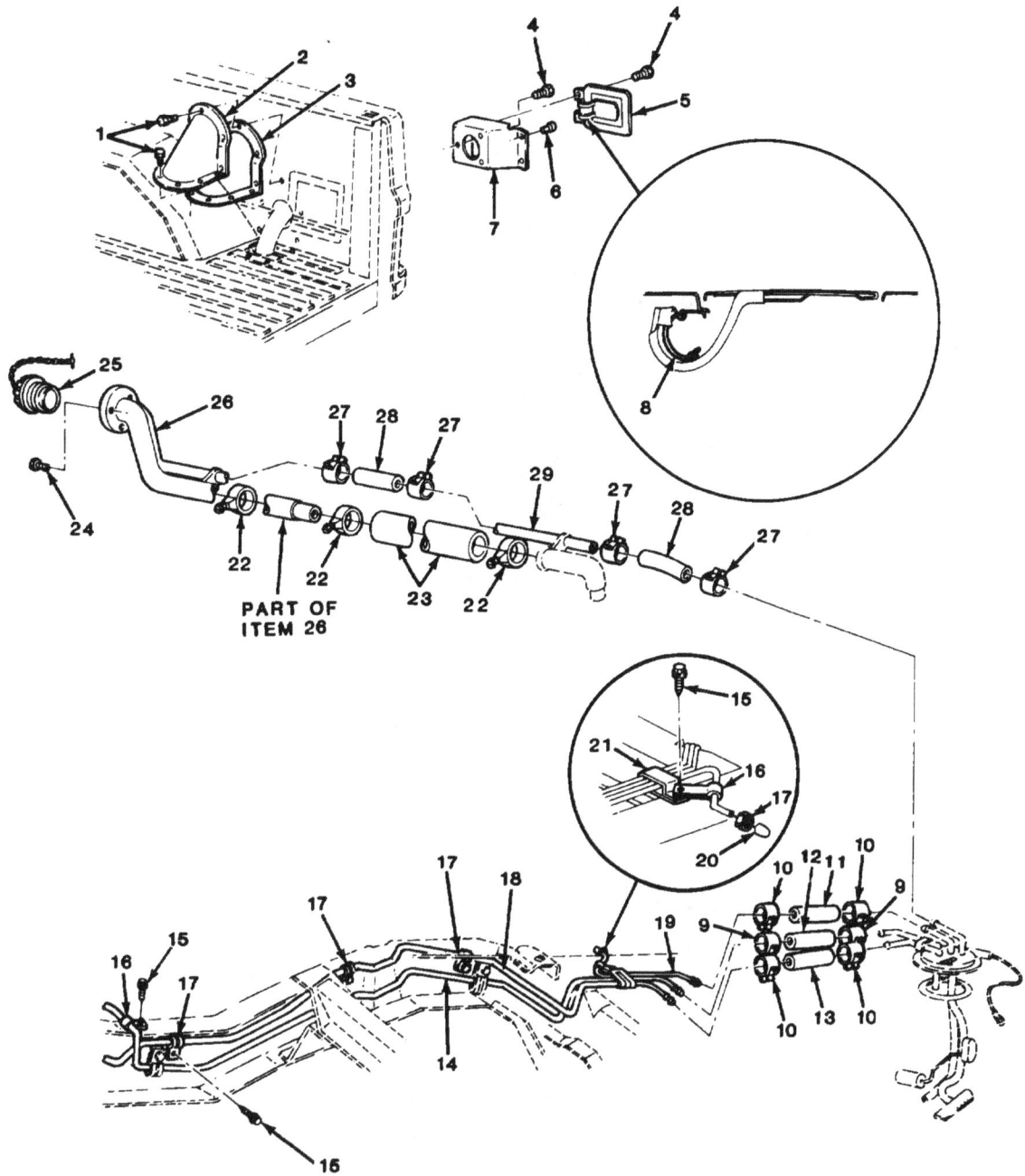

TA510827

FIGURE 20. REAR FUEL LINES AND RELATED PARTS (M1009).

(1) (2) (3) (4)	(5)	(6)
ITEM SMR PART		
NO CODE CAGEC NUMBER	DESCRIPTION AND USABLE ON CODES (UOC)	QTY

GROUP 0306 TANKS,LINES,FITTINGS, HEADERS

FIG. 20 REAR FUEL LINES AND RELATED PARTS (M1009)

1	PAOZZ 24617 9419327	SCREW	11
		UOC:209	
2	PAOZZ 11862 476916	COVER,FILLER NECK	1
		UOC:209	
3	PAOZZ 11862 476927	GASKET	1
		UOC:209	
4	PAOZZ 96906 MS51869-24	SCREW,CAP,SOCKET HE	5
		UOC:209	
5	PAOZZ 11862 14052026	DOOR,ACCESS	1
		UOC:209	
6	PAOZZ 11862 14010707	BUMPER,NONMETALLIC	2
		UOC:209	
7	PAOZZ 11862 14063329	HOUSING,FILLER PIPE	1
		UOC:209	
8	PAOZZ 11862 4813235	SPRING,FLAT	1
		UOC:209	
9	PAOZZ 19207 11608950-4	CLAMP,HOSE	3
		UOC:209	
10	PAOZZ 11862 25518880	CLAMP,HOSE	4
		UOC:209	
11	MOOZZ 11862 9439068	HOSE FUEL TANK DRAI (7.48"LG) MAKE FROM HOSE,P/N 9439104	1
		UOC:209	
12	MOOZZ 11862 9439128	HOSE FU FEED REAR (8.27"LG) MAKE FROM HOSE,P/N 9439162	1
		UOC:209	
13	MOOZZ 11862 9439010	HOSE,FUEL RET RR (7.48"LG) MAKE FROM HOSE,P/N 9439046	1
		UOC:209	
14	MOOZZ 11862 14018658	PIPE ASM-FUEL FEED (78.69"LG) MAKE FROM TUBE, P/N 3750950	1
		UOC:209	
15	PAOZZ 11862 11504447	SCREW,TAPPING,THREA	5
		UOC:209	
16	PAOZZ 11862 1638274	CLAMP,LOOP	2
		UOC:209	
17	PAOZZ 96906 MS21333-98	CLAMP,LOOP	3
		UOC:209	
18	MFOZZ 11862 14018647	PIPE FUEL RTN REAR (73.44"LG) MAKE FROM TUBE, P/N 603827	1
		UOC:209	
19	MFOZZ 11862 14063317	PIPE ASM F/T DRAIN (17.05"LG) MAKE FROM TUBE, P/N 1324714	1
		UOC:209	
20	PAOZZ 11862 14063319	CAP,FUEL TANK DRAIN	1
		UOC:209	

(1) (2) (3) (4)				(5)	(6)
ITEM	SMR	CAGEC	PART NUMBER	DESCRIPTION AND USABLE ON CODES (UOC)	QTY
NO	CODE				
21	PAOZZ	11862	14018630	CLAMP,LOOP	1
				UOC:209	
22	PAOZZ	96906	MS35842-13	CLAMP,HOSE	3
				UOC:209	
23	PAOZZ	11862	14049494	HOSE,NONMETALLIC (14.06"LG)	1
				UOC:209	
24	PAOZZ	11862	14026247	SCREW,ASSEMBLED WAS	3
				UOC:209	
25	PAOZZ	11862	14063326	CAP,FILLER OPENING	1
				UOC:209	
26	PAOZZ	11862	14063325	FILLER NECK	1
				UOC:209	
27	PAOZZ	96906	MS35842-11	CLAMP,HOSE	4
				UOC:209	
28	MOOZZ	11862	14041258	HOSE F/T FILL VENT (6.50"LG) MAKE FROM HOSE,P/N 9438383	2
				UOC:209	
29	PAOZZ	11862	468484	TUBE,METALLIC	1
				UOC:209	

END OF FIGURE

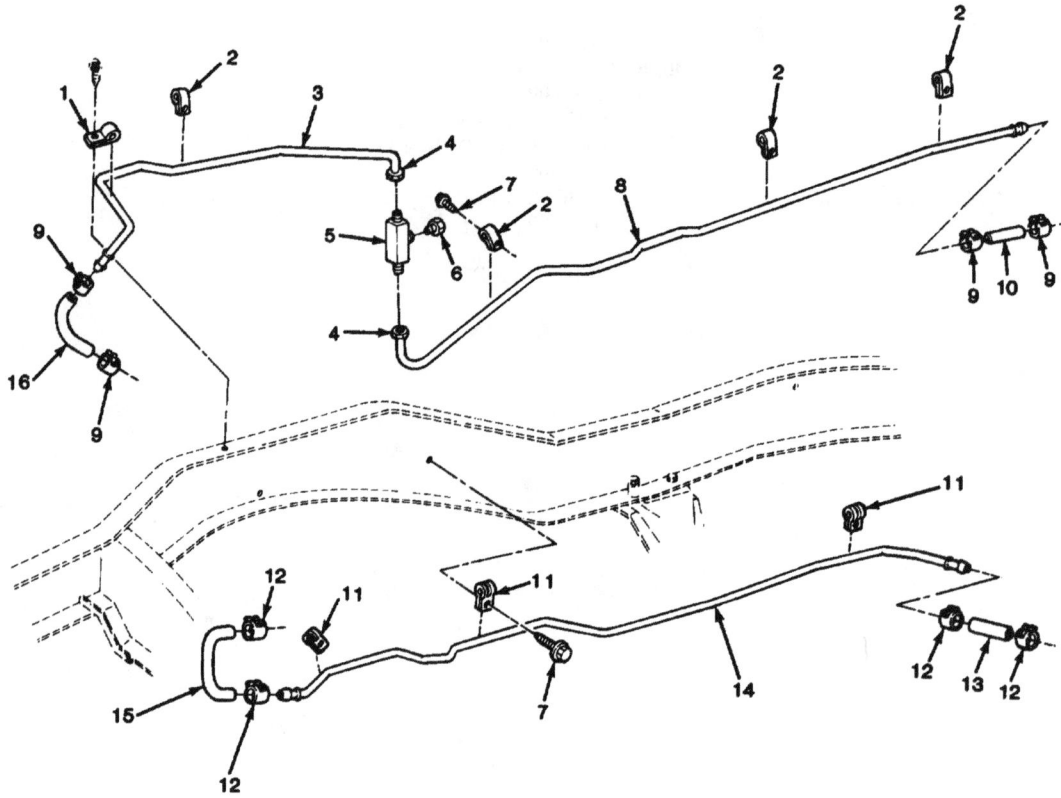

TA510828

FIGURE 21. FRONT FUEL LINES AND RELATED PARTS (M1009).

(1) (2) (3) (4) ITEM SMR PART NO CODE CAGEC NUMBER	(5) DESCRIPTION AND USABLE ON CODES (UOC)	(6) QTY
	GROUP 0306 TANKS,LINES,FITTINGS, HEADERS	
	FIG. 21 FRONT FUEL LINES AND RELATED PARTS (M1009)	
1 PAOZZ 11862 1638274	CLAMP,LOOP UOC:209	1
2 PAOZZ 96906 MS21333-111 CLAMP,LOOP	UOC:209	4
3 MOOZZ 11862 15599209	PIPE F/FEED FRONT (30.67"LG) MAKE FROM TUBE, P/N 3750950 UOC:209	1
4 PAOZZ 24617 142433	INVERTED NUT,TUBE C UOC:209	2
5 PAOZZ 72582 178917	TEE,PIPE TO TUBE UOC:209	1
6 PAOZZ 6N299 0917425	PLUG,PIPE UOC:209	1
7 PAOZZ 11862 11504447 SCREW,TAPPING,THREA	UOC:209	7
8 MOOZZ 11862 14063315	PIPE-FUEL FEED INTE (51.86"LG) MAKE FROM TUBE, P/N 3750950 UOC:209	1
9 PAOZZ 11862 477402	CLAMP,HOSE UOC:209	4
10 MOOZZ 11862 9439117	HOSE,NONMETALLIC (3.94:LG) MAKE FROM HOSE,P/N 9439162 UOC:209	1
11 PAOZZ 96906 MS21333-98 CLAMP,LOOP	UOC:209	3
12 PAOZZ 11862 25518880 CLAMP,HOSE	UOC:209	4
13 MOOZZ 11862 9439001	HOSE,FU,RET,INT (3.94"LG) MAKE FROM HOSE, P/N 9439043 UOC:209	1
14 MOOZZ 11862 474957	TUBE,BENT,METALLIC (71.45"LG) MAKE FROM TUBING-STEEL,P/N 603827 UOC:209	1
15 MOOZZ 11862 9438227	HOSE FUEL RTURN FRN (9.84"LG) MAKE FROM HOSE,P/N 9438257 UOC:209	1
16 PAOZZ 11862 14063391 HOSE,PREFORMED	UOC:209	1

END OF FIGURE

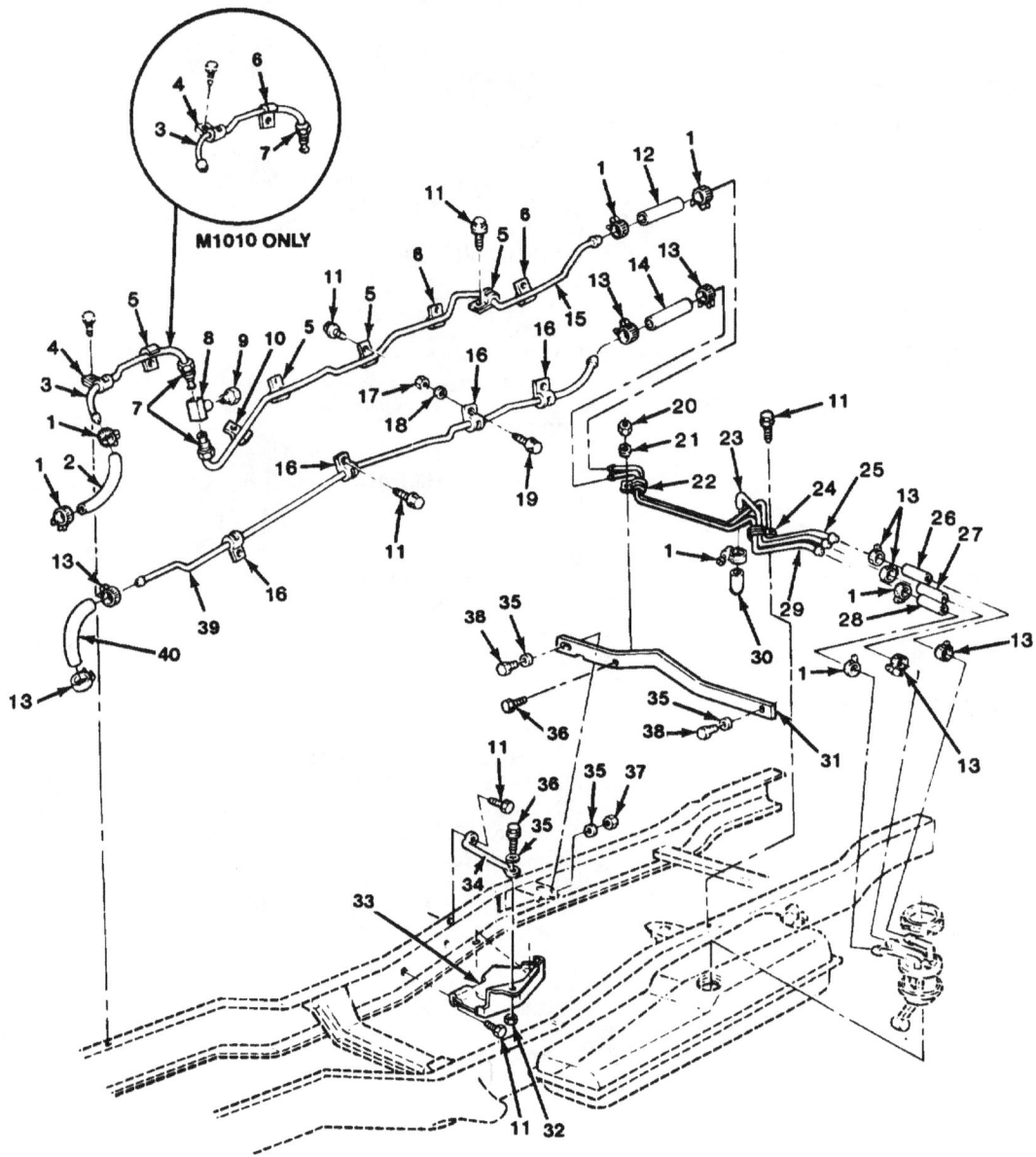

TA510829

FIGURE 22. FUEL LINES AND RELATED PARTS (ALL EXCEPT M1009).

(1) ITEM NO	(2) SMR CODE	(3) CAGEC	(4) PART NUMBER	(5) DESCRIPTION AND USABLE ON CODES (UOC)	(6) QTY
				GROUP 0306 TANKS,LINES,FITTINGS, HEADERS	
				FIG. 22 FUEL LINES AND RELATED PARTS (ALL EXCEPT M1009)	
1	PAOZZ	11862	477402	CLAMP,HOSE UOC:194,208,210,230,231,252,254,256	7
2	PAOZZ	11862	14063391	HOSE,PREFORMED UOC:194,208,210,230,231,252,254,256	1
3	MOOZZ	11862	15599209	PIPE F/FEED FRONT (30.67"LG) MAKE 1 FROM TUBE, P/N 3750950 UOC:194,208,210,230,231,252,254,256	1
4	PAOZZ	11862	1638274	CLAMP,LOOP UOC:194,208,210,230,231,252,256	1
5	PAOZZ	96906	MS21333-111	CLAMP,LOOP UOC:194,208,230,231,252,254,256	4
6	PAOZZ	96906	MS21333-111	CLAMP,LOOP UOC:210	3
7	PAOZZ	24617	142433	INVERTED NUT,TUBE C UOC:194,208,210,230,231,252,254,256	2
8	PAOZZ	72582	178917	TEE,PIPE TO TUBE UOC:194,208,210,230,231,252,254,256	1
9	PAOZZ	24617	444620	PLUG,PIPE UOC:194,208,230,231,252,254,256	1
10	PAOZZ	11862	25527423	CLAMP,LOOP UOC:210,230	1
11	PAOZZ	11862	11504447	SCREW,TAPPING,THREA UOC:194,208,210,230,231,252,254,256	12
12	MOOZZ	11862	9439117	HOSE,NONMETALLIC (3.94"LG) MAKE FROM HOSE,P/N 9439162 UOC:194,208,210,230,231,252,254,256	1
13	PAOZZ	11862	25518880	CLAMP,HOSE UOC:194,208,210,230,231,252,254,256	8
14	MOOZZ	11862	9439010	HOSE,FUEL RET RR (7.48"LG) MAKE FROM HOSE,P/N 9439046 UOC:194,208,210,230,231,252,254,256	1
15	MOOZZ	11862	14063314	PIPE-FUEL FEED INTE (57.14"LG) MAKE FROM TUBE, P/N 3750950 UOC:194,208,201,230,231,252,254,256	1
16	PAOZZ	96906	MS21333-45	CLAMP,LOOP UOC:194,20,8210,230,231,252,254,256	4
17	PAOZZ	96906	MS35691-406	NUT,PLAIN,HEXAGON UOC:194,208,210,230,231,252,254,256	1
18	PAOZZ	96906	MS35338-44	WASHER,LOCK UOC:194,208,210,230,231,252,254,256	1
9	PAOZZ	96906	MS90725-5	SCREW,CAP,HEXAGON H UOC:194,208,210,230,231,252,254,256	1
20	PAOZZ	96906	MS51967-6	NUT,PLAIN,HEXAGON UOC:194,208,210,230,231,252,254,256	1
21	PAOZZ	96906	MS35338-45	WASHER,LOCK UOC:194,208,210,230,231,252,254,256	1

(1) ITEM NO	(2) SMR CODE	(3) CAGEC	(4) PART NUMBER	(5) DESCRIPTION AND USABLE ON CODES (UOC)	(6) QTY
22	PAOZZ	11862	467524	CLAMP,LOOP UOC:194,208,210,230,231,252,254,256	1
23	MOOZZ	11862	14045605	PIPE ASM F/T DRAIN (12.25"LG) MAKE FROM TUBE, P/N 1324714 UOC:194,208,210,230,231,252,254,256	1
24	PAOZZ	11862	14034543	CLAMP,LOOP UOC:194,208,210,230,231,252,254,256	1
25	MFOZZ	11862	14061227	PIPE F/TRN HOSE RR (37.96"LG) MAKE FROM TUBE, P/N 603827 UOC:194,208,210,230,231,252,254,256	1
26	MOOZZ	11862	9439004	HOSE F/RETURN RR (5.12"LG) MAKE FROM HOSE,P/N 9439046 UOC:194,208,210,230,231,252,254,256	1
27	MOOZZ	11862	9439059	HOSE FUEL TANK DRAI (3.94"LG) MAKE FROM HOSE,P/N 9439104 UOC:194,208,210,230,231,252,254,256	1
28	MOOZZ	11862	9439120	HOSE F/FEED RR (5.12" LG) MAKE FROM HOSE,P/N 9439162 UOC:194,208,210,230,231,252,254,256	1
29	MOOZZ	11862	14061223	PIPE ASM-FUEL FEED (37.86"LG) MAKE FROM TUBE, P/N 3750950 UOC:194,208,210,230,231,252,254,256	1
30	PAOZZ	11862	14063319	CAP,FUEL TANK DRAIN UOC:194,208,210,230,231,252,254,256	1
31	PAOZZ	11862	467509	BRACKET,ANGLE UOC:194,208,210,230,231,252,254,256	1
32	PAOZZ	96906	MS17829-5C	NUT,SELF-LOCKING,HE UOC:194,208,210,230,231,252,254,256	1
33	PAOZZ	11862	15522392	SHIELD,FUEL HOSE UOC:194,208,210,230,231,252,254,256	1
34	PAOZZ	11862	14034546	BRACE,FUEL HOSE SHI UOC:194,208,210,230,231,252,254,256	1
35	PAOZZ	96906	MS27183-12	WASHER,FLAT UOC:194,208,210,230,231,252,254,256	5
36	PAOZZ	96906	MS90728-32	BOLT,MACHINE UOC:194,208,210,230,231,252,254,256	3
37	PAOZZ	24617	9422295	NUT,SELF-LOCKING,CO UOC:194,208,210,230,231,252,254,256	2
38	PAOZZ	96906	MS90728-33	BOLT,MACHINE UOC:194,208,210,230,231,252,254,256	2
39	MFOZZ	11862	15599999	PIPE FUEL RETRN FRT (74.60"LG) MAKE FROM TUBE, P/N 603827 UOC:194,208,210,230,231,252,254,256	1
40	MOOZZ	11862	9438227	HOSE FUEL RTURN FRN (9.84"LG)MAKE FROM HOSE,P/N 9438257 UOC:194,208,210,230,231,252,254,256	1

END OF FIGURE

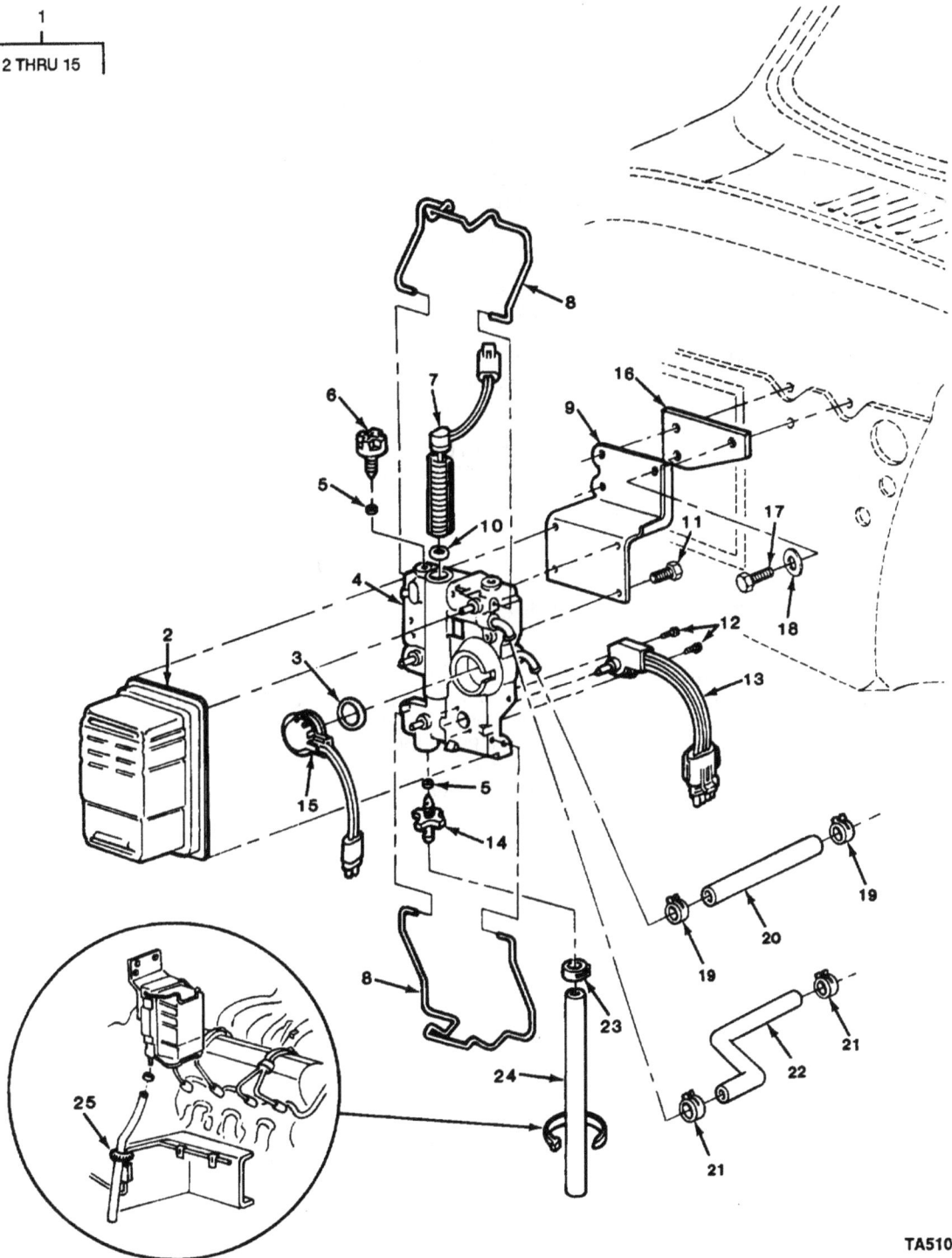

FIGURE 23. FUEL FILTER AND RELATED PARTS.

TA510830

(1) (2) (3) (4) ITEM SMR PART NO CODE CAGEC NUMBER	(5) DESCRIPTION AND USABLE ON CODES (UOC)	(6) QTY
	GROUP 0309 FUEL FILTERS	
	FIG. 23 FUEL FILTER AND RELATED PARTS	
1 PAOOO 84760 27290	FILTER,FLUID	1
2 PAOZZ 11862 14075347	.FILTER ELEMENT,FLUI	1
3 PAOZZ 84760 22591	.PACKING,PREFORMED	1
4 PAOFF 84760 24285	.HOUSING,FUEL FILTER	1
5 PAOZZ 84760 27820	.PACKING,PREFORMED	2
6 PAOZZ 84760 24267	.PLUG,VENT	1
7 PAOZZ 84760 29090	.HEATER ASSEMBLY,FUE	1
8 PAOZZ 84760 24265	.CLAMP,LOOP	2
9 PFOZZ 84760 24281	.BRACKET,ANGLE	1
10 PAOZZ 84760 15349	.PACKING,PREFORMED	1
11 PAOZZ 84760 24437	.SCREW,TAPPING,THREA	4
12 PAOZZ 84760 24322	.SCREW,TAPPING,THREA	2
13 PAOZZ 84760 27284	.SENSOR,FUEL FILTER	1
14 PAOZZ 84760 23796	.COCK,POPPET DRAIN	1
15 PAOZZ 61928 15596614	.SWITCH,PRESSURE	1
16 PFOZZ 11862 14043724	COVER,ACCESS	1
17 PAOZZ 96906 MS18154-58	SCREW,CAP,HEXAGON H	3
18 PAOZZ 96906 MS35338-46	WASHER,LOCK	3
19 PAOZZ 11862 25518880	CLAMP,HOSE	2
20 PAOZZ 11862 14063302	HOSE,PREFORMED	1
21 PAOZZ 19207 11608950-4	CLAMP,HOSE	2
22 PAOZZ 11862 14063301	HOSE,PREFORMED	1
23 PAOZZ 96906 MS35842-10	CLAMP,HOSE	1
24 MOOZZ 11862 9439092	HOSE,NONMETALLIC (25.59"LG) MAKE FROM HOSE,P/N 9439104	1
25 PAOZZ 11862 9785074	CLIP,SPRING TENSION	1

END OF FIGURE

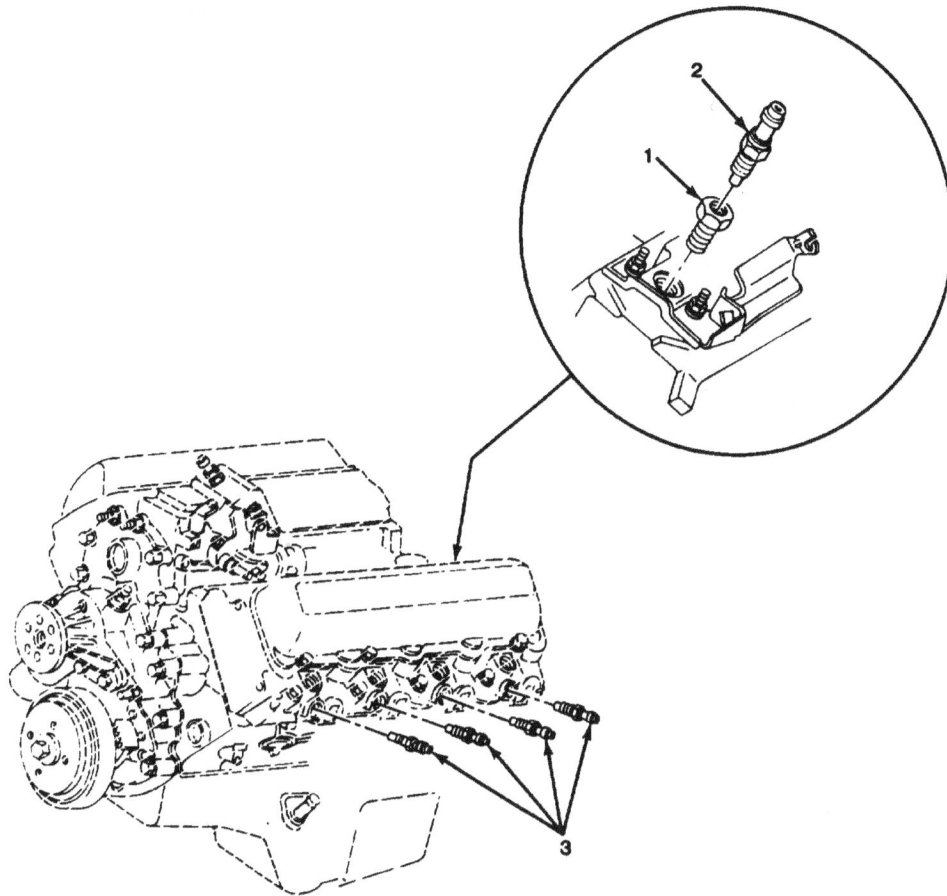

TA510831

FIGURE 24. GLOW PLUGS AND TEMPERATURE SENSOR.

(1) (2) (3) (4) ITEM SMR PART NO CODE CAGEC NUMBER	(5) DESCRIPTION AND USABLE ON CODES (UOC)	(6) QTY
	GROUP 0311 ENGINE STARTING AIDS	
	FIG. 24 GLOW PLUGS AND TEMPERATURE SENSOR	
1 PAOZZ 24617 444034 BUSHING,PIPE	UOC:208,209,210,230,231,252,254,256	1
2 PAOZZ 70040 10045847	SENSOR,TEMPERATURE INCLUDES ADAPTER FOR TWO TERMINAL SWITCH TO CONNECT TO ONE TERMINAL WIRING HARNESS	1
3 PAOZZ 11862 5613939	GLOW PLUG	8
	END OF FIGURE	

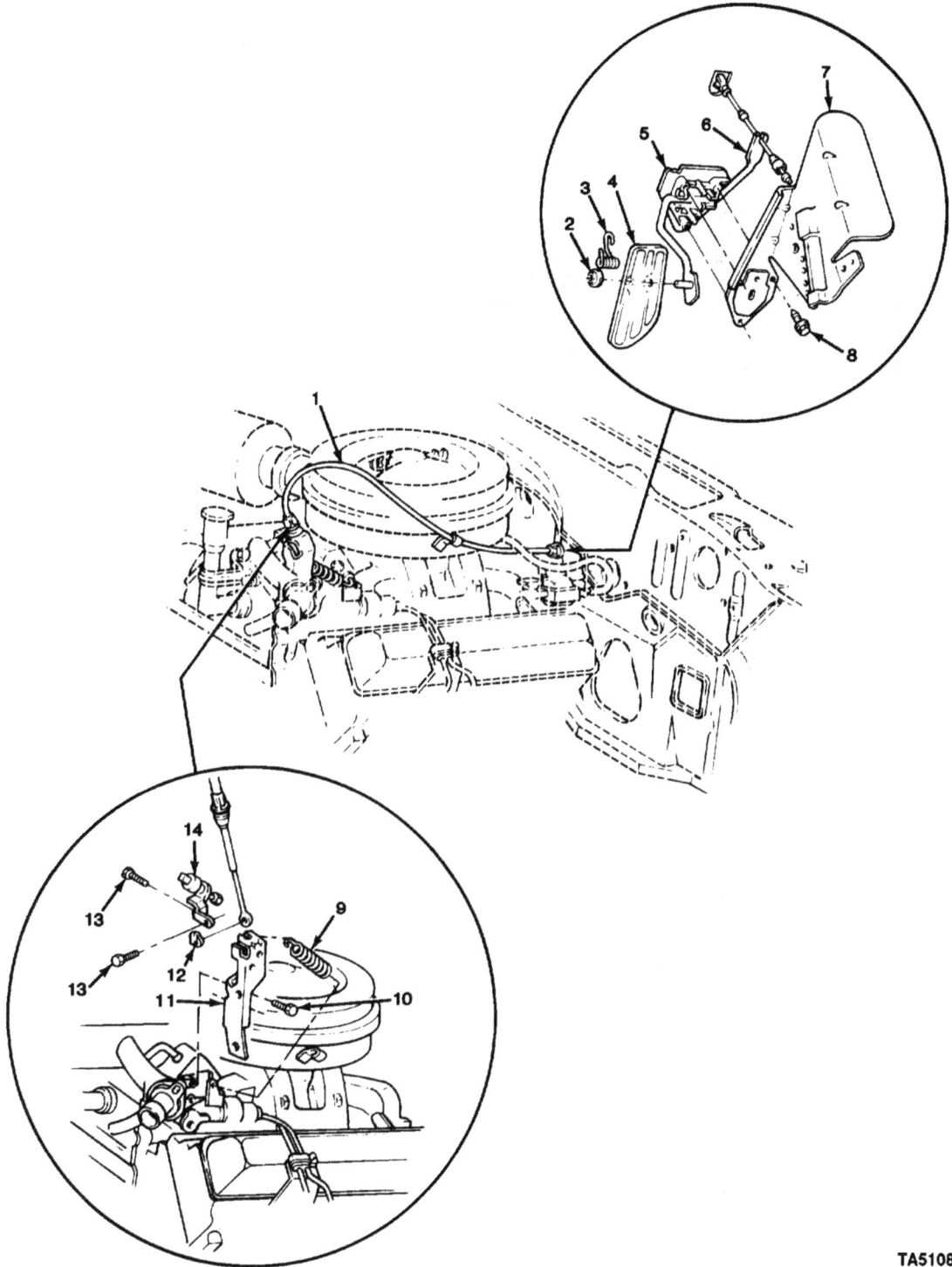

FIGURE 25. ACCELERATOR   LINKAGE AND RELATED PARTS.

TA510832

(1)	(2)	(3)	(4)	(5)	(6)
ITEM	SMR		PART		
NO	CODE	CAGEC	NUMBER	DESCRIPTION AND USABLE ON CODES (UOC)	QTY

GROUP 0312 ACCELERATOR,THROTTLE,
OR CHOKE CONTROLS

FIG. 25 ACCELERATOR LINKAGE AND
RELATED PARTS

1	PAOZZ	11862	14038644	CONTROL ASSEMBLY,PU	1
2	PAOZZ	73342	3909063	PUSH ON NUT	1
3	PAOZZ	11862	336989	SPRING,HELICAL,TORS	1
4	PAOZZ	11862	468234	PEDAL,CONTROL	1
5	PAOZZ	11862	3993087	SUPPORT,ACCELERATOR	1
6	PAOZZ	11862	15590123	LEVER,REMOTE CONTRO	1
7	PAOZZ	11862	342405	REINFORCEMENT,ACCEL	1
8	PAOZZ	24617	9422956	SCREW,TAPPING	3
9	PAOZZ	11862	14024997	SPRING,HELICAL,EXTE	1
10	PAOZZ	24617	11504986	BOLT,MACHINE	2
11	PAOZZ	11862	14038647	SUPPORT,ACCELERATOR	1
12	PAOZZ	11862	15567924	CLIP,RETAINING	1
13	PAOZZ	11862	11501095	BOLT,MACHINE	2
14	PAOZZ	11862	14066255	RELAY,ELECTROMAGNET	1

END OF FIGURE

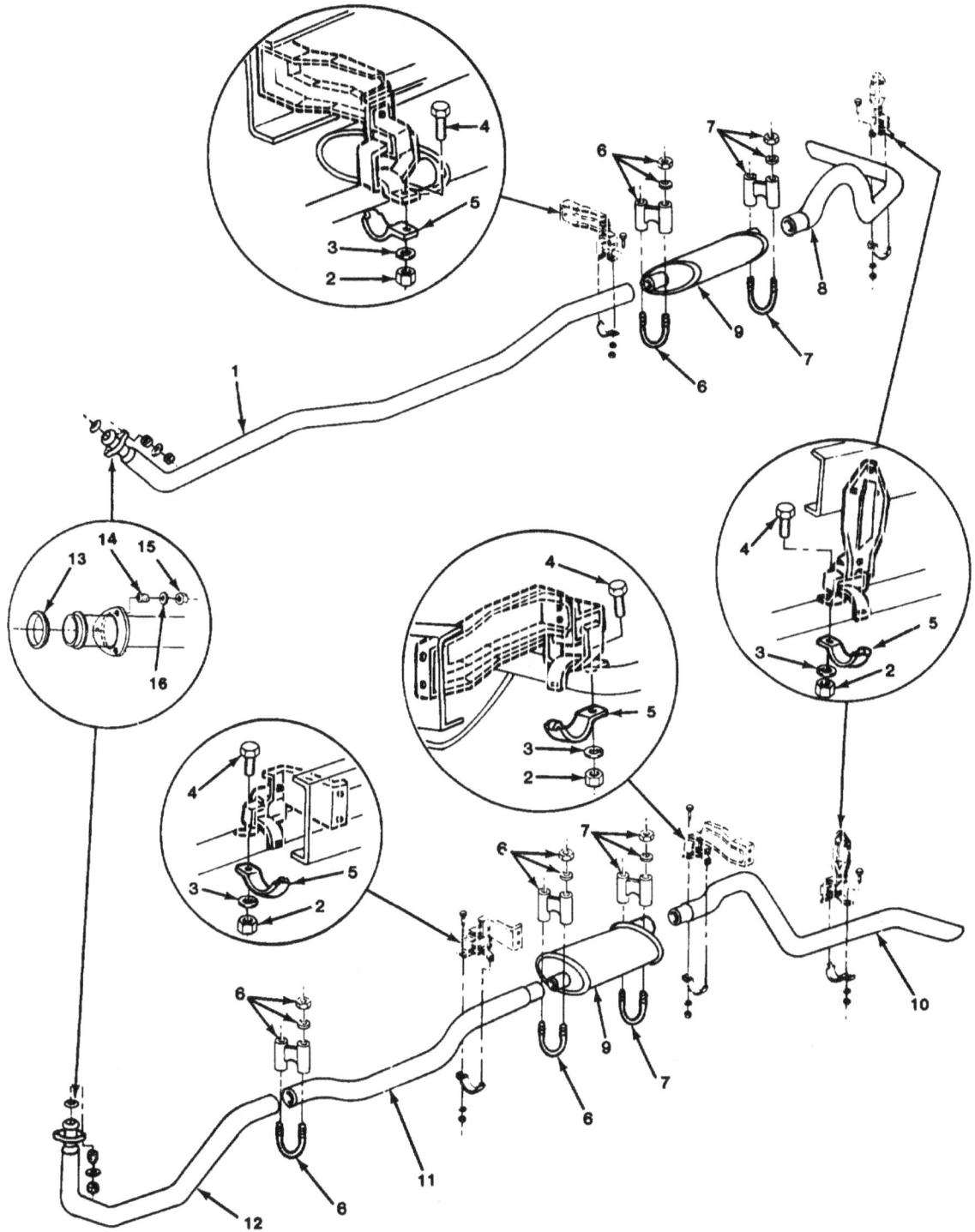

FIGURE 26. EXHAUST MUFFLERS AND PIPES (ALL EXCEPT M1009).

TA510833

(1) (2) (3) (4) ITEM SMR PART NO CODE CAGEC NUMBER	(5) DESCRIPTION AND USABLE ON CODES (UOC)	(6) QTY
	GROUP 04 EXHAUST SYSTEM	
	GROUP 0401 MUFFLER AND PIPES	
	FIG. 26 EXHAUST MUFFLERS AND PIPES (ALL EXCEPT M1009)	
1 PAOZZ 11862 14037856	PIPE,EXHAUST UOC:194,208,210,230,231,252,254,256	1
1 PAOZZ 11862 15595224	PIPE,EXHAUST APPLIES TO VEHICLES STARTING WITH VIN FF3000 UOC:194,208,210,230,231,252,254,256	1
2 PAOZZ 96906 MS51967-6 NUT,PLAIN,HEXAGON	UOC:194,208,210,230,231,252,254	5
3 PAOZZ 96906 MS35338-45 WASHER,LOCK	UOC:194,208,210,230,231,252,254,256	5
4 PAOZZ 96906 MS90728-37 BOLT,MACHINE	UOC:194,208,210,230,231,252,254	5
5 PAOZZ 11862 341160 STRAP,RETAINING	UOC:194,208,210,230,231,252,254,256	2
6 PAOZZ 96906 MS52150-31HE CLAMP,LOOP	UOC:194,208,210,230,231,252,254,256	3
7 PAOZZ 96906 MS52150-30HE CLAMP,LOOP	UOC:194,208,210,230,231,252,254,256	2
8 PAOZZ 79260 45823 PIPE,EXHAUST RIGHT REAR	UOC:194,208,210,230,231,252,254,256	1
9 PAOZZ 11862 14063795 MUFFLER,EXHAUST	UOC:194,208,21,230,231,252,254,256	2
9 PAOZZ 11862 14089132 MUFFLER,EXHAUST APPLIES TO VEHICLES STARTING WITH VIN FF3000 UOC:194,208,210,230,231,252,254,256		2
10 PAOZZ 11862 14029956 PIPE,EXHAUST LEFT REAR	UOC:194,208,210,230,231,252,254,256	1
11 PAOZZ 11862 14045525 PIPE,EXHAUST LEFT CENTER	UOC:194,208,210,230,231,252,254,256	1
11 PAOZZ 11862 15595271 PIPE,EXHAUST APPLIES TO VEHICLES STARTING WITH VIN FF3000 UOC:194,208,210,230,231,252,254,256		1
12 PAOZZ 11862 14045521 PIPE,EXHAUST LEFT FRONT	UOC:194,208,210,230,231,252,254,256	1
13 PAOZZ 11862 14072686 SEAL,EXHAUST PIPE	UOC:194,208,201,230,231,252,254,256	2
14 PAOZZ 11862 587575 SPRING,HELICAL,COMP	UOC:194,208,210,230,231,252,254,256	6
15 PAOZZ 80205 NAS1408A6 NUT,SELF-LOCKIGN,HE	UOC:194,208,210,230,231,252,254,256	6
16 PAOZZ 96906 MS27183-16 WASHER,FLAT	UOC:194,208,210,230,231,252,254,256	6

END OF FIGURE

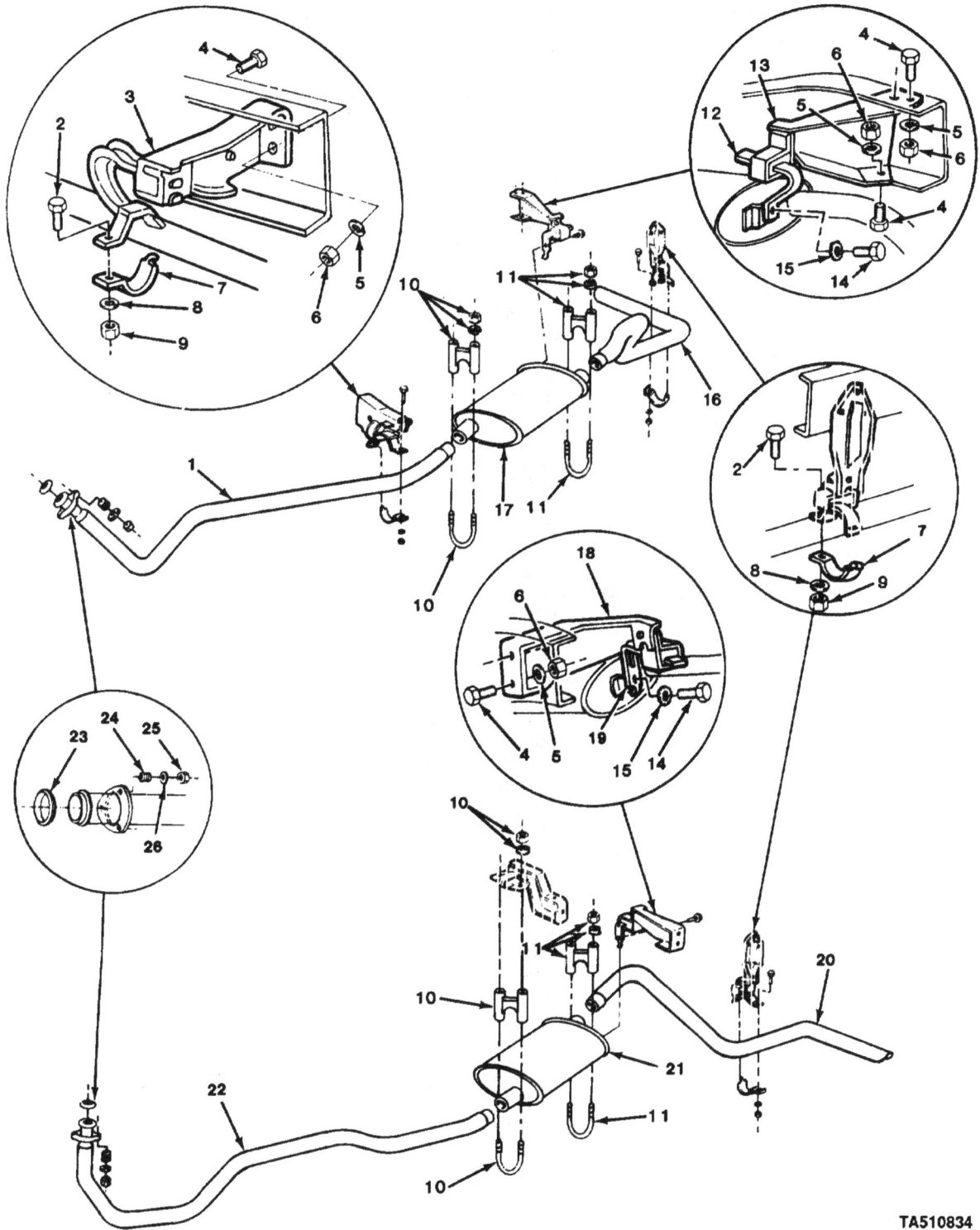

FIGURE 27. EXHAUST MUFFLERS AND PIPES (M1009).

TA510834

(1)	(2)	(3)	(4)	(5)	(6)
ITEM NO	SMR CODE	CAGEC	PART NUMBER	DESCRIPTION AND USABLE ON CODES (UOC)	QTY

GROUP 0401 MUFFLER AND PIPES

FIG. 27 EXHAUST MUFFLERS AND PIPES (M1009)

1	PAOZZ	11862	15595216	PIPE,EXHAUST (RIGHT) APPLIES TO VEHICLES BEFORE VIN FF136169 UOC:209	1
1	XDOZZ	11862	15599216	PIPE,EXHAUST (RIGHT) APPLIES TO VEHICLES STARTING WITH VIN FF136169 UOC:209	1
2	PAOZZ	96906	MS90728-37	BOLT,MACHINE UOC:209	3
3	PAOZZ	11862	14037808	BRACKET,PIPE UOC:209	1
4	PAOZZ	96906	MS18154-58	SCREW,CAP,HEXAGON H UOC:209	7
5	PAOZZ	96906	MS35338-46	WASHER,LOCK UOC:209	7
6	PAOZZ	96906	MS51967-8	NUT,PLAIN,HEXAGON UOC:209	7
7	PAOZZ	11862	341160	STRAP,RETAINING UOC:209	3
8	PAOZZ	96906	MS35338-45	WASHER,LOCK UOC:209	3
9	PAOZZ	96906	MS51967-6	NUT,PLAIN,HEXAGON UOC:209	3
10	PAOZZ	96906	MS52150-31HE	CLAMP,LOOP UOC:209	2
11	PAOZZ	96906	MS52150-30HE	CLAMP,LOOP UOC:209	2
12	PAOZZ	11862	14037836	HANGER,ENGINE EXHAU UOC:209	1
13	PAOZZ	11862	14037812	SUPPORT ASSEMBLY,MU UOC:209	1
14	PAOZZ	96906	MS90728-34	BOLT,MACHINE UOC:209	2
15	PAOZZ	96906	MS35335-34	WASHER,LOCK UOC:209	2
16	PAOZZ	11862	14044996	PIPE,EXHAUST RIGHT REAR UOC:209	1
17	PAOZZ	11862	14067430	MUFFLER AND INLET P RIGHT UOC:209	1
18	PAOZZ	11862	14034547	BRACKET,TAIL PIPE UOC:209	1
19	PAOZZ	11862	14024561	BRACKET,DOUBLE ANGL UOC:209	1
20	PAOZZ	11862	14044995	PIPE,EXHAUST LEFT REAR UOC:209	1
21	PAOZZ	11862	14067429	MUFFLER AND INLET P UOC:209	1
22	PAOZZ	11862	14067759	PIPE,EXHAUST (LEFT) APPLIES TO	1

(1) (2) (3) (4) ITEM SMR PART NO CODE CAGEC NUMBER	(5) DESCRIPTION AND USABLE ON CODES (UOC)	(6) QTY
	VEHICLES BEFORE VIN FF136169 UOC:209	
22 PAOZZ 11862 15599269	PIPE,EXHAUST (LEFT) APPLIES TO VEHICLES STARTING WITH VIN FF136169. UOC:209	1
23 PAOZZ 11862 14072686	SEAL,EXHAUST PIPE PART OF KIT P/N 15633467 UOC:209	2
24 PAOZZ 11862 587575	SPRING,HELICAL,COMP UOC:209	6
25 PAOZZ 80205 NAS1408A6	NUT,SELF-LOCKING,HE UOC:209	6
26 PAOZZ 96906 MS27183-16	WASHER,FLAT UOC:209	6

END OF FIGURE

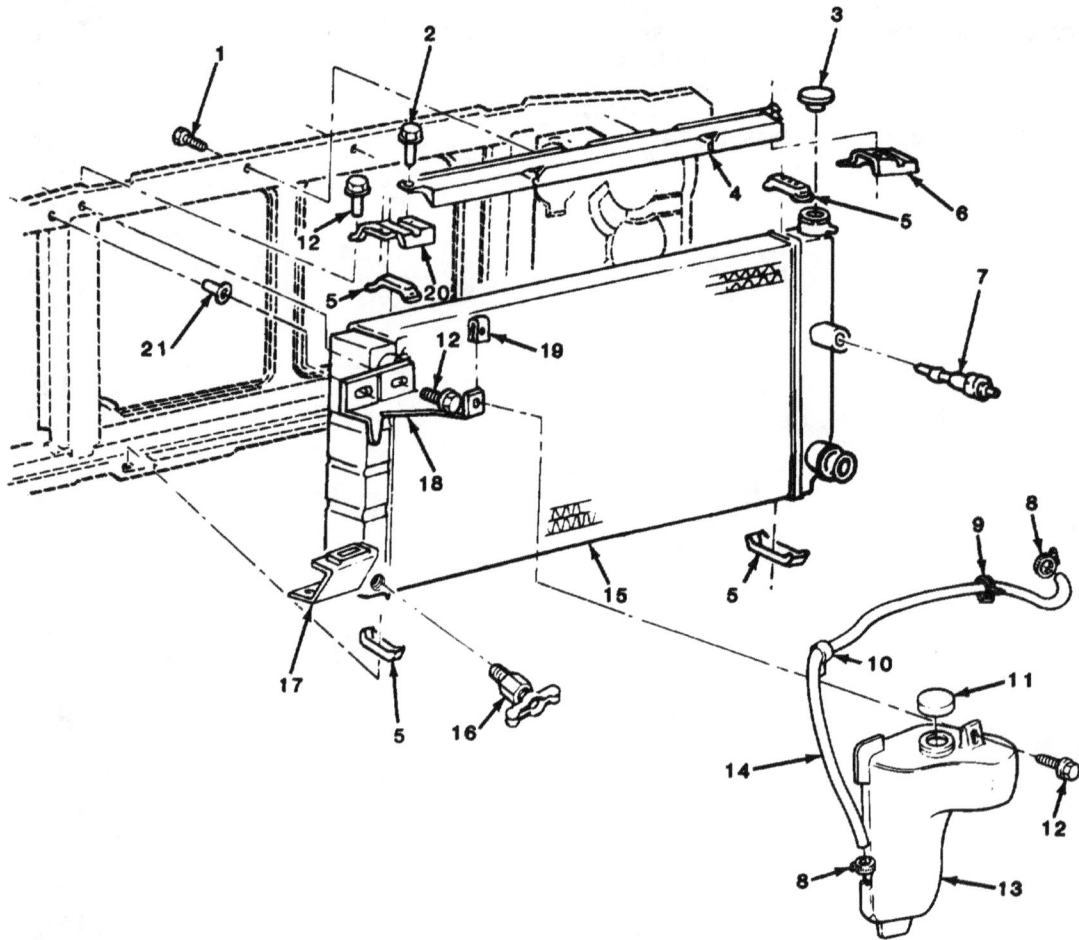

TA510835

FIGURE 28. RADIATOR, COOLANT RECOVERY RESERVOIR, AND RELATED PARTS.

(1) (2) (3) (4) ITEM SMR PART NO CODE CAGEC NUMBER	(5) DESCRIPTION AND USABLE ON CODES (UOC)	(6) QTY
	GROUP 05 COOLING SYSTEM	
	GROUP 0501 RADIATOR,EVAPORATIVE COOLER,OR HEAT EXHANGER	
	FIG. 28 RADIATOR, COOLANT RECOVERY RESERVOIR,AND RELATED PARTS	
1 PAOZZ 11862 11508566	SCREW,TAPPING,THREA	2
2 PAOZZ 11862 11504115	SCREW,TAPPING,THREA	2
3 PAOZZ 11862 6410785	CAP,FILLER OPENING	1
4 PFOZZ 11862 14039948	PANEL,RADIATOR	1
5 PAOZZ 11862 6264100	PAD,RADIATOR RETAIN	4
6 PFOZZ 11862 14039950	BRACKET,MOUNTING RA	1
7 PAOZZ 11862 358375	PROBE ASSEMBLY,TRAN	1
8 PAOZZ 96906 MS35842-10	CLAMP,HOSE	2
9 PAOZZ 11862 3816659	STRAP,LINE SUPPORTI	1
10 PAOZZ 96906 MS21333-114	CLAMP,LOOP	1
11 PAOZZ 11862 14011345	CAP ASSY	1
12 PAOZZ 11862 2014469	BOLT,ASSEMBLED WASH	7
13 PFOZZ 11862 14052221	TANK,RADIATOR,OVERF	1
14 MOOZZ 11862 14072430	HOSE COOLANT RES (59.00"LG) MAKE FROM HOSE,P/N 9438373	1
15 PAOFF 61928 3058966	RADIATOR,ENGINE COO	1
16 PAOZZ 11862 9437207	COCK,DRAIN	1
17 PFOZZ 11862 14072427	BRACKET,RESERVOIR,C	1
18 PFOZZ 11862 14072426	BRACKET,RESERVOIR,C	1
19 PAOZZ 24617 1494253	NUT,CLIP-ON	1
20 PFOZZ 11862 14039949	BRACKET,MOUNTING	1
21 PAOZZ 11862 3792287	NUT,PLAIN,BLIND RIV	1

END OF FIGURE

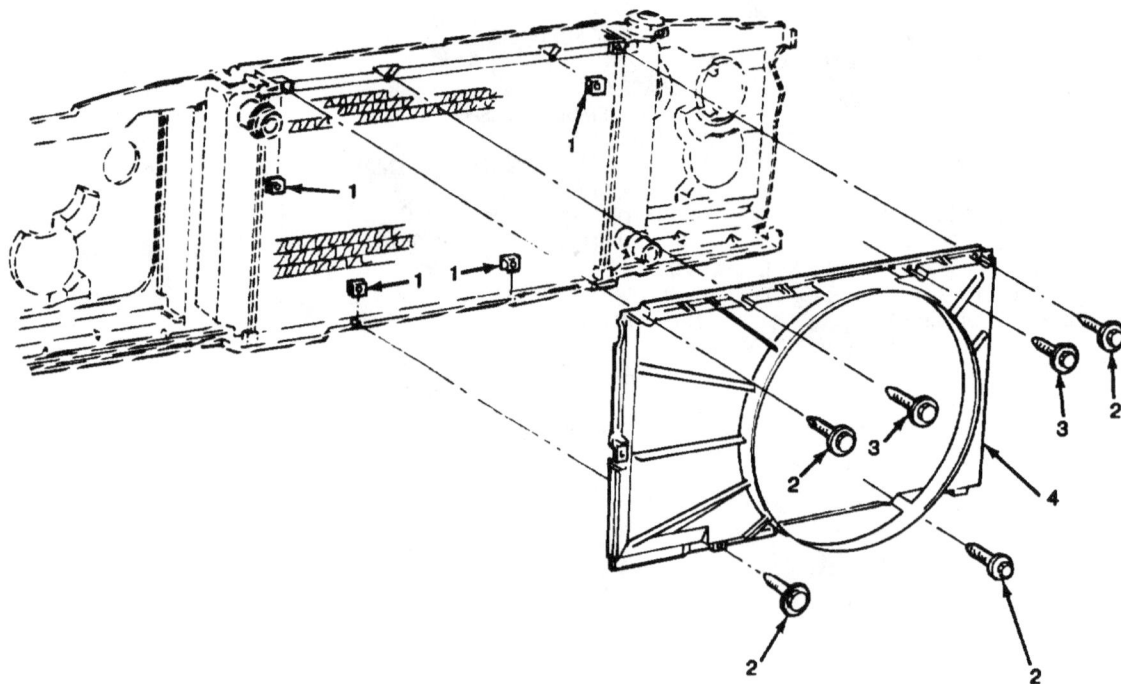

TA510836

FIGURE 29. RADIATOR SHROUD.

(1)	(2)	(3)	(4)	(5)	(6)
ITEM NO	SMR CODE	CAGEC	PART NUMBER	DESCRIPTION AND USABLE ON CODES (UOC)	QTY

GROUP 0502 COWLING,DEFLECTORS,AIR DUCTS,SHROUDS,ETC.

FIG. 29 RADIATOR SHROUD

1	PAOZZ	11862	3982098	NUT,PLAIN,HEXAGON	4
2	PAOZZ	11862	9440334	BOLT,ASSEMBLED WASH	4
3	PAOZZ	11862	11508566	SCREW,TAPPING,THREA	2
4	PFOZZ	11862	15522697	SHROUD,FAN,RADIATOR APPLIES TO VEHICLES STARTING WITH VIN FF136169 AND VIN FF300021	1

END OF FIGURE

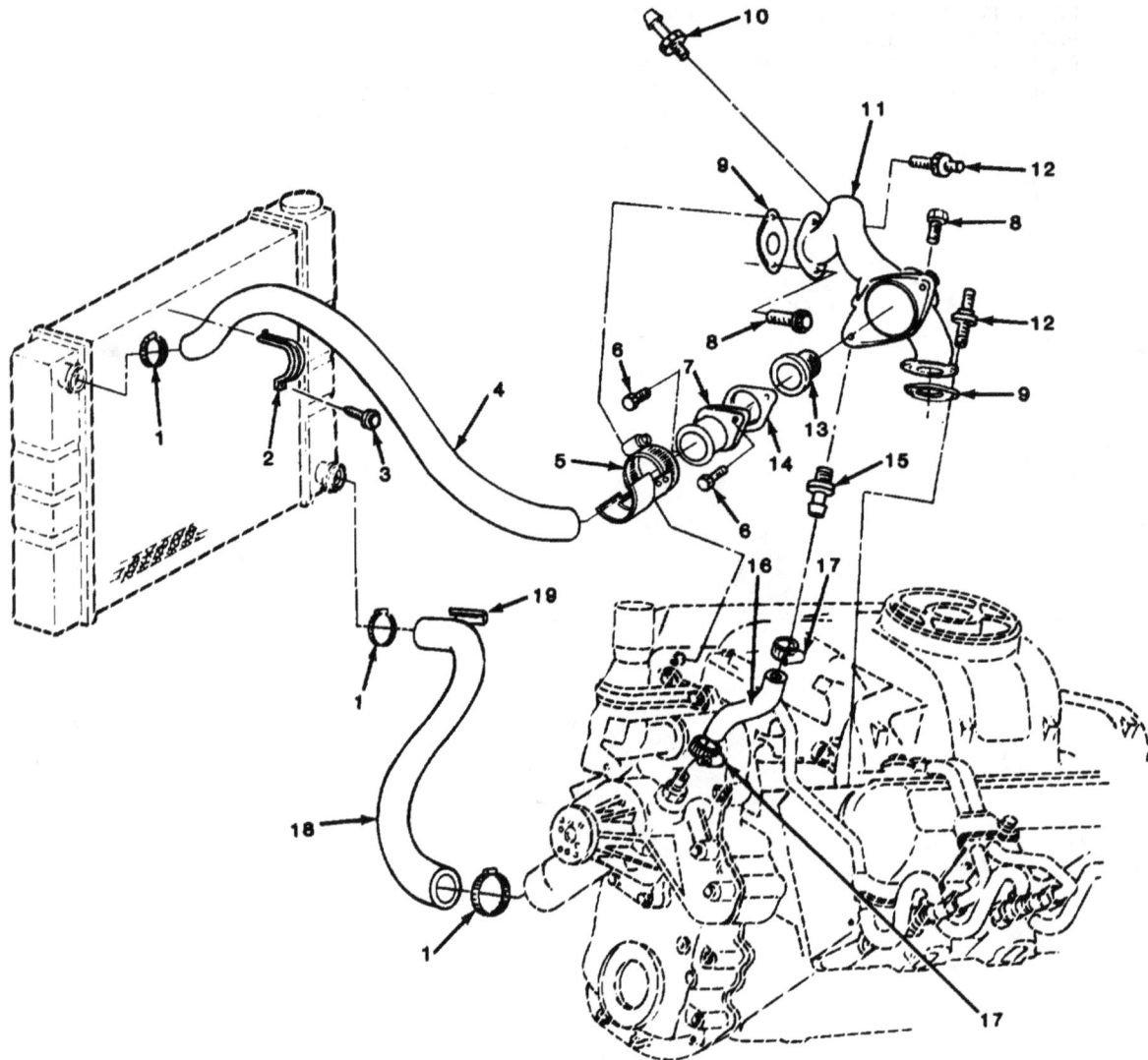

TA510837

FIGURE 30. ENGINE THERMOSTAT AND HOSES.

(1) (2) (3) (4)	(5)	(6)
ITEM SMR PART		
NO CODE CAGEC NUMBER	DESCRIPTION AND USABLE ON CODES (UOC)	QTY

GROUP 0503 WATER MANIFOLD,HEADERS,
THERMOSTATS,AND HOUSING GASKET

FIG. 30 ENGINE THERMOSTAT AND HOSES

1	PAOZZ	96906	MS35842-13	CLAMP,HOSE	3
				UOC:194,208,209,210,230,231,252,254	
2	PAOZZ	11862	14036779	STRAP,RETAINING	1
3	PAOZZ	11862	11508566	SCREW,TAPPING,THREA	1
4	PAOZZ	11862	14036744	HOSE,PREFORMED	1
5	PAOZZ	11862	14071983	CLAMP,HOSE	1
6	PAOZZ	11862	11513606	BOLT,MACHINE	2
7	PAOZZ	11862	14028918	FLANGE,WATER OUTLET	1
8	PAOZZ	11862	1635490	BOTL,SELF-LOCKING	2
9	PAOZZ	11862	23500846	GASKET PART OF KIT P/N 14067590	2
10	PAOZZ	11862	14067737	ELBOW,PIPE TO HOSE	1
				UOC:194,208,209,230,231,252,254,256	
10	PAOZZ	11862	14067727	ELBOW,PIPE TO HOSE	1
				UOC:210,230	
11	PAOZZ	11862	14028917	WATER OUTLET,ENGINE	1
12	PAOZZ	11862	22535073	STUD,CONTINUOUS THR	2
				UOC:194,208,209,210,230,231,252,254	
13	PAOZZ	11862	14077122	THERMOSTAT,FLOW CON	1
14	PAOZZ	11862	14028916	GASKET PART OFKIT P/N 14067590	1
15	PAOZZ	11862	354501	ADAPTER,STRAIGHT,PI	1
16	MOOZZ	11862	14033823	HOSE,NONMETALLIC (4.25"LG) MAKE	1
				FROM HOSE, P/N MS5213048203R	
17	PAOZZ	96906	MS35842-11	CLAMP,HOSE	2
18	PAOZZ	11862	14067763	HOSE,PREFORMED	1
19	PAOZZ	11862	14000217	GUARD,RADIATOR HOSE	1

END OF FIGURE

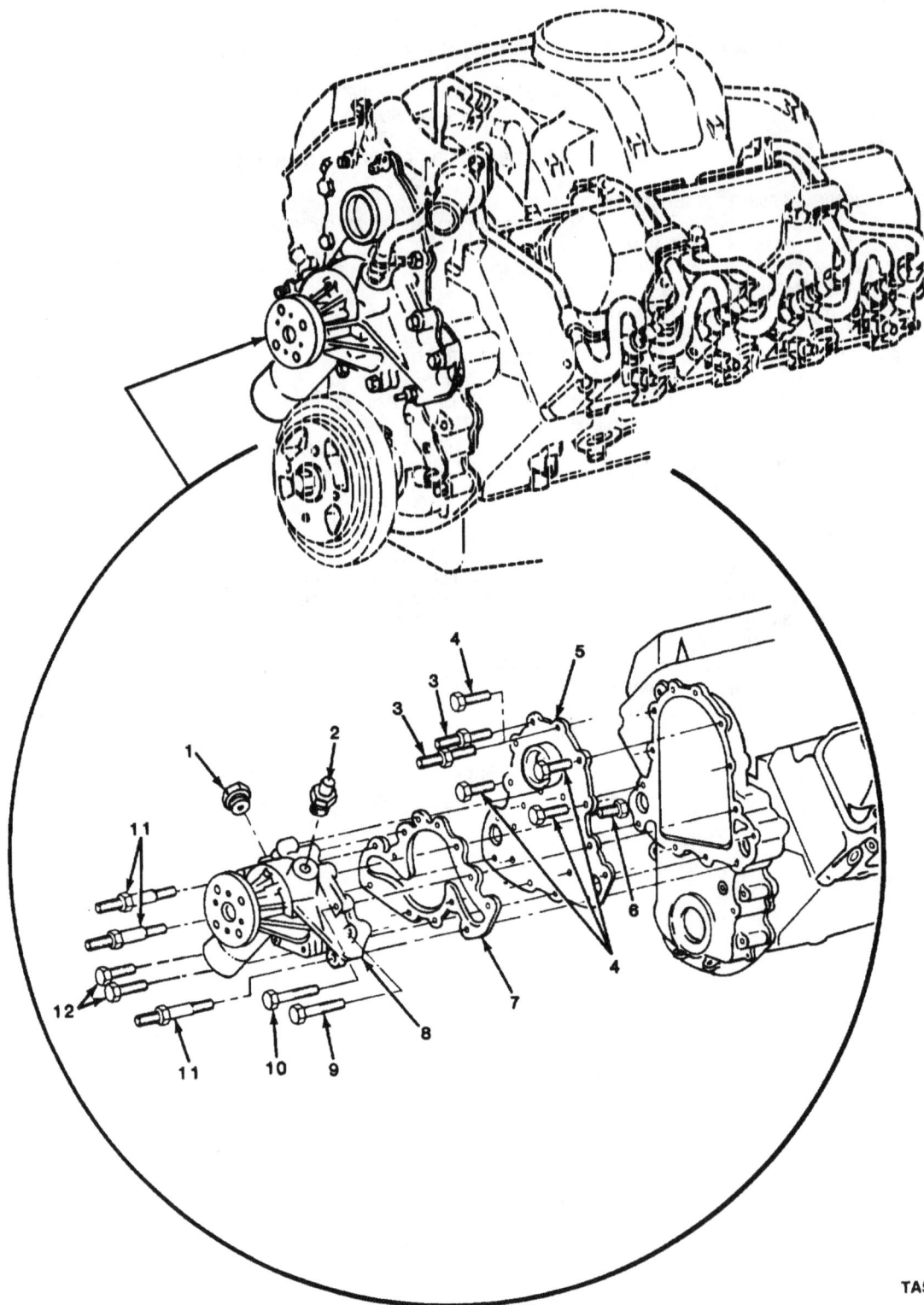

TA510838

FIGURE 31. WATER PUMP.

(1) ITEM NO	(2) SMR CODE	(3) CAGEC	(4) PART NUMBER	(5) DESCRIPTION AND USABLE ON CODES (UOC)	(6) QTY
				GROUP 0504 WATER PUMP	
				FIG. 31 WATER PUMP	
1	PAOZZ	30379	444789	PLUG,PIPE	1
2	PAOZZ	11862	354501	ADAPTER,STRAIGHT,PI	1
3	PAFZZ	11862	14045263	STUD,SHOULDERED	2
4	PAFZZ	11862	14060613	BOLT,MACHINE	4
5	PAFZZ	11862	14024208	PLATE ASSEMBLY,WATE	1
6	PAFZZ	24617	11500815	SCREW,CAP,HEXAGON H	7
7	PAFZZ	11862	14024209	GASKET PART OF KIT P/N 15633467 UOC:194,208,209,210,230,231,252,254	1
8	PAFZZ	11862	23500133	PUMP,COOLING SYSTEM	1
9	PAFZZ	11862	11504967	SCREW,CAP,HEXAGON H	1
10	PAFZZ	11862	11509202	SCREW,CAP,SOCKET HE	1
11	PAFZZ	11862	14071080	STUD,SHOULDERED	3
12	PAFZZ	11862	11500921	BOLT,MACHINE	2

END OF FIGURE

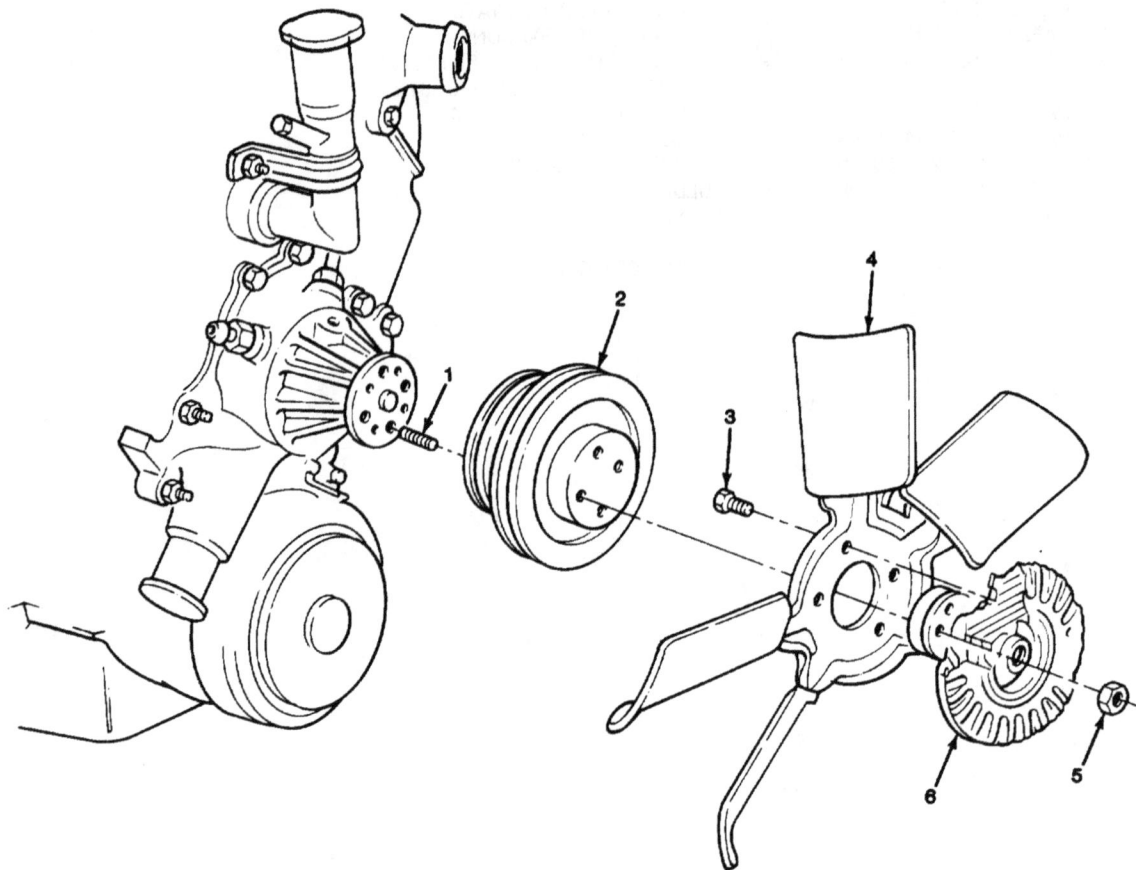

TA510839

FIGURE 32. ENGINE FAN ASSEMBLY.

(1) ITEM NO	(2) SMR CODE	(3) CAGEC	(4) PART NUMBER	(5) DESCRIPTION AND USABLE ON CODES (UOC)	(6) QTY
				GROUP 0505 FAN ASSEMBLY	
				FIG. 32 ENGINE FAN ASSEMBLY	
1	PAOZZ	11862	14061661	STUD,SHOULDERED	4
2	PAOZZ	11862	14067704	PULLEY,CONE	1
				UOC:194,208,209,230,231,252,254	
2	PAOZZ	11862	14067705	PULLEY,CONE	1
				UOC:210,230	
3	PAOZZ	11862	14020698	BOLT,ASSEMBLED WASH	4
4	PAOZZ	11862	14077928	IMPELLER,FAN,AXIAL	1
5	PAOZZ	96906	MS51967-6	NUT,PLAIN,HEXAGON	4
6	PAOZZ	11862	14032395	HUB,FAN CLUTCH	1

END OF FIGURE

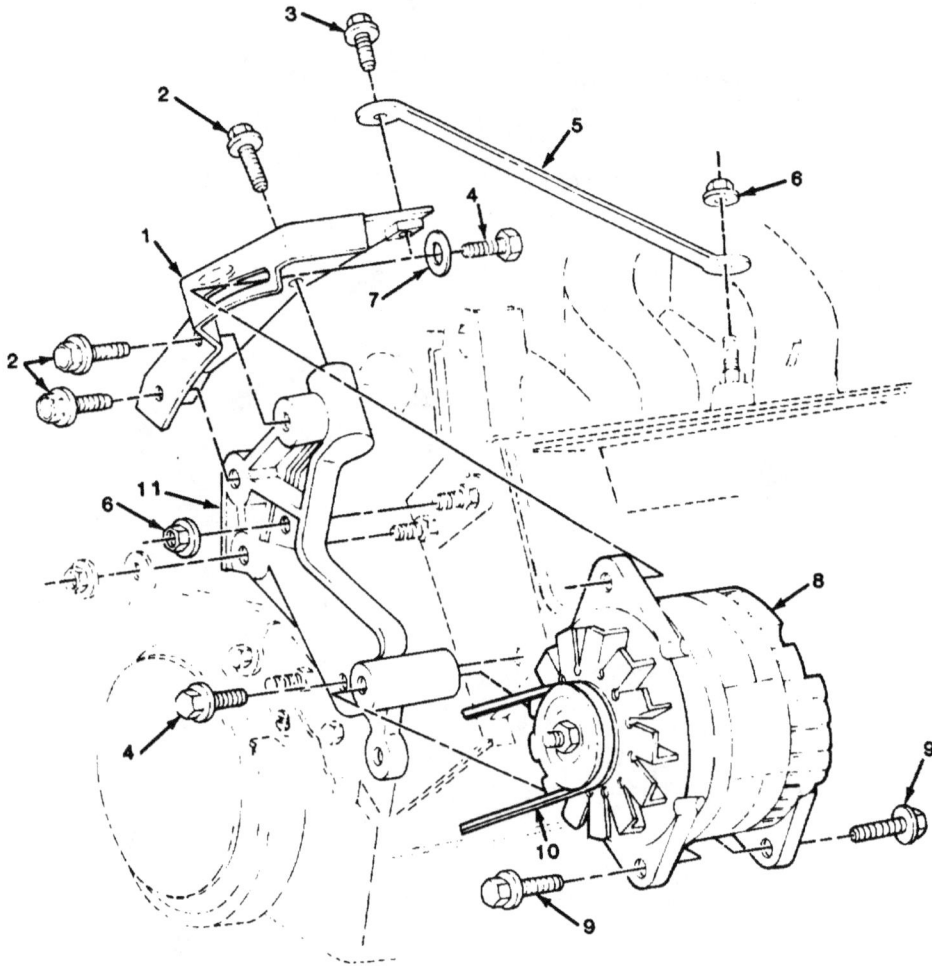

TA510840

FIGURE 33. ALTERNATOR AND MOUNTING HARDWARE, LEFT SIDE (ALL EXCEPT M1010).

(1) (2) (3) (4) ITEM SMR PART NO CODE CAGEC NUMBER	(5) DESCRIPTION AND USABLE ON CODES (UOC)	(6) QTY
	GROUP 06 ELECTRICAL SYSTEM	
	GROUP 0601 GENERATOR, ALTERNATOR	
	FIG. 33 ALTERNATOR AND MOUNTING HARDWARE, LEFT SIDE (ALL EXCEPT M1010)	
1 PAOZZ 11862 14077149	BRACKET,ENGINE ACCE UOC:194,208,209,230,231,252,254,256	1
2 PAOZZ 11862 1635450 BOLT,SELF-LOCKING	 UOC:194,208,209,230,231,252,254,256	3
3 PAOZZ 11862 11504595	SCREW,CAP,HEXAGON H UOC:194,208,209,230,231,252,254,256	1
4 PAOZZ 11862 1635490 BOLT,SELF-LOCKING	 UOC:194,208,209,230,231,252,254,256	2
5 PAOZZ 11862 14077151	BRACKET,DOUBLE ANGL UOC:194,208,209,230,231,252,254,256	1
6 PAOZZ 11862 11503643 NUT	 UOC:194,208,209,230,231,252,254,256	2
7 PAOZZ 96906 MS27183-16 WASHER,FLAT	 UOC:194,208,209,230,231,252,254,256	1
8 PAOFF 11862 1105500	GENERATOR,ENGINE AC FOR COMPONENT PARTS SEE FIG. 36 UOC:194,208,209,230,231,252,254,256	2
9 PAOZZ 11862 1610819 BOLT,MACHINE	 UOC:194,208,209,230,231,252,254,256	2
10 PAOZZ 20796 43-3226 BELT,V	 UOC:194,208,209,230,231,252,254,256	1
11 PAOZZ 11862 14077147	BRACKET,ENGINE ACCE UOC:194,208,209,230,231,252,254,256	1
	END OF FIGURE	

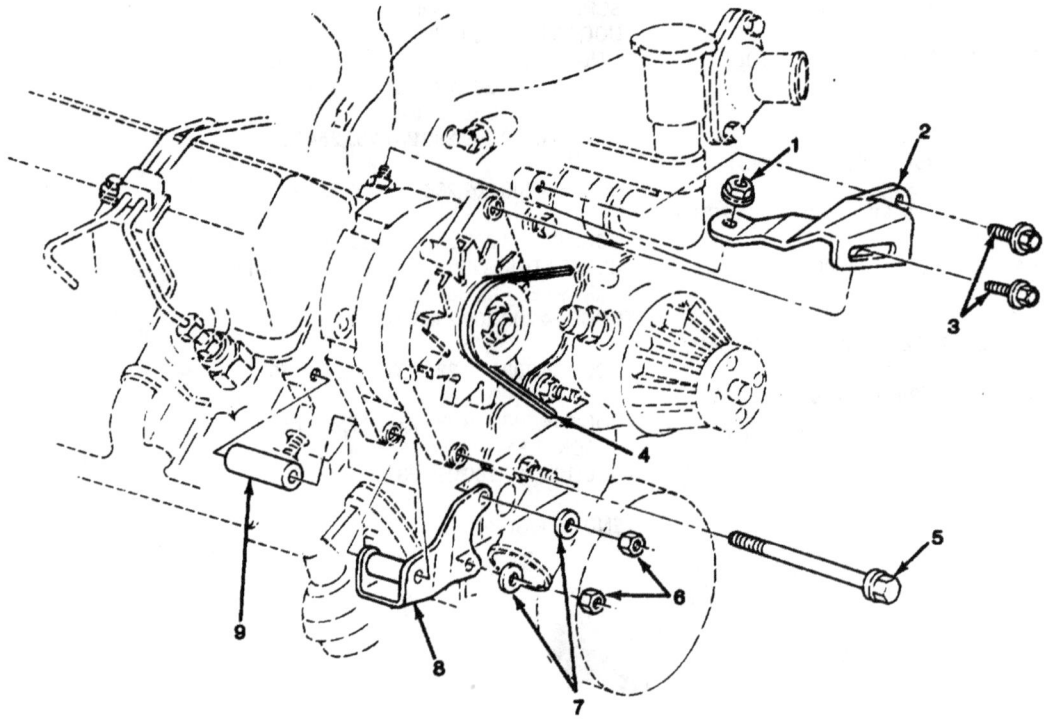

TA510841

FIGURE 34. ALTERNATOR AND MOUNTING HARDWARE, RIGHT SIDE (ALL EXCEPT M1010).

(1) (2) (3) (4)	(5)	(6)
ITEM SMR PART		
NO CODE CAGEC NUMBER	DESCRIPTION AND USABLE ON CODES (UOC)	QTY

GROUP 0601 GENERATOR, ALTERNATOR

FIG. 34 ALTERNATOR AND MOUNTING HARDWARE, RIGHT SIDE (ALL EXCEPT M1010)

1	PAOZZ	11862	14005953	SHIELD,EXPANSION	1
				UOC:194,208,209,230,231,252,254,256	
2	PAOZZ	11862	14067724	BRACKET,ENGINE ACCE	1
				UOC:194,208,209,230,231,252,254,256	
3	PAOZZ	11862	1635490	BOLT,SELF-LOCKING	2
				UOC:194,208,209,230,231,252,254,256	
4	PAOZZ	20796	42-6923	BELT,V	1
				UOC:194,208,209,230,231,252,254,256	
5	PAOZZ	11862	14067717	BOLT,MACHINE	1
				UOC:194,208,209,230,231,252,254,256	
6	PAOZZ	11862	11506101	NUT	2
				UOC:194,208,209,230,231,252,254,256	
7	PAOZZ	96906	MS27183-16	WASHER,FLAT	2
				UOC:194,208,209,230,231,252,254,256	
8	PAOZZ	11862	14067714	BRACKET,ENGINE ACCE	1
				UOC:194,208,209,230,231,252,254,256	
9	PAOZZ	11862	14067725	SPACER,SLEEVE	1
				UOC:194,208,209,230,231,252,254,256	

END OF FIGURE

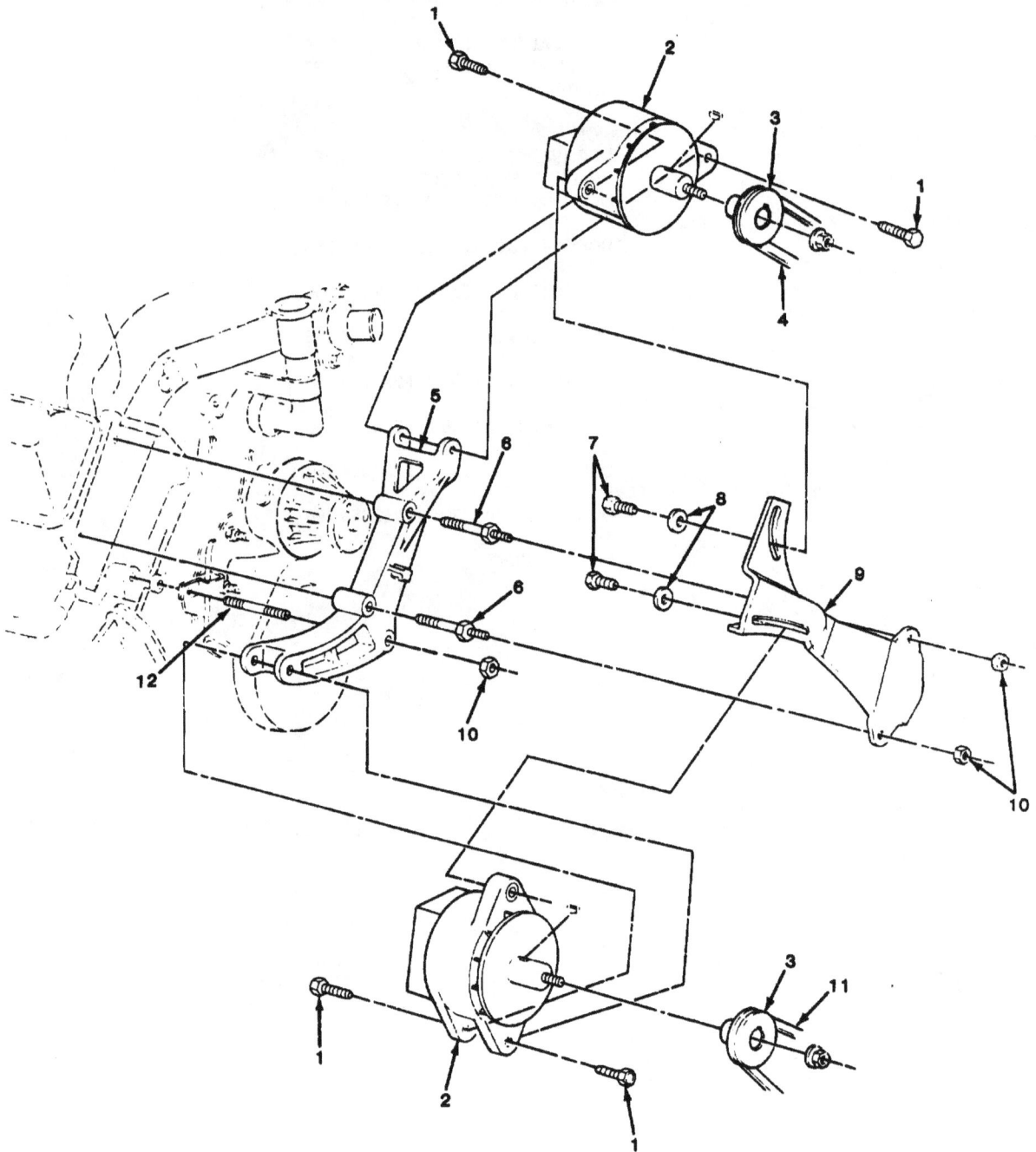

TA510842

FIGURE 35. ALTERNATOR AND MOUNTING HARDWARE (M1010),

(1) (2) (3) (4) ITEM SMR PART NO CODE CAGEC NUMBER	(5) DESCRIPTION AND USABLE ON CODES (UOC)	(6) QTY
	GROUP 0601 GENERATOR, ALTERNATOR	
	FIG. 35 ALTERNATOR AND MOUNTING HARDWARE (M1010)	
1 PAOZZ 11862 9440280	SCREW,CAP,HEXAGON H UOC:210,230	4
2 PAOZZ 35510 4629JA	GENERATOR,ENGINE AC FOR COMPONENT PARTS SEE FIG. 37 UOC:210,230	2
3 PAOZZ 11862 15599204 PULLEY,GROOVE	UOC:210,230	2
4 PAOZZ 20796 42-6919	BELT,V UOC:210,230	1
5 PAOZZ 11862 14067715	BRACKET,ENGINE ACCE UOC:210,230	1
6 PAOZZ 11862 14067718 STUD,RECESSED	UOC:210,230	2
7 PAOZZ 96906 MS90728-109	SCREW,CAP,HEXAGON H UOC:210,230	2
8 PAOZZ 11862 3954735	WASHER,FLAT UOC:210,230	2
9 PAOZZ 11862 14067721	BRACKET,ENGINE ACCE UOC:210,230	1
10 PAOZZ 11862 11506101 NUT	UOC:210,230	3
11 PAOZZ 20796 42-6921	BELT,V UOC:210,230	1
12 PAOZZ 11862 1623159	STUD,PLAIN UOC:210,230	1
	END OF FIGURE	

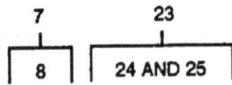

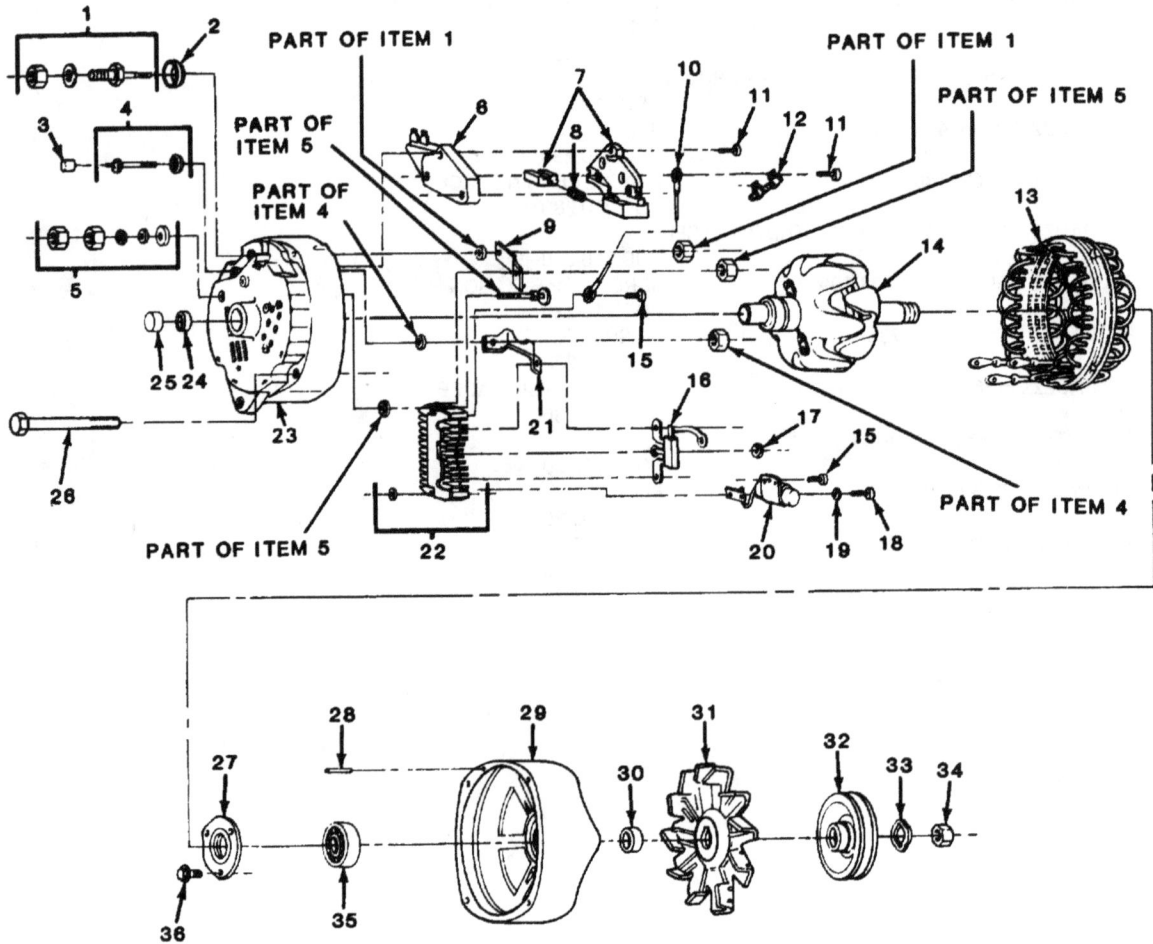

FIGURE 36. ALTERNATOR COMPONENT PARTS (ALL EXCEPT M1010.

TA510843

(1) ITEM NO	(2) SMR CODE	(3) CAGEC	(4) PART NUMBER	(5) DESCRIPTION AND USABLE ON CODES (UOC)	(6) QTY
				GROUP 0601 GENERATOR, ALTERNATOR	
				FIG. 36 ALTERNATOR COMPONENT PARTS (ALL EXCEPT M1010)	
1	PAFZZ	11862	1987808	PARTS KIT,ENGINE GE UOC:194,208,209,230,231,252,254,256	1
2	PAFZZ	16764	1876806	BOOT,DUST AND MOIST UOC:194,208,209,230,231,252,254,256	1
3	PAFZZ	16764	1970227	CAP,ELECTRICAL UOC:194,208,209,230,231,252,254,256	1
4	PAFZZ	11862	1852519	PARTS KIT,ENGINE GE UOC:194,208,209,230,231,252,254,256	1
5	PAFZZ	11862	1987809	PARTS KIT,ENGINE GE UOC:194,208,209,230,231,252,254,256	1
6	PAFZZ	11862	1116423	REGULAR,ENGINE GE UOC:194,208,209,230,231,252,254,256	1
7	PAFFF	16764	1876873	HOLDER,ELECTRICAL C UOC:194,208,209,230,231,252,254,256	1
8	PAFZZ	11862	1977357	.SPRING,HELICAL,COMP UOC:194,208,209,230,231,252,254,256	2
9	PAFZZ	16764	801815	PLATE,RETAINING,ELE UOC:194,208,209,230,231,252,254,256	1
10	PAFZZ	11862	1986433	LEAD,ELECTRICAL UOC:194,208,209,230,231,252,254,256	1
11	PAFZZ	16764	1876681	SETSCREW UOC:194,208,209,230,231,252,254,256	3
12	PAFZZ	11862	830478	RESISTOR,FIXED,WIRE UOC:194,208,209,230,231,252,254,256	1
13	PAFZZ	11862	1892941	STATOR,ENGINE GENER UOC:194,208,209,230,231,252,254,256	1
14	PAFZZ	11862	10499310	ROTOR UOC:194,208,209,230,231,252,254,256	1
15	PAFZZ	11862	447164	SCREW,TAPPING UOC:194,208,209,230,231,252,254,256	3
16	PAFZZ	11862	1977064	RECTIFIER,SEMICONDU UOC:194,208,209,230,231,252,254,256	1
17	PAFZZ	96906	MS35649-282	NUT,PLAIN,HEXAGON UOC:194,208,209,230,231,252,254,256	3
18	PAFZZ	11862	1986552	SCREW,MACHINE UOC:194,208,209,230,231,252,254,256	1
19	PAFZZ	11862	1970149	INSULATOR,WASHER UOC:194,208,209,230,231,252,254,256	1
20	PAFZZ	16764	1978146	CAPACITOR,FIXED,PAP UOC:194,208,209,230,231,252,254,256	1
21	PAFZZ	11862	1986428	CLAMP,CABLE,ELECTRI UOC:194,208,209,230,231,252,254,256	1
22	PAFZZ	16764	1875645	RECTIFIER,SEMICONDU UOC:194,208,209,230,231,252,254,256	1
23	PAFFF	11862	1986427	END BELL,ELECTRICAL UOC:194,208,209,230,231,252,254,256	1
24	PAFZZ	11862	1976143	.BUSHING,SLEEVE	1

(1) ITEM NO	(2) SMR CODE	(3) CAGEC	(4) PART NUMBER	(5) DESCRIPTION AND USABLE ON CODES (UOC)	(6) QTY
				UOC:194,208,209,230,231,252,254,256	
25	PAFZZ	16764	9436831	.BEARING,ROLLER,NEED	1
				UOC:194,208,209,230,231,252,254,256	
26	PAFZZ	16764	801810	BOLT	4
				UOC:194,208,209,230,231,252,254,256	
27	PAFZZ	11862	1971993	PLATE	1
				UOC:194,208,209,230,231,252,254,256	
28	PAFZZ	96906	MS16562-33	PIN,SPRING	1
				UOC:194,208,209,230,231,252,254,256	
29	XAFZZ	11862	1976049	END FRAME GEN DRV	1
				UOC:194,208,209,230,231,252,254,256	
30	PAFZZ	11862	1978059	COLLAR,BEARING	1
				UOC:194,208,209,230,231,252,254,256	
31	PAFZZ	11862	1978058	IMPELLER,FAN,AXIAL	1
				UOC;194,208,209,230,231,252,254,256	
32	PAFZZ	11862	1978068	PULLEY,FLAT	1
				UOC:194,208,209,230,231,252,254,256	
33	PAFZZ	16764	1941978	WASHER,SPRING TENSI	1
				UOC:194,208,209,230,231,252,254,256	
34	PAFZZ	24617	1915172	NUT,PLAIN,HEXAGON	1
				UOC:194,208,209,230,231,252,254,256	
35	PAFZZ	30760	AH068-31	BEARING,BALL,ANNULA	1
				UOC:194,208,209,230,231,252,254,256	
36	PAFZZ	16764	800549	SCREW RETAINER	3
				UOC:194,208,209,230,231,252,254,256	

END OF FIGURE

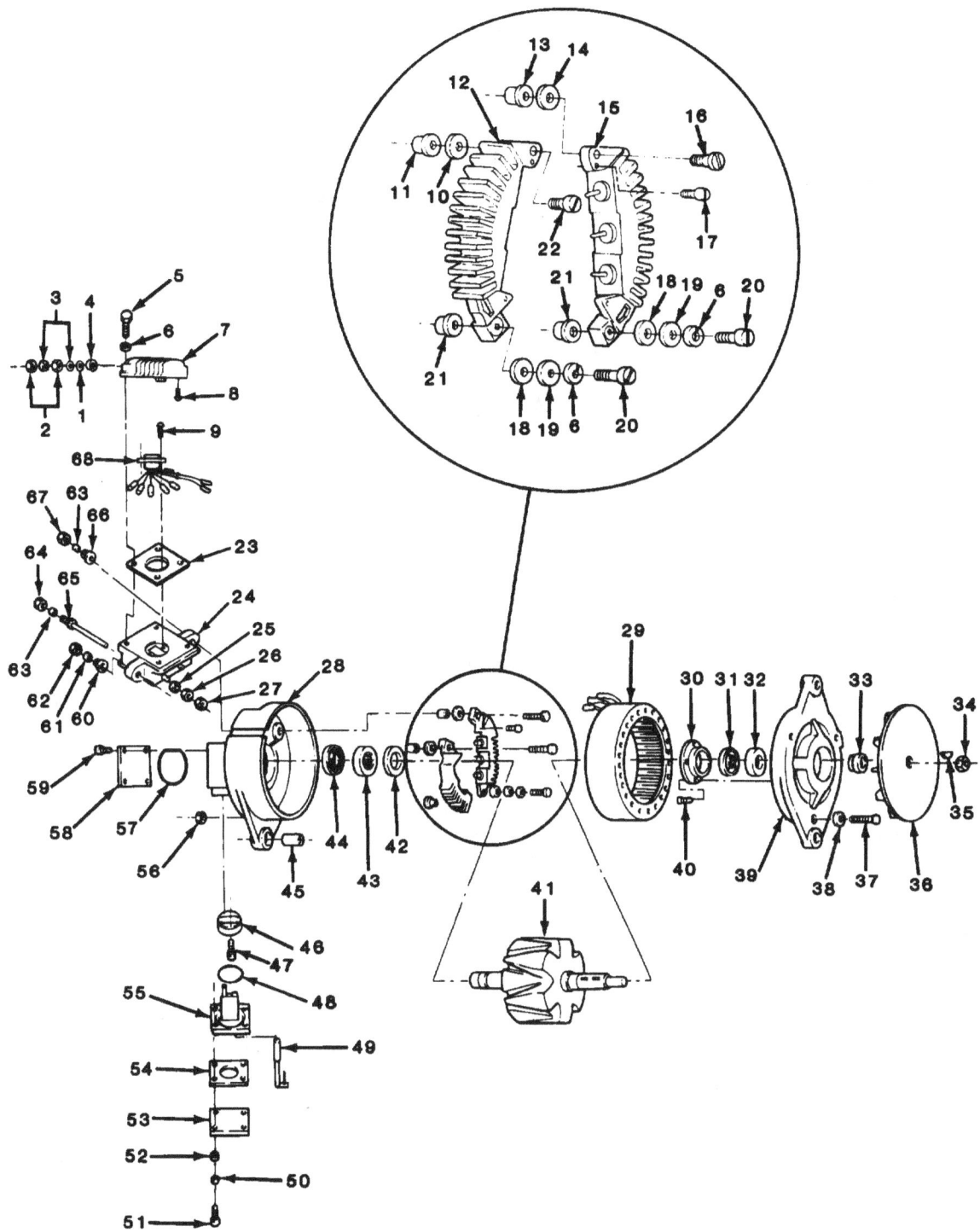

FIGURE 37. ALTERNATOR COMPONENT PARTS (M1010).

TA510844

(1) (2) (3) (4) ITEM SMR PART NO CODE CAGEC NUMBER	(5) DESCRIPTION AND USABLE ON CODES (UOC)	(6) QTY
	GROUP 0601 GENERATOR, ALTERNATOR	
	FIG. 37 ALTERNATOR COMPONENT PARTS (M1010)	
1 PAOZZ 35510 5413	WASHER,FLAT UOC:210	1
2 PAOZZ 35510 4340	NUT,PLAIN,HEXAGON UOC:210	2
3 PAOZZ 35510 31256	WASHER,LOCK UOC:210	2
4 PAOZZ 35510 7983	INSULATOR,WASHER UOC:210	1
5 PAOZZ 35510 99525	SCREW,MACHINE UOC:210	4
6 PAOZZ 35510 2434	WASHER,LOCK UOC:210	6
7 PAOZZ 35510 95300	REGULATOR,ENGINE GE UOC:210	1
8 PAOZZ 35510 99459	SCREW,ASSEMBLED WAS UOC:210	1
9 PAFZZ 35510 13771	SCREW,MACHINE UOC:210	2
10 PAFZZ 35510 73546	INSULATOR,BUSHING UOC:210	1
11 PAFZZ 35510 79071	BUSHING,SLEEVE UOC:210	1
12 PAFZZ 35510 71237	RECTIFIER,METALLIC UOC:210	1
13 PAFZZ 35510 79070	BUSHING,SLEEVE UOC:210	1
14 PAFZZ 35510 73545	INSULATOR,BUSHING UOC:210	1
15 PAFZZ 35510 71238	RECTIFIER,METALLIC UOC:210	1
16 PAFZZ 35510 79313	SCREW UOC:210	1
17 PAFZZ 35510 73543	SCREW,TAPPING,THREA UOC:210	3
18 PAFZZ 35510 52066	WASHER,FLAT UOC:210	2
19 PAFZZ 33510 2385	WASHER,FLAT UOC:210	2
20 PAFZZ 96906 MS35207-265 SCREW,MACHINE UOC:210		2
21 PAFZZ 35510 73547	BUSHING,SLEEVE UOC:210	2
22 PAFZZ 35510 78997	BOLT,SQUARE NECK UOC:210	1
23 PAFZZ 35510 99515	GASKET UOC:210	1
24 PAFZZ 35510 99478	HOLDER,VOLTAGE REGU	1

(1) ITEM NO	(2) SMR CODE	(3) CAGEC	(4) PART NUMBER	(5) DESCRIPTION AND USABLE ON CODES (UOC)	(6) QTY
				UOC:210	
25	PAFZZ	35510	79469	WASHER,FLAT	3
				UOC:210	
26	PAFZZ	35510	2313	NUT	3
				UOC:210	
27	PAFZZ	35510	31587	NUT,PLAIN,CASTELLAT	3
				UOC:210	
28	XAFZZ	35510	95279	HOUSING SRE	1
				UOC:210	
29	PAFZZ	35510	99789	STATOR ASSEMBLY,ALT	1
				UOC:210	
30	PAFZZ	35510	79486	PLATE,RETAINING,BEA	1
				UOC:210	
31	PAFZZ	35510	79405	SEAL	1
				UOC:210	
32	PAFZZ	35510	78895	BEARING	1
				UOC:210	
33	PAFZZ	35510	59225	WASHER,KEYWAY	1
				UOC:210	
34	PAFZZ	35510	76985	NUT,HEXAGON	1
				UOC:210	
35	PAFZZ	96906	MS35756-11	KEY,WOODRUFF	1
				UOC:210	
36	PAFZZ	35510	79035	HOLDER ASSEMBLY,FAN	1
				UOC:210	
37	PAFZZ	35510	95152	SCREW,CAP,HEXAGON H	3
				UOC:210	
38	PAFZZ	35510	79627	WASHER, GUARD	3
				UOC:210	
39	PAFZZ	35510	78983	HOUSING,ALTERNATOR	1
				UOC:210	
40	PAFZZ	35510	58754	SCREW,MACHINE	4
				UOC:210	
41	PAFZZ	35510	79517	ROTOR,GENERATOR	1
				UOC:210	
42	PAFZZ	35510	79404	SEAL,PLAIN,ENCASED	1
				UOC:210	
43	PAFZZ	35510	79406	BEARING,ROLLER	1
				UOC:210	
44	PAFZZ	35510	79403	SEAL,OIL	1
				UOC:210	
45	PAFZZ	35510	77916	BUSHING,SLEEVE	1
				UOC:210	
46	PAFZZ	35510	99508	ADAPTER,BRUSH HOLDE	1
				UOC:210	
47	PAFZZ	96906	MS35206-231	SCREW,MACHINE	2
				UOC:210	
48	PAFZZ	35510	99514	PACKING,PREFORMED	1
				UOC:210	
49	PAFZZ	35510	99516	BRUSH,ELECTRICAL CO	2
				UOC:210	
50	XDFZZ	35510	2435	LOCKWASHER	4

TM9-2320-289-34P

(1) (2) (3) (4) (5)						(6)
ITEM	SMR			PART		
NO	CODE	CAGEC	NUMBER		DESCRIPTION AND USABLE ON CODES (UOC)	QTY

					UOC:210	
51	PAFZZ	35510	99524		SCREW,CAP,HEXAGON H	4
					UOC:210	
52	PAFZZ	35510	36912	WASHER,FLAT		4
					UOC:210	
53	PAFZZ	35510	99517	COVER,BRUSH		1
					UOC:210	
54	PAFZZ	35510	99518	GASKET		1
					UOC:210	
55	PAFZZ	35510	99479		HOLDER,ELECTRICAL C	1
					UOC:210	
56	PAFZZ	35510	74732		NUT,FLANGE LOCK	3
					UOC:210	
57	PAFZZ	35510	99511	PACKING,PREFORMED		1
					UOC:210	
58	PAFZZ	35510	99512	COVER,ACCESS		1
					UOC:210	
59	PAFZZ	35510	75348		SCREW,CAP,HEXAGON H	4
					UOC:210	
60	PAFZZ	35510	73009		NUT TERMINAL	1
					UOC:210	
61	PAFZZ	35510	3231	WASHER,LOCK		1
					UOC:210	
62	PAFZZ	35510	2364	NUT		1
					UOC:210	
63	PAFZZ	35510	2523	WASHER,LOCK		4
					UOC:210	
64	PAFZZ	35510	3395	NUT,PLAIN,HEXAGON		3
					UOC:210	
65	PAFZZ	35510	99513	TERMINAL,STUD		3
					UOC:210	
66	PAFZZ	35510	59982		NUT TERMINAL	1
					UOC:210	
67	PAFZZ	35510	2771	NUT,PLAIN,HEXAGON		1
					UOC:210	
68	PAFZZ	35510	99481	CONNECTOR,PLUG,ELEC		1
					UOC:210	

END OF FIGURE

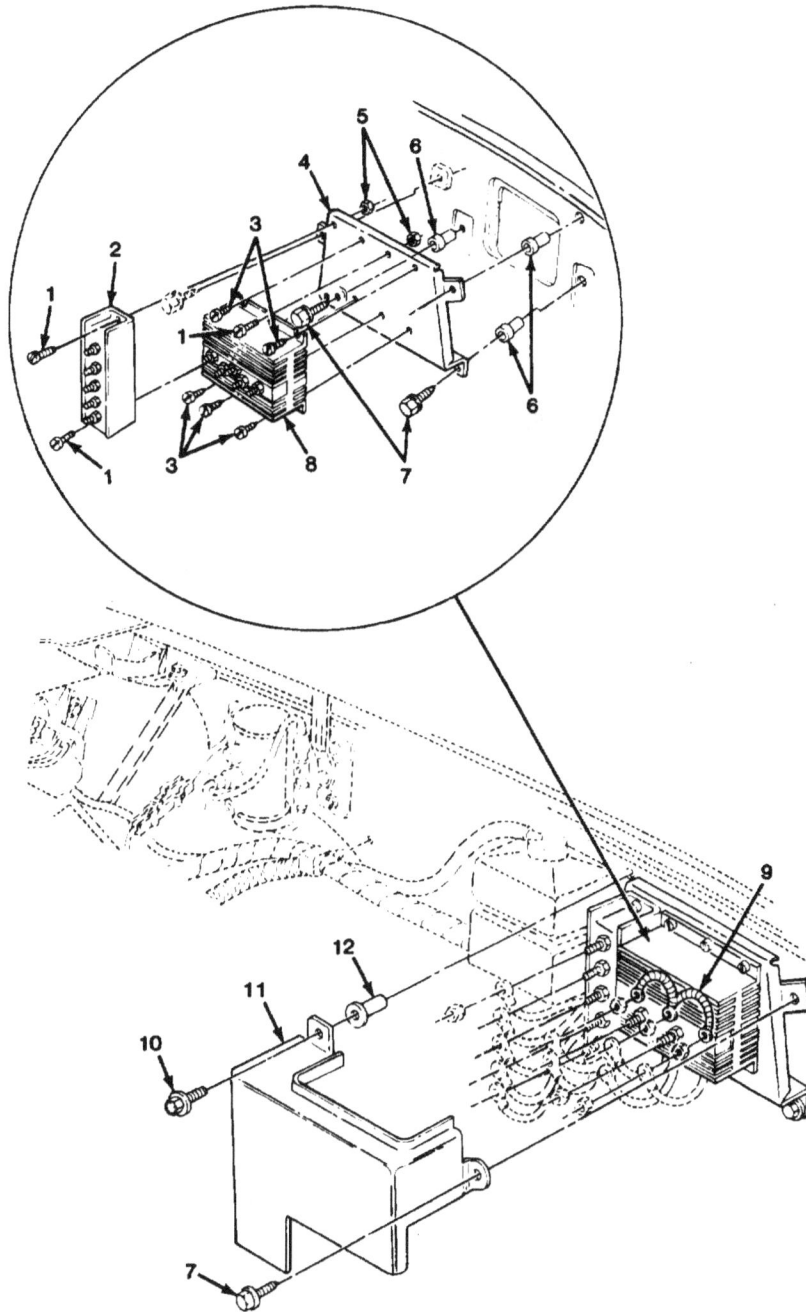

TA510845

FIGURE 38. VOLTAGE REGULATOR AND RELAY (M1010).

(1) (2) (3) (4) ITEM SMR NO CODE CAGEC	PART NUMBER	(5) DESCRIPTION AND USABLE ON CODES (UOC)	(6) QTY
		GROUP 0602 GENERATOR REGULATOR	
		FIG. 38 VOLTAGE REGULATOR AND RELAY (M1010)	
1 PAFZZ 96906	MS35206-243	SCREW,MACHINE UOC:210,230	3
2 PAFZZ 55156	40178	RELAY,ELECTROMAGNET UOC:210,230	1
3 PAFZZ 24617	9419320	SCREW,TAPPING,THREA UOC:210,230	5
4 PAFZZ 11862	14072428	BRACKET,MOUNTING UOC:210,230	1
5 PAFZZ 72582	271163	NUT,PLAIN,ASSEMBLED UOC:210,230	3
6 PAFZZ 11862	3792287	NUT,PLAIN,BLIND RIV UOC:210,230	3
7 PAFZZ 11862	2014469	BOLT,ASSEMBLED WASH UOC:210,230	3
8 PAFZZ 11862	14072341	REGULATOR,VOLTAGE UOC:210,230	1
9 PAFZZ 11862	15599251	SEMICONDUCTOR DEVIC UOC:210,230	1
10 PAFZZ 11862	9440033	BOLT,MACHINE UOC:210,230	1
11 PAFZZ 11862	15599949	COVER,VOLTAGE REGUL UOC:210,230	1
12 PAFZZ 11862	14005953	SHIELD,EXPANSION UOC:210,230	1

END OF FIGURE

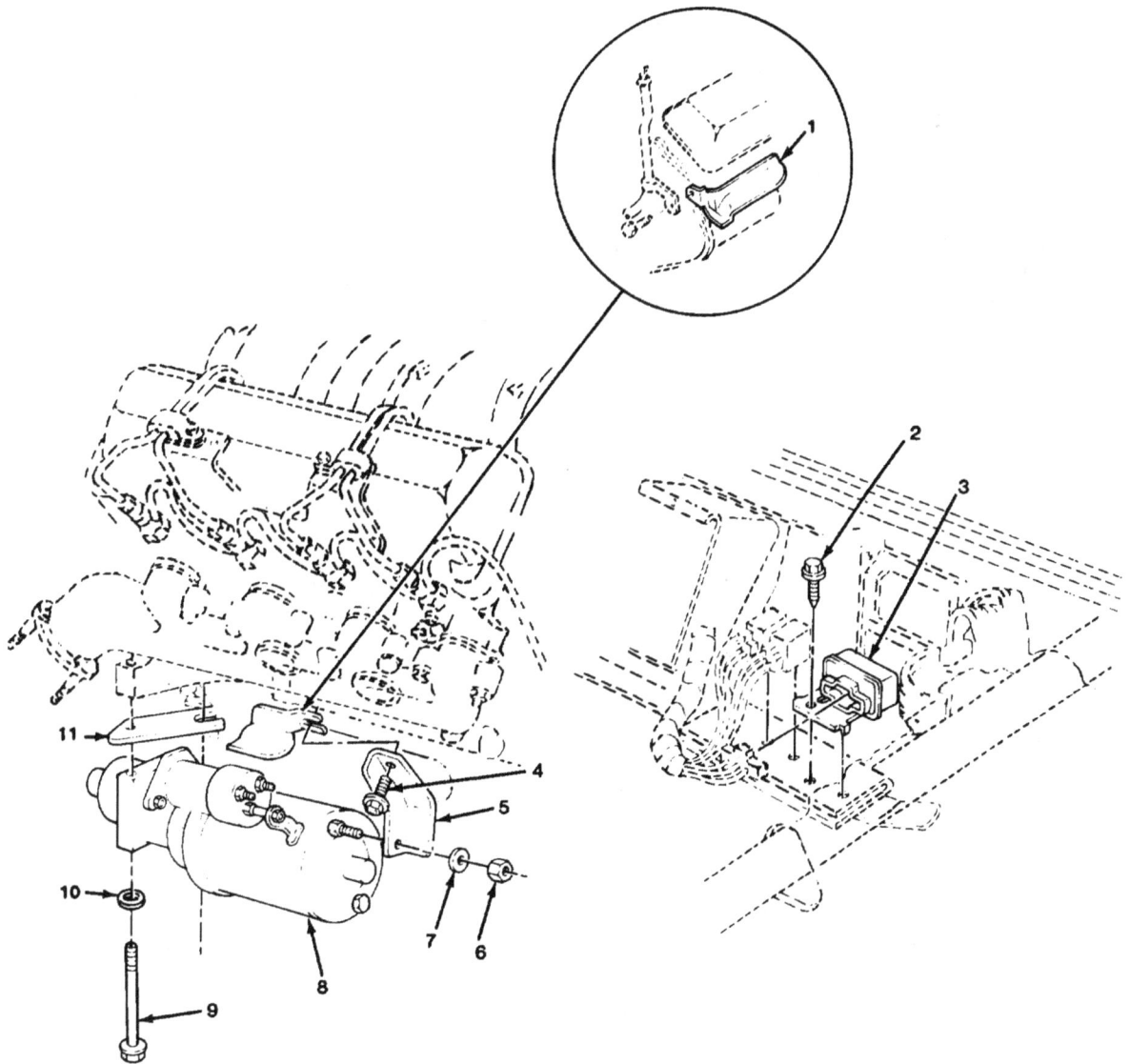

TA510846

FIGURE 39. STARTER MOTOR AND RELAY.

(1) (2) (3) (4)	(5)	(6)
ITEM SMR PART NO CODE CAGEC NUMBER	DESCRIPTION AND USABLE ON CODES (UOC)	QTY

GROUP 0603 STARTING MOTOR

FIG. 39 STARTER MOTOR AND RELAY

1	PAOZZ	11862	14066657	SHIELD,STARTER MOTO	1
2	PAOZZ	24617	9419663 SCREW,TAPPING,THREA	1
3	PAOZZ	11862	15591718 RELAY,ELECTROMAGNET	1
4	PAOZZ	11862	14060613 BOLT,MACHINE	1
5	PAOZZ	11862	14028931 BRACKET,ANGLE	1
6	PAOZZ	21450	131245 NUT,SELF-LOCKING,HE	1
7	PAOZZ	96906	MS27183-10 WASHER,FLAT	1
8	PAOZZ	16764	1113591 STARTER,ENGINE,ELEC FOR COMPONENT PARTS SEE FIG. 40 YZ R	1
9	PAOZZ	11862	15544950 BOLT,CLOSE TOLERANC	2
10	PAOZZ	11862	22521054 WASHER,FLAT	2
11	PAOZZ	11862	23500396 SPACER,PLATE (1.0 MM AS REQUIRED)	1

END OF FIGURE

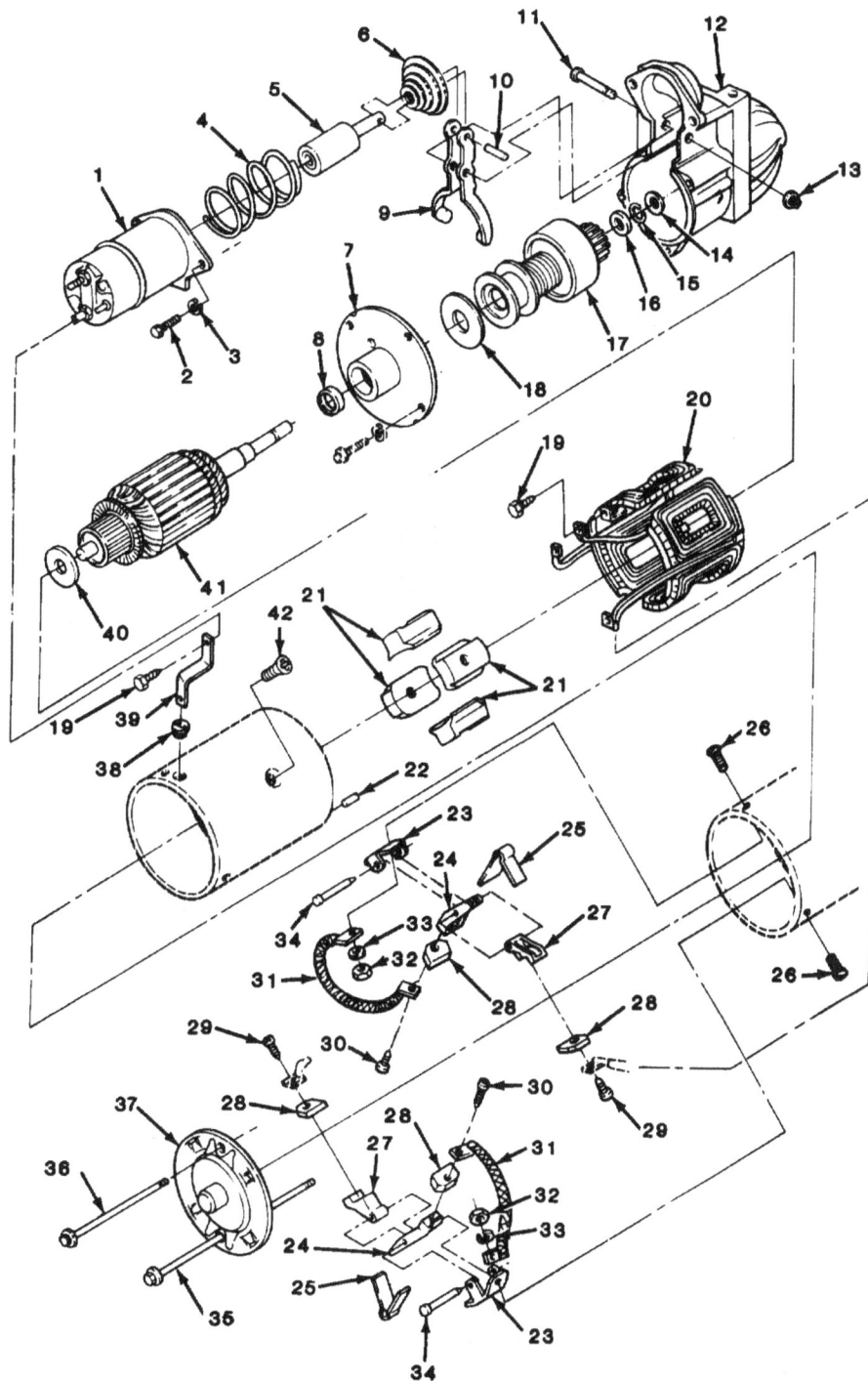

FIGURE 40. STARTER MOTOR COMPONENT PARTS.

TA510847

(1) ITEM NO	(2) SMR CODE	(3) CAGEC	(4) PART NUMBER	(5) DESCRIPTION AND USABLE ON CODES (UOC)	(6) QTY
				GROUP 0603 STARTER MOTOR	
				FIG. 40 STARTER MOTOR COMPONENT PARTS	
1	PAFZZ	11862	1114373	SOLENOID,ELECTRICAL	1
2	PAFZZ	96906	MS35265-79	SCREW,MACHINE	2
3	PAFZZ	96906	MS35340-44	WASHER,LOCK	2
4	PAFZZ	24617	1958679	SPRING,HELICAL,COMP	1
5	PAFZZ	11862	1941113	PLUNGER,SOLENOID	1
6	PAFZZ	16764	1970263	BOOT,DUST AND MOIST	1
7	PAFFF	11862	1976882	FLANGE,STARTER,MOTO	1
8	PAFZZ	16764	1894023	.BUSHING,SLEEVE	1
9	PAFZZ	11862	1951567	SHIFTER FORK	1
10	PAFZZ	11862	1987049	PIN,SPRING	1
11	PAFZZ	11862	1945804	PIN,GROOVED,HEADED	1
12	PAFZZ	16764	10496538	HOUSING,ENGINE DRIV	1
13	PAFZZ	96906	MS16624-1031	RING,RETAINING	1
14	PAFZZ	24617	1928023	WASHER,FLAT	1
15	PAFZZ	24617	1928022	RING,RETAINING	1
16	PAFZZ	16764	1928021	BEARING,SLEEVE	1
17	PAFZZ	16764	1893445	DRIVE,ENGINE,ELECTR	1
18	PAFZZ	24617	821453	WASHER,FLAT UOC:194,208,209,210,230,231,252	1
19	PAFZZ	16764	9427815	BOLT,SELF-LOCKING	2
20	PAFZZ	11862	1986473	WINDING,STARTER-GEN	1
21	PAFZZ	16764	1887021	POLE SHOE,STARTER	4
22	PAFZZ	11862	1976940	PIN,STRAIGHT,HEADLE	1
23	PAFZZ	11862	1876458	PARTS KIT,ELECTRICA INCLUDES ITEM # 29,30,31,32,33	2
24	PAFZZ	11862	1876359	HOLDER,ELECTRICAL C	2
25	PAFZZ	11862	1986019	CLIP,RETAINING	2
26	PAFZZ	96906	MS35206-263	SCREW,MACHINE	4
27	PAFZZ	11862	800091	HOLDER,ELECTRICAL C	2
28	PAFZZ	11862	1852890	BRUSH,ELECTRICAL CO	4
29	PAFZZ	11862	431615	SCREW	2
30	PAFZZ	24617	9414224	SCREW,TAPPING,THREA	2
31	PAFZZ	11862	1987254	LEAD,ELECTRICAL	2
32	PAFZZ	96906	MS35649-202	NUT,PLAIN,HEXAGON	4
33	PAFZZ	96906	MS35340-43	WASHER,LOCK	4
34	PAFZZ	16764	1966923	PIN	2
35	PAFZZ	16764	1960908	BOLT,SHOULDER	1
36	PAFZZ	11862	1975326	STUD,SHOULDERED UOC:194,208,209,210,230,231,252	1
37	PAFZZ	11862	1986464	END BELL,ELECTRICAL	1
38	PAFZZ	16764	1955946	GROMMET,NONMETALLIC	1
39	PAFZZ	11862	1876358	BRACKET,DOUBLE ANGL	1
40	PAFZZ	16764	1984076	WASHER	1
41	PAFZZ	11862	1986471	ARMATURE,MOTOR	1
42	PAFZZ	16764	1968396	SCREW,MACHINE	8

END OF FIGURE

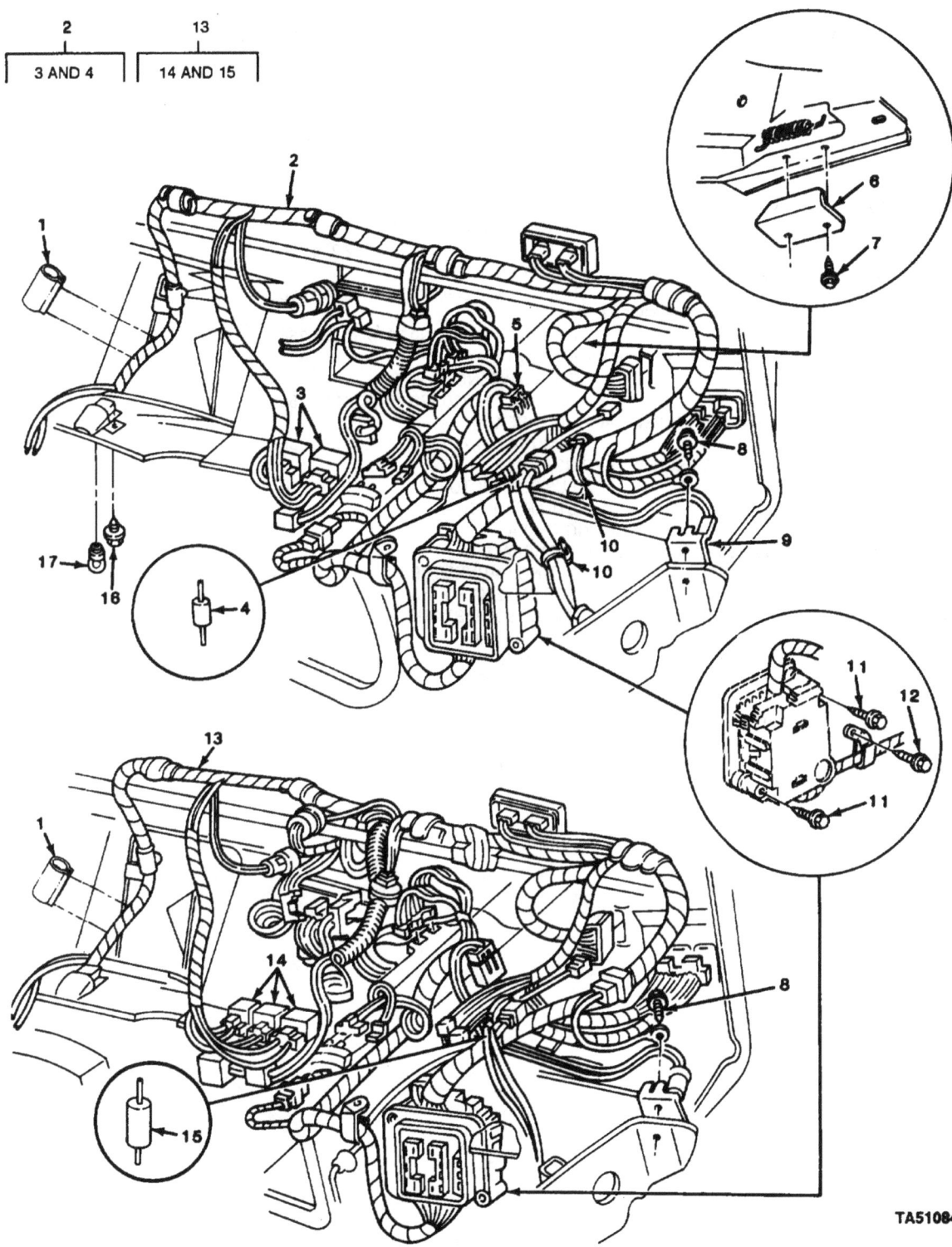

FIGURE 41. INSTRUMENT PANEL WIRING HARNESS.

TA510848

(1) (2) (3) (4) ITEM SMR PART NO CODE CAGEC NUMBER	(5) DESCRIPTION AND USABLE ON CODES (UOC)	(6) QTY
	GROUP 0607 INSTRUMENT OR ENGINE CONTROL PANEL	
	FIG. 41 INSTRUMENT PANEL WIRING HARNESS	
1 PAOZZ 11862 3999572	ADHESIVE,DRY MOUNTI	1
2 PAFFF 11862 12039309	WIRING HARNESS,BRAN UOC:194,208,209,230,231,252,254,256	1
3 PAFZZ 11862 12033862	.RELAY,SOLID STATE UOC:194,208,209,230,231,252,254,256	2
4 PAOZZ 11862 12006377	.RECTIFIER,CONNECTOR UOC:194,208,209,230,231,252,254,256	1
5 PAOZZ 11862 15591138	MODULE,INDICATOR,CO	1
6 PAFZZ 11862 329842	SHIELD,WIRING HARNE UOC:194,208,210,230,231,252,254,256	1
7 PAFZZ 96906 MS51850-86	SCREW,TAPPING,THREA UOC:194,208,210,230,231,252,254,256	2
8 PAFZZ 96906 MS51850-86	SCREW,TAPPING,THREA	1
9 PAFZZ 11862 459454	BUS,CONDUCTOR	1
10 PAFZZ 11862 3816659	STRAP,LINE SUPPORTI	2
11 PAFZZ 11862 9428613	SCREW,TAPPING,THREA	2
12 PAFZZ 11862 11508566	SCREW,TAPPING,THREA	1
13 PAFFF 11862 12039310	WIRING HARNESS,BRAN UOC:210,230	1
14 PAFZZ 11862 12033862	.RELAY,SOLID STATE UOC:210,230	3
15 PAFZZ 11862 12006377	.RECTIFIER,CONNECTOR UOC:210,230	1
16 PAFZZ 11862 11513932	SCREW,ASSEMBLED WAS	1
17 PAOZZ 08806 1003	LAMP,INCANDESCENT	1

END OF FIGURE

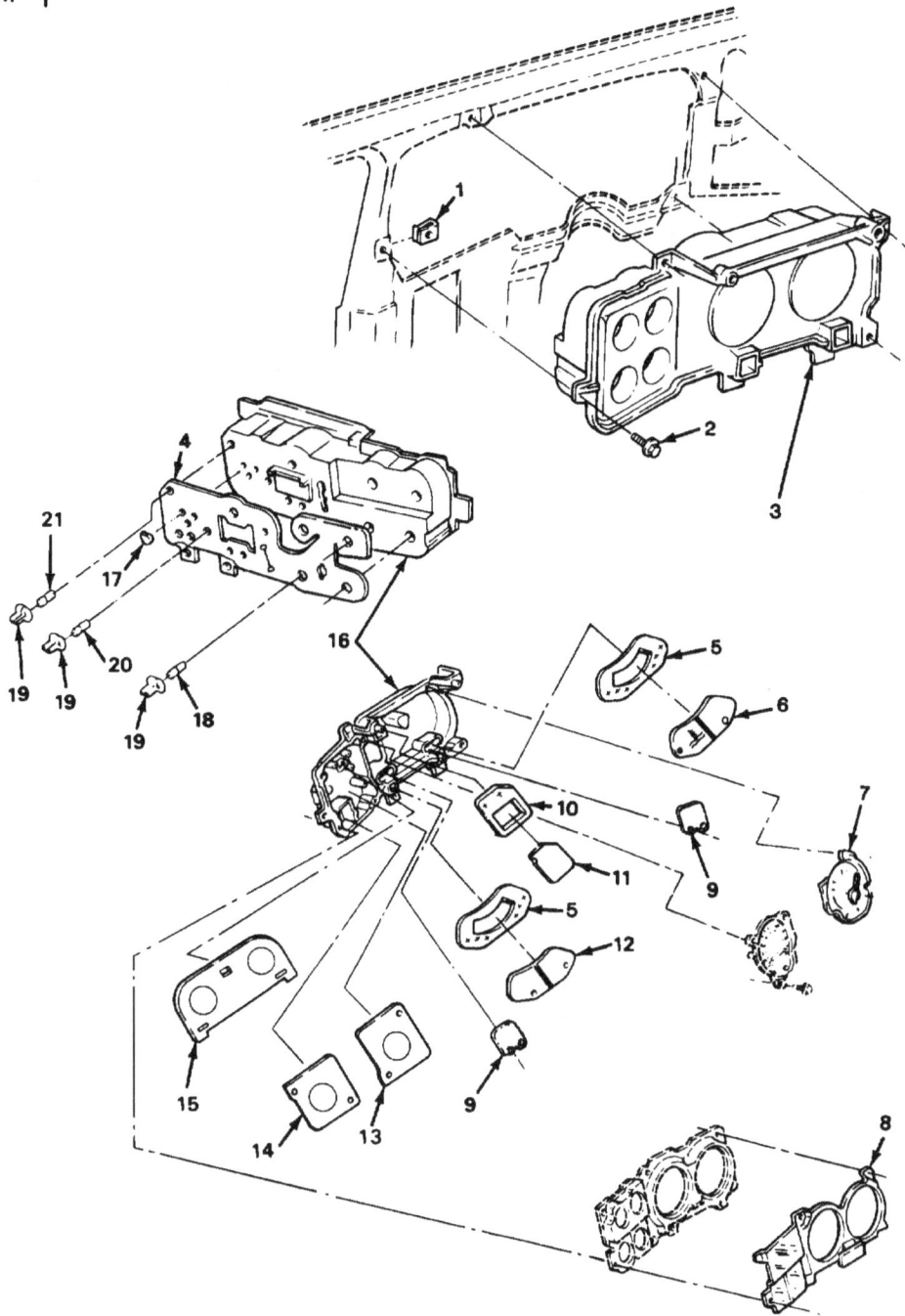

FIGURE 42. INSTRUMENT CLUSTER ASSEMBLY.

TA510849

(1) ITEM NO	(2) SMR CODE	(3) CAGEC	(4) PART NUMBER	(5) DESCRIPTION AND USABLE ON CODES (UOC)	(6) QTY
				GROUP 0607 INSTRUMENT OR ENGINE CONTROL PANEL	
				FIG. 42 INSTRUMENT CLUSTER ASSEMBLY	
1	PAOZZ	96906	MS90724-34	NUT,SHEET SPRING	4
2	PAOZZ	11862	11513932	SCREW,ASSEMBLED WAS	4
3	PAOOO	11862	25052807	PANEL,INSTRUMENT	1
4	PAOZZ	11862	25053623	.PRINTED WIRING BOAR	1
5	PAOZZ	11862	25015099	.GASKET	2
6	PAOZZ	11862	25022883	.LENS,LIGHT	1
7	PAOZZ	70040	6433429	.INDICATOR,LIQUID QU	1
8	PAOZZ	11862	25017376	.BEZEL,INSTRUMENT MO	1
9	PAOZZ	11862	6497476	.LENS,LIGHT	2
10	PAOZZ	11862	6497483	.GASKET	1
11	PAOZZ	11862	6497475	.FILTER,INDICATOR LI	1
12	PAOZZ	11862	25022884	.FILTER,INDICATOR LI UOC:194,208,209,210,230,231,252,254	1
13	PAOZZ	11862	25053501	.FILTER,INDICATOR LI	1
14	PAOZZ	11862	25053500	.FILTER,SIGNAL LIGHT	1
15	PAOZZ	11862	25076586	.FILTER,INDICATOR LI	1
16	PAOZZ	11862	25053622	.CASE,INSTRUMENT CLU	1
17	PAOZZ	11862	8986000	.CLIP,ELECTRICAL	3
18	PAOZZ	08806	194	LAMP,INCANDESCENT	1
19	PAOZZ	77060	2973932	LAMPHOLDER	14
20	PAOZZ	08806	194	LAMP,INCANDESCENT LEFT	1
21	PAOZZ	08806	168	LAMP,INCANDESCENT	12

END OF FIGURE

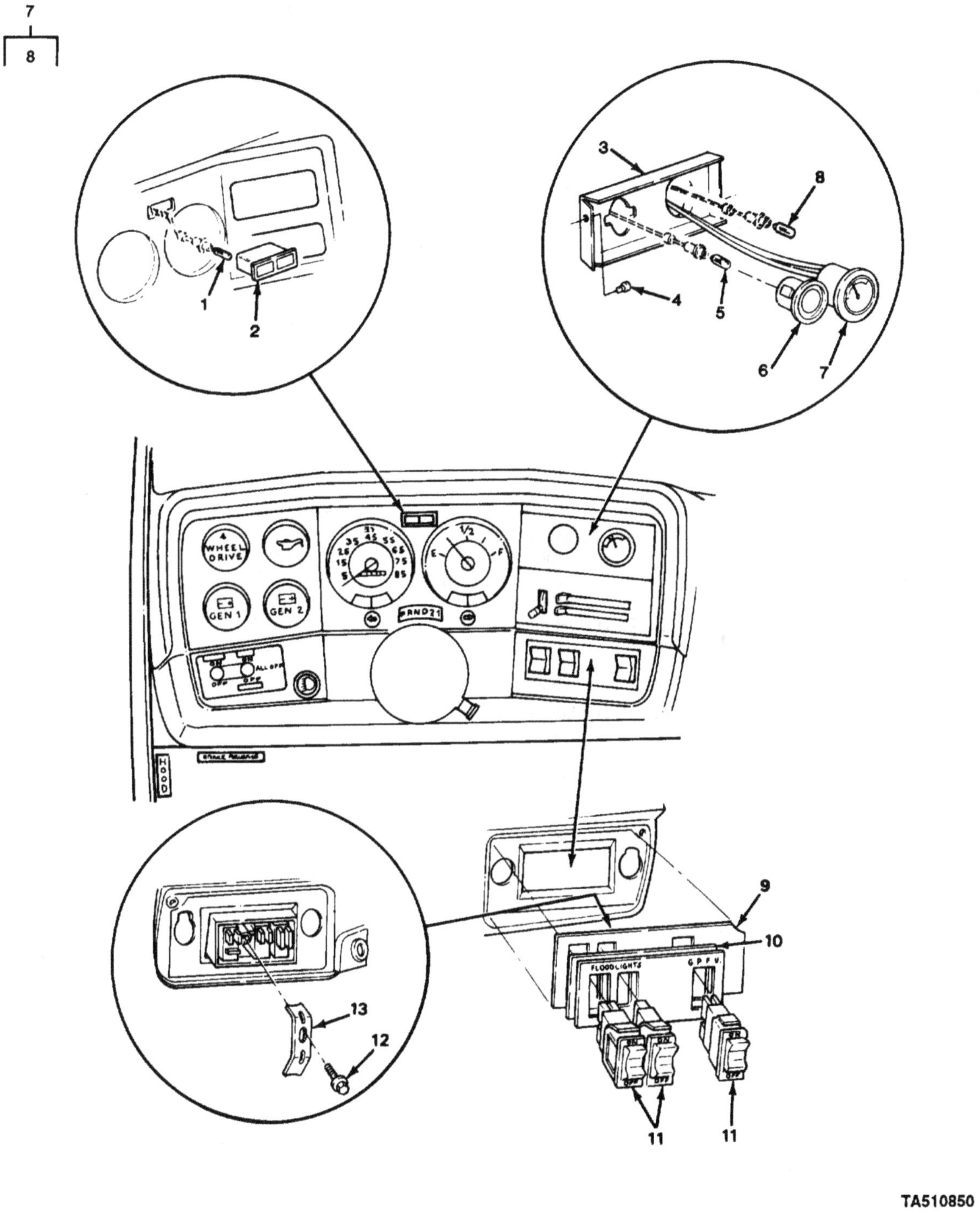

FIGURE 43. DOOR AJAR AND GLOW PLUG INDICATOR, VOLTMETER, GAS-PARTICULATE
FILTER UNIT AND FLOODLIGHT SWITCH.

TA510850

(1) (2) (3) (4)	(5)	(6)
ITEM SMR PART NO CODE CAGEC NUMBER	DESCRIPTION AND USABLE ON CODES (UOC)	QTY

GROUP 0607 INSTRUMENT OR ENGINE
CONTROL PANEL

FIG. 43 DOOR AJAR AND GLOW PLUG
INDICATOR, VOLTMETER,
GAS-PARTICULATE FILTER UNIT, AND
FLOODLIGHT SWITCH

Item	SMR	CAGEC	Part Number	Description	QTY
1	PAOZZ	08806	168	LAMP,INCANDESCENT	2
2	PAOZZ	11862	14066662	LENS,LIGHT	1
				UOC:194,208,209,210,230,231,252,254	
3	PAOZZ	11862	14072406	BRACKET,MOUNTING	1
				UOC:194,208,209,230,231,252,254,256	
3	PAOZZ	11862	14072409	BRACKET,MOUNTING	1
				UOC:210	
4	PAOZZ	11862	9441669	RIVET,BLIND	2
5	PAOZZ	08806	168	LAMP,INCANDESCENT	1
				UOC:210	
6	PAOZZ	11862	14075858	HOUSING,INDICATOR	1
				UOC:210	
7	PAOZZ	70040	6474942A	METER,SPECIAL SCALE	1
8	PAOZZ	08806	1445	.LAMP,INCANDESCENT VOLTMETER	1
9	PAOZZ	11862	14072410	COVER,ACCESSORY,SWI	1
				UOC:210	
10	PAOZZ	11862	14072448	COVER,ELECTRICAL SW	1
				UOC:210	
11	PAOZZ	11862	14072338	SWITCH,PUSH	3
				UOC:210	
12	PAOZZ	24617	9415163	SCREW,ASSEMBLED WAS	1
				UOC:210	
13	PAOZZ	11862	14072412	RETAINER,ELECTRICAL	1
				UOC:210	

END OF FIGURE

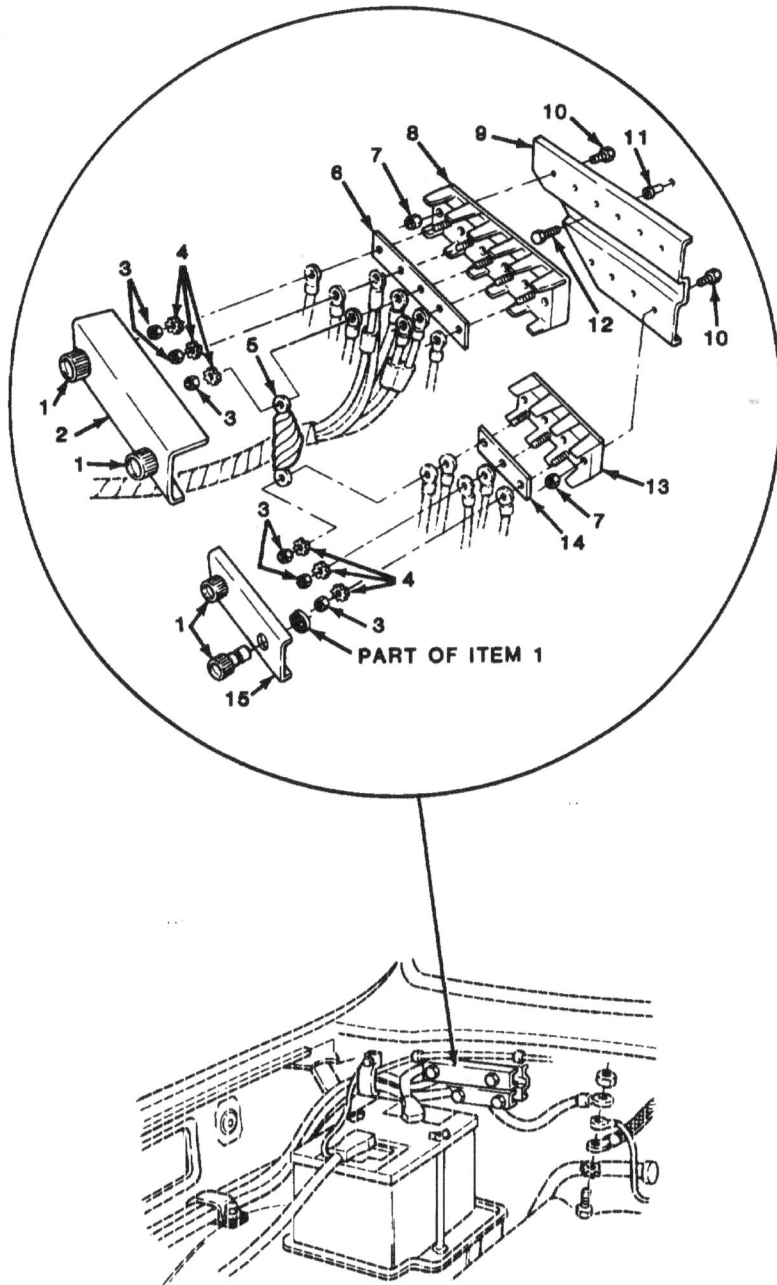

PART OF ITEM 1

TA510851

FIGURE 44. ACCESSORY WIRING TERMINAL BOARD AND COMPONENTS.

(1) (2) (3) (4) ITEM SMR PART NO CODE CAGEC NUMBER	(5) DESCRIPTION AND USABLE ON CODES (UOC)	(6) QTY
	GROUP 0608 MISCELLANEOUS ITEMS	
	FIG. 44 ACCESSORY WIRING TERMINAL BOARD AND COMPONENTS	
1 PAOZZ 96906 MS18029-24	NUT ASSEMBLY	4
2 MOOZZ 96906 MS18029-13L-5	BOARD,VEH ACC WRG MAKE FROM COVER, P/N MS18029-4S-8	1
3 PAOZZ 96906 MS35650-3314	NUT,PLAIN,HEXAGON	8
4 PAOZZ 96906 MS35335-34	WASHER,LOCK	8
5 PAOZZ 11862 15599225	CAPACITOR ASSEMBLY	1
6 PAOZZ 11862 14072336	BUS,CONDUCTOR	1
7 PAOZZ 96906 MS35649-262	NUT,PLAIN,HEXAGON	10
8 MOOZZ 96906 MS27212-4-5	TERMINAL BOARD MAKE FROM BOARD,P/N MS27212-4-8	1
9 PAOZZ 11862 14075900	TERMINAL BOARD BRAC	1
10 PAOZZ 96906 MS51849-33	SCREW,MACHINE	10
11 PAOZZ 11862 14005953	SHIELD,EXPANSION	2
12 PAOZZ 11862 9437702	BOLT,MACHINE	2
13 MOOZZ 96906 MS27212-4-3	BOARD,VEH ACC WRG MAKE FROM BOARD, P/N MS27212-4-8 UOC:194,208,209,210,230,231,252	1
14 MOOZZ 11862 14072337	CONNECTOR LINK (3.625" LG) MAKE FROM ITEM 6 (P/N 14072336)	1
15 MOOZZ 96906 MS18029-13S-3	COVER MAKE FROM COVER,P/N MS18029-13S-8	1

END OF FIGURE

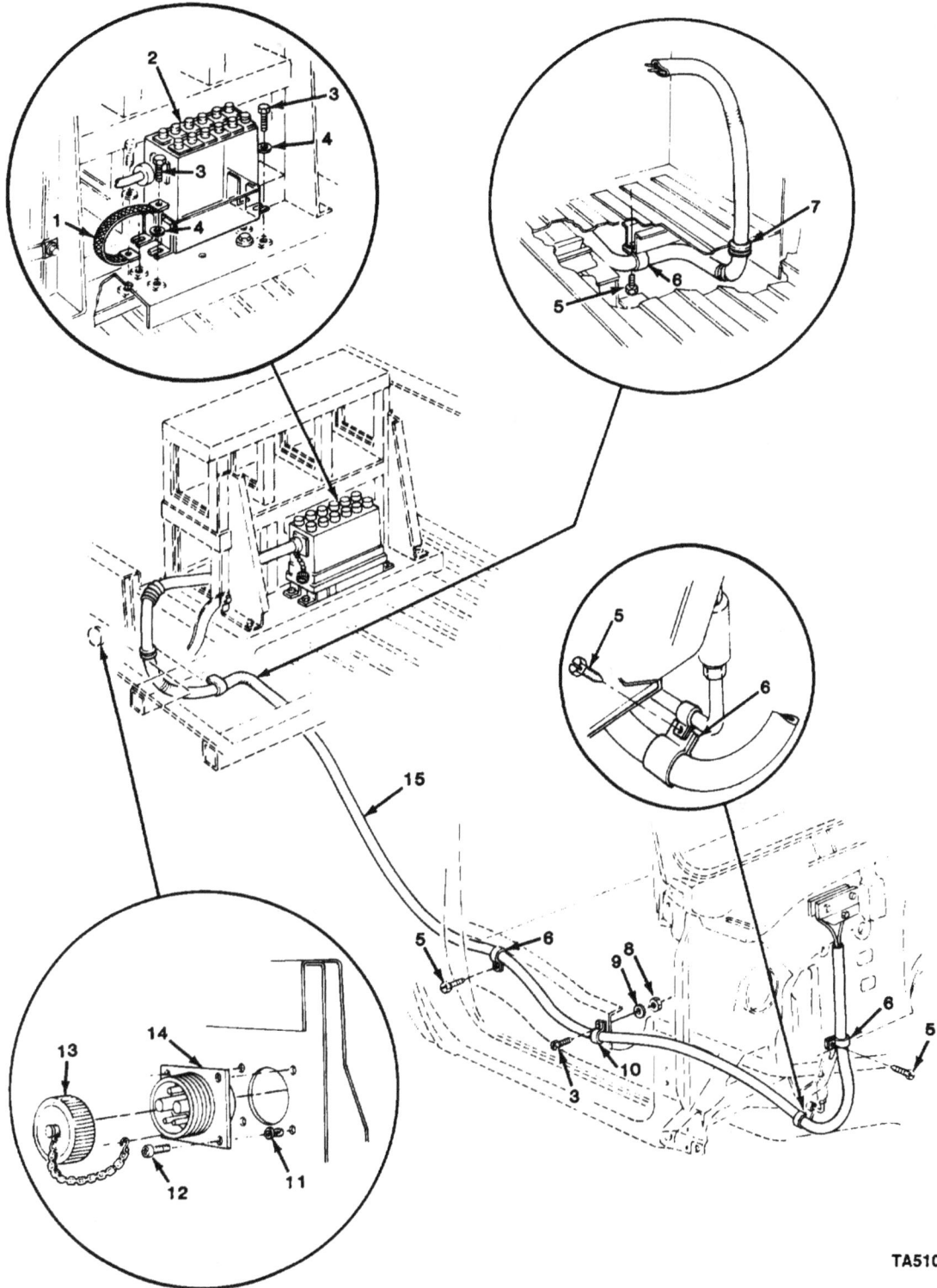

FIGURE 45. POWER JUNCTION BOX AND CABLE.

TA510852

(1) ITEM NO	(2) SMR CODE	(3) CAGEC	(4) PART NUMBER	(5) DESCRIPTION AND USABLE ON CODES (UOC)	(6) QTY
				GROUP 0608 MISCELLANEOUS ITEMS	
				FIG. 45 POWER JUNCTION BOX AND CABLE	
1	XDOZZ	80063	SC-B-75180-IV	LEAD, ELECTRICAL UOC:208,230	1
2	PAOFF	80063	SC-D-691391	INTERCONNECTING BOX FOR COMPONENT PARTS SEE TM11-5820-862-13&P UOC:194,208	1
3	PAOZZ	96906	MS90728-6	SCREW,CAP,HEXAGON H UOC:208,230	5
4	PAOZZ	96906	MS35335-33	WASHER,LOCK UOC:208,230	4
5	PAOZZ	11862	11504447	SCREW,TAPPING,THREA UOC:194,208,230,252,254,256	4
6	PAOZZ	11862	2043151	CLAMP,LOOP UOC:194,208,230,252,254,256	4
7	PAOZZ	11862	343124	GROMMET,NONMETALLIC UOC:208,230	1
8	PAOZZ	96906	MS51967-2	NUT,PLAIN,HEXAGON UOC:194,208	1
9	PAOZZ	96906	MS35338-44	WASHER,LOCK UOC:208,230	1
10	PAOZZ	96906	MS21333-116	CLAMP,LOOP UOC:208,230	1
11	PAOZZ	00613	C-5139-2	RECEPTACLE,TURNLOCK UOC:208,209	4
12	PAOZZ	96906	MS35206-267	SCREW,MACHINE UOC:208,230	4
13	PAOZZ	96906	MS25043-32DA	COVER,ELECTRICAL CO UOC:208,230	1
14	PAOZZ	19207	8701347	SHELL,ELECTRICAL CO UOC:208,230	1
15	PAOZZ	11862	12039254	CABLE ASSEMBLY,POWE UOC:208,230,252,254,256	1

END OF FIGURE

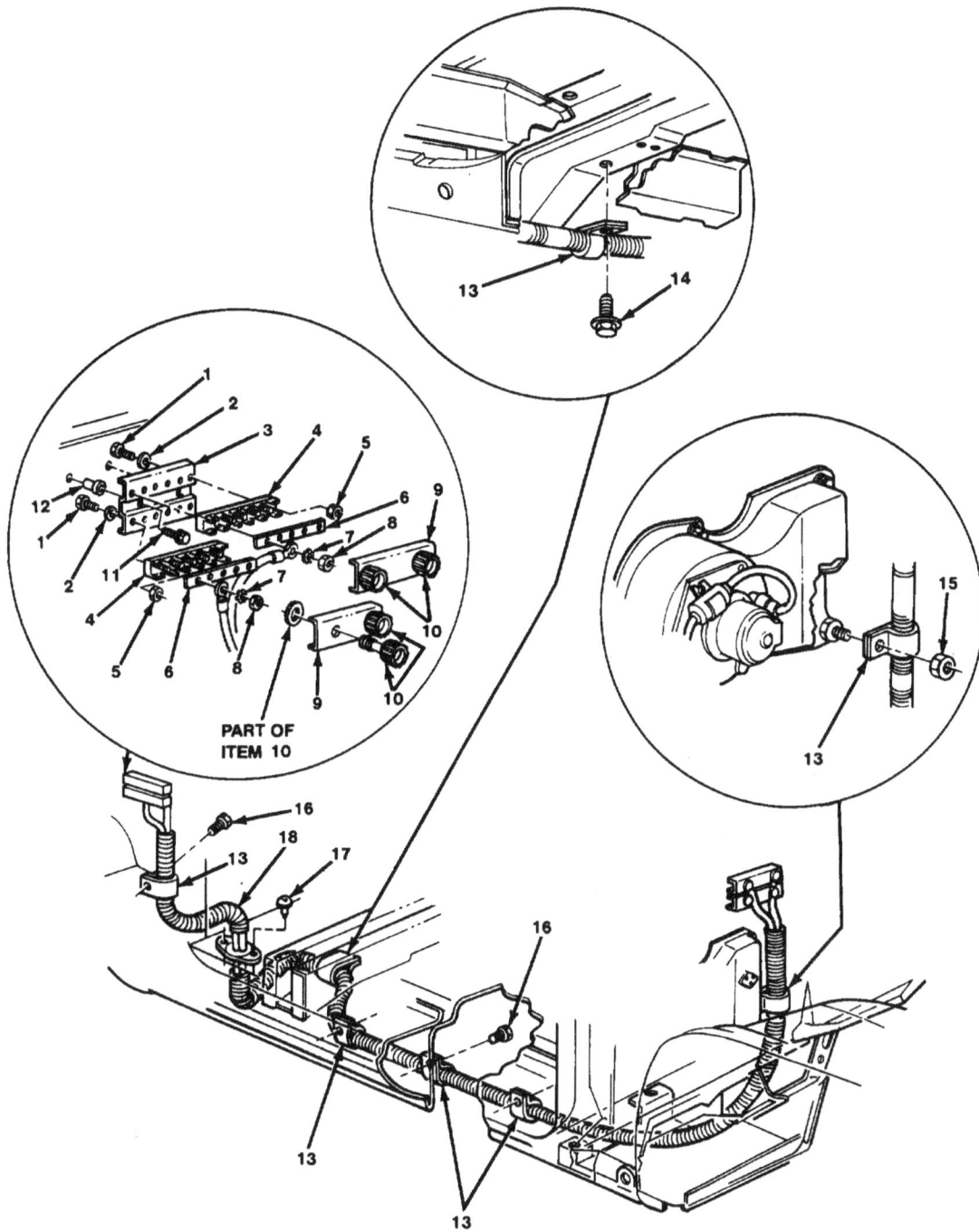

FIGURE 46. RADIO FEED HARNESS (M1009).

TA510853

(1) ITEM NO	(2) SMR CODE	(3) CAGEC	(4) PART NUMBER	(5) DESCRIPTION AND USABLE ON CODES (UOC)	(6) QTY
				GROUP 0608 MISCELLANEOUS ITEMS	
				FIG. 46 RADIO FEED HARNESS (M1009)	
1	PAOZZ	96906	MS51849-33	SCREW,MACHINE UOC:209	12
2	PAOZZ	96906	MS27183-6	WASHER,FLAT UOC:209	12
3	PAOZZ	11862	15599989	TERMINAL BOARD BRAC UOC:209	1
4	MOOZZ	96906	MS27212-4-5	TERMINAL BOARD MAKE FROM BOARD,P/N MS27212-4-8 UOC:209	2
5	PAOZZ	96906	MS35649-262	NUT,PLAIN,HEXAGON UOC:209	12
6	PAOZZ	11862	14072336	BUS,CONDUCTOR UOC:209	2
7	PAOZZ	96906	MS35335-34	WASHER,LOCK UOC:209	10
8	PAOZZ	96906	MS35650-3314	NUT,PLAIN,HEXAGON UOC:209	10
9	MOOZZ	96906	MS18029-13L-5	BOARD,VEH ACC WRG MAKE FROM COVER, P/N MS18029-4S-8 UOC:209	2
10	PAOZZ	96906	MS18029-24	NUT ASSEMBLY UOC:209	4
11	PAOZZ	11862	9437702	BOLT,MACHINE UOC:209	2
12	PAOZZ	11862	14005953	SHIELD,EXPANSION UOC:209	2
13	PAOZZ	11862	2043151	CLAMP,LOOP UOC:209	5
14	PAOZZ	11862	11508534	BOLT,MACHINE UOC:209	1
15	PAOZZ	24617	11506003	NUT,PLAIN,ASSEMBLED UOC:209	1
16	PAOZZ	11862	9436175	RIVET,BLIND UOC:209	4
17	PAOZZ	79846	ABA64LBA	RIVET,BLIND UOC:209	2
18	PAOZZ	11862	12044586	LEAD AND CONDUIT AS UOC:209	1

END OF FIGURE

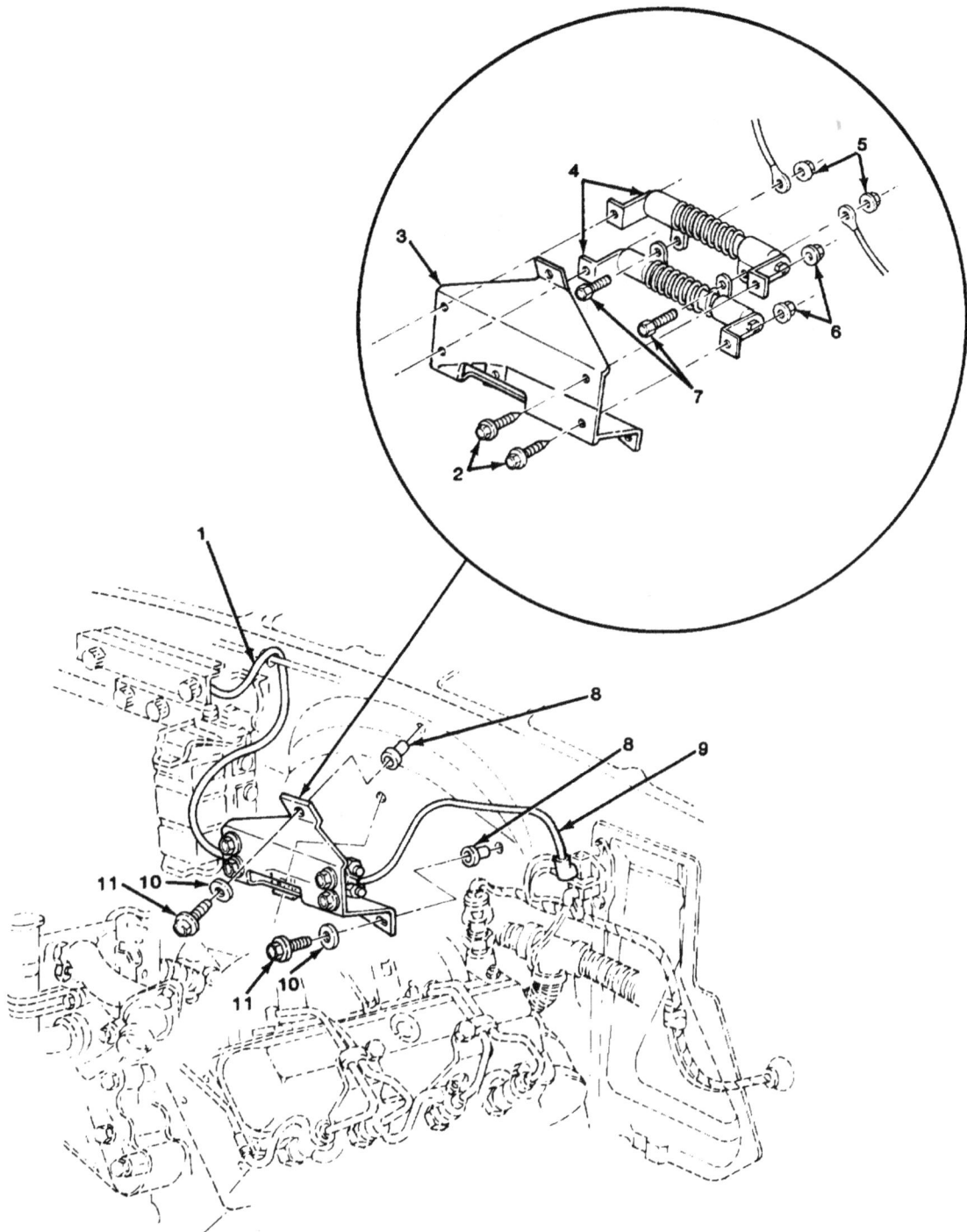

TA510854

FIGURE 47. BATTERY BOOSTER RESISTOR.

(1)	(2)	(3)	(4)	(5)	(6)
ITEM NO	SMR CODE	CAGEC	PART NUMBER	DESCRIPTION AND USABLE ON CODES (UOC)	QTY

GROUP 0608 MISCELLANEOUS ITEMS

FIG. 47 BATTERY BOOSTER RESISTOR

1	PAOZZ	77060	12039297	LEAD,IGNITION,ENGIN	1
2	PAOZZ	11862	9439770	BOLT,ASSEMBLED WASH	4
3	PAOZZ	11862	14076848	BRACKET,RESISTOR	1
4	PAOZZ	11862	14076847	RESISTOR,FIXED,WIRE	2
5	PAOZZ	72582	271163	NUT,PLAIN,ASSEMBLED	2
6	PAOZZ	24617	271172	NUT,SELF-LOCKING,AS	4
7	PAOZZ	96906	MS51849-55	SCREW, MACHINE	2
8	PAOZZ	11862	14005953	SHIELD,EXPANSION	3
9	PAOZZ	77060	12039298	LEAD,IGNITION,ENGIN	1
10	PAOZZ	96906	MS35335-33	WASHER,LOCK	3
11	PAOZZ	11862	9437702	BOLT,MACHINE	3

END OF FIGURE

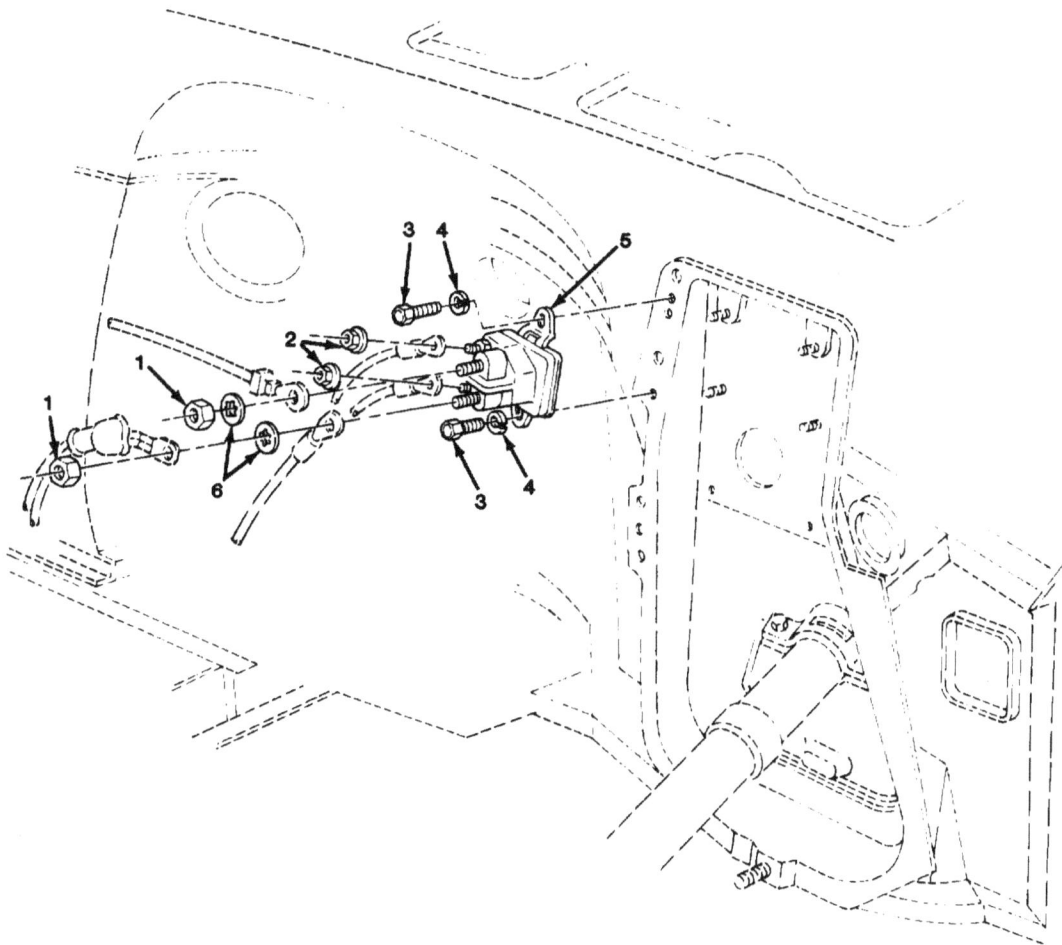

TA510855

FIGURE 48. GLOW PLUG RELAY.

(1) ITEM NO	(2) SMR CODE	(3) CAGEC	(4) PART NUMBER	(5) DESCRIPTION AND USABLE ON CODES (UOC)	(6) QTY
				GROUP 0608 MISCELLANEOUS ITEMS	
				FIG. 48 GLOW PLUG RELAY	
1	PAOZZ	96906	MS51968-5	NUT,PLAIN,HEXAGON	2
2	PAOZZ	96906	MS35649-205	NUT,PLAIN,HEXAGON	2
3	PAOZZ	11862	9440166	SCREW,TAPPING,THREA	2
4	PAOZZ	96906	MS35338-44	WASHER,LOCK	2
5	PAOZZ	72055	98226	RELAY,ELECTROMAGNET	1
6	PAOZZ	96906	MS35333-41	WASHER,LOCK	2

END OF FIGURE

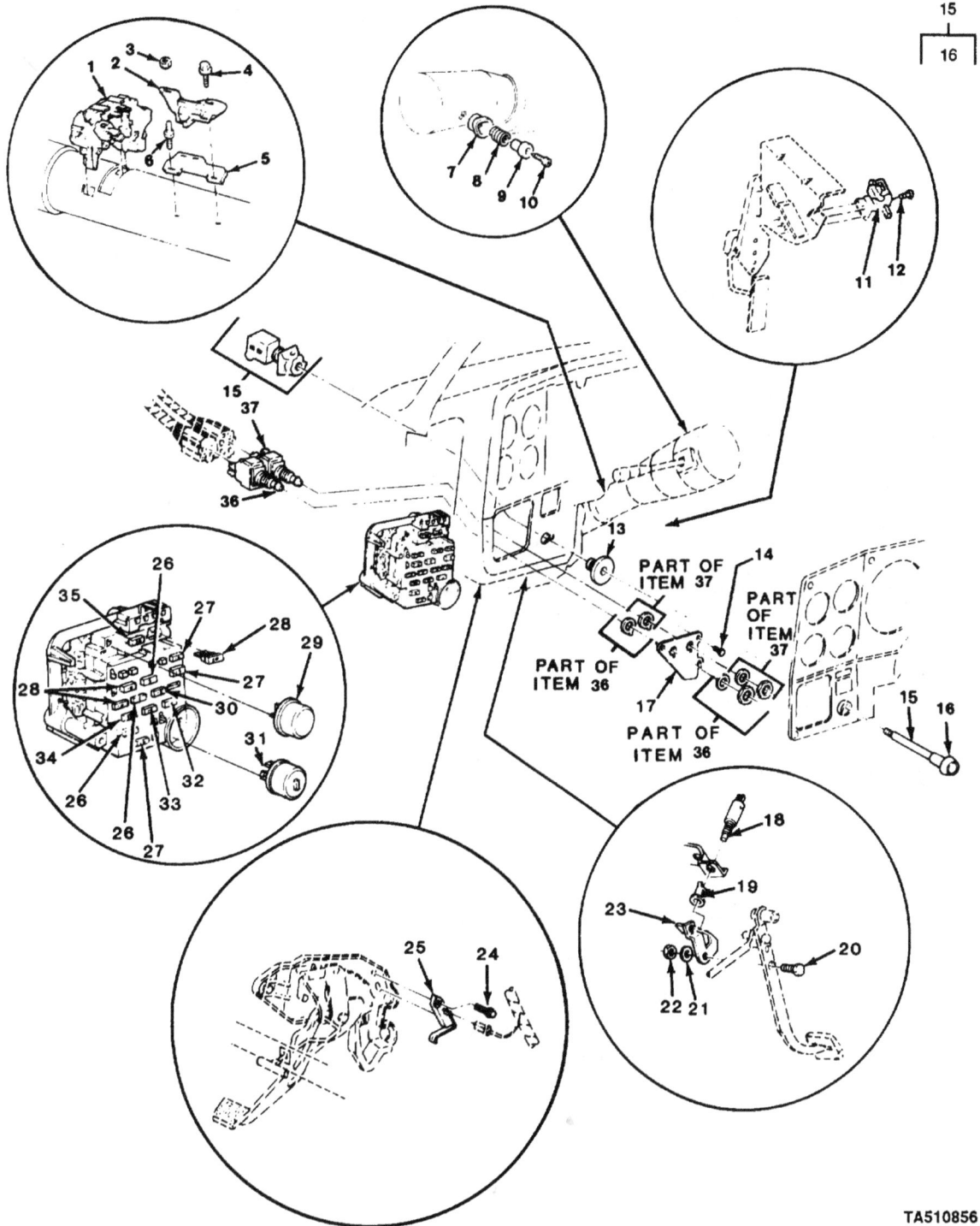

FIGURE 49. HEADLAMP, BLACKOUT LAMP AND TAILLAMP SWITCHES AND FUSES.

TA510856

(1) ITEM NO	(2) SMR CODE	(3) CAGEC	(4) PART NUMBER	(5) DESCRIPTION AND USABLE ON CODES (UOC)	(6) QTY
				GROUP 0608 MISCELLANEOUS ITEMS	
				FIG. 49 HEADLAMP, BLACKOUT LAMP, AND TAILLAMP SWITCHES AND FUSES	
1	PAOZZ	11862	22514861	SWITCH,SAFETY,NEUTR	1
2	PAFZZ	11862	26019661	SWITCH,ELECTRICAL	1
3	PAFZZ	96906	MS35649-202	NUT,PLAIN,HEXAGON	1
4	PAFZZ	11862	7806433	SCREW,CAP,HEXAGON H	1
5	PAFZZ	11862	1990115	SWITCH,ENGINE START	1
6	PAFZZ	11862	7846418	STUD	1
7	PAOZZ	11862	22507977	KNOB	1
8	PAOZZ	11862	556743	SPRING,HELICAL,COMP	1
9	PAOZZ	11862	556742	PUSH BUTTON	1
10	PAOZZ	11862	1640902	SCREW,TAPPING,THREA	1
11	PAOZZ	11862	1242101	SWITCH,DOWN SHIFT TRNS D/SHIFT	1
12	PAFZZ	96906	MS51850-86	SCREW,TAPPING,THREA	1
13	PAOZZ	11862	6258213	NUT,BEZEL	1
14	PAOZZ	11862	2477054	RIVET,BLIND	3
15	PAOOO	11862	1995217	SWITCH ASSEMBLY	1
16	PAOZZ	11862	469302	.KNOB	1
17	PAOZZ	11862	14072413	BRACKET,ELECTRICAL	1
18	PAOZZ	11862	1261219	SWITCH,PUSH	1
19	PAOZZ	11862	1361699	CLIP,RETAINING	1
20	PAOZZ	96906	MS90728-60	SCREW,CAP,HEXAGON H	1
21	PAOZZ	96906	MS35338-46	WASHER,LOCK	1
22	PAOZZ	96906	MS51967-8	NUT,PLAIN,HEXAGON	1
23	PAOZZ	11862	14040525	ACTUATOR,STOP LIGHT UOC:209	1
23	PAOZZ	11862	14000395	STRIKER,BRAKE PEDAL UOC:194,208,210,230,231,252,254,256	1
24	PAOZZ	11862	9418944	SCREW,TAPPING,THREA	1
25	PAOZZ	11862	477361	SWITCH,SENSITIVE	1
26	PAOZZ	11862	12004005	FUSE,INCLOSED LINK	3
27	PAOZZ	11862	12004007	FUSE,INCLOSED LINK	3
28	PAOZZ	11862	12004009	FUSE,INCLOSED LINK	3
29	PAOZZ	94988	552-12V	FLASHER,THERMAL	1
30	PAOZZ	11862	12004010	FUSE,INCLOSED LINK	1
31	PAOZZ	94988	224-12V	FLASHER,THERMAL	1
32	PAOZZ	11862	12004011	FUSE,INCLOSED LINK	1
33	PAOZZ	11862	12004008	FUSE,INCLOSED LINK	1
34	PAOZZ	11862	12004011	FUSE,INCLOSED LINK UOC:210	1
35	PAOZZ	11862	12004009	FUSE,INCLOSED LINK UOC:210	1
36	PAOZZ	11862	14072358	SWITCH,TOGGLE	1
37	PAOZZ	11862	14072339	SWITCH,TOGGLE	1

END OF FIGURE

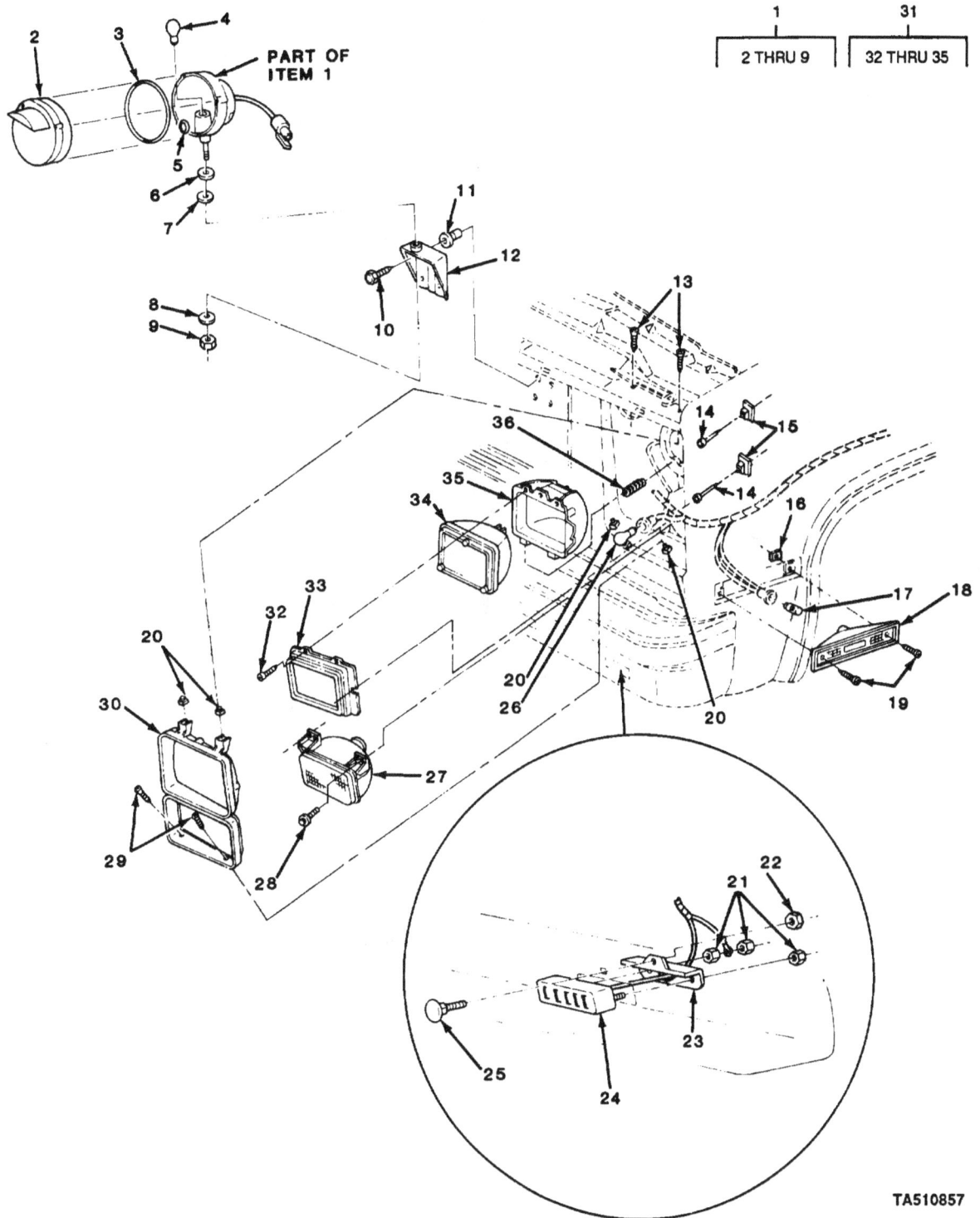

FIGURE 50. BLACKOUT HEADLAMP AND HEADLAMP ILLUMINATING COMPONENTS.

TA510857

(1) (2) (3) (4) ITEM SMR PART NO CODE CAGEC NUMBER	(5) DESCRIPTION AND USABLE ON CODES (UOC)	(6) QTY

GROUP 0609 LIGHTS

FIG. 50 BLACKOUT HEADLAMP AND
HEADLAMP ILLUMINATING COMPONENTS

1	PAOOO	11862	14072333	BLACKOUT LIGHT	1
2	PAOZZ	5A910	DC8218	.RETAINER,LENS	1
3	PAOZZ	34904	DC8226	.GASKET	1
4	PAOZZ	96906	MS35478-1073	.LAMP,INCANDESCENT	1
5	PAOZZ	5A910	DC8211	.GASKET	1
6	PAOZZ	5A910	DC8228	.WASHER,SADDLE	1
7	PAOZZ	19207	5294507	.WASHER,FINISHING	1
8	PAOZZ	96906	MS35338-46	.WASHER,LOCK	1
9	PAOZZ	96906	MS51967-8	.NUT,PLAIN,HEXAGON	1
10	PAOZZ	11862	9440334	BOLT,ASSEMBLED WASH	4
11	PAOZZ	11862	14005953	SHIELD,EXPANSION	4
12	PAOZZ	11862	14072431	BRACKET	1
13	PAOZZ	96906	MS51862-26	SCREW,TAPPING,THREA	4
14	PAOZZ	11862	9438150	SETSCREW	4
15	PAOZZ	11862	362379	NUT,HEADLIGHT MOUNT	4
16	PAOZZ	96906	MS90724-34	NUT,SHEET SPRING	4
17	PAOZZ	08806	194	LAMP,INCANDESCENT	2
18	PAOZZ	11862	915449	LENS,LIGHT LEFT	1
18	PAOZZ	11862	915450	LENS,LIGHT RIGHT	1
19	PAOZZ	11862	15605040	SCREW,TAPPING,THREA	4
20	PAOZZ	11862	347347	GROMMET,NONMETALLIC	8
21	PAOZZ	72582	271163	NUT,PLAIN,ASSEMBLED	6
22	PAOZZ	24617	11506003	NUT,PLAIN,ASSEMBLED	2
23	PAOZZ	11862	14072421	BRACKET,DOUBLE ANGL	2
24	PAOZZ	11862	15559316	LIGHT,MARKER,CLEARA	2
25	PAOZZ	24617	11503778	BOLT,SQUARE NECK	2
26	PAOZZ	08806	2057NA	LAMP,INCANDESCENT	2
27	PAOZZ	11862	915908	LENS,LIGHT	2
28	PAOZZ	96906	MS51871-4	SCREW,TAPPING,THREA	6
29	PAOZZ	11862	11504656	SCREW,TAPPING,THREA	4
30	PAOZZ	11862	14043873	RETAINER,LENS LEFT	1
30	PAOZZ	11862	14043874	RETAINER,LENS RIGHT	1
31	PAOOO	11862	16500591	HEADLIGHT	2
32	PAOZZ	11862	5966249	.SCREW	4
33	PAOZZ	11862	16501759	RING ASSEMBLY,RETAI UOC:194,208,209,210,230,231,252,254	1
34	PAOZZ	08806	H6054	.LAMP,INCANDESCENT	1
35	PAOZZ	11862	5968095	.HOUSING,LIGHT	1
36	PAOZZ	11862	459461	SPRING,HELICAL,EXTE	2

END OF FIGURE

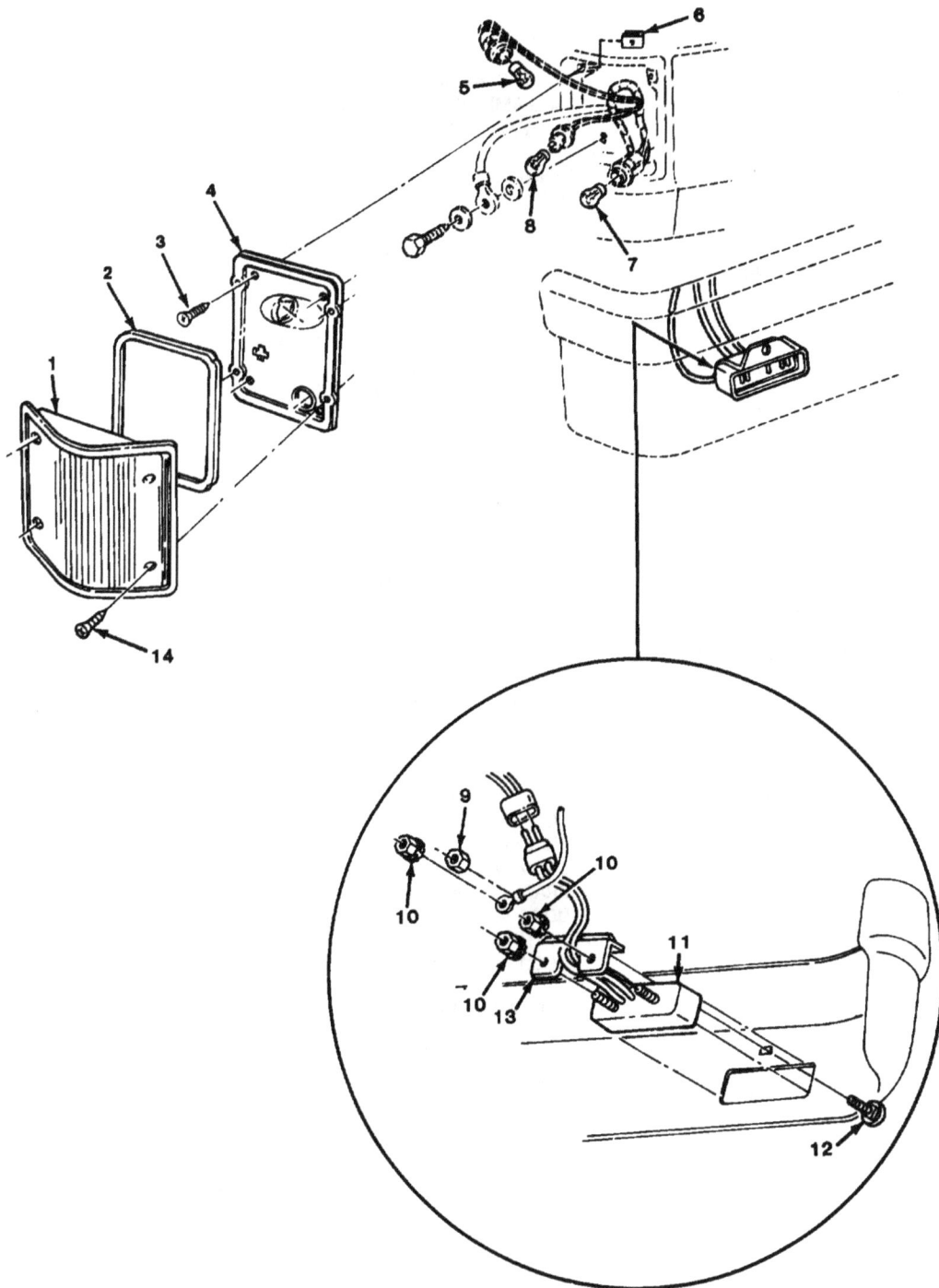

TA510858

FIGURE 51. REAR BLACKOUT LAMPS AND TAILLAMPS ILLUMINATING COMPONENTS
(ALL EXCEPT M1010 AND M1031).

(1) (2) (3) (4)	(5)	(6)
ITEM SMR PART		
NO CODE CAGEC NUMBER	DESCRIPTION AND USABLE ON CODES (UOC)	QTY

GROUP 0609 LIGHTS

FIG. 51 REAR BLACKOUT LAMPS AND
TAILLAMPS ILLUMINATING COMPONENTS
(ALL EXCEPT M1010 AND M1031)

ITEM NO	SMR CODE	CAGEC	PART NUMBER	DESCRIPTION AND USABLE ON CODES (UOC)	QTY
1	PAOZZ	11862	5965775	LENS,LIGHT LEFT UOC:194,208,209,230,252,254,256	1
1	PAOZZ	11862	5965776	LENS,LIGHT RIGHT UOC:194,208,209,230,252,254,256	1
2	PAOZZ	11862	5965748	GASKET UOC:194,208,209,230,252,254,256	2
3	PAOZZ	96906	MS51861-38	SCREW,TAPPING,THREA UOC:194,208,209,230,252,254,256	8
4	PAOZZ	11862	5965771	HOUSING,LIGHT LEFT UOC:194,208,209,230,252,254,256	1
4	PAOZZ	11862	5965772	HOUSING,TAIL LAMP RIGHT UOC:194,208,209,230,252,254,256	1
5	PAOZZ	08806	2057	LAMP,INCANDESCENT UOC:194,208,209,230,252,254,256	2
6	PAOZZ	96906	MS90724-34	NUT,SHEET SPRING UOC:194,208,209,230,252,254,256	8
7	PAOZZ	08806	1156	LAMP,INCANDESCENT UOC:194,208,209,230,252,254,256	2
8	PAOZZ	08806	168	LAMP,INCANDESCENT UOC:194,208,209,230,252,254,256	2
9	PAOZZ	24617	11506003	NUT,PLAIN,ASSEMBLED UOC:194,208,209,230,252,254,256	2
10	PAOZZ	72582	271163	NUT,PLAIN,ASSEMBLED UOC:194,208,209,230,252,254,256	6
11	PAOZZ	11862	15559312	LIGHT,MARKER,CLEARA UOC:194,208,209,230,252,254,256	2
12	PAOZZ	24617	11503778	BOLT,SQUARE NECK UOC:194,208,209,230,252	2
13	PAOZZ	11862	14072433	BRACKET,DOUBLE ANGL UOC:209	2
13	PAOZZ	11862	14072434	BRACKET,DOUBLE ANGL UOC:194,208,230,252,254,256	2
14	PAOZZ	11862	11504655	SCREW,TAPPING,THREA UOC:194,208,209,230,252,254,256	8

END OF FIGURE

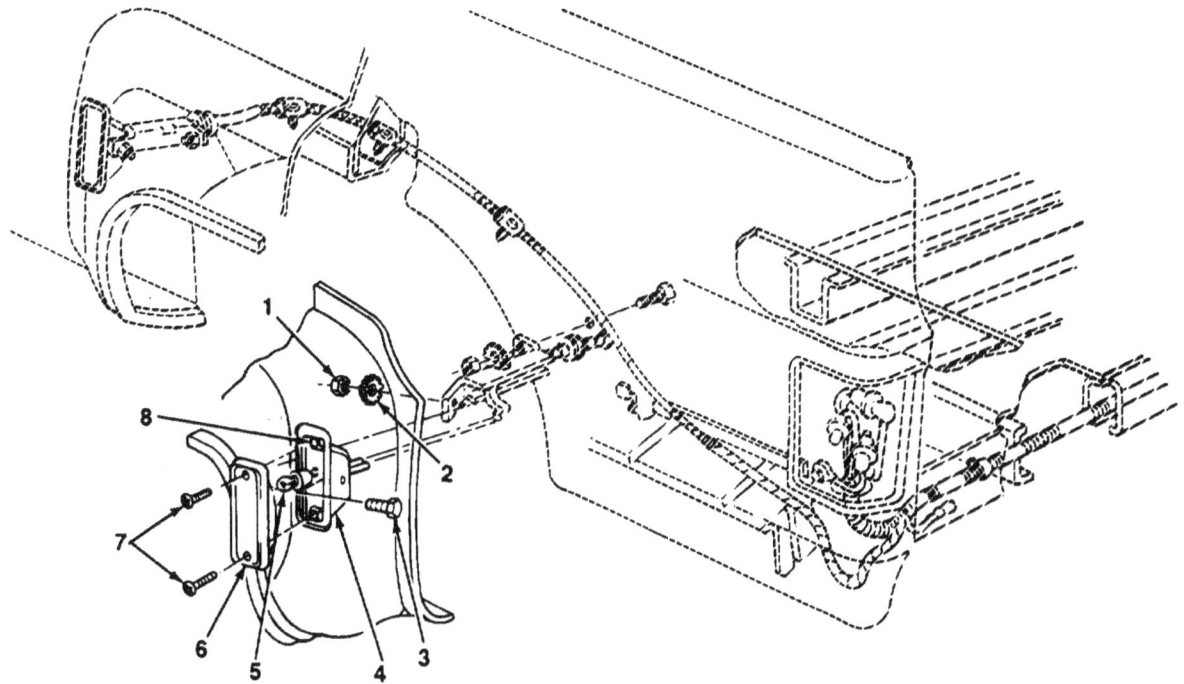

TA510859

FIGURE 52. REAR FENDER SIDE MARKERS (M1028A2 AND M1082A3).

(1) ITEM NO	(2) SMR CODE	(3) CAGEC	(4) PART NUMBER	(5) DESCRIPTION AND USABLE ON CODES (UOC)	(6) QTY
				GROUP 0609 LIGHTS	
				FIG. 52 REAR FENDER SIDE MARKERS (M1028A2 AND M1028A3)	
1	PAOZZ	96906	MS51967-2	NUT,PLAIN,HEXAGON UOC:254,256	2
2	PAOZZ	96906	MS35335-33	WASHER,LOCK UOC:254,256	2
3	PAOZZ	11862	9440033	BOLT,MACHINE UOC:254,256	2
4	PAOZZ	11862	330492	HOUSING,LIGHT UOC:254,256	4
5	PAOZZ	08806	168	LAMP,INCANDESCENT UOC:254,256	1
6	PAOZZ	11862	339885	LIGHT,MARKER,CLEARA UOC:254,256	2
6	PAOZZ	11862	339887	LENS,LIGHT UOC:254,256	2
7	PAOZZ	11862	11504655	SCREW,TAPPING,THREA UOC:254,256	2
8	PAOZZ	96906	MS90724-34	NUT,SHEET SPRING UOC:254,256	2

END OF FIGURE

7

8 THRU 11

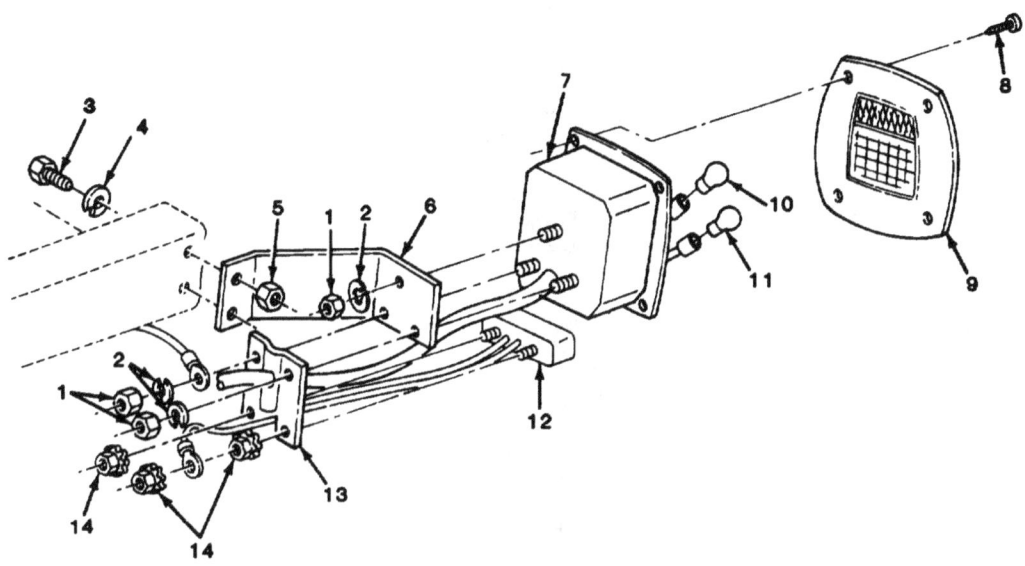

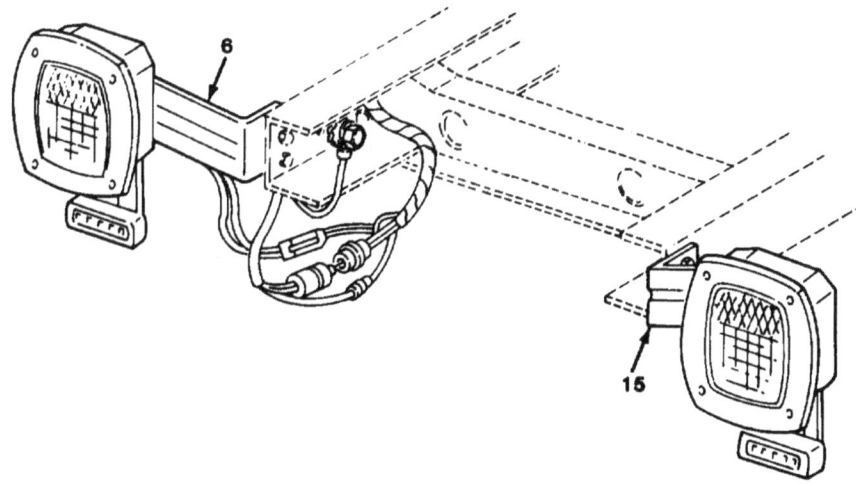

TA510860

FIGURE 53. REAR BLACKOUT LAMPS AND TAILLAMP ILLUMINATING COMPONENTS
(M1010 AND M1031).

(1) (2) (3) (4) ITEM SMR PART NO CODE CAGEC NUMBER	(5) DESCRIPTION AND USABLE ON CODES (UOC)	(6) QTY
	GROUP 0609 LIGHTS	
	FIG. 53 REAR BLACKOUT LAMPS AND TAILLAMP ILLUMINATING COMPONENTS (M1010 AND M1031)	
1 PAOZZ 96906 MS51967-2	NUT,PLAIN,HEXAGON UOC:210,231	6
2 PAOZZ 96906 MS35335-33	WASHER,LOCK UOC:210,231	6
3 PAOZZ 96906 MS90728-60	SCREW,CAP,HEXAGON H UOC:210,231	4
4 PAOZZ 96906 MS35340-46	WASHER,LOCK UOC:210,231	4
5 PAOZZ 96906 MS51967-8	NUT,PLAIN,HEXAGON UOC:210,231	4
6 PAOZZ 11862 370873	BRACKET,ANGLE UOC:210,231	1
7 PAOOO 11862 370867	STOP LIGHT-TAILLIGH LEFT UOC:210,231	1
7 PAOOO 11862 370868	STOP LIGHT-TAILLIGH RIGHT UOC:210,231	1
8 PAOZZ 11862 11504736	.SCREW,TAPPING,THREA UOC:210,231	4
9 PAOZZ 11862 475922	.LENS,LIGHT UOC:210,231	1
10 PAOZZ 08806 1156	.LAMP,INCANDESCENT UOC:210,231	1
11 PAOZZ 08806 1157	.LAMP,INCANDESCENT UOC:210,231	1
12 PAOZZ 11862 15559312	LIGHT,MARKER,CLEARA UOC:210,231	2
13 PAOZZ 11862 14072481	STRAP,RETAINING UOC:210,231	2
14 PAOZZ 72582 271163	NUT,PLAIN,ASSEMBLED UOC:210,231	6
15 PAOZZ 11862 370874	BRACE,LAMPHOLDER UOC:210,231	1

END OF FIGURE

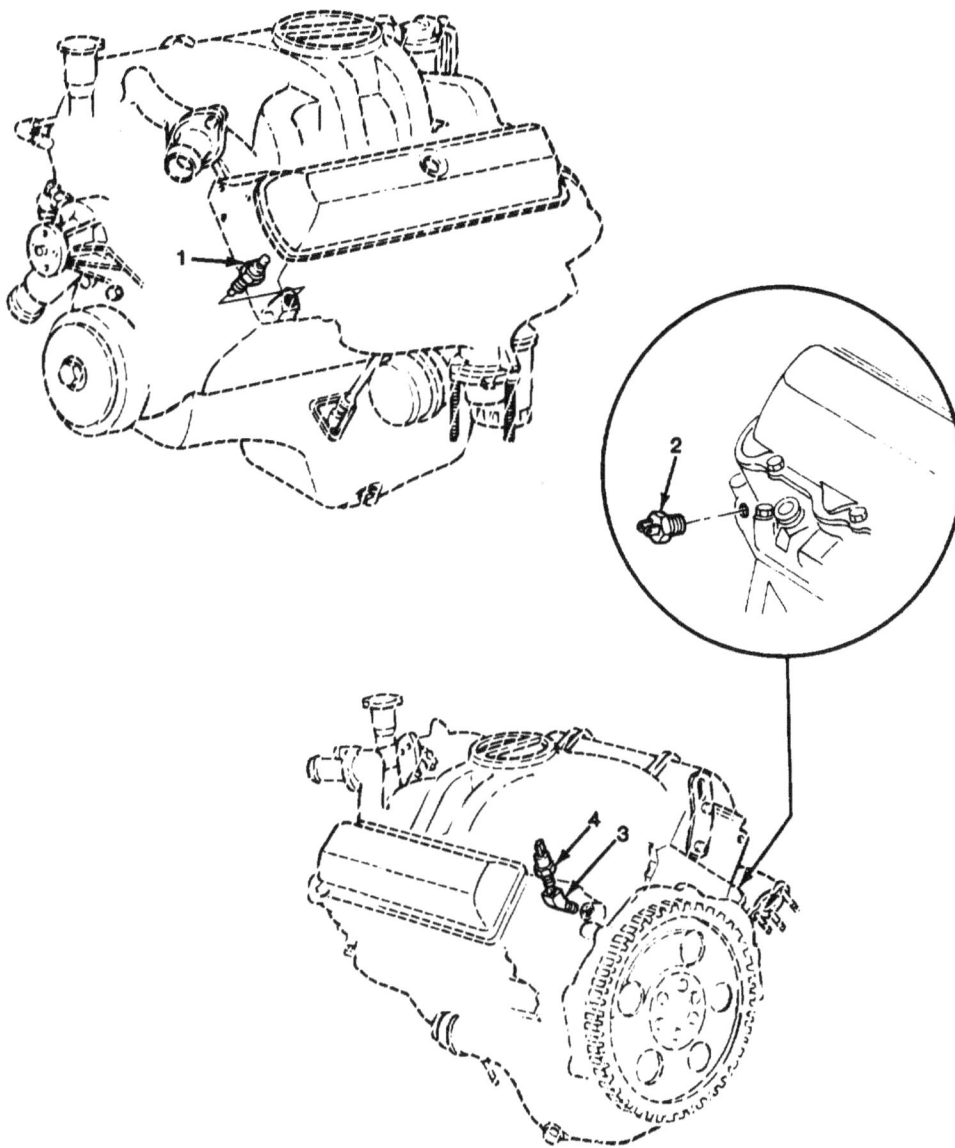

TA510861

FIGURE 54. ENGINE OIL PRESSURE AND WATER TEMPERATURE SENDING UNITS.

(1) (2) (3) (4) ITEM SMR NO CODE CAGEC	PART NUMBER	(5) DESCRIPTION AND USABLE ON CODES (UOC)	(6) QTY

GROUP 0610 SENDING UNITS AND WARNING
SWITCHES

FIG. 54 ENGINE OIL PRESSURE AND
WATER TEMPERATURE SENDING UNITS

1	PAOZZ 11862	25037177	TRANSMITTER,TEMPERA	1
2	PAOZZ 11862	14071047	SWITCH,COLD ADVANCE	1
3	PAOZZ 11862	14040817	ELBOW,PIPE	1
4	PAOZZ 11862	3815936	SWITCH,PRESSURE	1

END OF FIGURE

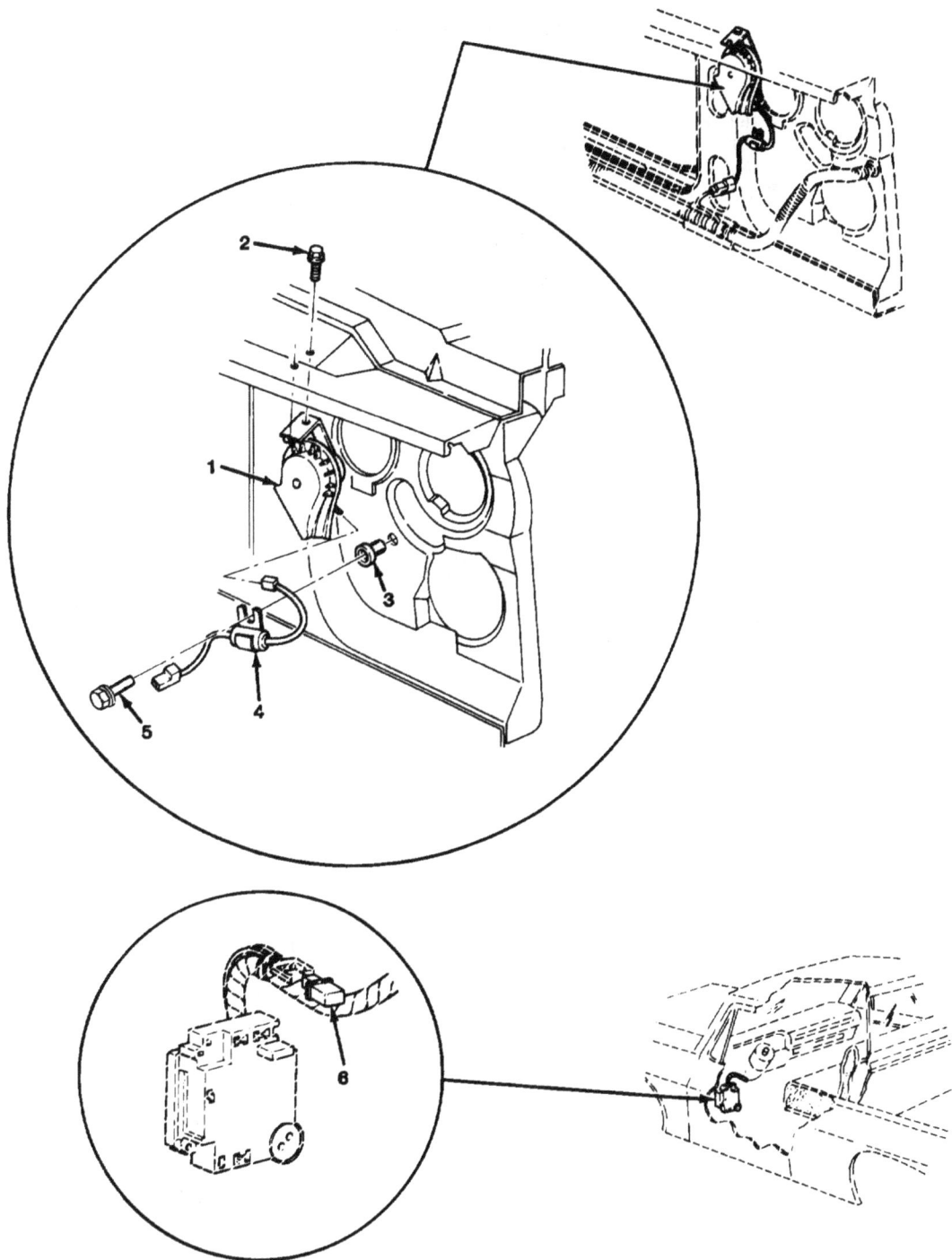

TA510862

FIGURE 55. TRUCK HORN AND RELATED PARTS.

(1)	(2)	(3)	(4)	(5)	(6)
ITEM NO	SMR CODE	CAGEC	PART NUMBER	DESCRIPTION AND USABLE ON CODES (UOC)	QTY

GROUP 0611 HORN, SIREN

FIG. 55 TRUCK HORN AND RELATED PARTS

1	PAOZZ	11862	1892163	HORN,ELECTRICAL	1
2	PAOZZ	11862	11501812	SCREW,TAPPING,THREA	1
3	PAOZZ	00613	C-5139-2	RECEPTACLE,TURNLOCK	1
4	PAOZZ	11862	15599224	CAPACITOR,FIXED,PAP	1
5	PAOZZ	11862	9437702	BOLT,MACHINE	1
6	PAOZZ	11862	25523703	RELAY,ELECTROMAGNET	1

END OF FIGURE

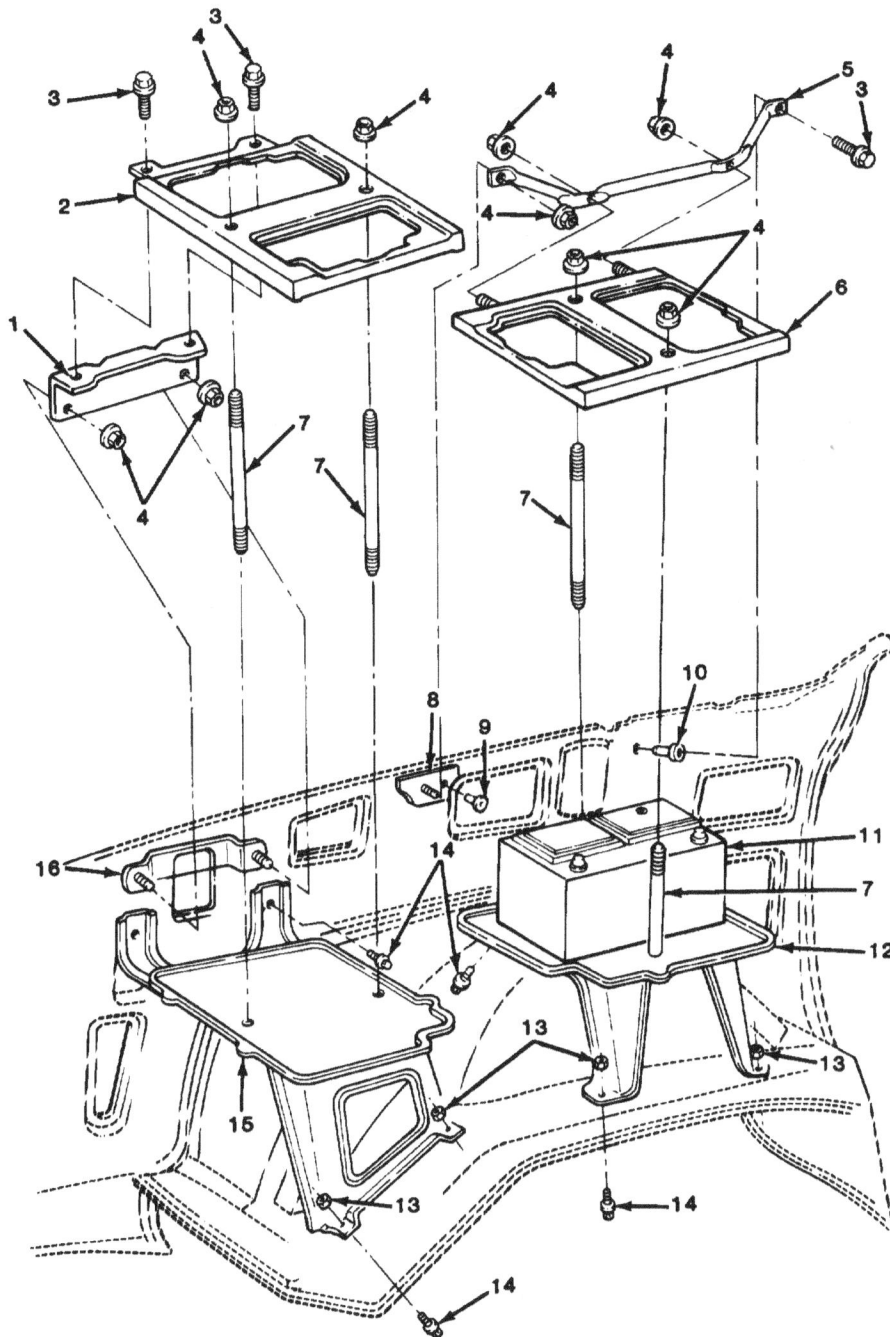

TA510863

FIGURE 56. BATTERIES, BATTERY TRAYS, AND RELATED PARTS.

(1) ITEM NO	(2) SMR CODE	(3) CAGEC	(4) PART NUMBER	(5) DESCRIPTION AND USABLE ON CODES (UOC)	(6) QTY
				GROUP 0612 BATTERIES, STORAGE	
				FIG. 56 BATTERIES, BATTERY TRAYS, AND RELATED PARTS	
1	PAOZZ	11862	15599900	BRACKET,ENGINE ACCE	1
2	PAOZZ	11862	14076856	RETAINER,BATTERY	1
3	PAOZZ	11862	9440334	BOLT,ASSEMBLED WASH	3
4	PAOZZ	24617	11506003	NUT,PLAIN,ASSEMBLED	9
5	PAOZZ	11862	14075896	RETAINER,BATTERY	1
6	PAOZZ	11862	14076857	RETAINER,BATTERY	1
7	PAOZZ	11862	14075894	STUD,PLAIN	4
8	PAOZZ	11862	15599902	RETAINER,BATTERY	1
9	PAOZZ	96906	MS20613-4P4	RIVET,SOLID	1
10	PAOZZ	11862	14005953	SHIELD,EXPANSION	1
11	PAOFA	81343	31-620	BATTERY,STORAGE OPTIONAL WITH P/N MS52149-1	2
11	PAOFA	96906	MS52149-1	BATTERY,STORAGE OPTIONAL WITH P/N 31-620 AND NOT APPLICABLE W/ WINTERIZED BATTERY BOX	2
12	PAOZZ	11862	14075389	TRAY,BATTERY	1
13	PAOZZ	11862	1359887	NUT,PLAIN,HEXAGON	4
14	PAOZZ	11862	2014469	BOLT,ASSEMBLED WASH	8
15	PAOZZ	11862	14075388	TRAY,BATTERY	1
16	PAOZZ	11862	15599901	RETAINER,BATTTERY	1

END OF FIGURE

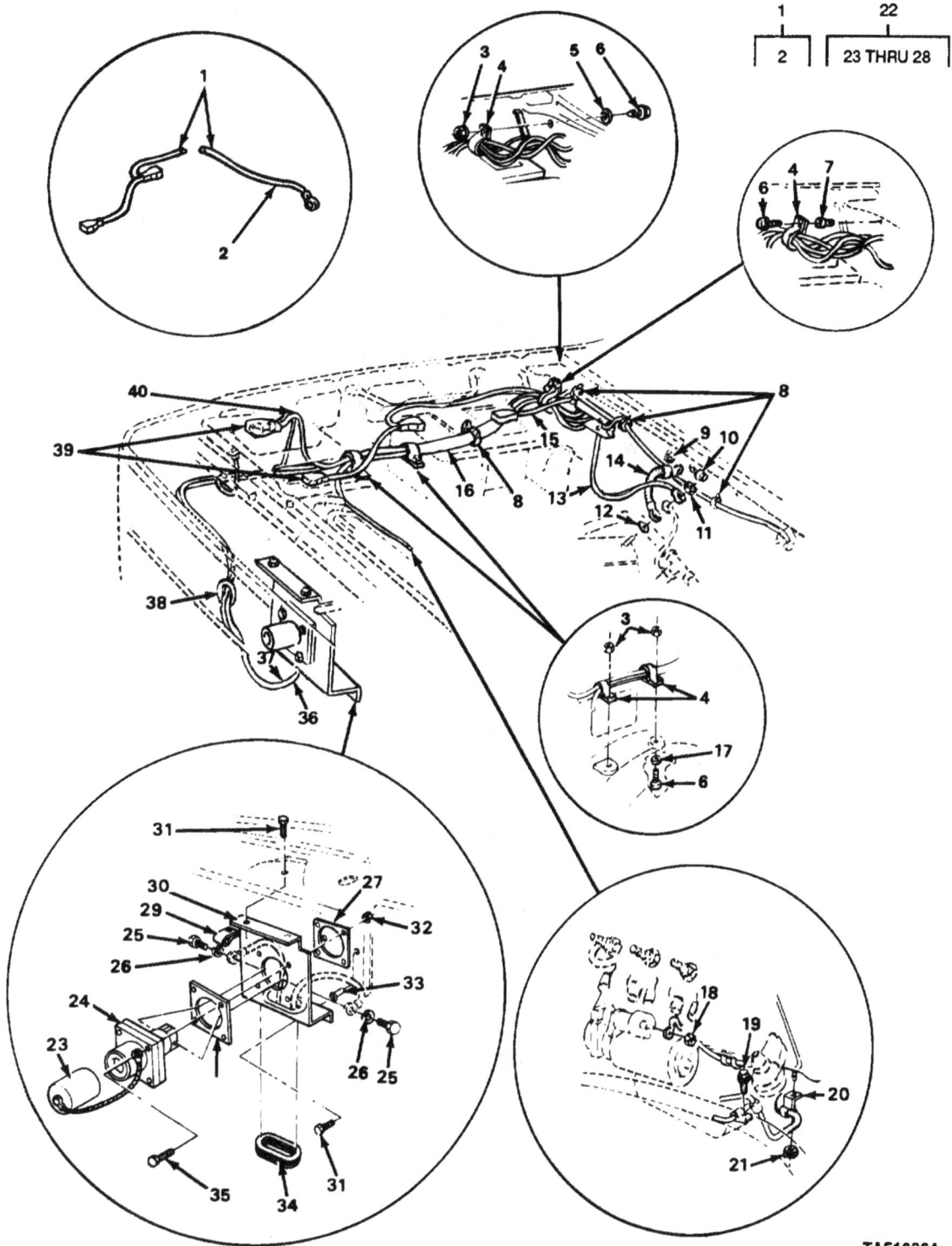

FIGURE 57. CABLES AND SLAVE CONNECTOR.

TA510864

(1) (2) (3) (4)	(5)	(6)
ITEM SMR PART NO CODE CAGEC NUMBER	DESCRIPTION AND USABLE ON CODES (UOC)	QTY

GROUP 0612 BATTERIES,STORAGE

FIG. 57 CABLES AND SLAVE CONNECTOR

1	PAOOO 11862 12039293	LEAD,STORAGE BATTER	1
2	MOOZZ 11862 FLW-12	.WIRE, FUSELINK (10 IN LG) MAKE FROM WIRE, P/N 6293923	1
3	PAOZZ 24617 271172	NUT,SELF-LOCKING,AS	3
4	PAOZZ 11862 2043151	CLAMP,LOOP UOC:208	3
5	PAOZZ 96906 MS27183-9	WASHER,FLAT	1
6	PAOZZ 96906 MS90728-6	SCREW,CAP,HEXAGON H UOC:194,208,209,210,230,231,252,254	3
7	PAOZZ 11862 14005953	SHIELD,EXPANSION	1
8	PAOZZ 11862 3816659	STRAP,LINE SUPPORTI	4
9	PAOZZ 96906 MS35335-34	WASHER,LOCK	1
10	PAOZZ 11862 11509371	SCREW,ASSEMBLED WAS	1
11	PAOZZ 11862 14005953	SHIELD,EXPANSION	1
12	PAOZZ 96906 MS35335-36	WASHER,LOCK	1
13	PAOZZ 11862 12039271	LEAD,ELECTRICAL	1
14	PAOZZ 11862 6287160	LEAD,STORAGE BATTER	1
15	PAOZZ 58499 6258	LEAD,STORAGE BATTER	1
16	PAOZZ 11862 12039272	LEAD,ELECTRICAL	1
17	PAOZZ 96906 MS27183-10	WASHER,FLAT	2
18	PAOZZ 11862 11504108	NUT,PLAIN,HEXAGON	1
19	PAOZZ 11862 11504447	SCREW,TAPPING,THREA UOC:208	1
20	PAOZZ 11862 14040813	CLIP,SPRING TENSION	1
21	PAOZZ 11862 11503739	NUT,PLAIN,HEXAGON	1
22	PAOOO 19207 11674728	CONNECTOR,RECEPTACL	1
23	PAOZZ 19207 11675004	.CAP,PROTECTIVE,DUST	1
24	PAOZZ 19207 11682345	.CONNECTOR,RECEPTACL	1
25	PAOZZ 96906 MS90727-57	.SCREW,CAP,HEXAGON H	2
26	PAOZZ 96906 MS35338-46	.WASHER,LOCK	2
27	PAOZZ 19207 11674730	.INSULATOR,PLATE	1
28	PAOZZ 19207 11674729	.GASKET	1
29	PAOZZ 11862 3982098	NUT,PLAIN,HEXAGON	2
30	PAOZZ 11862 14072432	BRACKET,DOUBLE ANGL	1
31	PAOZZ 11862 9440334	BOLT,ASSEMBLED WASH	4
32	PAOZZ 11862 11508446	NUT,PLAIN,EXTENDED	4
33	PAOZZ 11862 14005953	SHIELD,EXPANSION	2
34	PAOZZ 11862 3979756	GROMMET,NONMETALLIC	1
35	PAOZZ 11862 11508858	SCREW,ASSEMBLED WAS	4
36	PAOZZ 11862 12039257	LEAD,ELECTRICAL	1
37	PAOZZ 11862 12039267	LEAD,ELECTRICAL	1
38	MOOZZ 11862 8919163	CONDT BAT BST CBL (6.00" LG) MAKE FROM CONDUIT, P/N 8919355	1
39	PAOZZ 96906 MS75004-2	TERMINAL,LUG	1
39	PAOZZ 96906 MS75004-1	TERMINAL,LUG	1
40	PAOZZ 58499 6262	LEAD,STORAGE BATTER	1

END OF FIGURE

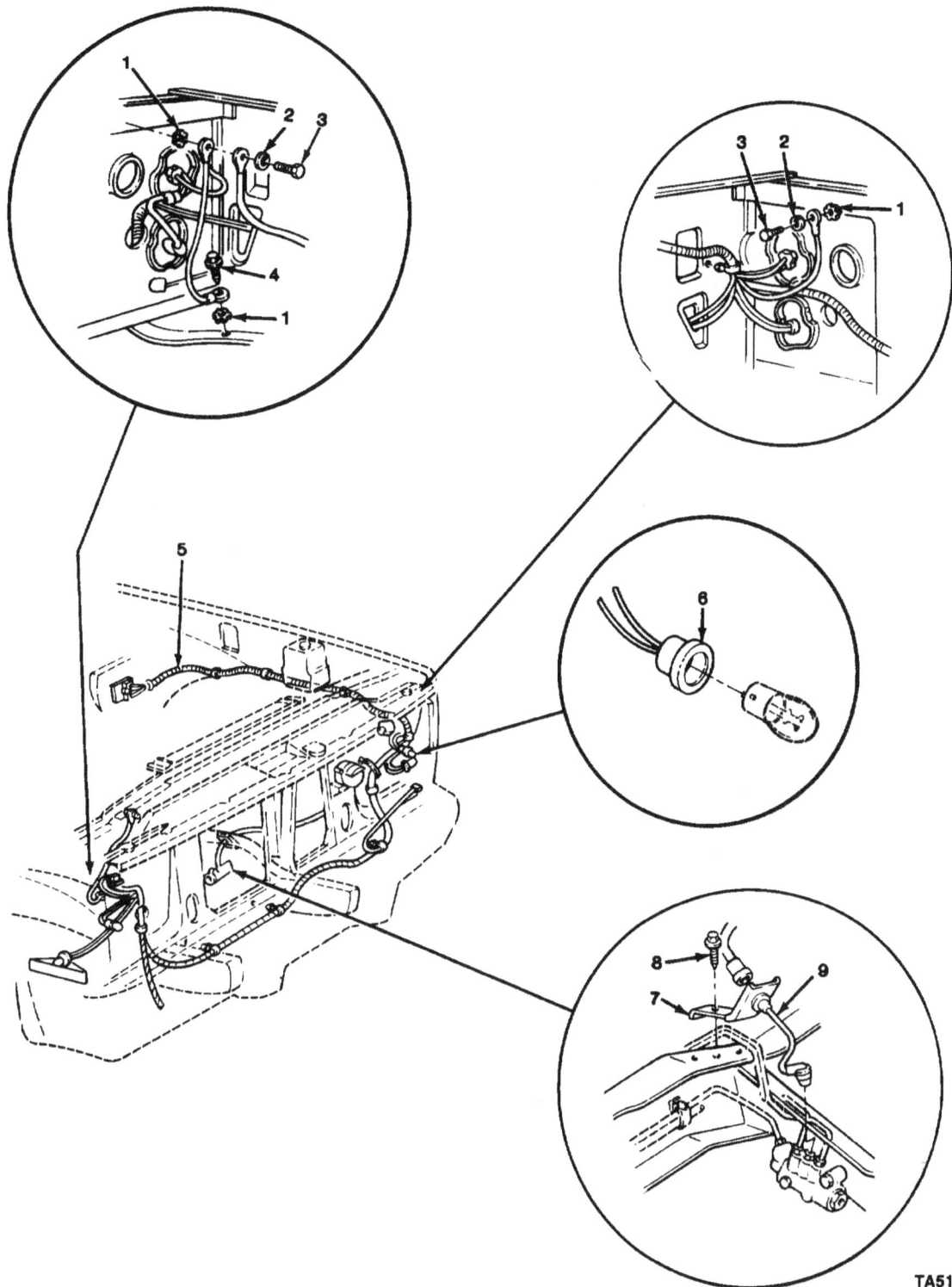

FIGURE 58. FRONT WIRING HARNESS.

TA510865

(1) (2) (3) (4) ITEM SMR NO CODE CAGEC	PART NUMBER	(5) DESCRIPTION AND USABLE ON CODES (UOC)	(6) QTY
		GROUP 0613 HULL OR CHASSIS WIRING HARNESS	
		FIG. 58 FRONT WIRING HARNESS	
1 PAFZZ 96906	MS35335-34	WASHER,LOCK	3
2 PAFZZ 96906	MS35340-45	WASHER,LOCK	2
3 PAFZZ 96906	MS90728-33	BOLT,MACHINE	2
4 PAFZZ 11862	11504447	SCREW,TAPPING,THREA	1
5 PAFZZ 11862	12039201	WIRING HARNESS,BRAN	1
6 PAOZZ 11862	12013813	LAMPHOLDER	2
7 PAOZZ 11862	469339	BRACKET,DOUBLE ANGL	1
8 PAOZZ 24617	9416187	SCREW,TAPPING,THREA	1
9 PAOZZ 11862	12001184	CABLE ASSEMBLY,SPEC	1

END OF FIGURE

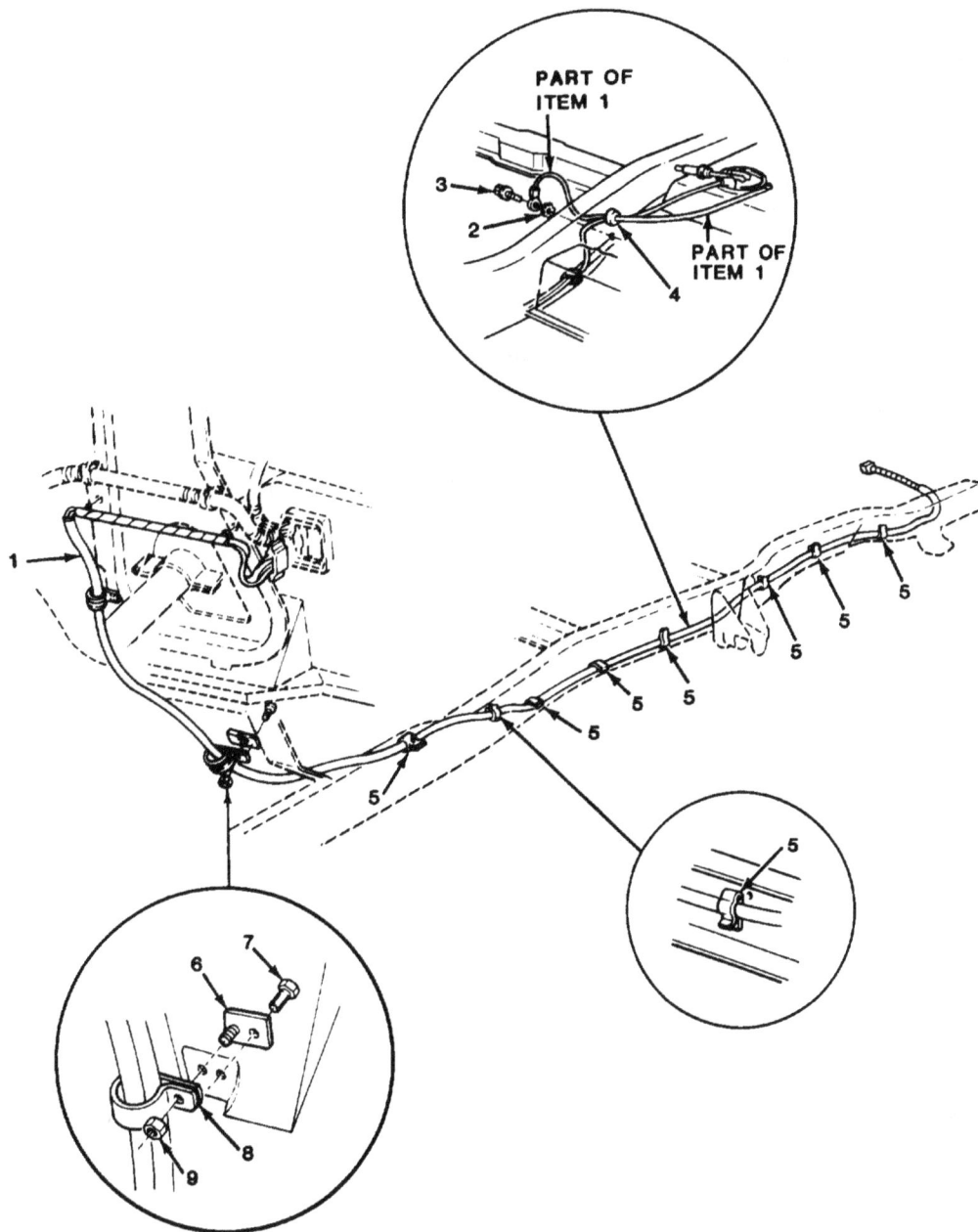

FIGURE 59. BODY AND FUEL SENDING WIRING HARNESSES (ALL EXCEPT M1009).

TA510866

(1) (2) (3) (4) ITEM SMR PART NO CODE CAGEC NUMBER	(5) DESCRIPTION AND USABLE ON CODES (UOC)	(6) QTY
	GROUP 0613 HULL OR CHASSIS WIRING HARNESS	
	FIG. 59 BODY AND FUEL SENDING WIRING HARNESSES (ALL EXCEPT M1009)	
1 PAFZZ 11862 12039434	WIRING HARNESS,BRAN UOC:194,208,210,230,231,252,254,256	1
2 PAFZZ 96906 MS35335-34 WASHER,LOCK	UOC:194,208,210,230,231,252,254,256	1
3 PAFZZ 11862 11504447 SCREW,TAPPING,THREA	UOC:194,208,210,230,231,252,254,256	1
4 PAFZZ 11862 3661804	GROMMET,NONMETALLIC UOC:194,208,210,230,231,252,254,256	1
5 PAFZZ 11862 9785074	CLIP,SPRING TENSION UOC:194,208,210,230,231,252,254,256	8
6 PAFZZ 11862 334963	PLATE,SPEEDOMETER,A UOC:194,208,210,230,231,252,254,256	1
7 PAFZZ 24617 9419327	SCREW UOC:194,208,210,230,231,252,254,256	1
8 PAFZZ 96906 MS21333-105 CLAMP,LOOP	UOC:194,208,210,230,231,252,254,256	1
9 PAFZZ 96906 MS35649-202 NUT,PLAIN,HEXAGON	UOC:194,208,210,230,231,252,254,256	1

END OF FIGURE

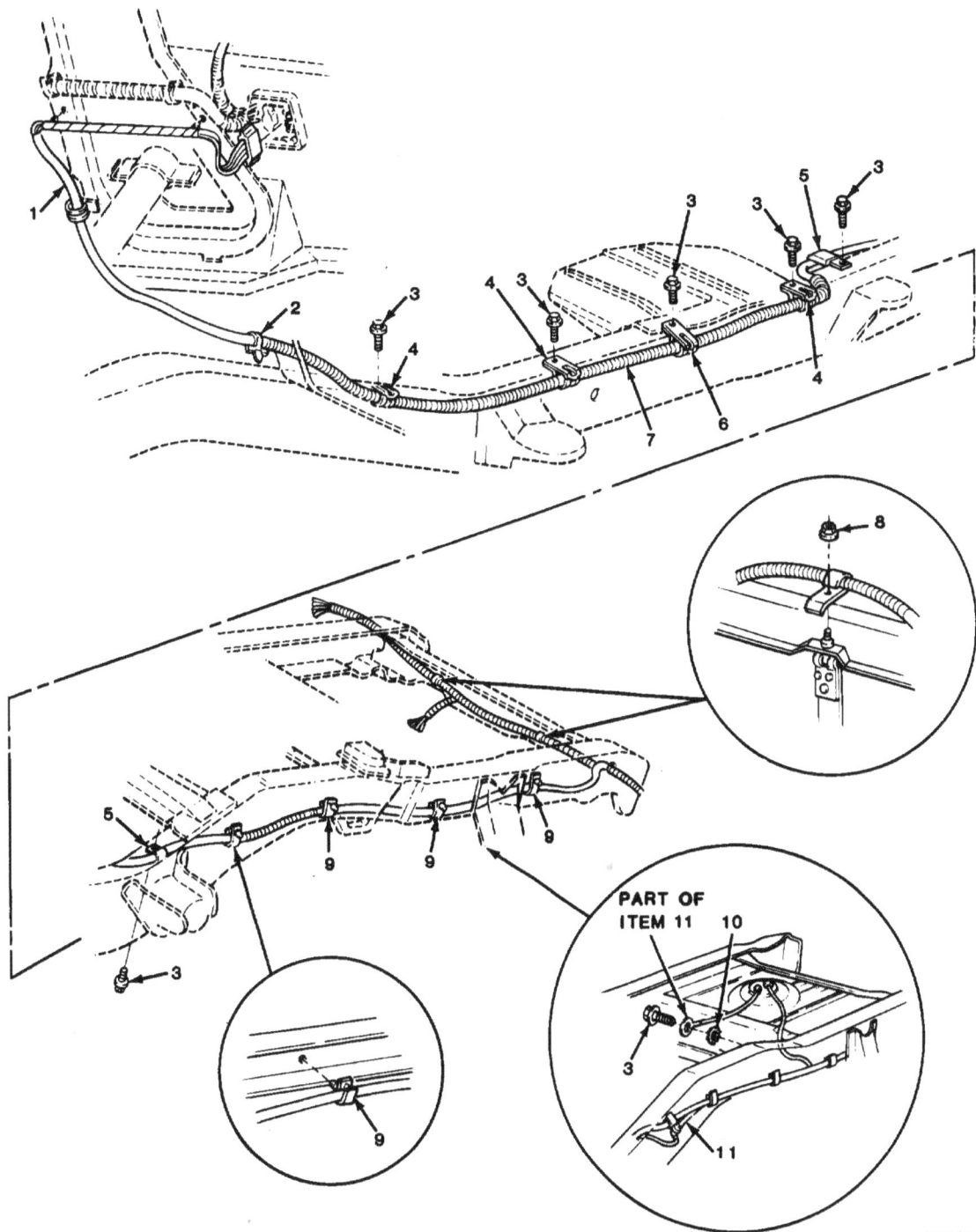

TA510867

FIGURE 60. BODY AND FUEL SENDING WIRING HARNESSES (M1009).

(1) (2) (3) (4) ITEM SMR PART NO CODE CAGEC NUMBER	(5) DESCRIPTION AND USABLE ON CODES (UOC)	(6) QTY
	GROUP 0613 HULL OR CHASSIS WIRING HARNESS	
	FIG. 60 BODY AND FUEL SENDING WIRING HARNESSES (M1009)	
1 PAFZZ 11862 12039206	WIRING HARNESS,BRAN UOC:209	1
2 PAFZZ 11862 3816659	STRAP,LINE SUPPORTI UOC:209	1
3 PAFZZ 11862 11504447 SCREW,TAPPING,THREA	UOC:209	7
4 PAFZZ 96906 MS21333-62 CLAMP,LOOP	UOC:209	3
5 PAFZZ 96906 MS21333-114 CLAMP,LOOP	UOC:209	2
6 PAFZZ 11862 3866187	HANGER,PIPE UOC:209	1
7 MFFZZ 11862 8917605	CLAMP,LOOP (54.00" LG) MAKE FROM CONDUIT, P/N 8919356 UOC:209	1
8 PAFZZ 96906 MS51967-12 NUT,PLAIN,HEXAGON	UOC:209	2
9 PAFZZ 11862 9785074	CLIP,SPRING TENSION UOC:209	4
10 PAFZZ 96906 MS35335-34 WASHER,LOCK	UOC:209	1
11 PAFZZ 11862 8906127	LEAD,ELECTRICAL UOC:209	1

END OF FIGURE

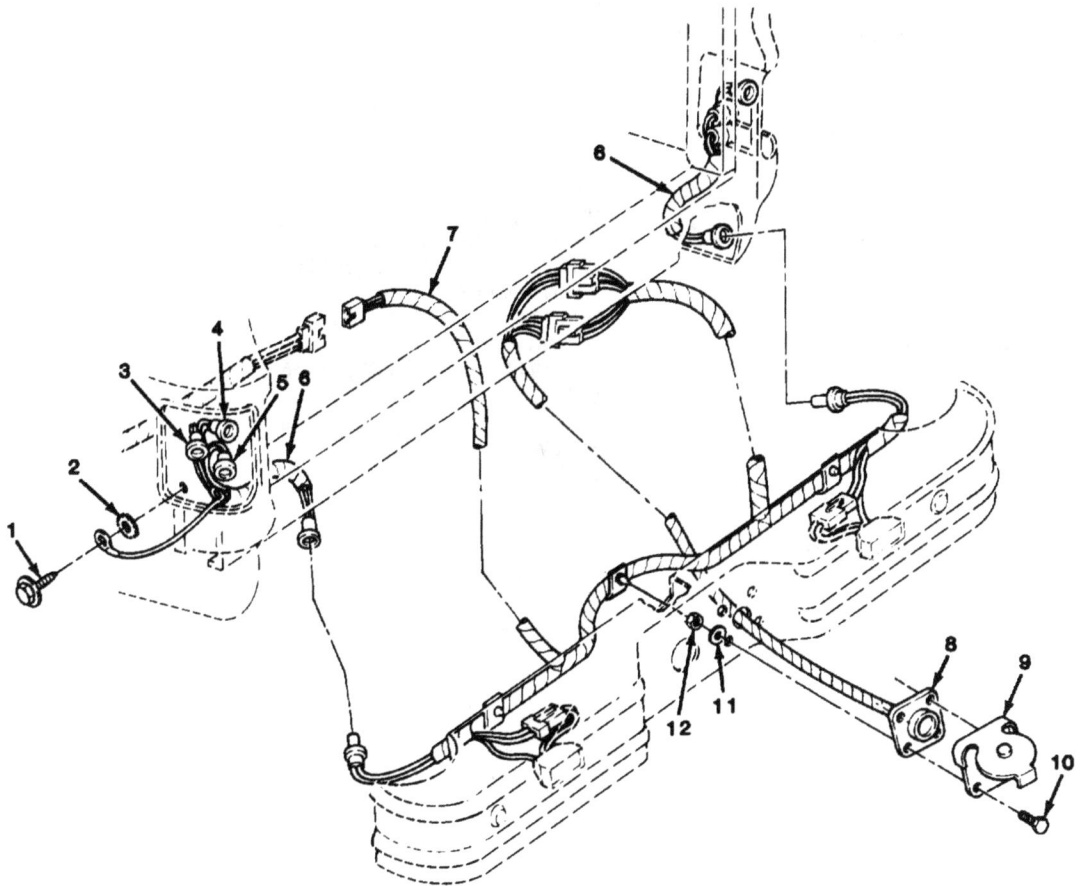

TA510868

FIGURE 61. BLACKOUT AND TAILLAMP WIRING HARNESS (ALL EXCEPT M1010
AND M1031) AND TRAILER WIRING HARNESS (ALL EXCEPT M1010).

(1) (2) (3) (4) ITEM SMR PART NO CODE CAGEC NUMBER	(5) DESCRIPTION AND USABLE ON CODES (UOC)	(6) QTY
	GROUP 0613 HULL OR CHASSIS WIRING HARNESS	
	FIG. 61 BLACKOUT AND TAILLAMP WIRING HARNESS (ALL EXCEPT M1010 AND M1031) AND TRAILER WIRING HARNESS (ALL EXCEPT M1010)	
1 PAOZZ 96906 MS51850-86	SCREW,TAPPING,THREA UOC:194,208,209,230,252,254,256	2
2 PAOZZ 96906 MS35335-33	WASHER,LOCK UOC:194,208,230,252,254,256	1
3 PAOZZ 11862 6298886	LAMPHOLDER UOC:194,208,209,2230,252,254,256	2
4 XDOZZ 11862 8914822	SOCKET,TAIL LAMP UOC:194,208,209,230,252,254,256	2
5 PAOZZ 11862 8909518	LAMPHOLDER UOC:194,208,209,230,252,254,256	2
6 PAOZZ 11862 8906150	WIRING HARNESS,BRAN UOC:194,208,209,230,252,254,256	2
7 PAOZZ 11862 12039205	WIRING HARNESS,BRAN UOC:194,208,230,252,254,256	1
8 PAOZZ 11862 12039208	WIRING HARNESS,BRAN UOC:194,208,209,230,252,254,256	1
9 PAOZZ 19207 7731428	COVER,ELECTRICAL CO UOC:194,208,209,230,252,254,256	1
10 PAOZZ 96906 MS90728-6	SCREW,CAP,HEXAGON H UOC:194,208,209,230,252,254,256	4
11 PAOZZ 96906 MS27183-10	WASHER,FLAT UOC:194,209,230,252,254,256	4
12 PAOZZ 96906 MS35691-406	NUT,PLAIN,HEXAGON UOC:194,208,209,230,252,254,256	4

END OF FIGURE

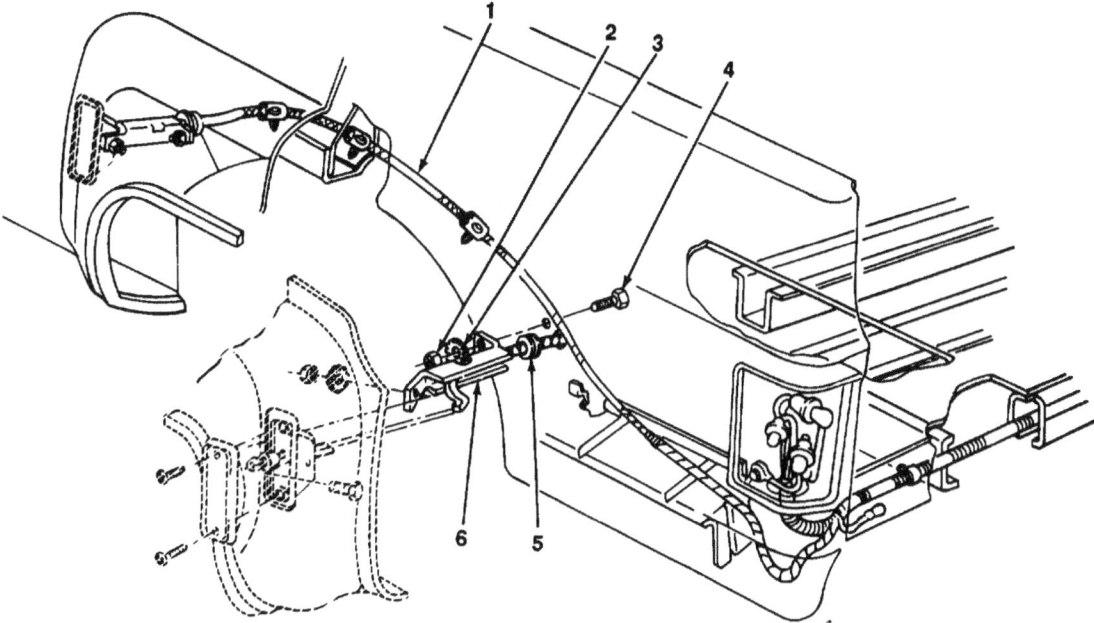

TA510869

FIGURE 62. REAR FENDER WIRING HARNESS (M1028A2 AND M1028A3).

(1) ITEM NO	(2) SMR CODE	(3) CAGEC	(4) PART NUMBER	(5) DESCRIPTION AND USABLE ON CODES (UOC)	(6) QTY
				GROUP 0613 HULL OR CHASSIS WIRING HARNESS	
				FIG. 62 REAR FENDER WIRING HARNESS (M1028A2 AND M1028A3)	
1	PAFZZ	11862	12096970	WIRING,HARNESS UOC:254,256	2
2	PAOZZ	96906	MS51967-2	NUT,PLAIN,HEXAGON UOC:254,256	2
3	PAOZZ	96906	MS35335-33	WASHER,LOCK UOC:254,256	2
4	PAOZZ	11862	9440033	BOLT,MACHINE UOC:254,256	2
5	PAOZZ	11862	15591130	GROMMET,NONMETALLIC UOC:254,256	8
6	PAFZZ	11862	329830	PROTECTOR,ELECTRICA UOC:254,256	4

END OF FIGURE

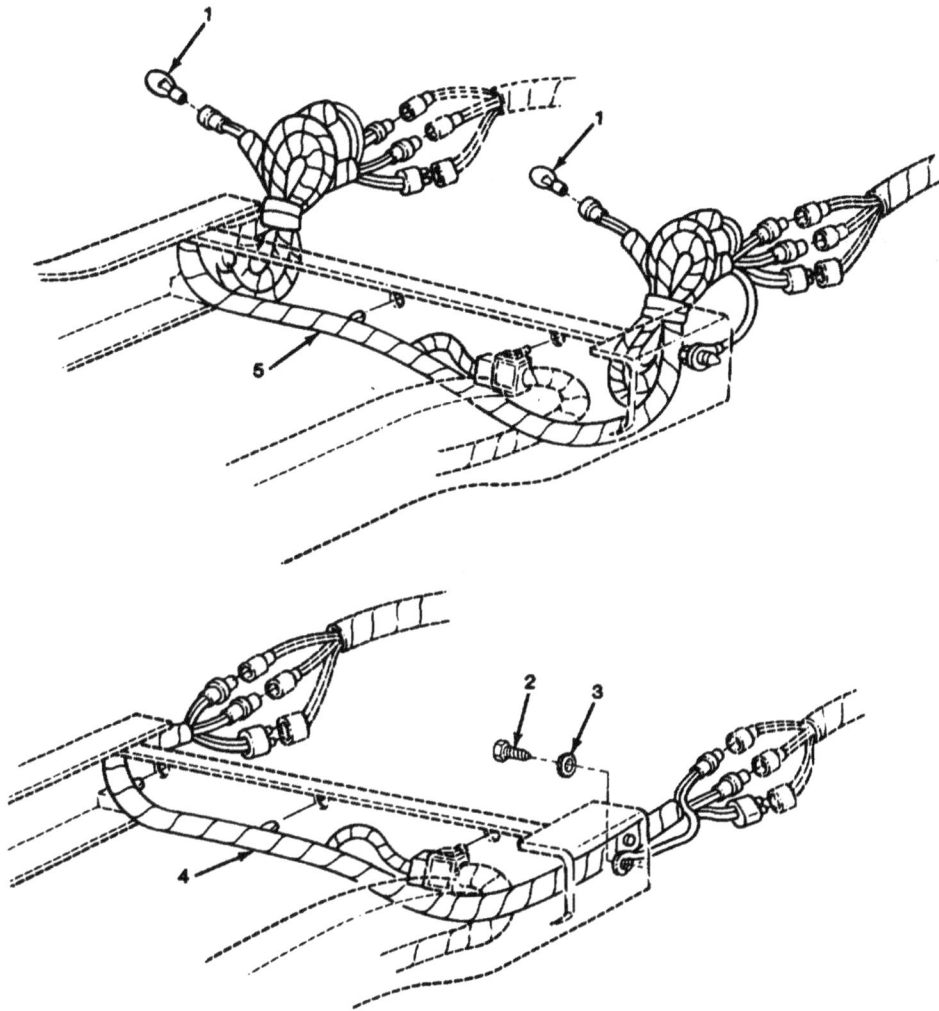

TA510870

FIGURE 63. BLACKOUT AND TAILLAMP WIRING HARNESSES (M1010 AND M1031).

(1) (2) (3) (4) ITEM SMR NO CODE CAGEC	PART NUMBER	(5) DESCRIPTION AND USABLE ON CODES (UOC)	(6) QTY
		GROUP 0613 HULL OR CHASSIS WIRING HARNESS	
		FIG. 63 BLACKOUT AND TAILLAMP WIRING HARNESSES (M1010 AND M1031)	
1 PAOZZ 08806	194	LAMP,INCANDESCENT UOC:210,231	2
2 PAOZZ 11862	11504447	SCREW,TAPPING,THREA UOC:210,231	1
3 PAOZZ 96906	MS35335-34	WASHER,LOCK UOC:210,231	1
4 PAFZZ 11862	12039204	WIRING HARNESS,BRAN UOC:231	1
5 PAFZZ 11862	12039203	WIRING HARNESS,BRAN UOC:210	1
		END OF FIGURE	

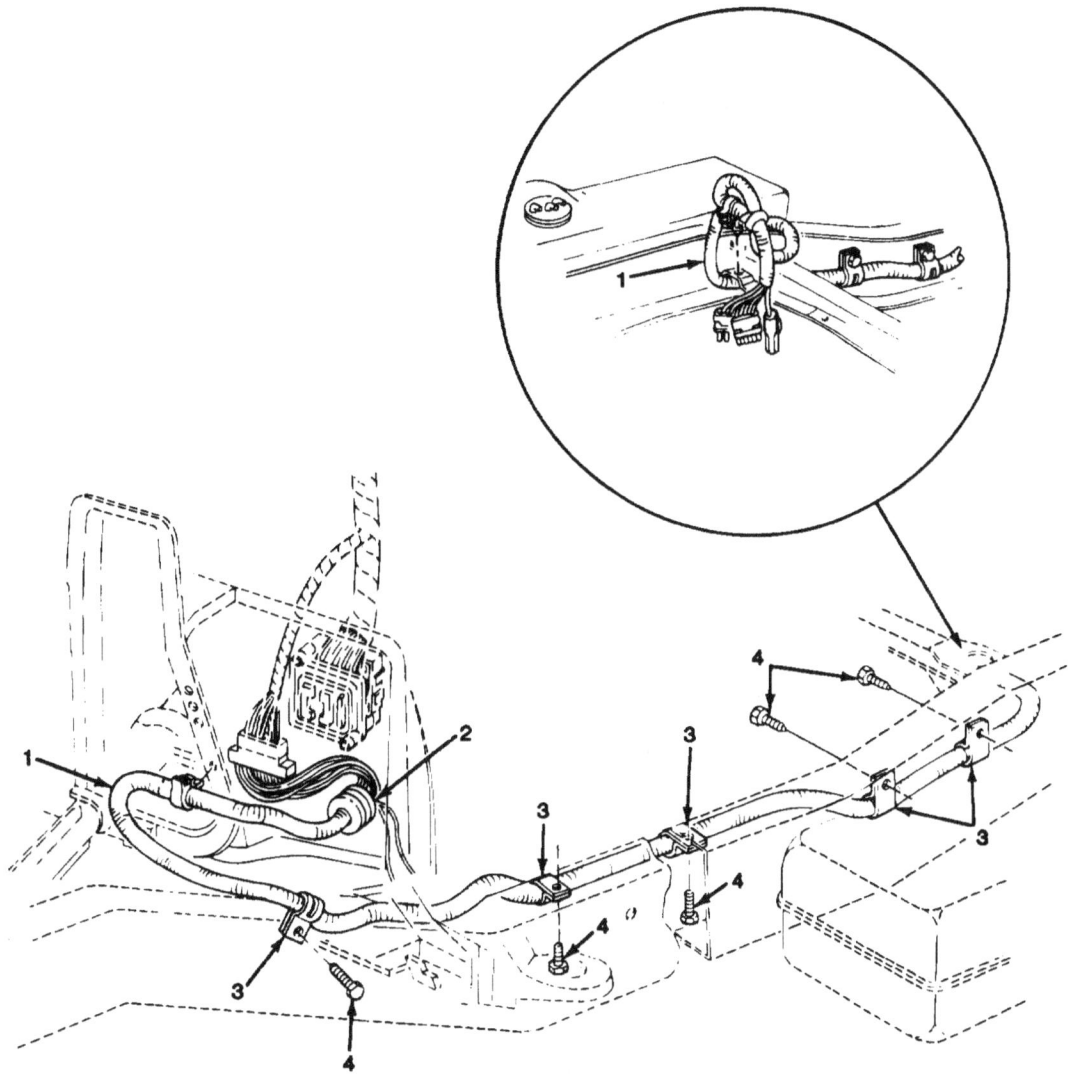

TA510871

FIGURE 64. INSTRUMENT PANEL EXTENSION WIRING HARNESS (M1010).

(1) (2) (3) (4) ITEM SMR PART NO CODE CAGEC NUMBER	(5) DESCRIPTION AND USABLE ON CODES (UOC)	(6) QTY
	GROUP 0613 HULL OR CHASSIS WIRING HARNESS	
	FIG. 64 INSTRUMENT PANEL EXTENSION WIRING HARNESS (M1010)	
1 PAFFF 77060 12039255	WIRING HARNESS,BRAN UOC:210	1
2 PAFZZ 11862 3918889	GROMMET,NONMETALLIC UOC:210	1
3 PAFZZ 96906 MS21333-114 CLAMP,LOOP 	 UOC:210	5
4 PAFZZ 11862 11504447 SCREW,TAPPING,THREA 	 UOC:210	5
	END OF FIGURE	

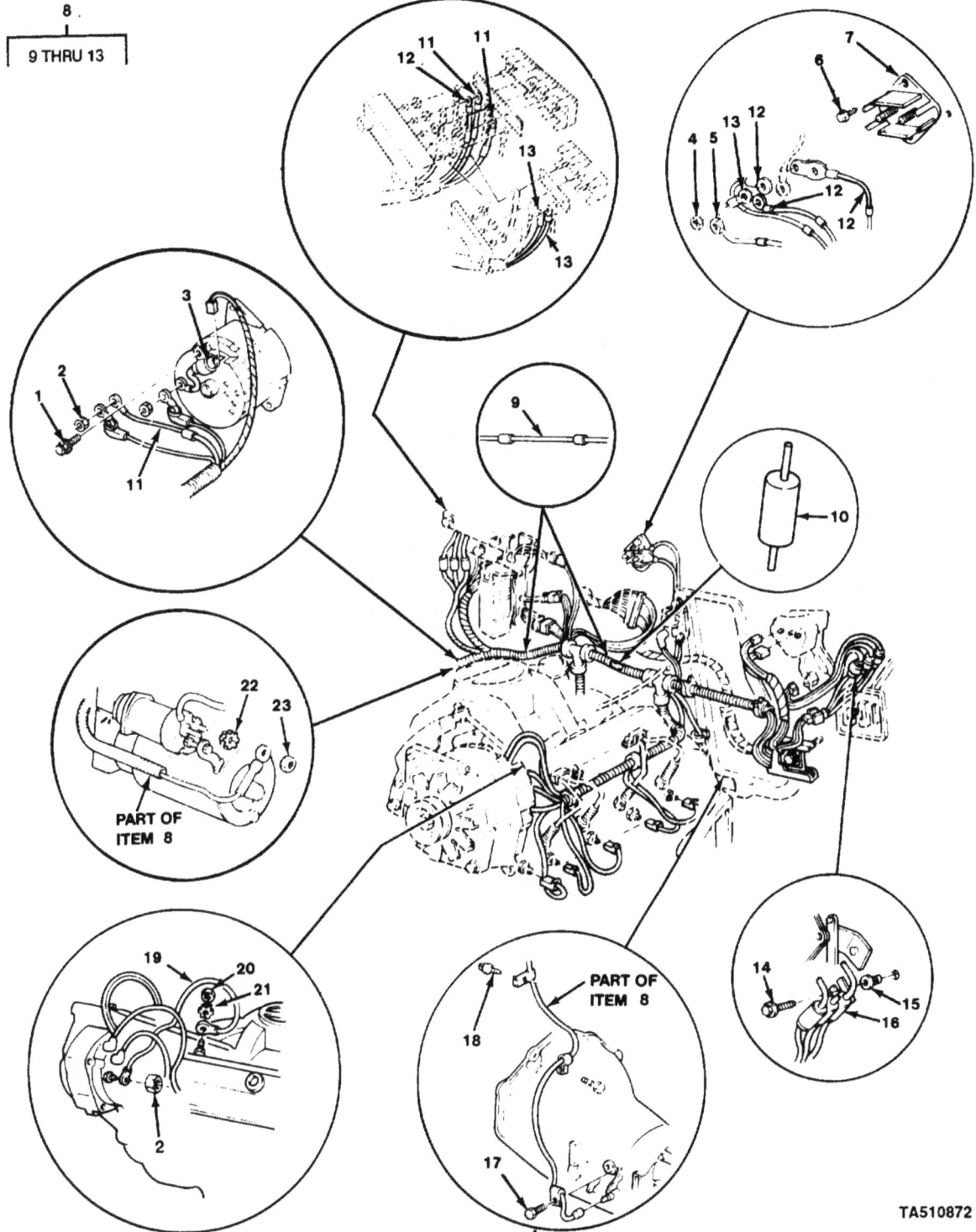

FIGURE 65. ENGINE WIRING HARNESS (ALL EXCEPT M1010).

TA510872

(1) (2) (3) (4) ITEM SMR PART NO CODE CAGEC NUMBER	(5) DESCRIPTION AND USABLE ON CODES (UOC)	(6) QTY
	GROUP 0613 HULL OR CHASSIS WIRING HARNESS	
	FIG. 65 ENGINE WIRING HARNESS (ALL EXCEPT M1010)	
1 PAOZZ 24617 11508687	BOLT,MACHINE UOC:194,208,209,230,231,252,254,256	1
2 PAOZZ 24617 11506003	NUT,PLAIN,ASSEMBLED UOC:194,208,209,230,231,252,254,256	2
3 PAOZZ 11862 1988380	CAPACITOR,FIXED,PAP UOC:194,208,209,230,231,252,254,256	1
4 PAOZZ 24617 271166	NUT,PLAIN,ASSEMBLED	1
5 PAOZZ 24617 271172	NUT,SELF-LOCKING,AS	1
6 PAOZZ 24617 1640810	SCREW,TAPPING,THREA	2
7 PAOZZ 11862 3996270	TERMINAL BOARD	1
8 PAFFF 11862 12039311	WIRING HARNESS,BRAN UOC:194,208,209,230,231,252,254,256	1
9 MFFZZ 11862 FLW-18	.WIRE,FUSELINK (10 IN LG) MAKE FROM WIRE, P/N 6293702 UOC:194,208,209,230,231,252,254,256	1
10 PAOZZ 11862 12006377	.RECTIFIER,CONNECTOR UOC:194,208,209,230,231,252,254,256	1
11 MFFZZ 11862 FLW-20	.WIRE,FUSELINK (10 IN LG) MAKE FROM WIRE, P/N 6292997 UOC:194,208,209,230,231,252,254,256	1
12 MFFZZ 11862 FLW-16	.WIRE,FUSELINK (10 IN LG) MAKE FROM WIRE, P/N 6292996 UOC:194,208,209,230,231,252,254,256	1
13 MFFZZ 11862 FLW-12	.WIRE, FUSELINK (10 IN LG) MAKE FROM WIRE, P/N 6293923 UOC:194,208,209,230,231,252,254,256	1
14 PAOZZ 11862 9437702	BOLT,MACHINE	1
15 PAOZZ 11862 14005953	SHIELD,EXPANSION	1
16 PAOZZ 11862 15599222	CAPACITOR ASSEMBLY	1
17 PAFZZ 24617 9419320	SCREW,TAPPING,THREA	1
18 PAFZZ 11862 9440166	SCREW,TAPPING,THREA	1
19 PAOZZ 11862 12039253	LEAD,ELECTRICAL UOC:194,208,209,230,231,252,254,256	1
20 PAOZZ 11862 14005953	SHIELD,EXPANSION UOC:194,208,209,210,230,231,252,254	1
21 PAOZZ 96906 MS35335-36	WASHER,LOCK	1
22 PAOZZ 96906 MS35335-33	WASHER,LOCK	1
23 PAFZZ 96906 MS35649-282	NUT,PLAIN,HEXAGON	1

END OF FIGURE

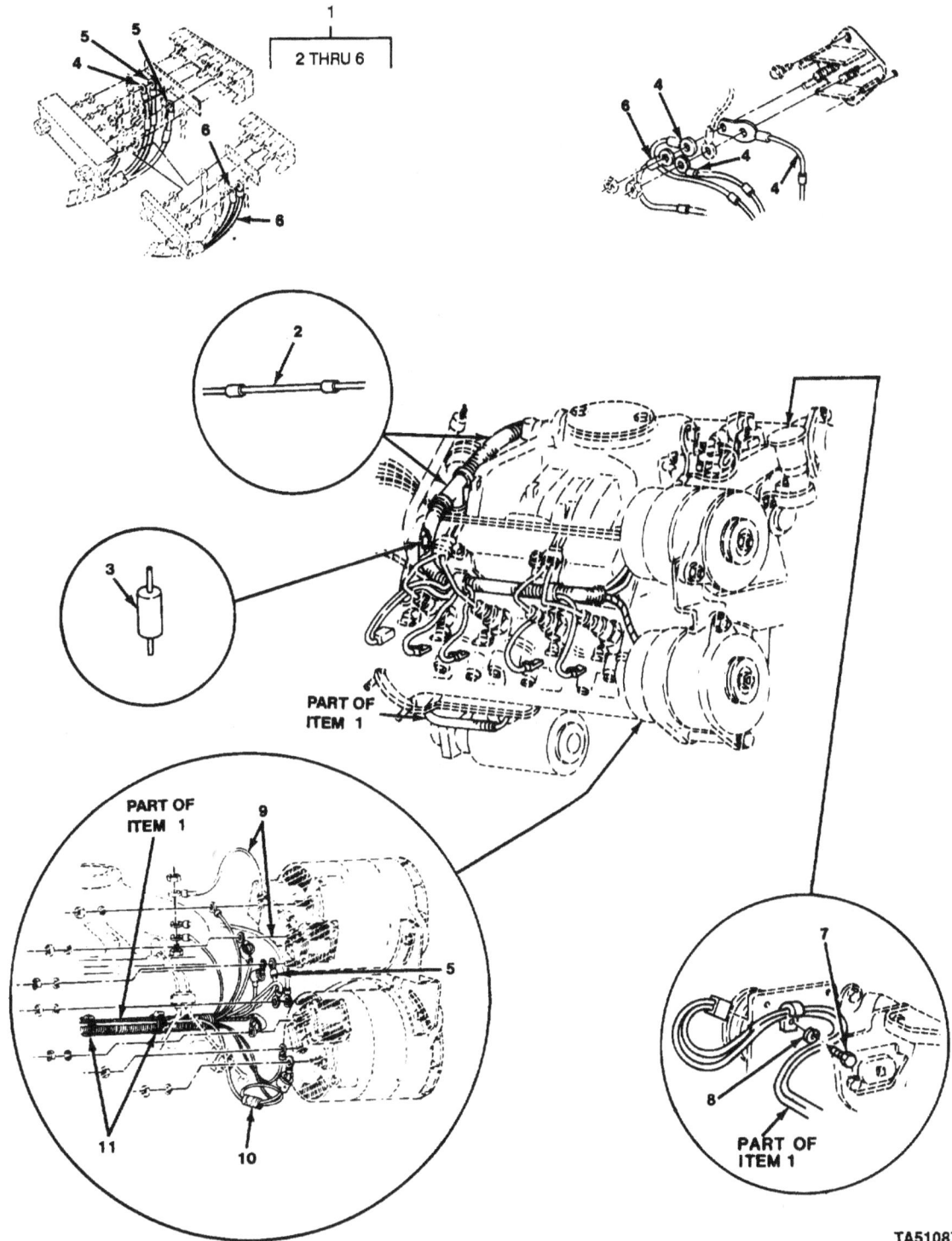

FIGURE 66. ENGINE WIRING HARNESS (M1010).

TA510873

(1) (2) (3) (4) ITEM SMR PART NO CODE CAGEC NUMBER	(5) DESCRIPTION AND USABLE ON CODES (UOC)	(6) QTY
	GROUP 0613 HULL OR CHASSIS WIRING HARNESS	
	FIG. 66 ENGINE WIRING HARNESS (M1010)	
1 PAFFF 11862 12044637	WIRING,HARNESS,BRAN UOC:210,230	1
2 MFFZZ 11862 FLW-18	.WIRE,FUSELINK (10" LG) MAKE FROM WIRE, P/N 6293702 UOC:210	1
3 PAOZZ 11862 12006377	.RECTIFIER,CONNECTOR UOC:210	3
4 MFFZZ 11862 FLW-16	.WIRE,FUSELINK (10" LG) MAKE FROM WIRE, P/N 6292996 UOC:210	1
5 MFFZZ 11862 FLW-20	.WIRE,FUSELINK (10" LG) MAKE FROM WIRE, P/N 6292997 UOC:210	3
6 MFFZZ 11862 FLW-12	.WIRE, FUSELINK (10" LG) MAKE FROM WIRE, P/N 6293923 UOC:210	1
7 PAOZZ 11862 11502656	BOLT,MACHINE UOC:210,230	1
8 PAOZZ 96906 MS35338-45	WASHER,LOCK UOC:210,230	1
9 PAOZZ 11862 12039253	LEAD,ELECTRICAL UOC:210	2
10 PAFZZ 11862 12044638	WIRING HARNESS,BRAN UOC:210,230	1
11 PAFZZ 11862 3816659	STRAP,LINE SUPPORTI UOC:210,230	2

END OF FIGURE

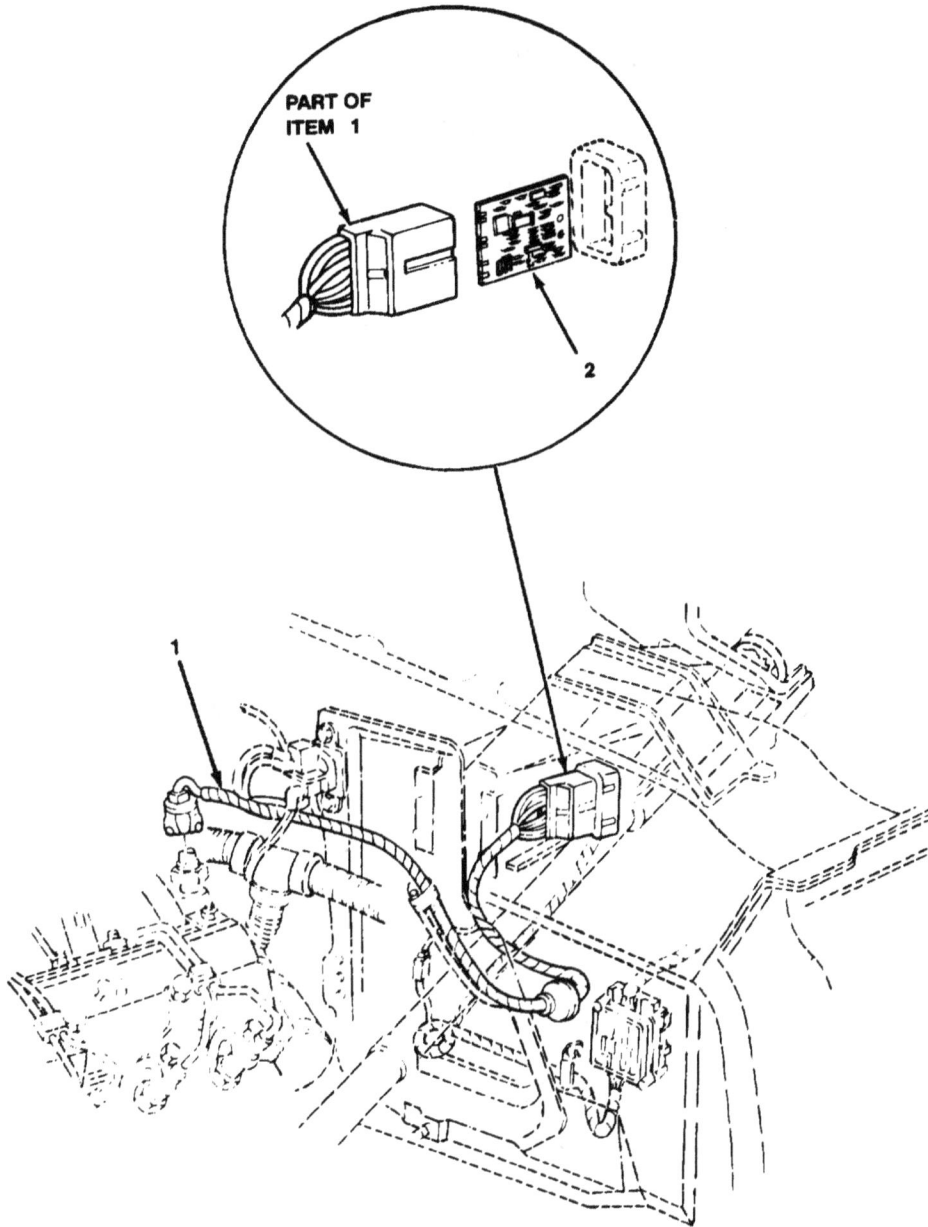

PART OF
ITEM 1

1

2

TA510874

FIGURE 67. GLOW PLUG WIRING HARNESS.

(1)	(2)	(3)	(4)	(5)	(6)
ITEM	SMR		PART		
NO	CODE	CAGEC	NUMBER	DESCRIPTION AND USABLE ON CODES (UOC)	QTY

GROUP 0613 HULL OR CHASSIS WIRING HARNESS

FIG. 67 GLOW PLUG WIRING HARNESS

| 1 | PAFFF | 11862 | 12039308 | WIRING HARNESS,BRAN | 1 |
| 2 | PAFZZ | 11862 | 12034592 | .MODULATOR ASSEMBLY | 1 |

END OF FIGURE

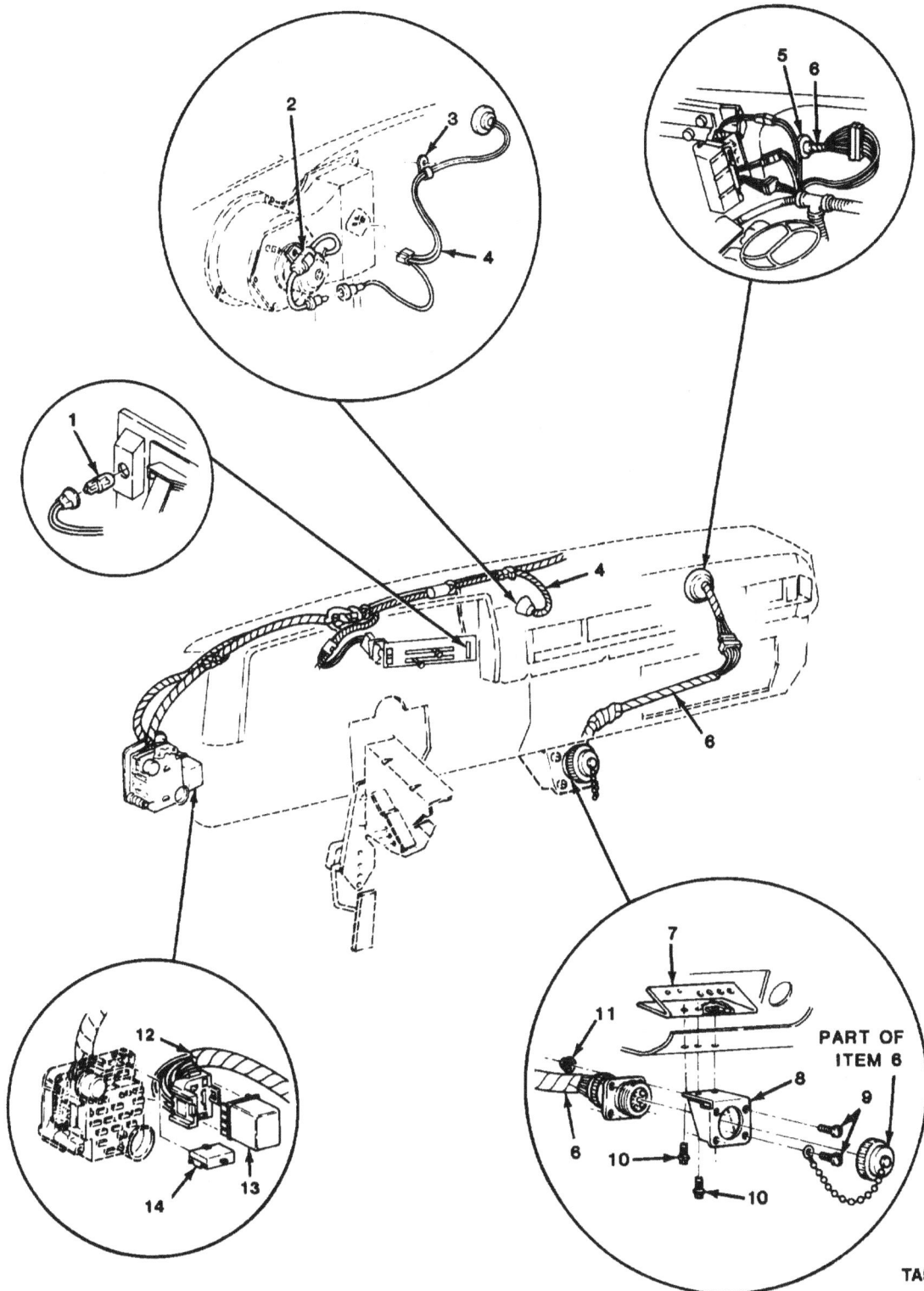

FIGURE 68. HEATER CONTROL, DIAGNOSTIC, AND DOOR ALARM WIRING HARNESSES.

TA510875

(1) (2) (3) (4) ITEM SMR PART NO CODE CAGEC NUMBER	(5) DESCRIPTION AND USABLE ON CODES (UOC)	(6) QTY
	GROUP 0613 HULL OR CHASSIS WIRING HARNESS	
	FIG. 68 HEATER CONTROL,DIAGNOSTIC, AND DOOR ALARM WIRING HARNESSES	
1 PAOZZ 08806 194	LAMP,INCANDESCENT UOC:194,208,209,210,230,231,252	1
2 PAOZZ 11862 15599223	CAPACITOR,FIXED,PAP	1
3 PAFZZ 11862 6270410	CLIP,SPRING TENSION	1
4 PAFZZ 11862 12031346	WIRING HARNESS,BRAN	1
5 PAFZZ 11862 3655180	GROMMET,NONMETALLIC	1
6 PAFZZ 11862 12039304	CABLE ASSEMBLY,SPEC UOC:194,208,209,230,231,252,254,256	1
6 PAFZZ 11862 12039305	WIRING HARNESS,BRAN UOC:210,230	1
7 PAOZZ 11862 14072340	BRACKET,ANGLE	1
8 PAOZZ 11862 14074480	BRACKET,ANGLE	1
9 PAOZZ 96906 MS35206-245	SCREW,MACHINE	4
10 PAOZZ 11862 1244067	BOLT,ASSEMBLED WASH	3
11 PAOZZ 72582 271163	NUT,PLAIN,ASSEMBLED	4
12 PAFZZ 11862 12039464	WIRING HARNESS,BRAN UOC:210	1
13 PAOZZ 11862 22529441	ALARM,DOOR ASSEMBLY UOC:210	1
14 PAOZZ 11862 1253637	BUZZER	1

END OF FIGURE

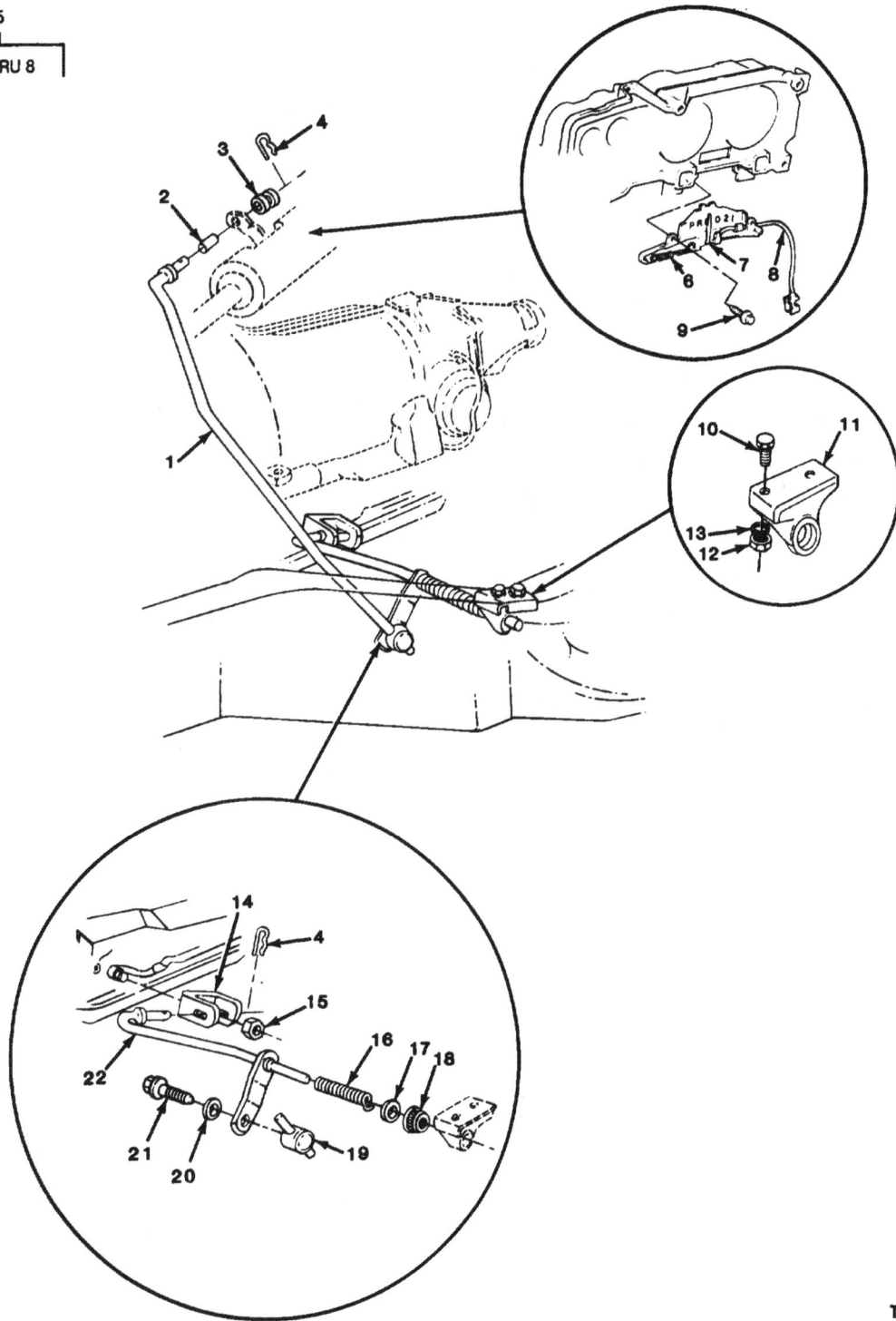

TA510876

FIGURE 69. SHIFTING LINKAGE.

TM9-2320-289-34P

(1) (2) (3) (4) ITEM SMR PART NO CODE CAGEC NUMBER	(5) DESCRIPTION AND USABLE ON CODES (UOC)	(6) QTY
	GROUP 07 TRANSMISSION	
	GROUP 0705 TRANSMISSION SHIFTING COMPONENTS	
	FIG. 69 SHIFTING LINKAGE	
1 PAFZZ 11862 6271387	ROD,RANGE SELECTOR	1
2 PAFZZ 11862 1394293	PARTS KIT,SHIFT LEV	1
3 PAFZZ 11862 6271391 GROMMET,NONMETALLIC		2
4 PAFZZ 11862 1244707 PIN,LOCK		2
5 PAOOO 11862 25078571	SHIFT INDICATOR,HYD	1
6 PAOZZ 11862 8985418	.SPRING,HELICAL,EXTE	1
7 PAOZZ 70040 25023641	.POINTER,SHIFT INDIC	1
8 PAOZZ 11862 25078578	.CABLE,SHIFT INDICAT	1
9 PAOZZ 96906 MS51850-86 SCREW,TAPPING,THREA		2
10 PAFZZ 96906 MS90728-32 BOLT,MACHINE		2
11 PFFZZ 11862 6271386 BRACKET,TRANSMISSIO		1
12 PAFZZ 96906 MS51967-6 NUT,PLAIN,HEXAGON		2
13 PAFZZ 96906 MS35338-45 WASHER,LOCK		2
14 PAFZZ 11862 9795470	LEVER,MANUAL CONTRO	1
15 PAFZZ 11862 14005953 SHIELD,EXPANSION		1
16 PAFZZ 11862 15538092 SPRING,HELICAL,COMP		1
17 PAFZZ 96906 MS27183-15 WASHER,FLAT		1
18 PAFZZ 11862 1377083 WASHER,FLAT		1
19 PAFZZ 11862 1385607 COLLAR,SHAFT		1
20 PAFZZ 11862 334533	WASHER,LOCK	1
21 PAFZZ 11862 1381477 BOLT,MACHINE		1
22 PAFZZ 11862 6271394	SHAFT ASSEMBLY,EQUA	1

END OF FIGURE

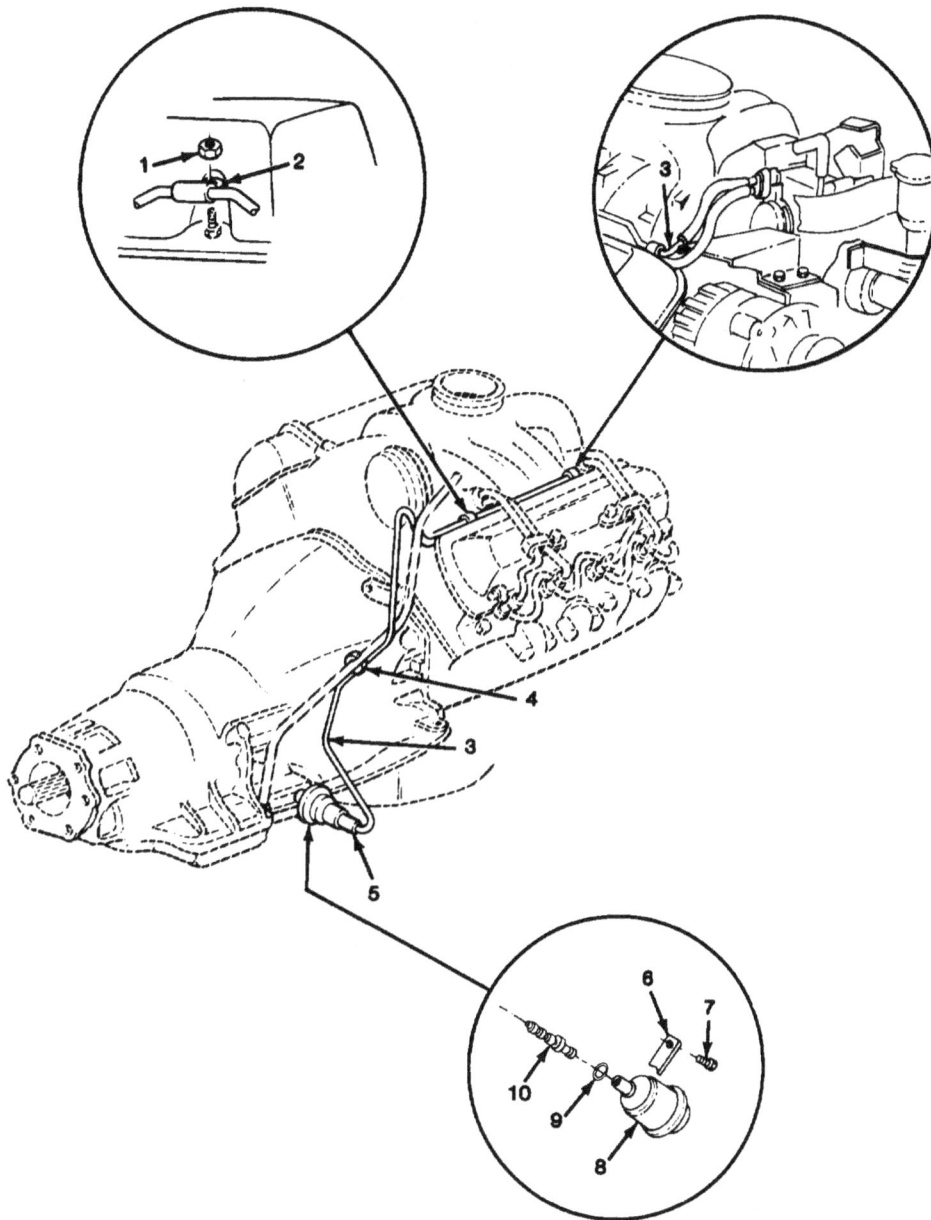

TA510877

FIGURE 70. VACUUM MODULATOR VALVE.

(1) (2) (3) (4) ITEM SMR PART NO CODE CAGEC NUMBER	(5) DESCRIPTION AND USABLE ON CODES (UOC)	(6) QTY

GROUP 0705 TRANSMISSION SHIFTING COMPONENTS

FIG. 70 VACUUM MODULATOR VALVE

1	PAOZZ	11862	22521550	NUT,PLAIN,HEXAGON	2
2	PAOZZ	11862	2044779	CLAMP	2
3	MFOZZ	11862	15599235	PIPE-TRNS VAC MOD (49.11"LG) MAKE FROM TUBE, P/N 603827	1
4	PAOZZ	11862	2043150	CLAMP,LOOP	1
5	MFOZZ	11862	326560	HOSE-TRNS VAL MOD P (1.30" LG) MAKE FROM HOSE, P/N 9439274	1
6	PAFZZ	73342	8627650	RETAINER,MODULATOR	1
7	PAFZZ	96906	MS90728-31	BOLT,MACHINE	1
8	PAFZZ	11862	359917	MODULATOR,TRANSMISSION	1
9	PAFZZ	11862	10054241	SEAL,SPEEDOMETER PART OF KT P/N 8625905	1
10	PAFZZ	11862	8623983	SLIDE,DIRECTIONAL C	1

END OF FIGURE

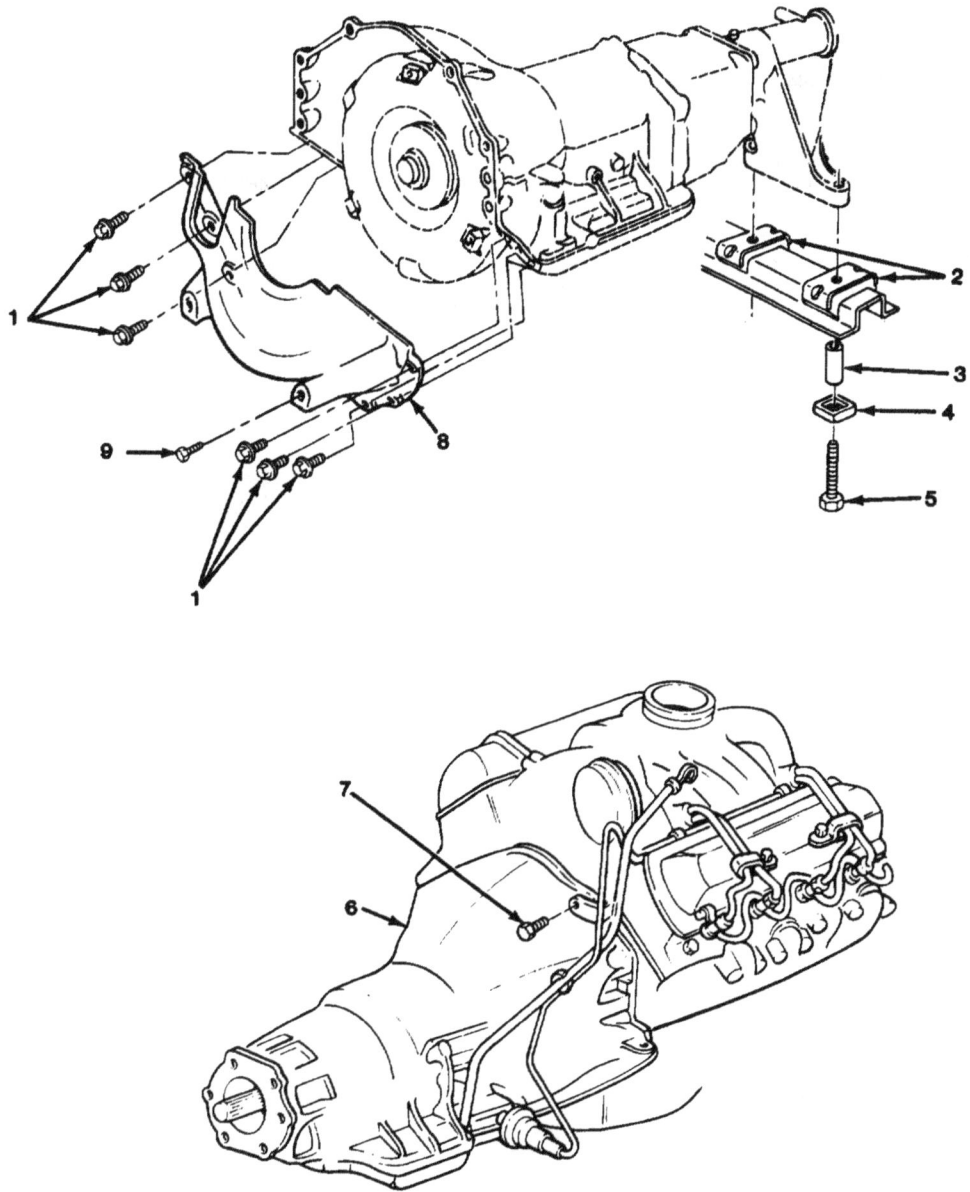

TA510878

FIGURE 71. TRANSMISSION ASSEMBLY AND MOUNTS.

(1) (2) (3) (4)				(5)	(6)
ITEM	SMR		PART		
NO	CODE	CAGEC	NUMBER	DESCRIPTION AND USABLE ON CODES (UOC)	QTY

GROUP 0710 TRANSMISSION ASSEMBLY AND
ASSOCIATED PARTS

FIG. 71 TRANSMISSION ASSEMBLY AND
MOUNTS

1	PAOZZ	11862	1635490	BOLT,SELF-LOCKING	6
2	PAFZZ	11862	3705444	MOUNT,RESILIENT	2
3	PAFZZ	11862	14009322	SPACER,SLEEVE	2
4	PAFZZ	11862	14009321	MOUNT,RESILIENT	2
5	PAFZZ	11862	14009324	SETSCREW	2
6	PAFHD	11862	8655111	TRANSMISSION,HYDRAU FOR COMPONENT PARTS SEE FIG'S 72, 73, 74, & 75	1
7	PAFZZ	96906	MS90728-63	SCREW,CAP,HEXAGON H	6
8	PAOZZ	11862	15522022	COVER,ACCESS	1
9	PAOZZ	96906	MS90728-83	SCREW,CAP,HEXAGON H UOC:194,208,209,210,230,231,252,254	2

END OF FIGURE

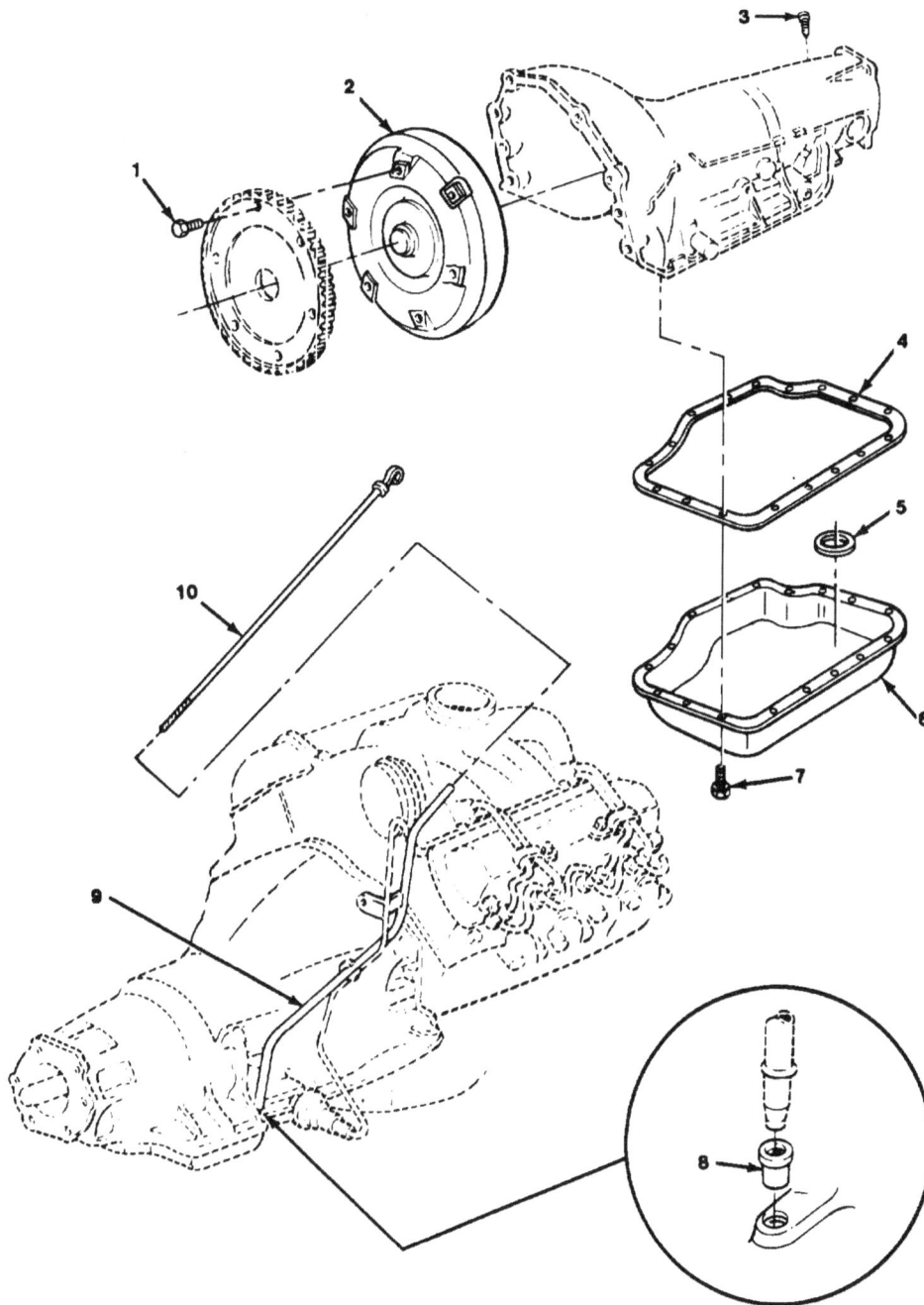

TA510879

FIGURE 72. TORQUE CONVERTER, OIL PAN, AND FILLER TUBE.

(1) (2) (3) (4)				(5)	(6)
ITEM SMR			PART		
NO CODE CAGEC NUMBER				DESCRIPTION AND USABLE ON CODES (UOC)	QTY

GROUP 0710 TRANSMISSION ASSEMBLY AND
ASSOCIATED PARTS

FIG. 72 TORQUE CONVERTER, OIL PAN,
AND FILLER TUBE

1	PAFZZ	11862	11508581		SCREW,CAP,HEXAGON H	6
2	PAFZZ	11862	8656942		TORQUE CONVERTER	1
3	PAFZZ	11862	8622361	SCREW,DRIVE		1
4	PAOZZ	11862	8655625	GASKET		1
5	PBOZZ	11862	3787240		MAGNET,CHIP COLLECT	1
6	PAOZZ	11862	8633203		OIL PAN	1
6	PAOZZ	11862	8655020		OIL PAN	1
7	PAOZZ	11862	9440224	BOLT,MACHINE		13
8	PAOZZ	11862	1259475		SEAL,OIL FILLER PIP	1
9	PAOZZ	11862	14045642	TUBE,BENT,METALLIC		1
				UOC:194,208,209,210,230,231,252,254	
10	PAOZZ	11862	334532		GAGE ROD-CAP,LIQUID	1

END OF FIGURE

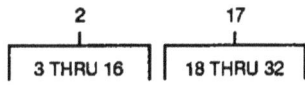

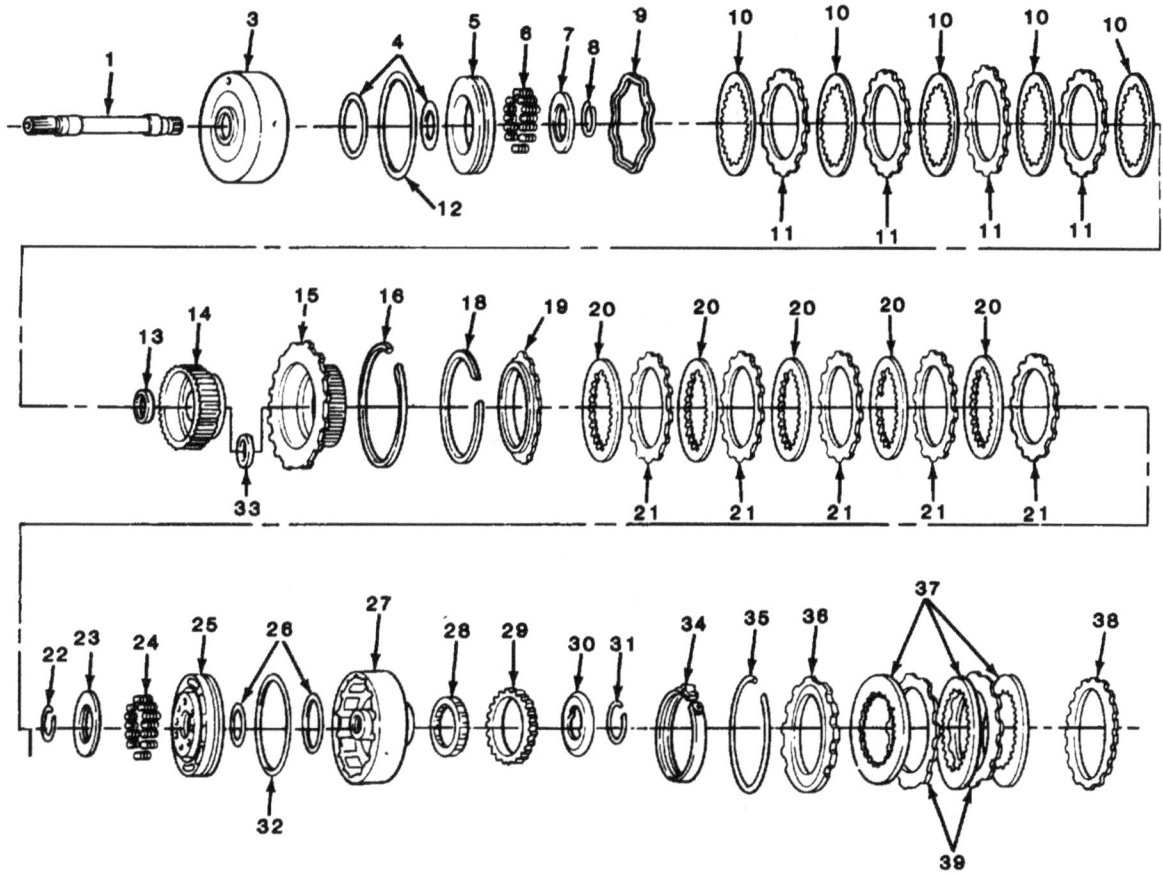

FIGURE 73. FORWARD AND DIRECT CLUTCH ASSEMBLIES.

TA510880

(1) ITEM NO	(2) SMR CODE	(3) CAGEC	(4) PART NUMBER	(5) DESCRIPTION AND USABLE ON CODES (UOC)	(6) QTY
				GROUP 0710 TRANSMISSION ASSEMBLY AND ASSOCIATED PARTS	
				FIG. 73 FORWARD AND DIRECT CLUTCH ASSEMBLIES	
1	PAHZZ	11862	8624772	SHAFT,STRAIGHT	1
2	AHHHH	11862	8627592	CLUTCH ASSEMBLY,FOR	1
3	PAHZZ	11862	8624908	.PARTS KIT,CLUTCH,HC	1
4	PAHZZ	23862	8623917	.SEAL,KIT PART OF KIT P/N 8629955	1
5	PAHZZ	11862	8627990	.PARTS KITS, TRANSMISS	1
6	PAHZZ	11862	8624101	.SPRING,HELICAL,COMP	16
7	PAHZZ	73342	8623104	.PLATE,RETAINING,BEA	2
8	PAHZZ	73342	8623105	.RING,SNAP PART OF KIT P/N 8623979	1
9	PAHZZ	11862	8623851	.DISK,CLUTCH	1
10	PAHZZ	11862	8655619	.DISK,CLUTCH PART OF KIT P/N 8629955	5
11	PAHZZ	11862	8623849	.DISK,CLUTCH	4
12	PAHZZ	11862	8625997	.PARTS KIT,CLUTCH SE	1
13	PAHZZ	11862	8624196	.BEARING,WASHER,THRU	1
14	PAHZZ	11862	8625717	.GEAR,SPUR	1
15	PAHZZ	73342	8625718	.HUB,CLUTCH DRIVING	1
16	PAHZZ	73342	8623112	.RING,RETAINING	1
17	AHHHH	11862	8626848	.CLUTCH ASM-DIR COMP	1
18	PAHZZ	73342	8623112	.RING,RETAINING	1
19	PAHZZ	73342	8623122	.PRESSURE PLATE ASSE	1
20	PAHZZ	11862	8655619	.DISK,CLUTCH	5
21	PAHZZ	11862	8625197	.DISK,CLUTCH	5
22	PAHZZ	73342	8623105	.RING,SNAP PART OF KIT P/N 8623979	1
23	PAHZZ	73342	8623104	.PLATE,RETAINING,BEA	1
24	PAHZZ	11862	8624101	.SPRING,HELICAL,COMP	14
25	PAHZZ	11862	8627989	.PARTS KIT,TRANSMISS	1
26	PAHZZ	23862	8623917	.SEAL,KIT PART OF KIT P/N 8629955	1
27	PAHZZ	11862	8627387	.HOUSING ASSEMBLY,CL	1
28	PAHZZ	11862	8625814	.ROLLER,CLUTCH ASSEM	1
29	PAHZZ	11862	8629487	.RING,BEARING,CUTER	1
30	PAHZZ	11862	8627334	.DISK,CLUTCH	1
31	PAHZZ	11862	8623120	.RING,RETAINING PART OF KIT P/N 8623979	1
32	PAHZZ	11862	8625997	.PARTS KIT,CLUTCH SE	1
33	PAHZZ	11862	8624781	BEARING,WASHER,THRU	1
34	PAHZZ	11862	12300852	BAND,CLUTCH,FRONT	1
35	PAHZZ	11862	8623149	RING,RETAINING	1
36	PAHZZ	11862	8623152	DISK,CLUTCH	1
37	PAHZZ	11862	8623151	DISK,CLUTCH PART OF KIT P/N 8629955	3
38	PAHZZ	11862	8624199	PLATE	1
39	PAHZZ	72590	8655621	PRESSURE PLATE ASSE	2

END OF FIGURE

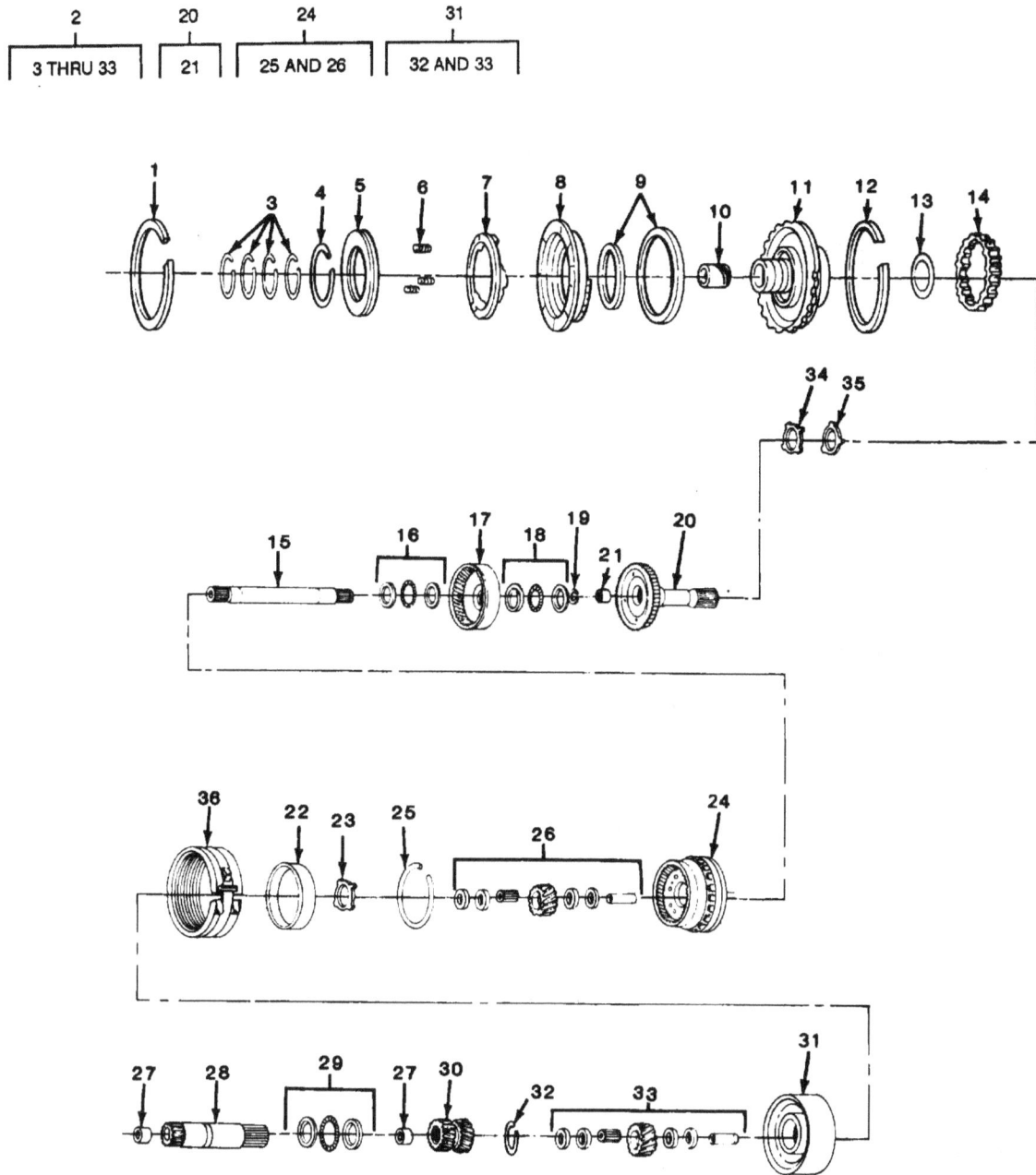

FIGURE 74. CENTER SUPPORT, REACTION CARRIER, AND OUTPUT
CARRIER ASSEMBLIES.

TA510881

(1) ITEM NO	(2) SMR CODE	(3) CAGEC	(4) PART NUMBER	(5) DESCRIPTION AND USABLE ON CODES (UOC)	(6) QTY
				GROUP 0710 TRANSMISSION ASSEMBLY AND ASSOCIATED PARTS	
				FIG. 74 CENTER SUPPORT, REACTION CARRIER, AND OUTPUT CARRIER ASSEMBLIES	
1	PAHZZ	11862	8623437	RING,RETAINING	1
2	AHHHH	11862	8633075	CARRIER,TRANSMISSION	1
3	PAHZZ	11862	8626356	.RETAINER,PACKING PART OF KIT P/N 8629955	4
4	PAHZZ	11862	8623153	.RING,RETAINING PART OF KIT P/N 8623979	1
5	PAHZZ	11862	8623146	.DISK,CLUTCH	1
6	PAHZZ	11862	8623145	.SPRING,HELICAL,COMP	3
7	PAHZZ	11862	8627385	.CLUTCH SPRING GUIDE	1
8	PAHZZ	11862	8626938	.PARTS KIT,CLUTCH DI	1
9	PAHZZ	11862	8623918	.PARTS KIT,CLUTCH,IN PART OF KIT P/N 8629955	1
10	PAHZZ	11862	8623942	.BUSHING,SLEEVE	1
11	PAHZZ	11862	8625990	.PARTS KIT,RACE ASSE	1
12	PAHZZ	11862	8626816	.RING,RETAINING	1
13	PAHZZ	11862	8624783	.BEARING,WASHER,THRU	1
14	PAHZZ	11862	8623842	.ROLLER ASSEMBLY,CLU	1
15	PAHZZ	11862	8625736	.SHAFT,STRAIGHT	1
16	PAHZZ	11862	8623921	.BEARING,ROLLER,THRU	1
17	PAHZZ	11862	8623202	.GEAR,INTERNAL	1
18	PAHZZ	11862	8623922	.BEARING,ROLLER,THRU	1
19	PAHZZ	11862	8625221	.RING,RETAINING PART OF KIT P/N 8623979	1
20	PAHHH	72590	8633059	.SHAFT AND BUSHING A	1
21	PAHZZ	11862	8653985	..BEARING,SLEEVE	1
22	PAHZZ	11862	8626112	.RING,RETAINING	1
23	PAHZZ	11862	8626372	.BEARING,WASHER,THRU	1
24	PAHHH	11862	8655059	.CARRIER,TRANSMISSIO	1
25	PAHZZ	11862	8626173	..RING,RETAINING	1
26	PAHZZ	11862	8625913	..PARTS KIT,PINION	4
27	PAHZZ	11862	8623944	.BEARING,SLEEVE	2
28	PAHZZ	11862	8626809	.SHAFT,STRAIGHT	1
29	PAHZZ	11862	8623920	.PARTS KIT,BEARING R	1
30	PAHZZ	11862	8626807	.GEAR CLUSTER	1
31	PAHHH	11862	8675627	.CARRIER ASSEMBLY,TR	1
32	PAHZZ	11862	8627657	..RING,RETAINING	1
33	PAHZZ	11862	8625913	..PARTS KIT,PINION	4
34	PAHZZ	11862	8626372	BEARING,WASHER,THRU	1
35	PAHZZ	11862	8625401	BEARING,WASHER,THRU (.074-.078)	1
35	PAHZZ	11862	8625402	BEARING,WASHER,THRU (.082-.086)	1
35	PAHZZ	11862	8625403	BEARING,WASHER,THRU (.090-.094)	1
35	PAHZZ	11862	8625404	BEARING,WASHER,THRU (.098-.102)	1
35	PAHZZ	11862	8625405	BEARING,WASHER,THRU (.106-110)	1
35	PAHZZ	11862	8625406	BEARING,WASHER,THRU (.114-.118)	1
36	PAHZZ	11862	8633257	BAND,CLUTCH BRAKE	1

END OF FIGURE

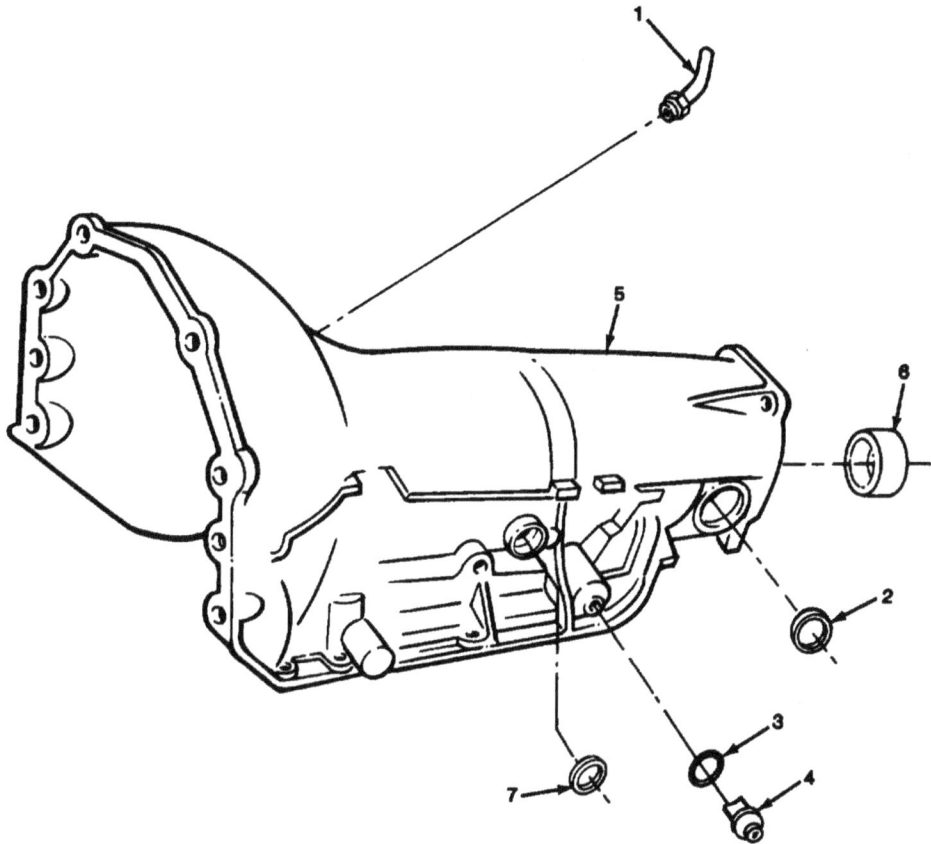

TA510882

FIGURE 75. TRANSMISSION CASE.

(1) (2) (3) (4) ITEM SMR PART NO CODE CAGEC NUMBER	(5) DESCRIPTION AND USABLE ON CODES (UOC)	(6) QTY

GROUP 0710 TRANSMISSION ASSEMBLY AND
ASSOCIATED PARTS

FIG. 75 TRANSMISSION CASE

1	PAHZZ	11862	8633408	TUBE,BENT,METALLIC	1
2	PAHZZ	11862	8623463	PLUG,PROTECTIVE,DUS	1
3	PAHZZ	11862	10054241	SEAL,SPEEDOMETER PART OF KIT P/N	1
				8625905	
4	PAHZZ	11862	8629503	CONNECTOR,PLUG,ELEC	1
5	XAHZZ	11862	8627996	HOUSING,MECHANICAL	1
6	PAHZZ	11862	8623941	.BEARING,SLEEVE	1
				UOC:194,208,209,210,230,231,252,254	
7	PAHZZ	11862	8611710	.PLUG,PROTECTIVE,DUS	1

END OF FIGURE

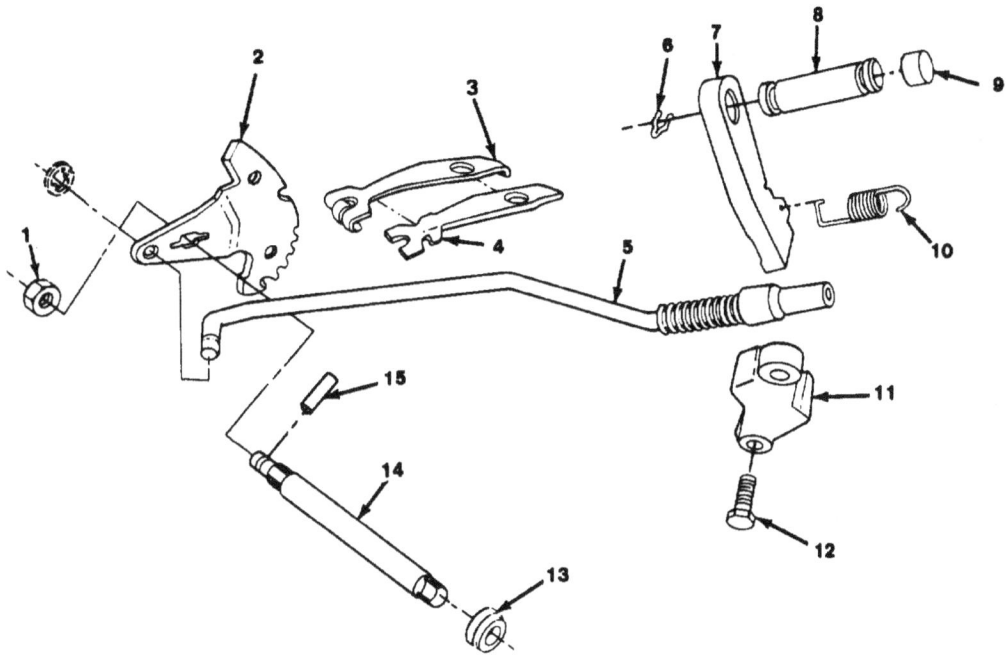

FIGURE 76. PARKING LOCK.

TA510883

(1)	(2)	(3)	(4)	(5)	(6)
ITEM NO	SMR CODE	CAGEC	PART NUMBER	DESCRIPTION AND USABLE ON CODES (UOC)	QTY

GROUP 0710 TRANSMISSION ASSEMBLY AND ASSOCIATED PARTS

FIG. 76 PARKING LOCK

1	PAHZZ	96906	MS51968-8	NUT,PLAIN,HEXAGON	1
2	PAHZZ	11862	8624336	DETENT PLATE	1
3	PAHHH	11862	8625773	ROLLER AND SPRING A	1
4	PAHZZ	11862	8633096	.SPRING,FLAT	1
5	PAHZZ	11862	8624091	ACTUATOR ASSEMBLY,P	1
6	PAHZZ	24617	451786	CLIP,RETAINING	1
7	PAHZZ	11862	8633060	PAWL	1
8	PAHZZ	11862	8624976	SHAFT AND PLUG ASSE	1
9	PAHZZ	11862	8620318	PLUG,EXPANSION	1
10	PAHZZ	11862	8623041	SPRING,HELICAL,EXTE	1
11	PAHZZ	11862	8624112	BRACKET,PARKING LOC	1
12	PAHZZ	96906	MS90728-31	BOLT,MACHINE	2
13	PAHZZ	11862	8657163	GASKET PART OF KIT P/N 8625905	1
14	PAHZZ	11862	8626979	PARTS KIT, MANUAL SH	1
15	PAHZZ	73342	23017556	PIN,STRAIGHT,HEADED	1

END OF FIGURE

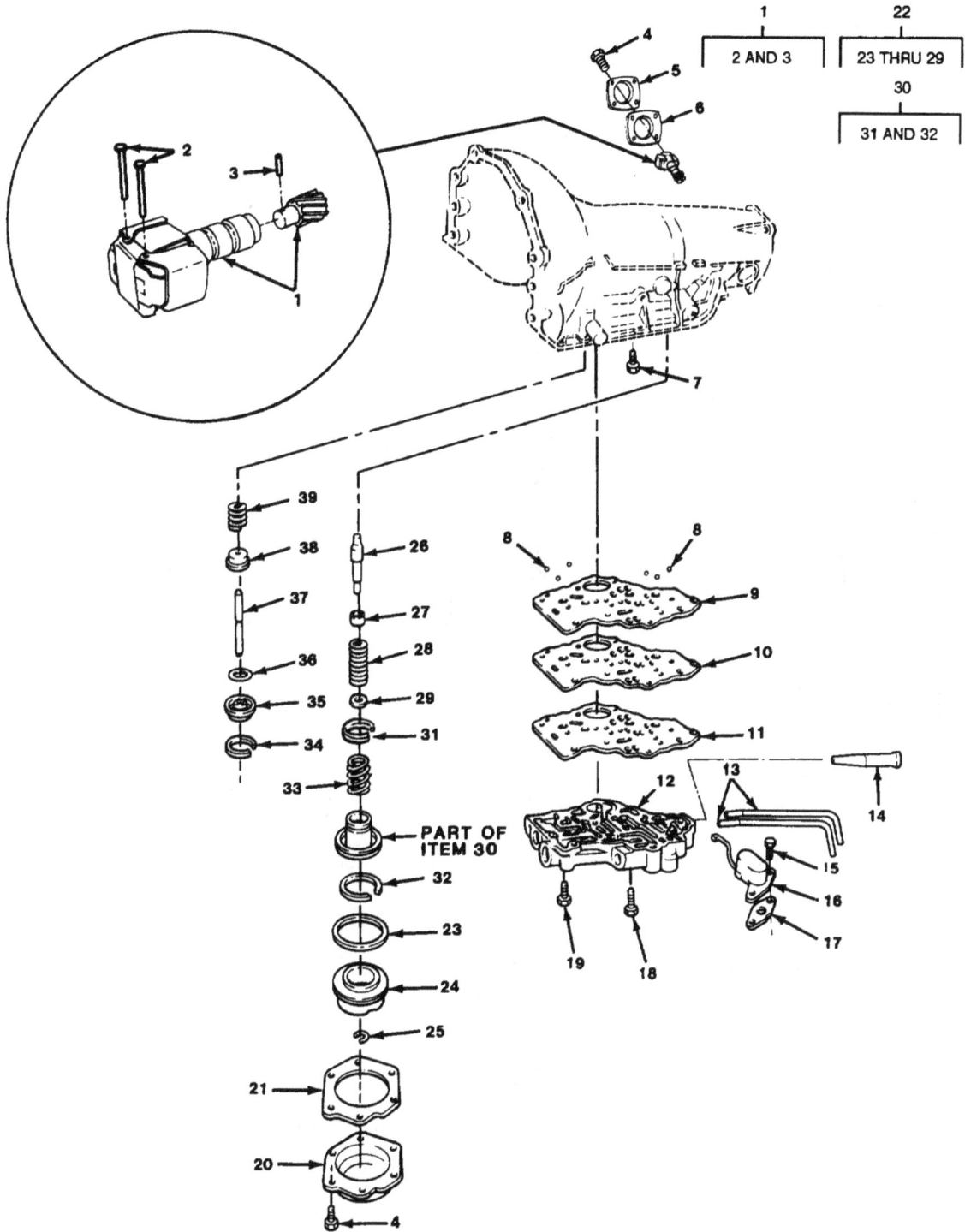

FIGURE 77. CONTROL VALVE ASSEMBLY GOVERNOR ASSEMBLY, AND SERVO VALVE UNIT.

TA510884

(1) (2) (3) (4)	(5)	(6)
ITEM SMR PART		
NO CODE CAGEC NUMBER	DESCRIPTION AND USABLE ON CODES (UOC)	QTY

GROUP 0714 SERVO UNIT

FIG. 77 CONTROL VALVE ASSEMBLY, GOVERNOR ASSEMBLY, AND SERVO VALVE UNIT

ITEM NO	SMR CODE	CAGEC	PART NUMBER	DESCRIPTION AND USABLE ON CODES (UOC)	QTY
1	PAFFF	11862	8629227	TRANSMISSION GOVERN	1
2	PAFZZ	73342	23017556	.PIN,STRAIGHT,HEADED PART OF KIT P/N 8629977	2
3	PAFZZ	96906	MS16562-35	.PIN,SPRING PART OF KIT P/N 8629977	1
4	PAFZZ	96906	MS90728-31	BOLT,MACHINE	10
5	PAFZZ	73342	8623262	COVER,GOVERNOR,TRAN	1
6	PAFZZ	11862	8623263	GASKET PART OF KIT P/N 8625905 PART OF KIT P/N 8629955	1
7	PAHZZ	11862	GM8670757	BOLT,INTERNALLY REL	1
8	PAHZZ	96906	MS19061-20007	BALL,BEARING	6
9	PAHZZ	11862	8670393	GASKET PART OF KIT P/N 8625905	1
10	PAHZZ	11862	8629796	PLATE,SPACER,VALVE	1
11	PAHZZ	11862	8623561	GASKET PART OF KIT P/N 8629955	1
12	PAHHH	11862	8633296	VALVE,LINEAR,DIRECT FOR COMPONENT PARTS SEE FIG 78	1
13	PAHZZ	11862	8623157	TUBE,BENT,METALLIC	2
14	PAFZZ	11862	8627509	STRAINER ELEMENT,SE	1
15	PAHZZ	96906	MS90725-5	SCREW,CAP,HEXAGON H	2
16	PAHZZ	11862	8629935	SOLENOID,ELECTRICAL UOC:194,208,209,210,230,231,252,254	1
17	PAHZZ	11862	8623741	GASKET PART OF KIT P/N 8625905	1
18	PAHZZ	11862	9430895	BOLT,MACHINE	8
19	PAHZZ	11862	9440338	BOLT,MACHINE	3
20	PAHZZ	11862	8624136	COVER,ACCESS	1
21	PAHZZ	11862	8623174	GASKET PART OF KIT P/N 8625905 PART OF KIT P/N 8629955	1
22	AHHHH	11862	8624145	PISTON ASM	1
23	PAHZZ	11862	8623430	.GASKET PART OF KIT P/N 8625905 PART OF KIT P/N 8629955	1
24	PAHZZ	11862	8624138	.PISTON,LINEAR ACTUA	1
25	PAHZZ	96906	MS16633-1031	.RING,RETAINING PART OF KIT P/N 8623979	1
26	PAHZZ	11862	8624139	.PIN,SHOULDER,HEADLE (3.300-3.306)	1
26	PAHZZ	11862	8624140	.PIN,SHOULDER,HEADLE (3.356-3.362)	1
26	PAHZZ	11862	8624141	.PIN,SHOULDER,HEADLE (3.412-3.418)	1
27	PAHZZ	11862	8623664	.SEAT,HELICAL COMPRE	1
28	PAHZZ	11862	8623666	.SPRING,HELICAL,COMP	1
29	PAHZZ	11862	8623744	.WASHER,FLAT	1
30	PAHHH	11862	8626902	PISTON AND RING ASS	1
31	PAHZZ	11862	8623671	.SEAL	1
32	PAHZZ	11862	8627153	.RING,WIPER	1
33	PAHZZ	11862	8624256	SPRING,HELICAL,COMP	1
34	PAHZZ	11862	8623131	SEAL RING,METAL PART OF KIT P/N 8629955	1
35	PAHZZ	11862	8626878	PISTON,LINEAR ACTUA	1
36	PAHZZ	11862	8626884	WASHER,SLOTTED	1

(1)	(2)	(3)	(4)	(5)		(6)
ITEM	SMR			PART		
NO	CODE	CAGEC	NUMBER		DESCRIPTION AND USABLE ON CODES (UOC)	QTY

37	PAHZZ	11862	8626879	PIN,GROOVED,HEADLES		1
38	PAHZZ	11862	8626881		RETAINER,HELICAL CO	1
39	PAHZZ	11862	8623489	SPRING,HELICAL,COMP		1

END OF FIGURE

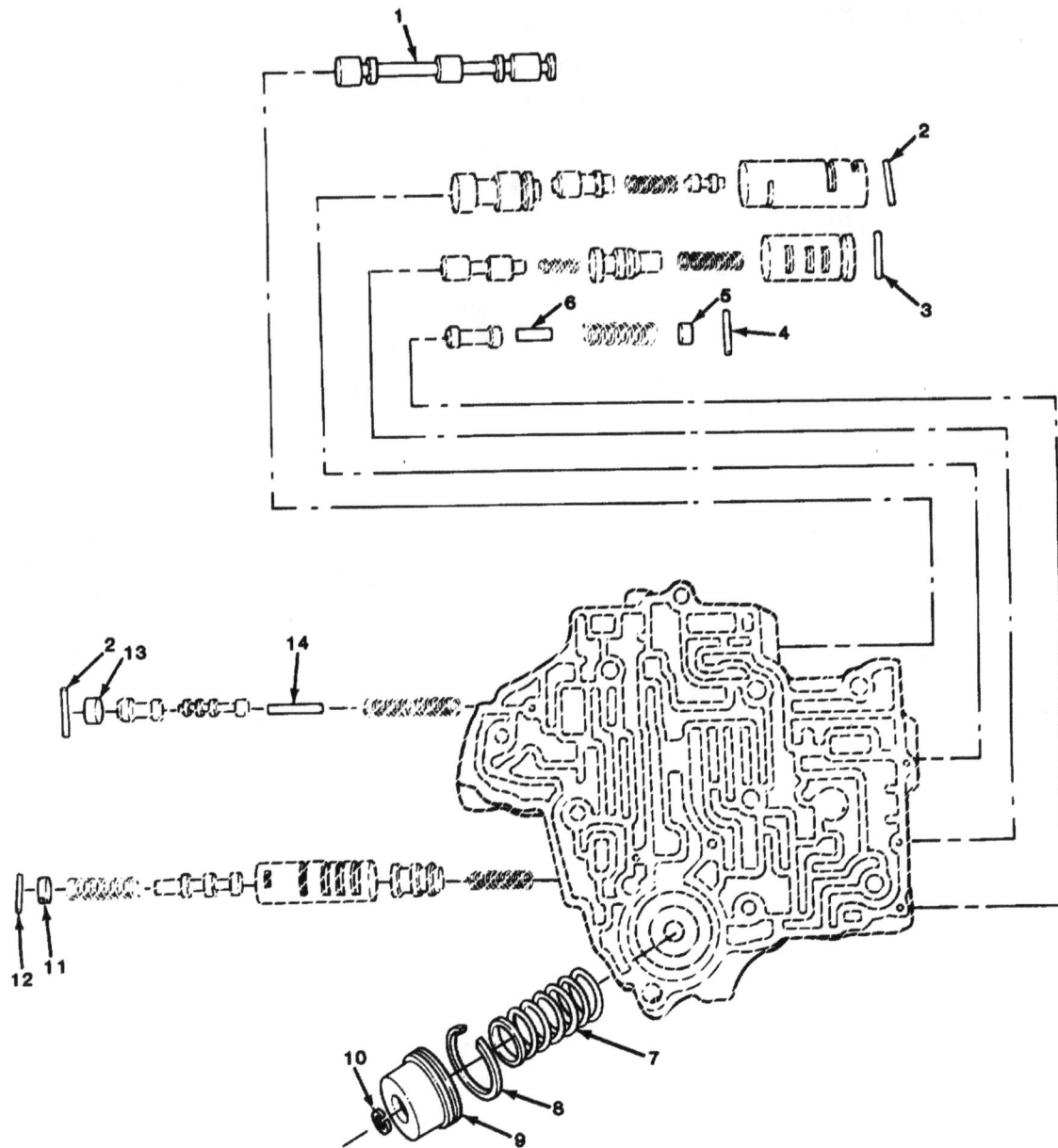

TA510885

FIGURE 78. CONTROL VALVE COMPONENT PARTS.

(1) ITEM NO	(2) SMR CODE	(3) CAGEC	(4) PART NUMBER	(5) DESCRIPTION AND USABLE ON CODES (UOC)	(6) QTY
				GROUP 0714 SERVO UNIT	
				FIG. 78 CONTROL VALVE COMPONENT PARTS	
1	PAHZZ	11862	8623592	VALVE,LINEAR,DIRECT	1
2	PAHZZ	11862	8624482	PIN,SPRING	2
3	PAHZZ	73342	8623078	PIN	1
4	PAHZZ	11862	8625324	PIN,STRAIGHT,HEADLE	1
5	PAHZZ	11862	8623298	PLUG,VALVE BORE	1
6	PAHZZ	11862	8622045	PIN,STRAIGHT,HEADLE	1
7	PAHZZ	11862	8623654	SPRING,HELICAL,COMP	1
8	PAHZZ	11862	8623653	SEAL RING,METAL	1
9	PAHZZ	11862	8629080	PISTON,ACCUMULATOR	1
10	PAHZZ	24617	9412978	RING,RETAINING PART OF KIT P/N 8623979	1
11	PAHZZ	11862	8625546	PLUG,VALVE BORE	1
12	PAHZZ	11862	8625544	PIN,GROOVED,HEADLES	1
13	PAHZZ	73342	8623295	PLUG	1
14	PAHZZ	11862	8623507	PIN,STRAIGHT,HEADLE	1

END OF FIGURE

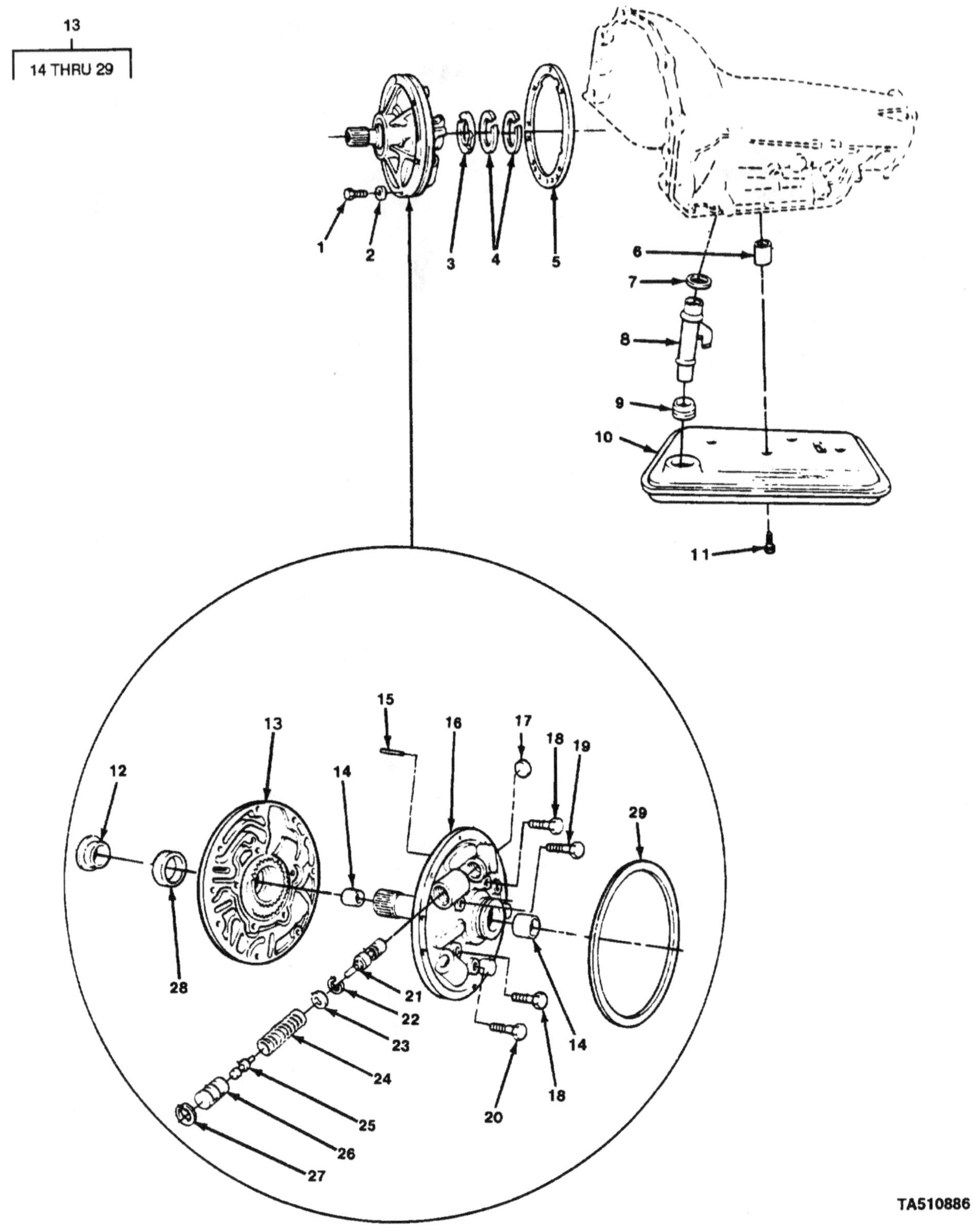

FIGURE 79. OIL PUMP ASSEMBLY AND TRANSMISSION FILTER.

TA510886

(1) (2) (3) (4) ITEM SMR PART NO CODE CAGEC NUMBER	(5) DESCRIPTION AND USABLE ON CODES (UOC)	(6) QTY

GROUP 0721 COOLERS, PUMPS, MOTORS

FIG. 79 OIL PUMP ASSEMBLY AND
TRANSMISSION FILTER

1	PAHZZ 11862 8655280		BOLT,MACHINE	6
2	PAHZZ 11862 8626281		WASHER,FLAT PART OF KIT P/N 8625905	6
3	PAHZZ 11862 8623301		BEARING,WASHER,THRU (.071-.075)	1
3	PAHZZ 11862 8623300		BEARING,WASHER,THRU (.060-.064)	1
3	PAHZZ 11862 8623303		BEARING,WASHER,THRU (.093-.097)	1
3	PAHZZ 11862 8623304		BEARING,WASHER,THRU (.104-.108)	1
3	PAHZZ 11862 8623305		BEARING,WASHER,THRU (.115-.119)	1
3	PAHZZ 11862 8623306		BEARING,WASHER,THRU (.126-.130)	1
3	PAHZZ 11862 8623302		BEARING,WASHER,THRU (.082-.086)	1
4	PAHZZ 11862 8626356		RETAINER,PACKING PART OF KIT P/N 8629955	2
5	PAHZZ 11862 8623978		GASKET PART OF KIT P/N 8625905	1
6	PAFZZ 11862 8629523		SPACER,SLEEVE	1
7	PAOZZ 73342 6771005		SEAL,NONMENTALLIC RO PART OF KIT P/N 8625905	1
8	PAOZZ 11862 8629526		TUBE,BENT,METALLIC	1
9	PAOZZ 11862 6437746		GROMMET,TRANSMISSIO	1
10	PAOZZ 11862 6259423		PARTS KIT,FLUID PRE	1
11	PAOZZ 11862 8633208		BOLT,SHOULDER	1
12	PAHZZ 11862 8626916		SEAL,PLAIN ENCASED	1
13	AHHHH 11862 8627644		PUMP ASM	1
14	PAHZZ 11862 8623944		.BEARING,SLEEVE	2
15	PAHZZ 73342 8623078		.PIN	1
16	PAHHH 11862 8625955		.COVER,ACCESS	1
17	PAHZZ 11862 8623292		.DISK,SOLID,PLAIN	1
18	PAHZZ 96906 MS90728-38	.BOLT,MACHINE		2
19	PAHZZ 96906 MS90728-39	.BOLT,MACHINE		1
20	PAHZZ 96906 MS90728-34	.BOLT,MACHINE		2
21	PAHZZ 11862 8623422		.VALVE,REGULATING,FL	1
22	PAHZZ 11862 8623368		.WASHER,SPLIT	1
23	PAHZZ 11862 8623039		.WASHER,KEY	1
24	PAHZZ 11862 8625648		.SPRING,HELICAL,COMP	1
25	KFHZZ 11862 8625650		.VALVE REG BOOST	1
26	KFHZZ 11862 8626885		.BUSHG REG BOOST VLV	1
27	PAHZZ 96906 MS16627-1087	.RING,RETAINING		1
28	PAHZZ 11862 8629961		.BUSHING	1
29	PAHZZ 73342 8626309		.OIL SEAL PART OF KIT P/N 8625905	1

END OF FIGURE

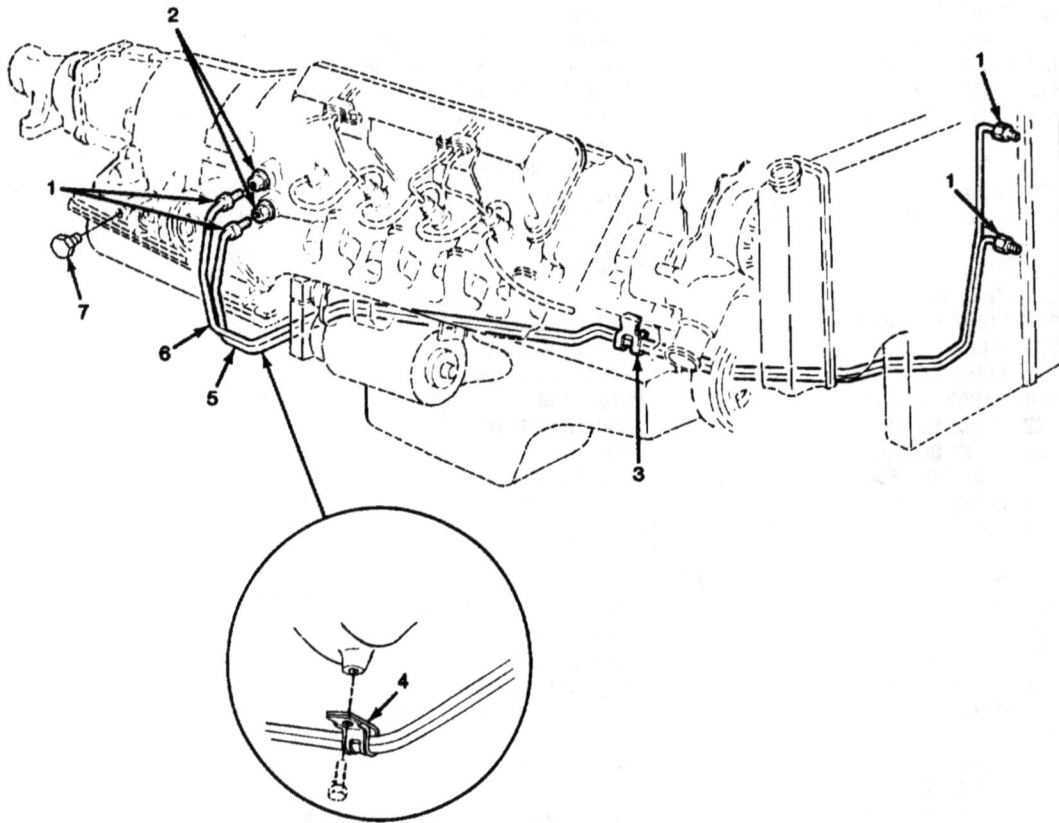

TA510887

FIGURE 80. TRANSMISSION OIL COOLER LINES.

(1) (2) (3) (4)	(5)	(6)
ITEM SMR PART		
NO CODE CAGEC NUMBER	DESCRIPTION AND USABLE ON CODES (UOC)	QTY

GROUP 0721 COOLERS, PUMPS, MOTORS

FIG. 80 TRANSMISSION OIL COOLER
LINES

1	PAOZZ 72582 137398		INVERTED NUT,TUBE C	4
2	PAOZZ 11862 8637742	ADAPTER,STRAIGHT,PI		2
3	PAOZZ 11862 3997718		CLIP,SPRING TENSION	1
4	PAOZZ 11862 15517986	CLAMP,LOOP		1
5	MOOZZ 11862 14045626		TUBE,METALLIC (73.95"LG) MAKE FROM TUBE, P/N 1324714	1
6	MOOZZ 11862 14045628		PIPE ASM-TRNS OIL C (80.64"LG) MAKE FROM TUBE, P/N 1324714	1
7	PAOZZ 14569 1007-2	PLUG,PIPE		1

END OF FIGURE

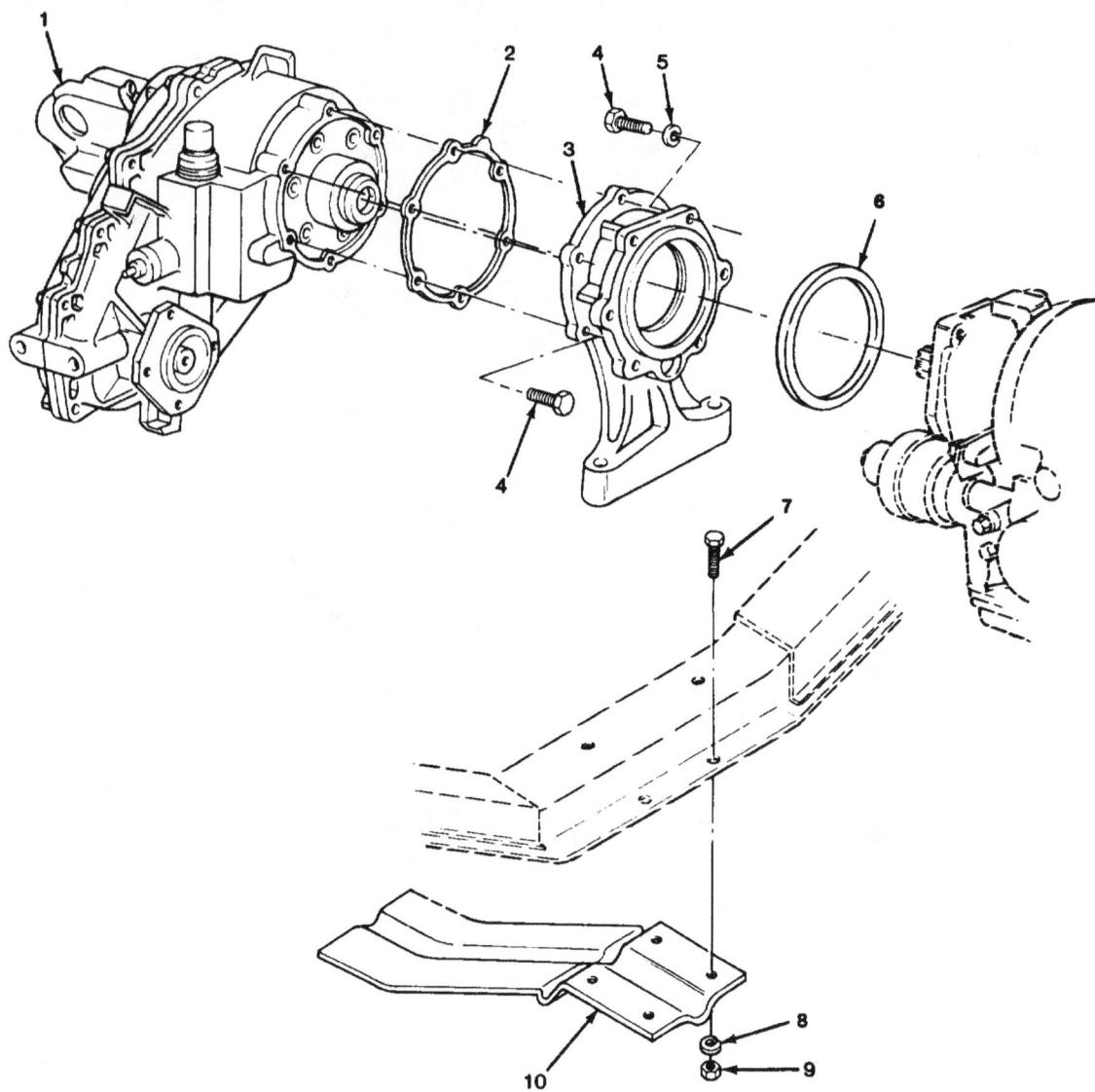

TA510888

FIGURE 81. TRANSFER CASE ASSEMBLY AND ADAPTER (MODEL 208).

(1) (2) (3) (4) ITEM SMR PART NO CODE CAGEC NUMBER	(5) DESCRIPTION AND USABLE ON CODES (UOC)	(6) QTY
	GROUP 08 TRANSFER, FINAL DRIVE, PLANETARY, AND DROP GEARBOX ASSEMBLIES	
	GROUP 0801 POWER TRANSFER, FINAL DRIVE, PLANETARY, OR DROP GEARBOX ASSEMBLIES FIG. 81 TRANSFER CASE ASSEMBLY AND ADAPTER (MODEL 208)	
1 PAFHD 11862 14067765	TRANSFER TRANSMISSION FOR COMPONENT PARTS SEE FIG 83 UOC:194,208,209,210,230	1
2 PAFZZ 11862 14020854 GASKET	UOC:194,208,209,210,230	1
3 PAFZZ 11862 14020861 HOUSING,MECHANICAL	UOC:194,208,209,210,230	1
4 PAFZZ 96906 MS90728-62 SCREW,CAP,HEXAGON H	UOC:194,208,209,210,230	12
5 PAFZZ 96906 MS35338-46 WASHER,LOCK	UOC:194,208,209,210,230	6
6 PAFZZ 11862 8623216 GASKET	UOC:194,208,209,210,230	1
7 PAOZZ 96906 MS90728-62 SCREW,CAP,HEXAGON,H	UOC:194,208,209,210,230	4
8 PAOZZ 96906 MS27183-14 WASHER,FLAT	UOC:194,208,209,210,230	4
9 PAOZZ 80205 NAS1408A6 NUT,SELF-LOCKING,HE	UOC:194,208,209,210,230	4
10 PAOZZ 11862 14067764 COVER,ACCESS	UOC:194,208,209,210,230,252,254,256	1

END OF FIGURE

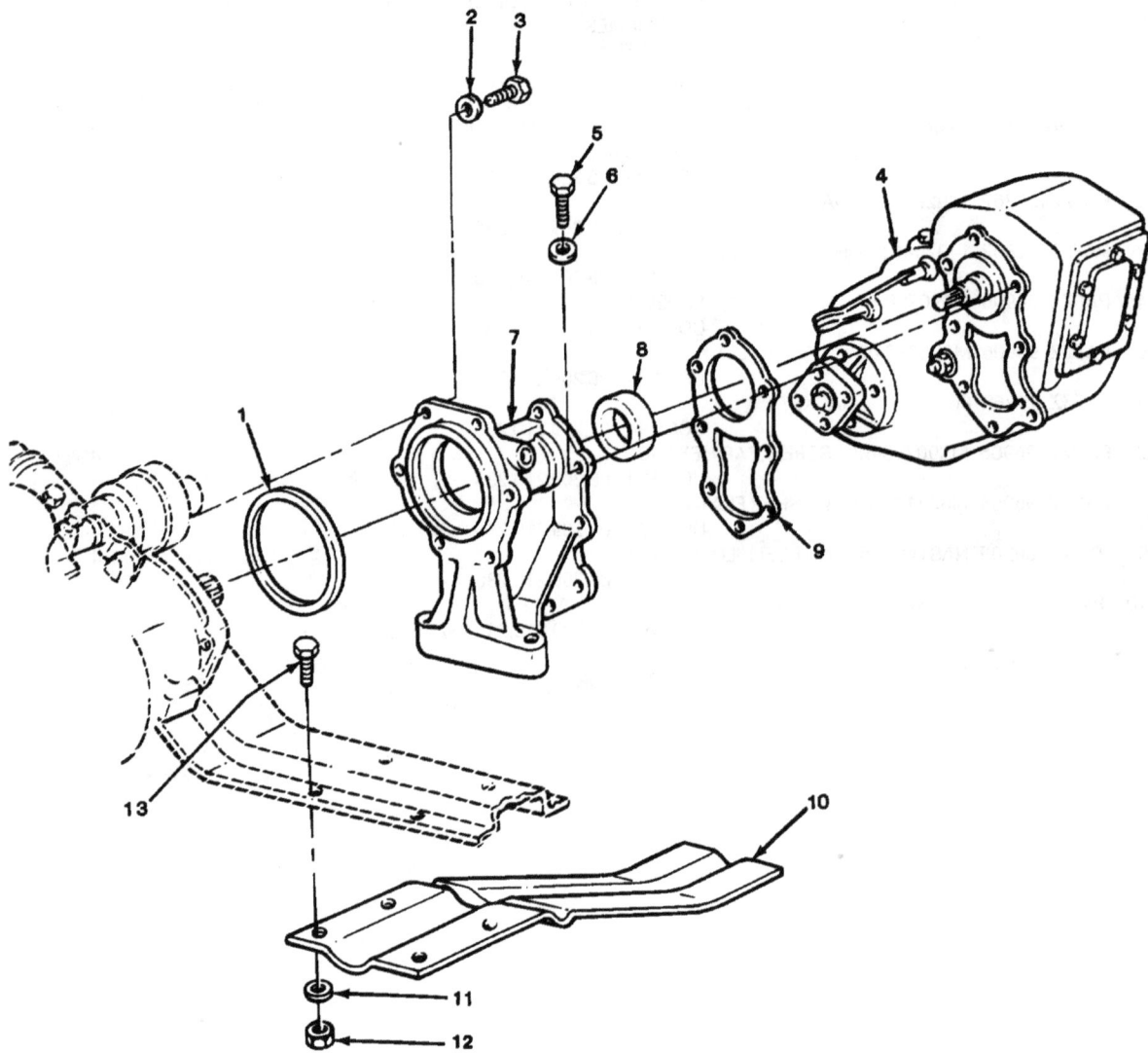

TA510889

FIGURE 82. TRANSFER CASE ASSEMBLY AND ADAPTER (MODEL 205).

(1) (2) (3) (4) ITEM SMR PART NO CODE CAGEC NUMBER	(5) DESCRIPTION AND USABLE ON CODES (UOC)	(6) QTY
	GROUP 0801 POWER TRANSFER, FINAL DRIVE, PLANETARY OR DROP GEARBOX ASSEMBLIES	
	FIG. 82 TRANSFER CASE ASSEMBLY AND ADAPTER (MODEL 205)	
1 PAFZZ 11862 8623216 GASKET		1
	UOC:231,252,254,256	
2 PAFZZ 96906 MS35338-46 WASHER,LOCK		6
	UOC:231,252,254,256	
3 PAFZZ 96906 MS90728-62 SCREW,CAP,HEXAGON H		6
	UOC:231,252,254,256	
4 PAFHD 11862 15599248 TRANSFER TRANSMISSI FOR COMPONENT PARTS SEE FIG 84		1
	UOC:231,252,254,256	
5 PAFZZ 96906 MS90728-62 .SCREW,CAP,HEXAGON H		8
	UOC:231,252,254,256	
6 PAFZZ 96906 MS35335-35 .WASHER,LOCK		8
	UOC:231,252,254,256	
7 PAFFF 11862 14022218 .ADAPTER,TRANSFER CA		1
	UOC:231,252,254,256	
8 PAFZZ 73680 29940-0445 ..SEAL,PLAIN ENCASED		1
	UOC:231,252,254,256	
9 PAFZZ 11862 14022219 .GASKET		1
	UOC:231,252,254,256	
10 PAOZZ 11862 14029158 PLATE,TRANSFER CASE		1
	UOC:231,252,254,256	
11 PAOZZ 96906 MS27183-14 WASHER,FLAT		4
	UOC:231,252,254,,256	
12 PAOZZ 80205 NAS1408A6 NUT,SELF-LOCKING,HE		4
	UOC:231,252,254,256	
13 PAOZZ 96906 MS90728-62 SCREW,CAP,HEXAGON H		4
	UOC:231,252,254,256	

END OF FIGURE

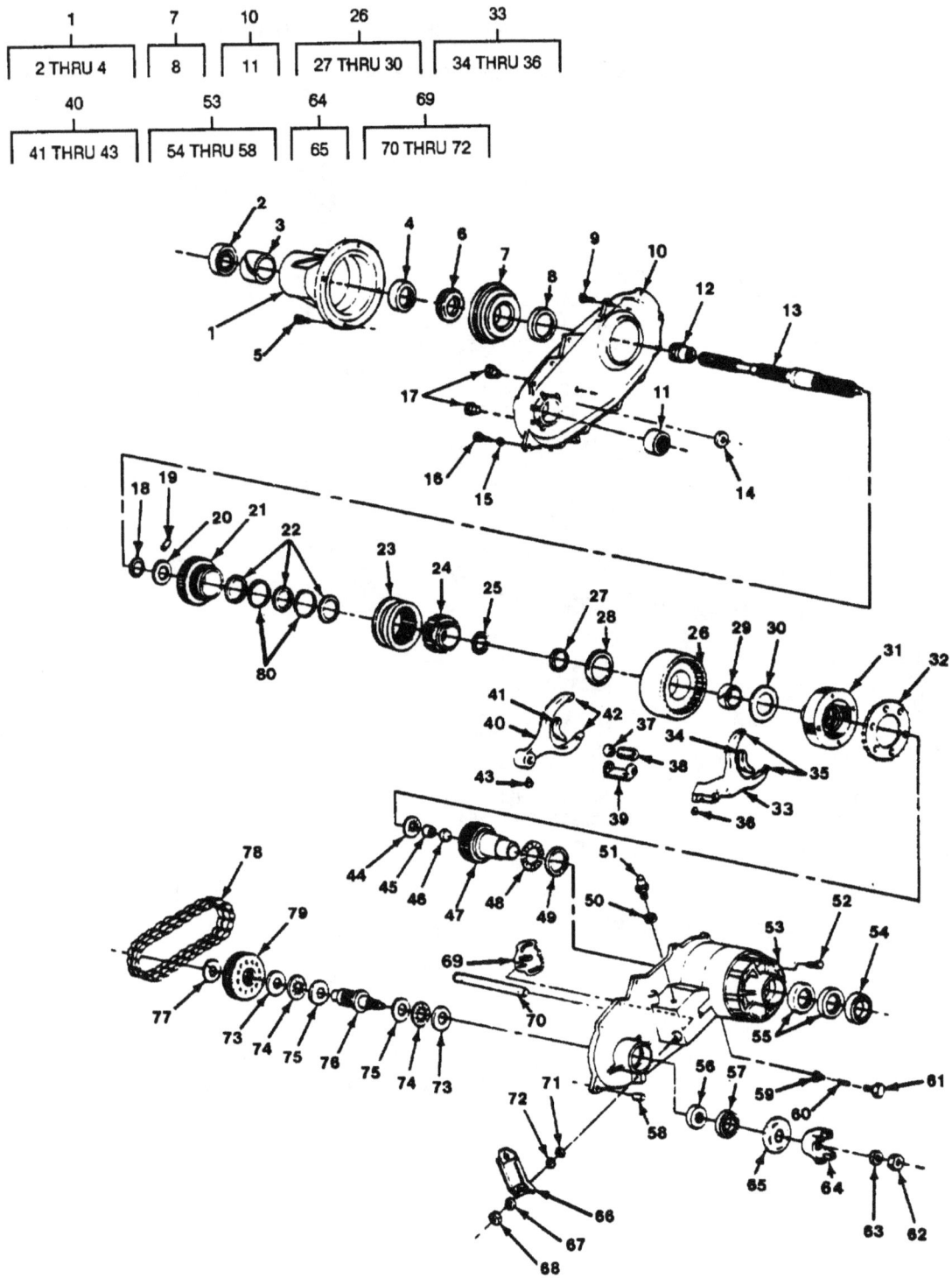

FIGURE 83. TRANSFER CASE COMPONENT PARTS (MODEL 208).

TA510890

(1) (2) (3) (4) ITEM SMR PART NO CODE CAGEC NUMBER	(5) DESCRIPTION AND USABLE ON CODES (UOC)	(6) QTY
	GROUP 0801 POWER TRANSFER, FINAL DRIVE, PLANETARY, OR DROP GEARBOX ASSEMBLIES	
	FIG. 83 TRANSFER CASE COMPONENT PARTS (MODEL 208)	
1 PAFFF 11862 14037990	HOUSING,MAIN SHAFT UOC:194,208,209,210,230	1
2 PAFZZ 76680 9449-K	.SEAL,CASE EXTENSION UOC:194,208,209,210,230	1
3 PAFZZ 11862 3978765	.BEARING,SLEEVE UOC:194,208,209,210,230	1
4 PAFZZ 21335 P207K	.BEARING,BALL,ANNULA UOC:194,208,209,210,230	1
5 PAFZZ 24617 11503428	BOLT,SELF-LOCKING UOC:194,208,209,210,230	4
6 PAFZZ 11862 376176	GEAR SET,SPUR,MATCH UOC:194,208,209,210,230	1
7 PAFFF 11862 14075205	HOUSING,PUMP UOC:194,208,209,210,230	1
8 PAFZZ 80201 22306	.SEAL,PLAIN ENCASED UOC:194,208,209,210,230	1
9 PAHZZ 24617 11503428	BOLT,SELF-LOCKING UOC:194,208,209,210,230	13
10 XAHHH 11862 14037983	HOUSING,MECHANICAL UOC:194,208,209,210,230	1
11 PAHZZ 96906 MS17131-42	.BEARING,ROLLER,NEED UOC:194,208,209,210,230	1
12 PAHZZ 11862 14037997	GEAR,HELICAL UOC:194,208,209,210,230	1
13 PAHZZ 11862 15594195	SHAFT,SHOULDERED (FIRST DESIGN APPLIES TO 1/CASE BEFORE 7-1-83 STAMPED ON THE T/CASE I.D. PLATE) UOC:194,208,209,210,230	1
13 PAHZZ 11862 15594195	SHAFT,SHOULDERED (SECOND DESIGN APPLIES TO 1/CASE AFTER 7-1-83 STAMPED ON THE T/CASE I.D. PLATE) UOC:194,208,209,210,230	1
14 PBHZZ 11862 3787240	MAGNET,CHIP COLLECT UOC:194,208,209,210,230	1
15 PAHZZ 11862 14037984	WASHER,FLAT UOC:194,208,209,210,230	2
16 PAHZZ 24617 11500831	BOLT,SELF-LOCKING UOC:194,208,209,210,230	2
17 PAOZZ 11862 14037987	PLUG,MACHINE THREAD UOC:194,208,209,210,230	2
18 PAHZZ 72800 XAN-225-H	RING,RETAINING UOC:194,208,209,210,230	1
19 PAHZZ 24617 443767	PIN,STRAIGHT,HEADLE (SECOND DESIGN APPLIES TO 1/CASE AFTER 7-1-83 STAMPED ON THE T/CASE I.D. PLATE	1

(1) ITEM NO	(2) SMR CODE	(3) CAGEC	(4) PART NUMBER	(5) DESCRIPTION AND USABLE ON CODES (UOC)	(6) QTY
20	PAHZZ	11862	14075212	UOC:194,208,209,210,230 WASHER,FLAT (FIRST DESIGN APPLIES TO T/CASE BEFORE 7-1-83 STAMPED ON THE T/CASE I.D. PLATE)	1
20	PAHZZ	11862	15594196	UOC:194,208,209,210,230 BEARING,WASHER,THRU (SECOND DESIGN APPLIES TO T/CASE AFTER 7-1-83 STAMPED ON THE T/CASE I.D. PLATE	1
21	PAHZZ	11862	14095623	UOC:194,208,209,210,230 SPROCKET WHEEL	1
22	PAHZZ	11862	14037976	UOC:194,208,209,210,230 SPACER,RING	3
23	PAHZZ	11862	14075211	UOC:194,208,209,210,230 SLEEVE,SYNCHRONIZER	1
24	PAHZZ	11862	14075210	UOC:194,208,209,210,230 GEAR,SPUR	1
25	PAHZZ	19207	8338687	UOC:194,208,209,210,230 RING,RETAINING	1
26	PAHHH	11862	14095676	UOC:194,208,209,210,230 GEAR,INTERNAL	1
27	PAHZZ	78553	XAN-262-H	UOC:194,208,209,210,230 .RING,RETAINING	1
28	PAHZZ	11862	14095678	UOC:194,208,209,210,230 .BEARING,WASHER,THRU	1
29	PAHZZ	11862	14044937	UOC:194,208,209,210,230 .BEARING,SLEEVE	1
30	PAHZZ	11862	14095677	UOC:194,208,209,210,230 .BEARING,WASHER,THRU	1
31	PAHZZ	11862	14071876	UOC:194,208,209,210,230 PLANETARY GEAR ASSE	1
32	PAHZZ	11862	14037949	UOCZ;194,208,209,210,230 .GEAR,SPUR	1
33	PAHHH	11862	14095679	UOC:194,208,209,210,230 SHIFTER FORK	1
34	PAHZZ	11862	14095680	UOC:194,208,209,210,230 .PAD,FORK	1
35	PAHZZ	11862	14037966	UOC:194,208,209,210,230 .PAD,FORK,TRANSFER	2
36	PAHZZ	11862	14037961	UOC:194,208,209,210,230 .PIN,STRAIGHT,HEADLE	1
37	PAHZZ	11862	14037967	UOC:194,208,209,210,230 RETAINER,HELICAL CO	1
38	PAHZZ	11862	14037968	UOC:194,208,209,210,230 SPRING,HELICAL,COMP	1
39	PAHZZ	11862	14037962	UOC:194,208,209,210,230 RETAINER,HELICAL CO	1
40	PAHHH	11862	14037963	UOC:194,208,209,210,230 SHIFTER FORK	1
41	PAHZZ	11862	14037965	UOC:194,208,209,210,230 .PAD,FORK,TRANSFER	1
42	PAHZZ	11862	14037966	UOC:194,208,209,210,230 .PAD,FORK,TRANSFER	2

(1) ITEM NO	(2) SMR CODE	(3) CAGEC	(4) PART NUMBER	(5) DESCRIPTION AND USABLE ON CODES (UOC)	(6) QTY
				UOC:194,208,209,210,230	
43	PAHZZ	11862	14037964	.PIN,STRAIGHT,HEADED	1
				UOC:194,208,209,210,230	
44	PAHZZ	60380	NTA2435	RETAINER AND ROLLER	1
				UOC:194,208,209,210,230	
45	PAHZZ	60380	B2012-0H	BEARING,ROLLER NEED	1
				UOC:194,208,209,210,230	
46	PAHZZ	24617	9427468	PLUG CUP,TRANSFER C	1
				UOC:194,208,209,210,230	
47	PAHZZ	11862	14037995	GEAR,SPUR	1
				UOC:194,208,209,210,230	
48	PAHZZ	60380	FNT5070	RETAINER AND ROLLER	1
				UOC:194,208,209,210,230	
49	PAHZZ	11862	14071875	SEAT,BEARING	1
				UOC:194,208,209,210,230	
50	PAOZZ	76760	99780	GASKET	1
				UOCl194,208,209,210,230	
51	PAOZZ	11862	14037986	SWITCH,PUSH	1
				UOC:194,208,209,210,230	
52	PAHZZ	11862	14075306	BOLT,SELF-LOCKING	6
				UOC:194,208,209,210,230	
53	XAHHH	11862	14037947	HOUSING,MECHANICAL	1
				UOC:194,208,209,210,230	
54	PAFZZ	73680	29940-0445	.SEAL,PLAIN ENCASED	1
				UOC:194,208,209,210,230	
55	PAHZZ	60380	F5020	.BEARING,ROLLER,NEED	2
				UOC:194,208,209,210,230	
56	PAHZZ	96906	MS17131-47	.BEARING,ROLLER,NEED	1
				UOC:194,208,209,210,230	
57	PAFZZ	07200	29940-0435	.SEAL,PLAIN ENCASED	1
				UOC:194,208,209,210,230	
58	PAHZZ	11862	14037948	.PIN,SPRING	2
				UOC:194,208,209,210,230	
59	PAHZZ	11862	14037956	PIN,SHOULDER,HEADLE	1
				UOC:194,208,209,210,230	
60	PAHZZ	11862	3967886	SPRING,HELICAL,COMP	1
				UOC:194,208,209,210,230	
61	PAHZZ	11862	14095597	BOLT,SELF-LOCKING	1
				UOC:194,208,209,210,230	
62	PAFZZ	11862	14037991	NUT,SELF-LOCKING,EX	1
				UOC:194,208,209,210,230	
63	PAFZZ	97271	6259149	WASHER,LOCK	1
				UOC:194,208,209,210,230	
64	PAFFF	11862	14037992	PRPOPELLER FLANGE AS	1
				UOC:194,208,209,210,230	
65	PAFZZ	11862	14037993	.DEFLECTOR,DIRT AND	1
				UOC:194,208,209,210,230	
66	PAHZZ	11862	14029126	LEVER,REMOTE CONTRO	1
				UOC:194,208,209,210,230	
67	PAFZZ	96906	MS27183-14	WASHER,FLAT	1
				UOC:194,208,209,210,230	
68	PAHZZ	96906	MS21045-6	NUT,SELF-LOCKING,HE	1

(1) ITEM NO	(2) SMR CODE	(3) CAGEC	(4) PART NUMBER	(5) DESCRIPTION AND USABLE ON CODES (UOC)	(6) QTY
				UOC:194,208,209,210,230	
69	PAHHH	11862	14037957	SHAFT AND SECTOR AS	1
				UOC:194,208,209,210,230	
70	PAHZZ	11862	14037959	.SHAFT,STRAIGHT	1
				UOC:194,208,209,210,230	
71	PAHZZ	02697	ARP583-116	.PACKING,PREFORMED	1
				UOC:194,208,209,210,230	
72	PAHZZ	11862	14037958	.RETAINER,PACKING	1
				UOC:194,208,209,210,230	
73	PAHZZ	11862	14071882	BEARING,WASHER,THRU	2
				UOC:194,208,209,210,230	
74	PAHZZ	60380	NTA2840	RETAINER AND ROLLER	2
				UOC:194,208,209,210,230	
75	PAHZZ	60380	TRA2840	SEAT,BEARING	2
				UOC:194,208,209,210,230	
76	PAHZZ	11862	14037979	SHAFT,SHOULDERED	1
				UOC:194,208,209,210,230	
77	PAHZZ	72800	XAN-250-H	RING,RETAINING	1
				UOC:194,208,209,210,230	
78	PAHZZ	55880	372380	CHAIN,SILENT	1
				UOC:194,208,209,210,230	
79	PAHZZ	11862	14037980	SPROCKET WHEEL	1
				UOC:194,208,209,210,230	
80	PAHZZ	60380	C10860	ROLLER,BEARING	120
				UOC:194,208,209,210,230	

END OF FIGURE

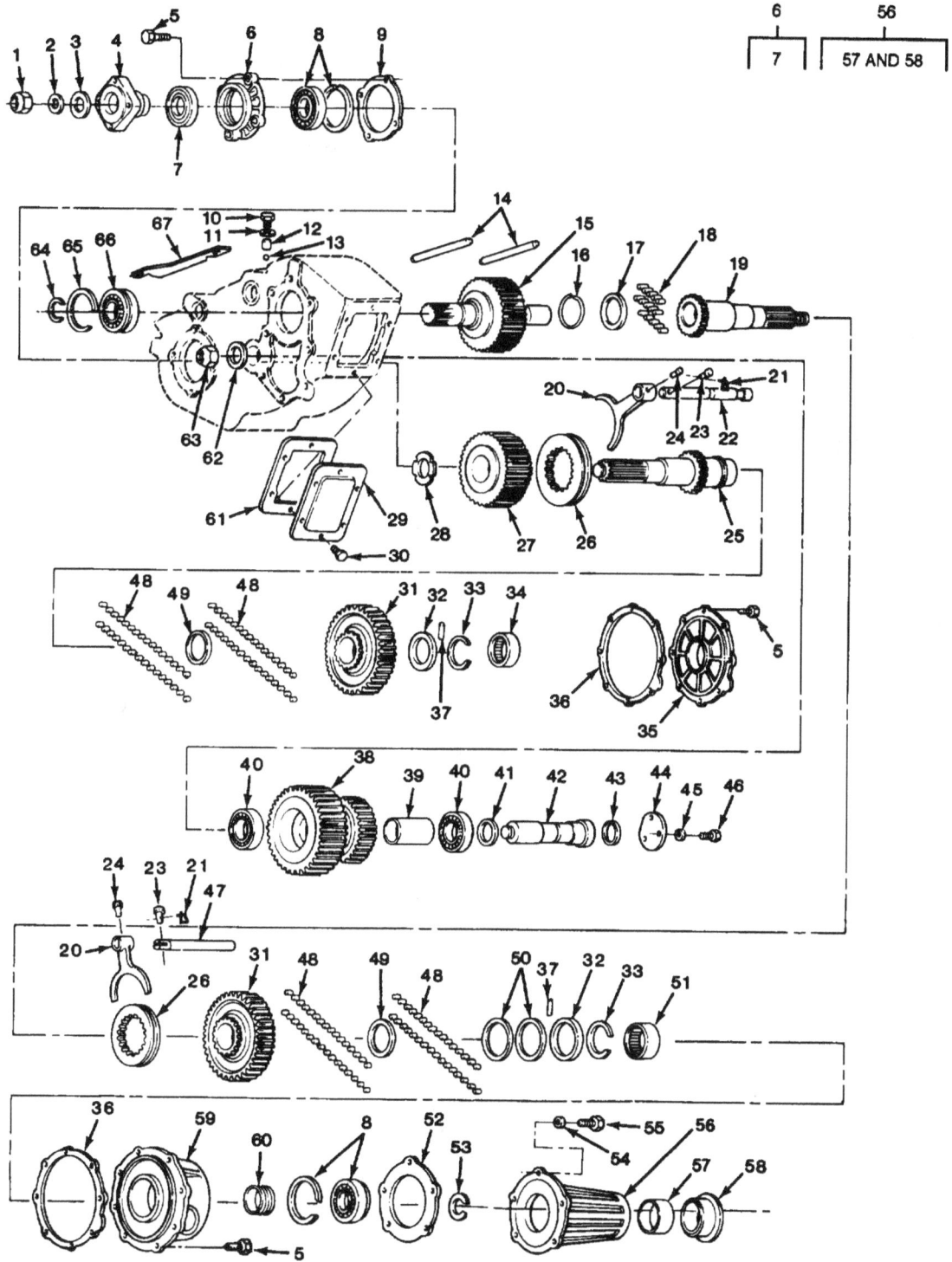

FIGURE 84. TRANSFER CASE COMPONENT PARTS (MODEL 205).

TA510891

(1) (2) (3) (4) ITEM SMR PART NO CODE CAGEC NUMBER	(5) DESCRIPTION AND USABLE ON CODES (UOC)	(6) QTY
	GROUP 0801 POWER TRANSFER, FINAL DRIVE, PLANETARY, OR DROP GEARBOX ASSEMBLIES	
	FIG. 84 TRANSFER CASE COMPONENT PARTS (MODEL 205)	
1 PAFZZ 11862 9422308	NUT,SELF-LOCKING,HE UOC:231,252,254,256	1
2 PAFZZ 11862 2423517	WASHER,FLAT UOC:231,252,254,256	1
3 PAFZZ 11862 14022212	GASKET UOC:231,252,254,256	1
4 PAFZZ 11862 460835	FLANGE,COMPANION,UN UOC:231,252,254,256	1
5 PAFZZ 11862 9437242	BOLT,MACHINE UOC:231,252,254,256	20
6 PAFFF 11862 465482	RETAINER,ASSEMBLY UOC:231,252,254,256	1
7 PAFZZ 11862 465483	.SEAL,PLAIN,ENCASED UOC:231,252,254,256	1
8 PAFZZ 43334 43307-P-S1568A BEARING,BALL,ANNULA	UOC:231,252,254,256	2
9 PAFZZ 11862 14037928	GASKET UOC:231,252,254,256	1
10 PAFZZ 11862 3975717	PLUG,MACHINE THREAD UOC:231,252,254,256	1
11 PAFZZ 11862 3967885	GASKET UOC:231,252,254,256	1
12 PAFZZ 11862 3967886	SPRING,HELICAL,COMP UOC:231,252,254,256	1
13 PAFZZ 24617 453610	BALL,BEARING UOC:231,252,254,256	1
14 PAHZZ 11862 3975716	PIN,STRAIGHT,HEADLE UOC:231,252,254,256	2
15 PAHZZ 11862 15594158	GEAR SHAFT,HELICAL-S UOC:231,252,254,256	1
16 PAHZZ 11862 471896	RING,RETAINING UOC:231,252,254,256	1
17 PAHZZ 11862 3967895	WASHER,FLAT UOC:231,252,254,256	1
18 PAHZZ 60380 QAR29523	ROLLER,BEARING UOC:231,252,254,256	15
19 PAHZZ 11862 14022216	SHAFT,AND CLUTCH HA UOC:231,252,254,256	1
20 PAHZZ 11862 3975715	FORK,CONTROL UOC:231,252,254,256	2
21 PAHZZ 11862 14022210	CLIP,SPRING TENSION UOC:231,252,254,256	1
22 PAHZZ 11862 14022221	SHAFT,STRAIGHT UOC:231,252,254,256	1
23 PAHZZ 24617 120690	PIN,COTTER	2

(1) ITEM NO	(2) SMR CODE	(3) CAGEC	(4) PART NUMBER	(5) DESCRIPTION AND USABLE ON CODES (UOC)	(6) QTY
				UOC:231,252,254,256	
24	PAHZZ	11862	9437415	PIN,SPRING	2
				UOC:231,252,254,256	
25	PAHZZ	11862	471727	SHAFT,AXLE,AUTOMOTI	1
				UOC:231,252,254,256	
26	PAHZZ	11862	3967879	CLUTCH,SLIDING SLEE	2
				UOC:231,252,254,256	
27	PAHZZ	11862	3975705	GEAR,HELICAL	1
				UOC:231,252,254,256	
28	PAHZZ	11862	3979450	BEARING,WASHER,THRU	1
				UOC:231,252,254,256	
29	PAHZZ	11862	6273213	COVER,ACCESS	1
				UOC:231,252,254,256	
30	PAHZZ	96906	MS90728-57	SCREW,CAP,HEXAGON H	6
				UOC:231,252,254,256	
31	PAHZZ	11862	3967882	GEAR,SPUR	2
				UOC:231,252,254,256	
32	PAHZZ	11862	3967877	BEARING,WASHER,THRU	2
				UOC:231,252,254,256	
33	PAHZZ	11862	361119	RING,RETAINING	2
				UOC:231,252,254,256	
34	PAHZZ	60380	M28161X0H	BEARING,ROLLER,NEED	1
				UOC:231,252,254,256	
35	PAHZZ	11862	3995881	RETAINER AND ROLLER	1
				UOC:231,252,254,256	
36	PAFZZ	11862	6259083	GASKET	2
				UOC:231,252,254,256	
37	PAHZZ	60380	AR-33761	PIN,STRAIGHT,HEADLE	2
				UOC:231,252,254,256	
38	PAHZZ	11862	3967884	GEAR CLUSTER	1
				UOC:231,252,254,256	
39	PAHZZ	11862	3967898	SPACER SLEEVE	1
				UOC:231,252,254,256	
40	PAHZZ	60038	2793-2720	BEARING,ROLLER,TAPE	2
				UOC:231,252,254,256	
41	PAHZZ	11862	3975698	SHIM	1
				UOC:231,252,254,256	
41	PAHZZ	11862	476781	SHIM	1
				UOC:231,252,254,256	
41	PAHZZ	11862	3975700	SHIM	1
				UOC:231,252,254,256	
41	PAHZZ	76760	A85344	SHIM	1
				UOC:231,252,254,256	
42	PAHZZ	11862	3975703	SHAFT,SHOULDERED	1
				UOC:231,252,254,256	
43	PAHZZ	11862	3975711	GASKET	1
				UOC:231,252,254,256	
44	PAHZZ	11862	3975710	COVER,ACCESS	1
				UOC:231,252,254,256	
45	PAHZZ	96906	MS35338-53	WASHER,LOCK	3
				UOC:231,252,254,256	
46	PAHZZ	96906	MS90728-32	BOLT,MACHINE	3

(1) (2) (3) (4)				(5)	(6)
ITEM	SMR		PART		
NO	CODE	CAGEC	NUMBER	DESCRIPTION AND USABLE ON CODES (UOC)	QTY

				UOC:231,252,254,256	
47	PAHZZ	11862	6259153	SHAFT,STRAIGHT	1
				UOC:231,252,254,256	
48	PAHZZ	23862	9419284	ROLLER,BEARING	128
				UOC:231,252,254,256	
49	PAHZZ	11862	3967883	SPACER,RING	2
				UOC:231,252,254,256	
50	PAHZZ	11862	3967881	WASHER,FLAT	2
				UOC:231,252,254,256	
51	PAHZZ	11862	9428004	BEARING,GEAR	1
				UOC:231,252,254,256	
52	PAFZZ	11862	3979453	GASKET	1
				UOC:231,252,254,256	
53	PAFZZ	11862	3709351	RING,RETAINING (.087")	1
				UOC:231,252,254,256	
53	PAFZZ	11862	3709353	RING,RETAINING (.093")	1
				UOC:231,252,254,256	
53	PAFZZ	11862	3709354	RING,RETAINING (.096")	1
				UOC:231,252,254,256	
53	PAFZZ	11862	3709352	RING,RETAINING (.090")	1
				UOC:231,252,254,256	
54	PAFZZ	96906	MS35335-35	WASHER,LOCK	5
				UOC:231,252,254,256	
55	PAFZZ	96906	MS90728-59	SCREW,CAP,HEXAGON H	5
				UOC:231,252,254,256	
56	PAFFF	11862	14022213	EXTENSION AND BUSHI	1
				UOC:231,252,254,256	
57	PAFZZ	11862	3978765	.BEARING,SLEEVE	1
				UOC:231,252,254,256	
58	PAFZZ	76680	9449-K	.SEAL,CASE EXTENSION	1
				UOC:231,252,254,256	
59	PFFZZ	11862	474071	RETAINER AND ROLLER	1
				UOC:231,252,254,256	
60	PAFZZ	11862	6273992	GEAR,HELICAL	1
				UOC:231,252,254,256	
61	PAHZZ	23862	6273214	GASKET	1
				UOC:231,252,254,256	
62	PAHZZ	11862	3967887	WASHER,FLAT	1
				UOC:231,252,254,256	
63	PAHZZ	11862	3967888	NUTK,SELF-LOCKING,HE	1
				UOC:231,252,254,256	
64	PAHZZ	11862	6259192	RING,RETAINING (.088")	1
				UOC:231,252,254,256	
64	PAHZZ	11862	6259193	RING,RETAINING (.091")	1
				UOC:231,252,254,256	
64	PAHZZ	11862	6259194	RING,RETAINING (.094")	1
				UOC:231,252,254,256	
64	PAHZZ	11862	6259195	RING,RETAINING (.097")	1
				UOC:231,252,254,256	
65	PAHZZ	11862	2600236	RING,RETAINING	1
				UOC:231,252,254,256	
66	PAHZZ	23862	9422715	BEARING,BALL,ANNULA	1

(1) (2) (3) (4) ITEM SMR PART NO CODE CAGEC NUMBER	(5) DESCRIPTION AND USABLE ON CODES (UOC)	(6) QTY
67 PAHZZ 11862 14069597	UOC:231,252,254,256 LEVER,REMOTE CONTRG UOC:231,252,254,256	1
	END OF FIGURE	

```
        3
   ┌─────────┐
   │ 4 THRU 6 │
   └─────────┘
```

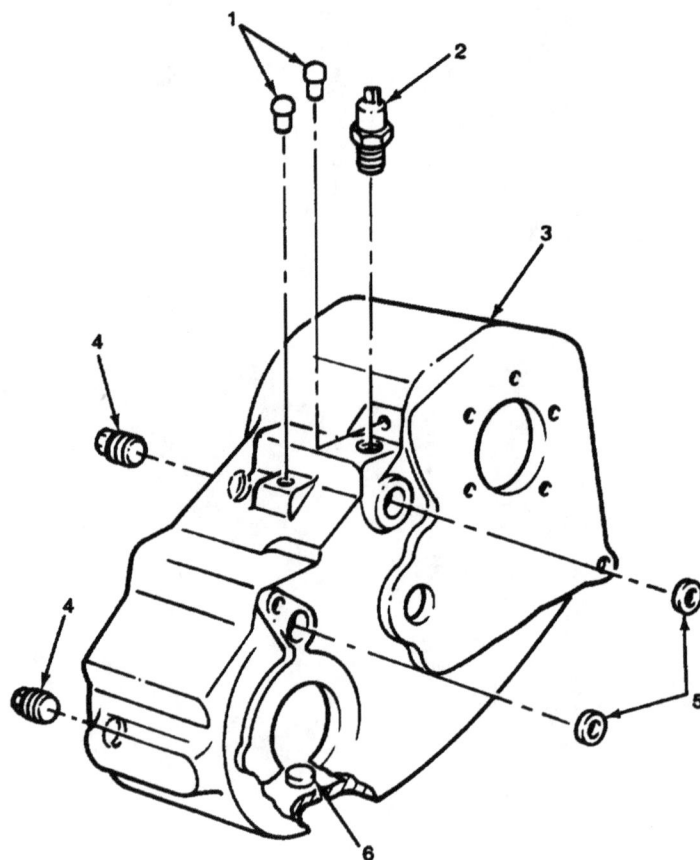

TA510892

FIGURE 85. TRANSFER CASE HOUSING (MODEL 205).

(1) ITEM NO	(2) SMR CODE	(3) CAGEC	(4) PART NUMBER	(5) DESCRIPTION AND USABLE ON CODES (UOC)	(6) QTY
				GROUP 0801 POWER TRANSFER, FINAL DRIVE, PLANETARY, AND DROP GEARBOX ASSEMBLIES	
				FIG. 85 TRANSFER CASE HOUSING (MODEL 205)	
1	PAFZZ	11862	14022217	PLUG,PROTECTIVE,DUS UOC:231,252,254,256	2
2	PAOZZ	11862	15594176	SWITCH,PUSH UOC:231,252,254,256	1
3	XAHHH	11862	14037909	HOUSING,MECHANICAL UOC:231,252,254,256	1
4	PAOZZ	89346	103868	.PLUG,PIPE UOC:231,252,254,256	2
5	PAFZZ	11862	3979626	.SEAL,PLAIN ENCASED UOC:231,252,254,256	2
6	PBHZZ	11862	3787240	.MAGNET,CHIP COLLECT UOC:231,252,254	1

END OF FIGURE

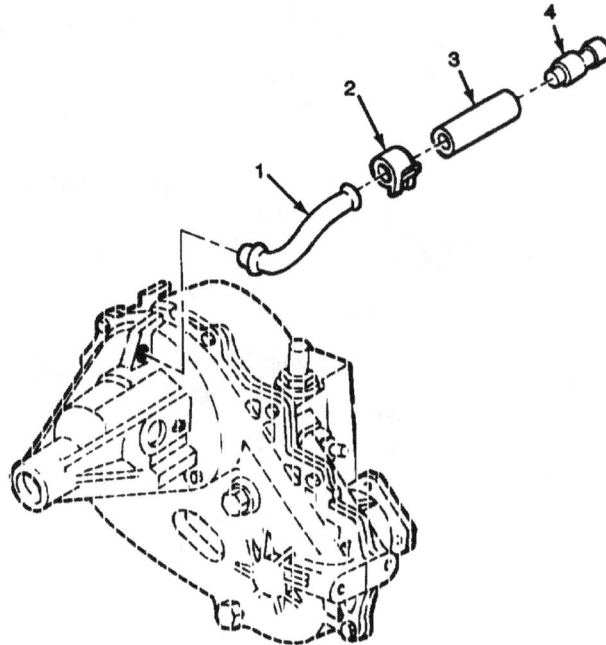

MODEL 208

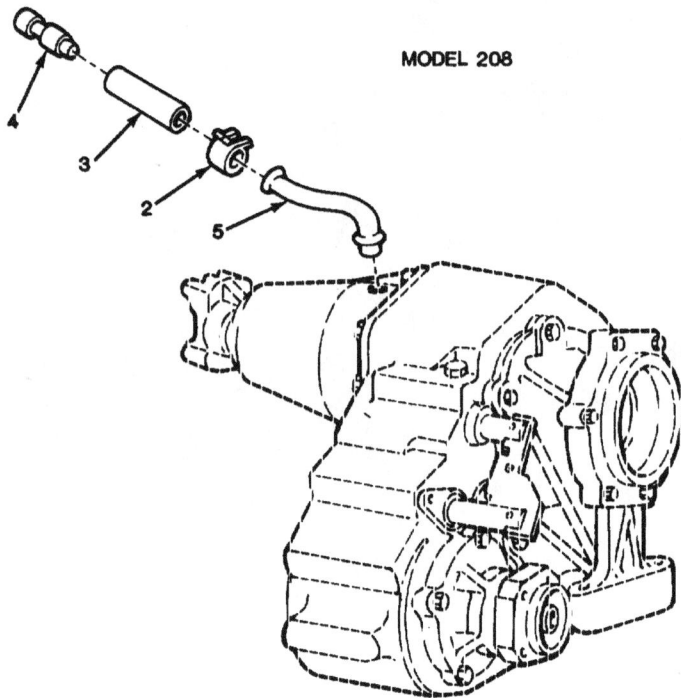

MODEL 205

TA510893

FIGURE 86. CASE VENT COMPONENT PARTS.

(1) (2) (3) (4) ITEM SMR NO CODE CAGEC	PART NUMBER	(5) DESCRIPTION AND USABLE ON CODES (UOC)	(6) QTY
		GROUP 0801 POWER TRANSFER, FINAL DRIVE, PLANETARY, OR DROP GEARBOX ASSEMBLIES	
		FIG. 86 CASE VENT COMPONENT PARTS	
1 PAOZZ 11862	14032995	TUBE ASSEMBLY,METAL UOC:194,208,209,210,230	1
2 PAOZZ 96906	MS35842-10	CLAMP,HOSE	1
3 MOOZZ 11862	9439048	HOSE,NONMETALLIC (1.77" LG) MAKE FROM HOSE, P/N 9439104	1
4 PAOZZ 11862	8640496	SHAFT,SHOULDERED	1
5 PAOZZ 11862	14032996	TUBE,BENT,METALLIC UOC:231,252,254,256	1

END OF FIGURE

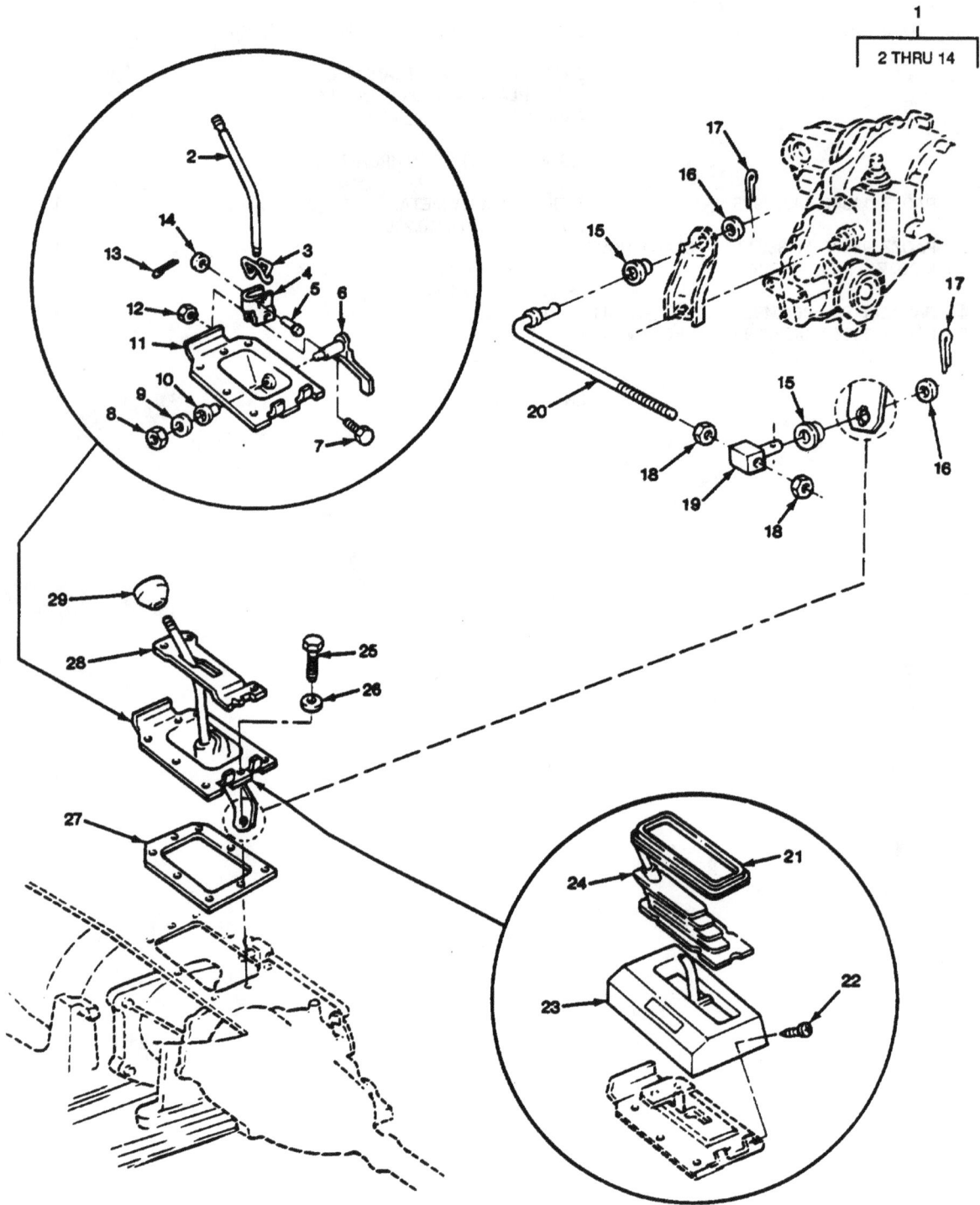

FIGURE 87. GEARSHIFT LEVER AND LINKAGE (MODEL 208).

TA510894

(1) (2) (3) (4)				(5)	(6)
ITEM SMR			PART		
NO	CODE	CAGEC	NUMBER	DESCRIPTION AND USABLE ON CODES (UOC)	QTY

GROUP 0803 GEARSHIFT,VACUUM
BOOSTER, AND CONTROLS

FIG. 87 GEARSHIFT LEVER AND LINKAGE
(MODEL 208)

1	ACOOO	11862	14063331	LVR A SM T/CASE CTL	1
				UOC:210	
1	ACOOO	11862	14071950	LVR A SM T/CASE CTL	1
				UOC:194,208,209,230	
2	PAOZZ	11862	14045658	.LEVER,MANUAL CONTRC	1
				UOC:194,208,209,230	
2	PAOZZ	11862	14063332	.ROD,SHIFT LINKAGE	1
				UOC:210	
3	PAOZZ	11862	14029108	.SPRING,HELICAL,TORS	1
				UOC:194,208,209,210,230	
4	PAOZZ	11862	14029107	.SHIFTER FORK	1
				UOC:194,208,209,210,230	
5	PAOZZ	11862	14029111	.PIN,STRAIGHT,HEADED	1
				UOC:194,208,209,210,230	
6	PAOZZ	11862	14037889	.LEVER,MANUAL CONTRC	1
				UOC:194,208,209,210,230	
7	PAOZZ	11862	14045504	.SCREW,CAP,HEXAGON H	2
				UOC:194,208,209,210,230	
8	PAOZZ	24617	9422295	.NUT,SELF-LOCKING,CC	1
				UOC:194,208,209,210,230	
9	PAOZZ	11862	466578	.WASHER,FLAT	1
				UOC:194,208,209,210,230	
10	PAOZZ	11862	3992925	.BUSHING,SLEEVE	2
				UOC:194,208,209,210,230	
11	PAOZZ	11862	14071952	.HOUSING,MECHANICAL	1
				UOC:194,208,209,210,230	
12	PAOZZ	21450	131245	.NUT,SELF-LOCKING,HE	2
				UOC:194,208,209,210,230	
13	PAOZZ	96906	MS24665-283	.PIN,COTTER	1
				UOC:194,208,209,210,230	
14	PAOZZ	96906	MS27183-12	.WASHER,FLAT	1
				UOC:194,208,209,210,230	
15	PAOZZ	11862	1234418	.GROMMET,NONMETALLIC	2
				UOC:194,208,209,210,230	
16	PAOZZ	11862	382105	.WASHER,FLAT	2
				UOC:194,208,209,210,230	
17	PAOZZ	89749	IF316	.PIN,COTTER	2
				UOC:194,208,209,210,230	
18	PAOZZ	96906	MS35691-29	.NUT,PLAIN,HEXAGON	2
				UOC:194,208,209,210,230	
19	PAOZZ	11862	718368	.SWIVEL,TRANSFER CAS	1
				UOC:194,208,209,210,230	
20	PAOZZ	11862	14029122	.PIN,STRAIGHT,HEADED	1
				UOC:194,208,209,210,230	
21	PAOZZ	11862	14029117	.PLATE,TRANSFER CASE	1
				UOC:194,208,209,210,230	

(1) (2) (3) (4)				(5)	(6)
ITEM	SMR		PART		
NO	CODE	CAGEC	NUMBER	DESCRIPTION AND USABLE ON CODES (UOC)	QTY
22	PAOZZ	11862	20365263	SCREW,TAPPING,THREA	4
				UOC:194,208,209,210,230	
23	PAOZZ	11862	14037893	LEVER,REMOTE CONTRO	1
				UOC:194,208,209,210,230	
24	PAOZZ	11862	14029116	BOOT,DUST AND MOIST	1
				UOC:194,208,209,210,230	
25	PAOZZ	96906	MS90728-6	SCREW,CAP,HEXAGON H	8
				UOC:194,208,209,210,230	
26	PAOZZ	96906	MS9549-10	WASHER,FLAT	8
				UOC:194,208,209,210,230	
27	PAOZZ	11862	14071954	GASKET	1
				UOC:194,208,209,210,230	
28	PAOZZ	11862	14045697	GUIDE,TRANSFER CASE	1
				UOC:194,208,209,210,230	
29	PAOZZ	11862	15588503	KNOB	1
				UOC:194,208,209,210,230	

END OF FIGURE

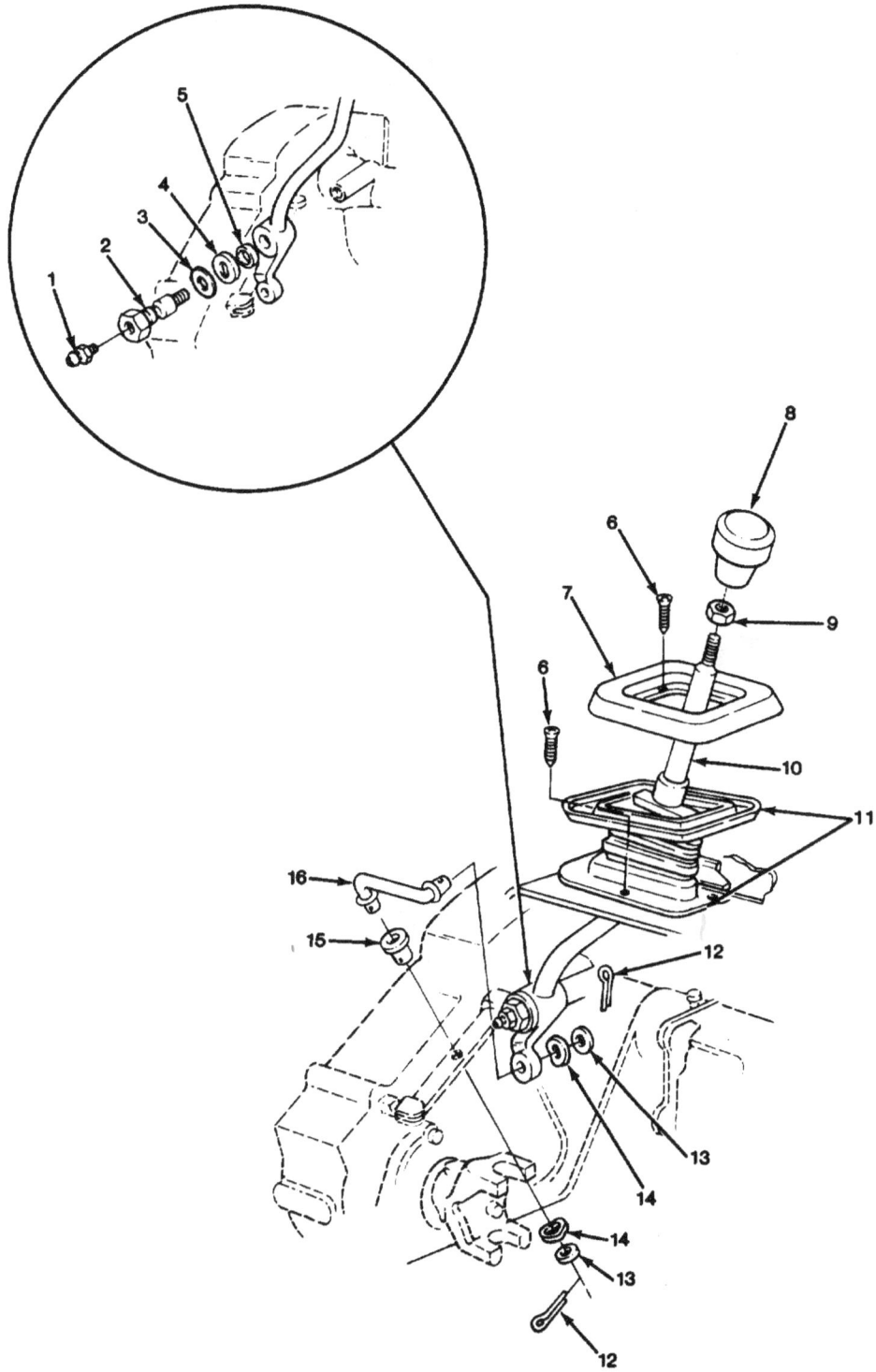

FIGURE 88. GEARSHIFT LEVER AND LINKAGE (MODEL 205).

TA510895

(1) (2) (3) (4)	(5)	(6)
ITEM SMR PART NO CODE CAGEC NUMBER	DESCRIPTION AND USABLE ON CODES (UOC)	QTY

GROUP 0803 GEARSHIFT,VACUUM
BOOSTER, AND CONTROLS

FIG. 88 GEARSHIFT LEVER AND LINKAGE
(MODEL 205)

ITEM NO	SMR CODE	CAGEC	PART NUMBER	DESCRIPTION AND USABLE ON CODES (UOC)	QTY
1	PAOZZ	11862	9417901	FITTING,LUBRICATION UOC:231,252,254,256	1
2	PAOZZ	11862	14009313	BOLT,FLUID PASSAGE UOC:231,252,254,256	1
3	PAOZZ	11862	4497001	WASHER,SPRING TENSI UOC:231,252,254,256	
4	PAOZZ	11862	3838153	WASHER,FLAT UOC:231,252,254,256	1
5	PAOZZ	11862	2423517	WASHER,FLAT UOC:231,252,254,256	1
6	PAOZZ	24617	9431995	SCREW,TAPPING,THREA UOC:231,252	8
7	PAOZZ	11862	14032789	PLATE,RETAINING,SHA UOC:231,252,254,256	1
8	PAOZZ	11862	15588504	KNOB UOC:231,252,254,256	1
9	PAOZZ	96906	MS35691-29	NUT,PLAIN,HEXAGON UOC:231,252,254,256	1
10	PAOZZ	11862	14055531	LEVER,MANUAL CONTRO UOC:231,252,254,256	1
11	PAOZZ	11862	14071955	BOOT,DUST AND MOIST UOC:231,252,254,256	1
12	PAOZZ	89749	IF316	PIN,COTTER UOC:231,252,254,256	2
13	PAOZZ	96906	MS27183-14	WASHER,FLAT UOC:231,252,254,256	2
14	PAOZZ	11862	3953987	WASHER,SPRING TENSI UOC:231,252,254,256	2
15	PAOZZ	11862	14049556	GROMMET,NONMETALLIC UOC:231,252,254,256	1
16	PAOZZ	11862	14054220	ROD,TRANSFER CASE UOC:231,252,254,256	1

END OF FIGURE

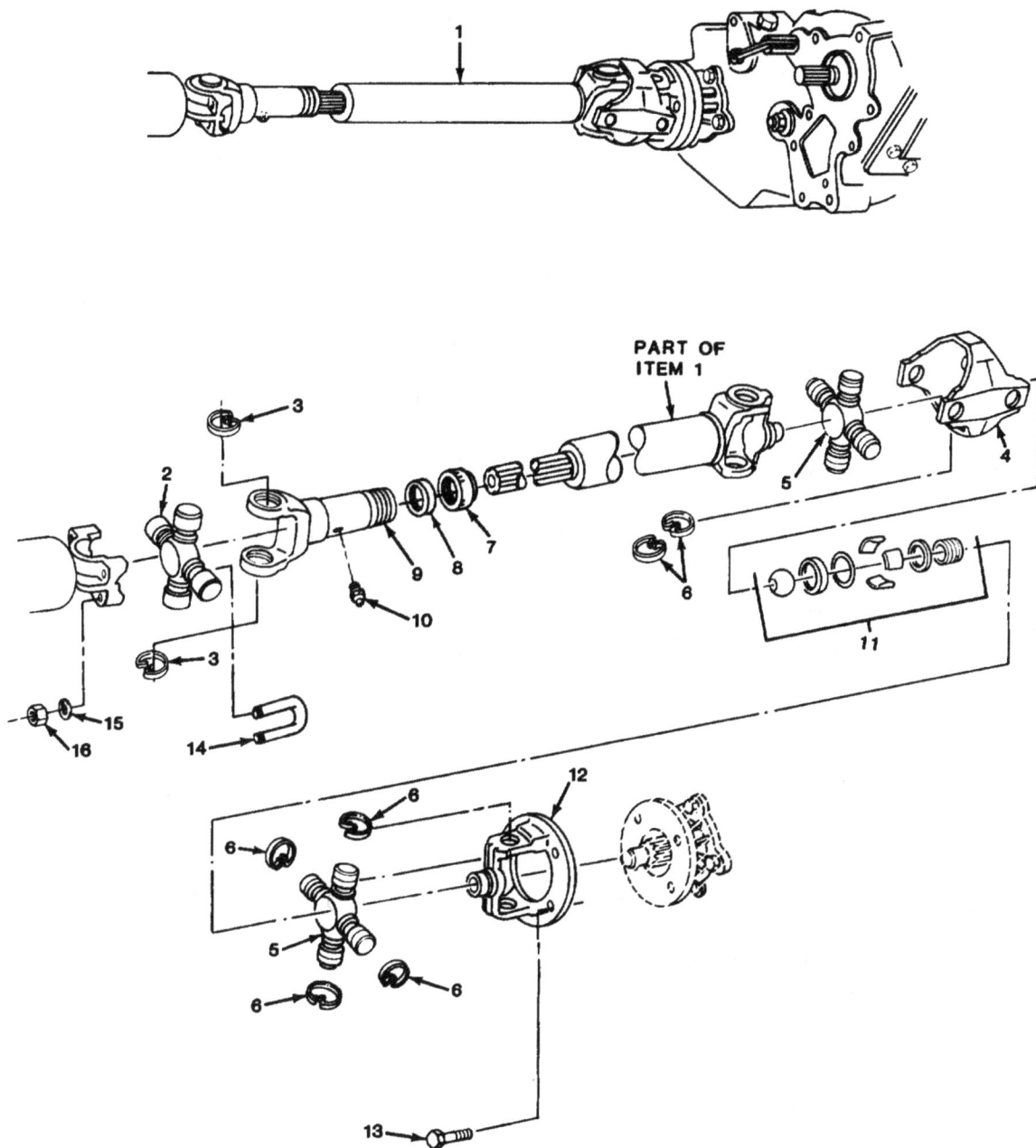

FIGURE 89. FRONT PROPELLER SHAFT ASSEMBLY (ALL EXCEPT M1009, FIRST DESIGN).

TA510896

(1) ITEM NO	(2) SMR CODE	(3) CAGEC	(4) PART NUMBER	(5) DESCRIPTION AND USABLE ON CODES (UOC)	(6) QTY
				GROUP 09 PROPELLER, PROPELLER SHAFTS, UNIVERSAL JOINTS, COUPLER, AND CLAMP ASSEMBLY	
				GROUP 0900 PROPELLER SHAFTS	
				FIG. 89 FRONT PROPELLER SHAFT ASSEMLBY (ALL EXCEPT M1009, FIRST DESIGN)	
1	PAOFF	11862	7845102	PROPELLER SHAFT (FIRST DESIGN APPLIES TO VEHICLES BEFORE VIN ENDING WITH 354457 UOC:194,208,210,230,231,252,254,256	1
2	PAOZZ	11862	386451	.PARTS KIT,UNIVERSAL INCLUDES ITEM #3 UOC:194,208,210,230,231,252,254,256	1
3	PAOZZ	11862	3721887	.RING,RETAINING UOC:194,208,210,230,231,252,254,256	2
4	PAFZZ	11862	7809057	.DOUBLE YOKE,UNIVERS UOC:194,208,210,230,231,252,254,256	1
5	PAOZZ	11862	7806140	.PARTS KIT,UNIVERSAL UOC:194,208,210,230,231,252,254,256	2
6	PAOZZ	11862	1456507	.RING,RETAINING UOC:194,208,210,230,231,252,254,256	8
7	PAOZZ	11862	458418	.CAP,DUST,PRCPELLER UOC:194,208,210,230,231,252,254,256	1
8	PAOZZ	11862	7827942	.PACKING,PERFORMED UOC:194,208,210,230,231,252,254,256	1
9	PAOZZ	11862	15596686	.YOKE,UNIVERSAL JOIN INCLUDES ITEM #10 UOC:194,208,210,230,231,252,254,256	1
10	PAOZZ	11862	9417901	.FITTING,LUBRICATION UOC:194,208,210,230,231,252,254,256	1
11	XDFZZ	11862	7815848	.KIT,JOINT BALL UOC:194,208,210,230,231,252,254,256	1
12	PAFZZ	11862	7815849	.YOKE,UNIVERSAL JOIN UOC:194,208,210,230,231,252,254,256	1
13	PACZZ	96906	MS90728-86	SCREW,CAP,HEXAGON H UOC:194,208,210,230,231,252,254,256	4
14	PAOZZ	72712	1358938	BOLT, U UOC:194,208,210,230,231,252,254,256	2
15	PAOZZ	96906	MS35338-45	WASHER,LOCK UOC:194,208,210,230,231,252,254,256	4
16	PAOZZ	96906	MS51967-6	NUT,PLAIN,HEXAGON UOC:194,208,210,230,231,252,254,256	4

END OF FIGURE

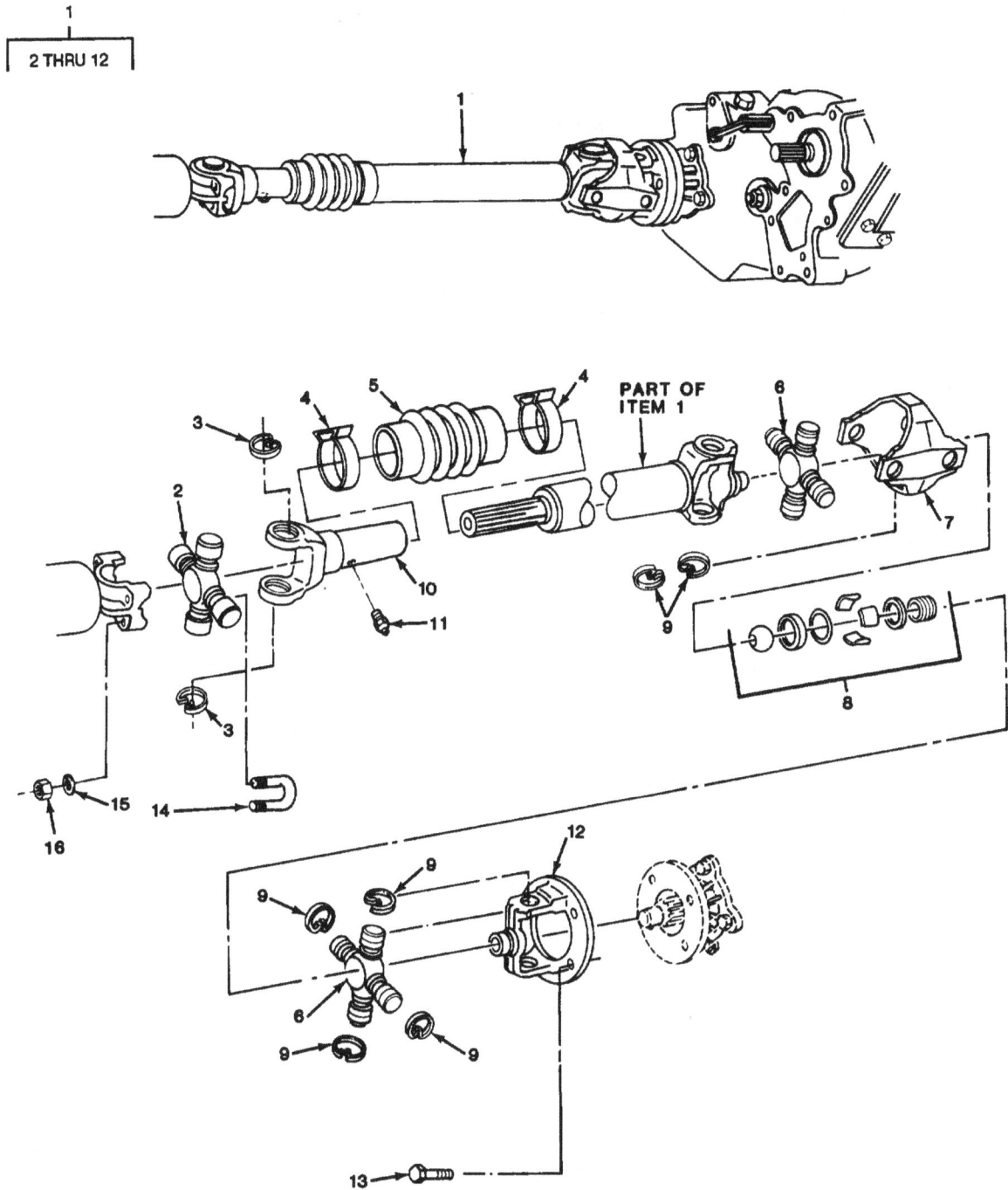

FIGURE 90. FRONT PROPELLER SHAFT ASSEMBLY (ALL EXCEPT M1009, SECOND DESIGN).

TA510897

(1)	(2)	(3)	(4)	(5)	(6)
ITEM NO	SMR CODE	CAGEC	PART NUMBER	DESCRIPTION AND USABLE ON CODES (UOC)	QTY

GROUP 0900 PROPELLER SHAFTS

FIG. 90 FRONT PROPELLER SHAFT ASSEMBLY (ALL EXCEPT M1009, SECOND DESIGN)

ITEM NO	SMR CODE	CAGEC	PART NUMBER	DESCRIPTION AND USABLE ON CODES (UOC)	QTY
1	PAOFF	11862	26013913	PROPELLER SHAFT WIT (SECOND DESIGN WITH 205 TRANSFER CASE APPLIES TO VEHICLES AFTER VIN ENDING WITH 354457) UOC:231,252,254,256	1
1	PAOFF	11862	7845102	PROPELLER SHAFT (SECOND DESIGN WITH 208 TRANSFER CASE APPLIES TO VEHICLES AFTER VIN ENDING WITH 354457) UOC:194,208,210,230	1
2	PAOZZ	11862	386451	.PARTS KIT,UNIVERSAL INCLUDES ITEM #3 UOC:194,208,210,230,231,252,254,256	1
3	PAOZZ	11862	3721887	.RING,RETAINING UOC:194,208,210,230,231,252,254,256	2
4	PAOZZ	11862	7840235	.CLIP,RETAINING UOC:194,208,210,230,231,252,254,256	2
5	PAOZZ	11862	7845127	.BOOT,DUST AND MOIST UOC:194,208,210,230,231,252,254,256	1
6	PAOZZ	11862	7806140	.PARTS KIT,UNIVERSAL INCLUDES ITEM #9 UOC:194,208,210,230,231,252,254,256	2
7	PAFZZ	11862	7809057	.DOUBLE YOKE,UNIVERS UOC:194,208,210,230,231,252,254,256	1
8	XDFZZ	11862	7815848	.KIT,JOINT BALL UOC:194,208,210,230,231,252,254,256	1
9	PAOZZ	11862	1456507	.RING,RETAINING UOC:194,208,210,230,231,252,254,256	8
10	PAOZZ	11862	7845119	.PROPELLER SHAFT INCLUDES ITEM #11 UOC:194,208,210,230,231,252,254,256	1
11	PAOZZ	11862	9417901	.FITTING,LUBRICATION UOC:194,208,210,230,231,252,254,256	1
12	PAFZZ	11862	7815849	.YOKE,UNIVERSAL JOIN UOC:194,208,210,230,231,252,254	1
13	PAOZZ	96906	MS90728-86	SCREW,CAP,HEXAGON H UOC:194,208,210,230,231,252,254,256	4
14	PAOZZ	11862	1358938	BOLT,U UOC:194,208,210,230,231,252,254,256	2
15	PAOZZ	96906	MS35338-45	WASHER,LOCK UOC:194,208,210,230,231,252,254,256	4
16	PAOZZ	96906	MS51967-6	NUT,PLAIN,HEXAGON UOC:194,208,210,230,231,252,254,256	4

END OF FIGURE

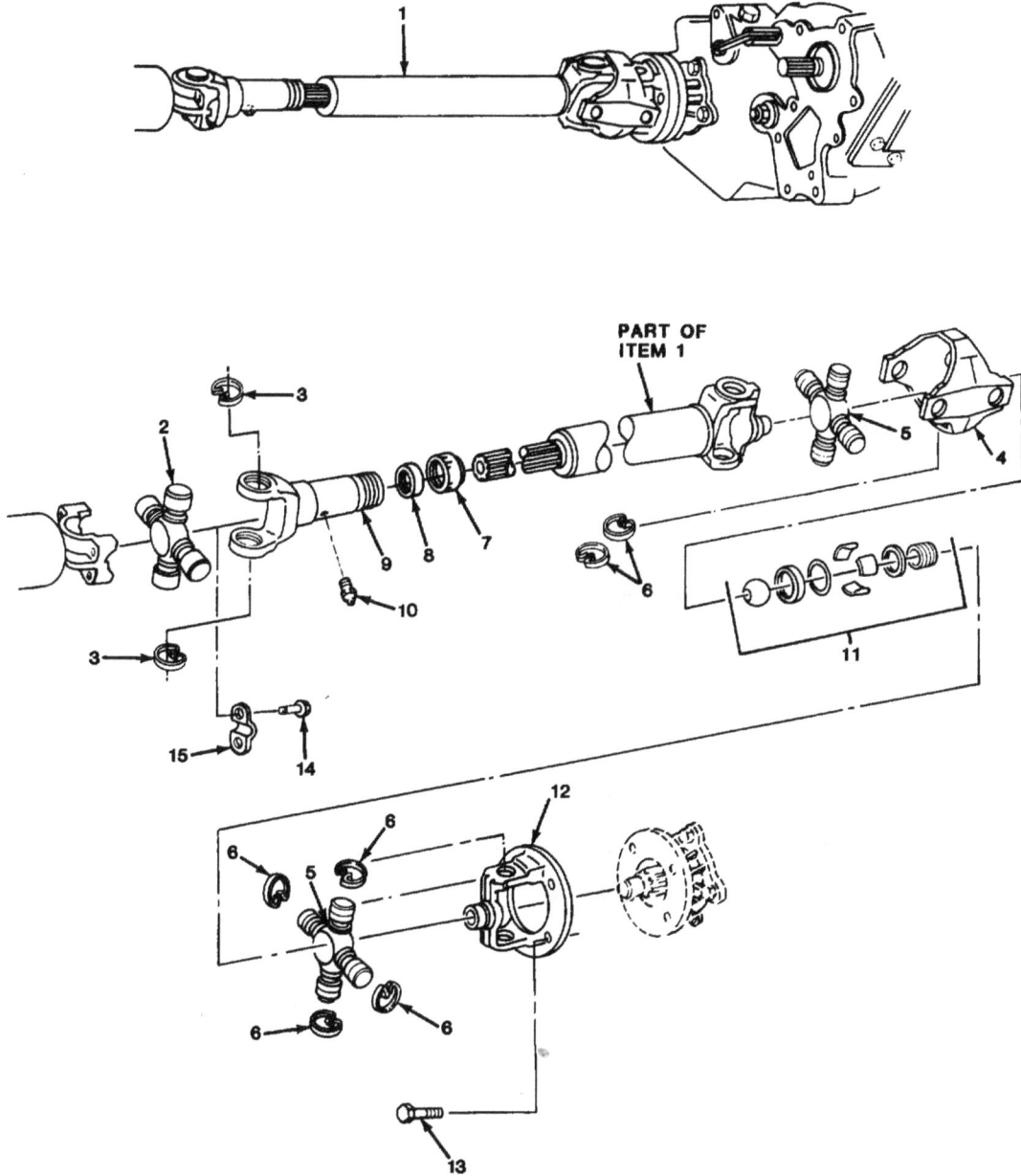

FIGURE 91. FRONT PROPELLER SHAFT ASSEMBLY (M1009, FIRST DESIGN).

TA510898

(1) ITEM NO	(2) SMR CODE	(3) CAGEC	(4) PART NUMBER	(5) DESCRIPTION AND USABLE ON CODES (UOC)	(6) QTY
				GROUP 0900 PROPELLER SHAFTS	
				FIG. 91 FRONT PROPELLER SHAFT ASSEMBLY (M1009, FIRST DESIGN)	
1	PAOFF	11862	7830927	PROPELLER SHAFT (FIRST DESIGN APPLIES TO VEHICLES BEFORE VIN ENDING WITH 168593 AND IS INTERCHANGEABLE WITH P/N 7845043 UOC:209	1
2	PAOZZ	11862	386451	.PARTS KIT,UNIVERSAL INCLUDES ITEM #3 UOC:209	1
3	PAOZZ	11862	3721887	.RING,RETAINING UOC:209	2
4	PAFZZ	118621	7809057	.DOUBLE YOKE,UNIVERS UOC:209	1
5	PAOZZ	11862	7806140	.PARTS KIT,UNIVERSAL INCLUDES ITEM #6 UOC:209	2
6	PAOZZ	11862	1456507	.RING,RETAINING UOC:209	8
7	PAOZZ	11862	458418	.CAP,DUST,PROPELLER UOC:209	1
8	PAOZZ	11862	7827942	.PACKING,PREFORMED UOC:209	1
9	PAOZZ	11862	15596686	.YOKE,UNIVERSAL JOIN UOC:209	1
10	PAOZZ	11862	9417901	.FITTING,LUBRICATION UOC:209	1
11	PAFZZ	86403	4049697	.PARTS KIT,UNIVERSAL UOC:209	1
12	PAOZZ	11862	7815849	.YOKE,UNIVERSAL JOIN UOC:209	1
13	PAOZZ	96906	MS90728-86	SCREW,CAP,HEXAGON H UOC:209	4
14	PAOZZ	11862	14018700	SCREW,CAP,HEXAGON H UOC:209	4
15	PAOZZ	11862	3882979	CLAMP,HUB UOC:209	2

END OF FIGURE

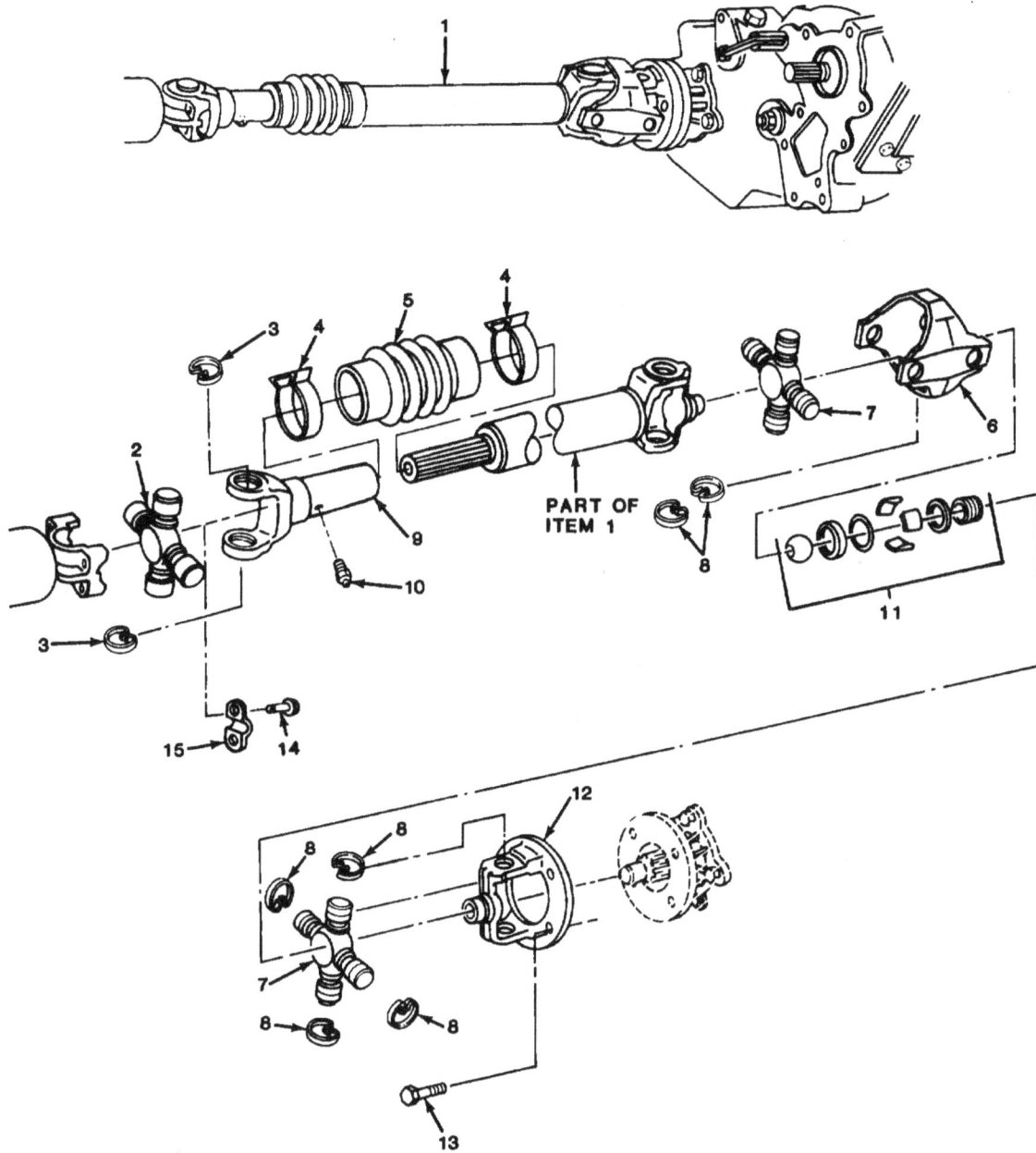

TA510899

FIGURE 92. FRONT PROPELLER SHAFT ASSEMBLY (M1009, SECOND DESIGN).

(1) (2) (3) (4) ITEM SMR NO CODE CAGEC	PART NUMBER	(5) DESCRIPTION AND USABLE ON CODES (UOC)	(6) QTY
		GROUP 0900 PROPELLER SHAFTS	
		FIG. 92 FRONT PROPELLER SHAFT ASSEMBLY (M1009, SECOND DESIGN)	
1 PAOFF 11862	26013911	PROPELLER SHAFT WIT (SECOND DESIGN APPLIES TO VEHICLES AFTER VIN ENDING WITH 168593 AND REPLACES P/N 7830927) UOC:209	1
2 PAOZZ 11862	386451	.PARTS KIT,UNIVERSAL INCLUDES ITEM #3 UOC:209	1
3 PAOZZ 11862	3721887	.RING,RETAINING UOC:209	2
4 PAOZZ 11862	7840235	.CLIP,RETAINING UOC:209	2
5 PAOZZ 11862	7845127	.BOOT,DUST AND MOIST UOC:209	1
6 PAFZZ 11862	7809057	.DOUBLE YOKE,UNIVERS UOC:209	1
7 PAOZZ 11862	7806140	.PARTS KIT,UNIVERSAL UOC:209	2
8 PAOZZ 11862	1456507	.RING,RETAINING UOC:209	8
9 PAOZZ 11862	7845119	.PROPELLER SHAFT INCLUDES ITEM #10 UOC:209	1
10 PAOZZ 11862	9417901	.FITTING,LUBRICATION UOC:209	1
11 XDFZZ 11862	7815848	.KIT,JOINT BALL UOC:209	1
12 PAOZZ 11862	7815849	.YOKE,UNIVERSAL JOIN UOC:209	1
13 PAOZZ 96906	MS90728-86	SCREW,CAP,HEXAGON H UOC:209	4
14 PAOZZ 11862	14018700	SCREW,CAP,HEXAGON H UOC:209	4
15 PAOZZ 11862	3882979	CLAMP,HUB UOC:209	2

END OF FIGURE

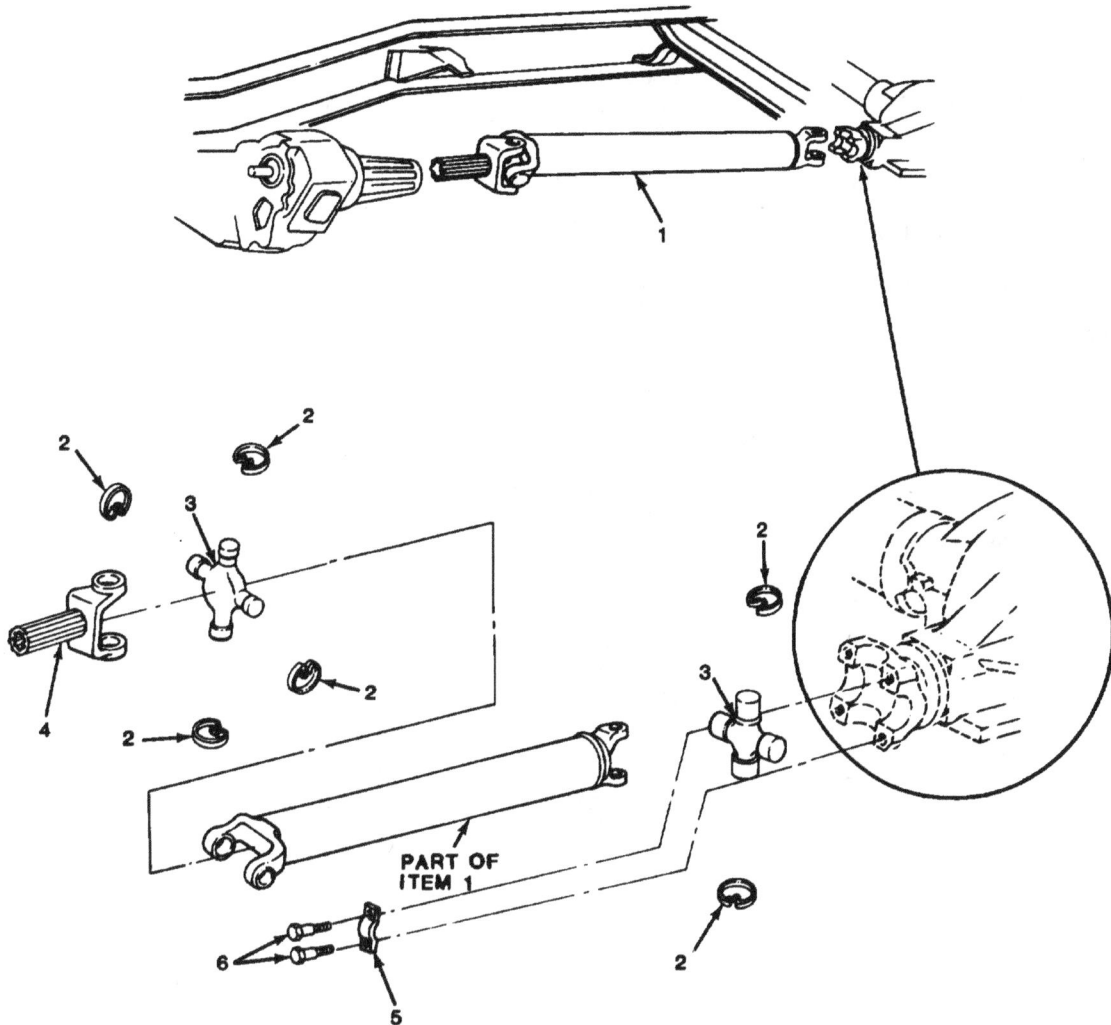

FIGURE 93. REAR PROPELLER SHAFT ASSEMBLY.

TA510900

(1) ITEM NO	(2) SMR CODE	(3) CAGEC	(4) PART NUMBER	(5) DESCRIPTION AND USABLE ON CODES (UOC)	(6) QTY
				GROUP 0900 PROPELLER SHAFTS	
				FIG. 93 REAR PROPELLER SHAFT ASSEMBLY	
1	PAOOO	11862	14067762	PROPELLER SHAFT WITH 208 TRANSFER CASE UOC:194,208,210,230	1
1	PAOOO	11862	7844074	PROPELLER SHAFT WIT UOC:209	1
1	PAOOO	11862	14020403	PROPELLER SHAFT WITH 205 TRANSFER CASE UOC:231,252	1
1	PAOZZ	97271	911105-5130	PROPELLER SHAFT WIT UOC:254,256	1
1	PFOOO	11862	14071980	PROPELLER SHAFT WIT UOC:254,256	1
2	PAOZZ	11862	1456507	.RING,RETAINING UOC:194,208,209,210,230,231,252	6
3	PAOZZ	11862	7806140	.PARTS KIT,UNIVERSAL INCLUDES ITEM #2 UOC:194,208,210,230,231,252	2
3	PAOZZ	11862	7806140	.PARTS KIT,UNIVERSAL INCLUDES ITEM #2 UOC:209	2
4	PAOZZ	11862	14029852	.SLIP YOKE ASSEMBLY UOC:194,208,210,230,231,252	1
4	PAOZZ	11862	7838665	.YOKE,UNIVERSAL JOIN UOC:209	1
5	PAOZZ	11862	3920486	STRAP,RETAINING UOC:194,208,210,230,231,252	2
5	PAOZZ	11862	7846740	STRAP,RETAINING UOC:209	2
5	PAOZZ	11862	14046907	BRACKET,VEHICULAR C UOC:254,256	2
6	PAOZZ	11862	14018700	SCREW,CAP,HEXAGON H UOC:194,208,209,210,230,231,252	4
6	PAOZZ	11862	458300	BOLT,SELF-LOCKING UOC:254,256	4

END OF FIGURE

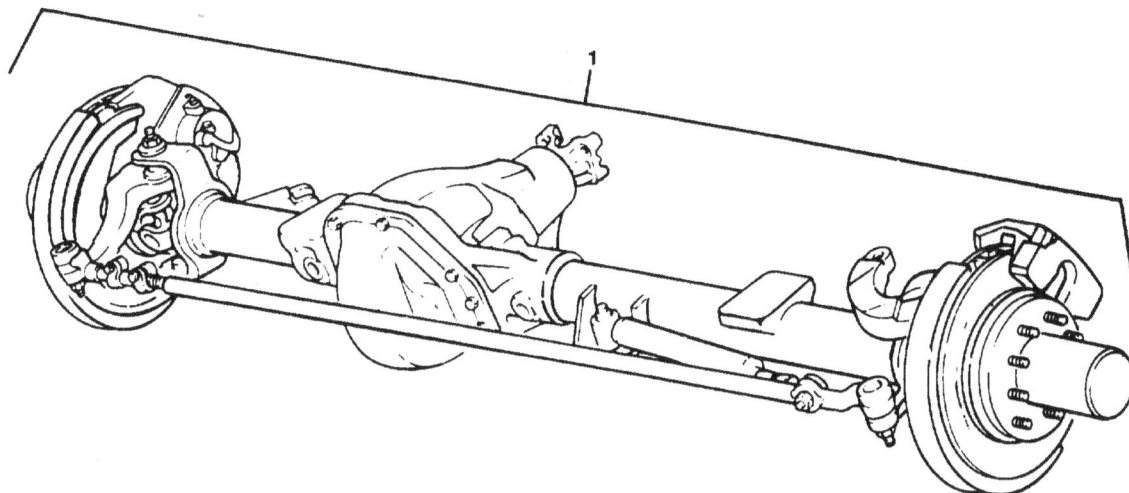

ALL EXCEPT M1009

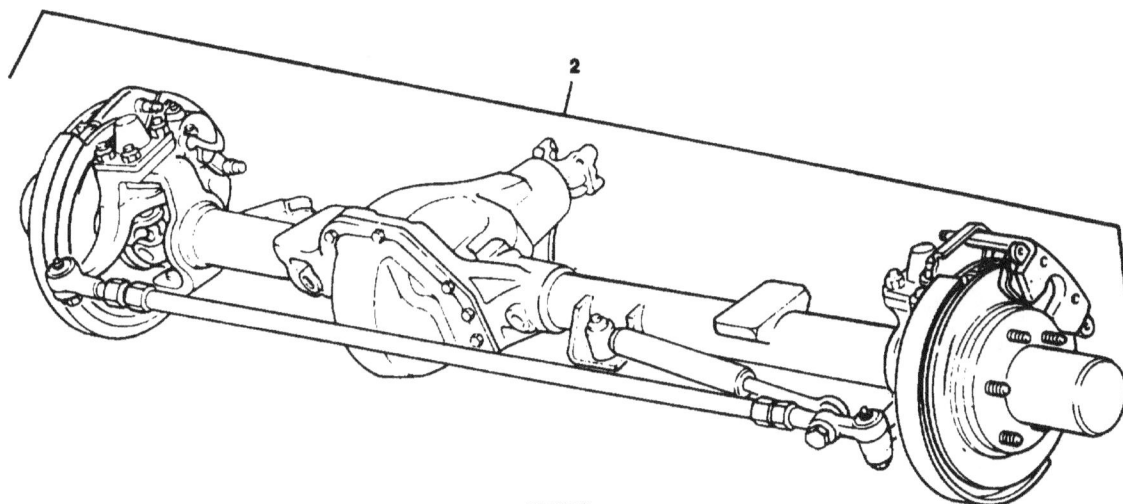

M1009

TA510901

FIGURE 94. FRONT AXLE.

(1)	(2)	(3)	(4)	(5)	(6)
ITEM NO	SMR CODE	CAGEC	PART NUMBER	DESCRIPTION AND USABLE ON CODES (UOC)	QTY

GROUP 10 FRONT AXLE

GROUP 1000 FRONT AXLE ASSEMBLY

FIG. 94 FRONT AXLE

1	PAFHD	11862	14072449	AXLE ASSEMBLY,AUTOM LESS LOOKING HUBS, FOR COMPONENT PARTS SEE FIG'S 96, 98, 101 UOC:194,208,210	1
1	PAFHD	11862	14072451	AXLE ASSEMBLY, AUTOM LESS LOCKING HUBS, FOR COMPONENT PARTS SEE FIG'S 96, 98, 101 UOC:230,231,252	1
2	PAFHD	11862	14063306	AXLE ASSEMBLY,AUTOM LESS LOOKING HUBS, FOR COMPONENT PARTS SEE FIG'S 97, 99, 100, 102 UOC:209	1
2	PAFHD	11862	15591700	AXLE ASSEMBLY,AUTOM LESS LOOKING HUBS, FOR COMPONENT PARTS SEE FIG'S 97, 99, 100, 102 JP R APPLIES TO VEHICLES WITH VIN ENDING F136169 JP R UOC:209	1

END OF FIGURE

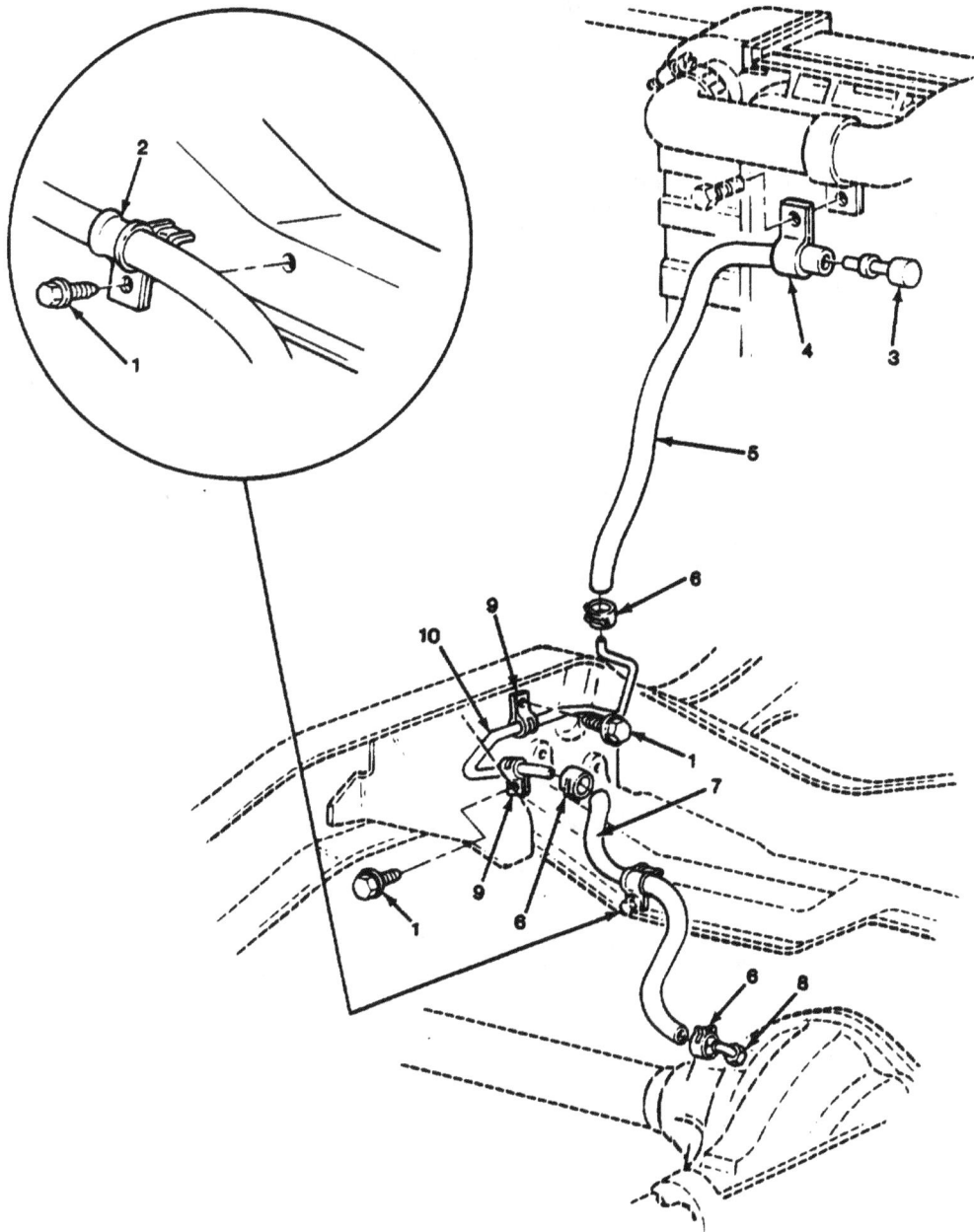

TA510902

FIGURE 95. FRONT AXLE BREATHER.

(1) ITEM NO	(2) SMR CODE	(3) CAGEC	(4) PART NUMBER	(5) DESCRIPTION AND USABLE ON CODES (UOC)	(6) QTY
				GROUP 1000 FRONT AXLE ASSEMBLY	
				FIG. 95 FRONT AXLE BREATHER	
1	PAOZZ	11862	11504447	SCREW,TAPPING,THREA	3
2	PAOZZ	96906	MS21333-112	CLAMP,LOOP	1
3	PAOZZ	11862	8640496	SHAFT,SHOULDERED	1
4	PAOZZ	96906	MS21333-48	CLAMP,LOOP	1
5	MOOZZ	11862	9439088	HOSE-FRT AX VENT (20.67" LG) MAKE FROM HOSE, P/N 9439104	1
6	PAOZZ	96906	MS35842-10	CLAMP,HOSE	3
7	MOOZZ	11862	9439091	HOSE-FRT AX VENT (23.62" LG) MAKE FROM HOSE, P/N 9439104	1
8	PAOZZ	11862	14056299	ELBOW,HOSE	1
				UOC:194,208,210,230,231,252,254,256	
9	PAOZZ	96906	MS21333-110	CLAMP,LOOP	2
10	MFOZZ	11862	14029235	PIPE-FRT AX VNT (23.82"LG) MAKE FROM TUBE, P/N 465246	1

END OF FIGURE

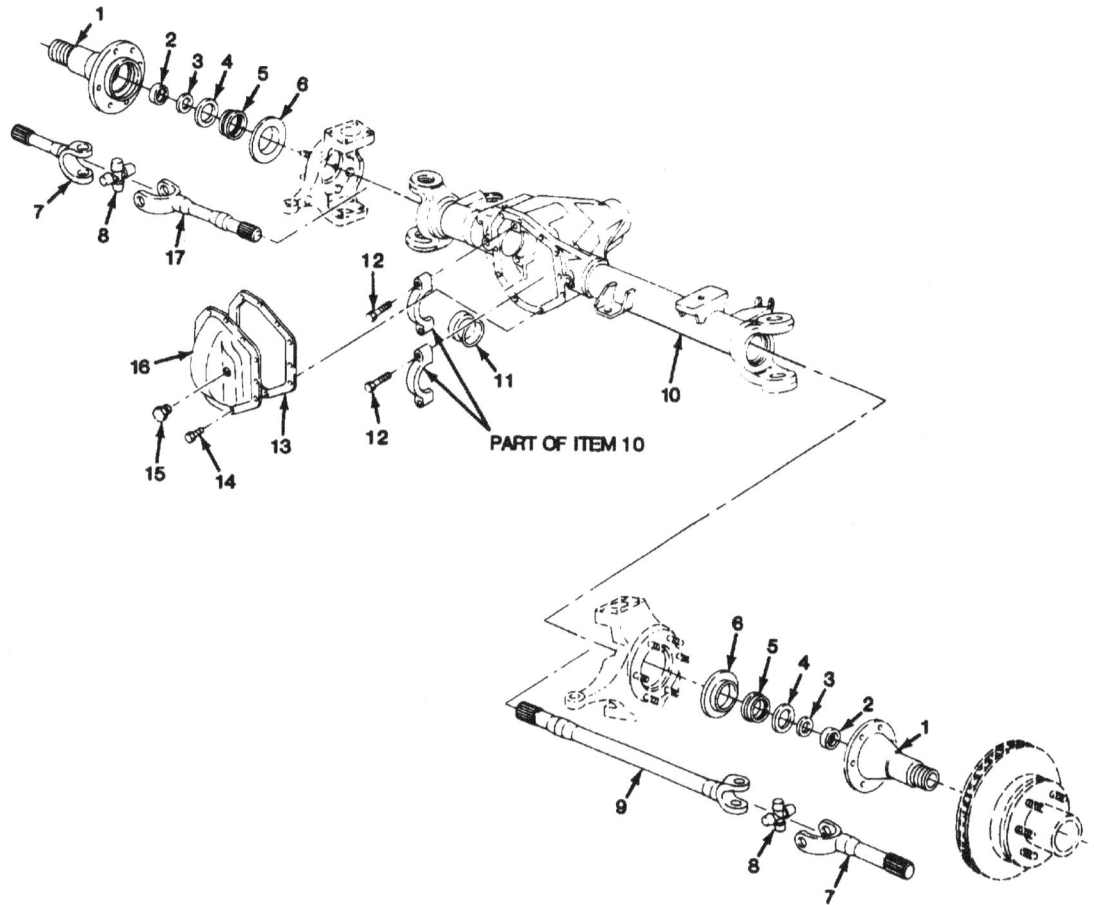

TA510903

FIGURE 96. FRONT AXLE COMPONENT PARTS AND SPINDLE ASSEMBLIES
(ALL EXCEPT M1009).

(1) ITEM NO	(2) SMR CODE	(3) CAGEC	(4) PART NUMBER	(5) DESCRIPTION AND USABLE ON CODES (UOC)	(6) QTY
				GROUP 1000 FRONT AXLE ASSEMBLY	
				FIG. 96 FRONT AXLE COMPONENT PARTS AND SPINDLE ASSEMBLIES (ALL EXCEPT M1009)	
1	PAOZZ	97271	700013L	PARTS KIT,VEHICULAR INCLUDES ITEMS 4,5 UOC:194,208,210,230,231,252,254,256	2
2	PAOZZ	11862	14009626	PARTS KIT,STEERING INCLUDES ITEMS 3,4,5 UOC:194,208,210,230,231,252,254,256	2
3	PAOZZ	97271	620062-B	SEAL,PLAIN ENCASED UOC:194,208,210,230,231,252,254,256	2
4	PAOZZ	97271	37312	SPACER,RING UOC:194,208,210,230,231,252,254,256	2
5	PAOZZ	11862	462811	SEAL,PLAIN ENCASED UOC:194,208,210,230,231,252,254,256	2
6	PAFZZ	97271	37308	DEFLECTOR,DIRT AND UOC:194,208,210,230,231,252,254,256	2
7	PAFZZ	97271	40955	YOKE,UNIVERSAL JOIN UOC:194,208,210,230,231,252,254,256	2
8	PAFZZ	23862	462809	SPIDER,UNIVERSAL JO UOC:194,208,210,230,231,252,254,256	2
9	PAFZZ	97271	660182-6	SHAFT,AXLE,AUTOMOTI UOC:194,208,210,230,231,252,254,256	1
10	XAHZZ	97271	71701-X	PARTS KIT,DRIVING A UOC:194,208,210,230,231,252,254,256	1
11	PAHZZ	23862	462857	SEAL,PLAIN ENCASED UOC:194,208,210,230,231,252,254,256	2
12	PAHZZ	96906	MS90728-117	SCREW,CAP,HEXAGON H UOC:194,208,210,230,231,252,254,256	4
13	PAFZZ	11862	6273951	GASKET UOC:194,208,210,230,231,252,254,256	1
14	PAFZZ	97271	34822	BOLT,SELF-LOCKING UOC:194,208,210,230,231,252,254,256	10
15	PAOZZ	72447	36472	PLUG,PIPE UOC:194,208,210,230,231,252,254,256	1
16	PFFZZ	97271	706968X	COVER,ACCESS INCLUDES ITEMS 13,14,15 UOC:194,208,210,230,231,252,254,256	1
17	PAFZZ	97271	660182-5	PROPELLER SHAFT UOC:194,208,210,230,231,252,254,256	1

END OF FIGURE

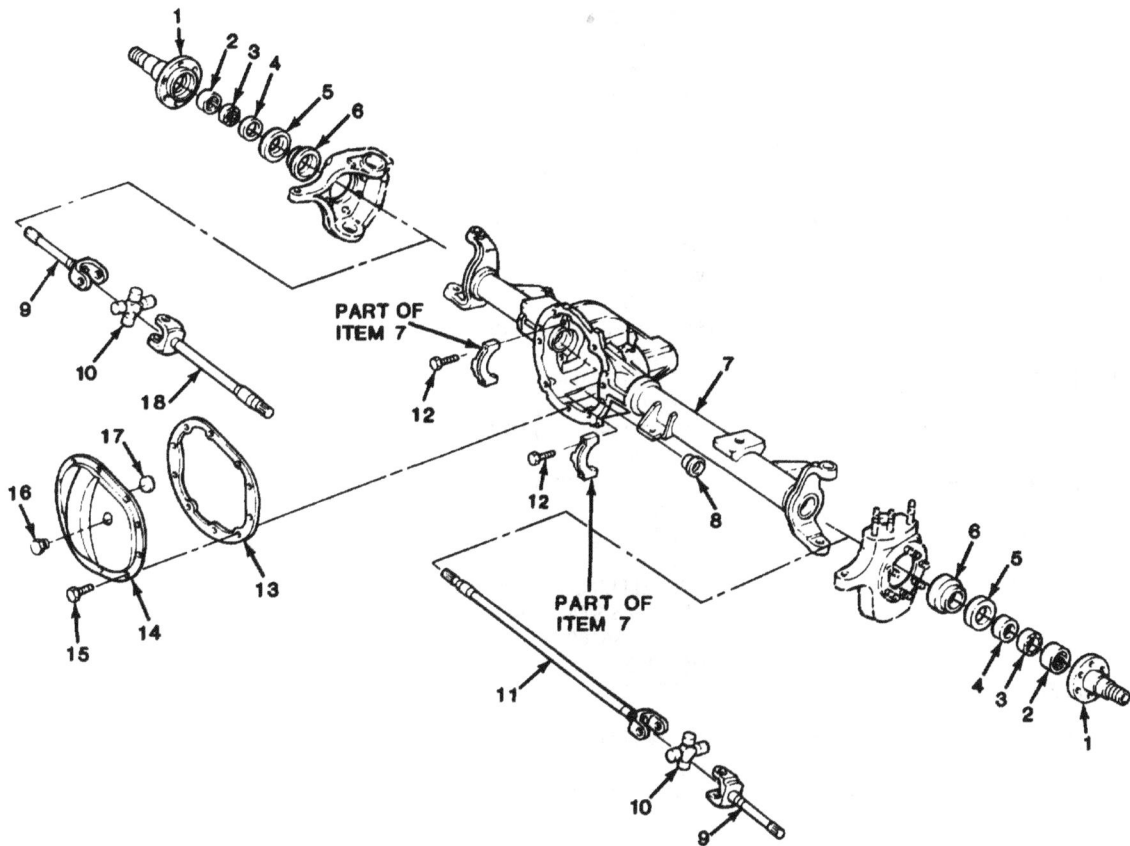

TA510904

FIGURE 97. FRONT AXLE COMPONENT PARTS AND SPINDLE ASSEMBLIES (M1009).

(1) (2) (3) (4)	(5)	(6)
ITEM SMR PART		
NO CODE CAGEC NUMBER	DESCRIPTION AND USABLE ON CODES (UOC)	QTY

GROUP 1000 FRONT AXLE ASSEMBLY

FIG. 97 FRONT AXLE COMPONENT PARTS AND SPINDLE ASSEMBLIES (M1009)

ITEM NO	SMR CODE	CAGEC	PART NUMBER	DESCRIPTION AND USABLE ON CODES (UOC)	QTY
1	PAOZZ	11862	14072919	SPINDLE,WHEEL,NONDR INCLUDES ITEMS 2,3 APPLIES TO 1984 MODELS ONLY UOC:209	2
1	PAOZZ	97271	29907X	SPINDLE,WHEEL,NONDR INCLUDES ITEMS 2,3,4 AND APPLIES TO VEHICLES STARTING WITH VIN ENDING WITH F136169 UOC:209	2
1	PAOZZ	11862	464039	SPINDLE,WHEEL,DRIVI INCLUDES ITEMS 2,3,4 UOC:209	2
2	PAOZZ	11862	3965121	PARTS KIT,BEARING R UOC:209	2
3	PAOZZ	11862	376855	SEAL,PLAIN ENCASED UOC:209	2
4	PAOZZ	11862	376852	WASHER,RECESSED UOC:209	2
5	PAOZZ	11862	376851	SEAL,PLAIN ENCASED UOC:209	2
6	PAFZZ	11862	6273934	DEFLECTOR,DIRT AND UOC:209	2
7	XAHZZ	11862	14095588	AXLE ASSEMBLY,AUTOM UOC:209	1
8	PAHZZ	97271	36352-1	SEAL,PLAIN ENCASED UOC:209	2
9	PAFZZ	11862	14072916	YOKE,UNIVERSAL JOIN UOC:209	2
9	PAFZZ	11862	15521884	PROPELLER SHAFT APPLIES TO VEHICLES STARTING WITH VIN ENDING WITH F136169 UOC:209	1
10	PAFZZ	97271	5-297X	PARTS KIT,UNIVERSAL UOC:209	2
11	PAFZZ	11862	458877	PROPELLER SHAFT UOC:209	1
12	PAHZZ	96906	MS90728-92	SCREW,CAP,HEXAGON H UOC:209	4
13	PAFZZ	11862	458860	GASKET UOC:209	1
14	PAFZZ	11862	14063308	COVER,ACCESS INCLUDES ITEMS 13,15, 16 UOC:209	1
15	PAFZZ	11862	10008936	BOLT,SELF-LOCKING UOC:209	10
16	PAOZZ	96906	MS51884-9	PLUG,PIPE UOC:209	1
17	PBFZZ	11862	3787240	MAGNET,CHIP COLLECT	1

(1) ITEM NO	(2) SMR CODE	(3) CAGEC	(4) PART NUMBER	(5) DESCRIPTION AND USABLE ON CODES (UOC)	(6) QTY
				UOC:209	
18	PAFZZ	11862	458878	SHAFT,AXLE,AUTOMOTI	1
				UOC:209	

END OF FIGURE

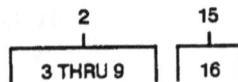

M1028 AND M1031 ONLY

ALL EXCEPT M1009, M1028,
AND M1031

PART OF
ITEM 15

TA510905

FIGURE 98. DIFFERENTIAL ASSEMBLY, RING, PINION, AND RELATED PARTS
(ALL EXCEPT M1009).

(1)	(2)	(3)	(4)	(5)	(6)
ITEM NO	SMR CODE	CAGEC	PART NUMBER	DESCRIPTION AND USABLE ON CODES (UOC)	QTY

GROUP 1002 DIFFERENTIAL

FIG. 98 DIFFERENTIAL ASSEMBLY, RING, PINION, AND RELATED PARTS (ALL EXCEPT M1009)

ITEM NO	SMR CODE	CAGEC	PART NUMBER	DESCRIPTION AND USABLE ON CODES (UOC)	QTY
1	PAHZZ	72447	30266	BOLT,MACHINE UOC:194,208,210,230,231,252,254,256	12
2	PAHHH	97271	73493-4X	CASE,DIFFERENTIAL UOC:230,231	1
3	PAHZZ	97271	500598-21	.PIN,STRAIGHT,HEADLE UOC:230,231	1
4	PAHZZ	97271	34367	.GEAR,BEVEL UOC:230,231	2
5	PAHZZ	97271	35830	.GEAR,BEVEL UOC:230,231	2
6	PAHZZ	97271	70696-4X	.HUB AND DISK,FRONT UOC:230,231	1
7	PAHZZ	97271	34730	.WASHER,CONCAVE UOC:230,231	2
8	PAHZZ	97271	30263	.SPIDER,DIFFERENTIAL UOC:230,231	1
9	PAHZZ	97271	35801	.CASE,DIFFERENTIAL UOC:230,231	1
10	PAHZZ	97271	25127-1X	GEAR ASSEMBLY, DIFFE UOC:194,208,210,230,231,252,254,256	1
11	XDHZZ	97271	30272-2	SHIM PNG BRG (.005) UOC:194,208,210,230,231,252,254,256	1
11	PAHZZ	97271	30272-3	SHIM (.010) UOC:194,208,210,230,231,252,254,256	1
11	PAHZZ	97271	30272-4	SPACER,RING (.030) UOC:194,208,210,230,231,252,254,256	1
11	PAHZZ	97271	30272-1	SHIM (.003) UOC:194,208,210,230,231,252,254,256	1
12	PAHZZ	97271	706045X	PARTS KIT,BEARING R UOC:194,208,210,230,231,252,254,256	1
13	PAFZZ	97271	30273	WASHER,FLAT UOC:194,208,210,230,231,252,254,256	1
14	PAFZZ	11862	14079089	SEAL UOC:194,208,210,230,231,252,254,256	1
15	PAFFF	97271	2-4-3801X	YOKE,UNIVERSSAL JOIN UOC:194,208,210,230,231,252,254,256	1
16	PAFZZ	97271	34592	.DEFLECTOR-PINION SE UOC:194,208,210,230,231,252,254,256	1
17	PAFZZ	72447	30275	WASHER,FLAT UOC:194,208,210,230,231,252,254,256	1
18	PAFZZ	72447	30271	NUT,SELF-LOCKING,HE UOC:194,208,210,230,231,252,254,256	1
19	PAHZZ	95019	30291-1	SHIM (.003) UOC:194,208,210,230,231,252,254,256	1
19	PAHZZ	95019	30291-2	SHIM (.005) UOC:194,208,210,230,231,252,254,256	1

(1) (2) (3) (4) ITEM SMR PART NO CODE CAGEC NUMBER	(5) DESCRIPTION AND USABLE ON CODES (UOC)	(6) QTY
19 PAHZZ 72447 30291-3	SPACER,RING (.010) UOC:194,208,210,230,231,252,254,256	1
20 PAHZZ 97271 706046X	CONE AND ROLLERS,TA UOC:194,208,210,230,231,252,254,256	1
21 PAHZZ 11862 462863	WASHER,FLAT UOC:194,208,210,230,231,252,254,256	1
22 PAHZZ 97271 706047X	PART KIT,BEARING R UOC:194,208,210,230,231,252,254,256	2
23 PAHZZ 97271 30276-1	SHIM (.003) UOC:194,208,210,230,231,252,254,256	1
23 PAHZZ 97271 30276-2	SHIM (.005) UOC:194,208,210,230,231,252,254,256	1
23 PAHZZ 97271 30276-3	SHIM (.010) UOC:194,208,210,230,231,252,254,256	1
23 PAHZZ 97271 30276-4	SPACER,RING (.030) UOC:194,208,210,231,252,254,256	1
24 PAHZZ 97271 500598-21	PIN,STRAIGHT,HEADLE UOC:194,208,210	1
25 PAHZZ 97271 34367	GEAR,BEVEL UOC:194,208,210	2
26 PAHZZ 97271 35625	GEAR,BEVEL UOC:194,208,210	2
27 PAHZZ 97271 34729	BEARING,WASHER,THRU UOC:194,208,210	2
28 PAHZZ 97271 34730	WASHER,CONCAVE UOC:194,208,210	2
29 PAHZZ 97271 30263	SPIDER,DIFFERENTIAL UOC:194,208,210	1
30 PAHZZ 95019 30258	CASE,DIFFERENTIAL G UOC:194,208,210	1

END OF FIGURE

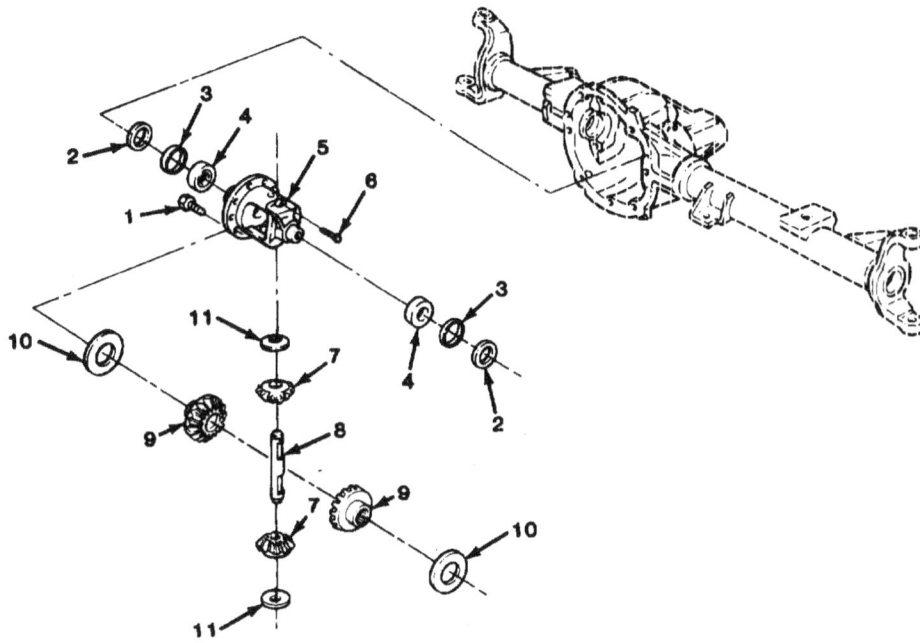

TA510906

FIGURE 99. DIFFERENTIAL COMPONENTS (M1009).

(1) ITEM NO	(2) SMR CODE	(3) CAGEC	(4) PART NUMBER	(5) DESCRIPTION AND USABLE ON CODES (UOC)	(6) QTY
				GROUP 1002 DIFFERENTIAL FIG. 99 DIFFERENTIAL COMPONENTS (M1009)	
1	PAHZZ	11862	14066913	BOLT,MACHINE UOC:209	10
2	PAHZZ	11862	3995791	SHIM (.040-.044) UOC:209	1
2	PAHZZ	11862	3995792	SHIM (.046-.050) UOC:209	1
2	PAHZZ	11862	3907088	WASHER ASSORTMENT (.052-.056) UOC:209	1
2	PAHZZ	11862	3907089	WASHER ASSORTMENT (.058-.062) UOC:209	1
2	PAHZZ	11862	3853759	SPACER SET,RING (.064-070) UOC:209	1
2	PAHZZ	11862	3853760	SHIM SET (.072-.078) UOC:209	1
2	PAHZZ	11862	3853761	WASHER ASSORTMENT (.080-.086) UOC:209	1
2	PAHZZ	11862	3853762	SHIM SET (.088-.094) UOC:209	1
2	PAHZZ	11862	3853912	SHIM (.096-.100) UOC:209	1
3	PAHZZ	11862	9775809	SPACER,RING UOC:209	2
4	PAHZZ	43334	LM501349/LM501314	BEARING,ROLLER,TAPE UOC:209	2
5	PAHZZ	11862	1252981	HOUSING,MECHANICAL UOC:209	1
6	PAHZZ	11862	14056196	SETSCREW UOC:209	1
7	PAHZZ	11862	6270977	GEAR,BEVEL UOC:209	2
8	PAHZZ	11862	14006401	SHAFT,STRAIGHT UOC:209	1
9	PAHZZ	11862	1397647	PARTS KIT,DRIVING A UOC:209	1
10	PAHZZ	11862	393578	WASHER,FLAT UOC:209	2
11	PAHZZ	11862	3984818	BEARING,WASHER,THRU UOC:209	2

END OF FIGURE

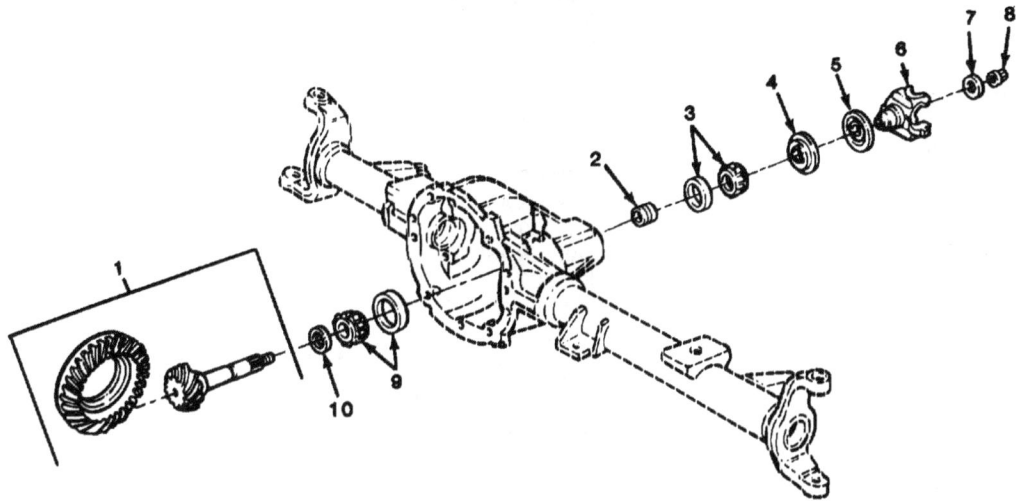

TA510907

FIGURE 100. RING AND PINION (M1009).

(1)	(2)	(3)	(4)	(5)	(6)
ITEM NO	SMR CODE	CAGEC	PART NUMBER	DESCRIPTION AND USABLE ON CODES (UOC)	QTY

GROUP 1002 DIFFERENTIAL

FIG. 100 RING AND PINION (M1009)

1	PAHZZ	11862	1258709	GEAR SET,BEVEL,MATCH UOC:2C9	1
2	PAHZZ	11862	1234726	SPACER,SLEEVE UOC:C9	1
3	PAFZZ	43334	M88048/M88010	BEARING,ROLLER,TAPE UOC:2C9	1
4	PAFZZ	11862	458859	SEAL,PLAIN ENCASED UOC:2C9	1
5	PAFZZ	48018	543319	DEFLECTOR,DIRT AND UOC:2C9	1
6	PAFZZ	11862	3988524	YOKE,UNIVERSAL JOIN UOC:209	1
7	PAFZZ	11862	517900	WASHER,FLAT UOC:209	1
8	PAFZZ	11862	1260823	NUT UOC:2C9	1
9	PAHZZ	43334	M802048/M802011	BEARING,ROLLER,TAPE UOC:2C9	1
10	PAHZZ	11862	1394892	WASHER ASSORTMENT (.020-.024) UOC:2C9	1
10	PAHZZ	11862	1394893	WASHER ASSORTMENT (.025-.029) UOC:2C9	1
10	PAHZZ	11862	1394894	WASHER ASSORTMENT (.030-.034) UOC:2C9	1
10	PAHZZ	11862	1394895	WASHER ASSORTMENT (.035-.039) UOC:2C9	1

END OF FIGURE

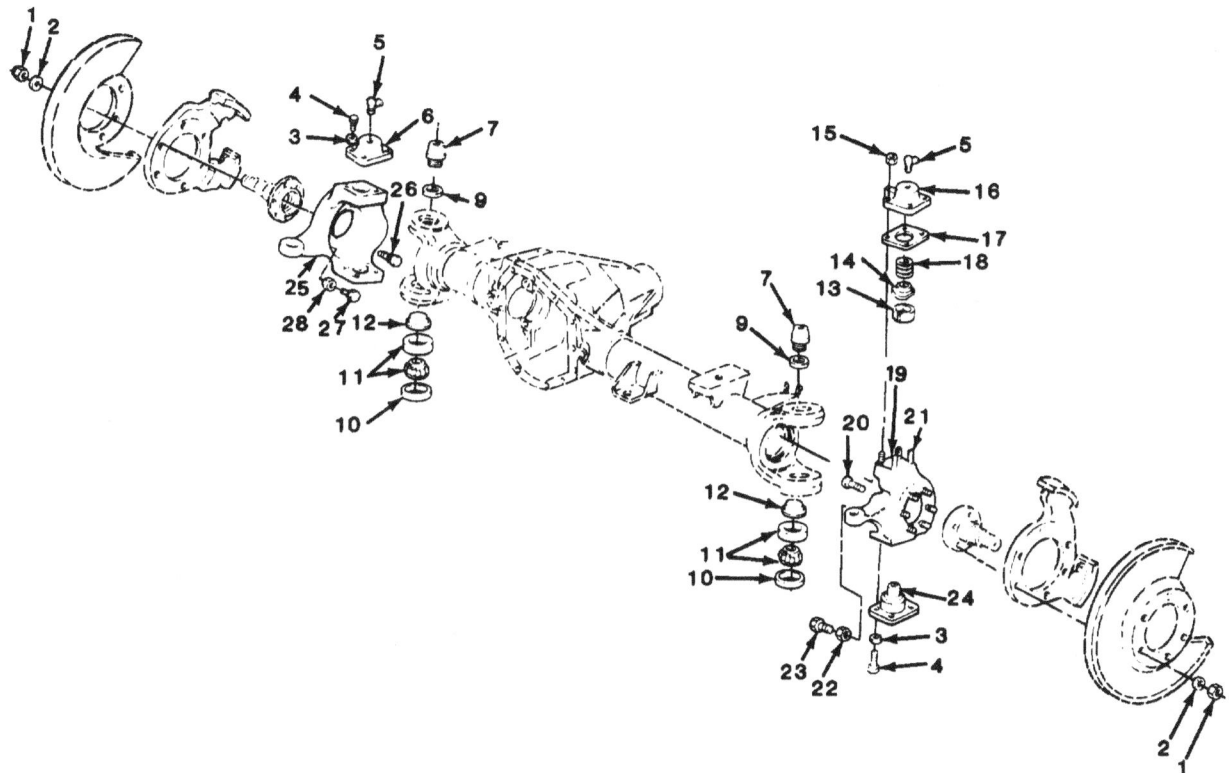

TA510908

FIGURE 101. STEERING KNUCKLE ASSEMBLY (ALL EXCEPT M1009).

(1) (2) (3) (4) ITEM SMR NO CODE CAGEC	PART NUMBER	(5) DESCRIPTION AND USABLE ON CODES (UOC)	(6) QTY
		GROUP 1004 STEERING AND LEANING WHEEL MECHANISM	
		FIG. 101 STEERING KNUCKLE ASSEMBLY (ALL EXCEPT M1009)	
1 PAOZZ 97271	36880	NUT,PLAIN UOC:194,208,210,230,231,252,254,256	12
2 PAOZZ 97271	38081-1	WASHER,LOCK UOC:194,208,210,230,231,252,254,256	12
3 PAOZZ 96906	MS35340-48	WASHER,LOCK UOC:194,208,210,230,231,252,254,256	12
4 PAFZZ 96906	MS90727-111	SCREW,CAP,HEXAGON H UOC:194,208,210,230,231,252,254,256	12
5 PAOZZ 24617	9411031	FITTING,LUBRICATION UOC:194,208,210,230,231,252,254,256	4
6 PAFZZ 97271	620132	CAP,STEERING KNUCKL UOC:194,208,210,230,231,252,254,256	1
7 PAFZZ 11862	462794	BOLT,INTERNAL WRENC UOC:194,208,210,230,231,252,254,256	2
8 PAFFF 97271	706395X	PARTS KIT,STEERING UOC:194,208,210,230,231,252,254,256	2
9 PAFZZ 97271	620058	.SEAL,PLAIN UOC:194,208,210,230,231,252,254,256	2
10 PAFZZ 80201	14700	.SEAL,PLAIN ENCASED UOC:194,208,210,230,231,252,254,256	2
11 PAFZZ 11862	9418355	.BEARING,ROLLER,TAPE UOC:194,208,210,230,231,252,254,256	2
12 PAFZZ 97271	37305	.CAP,PROTECTIVE,DUST UOC:194,208,210,230,231,252,254,256	2
13 PAFZZ 11862	462798	BUSHING UOC:194,208,210,230,231,252,254,256	2
14 PAFZZ 97271	620180	RETAINER,BUSHING,ST UOC:194,208,210,230,231,252,254,256	2
15 PAFZZ 97271	30875	NUT,SELF-LOCKING,CO UOC:194,208,210,230,231,252,254,256	4
16 PAFZZ 97271	38139	ARM,STEERING GEAR UOC:194,208,210,230,231,252,254,256	1
17 PAFZZ 97271	37307	GASKET UOC:194,208,210,230,231,252,254,256	2
18 PAFZZ 97271	37300	SPRING,HELICAL,COMP UOC:194,208,210,230,231,252,254,256	2
19 PAFFF 11862	462853	PARTS KIT,STEERING UOC:194,208,210,230,231,252,254,256	1
20 PAFZZ 97271	37879	.BOLT,RIBBED SHOULDE UOC:194,208,210,230,231,252,254,256	6
21 PAFZZ 9K937	37296	.STUD,PLAIN UOC:194,208,210,230,231,252,254,256	4
22 PAFZZ 97271	500381-3	.NUT,PLAIN,HEXAGON UOC:194,208,210,230,231,252,254,256	1
23 PAFZZ 97271	31026-1	.BOLT UCO:194,208,210,230,231,252,254,256	1

(1)	(2)	(3)	(4)	(5)	(6)
ITEM NO	SMR CODE	CAGEC	PART NUMBER	DESCRIPTION AND USABLE ON CODES (UOC)	QTY
24	PAFZZ	97271	37299	CAP,STEERING KNUCKL UOC:194,208,210,230,231,252,254,256	2
25	PAFFF	11862	15537219	SPINDLE,WHEEL,DRIVI UOC:194,208,210,230,231,252,254,256	1
26	PAFZZ	97271	37879	.BOLT,RIBBED SHOULDE UOC:194,208,210,230,231,252,254,256	6
27	PAFZZ	97271	31026-1	.BOLT UOC:194,208,210,230,231,252,254,256	1
28	PAFZZ	97271	500381-3	.NUT,PLAIN,HEXAGON UOC:194,208,210,230,231,252,254,256	1

END OF FIGURE

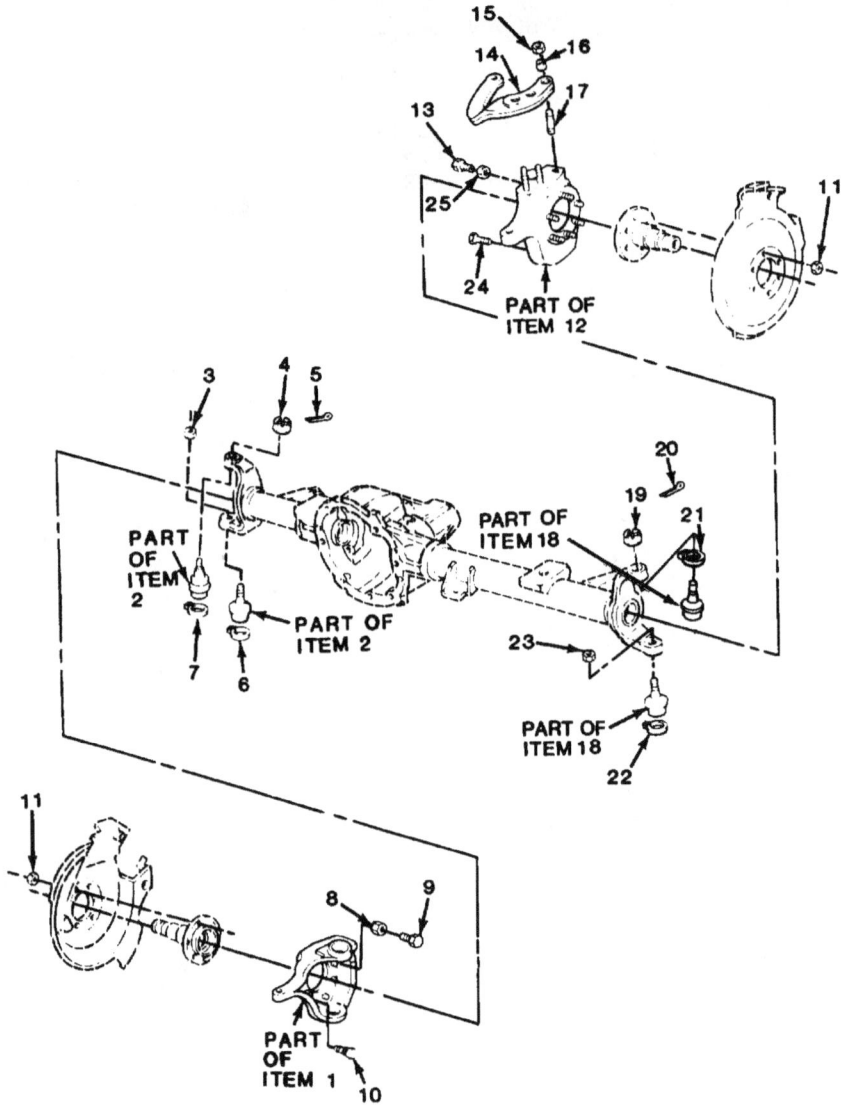

FIGURE 102. STEERING KNUCKLE ASSEMBLY (M1009).

TA510909

(1) (2) (3) (4) ITEM SMR NO CODE CAGEC	PART NUMBER	(5) DESCRIPTION AND USABLE ON CODES (UOC)	(6) QTY
		GROUP 1004 STEERING AND LEANING WHEEL MECHANISM	
		FIG. 102 STEERING KNUCKLE ASSEMBLY (M1009)	
1 PAFFF 11862	14013072	PARTS KIT,STEERING UOC:209	1
2 PAFFF 11862	14039008	.PARTS KIT,STEERING UOC:2C9	1
3 PAFZZ 24617	9422308	..NUT,SELF-LOCKING,HE UOC:2C9	1
4 PAFZZ 96906	MS35692-61	..NUT,PLAIN,SLOTTED,H UOC:2C9	1
5 PAFZZ 96906	MS24665-357	..PIN,COTTER UOC:209	1
6 PAFZZ 96906	MS16624-1175	..RING,RETAINIG UOC:2C9	1
7 PAFZZ 11862	3967845	..RING,EXTERNALLY THR UOC:209	1
8 PAOZZ 96906	MS35691-21	.NUT,PLAIN,HEXAGON UOC:2C9	1
9 PAOZZ 11862	3711876	.SCREW,CAP,HEXAGON H UOC:209	1
10 PAFZZ 11862	14008652	.BOLT,RIBBED NECK UOC:209	1
11 PAOZZ 11862	14056133	NUT,SELF-LOCKING,HE UOC:209	12
12 PAFFF 11862	14013071	SPINDLE,WHEEL,DRIVI UOC:209	1
13 PAOZZ 11862	3711876	.SCREW,CAP,HEXAGON H UOC:209	1
14 PAFZZ 11862	3979552	.ARM,STEERING GEAR UOC:209	1
15 PAFZZ 11862	9422303	.NUT,SELF-LOCKING,HE UOC:209	3
16 PAFZZ 11862	3965138	.BUSHING,TAPERED UOC:209	3
17 PAFZZ 11862	3965137	.STUD,PLAIN UOC:209	3
18 PAFFF 11862	14039008	.PARTS KIT,STEERING UOC:209	1
19 PAFZZ 96906	MS35692-61	..NUT,PLAIN,SLOTTED,H UOC:209	1
20 PAFZZ 96906	MS24665-357	..PIN,COTTER UOC:209	1
21 PAFZZ 11862	3967845	..RING,EXTERNALLY THR UOC:209	1
22 PAFZZ 96906	MS16624-1175	..RING,RETAINING UOC:209	1
23 PAFZZ 24617	9422308	..NUT,SELF-LOCKING,HE UOC:209	1

(1) (2) (3) (4)	(5)	(6)
ITEM SMR PART NO CODE CAGEC NUMBER	DESCRIPTION AND USABLE ON CODES (UOC)	QTY
24 PAFZZ 11862 14008652	.BOLT,RIBBED NECK UOC:209	6
25 PAOZZ 96906 MS35691-21	.NUT,PLAIN,HEXAGON UOC:209	1

END OF FIGURE

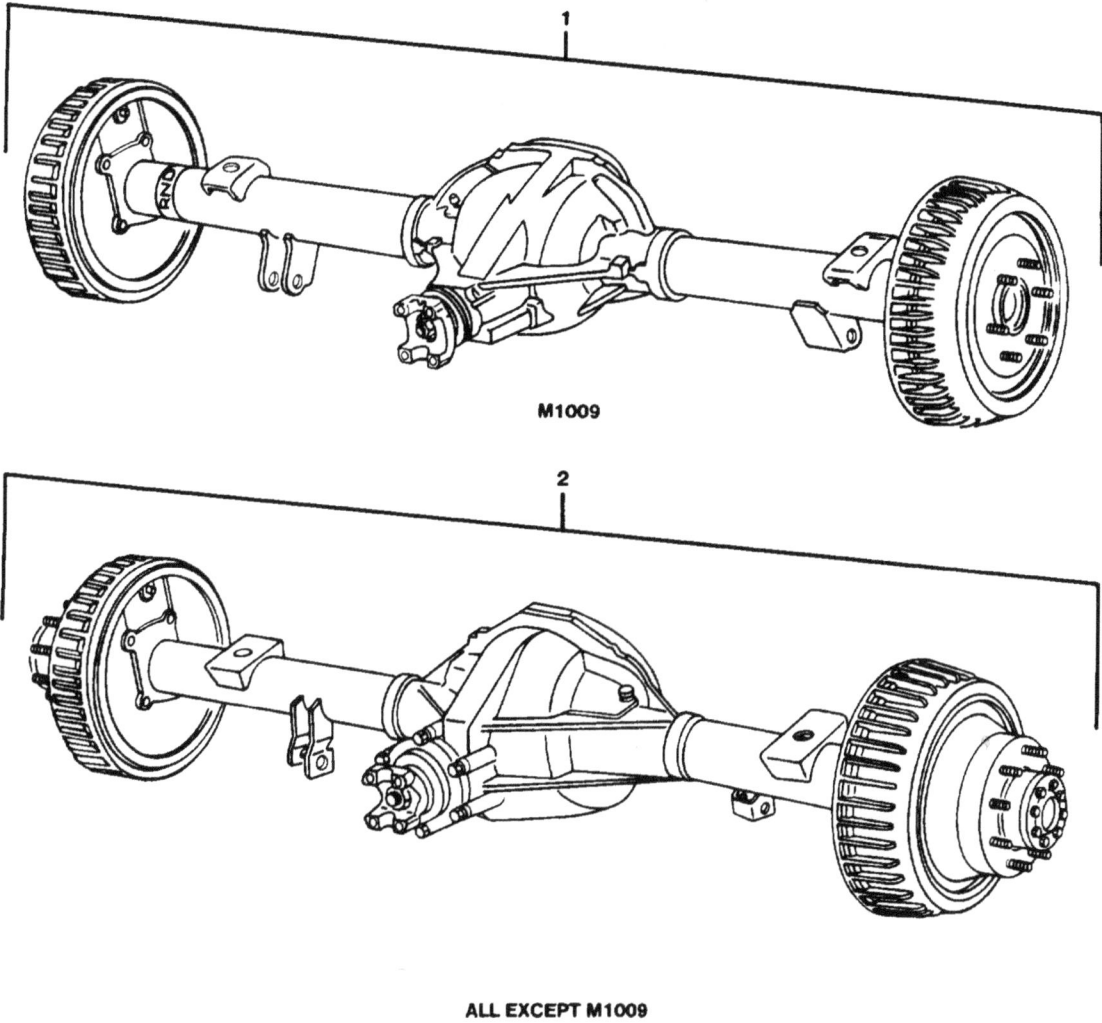

M1009

ALL EXCEPT M1009

TA510910

FIGURE 103. REAR AXLE.

TM9-2320-289-34P

(1) ITEM NO	(2) SMR CODE	(3) CAGEC	(4) PART NUMBER	(5) DESCRIPTION AND USABLE ON CODES (UOC)	(6) QTY
				GROUP 11 REAR AXLE	
				GROUP 1100 REAR AXLE ASSEMBLY	
				FIG.103 REAR AXLE	
1	PAFHD	11862	14055363	AXLE ASSEMBLY,AUTOM FOR COMPONENT PARTS SEE FIG'S 107, 109, 112,113 UOC:209	1
2	PAFHC	11862	15597787	AXLE ASSEMBLY,AUTOM APPLIES TO 85 MODEL WITH I.C. GHV STAMPED ON AA R TUBE, FOR COMPONENT PARTS SEE FIG'S 106,108,110 UOC:194,208,210,230,231,252,256	1
2	PAFHD	11862	15634657	AXLE ASSEMBLY,AUTOM UOC:254,256	1

END OF FIGURE

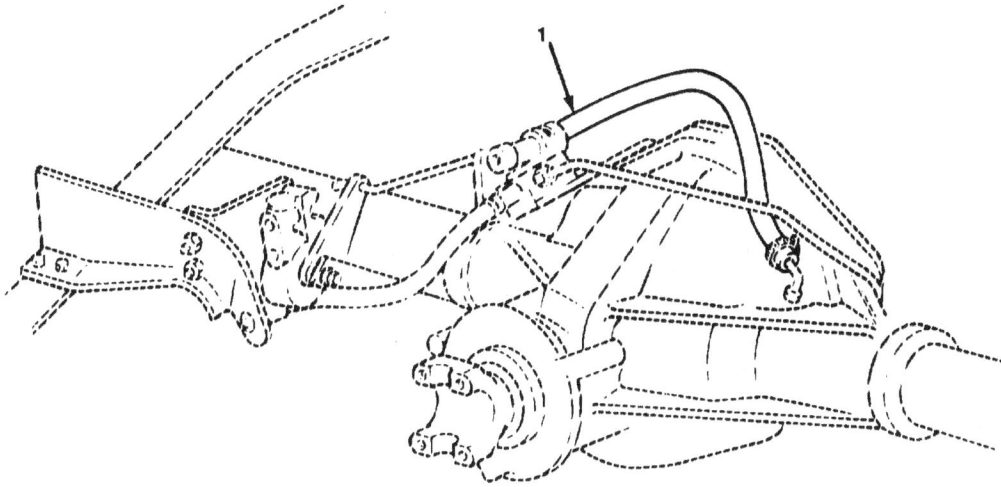

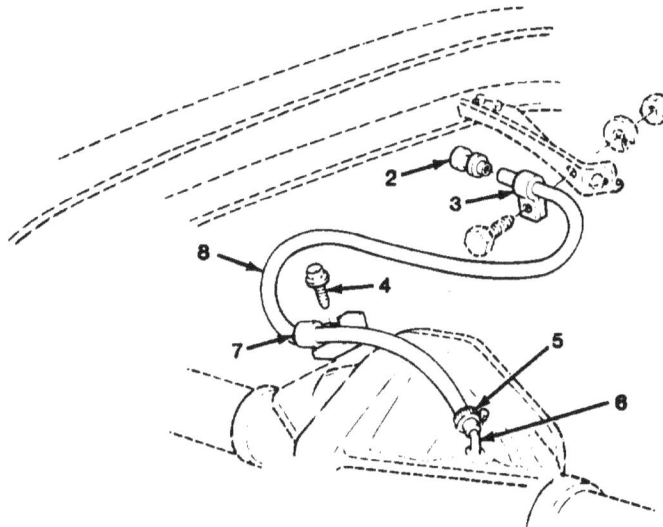

TA510911

FIGURE 104. REAR AXLE BREATHER (ALL EXCEPT M1009).

(1) (2) (3) (4) ITEM SMR PART NO CODE CAGEC NUMBER	(5) DESCRIPTION AND USABLE ON CODES (UOC)	(6) QTY
	GROUP 1100 REAR AXLE ASSEMBLY	
	FIG. 104 REAR AXLE BREATHER (ALL EXCEPT M1009)	
1 MOOZZ 11862 14040775	HOSE NON METALLIC (14.38" LG) MAKE FROM HOSE, P/N 9438315 UOC:210,231	1
2 PAOZZ 11862 8640496	SHAFT,SHOULDERED UOC:194,208,210,230,231,252	1
3 PAOZZ 11862 3866187	HANGER,PIPE UOC:210	1
4 PAOZZ 11862 11504447 SCREW,TAPPING,THREA	UOC:194,208,210,230,231,252	1
5 PAOZZ 96906 MS35842-10 CLAMP,HOSE	UOC:194,208,210,230,231,252	1
6 PAOZZ 11862 14056297 ELBOW,HOSE	UOC:194,208,210,230,231,252	1
7 PAOZZ 96906 MS21333-112 CLAMP,LOOP	UOC:194,208,210,230,231,252	1
8 MCOZZ 11862 474935	HOSE RR ZX VENT (24.00" LG) MAKE FROM HOSE, P/N 9438315 UOC:194,208,230,252	1
	END OF FIGURE	

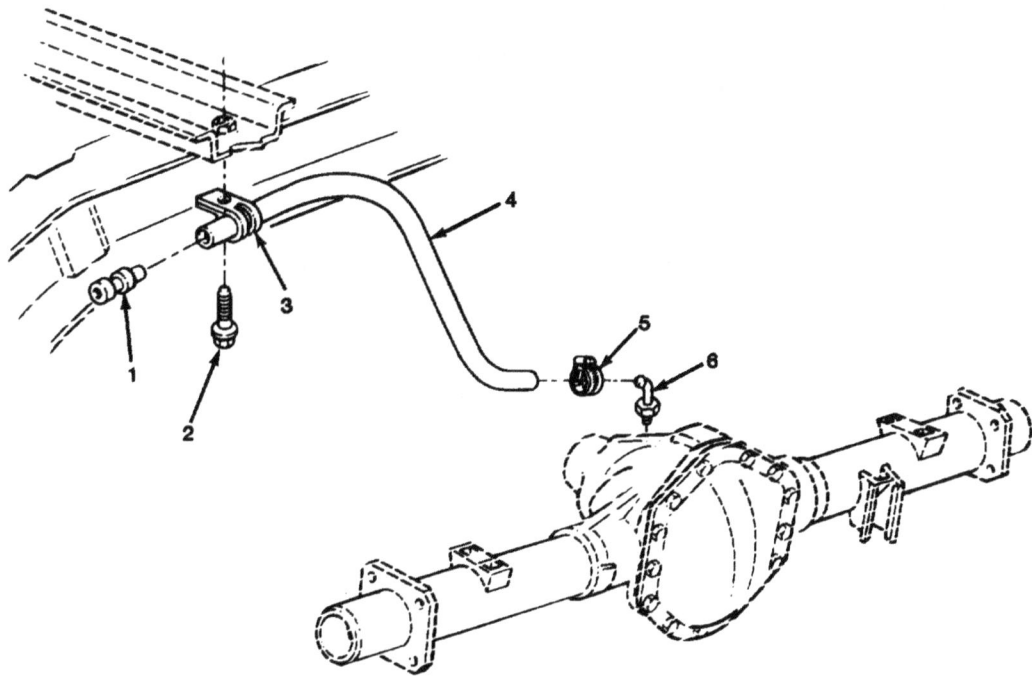

TA510912

FIGURE 105. REAR AXLE BREATHER (M1009).

(1) ITEM NO	(2) SMR CODE	(3) CAGEC	(4) PART NUMBER	(5) DESCRIPTION AND USABLE ON CODES (UOC)	(6) QTY
				GROUP 1100 REAR AXLE BREATHE (M1009)	
				FIG. 105 REAR AXLE BREATHER (M1009)	
1	PAOZZ	11862	8640496	SHAFT,SHOULDERED UOC:209	1
2	PAOZZ	11862	11504447	SCREW,TAPPING,THREA UOC:209	1
3	PAOZZ	96906	MS21333-112	CLAMP,LOOP UOC:209	1
4	MOOZZ	11862	474935	HOSE RR AX VENT (24.00" LG) MAKE FROM HOSE, P/N 9438315 UOC:209	1
5	PAOZZ	96906	MS35842-10	CLAMP,HOSE UOC:209	1
6	PAOZZ	11862	14072930	ELBOW,FLANGE TO HOS UOC:209	1

END OF FIGURE

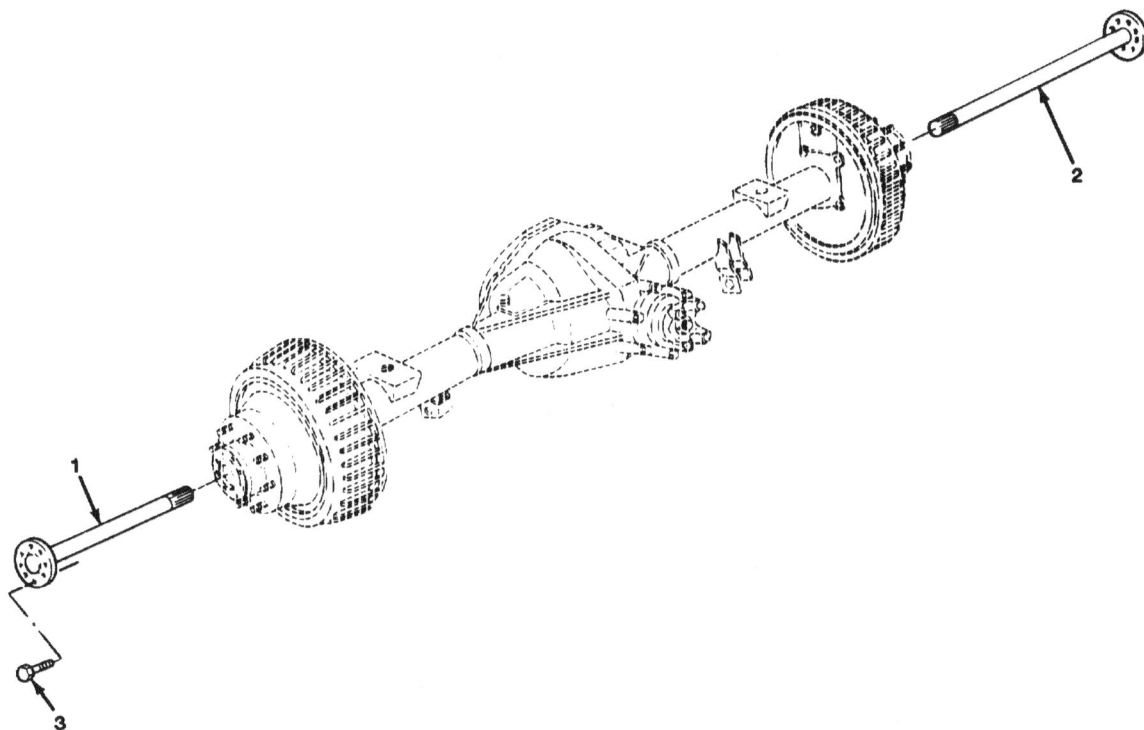

TA510913

FIGURE 106. AXLE SHAFTS (ALL EXCEPT M1009).

(1) ITEM NO	(2) SMR CODE	(3) CAGEC	(4) PART NUMBER	(5) DESCRIPTION AND USABLE ON CODES (UOC)	(6) QTY
				GROUP 1100 REAR AXLE ASSEMBLY	
				FIG. 106 AXLE SHAFTS (ALL EXCEPT M1009)	
1	PAOZZ	11862	3977384	SHAFT, AXLE,AUTOMOTI UOC:194,208,210,230,231,252	1
1	PAOZZ	11862	15599687	SHAFT,AXLE,AUTOMOTI UOC:254,256	1
2	PAOZZ	11862	3977383	SHAFT,AXLE,AUTOMOTI UOC:194,208,210,230,231,252	1
3	PAOZZ	11862	376869	BOLT,SELF-LOCKING UOC:194,208,210,230,231,252,254,256	16

END OF FIGURE

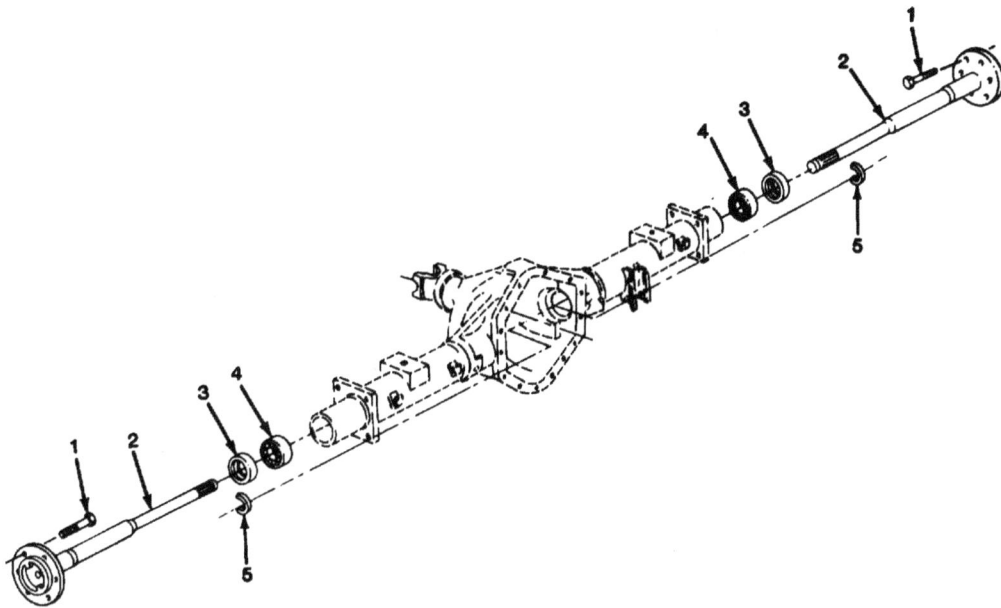

TA510914

FIGURE 107. REAR AXLE SHAFTS (M1009).

(1) (2) (3) (4) ITEM SMR PART NO CODE CAGEC NUMBER	(5) DESCRIPTION AND USABLE ON CODES (UOC)	(6) QTY
	GROUP 1100 REAR AXLE ASSEMBLY	
	FIG. 107 REAR AXLE ASSEMBLY	
	FIG. 107 REAR AXLE SHAFTS (M1009)	
1 PAFZZ 11862 9590271	BOLT,RIBBED SHOULDE UOC:209	12
2 PAFZZ 11862 14039547 SHAFT,AXLE,AUTOMOTI	UOC:209	2
3 PAFZZ 11862 14003417	SEAL,REAR AXLE UOC:209	2
4 PAFZZ 11862 7451809 BEARING,ROLLER,CYLI	UOC:209	2
5 PAFZZ 11862 3833322 RING,RETAINING	UOC:209	2

END OF FIGURE

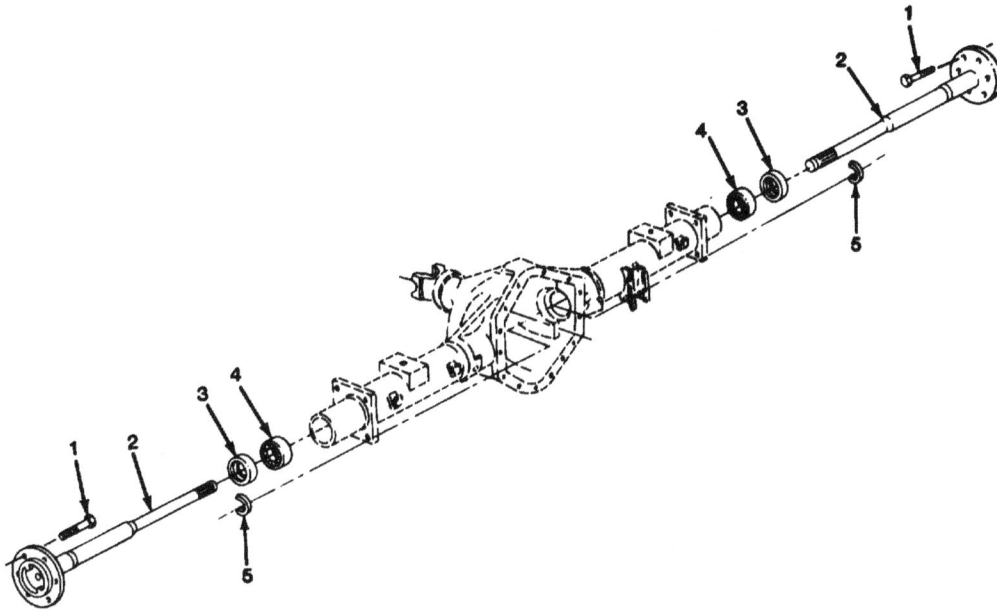

TA510914

FIGURE 107. REAR AXLE SHAFTS (M1009).

TM9-2320-289-34P

(1) ITEM NO	(2) SMR CODE	(3) CAGEC	(4) PART NUMBER	(5) DESCRIPTION AND USABLE ON CODES (UOC)	(6) QTY
				GROUP 1100 REAR AXLE ASSEMBLY	
				FIG. 107 REAR AXLE ASSEMBLY	
				FIG. 107 REAR AXLE SHAFTS (M1009)	
1	PAFZZ	11862	9590271	BOLT,RIBBED SHOULDE UOC:209	12
2	PAFZZ	11862	14039547	SHAFT,AXLE,AUTOMOTI UOC:209	2
3	PAFZZ	11862	14003417	SEAL,REAR AXLE UOC:209	2
4	PAFZZ	11862	7451809	BEARING,ROLLER,CYLI UOC:209	2
5	PAFZZ	11862	3833322	RING,RETAINING UOC:209	2

END OF FIGURE

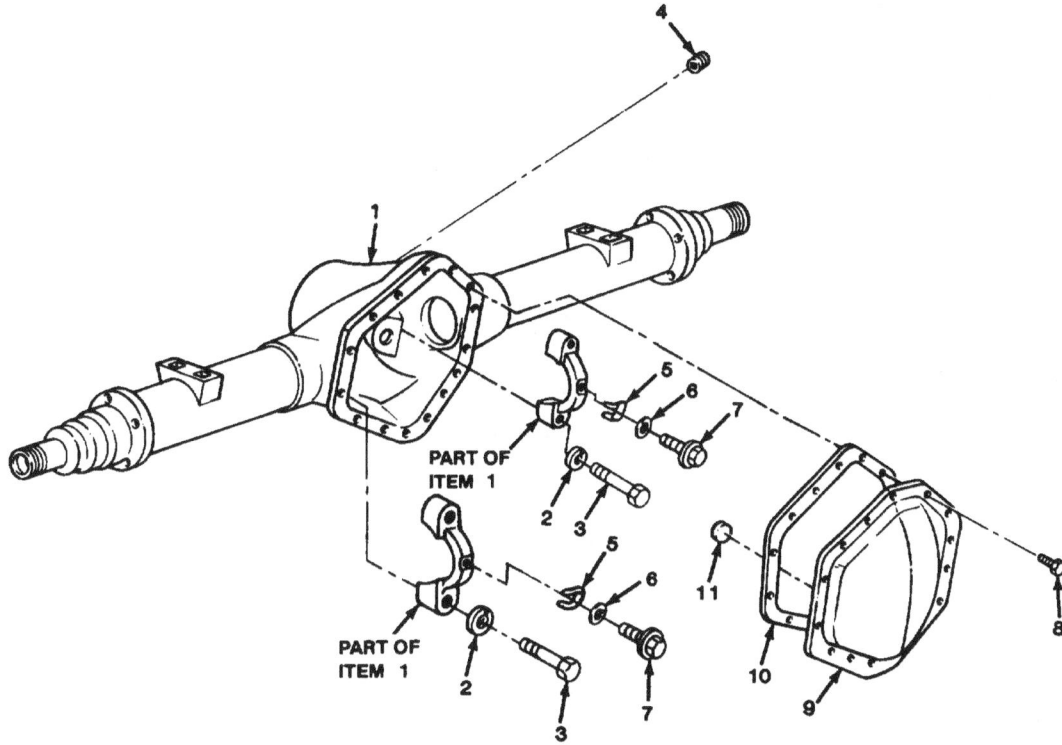

1

2 AND 3

FIGURE 108. REAR AXLE HOUSING (ALL EXCEPT M1009).

TA510915

(1) ITEM NO	(2) SMR CODE	(3) CAGEC	(4) PART NUMBER	(5) DESCRIPTION AND USABLE ON CODES (UOC)	(6) QTY
				GROUP 1101 HOUSING, BEAM, HOUSING COVERS, PLUGS, SEALS, ETC.	
				FIG. 108 REAR AXLE HOUSING (ALL EXCEPT M1009)	
1	XAHHH	11862	471777	DIFFERENTIAL AND DR APPLIES TO 84 MODEL WITH I.D. BMH STAMPED ON AXLE TUBE UOC:194,208,210,230,231,252	1
1	XAHHH	11862	15537684	DIFFERENTIAL CARRIE APPLIES TO 85 MODEL WITH I.D. GHV STAMPED ON AXLE TUBE UOC:194,208,210,230,231,252,254,256	1
1	XAHHH	11862	15537131	WLD ASM TUBE/HSG APPLIES TO 86 & 87 MODELS WITH HFF STAMP ON AXLE TUBE UOC:194,208,210,230,231,252,254,256	1
1	XAHHH	11862	15594111	HOUSING,AXLE UOC:254,256	1
2	PAHZZ	96906	MS35340-49	.WASHER,LOCK UOC:194,208,210,230,231,252	4
3	PAHZZ	96906	MS90728-143	.SCREW,CAP,HEXAGON H UOC:194,208,210,230,231,252	4
3	PAHZZ	11862	15594116	.BOLT UOC:254,256	4
4	PAOZZ	30379	444789	PLUG,PIPE UOC:194,208,210,230,231,252	1
5	PAHZZ	11862	3977326	CLIP,RETAINING UOC:194,208,210,230,231,252	2
6	PAHZZ	96906	MS35333-41	WASHER,LOCK UOC:194,208,210,230,231,252	2
7	PAHZZ	96906	MS90728-29	BOLT,MACHINE UOC:194,208,210,230,231,252	2
8	PAOZZ	11862	15521977	BOLT,SELF-LOCKING UOC:194,208,210,230,231,252	14
9	PAOZZ	11862	3977386	COVER,ACCESS UOC:194,208,210,230,252	1
9	PAOZZ	11862	14071884	COVER,TRANSMISSION UOC:254,256	1
10	PAOZZ	11862	3977387	GASKET UOC:194,208,210,230,231,252	1
11	PAOZZ	11862	3787240	MAGNET,CHIP COLLECT UOC:194,208,210,230,231,252	1

END OF FIGURE

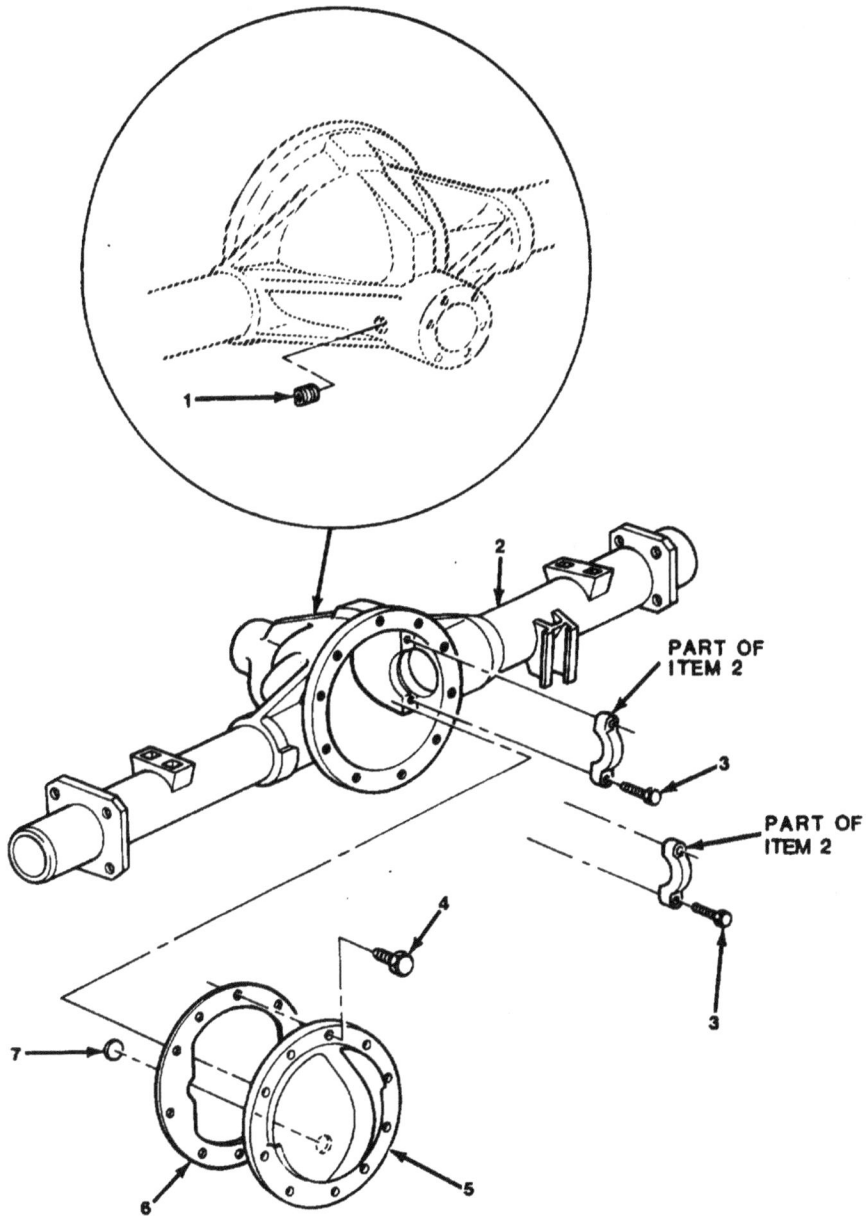

TA510916

FIGURE 109. REAR AXLE HOUSING (M1009).

SECTION II   TM9-2320-289-34P

(1) (2) (3) (4) ITEM SMR PART NO CODE CAGEC NUMBER	(5) DESCRIPTION AND USABLE ON CODES (UOC)	(6) QTY
	GROUP 1101 HOUSING,BEAM,HOUSING COVERS, PLUGS, SEASL, ETC.	
	FIG. 109 REAR AXLE HOUSING (M1009)	
1 PAOZZ 30379 444789	PLUG,PIPE UOC:209	1
2 XAHZZ 11862 15594126 HOUSING,DIFFERENTIA	UOC:209	1
3 PAHZZ 96906 MS90728-92	SCREW,CAP,HEXAGON H UOC:209	4
4 PAOZZ 11862 10008936 BOLT,SELF-LOCKING	UOC:209	10
5 PAOZZ 11862 1252415	COVER,GEAR CARRIER UOC:209	1
6 PAOZZ 11862 26016662 GASKET	UOC:209	1
7 PBOZZ 11862 3787240	MAGNET,CHIP COLLECT UOC:209	1
	END OF FIGURE	

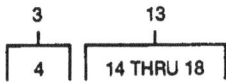

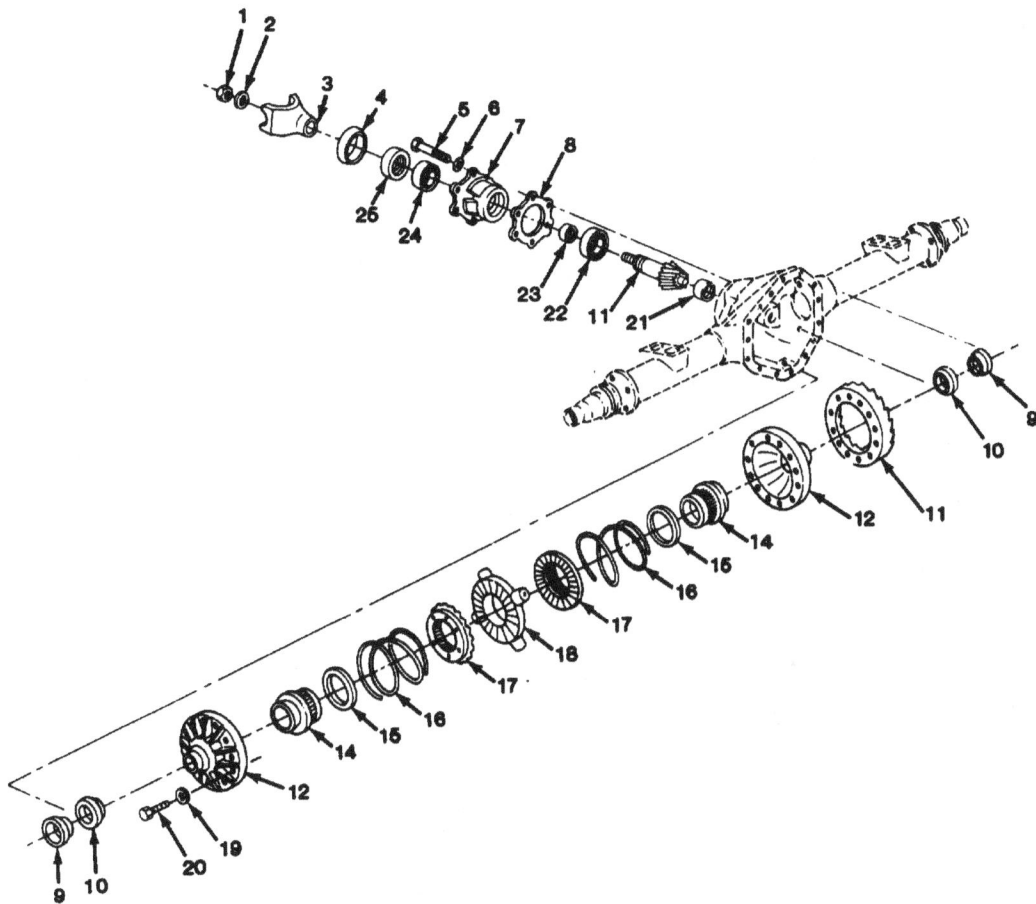

TA510917

FIGURE 110. DIFFERENTIAL LOCK ASSEMBLY RING, PINION, AND RELATED PARTS
(ALL EXCEPT M1009, M1028A2, AND M1028A3).

(1) (2) (3) (4) ITEM SMR PART NO CODE CAGEC NUMBER	(5) DESCRIPTION AND USABLE ON CODES (UOC)	(6) QTY
	GROUP 1102 DIFFERENTIAL	
	FIG. 110 DIFFERENTIAL LOCK ASSEMBLY, RING, PINION, AND RELATED PARTS (ALL EXCEPT M1009, M1028A2, AND M1028A3)	
1 PAFZZ 11862 644536	NUT,SELF-LOCKING,HE APPLIES TO BOTH 84 & 85 MODELS WITH I.D. B GHV STAMPED ON AXLE TURE UOC:194,208,210,230,231,252	1
1 PAFZZ 11862 15552844	NUT,SELF-LOCKING,EX APPLIES TO 86 & 87 MODELS WITH I.D. HFF STAMPED ON AXLE TUBE UOC:194,208,210,230,231,252,254,256	1
2 PAFZZ 11862 3977360	WASHER,FLAT APPLIES TO BOTH 84 & 85 MODELS WITH I.D. 8 GHV STAMPED ON AXLE TUBE UOC:194,208,210,230,231,252	1
2 PAFZZ 11862 2292300	WASHER,FLAT APPLIES TO 86 AND 87 MODELS WITH I.D. HFF STAMPED ON AXLE TUBE UOC:194,208,210,230,231,252,254,256	1
3 PAFFF 11862 26020811	PINION,FLANGE UOC:194,208,210,230,231,252	1
4 PAFZZ 11862 3977358	.DEFLECTOR,DIRT AND UOC:194,208,210,230,231,252	1
5 PAHZZ 96906 MS90728-86	SCREW,CAP,HEXAGON H UOC:194,208,210,230,231,252	6
6 PAHZZ 96906 MS35340-47	WASHER,LOCK UOC:194,208,210,230,231,252	6
7 PAHZZ 11862 3977354	RETAINER,OIL SEAL UOC:194,208,210,230,231,252	1
8 PAHZZ 11862 334362	SHIM (.006) UOC:194,208,210,230,231,252	1
8 PAHZZ 11862 334363	SHIM (.007) UOC:194,208,210,230,231,252	1
8 PAHZZ 11862 334364	SHIM (.008) UOC:194,208,210,230,231,252	1
8 PAHZZ 11862 334365	SHIM (.009) UOC:194,208,210,230,231,252	1
8 PAHZZ 11862 334366	SHIM (.010) UOC:194,208,210,230,231,252	1
8 PAHZZ 11862 334367	SHIM (.011) UOC:194,208,210,230,231,252	1
8 PAHZZ 11862 334368	SHIM (.012) UOC:194,208,210,230,231,252	1
8 PAHZZ 11862 334369	SHIM (.013) UOC:194,208,210,230,231,252	1
8 PAHZZ 11862 334370	SHIM (0.14) UOC:194,208,210,230,231,252	1
8 PAHZZ 11862 334370	SHIM (.015) UOC:194,208,210,230,231,252	1

(1) ITEM NO	(2) SMR CODE	(3) CAGEC	(4) PART NUMBER	(5) DESCRIPTION AND USABLE ON CODES (UOC)	(6) QTY
8	PAHZZ	11862	334372	SPACER,PLATE (.016) UOC:194,208,210,230,231,252	1
8	PAHZZ	11862	334373	SPACER,PLATE (.017) UOC:194,208,210,230,231,252	1
8	PAHZZ	11862	334374	SPACER,PLATE (.018) UOC:194,208,210,230,231,252	1
8	PAHZZ	11862	334375	SPACER,PLATE (.019) UOC:194,208,210,230,231,252	1
8	PAHZZ	11862	334376	SPACER,PLATE (.020) UOC:194,208,210,230,231,252	1
8	PAHZZ	11862	334377	SPACER,PLATE (.021) UOC:194,208,210,230,231,252	1
8	PAHZZ	11862	334378	SPACER,PLATE (.022) UOC:194,208,210,230,231,252	1
8	PAHZZ	11862	334379	SPACER,PLATE (.023) UOC:194,208,210,230,231,252	1
8	PAHZZ	11862	334380	SPACER,PLATE (.024) UOC:194,208,210,230,231,252	1
9	PAHZZ	11862	3977325	ADJUSTER,DIFFERENTI UOC:194,208,210,230,231,252	2
10	PAHZZ	60038	387AS-382A	BEARING,ROLLER,TAPE UOC:194,208,210,230,231,252	2
11	PAHZZ	11862	471873	PARTS KIT,DRIVING A UOC:194,208,210,230,231,252	1
12	PAHZZ	11862	6258340	HOUSING MECHANICAL UOC:194,208,210,230,231,252	1
13	PAHZZ	11862	14063392	SPIDER,DIFFERENTIAL UOC:194,208,210,230,231,252	1
14	PAHZZ	11862	14075355	.GEAR,SPUR UOC:194,208,210,230,231,252	2
15	PAHZZ	11862	14075354	.RETAINER,HELICAL CC UOC:194,208,210,230,231,252,254	2
16	PAHZZ	11862	14075353	.SPRING,HELICAL,COMP UOC:194,208,210,230,231,252,254	2
17	PAHZZ	11862	14075352	.CAM,CONTROL UOC:194,208,210,230,231,252,254	2
18	PAHZZ	11862	14075351	.SPIDER,DIFFERENTIAL UOC:194,208,210,230,231,252	1
19	PAHZZ	96906	MS35340-48	WASHER,LOCK UOC:194,208,210,230,231,252	12
20	PAHZZ	96906	MS90727-116	SCREW,CAP,HEXAGON H UOC:194,208,210,230,231,252,254,256	12
21	PAHZZ	11862	7451870	BEARING,ROLLER,CLY APPLIES TO 84 MODEL WITH I.D. BHM STAMPED ON AXLE TUBE UOC:194,208,210,230,231,252	1
21	PAHZZ	60380	DC57524	BEARING,ROLLER,NEED APPLIES TO 85 THRU 87 MODELS WITH I.D. GHV OR HFF STAMPED ON AXLE TUBE UOC:194,208,210,230,231,252,254,256	1
22	PAHZZ	11862	7451888	BEARING,ROLLER,TAPE UOC:194,208,210,230,231,252	1

(1) ITEM NO	(2) SMR CODE	(3) CAGEC	(4) PART NUMBER	(5) DESCRIPTION AND USABLE ON CODES (UOC)	(6) QTY
23	PAHZZ	11862	3977355	SPACER,SLEEVE UOC:194,208,210,230,231,252	1
24	PAHZZ	24617	7451155	BEARING,ROLLER,TAPE UOC:194,208,210,230,231,252	1
25	PAFZZ	11862	3977359	SEAL,PLAIN ENCASED UOC:194,208,210,230,231,252	1

END OF FIGURE

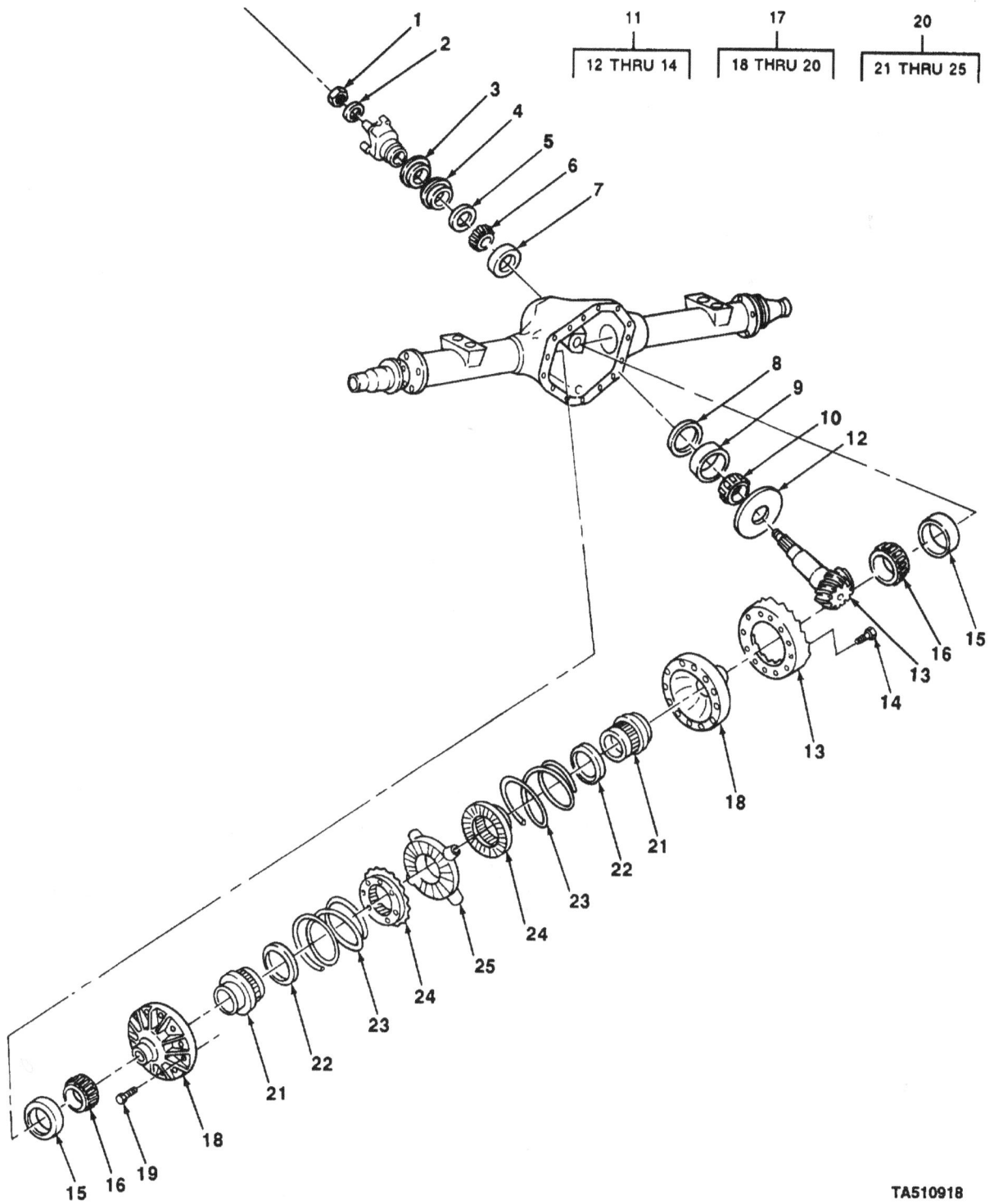

TA510918

FIGURE 111. DIFFERENTIAL LOCK ASSEMBLY, RING, PINION, AND RELATED
PARTS (M1028A2 AND M1028A3).

(1) ITEM NO	(2) SMR CODE	(3) CAGEC	(4) PART NUMBER	(5) DESCRIPTION AND USABLE ON CODES (UOC)	(6) QTY
				GROUP 1102 DIFFERENTIAL	
				FIG. 111 DIFFERENTIAL LOCK ASSEMBLY, RING, PINION, AND RELATED PARTS (M1028A2 AND M1028A3)	
1	PAFZZ	11862	2353021	NUT,SELF-LOCKING,HE UOC:254,256	1
2	PAFZZ	11862	2353015	WASHER,FLAT UOC:254,256	1
3	PAHZZ	97271	30276-4	SPACER,RING UOC:204,256	1
4	PAHZZ	97271	30276-1	SHIM UOC:254,256	1
4	PAHZZ	97271	30276-2	SHIM UOC:254,256	1
5	PAHZZ	97271	30276-3	SHIM UOC:254,256	1
6	PAOZZ	60038	HM88542	CONE AND ROLLERS,TA UOC:254,256	1
7	PAOZZ	60038	88510	CUP,TAPERED ROLLER UOC:254,256	1
8	KFHZZ	97271	30797-1	SMIM (.003) PART OF KIT P/N 15537086 UOC:254,256	1
8	DFHZZ	97271	30797-2	SHIM (.005) PART OF KIT P/N 15537086 UOC:254,256	1
8	KFHZZ	97271	30797-3	SHIM (.010) PART OF KIT P/N 15537086 UOC:254,256	1
8	KFHZZ	97271	34801-1	SHIM (.014) PART OF KIT P/N 15537086 UOC:254,256	1
8	KFHZZ	97271	34801-2	SHIM (.016) PART OF KIT P/N 15537086 UOC:254,256	1
8	KFHZZ	97271	34801-3	SHIM (.018) PART OF KIT P/N 15537086 UOC:254,256	1
8	LFJZZ	97271	34801-4	SHIM (.020) PART OF KIT P/N 15537086 UOC:254,256	1
8	KFHZZ	97271	34801-5	SHIM (.022) PART OF KIT P/N 15537086 UOC:254,256	1
8	KFHZZ	97271	34801-6	SHIM (.030) PART OF KIT P/N 15537086 UOC:254,256	1
8	KFHZZ	97271	34801-7	SHIM (.015) PART OF KIT P/N 15537086 UOC:254,256	1
8	KFHZZ	97271	34801-8	SHIM (.021) PART OF KIT P/N 15537086 UOC:254,256	1
8	KFHZZ	97271	34801-9	SHIM (.023) PART OF KIT P/N 15537086 UOC:254,256	1
8	KFHZZ	97271	34801-10	SHIM (.010) PART OF KIT P/N 15537086 UOC:254-256	1
9	PAOZZ	60038	HM807010	CUP,TAPERED ROLLER UOC:254,6256	1
10	PAOZZ	60038	HM807040	CONE AND ROLLERS,TA UOC:254,256	1

(1) ITEM NO	(2) SMR CODE	(3) CAGEC	(4) PART NUMBER	(5) DESCRIPTION AND USABLE ON CODES (UOC)	(6) QTY
11	PAHZZ	97271	72166 5X	PARTS KIT,DRIVING A UOC:254,256	1
12	PAHZZ	97271	701064 X	.SPACER, RING UOC:254,256	1
13	PAHZZ	97271	72166 X	.GEAR SET,BEVEL,MATC UOC:254,256	1
14	PAHZZ	97271	701056 1X	.PARTS KIT,DRIVING A UOC:254,256	1
15	PAOZZ	60038	453X	CUP,TAPERED ROLLER UOC:254,256	1
16	PAOZZ	60038	469	CONE AND ROLLERS,TA UOC:254,256	1
17	PAHZZ	11862	15637897	DIFFERENTIAL,DRIVIN UOC:254,256	1
18	PAHZZ	11862	15637902	.TRANSFER TRANSMISSI UOC:254,256	1
19	PAHZZ	96906	MS90728-90L	.BOLT,SELF-LOCKING UOC:254,256	12
20	PAHHH	11862	15637904	.SPIDER ASSEMBLY,DIF UOC:254,256	1
21	PAHZZ	11862	15637901	..GEAR CLUSTER UOC:254,256	2
22	PAHZZ	11862	14075354	..RETAINER,HELICAL CC UOC:254,256	1
23	PAHZZ	11862	14075353	..SPRING,HELICAL COMP UOC:254,256	2
24	PAHZZ	11862	14075352	..CAM,CONTROL UOC:254,256	2
25	PAHZZ	11862	15581505	..SPIDER,DIFFERENTIAL UOC:254,256	1

END OF FIGURE

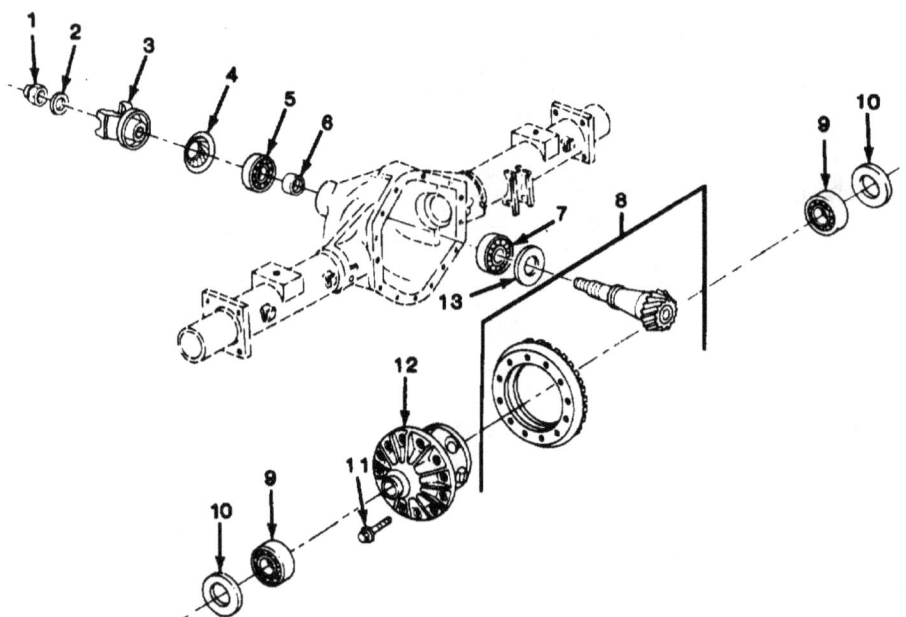

TA510919

FIGURE 112. DIFFERENTIAL LOCK ASSEMBLY, RING, PINION, AND RELATED
PARTS (M1009).

(1) (2) (3) (4) ITEM SMR PART NO CODE CAGEC NUMBER	(5) DESCRIPTION AND USABLE ON CODES (UOC)	(6) QTY
	GROUP 1102 DIFFERENTIAL	
	FIG. 112 DIFFERENTIAL LOCK ASSEMBLY, RING, PINION, AND RELATED PARTS (M1009)	
1 PAFZZ 11862 1260823	NUT UOC:209	1
2 PAFZZ 11862 517900	WASHER,FLAT UOC:209	1
3 PAFZZ 11862 1256654	FLANGE,COMPANION,UN UOC:209	1
4 PAFZZ 11862 1243465	SEAL,PLAIN ENCASED UOC:209	1
5 PAHZZ 4334 M88048/M88010	BEARING,ROLLER,TAPE UOC:209	1
6 PAHZZ 11862 1234726	SPACER,SLEEVE UOC:209	1
7 PAHZZ 24617 7451155	BEARING,ROLLER,TAPE UOC:209	1
8 PAHZZ 11862 1258709	GEAR SET,BEVEL,MATC UOC:209	1
9 PAHZZ 43334 LM501349/LM501314	BEARING,ROLLER,TAPE UOC:209	2
10 PAHZZ 11862 3995791	SHIM (.040"-.044") UOC:209	1
10 PAHZZ 11862 3995792	SHIM (.046"-.050") UOC:209	1
10 PAHZZ 11862 3907088	WASHER,ASSORTMENT (.052"-.062") UOC:209	1
10 PAHZZ 11862 3907089	WASHER ASSORTMENT (.058"-.062") UOC:209	1
10 PAHZZ 11862 3853759	SPACER SET,RING (.064"-.070") UOC:209	1
10 PAHZZ 11862 3853760	SHIM SET (.072"-.078") UOC:209	1
10 PAHZZ 11862 3853761	WASHER,ASSORTMENT (.080"-.086") UOC:209	1
10 PAHZZ 11862 3853762	SHIM SET (.088"-.094") UOC:209	1
10 PAHZZ 11862 3853912	SHIM (.086"-.100") UOC:209	1
10 PAHZZ 11862 9775809	SPACER,RING (.170"-.172") UOC:209	2
11 PAHZZ 11862 14066913	BOLT,MACHINE UOC:209	10
12 PAHZZ 11862 26004461	PARTS KIT, DRIVING A FOR COMPONENT PARTS SEE FIG 113 UOC:209	1
13 PAHZZ 11862 1394892	WASHER ASSORTMENT (.020-.024) UOC:209	1
13 PAHZZ 11862 1394893	WASHER ASSORTMENT (.025-.029)	1

(1) ITEM NO	(2) SMR CODE	(3) CAGEC	(4) (5) PART NUMBER	(6) DESCRIPTION AND USABLE ON CODES (UOC)	QTY
				UOC:209	
13	PAHZZ	11862	1394894	WASHER ASSORTMENT (.020-.034)	1
				UOC:209	
13	PAHZZ	11862	1394895	WASHER ASSORTMENT (.035-.039)	1
				UOC:209	

END OF FIGURE

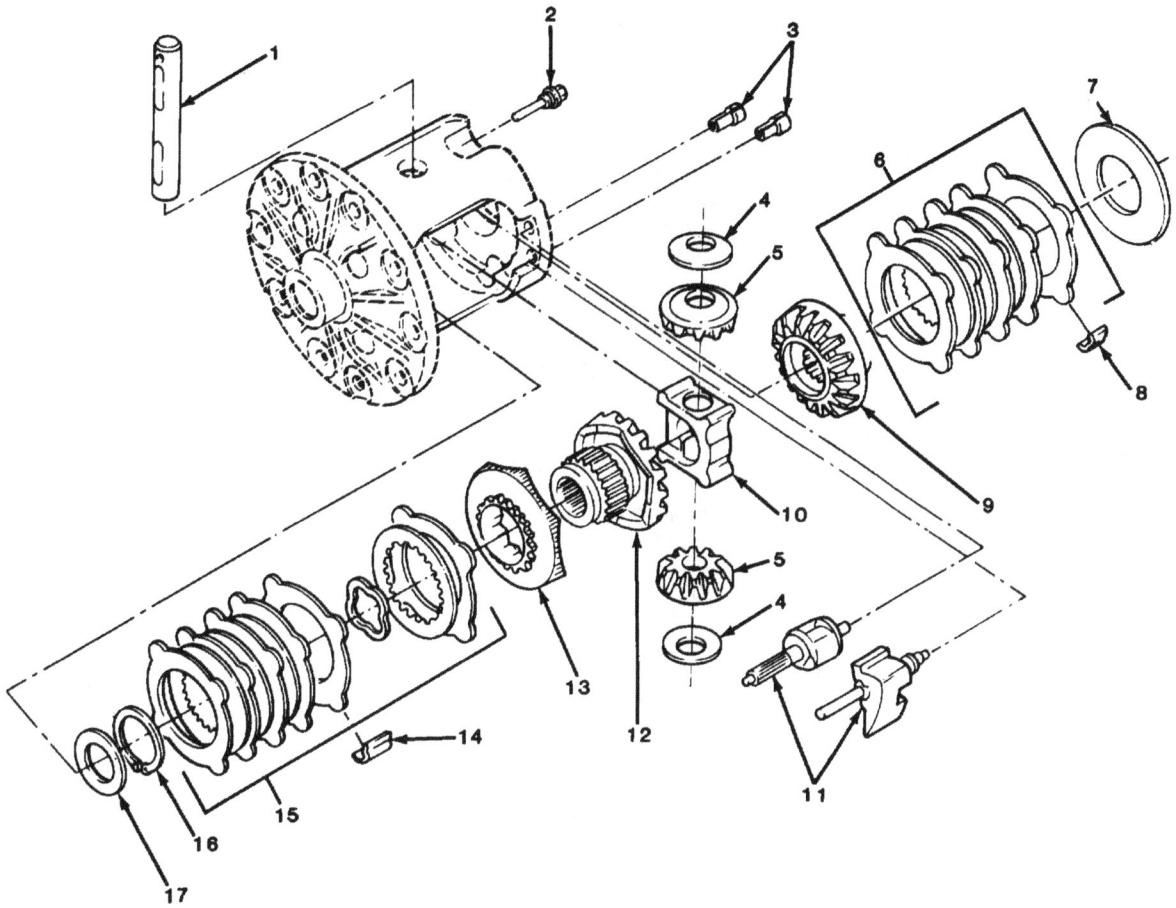

TA510920

FIGURE 113. DIFFERENTIAL LOCK ASSEMBLY COMPONENT PARTS (M1009).

(1) (2) (3) (4) ITEM SMR PART NO CODE CAGEC NUMBER	(5) DESCRIPTION AND USABLE ON CODES (UOC)	(6) QTY
	GROUP 1102 DIFFERENTIAL	
	FIG. 113 DIFFERENTIAL LOCK ASSEMBLY COMPONENT PARTS (M1009)	
1 PAHZZ 11862 14048412	.SHAFT,STRAIGHT UOC:209	1
2 PAHZZ 11862 14056196	.SET SCREW UOC:209	1
3 PAHZZ 11862 14048409	.BUSHING,TAPERED UOC:209	2
4 PAHZZ 11862 14048388	.WASHER,SPRING TENSI UOC:209	2
5 PAHZZ 11862 3995843	.GEAR,BEVEL UOC:209	2
6 PAHZZ 11862 14059083	.PARTS KIT,DRIVING A UOC:209	1
7 PAHZZ 11862 14063581	.SHIM (.010") UOC:209	1
7 PAHZZ 11862 14063582	.WASHER,FLAT (.015") UOC:209	1
7 PAHZZ 11862 14063583	.WASHER,FLAT (.020") UOC:209	1
7 PAHZZ 11862 14063584	.WASHER,FLAT (.025") UOC:209	1
7 PAHZZ 11862 14063585	.WASHER,FLAT (.030") UOC:209	1
7 PAHZZ 11862 14063586	.WASHER,FLAT (.035") UOC:209	1
7 PAHZZ 11862 14063587	.WASHER,FLAT (.040") UOC:209	1
7 PAHZZ 11862 14063588	.WASHER,FLAT (.045") UOC:209	1
8 PAHZZ 11862 462787	.CLIP,GUIDE UOC:209	4
9 PAHZZ 11862 15599620	.GEAR,HELICAL UOC:209	1
10 PAHZZ 11862 14048425	.BLOCK,SPACER,DIFFER (PURPLE, 1.322") UOC:209	1
10 PAHZZ 11862 14048426	.BLOCK,SPACER,DIFFER (WHITE, 1.326") UOC:209	1
10 PAHZZ 11862 14048427	.BLOCK,SPACER,DIFFFER (BROWN, 1.330") UOC:209	1
10 PAHZZ 11862 14048428	.BLOCK,SPACER DIFFER (YELLOW, 1.344") UOC:209	1
10 PAHZZ 11862 14048429	.BLOCK,SPACER DIFFER (ORANGE, 1.338") UOC:209	1

(1) (2) (3) (4) ITEM SMR PART NO CODE CAGEC NUMBER	(5) DESCRIPTION AND USABLE ON CODES (UOC)	(6) QTY
10 PAHZZ 11862 14048430	.BLOCK,SPACER,DIFFER (RED, 1.342") UOC:209	1
10 PAHZZ 11862 14048431	.BLOCK,SPACER,DIFFER (GREEN, 1.346") UOC:209	1
10 PAHZZ 11862 14048432	.BLOCK,SPACER,DIFFER (BLUE, 1.350") UOC:209	1
11 PAHZZ 11862 14048410	.PARTS KIT,GOVERNOR UOC:209	1
12 PAHZZ 11862 15599619	.GEAR,SPUR UOC:209	1
13 PAHZZ 11862 14048413	.GEAR,SPUR UOC:209	1
14 PAHZZ 11862 15599623	.CLIP,GUIDE UOC:209	4
15 PAHZZ 11862 26005078	.PARTS KIT,DRIVING A OUC:209	1
16 PAHZZ 24617 9414849	.RING,RETAINING UOC:209	1
17 PAHZZ 11862 14048417	.BEARING,WASHER,THRU (.027") UOC:209	1
17 PAHZZ 11862 14048416	.BEARING,WASHER,THRU (.022") UOC:209	1
17 PAHZZ 11862 14048418	.BEARING,WASHER,THRU (.032") UOC:209	1
17 PAHZZ 11862 14048419	.BEARING,WASHER,THRU (.040") UOC:209	1
17 XDHZZ 11862 14048420	.BEARING,WASHER,THRU (.040") UOC:209	1
17 PAHZZ 11862 14048421	.BEARING,WASHER,THRU (.044") UOC:209	1
17 PAHZZ 19954 EDS-98477-48	.BEARING,WASHER,THRU (.048") UOC:209	1
17 PAHZZ 11862 14048423	.BEARING,WASHER,THRU (.052") UOC:209	1

END OF FIGURE

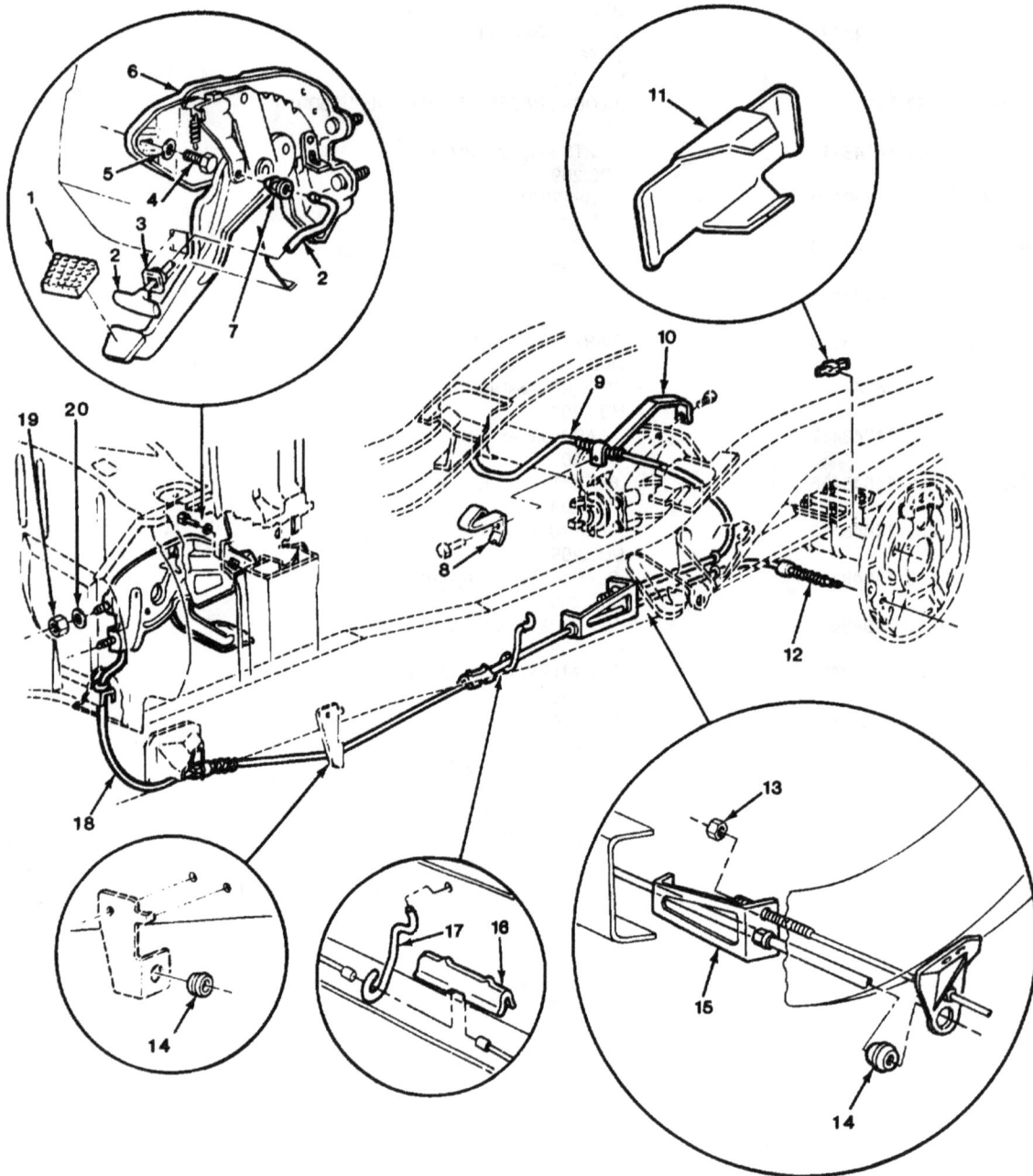

TA510921

FIGURE 114. PARKING BRAKE PEDAL AND RELEASE MECHANISM.

(1)	(2)	(3)	(4)	(5)	(6)
ITEM NO	SMR CODE	CAGEC	PART NUMBER	DESCRIPTION AND USABLE ON CODES (UOC)	QTY

GROUP 12 BRAKES

GROUP 1201 HANDBRAKES

FIG. 114 PARKING BRAKE PECAL AND RELEASE MECHANISM

1	PAOZZ	11862	3893181	PAD,PEDAL	1
2	PAOZZ	11862	334540	LEVER,MANUAL CONTROL	1
3	PAOZZ	11862	341990	GROMMET,NONMETALLIC	1
4	PAOZZ	96906	MS90728-33	BOLT,MACHINE	1
5	PAOZZ	96906	MS35340-45	WASHER,LOCK	1
6	PAOZZ	11862	14053591	LEVER,MANUAL CONTRO	1
7	PAOZZ	11862	334541	GROMMET,NONMETALLIC	1
8	PAOZZ	11862	14054122	CLIP,SPRING TENSION UOC:194,208,210,230,231,252,256	1
8	PAOZZ	11862	15557723	CLIP,SPRING TENSION UOC:254,256	1
9	PAOZZ	11862	14054174	CONTROL ASSEMBLY,PU UOC:209	1
9	PAOZZ	11862	14064664	CONTROL ASSEMBLY,PU UOC:194,208,210,230,231,252,254,256	1
10	PAOZZ	11862	14072697	CLIP,SPRING TENSION UOC:209	1
10	PAOZZ	11862	14054120	CLIP,SPRING TENSION UOC:254,256	1
11	PAOZZ	19207	7001423	PLUG,PROTECTIVE,DUS UOC:194,208,210,230,231,252,254,256	6
12	PAOZZ	11862	14054173	CONTROL ASSEMBLY,PU UOC:209	1
12	PAOZZ	11862	14064663	CONTROL ASSEMBLY,PU UOC:194,208,210,230,231,252,254,256	1
13	PAOZZ	96906	MS17829-5C	NUT,SELF-LOCKING,HE	1
14	PAOZZ	11862	15530620	GROMMET,NONMETALLIC	2
15	PAOZZ	11862	14072692	EQUALIZER,PARKING B	1
16	PAOZZ	11862	25516531	CLIP,RETAINING	1
17	PAOZZ	11862	368786	GUIDE,CABLE,PARKING UOC:209	1
18	PAOZZ	11862	14055591	CABLE ASSEMBLY,PARK UOC:209	1
18	PAOZZ	11862	14053593	CONTROL ASSEMBLY,PU UOC:194,208,210,230,231,252,254,256	1
19	PAOZZ	96906	MS51967-6	NUT,PLAIN,HEXAGON	2
20	PAOZZ	96906	MS27183-11	WASHER,FLAT	2

END OF FIGURE

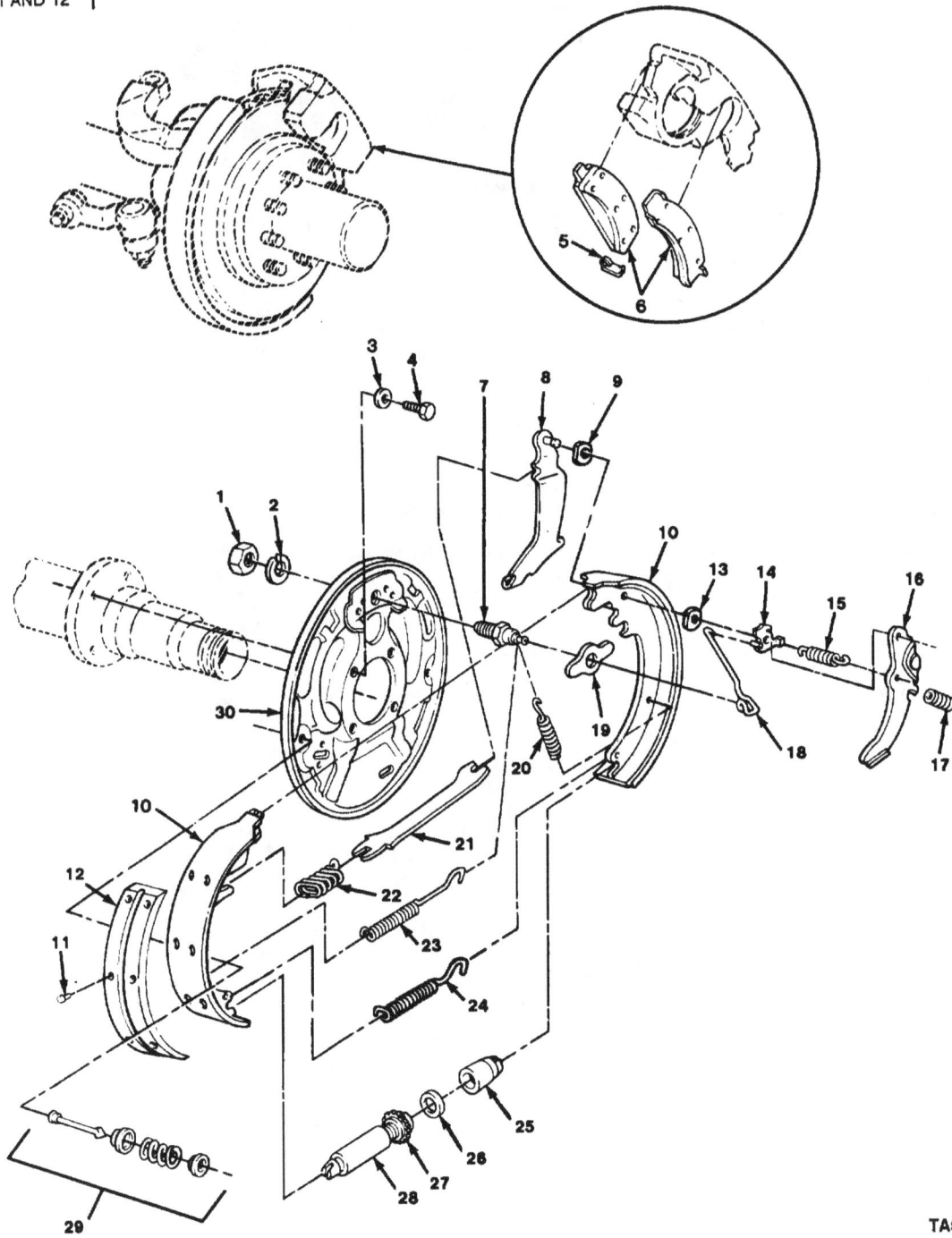

FIGURE 115. REAR BRAKESHOE ASSEMBLY COMPONENTS, FRONT BRAKE PADS,
AND RELATED PARTS (ALL EXCEPT M1009).

TA510922

(1) ITEM NO	(2) SMR CODE	(3) CAGEC	(4) PART NUMBER	(5) DESCRIPTION AND USABLE ON CODES (UOC)	(6) QTY
				GROUP 1202 SERVICE BRAKES	
				FIG.115 REAR BRAKESHOE ASSEMBLY COMPONENTS, FRONT BRAKE PADS, AND RELATED PARTS (ALL EXCEPT M1009)	
1	PAOZZ	96906	MS51968-24	NUT,PLAIN,HEXAGON UOC:194,208,210,230,252	2
2	PAOZZ	96906	MS35340-51	WASHER,LOCK UOC:194,208,210,230,252	2
3	PAOZZ	96906	MS35340-48	WASHER,LOCK UOC:194,208,210,230,231,252	8
4	PAOZZ	96906	MS90727-109	SCREW,CAP,HEXAGON H UOC:194,208,210,230,252	8
5	PAOZZ	14892	4150515	CLIP,BRAKE SHOE UOC:194,208,210,230,231,252,254,256	2
6	PAOZZ	14892	3368689	DISK BRAKE SHOE SET UOC:194,208,210,230,252,254,256	1
7	PAOZZ	11862	3856834	PIN,BRAKE SHOE ANCH UOC:194,208,210,230,231,252,254,256	2
8	PAOZZ	11862	334307	LEVER,MANUAL CONTRO UOC:194,208,210,230,231,252,254,256	1
8	PAOZZ	11862	334308	LEVER,MANUAL CONTRO UOC:194,208,210,230,231,252,254,256	1
9	PAOZZ	11862	5454797	WASHER,SPRING TENSI UOC:194,208,210,230,231,252,254,256	2
10	PAOFF	11862	372379	BRAKE SHOE SET,INTE UOC:194,208,210,230,231,252,254,256	1
11	PAFZZ	11862	344029	.RIVET,TUBULAR UOC:194,208,210,230,231,252,254,256	52
12	PAFZZ	23862	6260885	.BRAKE LINING KIT UOC:194,208,210,230,231,252,254,256	1
13	PAOZZ	96906	MS16633-1031	RING,RETAINING UOC:194,208,210,230,231,252	2
14	PAOZZ	11862	3856855	PIVOT,BRAKE SHOE AD UOC:194,208,210,230,231,252,254,256	1
14	PAOZZ	11862	3856856	PIVOT,BRAKE SHOE UOC:194,208,210,230,252,254,256	1
15	PAOZZ	11862	5461145	SPRING,HELICAL,EXTE UOC:194,208,210,230,231,252,254,256	2
16	PAOZZ	11862	357845	LEVER ASSEMBLY,BRAK UOC:194,208,210,230,231,252,254,256	1
16	PAOZZ	11862	357846	LEVER,LOCK-RELEASE UOC:194,208,210,230,231,252,254,256	1
17	PAOZZ	11862	5461984	SPRING,BRAKE SHOE UOC:194,208,210,230,231,252,254,256	2
18	PAOZZ	11862	3856857	CONNECTING LINK,RIG UOC:194,208,210,230,231,252,254,256	1
18	PAOZZ	11862	3856858	LIN,ANCHOR,BRAKE S UOC:194,208,210,230,231,252,254,256	1
19	PAOZZ	11862	18002428	NUT SPACER,PLATE UOC:194,208,210,230,231,252,254,256	2

(1) ITEM NO	(2) SMR CODE	(3) CAGEC	(4) PART NUMBER	(5) DESCRIPTION AND USABLE ON CODES (UOC)	(6) QTY
20	PAOZZ	11862	3856843	SPRING,HELICAL,EXTE UOC:194,208,210,230,231,252,254,256	2
21	PAOZZ	11862	3898059	STRU,PARKING BRAKE UOC:194,208,210,230,231,252,254,256	2
22	PAOZZ	11862	1312281	SPRING UOC:194,208,210,230,231,252,254,256	2
23	PAOZZ	11862	3767138	SPRING,HELICAL,EXTE UOC:194,208,210,230,231,252,254,256	2
24	PAOZZ	11862	3856850	SPRING,HELICAL,EXTE UOC:194,208,210,230,231,252,254,256	2
25	PAOZZ	11862	15522081	SOCKET,BRAKE SHOE UOC:194,208,210,230,231,252,254,256	2
26	PAOZZ	11862	3856849	SHIM UOC:194,208,210,230,231,252,254,256	2
27	PAOZZ	11862	15522077	ADJUSTER,SLACK,BRAK UOC:194,208,210,230,231,252,254,256	1
27	PAOZZ	11862	15522078	ADJUSTING SCREW ASS UOC:194,208,210,230,231,252,254,256	1
28	PAOZZ	11862	15522079	NUT,BRAKE SHOE ADJU LEFT UOC:194,208,210,230,231,252,254,256	1
28	PAOZZ	11862	15522080	ADJUSTING SCREW ASS RIGHT UOC:194,208,210,230,231,252,254,256	1
29	PAOZZ	11862	372249	PARTS KIT,BRAKE SHO UOC:194,208,210,230,231,252,254,256	2
30	PAOZZ	11862	14068905	PLATE,BACKING,BRAKE LEFT UOC:194,208,210,230,231,252,254,256	1
30	PAOZZ	11862	14068906	PLATE,BACKING,BRAKE FIGHT UOC:194,208,210,230,231,252,254,256	1

END OF FIGURE

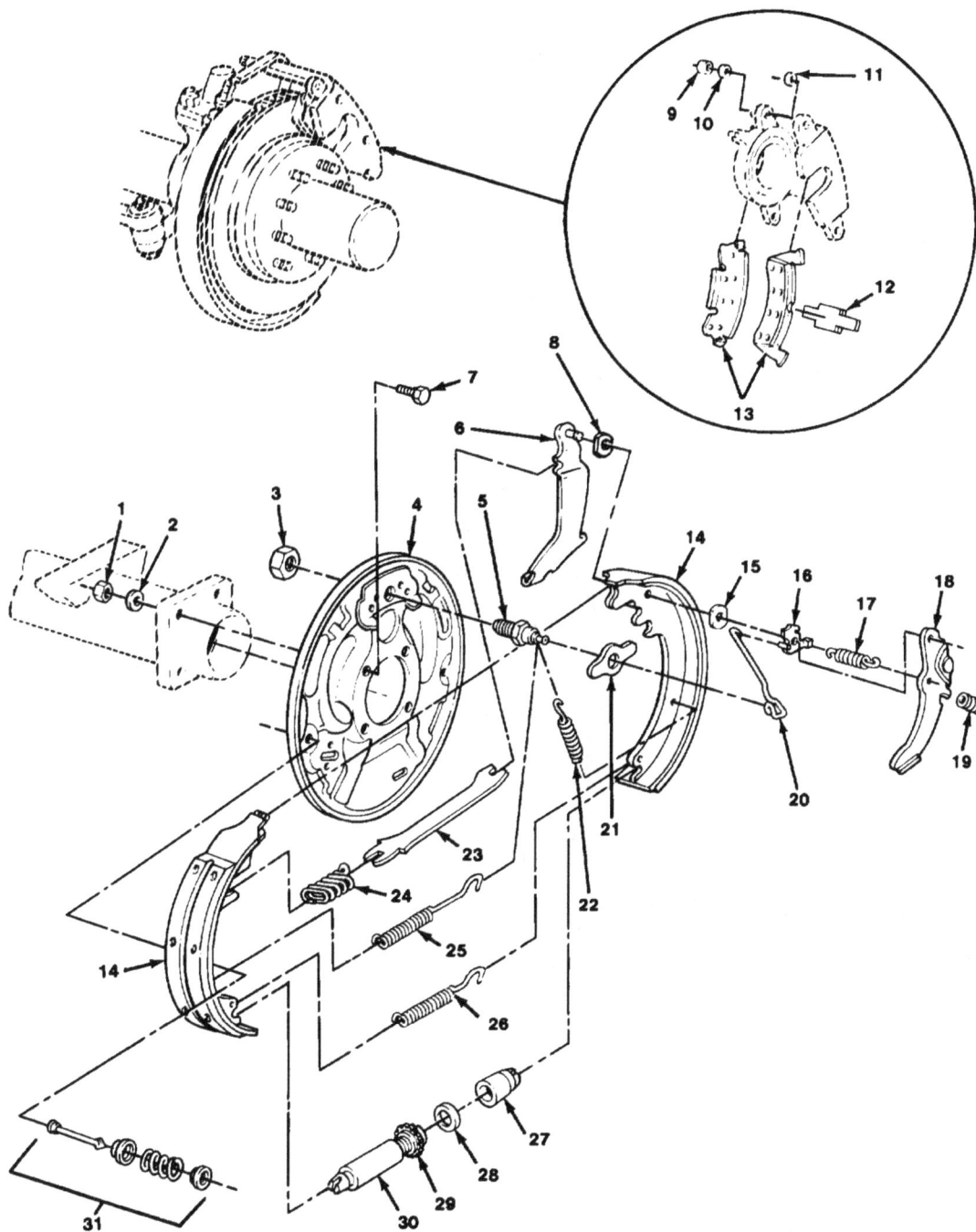

FIGURE 116, REAR BRAKESHOE ASSEMBLY COMPONENTS, FRONT BRAKE
PADS, AND RELATED PARTS (M1009).

TA510923

(1) ITEM NO	(2) SMR CODE	(3) CAGEC	(4) PART NUMBER	(5) DESCRIPTION AND USABLE ON CODES (UOC)	(6) QTY

GROUP 1202 SERVICE BRAKES

FIG 116 REAR BRASKESHOE ASSEMBLY COMPONENTS, FRONT BRAKE PADS, AND RELATED PARTS (M1009)

(1)	(2)	(3)	(4)	(5)	(6)
1	PAFZZ	96906	MS51968-9	NUT,PLAIN,HEXAGON UOC:209	8
2	PAFZZ	96906	MS35338-46	WASHER,LOCK UOC:209	8
3	PAOZZ	11862	3760300	NUT,PLAIN,ROUND UOC:209	2
4	PAFZZ	11862	14055279	PLATE ASSEMBLY,BRAK LEFT UOC:209	1
4	PAFZZ	11862	14055280	PLATE ASSEMBLY,BRAK FIGHT UOC:209	1
5	PAOZZ	11862	14009982	LINK,ANCHOR,BRAKE S UOC:209	2
6	PAOZZ	11862	15594177	PARTS KIT,PARKING B LEFT UOC:209	1
6	PAOZZ	11862	15594178	PARTS KIT,PARKING B RIGHT UOC:209	1
7	PAFZZ	96906	MS90727-60	SCREW,CAP,HEXAGON H UOC:209	8
8	PAOZZ	11862	5454797	WASHER,SPRING TENSI UOC:209	2
9	PAOZZ	11862	18004057	SPACER,SLEEVE UOC:209	4
10	PAOZZ	11862	18001032	BUSHING,NONMETALLIC UOC:209	4
11	PAOZZ	11862	5470497	PACKING,PREFORMED UOC:209	4
12	PAOZZ	11862	5469497	CLIP,SPRING TENSION UOC:209	2
13	PAOZZ	11862	12321435	DISK BRAKE SHOE SET UOC:209	1
14	PAOZZ	11862	1155445	BRAKE SHOE SET UOC:209	1
15	PAOZZ	96906	MS16633-1031	RING,RETAINING UOC:209	2
16	PAOZZ	11862	3856855	PIVOT,BRAKE SHOE AD LEFT UOC:209	1
16	PAOZZ	11862	3856856	PIVOT,BRAKE SHOE RIGHT UOC:209	1
17	PAOZZ	11862	5461145	SPRING,HELICAL,EXTE UOC:209	2
18	PAOZZ	11862	357890	LEVER,MANUAL CONTRO UOC:209	1
18	PAOZZ	11862	357889	LEVER,MANUAL CONTRO UOC:209	1
19	PAOZZ	11862	5461984	SPRING,BRAKE SHOE UOC:209	2

(1) (2) (3) (4) ITEM SMR PART NO CODE CAGEC NUMBER	(5) DESCRIPTION AND USABLE ON CODES (UOC)	(6) QTY
20 PAOZZ 11862 5461156	LIN,ANCHOR,BRAKE S RIGHT UOC:209	1
20 PAOZZ 11862 14055315	LINK,ADJUSTING BRAK LEFT UOC:209	1
21 PAOZZ 11862 18002428	NUT SPACER,PLATE UOC:209	2
22 PAOZZ 11862 3820163 SPRING,HELICAL,EXTE	 UOC:209	2
23 PAOZZ 11862 468661	STRUT,PARKING BRAKE UOC:209	2
24 PAOZZ 11862 1312281 SPRING	 UOC:209	2
25 PAOZZ 11862 3694822 SPRING,HELICAL,EXTE	 UOC:209	2
26 PAOZZ 11862 3887347 SPRING,HELICAL,EXTE	 UOC:209	2
27 PAOZZ 11862 468675 SCREW	 UOC:209	2
28 PAOZZ 11862 5462496 SHIM	 UOC:209	2
29 PAOZZ 11862 345943	ADJUSTING SCREW ASS LEFT UOC:209	1
29 PAOZZ 11862 345944	ADJUSTING SCREW ASS FIGHT UOC:209	1
30 PAOZZ 11862 468673	NUT,ADJUSTER BRAKE LET UOC:209	1
30 PAOZZ 11862 468674	NUT,ADJUSTER BRAKE FIGHT UOC:209	1
31 PAOZZ 11862 372249	PARTS KIT,BRAKE SHO UOC:209	2

END OF FIGURE

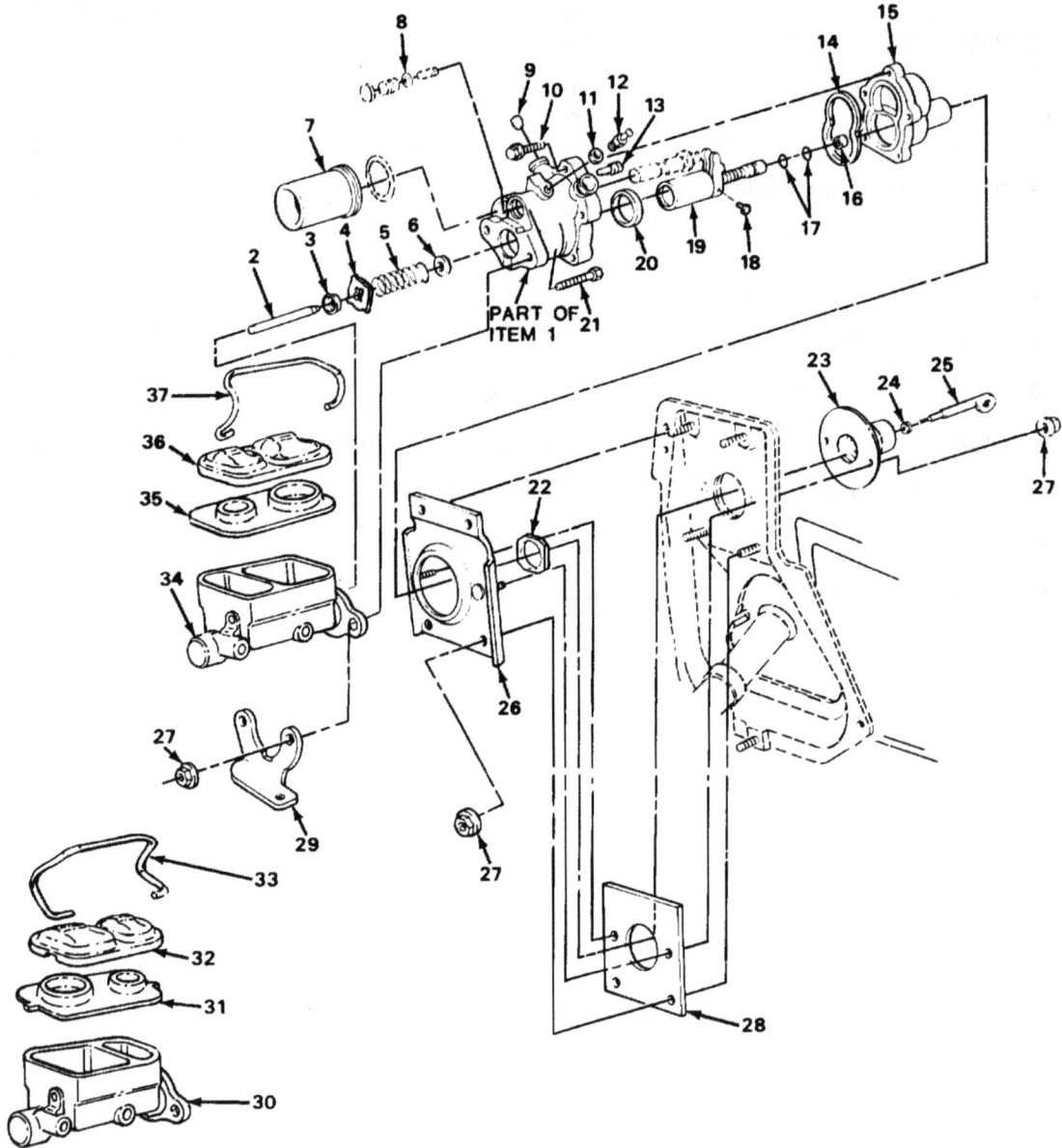

FIGURE 117. POWER BOOSTER AND MASTER CYLINDER ASSEMBLIES.

TA510924

(1) ITEM NO	(2) SMR CODE	(3) CAGEC	(4) PART NUMBER	(5) DESCRIPTION AND USABLE ON CODES (UOC)	(6) QTY
				GROUP 1204 HYDRAULIC BRAKE SYSTEM	
				FIG. 117 POWER BOOSTER AND MASTER CYLINDER ASSEMBLIES	
1	PAOZZ	14892	2770317	BRAKE BOOSTER ASSEM UOC:209	1
1	PAOZZ	14892	2770209	BOOSTER,HYDRAULLIC B UOC:194,208,210,230,231,252,254,256	1
2	XDFZZ	14892	129959	.PIN,STRAIGHT,HEADLE	1
3	XDFZZ	14892	129596	.BAFFLE,ROD SPRING	1
4	PAFZZ	14892	2771165	.RETAINER,HELICAL CO	1
5	XDFZZ	14892	129839	.SPRING,HELICAL,COMP	1
6	XDFZZ	14892	129497	.RETAINER	1
7	XDFZZ	14892	2770715	.PARTS KIT,BRAKE BOO	1
8	KFFZZ	14892	951965	.SEAL RING SPOOL PG PART OF KIT P/N 2770723 UOC:194,208,210,230,231,252,254,256	1
9	KFFZZ	14892	125666	.SEAT INSERT PART PART OF KIT P/N 2770723 UOC:194,208,210,230,231,252,254,256	2
10	XDFZZ	14892	2770746	.BOLT,EXTERNALLY REL	5
11	XDOZZ	14892	2770614	.PACKING,PREFORMED UOC:209	2
12	XAOZZ	14892	129494	.FITTING,BRAKE BOOST	1
13	XDFZZ	14892	2770720	.PARTS KIT,BRAKE BOO	1
14	KFFZZ	14892	2770532	.SEAL HOUSING PART OF KIT P/N 2770723	1
15	XDFZZ	14892	2770972	.COVER,BRAKE BOOSTER UOC:209	1
15	XDFZZ	14892	2770699	.COVER,ACCESS UOC:194,208,210,230,231,252,254,256	1
16	KFFZZ	14892	2770685	.O RING PART OF KIT P/N 2771098 UOC:209	1
16	KFFZZ	14892	2770685	.O RING PART OF KIT P/N 2770723	1
17	KFFZZ	14892	2770698	.SEAL PART OF KIT P/N 2770723 UOC:194,208,210,230,231,252,254,256	2
18	XDFZZ	14892	129472	.PIN,STRAIGHT,HEADED	1
19	XDFZZ	14892	2771247	.ROD,PISTON,LINEAR A UOC:194,208,210,230,231,252,254,256	1
20	KFFZZ	14892	129484	.PISTON SEAL PART OF KIT P/N 2770723	1
21	XDHZZ	14892	2771125	.BOLT,SHOULDER	2
22	XDFZZ	14892	129894	.NUT,PLAIN,RECTANGUL	1
23	PAFZZ	14892	2770771	.BOOT,DUST AND MOIST PART OF KIT P/N 2771098 UOC:209	1
24	PAFZZ	14892	129495	.BUSHING,NONMETALLIC PART OF KIT P/N 2770723 UOC:194,208,210,230,252,254,256	1
25	PAOZZ	11862	345683	.ROD,PEDAL CONTROL UOC:194,208,210,230,231,252,254,256	1
26	XDFZZ	14892	2770829	.BRACKET	1
27	PAOZZ	11862	11502812	NUT,SELF-LOCKING,EX	8

TM9-2320-289-34P

(1)	(2)	(3)	(4)	(5)	(6)
ITEM NO	SMR CODE	CAGEC	PART NUMBER	DESCRIPTION AND USABLE ON CODES (UOC)	QTY
28	PAOZZ	11862	14004810	GASKET	1
29	PAOZZ	11862	14045698	BRACKET,ANGLE	1
30	PAOOO	14892	2232073	CYLINDER ASSEMBLY,H UOC:194,208,210,230,231,252,254,256	1
31	PAOZZ	14892	2229044	.GASKET UOC:194,208,210,230,231,252,254,256	1
32	PBOZZ	14892	2232076	.COVER,MASTER CYLIND UOC:194,208,210,230,252,254,256	1
33	PBOZZ	14892	2229046	.HANDLE,BAIL UOC:194,208,210,230,231,252,254,256	2
34	PAOOO	14892	2232072	CYLINDER,HYDRAULIC UOC:209	1
35	PAOZZ	14892	2227168	.GASKET UOC:209	1
36	PBOZZ	14892	2232077	.COVER,RESERVOIR UOC:209	1
37	PBOZZ	14892	2229448	.CLIP,SPRING TENSION UOC:209	1

END OF FIGURE

(1) ITEM NO	(2) SMR CODE	(3) CAGEC	(4) PART NUMBER	(5) DESCRIPTION AND USABLE ON CODES (UOC)	(6) QTY
				GROUP 1204 HYDRAULIC BRAKE SYSTEM	
				FIG. 117 POWER BOOSTER AND MASTER CYLINDER ASSEMBLIES	
1	PAOZZ	14892	2770317	BRAKE BOOSTER ASSEM UOC:209	1
1	PAOZZ	14892	2770209	BOOSTER,HYDRAULLIC B UOC:194,208,210,230,231,252,254,256	1
2	XDFZZ	14892	129959	.PIN,STRAIGHT,HEADLE	1
3	XDFZZ	14892	129596	.BAFFLE,ROD SPRING	1
4	PAFZZ	14892	2771165	.RETAINER,HELICAL CO	1
5	XDFZZ	14892	129839	.SPRING,HELICAL,COMP	1
6	XDFZZ	14892	129497	.RETAINER	1
7	XDFZZ	14892	2770715	.PARTS KIT,BRAKE BOO	1
8	KFFZZ	14892	951965	.SEAL RING SPOOL PG PART OF KIT P/N 2770723 UOC:194,208,210,230,231,252,254,256	1
9	KFFZZ	14892	125666	.SEAT INSERT PART PART OF KIT P/N 2770723 UOC:194,208,210,230,231,252,254,256	2
10	XDFZZ	14892	2770746	.BOLT,EXTERNALLY REL	5
11	XDOZZ	14892	2770614	.PACKING,PREFORMED UOC:209	2
12	XAOZZ	14892	129494	.FITTING,BRAKE BOOST	1
13	XDFZZ	14892	2770720	.PARTS KIT,BRAKE BOO	1
14	KFFZZ	14892	2770532	.SEAL HOUSING PART OF KIT P/N 2770723	1
15	XDFZZ	14892	2770972	.COVER,BRAKE BOOSTER UOC:209	1
15	XDFZZ	14892	2770699	.COVER,ACCESS UOC:194,208,210,230,231,252,254,256	1
16	KFFZZ	14892	2770685	.O RING PART OF KIT P/N 2771098 UOC:209	1
16	KFFZZ	14892	2770685	.O RING PART OF KIT P/N 2770723	1
17	KFFZZ	14892	2770698	.SEAL PART OF KIT P/N 2770723 UOC:194,208,210,230,231,252,254,256	2
18	XDFZZ	14892	129472	.PIN,STRAIGHT,HEADED	1
19	XDFZZ	14892	2771247	.ROD,PISTON,LINEAR A UOC:194,208,210,230,231,252,254,256	1
20	KFFZZ	14892	129484	.PISTON SEAL PART OF KIT P/N 2770723	1
21	XDHZZ	14892	2771125	.BOLT,SHOULDER	2
22	XDFZZ	14892	129894	.NUT,PLAIN,RECTANGUL	1
23	PAFZZ	14892	2770771	.BOOT,DUST AND MOIST PART OF KIT P/N 2771098 UOC:209	1
24	PAFZZ	14892	129495	.BUSHING,NONMETALLIC PART OF KIT P/N 2770723 UOC:194,208,210,230,252,254,256	1
25	PAOZZ	11862	345683	.ROD,PEDAL CONTROL UOC:194,208,210,230,231,252,254,256	1
26	XDFZZ	14892	2770829	.BRACKET	1
27	PAOZZ	11862	11502812	NUT,SELF-LOCKING,EX	8

(1)	(2)	(3)	(4)	(5)	(6)
ITEM	SMR		PART		
NO	CODE	CAGEC	NUMBER	DESCRIPTION AND USABLE ON CODES (UOC)	QTY

28	PAOZZ	11862	14004810	GASKET	1
29	PAOZZ	11862	14045698	BRACKET,ANGLE	1
30	PAOOO	14892	2232073	CYLINDER ASSEMBLY,H	1
				UOC:194,208,210,230,231,252,254,256	
31	PAOZZ	14892	2229044	.GASKET	1
				UOC:194,208,210,230,231,252,254,256	
32	PBOZZ	14892	2232076	.COVER,MASTER CYLIND	1
				UOC:194,208,210,230,252,254,256	
33	PBOZZ	14892	2229046	.HANDLE,BAIL	2
				UOC:194,208,210,230,231,252,254,256	
34	PAOOO	14892	2232072	CYLINDER,HYDRAULIC	1
				UOC:209	
35	PAOZZ	14892	2227168	.GASKET	1
				UOC:209	
36	PBOZZ	14892	2232077	.COVER,RESERVOIR	1
				UOC:209	
37	PBOZZ	14892	2229448	.CLIP,SPRING TENSION	1
				UOC:209	

END OF FIGURE

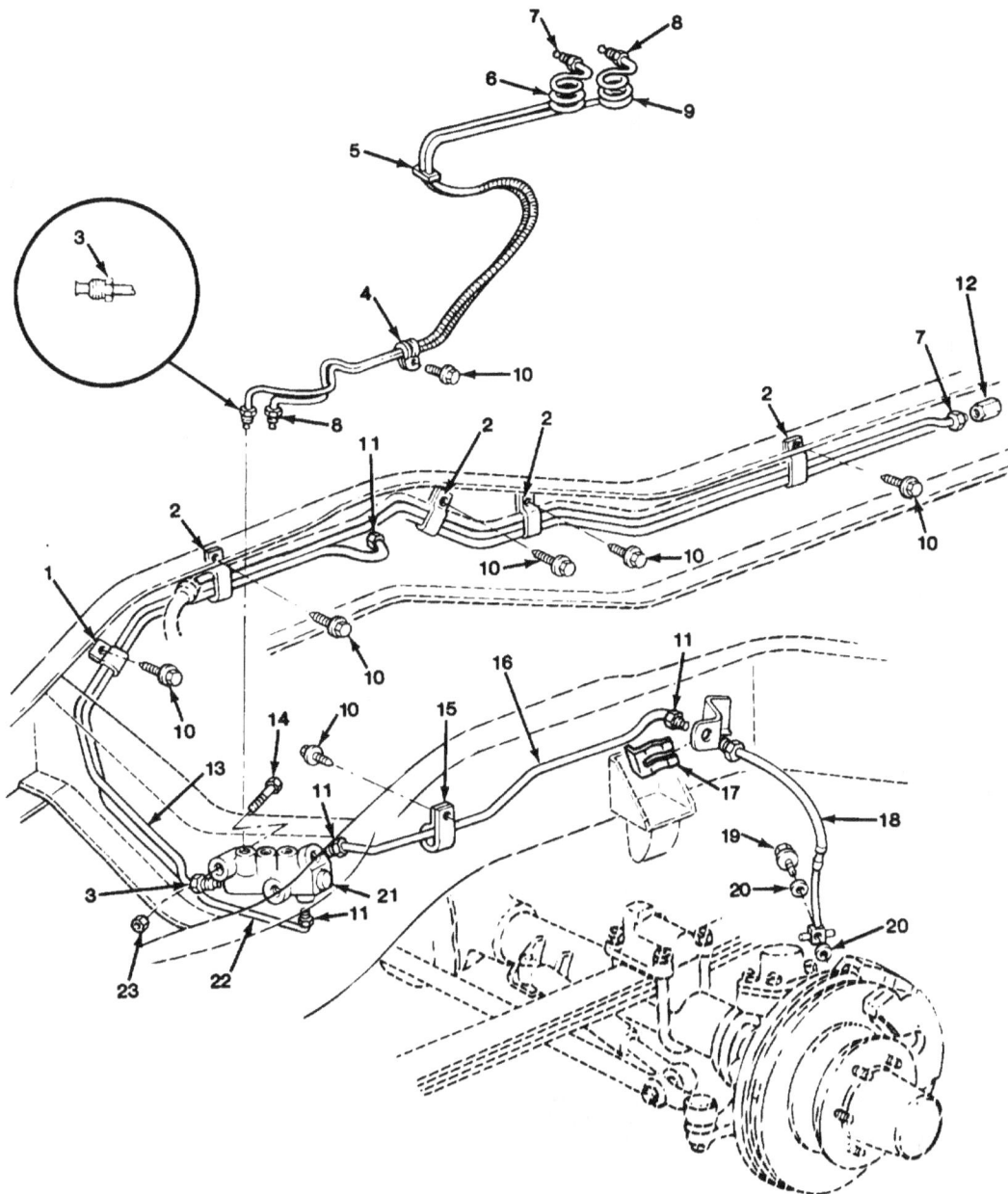

TA510925

FIGURE 118. FRONT BRAKE LINES AND RELATED PARTS.

(1) (2) (3) (4)				(5)	(6)
ITEM SMR			PART		
NO CODE CAGEC		NUMBER		DESCRIPTION AND USABLE ON CODES (UOC)	QTY

GROUP 1204 HYDRAULIC BRAKE SYSTEM

FIG. 118 FRONT BRAKE LINES AND
RELATED PARTS

ITEM NO	SMR CODE	CAGEC	PART NUMBER	DESCRIPTION	QTY
1	PAOZZ	11862	343438	CLAMP,LOOP	1
2	PAOZZ	11862	15599973 CLAMP,LOOP		4
3	PAOZZ	11862	9424955	INVERTED NUT,TUBE C	2
4	PAOZZ	11862	25527423 CLAMP,LOOP		1
5	PAOZZ	11862	3816659	STRAP,LINE SUPPORTI	1
6	MFOZZ	11862	15599260	PIPE ASM RR BRK CB (78.49" LG) MAKE FROM TUBE, P/N 603827 UOC:209	1
6	MFOZZ	11862	15599262	PIPE ASM RR BRK (75.25"LG) MAKE FROM TUBE,P/N 603827 UOC:194,208,210,230,231,252,254,256	1
7	PAOZZ	24617	137397	INVERTED NUT,TUBE C	2
8	PAOZZ	11862	15607227 NUT		2
9	MFOZZ	11862	15599259	PIPE ASM FRT BRK (72.68" LG) MAKE FROM PIPE, P/N 3696822 UOC:209	1
9	MFOZZ	11862	15599261	PIPE ASM FRT BRK CB (76.63" LG) MAKE FROM PIPE, P/N 3696822 UOC:194,208,210,230,231,252,254,256	1
10	PAOZZ	11862	11504447 SCREW,TAPPING,THREA		7
11	PAOZZ	24617	9432075	INVERTED NUT,TUBE C	4
12	PAOZZ	96906	MS51877-4 COUPLING,TUBE		1
13	MFOZZ	11862	14034586	PIPE ASM RR BRK (91.51" LG) MAKE FROM TUBE, P/N 603827 UOC:209	1
13	MFOZZ	11862	15522444	PIPE ASSY (89.68"LG) MAKE FROM TUBE, P/N 603827 UOC:194,208,210,230,231,252,254,256	1
14	PAOZZ	96906	MS90728-40 BOLT,MACHINE		2
15	PAOZZ	11862	342677	CLAMP,LOOP	1
16	MFOZZ	11862	14034571	PIPE ASM FRT BRK (29.43" LG) MAKE FROM PIPE, P/N 3696822 UOC:209	1
16	MFOZZ	11862	14054257	PIPE ASM FRT BRK LH (30.44"LG) MAKE FROM PIPE, P/N 3696822 UOC:194,208,210,230,231,252,254,256	1
17	PAOZZ	11862	22527167	SPRING,SPECIAL YOKE	2
18	PAOZZ	11862	14036723	HOSE ASSEMBLY,NONME LEFT UOC:209	1
18	PAOZZ	11862	14036736	HOSE ASSEMBLY,NONME RIGHT UOC:209	1
18	PAOZZ	11862	14054269	HOSE ASSEMBLY,NONME LEFT UOC:194,208,210,230,231,252,254,256	1
18	PAOZZ	11862	14054270	HOSE ASSEMBLY,NONME RIGHT UOC:194,208,210,230,252,254,256	1
19	PAOZZ	11862	14094948	BOLT,FLUID PASSAGE	2
20	PAOZZ	11862	14000172 WASHER,FLAT		4

(1) (2) (3) (4)				(5)	(6)
ITEM NO	SMR CODE	CAGEC	PART NUMBER	DESCRIPTION AND USABLE ON CODES (UOC)	QTY
21	PAOZZ	11862	25515635	VALVE,SAFETY RELIEF UOC:209	1
21	PAOZZ	11862	1257203	VALVE,BRAKE UOC:194,208,210,230,231,252,254,256	1
21	PAOZZ	11862	1257087	VALVE ASSEMBLY UOC:254,256	1
22	MFOZZ	11862	14034572	PIPE ASM FRT BRK (42.13"LG) MAKE FROM PIPE, P/N 3696822	1
23	PAOZZ	24617	9422295	NUT,SELF-LOCKING,CO	2

END OF FIGURE

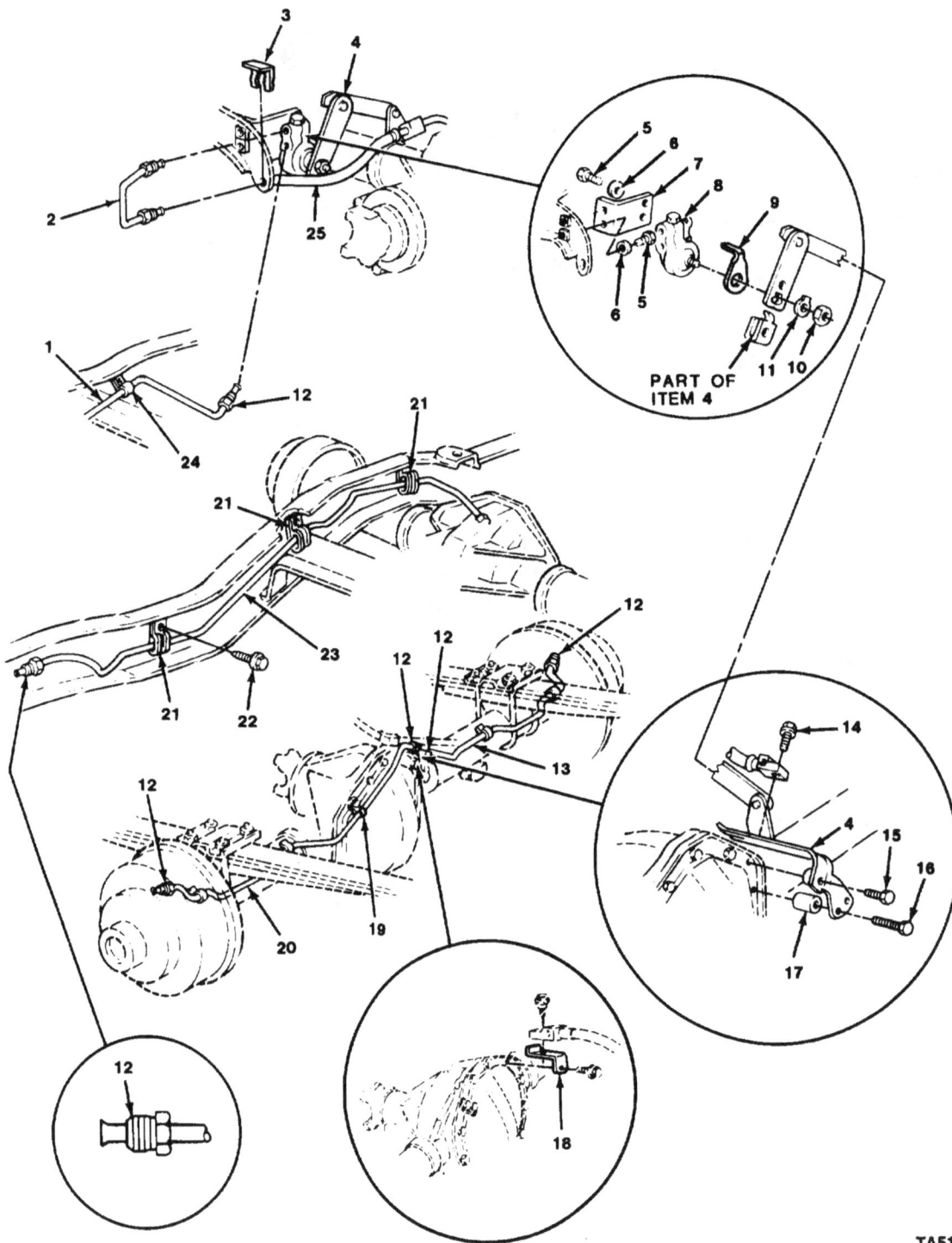

TA510926

FIGURE 119. REAR BRAKE LINES, WEIGHT PROPORTIONAL VALVE, AND RELATED PARTS.

(1) ITEM NO	(2) SMR CODE	(3) CAGEC	(4) PART NUMBER	(5) DESCRIPTION AND USABLE ON CODES (UOC)	(6) QTY
				GROUP 1204 HYDRAULIC BRAKE SYSTEM	
				FIG. 119 REAR BRAKE LINES, WEIGHT PROPORTIONAL VALVE, AND RELATED PARTS	
1	MFOZZ	11862	14036797	PIPE ASM RR BRK RR (78.68"LG) MAKE FROM TUBE, P/N 603827 UOC:194,208,210,230,231,252,254,256	1
2	PAOZZ	11862	14036792	TUBE ASSEMBLY,METAL UOC:194,208,230,231,252,254,256	1
3	PAOZZ	11862	22527167	SPRING,SPECIAL YOKE	1
4	PAOZZ	11862	15549248	LEVER ASSEMBLY UOC:194,208,210,230,231,252,254,256	1
5	PAOZZ	96906	MS90728-31	BOLT,MACHINE UOC:194,208,210,230,231,252,254,256	4
6	PAOZZ	96906	MS35340-45	WASHER,LOCK UOC:194,208,210,230,231,252,254,256	4
7	PAOZZ	11862	14036773	BRACKET,ANGLE UOC:194,208,210,230,231,252,254,256	1
8	PAOZZ	11862	15538215	VALVE,BRAKE,PROPORT UOC:194,208,210,230,231,252,254,256	1
9	PAOZZ	11862	14061396	GAGE,SETTING UOC:194,208,210,230,231,252,254,256	1
10	PAOZZ	96906	MS51967-6	NUT,PLAIN,HEXAGON UOC:194,208,210,230,252,254,256	1
11	PAOZZ	11862	14036775	BUSHING,PIPE UOC:194,208,210,230,231,252,254,256	1
12	PAOZZ	24617	9432075	INVERTED NUT,TUBE C	6
13	MFOZZ	11862	14036706	PIPE ASM RR BRK (31.29" LG) MAKE FROM PIPE, P/N 369822 UOC:209	1
13	MFOZZ	11862	14036712	PIPE ASM RR BRK (35.35"LG) MAKE FROM PIPE, P/N 369822 UOC:194,208,210,230,252,254,256	1
14	PAOZZ	11862	1239146	BOLT,SHOULDER	1
15	PAOZZ	96906	MS90728-63	SCREW,CAP,HEXAGON H UOC:194,208,210,230,231,252,254,256	1
16	PAOZZ	11862	9439637	SCREW,CAP,HEXAGON H UOC:194,208,210,230,231,252,254	1
17	PAOZZ	11862	14055556	SPACER,SLEEVE UOC:194,208,210,230,231,252,254,256	1
18	PAOZZ	11862	359816	BRACKET,DOUBLE ANGL UOC:209	1
19	PAOZZ	11862	331416	CLIP,SPRING TENSION	1
19	PAOZZ	11862	6259071	CLIP,SPRING TENSION UOC:254,256	1
20	MFOZZ	11862	14036705	PIPE ASM RR BRK C/O (38.59" LG) MAKE FROM PIPE, P/N 3696822 UOC:209	1
20	MFOZZ	11862	14036711	PIPE ASM RR BRK (41.54"LG) MAKE FROM PIPE, P/N 3696822	1

(1) ITEM NO	(2) SMR CODE	(3) CAGEC	(4) PART NUMBER	(5) DESCRIPTION AND USABLE ON CODES (UOC)	(6) QTY
				UOC:194,208,210,230,231,252,254,256	
21	PAOZZ	96906	MS21333-45	CLAMP,LOOP	3
				UOC:194,208,210,230,231,252,254,256	
21	PAOZZ	96906	MS21333-98	CLAMP,LOOP	3
				UOC:209	
22	PAOZZ	11862	11504447	SCREW,TAPPING,THREA	3
23	MFOZZ	11862	14034599	PIPE ASM RR BRK (73.51" LG) MAKE FROM TUBE, P/N 603827	1
				UOC:209	
24	PAOZZ	11862	15599973	CLAMP,LOOP	1
				UOC:194,208,210,230,231,252,254,256	
25	PAOZZ	11862	17981073	HOSE ASSEMBLY,NONME	1

END OF FIGURE

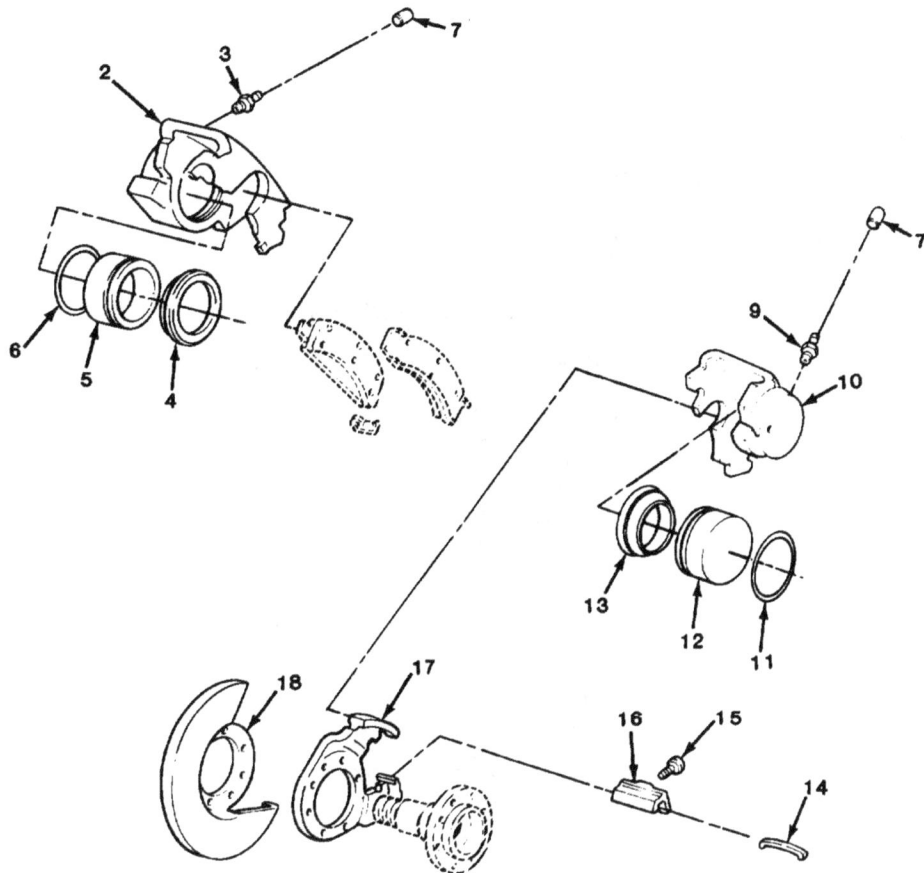

TA510927

FIGURE 120. FRONT BRAKE CALIPER ASSEMBLIES (ALL EXCEPT M1009).

(1) (2) (3) (4) ITEM SMR PART NO CODE CAGEC NUMBER	(5) DESCRIPTION AND USABLE ON CODES (UOC)	(6) QTY
	GROUP 1204 HYDRAULIC BRAKE SYSTEM	
	FIG. 120 FRONT BRAKE CALIPER ASSEMBLIES (ALL EXCEPT M1009)	
1 AFOFF 14892 2238739	CALIPER ASSEMB;LY,DI LEFT UOC:194,208,210,230,231,252,254,256	1
2 PAOFF 11862 14002543	.CALIPER ASSEMBLY,DI UOC:194,208,210,230,231,252,254,256	1
3 PAFZZ 14892 2238710	.VALVE,BLEEDER,HYDRA UOC:194,208,210,230,231,252,254,256	1
4 PAFZZ 11862 338269	.BOOT,DUST AND MOIST UOC:194,208,210,230,231,252,254,256	1
5 PAFZZ 14892 2230359	.PISTON,BRAKE CALIPE UOC:194,208,210,230,252,254,256	1
6 PAFZZ 11862 6259033	.SEAL UOC:194,208,210,230,231,252,254,256	1
7 PAOZZ 11862 5469581	CAP,PROTECTIVE,DUST UOC:194,208,210,230,231,252,254,256	2
8 AFOFF 14892 2238740	CALIPER ASSEMBLY,DI RIGHT UOC:194,208,210,230,231,252,254,256	1
9 PAFZZ 14892 2238710	.VALVE,BLEEDER,HYDRA UOC:194,208,210,230,231,252,254,256	1
10 PAOFF 14892 2238742	.CALIPER ASSEMBLY,DI UOC:194,208,210,230,231,252,254,256	1
11 PAFZZ 11862 6259033	.SEAL UOC:194,208,210,230,231,252,254,256	1
12 PAFZZ 14892 2230359	.PISTON,BRAKE CALIPE UOC:194,208,210,230,231,252,254,256	1
13 PAFZZ 11862 338269	.BOOT,DUST AND MOIST UOC:194,208,210,230,231,252,254,256	1
14 PAOZZ 14892 4150514	CLIP,SPRING TENSION UOC:194,208,210,230,231,252,254,256	2
15 PAOZZ 11862 331478	SCREW,SHOULDER UOC:194,208,210,230,231,252,254,256	2
16 PAOZZ 11862 14023439	KEY,CALIPER SUPPORT UOC:194,208,210,230,231,252,254,256	2
17 PAOZZ 14892 3203466	CALIPER,DISC BRAKE RIGHT UOC:194,208,210,230,231,252,254,256	1
17 PAOZZ 14894 3203465	BRACKET,BRAKE SHOE LEFT UOC:194,208,210,230,231,252,254,256	1
18 PAOZZ 97271 38001	SHIELD,BRAKE DISK LEFT UOC:194,208,210,230,231,252,254,256	1
18 PAOZZ 97271 38000	SHIELD,BRAKE DISK RIGHT UOC:194,208,210,230,231,252,254,256	1

END OF FIGURE

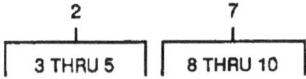

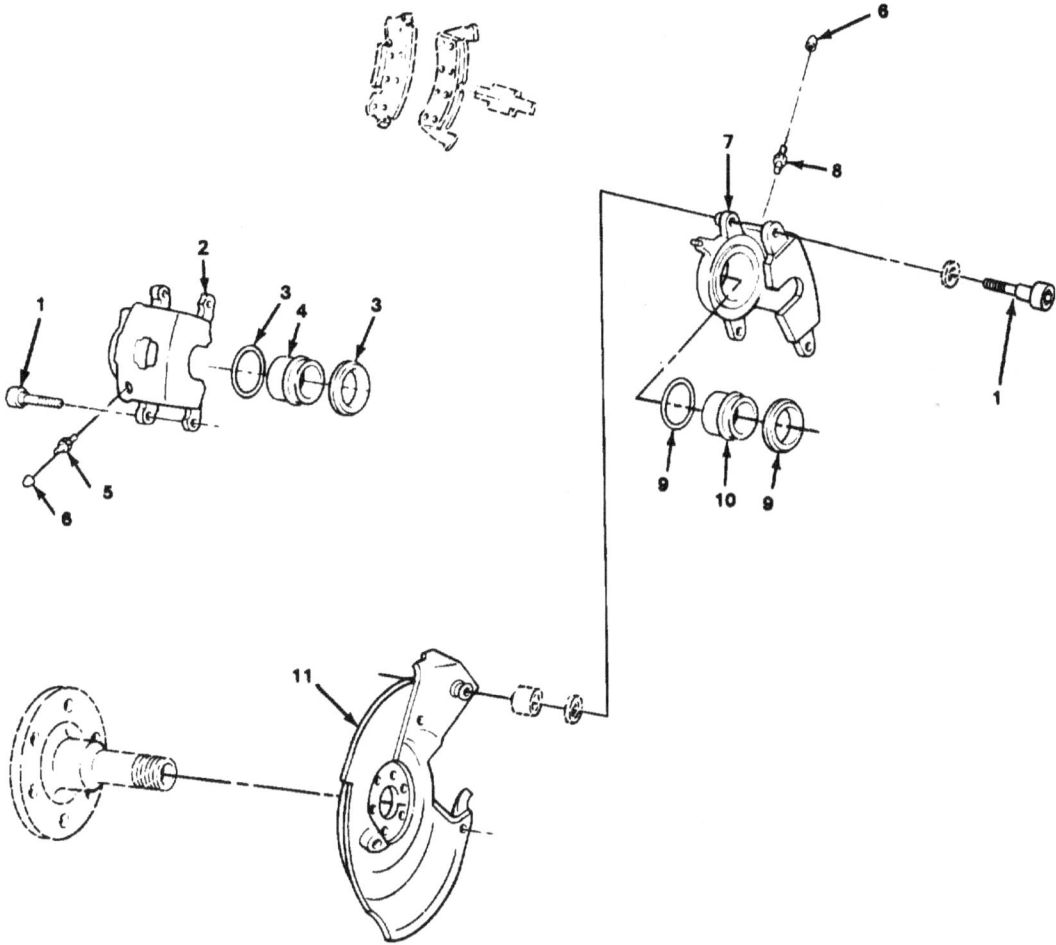

TA510928

FIGURE 121. FRONT BRAKE CALIPER ASSEMBLIES (M1009).

(1)	(2)	(3)	(4)	(5)	(6)
ITEM NO	SMR CODE	CAGEC	PART NUMBER	DESCRIPTION AND USABLE ON CODES (UOC)	QTY

GROUP 1204 HYDRAULIC BRAKE SYSTEM

FIG. 121 FRONT BRAKE CALIPER
ASSEMBLIES (M1009)

Item	SMR	CAGEC	Part Number	Description	QTY
1	PAOZZ	11862	5468226	BOLT UOC:209	4
2	PAOFF	11862	18015381	CALIPER,DISC BRAKE LEFT HAND UOC:209	1
3	PAFZZ	11862	5468767	.BOOT,DUST AND MOIST UOC:209	1
4	PAFZZ	11862	2622207	.PISTON,CALIPER,DISK UOC:209	1
5	PAOZZ	11862	18013395	.VALVE,SAFETY RELIEF UOC:209	1
6	PAOZZ	11862	5469581	CAP,PROTECTIVE,DUST UOC:209	2
7	PAOFF	11862	18007952	CALIPER,DISC BRAKE RIGHT HAND UOC:209	1
8	PAOZZ	11862	18013395	.VALVE,SAFETY RELIEF UOC:209	1
9	PAFZZ	11862	5468767	.BOOT,DUST AND MOIST UOC:209	1
10	PAFZZ	11862	2622207	.PISTON,CALIPER,DISK UOC:209	1
11	PAOZZ	11862	14023429	PLATE,BACKING,BRAKE LEFT UOC:209	1
11	PAOZZ	11862	14023430	PLATE,BACKING,BRAKE RIGHT UOC:209	1

END OF LIST

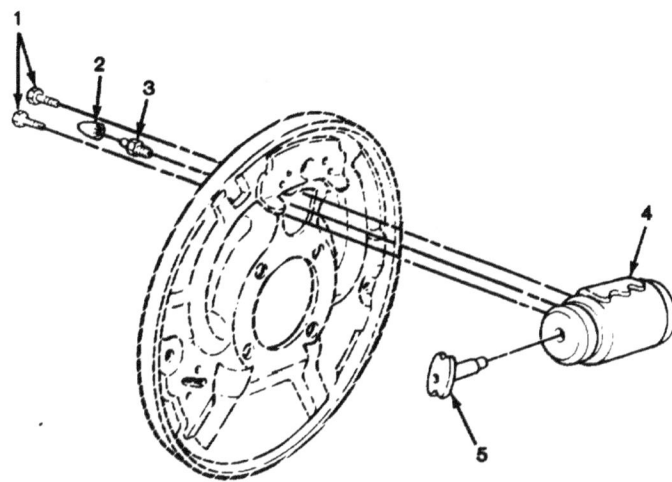

TA510929

FIGURE 122. REAR BRAKE WHEEL CYLINDER.

(1) (2) (3) (4) ITEM SMR PART NO CODE CAGEC NUMBER	(5) DESCRIPTION AND USABLE ON CODES (UOC)	(6) QTY
	GROUP 1204 HYDRAULIC BRAKE SYSTEM	
	FIG. 122 REAR BRAKE WHEEL CYLINDER	
1 PAOZZ 24617 456697 BOLT,MAHCINE		4
2 PAOZZ 11862 5469581 CAP,PROTECTIVE,DUST		2
3 PAOZZ 11862 18003151 BLEEDER VALVE,HYDRA		2
4 PAOZZ 11862 18004890 CYLINDER ASSEMBLY,H UOC:209		2
4 PAOZZ 11862 18004794 CYLINDER,HYDRAULIC UOC:194,208,210,230,231,252,254,256		2
5 PAOZZ 11862 2622667 LINK,WHEEL CYLINDER		4
	END OF FIGURE	

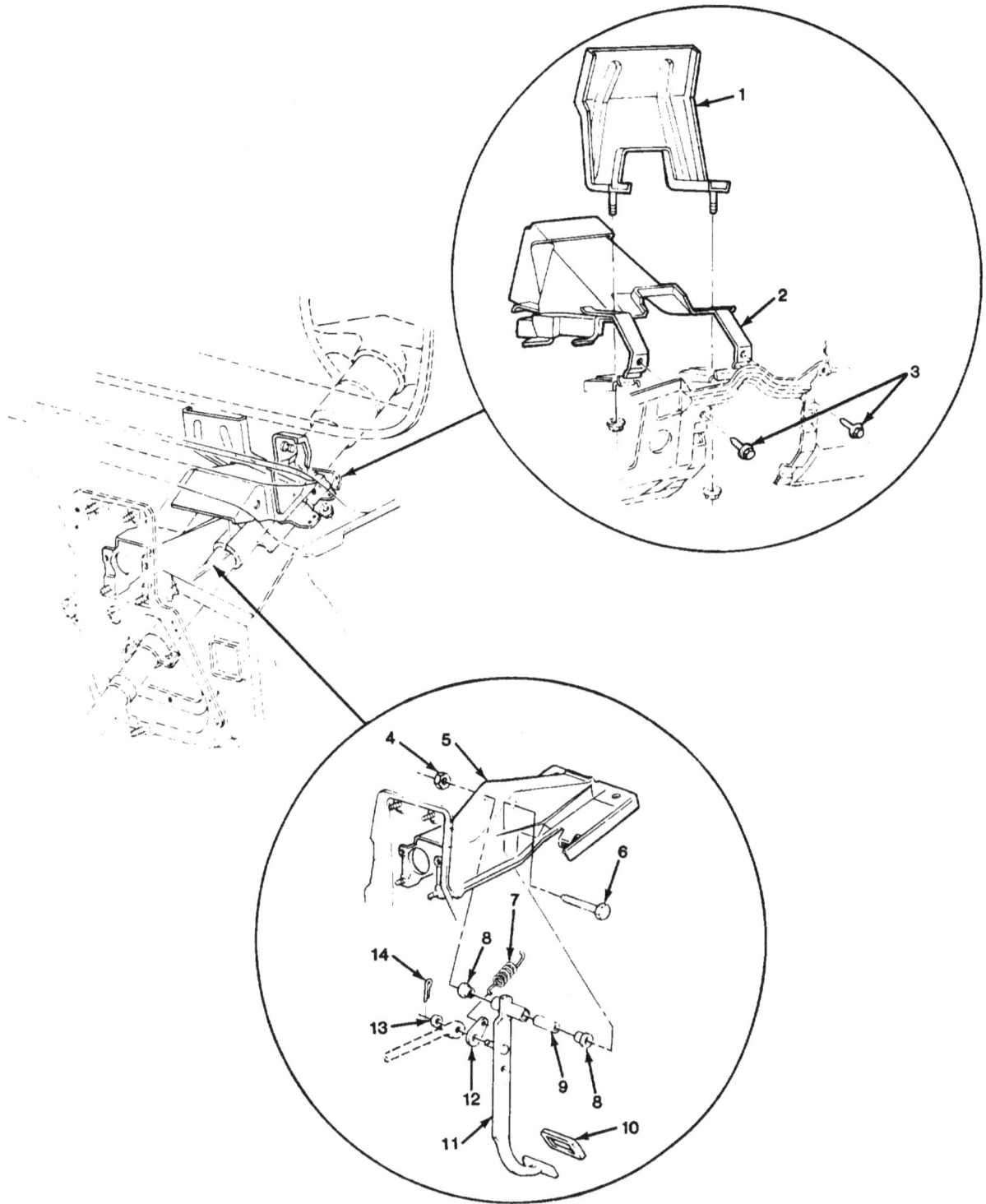

FIGURE 123. BRAKE PEDAL, MOUNTING BRACKETS, AND RELATED PARTS.

TA510930

(1) (2) (3) (4) ITEM SMR PART NO CODE CAGEC NUMBER	(5) DESCRIPTION AND USABLE ON CODES (UOC)	(6) QTY

GROUP 1206 MECHANICAL BRAKE SYSTEM

FIG. 123 BRAKE PEDAL, MOUNTING
BRACKET, AND RELATED PARTS

ITEM NO	SMR CODE	CAGEC	PART NUMBER	DESCRIPTION	QTY
1	PAFZZ	11862	467282	BRACE ASSEMBLY,BRAK	1
2	PAFZZ	11862	334146	BRACKET,DOUBLE ANGL	1
3	PAFZZ	11862	11509135	SCREW,ASSEMBLED WAS	2
4	PAOZZ	11862	9422299	NUT,SELF-LOCKING,HE	1
5	PBOZZ	11862	15593849	BRACKET,BRAKE PEDAL	1
6	PAOZZ	96906	MS18154-96	SCREW,CAP,HEXAGON H	1
7	PAOZZ	11862	3850084	SPRING,HELICAL,EXTE	1
8	PAOZZ	11862	346381	BUSHING,SLEEVE	2
9	PAOZZ	11862	6264951	SPACER,SLEEVE	1
10	PAOZZ	11862	15522095	COVER,BRAKE PEDAL P	1
11	PAOZZ	11862	355561	PEDAL ASSEMBLY,BRAK	1
12	PAOZZ	11862	336926	CONNECTING LINK,RIG	1
13	PAOZZ	11862	3702807	WASHER,FLAT	1
14	PAOZZ	11862	1244707	PIN,LOCK	1

END OF FIGURE

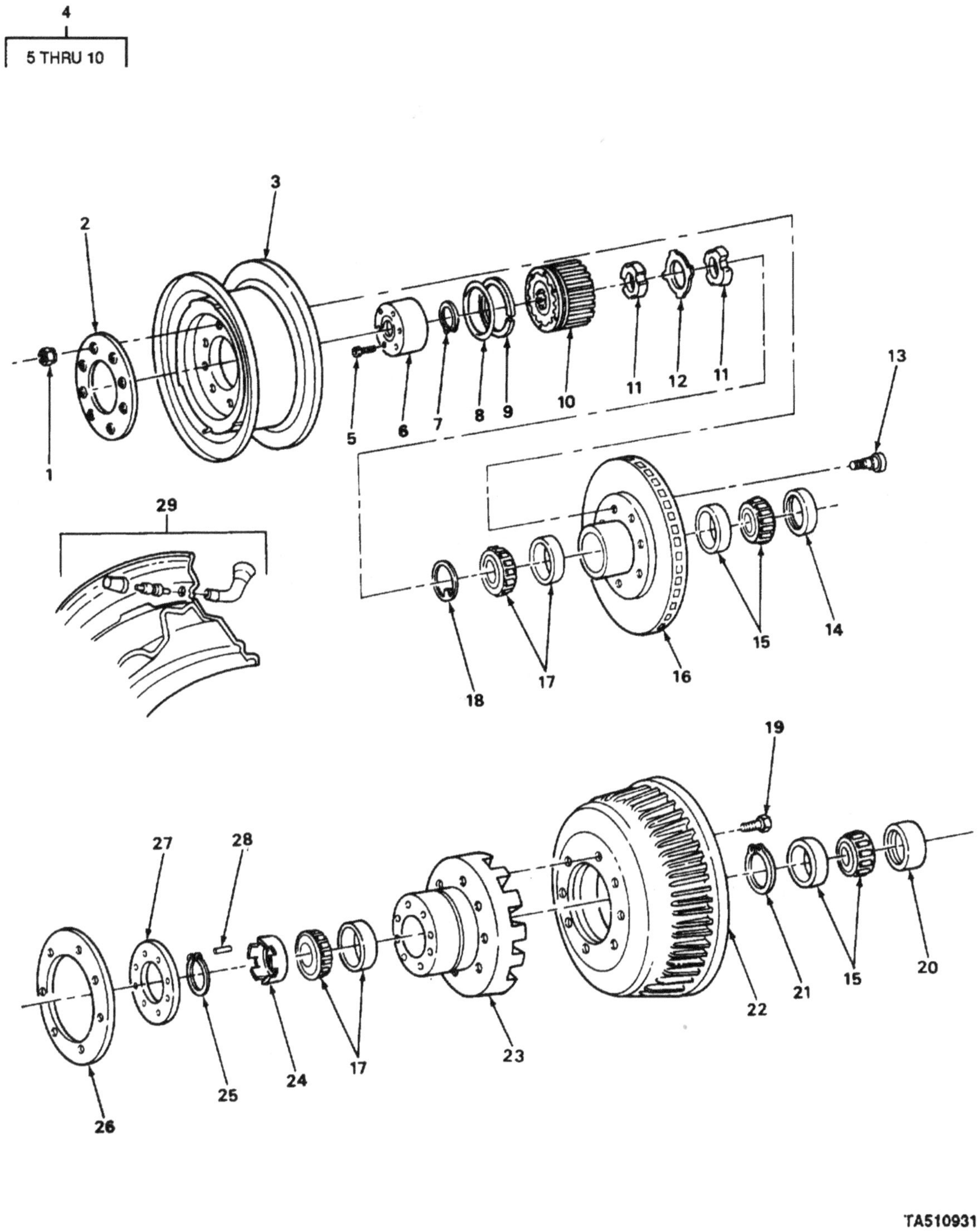

FIGURE 124. FRONT HUB, ROTOR, REAR HUB, BRAKEDRUM, AND
RELATED PARTS (ALL EXCEPT M1009).

TA510931

(1) ITEM NO	(2) (3) SMR CODE CAGEC	(4) PART NUMBER	(5) DESCRIPTION AND USABLE ON CODES (UOC)	(6) QTY
			GROUP 13 WHEELS AND TRUCKS	
			GROUP 1311 WHEEL ASSEMBLY	
			FIG. 124 FRONT HUB, ROTOR, REAR HUB, BRAKEDOWN, AND RELATED PARTS (ALL EXCEPT M1009)	
1	PAOZZ 11862	3978901	NUT,PLAIN,CONE SEAT UOC:194,208,210,230,231,252,254,256	32
1	PAOZZ 23862	334387	NUT UOC:254,256	32
2	PAOZZ 11862	472536	RING,WHEEL CLAMPING UOC:254,256	4
3	PAOZZ 11862	14035374	WHEEL,PNEUMATIC TIR	5
3	PAOZZ 11862	15668598	WHEEL,PNEUMATIC TIR UOC:254,256	1
4	PAOOO 27647	25113	LOCK HUB,FRONT UOC:194,208,210,230,231,252,254,256	2
5	PAOZZ 27647	9477	.SCREW PART OF KIT P/N 15528 UOC:194,208,210,230,231,252,254,256	6
6	PAOZZ 27647	15149	.HUB CAP,WHEEL UOC:194,208,210,230,231,252,254,256	1
7	PAOZZ 96906	MS16624-1131	.RING,RETAINING PART OF KIT P/N 15528 UOC:194,208,210,230,231,252,254,256	1
8	KFOZZ 27647	9952	.SEAL "O" RING PART OF KIT P/N 15528 UOC:194,208,210,230,231,252,254,256	1
9	PAOZZ 11862	14070396	.RING,RETAINING PART OF KIT P/N 15528 UOC:194,208,210,230,231,252,254,256	1
10	PAOZZ 27647	13109	.HUB BODY ASSEMBLY,F UOC:194,208,210,230,231,252,254,256	1
11	PAOZZ 11862	14050679	DISC,OUTER LOCKNUT UOC:194,208,210,230,231,252,254,256	4
11	PAOZZ 11862	15582233	DISC,INNER LOCKNUT UOC:194,208,210,230,231,252,254,256	4
12	PAOZZ 11862	14038051	DISC,LOCK RING UOC:194,208,210,230,252,254,256	2
13	PAOZZ 09386	102007	BOLT UOC:194,208,210,230,231,252,254,256	16
13	XDOZZ 11862	15634658	BOLT UOC:254,256	8
14	PAOZZ 80201	27467	SEAL,PLAIN UOC:194,208,210,230,231,252,254,256	2
14	PAOZZ 80201	27467	SEAL,PLAIN UOC:254,256	1
15	PAOZZ 60038	387AS-382A	BEARING,ROLLER,TAPE UOC:194,208,210,230,231,252,254,256	4
16	PAOFF 09386	SR104396	HUB,WHEEL,VEHICULAR INCLUDES ITEM 13 UOC:194,208,210,230,231,252,256	2

(1) ITEM NO	(2) SMR CODE	(3) CAGEC	(4) PART NUMBER	(5) DESCRIPTION AND USABLE ON CODES (UOC)	(6) QTY
16	PAOZZ	09386	104192	ROTOR AND HUB UOC:254,256	2
17	PAOZZ	43334	LM104949LM104911	BEARING,ROLLER,TAPE UOC:194,208,210,230,231,252,254,256	4
18	PAOZZ	97271	33734	WASHER UOC:194,208,210,230,231,252,254,256	2
19	PAOZZ	11862	3988538	SCREW,CAP,HEXAGON H UOC:194,208,210,230,231,252	16
20	PAOZZ	11862	469694	SEAL ASSY UOC:194,208,210,230,231,252,254,256	2
21	PAOZZ	11862	474309	RING,RETAINING UOC:194,208,210,230,231,252	2
22	PAOFF	11862	6260830	BRAKE DRUM UOC:194,208,210,230,231,252	2
23	PAOZZ	11862	3977397	HUB,WHEEL,VEHICULAR UOC:194,208,210,230,231,252	2
23	PAOFF	11862	15634663	HUB,WHEEL,VEHICULAR INCLUDES DRUM UOC:254,256	2
24	PAOZZ	11862	341509	NUT,PLAIN,ROUND UOC:194,208,210,230,231,252	2
25	PAOZZ	11862	341511	REAR WHEEL HUB RETA UOC:194,208,210,230,231,252	2
26	PAOZZ	11862	15634661	SPACER,PLATE UOC:254,256	2
27	PAOZZ	24617	327739	GASKET UOC:194,208,210,230,231,252	2
28	PAOZZ	11862	341510	KEY,MACHINE UOC:194,208,210,230,231,252	2
29	PAOZZ	11862	273487	VALVE,PNEUMATIC TIR UOC:194,208,210,230,231,252,254,256	5
29	PAOZZ	11862	9591270	VALVE,PNEUMATIC TIR UOC:254,256	5
29	PAOZZ	6V6625	30-600	VALVE,PNEUMATIC TIR USED ON WHEELS WITH PIN 9591597 UOC:194,208,210,230,231,252,254,256	5

END OF FIGURE

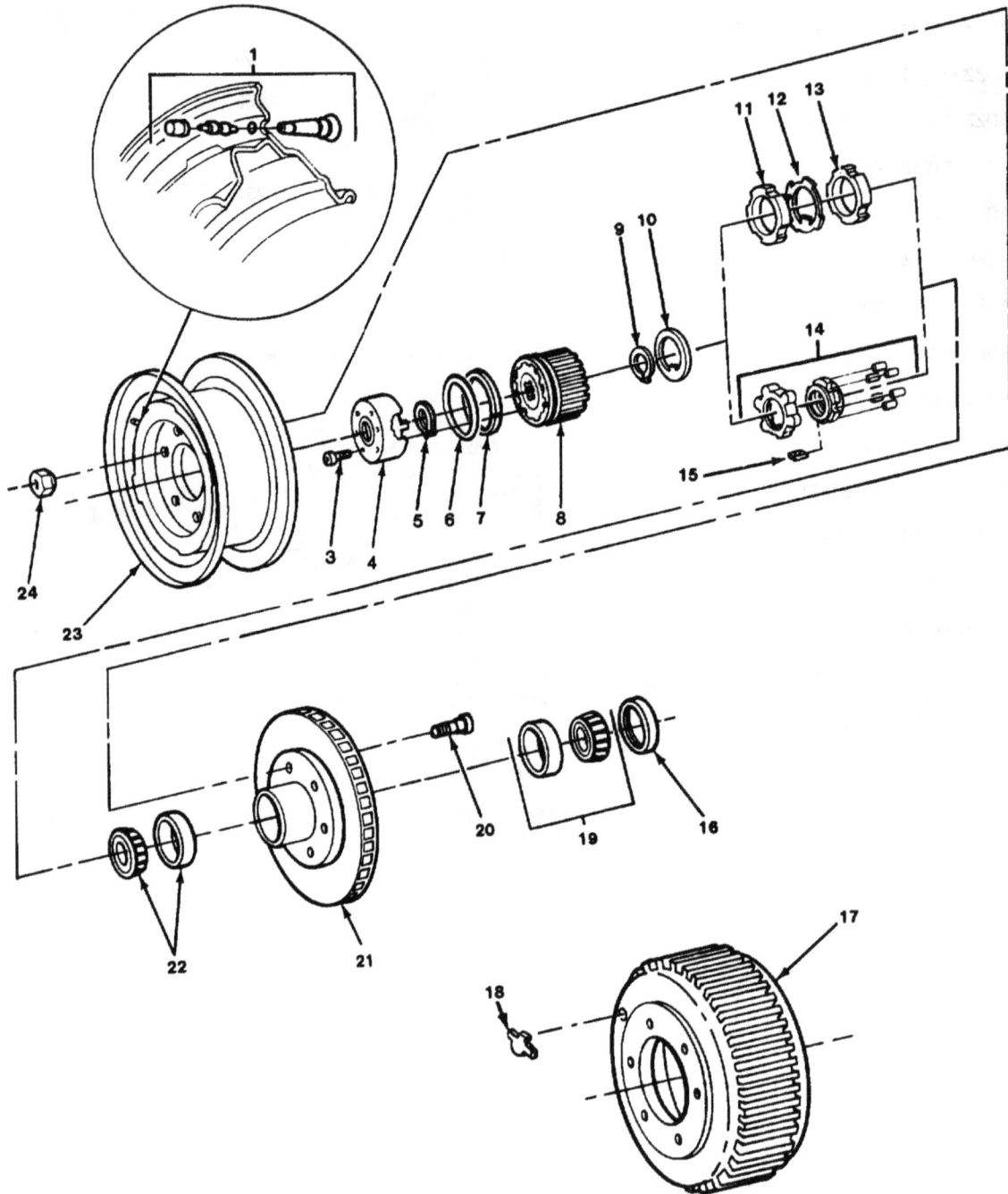

FIGURE 125. FRONT HUB, ROTOR, BRAKEDRUM, AND RELATED PARTS (M1009).

TA510932

(1)	(2)	(3)	(4)	(5)	(6)
ITEM NO	SMR CODE	CAGEC	PART NUMBER	DESCRIPTION AND USABLE ON CODES (UOC)	QTY

GROUP 1311 WHEEL ASSEMBLY

FIG. 125 FRONT HUB, ROTOR,
BRAKEDOWN, AND RELATED PARTS (M1009)

1	PAOZZ	17875	T14R	VALVE,PNEUMATIC TIR UOC:209	5
2	PAOOO	27647	M257	LOCK,DRIVE SHAFT,DI UOC:209	2
3	PAOZZ	27647	9477	.SCREW PART OF KIT P/N 11967 UOC:209	6
4	PAOZZ	27647	15147	.HUB CAP ASSEMBLY UOC:209	1
5	PAOZZ	96906	MS16624-1125	.RING,RETAINING PART OF KIT P/N 11967 UOC:209	2
6	PAOZZ	27647	13446	.PACKING,PREFORMED UOC:209	1
7	PAOZZ	11862	14070348	.RING,RETAINING PART OF KIT P/N 11967 UOC:209	1
8	PAOZZ	27647	13113	.HUB,BODY UOC:209	1
9	PAOZZ	96906	MS16624-1125	RING,RETAINING UOC:209	2
10	PAOZZ	11862	14072921	SPACER,RING UOC:209	2
11	PAOZZ	11862	14034413	NUT,PLAIN,ROUND APPLIES TO VEHICLES STARTING WITH VIN ENDING WITH F136169 UOC:209	2
12	PAOZZ	79410	17-01-014-001	WASHER,KEY APPLIES TO VEHICLES STARTING WITH VIN ENDING WITH F136169 UOC:209	2
13	PAOZZ	11862	14034410	NUT,PLAIN,ROUND APPLIES TO VEHICLES STARTING WITH VIN ENDING WITH F136169 UOC:209	2
14	PAOZZ	76445	40424	NUT,PLAIN,ROUND APPLIES TO 84 MODEL UOC:209	2
15	XDOZZ	11862	14072927	KEY,MACHINE APPLIES TO 84 MODEL UOC:209	2
16	PAOZZ	11862	6273948	SEAL,PLAIN ENCASED UOC:209	2
17	PAOFF	11862	14070352	BRAKE DRUM UOC:209	2
18	PAOZZ	19207	6262328	PLUG,PROTECTIVE,DUS UOC:209	2
19	PAOZZ	11862	7455617	BEARING,ROLLER,TAPE UOC:209	2
20	PAOZZ	09386	96735	BOLT,RIBBED SHOULDE	12

(1) (2) (3) (4)				(5)	(6)
ITEM NO	SMR CODE	CAGEC	PART NUMBER	DESCRIPTION AND USABLE ON CODES (UOC)	QTY
				UOC:209	
21	PAOFF	11862	14026765	ROTOR ASEMBLY,DISC INCLUDES ITEM 20	2
				UOC:209	
22	PAOZZ	43334	LM501349-LM501310	BEARING,ROLLER,TAPE	2
				UOC:209	
23	PAOZZ	11862	14063307	RIM,WHEEL,PNEUMATIC	5
				UOC:209	
24	PAOZZ	11862	358501	NUT,PLAIN,CONE SEAT	24
				UOC:209	

END OF FIGURE

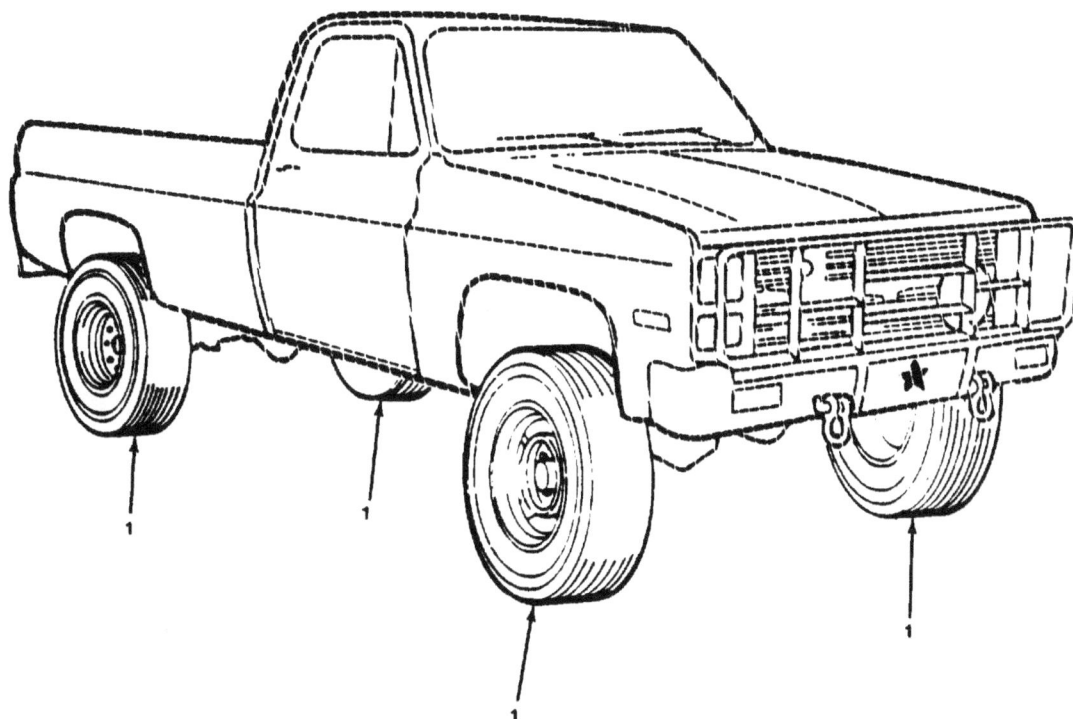

TA510933

FIGURE 126. TIRES.

| (1) (2) (3) (4) | (5) | (6) |
ITEM SMR PART NO CODE CAGEC NUMBER	DESCRIPTION AND USABLE ON CODES (UOC)	QTY
	GROUP 1313 TIRES, TUBES, TIRE CHAINS	
	FIG. 126 TIRES	
1 PAOOO 22337 212-776	TIRE,PNEUMATIC UOC:209	5
1 PAOOO 81348 GP2A/LT235/85R16 /E/LTAW	TIRE,PNEUMATIC UOC:194,208,210,230,231,252,254,256	5
	END OF FIGURE	

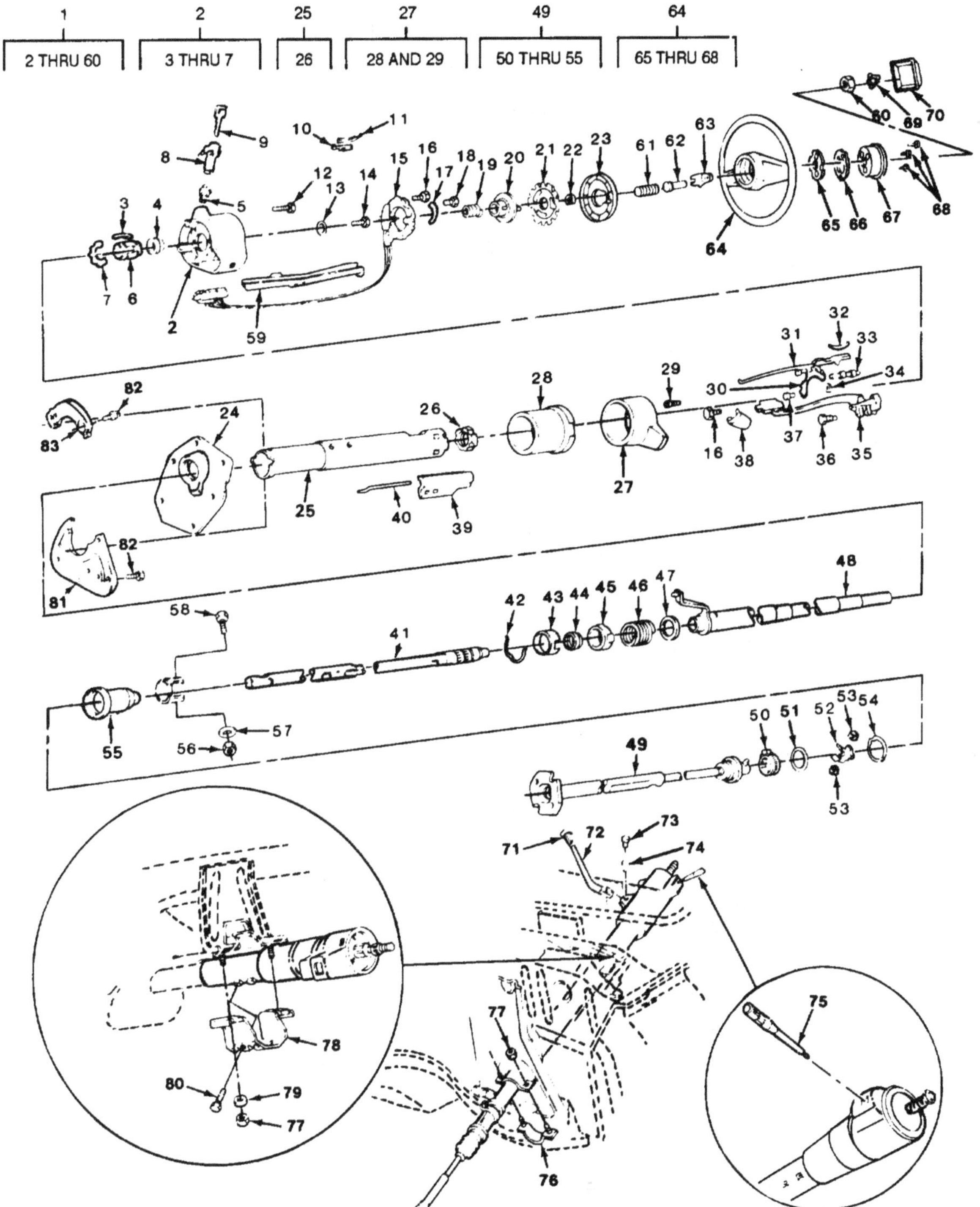

FIGURE 127. STEERING WHEEL AND COLUMN.

TA510934

TM9-2320-289-34P

(1) (2) (3) (4)   (5)   (6)
ITEM SMR   PART
NO   CODE CAGEC NUMBER   DESCRIPTION AND USABLE ON CODES (UOC)   QTY

GROUP 14 STEERING

GROUP 1401 MECHANICAL STEERING GEAR
ASSEMBLY

FIG. 127 STEERING WHEEL AND COLUMN

1	PAFHH	11862	7843369	COLUMN ASSEMBLY,STE	1
2	PAFFF	11862	7837032	.HOUSING,STEERING CO	1
3	PAFZZ	11862	7832984	..HORN CONTACT CONTRO	1
4	PAFZZ	11862	7819517	..BEARING,SLEEVE	1
5	PAFZZ	11862	7812526	..ACTUATOR,IGNITION S	1
6	PAFZZ	11862	7832808	..BUSHING,SLEEVE	1
7	PAFZZ	11862	7832311	..RING,RETAINING	1
8	PAFZZ	11862	7843294	.SWITCH,LOCK,IGNITIO	1
9	PAFZZ	11862	1154611	.KEY BLANK	1
10	PAFZZ	11862	7830375	.CLIP,BUZZER	1
11	PAFZZ	11862	7804414	.SWITCH,BUZZER,IGNIT	1
12	PAFZZ	11862	7830377	.SCREW	1
13	PAFZZ	11862	7800580	.WASHER,FLAT	1
14	PAFZZ	11862	7806867	.SCREW	4
15	PAFZZ	11862	1997983	.CONTROL,DIRECTIONAL	1
16	PAFZZ	11862	7809128	.SCREW,TAPPING,THREA	4
17	PAFZZ	11862	7843400	.ARM,DIRECTIONAL SIG	1
18	PAFZZ	11862	11504678	.SCREW,CAP,HEXAGON H	1
19	PAFZZ	11862	7819898	.SPRING,HELICAL,COMP	1
20	PAFZZ	11862	7812136	.CAM,CONTROL	1
21	PAFZZ	11862	7837185	.LOCK RING,SHAFT,STE	1
22	PAFZZ	52788	5694191	.RING,RETAINING	1
23	PAFZZ	11862	7819738	.COVER,SHAFT	1
24	PFFZZ	11861	7814387	.SEAL,RUBBER SPECIAL	1
25	PAFFF	11862	7842684	.HOUSING,STEERING CO	1
26	PAFZZ	11862	7827111	..BEARING,BOWL,LOWER	1
27	PFFFF	11862	7836172	.BOWL,GEARING SHIFT	1
28	PFFZZ	11862	7827077	..HOUSING,STEERING CO	1
29	PAFZZ	11862	7837378	..SPRING,HELICAL,COMP	1
30	PAFZZ	11862	7845130	.LEVER,GATE	1
				UOC:194,208,209,210,230,231,252,254	
31	PAFZZ	11862	7843368	.RACK,ROD,STEERING	1
32	PAFZZ	11862	7804410	.SPRING,FLAT	1
				UOC:2008,209,210,230,231,252,254,256	
33	PAFZZ	11862	7806185	.BOLT,SPRING,LOCK	1
34	PAFZZ	11862	7805950	.BEARING,WASHER,THR	1
35	PAFZZ	11862	7840274	.SWITCH,ELECTRICAL	1
36	PAFZZ	11862	7830384	.PIN,SHOULDER,HEADED	1
37	PAFZZ	11862	7830314	.SCREW,MACHINE	2
				UOC:194,208,209,210,230,252,254	
38	PAFZZ	11862	7830209	.COVER,STEERING COLU	1
39	PAFZZ	11862	7830335	.PROTECTOR,WIRING	1
40	PAFZZ	11862	7843467	.ROD,SWITCH,ACTUATOR	1
41	PAFZZ	11862	7831538	.SHAFT,STRAIGHT	1
42	PAFZZ	11862	7804439	.CLIP,RETAINING	1

(1)	(2)	(3)	(4)	(5)	(6)
ITEM NO	SMR CODE	CAGEC	PART NUMBER	DESCRIPTION AND USABLE ON CODES (UOC)	QTY
43	PAFZZ	11862	7804440	.RETAINER,BEARING AD UOC:194,208,209,210,230,231,252,254	1
44	PAFZZ	11862	7805700	.BEARING	1
45	PAFZZ	11862	7805822	.ADAPTER,SHIFT,LOWER	1
46	PAFZZ	11862	7810567	.SPRING,HELICAL,COMP	1
47	PAFZZ	11862	7832112	.BEARING,WASHER,THRU	1
48	PAFZZ	11862	7842688	.TUBE,SHIFT	1
49	PAFFF	11862	7831570	.SHAFT ASSEMBLY,FLEX	1
50	PAFZZ	96906	MS35842-11	..CLAMP,HOSE	1
51	PAFZZ	11862	7817454	..WASHER,FLAT	1
52	PAFZZ	11862	7809409	..SPRING,STEERING SHA	1
53	PAFZZ	11862	5671921	..BEARING,UNIVERSAL J	2
54	PAFZZ	11862	7809408	..RETAINER,PACKING	1
55	PAFZZ	11862	7831571	..COUPLING,SHAFT,RIGI	1
56	PAFZZ	73342	11501033	.NUT	1
57	PAFZZ	96906	MS27183-16	.WASHER,FLAT	1
58	PAFZZ	11862	7832907	.BOLT,SQUARE NECK	1
59	PAFZZ	11862	7828271	.COVER,ACCESS	1
60	PAOZZ	11862	7846970	.NUT,PLAIN,HEXAGON	1
61	PAOZZ	11862	404234	SPRING,HELICAL,COMP	1
62	PAOZZ	11862	474102	PARTS KIT,HORN BUTT INCLUDES ITEMS 61 & 63	1
63	PAOZZ	11862	409190	INSULATOR,HORN CONT	1
64	PAOOO	11862	9762199	STEERING WHEEL	1
65	PAOZZ	24617	9749363	.SPRING	1
66	PAOZZ	11862	9754764	.SPACER,RING	1
67	PAOZZ	11862	17983936	.HORN BUTTON,VEHICLE	1
68	PAOZZ	11862	9767270	.SCREW,CAP,SOCKET HE	3
69	PAOZZ	11862	419454	RING,RETAINING	1
70	PAOZZ	11862	17987489	HORN BUTTON,VEHICLE	1
71	PAOZZ	11862	14049351	KNOB	1
71	PAOZZ	11862	470205	KNOB APPLIES TO VEHICLES STARTING WITH VIN FF136169 AND VIN FF300021	1
72	PAOZZ	11862	14034728	LEVER,MANUAL CONTRO	1
72	PAOZZ	11862	14081915	LEVER,MANUAL CONTRO	1
73	PAOZZ	11862	3793014	PIN,STRAIGHT,HEADED	1
74	PAOZZ	11862	1604854	BUSHING,SLEEVE APPLIES TO VEHICLES STARTING WITH VIN FF136169 AND VIN FF300021	1
75	PAOZZ	11862	22510143	LEVER,MANUAL CONTRO	1
76	PAFZZ	11862	467283	CLAMP,LOOP	1
77	PAFZZ	73342	11501033	NUT	4
78	PFFZZ	11862	7809925	SUPPORT,STEERING CO	1
79	PAFZZ	96906	MS27183-16	WASHER,FLAT	2
80	PAFZZ	24617	11504270	BOLT,MACHINE	4
81	PAFZZ	11862	14027745	COVER,ACCESS	1
82	PAFZZ	11862	11508566	SCREW,TAPPING,THREA	5
83	PAFZZ	11862	14027746	COVER,STEERING COLU	1

END OF FIGURE

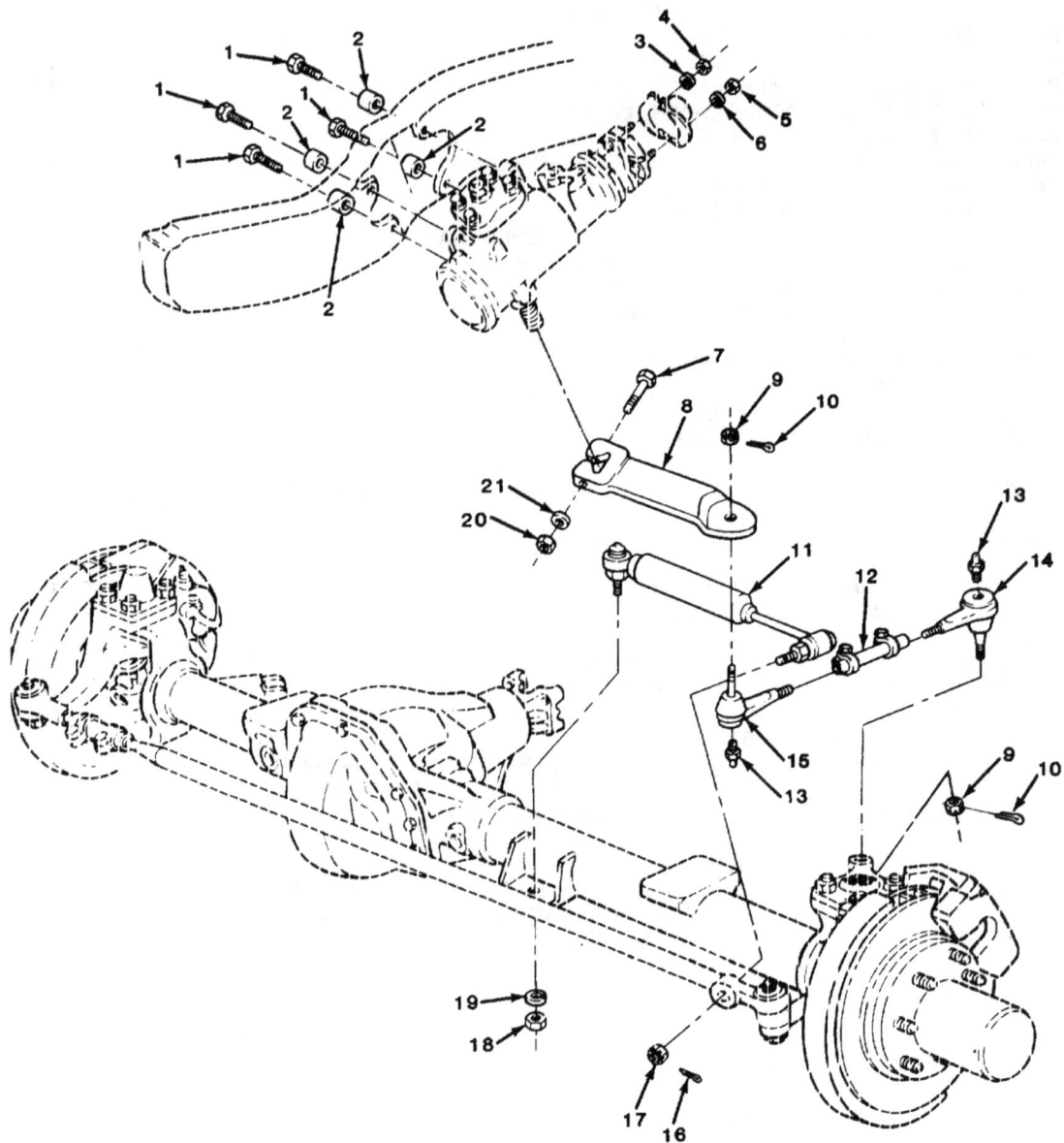

TA510935

FIGURE 128. PITMAN ARM, ABSORBER, AND RELATED PARTS.

(1) (2) (3) (4) ITEM SMR PART NO CODE CAGEC NUMBER	(5) DESCRIPTION AND USABLE ON CODES (UOC)	(6) QTY
	GROUP 1401 MECHANICAL STEERING GEAR ASSEMBLY	
	FIG. 128 PITMAN ARM, ABSORBER, AND RELATED PARTS	
1 PAOZZ 11862 343178	SCREW,CAP,HEXAGON H	4
2 PAOZZ 11862 343179	SPACER,SLEEVE	4
3 PAOZZ 96906 MS35338-45 WASHER,LOCK		1
4 PAOZZ 96906 MS51968-6 NUT,PLAIN,HEXAGON		1
5 PAOZZ 96906 MS51968-9 NUT,PLAIN,HEXAGON		1
6 PAOZZ 96906 MS35338-46 WASHER,LOCK		1
7 PAOZZ 96906 MS90727-145	SCREW,CAP,HEXAGON H	1
8 PAOZZ 11862 14064660	ARM,STEERING GEAR	1
9 PAOZZ 96906 MS35692-53 NUT,PLAIN,SLOTTED,H		2
10 PAOZZ 96906 MS24665-357 PIN,COTTER		2
11 PAOZZ 11862 4993563	CYLINDER ASSEMBLY,A	1
12 PAOZZ 11862 14007644	SLEEVE ASSEMBLY,CON	1
13 PAOZZ 11862 9417901	FITTING,LUBRICATION	2
14 PAOZZ 11862 362297	TIE ROD END,STEERIN	1
15 PAOZZ 11862 362298	TIE ROD END,STEERIN	1
16 PAOZZ 89749 IF316	PIN,COTTER	1
17 PAOZZ 96906 MS35692-37 NUT,PLAIN,SLOTTED,H		1
18 PAOZZ 96906 MS35690-824 NUT,PLAIN,HEXAGON		1
19 PAOZZ 96906 MS35338-48 WASHER,LOCK		1
20 PAOZZ 11862 9422303	NUT,SELF-LOCKING,HE	1
21 PAOZZ 96906 MS27183-20 WASHER,FLAT		1

END OF FIGURE

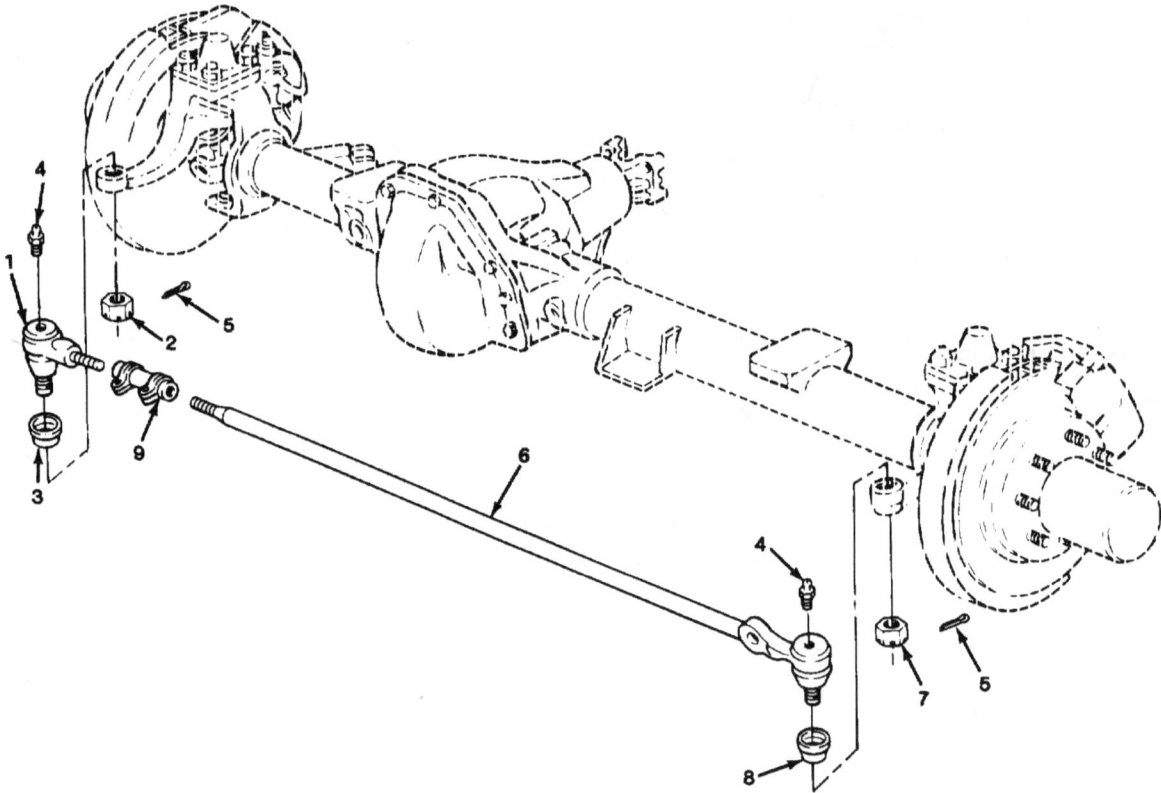

TA510936

FIGURE 129. TIE-ROD ASSEMBLY (ALL EXCEPT M1009).

TM9-2320-289-34P

(1) ITEM NO	(2) SMR CODE	(3) CAGEC	(4) PART NUMBER	(5) DESCRIPTION AND USABLE ON CODES (UOC)	(6) QTY
				GROUP 1401 MECHANICAL STEERING GEAR ASSEMBLY	
				FIG. 129 TIE-ROD ASSEMBLY (ALL EXCEPT M1009)	
1	PAOOO	72210	D23980-J	TIE ROD END,STEERIN RIGHT UOC:194,208,210,230,231,252,254,256	1
2	PAOZZ	24617	125384	.NUT,PLAIN,SLOTTED,H UOC:194,208,210,230,231,252,254,256	1
3	PAOZZ	11862	6259074	.PACKING,PREFORMED UOC:194,208,210,230,231,252,254,256	1
4	PAOZZ	11862	9417901	FITTING,LUBRICATION UOC:194,208,210,230,231,252,254,256	2
5	PAOZZ	96906	MS24665-425	PIN,COTTER UOC:194,208,210,230,231,252,254,256	2
6	PAOOO	72210	24004-J	TIE ROD END,STEERING LEFT UOC:194,208,210,230,231,252,254,256	1
7	PAOZZ	24617	125384	.NUT,PLAIN,SLOTTED,H UOC:194,208,210,230,231,252,254,256	1
8	PAOZZ	97271	S19587-T	.PACKING,PREFORMED UOC:194,208,210,230,231,252,254,256	1
9	PAOZZ	72210	S23981-H	ADJUSTING SLEEVE,TI UOC:194,208,210,230,231,252,254,256	1

END OF FIGURE

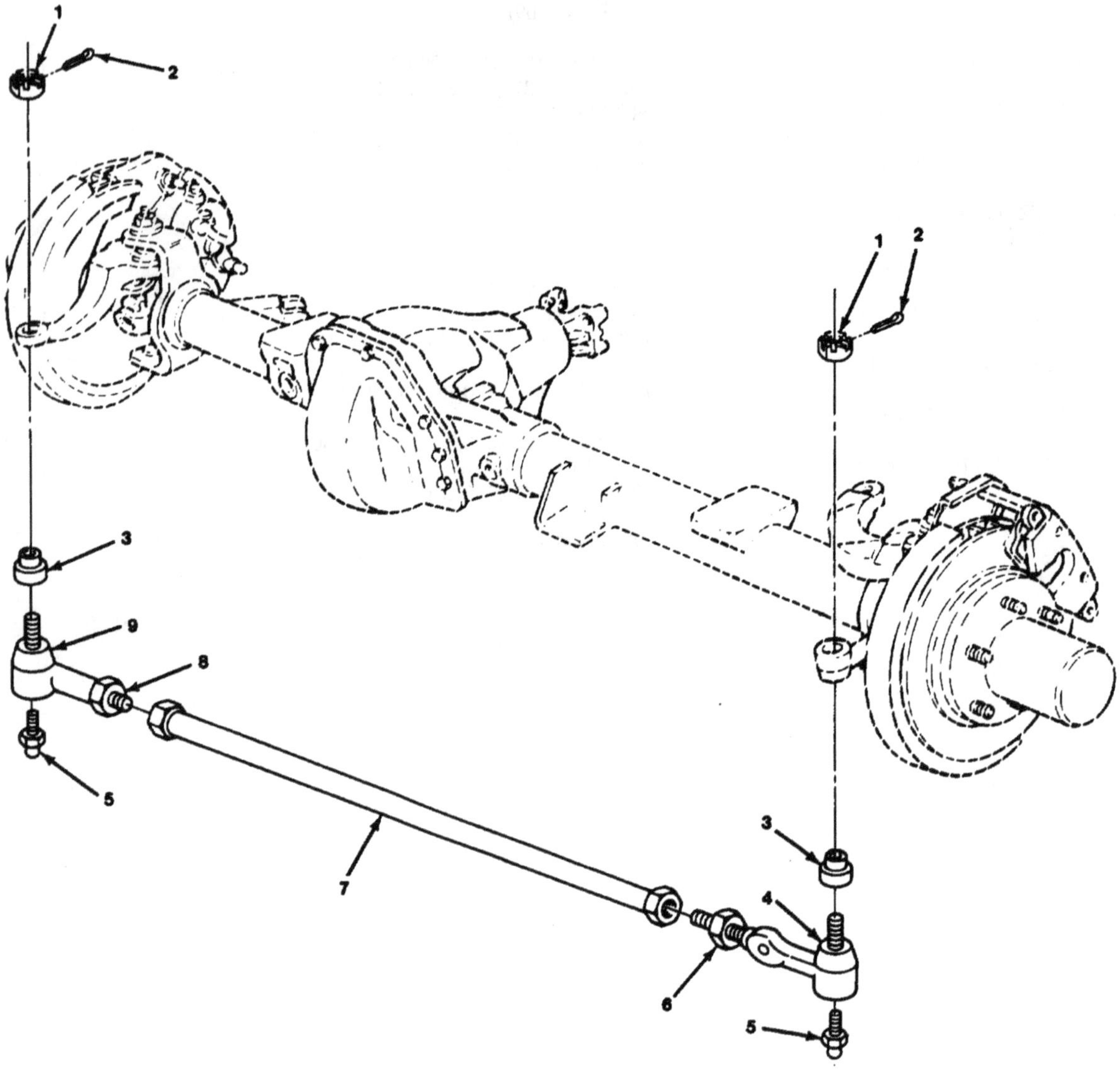

TA510937

FIGURE 130. TIE-ROD ASSEMBLY (M1009).

(1) ITEM NO	(2) SMR CODE	(3) CAGEC	(4) PART NUMBER	(5) DESCRIPTION AND USABLE ON CODES (UOC)	(6) QTY
				GROUP 1401 MECHANICAL STEERING GEAR ASSEMBLY	
				FIG. 130 TIE-ROD ASSEMBLY (M1009)	
1	PAOZZ	11862	467117	NUT,PLAIN,SLOTED,H UOC:209	2
2	PAOZZ	11862	11514337	PIN,COTTER UOC:209	2
3	PAOZZ	11862	6259074	PACKING,PREFORMED UOC:209	2
4	PAOZZ	11862	14026803	TIE ROD END,STEERIN LEFT UOC:209	1
5	PAOZZ	11862	9417901	FITTING,LUBRICATION UOC:209	2
6	PAOZZ	11862	14026805	NUT,PLAIN,HEXAGON UOC:209	1
7	PAOZZ	11862	14026804	TIE ROD,STEERING UOC:209	1
8	PAOZZ	30076	160635	NUT,PLAIN,HEXAGON UOC:209	1
9	PAOZZ	11862	14026802	TIE ROD END,STEERIN RIGHT UOC:209	1

END OF FIGURE

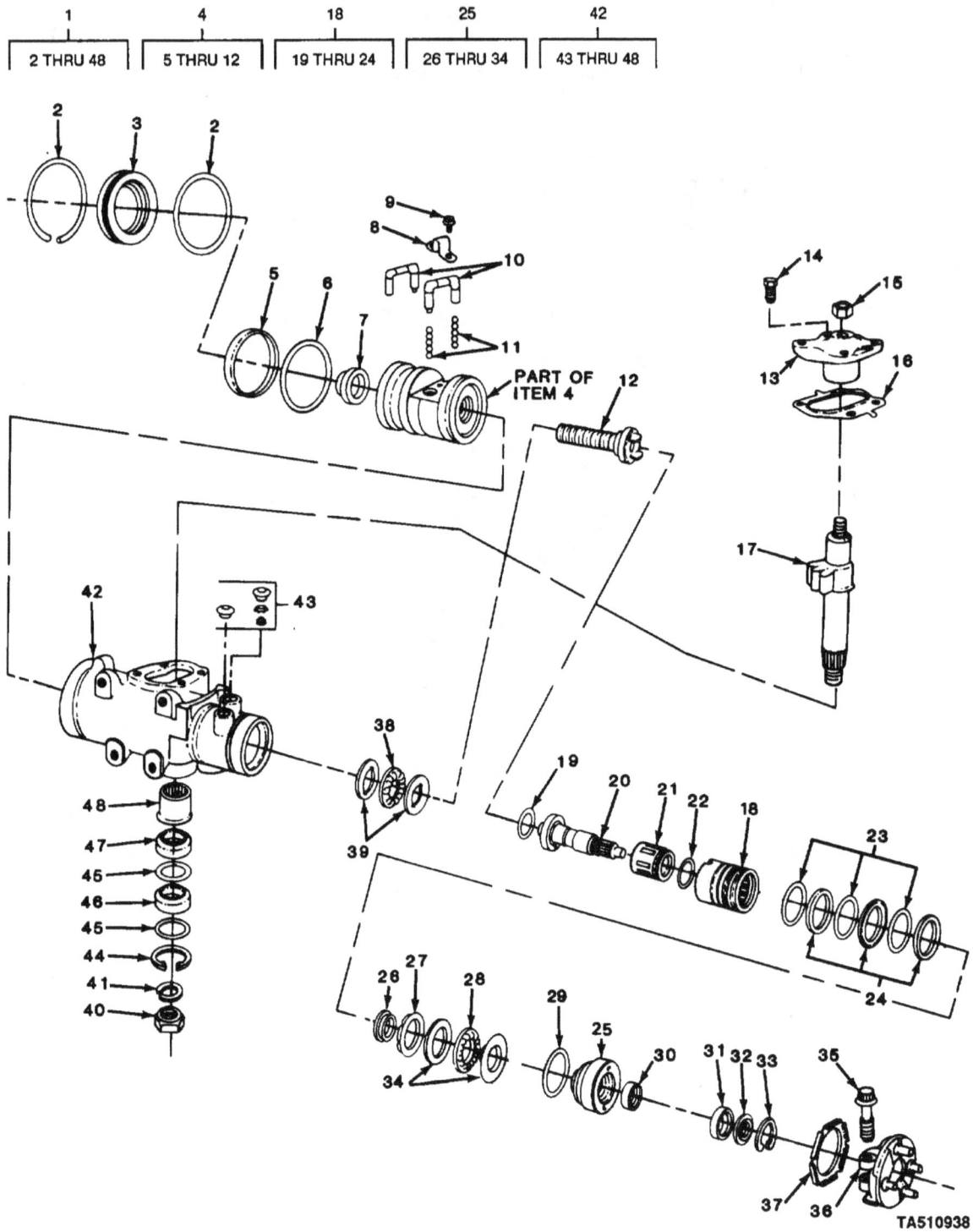

FIGURE 131. POWER STEERING GEAR AND COMPONENT PARTS.

(1) ITEM NO	(2) SMR CODE	(3) CAGEC	(4) PART NUMBER	(5) DESCRIPTION AND USABLE ON CODES (UOC)	(6) QTY
				GROUP 1407 POWER STEERING GEAR ASSEMBLY	
				FIG. 131 POWER STEERING GEAR AND COMPONENT PARTS	
1	PAOHH	11862	7846959	STEERING GEAR	1
2	KFHZZ	11862	5686814	.RING PART OF KIT P/N 7817485	1
3	PAHZZ	11862	5686815	.PLUG,HOUSING END	1
4	XAHHH	11862	7817529	.CYLINDER ASSEMBLY,A INCLUDES ITEM #14	1
5	PAHZZ	11862	5687958	..GASKET PART OF KIT P/N 7817487	1
6	PAHZZ	11862	7814503	..PACKING,PREFORMED PART OF KIT P/N 7817487	1
7	PAHZZ	11862	7817355	..PLUG,RACK-PISTON E	1
8	PAHZZ	11862	5686539	..STRAP,RETAINING	1
9	PAHZZ	11862	5696151	..SCREW ASSEMBLED	2
10	PFHZZ	11862	5695774	..GUIDE,BALL,POWER ST	2
11	PAHZZ	11862	5695513	..PARTS KIT,STEERING	1
12	XAHZZ	11862	5695777	..WORM-RACK & PSTN GR	1
13	PAHZZ	11862	7817484	.PARTS KIT,COVER INCLUDES ITEMS 14, 15,16	1
14	PAHZZ	11862	7816516	.BOLT,MACHINE	4
15	PAHZZ	11862	5687973	.NUT,SELF-LOCKING,SI	1
16	PAHZZ	11862	7817486	.SEAL KIT	1
17	XAHZZ	11862	7818809	.PARTS KIT,STEERING	1
18	XAHHH	11862	7846626	.VALVE,STEERING SELE	1
19	KFHZZ	11862	5672489	..SEAL-P/S GR VLV ASM PART OF KIT P/N 5687182	1
20	XAHZZ	11862	7832422	..SHAFT-STUB T/S	1
21	XDHZZ	11862	7807845	..SEAL SPOOL-VLV P/S	1
22	KFHZZ	11862	7819986	..SEAL-P/S SPOOL VLV PART OF KIT P/N 5687182	1
23	KFHZZ	11862	5686550	..SEAL-P/S GR VLV ASM PART OF KIT P/N 5687182	1
24	PAHZZ	11862	5686611	..PACKING,PREFORMED PART OF KIT P/N 5687182	1
25	XAHZZ	52788	7817725	.PARTS KIT,STEERING	1
26	XAHZZ	11862	7832429	..RETAIN-P/S PLUG KIT	1
27	XAHZZ	11862	7819849	..SPACER-P/S PLUG KIT	1
28	PAHZZ	11862	7832729	..BEARING,ROLLER,THRU INCLUDES ITEMS 26,27,34	1
29	PAHZZ	11862	5686527	..PACKING,PREFORMED PART OF KIT P/N 7828565	1
30	PAHZZ	11862	7828012	..BUSHING,SLEEVE	1
31	KFHZZ	11862	7832426	..SEAL-P/S GR STUB PART OF KIT P/N 7828565	1
32	KFHZZ	11862	7832427	..SEAL-P/S GR STUB PART OF KIT P/N 7828565	1
33	KFHZZ	11862	7832428	..RING-STUB SHF RET PART OF KIT P/N 7828565	1
34	XAHZZ	11862	5687209	..RACE-P/S PLUG KIT	2

(1) ITEM NO	(2) SMR CODE	(3) CAGEC	(4) PART NUMBER	(5) DESCRIPTION AND USABLE ON CODES (UOC)	(6) QTY
35	PAHZZ	11862	7807271	.BOLT,MACHINE	1
36	PAHZZ	11862	7826542	.FLANGE,STEERING,POW	1
37	PAHZZ	11862	7826012	.NUT	1
38	PAHZZ	11862	7826850	.BEARING ASSEMBLY INCLUDES ITEM 39	1
39	XAHZZ	11862	7816509	.RACE-LWR BRG P/S	2
40	PAHZZ	11862	5667628	.NUT,PLAIN,HEXAGON	1
41	PAHZZ	11862	5697702	.WASHER,LOCK	1
42	XAHHH	11862	7834333	.PARTS KIT,POWER STE	1
43	PAHZZ	11862	7834284	..VALVE,SAFETY RELIEF	1
44	KFHZZ	11862	5682898	..RING-PIT SHF SNAP PART OF KIT P/N 7826470	1
45	KFHZZ	11862	5690517	..WASHER-SPECIAL PART OF KIT P/N 7826470	2
46	KFHZZ	11862	7826470	..SEAL,PITMAN SHAFT PART OF KIT P/N 7826470	1
47	KFHZZ	11862	5683373	..SEAL-PITMAN SHF PART OF KIT P/N 7826470	1
48	PAHZZ	11862	5697804	..BEARING,ROLLER,NEED	1

END OF FIGURE

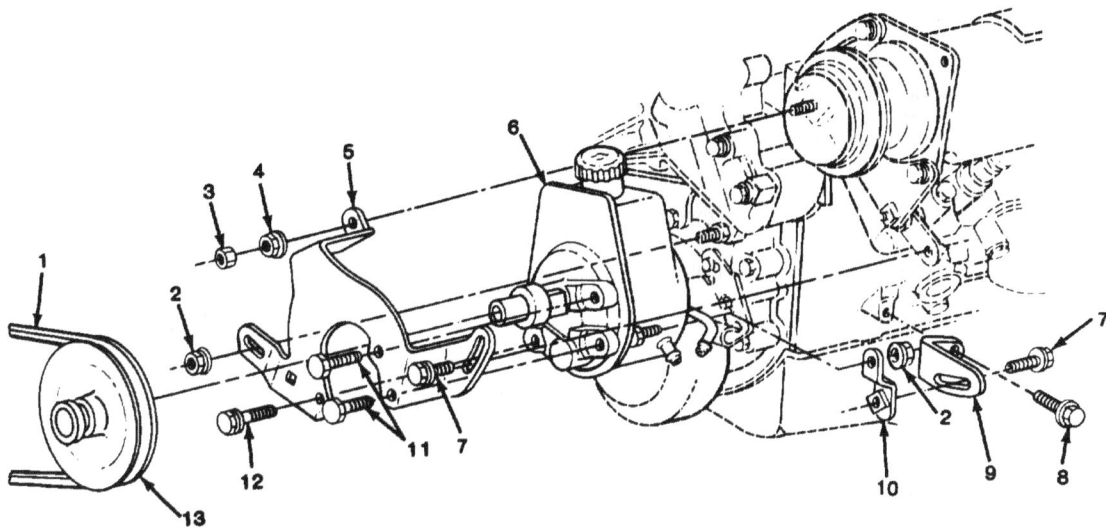

TA510939

FIGURE 132. HYDRAULIC PUMP.

(1) (2) (3) (4) ITEM SMR PART NO CODE CAGEC NUMBER	(5) DESCRIPTION AND USABLE ON CODES (UOC)	(6) QTY
	GROUP 1410 HYDRAULIC PUMP OR FLUID MOTOR ASSEMBLY	
	FIG. 132 HYDRAULIC PUMP	
1 PAOZZ 20796 42-5023 BELT,V		1
2 PAOZZ 11862 11506101 NUT		2
3 PAOZZ 11862 11503643 NUT		1
	UOC:194,208,209,230,231,252,254,256	
4 PAOZZ 73342 11501033 NUT		1
	UOC:210	
5 PAOZZ 11862 14033879	BRACKET,POWER STEER	1
6 PAOFH 52788 7838936	PUMP ASSEMBLY,POWER FOR COMPONENT PARTS SEE FIG 133	1
7 PAOZZ 11862 1635490 BOLT,SELF-LOCKING		2
8 PAOZZ 11862 11504595	SCREW,CAP,HEXAGON H	1
9 PAOZZ 11862 14033880 BRACKET,ANGLE		1
10 PAOZZ 11862 14033881	BRACKET,DOUBLE ANGL	1
11 PAOZZ 11862 11504512 BOLT,MACHINE		2
12 PAOZZ 11862 11505299 BOLT,MACHINE		1
13 PAOZZ 11862 14023174 PULLEY,GROOVE		1
	UOC:210,230	
13 PAOZZ 11862 14067701 PULLEY,CONE		1
	UOC:194,208,209,230,231,252,254,256	

END OF FIGURE

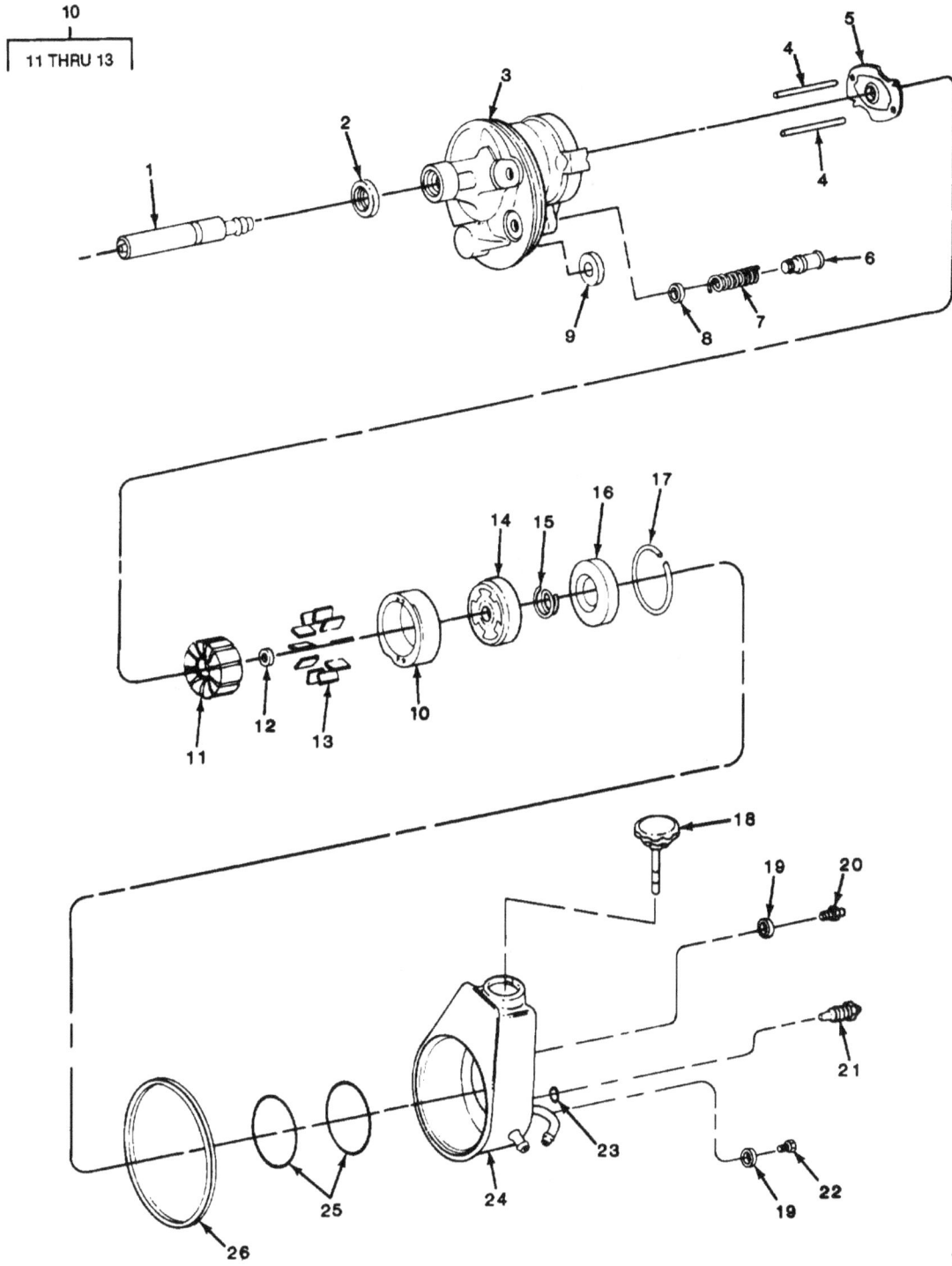

FIGURE 133. HYDRAULIC PUMP COMPONENT PARTS.

TA510940

(1)	(2)	(3)	(4)	(5)		(6)
ITEM	SMR			PART		
NO	CODE	CAGEC	NUMBER		DESCRIPTION AND USABLE ON CODES (UOC)	QTY

GROUP 1410 HYDAULIC PUMP OR FLUID
MOTOR ASSEMBLY

FIG. 133 HYDRAULIC PUMP COMPONENT
PARTS

1	XDFZZ	11862	7837321		.SHAFT,SHOULDERED INCLUDES ITEM 12	1
2	PAFZZ	11862	7808195		.SEAL,PLAIN ENCASED PART OF KIT P/N 5688044	1
					UOC:194,208,209,210,231,252,254,256	
3	XAFZZ	11862	7830236		.HOUSING,LIQUID PUMP	1
4	PAFZZ	11862	5689357	.PIN		2
5	XDFZZ	11862	7836369	.PLATE,THRUST		1
6	PAFZZ	11862	7809232		.VALVE,SAFETY RELIEF	1
7	PAFZZ	11862	5688037	.SPRING,HELICAL,COMP		1
8	PAFZZ	11862	5688035		.PACKING,PREFORMED PART OF KIT P/N 5688044	1
9	PAFZZ	11862	5692682		.MAGNET,PERMANENT PART OF KIT P/N 5688044	1
10	XDFZZ	11862	7837322		.PARTS KIT,POWER STE	1
11	XAFZZ	11862	7839341		..ROTOR P/S PMP	1
12	XDFZZ	11862	7837284	..RING,RETAINING		1
13	XAFZZ	11862	7825561		..VANE ROTOR P/S PMP	1
14	XDFZZ	11862	7839669	.PLATE,PRESSURE		1
15	XDFZZ	11862	7839667	.SPRING,FLAT		1
16	XDFZZ	11862	5689358	.PLATE,MENDING		1
17	XDFZZ	11862	5688014	.RING,RETAINING		1
18	PAOZZ	11862	7834183		.GAGE ROD-CAP,LIQUID	1
19	PAFZZ	11862	7814503		.PACKING,PREFORMED PART OF KIT P/N 5688044	2
20	PAFZZ	11862	7831388	.STUD		1
21	PAFZZ	11862	7830913	.CONNECTOR,FLUID,PUM		1
22	PAFZZ	11862	7830239		.SCREW,CAP,HEXAGON H	1
23	PAFZZ	48018	5688049		.GASKET PART OF KIT P/N 5688044	1
24	XDFZZ	11862	7839499		.RESERVOIR,PUMP ASSE	1
25	KFFZZ	11862	5688109		.SEAL P/S HSG END PART OF KIT P/N 5688044	2
26	PAFZZ	11862	5688015		.PACKING,PREFORMED PART OF KIT P/N 5688044	1

END OF FIGURE

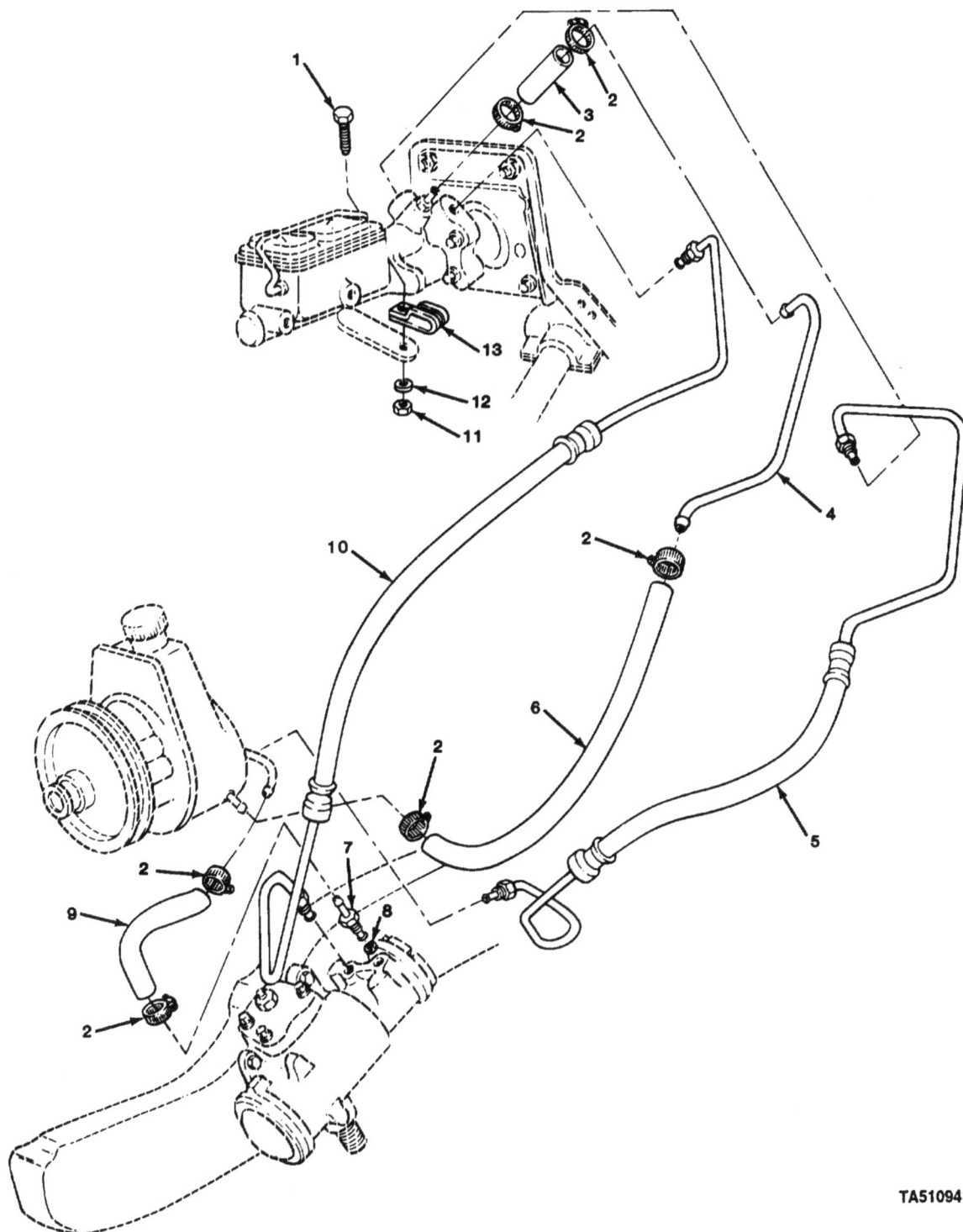

FIGURE 134. HYDRAULIC STEERING LINES.

TA510941

(1) (2) (3) (4) ITEM SMR PART NO CODE CAGEC NUMBER	(5) DESCRIPTION AND USABLE ON CODES (UOC)	(6) QTY
	GROUP 1411 HOSES, LINES, FITTINGS	
	FIG. 134 HYDRAULIC STEERING LINES	
1 PAOZZ 96906 MS90728-32 BOLT,MACHINE		1
2 PAOZZ 11862 25518880 CLAMP,HOSE		6
3 MOOZZ 11862 350371	HOSE P/S BSTR RTN (1.75" LG) MAKE FROM HOSE, P/N 7828506	1
4 MFOZZ 11862 14040735	PIPE P/S BSTR RTN (15.90"LG) MAKE FROM TUBE,P/N 603827	1
5 PAOZZ 11862 7838941	HOSE ASSEMBLY,NONME	1
6 MOOZZ 11862 1488565	HOSE P/S BSTR RTN (20.50" LG) MAKE FROM HOSE, P/N 7828506	1
7 PAOZZ 11862 22514738	TUBE AND ADAPTER AS	1
8 PAOZZ 11862 7829923 PACKING,PREFORMED		1
9 MOOZZ 11862 3773687	HOSE P/S GR CTLT (13.00" LG) MAKE FROM HOSE, P/N 7828506	1
10 PAOZZ 11862 7838942	HOSE ASSEMBLY,NONME	1
11 PAOZZ 96906 MS51967-6 NUT,PLAIN,HEXAGON		1
12 PAOZZ 96906 MS35340-45 WASHER,LOCK		1
13 PAOZZ 11862 338696	CLAMP,LOOP	1

END OF FIGURE

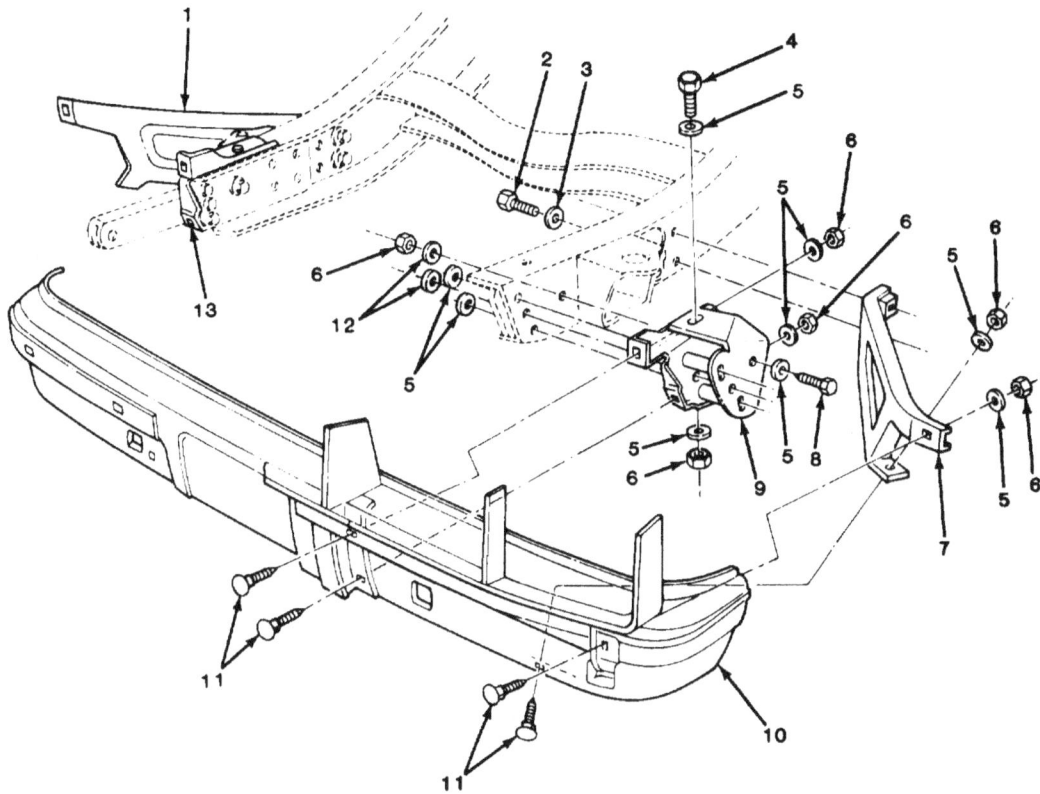

TA510942

FIGURE 135. FRONT BUMPER AND RELATED PARTS.

(1) (2) (3) (4)	(5)	(6)
ITEM SMR PART NO CODE CAGEC NUMBER	DESCRIPTION AND USABLE ON CODES (UOC)	QTY

GROUP 15 FRAME, TOWING ATTACHMENTS,
DRAWBARS, AND ARTICULATION SYSTEMS

GROUP 1501 FRAME ASSEMBLY

FIG. 135 FRONT BUMPER AND RELATED
PARTS

1	PAOZZ	11862	14045516	BRACE,FENDER	1
2	PAOZZ	96906	MS90728-87	SCREW,CAP,HEXAGON H	4
3	PAOZZ	11862	3790768	WASHER,FLAT	4
4	PAOZZ	96906	MS90728-111	SCREW,CAP,HEXAGON H	2
5	PAOZZ	96906	MS27183-18	WASHER,FLAT	24
6	PAOZZ	96906	MS51967-14	NUT,PLAIN,HEXAGON	18
7	PAOZZ	11862	14045515	BRACE,FENDER	1
8	PAOZZ	96906	MS90728-119	SCREW,CAP,HEXAGON H	8
9	PAOZZ	11862	14067791	BRACE,FENDER LEFT	1
10	PFOZZ	11862	14072422	BUMPER,VEHICULAR	1
11	PAOZZ	11862	14072425	BOLT,SQUARE NECK	8
12	PAOZZ	96906	MS27183-19	WASHER,FLAT	4
13	PAOZZ	11862	14067792	BRACKET ASSEMBLY,BU RIGHT	1

END OF FIGURE

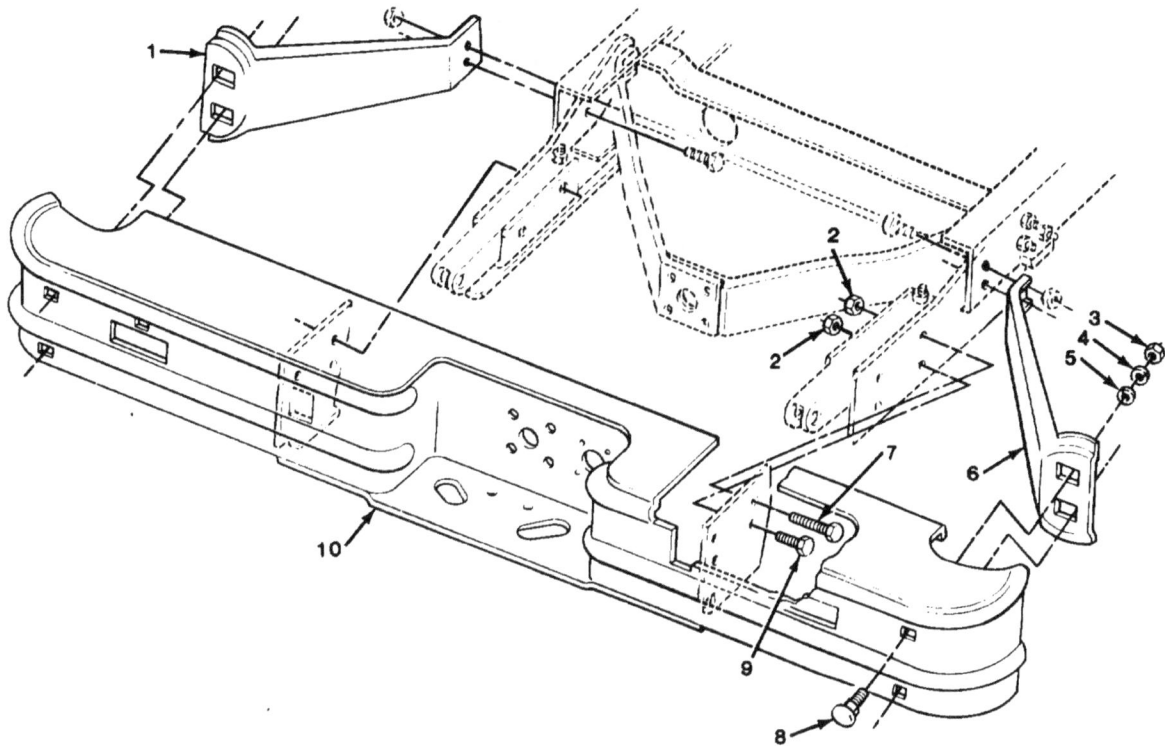

TA510943

FIGURE 136. REAR BUMPER AND MOUNTING PARTS (ALL EXCEPT M1009 AND M1010).

(1) (2) (3) (4) ITEM SMR NO CODE CAGEC	PART NUMBER	(5) DESCRIPTION AND USABLE ON CODES (UOC)	(6) QTY
		GROUP 1501 FRAME ASSEMBLY	
		FIG. 136 REAR BUMPER AND MOUNTING PARTS (ALL EXCEPT M1009 AND M10101	
1 PAOZZ 11862	14014799	BRACE,FENDER UOC:194,208,230,231,252,254,256	1
2 PAOZZ 24617	9422301	NUT,SELF-LOCKING,HE UOC:194,208,230,231,252,254,256	8
3 PAOZZ 96906	MS51967-14	NUT,PLAIN,HEXAGON UOC:194,208,230,231,252,254,256	4
4 PAOZZ 96906	MS35340-48	WASHER,LOCK UOC:194,208,230,231,252,254,256	4
5 PAOZZ 96906	MS27183-18	WASHER,FLAT UOC:194,208,230,231,252,254,256	4
6 PAOZZ 11862	14014800	BRACE,FENDER UOC:194,208,230,231,252,254,256	1
7 PAOZZ 96906	MS90728-119	SCREW,CAP,HEXAGON H UOC:194,208,230,231,252,254,256	4
8 PAOZZ 11862	14072425	BOLT,SQUARE NECK UOC:194,208,230,231,252,254,256	4
9 PAOZZ 96906	MS90728-111	SCREW,CAP,HEXAGON H UOC:194,208,230,231,252,254,256	4
10 PFOZZ 11862	14072435	BUMPER,VEHICULAR UOC:194,208,230,231,252,254,256	1

END OF FIGURE

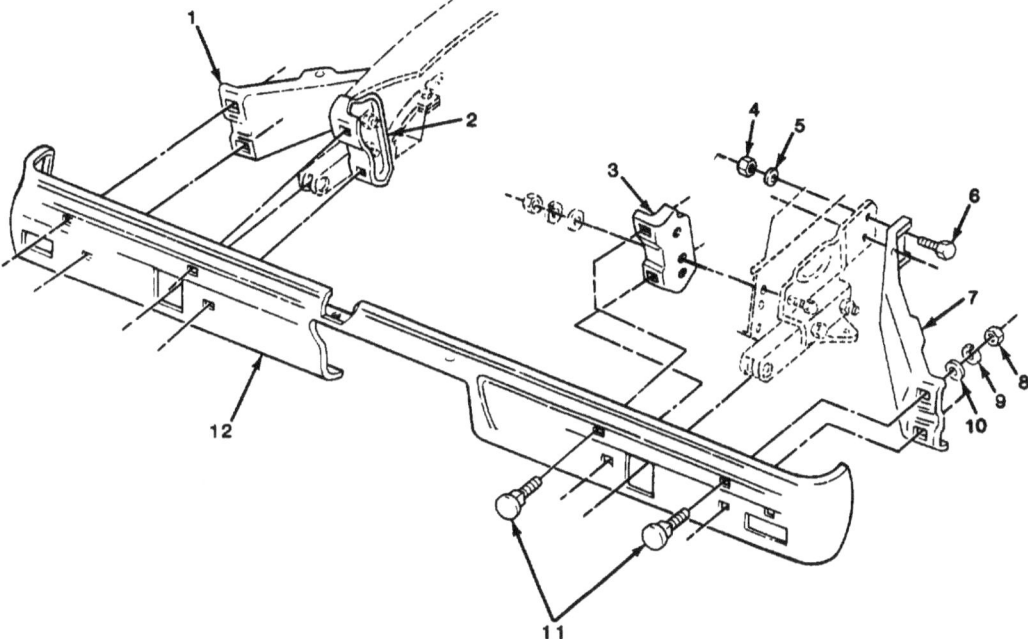

TA510944

FIGURE 137. REAR BUMPER AND MOUNTING PARTS (M1009).

(1) ITEM NO	(2) SMR CODE	(3) CAGEC	(4) PART NUMBER	(5) DESCRIPTION AND USABLE ON CODES (UOC)	(6) QTY
				GROUP 1501 FRAME ASSEMBLY	
				FIG. 137 REAR BUMPER AND MOUNTING PARTS (M1009)	
1	PAOZZ	11862	14067783	BRACE,FENDER UOC:209	1
2	PAOZZ	11862	14021357	BRACE,FENDER UOC:209	1
3	PAOZZ	11862	14021358	BRACE,FENDER UOC:209	1
4	PAOZZ	96906	MS51967-12	NUT,PLAIN,HEXAGON UOC:209	4
5	PAOZZ	96906	MS35340-47	WASHER,LOCK UOC:209	4
6	PAOZZ	96906	MS90728-87	SCREW,CAP,HEXAGON H UOC:209	4
7	PAOZZ	11862	14067784	BRACE,FENDER UOC:209	1
8	PAOZZ	96906	MS51967-14	NUT,PLAIN,HEXAGON UOC:209	8
9	PAOZZ	96906	MS35340-48	WASHER,LOCK UOC:209	8
10	PAOZZ	96906	MS27183-18	WASHER,FLAT UOC:209	8
11	PAOZZ	11862	14072425	BOLT,SQUARE NECK UOC:209	8
12	PFOZZ	11862	14072436	BUMPER,VEHICULAR UOC:209	1

END OF FIGURE

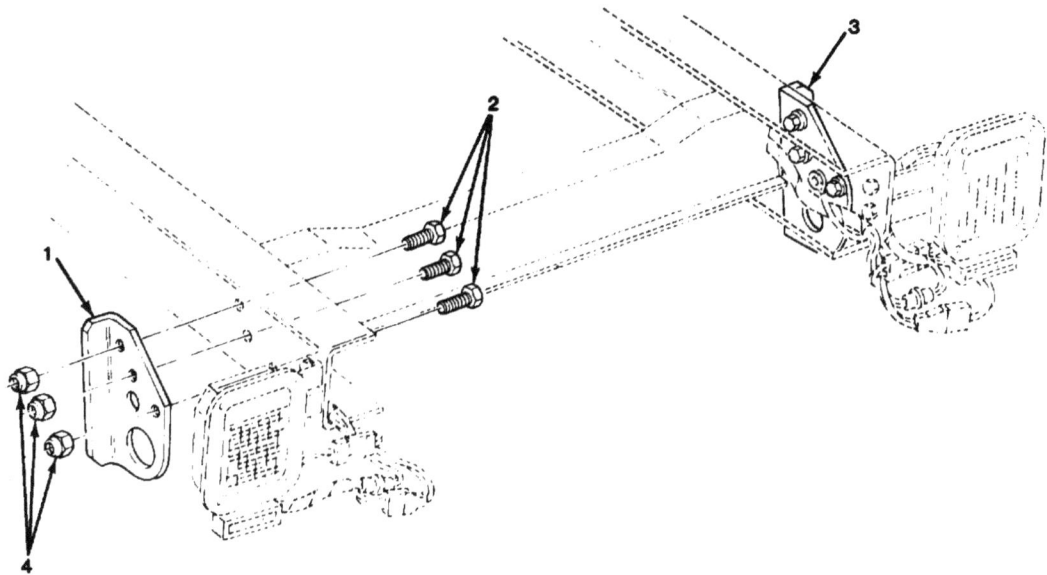

TA510945

FIGURE 138. REAR TIE-DOWNS (M1031).

(1) (2) (3) (4)				(5)	(6)
ITEM SMR			PART	DESCRIPTION AND USABLE ON CODES (UOC)	QTY
NO	CODE	CAGEC	NUMBER		

GROUP 1501 FRAME ASSEMBLY

FIG. 138 REAR TIE-DOWNS (M1031)

1	PAOZZ	11862	15599219	BRACKET,ANGLE LEFT UOC:231	1
2	PAOZZ	96906	MS90728-111	SCREW,CAP,HEXAGON H UOC:231	6
3	PAOZZ	11862	15599220	BRACKET,ANGLE UOC:231	1
4	PAOZZ	24617	9422301	NUT,SELF-LOCKING,HE UOC:231	6

END OF FIGURE

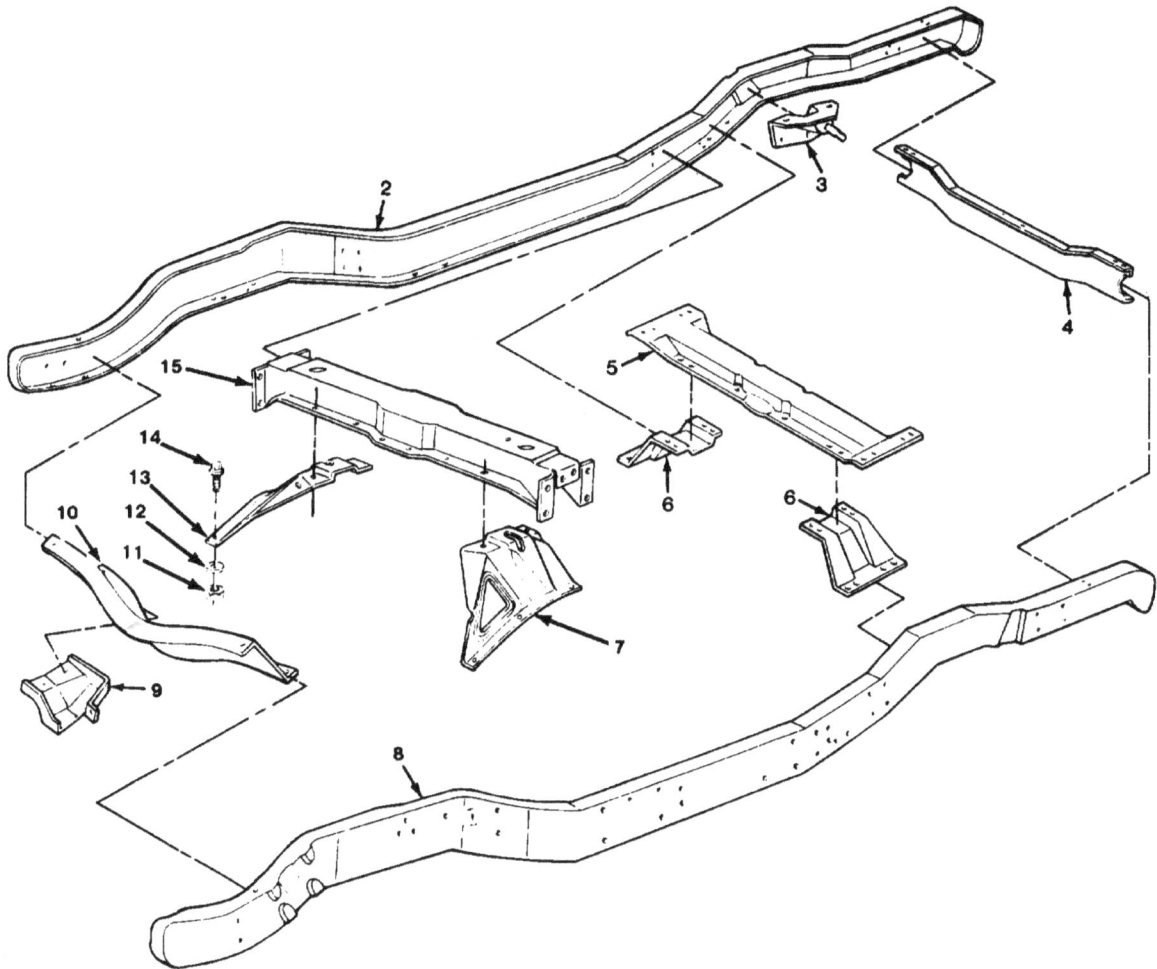

FIGURE 139. FRAME ASSEMBLY COMPONENTS (ALL EXCEPT M1009).

TA510946

(1) (2) (3) (4) ITEM SMR NO CODE CAGEC	PART NUMBER	(5) DESCRIPTION AND USABLE ON CODES (UOC)	(6) QTY
		GROUP 1501 FRAME ASSEMBLY	
		FIG. 139 FRAME ASSEMBLY COMPONENTS (ALL EXCEPT M1009)	
1 PAHHH 11862	15589799	FRAME,STRUCTURAL, VE UOC:208,210,230,231,252,254,256,	2
2 PFHZZ 11862	14072640	.SIDE MEMEBER FRAME, R UOC:194,208,210,230,231,252,254,256	1
3 PFHZZ 11862	334666	.BRACKET,EYE,NONROTA UOC:194,208,210,310,252,254,256	2
4 PFHZZ 86403	4086641	.PART KIT,STEERING UOC:194,208,210,230,231,252,254,256	1
5 PFHZZ 11862	15538347	.FRAME SECTION,STRUC UOC:194,208,210,230,231,252,254,256	1
6 PFHZZ 11862	14041286	.BRACE,REAR SPRING UOC:194,208,210,230,231,252,254,256	2
7 PFFZZ 11862	15593979	.BRACE,PROPELLER SHA APPLIES TO VEHICLES STARTING WITH VIN FF3000 UOC:194,208,210,230,231,252,254,256	1
8 PFHZZ 11862	14072639	.SIDE MEMEBER FRAME,L UOC:194,208,210,230,231,252,254,256	1
9 PFHZZ 11862	328059	.BRACKET,MOUNTING LEFT UOC:194,208,210,230,231,252,254,256	1
9 PFHZZ 11862	328060	.BRACKET,MOUNTING RIGHT UOC:194,208,210,230,231,252,254,256	1
10 PFHZZ 11862	14024587	.FRAME SECTION, STRUC UOC:194,208,210,230,231,252,254,256	1
11 PAFZZ 80205	NAS1408A6	.NUT,SELF-LOCKING,HE APPLIES TO VEHICLES STARTING WITH VIN FF3000 UOC:194,208,210,231,252,254,256	10
12 PAFZZ 96906	MS27183-16	.WASHER,FLAT APPLIES TO VEHICLES STARTING WITH VIN FF3000 UOC:194,208,210,230,231,252,254,256	10
13 PFFZZ 11862	15593980	.FRAME SECTION,STRUC UOC:194,208,210,230,231,252,254,256	1
14 PAFZZ 11862	9424320	.SCREW,CAP,HEXAGON H APPLIES TO VEHICLES STARTING WITH VIN FF3000 UOC:194,208,210,230,231,252,254,256	10
15 PFHZZ 11862	15593955	.FRAME SECTION,STRUC APPLIES TO VEHICLES STARTING WITH VIN FF3000 UOC:194,208,210,230,231,252,254,256	1

END OF FIGURE

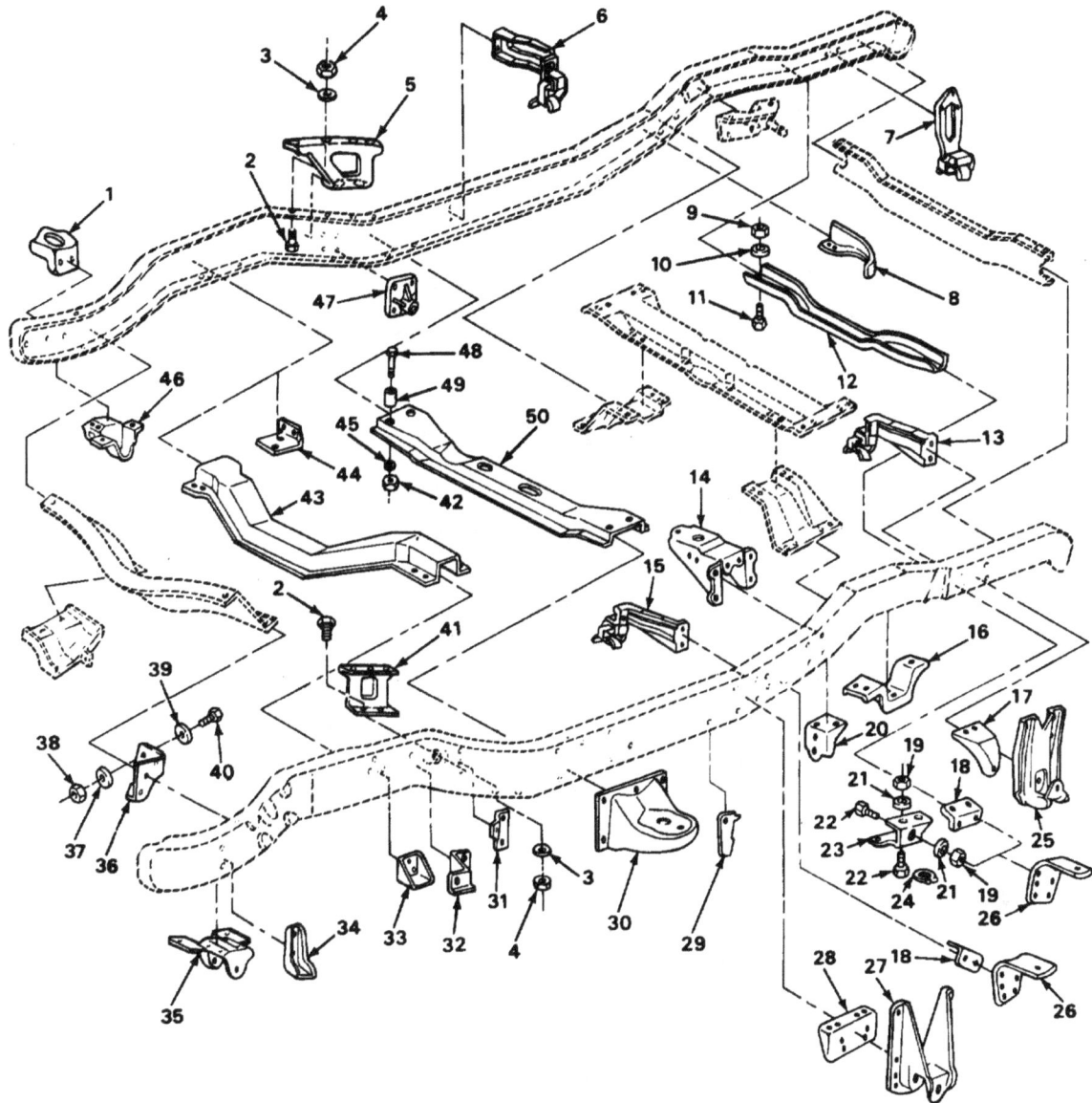

TA510947

FIGURE 140. FRAME ASSEMBLY RELATED PARTS (ALL EXCEPT M1009).

(1)	(2)	(3)	(4)	(5)	(6)
ITEM NO	SMR CODE	CAGEC	PART NUMBER	DESCRIPTION AND USABLE ON CODES (UOC)	QTY

GROUP 1501 FRAME ASSEMBLY

FIG. 140 FRAME ASSEMBLY RELATED PARTS (ALL EXCEPT M1009)

1	PFHZZ	11862	14029193	BRACKET	2
				UOC:194,208,210,230,231,252,254,256	
2	PAOZZ	11862	9424320	SCREW,CAP,HEXAGON H	10
				UOC:194,208,210,230,231,252,254,256	
3	PAOZZ	96906	MS27183-16	WASHER,FLAT	10
				UOC:194,208,210,230,231,252,254,256	
4	PAOZZ	80205	NAS1408A6	NUT,SELF-LOCKING,HE	10
				UOC:194,208,210,230,231,252,254,256	
5	PAOZZ	11862	15599998	BRACKET, SIDE MEMEBER RIGHT	1
				UOC:194,208,210,230,231,252,254,256	
6	PAOZZ	11862	14022538	BRACKET,PIPE	1
				UOC:194,208,210,230,231,252,254,256	
7	PAOZZ	11862	14034504	BRACKET,PIPE RIGHT	1
				UOC:194,208,210,230,231,252,254,256	
7	PAOZZ	11862	14034503	HANGER,PIPE LEFT	1
				UOC:194,208,210,230,231,252,254,256	
8	PFHZZ	11862	14036789	BRACKET,MOUNTING	1
				UOC:194,208,210,230,231,252,254,256	
9	PAOZZ	80205	NAS1408A6	NUT,SELF-LOCKING,HE	1
				UOC:194,208,210,230,231,252,254,256	
10	PAOZZ	11862	3914674	WASHER,FLAT	1
				UOC:194,208,210,230,231,252,254,256	
11	PAOZZ	96906	MS90728-59	SCREW,CAP,HEXAGON H	1
				UOC:194,208,210,230,231,252,254,256	
12	PAOZZ	11862	15595211	BAR,SPARE WHEEL	1
				UOC:194,208,210,230,231,252,254,256	
13	PAOZZ	11862	14034501	BRACKET,PIPE	1
				UOC:194,208,210,230,231,252,254,256	
14	PAHZZ	11862	14024567	BRACKET	2
				UOC:194,208,210,230,231,252,254,256	
15	PAOZZ	11862	14022537	BRACKET,PIPE	1
				UOC:194,208,210,230,231,252,254,256	
16	PFHZZ	11862	14000214	BRACKET,REAR AXLE B	2
				UOC:194,208,210,230,231,252,254,256	
17	PFHZZ	11862	14000212	HANGER,SPRING	2
				UOC:194,208,210,230,231,252,254,256	
18	PFHZZ	11862	14000213	BRACKET,ANGLE	4
				UOC:230,252,254,256	
19	PAOZZ	96906	MS51967-8	NUT,PLAIN,HEXAGON	2
				UOC:254,256	
20	PFHZZ	11862	14071973	BRACKET,ANGLE	1
				UOC:194,208,210,230,231,252,254,256	
21	PAOZZ	96906	MS35338-46	WASHER,LOCK	2
				UOC:254,256	
22	PAOZZ	96906	MS90728-60	SCREW,CAP,HEXAGON H	3
				UOC:254,256	
23	PAOZZ	11862	14001067	BRACKET,DOUBLE ANGL	2

(1) ITEM NO	(2) SMR CODE	(3) CAGEC	(4) PART NUMBER	(5) DESCRIPTION AND USABLE ON CODES (UOC)	(6) QTY
				UOC:254,256	
23	PAOZZ	11862	14001068	BRACKET,DOUBLE ANGL	2
				UOC:254,256	
24	PAOZZ	11862	337957	RETAINER,PACKING	2
				UOC:254,256	
25	PFHZZ	11862	14000211	HANGER,SPRING,VEHIC	2
				UOC:194,208,210,230,231,252,254,256	
26	PFHZZ	11862	328200	BRACKET,ANGLE	4
				UOC:230,252,254,256	
27	PFHZZ	11862	14000209	HANGER,SPRING,REAR	2
				UOC:194,208,210,230,231,252,254,256	
28	PFHZZ	11862	14041287	HANGER,SPRING LEFT	1
				UOC:194,208,210,230,231,252,254,256	
28	PFHZZ	11862	14041288	HANGER,SPRING RIGHT	1
				UOC:194,208,210,230,231,252,254,256	
29	PFHZZ	11862	14053598	BRACKET,ANGLE	1
				UOC:194,208,210,230,231,252,254,256	
30	PFHZZ	11862	14040692	BRACKET,MOUNTING	2
				UOC:194,208,210,230,231,252,254,256	
31	PFHZZ	11862	3943493	BRACKET,DOUBLE ANGL LEFT	1
				UOC:194,208,210,230,231,252,254,256	
31	PFHZZ	11862	3943494	BRACKET,DOUBLE ANGL RIGHT	1
				UOC:194,208,210,230,231,252,254,256	
32	PFHZZ	11862	14054259	BRACKET,DOUBLE ANGL LEFT	1
				UOC:194,208,210,230,231,252,254,256	
32	PFHZZ	11862	14036720	BRACKET RIGHT	1
				UOC:194,208,210,230,231,252,254,256	
33	PFHZZ	11862	460397	BRACKET	2
				UOC:194,208,210,230,231,252,254,256	
34	PFHZZ	11862	14029194	BRACKET,ANGLE	2
				UOC:194,208,210,230,231,252,254,256	
35	PFHZZ	11862	326439	HANGER,FRONT SPRING	1
				UOC:194,208,210,230,231,252,254,256	
36	PFHZZ	11862	14024593	BRACKET,FRAME	1
				UOC:194,208,210,230,231,252,254,256	
37	PAHZZ	96906	MS35338-46	WASHER,LOCK	2
				UOC:194,208,210,230,231,252,254,256	
38	PAHZZ	96906	MS51967-8	NUT,PLAIN,HEXAGON	2
				UOC:194,208,210,230,231,252,254,256	
39	PAHZZ	96906	MS27183-15	WASHER,FLAT	2
				UOC:194,208,210,230,231,252,254,256	
40	PAHZZ	96906	MS90728-61	SCREW,CAP,HEXAGON H	2
				UOC:194,208,210,230,231,252,254,256	
41	PAOZZ	11862	15599997	BRACKET,SIDE MEMEBER LEFT	1
				UOC:194,208,210,230,231,252,254,256	
42	PAHZZ	96906	MS51967-12	NUT,PLAIN,HEXAGON	4
				UOC:194,208,210,230,231,252,254,256	
43	PFHZZ	11862	14072695	FRAME SECTION,STRUC	1
				UOC:194,208,210,230,231,252,254,256,	
44	PFHZZ	11862	14072690	SUPPORT,ENGINE,FRON	1
				UOC:194,208,210,230,231,252,254,256	
45	PAHZZ	96906	MS27183-16	WASHER,FLAT	4

(1) ITEM NO	(2) SMR CODE	(3) CAGEC	(4) PART NUMBER	(5) DESCRIPTION AND USABLE ON CODES (UOC)	(6) QTY
				UOC:194,208,210,230,231,252,254,256	
46	PFHZZ	11862	326440	HANGER,FRONT SPRING	1
				UOC:194,208,210,230,231,252,254,256	
47	PFHZZ	11862	6271344	HANGER,SPRING	2
				UOC:194,208,210,230,231,252,254,256	
48	PAHZZ	11862	343178	SCREW,CAP,HEXAGON H	4
				UOC:194,208,210,230,231,252,254,256	
49	PAHZZ	11862	343179	SPACER,SLEEVE	4
				UOC:194,208,210,230,231,252,254,256	
50	PFHZZ	11862	14024588	BRACET,TRANSMISSIO	1
				UOC:194,208,210,230,231,252,254,256	

END OF FIGURE

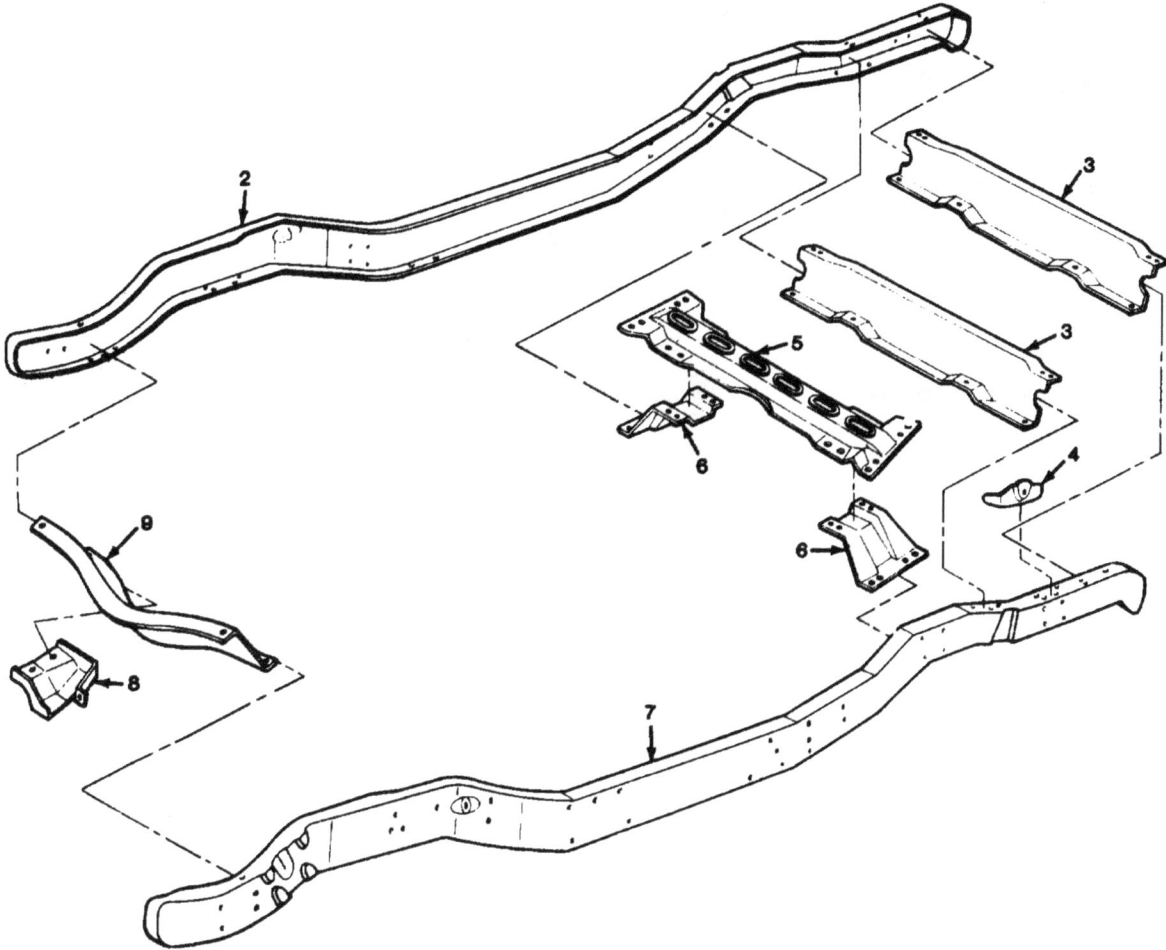

TA510948

FIGURE 141. FRAME ASSEMBLY COMPONENTS (M1009).

(1) ITEM NO	(2) SMR CODE	(3) CAGEC	(4) PART NUMBER	(5) DESCRIPTION AND USABLE ON CODES (UOC)	(6) QTY
				GROUP 1501 FRAME ASSEMBLY	
				FIG. 141 FRAME ASSEMBLY COMPONENTS (M1009)	
1	PAHHH	11862	15589794	FRAME,STRUCTURAL,VE UOC:209	1
2	PAHZZ	11862	14072632	.SIDE MEMBER FRAME,R UOC:209	1
3	PFHZZ	11862	14067782	.FRAME SECTION,STRUC UOC:209	2
4	PAHZZ	11862	6271341	.BRACKET,EYE,NONROTA UOC:209	1
5	PFHZZ	11862	14024577	.FRAME SECTION,STRUC UOC:209	1
6	PFHZZ	11862	14041286	.BRACE,REAR SPRING UOC:209	2
7	PAHZZ	11862	14072631	.SIDE MEMBER,FRAME,L UOC:209	1
8	PAHZZ	11862	328059	.BRACKER,MOUNTING LEFT UOC:209	1
8	PFHZZ	11862	328060	.BRACKET,MOUNTING RIGHT UOC:209	1
9	PFHZZ	11862	14024587	.FRAME SECTION,STRUC UOC:209	1

END OF FIGURE

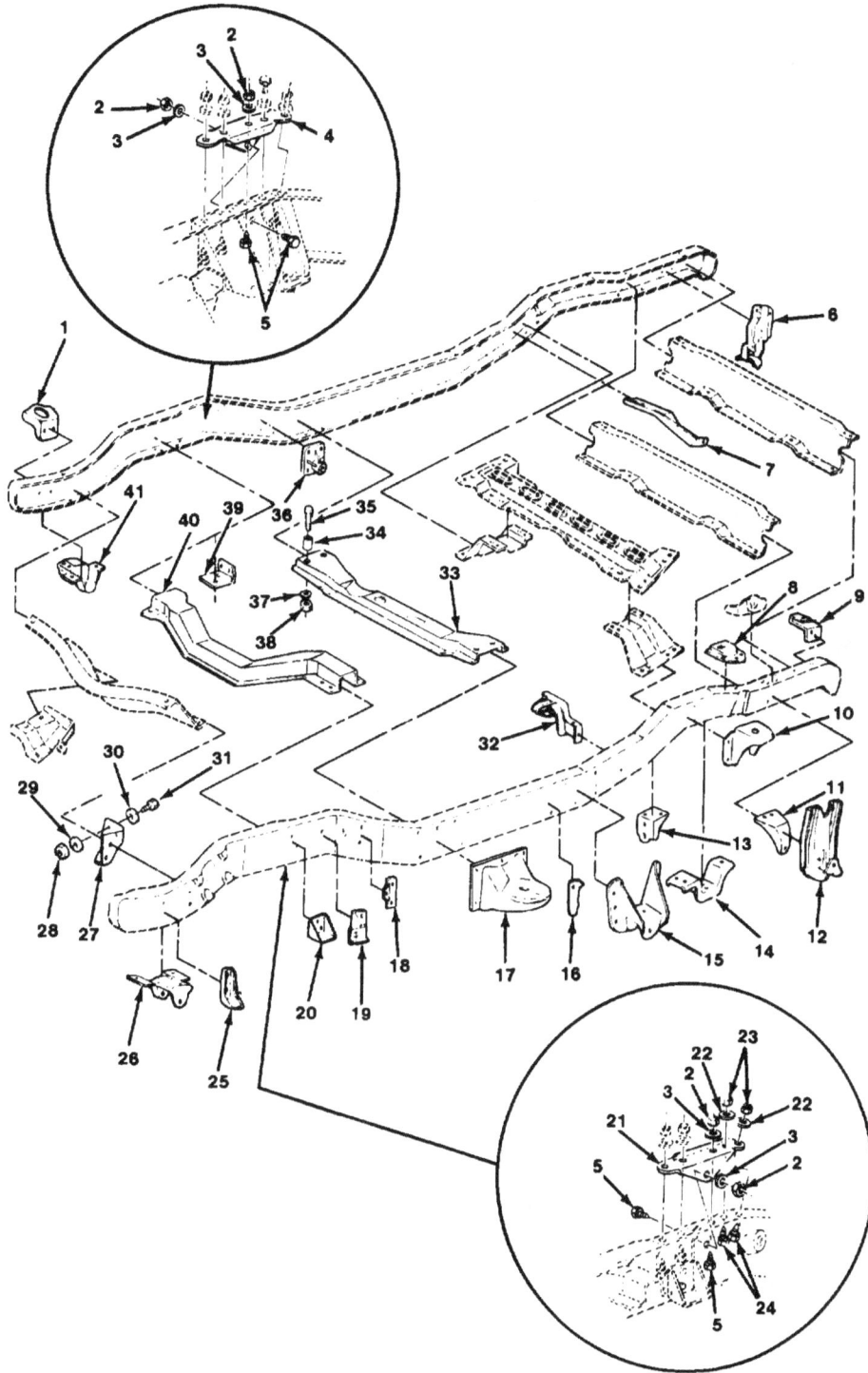

FIGURE 142. FRAME ASSEMBLY RELATED PARTS (M1009).

TA510949

(1) ITEM NO	(2) SMR CODE	(3) CAGEC	(4) PART NUMBER	(5) DESCRIPTION AND USABLE ON CODES (UOC)	(6) QTY
				GROUP 1501 FRAME ASSEMBLY FIG. 142 FRAME ASSEMBLY RELATED PARTS (M1009)	
1	PFHZZ	11862	14029193	BRACKET UOC:209	2
2	PAOZZ	11862	9422299	NUT,SELF-LOCKING,HE UOC:209	8
3	PAOZZ	24617	3990160	WASHER,FLAT UOC:209	8
4	XBOZZ	11862	TX001488	BRACKET,RH UOC:209	1
5	PAOZZ	11862	9440344	BOLT,MACHING UOC:209	8
6	PAOZZ	11862	14020478	SUPPORT ASSEMBLY,MU RIGHT UOC:209	1
6	PAOZZ	11862	14034513	SUPPORT,MUFFLER AND LEFT UOC:209	1
6	PFHZZ	11862	15593907	HANGER,ENGINE EXHAU LEFT, APPLIES TO VEHICLES STARTING WITH VIN 136169 UOC:209	1
6	XDHZZ	11862	15593908	HANGER,ENGINE EXHAU APPLIES TO VEHICLES STARTING WITH VIN FF136169 UOC:209	1
7	PFHZZ	11862	338699	BRACKET,MOUNTING UOC:209	1
8	PFHZZ	11862	326418	BRACKET,BODY RIGHT UOC:209	1
8	PFHZZ	11862	6274179	BRACKET,BODY LEFT UOC:209	1
9	PFHZZ	11862	14043889	SUPPORT ASSEMBLY,CU LEFT UOC:209	1
9	PFHZZ	11862	14043890	SUPPORT,BODY HOLD D RIGHT UOC:209	1
10	PFHZZ	11862	14029286	BRACKET,BODY UOC:209	2
11	PFHZZ	11862	6270214	HANGER,SPRING,VEHIC UOC:209	2
12	PFHZZ	11862	6270213	HANGER,SPRING,VEHIC UOC:209	2
13	PAHZZ	11862	14055592	BRACKET,ANGLE UOC:209	1
14	PFHZZ	11862	327260	BRACKET,REAR BUMPER UOC:209	2
15	PFHZZ	11862	460373	HANGER,SPRING,VEHIC LEFT UOC:209	1
15	PFHZZ	11862	460374	HANGER,SPRING,VEHIC RIGHT UOC:209	1
16	PFHZZ	11862	14053598	BRACKET,ANGLE UOC:209	1
17	PFHZZ	11862	14040692	BRACKET,MOUNTING UOC:209	2

(1) ITEM NO	(2) (3) (4) SMR CODE CAGEC PART NUMBER	(5) DESCRIPTION AND USABLE ON CODES (UOC)	(6) QTY
18	PFHZZ 11862 3943493	BRACKET,DOUBLE ANGL LEFT UOC:209	1
18	PFHZZ 11862 3943494	BRACKET,DOUBLE ANGL RIGHT UOC:209	1
19	PFHZZ 11862 14036719	BRACKET,DOUBLE ANGL LEFT UOC:209	1
19	PFHZZ 11862 14036720	BRACKET RIGHT UOC:209	1
20	PFHZZ 11862 460397	BRACKET UOC:209	2
21	XBOZZ 11862 TX001487	BRACKET,LH UOC:209	1
22	PAOZZ 96906 MS35340-46	WASHER,LOCK UOC:209	4
23	PAOZZ 11862 9418931	NUT,PLAIN,HEXAGON UOC:209	3
24	PAOZZ 96906 MS90728-63	SCREW,CAP,HEXAGON H UOC:209	2
25	PFHZZ 11862 14029194	BRACKET,ANGLE UOC:209	2
26	PFHZZ 11862 326439	HANGER,FRONT SPRING UOC:209	1
27	PFHZZ 11862 14024593	BRACKET,FRAME UOC:209	1
28	PAHZZ 96906 MS51967-5	NUT,PLAIN,HEXAGON UOC:209	2
29	PAHZZ 96906 MS35338-46	WASHER,LOCK UOC:209	2
30	PAHZZ 96906 MS27183-15	WASHER,FLAT UOC:209	2
31	PAHZZ 96906 MS90728-61	SCREW,CAP,HEXAGON H UOC:209	2
32	PAOZZ 11862 14007142	HANGER,ENGINE EXHAU UOC:209	1
33	PFHZZ 11862 14034558	BRACKET,TRANSMISSIO UOC:209	1
34	PAHZZ 11862 343179	SPACER,SLEEVE UOC:209	4
35	PAHZZ 11862 343178	SCREW,CAP,HEXAGON H UOC:209	4
36	PFHZZ 11862 6271344	HANGER,SPRING UOC:209	2
37	PAHZZ 96906 MS27183-16	WASHER,FLAT UOC:209	4
38	PAHZZ 96906 MS51967-12	NUT,PLAIN,HEXAGON UOC:209	4
39	PFHZZ 11862 14072690	SUPPORT,ENGINE,FRON UOC:209	1
40	PFHZZ 11862 14072695	FRAME SECTION,STRUC UOC:209	1
41	PFHZZ 11862 326440	HANGER,FRONT SPRING UOC:209	1

END OF FIGURE

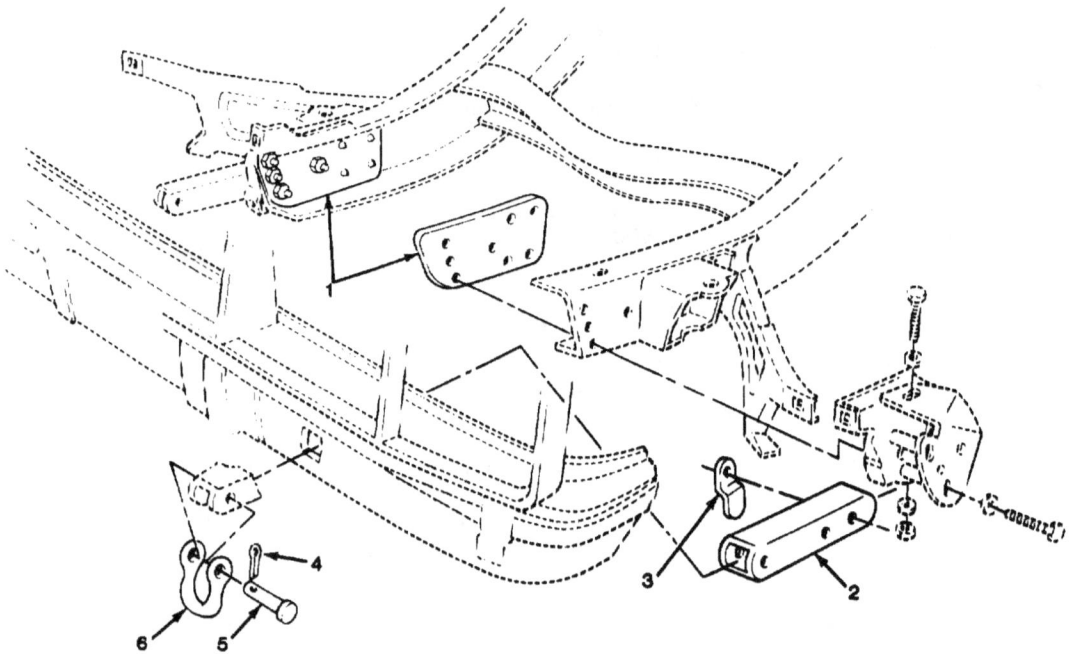

TA510950

FIGURE 143. CLEVISES AND SUPPORTS.

(1) (2) (3) (4)				(5)	(6)
ITEM	SMR		PART		
NO	CODE	CAGEC	NUMBER	DESCRIPTION AND USABLE ON CODES (UOC)	QTY

GROUP 1503 PINTLES AND TOWING ATTACHMENTS

FIG. 143 CLEVISES AND SUPPORTS

1	PAOZZ	11862	14067790	PLATE,TIE-DOWN	2
2	PAOZZ	11862	14067795	EXTENSION,TIE DOWN	2
3	PAOZZ	11862	14067796	BRACKET,DOUBLE ANGL	2
4	PAOZZ	96906	MS24665-497	PIN,COTTER	2
5	PAOZZ	11862	14067794	PIN,STRAIGHT,HEADED	2
6	PAOZZ	19207	7358030	SHACKLE	2

END OF FIGURE

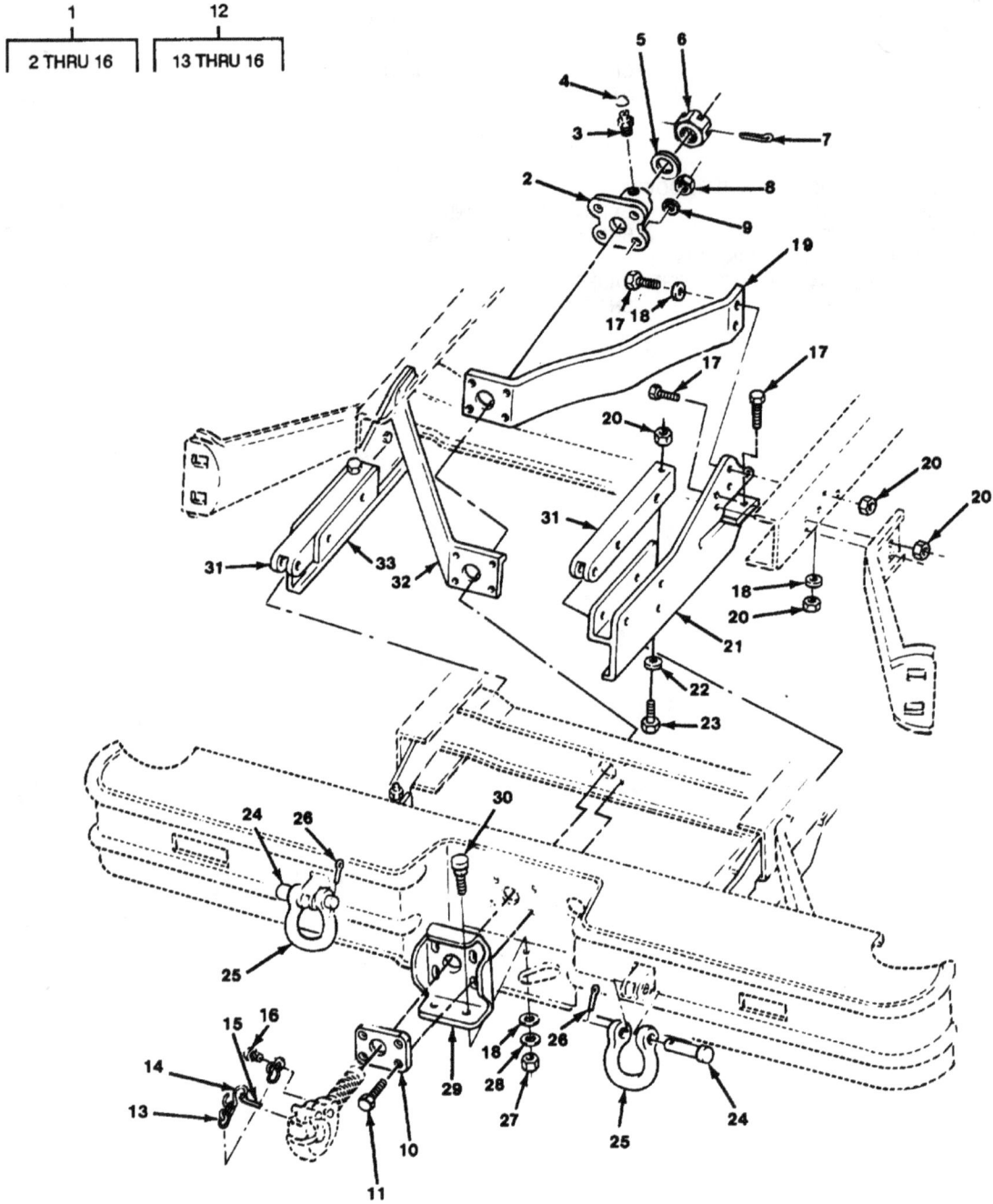

TA510951

FIGURE 144. PINTLE ASSEMBLY AND CLEW (M1008, M1028, AND M1028A1).

(1)	(2)	(3)	(4)	(5)	(6)
ITEM NO	SMR CODE	CAGEC	PART NUMBER	DESCRIPTION AND USABLE ON CODES (UOC)	QTY

GROUP 1503 PINTLES AND TOWING ATTACHMENTS

FIG. 144 PINTLE ASSEMBLY AND CLEVIS (M1008, M1028, AND M1028A1)

ITEM NO	SMR CODE	CAGEC	PART NUMBER	DESCRIPTION AND USABLE ON CODES (UOC)	QTY
1	PAOOO	96906	MS51335-1	PINTLE ASSEMBLY,TOW UOC:194,208,230,252	1
2	PAOZZ	74410	XA-T-61-SR	.COLLAR,SHAFT UOC:194,208,230,252,254,256	1
3	PAOZZ	96906	MS15001-1	.FITTING,LUBRICATION UOC:194,208,230,252,254,256	1
4	PAOZZ	74410	XB-767-10	.CAP,PROTECTIVE,DUST UOC:194,208,230,252,254,256	1
5	PAOZZ	74410	XA-T-88	.WASHER,FLAT UOC:194,208,230,252,254,256	1
6	PAOZZ	96906	MS35692-94	.NUT,PLAIN,SLOTTED,H UOC:194,208,230,252,254,256	1
7	PAOZZ	96906	MS24665-628	.PIN,COTTER UOC:194,208,230,252,254,256	1
8	XDOZZ	74410	XB-766	.NUT,PLAIN,HEXAGON UOC:194,208,230,252,254,256	4
9	PAOZZ	74410	XB-T-45-1	.WASHER,LOCK UOC:194,208,230,252,254,256	4
10	PAOZZ	74410	XA-T-61-SF	.BRACKET,PINTLE HOOK UOC:194,208,230,252,254,256	1
11	PAOZZ	96906	MS90727-117	.SCREW,CAP,HEXAGON H UOC:194,208,230,252,254,256	4
12	PAOOO	96906	MS51335-1	.PINTLE ASSEMBLY,TOW UOC:194,208,230,252,254,256	1
13	MOOZZ	74410	XX123	..CHAIN (5" LG) MAKE FROM CHAIN, P/N RRC271BTY2CLDIA072 UOC:194,208,230,252,254,256	1
14	PAOZZ	96906	MS87006-53	..HOOK,CHAIN,S UOC:194,208,230,252,254,256	1
15	PAOZZ	80020	36344N24	..PIN,COTTER UOC:194,208,230,252,254,256	1
16	PAOZZ	96906	MS21318-47	..SCREW,DRIVE UOC:194,208,230,252,254,256	1
17	PAOZZ	96906	MS90728-113	SCREW,CAP,HEXAGON H UOC:194,208,230,252	12
18	PAOZZ	96906	MS27183-18	WASHER,FLAT UOC:194,208,230,252	8
19	PAOZZ	11862	14067786	BAR,REINFORCEMENT,P UOC:194,208,230,252,254,256	1
20	PAOZZ	24617	9422301	NUT,SELF-LOCKING,HE UOC:194,208,230,252	14
21	PAOZZ	11862	14067780	BRACKET,SPECIAL UOC:194,208,230,252,254,256	1
22	PAOZZ	96906	MS27183-19	WASHER,FLAT UOC:194,208,230,252	2
23	PAOZZ	96906	MS90728-119	SCREW,CAP,HEXAGON H	2

(1) ITEM NO	(2) SMR CODE	(3) CAGEC	(4) PART NUMBER	(5) DESCRIPTION AND USABLE ON CODES (UOC)	(6) QTY
				UOC:194,208,230,252	
24	PAOZZ	11862	14067794	PIN,STRAIGHT,HEADED	2
				UOC:194,208,230,252	
25	PAOZZ	11862	14067793	SHACKLE	2
				UOC:194,208,230,252	
26	PAOZZ	96906	MS24665-497	PIN,COTTER	2
				UOC:194,208,230,252	
27	PAOZZ	96906	MS51967-14	NUT,PLAIN,HEXAGON	2
				UOC:194,208,230,252,254,256	
28	PAOZZ	96906	MS35340-48	WASHER,LOCK	2
				UOC:194,208,230,252,254,256	
29	PAOZZ	11862	14078806	REINFORCEMENT,BUMPE	1
				UOC:194,208,230,252,254,256	
30	PAOZZ	11862	14072425	BOLT,SQUARE NECK	2
				UOC:194,208,230,252,254,256	
31	PAOZZ	11862	14067787	EXTENSION,TIE DOWN	2
				UOC:194,208,230,252,254,256	
32	PAOZZ	11862	14067785	BRACKET,TOW HOOK	1
				UOC:194,208,230,252,254,256	
33	PAOZZ	11862	14067779	BRACKET,SPECIAL	1
				UOC:194,208,230,252,254,256	

END OF FIGURE

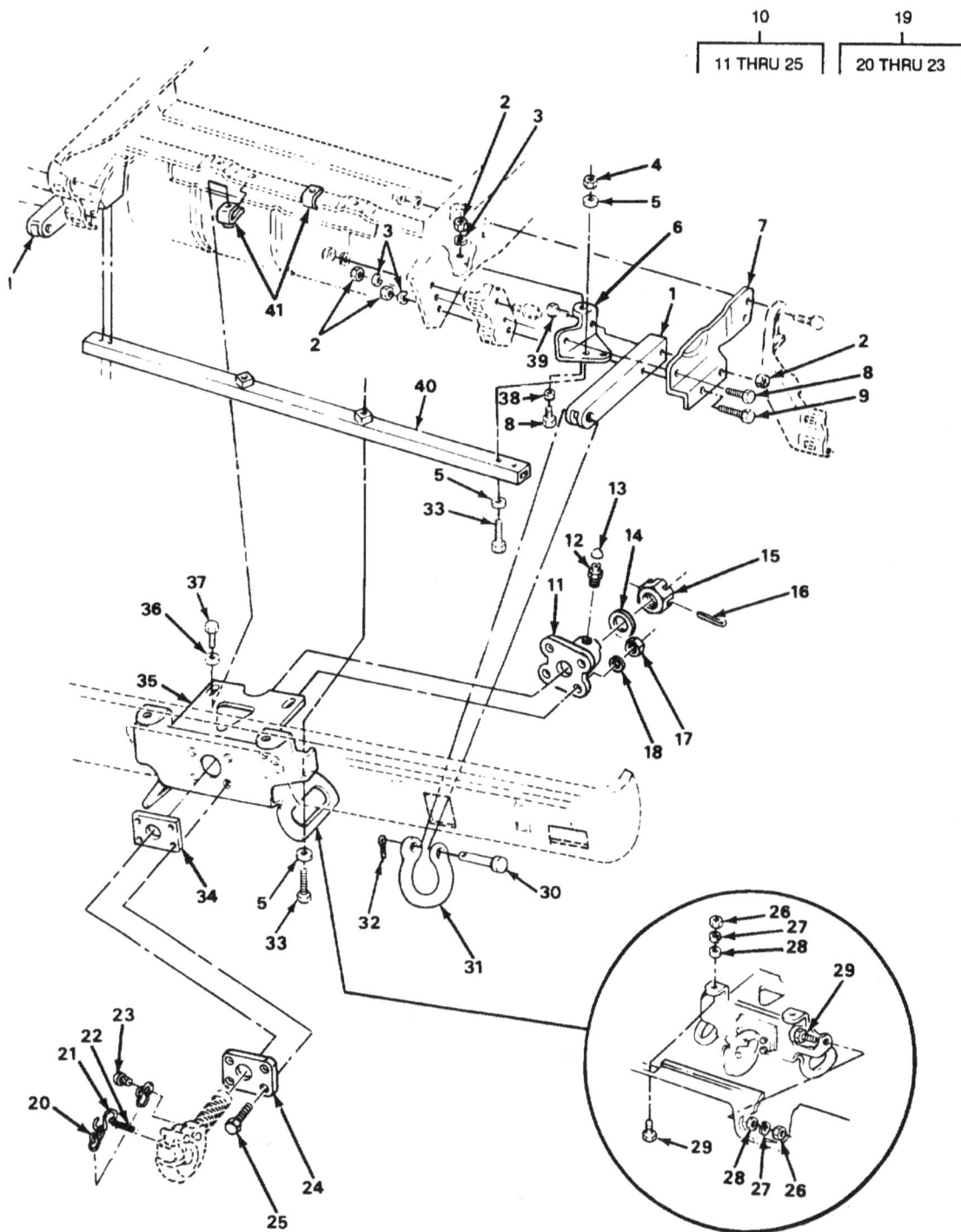

FIGURE 145. PINTLE ASSEMBLY AND CLEVIS (M1009).

TA510952

(1) ITEM NO	(2) SMR CODE	(3) CAGEC	(4) PART NUMBER	(5) DESCRIPTION AND USABLE ON CODES (UOC)	(6) QTY
				GROUP 1503 PINTLES AND TOWING ATTACHMENTS	
				FIG. 145 PINTLE ASSEMBLY AND CLEVIS (M1009)	
1	PAOZZ	11862	14067770	EXTENSION,TIE DOWN UOC:209	2
2	PAOZZ	96906	MS51967-14	NUT,PLAIN,HEXAGON UOC:209	10
3	PAOZZ	96906	MS35340-48	WASHER,LOCK UOC:209	6
4	PAOZZ	96906	MS51967-12	NUT,PLAIN,HEXAGON UOC:209	4
5	PAOZZ	11862	3790768	WASHER,FLAT UOC:209	10
6	PAOZZ	11862	14067773	BRACKET,REINFORCEME LEFT UOC:209	1
6	PAOZZ	11862	14067774	BRACKET,TOW HOOK RIGHT UOC:209	1
7	PAOZZ	11862	14067771	BRACKET,DOUBLE ANGL LEFT UOC:209	1
7	PAOZZ	11862	14067772	BRACKET,DOUBLE ANGL RIGHT UOC:209	1
8	PAOZZ	96906	MS90728-111	SCREW,CAP,HEXAGON H UOC:209	4
9	PAOZZ	96906	MS90728-119	SCREW,CAP,HEXAGON H UOC:209	2
10	PAOOO	96906	MS51335-1	PINTLE ASSEMBLY,TOW UOC:209	1
11	PAOZZ	74410	XA-T-61-SR	.COLLAR,SHAFT UOC:209	1
12	PAOZZ	96906	MS15001-1	.FITTING,LUBRICATION UOC:209	1
13	PAOZZ	74410	XB-767-10	.CAP,PROTECTIVE,DUST UOC:209	1
14	PAOZZ	74410	XA-T-88	.WASHER,FLAT UOC:209	1
15	PAOZZ	96906	MS35692-94	.NUT,PLAIN,SLOTTED,H UOC:209	1
16	PAOZZ	96906	MS24665-628	.PIN,COTTER UOC:209	1
17	XDOZZ	74410	XB-766	.NUT,PLAIN,HEXAGON UOC:209	4
18	PAOZZ	74410	XB-T-45-1	.WASHER,LOCK UOC:209	4
19	PAOOO	96906	MS51335-1	.PINTLE ASSEMBLY,TOW UOC:209	1
20	MOOZZ	74410	XX123	..CHAIN (5" LG) MAKE FROM CHAIN, P/N RRC271BTY2CLDIA072 UOC:209	1
21	PAOZZ	96906	MS87006-53	..HOOK,CHAIN,S	1

(1) ITEM NO	(2) SMR CODE	(3) CAGEC	(4) PART NUMBER	(5) DESCRIPTION AND USABLE ON CODES (UOC)	(6) QTY
				UOC:209	
22	PAOZZ	80020	36344N24	..PIN,COTTER	1
				UOC:209	
23	PAOZZ	96906	MS21318-47	..SCREW,DRIVE	1
				UOC:209	
24	PAOZZ	74410	XA-T-61-SF	.BRACKET,PINTLE HOOK	1
				UOC:209	
25	PAOZZ	96906	MS90727-117	.SCREW,CAP,HEXAGON H	4
				UOC:209	
26	PAOZZ	96906	MS51967-8	NUT,PLAIN,HEXAGON	4
				UOC:209	
27	PAOZZ	96906	MS35338-46	WASHER,LOCK	4
				UOC:209	
28	PAOZZ	96906	MS27183-14	WASHER,FLAT	4
				UOC:209	
29	PAOZZ	96906	MS90728-60	SCREW,CAP,HEXAGON H	4
				UOC:209	
30	PAOZZ	11862	14067794	PIN,STRAIGHT,HEADED	2
				UOC:209	
31	PAOZZ	11862	14067793	SHACKLE	2
				UOC:209	
32	PAOZZ	96906	MS24665-497	PIN,COTTER	2
				UOC:209	
33	PAOZZ	96906	MS90728-92	SCREW,CAP,HEXAGON H	6
				UOC:209	
34	PAOZZ	11862	14067789	SPACER,PLATE	2
				UOC:209	
35	PAOZZ	11862	14067776	BRACKET,TOW HOOK	1
				UOC:209	
36	PAOZZ	96906	MS27183-15	WASHER,FLAT	2
				UOC:209	
37	PAOZZ	96906	MS90728-87	SCREW,CAP,HEXAGON H	2
				UOC:209	
38	PAOZZ	96906	MS27183-18	WASHER,FLAT	2
				UOC:209	
39	PAOZZ	96906	MS90728-117	SCREW,CAP,HEXAGON H	2
				UOC:209	
40	PAOZZ	11862	14067775	BRACKET,TOW HOOK	1
				UOC:209	
41	PAOZZ	11862	480567	NUT,CLIP-ON	2
				UOC:209	

END OF FIGURE

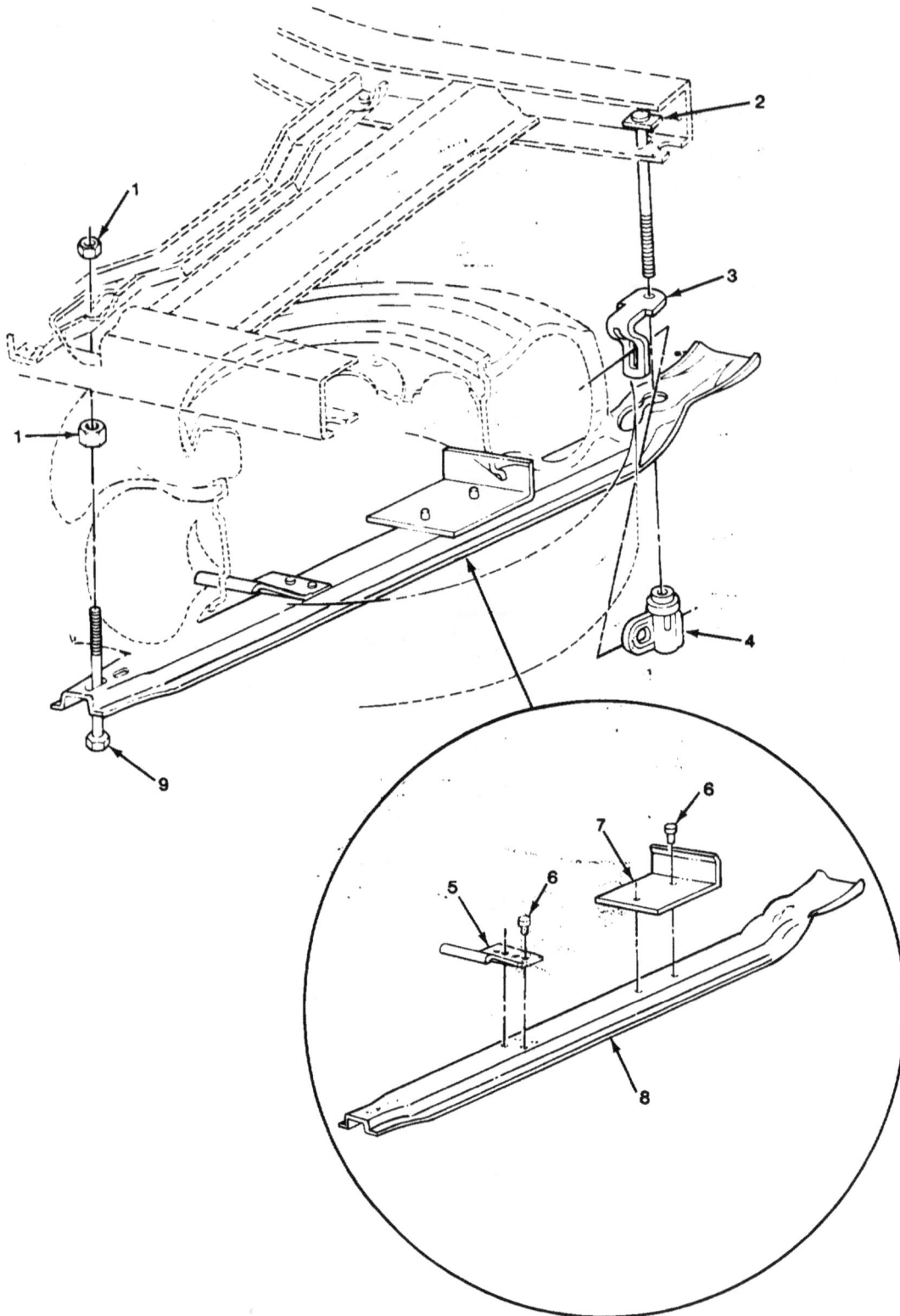

FIGURE 146. SPARE WHEEL CARRIER (ALL EXCEPT M1009).

TA510953

(1)	(2)	(3)	(4)	(5)	(6)
ITEM NO	SMR CODE	CAGEC	PART NUMBER	DESCRIPTION AND USABLE ON CODES (UOC)	QTY

GROUP 1504 SPARE WHEEL CARRIER AND TIRE LOCK

FIG. 146 SPARE WHEEL CARRIER (ALL EXCEPT M1009)

1	PAOZZ	96906	MS51967-14	NUT,PLAIN,HEXAGON UOC:194,208,210,230,231,252,254,256	2
2	PAOZZ	11862	3991022	BOLT,MACHINE UOC:194,208,210,230,231,252,254,256	1
3	PAOZZ	11862	15599915	HOLD DOWN,SPARE TIR UOC:194,208,210,230,231,252,254,256	1
4	PAOZZ	11862	6274031	NUT,PLAIN,HEXAGON UOC:194,208,210,230,231,252,254,256	1
5	PAOZZ	11862	350037	PILOT,WHEEL CARRIER UOC:194,208,210,230,231,252,254,256	1
6	PAOZZ	24617	189448	RIVET,SOLID UOC:194,208,210,230,231,252,254,256	4
7	PAOZZ	11862	6274036	PILOT,WHEEL CARRIER UOC:194,208,210,230,231,252,254,256	1
8	PAOZZ	11862	350036	STRAP,SPARE WHEEL UOC:194,208,210,230,231,252,254,256	1
9	PAOZZ	11862	371603	SCREW,CAP,HEXAGON H UOC:194,208,210,230,231,252,254,256	1

END OF FIGURE

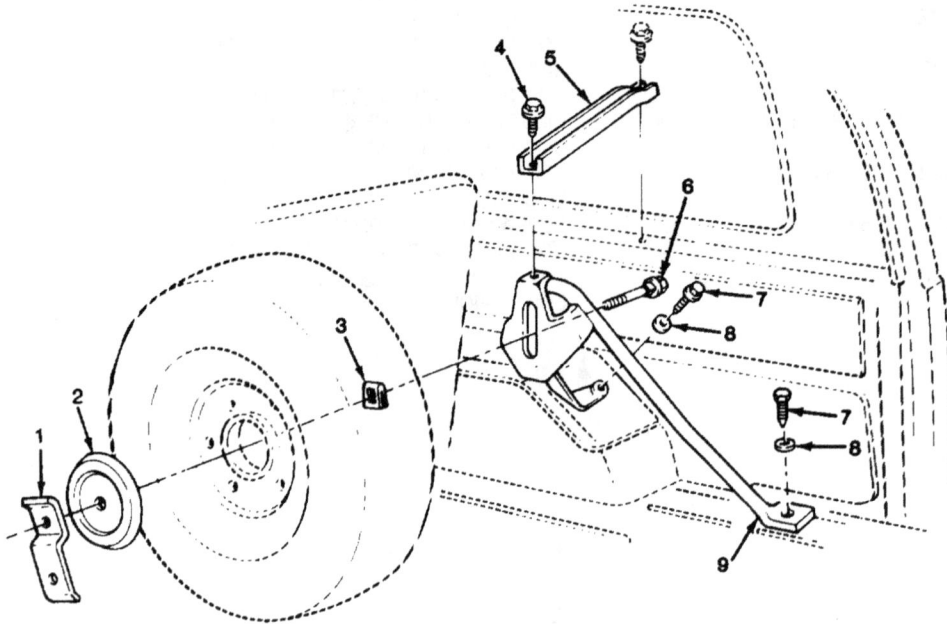

TA510954

FIGURE 147. SPARE WHEEL CARRIER (M1009).

(1) (2) (3) (4)			(5)	(6)
ITEM SMR PART				
NO CODE CAGEC NUMBER			DESCRIPTION AND USABLE ON CODES (UOC)	QTY

GROUP 1504 SPARE WHEEL CARRIER AND
TIRE LOCK

FIG. 147 SPARE WHEEL CARRIER (M1009)

1	PAOZZ 11862 330046		RETAINER,SPARE TIRE	1
			UOC:209	
2	PAOZZ 11862 14027926		PLATE,RESILIENT MOU	1
			UOC:209	
3	PAOZZ 11862 3725668		PUSH ON NUT	1
			UOC:209	
4	PAOZZ 24617 443945		SCREW	2
			UOC:209	
5	PAOZZ 11862 14007545		BRACE,SPARE WHEEL S	1
			UOC:209	
6	PAOZZ 11862 15599432		BOLT,SQUARE NECK	1
			UOC:209	
7	PAOZZ 11862 3954730	SETSCREW		2
			UOC:209	
8	PAOZZ 96906 MS27183-18	WASHER,FLAT		2
			UOC:209	
9	PAOZZ 11862 343951		SUPPORT ASSEMBLY,WH	1
			UOC:209	

END OF FIGURE

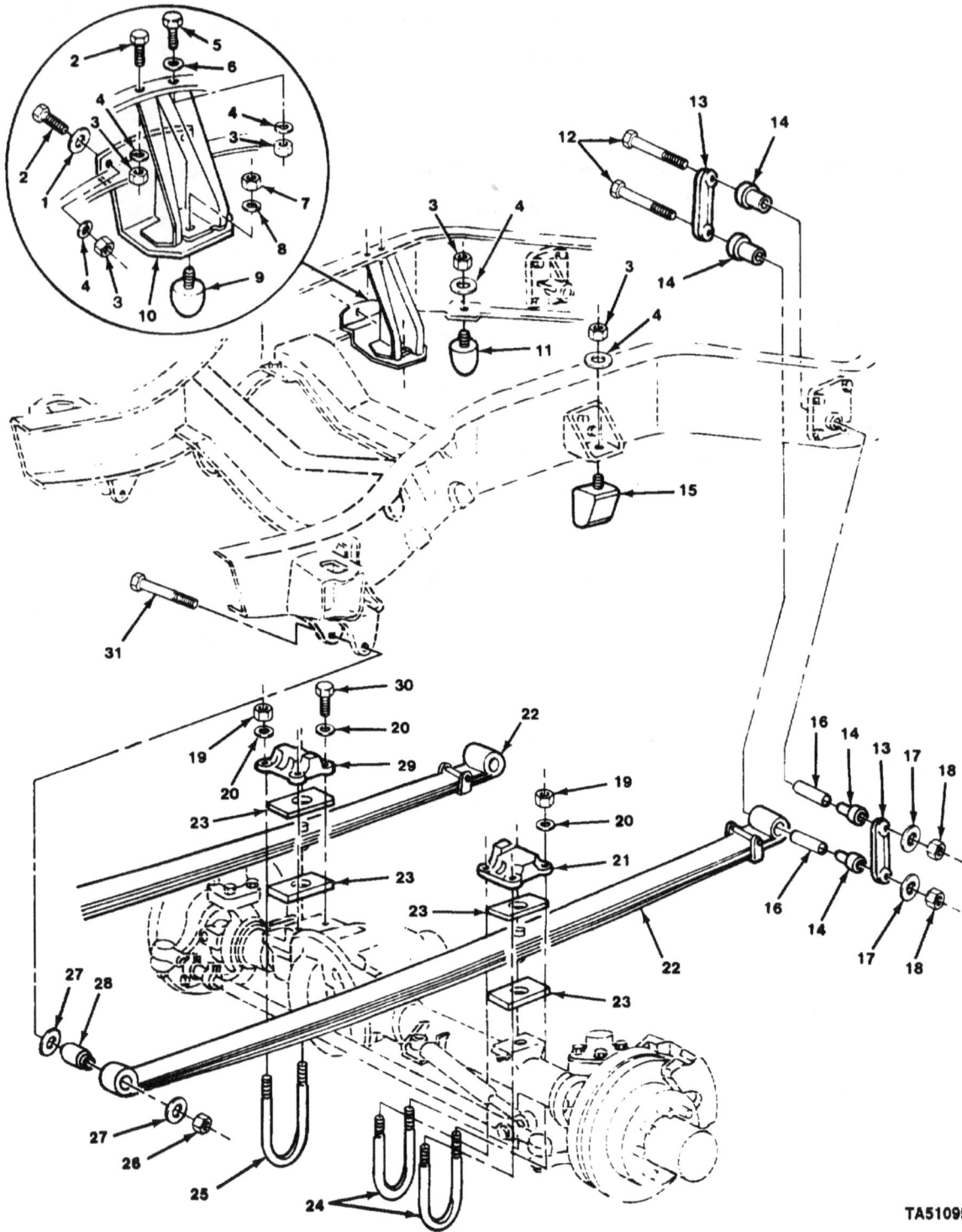

TA510955

FIGURE 148. FRONT SPRINGS AND RELATED PARTS (ALL EXCEPT M1009).

(1) ITEM NO	(2) SMR CODE	(3) CAGEC	(4) PART NUMBER	(5) DESCRIPTION AND USABLE ON CODES (UOC)	(6) QTY
				GROUP 16 SPRINGS AND SHOCK ABSORBERS	
				GROUP 1601 SPRINGS	
				FIG. 148 FRONT SPRINGS AND RELATED PARTS (ALL EXCEPT M1009)	
1	PAOZZ	96906	MS27183-14	WASHER,FLAT	2
				UOC:194,208,210,230,231,252,254	
2	PAOZZ	96906	MS90728-60	SCREW,CAP,HEXAGON H	4
				UOC:194,208,210,230,231,252,254,256	
3	PAOZZ	96906	MS51967-8	NUT,PLAIN,HEXAGON	8
				UOC:194,208,210,230,231,252,254,256	
4	PAOZZ	96906	MS35338-46	WASHER,LOCK	10
				UOC:194,208,210,230,231,252,254,256	
5	PAOZZ	96906	MS90728-62	SCREW,CAP,HEXAGON H	1
				UOC:194,208,210,230,231,252,254,256	
6	PAOZZ	96906	MS27183-16	WASHER,FLAT	1
				UOC:194,208,210,230,231,252,254,256	
7	PAOZZ	96906	MS51967-14	NUT,PLAIN,HEXAGON	1
				UOC:194,208,210,230,231,252,254	
8	PAOZZ	96906	MS35340-48	WASHER,LOCK	1
				UOC:194,208,210,230,231,252,254,256	
9	PAOZZ	11862	14029200	BUMPER,NONMETALLIC	1
				UOC:194,208,210,230,231,252,254,256	
10	PAOZZ	11862	14045654	BRACKET ASSEMBLY,FR	1
				UOC:194,208,210,230,231,252,254,256	
11	PAOZZ	11862	359878	BUMPER ASSEMBLY,FRO	2
				UOC:194,208,210,230,231,252	
12	PAFZZ	96906	MS90727-103	SCREW,CAP,HEXAGON H	4
				UOC:194,208,210,230,231,252,254,256	
13	PAFZZ	11862	3975202	PARTS KIT,LEAF SPRI	4
				UOC:194,208,210,230,231,252,254,256	
14	PAFZZ	11862	363913	BUSHING,NONMETALLIC	8
				UOC:194,208,210,230,231,252,254,256	
15	PAOZZ	11862	359877	BUMPER ASSEMBLY,FRO	2
				UOC:194,208,210,230,231,252,254,256	
16	PAFZZ	11862	15571581	SPACER,SLEEVE	4
				UOC:194,208,210,230,231,252,254,256	
17	PAFZZ	11862	3790768	WASHER,FLAT	4
				UOC:194,208,210,230,231,252,254,256	
18	PAFZZ	80205	NAS1409A7	NUT,SELF-LOCKING,HE	4
				UOC:194,208,210,230,231,252,254,256	
19	PAFZZ	96906	MS51968-20	NUT,PLAIN,HEXAGON	6
				UOC:194,208,210,230,231,252,254,256	
20	PAFZZ	96906	MS27183-21	WASHER,FLAT	8
				UOC:194,208,210,230,231,252,254,256	
21	PAFZZ	11862	370055	PLATE,WEAR,LEAF SPR	1
				UOC:194,208,210,230,231,252,254,256	
22	PAFZZ	11862	460354	SPRING ASSEMBLY,LEA	2
				UOC:194,208,210,230,231,252,254,256	
23	PAFZZ	11862	3970988	WASHER,FLAT	4

(1) ITEM NO	(2) SMR CODE	(3) CAGEC	(4) PART NUMBER	(5) DESCRIPTION AND USABLE ON CODES (UOC)	(6) QTY
				UOC:194,208,210,230,231,252,254,256	
24	PAFZZ	11862	370053	BOLT,	2
				UOC:194,208,210,230,231,252,254,256	
25	PAFZZ	11862	370054	BOLT,U	1
				UOC:194,208,210,230,231,252,254,256	
26	PAFZZ	11862	9422303	NUT,SELF-LOCKING,HE	2
				UOC:194,208,210,231,252,254,256	
27	PAFZZ	96906	MS27183-19	WASHER,FLAT	4
				UOC:194,208,210,230,252	
28	PAFZZ	11862	3887751	BUSHING	2
				UOC:194,208,210,230,231,252,254,256	
29	PAFZZ	11862	370056	PLATE,WEAR,LEAF SPR	1
				UOC:194,208,210,230,231,252,254,256	
30	PAFZZ	11862	460340	SETSCREW	2
				UOC:194,208,210,231,252,254,256	
31	PAFZZ	96906	MS90728-152	SCREW,CAP,HEXAGON H	2
				UOC:194,208,210,230,231,252,254,256	

END OF FIGURE

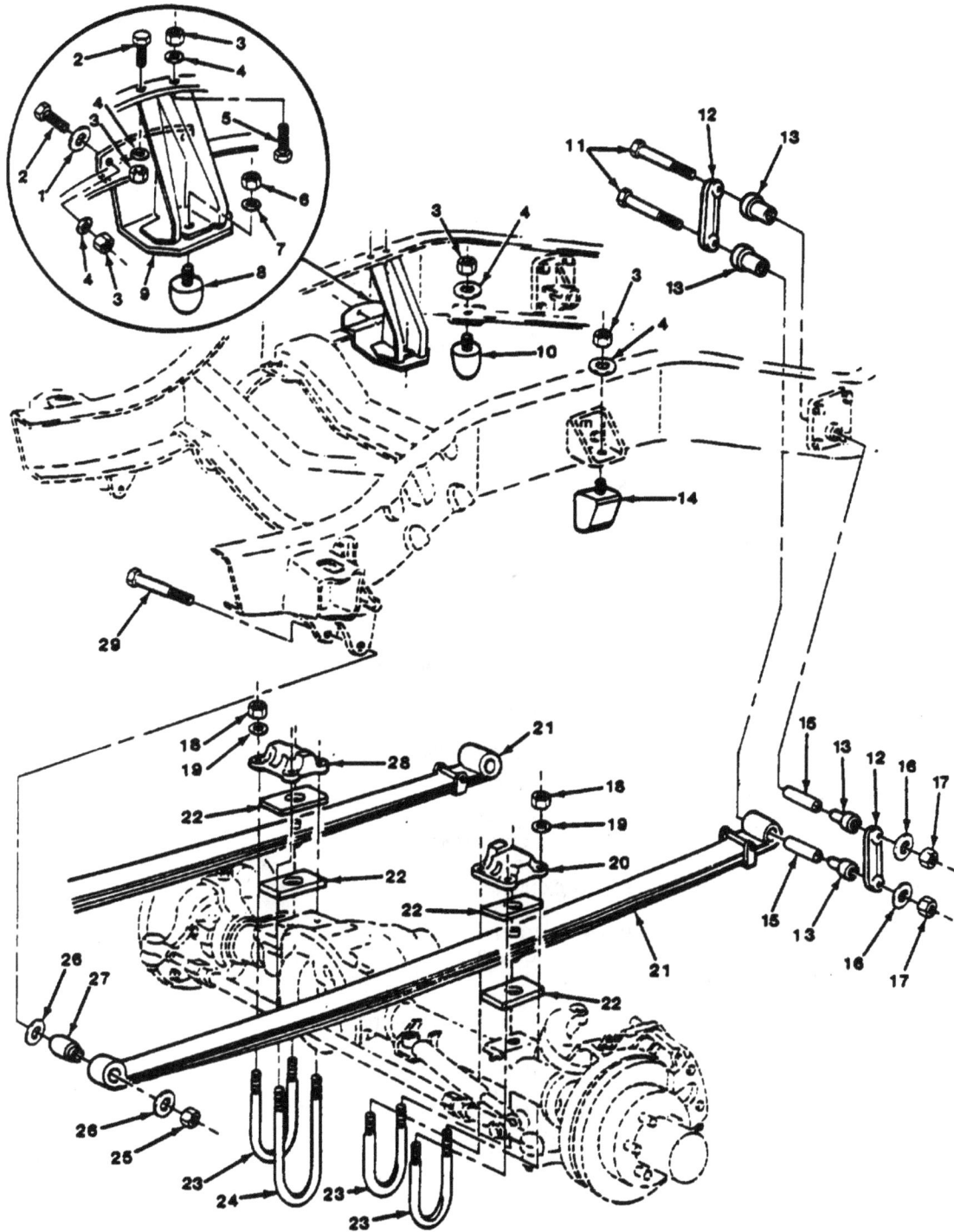

FIGURE 149. FRONT SPRINGS AND RELATED PARTS (M1009).

TA510956

(1) ITEM NO	(2) SMR CODE	(3) CAGEC	(4) PART NUMBER	(5) DESCRIPTION AND USABLE ON CODES (UOC)	(6) QTY
				GROUP 1601 SPRINGS	
				FIG. 149 FRONT SPRINGS AND RELATED PARTS (M1009)	
1	PAOZZ	96906	MS27183-14	WASHER,FLAT UOC:209	2
2	PAOZZ	96906	MS90728-60	SCREW,CAP,HEXAGON H UOC:209	3
3	PAOZZ	96906	MS51967-8	NUT,PLAIN,HEXAGON UOC:209	8
4	PAOZZ	96906	MS35338-46	WASHER,LOCK UOC:209	8
5	PAOZZ	96906	MS90728-63	SCREW,CAP,HEXAGON H UOC:209	1
6	PAOZZ	96906	MS51967-14	NUT,PLAIN,HEXAGON UOC:209	1
7	PAOZZ	96906	MS35340-48	WASHER,LOCK UOC:209	1
8	PAOZZ	11862	14029200	BUMPER,NONMETALLIC UOC:209	1
9	PAOZZ	11862	15522381	BRACKET,BUMPER UOC:209	1
10	PAOZZ	11862	359878	BUMPER ASSEMBLY,FRO UOC:209	2
11	PAFZZ	96906	MS90727-103	SCREW,CAP,HEXAGON H UOC:209	4
12	PAFZZ	11862	3975202	PARTS KIT,LEAF SPRI UOC:209	4
13	PAFZZ	11862	363913	BUSHING,NONMETALLIC UOC:209	8
14	PAOZZ	11862	359877	BUMPER ASSEMBLY,FRO UOC:209	2
15	PAFAZZ	11862	15571581	SPACER,SLEEVE UOC:209	4
16	PAFZZ	11862	3790768	WASHER,FLAT UOC:209	4
17	PAFZZ	80205	NAS1409A7	NUT,SELF-LOCKING,HE UOC:209	4
18	PAFZZ	96906	MS51968-20	NUT,PLAIN,HEXAGON UOC:209	8
19	PAFZZ	96906	MS27183-21	WASHER,FLAT UOC:209	8
20	PAFZZ	11862	379393	PLATE,WEAR,LEAF SPR UOC:209	1
21	PAFZZ	11862	460354	SPRING ASSEMBLY,LEA UOC:209	2
22	PAFZZ	11862	3970988	WASHER,FLAT UOC:209	4
23	PAFZZ	11862	379397	BOLT,U UOC:209	3
24	PAFZZ	11862	379398	BOLT,U	1

(1)	(2)	(3)	(4)	(5)	(6)
ITEM NO	SMR CODE	CAGEC	PART NUMBER	DESCRIPTION AND USABLE ON CODES (UOC)	QTY
				UOC:209	
25	PAFZZ	24617	443342	NUT,SELF-LOCKING,HE	2
				UOC:209	
26	PAFZZ	96906	MS27183-19	WASHER,FLAT	4
				UOC:209	
27	PAFZZ	11862	3887751	BUSHING	2
				UOC:209	
28	PAFZZ	11862	379394	PLATE,WEAR,LEAR SPR	1
				UOC:209	
29	PAFZZ	96906	MS90728-152	SCREW,CAP,HEXAGON H	2
				UOC:209	

END OF FIGURE

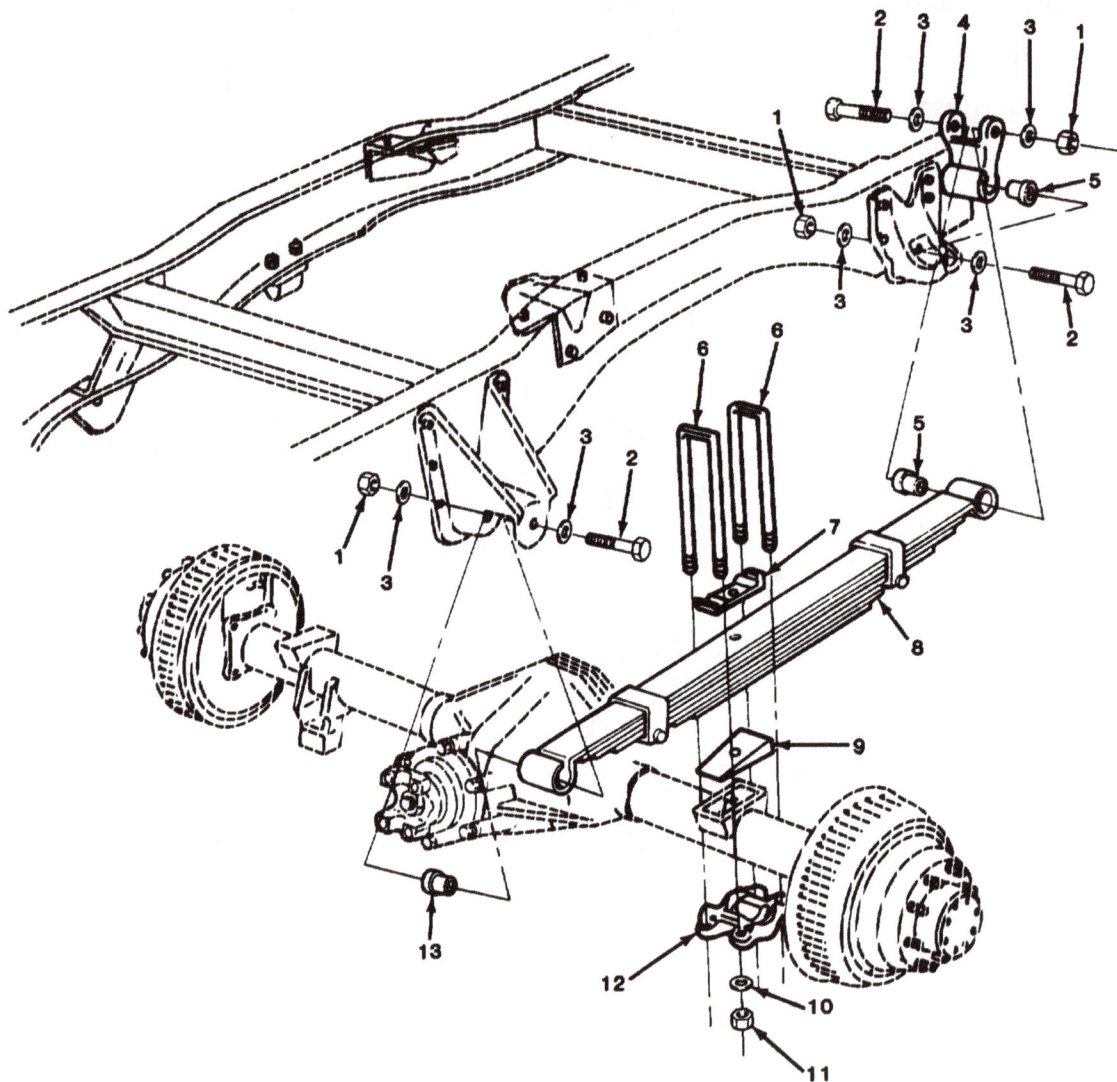

TA510957

FIGURE 150. REAR SPRINGS AND RELATED PARTS (ALL EXCEPT M1009, M1028, M1028A1, M1028A2, AND M1028A3).

(1) (2) (3) (4) ITEM SMR PART NO CODE CAGEC NUMBER	(5) DESCRIPTION AND USABLE ON CODES (UOC)	(6) QTY
	GROUP 1601 SPRINGS	
	FIG. 150 REAR SPRINGS AND RELATED PARTS (ALL EXCEPT M1009,M1028, M1028A1,M1028A2,AND M1028A3)	
1 PAFZZ 24617 443342	NUT,SELF-LOCKING,HE UOC:194,208,210,231	6
2 PAFZZ 96906 MS90728-150	SCREW,CAP,HEXAGON H UOC:194,208,210,231	6
3 PAFZZ 96906 MS27183-19 WASHER,FLAT	UOC:194,208,210,231	12
4 PAFZZ 11862 14022597	SHACKLE,LEAF SPRING UOC:194,208,210,231	2
5 PAFZZ 11862 468481	BUSHING,SLEEVE UOC:194,208,210,231	4
6 PAFZZ 11862 471667	BOLT,U UOC:194,208,210,231	4
7 PAFZZ 11862 337520	SPACER,REAR SPRING UOC:194,208,210,231	2
8 PAFZZ 11862 14071877	SPRING ASSEMBLY,LEA UOC:194,208,210,231	2
9 PAFZZ 11862 488369	SHIM SET UOC:194,208,210,231	2
10 PAFZZ 96906 MS27183-21 WASHER,FLAT	UOC:194,208,210,231	8
11 PAFZZ 96906 MS51968-20 NUT,PLAIN,HEXAGON	UOC:194,208,210,231	8
12 PAFZZ 11862 362275	SADDLE,LEAF SPRING UOC:194,208,210,231	2
13 PAFZZ 11862 350544	BUSHING,SLEEVE UOC:194,208,210,231	2
	END OF FIGURE	

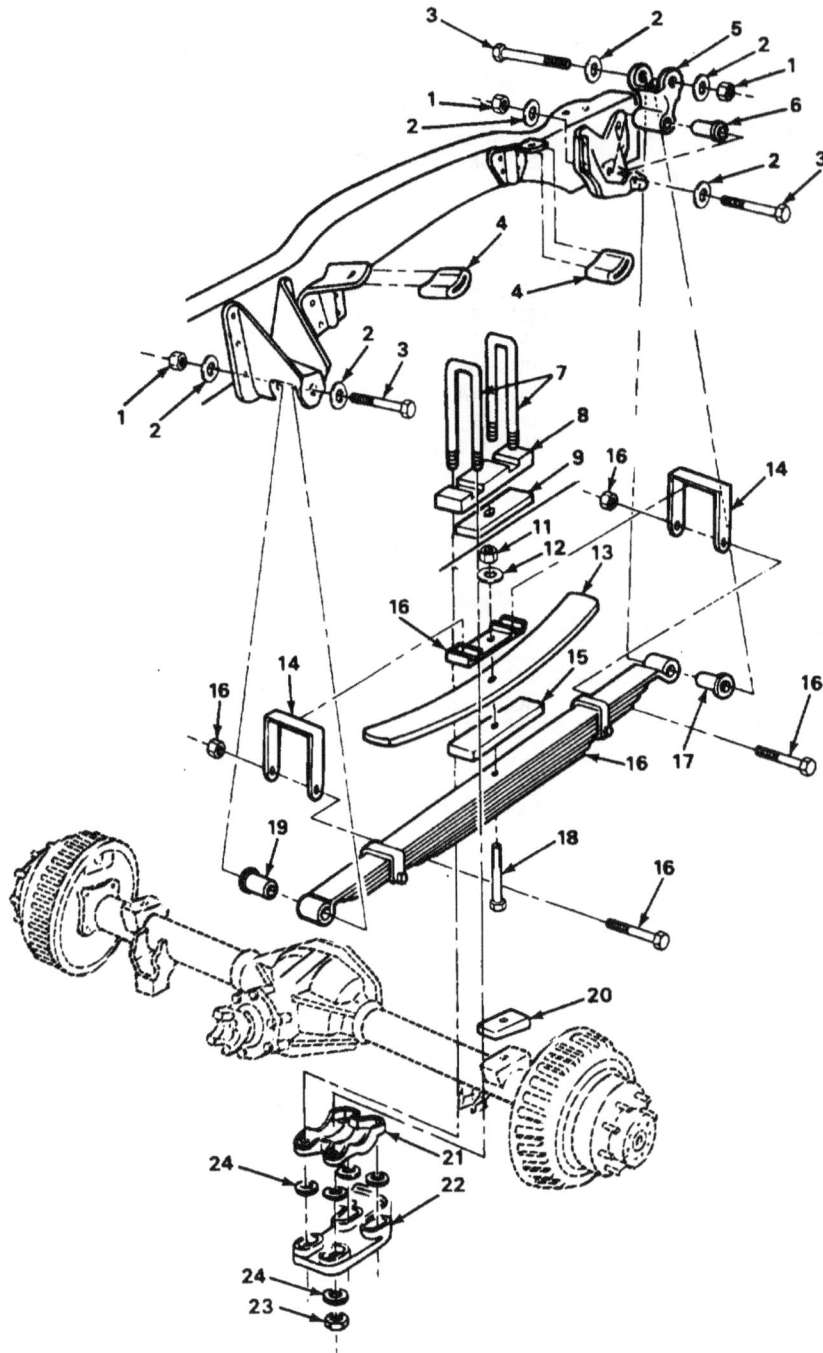

FIGURE 151. REAR SPRINGS (M1028, M1028A1, M1028A2, AND M1028A3).

TA510958

(1) (2) (3) (4) ITEM SMR PART NO CODE CAGEC NUMBER	(5) DESCRIPTION AND USABLE ON CODES (UOC)	(6) QTY
	GROUP 1601 SPRINGS	
	FIG. 151 REAR SPRINGS AND RELATED PARTS (M1028,M1028A1,M1028A2,AND M1028A3)	
1 PAFZZ 24617 443342	NUT,SELF-LOCKING,HE UOC:230,252,254,256	6
2 PAFZZ 96906 MS27183-19 WASHER,FLAT	UOC:230,252,254,256	12
3 PAFZZ 96906 MS90728-150	SCREW,CAP,HEXAGON H UOC:230,252,254,256	6
4 PAFZZ 11862 328169	BUMPER,NONMETALLIC UOC:230,252	4
5 PAFZZ 11862 14022597	SHACKLE,LEAF SPRING UOC:230,252	2
6 PAFZZ 11862 468481	BUSHING,SLEEVE UOC:230,252,254,256	2
7 PAFZZ 11862 471665	BOLT,U UOC:230,252	4
7 PAFZZ 11862 471663	BOLT,U UOC:254,256	4
8 PAFZZ 11862 471657	SPACER SPECIAL UOC:230,252	2
8 PAFZZ 11862 471658	SPACER,PLATE UOC:254,256	2
9 PAFZZ 11862 3970988	WASHER,FLAT UOC:230,252	2
9 PAFZZ 11862 327225	SPACER,PLATE UOC:254,256	2
10 AFFFF 11862 14067752	SPRING ASSEMBLY,LEA UOC:230,252	2
11 PAFZZ 96906 MS51968-8 .NUT,PLAIN,HEXAGON	UOC:230,252	1
12 PAFZZ 96906 MS27183-13 .WASHER,FLAT	UOC:230,252	1
13 PAFZZ 11862 350726	.LEAF,SPRING UOC:230,252	1
14 PAFZZ 11862 14067756	.STRAP,RETAINING UOC:230,254,256	2
15 PAFZZ 11862 6271322	.SPACER,PLATE UOC:230,252	1
16 PAFZZ 11862 14071877	.SPRING ASSEMBLY,LEA UOC:230,252	1
17 PAFZZ 11862 468481	.BUSHING,SLEEVE UOC:230,252	1
18 PAFZZ 11862 3717658 N.D.	.SCREW,CAP,SOCKET HE UOC:230,252	1
19 PAFZZ 11862 350544	.BUSHING,SLEEVE UOC:230,252	1
20 PAFZZ 11862 488369	SHIM SET UOC:230,252	2

TM9-2320-289-34P

(1) ITEM NO	(2) SMR CODE	(3) CAGEC	(4) PART NUMBER	(5) DESCRIPTION AND USABLE ON CODES (UOC)	(6) QTY
21	PAFZZ	11862	362275	SADDLE,LEAF SPRING UOC:230,252	2
21	PAFZZ	11862	464718	HANGER,SPRING,VEHIC UOC:254,256	1
22	PAFZZ	11862	14033573	BRACKET,VEHICULAR C UOC:254,256	2
22	PAFZZ	11862	14033574	BRACKET,VEHICULAR C UOC:254,256	2
23	PAFZZ	96906	MS51968-20	NUT,PLAIN,HEXAGON UOC:230,252	8
24	PAFZZ	96906	MS27183-21	WASHER,FLAT UOC:230,252,254,256	8
24	PAFZZ	11862	474978	WASHER,FLAT UOC:254,256	16

END OF FIGURE

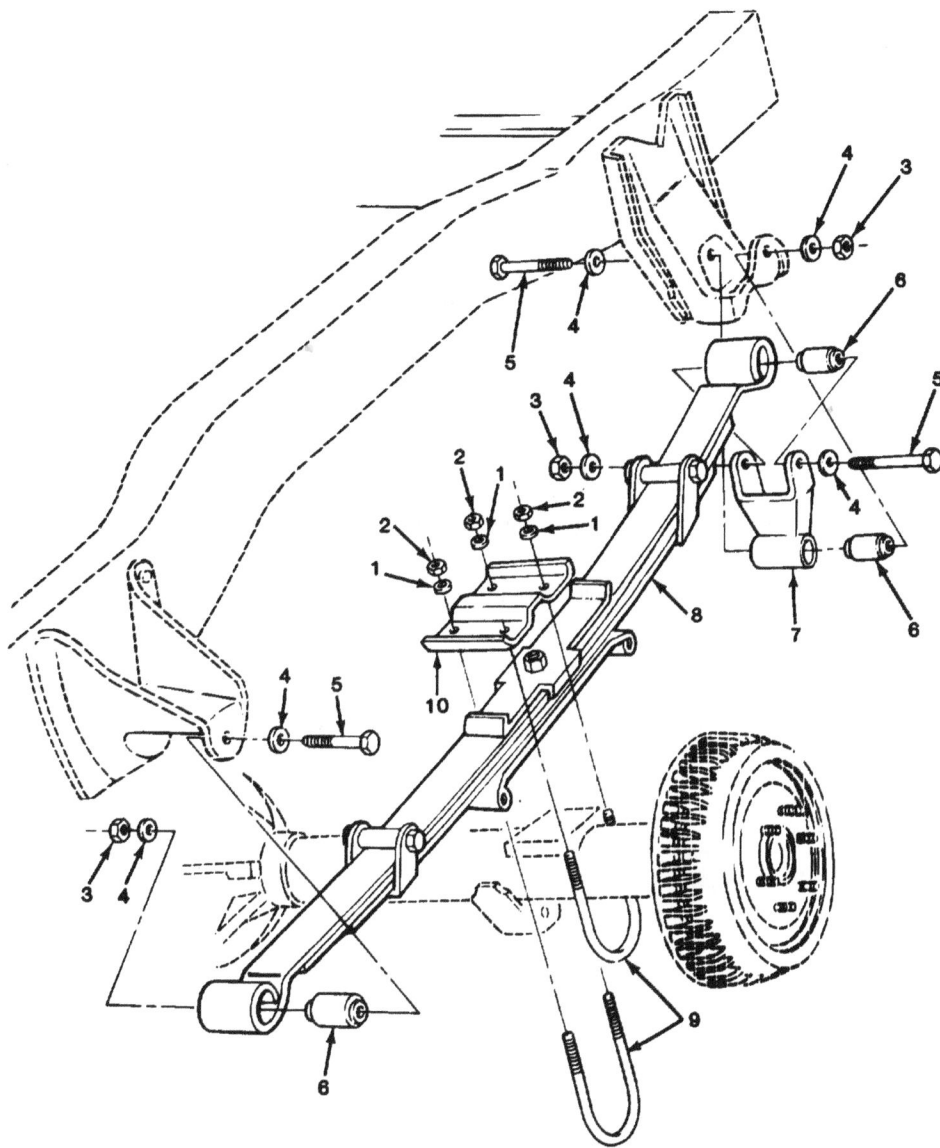

TA510959

FIGURE 152. REAR SPRINGS AND RELATED PARTS (M1009).

(1)	(2)	(3)	(4)	(5)	(6)
ITEM NO	SMR CODE	CAGEC	PART NUMBER	DESCRIPTION AND USABLE ON CODES (UOC)	QTY

GROUP 1601 SPRINGS

FIG. 152 REAR SPRINGS AND RELATED PARTS (M1009)

1	PAFZZ	96906	MS27183-21	WASHER,FLAT UOC:209	8
2	PAFZZ	96906	MS51968-20	NUT,PLAIN,HEXAGON UOC:209	8
3	PAFZZ	24617	443342	NUT,SELF-LOCKING,HE UOC:209	6
4	PAFZZ	96906	MS27183-19	WASHER,FLAT UOC:209	12
5	PAFZZ	96906	MS90728-150	SCREW,CAP,HEXAGON H UOC:209	6
6	PAFZZ	11862	468481	BUSHING,SLEEVE UOC:209	6
7	PAFZZ	11862	14022597	SHACKLE,LEAF SPRING UOC:209	2
8	PAFZZ	11862	14067753	SPRING ASSEMBLY,LEA UOC:209	2
9	PAFZZ	11862	326489	BOLT,U UOC:209	4
10	PAFZZ	11862	6263945	PLATE,WEAR,LEAF SPR UOC:209	2

END OF FIGURE

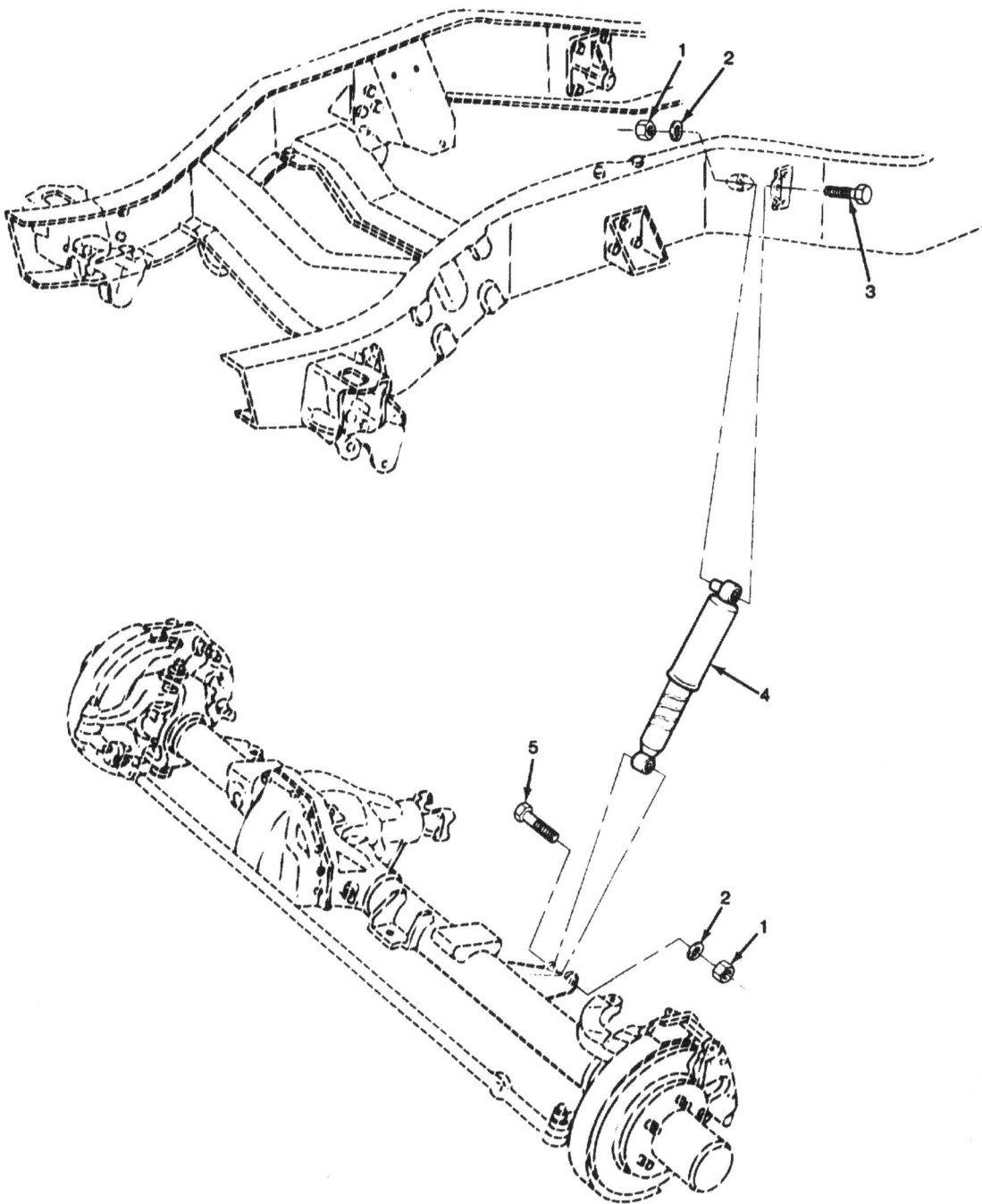

TA510960

FIGURE 153. FRONT SHOCK ABSORBERS.

(1) (2) (3) (4)	(5)	(6)
ITEM SMR PART	DESCRIPTION AND USABLE ON CODES (UOC)	QTY
NO CODE CAGEC NUMBER		

GROUP 1604 SHOCK ABSORBER EQUIPMENT

FIG. 153 FRONT SHOCK ABSORBERS

1	PAOZZ 96906 MS51967-14	NUT,PLAIN,HEXAGON	4
2	PAOZZ 96906 MS35340-48	WASHER,LOCK	4
3	PAOZZ 96906 MS90728-121	SCREW,CAP,HEXAGON H	2
4	PAOZZ 11862 3187843	SHOCK ABSORBER,DIRE	2
		UOC:209	
4	PAOZZ 11862 3187846	SHOCK ABSORBER,DIRE	2
		UOC:194,208,210,230,231,252,254,256	
5	PAOZZ 96906 MS90728-119	SCREW,CAP,HEXAGON H	2

END OF FIGURE

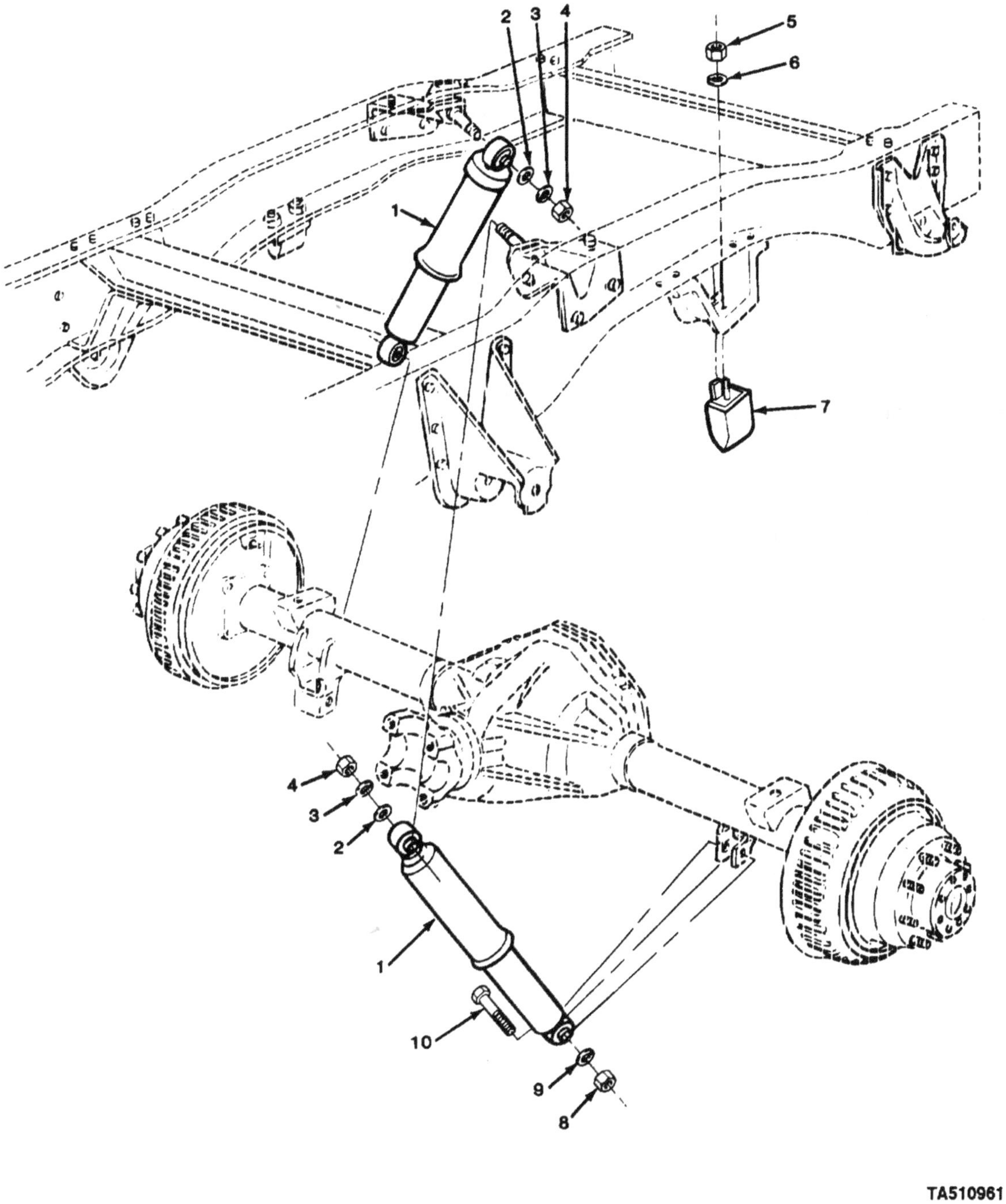

TA510961

FIGURE 154. REAR SHOCK ABSORBERS (ALL EXCEPT M1009).

(1)	(2)	(3)	(4)	(5)	(6)
ITEM NO	SMR CODE	CAGEC	PART NUMBER	DESCRIPTION AND USABLE ON CODES (UOC)	QTY

GROUP 1604 SHOCK ABSORBER EQUIPMENT

FIG. 154 REAR SHOCK ABSORBERS (ALL EXCEPT M1009)

1	PAOZZ	11862	3187845	SHOCK ABSORBER,DIRE UOC:194,208,210,230,231,252,254,256	2
2	PAOZZ	96906	MS27183-18	WASHER,FLAT UOC:194,208,210,230,231,252,254,256	2
3	PAOZZ	96906	MS35340-49	WASHER,LOCK UOC:194,208,210,230,231,252,254,256	2
4	PAOZZ	96906	MS51967-14	NUT,PLAIN,HEXAGON UOC:194,208,210,230,231,252,254,256	2
5	PAOZZ	96906	MS51967-8	NUT,PLAIN,HEXAGON UOC:194,208,210,230,231,252,254,256	2
6	PAOZZ	96906	MS35338-46	WASHER,LOCK UOC:194,208,210,230,231,252,254,256	2
7	PAOZZ	11862	3764438	BUMPER,RUBBER ASSEM UOC:194,208,210,230,231,252,254,256	2
8	PAOZZ	96906	MS51967-17	NUT,PLAIN,HEXAGON UOC:194,208,210,230,231,252,254,256	2
9	PAOZZ	96906	MS35338-49	WASHR,LOCK UOC:194,208,210,230,231,252,254,256	2
10	PAOZZ	11862	9419138	SCREW,CAP,HEXAGON H UOC:194,208,210,230,231,252,254,256	2

END OF FIGURE

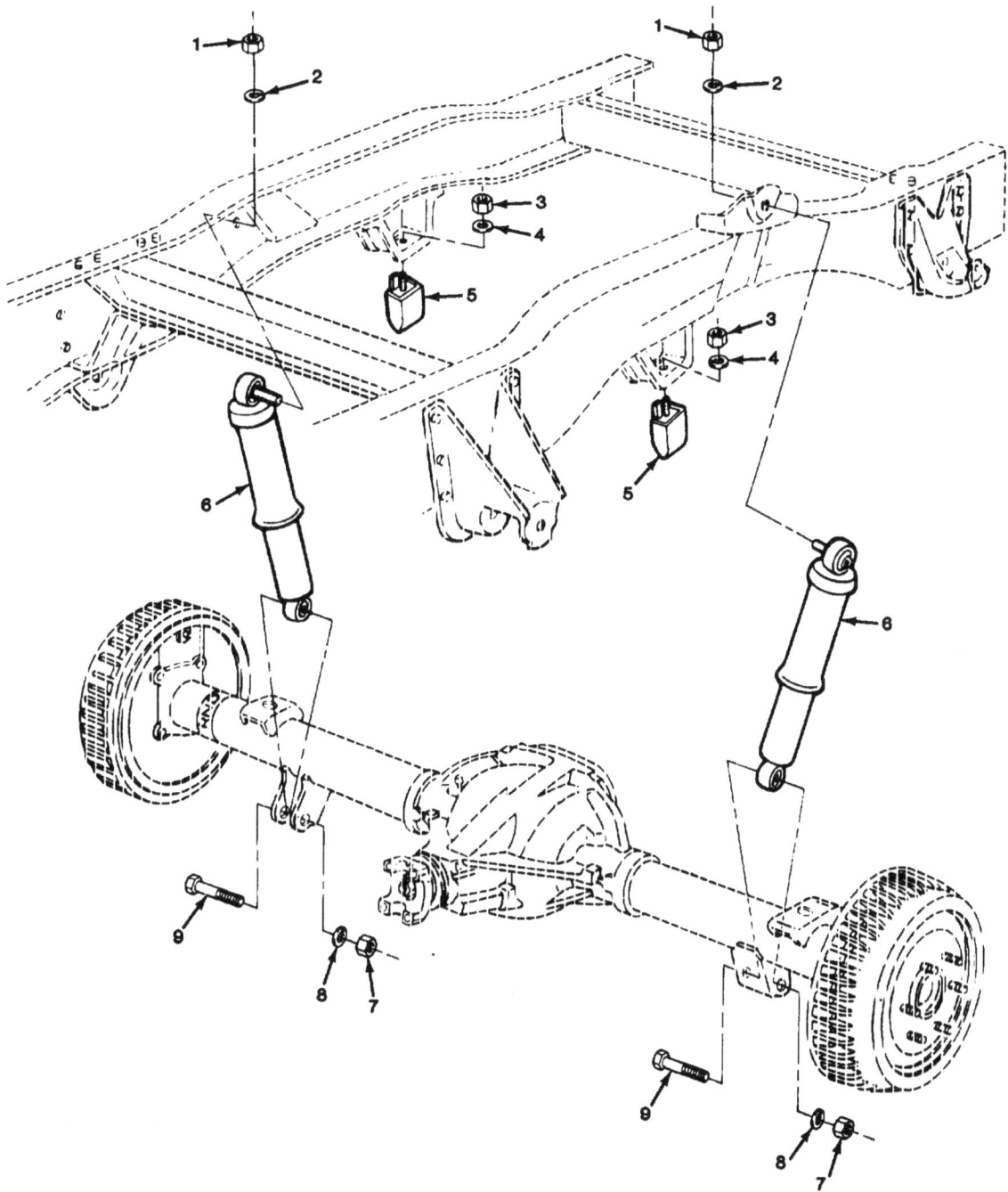

TA510962

FIGURE 155. REAR SHOCK ABSORBERS (M1009).

(1)	(2)	(3)	(4)	(5)	(6)
ITEM NO	SMR CODE	CAGEC	PART NUMBER	DESCRIPTION AND USABLE ON CODES (UOC)	QTY

GROUP 1604 SHOCK ABSORBER EQUIPMENT

FIG. 155 REAR SHOCK ABSORBERS
(M1009)

1	PAOZZ	96906	MS51968-20	NUT,PLAIN,HEXAGON UOC:209	2
2	PAOZZ	96906	MS35340-50	WASHER,LOCK UOC:209	2
3	PAOZZ	96906	MS51967-8	NUT,PLAIN,HEXAGON UOC:209	2
4	PAOZZ	96906	MS35338-46	WASHER,LOCK UOC:209	2
5	PAOZZ	11862	3764438	BUMPER,RUBBER ASSEM UOC:209	2
6	PAOZZ	11862	3187844	SHOCK ABSORBER,LEVE UOC:209	2
7	PAOZZ	96906	MS51967-17	NUT,PLAIN,HEXAGON UOC:209	2
8	PAOZZ	96906	MS35338-49	WASHER,LOCK UOC:209	2
9	PAOZZ	11862	9419138	SCREW,CAP,HEXAGON H UOC:209	2

END OF FIGURE

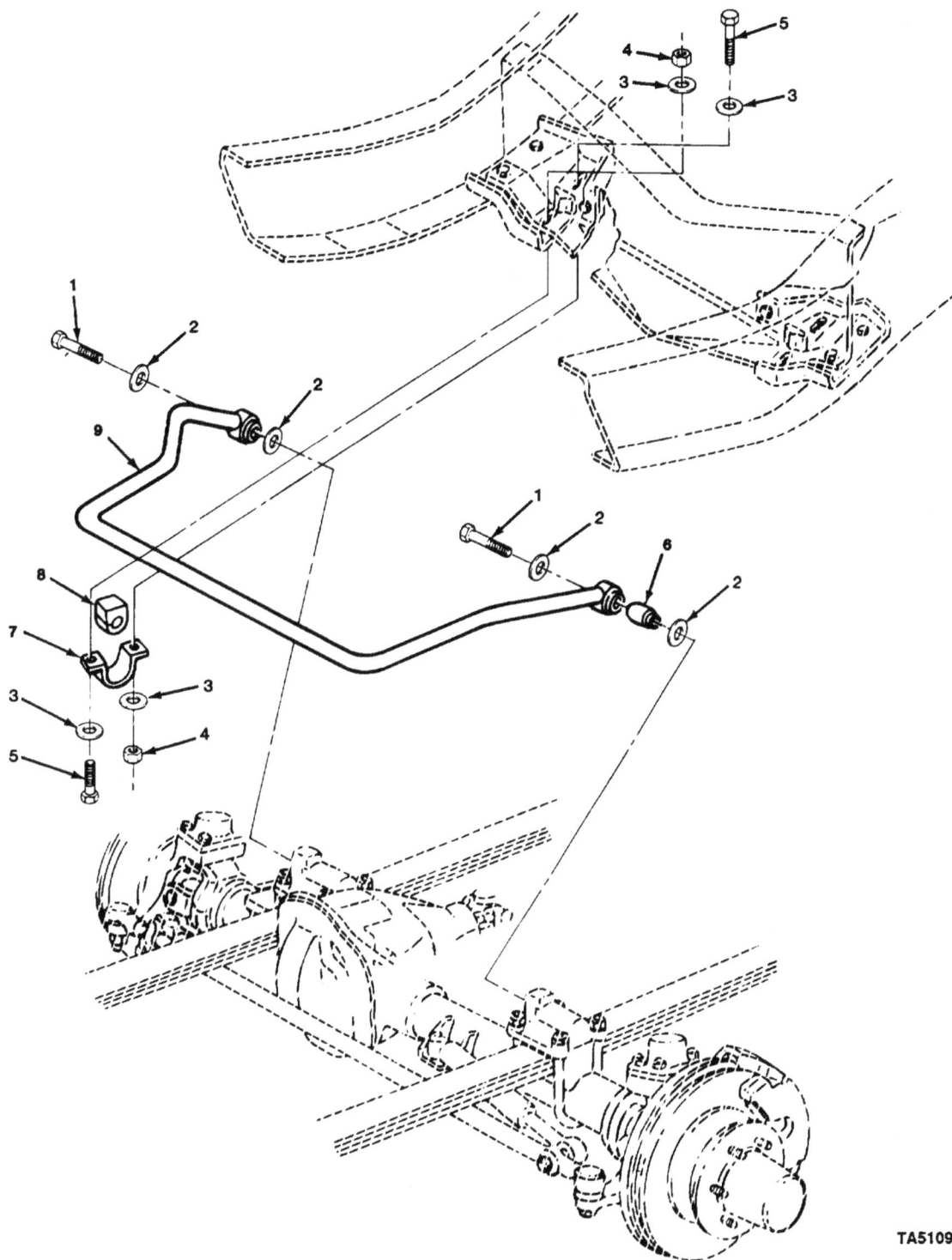

FIGURE 156. FRONT STABILIZER BAR.

TA510963

(1) ITEM NO	(2) SMR CODE	(3) CAGEC	(4) PART NUMBER	(5) DESCRIPTION AND USABLE ON CODES (UOC)	(6) QTY
				GROUP 1605 TORQUE,RADIUS,AND STABILIZER RODS	
				FIG. 156 FRONT STABILIZER BAR	
1	PAOZZ	11862	328130	BOLT,SELF-LOCKING	2
2	PAOZZ	11862	328131	WASHER,FLAT	4
3	PAOZZ	24617	3990160	WASHER,FLAT	8
4	PAOZZ	11862	9422299	NUT,SELF-LOCKING,HE	4
5	PAOZZ	96906	MS90728-89	SCREW,CAP,HEXAGON H	4
6	PAOZZ	11862	328128	BUSHING,SLEEVE	2
7	PAOZZ	11862	14015726	STRAP,RETAINING	2
8	PAOZZ	11862	14015724	BUSHING,NONMETALLIC	2
9	PAOFF	11862	328132	SHAFT ASSEMBLY,FRON	1

END OF FIGURE

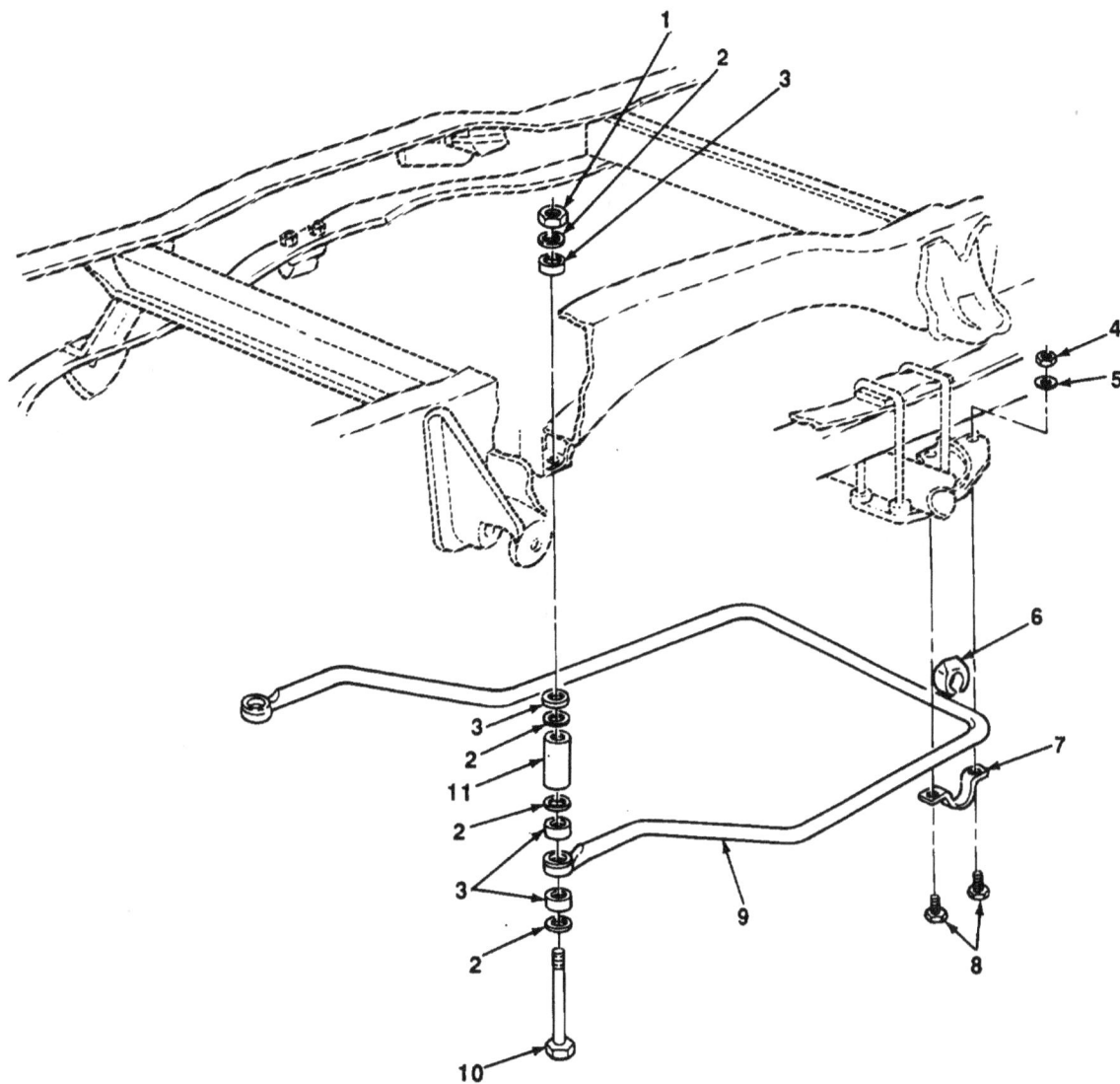

TA510964

FIGURE 157. REAR STABILIZER BAR (M1028A2 AND M1028A3).

(1) ITEM NO	(2) SMR CODE	(3) CAGEC	(4) PART NUMBER	(5) DESCRIPTION AND USABLE ON CODES (UOC)	(6) QTY
				GROUP 1605 TORQUE,RADIUS,AND STABILIZER RODS	
				FIG. 157 REAR STABILIZER BAR (M1028A2 AND M1028A3)	
1	KFOZZ	11862	1365065	NUT,SPECIAL PART OF KIT P/N 6258545 UOC:254,256	2
2	KFOZZ	11862	3993729	RETAINER,WASHER SPE PART OF KIT P/N 6258545 UOC:254,256	4
3	PAOZZ	11862	6270704	GROMMET STABIL LINK PART OF KIT P/N 6258545 UOC:254,256	4
4	PAOZZ	96906	MS51967-8	NUT,PLAIN,HEXAGON UOC:254,256	4
5	PAOZZ	96906	MS35338-46	WASHER,LOCK UOC:254,256	2
6	PAOZZ	11862	404062	BUSHING,SLEEVE UOC:254,256	2
7	PAOZZ	11862	406887	BRACKET,VEHICULAR C UOC:254,256	2
8	PAOZZ	96906	MS90728-65	SCREW,CAP,HEXAGON H UOC:254,256	2
9	PAOZZ	11862	328107	BAR,STABILIZER UOC:254,256	1
10	KFOZZ	11862	328111	BOLT,STAB,LINK PART OF KIT P/N 6258545 UOC:254,256	2
11	KFOZZ	11862	328108	SPACER,STAB,LINK PART OF KIT P/N 6258545 UOC:254,256	2

END OF FIGURE

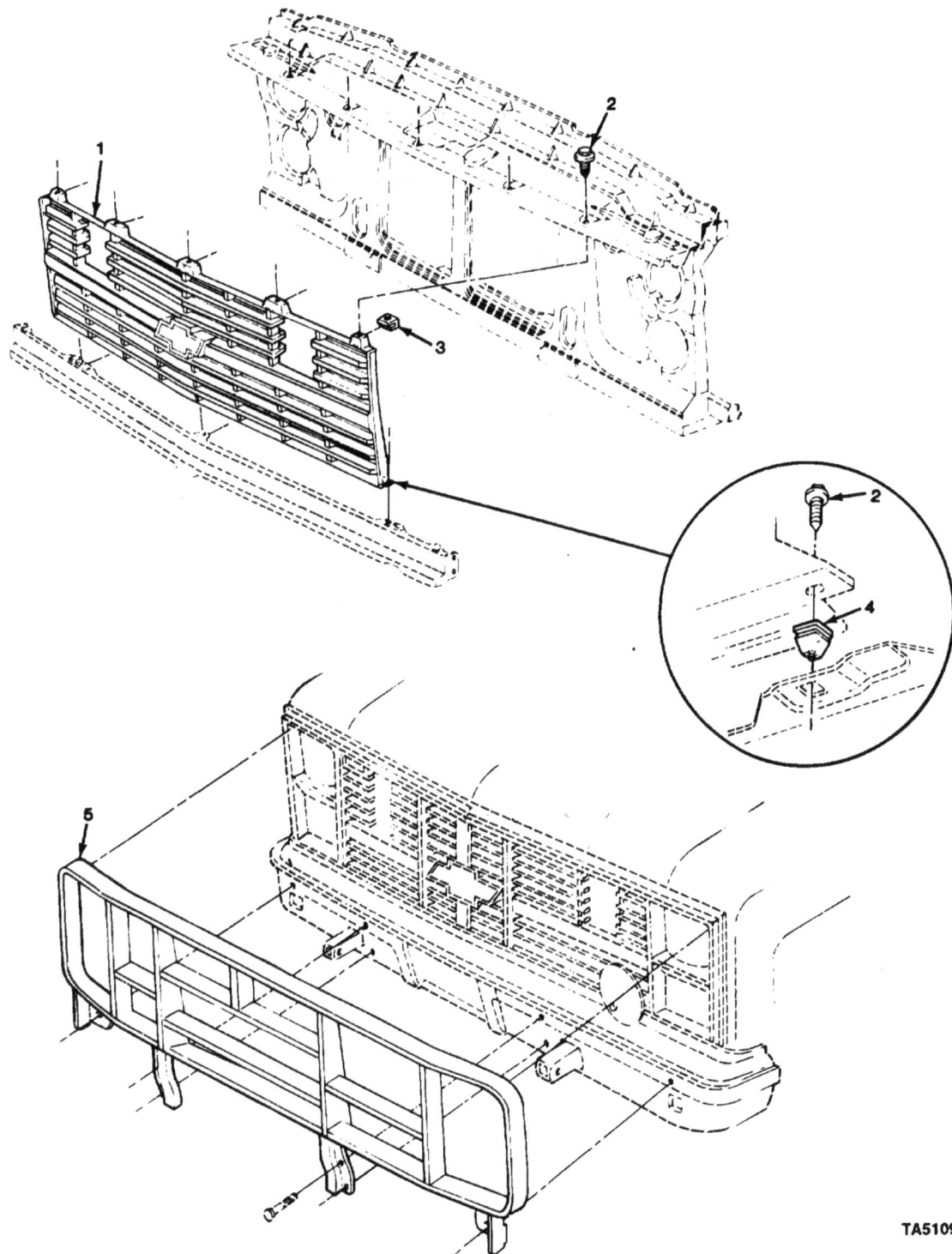

TA510965

FIGURE 158. BRUSH GUARD AND GRILLE.

(1) ITEM NO	(2) SMR CODE	(3) CAGEC	(4) PART NUMBER	(5) DESCRIPTION AND USABLE ON CODES (UOC)	(6) QTY
				GROUP 18 BODY, CAB, HOOD, AND HULL	
				GROUP 1801 BODY, CAB, HOOD, AND HULL ASSEMBLIES	
				FIG. 158 BRUSH GUARD AND GRILLE	
1	PFOZZ	11862	15554915	GRILLE,RADIATOR,VEH	1
2	PAOZZ	24617	11503395	SCREW,ASSEMBLED WAS	8
3	PAOZZ	96906	MS90724-34	NUT,SHEET SPRING	5
4	PAOZZ	11862	347347	GROMMET,NONMETALLIC	3
5	PFOZZ	11862	14072488	GUARD,GRILLE,RADIAT	1
				END OF FIGURE	

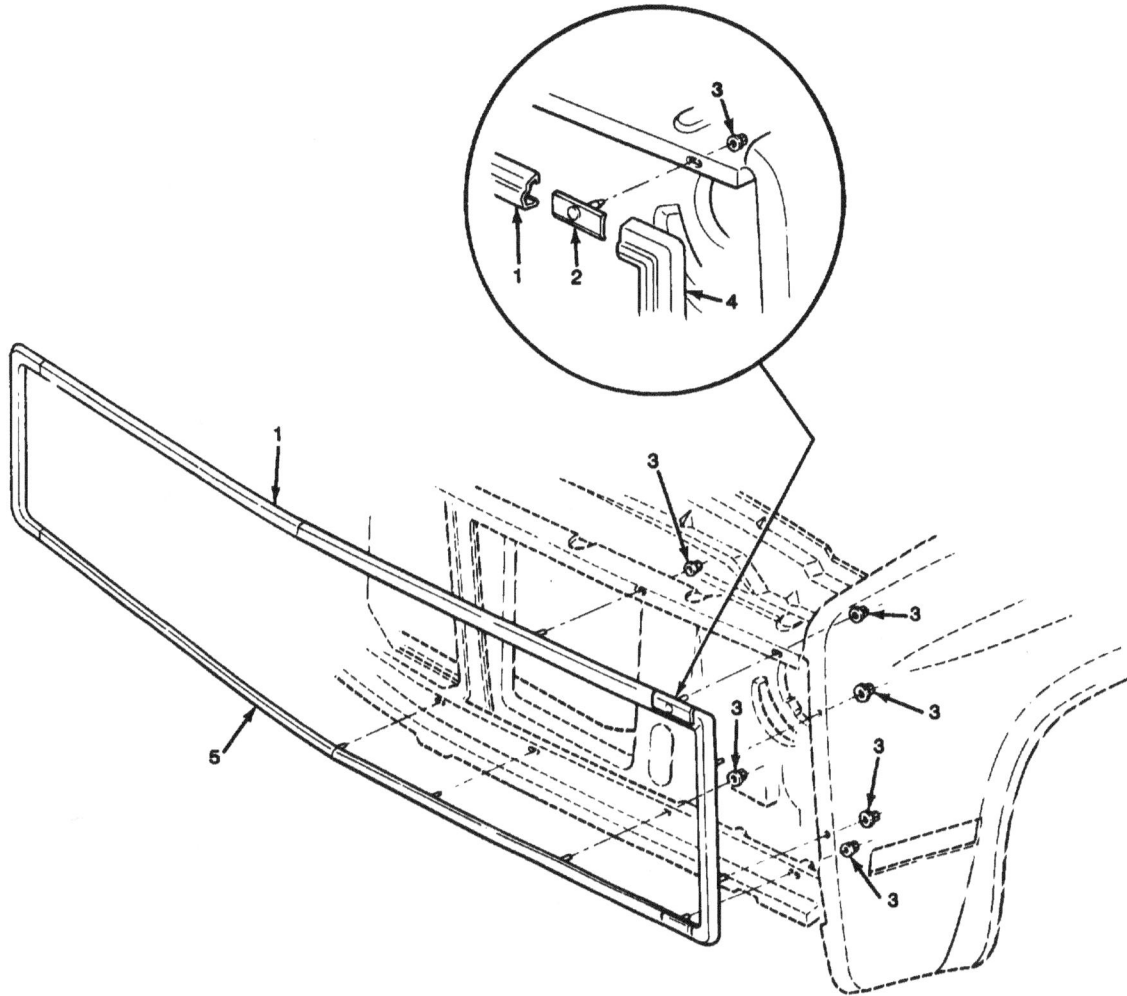

TA510966

FIGURE 159. FRONT MOLDINGS.

(1)	(2)	(3)	(4)	(5)	(6)
ITEM NO	SMR CODE	CAGEC	PART NUMBER	DESCRIPTION AND USABLE ON CODES (UOC)	QTY

GROUP 1801 BODY, CAB, HOOD, AND HULL ASSEMBLIES

FIG. 159 FRONT MOLDINGS

1	PAOZZ	11862	15599285	MOLDING ASSEMBLY,FR	1
2	PAOZZ	11862	14072850	RETAINER,MOLDING	4
3	PAOZZ	24617	9420621	PUSH ON NUT	18
4	PAOZZ	11862	15599283	MOLDING,METAL LEFT	1
4	PAOZZ	11862	15599284	MOLDING,METAL RIGHT	1
5	PAOZZ	11862	15599286	MOLDING,METAL	1

END OF FIGURE

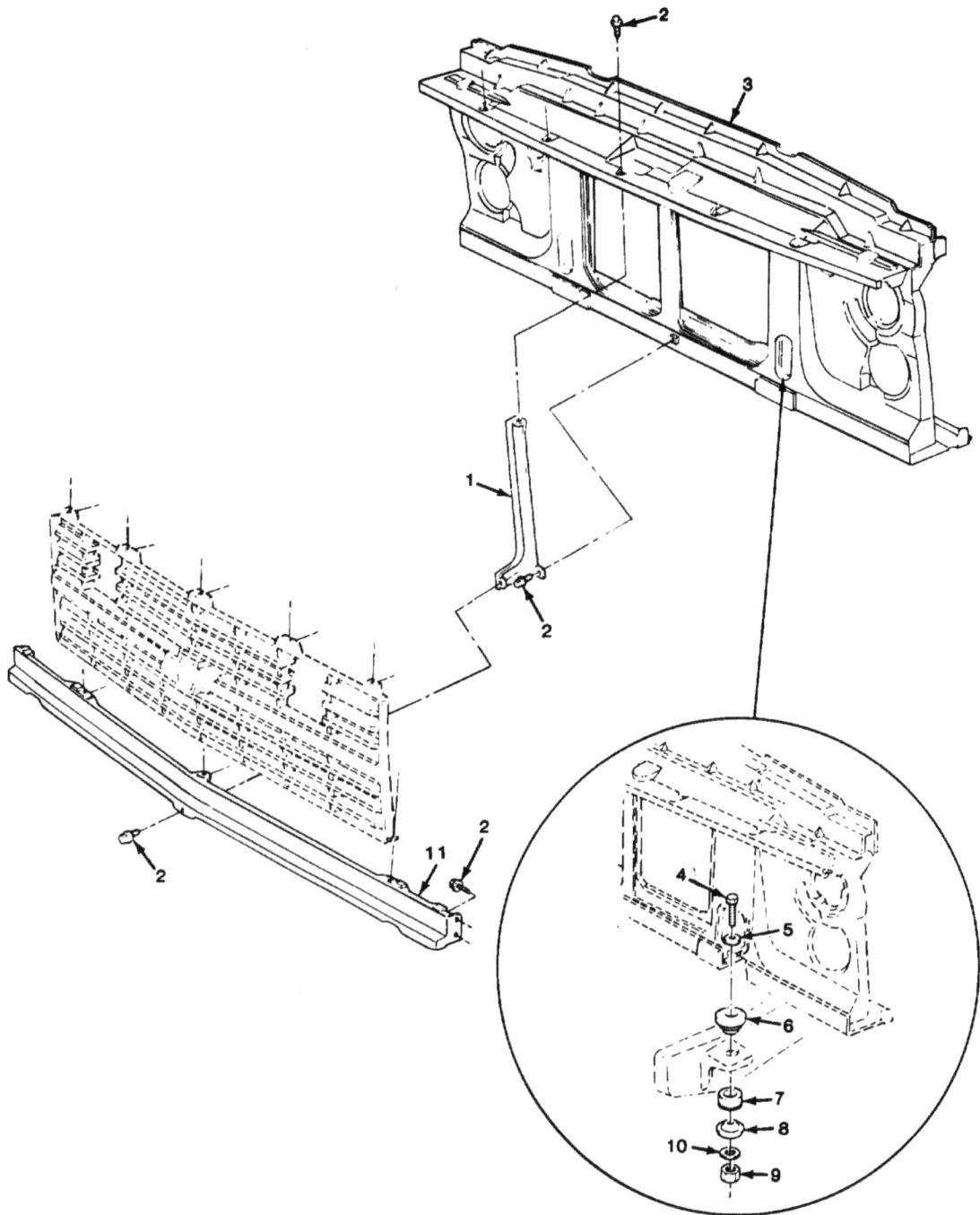

FIGURE 160. FRONT SUPPORT PANELS.

TA510967

(1) ITEM NO	(2) SMR CODE	(3) CAGEC	(4) PART NUMBER	(5) DESCRIPTION AND USABLE ON CODES (UOC)	(6) QTY
				GROUP 1801 BODY, CAB, HOOD, AND HULL ASSEMBLIES	
				FIG. 160 FRONT SUPPORT PANELS	
1	PAOZZ	11862	14021243	SUPPORT ASSEMBLY,HO	1
2	PAOZZ	11862	2014469	BOLT,ASSEMBLED WASH	7
3	PAFZZ	11862	15598787	SUPPORT ASSEMBLY,RA	1
4	PAFZZ	96906	MS90728-127	SCREW,CAP,HEXAGON H	2
5	PAFZZ	11862	14049809	WASHER,FLAT	2
6	PAFZZ	11862	15597600	RETAINER,BOLT CUSHI UPPER	2
7	PAFZZ	11862	15597629	BUSHING,RUBBER LOWER	2
8	PAFZZ	11862	14027472	RETAINER,BODY BOLT	2
9	PAFZZ	96906	MS51967-14	NUT,PLAIN,HEXAGON	2
10	PAFZZ	96906	MS27183-18	WASHER,FLAT	2
11	PAOZZ	11862	14043880	PANEL,FRONT END	1

END OF FIGURE

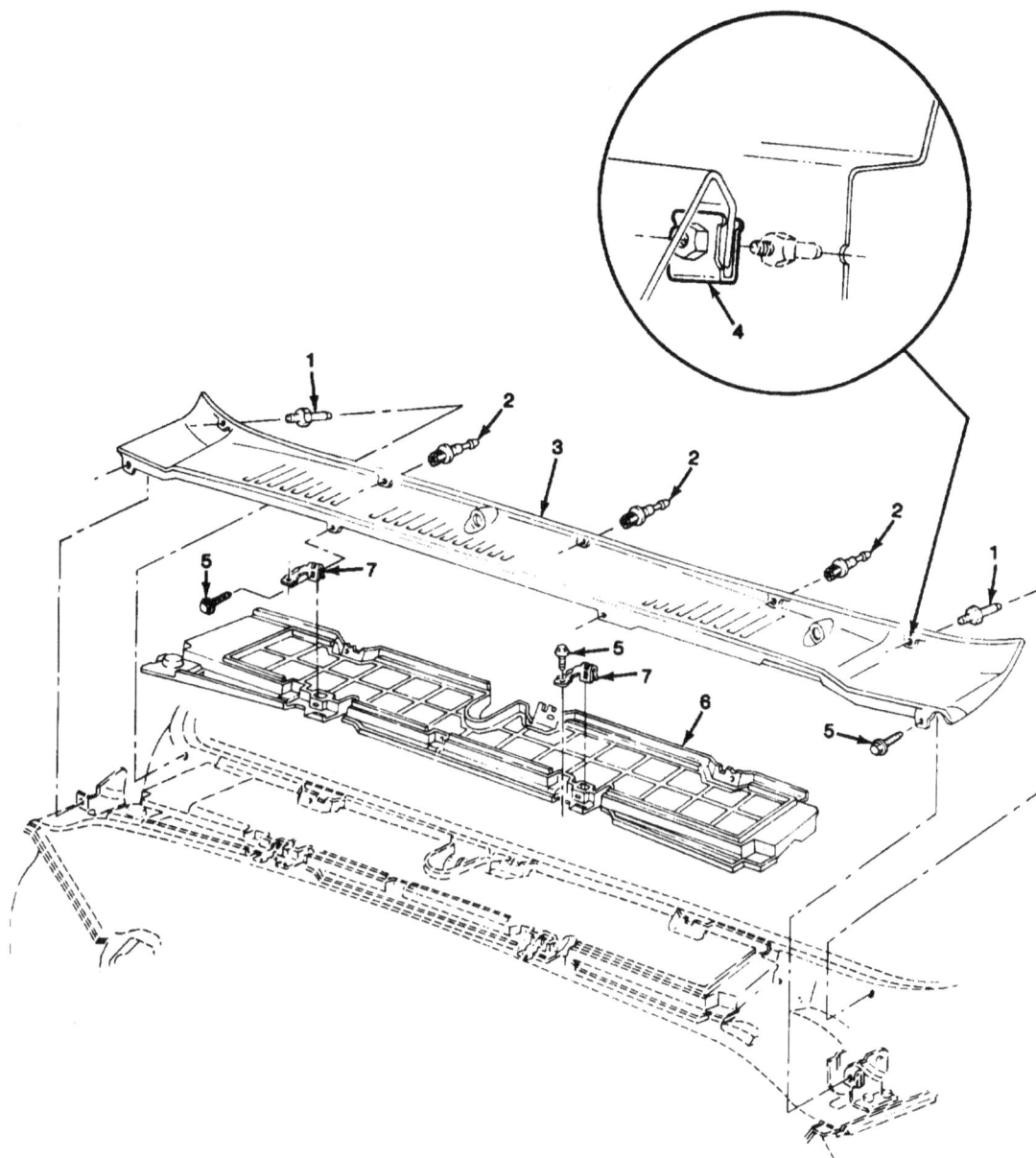

TA510968

FIGURE 161. COWL TOP VENTILATOR PANEL.

(1)	(2)	(3)	(4)	(5)	(6)
ITEM	SMR		PART		
NO	CODE	CAGEC	NUMBER	DESCRIPTION AND USABLE ON CODES (UOC)	QTY

GROUP 1801 BODY, CAB, HOOD, AND HULL
ASSEMBLIES

FIG. 161 COWL TOP VENTILATOR PANEL

1	PAOZZ	11862	15598770	RETAINER,COWL,TOP V	2
2	PAOZZ	11862	15598769	RETAINER,AIR VENT,T	3
3	PAOZZ	11862	15598708	PANEL,AIR INLET GRI	1
4	PAOZZ	11862	3982098	NUT,PLAIN,HEXAGON	2
5	PAOZZ	11862	14026247	SCREW,ASSEMBLED WAS	6
6	PAOZZ	11862	15598709	PLENUM,AIR	1
7	PAOZZ	11862	14027555	BRACKET	2

END OF FIGURE

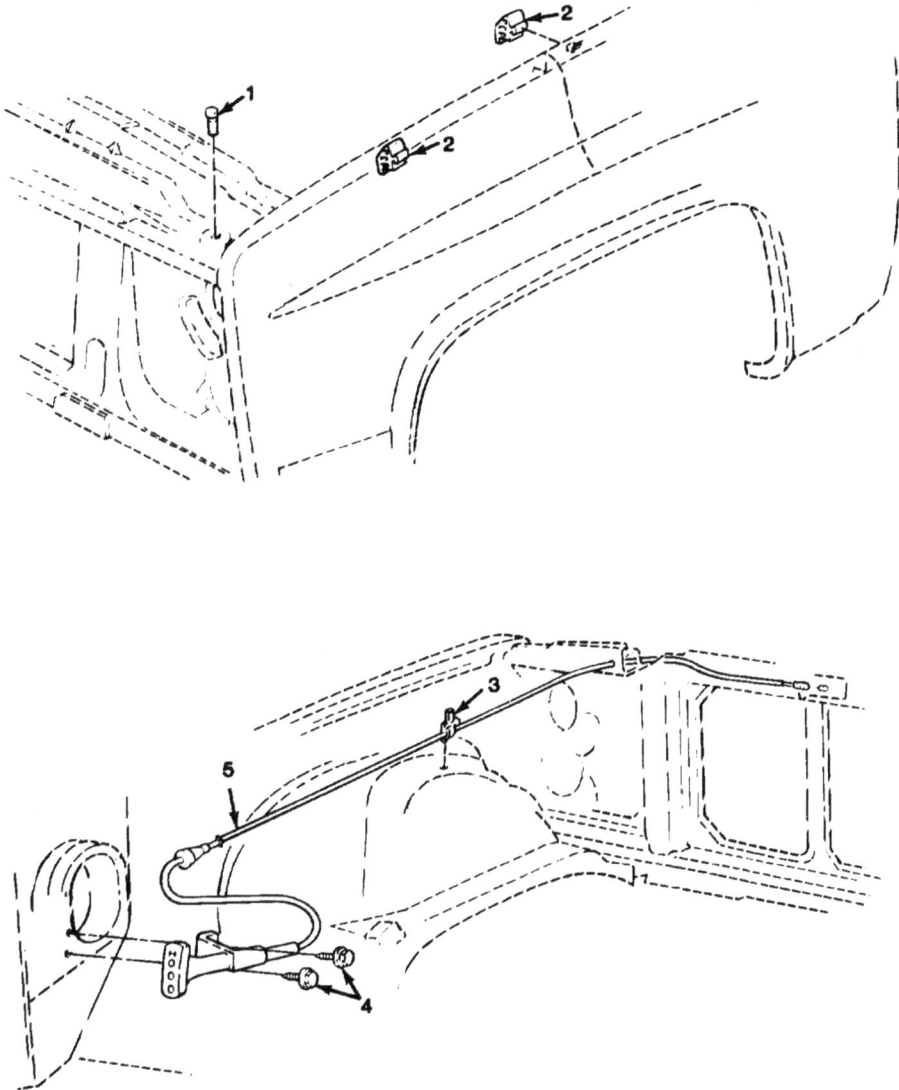

TA510969

FIGURE 162. FRONT HOOD LATCH RELEASE CABLE.

(1)	(2)	(3)	(4)	(5)	(6)
ITEM NO	SMR CODE	CAGEC	PART NUMBER	DESCRIPTION AND USABLE ON CODES (UOC)	QTY

GROUP 1801 BODY, CAB, HOOD, AND HULL ASSEMBLIES

FIG. 162 FRONT HOOD LATCH RELEASE CABLE

1	PAOZZ	11862	472450	BUMPER,NONMETALLIC	2
2	PAOZZ	11862	6262054	BUMPER,NONMETALLIC	4
3	PAOZZ	11862	3816659	STRAP,LINE SUPPORTI	1
4	PAOZZ	24617	11503396	SCREW,ASSEMBLED WAS	2
5	PAOZZ	11862	14039963	CONTROL ASSEMBLY,PU	1

END OF FIGURE

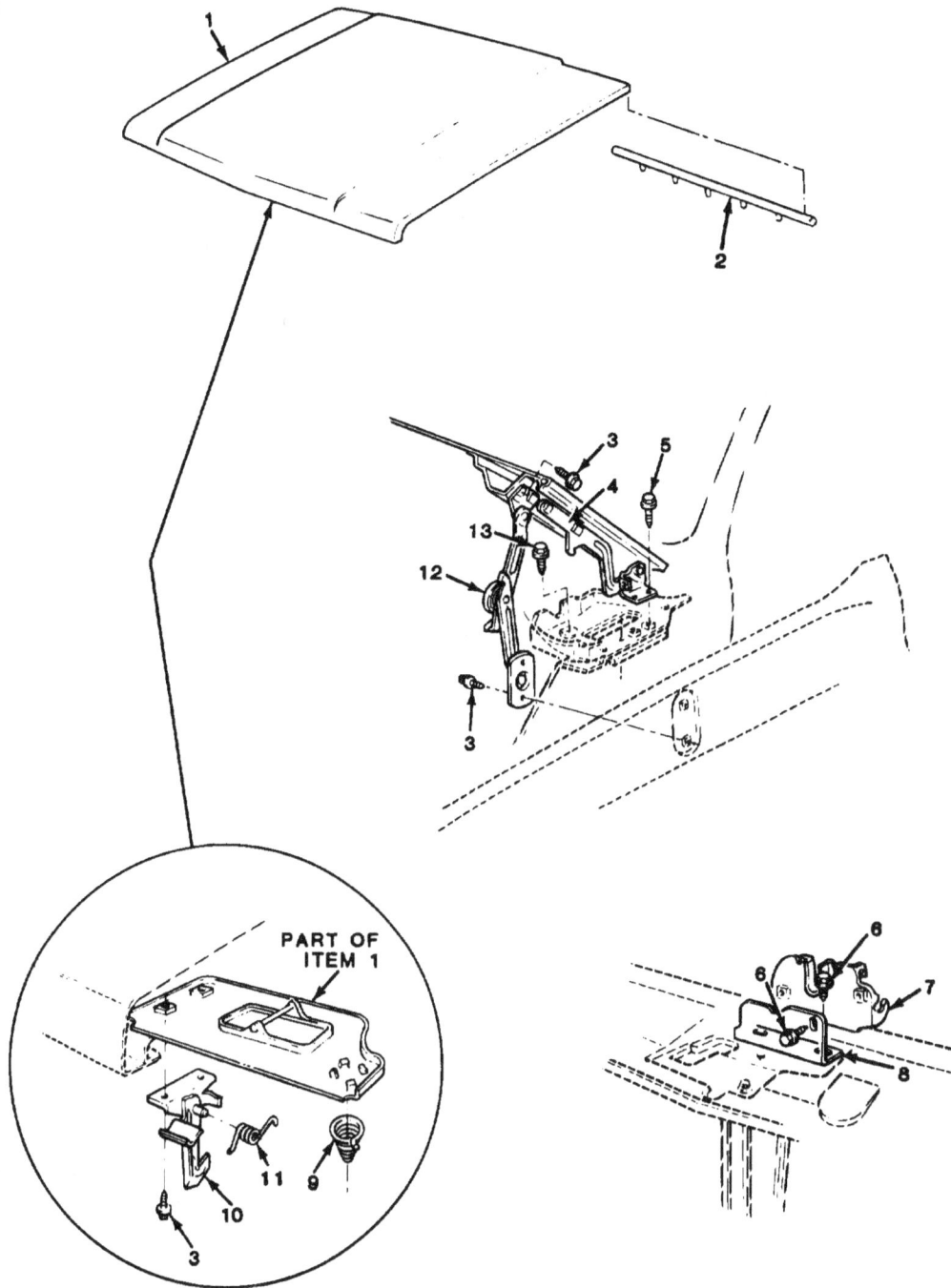

FIGURE 163. FRONT HOOD, LATCH, AND HINGES.

TA510970

TM9-2320-289-34P

(1) (2) (3) (4) ITEM SMR PART NO CODE CAGEC NUMBER	(5) DESCRIPTION AND USABLE ON CODES (UOC)	(6) QTY
	GROUP 1801 BODY, CAB, HOOD, AND HULL ASSEMBLIES	
	FIG. 163 FRONT HOOD, LATCH, AND HINGES	
1 PAOZZ 11862 15571640	HOOD,ENGINE COMPART	1
2 PAOZZ 11862 14018523	SEAL,NONMETALLIC SP	1
3 PAOZZ 11862 2014469	BOLT,ASSEMBLED WASH	14
4 PAOZZ 11862 14043823	HINGE,HOOD,VEHICULA LEFT	1
4 PAOZZ 11862 14043824	HINGE,HOOD,VEHICULA RIGHT	1
5 PAOZZ 96906 MS90728-59	SCREW,CAP,HEXAGON H	4
6 PAOZZ 11862 11508164	SCREW,ASSEMBLED WAS	4
7 PAOZZ 11862 14070703	LATCH HOOD VEHICULA	1
8 PAOZZ 11862 14018531 BRACKET,ANGLE		1
9 PAOZZ 11862 14018532 SPRING,HELICAL,COMP		1
10 PAOOO 11862 14018526	LATCH SET, MORTISE	1
11 PAOZZ 11862 14018529 .SPRING		1
12 PAOZZ 11862 14021253	HINGE,HOOD,VEHICULA LEFT	1
12 PAOZZ 11862 14021254	HINGE,HOOD,VEHICULA RIGHT	1
13 PAOZZ 11862 11508566 SCREW,TAPPING,THREA		4

END OF FIGURE

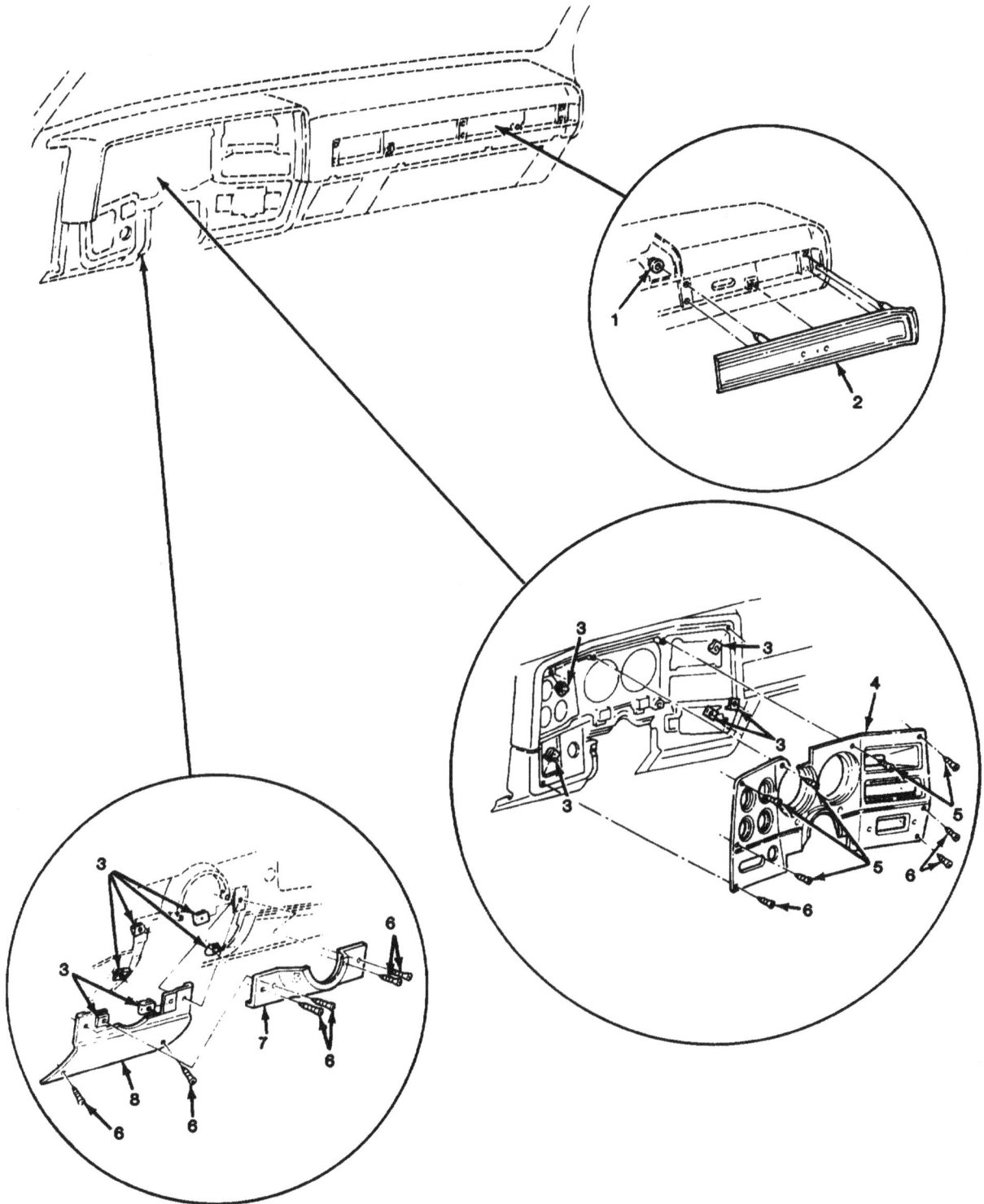

TA510971

FIGURE 164. INSTRUMENT PANEL TRIM.

(1) ITEM NO	(2) SMR CODE	(3) CAGEC	(4) PART NUMBER	(5) DESCRIPTION AND USABLE ON CODES (UOC)	(6) QTY

GROUP 1801 BODY, CAB, HOOD, AND HULL ASSEMBLIES

FIG. 164 INSTRUMENT PANEL TRIM

1	PAOZZ	11862	11501047	NUT	5
2	PFOZZ	11862	14023039	PANEL,INSTRUMENT,PA	1
3	PAOZZ	96906	MS90724-34	NUT,SHEET SPRING	11
4	PAOZZ	11862	14072405	PLATE,INSTRUMENT PA	1
5	PAOZZ	24617	11501153	SCREW,ASSEMBLED WAS	5
6	PAOZZ	24617	11501151	SCREW,TAPPING,THREA	9
7	PAOZZ	11862	14023008	SUPPORT,STEERING CO	1
8	PAOZZ	11862	6274970	COVER,PANEL	1

END OF FIGURE

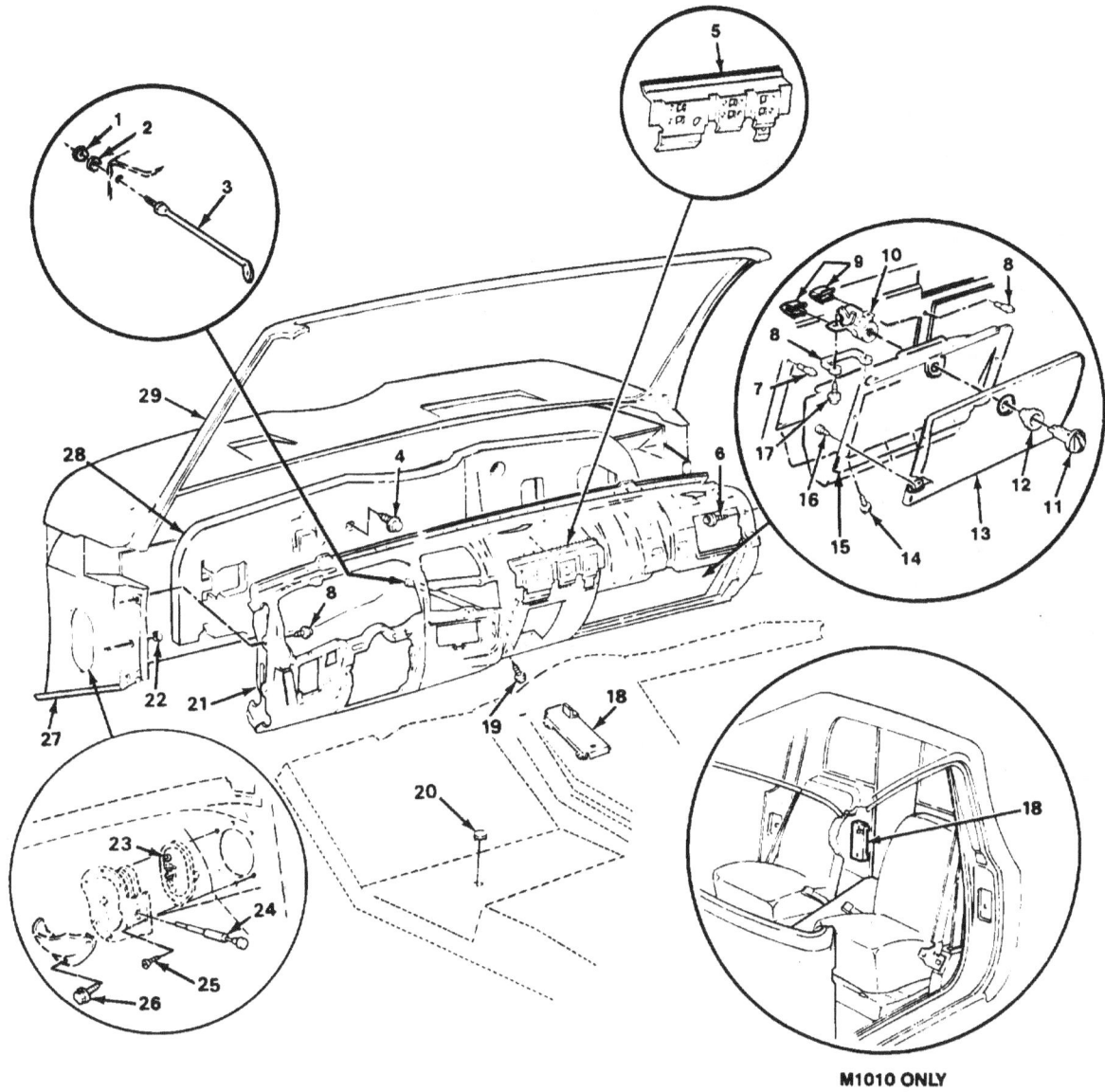

FIGURE 165. INSTRUMENT PANEL COMPONENTS.

M1010 ONLY

TA510972

(1) ITEM NO	(2) SMR CODE	(3) CAGEC	(4) PART NUMBER	(5) DESCRIPTION AND USABLE ON CODES (UOC)	(6) QTY

GROUP 1801 BODY, CAB, HOOD, AND HULL ASSEMBLIES

FIG. 165 INSTRUMENT PANEL COMPONENTS

(1)	(2)	(3)	(4)	(5)	(6)
1	PAFZZ	96906	MS51967-2	NUT,PLAIN,HEXAGON	1
2	PAFZZ	96906	MS27183-11	WASHER,FLAT	1
3	PAFZZ	11862	351789	BRACE	1
4	PAFZZ	24617	9422956	SCREW,TAPPING	2
5	PAFZZ	11862	14072414	PANEL,INSTRUMENT,RE	1
6	PAFZZ	11862	2014469	BOLT,ASSEMBLED WASH	4
7	PAOZZ	11862	3765243	BUMPER,NONMETALLIC	2
8	PAOZZ	11862	6274550	STRIKE,CATCH	1
				UOC:194,208,209,210,230,231,252	
9	PAOZZ	96906	MS90724-34	NUT,SHEET SPRING	2
10	PAOZZ	11862	14044471	CYLINDER,LOCK,VEHIC	1
11	PAOZZ	11862	6260421	CYLINDER,LOCK,VEHIC	1
12	PAOZZ	11862	3957093	MOLDING,METAL	1
13	PAOZZ	11862	6264131	DOOR,ACCESS	1
14	PAOZZ	11862	11509135	SCREW,ASSEMBLED WAS	4
15	PAOZZ	11862	343915	GLOVE BOX	1
16	PAOZZ	24617	9420408	SCREW,TAPPING,THREA	4
17	PAOZZ	24617	9415163	SCREW,ASSEMBLED WAS	2
18	PAFZZ	11862	15599270	BRACKET,FIRE EXTING	1
19	PAFZZ	11862	9420818	SCREW	1
20	PAFZZ	11862	14049810	PLUG,PROTECTIVE,DUS	2
21	PBFFF	11862	15590559	PANEL,INSTRUMENT	1
22	PAFZZ	11862	14000077	BUTTON,PLUG	2
23	PAFZZ	11862	11501937	NUT	2
24	PFFZZ	11862	466717	CONTROL ASSEMBLY,PU	2
25	PAFZZ	11862	11503537	SCREW,ASSEMBLED WAS	2
26	PAFZZ	11862	11509121	SCREW,TAPPING,THREA	12
27	PFFZZ	11862	15593569	PANEL,DASH AND COWL	1
28	PFFZZ	11862	15593831	INSTRUMENT PANEL,PA	1
29	PAOZZ	11862	15598706	FRAME,WINDOW,VEHICU	1

END OF FIGURE

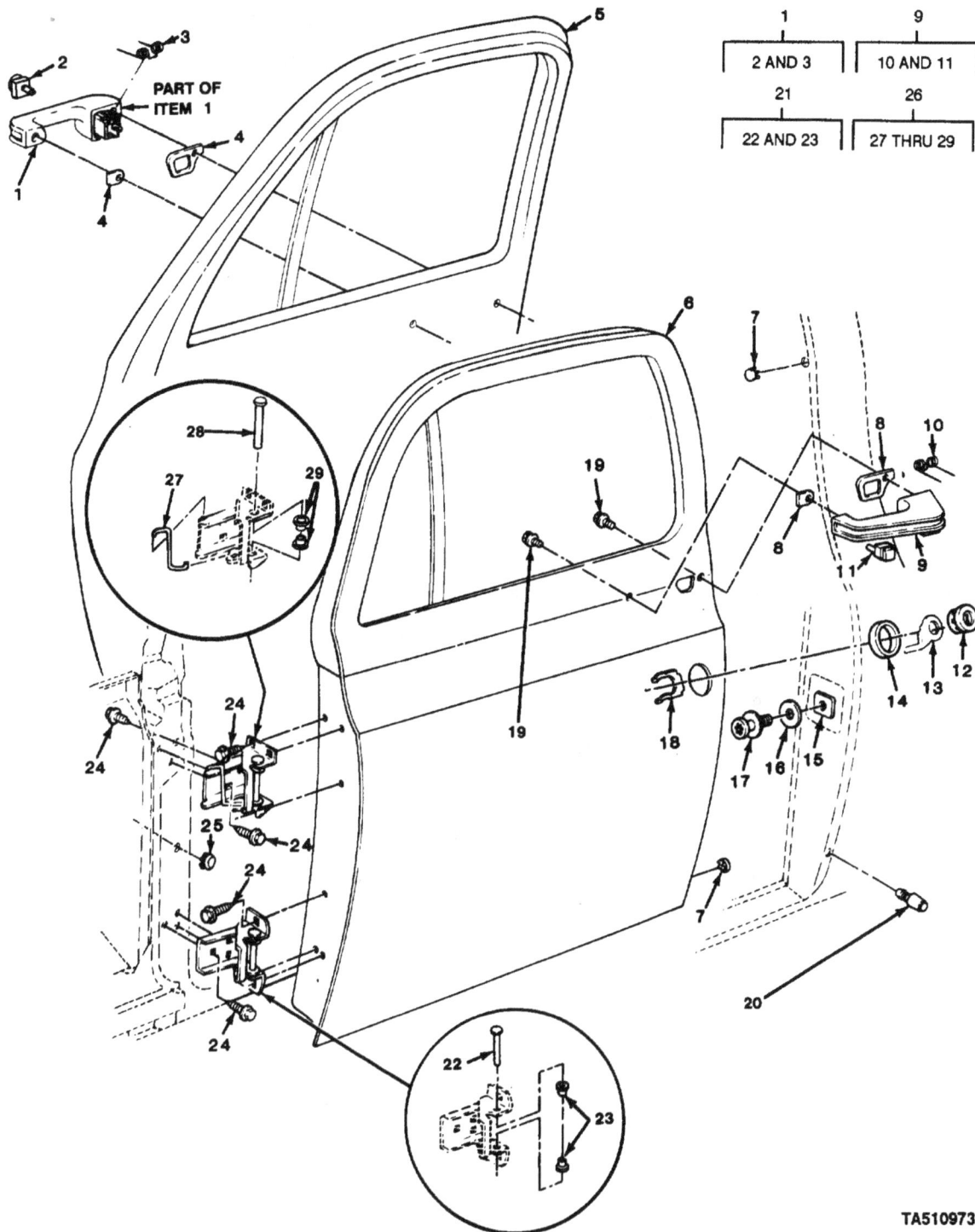

FIGURE 166. SIDE DOORS AND COMPONENTS.

TA510973

(1) ITEM NO	(2) SMR CODE	(3) CAGEC	(4) PART NUMBER	(5) DESCRIPTION AND USABLE ON CODES (UOC)	(6) QTY
				GROUP 1801 BODY, CAB, HOOD, AND HULL ASSEMBLIES	
				FIG. 166 SIDE DOORS AND COMPONENTS	
1	PAOOO	11862	14072494	HANDLE,DOOR	1
2	PAOZZ	11862	327062	.BUTTON,DOOE HANDLE	1
3	PAOZZ	11862	327065	.SPRING,HELICAL,TORS	1
4	PAOZZ	11862	6258562	GASKET SET	2
5	PAOZZ	11862	15593230	DOOR ASSEMBLY,CAB RIGHT	1
6	PAOZZ	11862	15571643	DOOR,VEHICULAR LEFT	1
7	PAOZZ	11862	4410574	PLUG,PROTECTIVE,DUS	4
8	PAOZZ	11862	6258561	GASKET SET	2
9	PAOOO	11862	14072493	HANDLE,DOOR	1
10	PAOZZ	11862	327065	.SPRING,HELICAL,TORS	1
11	PAOZZ	11862	327062	.BUTTON,DOOR HANDLE	1
12	PAOZZ	11862	15599677	CYLINDER,LOCK,VEHIC	2
13	PAOZZ	11862	7040173	PAW LEFT	1
13	PAOZZ	11862	7040174	PAWL RIGHT	1
14	PAOZZ	11862	4587931	GASKET	2
15	XDOZZ	11862	6272627	STRIKE,CATCH	2
16	PAOZZ	11862	1260895	WASHER,FLAT	2
17	PAOZZ	11862	9601750	BOLT,SHOULDER	2
18	PAOZZ	11862	9728247	RETAINER	2
19	PAOZZ	11862	9439771	BOLT,ASSEMBLED WASH	4
20	PAOZZ	11862	3900684	BUMPER,NONMETALLIC	4
21	PAOOO	11862	14000091	HINGE,ACCESS DOOR LEFT	1
21	PAOOO	11862	14000092	HINGE,DOOR VEHICULA RIGHT	1
22	PAOZZ	11862	6271989	.PIN,STRAIGHT,HEADED	1
23	PAOZZ	11862	9721917	.BEARING,SLEEVE	2
24	PAOZZ	11862	9438916	SCREW,TAPPING,THREA	24
25	PAOZZ	11862	3944769	CAP-PLUG,PROTECTIVE	2
26	PAOOO	11862	14000093	HINGE,ACCESS DOOR LEFT	1
26	PAOOO	11862	14000094	HINGE RIGHT	1
27	PAOZZ	11862	327960	.ROD,HINGE TORQUE RH	1
27	PAOZZ	11862	327959	.ROD,HINGE TORQUE LH	1
28	PAOZZ	11862	6271989	.PIN,STRAIGHT,HEADED	1
29	PAOZZ	11862	9721917	.BEARING,SLEEVE	2

END OF FIGURE

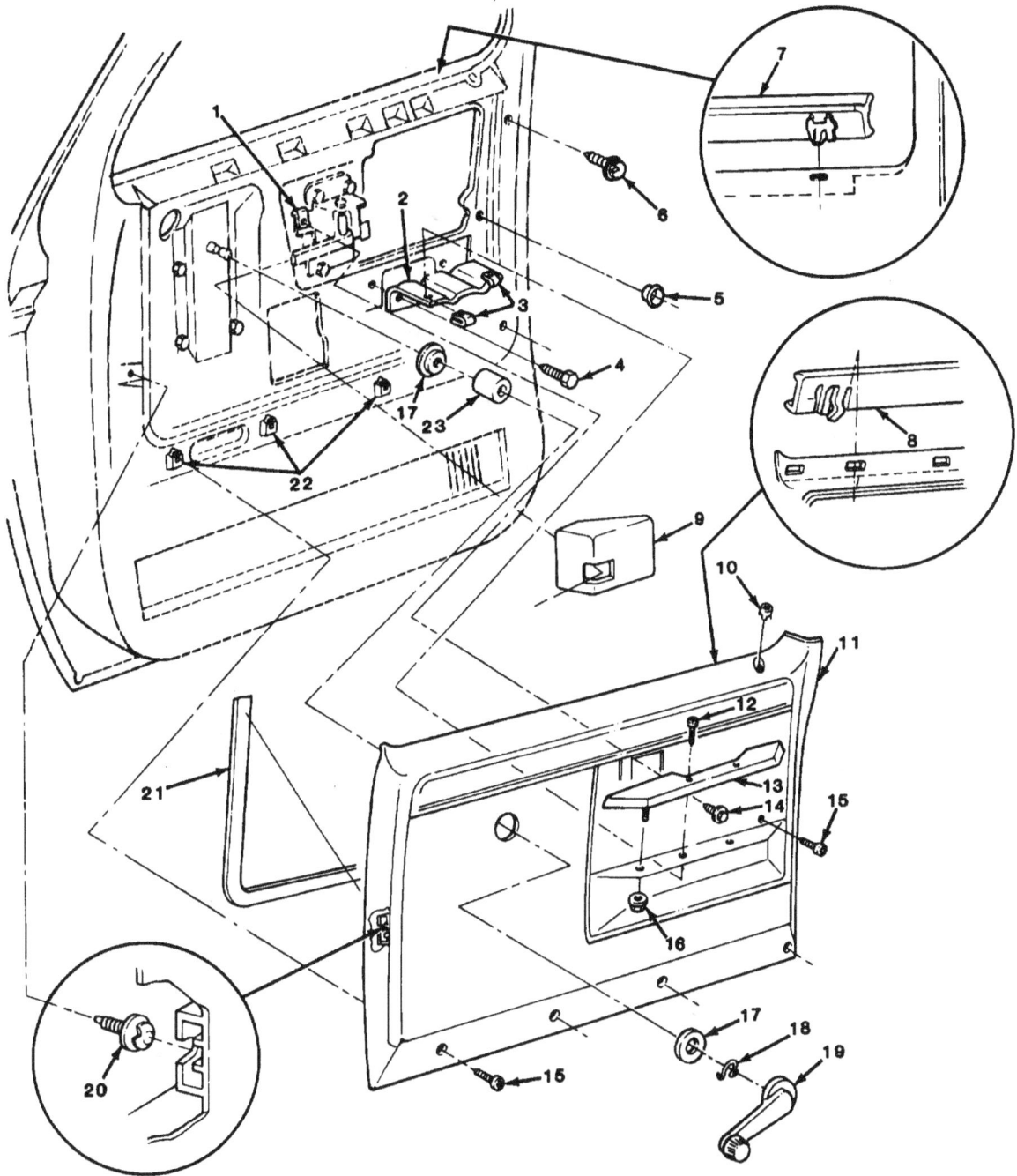

TA510974

FIGURE 167. INSIDE DOOR TRIM PANEL AND RELATED PARTS.

(1)	(2)	(3)	(4)	(5)	(6)
ITEM NO	SMR CODE	CAGEC	PART NUMBER	DESCRIPTION AND USABLE ON CODES (UOC)	QTY

GROUP 1801 BODY, CAB, HOOD, AND HULL
ASSEMBLIES

FIG. 167 INSIDE DOOR TRIM PANEL AND
RELATED PARTS

1	PAOZZ	96906	MS90724-34	NUT,SHEET SPRING	2
2	PAOZZ	11862	362433	BRACKET,ANGLE LEFT	1
2	PAOZZ	11862	362434	BRACKET,ANGLE RIGHT	1
3	PAOZZ	96906	MS90724-40	NUT,SHEET SPRING	4
4	PAOZZ	11862	9439770	BOLT,ASSEMBLED WASH	4
5	PAOZZ	11862	1355003	NUT	4
6	PAOZZ	11862	20696927	FASTENER,SPRING TEN	2
7	PAOZZ	11862	14027775	SEAL,RUBBER SPECIAL LEFT	1
7	PAOZZ	11862	14027776	SEAL,RUBBER SPECIAL OUTER RIGHT	1
8	PAOZZ	11862	15569071	SEAL	1
8	PAOZZ	11862	15569072	SEAL,NONMETALLIC SP INNER RIGHT	1
9	PAOZZ	11862	14026383	SEAL,NONMETALLIC SP LEFT	1
9	PAOZZ	11862	14026384	SEAL,NONMETALLIC SP RIGHT	1
10	PAOZZ	11862	330485	GROMMET	2
11	PAOZZ	11862	15597667	PANEL,BODY,VEHICULA LEFT	1
11	PAOZZ	11862	15597668	PANEL,BODY,VEHICULA RIGHT	1
12	PAOZZ	24617	9414724	SCREW,TAPPING,THREA	4
13	PAOZZ	11862	14026409	PAD,CUSHIONING LEFT	1
13	PAOZZ	11862	14026410	PAD,CUSHIONING RIGHT	1
14	PAOZZ	11862	11503537	SCREW,ASSEMBLED WAS	2
15	PAOZZ	24617	1640810	SCREW,TAPPING,THREA	10
16	PAOZZ	24617	9420621	PUSH ON NUT	2
17	PAOZZ	11862	14010954	ESCUTCHEON PLATE	4
18	PAOZZ	11862	4168122	RETAINER,SPECIAL	2
19	PAOZZ	11862	14030586	HANDLE,WINDOW REGUL	2
20	PAOZZ	11862	363137	FASTENER,SPRING TEN	2
21	MOOZZ	11862	15590415	SEAL,FRONT DOOR (46.5" X 0.36" X 0.50") MAKE FROM SEAL, P/N 363139	2
22	PAOZZ	11862	15590422	RETAINER,DOOR TRIM	6
23	PAOZZ	11862	364372	BUSHING,NONMETALLIC	2

END OF FIGURE

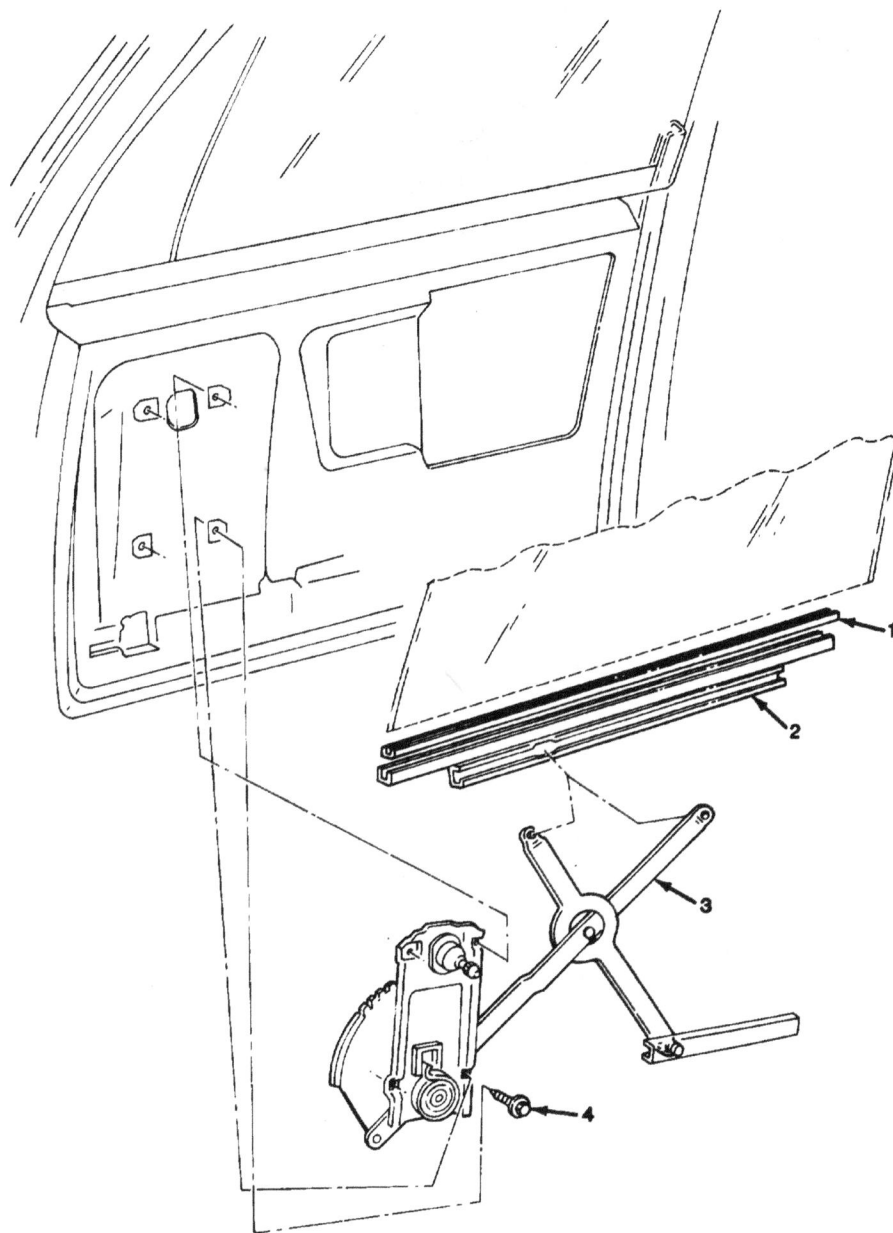

TA510975

FIGURE 168. SIDE DOOR WINDOW REGULATOR.

(1) (2) (3) (4) ITEM SMR PART NO CODE CAGEC NUMBER	(5) DESCRIPTION AND USABLE ON CODES (UOC)	(6) QTY
	GROUP 1801 BODY, CAB, HOOD, AND HULL ASSEMBLIES	
	FIG. 168 SIDE DOOR WINDOW REGULATOR	
1 MOOZZ 11862 365953	FILLER,GLASS CHANNE (25.62" LG) MAKE FROM FILLER, P/N 370389	1
2 PAOZZ 11862 15590401	CHANNEL,LIFT,VEHIVL LEFT	1
2 PAOZZ 11862 15590402	CHANNEL,LIFT,VEHICL RIGHT	1
3 PAOZZ 11862 14027431	REGULATOR,VEHICLE W LH	1
3 PAOZZ 11862 14027432	REGULATOR,VEHICLE W RH	1
4 PAOZZ 11862 9439770	BOLT,ASSEMBLED WASH	8

END OF FIGURE

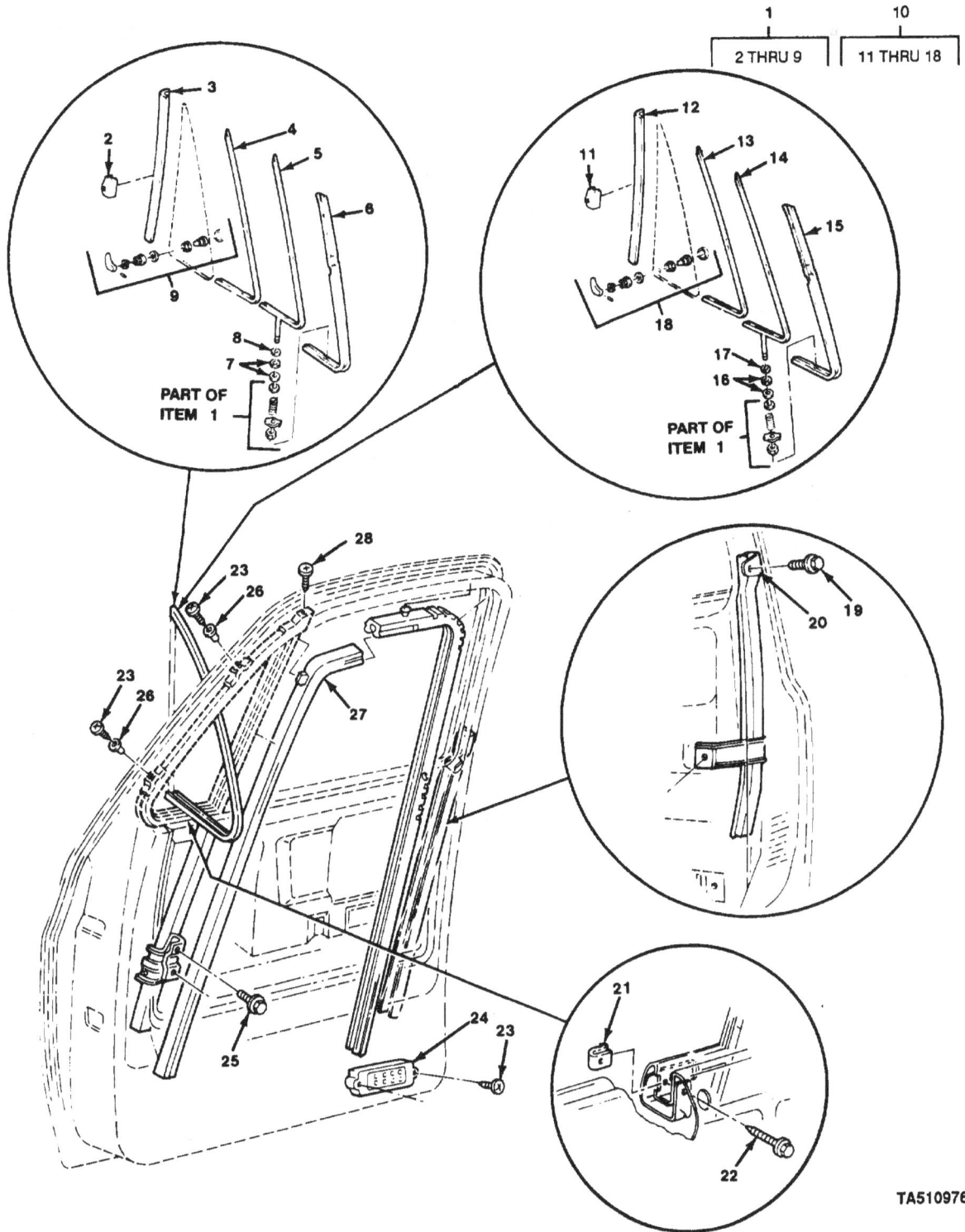

FIGURE 169. SIDE DOOR COMPONENTS.

TA510976

(1)	(2)	(3)	(4)	(5)	(6)
ITEM NO	SMR CODE	CAGEC	PART NUMBER	DESCRIPTION AND USABLE ON CODES (UOC)	QTY

GROUP 1801 BODY, CAB, HOOD, AND HULL
ASSEMBLIES

FIG. 169 SIDE DOOR COMPONENTS

1	PAOOO	11862	15635685	VENT WING ASSEMBLY RIGHT	1
2	PAOZZ	11862	20264729	.STRIKE,CATCH	1
3	PAOZZ	11862	20264731	.SEAL,NONMETALLIC SP	1
4	MOOZZ	11862	365953-1	.FILLER,GLASS CHANNE (25.10" LG.) MAKE FROM RUBBER, P/N 370390	1
5	PAOZZ	11862	20264737	.CHANNEL,LIFT,VEHICL	1
6	PAOZZ	11862	20354946	.WEATHER STRIP	1
7	PAOZZ	11862	365443	.WASHER,FLAT	2
8	PAOZZ	11862	3762400	.STOP,DOOR VENTILATC	1
9	PAOZZ	11862	15617126	.PARTS KIT,VEHICULAR	1
10	PAOOO	11862	15635684	VENTILATOR,AIR CIRC LEFT	1
11	PAOZZ	11862	20264728	.STRIKE,CATCH	1
12	PAOZZ	11862	20264730	.SEAL,RUBBER SPECIAL	1
13	MOOZZ	11862	365953-1	.FILLER,GLASS CHANNE (25.10" LG.) MAKE FROM RUBBER, P/N 370390	1
14	PAOZZ	11862	20264736	.VENTILATOR,AIR CIRC	1
15	PAOZZ	11862	20354945	.WEATHER STRIP	1
16	PAOZZ	11862	365443	.WASHER,FLAT	2
17	PAOZZ	11862	3762400	.STOP,DOOR VENTILATO	1
18	PAOZZ	11862	12300197	.PARTS KIT,VEHICULAR	1
19	PAOZZ	11862	9439770	BOLT,ASSEMBLED WASH	4
20	PAOZZ	11862	337715	CHANNEL,LIFT,VEHICL LEFT	1
20	PAOZZ	11862	337716	CHANNEL,LIFT,VEHICL RIGHT	1
21	PAOZZ	96906	MS90724-34	NUT,SHEET SPRING	2
22	PAOZZ	11862	15531547	SCREW,TAPPING,THREA	2
23	PAOZZ	11862	11501149	SCREW,TAPPING,THREA	8
24	PAOZZ	11862	461610	VALVE ASSEMBLY,DOOR	2
25	PAOZZ	11862	9439770	BOLT,ASSEMBLED WASH	4
26	PAOZZ	11862	14013789	SPACER,SLEEVE	4
27	PAOZZ	11862	14027777	NONMETALLIC SPECIAL LEFT	1
27	PAOZZ	11862	14027778	NONMETALLIC SPECIAL RIGHT	1
28	PAOZZ	11862	15590443	SCREW,ASSEMBLED WAS	2

END OF FIGURE

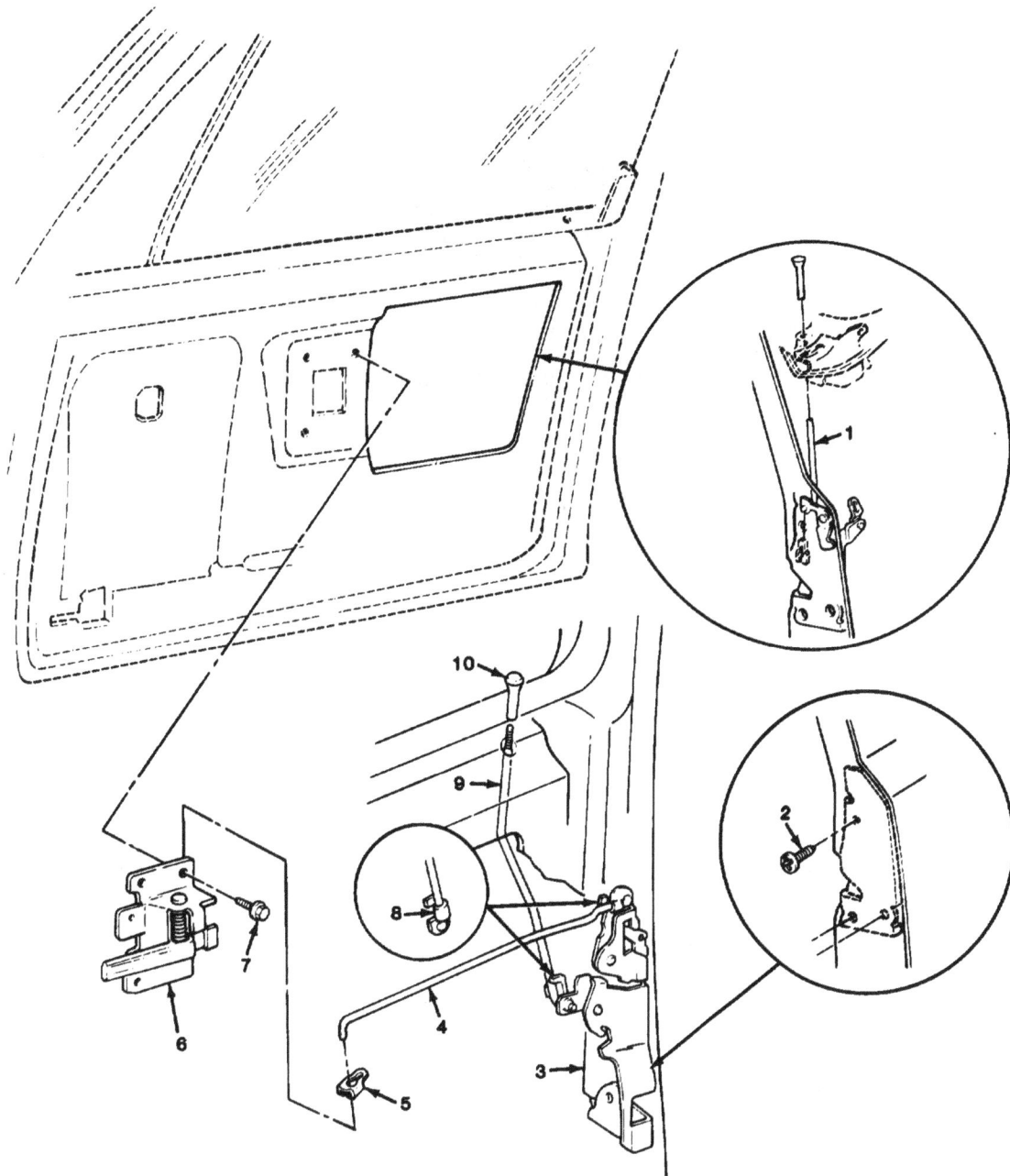

TA510977

FIGURE 170. SIDE DOOR LOCK ASSEMBLY

(1) (2) (3) (4) ITEM SMR NO CODE CAGEC	PART NUMBER	(5) DESCRIPTION AND USABLE ON CODES (UOC)	(6) QTY
		GROUP 1801 BODY, CAB, HOOD, AND HULL ASSEMBLIES	
		FIG 170 SIDE DOOR LOCK ASSEMBLY	
1 PAOZZ 11862	327067	CONNECTING LINK,RIG	2
2 PAOZZ 11862	9439772	SCREW,SELF-LOCKING	6
3 PAOZZ 11862	14039763	LOCK ASSEMBLY,VEHIC LEFT	1
3 PAOZZ 11862	14039764	LOCK ASSEMBLY,VEHIC RIGHT	1
4 PAOZZ 11862	14039765	CONNECTING LINK,RIG LEFT	1
4 PAOZZ 11862	14039766	CONNECTING LINK,RIG RIGHT	1
5 PAOZZ 11862	375180	CLIP,SPRING TENSION	2
6 PAOZZ 11862	15597653	CONTROL ASSEMBLY,RE LEFT	1
6 PAOZZ 11862	15597654	CONTROL ASSEMBLY,LO RIGHT	1
7 PAOZZ 11862	9439770	BOLT,ASSEMBLED WASH	6
8 PAOZZ 11862	15545178	CLIP,RETAINING	4
9 PAOZZ 11862	14039767	CONNECTING LINK,RIG LEFT	1
9 PAOZZ 11862	14039768	CONNECTING LINK,RIG RIGHT	1
10 PAOZZ 11862	7591126	KNOB	2

END OF FIGURE

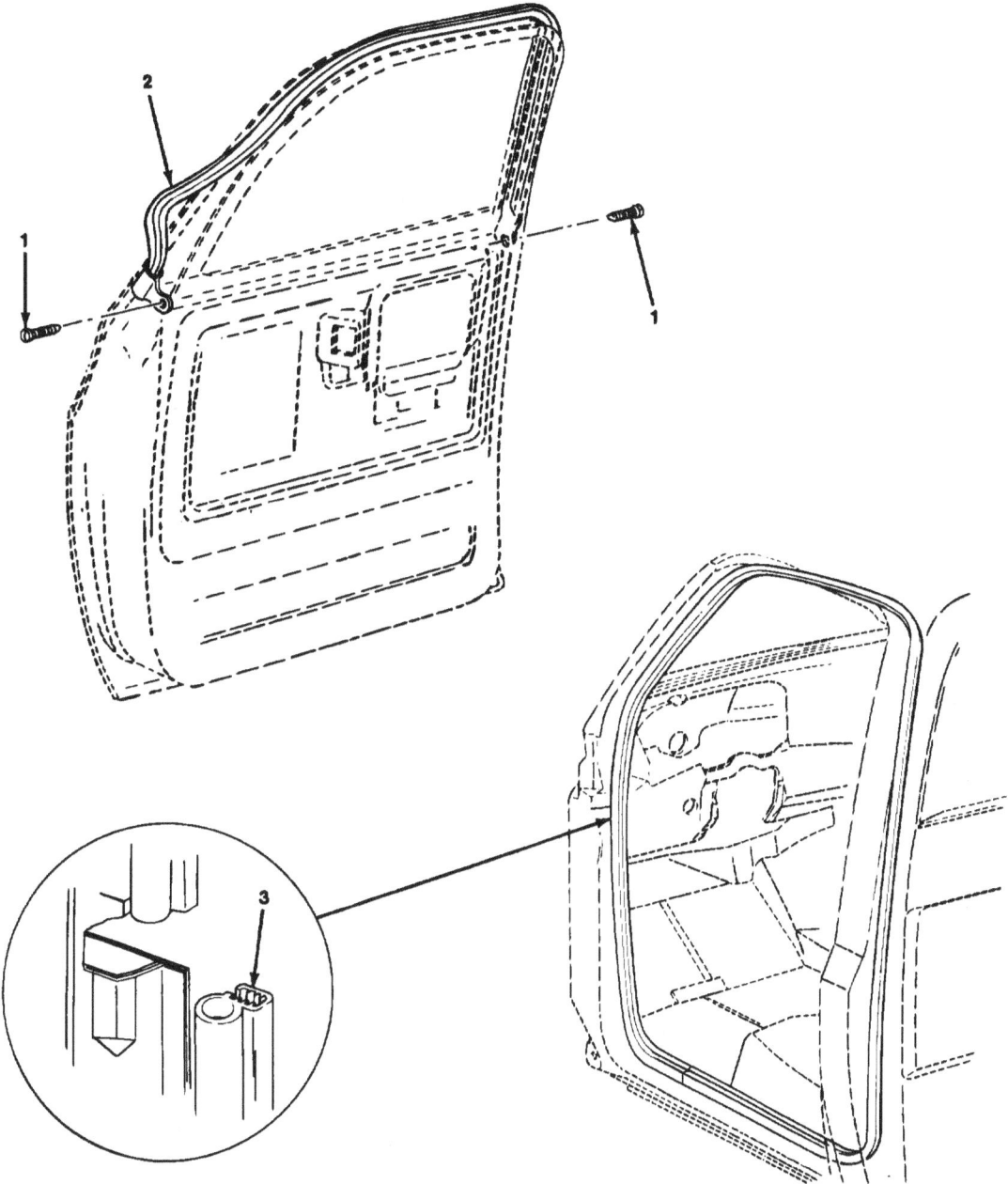

TA510978

FIGURE 171. SIDE DOOR WEATHERSTRIPS.

(1) (2) (3) (4) ITEM SMR PART NO CODE CAGEC NUMBER	(5) DESCRIPTION AND USABLE ON CODES (UOC)	(6) QTY
	GROUP 1801 BODY, CAB, HOOD, AND HULL ASSEMBLIES	
	FIG. 171 SIDE DOOR WEATHERSTRIPS	
1 PAOZZ 11862 14022885	SCREW,TAPPING,THREA UOC:209	4
2 PAOZZ 11862 14016511	MOLDING,FRAME LEFT UOC:209	1
2 PAOZZ 11862 14016512	MOLDING,FRAME RIGHT UOC:209	1
3 PAOZZ 11862 15522764	WEATHER STRIP	2
	END OF FIGURE	

TM9-2320-289-34P

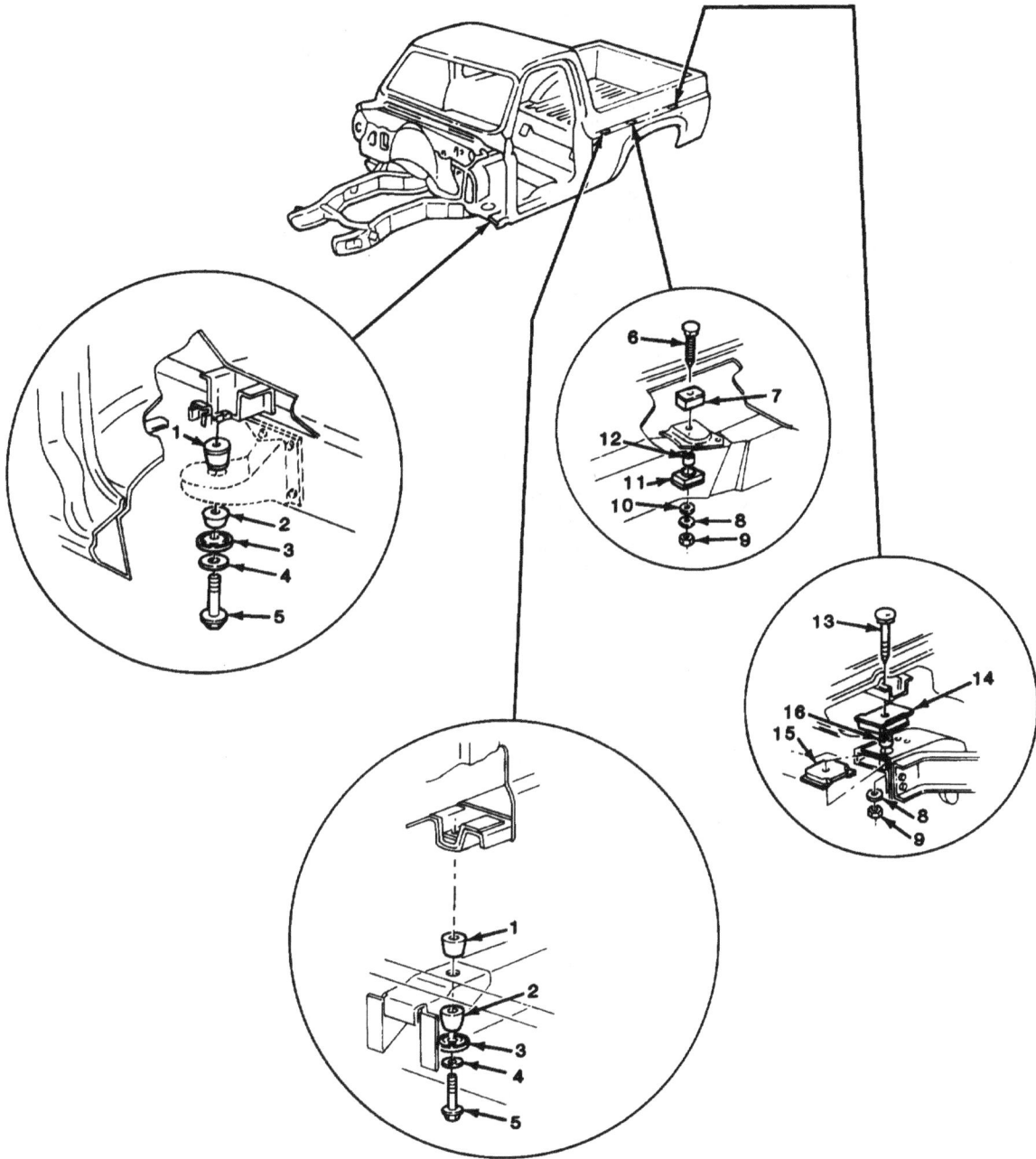

TA510979

FIGURE 172. BODY CAB MOUNTING (M1009).

(1) (2) (3) (4)	(5)	(6)
ITEM SMR PART		
NO CODE CAGEC NUMBER	DESCRIPTION AND USABLE ON CODES (UOC)	QTY

GROUP 1801 BODY, CAB, HOOD, AND HULL
ASSEMBLIES

FIG. 172 BODY CAB MOUNTING (M1009)

1	PAFZZ 11862 15597600		RETAINER,BOLT CUSHI UPPER	4
2	PAFZZ 11862 15597629		BUSHING,RUBBER LOWER	4
3	PAFZZ 11862 14027472		RETAINER,BODY BOLT	2
4	PAFZZ 96906 MS27183-20	WASHER,FLAT		4
5	PAFZZ 96906 MS90728-121		SCREW,CAP,HEXAGON H	4
6	PAFZZ 11862 14014389		BOLT,SQUARE NECK	2
			UOC:209	
7	PFFZZ 11862 3946246		MOUNT,RESLILIENT UPPER	2
			UOC:209	
8	PAFZZ 96906 MS27183-18	WASHER,FLAT		4
			UOC:209	
9	PAFZZ 96906 MS51967-14	NUT,PLAIN,HEXAGON		4
			UOC:209	
10	PAFZZ 11862 2387523		WASHER,FLAT	2
			UOC:209	
11	PFFZZ 11862 3909351		BUMPER,NONMETALLIC LOWER	2
			UOC:209	
12	PAFZZ 11862 3946247		SPACER,SLEEVE	2
			UOC:209	
13	PAFZZ 11862 14013710		BOLT,SQUARE NECK	2
			UOC:209	
14	PAFZZ 11862 337751		MOUNT,RESILIENT	2
			UOC:209	
15	PFFZZ 1862 327100		MOUNT,RESILIENT	2
			UOC:209	
16	PAFZZ 11862 337773		SPACER,SLEEVE	2
			UOC:209	

END OF FIGURE

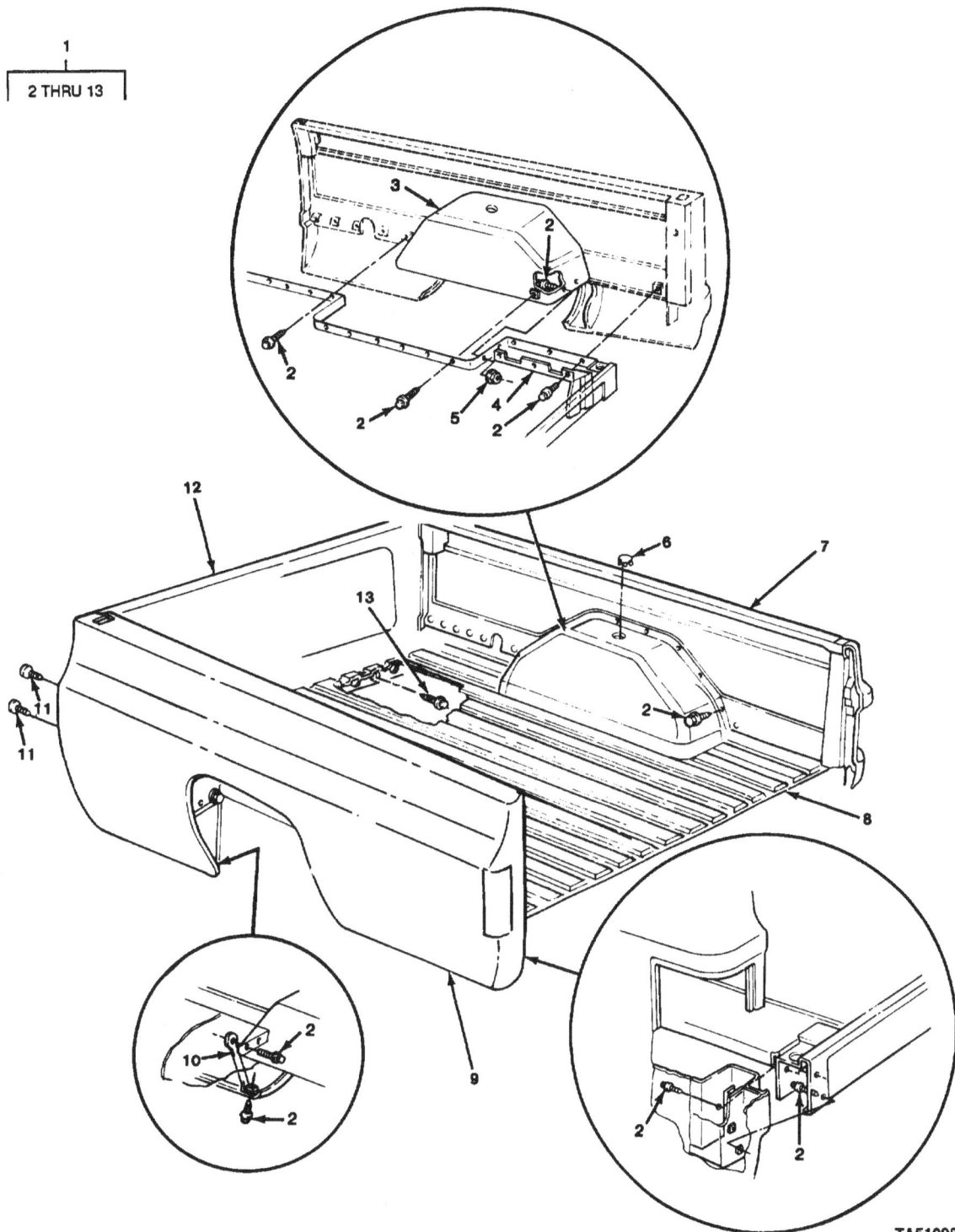

FIGURE 173. CARGO BOX AND COMPONENTS (M1008, M1008A1, M1028, M1028A1, M1028A2, AND M1028A3).

TA510980

(1) (2) (3) (4) ITEM SMR PART NO CODE CAGEC NUMBER	(5) DESCRIPTION AND USABLE ON CODES (UOC)	(6) QTY
	GROUP 1801 BODY, CAB, HOOD, AND HULL ASSEMBLIES	
	FIG. 173 CARGO BOX AND COMPONENTS (M1008, M1008A1, M1028, M1028A1, M1028A2, AND M1028A3)	
1 PBFFF 11862 15596628	BODY,CARGO TRUCK UOC:194,208,230,252,254,256	1
2 PAFZZ 96906 MS51869-24	.SCREW,CAP,SOCKET HE UOC:194,208,230,252,254,256	66
3 PAFZZ 11862 14027451	.PANEL,WHEEL HOUSE UOC:194,208,230,252,254,256	2
4 PAFZZ 11862 467913	.SHIELD,BRAKE DISK LEFT UOC:194,208,230,252	1
4 PAFZZ 11862 15593254	.SHIELD,BRAKE DISK RIGHT UOC:194,208,230,252,254,256	1
5 PAFZZ 96906 MS51967-6	.NUT,PLAIN,HEXAGON UOC:194,208,230,252,254,256	4
6 PAOZZ 11862 14049810	.PLUG,PROTECTIVE,DUS UOC:194,208,230,252,254,256	2
7 PAFZZ 11862 15628618	.PANEL,BODY,VEHICULA RIGHT UOC:194,208,230,252,254,256	1
8 XAFZZ 11862 15596994	.PLATFORM UOC:194,208,230,252,254,256	1
9 PAFZZ 11862 15597621	.FRAME SECTION,STRUC LEFT UOC:194,208,230,252,254,256	1
10 PAFZZ 11862 330438	.BRACE ASSEMBLY,PANE UOC:194,208,230,252,254,256	2
11 PAFZZ 11862 480534	.RETAINER,PANEL HOLE UOC:194,208,230,252,254,256	4
12 PBFZZ 11862 14049881	.PANEL,BODY,VEHICULA UOC:194,208,230,252,254,256	1
13 PAFZZ 11862 2014469	.BOLT,ASSEMBLED WASH UOC:194,208,230,252,254,256	18

END OF FIGURE

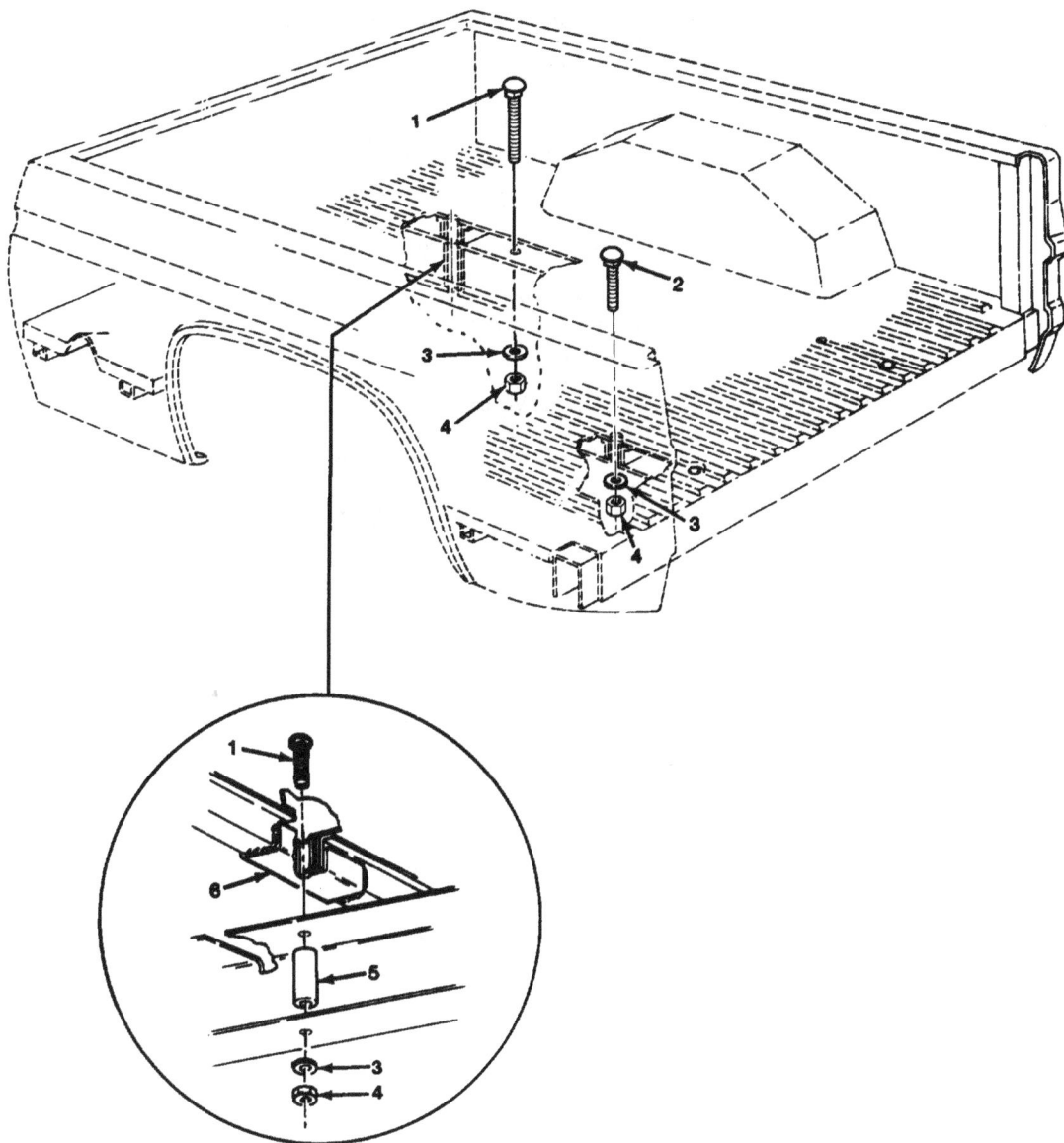

TA510981

FIGURE 174. CARGO BOX MOUNTING (M1008, M1008A1 ,M1028, M1028A1,
M1028A2, AND M1028A3).

(1) ITEM NO	(2) SMR CODE	(3) CAGEC	(4) PART NUMBER	(5) DESCRIPTION AND USABLE ON CODES (UOC)	(6) QTY
				GROUP 1801 BODY, CAB, HOOD, AND HULL ASSEMBLIES	
				FIG. 174 CARGO BOX MOUNTING (M1008, M1008A1, M1028, M1028A1, M1028A2, AND M1028A3)	
1	PAOZZ	11862	14039924	BOLT,SQUARE NECK UOC:194,208,230,231,252,254,256	4
2	PAOZZ	11862	15599971	BOLT,SQUARE NECK UOC:194,208,230,231,252,254,256	4
3	PAOZZ	96906	MS27183-18	WASHER,FLAT UOC:194,208,230,231,252,254,256	8
4	PAOZZ	96906	MS51967-14	NUT,PLAIN,HEXAGON UOC:194,208,230,231,252,254,256	8
5	PAOZZ	11862	14076894	SPACER,PLATE USE WITH BOLT, P/N 15599971 UOC:194,208,230,252,254,256	4
5	PAOZZ	11862	15599250	SPACER,SLEEVE USE WITH BOLT, P/N 14039924 UOC:194,208,230,231,252,254,256	4
6	PAOZZ	11862	14076887	STRAP,RETAINING UOC:194,208,230,231,252,254,256	6

END OF FIGURE

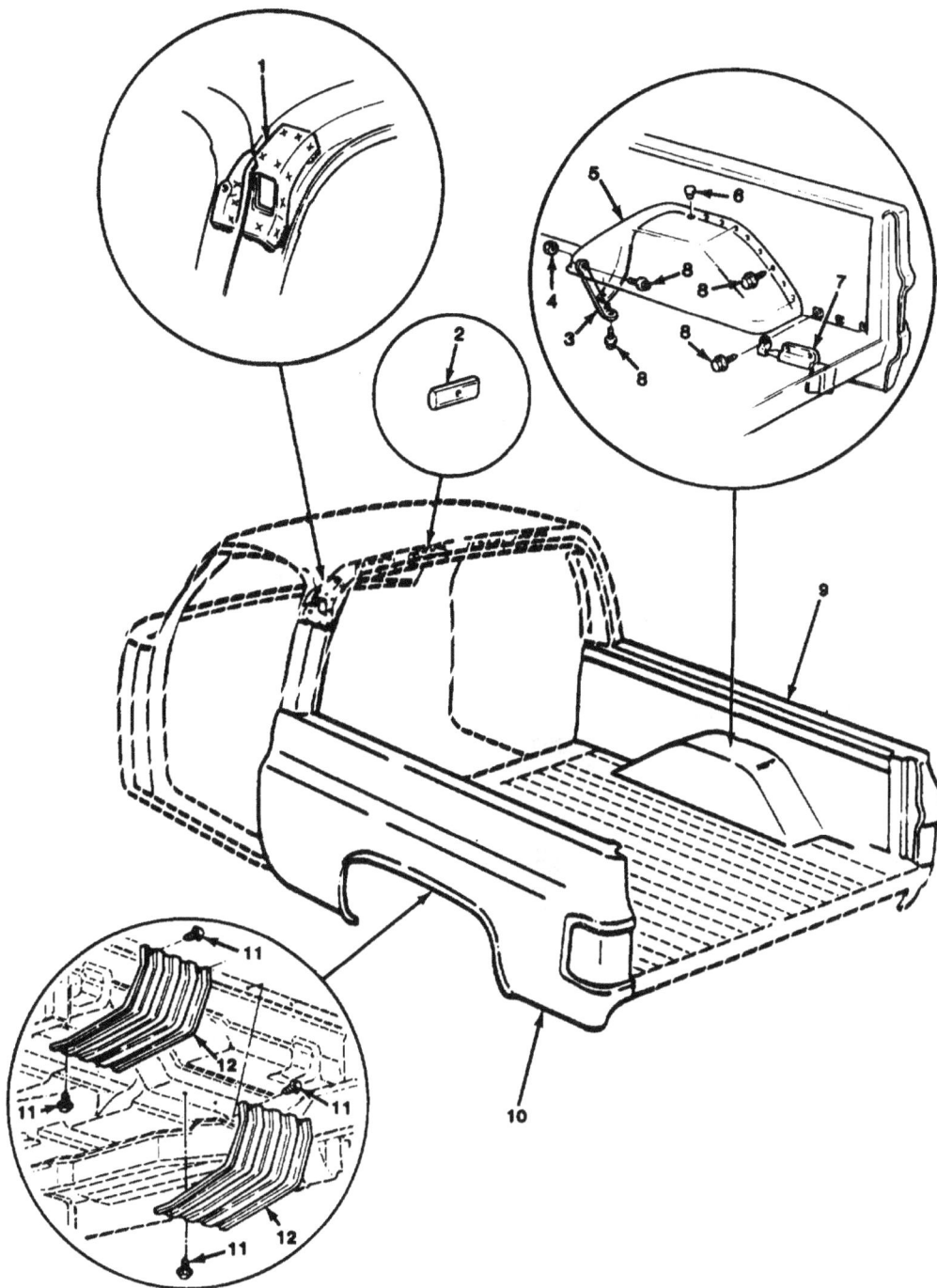

TA510982

FIGURE 175. BODY CARGO AND COMPONENTS (M1009).

(1) (2) (3) (4)	(5)	(6)
ITEM SMR PART		
NO CODE CAGEC NUMBER	DESCRIPTION AND USABLE ON CODES (UOC)	QTY

GROUP 1801 BODY, CAB, HOOD, AND HULL
ASSEMBLIES

FIG. 175 MBODY CARGO AND COMPONENTS
(M1009)

1	PAFZZ	11862	358543	REINFORCEMENT ASSEM UOC:209	1
2	PAFZZ	11862	357445	NUT,PLAIN,RECTANGUL UOC:209	4
3	PAOZZ	11862	330438	BRACE ASSEMBLY,PANE UOC:209	2
4	PAOZZ	96906	MS51967-6	NUT,PLAIN,HEXAGON UOC:209	2
5	PAOZZ	11862	14027451	PANEL,WHEEL HOUSE LEFT UOC:209	1
5	PAOZZ	11862	14027454	LINER SECTION,VEHIC RIGHT UOC:209	1
6	PAOZZ	11862	14049810	PLUG,PROTECTIVE,DUS UOC:209	2
7	PAOZZ	11862	467911	SHIELD,PANEL STONE LEFT UOC:209	1
7	PAOZZ	11862	467912	SHIELD,PANEL STONE RIGHT UOC:209	1
8	PAOZZ	11862	2014469	BOLT,ASSEMBLED WASH UOC:209	36
9	PAFZZ	11862	14073962	FRAME SECTION,STRUC RIGHT UOC:209	1
9	PAFZZ	11862	14025874	FENDER,VEHICULAR UOC:209	1
10	PAFZZ	11862	14026209	PANEL,BODY,SIDE LEFT UOC:209	1
10	PAFZZ	11862	14025873	PANEL,BODY,VEHICULA UOC:209	1
11	PAOZZ	11862	11504447	SCREW,TAPPING,THREA UOC:209	8
12	PAOZZ	11862	471089	GUARD,MUFFLER-EXHAU UOC:209	2

END OF FIGURE

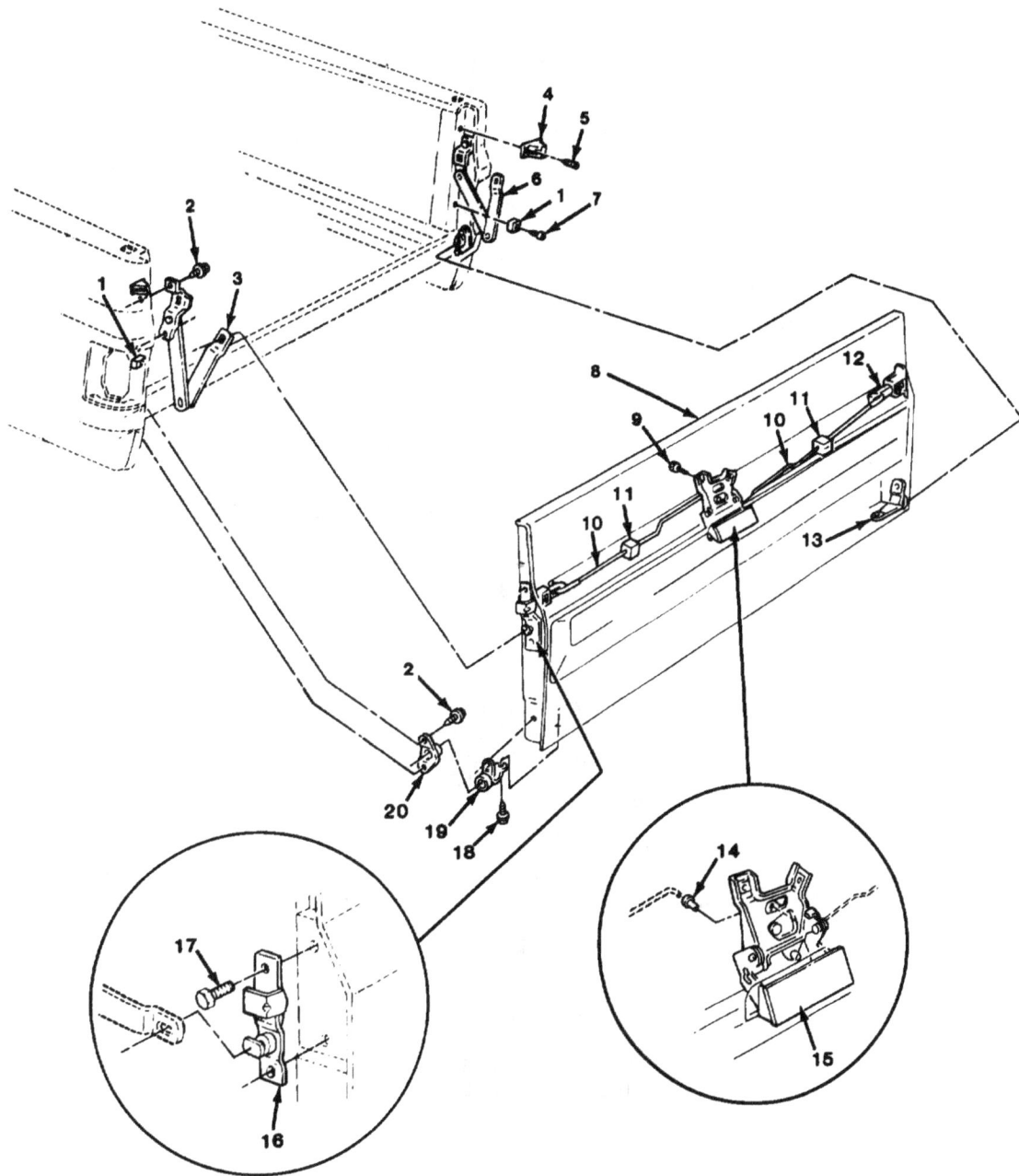

TA510983

FIGURE 176. CARGO ENDGATE AND COMPONENTS (M1008, M1008A1, M1028, M1028A1, M1028A2, AND M1028A3).

(1)	(2)	(3)	(4)	(5)	(6)
ITEM NO	SMR CODE	CAGEC	PART NUMBER	DESCRIPTION AND USABLE ON CODES (UOC)	QTY

GROUP 1801 BODY, CAB, HOOD, AND HULL ASSEMBLIES

FIG. 176 CARGO ENDGATE AND COMPONENTS (M1008, M1008A1, M1028, M1028A1,M1028A2,AND M1028A3)

1	PAOZZ	11862	6262029	BUMPER,END GATE LIN UOC:194,208,230,252,254,256	2
2	PAOZZ	96906	MS90728-59	SCREW,CAP,HEXAGON H UOC:194,208,230,252,254,256	8
3	PAOZZ	11862	458025	STRIKER PLATE AND L UOC:194,208,230,252,254,256	1
4	PAOZZ	11862	14032787	BUMPER,NONMETALLIC UOC:194,208,230,252,254,256	2
5	PAOZZ	96906	MS51869-24	SCREW,CAP,SOCKET HE UOC:194,208,230,252,254,256	2
6	PAOZZ	11862	458026	STRIKER PLATE ASSEM UOC:194,208,230,252,254,256	1
7	PAOZZ	96906	MS51861-45	SCREW,TAPPING,THREA UOC:194,208,230,252,254,256	2
8	PAOZZ	11862	15628608	TAILGATE,VEHICLE BO UOC:194,208,230,252,254,256	1
9	PAOZZ	11862	9431663	SCREW,ASSEMBLED WAS UOC:194,208,230,252,254,256	3
10	PAOZZ	11862	14021389	ROD,TAILGATE LATCH UOC:194,208,230,252,254,256	2
11	PAOZZ	11862	14007449	BUMPER,NONMETALLIC UOC:194,208,230,252,254,256	2
12	PAOZZ	11862	15594644	LATCH,END GATE UOC:194,208,230,252,254,256	1
13	PAOZZ	11862	15596976	TRUNION ASSEMBLY RIGHT UOC:194,208,230,252,256	1
14	PAOZZ	11862	3889864	CLIP,SPLIT TUBULAR UOC:194,208,230,252,254,256	2
15	PAOZZ	11862	14021275	HANDLE,DOOR UOC:194,208,230,252,254,256	1
16	PAOZZ	11862	15594643	LATCH,DOOR,VEHICULA UOC:194,208,230,252,254,256	1
17	PAOZZ	24617	9414712	SCREW,TAPPING,THREA UOC:194,208,230,252,254,256	4
18	PAOZZ	11862	2014469	BOLT,ASSEMBLED WASH UOC:194,208,230,252,254,256	8
19	PAOZZ	11862	15596975	BUSHING ASSEMBLY,EN LEFT UOC:194,208,230,252,254,256	1
20	PAOZZ	11862	14021385	TRUNNION ASSEMBLY LEFT UOC:194,208,230,252,254,256	1
20	PAOZZ	11862	14021386	TRUNNION ASSEMBLY,E RIGHT UOC:194,208,230,252,254,256	1

END OF FIGURE

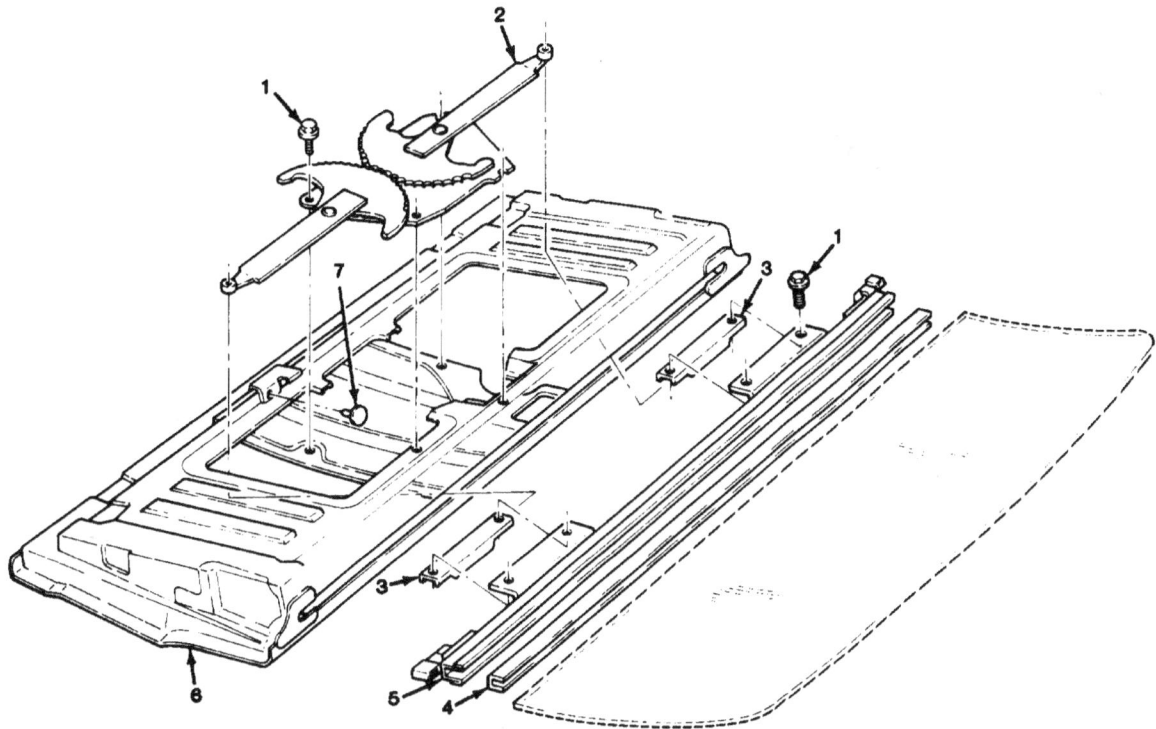

TA510984

FIGURE 177. ENDGATE AND COMPONENTS (M1009).

(1) (2) (3) (4) ITEM SMR PART NO CODE CAGEC NUMBER	(5) DESCRIPTION AND USABLE ON CODES (UOC)	(6) QTY
	GROUP 1801 BODY, CAB, HOOD, AND HULL ASSEMBLIES	
	FIG. 177 ENDGATE AND COMPONENTS (M1009)	
1 PAOZZ 11862 1244067	BOLT,ASSEMBLED WASH UOC:209	8
2 PAOOO 11862 15641780 REGULATOR,WINDOW,VE	UOC:209	1
2 PAOZZ 11862 4158246 ROLLER,LINEAR-ROTAR	UOC:209	2
3 PAOZZ 11862 20171141 CHANNEL,LIFT,VEHICL	UOC:209	2
4 MOOZZ 11862 14027542	FILLER E/GATE SASH (60.00"LG) ,MAKE FROM RUBBER, P/N 370390 UOC:209	1
5 PAOZZ 11862 340053 CHANNEL,LIFT,VEHICL	UOC:209	1
6 PAOZZ 11862 15614462	TAILGATE,VEHICLE BO UOC:209	1
7 PAOZZ 11862 8785295 BUMPER,NONMETALLIC	UOC:209	1

END OF FIGURE

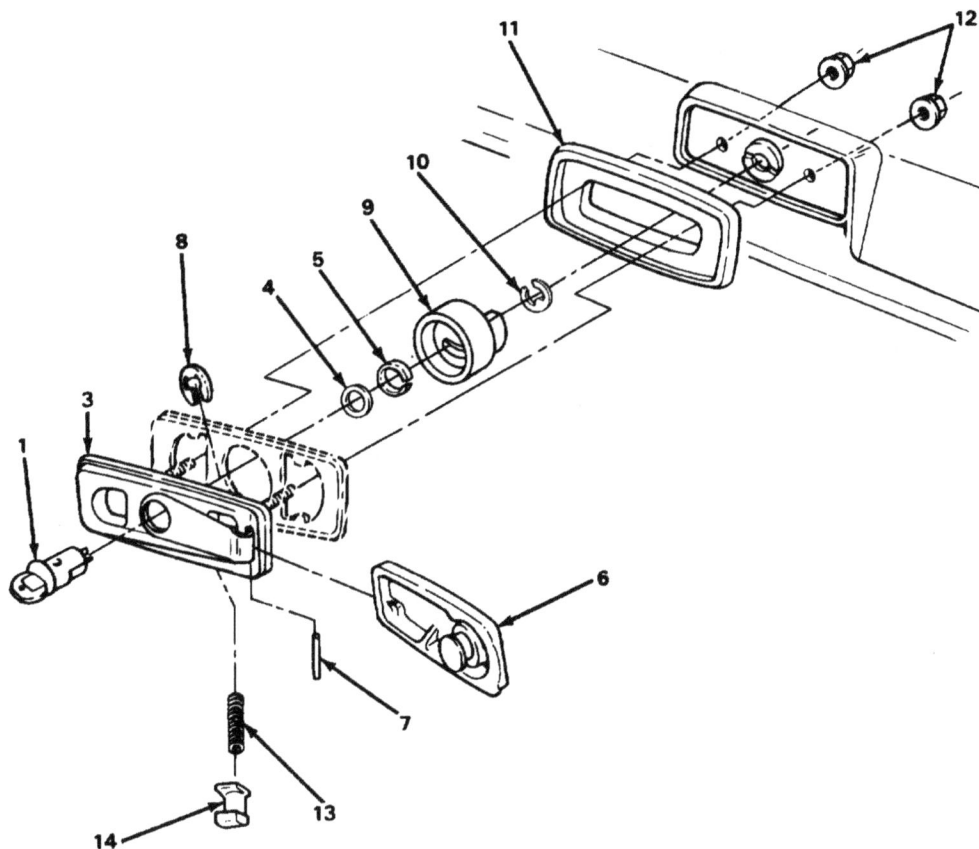

TA510985

FIGURE 178. ENDGATE WINDOW HANDLE ASSEMBLY (M1009).

(1) (2) (3) (4) ITEM SMR PART NO CODE CAGEC NUMBER	(5) DESCRIPTION AND USABLE ON CODES (UOC)	(6) QTY
	GROUP 1801 BODY, CAB, HOOD, AND HULL ASSEMBLIES	
	FIG. 178 ENDGATE WINDOW HANDLE ASSEMBLY (M1009)	
1 PAOZZ 11862 15599678	CYLINDER,LOCK,VEHIC UOC:209	1
2 PAOZZ 11862 14072497	HANDLE ASSEMBLY,WIN UOC:209	1
3 PAOZZ 11862 14072499	.LATCH,DOOR,VEHICULA UOC:209	1
4 PAOZZ 11862 5713274	.WASHER,SPRING TENSI UOC:209	1
5 PAOZZ 11862 5713276	.RING,RETAINING UOC:209	1
6 PAOZZ 11862 9703344	.HANDLE,DOOR UOC:209	1
7 PAOZZ 11862 5717887	.PIN,STRAIGHT,HEADLE UOC:209	1
8 PAOZZ 96906 MS16633-1021	RING,RETAINING UOC:209	1
9 PAOZZ 11862 9702916	CLUTCH,WINDOW REGUL UOC:209	1
10 PAOZZ 96906 MS16624-1024	RING,RETAINING UOC:209	1
11 PAOZZ 11862 327015	GASKET UOC:209	1
12 PAOZZ 24617 271172	NUT,SELF-LOCKING,AS UOC:209	2
13 PAOZZ 11862 4303911	SPRING,HELICAL,COMP UOC:209	1
14 PAOZZ 11862 5713268	PAWL UOC:209	1

END OF FIGURE

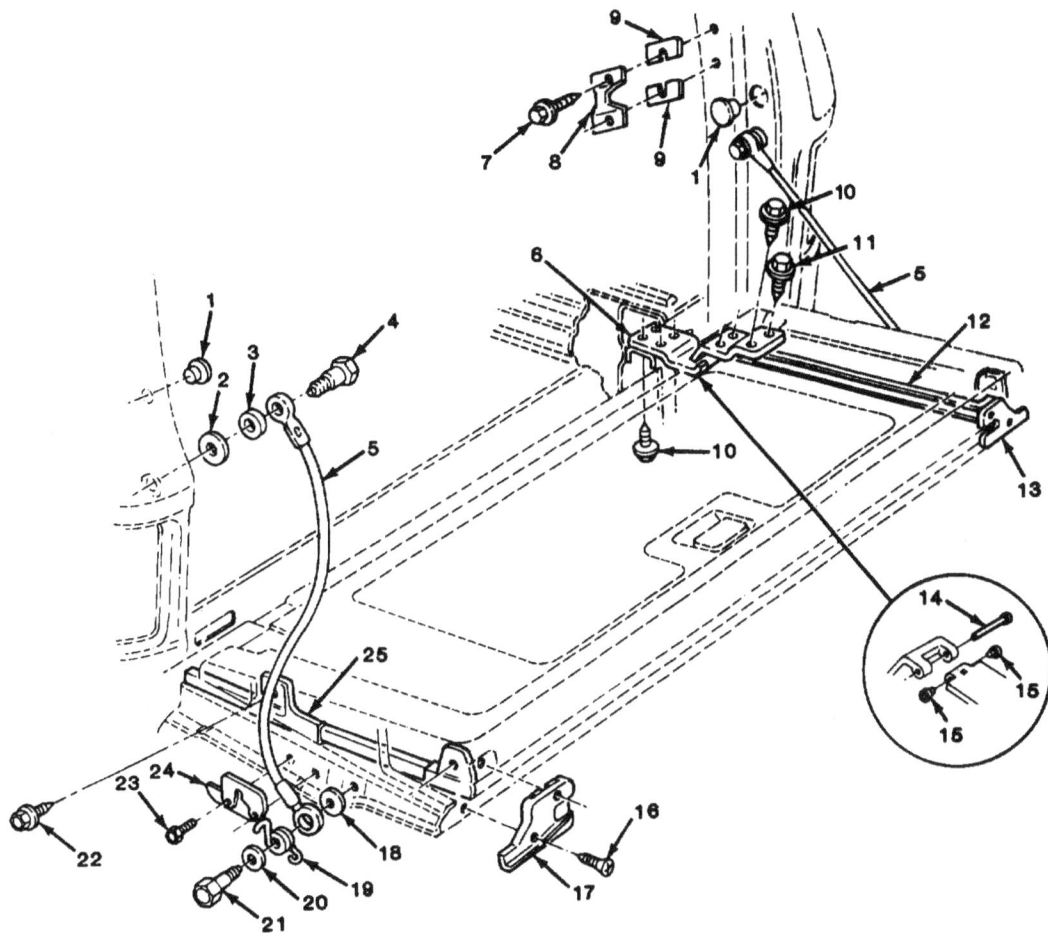

TA510986

FIGURE 179. ENDGATE HINGES AND RELATED PARTS (M1009).

(1) ITEM NO	(2) SMR CODE	(3) CAGEC	(4) PART NUMBER	(5) DESCRIPTION AND USABLE ON CODES (UOC)	(6) QTY
				GROUP 1801 BODY, CAB, HOOD, AND HULL ASSEMBLIES	
				FIG. 179 ENDGATE HINGES AND RELATED PARTS (M1009)	
1	PAOZZ	11862	14050440	PLUG,PROTECTIVE,DUS UOC:209	2
2	PAOZZ	11862	1260895	WASHER,FLAT UOC:209	2
3	PAOZZ	11862	14021292	SPACER,SLEEVE UOC:209	2
4	PAOZZ	11862	7740374	BOLT,SHOULDER UOC:209	2
5	PAOZZ	11862	6274850	WIRE ROPE ASSEMBLY UOC:209	2
6	PAOZZ	11862	337788	HINGE,BUTT UOC:209	2
7	PAOZZ	11862	2014469	BOLT,ASSEMBLED WASH UOC:209	4
8	PAOZZ	11862	6274847	STRIKE,CATCH LEFT UOC:209	1
8	PAOZZ	11862	6274848	STRIKER,END GATE RIGHT UOC:209	1
9	PAOZZ	11862	6274849	SPACER,PLATE UOC:209	1
10	XDOZZ	11862	9438039	BOLT,ASSEMBLED WASH UOC:209	14
11	PAOZZ	11862	9437242	BOLT,MACHINE UOC:209	2
12	PAOZZ	11862	6274836	GUIDE,WINDOW,VEHICU UOC:209	1
13	PAOZZ	11862	15522708	COP ASSEMBLY,TAILGA RIGHT UOC:209	1
14	PAOZZ	11862	6271989	PIN,STRAIGHT,HEADED UOC:209	2
15	PAOZZ	11862	335524	BUSHING,SLEEVE UOC:209	4
16	PAOZZ	96906	MS51862-35	SCREW,TAPPING,THREA UOC:209	4
17	PAOZZ	11862	15522707	CAP ASSEMBLY,TAILGA LEFT UOC:209	1
18	PAOZZ	96906	MS27183-16	WASHER,FLAT UOC:209	2
19	PAOZZ	11862	14021315	SPRING,HELICAL,TORS LEFT UOC:209	1
19	PAOZZ	11862	14021316	SPRING,HELICAL,TORS RIGHT UOC:209	1
20	PAOZZ	96906	MS27183-19	WASHER,FLAT UOC:209	2
21	PAOZZ	11862	14021317	BOLT,SHOULDER UOC:209	2

(1) ITEM NO	(2) SMR CODE	(3) CAGEC	(4) PART NUMBER	(5) DESCRIPTION AND USABLE ON CODES (UOC)	(6) QTY
22	PAOZZ	11862	9439771	BOLT,ASSEMBLED WASH UOC:209	4
23	PAOZZ	24617	9414712	SCREW,TAPPING,THREA UOC:209	2
24	PAOZZ	11862	334132	GUIDE,END GATE SUPP UOC:209	2
25	PAOZZ	11862	470949	GUIDE,WINDOW,VEHICU UOC:209	1

END OF FIGURE

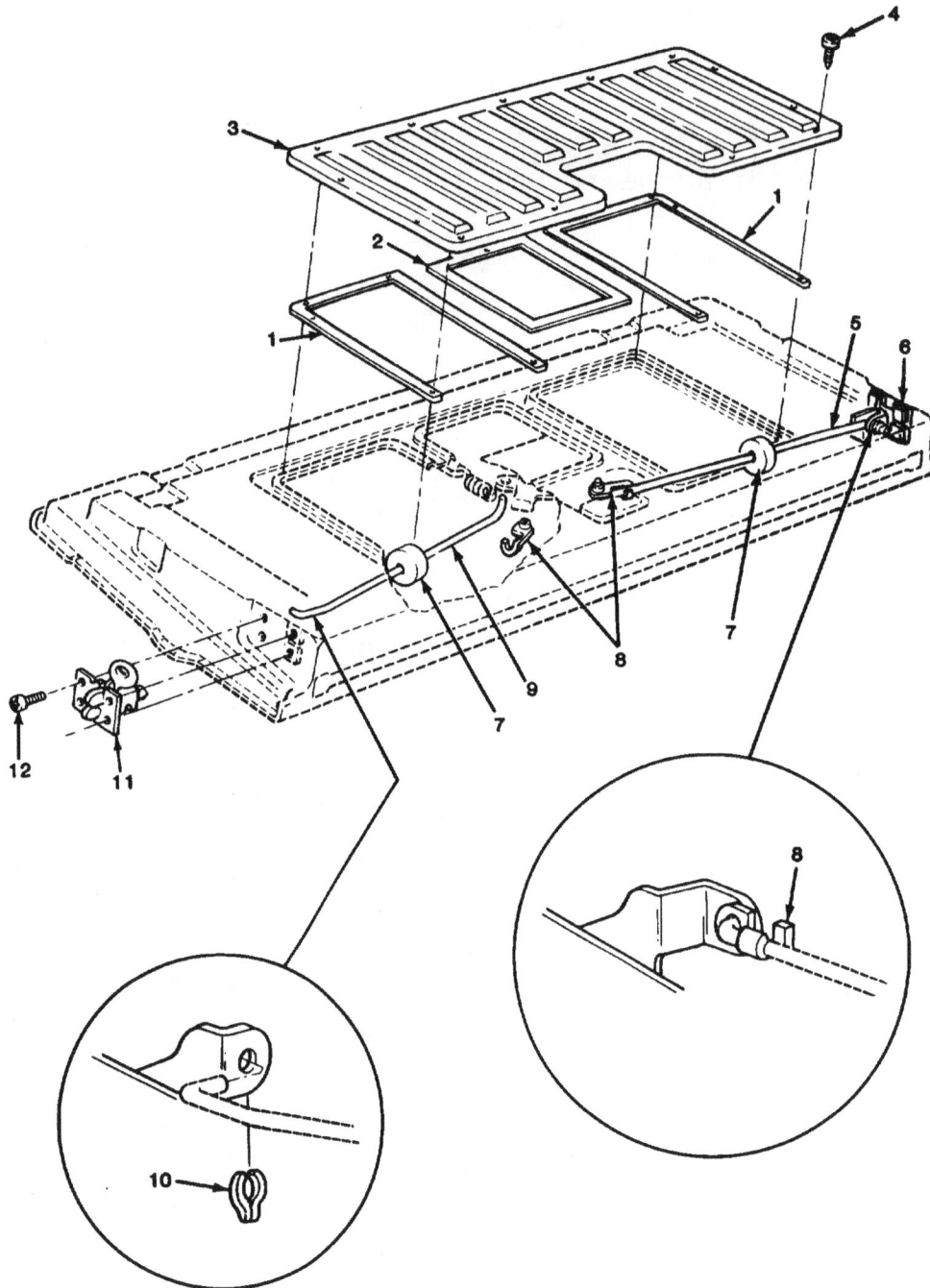

TA510987

FIGURE 180. ENDGATE LATCHES AND RELATED PARTS (M1009).

(1)	(2)	(3)	(4)	(5)	(6)
ITEM NO	SMR CODE	CAGEC	PART NUMBER	DESCRIPTION AND USABLE ON CODES (UOC)	QTY

GROUP 1801 BODY, CAB, HOOD, AND HULL ASSEMBLIES

FIG. 180 ENDGATE LATCHES AND RELATED PARTS (M1009)

1	PAOZZ	11862	15593570	SEAL	2
				UOC:209	
2	PAOZZ	11862	15593571	GASKET	1
				UOC:209	
3	PAOZZ	11862	6274890	COVER,ACCESS	1
				UOC:209	
4	PAOZZ	24617	9419327	SCREW	16
				UOC:209	
5	PAOZZ	11862	14039716	CONNECTING LINK RIG RIGHT	1
				UOC:209	
6	PAOZZ	11862	14039710	LATCH ASSEMBLY,END RIGHT	1
				UOC:209	
7	PAOZZ	11862	14007449	BUMPER,NONMETALLIC	2
				UOC:209	
8	PAOZZ	11862	15545178	CLIP,RETAINING	3
				UOC:209	
9	PAOZZ	11862	473917	ROD,LATCH,END GATE, LEFT	1
				UOC:209	
10	PAOZZ	11862	3905674	CLIP,SPRING TENSION	1
				UOC:209	
11	PAOZZ	11862	473995	LATCH ASSEMBLY,END LEFT	1
				UOC:209	
12	PAOZZ	11862	9439772	SCREW,SELF-LOCKING	8
				UOC:209	

END OF FIGURE

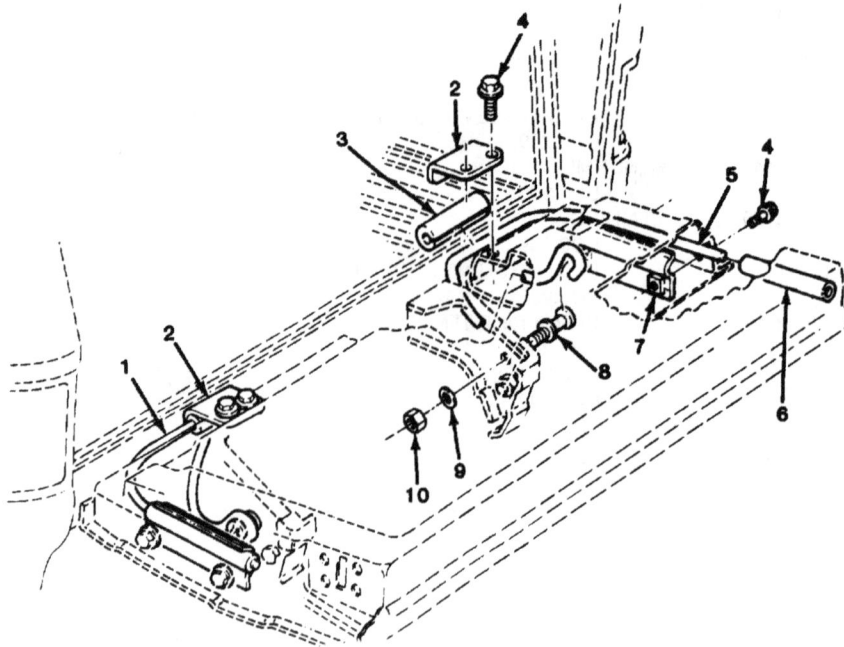

TA510988

FIGURE 181. ENDGATE TORQUE RODS (M1009).

(1) ITEM NO	(2) SMR CODE	(3) CAGEC	(4) PART NUMBER	(5) DESCRIPTION AND USABLE ON CODES (UOC)	(6) QTY
				GROUP 1801 BODY, CAB, HOOD, AND HULL ASSEMBLIES	
				FIG. 181 ENDGATE TORQUE RODS (M1009)	
1	PAOZZ	11862	334101	ROD,TAILGATE,VEHICU LEFT UOC:209	1
2	PAOZZ	11862	6274854	BRACKET,ANGLE UOC:209	2
3	PAOZZ	22593	30103	GASKET LEFT UOC:209	1
3	PAOZZ	11862	334104	SEAL,NONMETALLIC SP RIGHT UOC:209	1
4	PAOZZ	11862	9439771	BOLT,ASSEMBLED WASH UOC:209	4
5	PAOZZ	11862	334102	ROD,END GATE,VEHICU RIGHT UOC:209	1
6	PAOZZ	11862	6274853	RUBBER ROUND SECTION UOC:209	2
7	PAOZZ	11862	337899	BRACKET,DOUBLE ANGL UOC:209	2
8	PAOZZ	11862	473894	STUD ASSEMBLY,ROD,T UOC:209	2
9	PAOZZ	96906	MS35340-48	WASHER,LOCK UOC:209	2
10	PAOZZ	96906	MS51967-14	NUT,PLAIN,HEXAGON UOC:209	2

END OF FIGURE

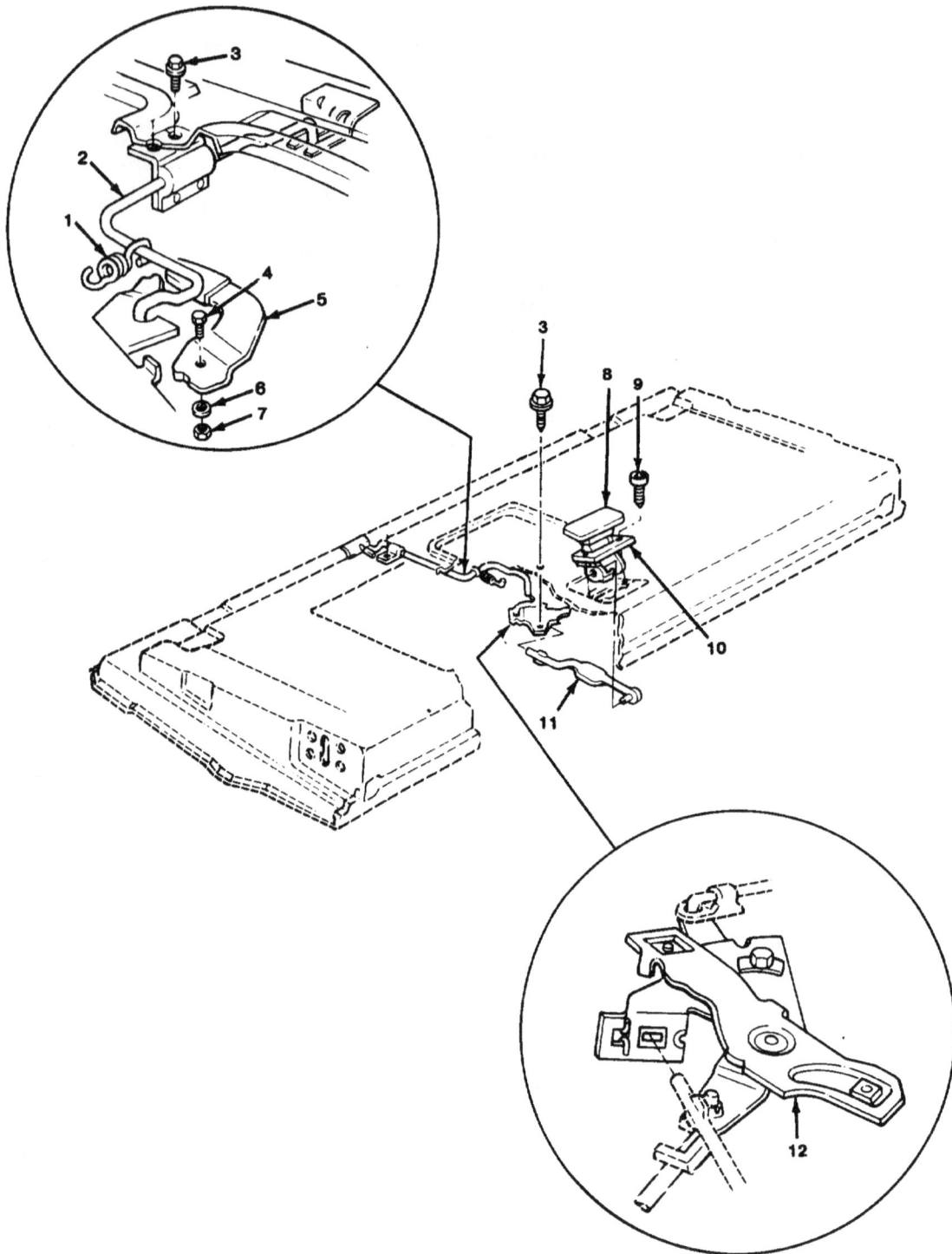

TA510989

FIGURE 182. ENDGATE LATCH HANDLE (M1009).

(1) (2) (3) (4)			(5)	(6)
ITEM SMR	PART			
NO CODE CAGEC NUMBER			DESCRIPTION AND USABLE ON CODES (UOC)	QTY

GROUP 1801 BODY, CAB, HOOD, AND HULL
ASSEMBLIES

FIG. 182 ENDGATE LATCH HANDLE
(M1009)

1	PAOZZ 11862 15593866	SPRING,HELICAL,EXTE		1
		UOC:209		
2	PAOZZ 11862 331638	ROD ASSEMBLY,LATCH		1
		UOC:209		
3	PAOZZ 11862 1244067	BOLT,ASSEMBLED WASH		5
		UOC:209		
4	PAOZZ 96906 MS90728-6	SCREW,CAP,HEXAGON H		1
		UOC:209		
5	PAOZZ 11862 335452	LEVER,MANUAL CONTRO		1
		UOC:209		
6	PAOZZ 96906 MS35338-44	WASHER,LOCK		1
		UOC:209		
7	PAOZZ 96906 MS35691-406	NUT,PLAIN,HEXAGON		1
		UOC:209		
8	PAOZZ 11862 4495180	HANDLE ASSEMBLY,TAI		1
		UOC:209		
9	PAOZZ 11862 8782501	SCREW,TAPPING,THREA		2
		UOC:209		
10	PAOZZ 11862 470993	GASKET		1
		UOC:209		
11	PAOZZ 11862 8742340	CONNECTING LINK,RIG		1
		UOC:209		
12	PAOZZ 11862 14039712	LEVER,REMOTE CONTRO		1
		UOC:209		

END OF FIGURE

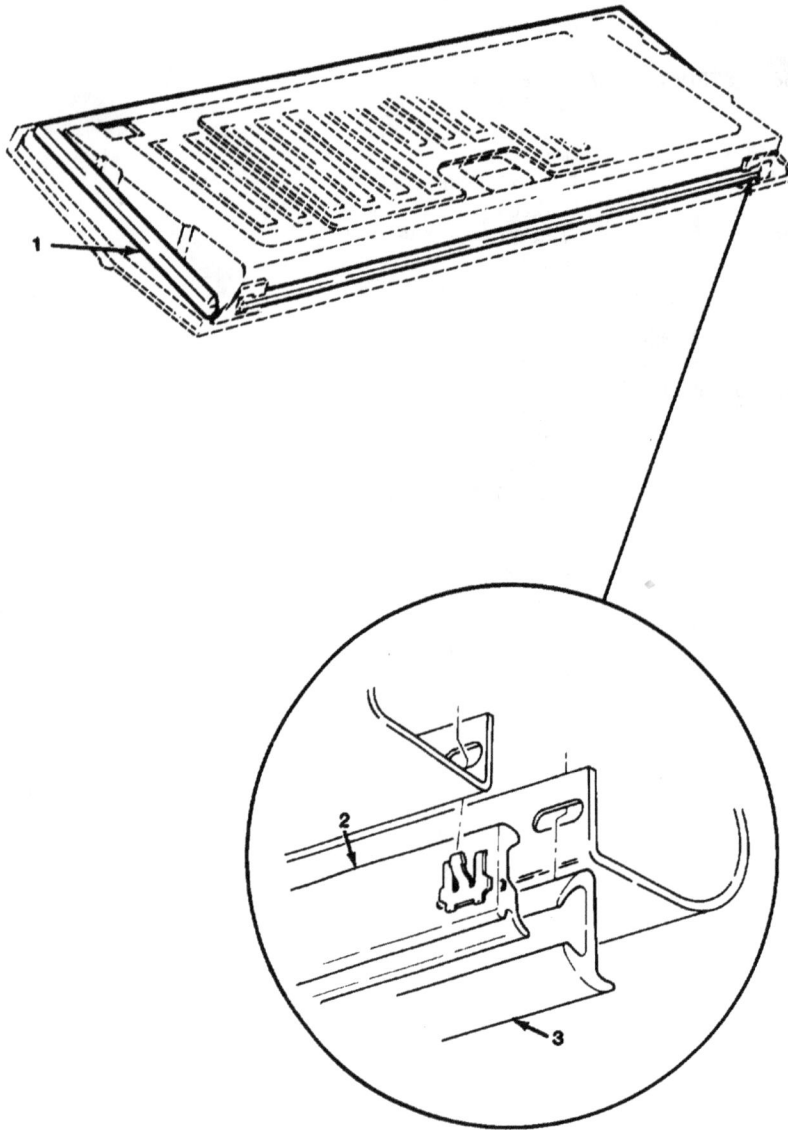

TA510990

FIGURE 183. ENDGATE WEATHERSTRIPS (M1009).

(1) (2) (3) (4) (5) ITEM SMR NO CODE CAGEC NUMBER	PART	DESCRIPTION AND USABLE ON CODES (UOC)	(6) QTY
		GROUP 1801 BODY, CAB, HOOD, AND HULL ASSEMBLIES	
		FIG. 183 ENDGATE WEATHERSTRIPS (M1009)	
1 PAOZZ 11862 326934		SEAL,NONMETALLIC SP UOC:209	1
2 PAOZZ 11862 327006		SEAL,NONMETALLIC SP UOC:209	1
3 PAOZZ 11862 327005		SEAL,NONMETALLIC SP UOC:209	1
		END OF FIGURE	

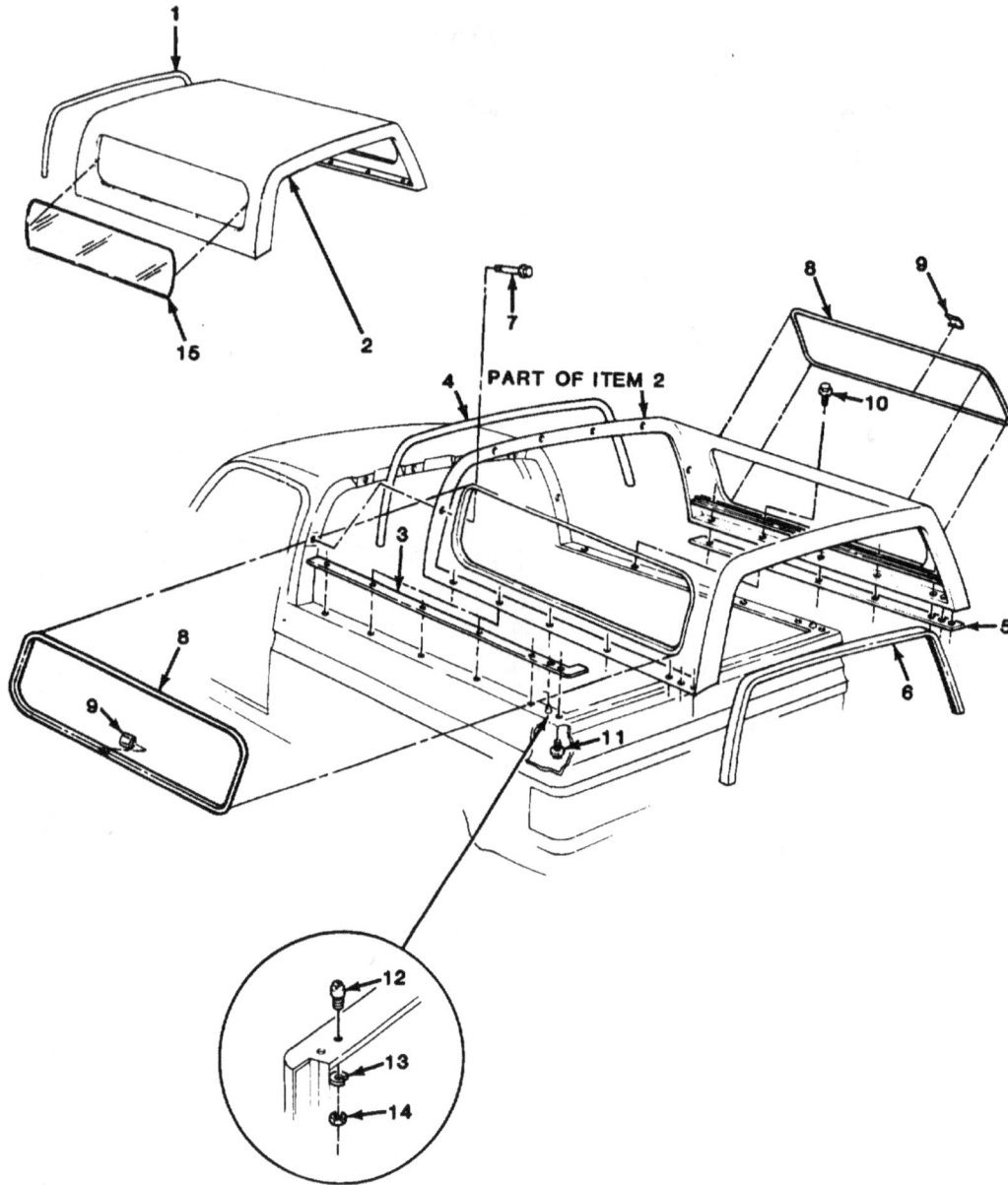

FIGURE 184. TOP ASSEMBLY AND COMPONENTS (M1009).

TA510991

(1) (2) (3) (4) ITEM SMR PART NO CODE CAGEC NUMBER	(5) DESCRIPTION AND USABLE ON CODES (UOC)	(6) QTY
	GROUP 1801 BODY, CAB, HOOD, AND HULL ASSEMBLIES	
	FIG. 184 TOP ASSEMBLY AND COMPONENTS (M1009)	
1 PAFZZ 11862 357490	SEAL,NONMETALLIC SP UOC:209	1
2 PAFZZ 11862 15565342	TOP ASSEMBLY,TRUCK UOC:209	1
3 PAFZZ 11862 358589	.SEAL,RUBBER SPECIAL UOC:209	1
4 PAFZZ 11862 358590	.SEAL,NONMETALLIC SP UOC:209	1
5 PAFZZ 11862 15590455	.SEAL,NONMETALLIC SP UOC:209	1
6 PAFZZ 11862 358593	.SEAL,NONMETALLIC SP UOC:209	1
7 PAFZZ 11862 358549	BOLT,ASSEMBLED WASH UOC:209	6
8 PAFZZ 11862 466157	RUBBER STRIP UOC:209	2
9 PAFZZ 11862 334150	MOLDING,METAL UOC:209	2
10 PAOZZ 11862 458985	SCREW,TAPPING,THREA UOC:209	10
11 PAFZZ 11862 407217	SCREW,ASSEMBLED WAS UOC:209	2
12 PAFZZ 11862 340034	PIN,SHOULDER,HEADLE UOC:209	2
13 PAFZZ 96906 MS35338-45	WASHER,LOCK UOC:209	2
14 PAFZZ 96906 MS51967-6	NUT,PLAIN,HEXAGON UOC:209	2
15 PAFZZ 11862 475968	SEAL,NONMETALLIC SP UOC:209	2

END OF FIGURE

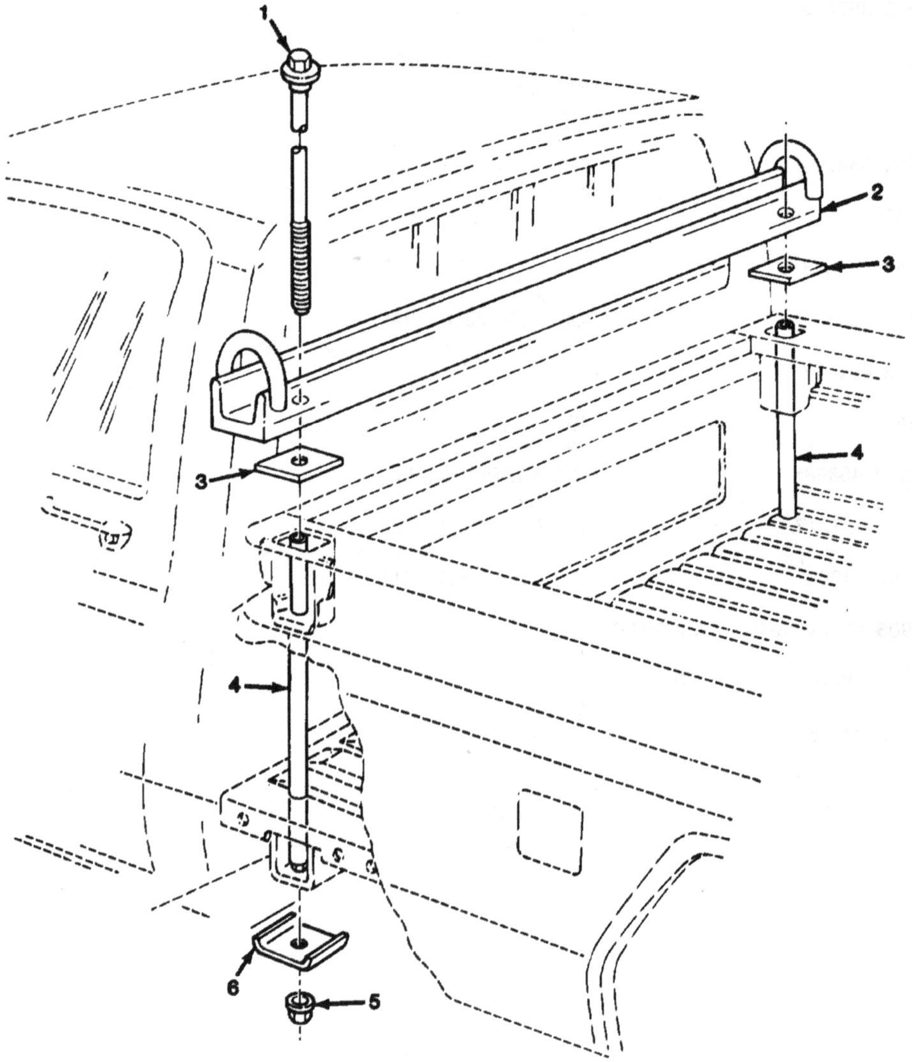

TA510992

FIGURE 185. FRONT TIE-DOWN BRACKETS (M1028,M1028A1,M1028A2, AND M1028A3).

(1) (2) (3) (4) ITEM SMR PART NO CODE CAGEC NUMBER	(5) DESCRIPTION AND USABLE ON CODES (UOC)	(6) QTY
	GROUP 1801 BODY, CAB, HOOD, AND HULL ASSEMBLIES	
	FIG. 185 FRONT TIE-DOWN BRACKETS (M1028, M1028A1, M1028A2, AND M1028A3)	
1 PAOZZ 11862 15599929	BOLT,EXTERNALLY REL UOC:230,252,254,256	1
2 PAOZZ 11862 14072452 CLAMP,LOOP	UOC:230,252,254,256	1
3 PAOZZ 11862 14072460	PAD,BRACKET SUPPORT UOC:230,252,254,256	2
4 PAOZZ 11862 14072454 SPACER,SLEEVE	UOC:230,252,254,256	1
5 PAOZZ 11862 9439757 NUT,PLAIN,EXTENDED	UOC:230,252,254,256	2
6 PAOZZ 11862 14072459 PLATE,REINFORCEMENT	UOC:230,252,254,256	2

END OF FIGURE

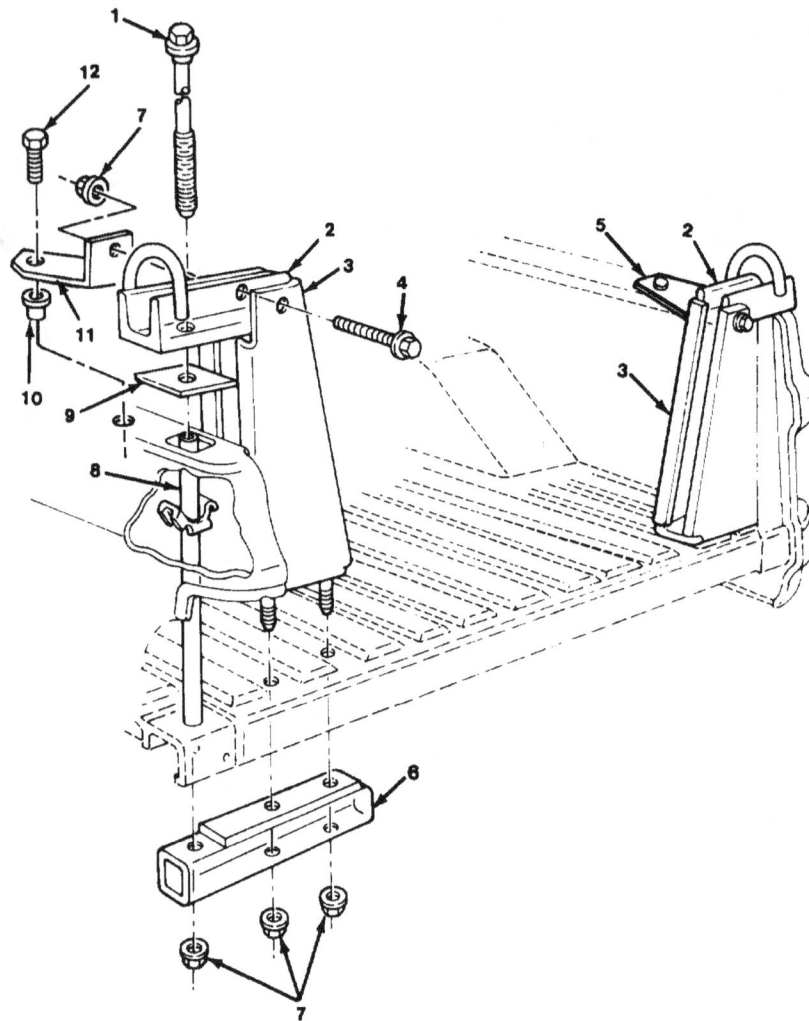

TA510993

FIGURE 186. REAR TIE-DOWN BRACKETS (M1028,M1028A1,M1028A2, AND M1028A3).

(1) (2) (3) (4) ITEM SMR PART NO CODE CAGEC NUMBER	(5) DESCRIPTION AND USABLE ON CODES (UOC)	(6) QTY
	GROUP 1801 BODY, CAB, HOOD, AND HULL ASSEMBLIES	
	FIG. 186 REAR TIE-DOWN BRACKETS (M1028,M1028A1,M1028A2, AND M1028A3)	
1 PAOZZ 11862 15599929	BOLT,EXTERNALLY REL UOC:230,252,254,256	2
2 PAOZZ 11862 15591705	BRACKET,TIE DOWN UOC:230,252,254,256	2
3 PAOZZ 23862 15591706	BRACKET,TOW HOOK RIGHT UOC:230,252,254,256	1
3 PAOZZ 11862 15591709	BRACKET,TIE DOWN LEFT UOC:230,252,254,256	1
4 PAOZZ 11862 14072458 SETSCREW	 UOC:230,252,254,256	2
5 PAOZZ 1T998 15591710	BRACKET,ANGLE RIGHT UOC:230,252,254,256	1
6 PAOZZ 11862 14072455 REINFORCEMENT,REAR	 UOC:230,252,254,256	2
7 PAOZZ 96906 MS51967-14 NUT,PLAIN,HEXAGON	 UOC:230,252,254,256	8
8 PAOZZ 11862 14072454 SPACER,SLEEVE	 UOC:230,252,254,256	2
9 PAOZZ 11862 14072460	PAD,BRACKET SUPPORT UOC:230,252,254,256	2
10 PAOZZ 11862 3792287	NUT,PLAIN,BLIND RIV UOC:230,252,254,256	2
11 PAOZZ 11862 15591707	BRACKET,ANGLE LEFT UOC:230,252,254	1
12 PAOZZ 11862 9440300	SCREW,CAP,HEXAGON H UOC:230,252,254,256	2

END OF FIGURE

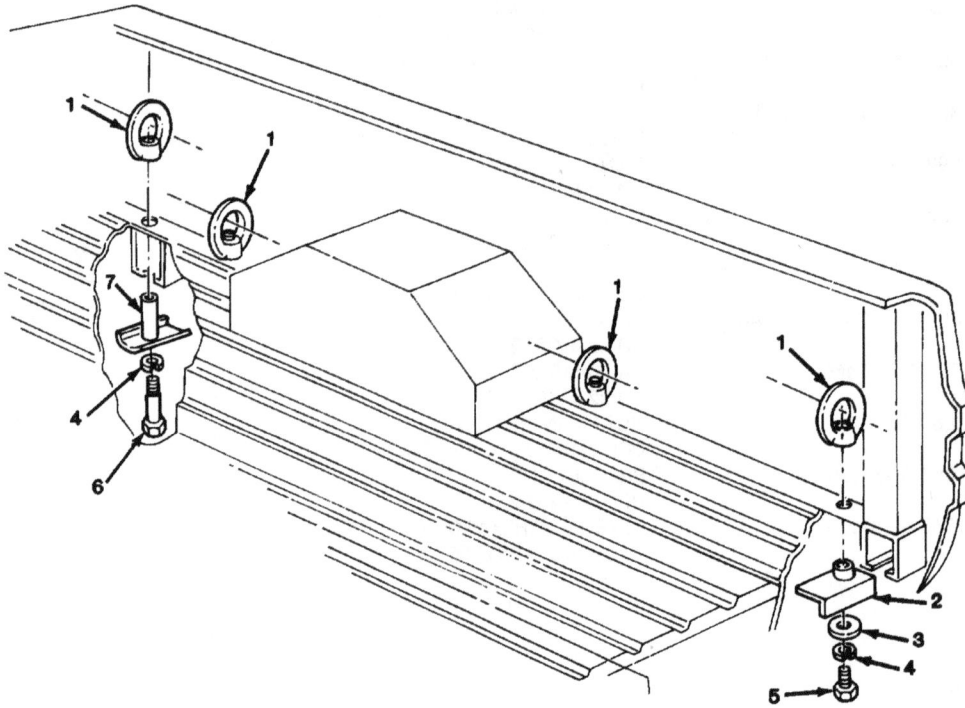

TA510994

FIGURE 187. CARGO TIE-DOWNS (M1008 AND M1008A1).

(1)	(2)	(3)	(4)	(5)	(6)
ITEM NO	SMR CODE	CAGEC	PART NUMBER	DESCRIPTION AND USABLE ON CODES (UOC)	QTY

GROUP 1801 BODY, CAB, HOOD, AND HULL ASSEMBLIES

FIG. 187 CARGO TIE-DOWNS (M1008 AND M1008A1)

1	PAOZZ	11862	14072307	NUT,EYE UOC:194,208	8
2	PAOZZ	11862	14072305	STRAP,RETAINING UOC:194,208	2
3	PAOZZ	11862	94009398	WASHER,FLAT UOC:194,208	2
4	PAOZZ	96906	MS35340-48	WASHER,LOCK UOC:194,208	8
5	PAOZZ	96906	MS90728-114	SCREW,CAP,HEXAGON H UOC:194,208	2
6	PAOZZ	96906	MS90728-125	SCREW,CAP,HEXAGON H UOC:194,208	6
7	PAOZZ	11862	14072306	REINFORCEMENT,CARGO UOC:194,208	6

END OF FIGURE

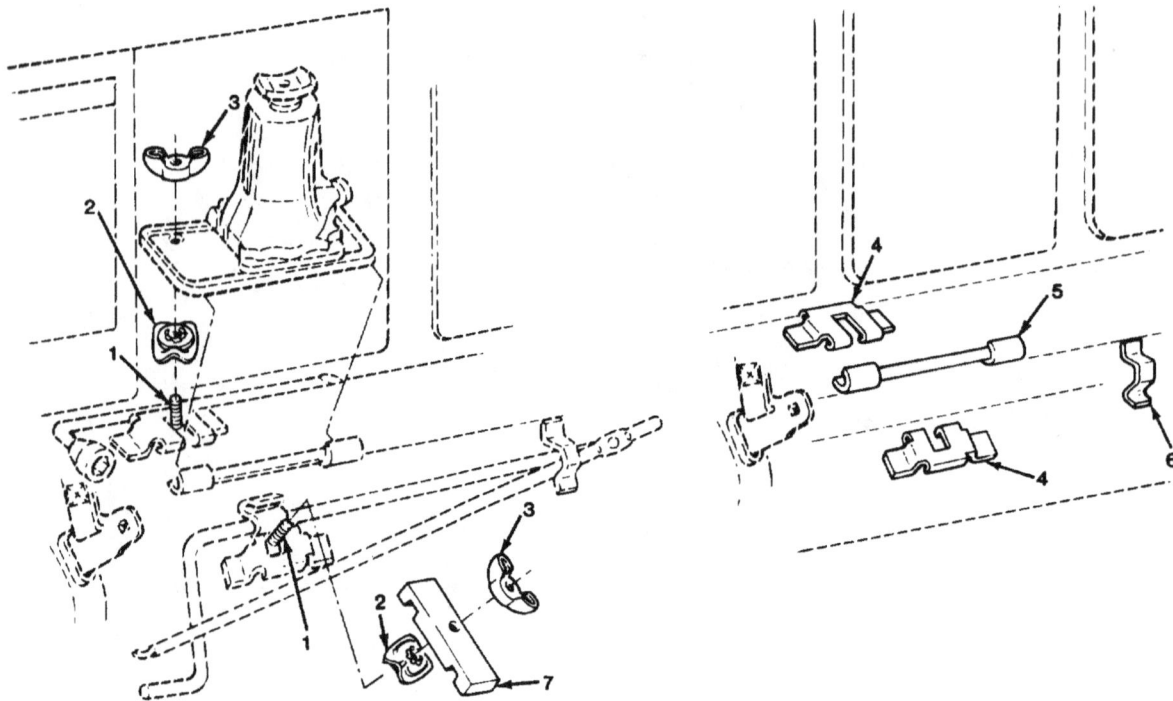

ALL EXCEPT M1009 AND M1010

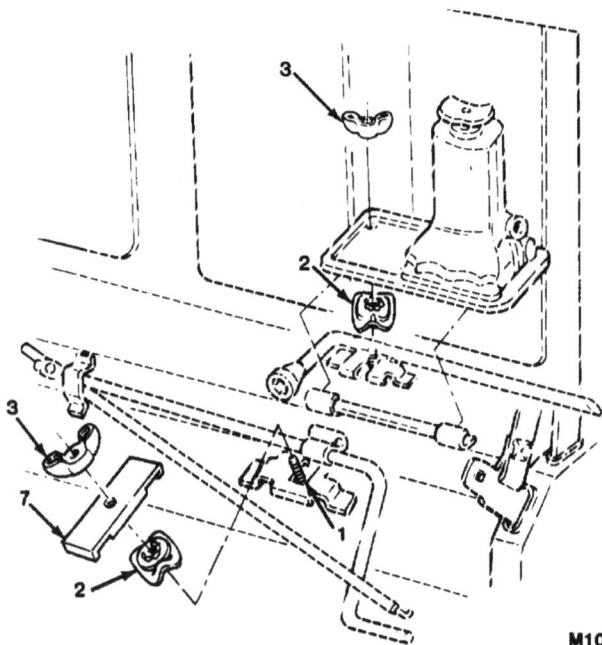

M1010 ONLY

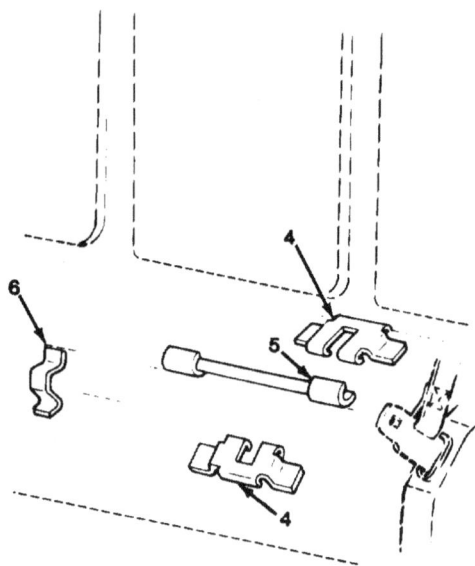

TA510995

FIGURE 188. JACK STOWAGE COMPONENTS (ALL EXCEPT M1009).

(1) (2) (3) (4) ITEM SMR PART NO CODE CAGEC NUMBER	(5) DESCRIPTION AND USABLE ON CODES (UOC)	(6) QTY
	GROUP 1801 BODY, CAB, HOOD, AND HULL ASSEMBLIES	
	FIG. 188 JACK STOWAGE COMPONENTS (ALL EXCEPT M1009)	
1 PAOZZ 11862 14032814	BOLT,SQUARE NECK UOC:194,208,210,230,231,252,254,256	2
2 PAOZZ 24617 9409613	PUSH ON NUT UOC:194,208,210,230,231,252,254,256	2
3 PAOZZ 96906 MS35425-72	NUT,PLAIN,WING UOC:194,208,210,230,231,252,254,256	2
4 PFFZZ 11862 14074431	BRACKET,MOUNTING UOC:194,208,210,230,231,252,254,256	2
5 PFFZZ 11862 14074443	BRACKET,MOUNTING UOC:194,208,210,230,231,252,254,256	1
6 PFFZZ 11862 14074440	BRACKET,MOUNTING UOC:194,208,210,230,231,252,254,256	1
7 PAOZZ 11862 467299	STRAP,RETAINING UOC:194,208,210,230,231,252,254,256	1

END OF FIGURE

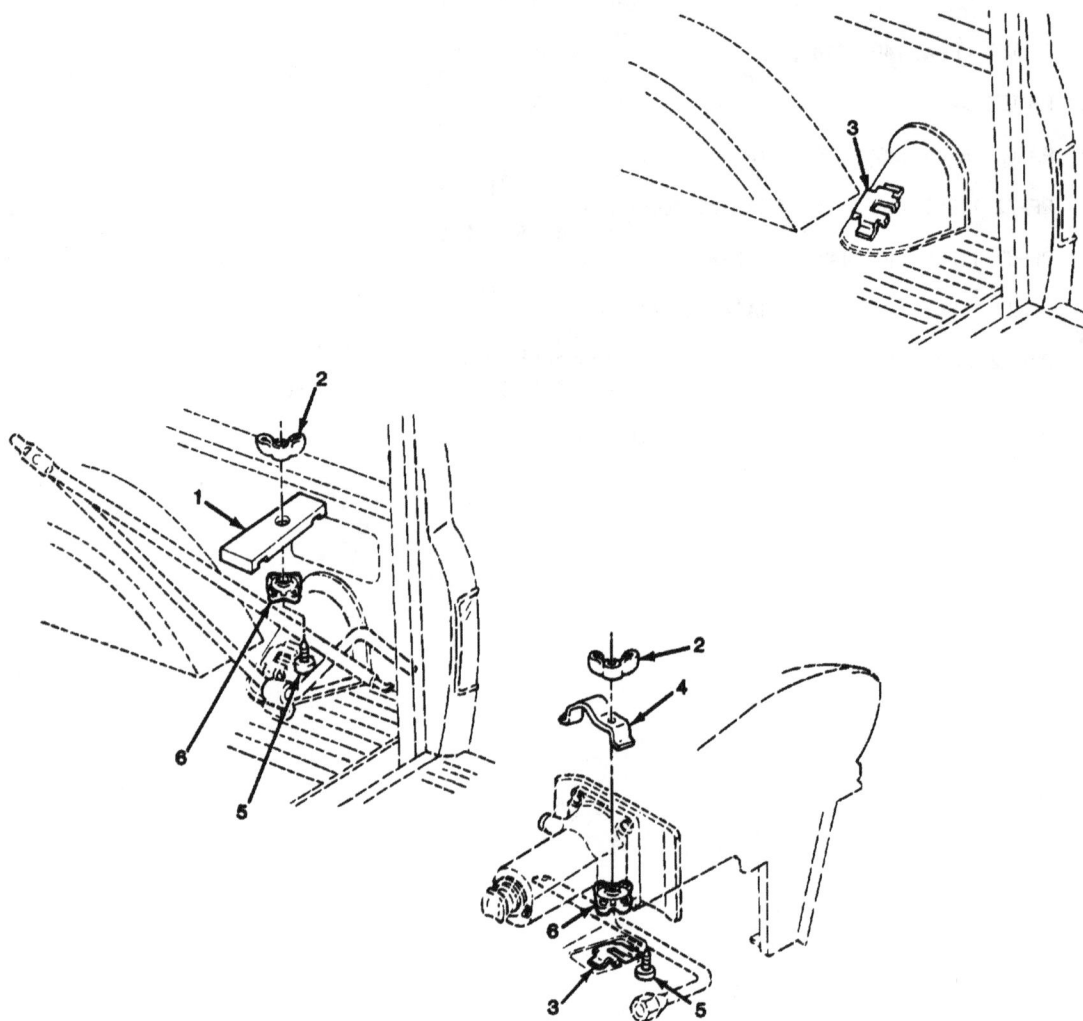

TA510996

FIGURE 189. JACK STOWAGE COMPONENTS (M1009).

(1) (2) (3) (4) ITEM SMR PART NO CODE CAGEC NUMBER	(5) DESCRIPTION AND USABLE ON CODES (UOC)	(6) QTY
	GROUP 1801 BODY, CAB, HOOD, AND HULL ASSEMBLIES	
	FIG. 189 JACK STOWAGE COMPONENTS (M1009)	
1 PAOZZ 11862 467299	STRAP,RETAINING UOC:209	1
2 PAOZZ 96906 MS35425-72 NUT,PLAIN,WING	UOC:209	2
3 PFFZZ 11862 14074431 BRACKET,MOUNTING	UOC:209	2
4 PAOZZ 11862 14032812 RETAINER,JACK	UOC:209	1
5 PAOZZ 11862 14032814	BOLT,SQUARE NECK UOC:209	2
6 PAOZZ 24617 9409613	PUSH ON NUT UOC:209	2
	END OF FIGURE	

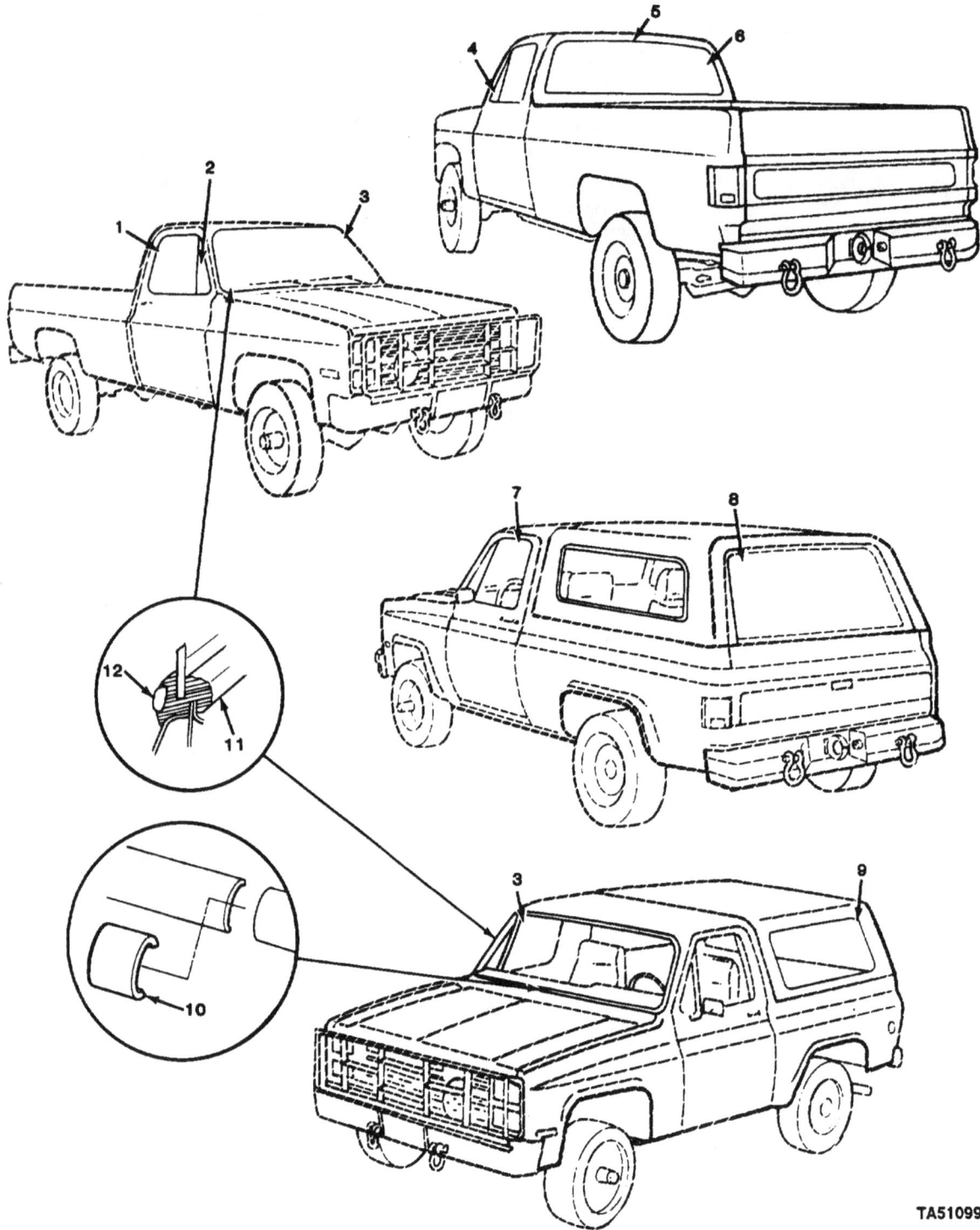

FIGURE 190. WINDSHIELD AND WINDOW GLASS.

TA510997

(1) (2) (3) (4) ITEM SMR PART NO CODE CAGEC NUMBER	(5) DESCRIPTION AND USABLE ON CODES (UOC)	(6) QTY
	GROUP 1802 FENDERS, RUNNING BOARDS WITH MOUNTING AND ATTACHING PARTS, OUTRIGGERS, WINDSHIELD, GLASS, ETC.	
	FIG. 190 WINDSHIELD AND WINDOW GLASS	
1 PAOZZ 75829 14022842	WINDOW,VEHICULAR RIGHT	1
2 PAOZZ 11862 20264744	WINDOW,VEHICULAR RIGHT	1
3 PAFZZ 75829 14018597	WINDOW,VEHICULAR	1
4 PAOZZ 11862 20264743	WINDOW,VEHICULAR LEFT	1
5 PAFZZ 11862 15590421	SEAL,NONMETALLIC SP UOC:194,208,230,231,252,254,256	1
6 PAFZZ 75829 363107	WINDOW,VEHICULAR UOC:194,208,230,231,252,254,256	1
7 PAOZZ 75829 14022841	WINDOW,VEHICULAR LEFT	1
8 PAOZZ 75829 14076899	WINDOW,VEHICULAR UOC:209	1
9 PAFZZ 75829 14076859	GLASS,VEHICULAR LEFT UOC:209	1
9 PAFZZ 75829 14076860	GLASS,VEHICULAR RIGHT UOC:209	1
10 PAFZZ 11862 334150	MOLDING,METAL	1
11 PAFZZ 11862 471009	WEATHERSTRIP,GLASS	1
12 PAFZZ 11862 466157	RUBBER STRIP	1

END OF FIGURE

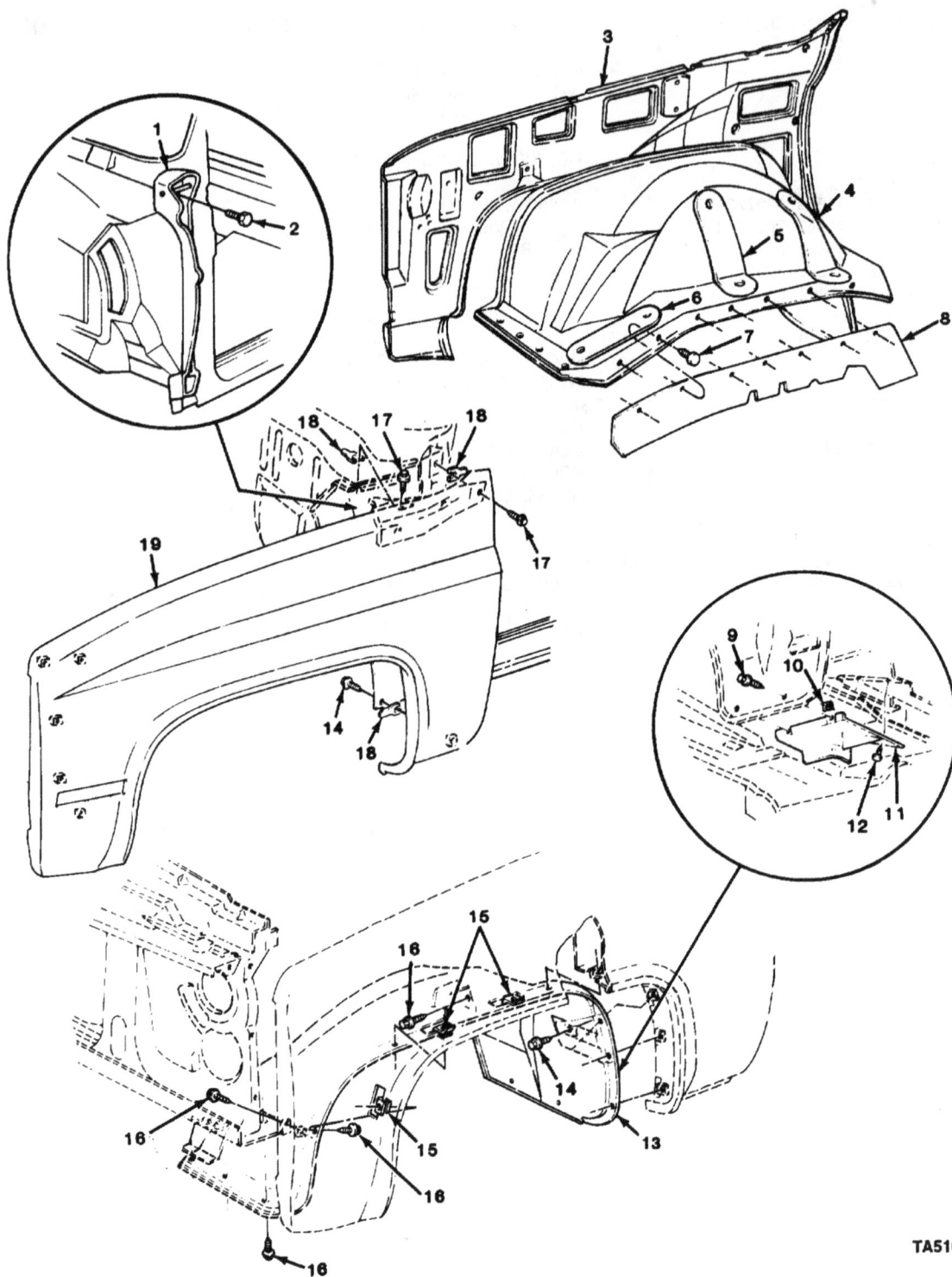

FIGURE 191. FRONT FENDERS, INNER WHEEL PANELS, AND RELATED PARTS.

TA510998

(1) (2) (3) (4) ITEM SMR PART NO CODE CAGEC NUMBER	(5) DESCRIPTION AND USABLE ON CODES (UOC)	(6) QTY
	GROUP 1802 FENDERS, RUNNING BOARDS WITH MOUNTING AND ATTACHING PARTS OUTRIGGERS, WINDSHIELD, GLASS, ETC.	
	FIG. 191 FRONT FENDERS, INNER WHEEL PANELS, AND RELATED PARTS	
1 PAFZZ 11862 14022831	COWL ASSEMBLY,VEHIC LEFT	1
1 PAFZZ 11862 14022832	COWL ASSEMBLY,VEHIC RIGHT	1
2 PAFZZ 24617 11503395	SCREW,ASSEMBLED WAS	8
3 PAFZZ 11862 15522752	FENDER,VEHICULAR RIGHT	1
4 PAFZZ 11862 14075827	BRACE,FRONT FENDER RIGHT	1
5 PAFZZ 11862 14075829	BRACKET,ANGLE RIGHT	1
6 PAFZZ 11862 14075839	BRACE,FRONT FENDER RIGHT	1
7 PAFZZ 11862 411700	FASTENER,SPRING TEN	14
8 PAFZZ 11862 14027795	SHIELD,WHEEL HOUSE PANEL SIDE SPLASH LEFT	1
8 P·AFZZ 11862 14027796	SHIELD,WHEEL HOUSE PANEL SIDE SPLASH RIGHT	1
9 PAFZZ 11862 335551	SCREW,ASSEMBLED WAS	4
10 PAFZZ 96906 MS90724-40	NUT,SHEET SPRING	4
11 PAFZZ 11862 14027645	SHIELD,LOWER REAR SPLASH,LEFT	1
11 PAFZZ 11862 14027646	SHIELD,LOWER REAR SPLASH,RIGHT	1
12 PAFZZ 11862 480534	RETAINER,PANEL HOLE	4
13 PAFZZ 11862 15594889	PANEL,WHEEL HOUSE LEFT	1
13 PAFZZ 11862 15594890	PANEL,WHEEL HOUSE RIGHT	1
14 PAFZZ 11862 407217	SCREW,ASSEMBLED WAS	4
15 PAFZZ 24617 1494253 NUT,CLIP-ON		14
16 PAFZZ 11862 2014469	BOLT,ASSEMBLED WASH	30
17 PAFZZ 11862 334195	BOLT,ASSEMBLED WASH	6
18 PAFZZ 11862 345819	SHIM	6
19 PAFZZ 11862 15614467	FENDER,VEHICULAR LEFT	1

END OF FIGURE

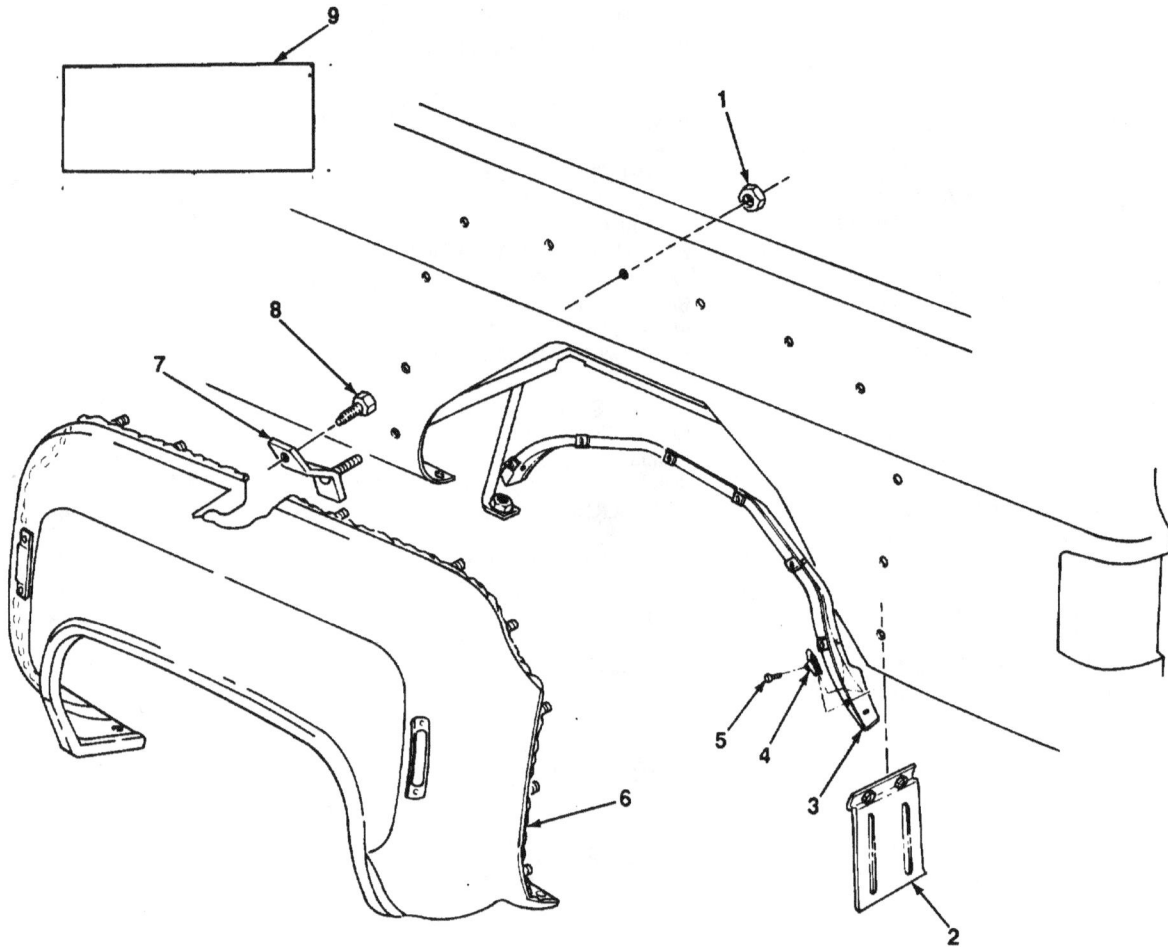

TA510999

FIGURE 192. REAR FENDER AND RELATED PARTS (M1028A2 AND M1028A3).

(1) ITEM NO	(2) SMR CODE	(3) CAGEC	(4) PART NUMBER	(5) DESCRIPTION AND USABLE ON CODES (UOC)	(6) QTY
				GROUP 1802 FENDER, RUNNING BOARDS WITH MOUNTING AND ATTACHING PARTS, OUTRIGGERS, WINDSHIELDS, GLASS, ETC.	
				FIG. 192 REAR FENDER AND RELATED PARTS (M1028A2 AND M1028A3)	
1	PAOZZ	11862	1359887	NUT,PLAIN,HEXAGON UOC:254,256	2
2	PAFZZ	11862	330489	PANEL,BODY,VEHICULA UOC:254,256	1
2	PAFZZ	11862	330490	PANEL,BODY,VEHICULA UOC:254,256	1
3	PAFZZ	11862	330393	LINER,VEHICULAR FEN UOC:254,256	1
3	PAFZZ	11862	330394	LINER,VEHICULAR FEN UOC:254,256	1
4	PAOZZ	24617	1494253	NUT,CLIP-ON UOC:254,256	2
5	PAOZZ	96906	MS51869-24	SCREW,CAP,SOCKET HE UOC:254,256	2
6	PAFZZ	11862	15606405	FENDER,VEHICULAR UOC:254,256	1
6	PAFZZ	11862	15606406	FENDER,VEHICULAR UOC:254,256	1
7	PAFZZ	11862	15606404	BRACE,FENDER UOC:254,256	12
8	PAFZZ	11862	15606407	SCREW,TAPPING UOC:254,256	12
9	XODZZ	11862	TX015963	TEMPL,RH,LH,QTR CT UOC:254,256	2

END OF FIGURE

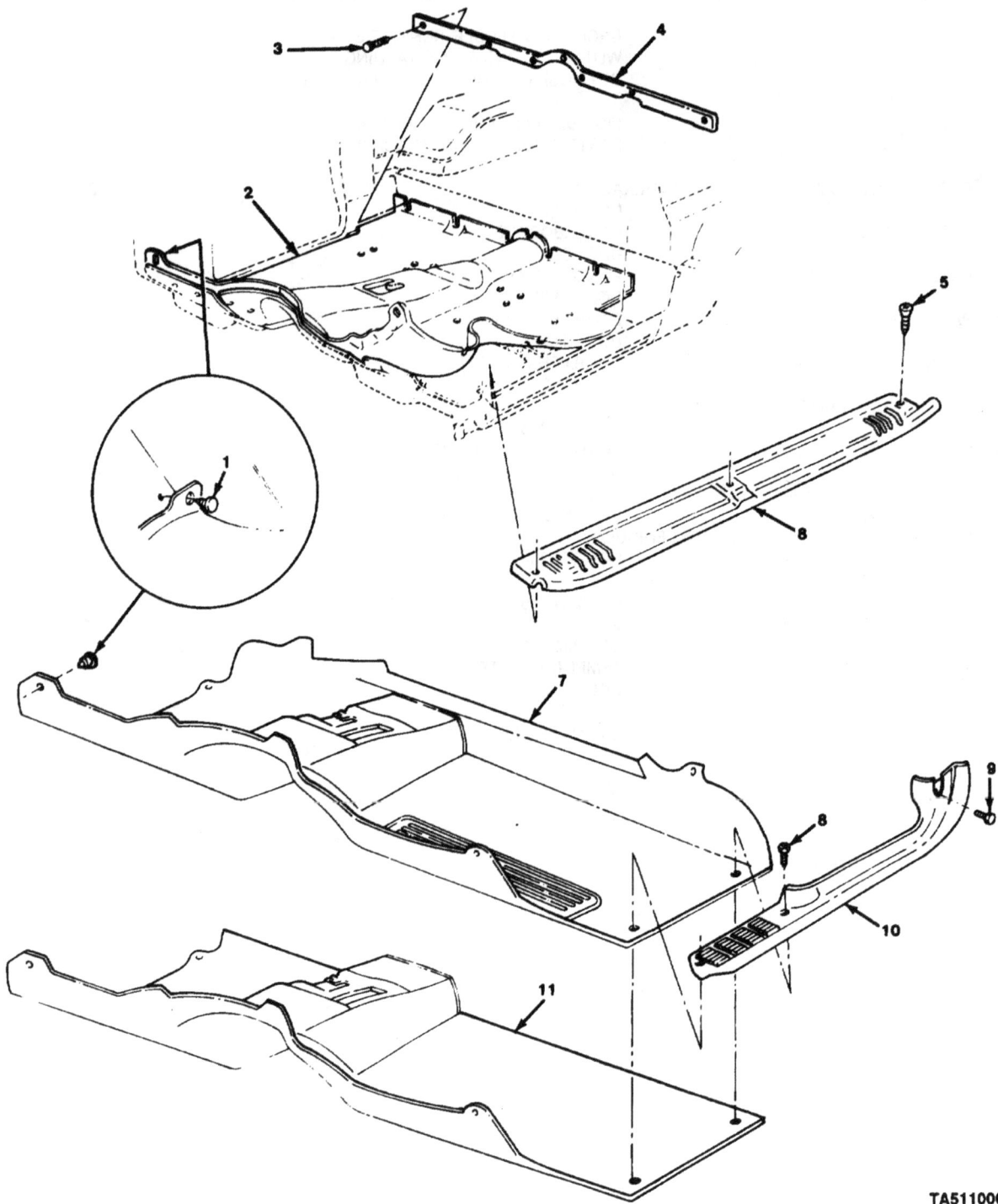

TA511000

FIGURE 193. FLOORMATS, INSULATORS, AND RELATED PARTS.

(1) (2) (3) (4) ITEM SMR NO CODE CAGEC	PART NUMBER	(5) DESCRIPTION AND USABLE ON CODES (UOC)	(6) QTY
		GROUP 1805 FLOOR, SUBFLOORS, AND RELATED COMPONENTS	
		FIG. 193 FLOORMATS, INSULATORS, AND RELATED PARTS	
1 PAOZZ 11862	20030401	FASTENER,SPRING TEN	2
2 PFOZZ 11862	15591702	MAT,FLOOR UOC:209	1
3 PAOZZ 11862	11500668	SCREW,ASSEMBLED WAS UOC:209	7
4 PBOZZ 11862	474022	FRAME SECTION,STRUC UOC:209	1
5 PAOZZ 96906	MS51862-26	SCREW,TAPPING,THREA UOC:209	6
6 PAOZZ 11862	467247	PLATE,DOOR,LEFT UOC:209	1
6 PAOZZ 11862	467248	TREAD,METALLIC,NONS RIGHT UOC:209	1
7 PAOZZ 11862	15594983	MAT,FLOOR UOC:194,208,210,230,231,252,254,256	1
8 PAOZZ 11862	11502634	SCREW,TAPPING,THREA UOC:194,208,210,230,231,252,254,256	4
9 PAOZZ 24617	11503606	SCREW,TAPPING,THREA UOC:194,208,210,230,231,252,254,256	2
10 PAOZZ 11862	15594895	PLATE,DOOR,KICK LEFT UOC:194,208,210,230,231,252,254,256	1
10 PAOZZ 11862	15594896	MOLDING,METAL RIGHT UOC:194,208,210,230,231,252,254,256	1
11 PAOZZ 11862	462233	IINSULATOR,FLOOR,VEH	1

END OF FIGURE

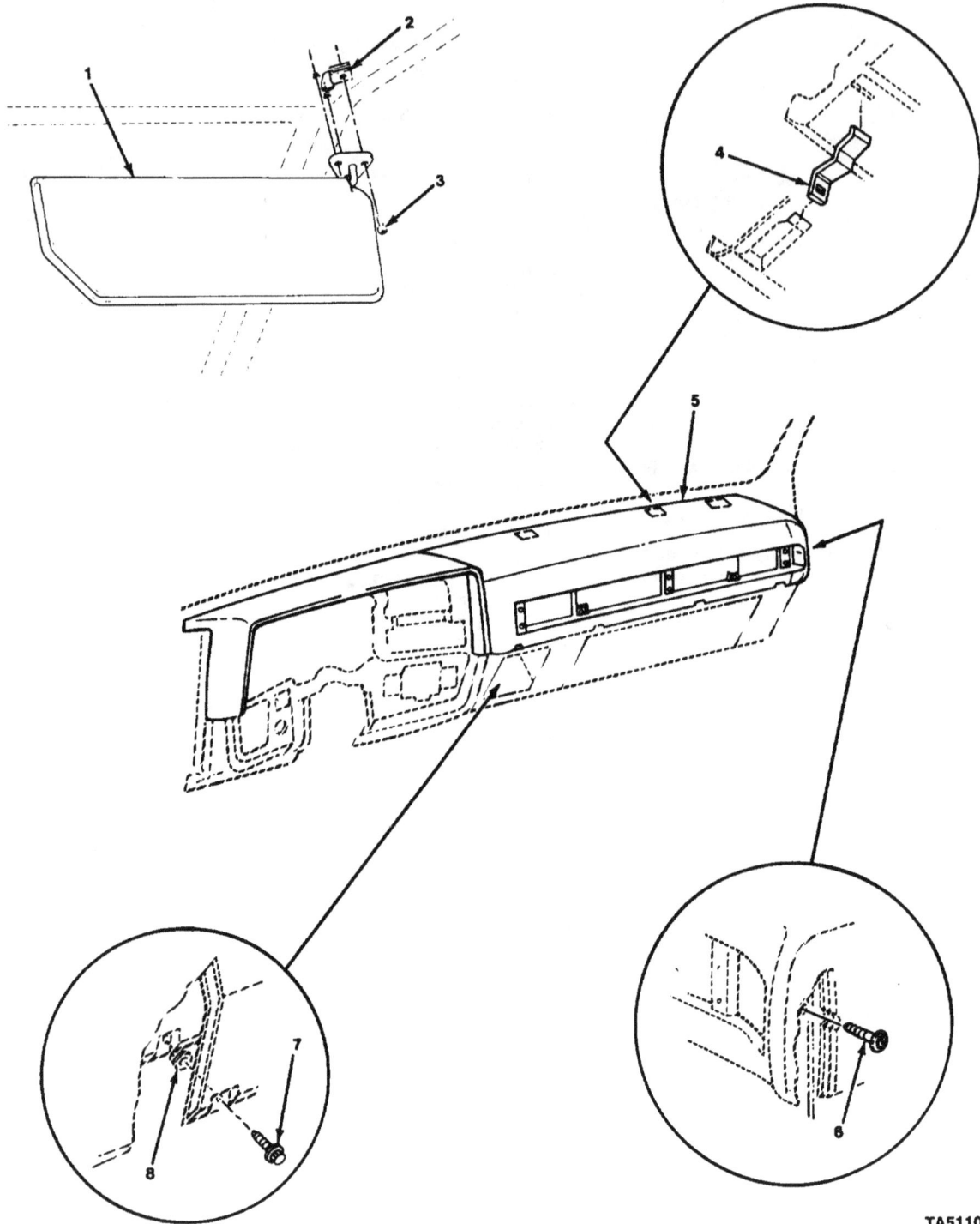

FIGURE 194. SUNVISOR AND INSTRUMENT PANEL PAD.

TA511001

(1)	(2)	(3)	(4)	(5)	(6)
ITEM NO	SMR CODE	CAGEC	PART NUMBER	DESCRIPTION AND USABLE ON CODES (UOC)	QTY

GROUP 1806 UPHOLSTERY, SEATS, AND CARPETS

FIG. 194 SUNVISOR AND INSTRUMENT PANEL PAD

ITEM NO	SMR CODE	CAGEC	PART NUMBER	DESCRIPTION AND USABLE ON CODES (UOC)	QTY
1	PAOZZ	11862	14013753	VISOR,SUN,VEHICLE LEFT	1
1	PAOZZ	11862	14013754	VISOR,SUN,VEHICLE RIGHT	1
2	PAOZZ	96906	MS90724-34	NUT,SHEET SPRING	6
3	PAOZZ	11862	11501149	SCREW,TAQPPING,THREA	6
4	PFOZZ	11862	14031893	CLIP,SPRING TENSION	6
5	PAOZZ	11862	15646949	PAD ASSEMBLY INSTRU	1
6	PAOZZ	11862	14044340	SCREW,ASSEMBLED WAS	1
7	PAOZZ	24617	11503395	SCREW,ASSEMBLED WAS	4
8	PAOZZ	11862	347347	GROMMET,NONMETALLIC	4

END OF FIGURE

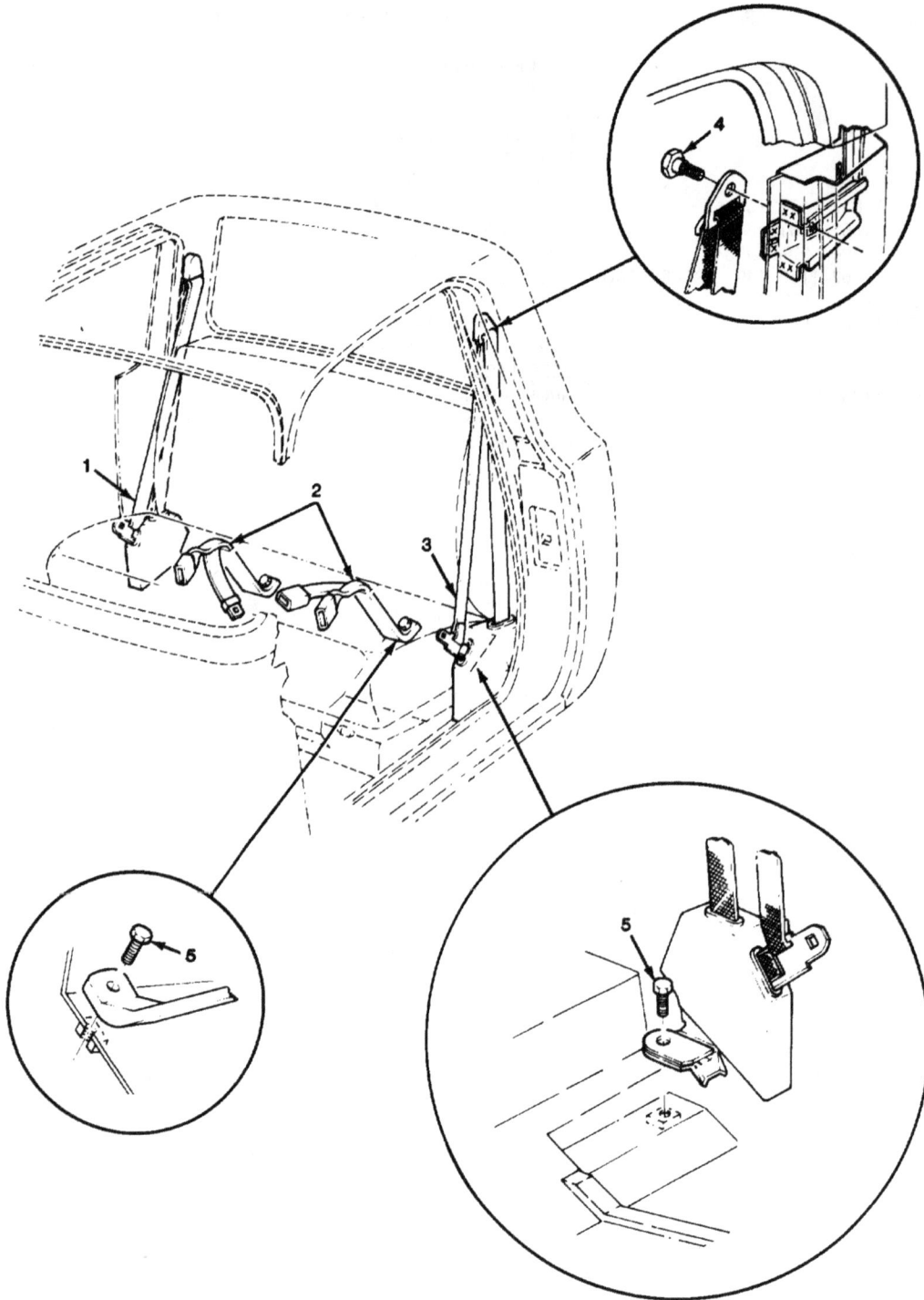

TA511002

FIGURE 195. SEATBELTS AND MOUNTING PARTS (ALL EXCEPT M1009 AND M1010).

(1) (2) (3) (4) ITEM SMR NO CODE CAGEC	PART NUMBER	(5) DESCRIPTION AND USABLE ON CODES (UOC)	(6) QTY
		GROUP 1806 UPHOLSTERY, SEATS, AND CARPETS	
		FIG. 195 SEATBELTS AND MOUNTING PARTS (ALL EXCEPT M1009 AND M1010)	
1 PAOZZ 11862	14079058	BELT,VEHICULAR SAFE RIGHT UOC:194,208,230,231,252	1
2 PAOZZ 11862	14079056	BELT,VEHICULAR SAFE UOC:194,208,230,231,252,254,256	1
3 PAOZZ 11862	14079057	BELT,VEHICULAR SAFE LEFT UOC:194,208,230,231,252,254,256	1
4 PAOZZ 11862	342221	BOLT,SHOULDER UOC:194,208,230,231,252,254,256	2
5 PAOZZ 11862	471083	BOLT,MACHINE UOC:194,208,230,231,252,254,256	4

END OF FIGURE

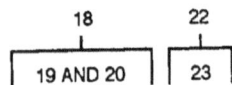

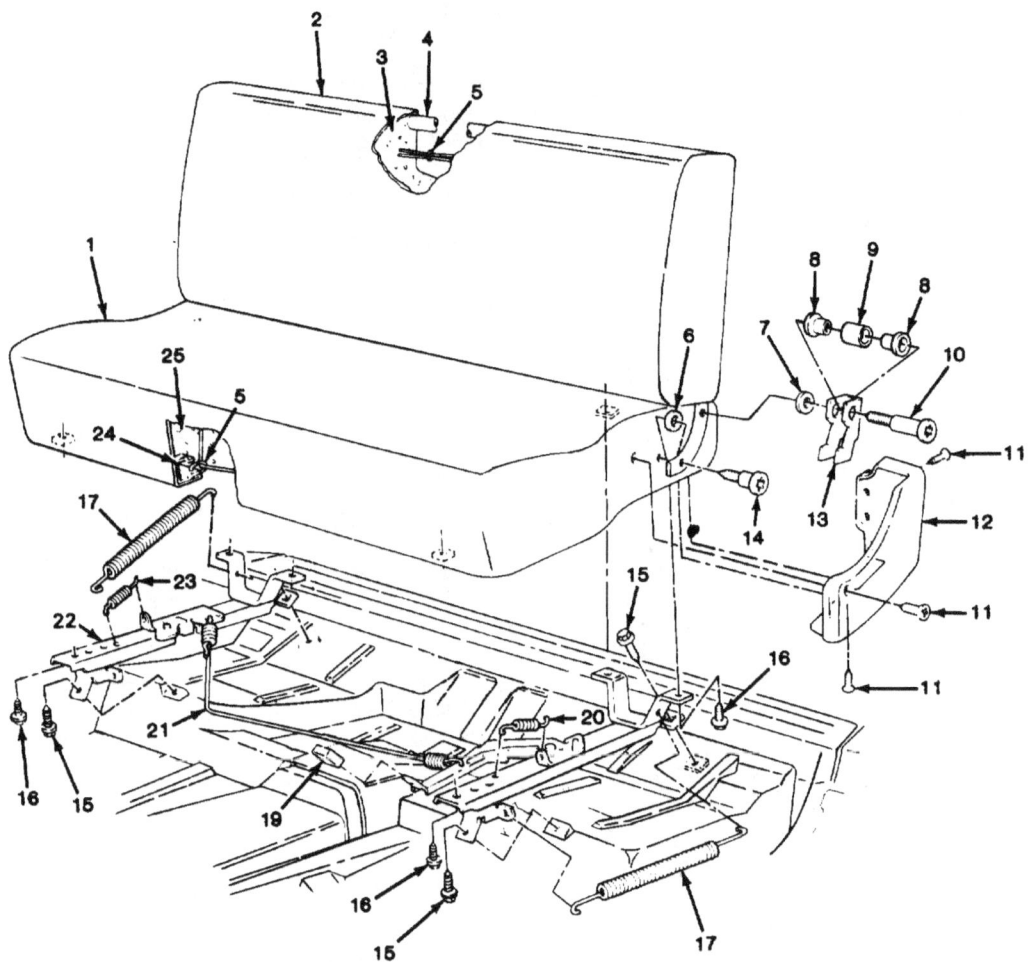

TA511003

FIGURE 196. BENCH SEAT AND COMPONENTS (ALL EXCEPT M1009 AND M1010).

(1) ITEM NO	(2) SMR CODE	(3) CAGEC	(4) PART NUMBER	(5) DESCRIPTION AND USABLE ON CODES (UOC)	(6) QTY
				GROUP 1806 UPHOLSTERY, SEATS, AND CARPETS	
				FIG. 196 BENCH SEAT AND COMPONENTS (ALL EXCEPT M1009 AND M1010)	
1	XDFZZ	11862	14070448	CUSHION,SEAT,VEHICU APPLIES TO FIRST DESIGN UOC:194,208,230,231,252,254,256	1
1	PAFZZ	11862	15594722	SEAT,VEHICULAR APPLIES TO SECOND DESIGN WHICH IS IDENTIFIED BY RMC STICKER ON GLOVEBOX DOOR UOC:194,208,230,231,252,254,256	1
2	PAFZZ	11862	14070453	TRIM ASSEMBLY,SEAT APPLIES TO FIRST DESIGN UOC:194,208,230,231,252,254,256	1
2	PAFZZ	11862	15594727	TRIM ASSEMBLY,SEAT APPLIES TO SECOND DESIGN WHICH IS INDENTIFIED BY RMC STICKER ON GLOVEBOX DOOR UOC:194,208,230,231,252,254,256	1
3	PAFZZ	11862	14021208	CUSHION,SEAT BACK,V UOC:194,208,230,231,252,254,256	1
4	PAFZZ	11862	14021206	FRAME ASSEMBLY,SEAT APPLIES TO FIRST DESIGN UOC:194,208,230,231,252,254,256	1
4	PAFZZ	11862	15598767	CUSHION,SEAT BACK,V INCLUDES ITEMS 3.5 AND APPLIES TO SECOND DESIGN WHICH IS IDENTIFIED BY RMC STICKER ON GLOVEBOX DOOR UOC:194,208,230,231,252,254,256	1
5	PAFZZ	11862	4876466	CLIP,SPRING TENSION UOC:194,208,230,231,252,254,256	51
6	PAOZZ	11862	343978	SHIM UOC:194,208,230,231,252,254,256	2
7	PAOZZ	11862	3914674	WASHER,FLAT UOC:194,208,230,231,252,254,256	2
8	PAOZZ	11862	14066195	BUSHING,SLEEVE UOC:194,208,230,231,252,254,256	4
9	PAOZZ	11862	14066196	BUSHING,NONMETALLIC UOC:194,208,230,231,252,254,256	2
10	PAOZZ	11862	14021211	SCREW,SHOULDER UOC:194,208,230,231,252,254,256	2
11	PAOZZ	24617	9417325	SCREW,TAPPING,THREA UOC:194,208,230,231,252,254,256	10
12	PAOZZ	11862	14037059	COVER,TRIM SEAT LEFT UOC:194,208,230,231,252,254,256	1
12	PAOZZ	11862	14037060	COVER,TRIM SEAT RIGHT UOC:194,208,230,231,252,254,256	1
13	PAOZZ	11862	14021209	CATCH,SEAT FRAME LEFT UOC:194,208,230,231,252,254,256	1
13	PAOZZ	11862	14021210	CATCH,SEAT FRAME RIGHT UOC:194,208,230,231,252,254,256	1

(1)	(2)	(3)	(4)	(5)	(6)
ITEM NO	SMR CODE	CAGEC	PART NUMBER	DESCRIPTION AND USABLE ON CODES (UOC)	QTY
14	PAOZZ	11862	14021213	SCREW,SHOULDER UOC:194,208,230,231,252,254,256	2
15	PAOZZ	11862	9438916	SCREW,TAPPING,THREA UOC:194,208,230,231,252,254,256	4
16	PAOZZ	11862	2014469	BOLT,ASSEMBLED WASH UOC:194,208,230,231,252,254,256	6
17	PAOZZ	11862	20056525	SPRING,HELICAL,EXTE UOC:194,208,230,231,252,254,256	2
18	PAOOO	11862	14022777	LEVER,MANUAL CONTRO LEFT UOC:194,208,230,231,252,254,256	1
19	PAOZZ	11862	465536	.KNOB UOC:194,208,230,231,252,254,256	1
20	PAOZZ	11862	329457	.SPRING,HELICAL,EXTE UOC:194,208,230,231,252,254,256	1
21	PAOZZ	11862	14022786	CLIP ASSEMBLY UOC:194,208,230,231,252,254,256	1
22	PAOOO	11862	14022778	LEVER,MANUAL CONTRO RIGHT UOC:194,208,230,231,252,254,256	1
23	PAOZZ	11862	329457	.SPRING,HELICAL,EXTE UOC:194,208,230,231,252,254,256	1
24	PAFZZ	11862	14021204	FRAME ASSEMBLY,SEAT UOC:194,208,230,231,252,254,256	1
25	PAFZZ	11862	14021207	CUSHION,SEAT,VEHICU APPLIES TO FIRST DESIGN UOC:194,208,230,231,252,254,256	1
25	PAFZZ	11862	15598766	CUSHION,SEAT,BACK,V APPLIES TO SECOND DESIGN WHICH IS INDENTIFIED BY RMC STICKER ON GLOVEBOX DOOR UOC:194,208,230,231,252,254,256	1

END OF FIGURE

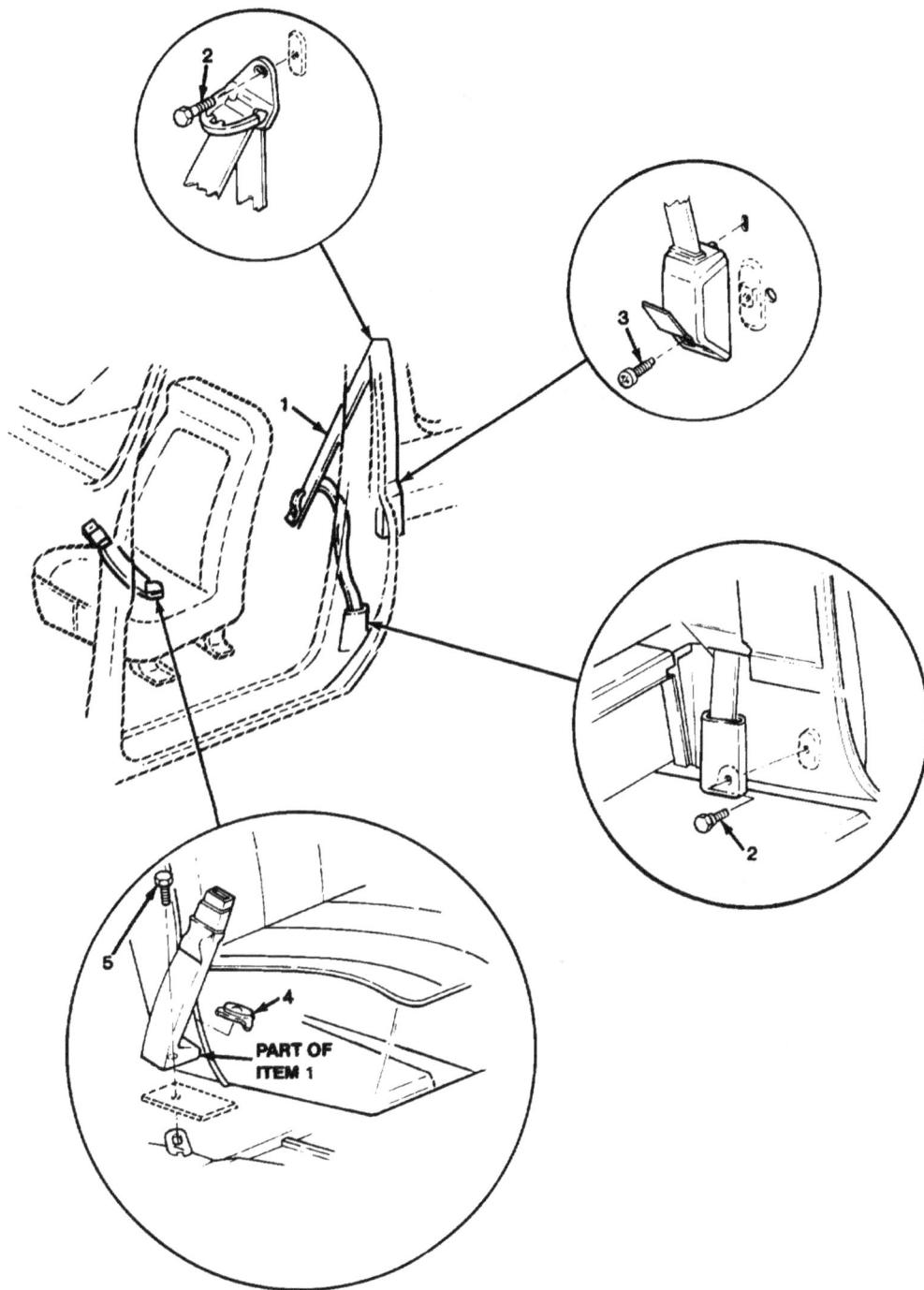

TA511004

FIGURE 197. DRIVER'S SEATBELT AND MOUNTING PARTS (M1009).

(1) (2) (3) (4) ITEM SMR PART NO CODE CAGEC NUMBER	(5) DESCRIPTION AND USABLE ON CODES (UOC)	(6) QTY
	GROUP 1806 UPHOLSTERY,SEATS, AND CARPETS	
	FIG. 197 DRIVER'S SEATBELT AND MOUNTING PARTS (M1009)	
1 PAOZZ 11862 15591247	BELT,VEHICULAR SAFE UOC:209	1
2 PAOZZ 11862 342221	BOLT,SHOULDER UOC:209	2
3 PAOZZ 11862 3954730	SETSCREW UOC:209	1
4 PAOZZ 11862 1731168	PLUG,PROTECTIVE,DUS UOC:209	1
5 PAOZZ 11862 471083	BOLT,MACHINE UOC:209	1
	END OF FIGURE	

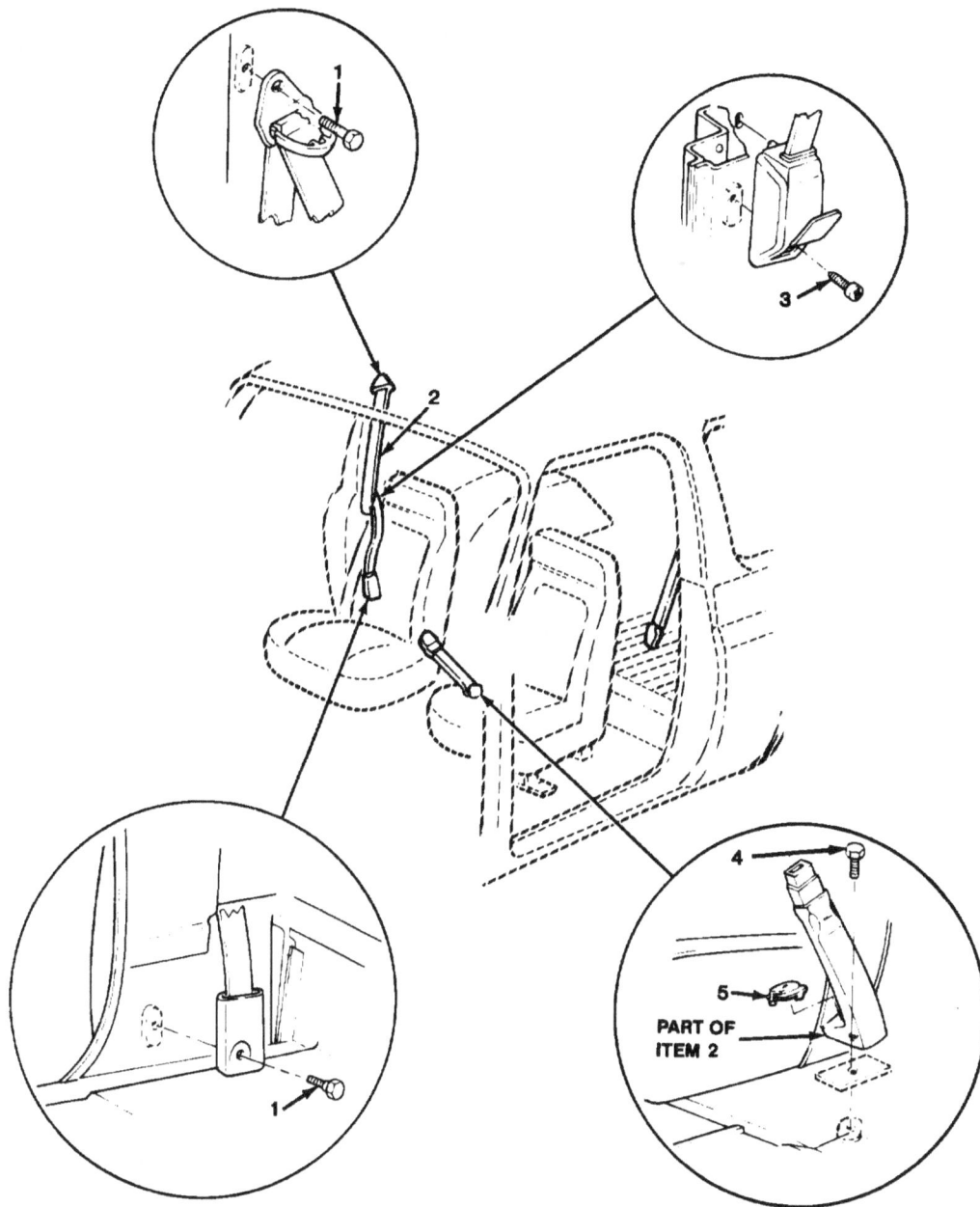

TA511005

FIGURE 198. PASSENGER'S SEATBELT AND MOUNTING PARTS (M1009).

(1) (2) (3) (4) ITEM SMR PART NO CODE CAGEC NUMBER	(5) DESCRIPTION AND USABLE ON CODES (UOC)	(6) QTY
	GROUP 1806 UPHOLSTERY,SEATS,AND CARPETS	
	FIG. 198 PASSENGER'S SEATBELT AND MOUNTING PARTS (M1009)	
1 PAOZZ 11862 342221 BOLT,SHOULDER	UOC:209	2
2 PAOZZ 11862 15591248 BELT,VEHICULAR SAFE	UOC:209	1
3 PAOZZ 11862 3954730 SETSCREW	UOC:209	1
4 PAOZZ 11862 471083 BOLT,MACHINE	UOC:209	1
5 PAOZZ 11862 1731168 PLUG,PROTECTIVE,DUS	UOC:209	1
	END OF FIGURE	

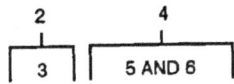

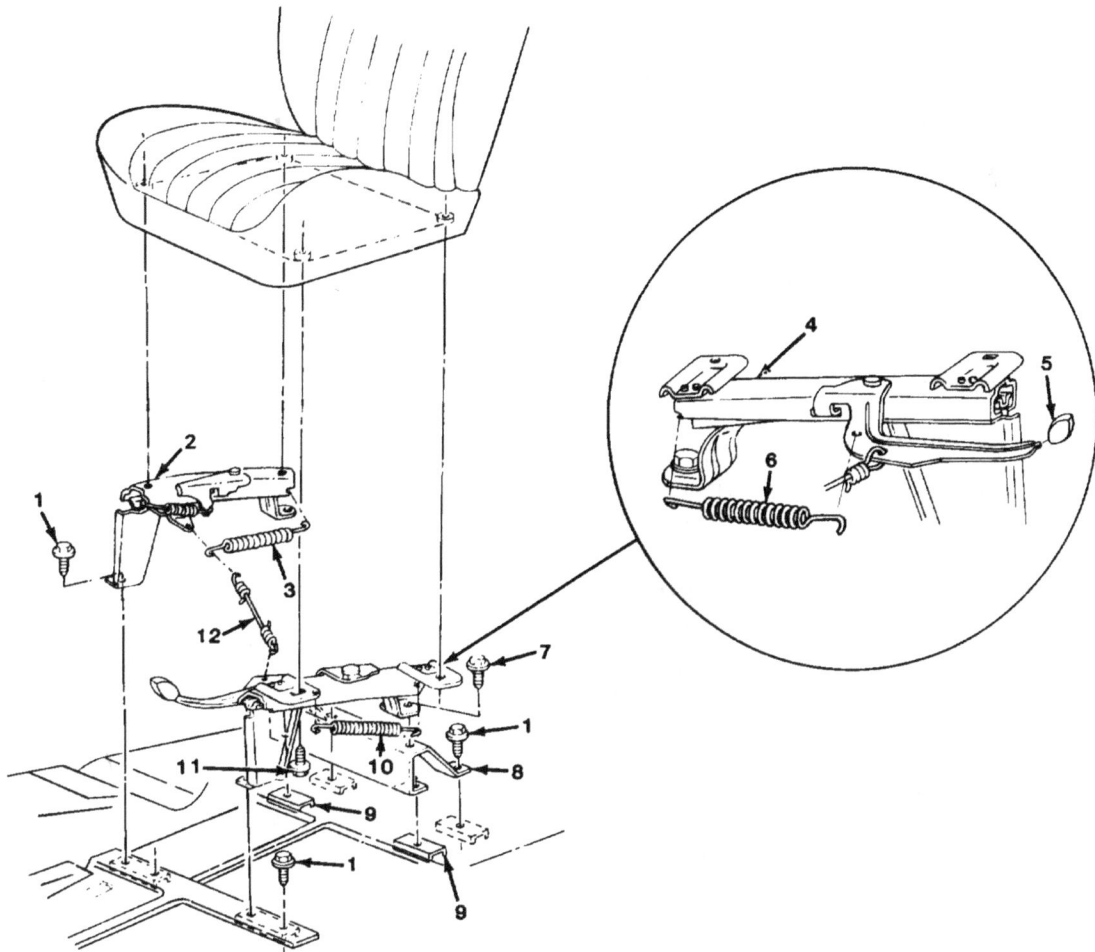

TA511006

FIGURE 199. DRIVER'S SEAT ADJUSTER ASSEMBLY (M1009).

(1) ITEM NO	(2) SMR CODE	(3) CAGEC	(4) PART NUMBER	(5) DESCRIPTION AND USABLE ON CODES (UOC)	(6) QTY
				GROUP 1806 UPHOLSTERY,SEATS,AND CARPETS	
				FIG. 199 DRIVER'S SEAT ADJUSTER ASSEMBLY (M1009)	
1	PAOZZ	11862	9438916	SCREW,TAPPING,THREA UOC:209	8
2	PAOZZ	11862	14075823	ADJUSTER ASSEMBLY,S UOC:209	1
3	PAOZZ	11862	9711038	.SPRING,HELICAL,EXTE UOC:209	1
4	PAOZZ	11862	14075824	ADJUSTER ASSEMBLY,S UOC:209	1
5	PAOZZ	11862	9834636	.HANDLE,MANUAL CONTR UOC:209	1
6	PAOZZ	11862	20351007	.SPRING,HELICAL,EXTE UOC:209	1
7	PAOZZ	11862	334195	BOLT,ASSEMBLED WASH UOC:209	1
8	PAOZZ	11862	14075821	BRACKET,SEAT ADJUST UOC:209	1
9	PAFZZ	11862	467983	PLATE,ANCHOR,SEAT UOC:209	2
10	PAOZZ	11862	20056525	SPRING,HELICAL,EXTE UOC:209	1
11	PAOZZ	11862	14060613	BOLT,MACHINE UOC:209	4
12	PAOZZ	11862	20243999	WIRE,SEAT ADJUSTMEN UOC:209	1

END OF FIGURE

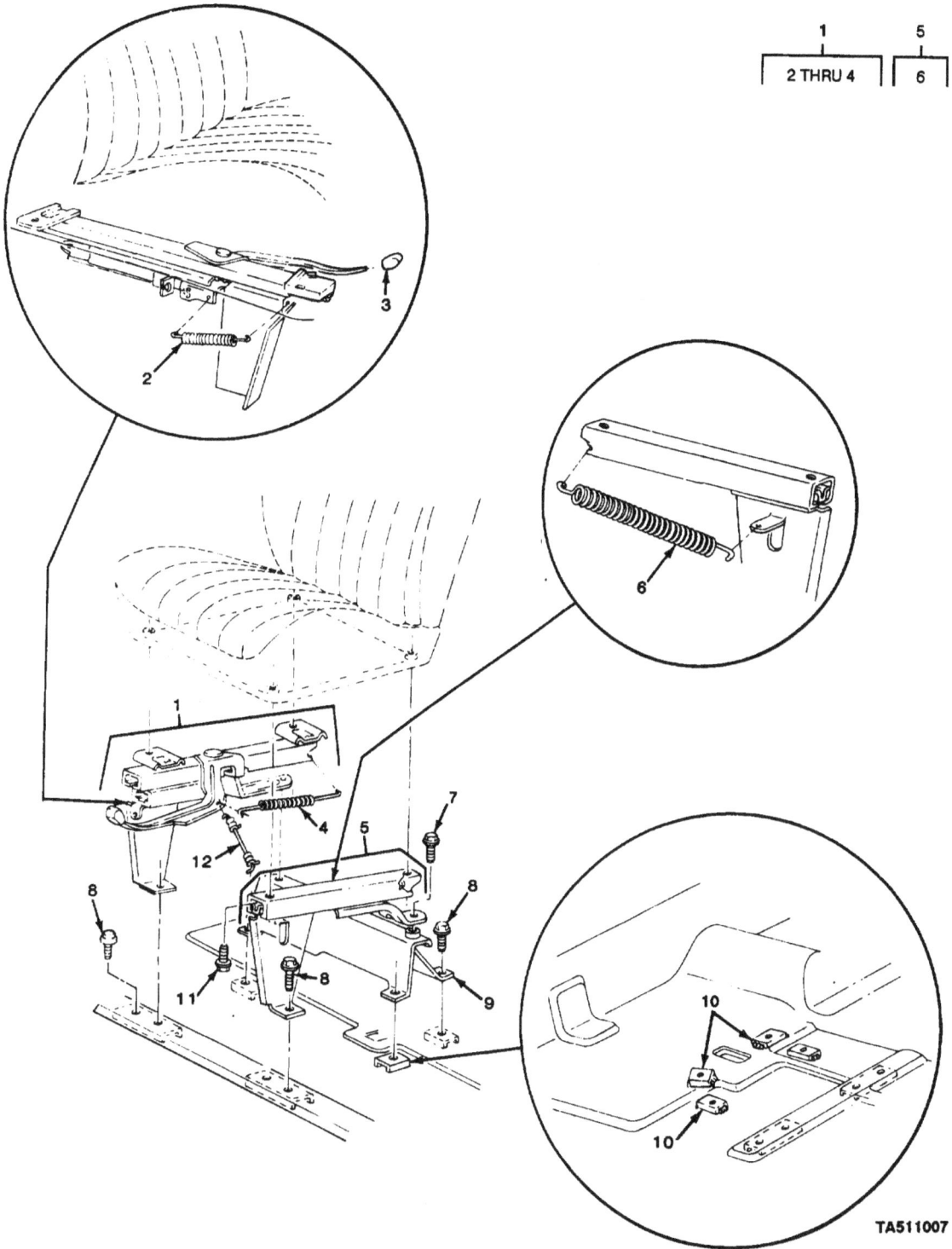

FIGURE 200. PASSENGER'S SEAT ADJUSTER ASSEMBLY (M1009).

(1) (2) (3) (4) ITEM SMR PART NO CODE CAGEC NUMBER	(5) DESCRIPTION AND USABLE ON CODES (UOC)	(6) QTY
	GROUP 1806 UPHOLSTERY,SEATS, AND CARPETS	
	FIG. 200 PASSENGER'S SEAT ADJUSTER ASSEMBLY (M1009)	
1 PAOZZ 11862 15599975	ADJUSTER ASSEMBLY,S UOC:209	1
2 PAOZZ 11862 9645073	.SPRING,HELICAL,EXTE UOC:209	1
3 PAOZZ 11862 9834636	.HANDLE,MANUEL CONTR UOC:209	
4 PAOZZ 11862 9826897	.SPRING,HELICAL,EXTE UOC:209	1
5 PAOOO 11862 15599976	SUPPORT,SEAT ADJUST UOC:209	1
6 PAOZZ 11862 20056525	.SPRING,HELICAL,EXTE UOC:209	1
7 PAOZZ 11862 334195	BOLT,ASSEMBLED WASH UOC:209	2
8 PAOZZ 11862 9438916	SCREW,TAPPING,THREA UOC:209	8
9 PAOZZ 11862 14075822	BRACKET,ADJUSTMENT UOC:209	1
10 PAFZZ 11862 467983	PLATE,ANCHOR,SEAT UOC:209	3
11 PAOZZ 11862 14060613	BOLT,MACHINE UOC:209	4
12 PAOZZ 11862 20243999	WIRE,SEAT ADJUSTMEN UOC:209	1

END OF FIGURE

```
        1
   ┌─────────────┐
   │  2 THRU 14  │
   └─────────────┘
```

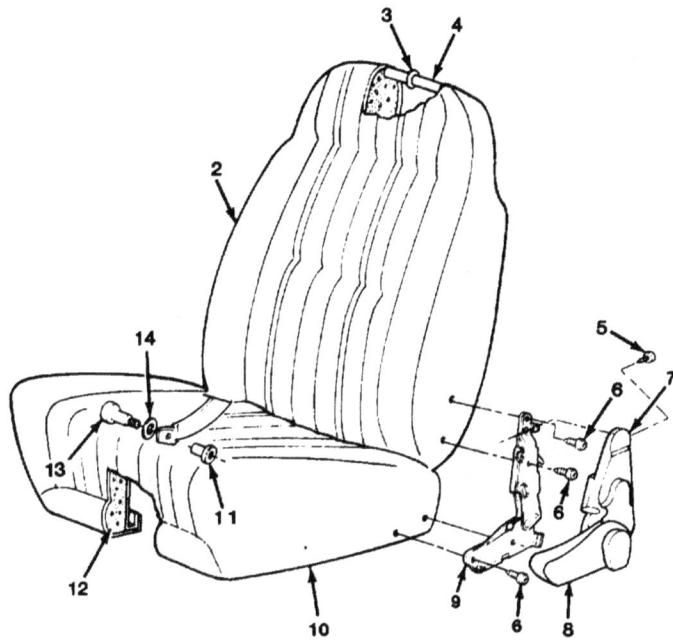

TA511008

FIGURE 201. DRIVER'S SEAT ASSEMBLY AND COMPONENTS (M1009).

(1) ITEM NO	(2) SMR CODE	(3) CAGEC	(4) PART NUMBER	(5) DESCRIPTION AND USABLE ON CODES (UOC)	(6) QTY
				GROUP 1806 UPHOLSTERY,SEATS,AND CARPETS	
				FIG. 201 DRIVER'S SEAT ASSEMBLY AND COMPONENTS (M1009)	
1	PAOFF	11862	14075820	SEAT,VEHICULAR UOC:209	1
2	PAFZZ	11862	14075817	.COVER,SEAT,VEHICULA UOC:209	1
3	PAFZZ	11862	4876466	.CLIP,SPRING TENSION UOC:209	51
4	PAFZZ	11862	14071555	.CUSHION,SEAT BACK,V UOC:209	1
5	PAOZZ	11862	11503537	.SCREW,ASSEMBLED WAS UOC:209	4
6	PAOZZ	11862	20025648	.SCREW,TAPPING,THREA UOC:209	4
7	PAOZZ	11862	20293843	.COVER,RECLINING HIN UOC:209	1
8	PAOZZ	11862	20289493	.COVER,SEAT LOCK UOC:209	1
9	PAOZZ	11862	16604537	.LOCK,SEAT,HINGE UOC:209	1
10	PAOZZ	11862	14075818	.COVER,SEAT,VEHICULA UOC:209	1
11	PAOZZ	11862	14059238	.BUSHING,SLEEVE UOC:209	2
12	PAFZZ	11862	14071461	.CUSHION,SEAT,VEHICU UOC:209	1
13	PAOZZ	11862	14056723	.SCREW,SHOULDER UOC:209	2
14	PAOZZ	11862	1727059	.WASHER,FLAT UOC:209	2

END OF FIGURE

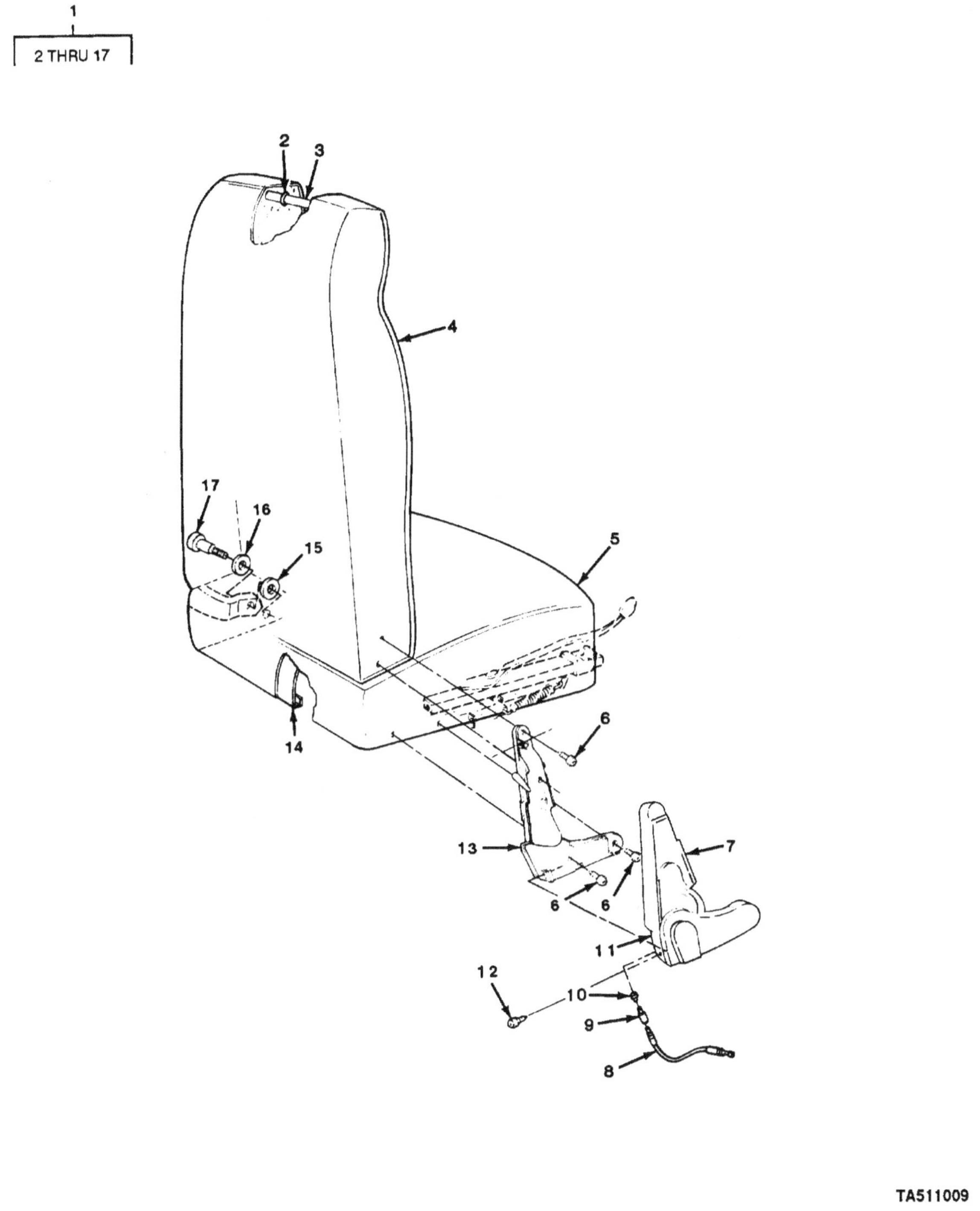

TA511009

FIGURE 202. PASSENGER'S SEAT ASSEMBLY AND COMPONENTS (M1009).

(1)	(2)	(3)	(4)	(5)	(6)
ITEM NO	SMR CODE	CAGEC	PART NUMBER	DESCRIPTION AND USABLE ON CODES (UOC)	QTY

GROUP 1806 UPHOLSTERY,SEATS,AND CARPETS

FIG. 202 PASSENGER'S SEAT ASSEMBLY AND COMPONENTS (M1009)

1	PAOFF	11862	14075819	SEAT,VEHICULAR UOC:209	1
2	PAFZZ	11862	4876466	.CLIP,SPRING TENSION UOC:209	51
3	PAFZZ	11862	14071556	.CUSHION,SEAT BACK,V UOC:209	1
4	PAFZZ	11862	14075817	.COVER,SEAT,VEHICULA UOC:209	1
5	PAFZZ	11862	14075818	.COVER,SEAT,VEHICULA UOC:209	1
6	PAOZZ	11862	20025648	.SCREW,TAPPING,THREA UOC:209	4
7	PAOZZ	11862	20293842	.COVER,SEAT BACK HIN UOC:209	1
8	PAOZZ	11862	20369919	.CABLE,PASSENGER SEA UOC:209	1
9	PAOZZ	11862	20369920	.LOCK,BELT RELEASE,V UOC:209	1
10	PAOZZ	11862	20573776	.GROMMET,NONMETALLIC UOC:209	1
11	PAOZZ	11862	15627452	.COVER,SEAT,LOCK UOC:209	1
12	PAOZZ	11862	11503537	.SCREW,ASSEMBLED WAS UOC:209	4
13	PAOZZ	11862	20410901	.LOCK,SEAT,BACK UOC:209	1
14	PAFZZ	11862	14071461	.CUSHION,SEAT,VEHICU UOC:209	1
15	PAOZZ	11862	14059238	.BUSHING,SLEEVE UOC:209	2
16	PAOZZ	11862	1727059	.WASHER,FLAT UOC:209	2
17	PAOZZ	11862	14056723	.SCREW,SHOULDER UOC:209	2

END OF FIGURE

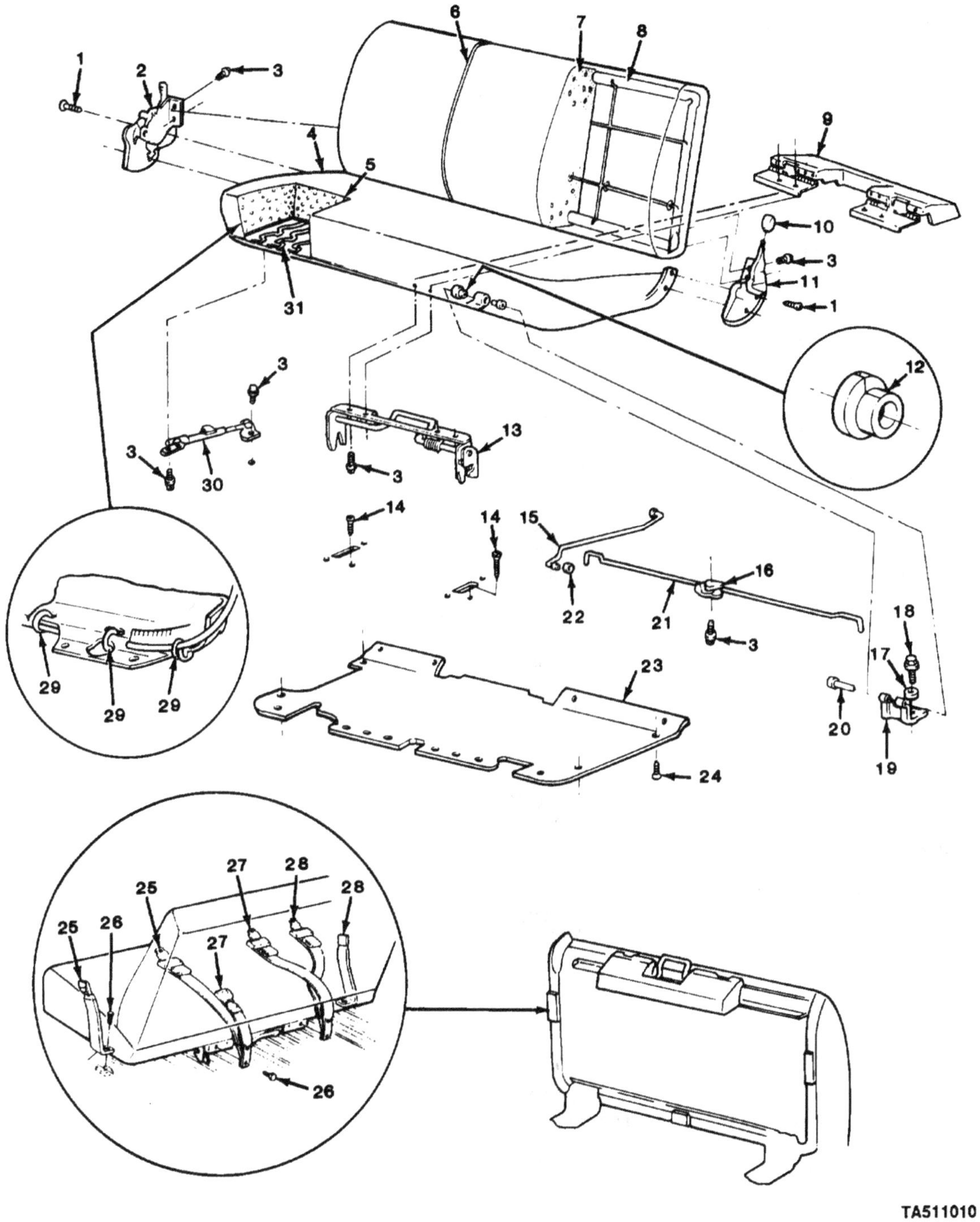

FIGURE 203. REAR BENCH SEAT ASSEMBLY AND COMPONENTS (M1009).

TA511010

(1) ITEM NO	(2) SMR CODE	(3) CAGEC	(4) PART NUMBER	(5) DESCRIPTION AND USABLE ON CODES (UOC)	(6) QTY
				GROUP 1806 UPHOLSTERY,SEATS,AND CARPETS	
				FIG. 203 REAR BENCH SEAT ASSEMBLY AND COMPONENTS (M1009)	
1	PAOZZ	11862	14013724	SCREW,TAPPING,THREA UOC:209	4
2	PAOZZ	11862	15569010	HINGE AND LATCH,SEA RIGHT UOC:209	1
3	PAOZZ	11862	2014469	BOLT,ASSEMBLED WASH UOC:209	10
4	PAFZZ	11862	14075815	CUSHION,SEAT,VEHICU UOC:209	1
5	PAFZZ	11862	471020	CUSHION,SEAT,VEHICU UOC:209	1
6	PAFZZ	11862	14075816	TRIM,REAR SEAT UOC:209	1
7	PAFFF	11862	471021	CUSHION,SEAT BACK,V UOC:209	1
8	PAFZZ	11862	14027798	FRAME,SEAT,REAR BAC UOC:209	1
9	PAOZZ	11862	14014293	COVER,LATCH SEAT UOC:209	1
10	PAOZZ	11862	473187	KNOB UOC:209	1
11	PAOZZ	11862	15569009	HINGE,SEAT,VEHICULA LEFT UOC:209	1
12	PAOZZ	11862	335524	BUSHING,SLEEVE UOC:209	4
13	PAOZZ	11862	470984	LATCH AND GLIDE ASS UOC:209	1
14	PAOZZ	24617	9419327	SCREW UOC:209	2
15	PAOZZ	11862	473197	ROD ASSEMBLY,SEAT L UOC:209	1
16	PAOZZ	11862	14027799	CLAMP,RIM CLENCHING UOC:209	1
17	PAOZZ	96906	MS35340-48	WASHER,LOCK UOC:209	2
18	PAOZZ	24617	9432194	SCREW,CAP,HEXAGON H UOC:209	2
19	PAOZZ	11862	471018	HINGE,LOWER FRONT,R UOC:209	2
20	PAOZZ	11862	470974	PIN,STRAIGHT,HEADED UOC:209	2
21	PAOZZ	11862	471010	LEVER,MANUAL CONTRO UOC:209	1
22	PAOZZ	96906	MS16633-1015	RING,RETAINING UOC:209	2
23	PAOZZ	11862	14074479	PANEL,SEAT CUSHION UOC:209	1

(1)	(2)	(3)	(4)	(5)	(6)
ITEM NO	SMR CODE	CAGEC	PART NUMBER	DESCRIPTION AND USABLE ON CODES (UOC)	QTY
24	PAOZZ	11862	14069636	SCREW,ASSEMBLED WAS UOC:209	16
25	PAOZZ	11862	14075375	BELT,VEHICULAR SAFE LEFT UOC:209	1
26	PAOZZ	11862	3954730	SETSCREW UOC:209	4
27	PAOZZ	11862	14075374	BELT,VEHICULAR SAFE CENTER UOC:209	1
28	PAOZZ	11862	14075376	BELT,VEHICULAR SAFE RIGHT UOC:209	1
29	PAFZZ	11862	4876466	CLIP,SPRING TENSION UOC:209	31
30	PAOZZ	11862	471079	STRUT,STOWAGE,REAR UOC:209	1
31	PAFZZ	11862	14027797	FRAME ASSEMBLY,REAR UOC:209	1

END OF FIGURE

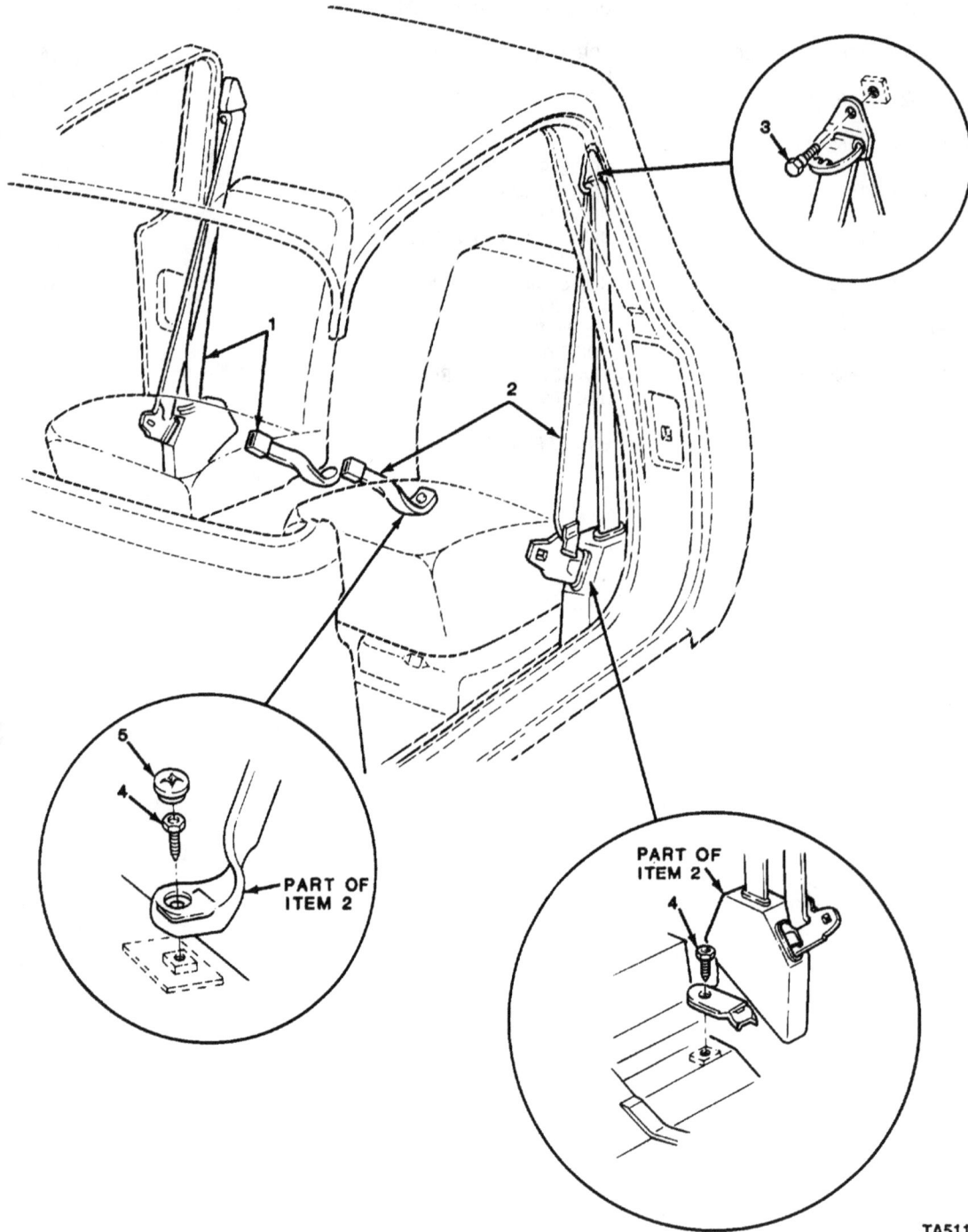

FIGURE 204. SEATBELTS AND MOUNTING PARTS (M1010).

TA511011

(1) ITEM NO	(2) SMR CODE	(3) CAGEC	(4) PART NUMBER	(5) DESCRIPTION AND USABLE ON CODES (UOC)	(6) QTY
				GROUP 1806 UPHOLSTERY,SEATS,AND CARPETS	
				FIG. 204 SEATBELTS AND MOUNTING PARTS (M1010)	
1	PAOZZ	11862	15577694	BELT,VEHICULAR SAFE RIGHT UOC:210	1
2	PAOZZ	11862	14075379	BELT,VEHICULAR SAFE LEFT UOC:210	1
3	PAOZZ	11862	342221	BOLT,SHOULDER UOC:210	2
4	PAOZZ	11862	471083	BOLT,MACHINE UOC:210	4
5	PAOZZ	11862	1731168	PLUG,PROTECTIVE,DUS UOC:210	2

END OF FIGURE

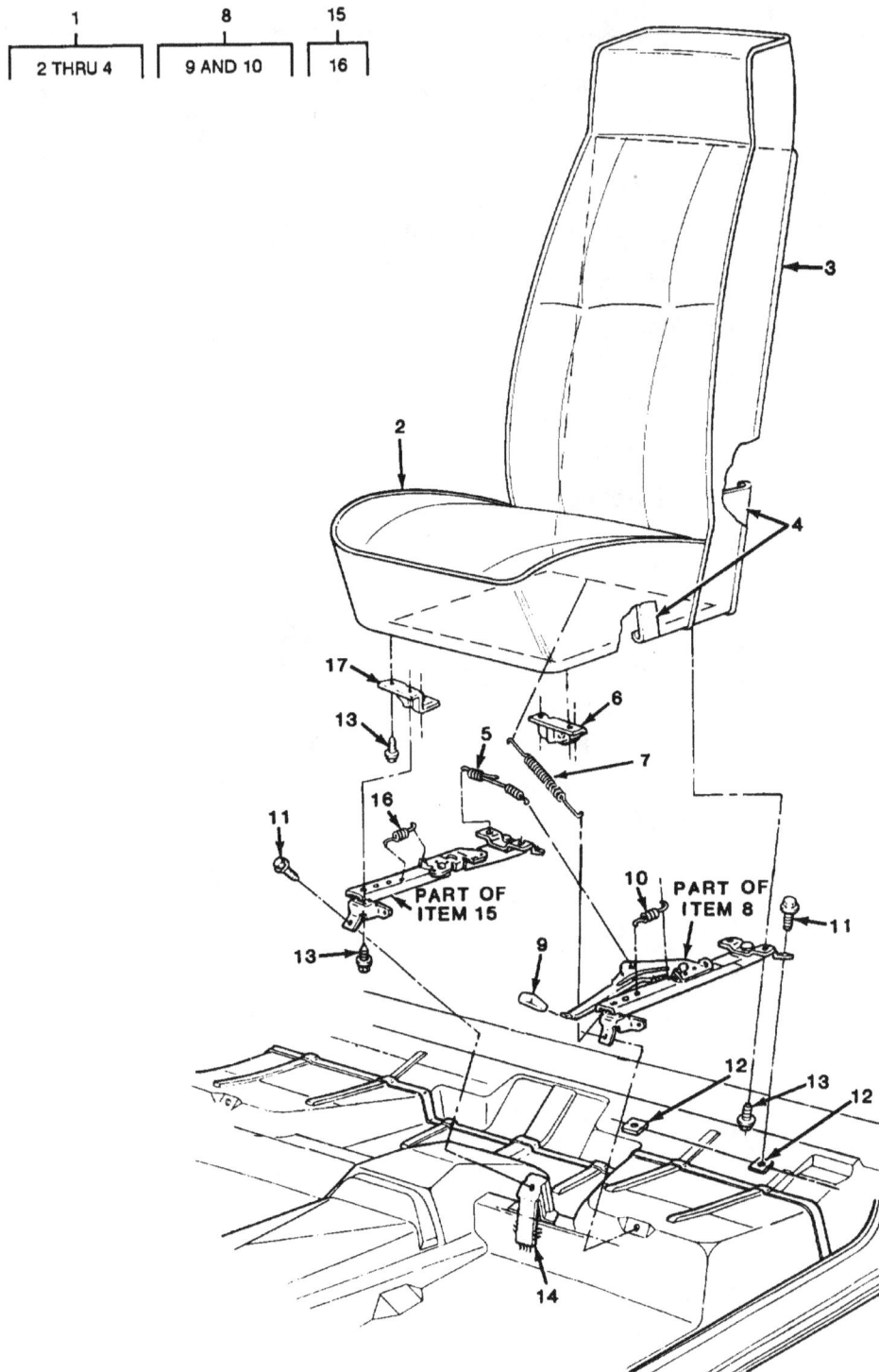

FIGURE 205. SEAT ASSEMBLY, DRIVER'S SEAT, ADJUSTER ASSEMBLY
AND RELATED PARTS (M1010).

TA511012

(1) (2) (3) (4) ITEM SMR PART NO CODE CAGEC NUMBER	(5) DESCRIPTION AND USABLE ON CODES (UOC)	(6) QTY
	GROUP 1806 UPHOLSTERY,SEATS,AND CARPETS	
	FIG. 205 SEAT ASSEMBLY,DRIVER'S SEAT,ADJUSTER ASSEMBLY,AND RELATED PARTS (M1010)	
1 PAOFF 11862 14075815	CUSHION,SEAT,VEHICU UOC:210	2
2 PAFZZ 11862 14075372	.TRIM ASSEMBLY,SEAT UOC:210	1
3 PAFZZ 11862 14075373	.TRIM ASSEMBLY,SEAT UOC:210	1
4 PAFZZ 11862 14067457	.SPRING,FRAME UOC:210	1
5 PAOZZ 11862 14025679	WIRE ASSEMBLY,LOCK, UOC:210	1
6 PAOZZ 11862 14025673	BRACKET,DOUBLE ANGL UOC:210	1
7 PAOZZ 11862 20056525	SPRING,HELICAL,EXTE UOC:210	1
8 PAOOO 11862 14023889	LEVER,MANUAL CONTROL LEFT UOC:210	1
9 PAOZZ 11862 465536	.KNOB UOC:210	1
10 PAOZZ 11862 329457	.SPRING,HELICAL,EXTE UOC:210	1
11 PAOZZ 11862 9438916	SCREW,TAPPING,THREA UOC:210	4
12 PAFZZ 11862 335435	PLATE ASSEMBLY,SEAT UOC:210	2
13 PAOZZ 11862 2014469	BOLT,ASSEMBLED WASH UOC:210	10
14 PAFZZ 11862 14025677	BRACKET,SEAT ADJUST UOC:210	1
15 PAOOO 11862 14023890	LEVER,MANUAL CONTRO RIGHT UOC:210	1
16 PAOZZ 11862 329457	.SPRING,HELICAL,EXTE UOC:210	1
17 PAOZZ 11862 14025674	BRACKET,DOUBLE ANGL UOC:210	1
	END OF FIGURE	

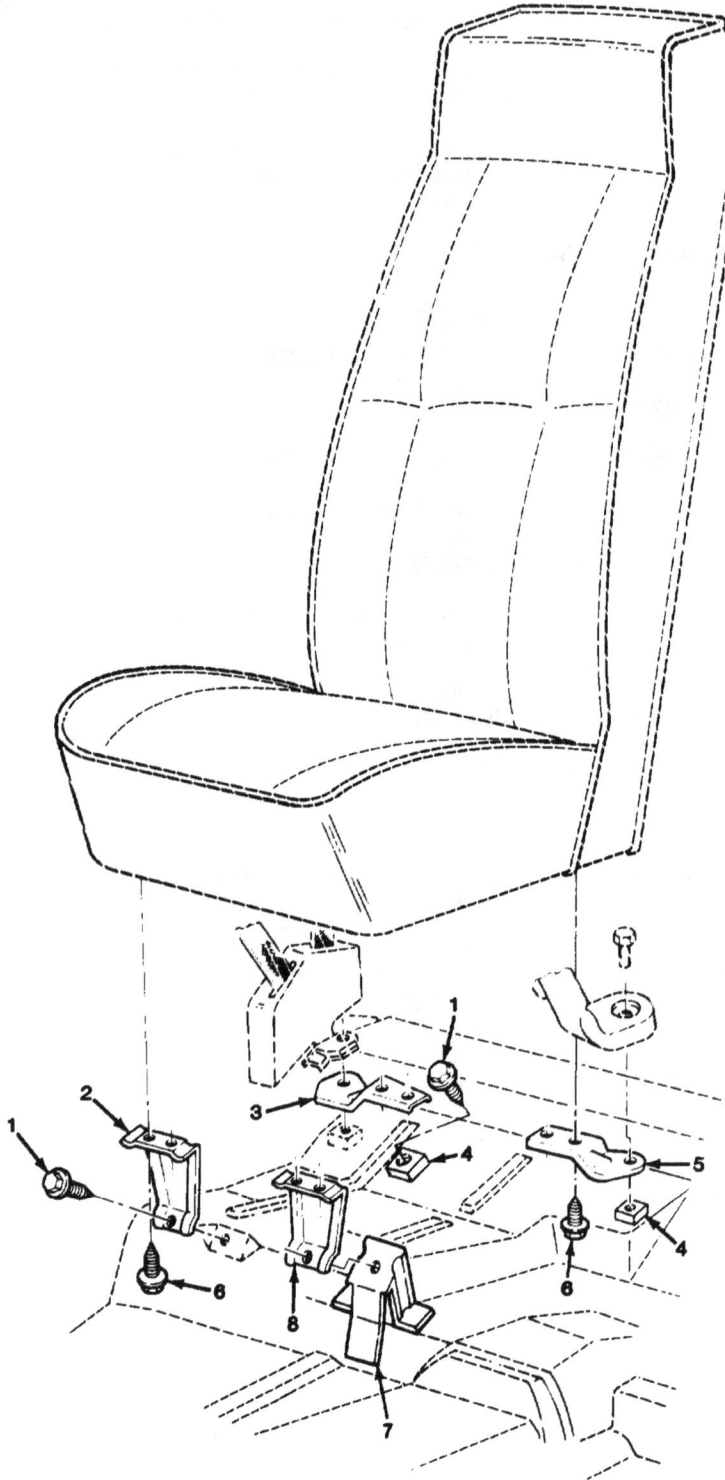

FIGURE 206. SEAT MOUNTING PARTS (M1010).

TA511013

TM9-2320-289-34P

(1) (2) (3) (4) ITEM SMR PART NO CODE CAGEC NUMBER	(5) DESCRIPTION AND USABLE ON CODES (UOC)	(6) QTY
	GROUP 1806 UPHOLSTERY,SEATS,AND CARPETS	
	FIG. 206 SEAT MOUNTING PARTS (M1010)	
1 PAOZZ 11862 9438916	SCREW,TAPPING,THREA UOC:210	3
2 PAOZZ 11862 14027346	SUPPORT,SEAT FRAME RIGHT UOC:210	1
3 PAOZZ 11862 14027348	SUPPORT,SEAT,VEHICU RIGHT UOC:210	1
4 PAFZZ 11862 335435	PLATE ASSEMBLY,SEAT UOC:210	2
5 PAOZZ 11862 14027347	SUPPORT,SEAT,VEHICU LEFT UOC:210	1
6 PAOZZ 11862 2014469	BOLT,ASSEMBLED WASH UOC:210	8
7 PAFZZ 11862 14025677	BRACKET,SEAT ADJUST UOC:210	1
8 PAOZZ 11862 14027345	SUPPORT,SEAT,VEHICU LEFT UOC:210	1

END OF FIGURE

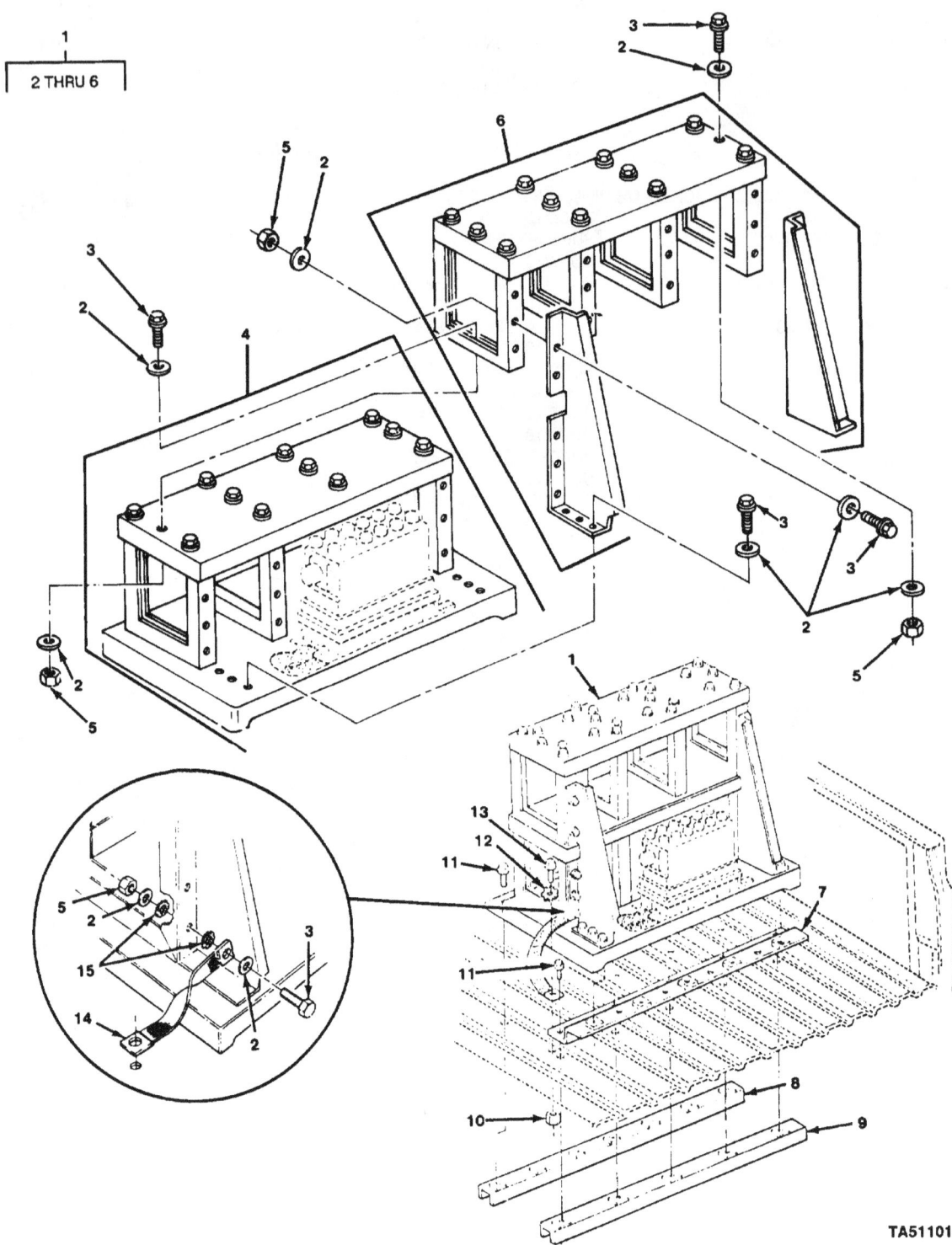

FIGURE 207. COMMUNICATIONS RACK ASSEMBLY AND RELATED PARTS (M1028A1).

TA511014

(1)	(2)	(3)	(4)	(5)	(6)
ITEM NO	SMR CODE	CAGEC	PART NUMBER	DESCRIPTION AND USABLE ON CODES (UOC)	QTY

GROUP 1808 STOWAGE RACKS,BOXES,
STRAPS,CARRYING CASES,CABLE REELS,
HOSE REELS, ETC.

FIG. 207 COMMUNICATIONS RACK
ASSEMBLY AND RELATED PARTS (M1028A1)

1	PAOOO	80063	SC-D-866091	RACK,ELECTRICAL EQU UOC:208	1
2	PAOZZ	96906	MS35338-139	.WASHER,LOCK UOC:208	110
3	PAOZZ	96906	MS90725-6	.SCREW,CAP,HEXAGON H UOC:208	61
4	PAOZZ	80063	DL-SC-B-691368	.RACK ASSEMBLY UOC:194,208	1
5	PAOZZ	96906	MS51971-1	.NUT,PLAIN,HEXAGON UOC:208	49
6	PAOZZ	80063	SC-D-691375	.RACK,ELECTRICAL EQU UOC:194,208	1
7	PAOZZ	80063	SC-C-691545	BRACKET,DOUBLE ANGL UOC:208	1
8	PAOZZ	11862	15599959	STRAP,RETAINING UOC:208	1
9	PAOZZ	11862	15599958	STRAP,RETAINING UOC:208	1
10	PAOZZ	24617	271172	NUT,SELF-LOCKING,AS UOC:208	1
11	PAOZZ	11862	9440334	BOLT,ASSEMBLED WASH UOC:208	12
12	PAOZZ	96906	MS35335-33	WASHER,LOCK UOC:208	1
13	PAOZZ	96906	MS90728-6	SCREW,CAP,HEXAGON H UOC:208	1
14	XDOZZ	80063	SC-B-75180-IV	LEAD,ELECTRICAL UOC:208	1
15	PAOZZ	96906	MS45904-68	WASHER,LOCK UOC:208	2

END OF FIGURE

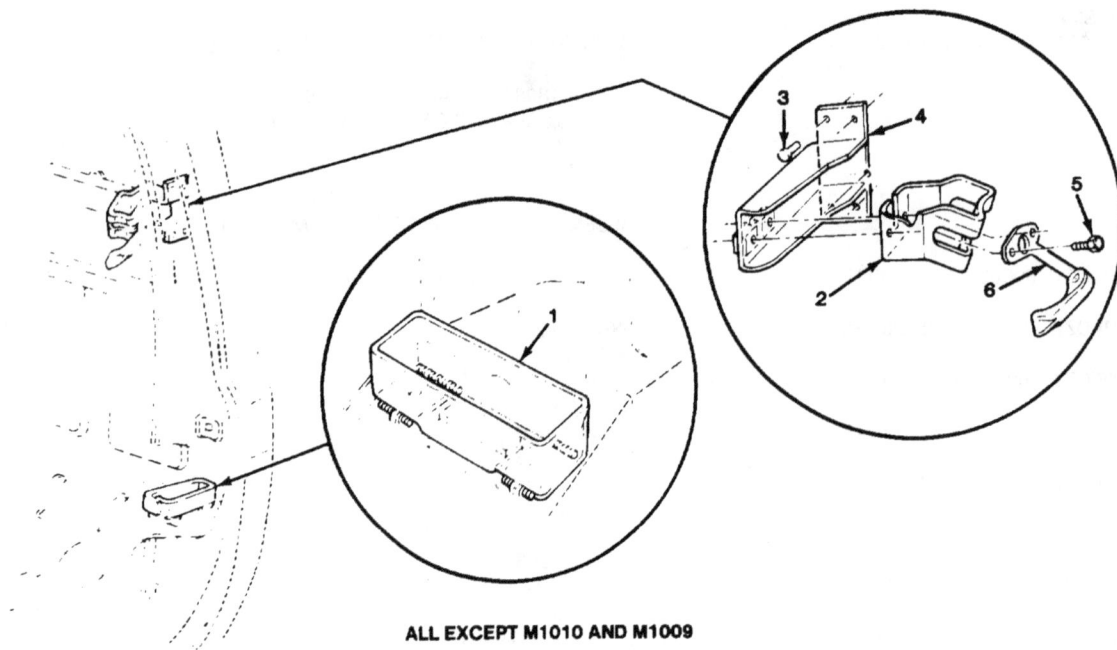

ALL EXCEPT M1010 AND M1009

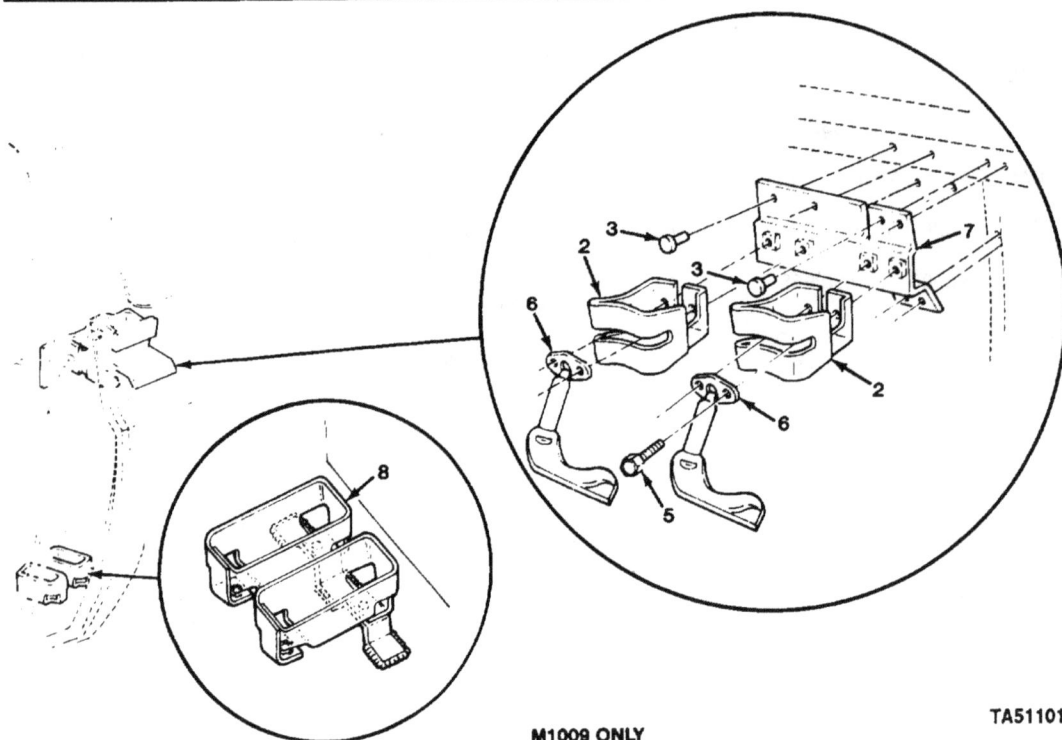

M1009 ONLY

TA511015

FIGURE 208. WEAPONS MOUNT (ALL EXCEPT M1010).

(1) (2) (3) (4)	(5)	(6)
ITEM SMR PART NO CODE CAGEC NUMBER	DESCRIPTION AND USABLE ON CODES (UOC)	QTY

GROUP 1808 STOWAGE RACKS,BOXES,
STRAPS,CARRYING CASES,CABLE REELS,
HOSE REELS,ETC.

FIG. 208 WEAPONS MOUNT (ALL EXCEPT
M1010)

1	PAOZZ 11852 14074435		SUPPORT,WEAPON MOUN UOC:194,208,230,231,252,254,256	2
2	PAOZZ 11862 14074437	BRACKET,MOUNTING,CA	UOC:194,208,209,230,231,252,254,256,	2
3	PAOZZ 11862 2477054	RIVET,BLIND	UOC:194,208,209,230,231,252,254,256	8
4	PAOZZ 11862 14074441		BRACKET,WEAPON MOUN LEFT UOC:194,208,230,231,252,254,256	1
4	PAOZZ 11862 14074442		BRACKET,WEAPON MOUN RIGHT UOC:194,208,230,231,252,254,256	1
5	PAOZZ 11862 9440033	BOLT,MACHINE	UOC:194,208,209,230,231,252,254,256	4
6	PAOZZ 11862 14074438		LATCH ASSEMBLY,WEAP UOC:194,208,209,230,231,252,254,256	2
7	PAOZZ 11862 15599965		BRACKET,MOUNTING LEFT UOC:209	1
8	PAOZZ 11862 15599271	SUPPORT,WEAPON	UOC:209	1

END OF FIGURE

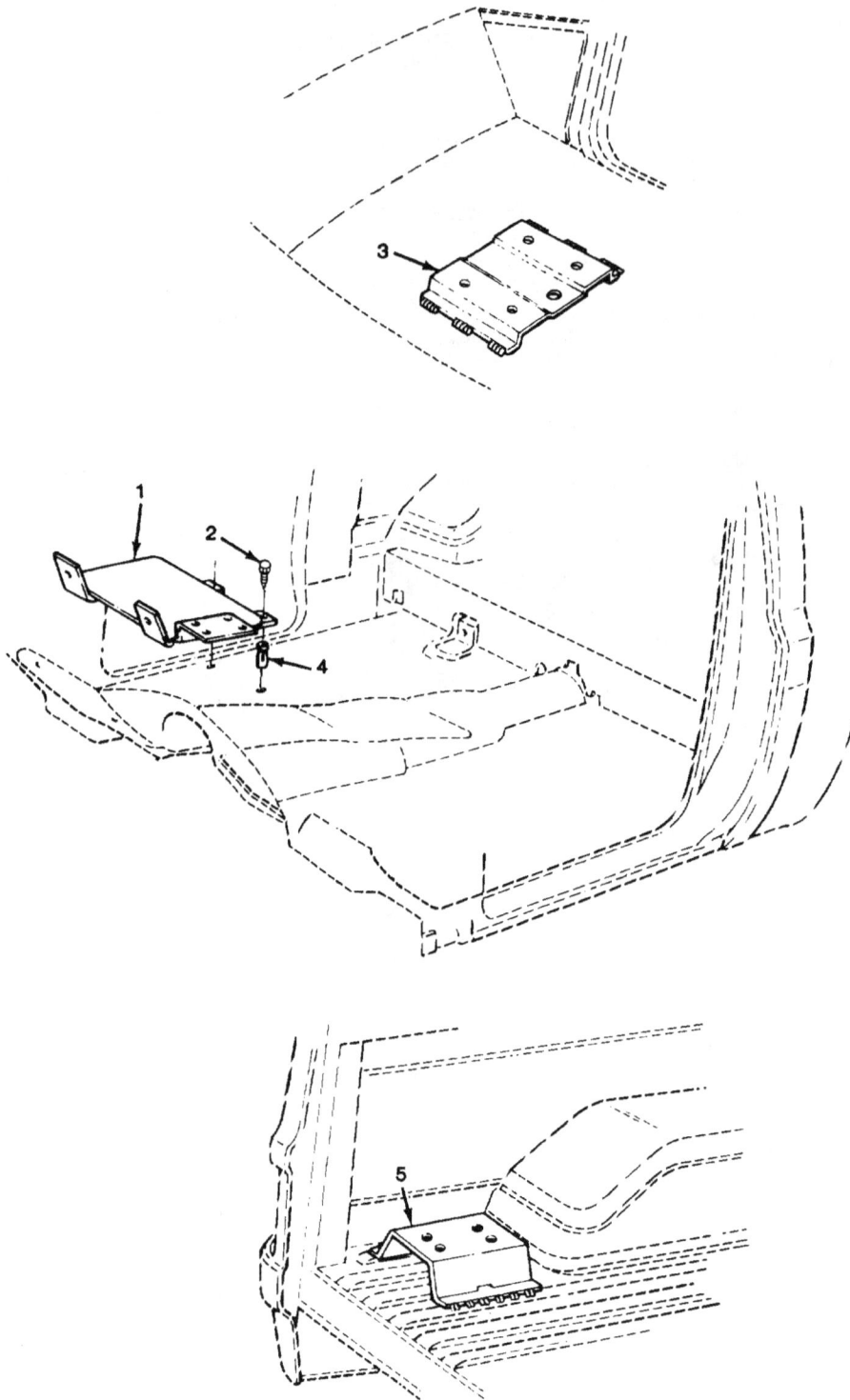

TA511016

FIGURE 209. BRACKETS AND MOUNTING PARTS.

(1)	(2)	(3)	(4)	(5)	(6)
ITEM NO	SMR CODE	CAGEC	PART NUMBER	DESCRIPTION AND USABLE ON CODES (UOC)	QTY

GROUP 1808 STOWAGE RACKS, BOXES,
STRAPS, CARRYING CASES, CABLE REELS,
HOSE REELS, ETC.

FIG. 209 BRACKETS AND MOUNTING PARTS

1	PAOZZ	11852	14074439	BRACKET ASSEMBLY,CH	1
2	PAOZZ	24617	9421432	SCREW,MACHINE	4
3	PAFZZ	11862	14074432	BRACKET ASSEMBLY,DE UOC:194,208,210,230,231,252,254,256	1
4	PAOZZ	11862	14005953	SHIELD,EXPANSION	4
5	PAFZZ	11862	15599961	BRACKET,ANGLE UOC:209	1

END OF FIGURE

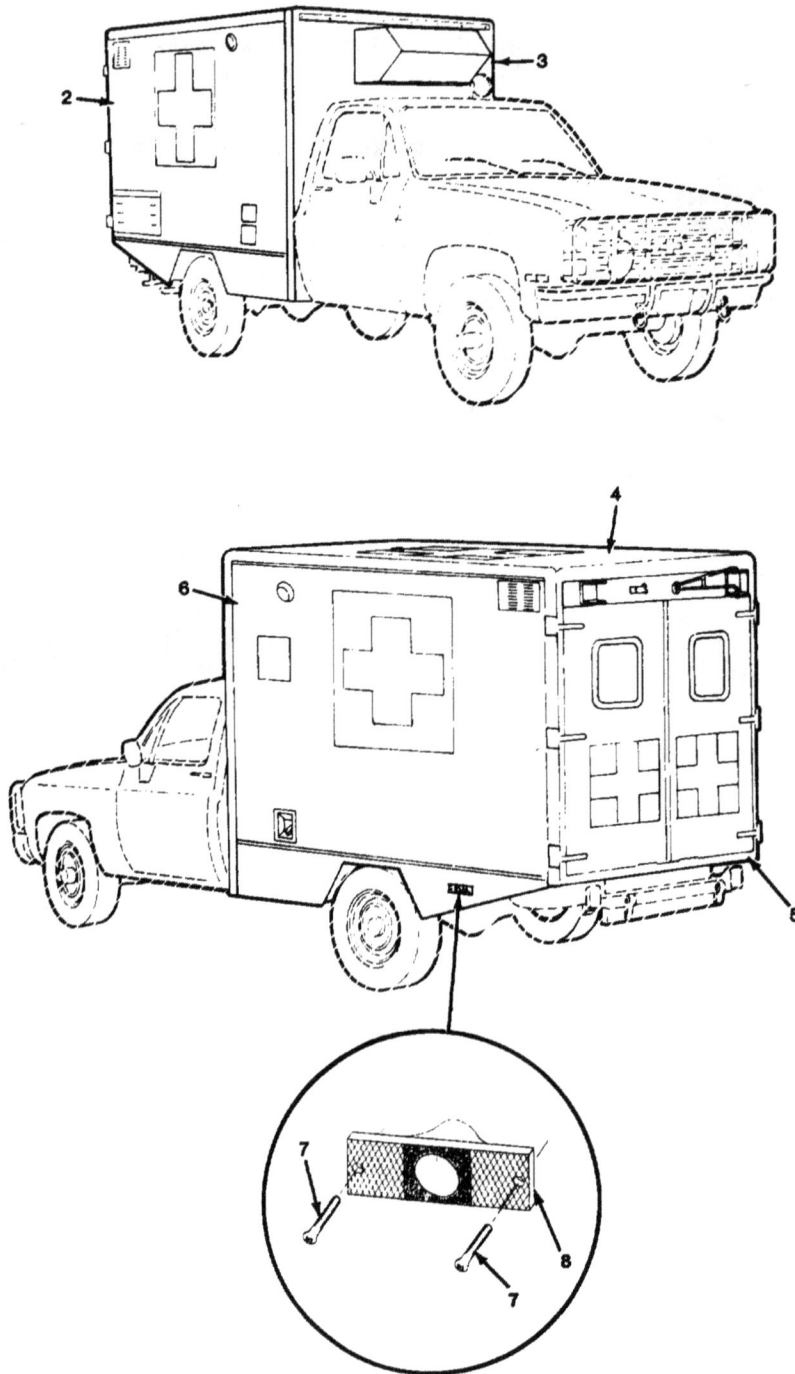

FIGURE 210. BODY ASSEMBLY (M1010).

TA511017

(1) (2) (3) (4) ITEM SMR　　　PART NO CODE CAGEC NUMBER	(5) DESCRIPTION AND USABLE ON CODES (UOC)	(6) QTY
	GROUP 1812 SPECIAL PURPOSES BODIES	
	FIG. 210 BODY ASSEMBLY (M1010)	
1　XAFFF 25022 99-4361-0	AMBULANCE BODY UOC:210	1
2　XAFFF 25022 99-4317-0	.BODYSIDE FRAME RH FOR COMPONENT PARTS SEE FIG'S 212,219 UOC:210	1
3　XAFFF 25022 99-4276-0	.FRONT FRAME ASM FOR COMPONENT PARTS SEE FIG 227 UOC:210	1
4　XAFFF 25022 99-4290-0	.ROOF ASSEMBLY FOR COMPONENT PARTS SEE FIG 213 UOC:210	1
5　XAFFF 25022 99-4299-0	.REAR FRAME ASM FOR COMPONENT PARTS SEE FIG 214 UOC:210	1
6　XAFFF 25022 99-4342-0	.BODYSIDE FRAME LH FOR COMPONENT PARTS SEE FIG'S 211,218 UOC:210	1
7　PAOZZ 25022 07-1016	.SCREW,TAPPING THRD UOC:210	4
8　PAOZZ 13548 99012R	.LENS,LIGHT RED MARKER LIGHT UOC:210	2

END OF FIGURE

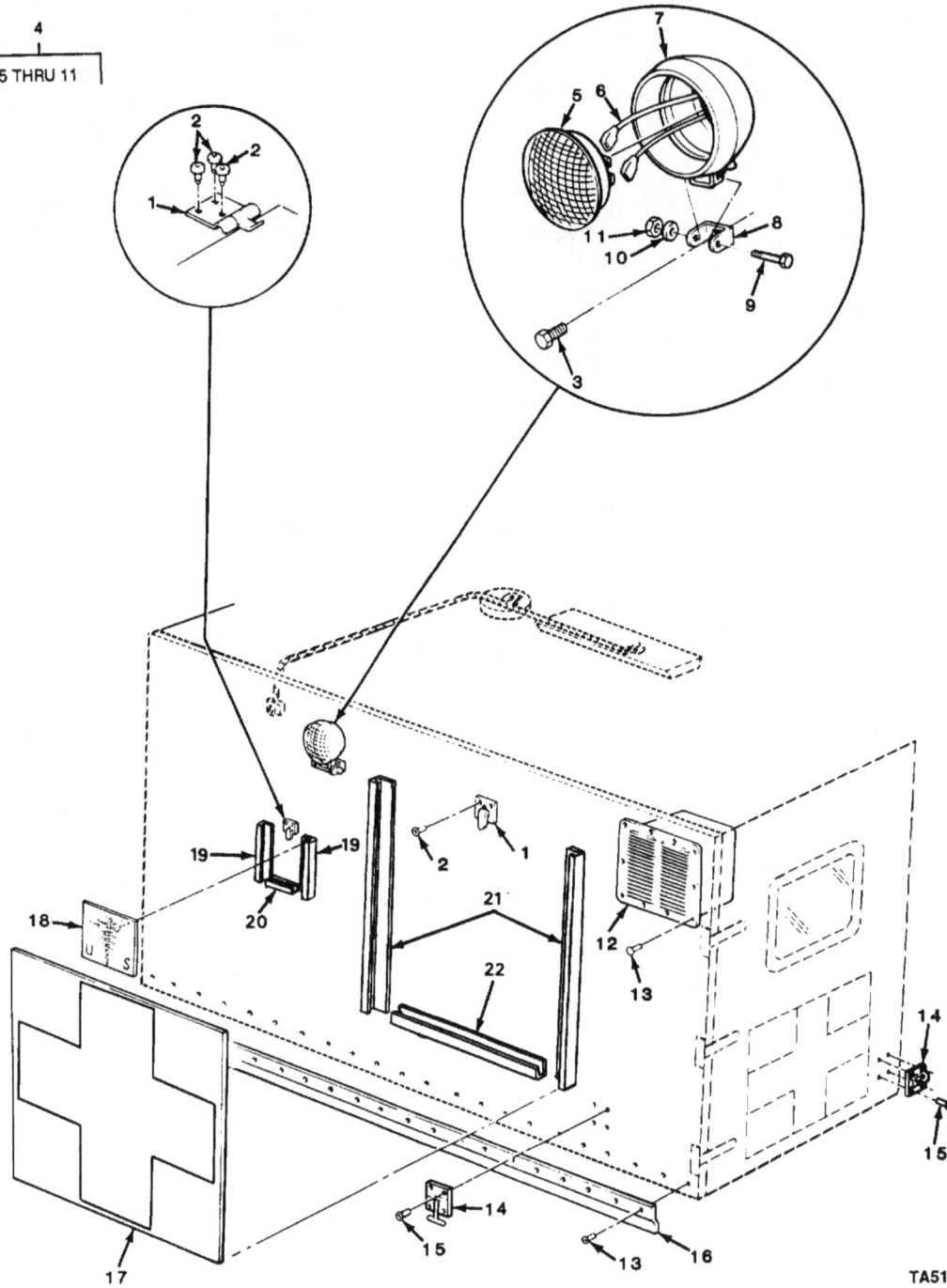

FIGURE 211. LEFT SIDE OUTER PANEL COMPONENT PARTS (M1010)

TA511018

(1)	(2)	(3)	(4)	(5)	(6)
ITEM NO	SMR CODE	CAGEC	PART NUMBER	DESCRIPTION AND USABLE ON CODES (UOC)	QTY

GROUP 1812 SPECIAL PURPOSE BODIES

FIG. 211 LEFT SIDE OUTER PANEL
COMPONENT PARTS (M1010)

Item	SMR	CAGEC	Part Number	Description	QTY
1	PAOZZ	25022	09-0954	CLIP,SPRING TENSION UOC:210	2
2	PAOZZ	96906	MS20600-B4W3	RIVET,BLIND UOC:210	6
3	PAOZZ	96906	MS90728-60	SCREW,CAP,HEXAGON H UOC:210	1
4	PAOOO	25022	19-0817	FLOODLIGHT,ELECTRIC UOC:210	1
5	PAOZZ	08806	4411	.LAMP,INCANDESCENT UOC:210	1
6	PAOZZ	25022	19-0884	.CABLE ASSEMBLY,SPEC UOC:210	2
7	PAOZZ	25022	19-0882	.HOUSING,FLOODLIGHT UOC:210	1
8	PAOZZ	25022	19-0885	.BRACKET,FLOODLIGHT UOC:210	1
9	PAOZZ	96906	MS90728-40	.BOLT,MACHINE UOC:210	1
10	PAOZZ	96906	MS35338-45	.WASHER,LOCK UOC:210	1
11	PAOZZ	96906	MS51967-5	.NUT,PLAIN,HEXAGON UOC:210	1
12	PAOZZ	25022	31-0030	VENTILATOR,AIR CIRC UOC:210	1
13	PAOZZ	96906	MS20600B6W4	RIVET,BLIND UOC:210	V
14	PAOZZ	25022	09-0093	LATCH SET,RIM UOC:210	1
15	PAOZZ	96906	MS21141-U0604	RIVET,BLIND UOC:210	8
16	PAOZZ	25022	99-4308-1	MOLDING,METAL UOC:210	1
17	PAOZZ	25022	23-0196	PLATE,DESIGNATION UOC:210	1
18	PAOZZ	25022	23-0193	PLATE,DESIGNATION UOC:210	1
19	PAOZZ	25022	99-4582-1	BRACKET,MOUNTING UOC:210	2
20	PAOZZ	25022	99-4581-1	BRACKET,MOUNTING UOC:210	1
21	PAOZZ	25022	99-4580-1	BRACKET,MOUNTING UOC:210	2
22	PAOZZ	25022	99-4579-1	BRACKET,MOUNTING UOC:210	1

END OF FIGURE

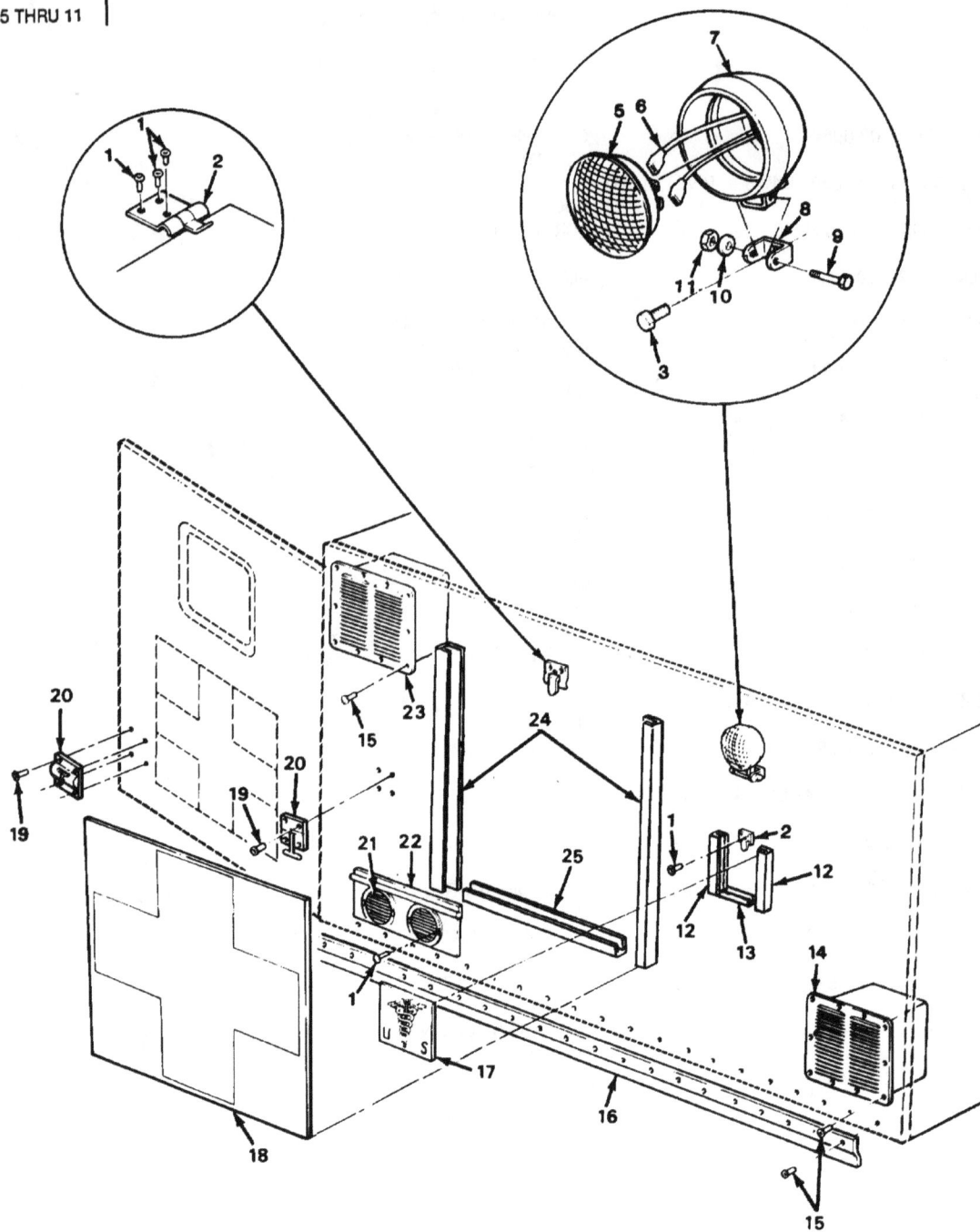

TA511019

FIGURE 212. RIGHT SIDE OUTER PANEL COMPONENT PARTS (M1010).

(1) (2) (3) (4)	(5)	(6)
ITEM SMR PART		
NO CODE CAGEC NUMBER	DESCRIPTION AND USABLE ON CODES (UOC)	QTY

GROUP 1812 SPECIAL PURPOSE BODIES

FIG. 212 RIGHT SIDE OUTER PANEL
COMPONENT PARTS (M1010)

ITEM NO	SMR CODE	CAGEC	PART NUMBER	DESCRIPTION AND USABLE ON CODES (UOC)	QTY
1	PAOZZ	96906	MS20600-B4W3	RIVET,BLIND	10
				UOC:210	
2	PAOZZ	25022	09-0954	CLIP,SPRING TENSION	2
				UOC:210	
3	PAOZZ	96906	MS90728-60	SCREW,CAP,HEXAGON H	1
				UOC:210	
4	PAOOO	25022	19-0817	FLOODLIGHT,ELECTRIC	1
				UOC:210	
5	PAOZZ	08806	4411	.LAMP,INCANDESCENT	1
				UOC:210	
6	PAOZZ	25022	19-0884	.CABLE ASSEMBLY,SPEC	2
				UOC:210	
7	PAOZZ	25022	19-0882	.HOUSING,FLOODLIGHT	1
				UOC:210	
8	PAOZZ	25022	19-0885	.BRACKET,FLOODLIGHT	1
				UOC:210	
9	PAOZZ	96906	MS90728-40	.BOLT,MACHINE	1
				UOC:210	
10	PAOZZ	96906	MS35338-45	.WASHER,LOCK	1
				UOC:210	
11	PAOZZ	96906	MS51967-5	.NUT,PLAIN,HEXAGON	1
				UOC:210	
12	PAOZZ	25022	99-4582-1	BRAKET,MOUNTING	2
				UOC:210	
13	PAOZZ	25022	99-4581-1	BRACKET,MOUNTING	1
				UOC:210	
14	PAOZZ	99688	78404	VENTILATOR,AIR CIRC	1
				UOC:210	
15	PAOZZ	96906	MS20600B6W4	RIVET,BLIND	V
				UOC:210	
16	PAOZZ	25022	99-4308-1	MOLDING,METAL	1
				UOC:210	
17	PAOZZ	25022	23-0193	PLATE,DESIGNATION	1
				UOC:210	
18	PAOZZ	25022	23-0196	PLATE,DESIGNATION	1
				UOC:210	
19	PAOZZ	96906	MS21141-U0604	RIVET,BLIND	8
				UOC:210	
20	PAOZZ	25022	09-0093	LATCH SET,RIM	1
				UOC:210	
21	PAOZZ	25022	22-0881	PLUG	1
				UOC:210	
22	PAOZZ	25022	99-4888-0	COVER,ACCESS	1
				UOC:210	
23	PAOZZ	25022	31-0030	VENTILATOR,AIR CIRC	1
				UOC:210	
24	PAOZZ	25022	99-4580-1	BRACKET,MOUNTING	2

(1) ITEM NO	(2) SMR CODE	(3) CAGEC	(4) PART NUMBER	(5) DESCRIPTION AND USABLE ON CODES (UOC)	(6) QTY
				UOC:210	
25	PAOZZ	25022	99-4579-1	BRACKET,MOUNTING	1
				UOC:210	
				END OF FIGURE	

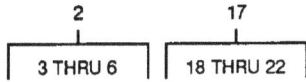

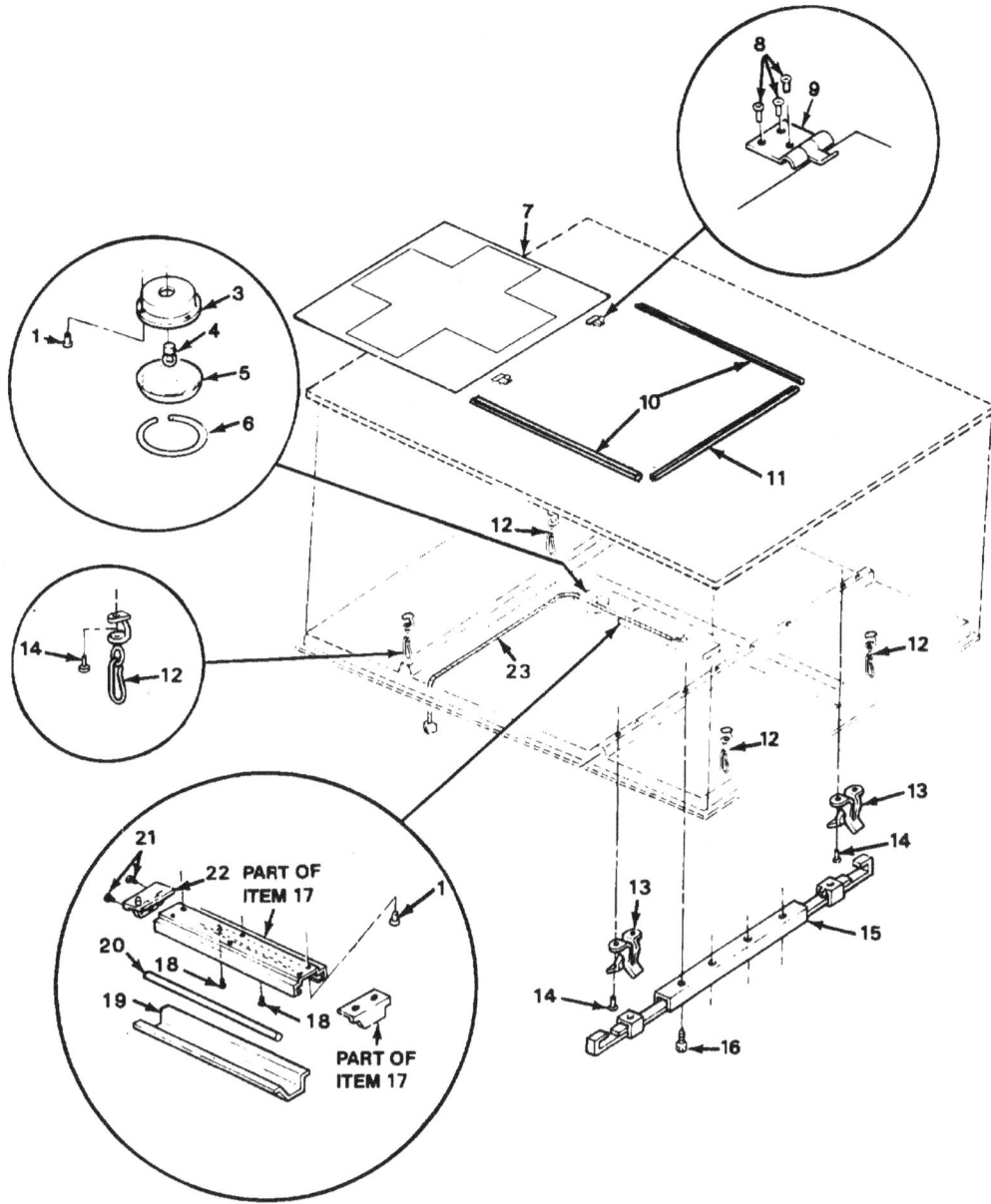

FIGURE 213. ROOF PANEL COMPONENT PARTS (M1010).

TA511020

(1) (2) (3) (4)	(5)	(6)
ITEM SMR PART		
NO CODE CAGEC NUMBER	DESCRIPTION AND USABLE ON CODES (UOC)	QTY

GROUP 1812 SPECIAL PURPOSE BODIES

FIG. 213 ROOF PANEL COMPONENT PARTS
(M1010)

1	PAOZZ	96906	MS20600B6W4	RIVET,BLIND	9
				UOC:210	
2	AOOOO	25022	19-0445	LIGHT,DOME	1
				UOC:210	
3	PAOZZ	25022	19-0893	.LIGHT,DOME	1
				UOC:210	
4	PAOZZ	08806	1157	.LAMP,INCANDESCENT	1
				UOC:210	
5	PAOZZ	25022	19-0861	.LENS,LIGHT	1
				UOC:210	
6	PAOZZ	25022	19-0969	.RING,RETAINING	1
				UOC:210	
7	PAOZZ	25022	23-0198	PLATE,DESIGNATION	1
				UOC:210	
8	PAOZZ	96906	MS20600-B4W3	RIVET,BLIND	6
				UOC:210	
9	PAOZZ	25022	09-0954	CLIP,SPRING TENSION	2
				UOC:210	
10	PAOZZ	25022	99-4577-1	BRACKET,MOUNTING	2
				UOC:210	
11	PAOZZ	25022	99-4578-1	BRACKET,MOUNTING	1
				UOC:210	
12	PAOZZ	25022	99-1088-0	SWIVEL,LINK AND LIN	4
				UOC:210	
13	PAOZZ	25022	09-0959	CLIP,SPRING TENSION	2
				UOC:210	
14	PAOZZ	96906	MS20600-B4W4	RIVET,BLIND	8
				UOC:210	
15	PAOZZ	25022	22-0875	HANGER,UPPER LITTER	1
				UOC:210	
16	PAOZZ	96906	MS90728-29	BOLT,MACHINE	4
				UOC:210	
17	PAOOO	25022	19-0818	LIGHT,DOME	1
				UOC:210	
18	PAOZZ	96906	MS35206-243	.SCREW,MACHINE	2
				UOC:210	
19	PAOZZ	25022	51-1942	.LENS,LIGHT	1
				UOC:210	
20	PAOZZ	08805	F48T12/CW/WM	.LAMP,FLOURESCRENT	1
				UOC:210	
21	PAOZZ	24617	9414238	.SCREW	2
				UOC:210	
22	PAOZZ	25022	51-1991	.INVERTER,POWER,STAT	1
				UOC:210	
23	PAOZZ	25022	19-0820	WIRING HARNESS,BRAN	1
				UOC:210	

END OF FIGURE

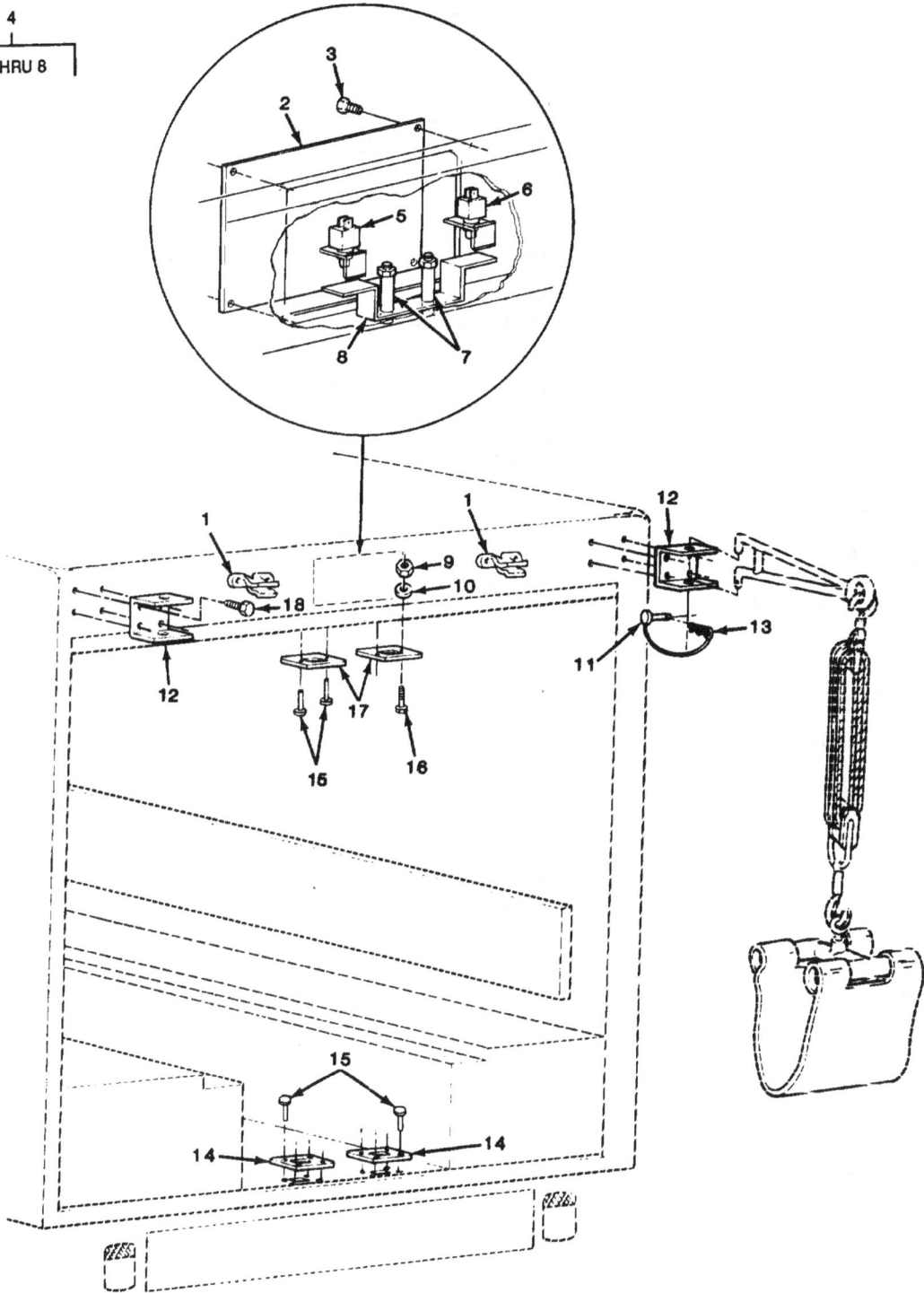

FIGURE 214. REAR FRAME COMPONENT PARTS (M1010).

TA511021

(1) ITEM NO	(2) SMR CODE	(3) CAGEC	(4) PART NUMBER	(5) DESCRIPTION AND USABLE ON CODES (UOC)	(6) QTY
				GROUP 1812 SPECIAL PURPOSE BODIES	
				FIG. 214 REAR FRAME COMPONENT PARTS (M1010)	
1	PAOZZ	25022	09-0959	CLIP,SPRING TENSION UOC:210	2
2	PAOZZ	25022	99-4356-1	PLATE,MENDING UOC:210	1
3	PAOZZ	24617	9414714	SCREW,MACHINE UOC:210	4
4	PAOOO	25022	99-4256-0	SWITCH,TOGGLE UOC:210	1
5	PAOZZ	25022	51-1902	.SWITCH,PUSH UOC:210	1
6	PAOZZ	25022	51-1901	.SWITCH,PUSH UOC:210	1
7	PAOZZ	25022	06-0075	.TUBE,METALLIC UOC:210	2
8	PAOZZ	25022	99-4256-2	.BRACKET,HEATER UOC:210	1
9	PAOZZ	96906	MS35649-202	NUT,PLAIN,HEXAGON UOC:210	2
10	PAOZZ	96906	MS35338-43	WASHER,LOCK UOC:210	2
11	PAOZZ	11862	3929059	PIN,LOCK UOC:210	1
12	PAOZZ	25022	99-4315-1	BRACKET UOC:210	2
13	PAOZZ	25022	09-0963	LANYARD,VEHICULAR UOC:210	1
14	PAOZZ	25022	51-1305	FRAME SECTION,STRUC UOC:210	2
15	PAOZZ	96906	MS21141-U0604	RIVET,BLIND UOC:210	10
16	PAOZZ	96906	MS51957-71	SCREW,MACHINE UOC:210	2
17	PAOZZ	25022	51-1304	STRIKE,CATCH UOC:210	2
18	PAOZZ	96906	MS90728-36	BOLT,MACHINE UOC:210	8

END OF FIGURE

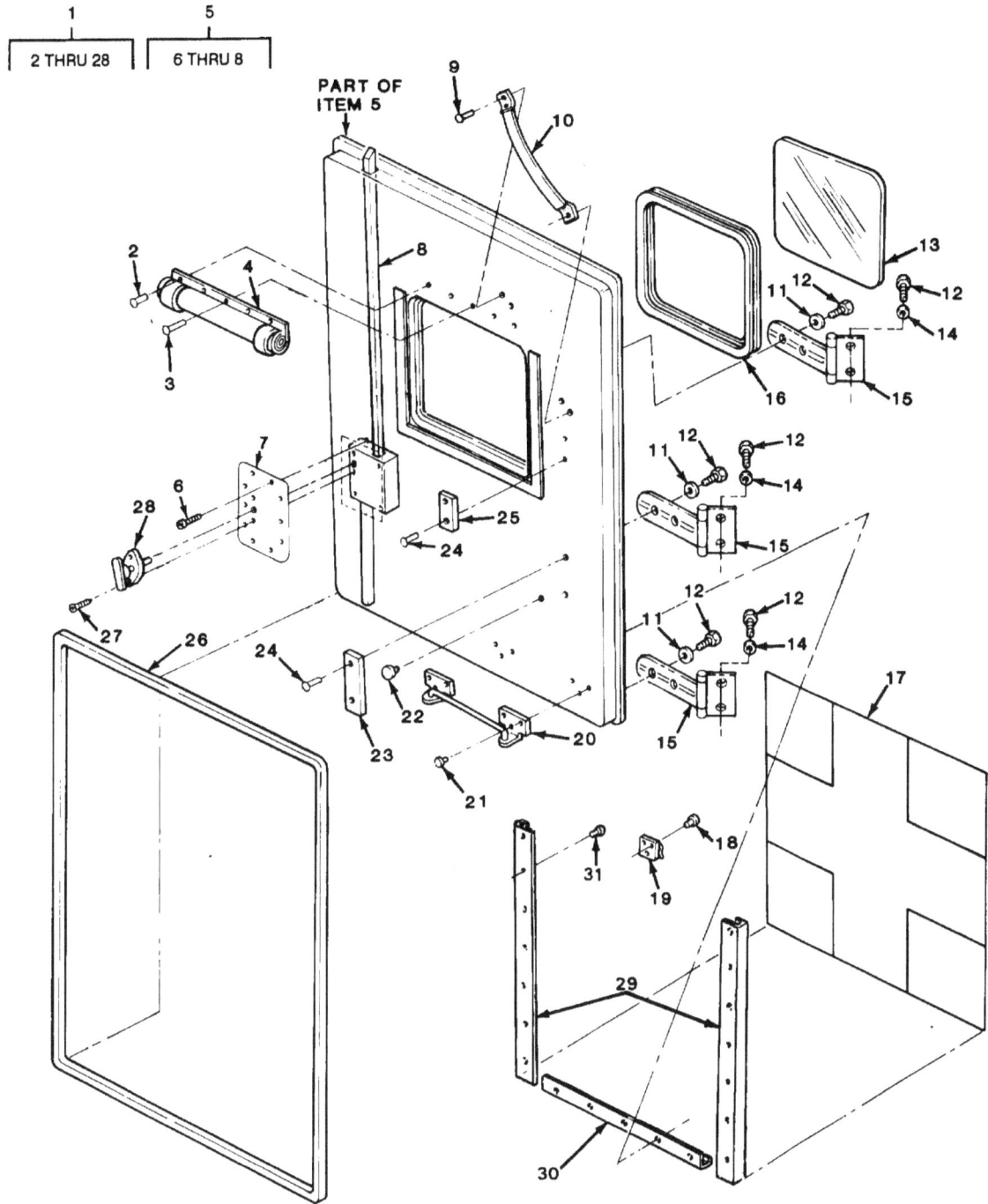

FIGURE 215. LEFT REAR DOOR ASSEMBLY (M1010).

TA511022

(1) ITEM NO	(2) SMR CODE	(3) CAGEC	(4) PART NUMBER	(5) DESCRIPTION AND USABLE ON CODES (UOC)	(6) QTY
				GROUP 1812 SPECIAL PURPOSE BODIES	
				FIG. 215 LEFT REAR DOOR ASSEMBLY (M1010)	
1	XDOFF	25022	99-1265-01	DOOR,METAL,SWINGING UOC:210	1
2	PAOZZ	96906	MS20600-B6W5	.RIVET,BLIND UOC:210	4
3	PAOZZ	96906	MS20600-B4W4	.RIVET,BLIND UOC:210	1
4	PAOZZ	25022	51-2294	.CURTAIN,BLACKOUT UOC:210	1
5	PAOOO	25022	13-1105	.DOOR,METAL,SWINGING INCLUDES ITEMS 11,12,15 UOC:210	1
6	PAOZZ	24617	447143	..SCREW,TAPPING,THREA UOC:210	10
7	PAOZZ	25022	51-2204	..PANEL,ACCESS DOOR UOC:210	1
8	PAOZZ	25022	51-0999	..BAR,LOCK,DOOR,VEHIC UOC:210	1
9	PAOZZ	96906	MS20600MP6W10	.RIVET,BLIND UOC:210	4
10	PAOZZ	25022	99-4675-1	.STRAP UOC:210	1
11	PAOZZ	96906	MS27183-13	.WASHER,FLAT UOC:210	6
12	PAOZZ	96906	MS90728-60	.SCREW,CAP,HEXAGON H UOC:210	12
13	PAOZZ	25022	51-1603	.WINDOW,OBSERVATION UOC:210	1
14	PAOZZ	96906	MS35338-46	.WASHER,LOCK UOC:210	6
15	PAOZZ	25022	51-0998	.HINGE,TEE UOC:210	3
16	PAOZZ	25022	51-1721	.FRAME,WINDOW DOOR,V UOC:210	1
17	PAOZZ	25022	23-0200	.PLATE,DESIGNATION UOC:210	1
18	PAOZZ	96906	MS20600-B4W3	.RIVET,BLIND UOC:210	3
19	PAOZZ	25022	09-0954	.CLIP,SPRING TENSION UOC:210	1
20	PAOZZ	25022	52-0914	.HOLDER,BACK,DOOR,VE UOC:210	1
21	PAOZZ	96906	MS21141-U0604	.RIVET,BLIND UOC:210	6
22	PAOZZ	25022	17-0209	.BUMPER,VEHICULAR UOC:210	1
23	PAOZZ	25022	17-0178	.BUMPR UOC:210	1

(1) ITEM NO	(2) SMR CODE	(3) CAGEC	(4) PART NUMBER	(5) DESCRIPTION AND USABLE ON CODES (UOC)	(6) QTY
24	PAOZZ	96906	MS24662-153	.RIVET,BLIND UOC:210	4
25	PAOZZ	25022	17-0177	.BUMPER UOC:210	1
26	PAOZZ	25022	17-0202	.WEATHER STRIP UOC:210	1
27	PAOZZ	24617	165079	.SCREW,MACHINE UOC:210	2
28	PAOZZ	25022	51-0988	.HANDLE,DOOR UOC:210	1
29	PAOZZ	25022	99-4583-1	BRACKET,MOUNTING UOC:210	2
30	PAOZZ	25022	99-4584-1	BRACKET,MOUNTING UOC:210	1
31	PAOZZ	96906	MS20600B6W4	RIVET,BLIND UOC:210	V

END OF FIGURE

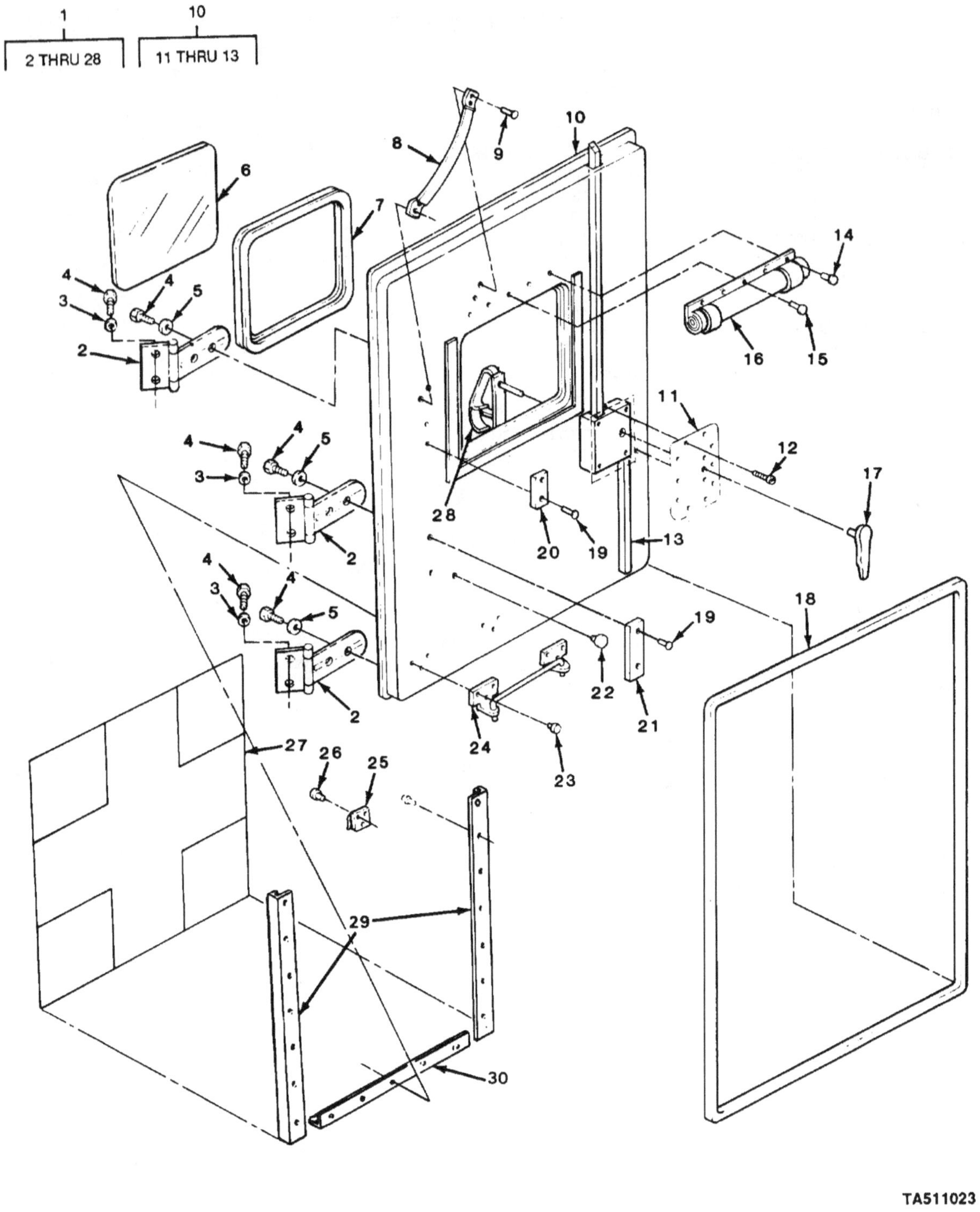

FIGURE 216. RIGHT REAR DOOR ASSEMBLY (M1010).

TA511023

(1) ITEM NO	(2) SMR CODE	(3) CAGEC	(4) PART NUMBER	(5) DESCRIPTION AND USABLE ON CODES (UOC)	(6) QTY
				GROUP 1812 SPECIAL PURPOSE BODIES	
				FIG. 216 RIGHT REAR DOOR ASSEMBLY (M1010)	
1	XDOFF	25022	99-1265-02	DOOR,METAL,SWINGING UOC:210	1
2	PAOZZ	25022	51-0998	.HINGE,TEE UOC:210	3
3	PAOZZ	96906	MS35338-46	.WASHER,LOCK UOC:210	6
4	PAOZZ	96906	MS90728-60	.SCREW,CAP,HEXAGON H UOC:210	12
5	PAOZZ	96906	MS27183-13	.WASHER,FLAT UOC:210	6
6	PAOZZ	25022	51-1603	.WINDOW,OBSERVATION UOC:210	1
7	PAOZZ	25022	51-1721	.FRAME,WINDOW DOOR,V UOC:210	1
8	PAOZZ	25022	99-4675-1	.STRAP UOC:210	1
9	PAOZZ	96906	MS20600MP6W10	.RIVET,BLIND UOC:210	4
10	PAOOO	25022	13-1104	.DOOR,METAL,SWINGING INCLUDES ITEMS 2,4,5 UOC:210	1
11	PAOZZ	25022	51-2204	..PANEL,ACCESS DOOR UOC:210	1
12	PAOZZ	24617	447143	..SCREW,TAPPING,THREA UOC:210	10
13	PAOZZ	25022	51-0999	..BAR,LOCK,DOOR,VEHIC UOC:210	1
14	PAOZZ	96906	MS20600-B6W5	.RIVET,BLIND UOC:210	4
15	PAOZZ	96906	MS20600-B4W4	.RIVET,BLIND UOC:210	1
16	PAOZZ	25022	51-2294	.CURTAIN,BLACKOUT UOC:210	1
17	PAOZZ	25022	51-0989	.HANDLE,DOOR UOC:210	1
18	PAOZZ	25022	17-0201	.WEATHER STRIP UOC:210	1
19	PAOZZ	96906	MS24662-153	.RIVET,BLIND UOC:210	4
20	PAOZZ	25022	17-0177	.BUMPER UOC:210	1
21	PAOZZ	25022	17-0178	.BUMPR UOC:210	1
22	PAOZZ	25022	17-0209	.BUMPER,VEHICULAR UOC:210	1
23	PAOZZ	96906	MS21141-UO604	.RIVET,BLIND UOC:210	6

(1)	(2)	(3)	(4)	(5)	(6)
ITEM NO	SMR CODE	CAGEC	PART NUMBER	DESCRIPTION AND USABLE ON CODES (UOC)	QTY
24	PAOZZ	25022	52-0914	.HOLDER,BACK,DOOR,VE UOC:210	1
25	PAOZZ	25022	09-0954	.CLIP,SPRING TENSION UOC:210	1
26	PAOZZ	96906	MS20600-B4W3	.RIVET,BLIND UOC:210	3
27	PAOZZ	25022	23-0200	.PLATE,DESIGNATION UOC:210	1
28	PAOZZ	25022	99-4680-1	.HANDLE,BAIL UOC:210	1
29	PAOZZ	25022	99-4583-1	BRACKET,MOUNTING UOC:210	2
30	PAOZZ	25022	99-4584-1	BRACKET,MOUNTING UOC:210	1
31	PAOZZ	96906	MS20600B6W4	RIVET,BLIND UOC:210	V

END OF FIGURE

11
12 THRU 19

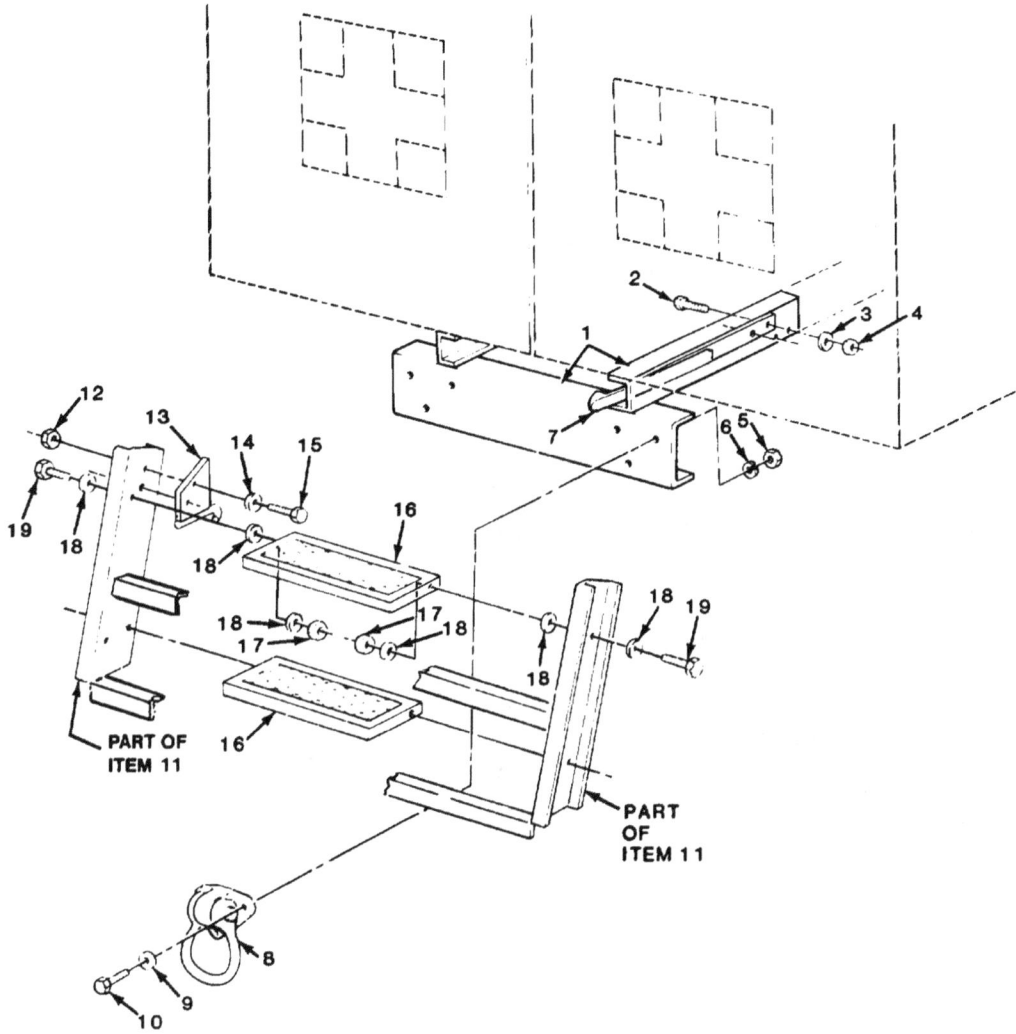

FIGURE 217. REAR BUMPER AND STEP ASSEMBLY (M1010).

TA511024

(1) (2) (3) (4)	(5)	(6)
ITEM SMR PART		
NO CODE CAGEC NUMBER	DESCRIPTION AND USABLE ON CODES (UOC)	QTY

GROUP 1812 SPECIAL PURPOSE BODIES

FIG. 217 REAR BUMPER AND STEP
ASSEMBLY (M1010)

Item	SMR	CAGEC	Part Number	Description	Qty
1	XDFZZ	25022	99-4358-0	BUMPER,VEHICULAR UOC:210	1
2	PAOZZ	96906	MS90728-8	SCREW,CAP,HEXAGON H UOC:210	2
3	PAOZZ	96906	MS27183-9	WASHER,FLAT UOC:210	4
4	PAOZZ	96906	MS35691-1	NUT,PLAIN,HEXAGON UOC:210	2
5	PAOZZ	96906	MS51967-14	NUT,PLAIN,HEXAGON UOC:210	6
6	PAOZZ	96906	MS35340-48	WASHER,LOCK UOC:210	6
7	PAOZZ	25022	09-0955	RETAINER,SPRING UOC:210	1
8	PAOZZ	74410	TH-0681	RING,HITCH UOC:210	2
9	PAOZZ	96906	MS27183-17	WASHER,FLAT UOC:210	6
10	PAOZZ	96906	MS90728-114	SCREW,CAP,HEXAGON H UOC:210	6
11	PAOOO	25022	99-4340-0	LADDER,VEHICLE BOAR UOC:210	1
12	PAOZZ	96906	MS35691-1	.NUT,PLAIN,HEXAGON UOC:210	2
13	PAOZZ	25022	99-4469-0	.HOOK,LADDER,VEHICLE UOC:210	1
14	PAOZZ	96906	MS27183-9	.WASHER,FLAT UOC:210	2
15	PAOZZ	96906	MS90728-8	.SCREW,CAP,HEXAGON H UOC:210	2
16	PAOZZ	25022	99-4456-0	.STEP,VEHICULAR UOC:210	2
17	PAOZZ	96906	MS35691-9	.NUT,PLAIN,HEXAGON UOC:210	4
18	PAOZZ	96906	MS27183-11	.WASHER,FLAT UOC:210	12
19	PAOZZ	96906	MS90728-34	.BOLT,MACHINE UOC:210	4

END OF FIGURE

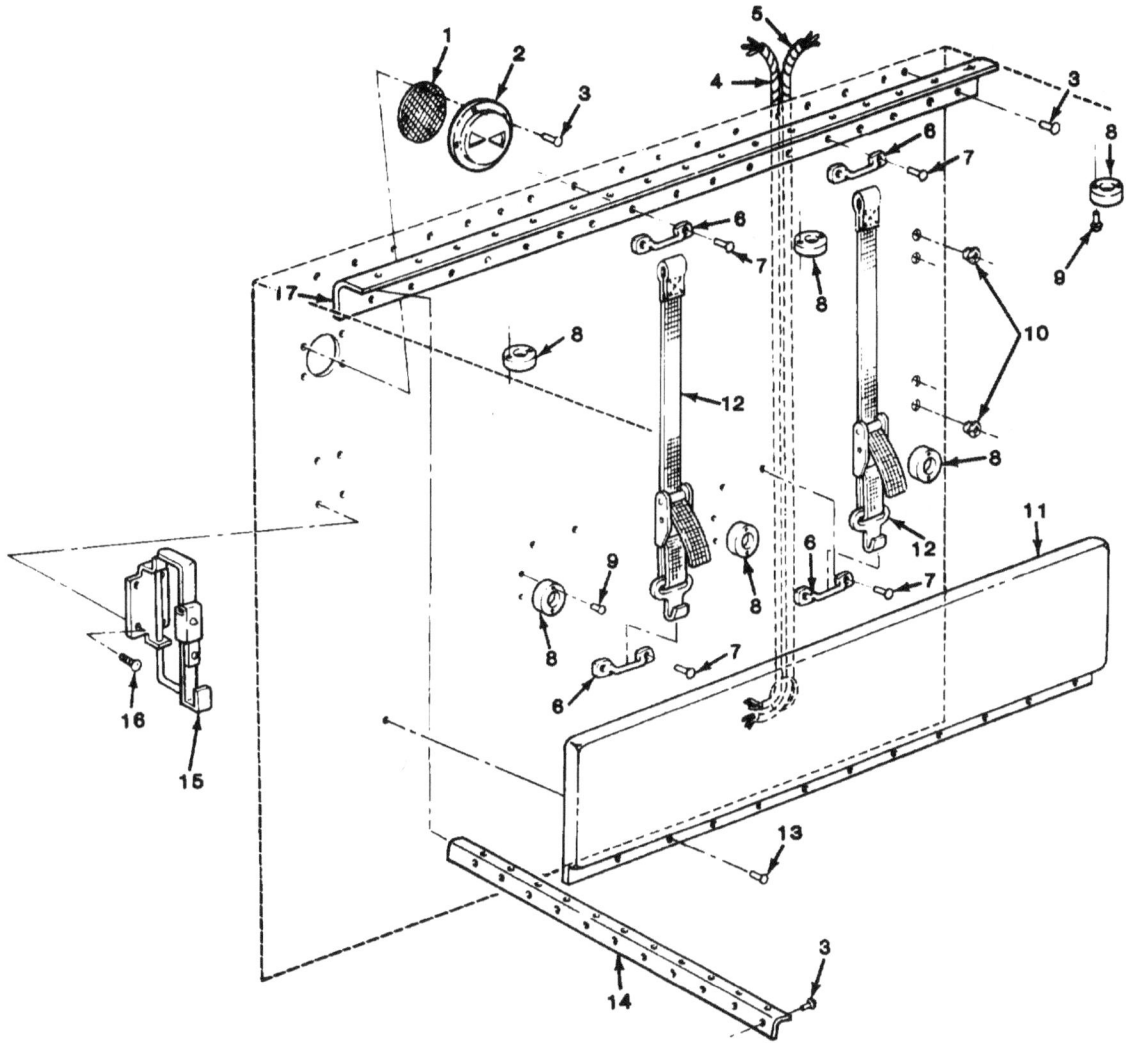

TA511025

FIGURE 218. LEFT SIDE INNER PANEL RELATED PARTS (M1010).

(1) (2) (3) (4) ITEM SMR PART NO CODE CAGEC NUMBER	(5) DESCRIPTION AND USABLE ON CODES (UOC)	(6) QTY
	GROUP 1812 SPECIAL PURPOSE BODIES	
	FIG. 218 LEFT SIDE INNER PANEL RELATED PARTS (M1010)	
1 PAOZZ 25022 22-0832	STRAINER ELEMENT,SE UOC:210	1
2 PAOZZ 25022 22-0884	VENTILATOR,AIR CIRC INCLUDES ITEM 1 UOC:210	1
3 PAOZZ 96906 MS20600B6W4 RIVET,BLIND	UOC:210	51
4 PAFZZ 25022 19-0819	WIRING HARNESS,BRAN UOC:210	1
5 PAFZZ 25022 19-0831	WIRING HARNESS,BRAN UOC:210	1
6 PAOZZ 25022 51-0979	STRAP,RETAINING UOC:210	4
7 PAOZZ 96906 MS21141-U0604 RIVET,BLIND	UOC:210	8
8 PAOZZ 25022 19-0942	BASE,LAMPHOLDER UOC:210	6
9 PAOZZ 96906 MS20600-B6W10 RIVET,BLIND	UOC:210	12
10 PAOZZ 25022 19-0849	STRAIN RELIEF UOC:210	4
11 PAOZZ 25022 22-0861	PAD,CUSHIONING UOC:210	1
12 PAOZZ 25022 22-0877	STRAP,RETAINING UOC:210	2
13 PAOZZ 96906 MS24662-155 RIVET,BLIND	UOC:210	14
14 PAOZZ 25022 99-4420-2	MOLDING,FRONT UOC:210	1
15 PAOZZ 25022 09-0965	BRACKET,PIVOT UOC:210	1
16 PAOZZ 24617 171108	SCREW,TAPPING,THREA UOC:210	4
17 PAOZZ 25022 99-4420-1	MOLDING,METAL UOC:210	1
	END OF FIGURE	

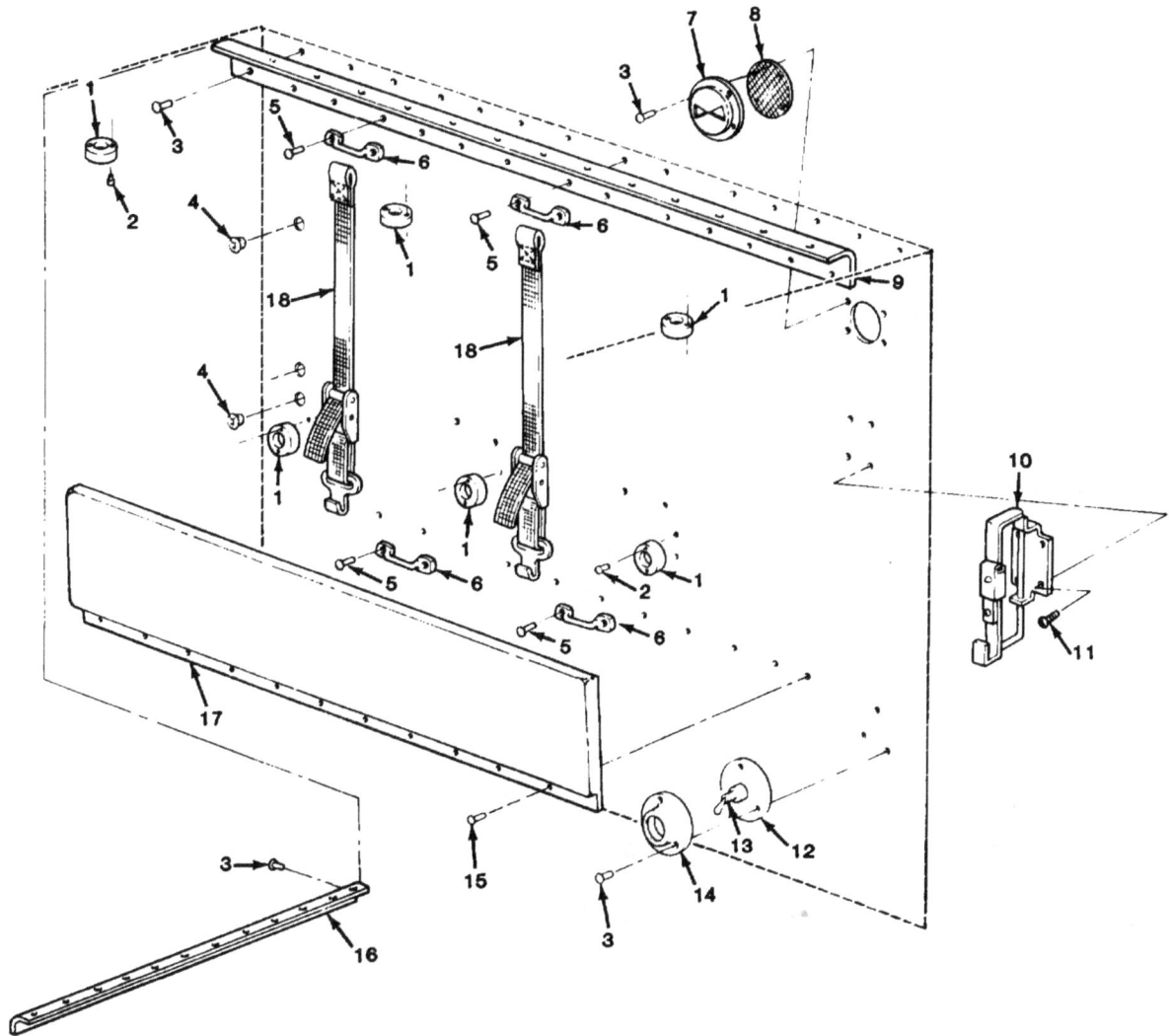

TA511026

FIGURE 219. RIGHT SIDE INNER PANEL RELATED PARTS (M1010).

(1) ITEM NO	(2) SMR CODE	(3) CAGEC	(4) PART NUMBER	(5) DESCRIPTION AND USABLE ON CODES (UOC)	(6) QTY
				GROUP 1812 SPECIAL PURPOSE BODIES	
				FIG. 219 RIGHT SIDE INNER PANEL RELATED PARTS (M1010)	
1	PAOZZ	25022	19-0942	BASE,LAMPHOLDER UOC:210	6
2	PAOZZ	96906	MS20600-B6W10 RIVET,BLIND UOC:210		12
3	PAOZZ	96906	MS20600B6W4 RIVET,BLIND UOC:210		54
4	PAOZZ	25022	19-0849	STRAIN RELIEF UOC:210	3
5	PAOZZ	96906	MS21141-U0604 RIVET,BLIND UOC:210		8
6	PAOZZ	25022	51-0979	STRAP,RETAINING UOC:210	4
7	PAOZZ	25022	22-0884	VENTILATOR,AIR CIRC INCLUDES ITEM #8 UOC:210	1
8	PAOZZ	25022	22-0832	STRAINER ELEMENT,SE UOC:210	1
9	PAOZZ	25022	99-4420-1	MOLDING,METAL UOC:210	1
10	PAOZZ	25022	09-0965	BRACKET,PIVOT UOC:210	1
11	PAOZZ	24617	171108	SCREW,TAPPING,THREA UOC:210	4
12	PAOZZ	25022	37-0020	PLATE,SWITCH UOC:210	1
13	PAOZZ	25022	51-1959	SWITCH,TOGGLE UOC:210	1
14	PAOZZ	25022	37-0019	COVER,ELECTRICAL SW UOC:210	1
15	PAOZZ	96906	MS24662-155 RIVET,BLIND UOC:210		14
16	PAOZZ	25022	99-4420-2	MOLDING,FRONT UOC:210	1
17	PAOZZ	25022	22-0861	PAD,CUSHIONING UOC:210	1
18	PAOZZ	25022	22-0877	STRAP,RETAINING UOC:210,230	2

END OF FIGURE

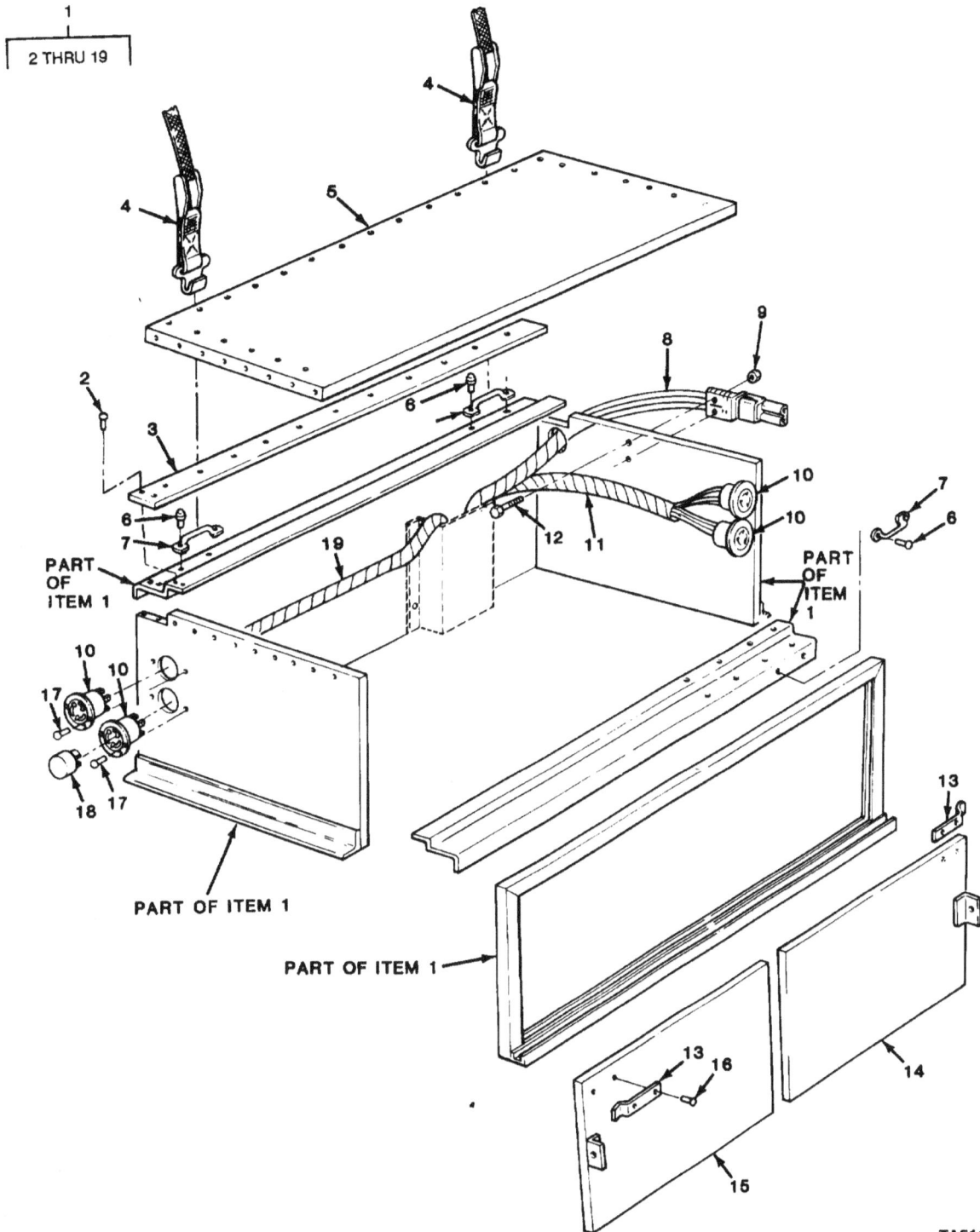

FIGURE 220. LEFT SIDE STOWAGE BOX ASSEMBLY AND COMPONENT PARTS (M1010).

TA511027

(1) ITEM NO	(2) SMR CODE	(3) CAGEC	(4) PART NUMBER	(5) DESCRIPTION AND USABLE ON CODES (UOC)	(6) QTY
				GROUP 1812 SPECIAL PURPOSE BODIES	
				FIG. 220 LEFT SIDE STOWAGE BOX ASSEMBLY AND COMPONENT PARTS (M1010)	
1	XAOFF	25022	99-4447-0	BOX,VEHICULAR ACCES UOC:210	1
2	PAOZZ	96906	MS20601-B6W6	.RIVET,BLIND UOC:210	30
3	PAOZZ	25022	51-2297	.PLASTIC STRIP UOC:210	2
4	PAOZZ	25022	22-0880	.STRAP,WEBBING,LITTE UOC:210	4
5	XAOZZ	25022	22-0867	.LITTER,VEHICULAR MC UOC:210	1
6	PAOZZ	96906	MS21141-U0604	.RIVET,BLIND UOC:210	8
7	PAOZZ	25022	51-0979	.STRAP,RETAINING UOC:210	4
8	PAFZZ	25022	19-0837	.LEAD ASSEMBLY,ELECT UOC:210	1
9	PAOZZ	96906	MS35691-1	.NUT,PLAIN,HEXAGON UOC:210	2
10	PAFZZ	25022	19-0843	.CYLINDER,LOCK,VEHIC UOC:210	4
11	PAFZZ	25022	19-0881	.WIRING HARNESS UOC:210	1
12	PAOZZ	96906	MS90728-10	.SCREW,CAP,HEXAGON H UOC:210	2
13	PAOZZ	25022	51-0991	.CLIP,SPRING TENSION UOC:210	2
14	PAOZZ	25022	99-4894-02	.DOOR,HATCH,VEHICLE UOC:210	1
15	PAOZZ	25022	99-4894-01	.DOOR,HATCH,VEHICLE UOC:210	1
16	PAOZZ	96906	MS20600B6W4	.RIVET,BLIND UOC:210	4
17	PAOZZ	96906	MS20600-B4W4	.RIVET,BLIND UOC:210	8
18	PAOZZ	81348	WC596/9-1	.CONNECTOR,PLUG,ELEC UOC:210	4
19	PAFZZ	25022	19-0880	.WIRING HARNESS UOC:210	1

END OF FIGURE

1

2 THRU 11

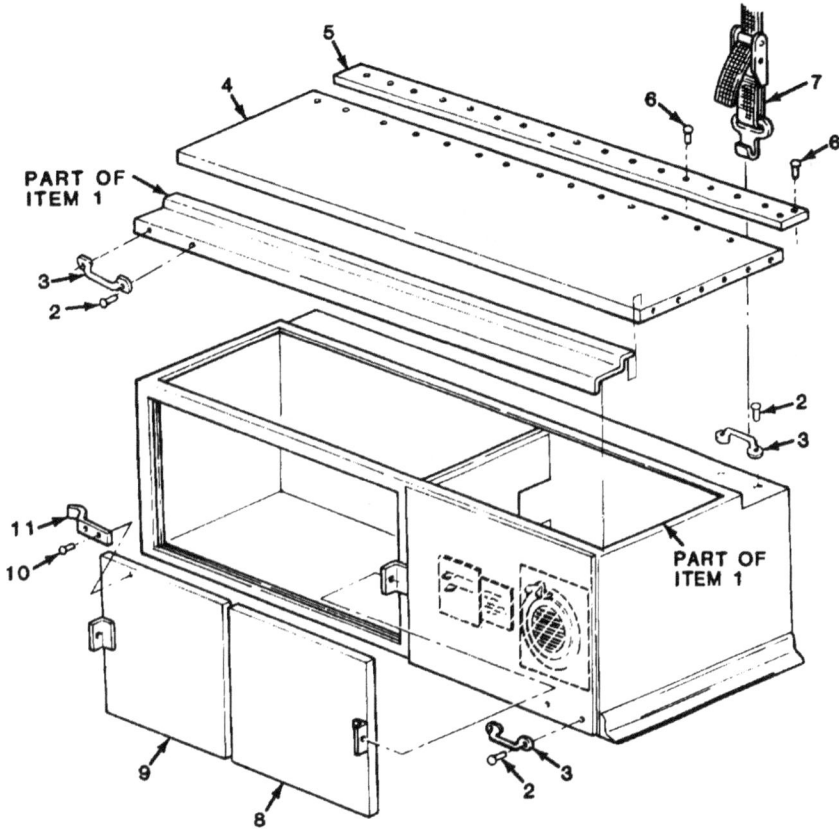

FIGURE 221. RIGHT SIDE STOWAGE BOX ASSEMBLY AND COMPONENT PARTS (M1010).

(1) ITEM NO	(2) SMR CODE	(3) CAGEC	(4) PART NUMBER	(5) DESCRIPTION AND USABLE ON CODES (UOC)	(6) QTY
				GROUP 1812 SPECIAL PURPOSE BODIES	
				FIG. 221 RIGHT SIDE STOWAGE BOX ASSEMBLY AND COMPONENT PARTS (M1010)	
1	XAFFF	25022	99-4325-0	BOX,VEHICULAR ACCES UOC:210	1
2	PAOZZ	96906	MS21141-U0604	.RIVET,BLIND UOC:210	12
3	PAOZZ	25022	51-0979	.STRAP,RETAINING UOC:210	6
4	XAFZZ	25022	22-0866	.LITTER,VEHICULAR UOC:210	1
5	PAOZZ	25022	51-2297	.PLASTIC STRIP UOC:210	2
6	PAOZZ	96906	MS20601-B6W6	.RIVET,BLIND UOC:210	30
7	PAOZZ	25022	22-0880	.STRAP,WEBBING,LITTE UOC:210	4
8	PAOZZ	25022	99-4893-01	.DOOR,HATCH,VEHICLE UOC:210	1
9	PAOZZ	25022	99-4893-02	.DOOR,HATCH,VEHICLE UOC:210	1
10	PAOZZ	96906	MS20600B6W4	.RIVET,BLIND UOC:210	4
11	PAOZZ	25022	51-0991	.CLIP,SPRING TENSION UOC:210	2

END OF FIGURE

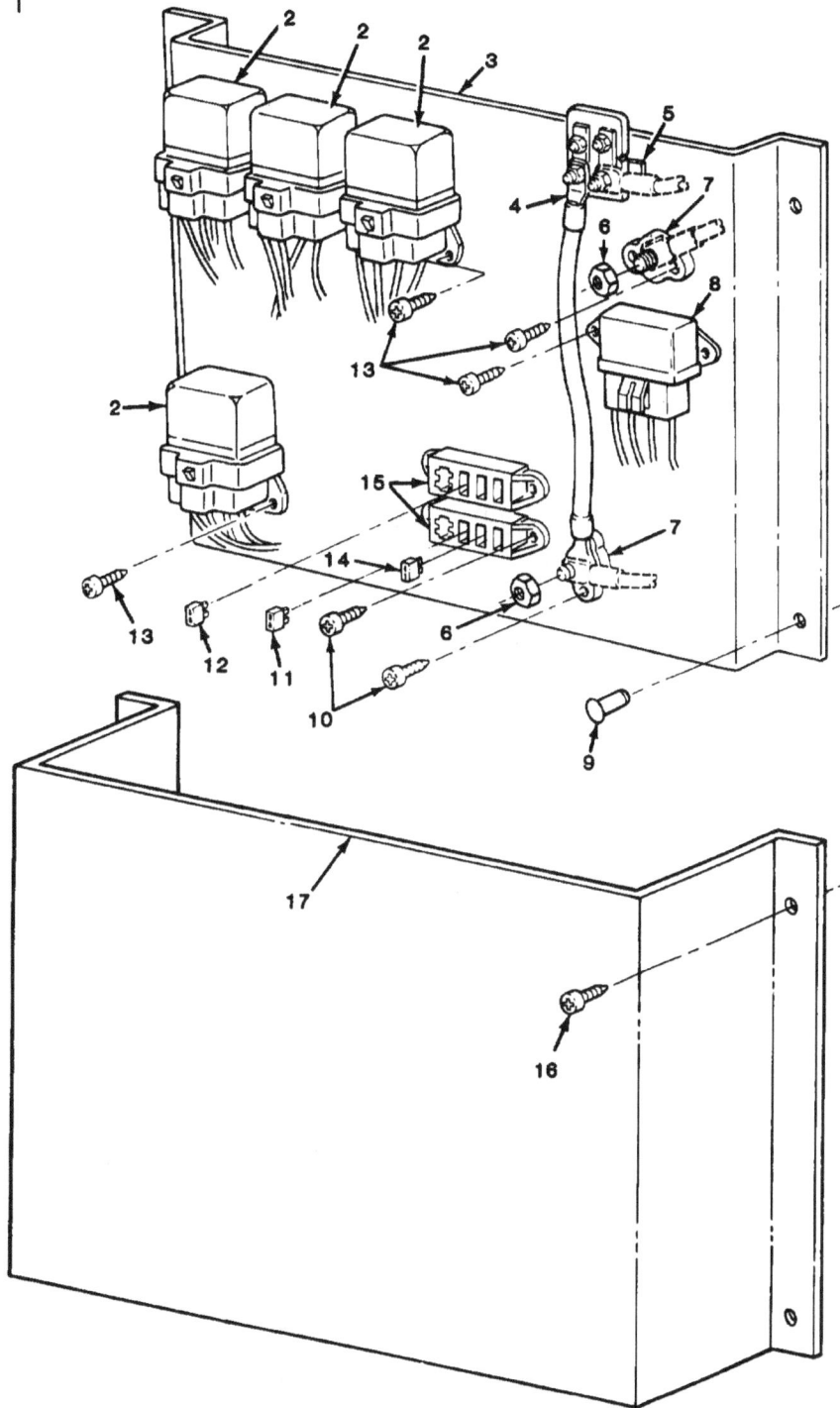

FIGURE 222. RELAY PANEL ASSEMBLY COMPONENTS AND RELATED PARTS (M1010).

TA511029

(1) ITEM NO	(2) SMR CODE	(3) CAGEC	(4) PART NUMBER	(5) DESCRIPTION AND USABLE ON CODES (UOC)	(6) QTY
				GROUP 1812 SPECIAL PURPOSE BODIES	
				FIG. 222 RELAY PANEL ASSEMBLY COMPONENTS AND RELATED PARTS (M1010)	
1	PAFFF	25022	19-0833	POWER SUPPLY SUBASS UOC:210	1
2	PAOZZ	25022	19-0875	.RELAY,ELECTROMAGNET UOC:210	4
3	PAFZZ	25022	99-4519-1	.RELAY UOC:210	1
4	PAOZZ	25022	19-0910	.CABLE ASSEMBLY,SPEC UOC:210	1
5	PAOZZ	25022	19-0909	.CIRCUIT BREAKER UOC:210	1
6	PAOZZ	25022	07-01-01	.NUT,PLAIN,HEXAGON UOC:210	2
7	PAOZZ	25022	19-0877	.TERMINAL,LUG UOC:210	2
8	PAOZZ	25022	19-0876	.RELAY SUBASSEMBLY UOC:210	1
9	PAFZZ	96906	MS20600B6W4	.RIVET,BLIND UOC:210	4
10	PAOZZ	24617	9426623	.SCREW,TAPPING,THREA UOC:210	8
11	PAOZZ	11862	12004011	.FUSE,INCLOSED LINK UOC:210	2
12	PAOZZ	11862	12004008	.FUSE,INCLOSED LINK UOC:210	4
13	PAOZZ	24617	9418719	.SCREW UOC:210	12
14	PAOZZ	11862	12004010	.FUSE,INCLOSED LINK UOC:210	1
15	PAFZZ	25022	19-0878	.FUSE,CARTRIDGE UOC:210	2
16	PAOZZ	24617	9414714	SCREW,MACHINE UOC:210	4
17	PAOZZ	25022	99-4360-1	COVER,PANEL UOC:210	1

END OF FIGURE

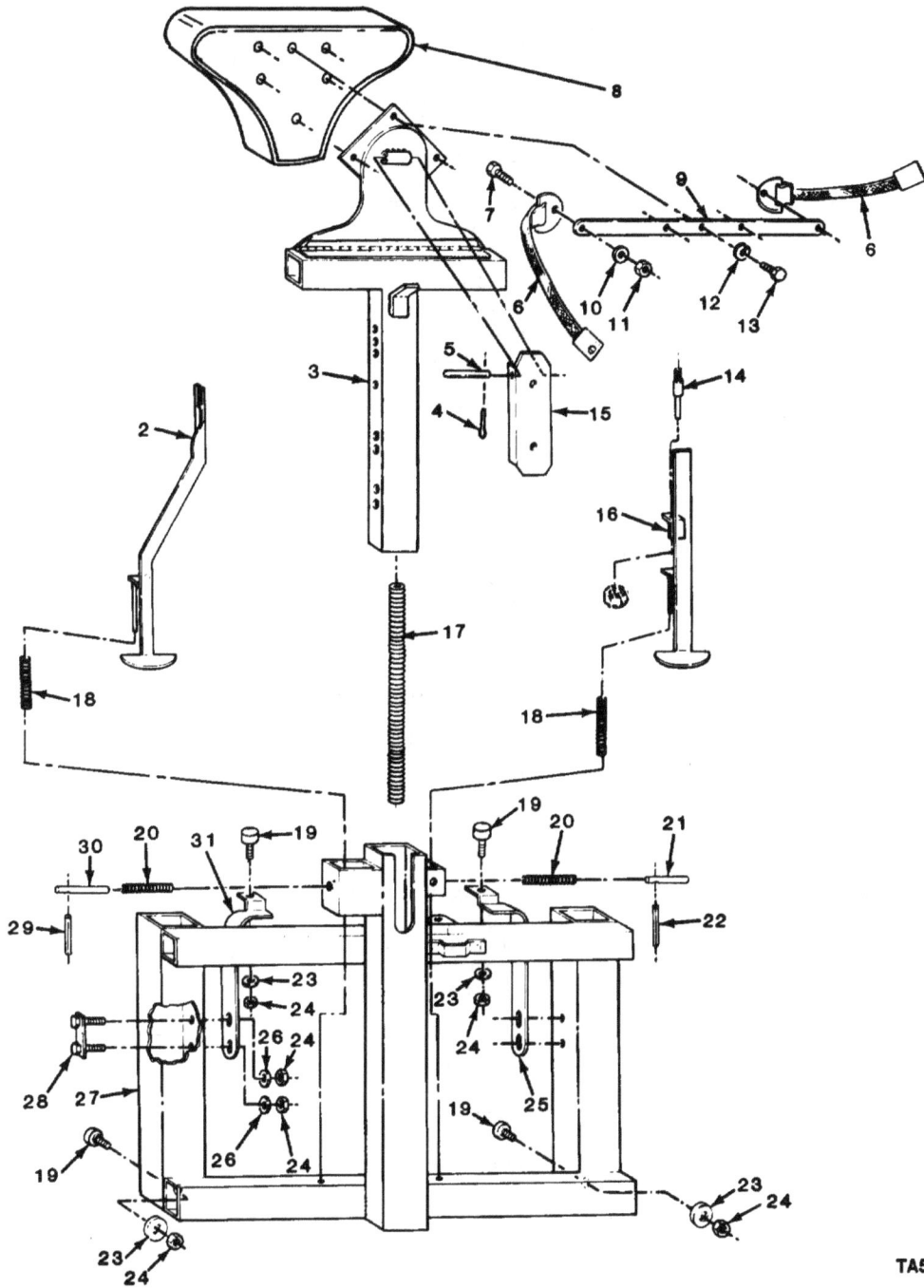

FIGURE 223. ATTENDANT'S SEAT ASSEMBLY AND COMPONENT PARTS (M1010).

TA511030

(1)(2)(3)(4) ITEM SMR PART NO CODE CAGEC NUMBER	(5) DESCRIPTION AND USABLE ON CODES (UOC)	(6) QTY
	GROUP 1812 SPECIAL PURPOSE BODIES	
	FIG. 223 ATTENDANT'S SEAT ASSEMBLY AND COMPONENT PARTS (M1010)	
1 PAOOO 25022 36-0039	SEAT,VEHICULAR UOC:210	1
2 PAOZZ 25022 36-0045	.ARM,ADJUSTING,VEHIC UOC:210	1
3 PFOZZ 25022 36-0044	.STEM,SEAT VEHICULAR UOC:210	1
4 PAOZZ 96906 MS24665-134	.PIN,COTTER UOC:210	1
5 PAOZZ 25022 36-0056	.PIN UOC:210	1
6 PAOZZ 25022 36-0055	.BELT,VEHICULAR SAFE UOC:210	1
7 PAOZZ 96906 MS90727-85	.SCREW,CAP,HEXAGON H UOC:210	2
8 PAOZZ 25022 36-0054	.SEAT,VEHICULAR UOC:210	1
9 PAOZZ 25022 36-0070	.BRACKET,SEAT BELT UOC:210	1
10 PAOZZ 96906 MS27183-15	.WASHER,FLAT UOC:210	2
11 PAOZZ 96906 MS35691-29	.NUT,PLAIN,HEXAGON UOC:210	2
12 PAOZZ 96906 MS35338-47	.WASHER,LOCK UOC:210	6
13 PAOZZ 96906 MS90728-3	.SCREW,CAP,HEXAGON H UOC:210	6
14 PAOZZ 25022 36-0071	.PIN,SHOULDER,HEADLE UOC:210	1
15 PAOZZ 25022 36-0047	.BRACKET,SEAT SUPPOR UOC:210	1
16 PAOZZ 25022 36-0046	.ARM,ADJUSTING,VEHIC UOC:210	1
17 PAOZZ 25022 36-0048	.SPRING,HELICAL,COMP UOC:210	1
18 PAOZZ 25022 36-0049	.SPRING,HELICAL,COMP UOC:210	2
19 PAOZZ 25022 09-0056	.ROLLER,SEAT UOC:210	4
20 PAOZZ 25022 36-0050	.SPRING,HELICAL,COMP UOC:210	2
21 PAOZZ 25022 36-0075	.PIN UOC:210	1
22 PAOZZ 96906 MS16562-147	.PIN,SPRING UOC:210	1
23 PAOZZ 96906 MS35338-46	.WASHER,LOCK UOC:210	8
24 PAOZZ 96906 MS51968-8	.NUT,PLAIN,HEXAGON	8

(1) ITEM NO	(2) SMR CODE	(3) CAGEC	(4) PART NUMBER	(5) DESCRIPTION AND USABLE ON CODES (UOC)	(6) QTY
				UOC:210	
25	PAOZZ	25022	36-0074	.BRACKET,ANGLE RIGHT	1
				UOC:210	
26	PAOZZ	96906	MS27183-13	.WASHER,FLAT	4
				UOC:210	
27	PFOZZ	25022	36-0043	.BASE,SEAT VEHICULAR	1
				UOC:210	
28	PAOZZ	25022	36-0073	.PIN,SUPPORT	2
				UOC:210	
29	PAOZZ	96906	MS16562-50	.PIN,SPRING	1
				UOC:210	
30	PAOZZ	25022	36-0051	.PIN	1
				UOC:210	
31	PAOZZ	25022	36-0072	.BRACKET,ANGLE LEFT	1
				UOC:210	

END OF FIGURE

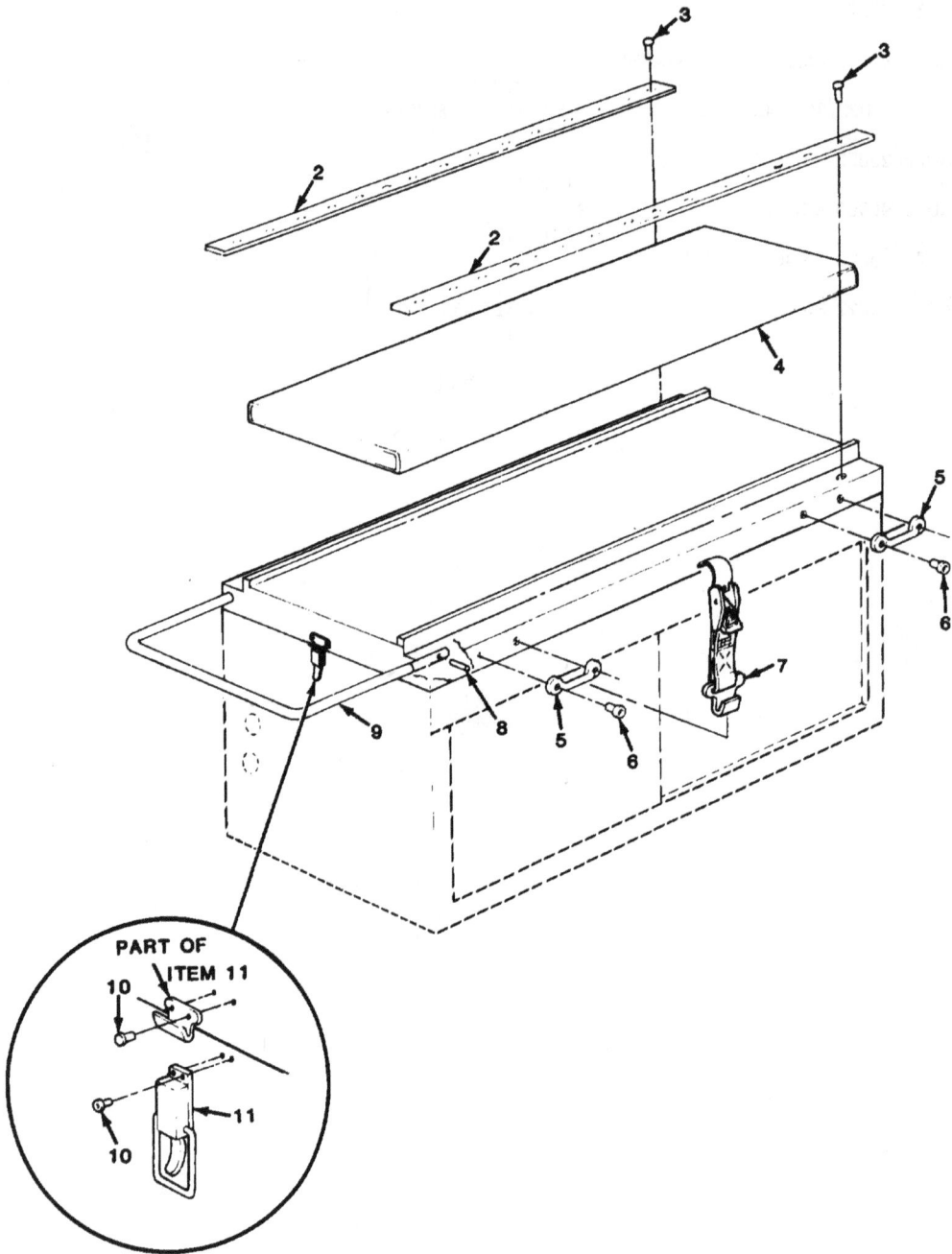

TA511031

FIGURE 224. UPPER LITTER ASSEMBLY AND COMPONENT PARTS (M1010).

(1) (2) (3) (4) ITEM SMR NO CODE CAGEC	PART NUMBER		(5) DESCRIPTION AND USABLE ON CODES (UOC)	(6) QTY
			GROUP 1812 SPECIAL PURPOSE BODIES	
			FIG. 224 UPPER LITTER ASSEMBLY AND COMPONENT PARTS (M1010)	
1	PAOFF	25022 22-0865	LITTER,VEHICULAR UOC:210	2
2	PAOZZ	25022 51-2297	.PLASTIC STRIP UOC:210	2
3	PAOZZ	96906 MS20601-B6W6	.RIVET,BLIND UOC:210	30
4	PAOZZ	25022 22-0860	.CUSHION,SEAT,VEHICU UOC:210	1
5	PAOZZ	25022 51-0979	.STRAP,RETAINING UOC:210	2
6	PAOZZ	96906 MS21141-U0607	.RIVET,BLIND UOC:210	4
7	PAOZZ	25022 37-0051	.STRAP,WEBBING UOC:210	4
8	PAOZZ	96906 MS16562-35	.PIN,SPRING UOC:210	2
9	PAOZZ	25022 36-0066	.HANDLE,LITTER,TELES UOC:210	1
10	PAOZZ	96906 MS21141-U0604	.RIVET,BLIND UOC:210	4
11	PAOZZ	25022 51-0985	.CLASP,LITTER HOLD D UOC:210	1

END OF FIGURE

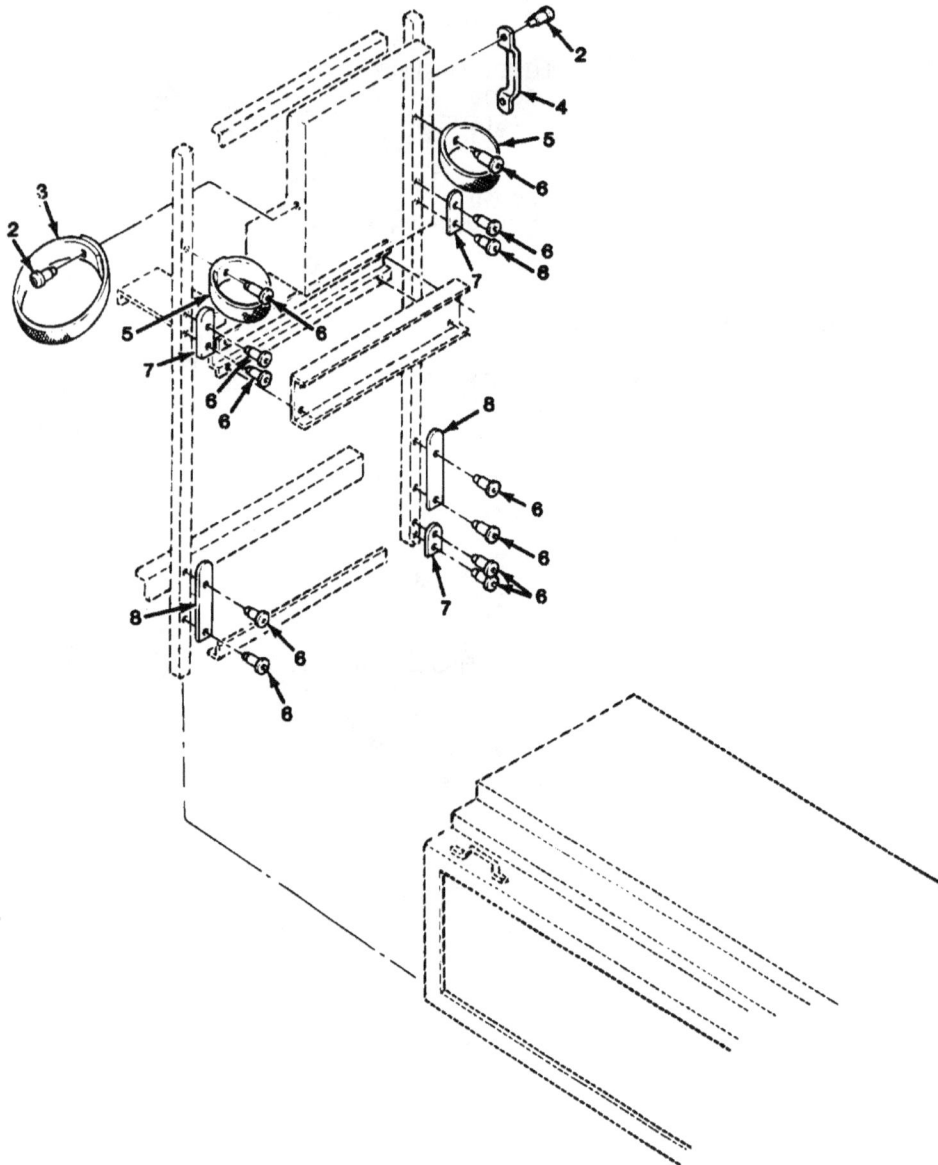

TA511032

FIGURE 225. FRONT HALF-PARTITION ASSEMBLY (M1010).

(1)	(2)	(3)	(4)	(5)	(6)
ITEM NO	SMR CODE	CAGEC	PART NUMBER	DESCRIPTION AND USABLE ON CODES (UOC)	QTY

GROUP 1812 SPECIAL PURPOSE BODIES

FIG. 225 FRONT HALF-PARTITION
ASSEMBLY (M1010)

1	XAFFF	25022	99-4593-0	HALF PARTITION UOC:210	2
2	PAOZZ	96906	MS21141-U0604	.RIVET,BLIND UOC:210	3
3	PAOZZ	25022	22-0876	.STRAP,WEBBING UOC:210	1
4	PAOZZ	25022	51-0979	.STRAP,RETAINING UOC:210	1
5	PAOZZ	25022	22-0878	.STRAP,WEBBING UOC:210	2
6	PAOZZ	96906	MS24662-153	.RIVET,BLIND UOC:210	12
7	PAOZZ	25022	17-0177	.BUMPER UOC:210	3
8	PAOZZ	25022	17-0178	.BUMPR UOC:210	2

END OF FIGURE

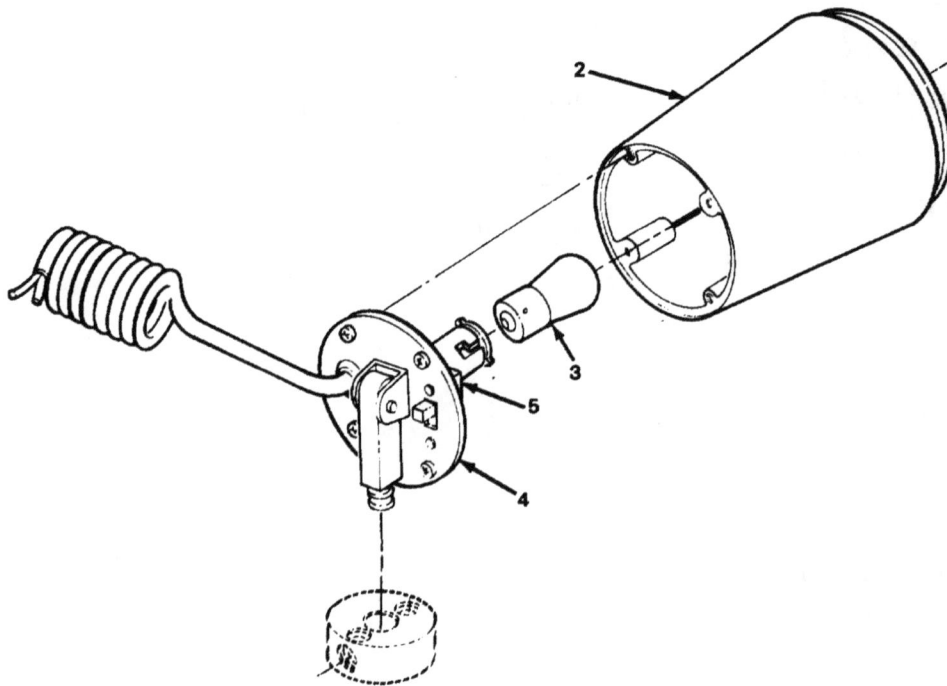

```
    1           4
 ┌──┴──┐      ┌─┴─┐
 │2 THRU 5│    │ 5 │
 └─────────┘   └───┘
```

TA511033

FIGURE 226. INTERIOR FOCUS LIGHT (M1010).

(1)	(2)	(3)	(4)	(5)	(6)
ITEM NO	SMR CODE	CAGEC	PART NUMBER	DESCRIPTION AND USABLE ON CODES (UOC)	QTY

GROUP 1812 SPECIAL PURPOSE BODIES

FIG. 226 INTERIOR FOCUS LIGHT
(M1010)

1	PAOOO	25022	19-0816	FIXTURE,LIGHTING	4
				UOC:210	
2	PAOZZ	25022	19-0940	.LENS,LIGHT	1
				UOC:210	
3	PAOZZ	08806	1003	.LAMP,INCANDESCENT	1
				UOC:210	
4	PAOOO	25022	19-0941	.LIGHT,EXTENSION	1
				UOC:210	
5	XDOZZ	25022	19-0941-1	..SWITCH,TOGGLE	1
				UOC:210	

END OF FIGURE

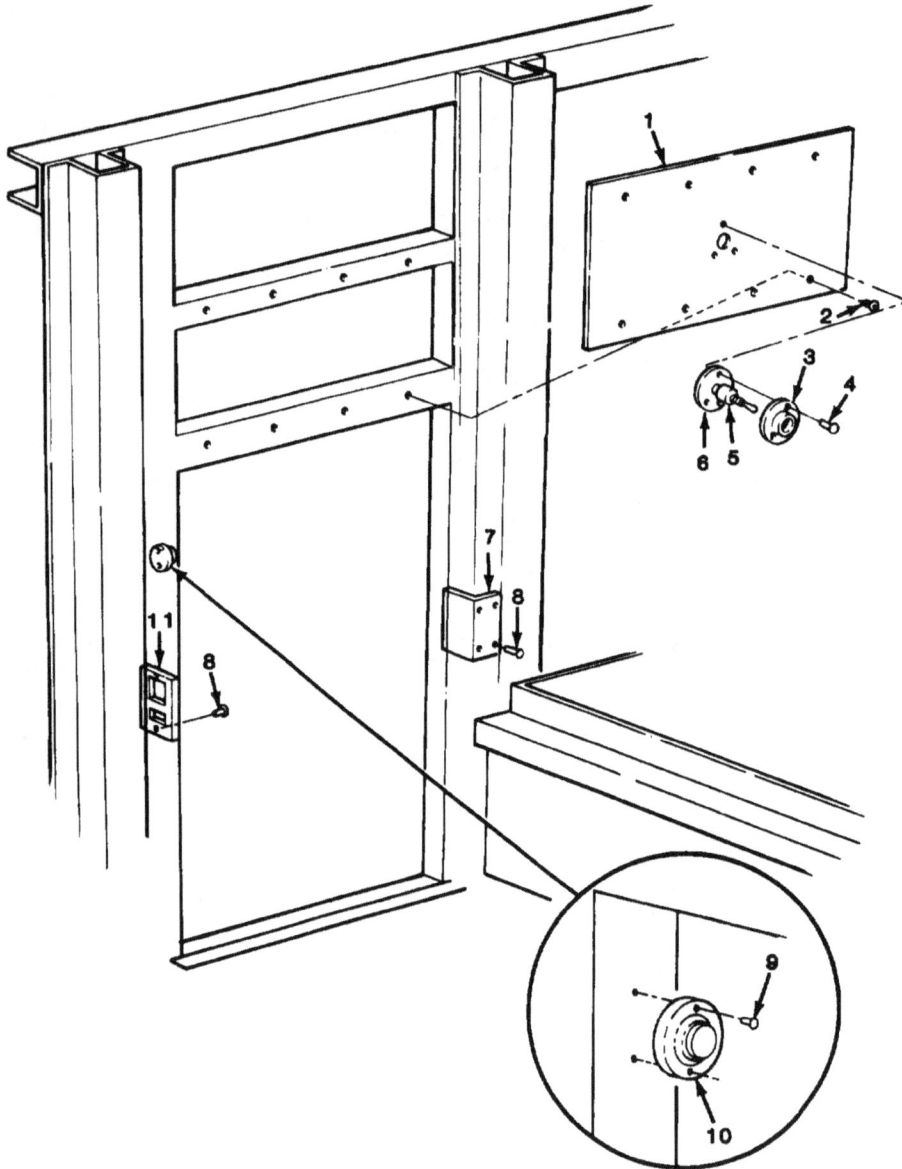

TA511034

FIGURE 227. FRONT FRAME ASSEMBLY COMPONENTS (M1010).

(1) ITEM NO	(2) SMR CODE	(3) CAGEC	(4) PART NUMBER	(5) DESCRIPTION AND USABLE ON CODES (UOC)	(6) QTY
				GROUP 1812 SPECIAL PURPOSE BODIES	
				FIG. 227 FRONT FRAME ASSEMBLY COMPONENTS (M1010)	
1	PAOZZ	25022	99-4282-1	PANEL,SWITCH MOUNTI UOC:210	1
2	PAOZZ	24617	9414714	SCREW,MACHINE UOC:210	8
3	PAOZZ	25022	37-0019	COVER,ELECTRICAL SW UOC:210	1
4	PAOZZ	96906	MS20600B6W4	RIVET,BLIND UOC:210	3
5	PAOZZ	25022	51-1959	SWITCH,TOGGLE UOC:210	1
6	PAOZZ	25022	37-0020	PLATE,SWITCH UOC:210	1
7	PAOZZ	25022	99-4408-1	STRIKER,STOP UOC:210	1
8	PAOZZ	96906	MS21141-U0604	RIVET,BLIND UOC:210	6
9	PAOZZ	96906	MS20600-B4W4	RIVET,BLIND UOC:210	2
10	PAOZZ	25022	51-1926	SWITCH,SENSITIVE UOC:210	1
11	PAOZZ	25022	51-1304	STRIKE,CATCH UOC:210	1

END OF FIGURE

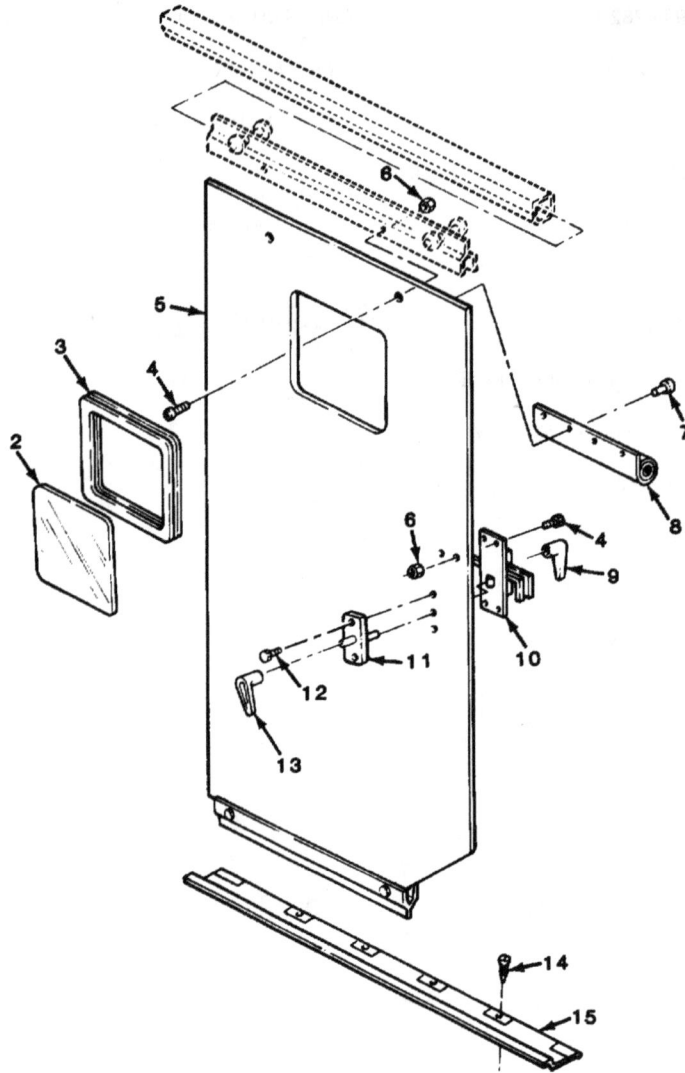

TA511035

FIGURE 228. INTERIOR SLIDING DOOR ASSEMBLY AND COMPONENT PARTS (M1010).

(1) ITEM NO	(2) SMR CODE	(3) CAGEC	(4) PART NUMBER	(5) DESCRIPTION AND USABLE ON CODES (UOC)	(6) QTY
				GROUP 1812 SPECIAL PURPOSE BODIES	
				FIG. 228 INTERIOR SLIDING DOOR ASSEMBLY AND COMPONENT PARTS (M1010)	
1	PAOOO	25022	99-4324-0	DOOR,VEHICULAR UOC:210	1
2	PAOZZ	25022	51-1604	.WINDOW,OBSERVATION UOC:210	1
3	PAOZZ	25022	51-1724	.SEAL,RUBBER SPECIAL UOC:210	1
4	PAOZZ	21450	126281	.BOLT,SQUARE NECK UOC:210	6
5	PAOOO	25022	22-0864	.DOOR,VEHICULAR UOC:210	1
6	PAOZZ	88044	AN365-1024A	.NUT,SELF-LOCKING,HE UOC:210	6
7	PAOZZ	96906	MS24662-153	.RIVET,BLIND UOC:210	4
8	PAOZZ	25022	22-0873	.CURTAIN,BLACKOUT UOC:210	1
9	PAOZZ	25022	09-0951	.HANDLE,DOOR UOC:210	1
10	PAOZZ	25022	09-0952	.LOCK,DOOR,VEHICULAR UOC:210	1
11	PAOZZ	25022	99-4683-1	.PIN UOC:210	1
12	PAOZZ	96906	MS21207-10-10	.SCREW,TAPPING,THREA UOC:210	2
13	PAOZZ	25022	08-0475	.HANDLE,DOOR UOC:210	1
14	PAOZZ	24617	163881	RIVET UOC:210	4
15	PFOZZ	25022	99-4257-1	TRACK,SLIDING DOOR UOC:210	1

END OF FIGURE

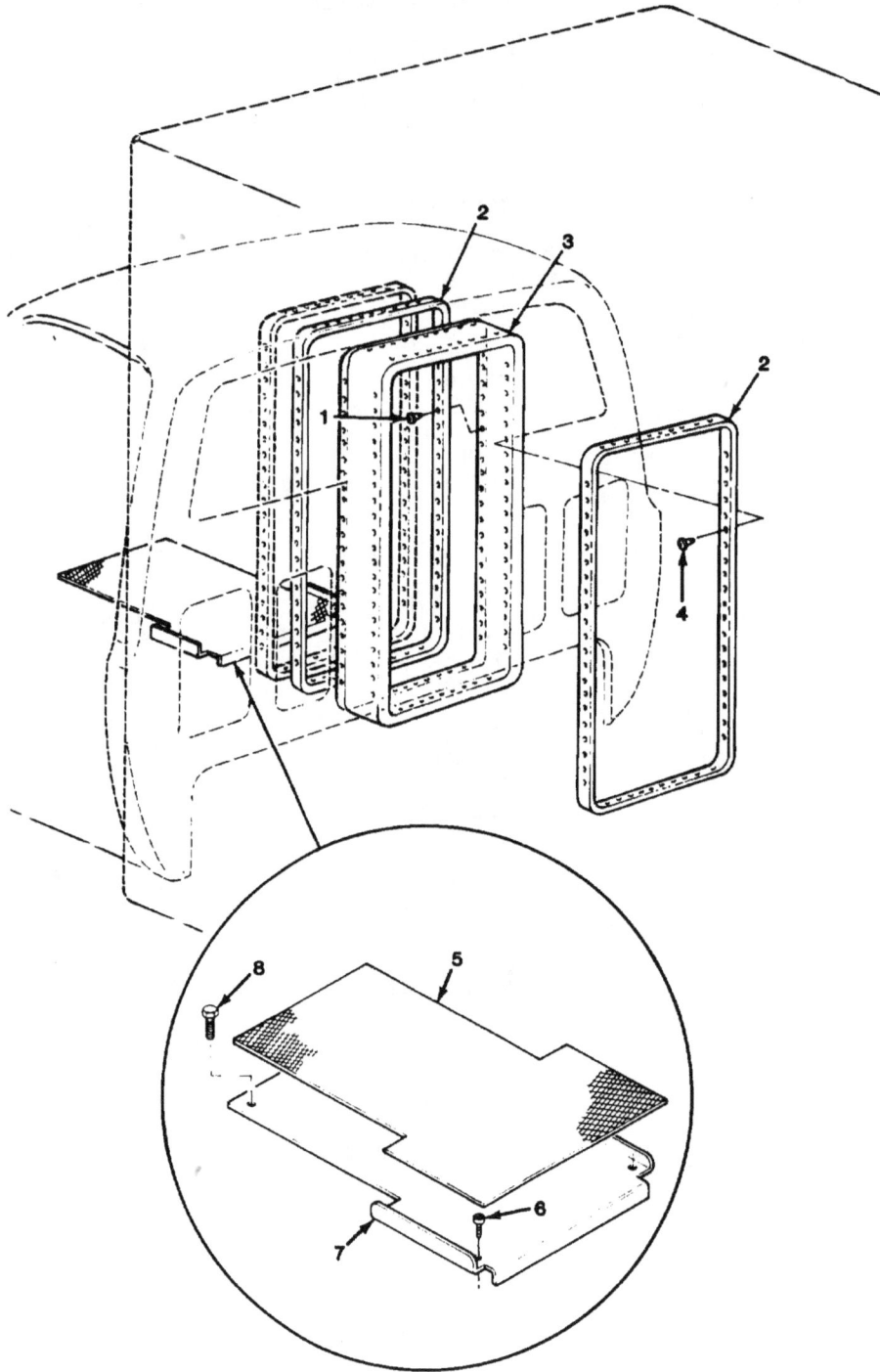

TA511036

FIGURE 229. PULLMAN COLLAR AND RELATED PARTS (M1010).

(1) (2) (3) (4) ITEM SMR PART NO CODE CAGEC NUMBER	(5) DESCRIPTION AND USABLE ON CODES (UOC)	(6) QTY
	GROUP 1812 SPECIAL PURPOSE BODIES	
	FIG. 228 INTERIOR SLIDING DOOR ASSEMBLY AND COMPONENT PARTS (M1010)	
1 PAOOO 25022 99-4324-0	DOOR,VEHICULAR UOC:210	1
2 PAOZZ 25022 51-1604	.WINDOW,OBSERVATION UOC:210	1
3 PAOZZ 25022 51-1724	.SEAL,RUBBER SPECIAL UOC:210	1
4 PAOZZ 21450 126281	.BOLT,SQUARE NECK UOC:210	6
5 PAOOO 25022 22-0864	.DOOR,VEHICULAR UOC:210	1
6 PAOZZ 88044 AN365-1024A	.NUT,SELF-LOCKING,HE UOC:210	6
7 PAOZZ 96906 MS24662-153	.RIVET,BLIND UOC:210	4
8 PAOZZ 25022 22-0873	.CURTAIN,BLACKOUT UOC:210	1
9 PAOZZ 25022 09-0951	.HANDLE,DOOR UOC:210	1
10 PAOZZ 25022 09-0952	.LOCK,DOOR,VEHICULAR UOC:210	1
11 PAOZZ 25022 99-4683-1	.PIN UOC:210	1
12 PAOZZ 96906 MS21207-10-10	.SCREW,TAPPING,THREA UOC:210	2
13 PAOZZ 25022 08-0475	.HANDLE,DOOR UOC:210	1
14 PAOZZ 24617 163881	RIVET UOC:210	4
15 PFOZZ 25022 99-4257-1	TRACK,SLIDING DOOR UOC:210	1

END OF FIGURE

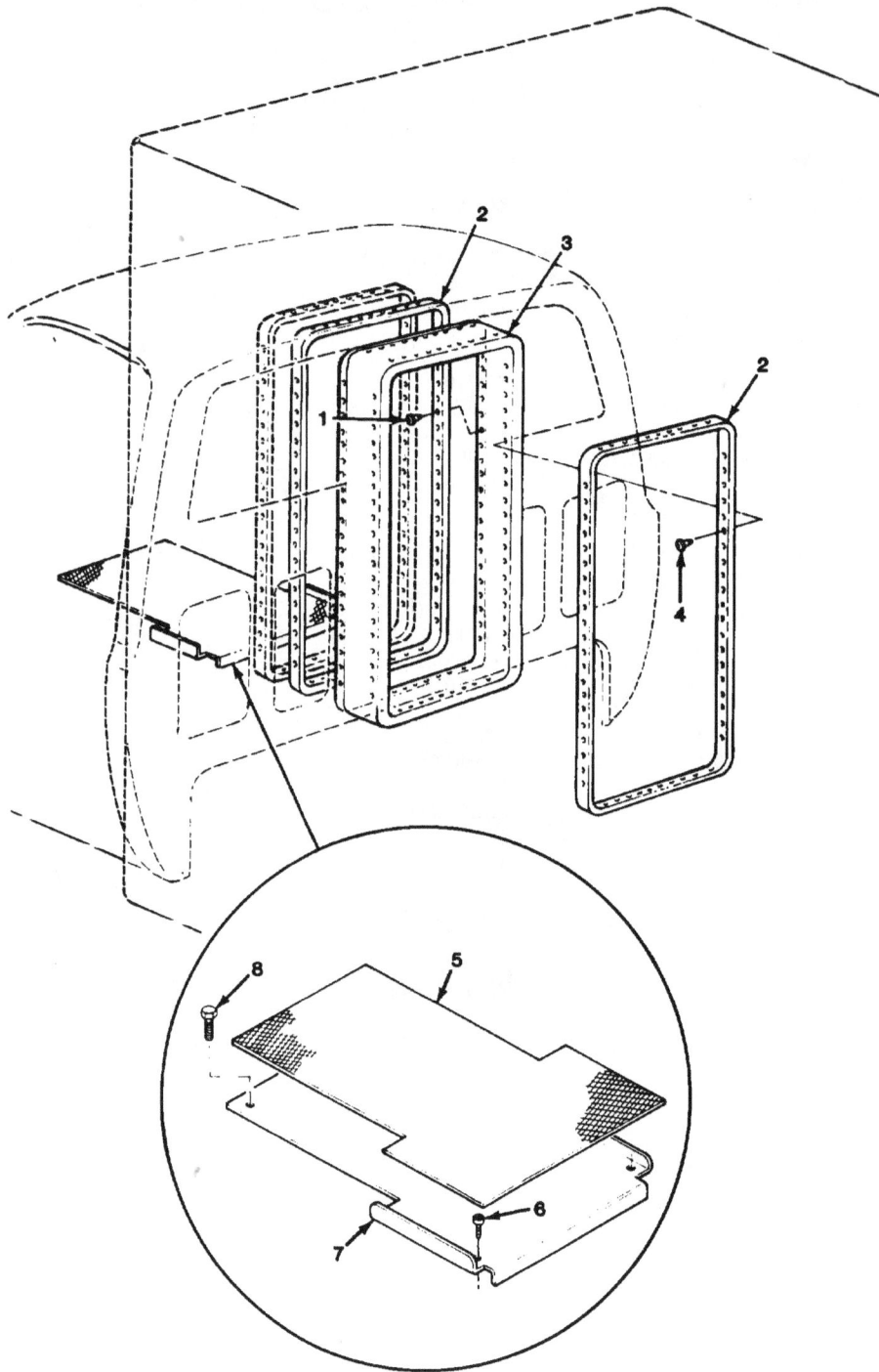

TA511036

FIGURE 229. PULLMAN COLLAR AND RELATED PARTS (M1010).

TM9-2320-289-34P

(1) (2) (3) (4) ITEM SMR PART NO CODE CAGEC NUMBER	(5) DESCRIPTION AND USABLE ON CODES (UOC)	(6) QTY
	GROUP 1812 SPECIAL PURPOSE BODIES	
	FIG. 229 PULLMAN COLLAR AND RELATED PARTS (M1010)	
1 PAFZZ 96906 MS20600B6W8 RIVET,BLIND		60
	UOC:210	
2 PAFZZ 25022 99-4530-1	PULLMAN COLLAR,CAB	2
	UOC:210	
3 PAFZZ 25022 22-0858	PULLMAN,VEHICULAR	1
	UOC:210	
4 PAFZZ 96906 MS20600-B6W12 RIVET,BLIND		60
	UOC:210	
5 PAOZZ 25022 17-0195	MAT,FLOOR	1
	UOC:210	
6 PAOZZ 96906 MS24627-67	SCREW,TAPPING	2
	UOC:210	
7 PAOZZ 25022 99-4892-0	RAMP,CAB	1
	UOC:210	
8 PAOZZ 96906 MS90728-8	SCREW,CAP,HEXAGON H	2
	UOC:210	
	END OF FIGURE	

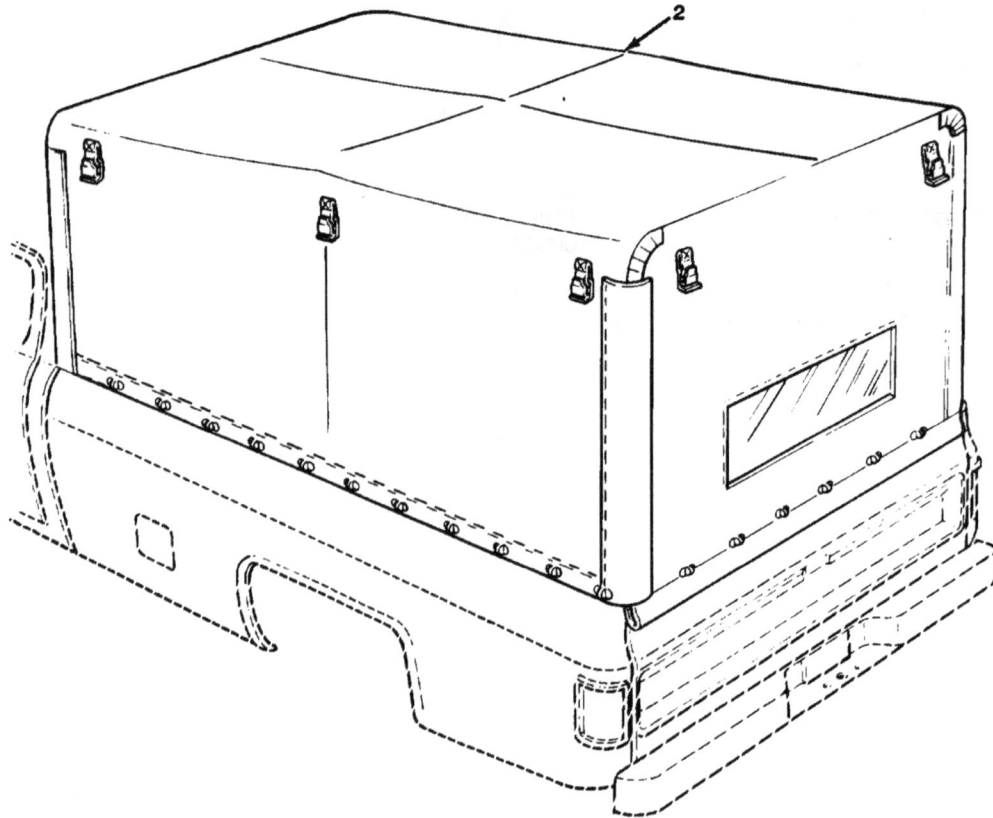

TA511037

FIGURE 230. CARGO COVER ASSEMBLY (M1008 AND M1008A1).

(1) (2) (3) (4) ITEM SMR PART NO CODE CAGEC NUMBER	(5) DESCRIPTION AND USABLE ON CODES (UOC)	(6) QTY
	GROUP 22 BODY, CHASSIS, AND HULL ACCESSORY ITEMS	
	GROUP 2201 CANVAS, RUBBER, OR PLASTIC ITEMS	
	FIG. 230 CARGO COVER ASSEMBLY (M1008 AND M1008A1)	
1 PDOFF 11862 14072479	PARTS KIT,CARGO COV FOR COMPONENT PARTS SEE FIG'S 231,232 UOC:194,208	1
2 PAOFF 11862 14072475	.TOP ASSEMBLY,TRUCK UOC:194,208	1
	END OF FIGURE	

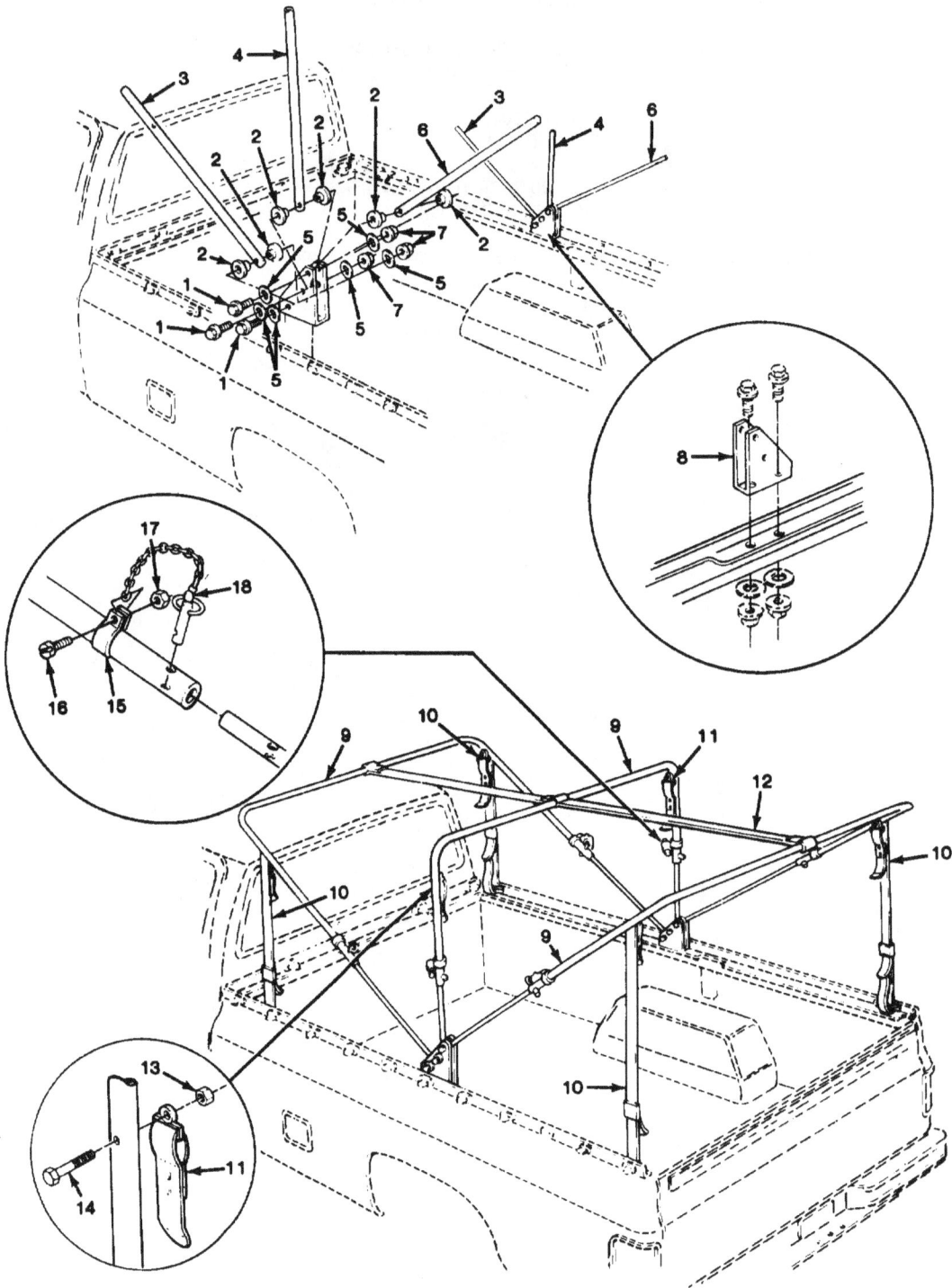

FIGURE 231. COVER FRAME COMPONENTS (M1008 AND M1008A1).

TA511038

(1)	(2)	(3)	(4)		(5)	(6)
ITEM	SMR		PART			
NO	CODE	CAGEC	NUMBER		DESCRIPTION AND USABLE ON CODES (UOC)	QTY

GROUP 2201 CANVAS, RUBBER, OR
PLASTIC ITEMS

FIG. 231 COVER FRAME COMPONENTS
(M1008 AND M1008A1)

1	PAOZZ	96906	MS90728-39	BOLT,MACHINE UOC:194,208	6
2	PAOZZ	11862	15599919	BUSHING,SLEEVE UOC:194,208	12
3	MOOZZ	11862	14072471	STRUT-C/CVR (33.75" LG) MAKE FROM STRUT, P/N 15599621 UOC:194,208	2
4	MOOZZ	11862	14072472	STRUT-C/CVR RF BOW (22.50" LG) MAKE FROM STRUT, P/N 15599621 UOC:194,208	2
5	PAOZZ	96906	MS27183-12	WASHER,FLAT UOC:194,208	12
6	MOOZZ	11862	14072473	STRUT-C/CVR RF BOW (32.34" LG) MAKE FROM STRUT, P/N 15599621 UOC:194,208	2
7	PAOZZ	96906	MS51943-33	NUT,SELF-LOCKING,HE UOC:194,208	6
8	PAOZZ	11862	14072468	BRACKET,BOW HINGE,V UOC:194,208	2
9	PAOZZ	11862	14072470	BOW,VEHICULAR TOP UOC:194,208	3
10	PAOZZ	19207	12255559	STRAP,WEBBING UOC:194,208	4
11	PAOZZ	19207	12255567	STRAP,WEBBING UOC:194,208	2
12	PAOZZ	11862	14072469	STRAP,VEHICULAR TOP UOC:194,208	1
13	PAOZZ	72582	9419454	NUT,SELF-LOCKING,HE UOC:194,208	6
14	PAOZZ	24617	157456	SCREW,MACHINE UOC:194,208	6
15	PAOZZ	19207	12255561	CLAMP,LOOP UOC:194,208	6
16	PAOZZ	96906	MS51957-30	SCREW,MACHINE UOC:194,208	6
17	PAOZZ	24617	9423530	NUT,PLAIN,ASSEMBLED UOC:194,208	6
18	PAOZZ	19207	11669126-1	PIN,QUICK RELEASE UOC:194,208	6

END OF FIGURE

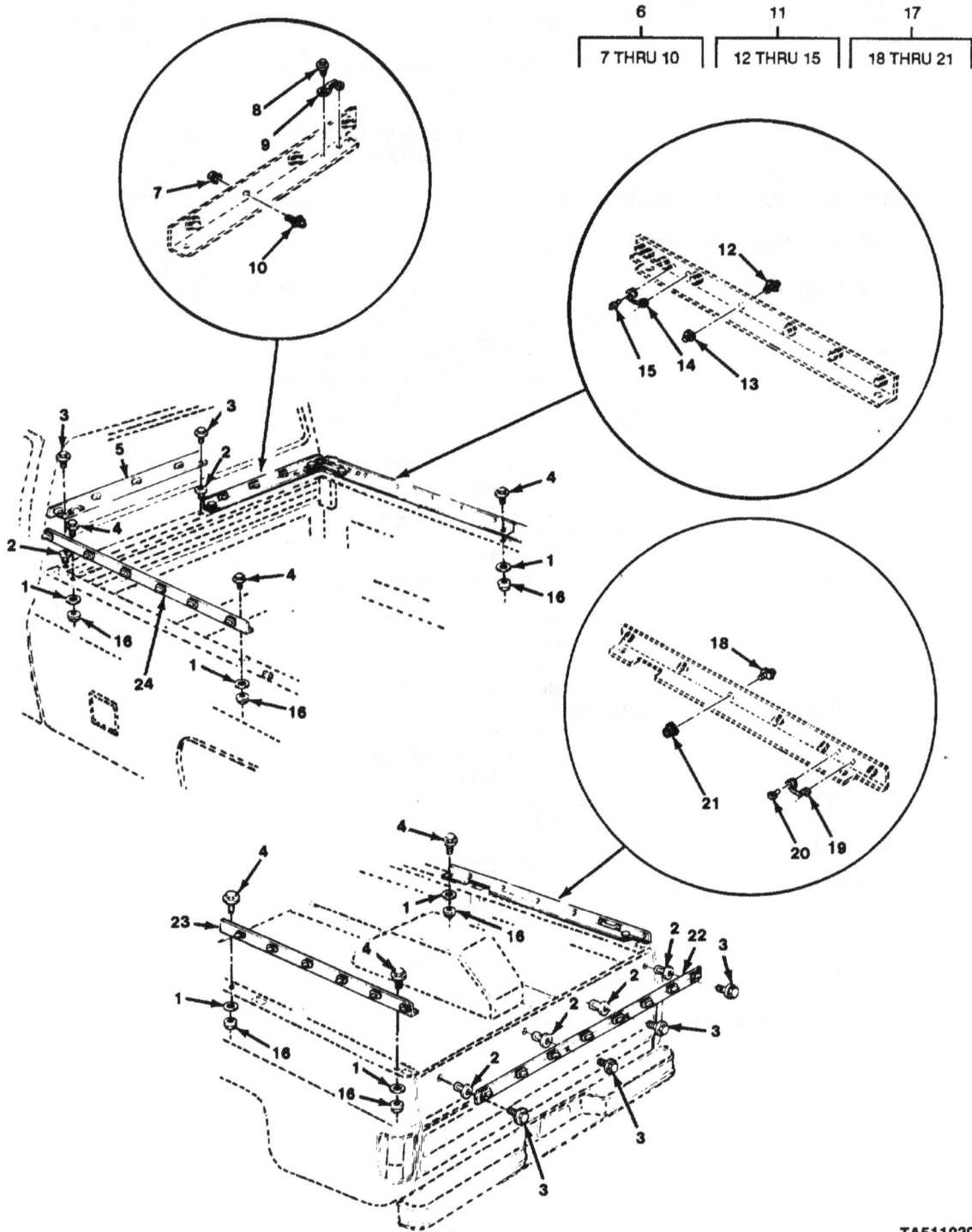

FIGURE 232. CARGO COVER MOUNTING PARTS (M1008 AND M1008A1).

TA511039

(1) ITEM NO	(2) SMR CODE	(3) CAGEC	(4) PART NUMBER	(5) DESCRIPTION AND USABLE ON CODES (UOC)	(6) QTY
				GROUP 2201 CANVAS, RUBBER, OR PLASTIC ITEMS	
				FIG. 232 CARGO COVER MOUNTING PARTS (M1008 AND M1008A1)	
1	PAOZZ	96906	MS27183-14	WASHER,FLAT UOC:194,208	8
2	PAOZZ	11862	14005953	SHEILD,EXPANSION UOC:194,208	8
3	PAOZZ	96906	MS90728-8	SCREW,CAP,HEXAGON H UOC:194,208	8
4	PAOZZ	24617	9415153	SCREW,ASSEMBLED WAS UOC:194,208	8
5	PAOZZ	11862	14072461	RAIL,BOW RECEPTACLE LEFT UOC:194,208	1
6	PAOZZ	11862	14072462	RAIL,CARGO COVER RIGHT UOC:194,208	1
7	PAOZZ	24617	271172	.NUT,SELF-LOCKING,AS UOC:194,208	4
9	PAOZZ	96906	MS51939-2	.LOOP,STRAP FASTENER UOC:194,208	1
10	PAOZZ	19207	12255564-2	.STUD,TURNBUTTON FAS UOC:194,208	4
11	PAOOO	11862	14072464	RAIL,CARGO COVER RIGHT UOC:194,208	1
12	PAOZZ	19207	12255564-2	.STUD,TURNBUTTON FAS UOC:194,208	6
13	PAOZZ	24617	271172	.NUT,SELF-LOCKING,AS UOC:194,208	6
14	PAOZZ	96906	MS51939-2	.LOOP,STRAP FASTENER UOC:194,208	25
15	PAOZZ	96906	MS21141-U0604	.RIVET,BLIND UOC:194,208	4
16	PAOZZ	24617	456004	NUT,SELF-LOCKING,HE UOC:194,208	8
17	PAOOO	11862	14072466	RAIL,CARGO COVER RIGHT UOC:194,208	1
18	PAOZZ	19207	12255564-2	.STUD,TURNBUTTON FAS UOC:194,208	6
19	PAOZZ	96906	MS51939-2	.LOOP,STRAP FASTENER UOC:194,208	1
20	PAOZZ	96906	MS21141-U0604	.RIVET,BLIND UOC:194,208	2
21	PAOZZ	24617	271172	.NUT,SELF-LOCKING,AS UOC:194,208	6
22	PAOZZ	11862	14072467	RAIL,CARGO COVER UOC:194,208	1
23	PAOOO	11862	14072465	RAIL,CARGO COVER UOC:194,208	1

(1) ITEM NO	(2) SMR CODE	(3) CAGEC	(4) PART NUMBER	(5) DESCRIPTION AND USABLE ON CODES (UOC)	(6) QTY
24	PAOOO	11862	14072463	RAIL,CARGO COVER UOC:194,208	1

END OF FIGURE

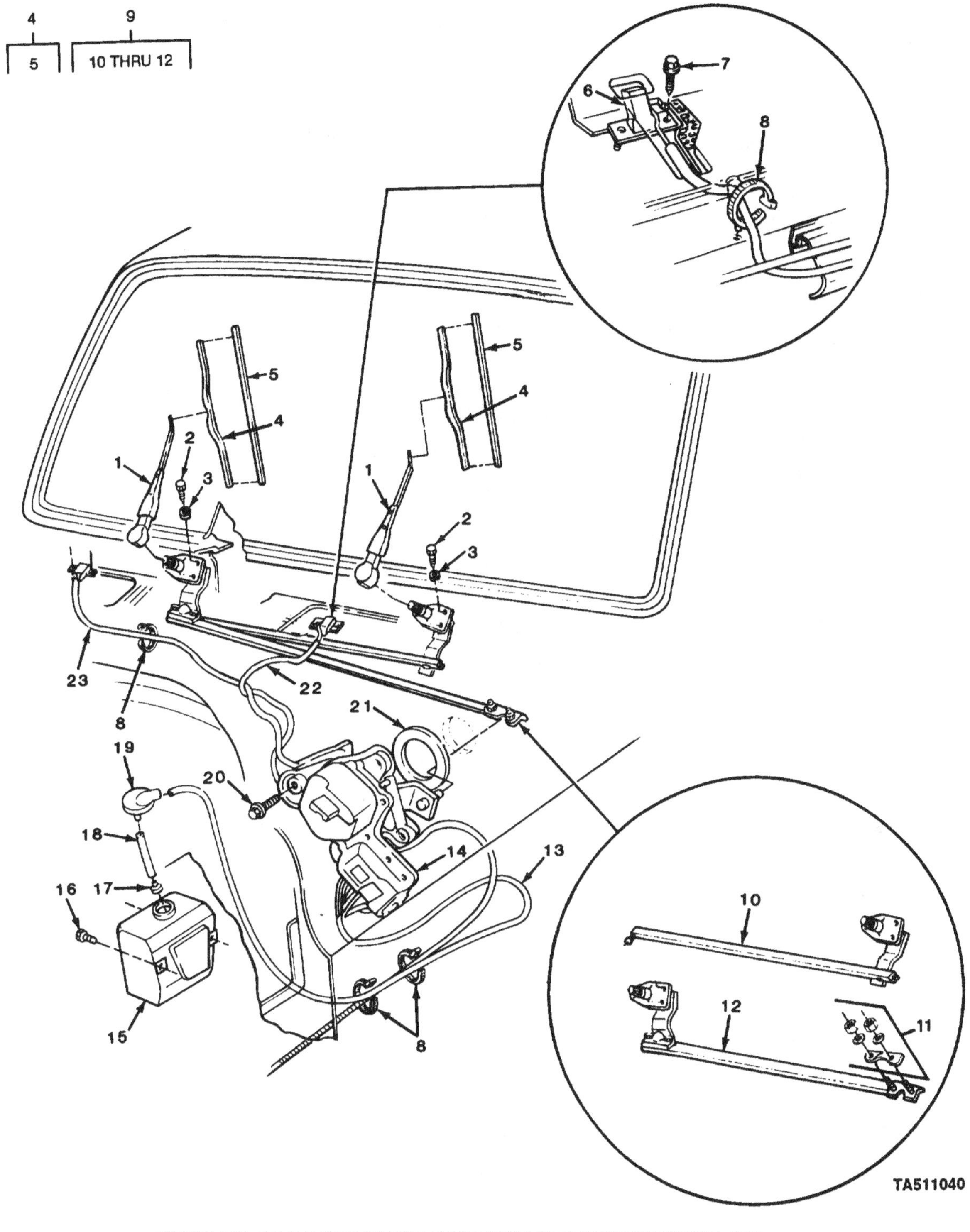

FIGURE 233. WINDSHIELD WIPER MOTOR AND PUMP ASSEMBLY WIPER ARM
LINK ASSEMBLY, WIPER BLADE ASSEMBLY, AND RELATED PARTS.

TA511040

(1) (2) (3) (4) ITEM SMR PART NO CODE CAGEC NUMBER	(5) DESCRIPTION AND USABLE ON CODES (UOC)	(6) QTY
	GROUP 2202 ACCESSORY ITEMS	
	FIG. 233 WINDSHIELD WIPER MOTOR AND PUMP ASSEMBLY, WIPER ARM LINK ASSEMBLY, WIPER BLADE ASSEMBLY, AND RELATED PARTS	
1 PAOZZ 11862 15591703	ARM,WINDSHIELD WIPE	2
2 PAOZZ 24617 9421985 SCREW,TAPPING,THREA		6
3 PAOZZ 96906 MS35335-34 WASHER,LOCK		6
4 PAOOO 11862 15591704	BLADE,WINDSHIELD WI	2
5 XAOZZ 11862 14044931	.REFILL BLADE,WIPER	25
6 PAOZZ 11862 14076838	NOZZLE ASSEMBLY,WIN	25
7 PAOZZ 24617 9419163	SCREW,CAP,HEXAGON H	2
8 PAOZZ 11862 3816659	STRAP,LINE SUPPORTI	6
9 AOOOO 11862 22021655	TRANS & LINK ASM W/	1
10 PAOZZ 11862 22029629	.LINK,WIPER,WINDSHIELD LEFT	1
11 PAOZZ 11862 4918562	.PARTS KIT,WINDSHIELD	1
12 PAOZZ 11862 22029630	.WIPER ASSEMBLY,WIND RIGHT	1
13 MOOZZ 11862 3782730	HOSE ASM PMP (46.00" LG) MAKE FROM HOSE, P/N 3987364	1
14 PAOFF 72560 22048352	MOTOR,WINDSHIELD WI FOR COMPONENT PARTS SEE FIG 234	1
15 PAOZZ 11862 3990892	JAR ASSEMBLY,WINDSH	1
16 PAOZZ 96906 MS51871-4 SCREW,TAPPING,THREA		2
17 PAOZZ 11862 3986821 STRAINER,SUCTION		1
18 MOOZZ 11862 329198	HOSE (CUT TO 7.0") MAKE FROM HOSE, P/N 3987364	1
19 PAOZZ 11862 3798372	CAP,FILLER OPENING	1
20 PAOZZ 11862 3824124	BOLT,ASSEMBLED WASH	3
21 PAOZZ 11862 20489125 GASKET		1
22 MOOZZ 11862 337714	HOSE ASM NOZZLE L/H (16.00" LG) MAKE FROM HOSE,P/N 1359744	1
23 MOOZZ 11862 3782732	HOSE ASM NOZZLE R/H (36.00" LG) MAKE FROM P/N 1359744	1
	END OF FIGURE	

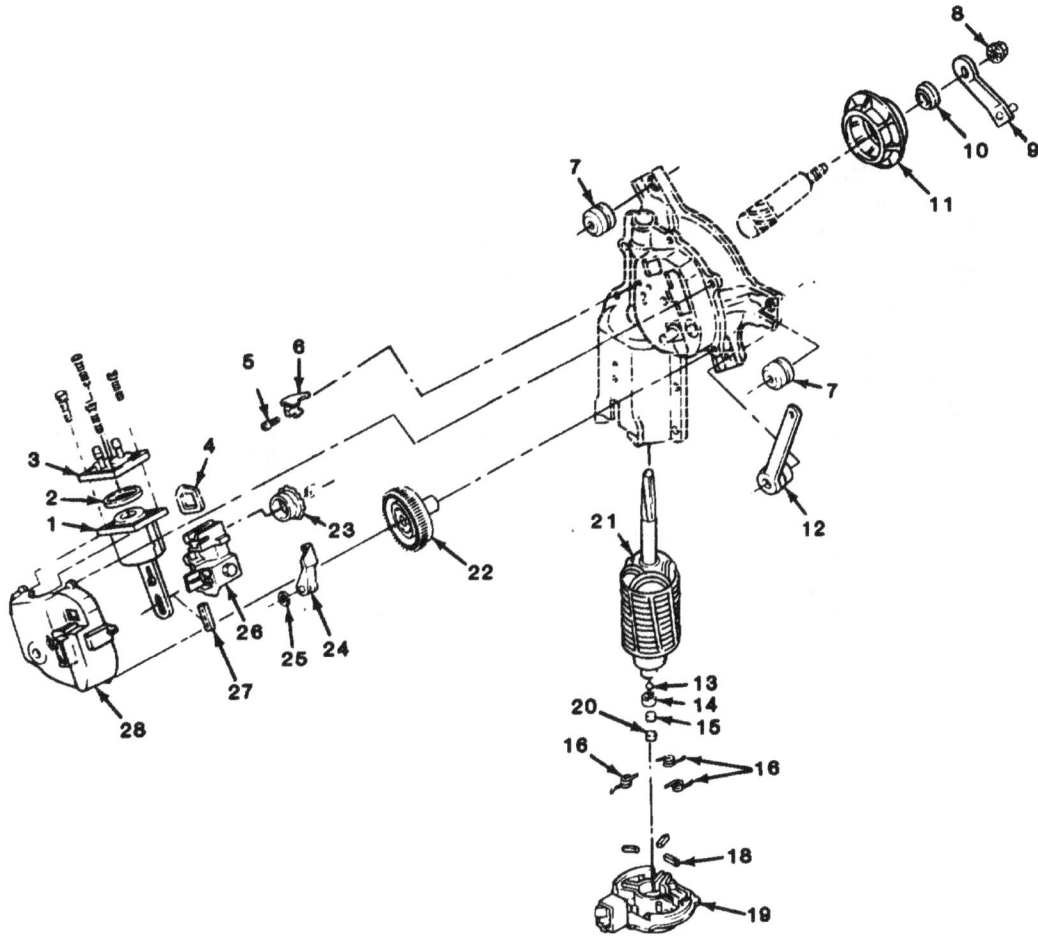

FIGURE 234. WIPER MOTOR COMPONENT PARTS.

TA511041

(1) (2) (3) (4) ITEM SMR PART NO CODE CAGEC NUMBER	(5) DESCRIPTION AND USABLE ON CODES (UOC)	(6) QTY
	GROUP 2202 ACCESSORY ITEMS	
	FIG. 234 WIPER MOTOR COMPONENT PARTS	
1 PAFZZ 11862 22048368	PISTON AND HOUSING	1
2 PAFZZ 11862 22048303	PACKING,PREFORMED	1
3 PAFZZ 11862 4914356	PARTS KIT,WINDSHIEL	1
4 XFOZZ 11862 22054154	GASKET	1
5 PAFZZ 11862 22049837	SCREW,MACHINE	1
6 PAFZZ 11862 22049511	SPRING,WINDSHIELD W	1
7 PAOZZ 11862 22049531	SPACER AND GROMMET	3
8 PAOZZ 11862 22054153	NUT,SELF-LOCKING,EX	1
9 PAOZZ 11862 22038927	LEVER,REMOTE CONTRO	1
10 PAOZZ 11862 22054156	GASKET	1
11 PAFZZ 11862 22054155	SPACER,SEAL,WINDSHI	1
12 PAOZZ 11862 22021679	GROMMET	1
13 PAFZZ 96906 MS19061-20005 BALL,BEARING		1
14 PAFZZ 11862 22048281	FELT,MECHANICAL,PRE	1
15 PAFZZ 11862 22048287	DISK,SOLID,PLAIN	1
16 PAFZZ 11862 22009291	SPRING,HELICAL,TORS	3
17 PAFFF 11862 22038909	PARTS KIT,WINDSHEIL	1
18 PAFZZ 16748 22009290	.BRUSH,ELECTRICAL CC	3
19 PAFZZ 11862 22038924	.END BALL,ELECTRICAL	1
20 PAFZZ 11862 22049534	DISK,RUPTURABLE	1
21 PAFZZ 11862 22038925	ARMATURE,MOTOR	1
22 XAFZZ 11862 22021690	GEAR W/SHAFT W/S WI	1
23 PFFZZ 11862 4939314	RATCHET WHEEL	1
24 PAFZZ 11862 4918446	PARTS KIT,WINDSHIEL	1
25 PAFZZ 96906 MS16633-1018 RING,RETAINING		1
26 PAFZZ 11862 22038931	SWITCH ASSEMBLY	1
27 XDFZZ 11862 22054158	GASKET	1
28 PAOZZ 11862 22038928	COVER,WASHER,WINDSH	1

END OF FIGURE

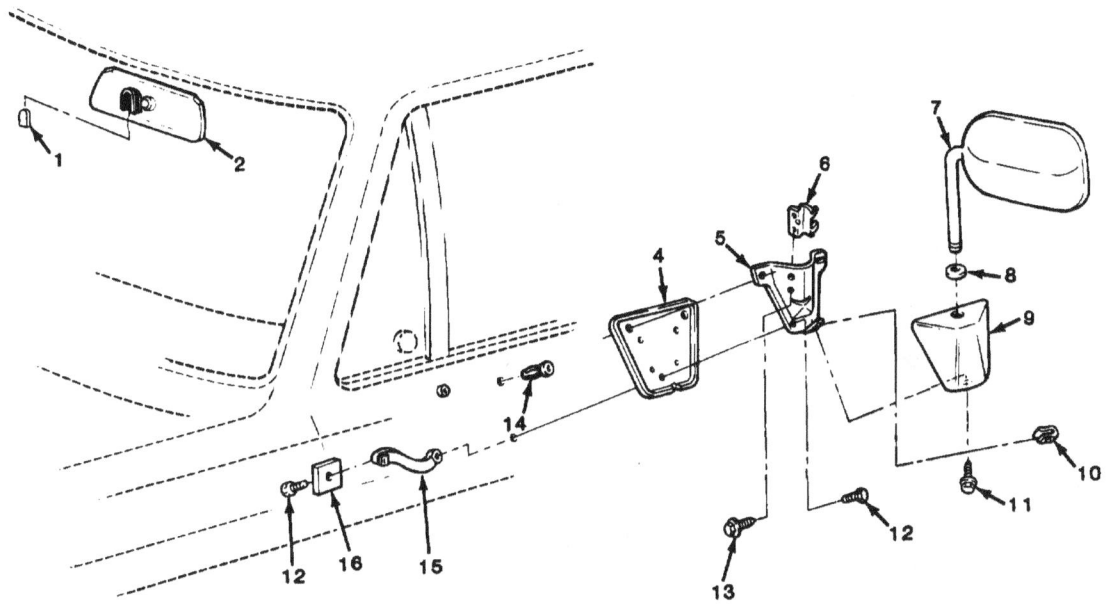

FIGURE 235. REARVIEW MIRRORS.

TA511042

(1) (2) (3) (4) ITEM SMR PART NO CODE CAGEC NUMBER	(5) DESCRIPTION AND USABLE ON CODES (UOC)	(6) QTY
	GROUP 2202 ACCESSORY ITEMS	
	FIG. 235 REARVIEW MIRRORS	
1 PAOZZ 11862 9831062	SUPPORT,REARVIEW MI INCLUDES GLUE UOC:194,208,209,231	1
2 PAOZZ 11862 918656	MIRROR ASSEMBLY,REA UOC:194,208,209,231	1
3 AOOOO 11862 14072485	MIRROR ASSEMBLY,REA LEFT UOC:194,208,209,210,230,231,252	1
3 AOOOO 11862 14072486	MIRROR ASSEMBLY,REA RIGHT UOC:194,208,209,210,230,231,252	1
4 PAOZZ 11862 14008191 .GASKET	 UOC:194,208,209,210,230,231,252	1
5 PAOZZ 11862 14007429 .BRACKET,MIRROR	 UOC:194,208,209,210,230,231,252	1
6 PAOZZ 11862 14007430	.CLAMP,REAR VIEW MIR UOC:194,208,209,210,230,231,252	1
7 PAOZZ 11862 14072487	.MIRROR ASSEMBLY,REA UOC:194,208,209,210,230,231,252	1
7 PAOZZ 11862 15634659	.MIRROR ASSEMBLY,REA UOC:254,256	1
7 PAOZZ 11862 15634660	.MIRROR ASSEMBLY,REA UOC:254,256	1
8 PAOZZ 11862 14007435 .GROMMET,NONMETALLIC	 UOC:194,208,209,210,230,231,252	1
9 PAOZZ 11862 14072489	.COVER,MIRROR,REAR V LEFT UOC:194,208,209,210,230,231,252	1
10 PAOZZ 11862 52351724	NUT,SHEET SPRING	2
11 PAOZZ 11862 15554576	SCREW,ASSEMBLED WAS	2
12 PAOZZ 11862 14007511	BOLT,ASSEMBLED WASH	8
13 PAOZZ 24617 9426277 BOLT		4
14 PAOZZ 11862 14005953 SHIELD,EXPANSION	 UOC:194,208,209,210,230,231,252	4
15 PAOZZ 11862 14007541	ARM,REARVIEW MIRROR LEFT UOC:194,208,209,210,230,231,252	1
15 PAOZZ 11862 14007542	ARM,REARVIEW MIRROR RIGHT UOC:194,208,209,210,230,231,252	1
16 PAOZZ 11862 14007539 WASHER,FLAT		2

END OF FIGURE

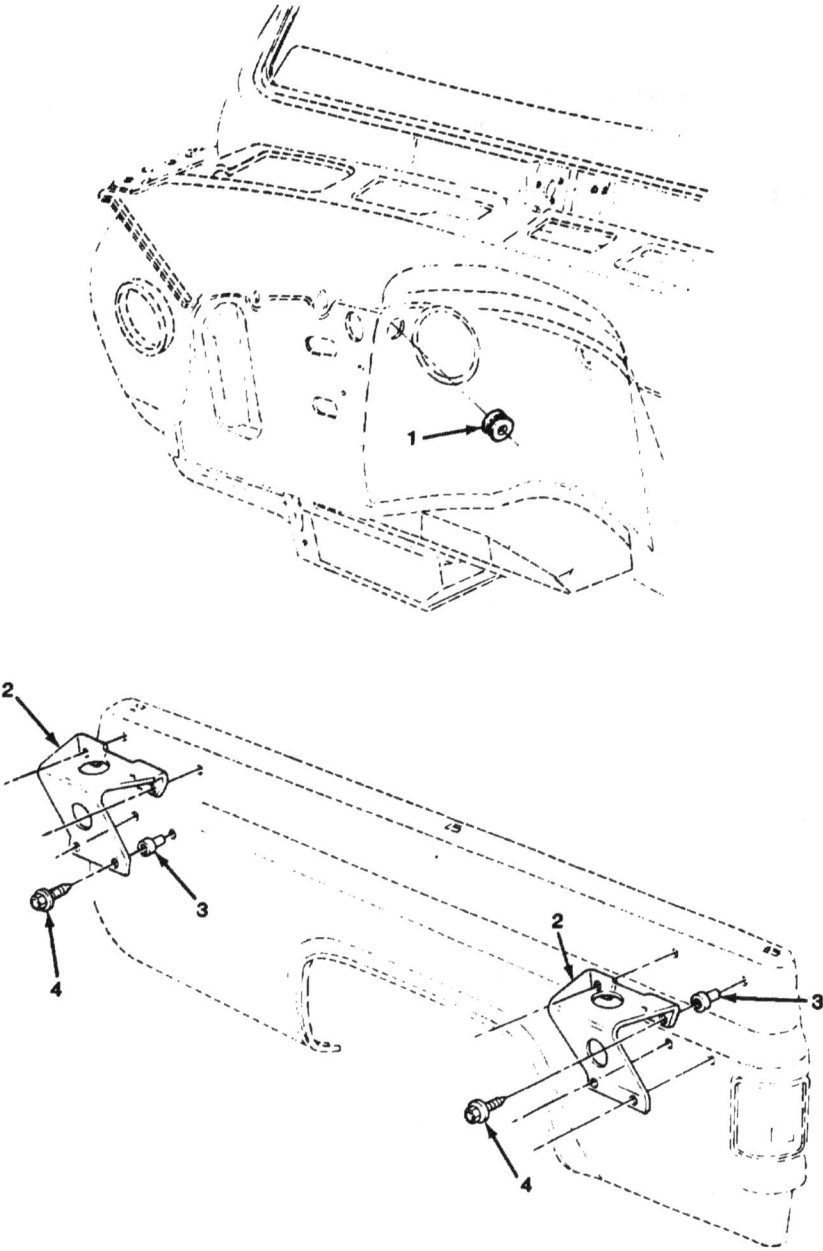

TA511043

FIGURE 236. ANTENNA MOUNTING (ALL EXCEPT M1009 AND M1010).

(1) (2) (3) (4) ITEM SMR PART NO CODE CAGEC NUMBER	(5) DESCRIPTION AND USABLE ON CODES (UOC)	(6) QTY
	GROUP 2202 ACCESSORY ITEMS	
	FIG. 236 ANTENNA MOUNTING (ALL EXCEPT M1009 AND M1010)	
1 PAOZZ 11862 14076298	GROMMET,NONMETALLIC UOC:194,208,230,252,254,256	1
2 PAOZZ 11862 14072445	BRACKET,ANGLE UOC:194,208,230,252,254,256	4
3 PAOZZ 11862 15599994	NUT,PLAIN,BLIND RIV UOC:194,208,230,231,252,254,256,	16
4 PAOZZ 11862 9440300	SCREW,CAP,HEXAGON H UOC:194,208,230,252,254,256,	16
	END OF FIGURE	

TM9-2320-289-34P

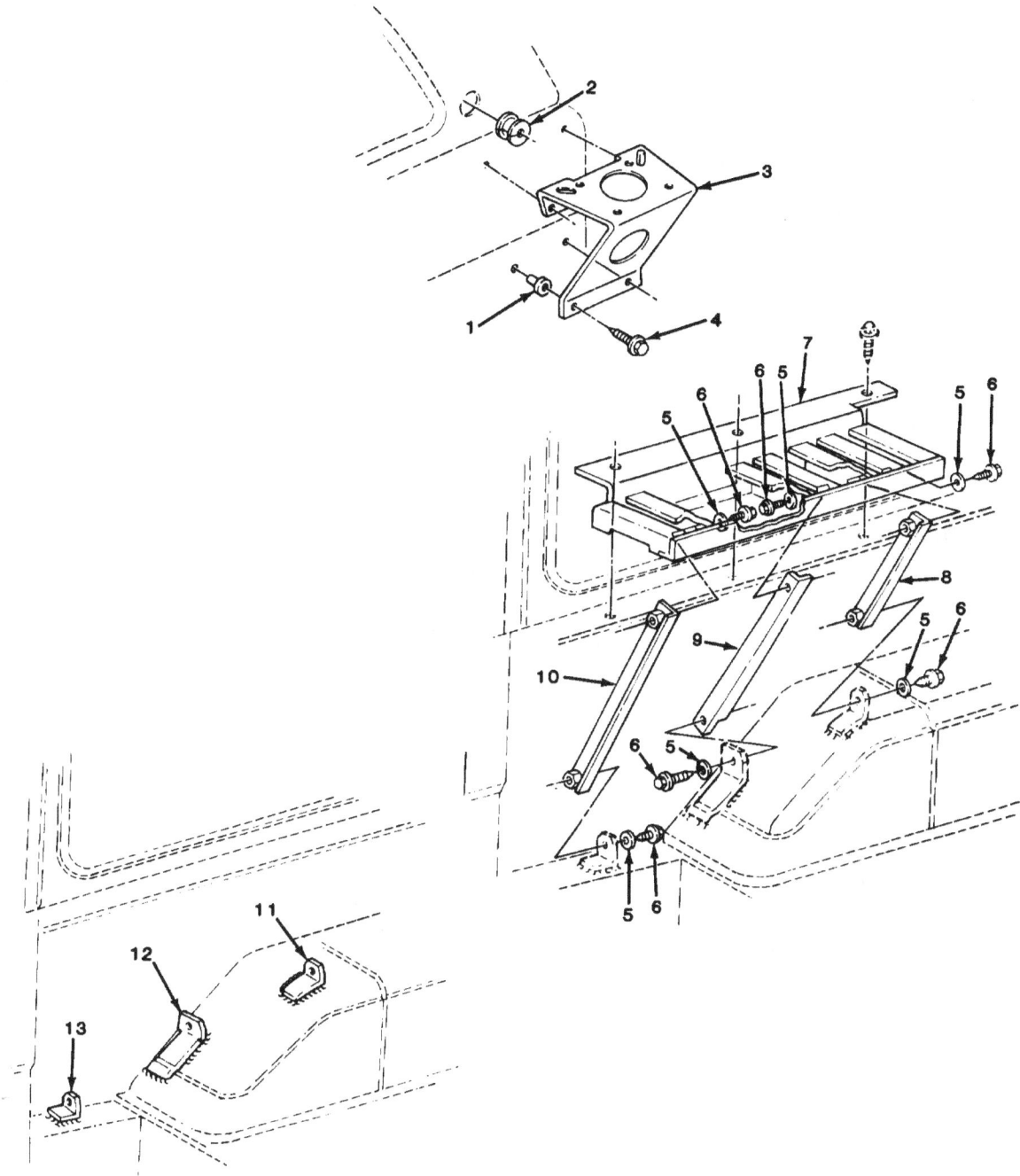

TA511044

FIGURE 237. RADIO AND ANTENNA BRACKETS (M1009).

(1) (2) (3) (4) ITEM SMR PART NO CODE CAGEC NUMBER	(5) DESCRIPTION AND USABLE ON CODES (UOC)	(6) QTY
	GROUP 2202 ACCESSORY ITEMS	
	FIG. 237 RADIO AND ANTENNA BRACKETS (M1009)	
1 PAOZZ 11862 15599994	NUT,PLAIN,BLIND RIV UOC:209	8
2 PAOZZ 11862 15599990 GROMMET,NONMETALLIC	UOC:209	2
3 PAOZZ 11862 14072445 BRACKET,ANGLE	UOC:209	2
4 PAOZZ 11862 9440300	SCREW,CAP,HEXAGON H UOC:209	8
5 PAOZZ 96906 MS35340-45 WASHER,LOCK	UOC:209	6
6 PAOZZ 96906 MS90728-34 BOLT,MACHINE	UOC:209	6
7 PAOZZ 11862 15599962 BRACKET	UOC:209	1
8 PAOZZ 11862 14072441 BRACKET,ANGLE	UOC:209	1
9 PAOZZ 11862 14072440 BRACKET,ANGLE	UOC:209	1
10 PAOZZ 11862 14072439 BRACKET,ANGLE	UOC:209	1
11 PAOZZ 11862 15599964 BRACKET,ANGLE	UOC:209	1
12 PAOZZ 11862 15599963	BRACKET,DOUBLE ANGL UOC:209	1
13 PAOZZ 11862 14072442 BRACKET,ANGLE	UOC:209	1

END OF FIGURE

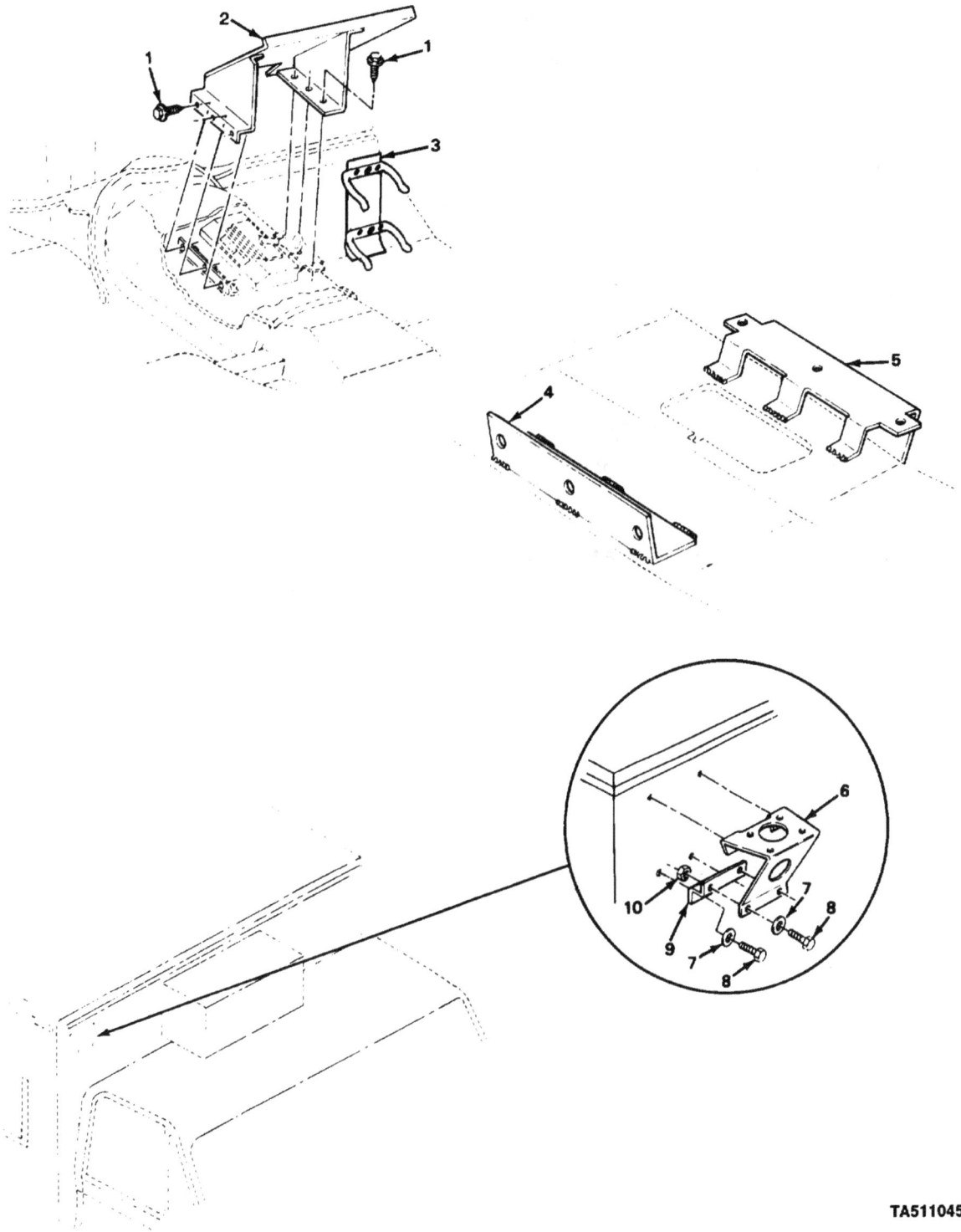

FIGURE 238. RADIO, FIRE EXTINGUISHER, AND ANTENNA BRACKETS (M1010).

TA511045

(1) (2) (3) (4)	(5)	(6)
ITEM SMR PART NO CODE CAGEC NUMBER	DESCRIPTION AND USABLE ON CODES (UOC)	QTY

GROUP 2202 ACCESSORY ITEMS

FIG. 238 RADIO, FIRE EXTINGUISHER,
AND ANTENNA BRACKETS (M1010)

1	PAOZZ	24617	9425117	BOLT,MACHINE UOC:210	6
2	PAOZZ	11862	14075846	BOLT,DOUBLE ANGL UOC:210	1
3	PAOZZ	19207	7357009	BRACKET,FIRE EXTING UOC:210	1
4	PAFZZ	11862	14075847	BRACKET,ANGLE UOC:210	1
5	PAFZZ	11862	14075848	BRACKET,ANGLE UOC:210	1
6	PAOZZ	25022	99-4602-1	BRACKET,ANGLE UOC:210	1
7	PAOZZ	96906	MS35338-45	WASHER,LOCK UOC:210	6
8	PAOZZ	96906	MS90728-34	BOLT,MACHINE UOC:210	6
9	PAOZZ	25022	99-4341-1	BRACKET,DOUBLE ANGL UOC:210	1
10	PAOZZ	96906	MS35691-9	NUT,PLAIN,HEXAGON UOC:210	2

END OF FIGURE

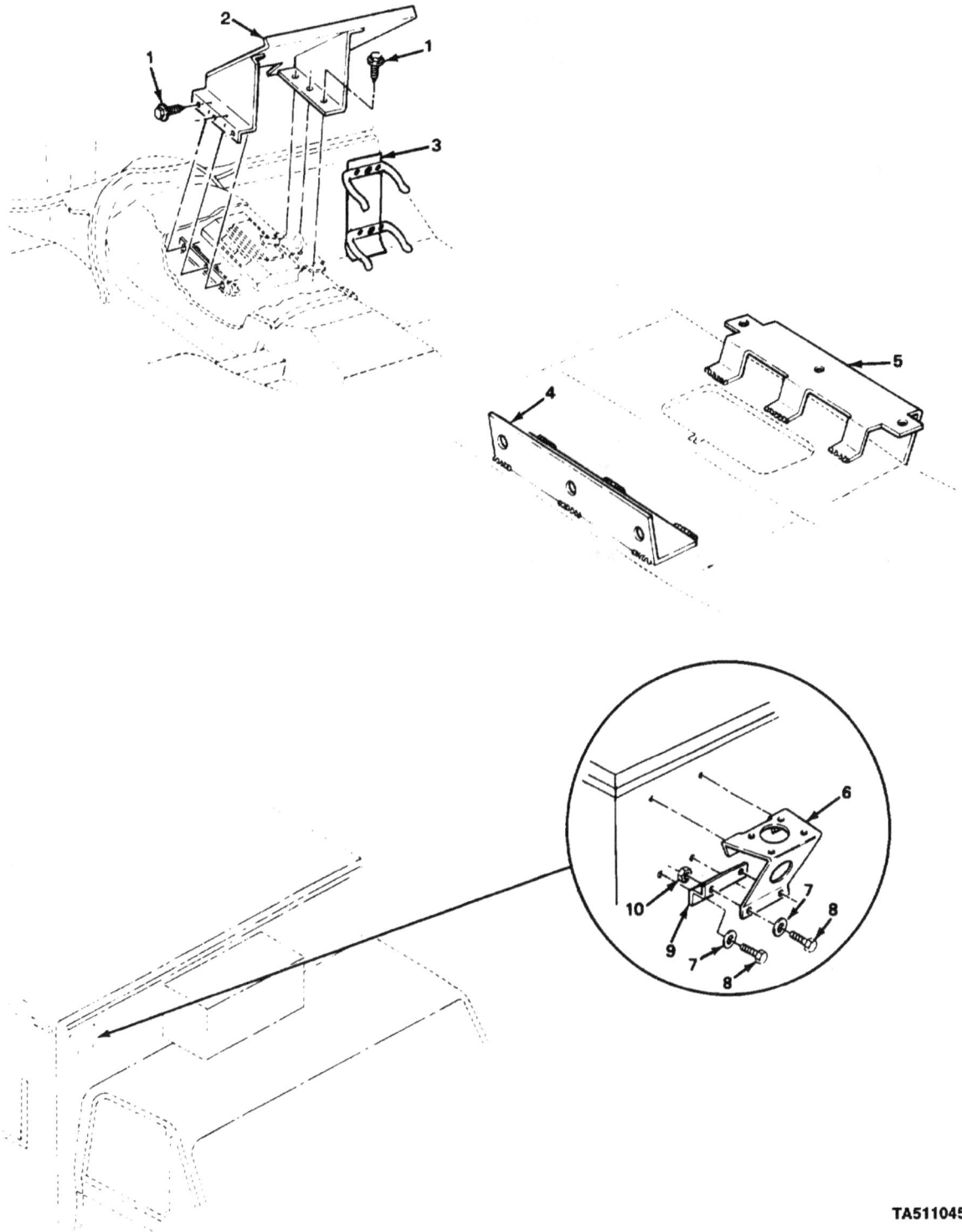

TA511045

FIGURE 238. RADIO, FIRE EXTINGUISHER, AND ANTENNA BRACKETS (M1010).

(1) ITEM NO	(2) SMR CODE	(3) CAGEC	(4) PART NUMBER	(5) DESCRIPTION AND USABLE ON CODES (UOC)	(6) QTY
				GROUP 2202 ACCESSORY ITEMS	
				FIG. 238 RADIO, FIRE EXTINGUISHER, AND ANTENNA BRACKETS (M1010)	
1	PAOZZ	24617	9425117	BOLT,MACHINE UOC:210	6
2	PAOZZ	11862	14075846	BOLT,DOUBLE ANGL UOC:210	1
3	PAOZZ	19207	7357009	BRACKET,FIRE EXTING UOC:210	1
4	PAFZZ	11862	14075847	BRACKET,ANGLE UOC:210	1
5	PAFZZ	11862	14075848	BRACKET,ANGLE UOC:210	1
6	PAOZZ	25022	99-4602-1	BRACKET,ANGLE UOC:210	1
7	PAOZZ	96906	MS35338-45	WASHER,LOCK UOC:210	6
8	PAOZZ	96906	MS90728-34	BOLT,MACHINE UOC:210	6
9	PAOZZ	25022	99-4341-1	BRACKET,DOUBLE ANGL UOC:210	1
10	PAOZZ	96906	MS35691-9	NUT,PLAIN,HEXAGON UOC:210	2

END OF FIGURE

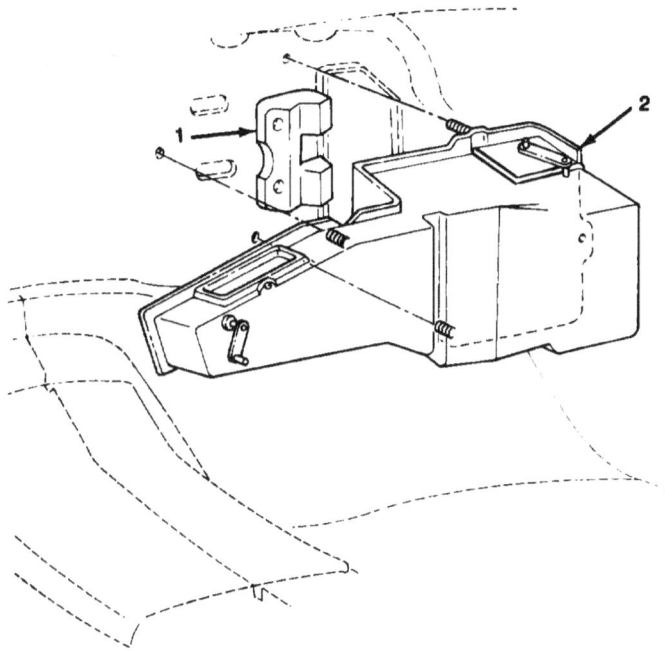

TA511046

FIGURE 239. HEATER ASSEMBLY.

(1) ITEM NO	(2) SMR CODE	(3) CAGEC	(5) PART NUMBER	DESCRIPTION AND USABLE ON CODES (UOC)	(6) QTY
				GROUP 2207 WINTERIZATION EQUIPMENT	
				FIG. 239 HEATER ASSEMBLY	
1	PAOZZ	11862	6273325	SEAL,RUBBER SPECIAL	1
2	AOOOO	11862	3054308	HEATER ASM FOR COMPONENT PARTS SEE FIG 240	1
				END OF FIGURE	

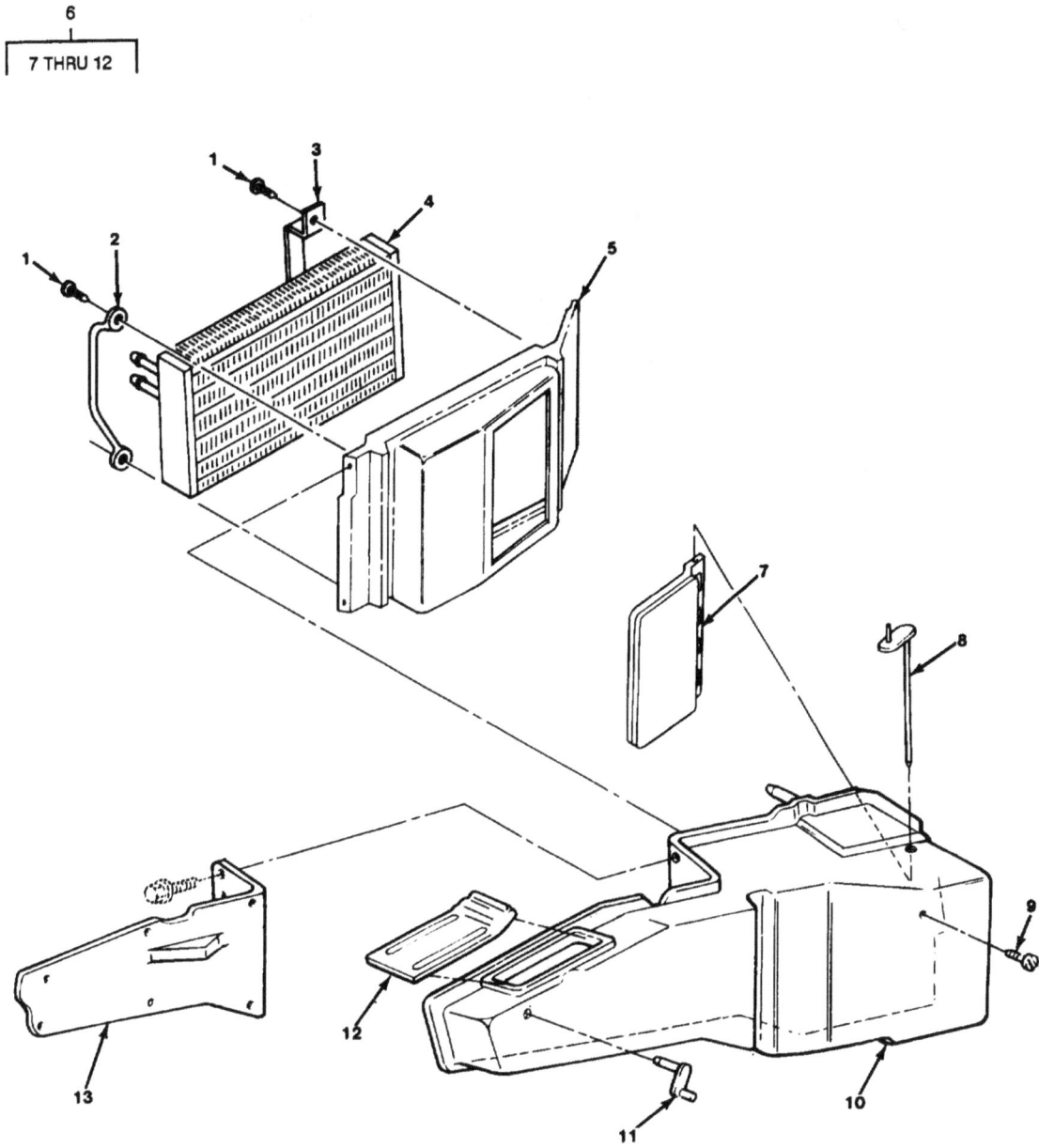

TA511047

FIGURE 240. HEATER ASSEMBLY COMPONENTS.

(1) (2) (3) (4)	(5)	(6)
ITEM SMR PART		
NO CODE CAGEC NUMBER	DESCRIPTION AND USABLE ON CODES (UOC)	QTY

GROUP 2207 WINTERIZATION EQUIPMENT

FIG. 240 HEATER ASSEMBLY COMPONENTS

1	PAOZZ 24617 9419699	BOLT,MACHINE		4
		UOC:194,208,209,210,230,231,252,254		
2	PAOZZ 11862 3025501	STRAP,RETAINING		1
3	PAOZZ 11862 3024673	STRAP,RETAINING		1
4	PAOZZ 11862 3027247	CORE,HEATER ASSEMBL		1
5	PAOZZ 11862 3048083	SHROUD,VALVE SEAT		1
6	PAOZZ 11862 3054315	HOUSING,HEATER COMP		1
7	PAOZZ 11862 3024867	.VALVE,HEATER CONTRO		1
8	PAOZZ 11862 3030075	.SHAFT AND LEVER ASS		1
9	PAOZZ 24617 11500999	.SCREW,ASSEMBLED WAS		1
10	PAOZZ 27462 3054316	.HOUSING ASSEMBLY,HE		1
11	PAOZZ 11862 3030072	.SHAFT AND LEVER ASS		1
12	PAOZZ 11862 3027308	.VALVE,DEFROSTER		1
13	PAOZZ 11862 3048067	PLATE AND BOLT ASSE		1

END OF FIGURE

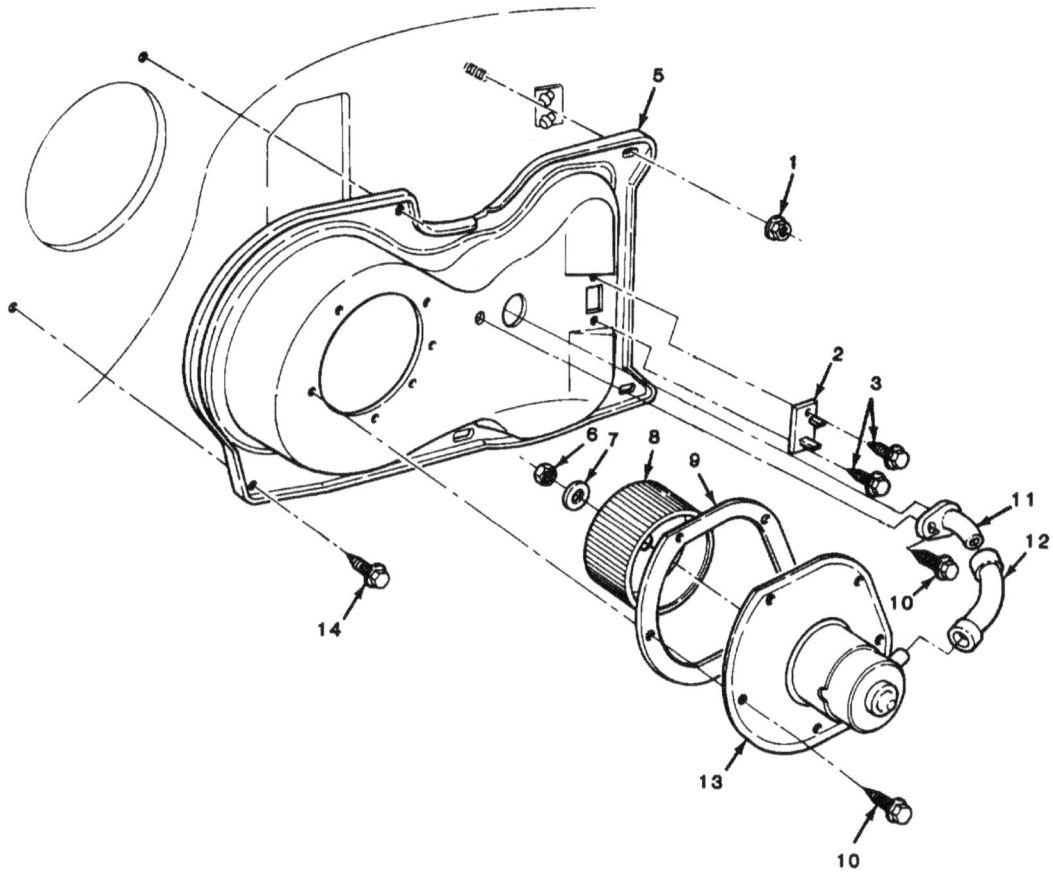

FIGURE 241. HEATER BLOWER MOTOR ASSEMBLY AND COMPONENT PARTS.

TA511048

(1) ITEM NO	(2) SMR CODE	(3) CAGEC	(4) PART NUMBER	(5) DESCRIPTION AND USABLE ON CODES (UOC)	(6) QTY

GROUP 2207 HEATER BLOWER MOTOR

FIG. 241 HEATER BLOWER MOTOR
ASSEMBLY AND COMPONENT PARTS

1	PAOZZ	24617	11506003	NUT,PLAIN,ASSEMBLED	3
2	PAOZZ	11862	500890	RESISTOR,FIXED,WIRE	1
3	PAOZZ	11862	11513932	SCREW,ASSEMBLED WAS	2
4	AOOOO	11862	3058097	BLOWER ASM	1
5	PAOZZ	11862	3029730	.CASE ASSEMBLY,BLOWE	1
6	PAOZZ	11862	3042351	.NUT,SHEET SPRING	1
7	PAOZZ	11862	3015545	.WASHER,FAN SUPPORT	1
8	PAOZZ	11862	3037550	.IMPELLER,FAN,CENTRI	1
9	PAOZZ	11862	3039873	.GASKET	1
10	PAOZZ	11862	11500742	.SCREW,TAPPING, THREA	6
11	PAOZZ	11862	3013475	.ELBOW,TUBE	1
12	PAOZZ	11862	3036927	.HOSE,PREFORMED	1
13	PAOZZ	11862	22020945	.MOTOR,DIRECT CURREN	1
14	PAOZZ	24617	11500997	SCREW,ASSEMBLED WAS	2

END OF FIGURE

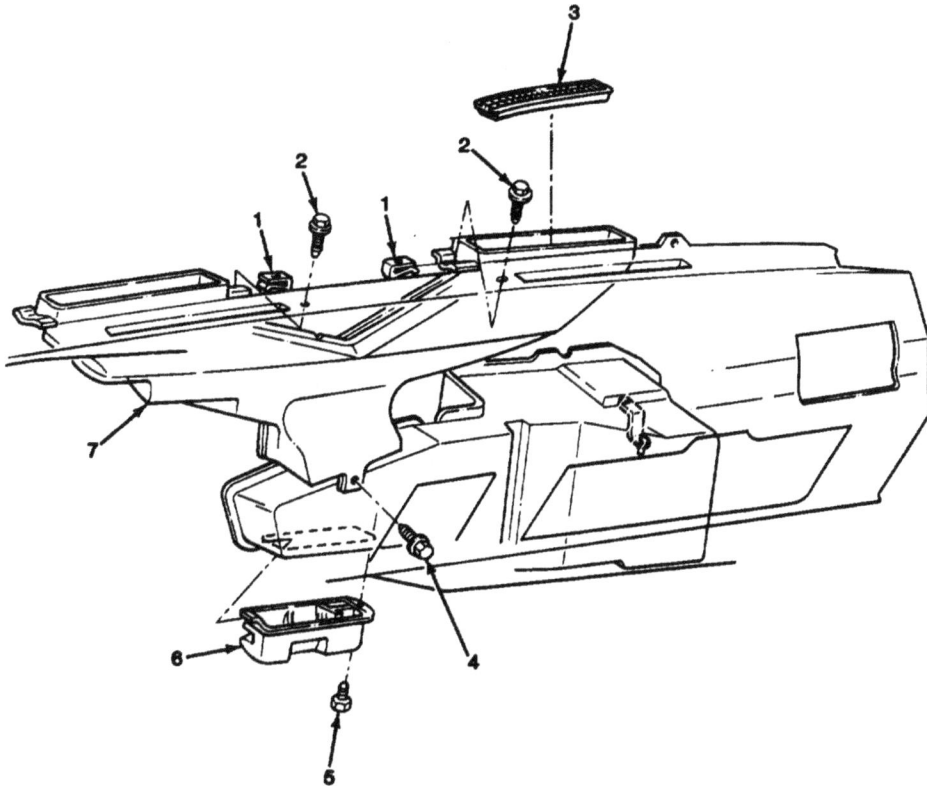

TA511049

FIGURE 242. HEATER ASSEMBLY AIR DUCTS.

(1) (2) (3) (4) ITEM SMR PART NO CODE CAGEC NUMBER	(5) DESCRIPTION AND USABLE ON CODES (UOC)	(6) QTY
	GROUP 2207 WINTERIZATION EQUIPMENT	
	FIG. 242 HEATER ASSEMBLY AIR DUCTS	
1 PAFZZ 11862 52351724	NUT,SHEET SPRING	2
2 PAFZZ 24617 9415163	SCREW,ASSEMBLED WAS	2
3 PAFZZ 11862 14064924 GRILLE,DEFROSTER		2
4 PAOZZ 24617 9440025 SCREW,TAPPING,THREA		1
5 PAOZZ 24617 11503396	SCREW,ASSEMBLED WAS	1
6 PAOZZ 11862 14013122	DUCT ASSEMBLY,HEATI RIGHT	1
7 PAFZZ 11862 14074319	NOZZLE ASSEMBLY,WIN	1

END OF FIGURE

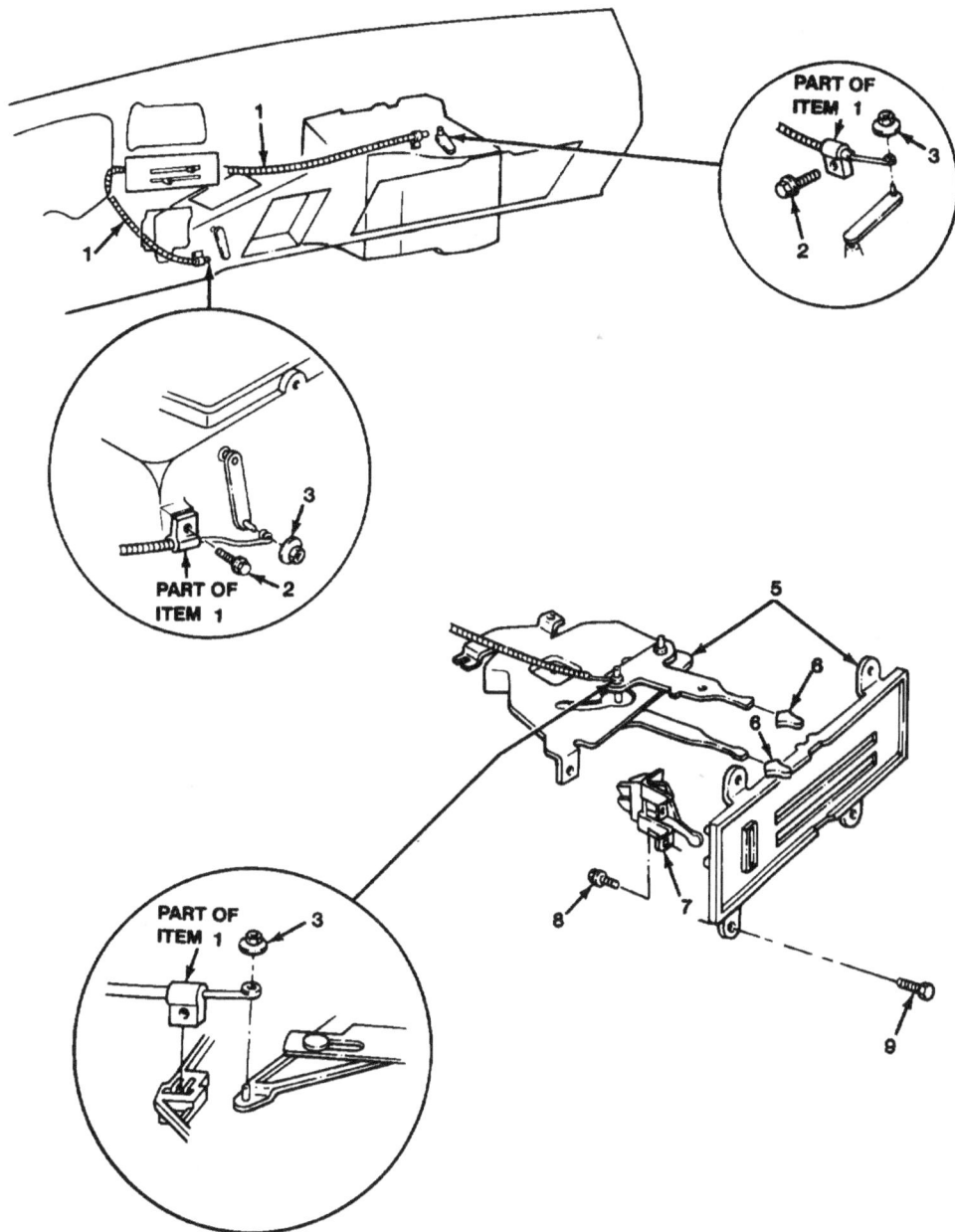

TA511050

FIGURE 243. HEATER CABLE AND CONTROL ASSEMBLY.

(1) ITEM NO	(2) SMR CODE	(3) CAGEC	(4) PART NUMBER	(5) DESCRIPTION AND USABLE ON CODES (UOC)	(6) QTY
				GROUP 2207 WINTERIZATION EQUIPMENT	
				FIG. 243 HEATER CABLE AND CONTROL ASSEMBLY	
1	PAOZZ	11862	6258364	CONTROL ASSEMBLY,PU	2
2	PAOZZ	96906	MS51850-86	SCREW,TAPPING,THREA	2
3	PAOZZ	11862	11501937	NUT	4
4	PAOOO	11862	16034561	CONTROL,HEATER	1
5	PAOZZ	11862	1226174	.PARTS KIT,HEATER CC	1
6	PAOZZ	11862	336406	.KNOB	2
7	PAOZZ	16758	16015256	.SWITCH,LEVER	1
8	PAOZZ	24617	9419663	.SCREW,TAPPING,THREA	2
9	PAOZZ	24617	9415163	SCREW,ASSEMBLED WAS	4

END OF FIGURE

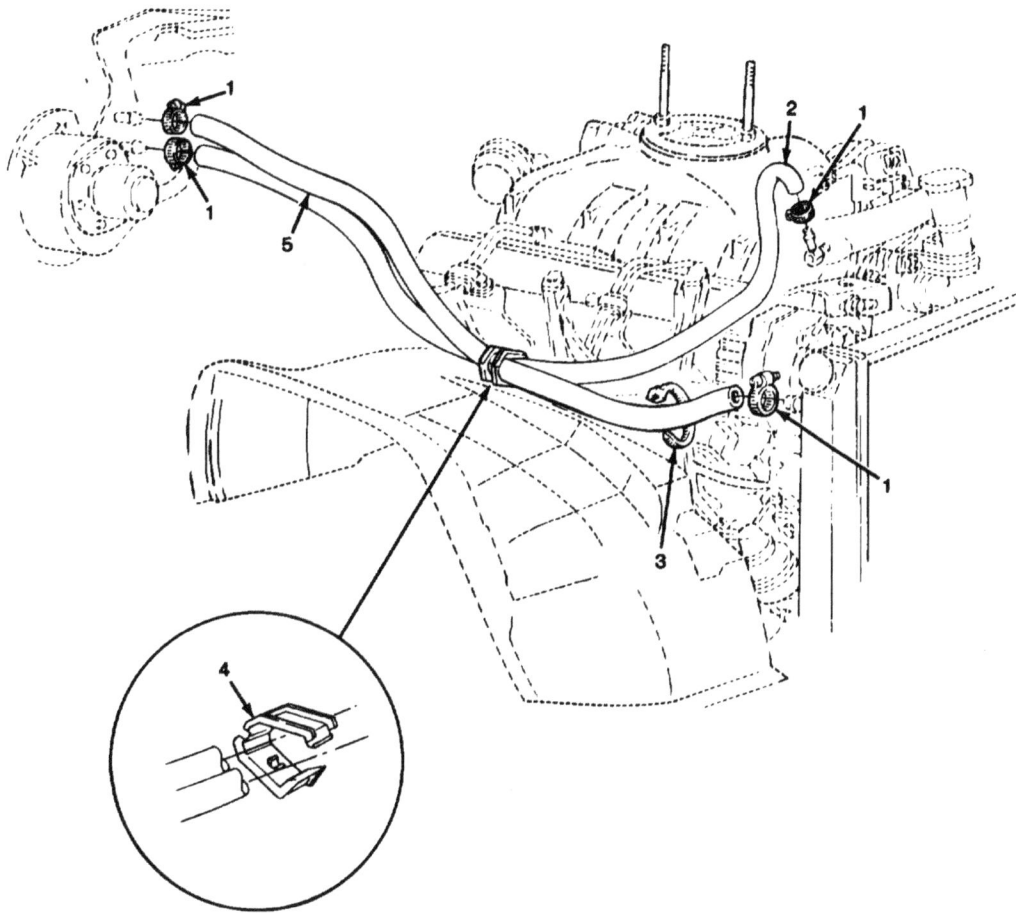

TA511051

FIGURE 244. HEATER HOSES.

(1)	(2)	(3)	(4)	(5)	(6)
ITEM NO	SMR CODE	CAGEC	PART NUMBER	DESCRIPTION AND USABLE ON CODES (UOC)	QTY

GROUP 2207 WINTERIZATION EQUIPMENT

FIG. 244 HEATER HOSES

1	PAOZZ	96906	MS35842-11	CLAMP,HOSE	4
2	MOOZZ	11862	487425	HOSE HEATER INLET (48.00"LG) MAKE FROM HOSE,P/N 482420 UOC:210	1
2	MOOZZ	11862	482995	HOSE,NONMETALLIC (42.00"LG) MAKE FROM HOSE,P/N MS521304B203R UOC:194,208,209,230,231,252,254,256	1
3	PAOZZ	11862	11509088	STRAP,LINE SUPPORTI UOC:210,230	1
4	PAOZZ	11862	3825416	CLAMP,LOOP	1
5	MOOZZ	11862	10012288-45	HOSE,PREFORMED (44.75"LG) MAKE FROM HOSE,P/N 482420	1

END OF FIGURE

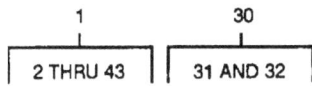

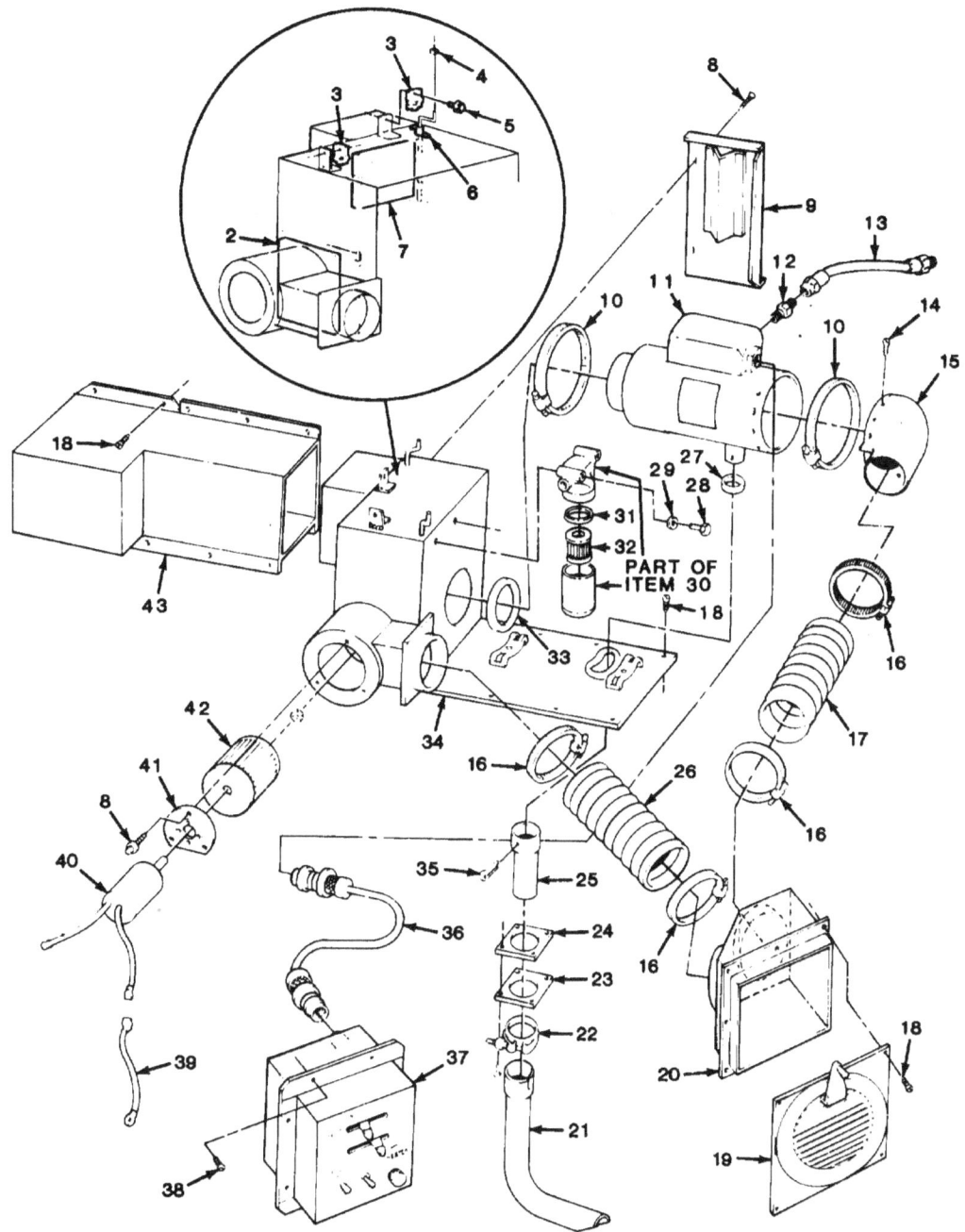

TA511052

FIGURE 245. PERSONNEL HEATER ASSEMBLY (M1010).

(1) ITEM NO	(2) SMR CODE	(3) CAGEC	(4) PART NUMBER	(5) DESCRIPTION AND USABLE ON CODES (UOC)	(6) QTY
				GROUP 2207 WINTERIZATION EQUIPMENT	
				FIG. 245 PERSONNEL HEATER ASSEMBLY (M1010)	
1	AOOOO	25022	22-0871	PARTS KIT,MANIFOLD UOC:210	1
2	PAOZZ	99688	58636	.GASKET UOC:210	1
3	PAOZZ	99688	58659	.CLIP,SPRING TENSION .UOC:210	2
4	PAOZZ	99688	62229	.PUSH ON NUT .UOC:210	2
5	PAOZZ	96906	MS51861-24	.SCREW,TAPPING THREA .UOC:210	2
6	PAOZZ	99688	58627	.SPRING,HELICAL,TORS UOC:210	1
7	PAOZZ	99688	58637	.GASKET UOC:210	1
8	PAOZZ	24617	9421073	.SCREW,TAPPING,THREA UOC:210	7
9	PAOZZ	99688	85941	.COVER,DOOR UOC:210	1
10	PAOZZ	96906	MS35842-16	.CLAMP,HOSE UOC:210	2
11	PAOFF	78385	10530B	.HEATER,VEHICULAR,CO FOR COMPONENT PARTS SEE FIG'S 246,248,249,250 .UOC:210	1
12	PAOZZ	72923	248X4	ADAPTER,STRAIGHT,PI .UOC:210	1
13	PAOZZ	99688	58625	HOSE ASSEMBLY,NONME .UOC:210	1
14	PAOZZ	96906	MS35206-247	.SCREW,MACHINE UOC:210	4
15	PAOZZ	99688	58623	.ADAPTER,AIR CONDITI UOC:210	1
16	PAOZZ	96906	MS35842-15	.CLAMP,HOSE UOC:210	4
16	PAOZZ	96906	MS35842-15	.CLAMP,HOSE UOC:210	4
17	MOOZZ	99688	58628	.HOSE,AIR DUCT (CUT TO 16.0" LG) MAKE FROM HOSE,P/N 319029 UOC:210	1
18	PAOZZ	96906	MS21141-U0604	.RIVET,BLIND UOC:210	8
19	PFOZZ	99688	784030	.VENTILATOR,AIR CIRC UOC:210	1
20	XDOZZ	99688	78405	.VENT,HEATER UOC:210	1
21	PAOZZ	99688	58649	.PIPE,EXHAUST UOC:210	1
22	PAOZZ	99688	58620	.CLAMP,HOSE UOC:210	1
23	PAOZZ	99688	46090	.COLLAR,AIR CONDITION	1

(1) ITEM NO	(2) SMR CODE	(3) CAGEC	(4) PART NUMBER	(5) DESCRIPTION AND USABLE ON CODES (UOC)	(6) QTY
				UOC:210	
24	PAOZZ	99688	58635	.GASKET	1
				UOC:210	
25	PAOZZ	99688	58651	.CONNECTOR,EXHAUST P	1
				UOC:210	
26	MOOZZ	99688	58629	.HOSE,AIR DUCT (CUT TO 23.0" LG) MAKE FROM HOSE, P/N 319029	1
				UOC:210	
27	PAOZZ	99688	58632	.GASKET	1
				UOC:210	
28	PAOZZ	96906	MS51849-70	.SCREW,MACHINE	2
				UOC:210	
29	PAOZZ	96906	MS27183-42	.WASHER,FLAT	2
				UOC:210	
30	PAOOO	96906	MS51085-1	.FILTER,FLUID	1
				UOC:210	
31	PAOZZ	96906	MS29513-125	..PACKING,PREFORMED	1
				UOC:210	
32	PAOZZ	90005	26422-B	..FILTER,ELEMENT,FLUI	1
				UOC:210	
33	POOZZ	99688	58624	.GASKET	1
				UOC:210	
34	PFOZZ	99688	85942	.BASE,HEATER	1
				UOC:210	
35	PAOZZ	96906	MS24665-446	.PIN,COTTER	1
				UOC:210	
36	PAOZZ	99688	65148	.WIRING HARNESS,BRAN	1
				UOC:210	
37	PAOFF	99588	859430	.PANEL,HEATER CONTRO FOR COMPNENT PARTS SEE FIG 247	1
				UOC:210	
38	PAOZZ	96906	MS20600B6W4	.RIVET,BLIND	28
				UOC:210	
39	PAOZZ	99688	65149	.LEAD,ELECTRICAL	1
				UOC:210	
40	PAOZZ	0A7R8	PV23135R	.MOTOR,DIRECT CURREN 24 VOLT D.C	1
				UOC:210	
41	PAOZZ	99688	46023	.PLATE,MOUNTING,GROM	1
				UOC:210	
42	PAOZZ	99688	58634	.IMPELLER,FAN,CENTRI	1
				UOC:210	
43	PFOZZ	99688	78406	.DUCT,FRESH AIR	1
				UOC:210	

END OF FIGURE

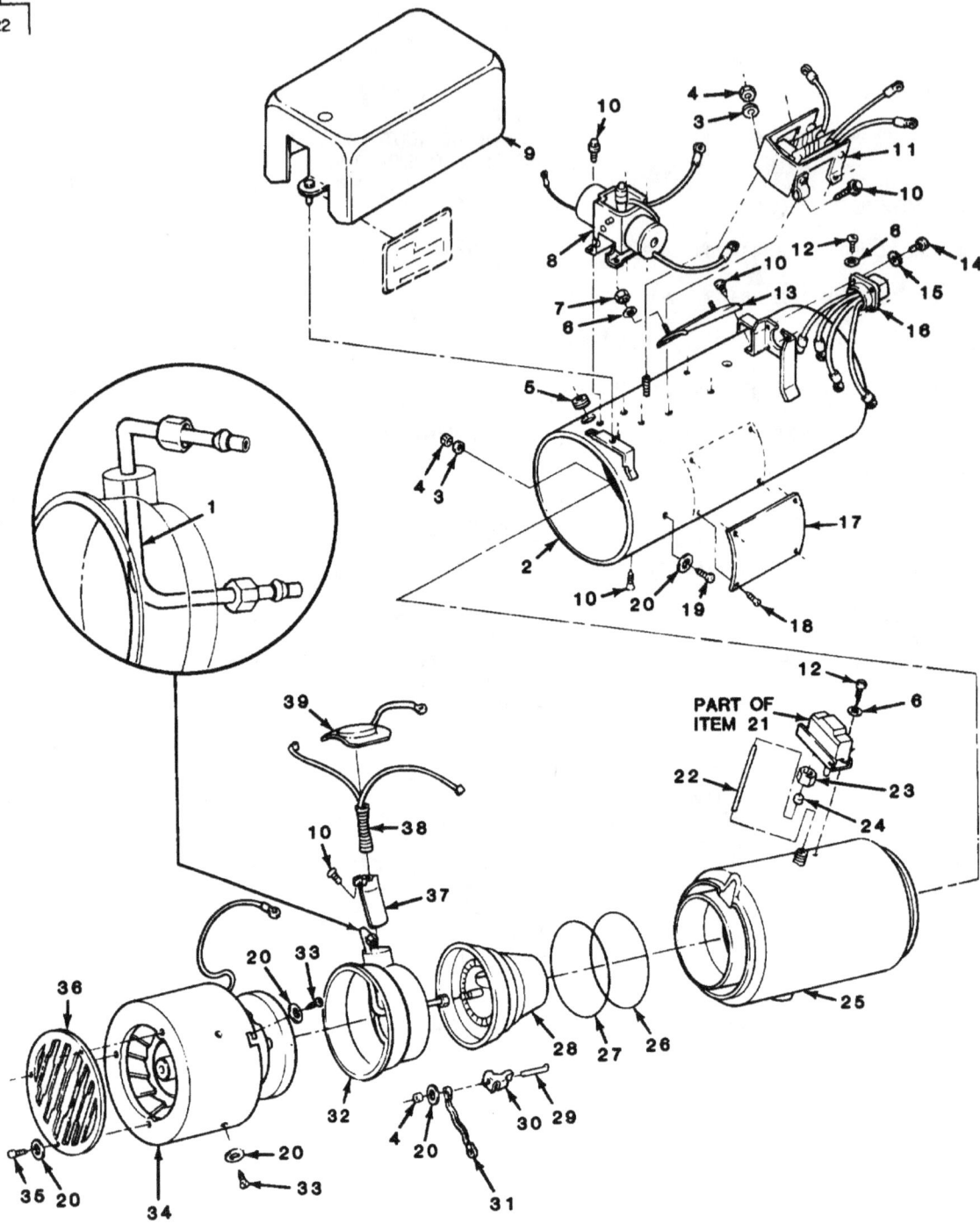

FIGURE 246. WARM AIR HEATER COMPONENT PARTS (M1010).

TA511053

(1) (2) (3) (4) ITEM SMR PART NO CODE CAGEC NUMBER	(5) DESCRIPTION AND USABLE ON CODES (UOC)	(6) QTY
	GROUP 2207 WINTERIZATION EQUIPMENT	
	FIG. 246 WARM AIR HEATER COMPONENT PARTS (M1010)	
1 PAFZZ 78385 704363	TUBE ASSEMBLY,METAL UOC:210	1
2 XAFZZ 78385 G-704213	HOUSING HEATER UOC:210	1
3 PAFZZ 96906 MS45904-57	WASHER,LOCK UOC:210	3
4 PAFZZ 96906 MS35649-282	NUT,PLAIN,HEXAGON PART OF KIT P/N 5704051 PART OF KIT P/N 5704052 UOC:210	6
5 PAFZZ 96906 MS35489-5	GROMMET,NONMETALLIC UOC:210	1
6 PAFZZ 96906 MS35335-30	WASHER,LOCK UOC:210	8
7 PAFZZ 96906 MS35649-262	NUT,PLAIN,HEXAGON UOC:210	2
8 PAFZZ 19207 11663058	VALVE ASSEMBLY,FUEL FOR COMPONENT PARTS SEE FIG 248 UOC:210	1
9 XAFZZ 78385 G-704183	GAURD UOC:210	1
10 PAFZZ 78385 719675	SCREW,MACHINE UOC:210	13
11 PAFZZ 19207 11663061	RESISTOR,IGNITION C UOC:210	1
12 PAFZZ 96906 MS35206-226	SCREW,MACHINE UOC:210	6
13 PAFZZ 19207 10948233	SWITCH,THERMOSTATIC UOC:210	1
14 PAFZZ 96906 MS51863-22	SCREW,TAPPING,THREA UOC:210	4
15 PAFZZ 96906 MS35333-37	WASHER,LOCK UOC:210	4
16 PAFZZ 78385 G-704234	CONNECTOR,RECEPTACL UOC:210	1
17 XAFZZ 78385 704501	PLATE,IDENTIFICATIO UOC:210	1
18 PAFZZ 78385 488755	RIVET,BLIND UOC:210	4
19 PAFZZ 96906 MS35206-243	SCREW,MACHINE UOC:210	1
20 PAFZZ 96906 MS35335-31	WASHER,LOCK PART OF KIT P/N 5704051 PART OF KIT P/N 5704052 UOC:210	14
21 PAFZZ 19207 11663057	SWITCH,THERMOSTATIC UOC:210	1
22 PAFZZ 78385 704225	.ROD,NONEXPANSIVE UOC:210	1

(1)	(2)	(3)	(4)	(5)	(6)
ITEM NO	SMR CODE	CAGEC	PART NUMBER	DESCRIPTION AND USABLE ON CODES (UOC)	QTY
23	PAFZZ	79470	1461-4	NUT UOC:210	1
24	XDFZZ	21450	114628	SLEEVE,COMPRESSION, UOC:210	1
25	XAFZZ	78385	G-704232	EXCHANGER HEAT UOC:210	1
26	PAFZZ	78385	702903	GASKET PART OF KIT P/N 5704051 PART OF KIT P/N 5704052 UOC:210	1
27	PAFZZ	19207	11588688	PACKING,PREFORMED PART OF KIT P/N 5704051 PART OF KIT P/N 5704052 UOC:210	1
28	KFFZZ	78385	G-704195	BURNER ASSEMBLY FOR COMPONENT PARTS SEE FIG 250 PART OF KIT P/N 5704051 UOC:210	1
29	KFFZZ	78385	703547	BOLT-HOOK PART OF KIT P/N 5704051 PART OF KIT P/N 57040452 UOC:210	4
30	KFFZZ	78385	703546	CLAMP,RIM CLENCHING PART OF KIT P/N 4 5704051 PART OF KIT P/N 5704052 UOC:210	4
31	XAFFZZ	78385	G-704373	LEAD GROUND UOC:210	1
32	XAFZZ	78385	G704177	HOUSING,BAFFLE,HEAT UOC:210	1
33	PAFZZ	96906	MS51863-32	SCREW,TAPPING,THREA UOC:210	4
34	PAFFF	78385	G-704554	BLOWER ASSEMBLY FOR COMPONENT PARTS SEE FIG 249 UOC:210	1
35	PAFZZ	96906	MS35206-241	SCREW,MACHINE UOC:210	3
36	XAFZZ	78385	488773	VENTILATOR,AIR CIRC UOC:210	1
37	XAFZZ	78385	G-704288-1	IGNITER TUBE UOC:210	1
38	PAFZZF	16236	CS-4520-SV-0705	IGNITER,SPARK,FUEL UOC:210	1
39	XAFZZ	78385	G-704293	COVER ASSY, HATCH UOC:210	

END OF FIGURE

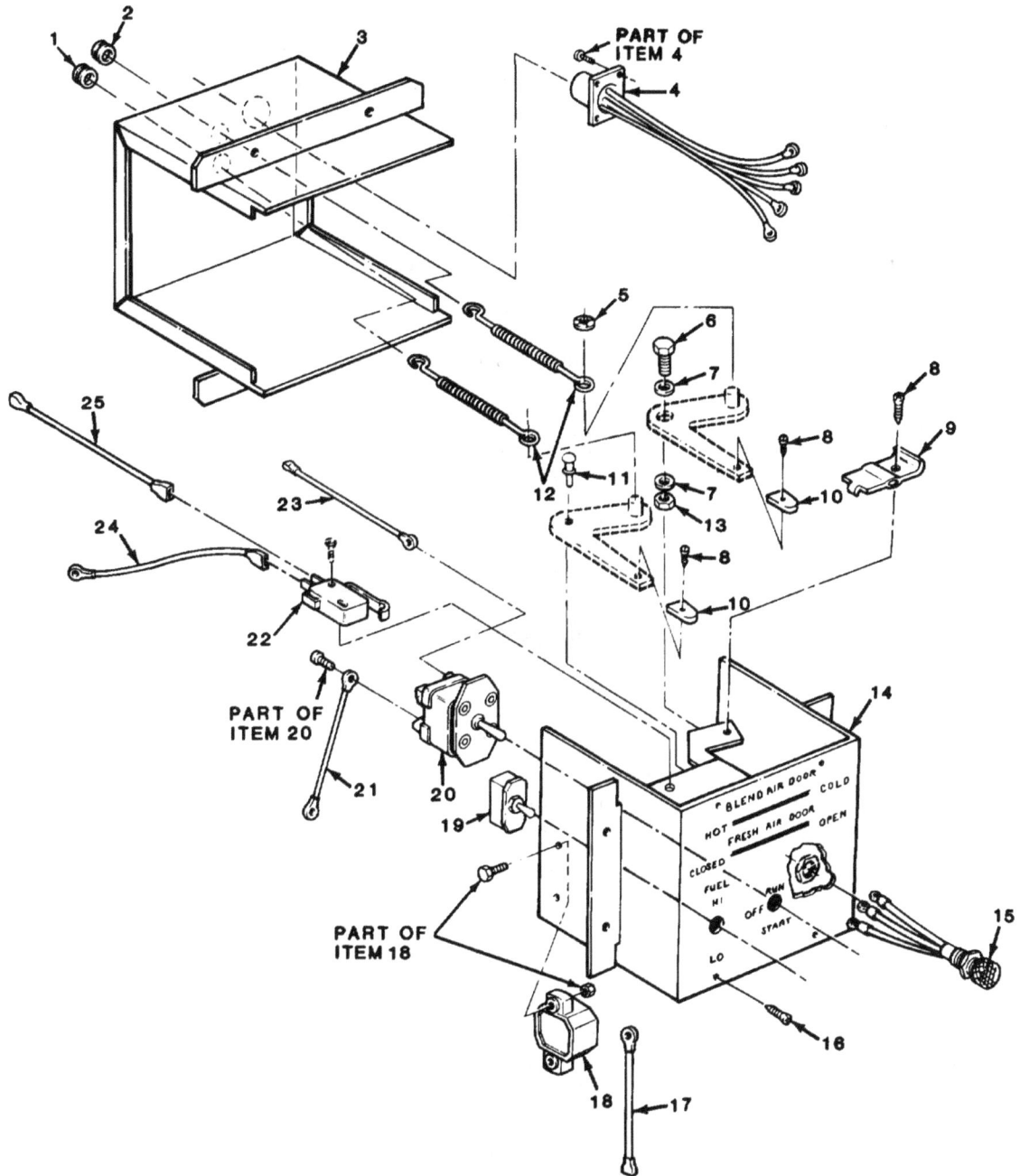

TA511054

FIGURE 247. PERSONNEL HEATER CONTROL COMPONENTS (M1010).

(1) (2) (3) (4) ITEM SMR PART NO CODE CAGEC NUMBER	(5) DESCRIPTION AND USABLE ON CODES (UOC)	(6) QTY
	GROUP 2207 WINTERIZATION EQUIPMENT	
	FIG. 247 PERSONNEL HEATER CONTROL COMPONENTS (M1010)	
1 PAFZZ 99688 58660	GROMMET,NONMETALLIC UOC:210	1
2 PAFZZ 99688 50090	GROMMET,NONMETALLIC UOC:210	1
3 PFFZZ 99688 859450	BOX ASSEMBLY,HEATER UOC:210	1
4 PAFZZ 24234 221968	CABLE ASSEMBLY,POWE UOC:210	1
5 PAFZZ 99688 62229	PUSH ON NUT UOC:210	2
6 PAFZZ 96906 MS51869-26	SCREW,TAPPING,THREA UOC:210	1
7 PAFZZ 99688 62220	WASHER,SPRING TENSI UOC:210	2
8 PAFZZ 96906 MS51861-24	SCREW,TAPPING,THREA UOC:210	4
9 PAFZZ 99688 58659	CLIP,SPRING TENSION UOC:210	2
10 PAFZZ 99688 58668	KNOB UOC:210	2
11 PAFZZ 99688 58661	FASTENER,SNAP UOC:210	1
12 PAFZZ 99688 58643	WIRE ROPE ASSEMBLY UOC:210	2
13 PAFZZ 99688 62219	NUT,SELF-LOCKING,HE UOC:210	1
14 PAFZZ 99688 85944	COVER ASSEMBLY,HEAT UOC:210	1
15 PAFZZ 96906 MS25331-4-313S LIGHT,INDICATOR UOC:210		1
16 PAFZZ 24617 9421073	SCREW,TAPPING,THREA UOC:210	8
17 PAFZZ 99688 65151	LEAD,ELECTRICAL UOC:210	1
18 PAFZZ 82647 PDA-15	CIRCUIT BREAKER UOC:210	1
19 PAFZZ 96906 MS35058-22	SWITCH,TOGGLE UOC:210	1
20 PAFZZ 91929 32TS1-5	SWITCH,TOGGLE UOC:210	1
21 PAFZZ 99688 65153	LEAD,ELECTRICAL UOC:210	1
22 PAFZZ 99688 65171	SWITCH,SENSITIVE UOC:210	1
23 PAFZZ 99688 65155	LEAD,ELECTRICAL UOC:210	1
24 PAFZZ 99688 65152	LEAD,ELECTRICAL	1

(1) ITEM NO	(2) SMR CODE	(3) CAGEC	(4) PART NUMBER	(5) DESCRIPTION AND USABLE ON CODES (UOC)	(6) QTY
				UOC:210	
25	PAFZZ	99688	65154	LEAD,ELECTRICAL	1
				UOC:210	

END OF FIGURE

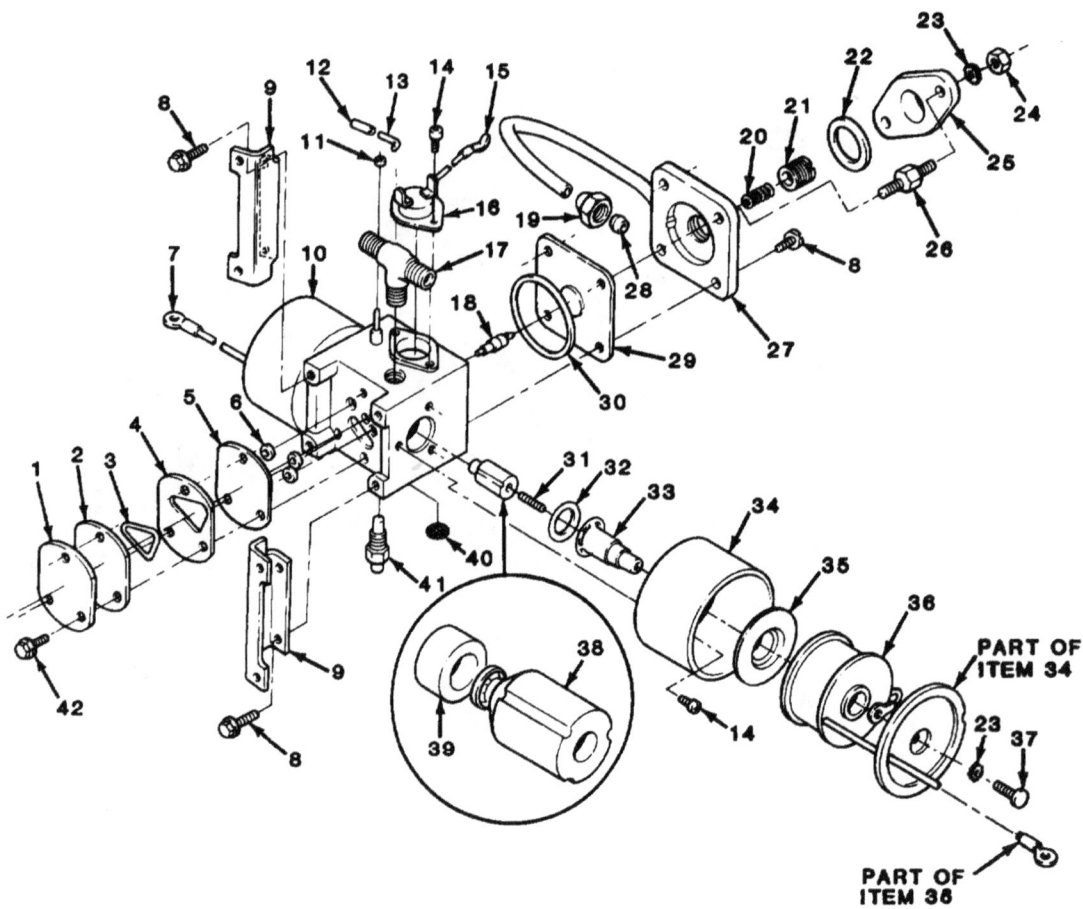

TA511055

FIGURE 248. WARM AIR HEATER VALVE ASSEMBLY COMPONENT PARTS (M1010).

(1) (2) (3) (4) ITEM SMR PART NO CODE CAGEC NUMBER	(5) DESCRIPTION AND USABLE ON CODES (UOC)	(6) QTY
	GROUP 2207 WINTERIZATION EQUIPMENT	
	FIG. 248 WARM AIR HEATER VALVE ASSEMBLY COMPONENT PARTS (M1010)	
1 XAFZZ 78385 704396	PLATE,OUTER UOC:210	1
2 XAFZZ 78385 704395	PLATE,INSULATOR UOC:210	1
3 KFFZZ 78385 704432	PACKING,PREFORMED PART OF KIT P/N 5704064 UOC:210	1
4 XAFZZ 78385 704402	PLATE,THREE HOLE UOC:210	1
5 XDFZZ 78385 704390	PLATE,RESTRICTION,F UOC:210	1
6 PAFZZ 78385 704406	PACKING,PREFORMED PART OF KIT P/N 5704064 UOC:210	3
7 PAFZZ 78385 G-700637-101	PARTS KIT,SCLENOID UOC:210	1
8 PAFZZ 24617 9415319	SCREW,CAKP,HECAGON H UOC:210	6
9 XAFZZ 78385 705311	BRACKET,MOUNTING UOC:210	2
10 XDFZZ 78385 704391-1	BODY,VALVE UOC:210	1
11 XAFZZ 78385 704789	INSULATOR,WASHER UOC:210	2
12 XAFZZ 78385 735447-5	INSULATION SLEEVING UOC:210	1
13 XAFZZ 78385 704447	WIRE UOC:210	1
14 PAFZZ 96906 MS51863-21	SCREW,TAPPING,THREA UOC:210	8
15 XDFZZ 78385 G704676	CABLE ASSEMBLY UOC:210	1
16 PAFZZ 78385 704401	SWITCH,THERMOSTATIC UOC:210	1
17 PAFZZ 96906 MS39173-2	TEE,PIPE TO TUBE UOC:210	1
18 KFFZZ 78385 702942	CORE VALVE PART OF KIT P/N 5704064 UOC:210	1
19 PAFZZ 35211 RL984A	NUT,TUBE COUPLING UOC:210	1
20 PAFZZ 78385 704434	SPRING,HELICAL,COMP UOC:210	1
21 PAFZZ 78385 704397	SCREW UOC:210	1
22 PAFZZ 78385 705344	GASKET UOC:210	1
23 PAFZZ 96906 MS35335-31	WASHER,LOCK	4

(1) (2) (3) (4)	(5)	(6)
ITEM SMR PART		
NO CODE CAGEC NUMBER	DESCRIPTION AND USABLE ON CODES (UOC)	QTY

	UOC:210	
24 PAFZZ 96906 MS35649-282	NUT,PLAIN,HEXAGON	2
	UOC:210	
25 XAFZZ 78385 705347	RETAINER,PACKING	1
	UOC:210	
26 PFFZZ 78385 705345	STUD,PLAIN	2
	UOC:210	
27 XAFZZ 78385 G705349	CAP AND BYPASS ASSE	1
	UOC:210	
28 PAFZZ 79470 60X3	SLEEVE,COMPRESSION	1
	UOC:210	
29 KFFZZ 78385 G-704398	DIAPHRAGM PART OF KIT P/N 5704064	1
	UOC:210	
30 KFFZZ 78385 704580	PACKING,PREFORMED PART OF KIT P/N 5704064	1
	UOC:210	
31 PAFZZ 78385 476339	SPRING,HELICAL,COMP	2
	UOC:210	
32 KFFZZ 78385 474669	GASKET PART OF KIT P/N 5704064	2
	UOC:210	
33 XDFZZ 78385 476220	BUSHING,SLEEVE	2
	UOC:210	
34 XAFZZ 78385 484044	HOUSING,HEATER SOLE	2
	UOC:210	
35 PAFZZ 78385 476229	WASHER,FLAT	2
	UOC:210	
36 PAFZZ 78385 G-700637-100	PARTS KIT,SOLENOID	1
	UOC:210	
37 PAFZZ 24617 455956	SCREW,MACHINE	2
	UOC:210	
38 XAFZZ 78385 701328	PLUNGER,SOLENOID	2
	UOC:210	
39 KFFZZ 78385 735411	SEAL,PLAIN PART OF KIT P/N 5704064	2
	UOC:210	
40 KFFZZ 78385 704898	SCREEN,INLET PART OF KIT P/N 5704064	1
	UOC:210	
41 PAFZZ 78385 704579	STEM,NEEDLE VALVE	1
	UOC:210	
42 PAFZZ 24617 9419950	SCREW,TAPPING	3
	UOC:210	

END OF FIGURE

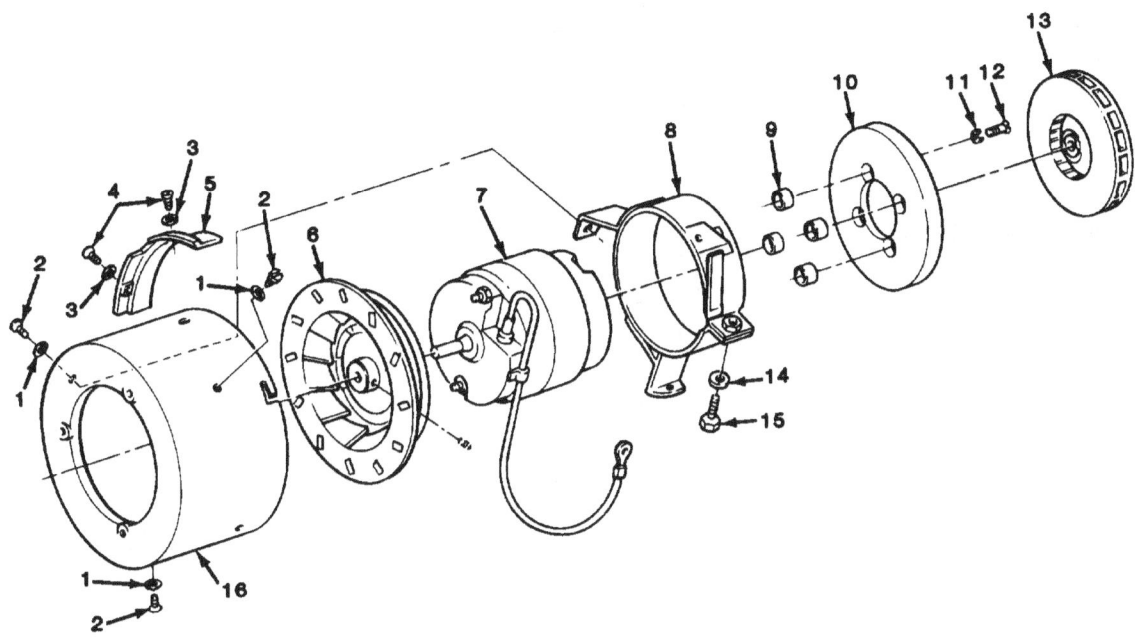

TA511056

FIGURE 249. WARM AIR HEATER BLOWER ASSEMBLY COMPONENT PARTS (M1010).

(1)	(2)	(3)	(4)	(5)	(6)
ITEM	SMR		PART		
NO	CODE	CAGEC	NUMBER	DESCRIPTION AND USABLE ON CODES (UOC)	QTY

GROUP 2207 WINTERIZATION EQUIPMENT

FIG. 249 WARM AIR HEATER BLOWER
ASSEMBLY COMPONENT PARTS (M1010)

Item	SMR Code	CAGEC	Part Number	Description	QTY
1	PAFZZ	96906	MS35335-32	WASHER,LOCK UOC:210	3
2	PAFZZ	96906	MS35207-261	SCREW,MACHINE UOC:210	3
3	PAFZZ	96906	MS35335-31	WASHER,LOCK UOC:210	2
4	PAFZZ	96906	MS35206-241	SCREW,MACHINE UOC:210	2
5	XAFZZ	78385	G-704553	BAFFLE VENT UOC:210	1
6	XAFZZ	78385	G-720122	WHEEL BLOWER UOC:210	1
7	PAFZZ	19207	8720780-1	MOTOR,DIRECT CURREN UOC:210	1
8	XAFZZ	78385	G-700038	BRACKET ASSEMBLY UOC:210	1
9	PAFZZ	78385	702129	SPACER UOC:210	4
10	PAFZZ	78385	488935	HEADER,SECONDARY BL UOC:210	1
11	PAFZZ	96906	MS35790-9	WASHER,LOCK UOC:210	4
12	PAFZZ	96906	MS35190-254	SCREW,MACHINE UOC:210	4
13	XAFZZ	78385	G-488934	WHEEL BLOWER UOC:210	1
14	PAFZZ	78385	736861	WASHER,LOCK UOC:210	1
15	PAFZZ	96906	MS90728-8	SCREW,CAP,HEXAGON H UOC:210	1
16	XAFZZ	78385	704408	HOUSING,BLOWER,HEAT UOC:210	1

END OF FIGURE

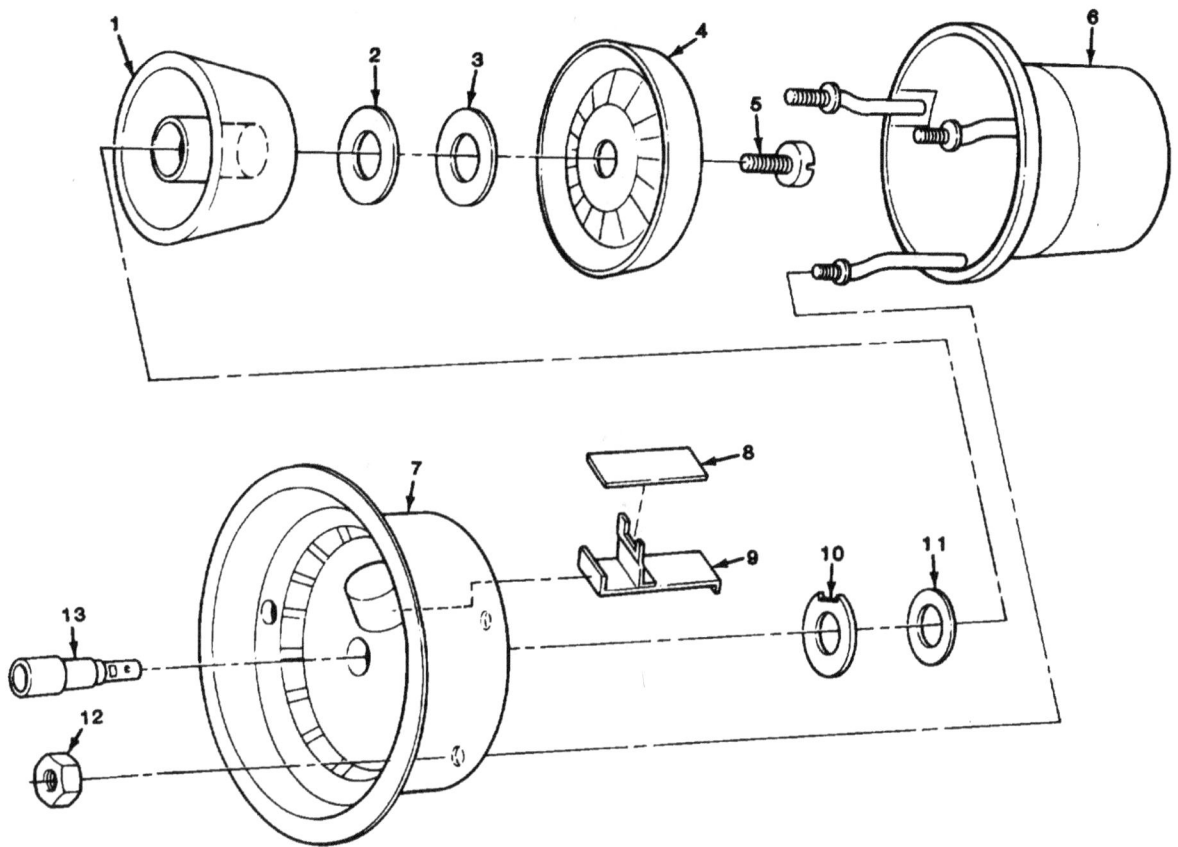

TA511057

FIGURE 250. WARM AIR HEATER BURNER ASSEMBLY COMPONENT PARTS (M1010).

(1) (2) (3) (4) ITEM SMR PART NO CODE CAGEC NUMBER	(5) DESCRIPTION AND USABLE ON CODES (UOC)	(6) QTY
	GROUP 2207 WINTERIZATION EQUIPMENT	
	FIG. 250 WARM AIR HEATER BURNER ASSEMBLY COMPONENT PARTS (M1010)	
1 KFFZZ 78385 704192	VAPORIZER PART OF KIT P/N 5704052 UOC:210	1
2 KFFZZ 78385 704678	WASHER,FLAT PART OF KIT P/N 5704052 UOC:210	1
3 PAFZZ 96906 MS14151-6	WASHER,FLAT PART OF KIT P/N 5704052 UOC:210	1
4 XAFZZ 78385 G-704284	HDR PLATE BUSHG FUL UOC:210	1
5 PAFZZ 78385 705117	SCREW,MACHINE PART OF KIT P/N 5704052 UOC:210	1
6 XAFZZ 78385 G-704196	BURNER CAP UOC:210	1
7 XAFZZ 78385 704181	SHIELD,FUEL VAPOR UOC:210	1
8 KFFZZ 78385 704371-1	WICK PART OF KIT P/N 5704052 UOC:210	1
9 PAFZZ 78385 704285	CLAMP,LOOP UOC:210	1
10 XAFZZ 78385 704189	WASHER UOC:210	1
11 KFFZZ 78385 704191	WASHER,FIBER PART OF KIT P/N 5704052 UOC:210	1
12 PAFZZ 96906 MS21043-08	NUT,SELF-LOCKING,EX UOC:210	3
13 XAFZZ 78385 704283	BUSHING,SLEEVE UOC:210	1

END OF FIGURE

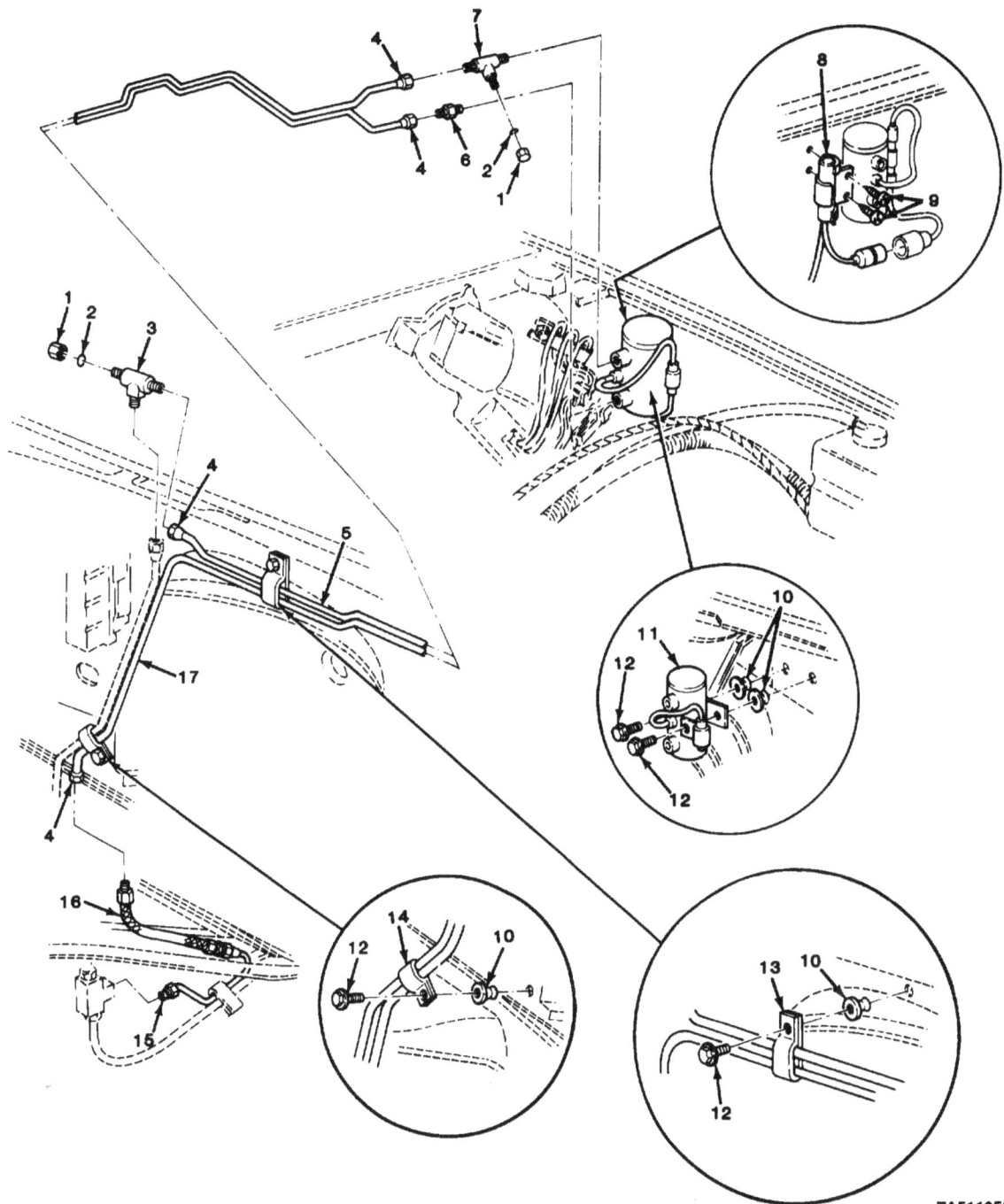

FIGURE 251. PERSONNEL HEATER FUEL PUMP, LINES, AND RELATED PARTS (M1010).

TA511058

(1) (2) (3) (4)	(5)	(6)
ITEM SMR PART		
NO CODE CAGEC NUMBER	DESCRIPTION AND USABLE ON CODES (UOC)	QTY

GROUP 2207 WINTERIZATION EQUIPMENT

FIG. 251 PERSONNEL HEATER FUEL PUMP,
LINES, AND RELATED PARTS (M1010)

1	PAOZZ 24617 9409761	NUT,PLAIN,CAP		2
		UOC:210,230		
2	PAOZZ 11862 9409754	WASHER,FINISHING		2
		UOC:210,230		
3	PAOZZ 93061 144F-5	TEE,TUBE		1
		UOC:210,230		
4	PAFZZ 81343	5 010111B NUT,TUBE COUPLING		4
		UOC:210		
5	MFOZZ 11862 14072371	PIPE ASM-AUX HTR (44.45"LG) MAKE		1
		FROM TUBE, P/N 1324714		
		UOC:210,230		
6	PAOZZ 81343 SAEJ513	ADAPTER,STRAIGHT,PI		1
		UOC:210,230		
7	PAOZZ 24617 140642	TEE,PIPE TO TUBE		1
		UOC:210,230		
8	PAOZZ 11862 14072347	SWITCH,HEATER FUEL		1
		UOC:210,230		
9	PAOZZ 24617 9415163	SCREW,ASSEMBLED WAS		2
		UOC:210,230		
10	PAOZZ 11862 14005953	SHIELD,EXPANSION		4
		UOC:210,230		
11	PAOZZ 96906 MS51321-1	PUMP,FUEL,ELECTRICA		1
		UOC:210,230		
12	PAOZZ 11862 14007511	BOLT,ASSEMBLED WASH		4
		UOC:210,230		
13	PAOZZ 11862 343444	CLAMP,LOOP		1
		UOC:210,230		
14	PAOZZ 11862 14004512	CLAMP,LOOP		1
		UOC:210,230		
15	PAOZZ 16764 110200	ADAPTER,STRAIGHT,PI		1
		UOC:210,230		
16	PAOZZ 11862 14072368	HOSE ASSEMBLY,NONME		1
		UOC:210,230		
17	MFOZZ 11862 14072369	PIPE ASM-AUX HTR (65.22"LG) MAKE		1
		FROM TUBE, P/N 1324714		
		UOC:210,230		

END OF FIGURE

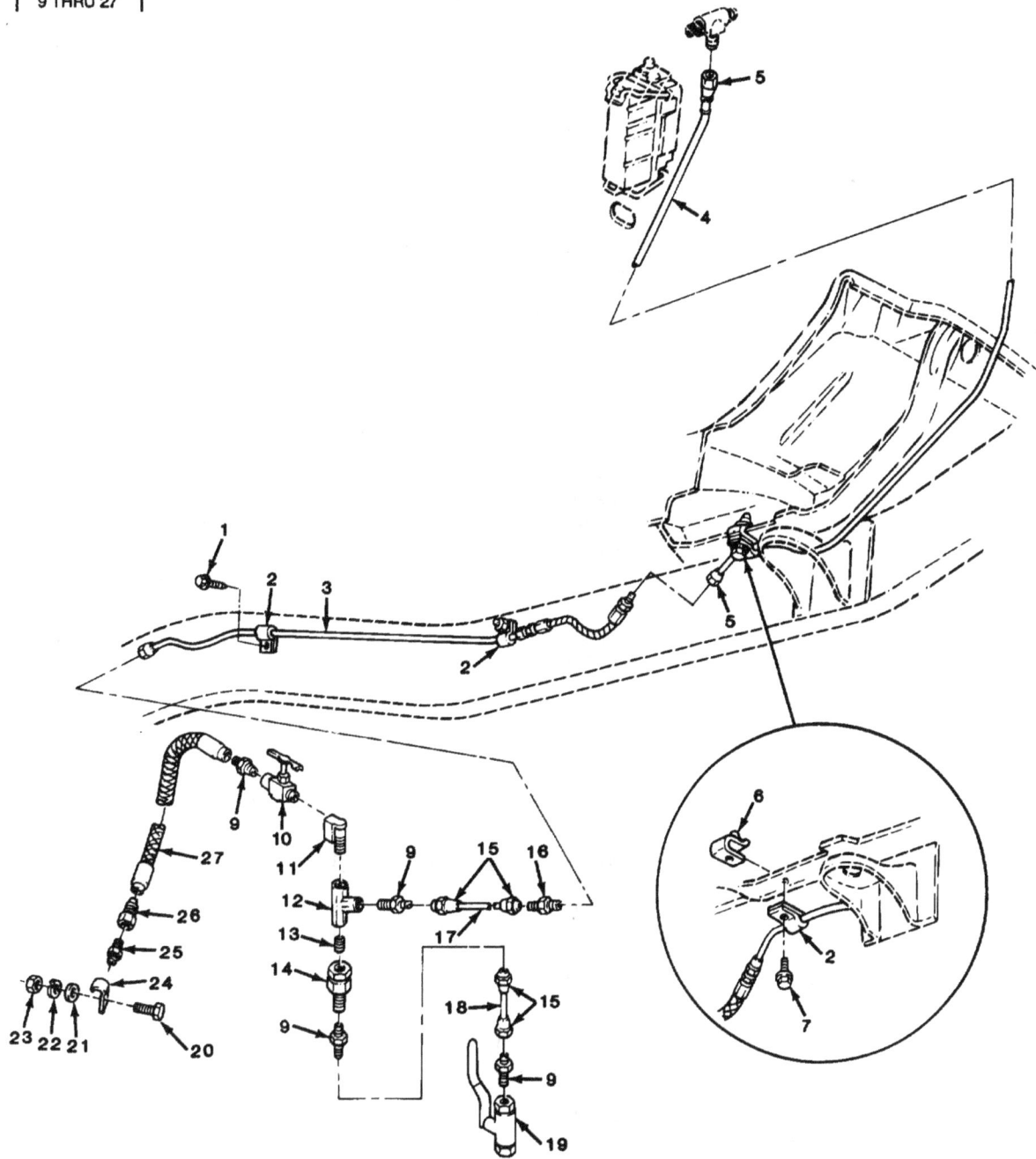

FIGURE 252. PERSONNEL HEATER FUEL LINES AND RELATED PARTS (M1010).

TA511059

(1) ITEM NO	(2) SMR CODE	(3) CAGEC	(4) PART NUMBER	(5) DESCRIPTION AND USABLE ON CODES (UOC)	(6) QTY
				GROUP 2207 WINTERIZATION EQUIPMENT	
				FIG. 252 PERSONNEL HEATER FUEL LINES AND RELATED PARTS (M1010)	
1	PAOZZ	11862	11504447	SCREW,TAPPING,THREA UOC:210,230	1
2	PAOZZ	96906	MS21333-110	CLAMP,LOOP UOC:210,230	3
3	PAOZZ	11862	14063324	HOSE ASSEMBLY,NONME UOC:210,230	1
4	MFFZZ	11862	14063323	TUBE ASSEMBLY,METAL (45.74"LG) MAKE FROM TUBE, P/N 1324714 UOC:210,230	1
5	PAFZZ	81343	5 010111B	NUT,TUBE COUPLING UOC:210	2
6	PAOZZ	24617	1494253	NUT,CLIP-ON UOC:210,230	1
7	PAOZZ	96906	MS90728-31	BOLT,MACHINE UOC:210,230	1
8	PAFFF	9W635	MBH-V	PARTS KIT,HEATER UOC:210	1
9	PAFZZ	96906	MS39158-23	.ADAPTER,STRAIGHT,PI UOC:210	4
10	PAFZZ	93061	NV108P-4	.COCK,SHUTOFF,SCREW UOC:210	1
11	PAFZZ	93061	2202P-4-4	.ELBOW,PIPE UOC:210	1
12	PAFZZ	30780	1203P-4	.TEE,PIPE UOC:210	1
13	PAFZZ	93061	215PN-4	.NIPPLE,PIPE UOC:210	1
14	PAFZZ	93061	207ACBHS-4	.COUPLING,PIPE UOC:210	1
15	PAFZZ	81343	5 010111B	.NUT,TUBE COUPLING UOC:210	4
16	XDFZZ	93061	42F-5	.NIPPLE,TUBE UOC:210	1
17	MFFZZ	9W635	31-0040	.PIPE (96" LG) MAKE FROM TUBE, P/N 1324714 UOC:210	1
18	MFFZZ	9W635	31-0040A	.PIPE (3.0"LG) MAKE FROM TUBE, P/N 1324714 UOC:210	1
19	PAFZZ	79227	B-6000-1/4IN	.VALVE,BALL UOC:210	1
20	PAFZZ	96906	MS90728-8	.SCREW,CAP,HEXAGON H UOC:210	4
21	PAFZZ	96906	MS27183-9	.WASHER,FLAT UOC:210	4
22	PAFZZ	96906	MS35338-47	.WASHER,LOCK UOC:210	4

(1)	(2)	(3)	(4)	(5)	(6)
ITEM NO	SMR CODE	CAGEC	PART NUMBER	DESCRIPTION AND USABLE ON CODES (UOC)	QTY
23	PAFZZ	96906	MS35691-1	.NUT,PLAIN,HEXAGON UOC:210	4
24	PAFZZ	17773	11176106-5	.CLAMP,LOOP UOC:210	4
25	PAFZZ	81343	SAEJ513	.ADAPTER,STRAIGHT,PI UOC:210	1
26	XDFZZ	30780	260-6-6	.CONNECTOR SWIVEL UOC:210	1
27	PAFZZ	9W635	500-4	.HOSE ASSEMBLY,NONME UOC:210	1

END OF FIGURE

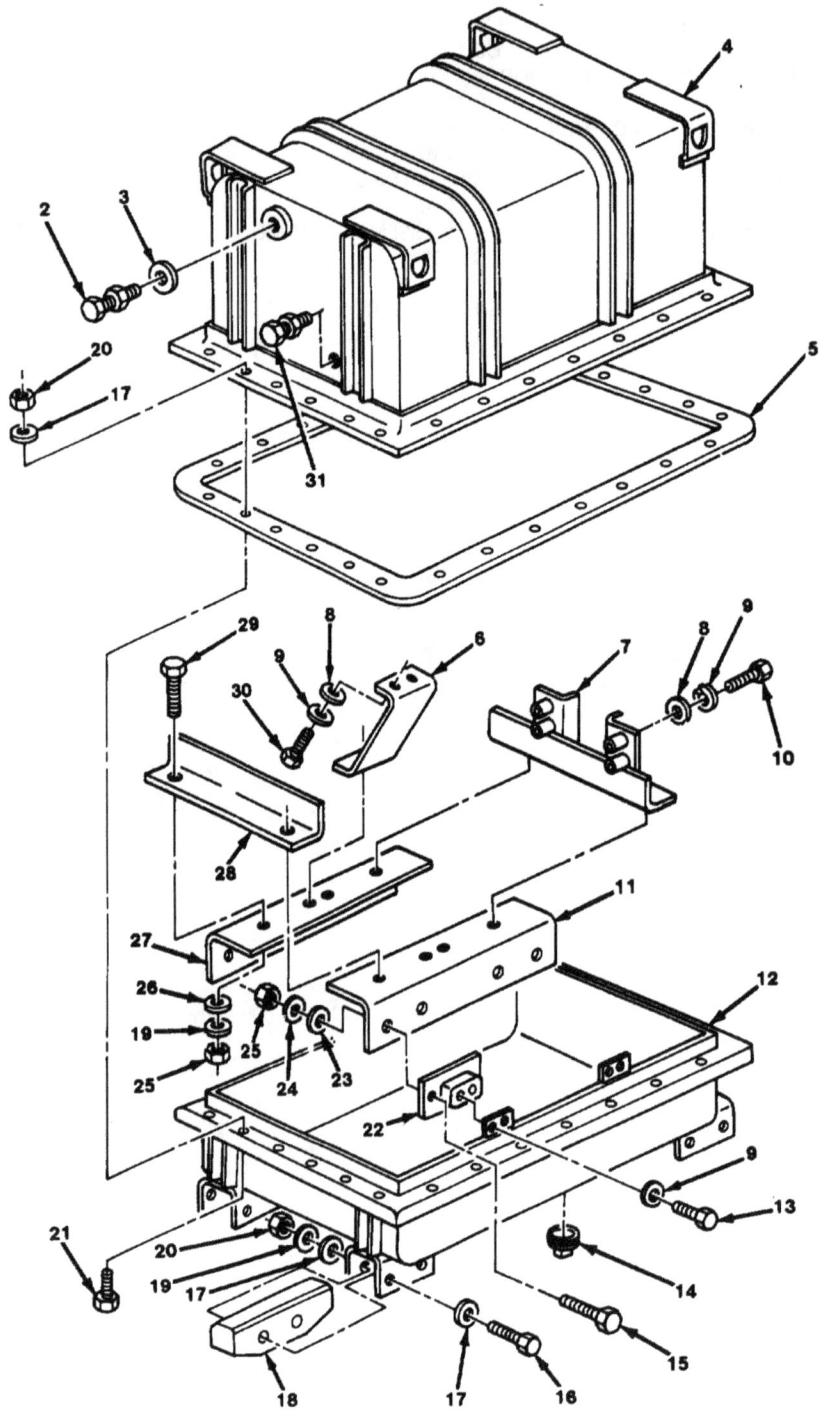

FIGURE 253. ENGINE SHIPPING CONTAINER ASSEMBLY.

TA511060

(1) (2) (3) (4)	(5)	(6)
ITEM SMR　　　　　　PART		
NO CODE CAGEC NUMBER	DESCRIPTION AND USABLE ON CODES (UOC)	QTY

GROUP 33 SPECIAL PURPOSE KITS

GROUP 3301 REUSABLE SHIPPING CONTAINERS

FIG. 253 ENGINE SHIPPING CONTAINER ASSEMBLY

ITEM NO	SMR CODE	CAGEC	PART NUMBER	DESCRIPTION	QTY
1	PFFFF	19207	12338064	SHIPPING AND STORAG	1
2	PAFZZ	19207	12338078	.BREATHER	1
3	PAFZZ	10001	2642880	.VALVE,BREATHER	1
4	XAFZZ	19207	12338066	.CONTAINER,UPPER	1
5	PAFZZ	19207	12338073	.RUBBER ROUND SECTIO	1
6	PFFZZ	19207	12338068	.BRACKET,DOUBLE ANGL	2
7	PFFZZ	19207	12338067	.BRACKET,MOUNTING	1
8	PAFZZ	96906	MS27183-15	.WASHER,FLAT	8
9	PAFZZ	96906	MS35338-46	.WASHER,LOCK	16
10	PAFZZ	96906	MS90725-69	.SCREW,CAP,HEXAGON H	4
11	PFFZZ	19207	12338074	.BRACKET,ANGLE LEFT	1
12	XAFZZ	19207	12338065	.CONTAINER,LOWER	1
13	PAFZZ	96906	MS90727-57	.SCREW,CAP,HEXAGON H	8
14	PAFZZ	96906	MS20913-6S	.PLUG,PIPE	1
15	PAFZZ	96906	MS90728-35	.BOLT,MACHINE	8
16	PAFZZ	96906	MS90728-127	.SCREW,CAP,HEXAGON H	8
17	PAFZZ	96906	MS27183-18	.WASHER,FLAT	34
18	PFFZZ	19207	12338071	.RUNNER,METAL	4
19	PAFZZ	96906	MS35338-48	.WASHER,LOCK	24
20	PAFZZ	96906	MS51967-14	.NUT,PLAIN,HEXAGON	42
21	PAFZZ	96906	MS90728-110	.SCREW,CAP,HEXAGON H	26
22	PFFZZ	19207	12338070	.MOUNT,RESILIENT	4
23	PAFZZ	96906	MS27183-12	.WASHER,FLAT	8
24	PAFZZ	96906	MS35338-45	.WASHER,LOCK	8
25	PAFZZ	96906	MS51967-5	.NUT,PLAIN,HEXAGON	8
26	PAFZZ	96906	MS27183-19	.WASHER,FLAT	8
27	PFFZZ	19207	12338075	.BRACKET,ANGLE	1
28	PFFZZ	19207	12338076	.BRACKET,ANGLE	1
29	PAFZZ	96906	MS90728-113	.SCREW,CAP,HEXAGON H	8
30	PAFZZ	96906	MS90728-62	.SCREW,CAP,HEXAGON H	4
31	PAFZZ	19207	10906697	.INDICATOR,HUMIDITY	1

END OF FIGURE

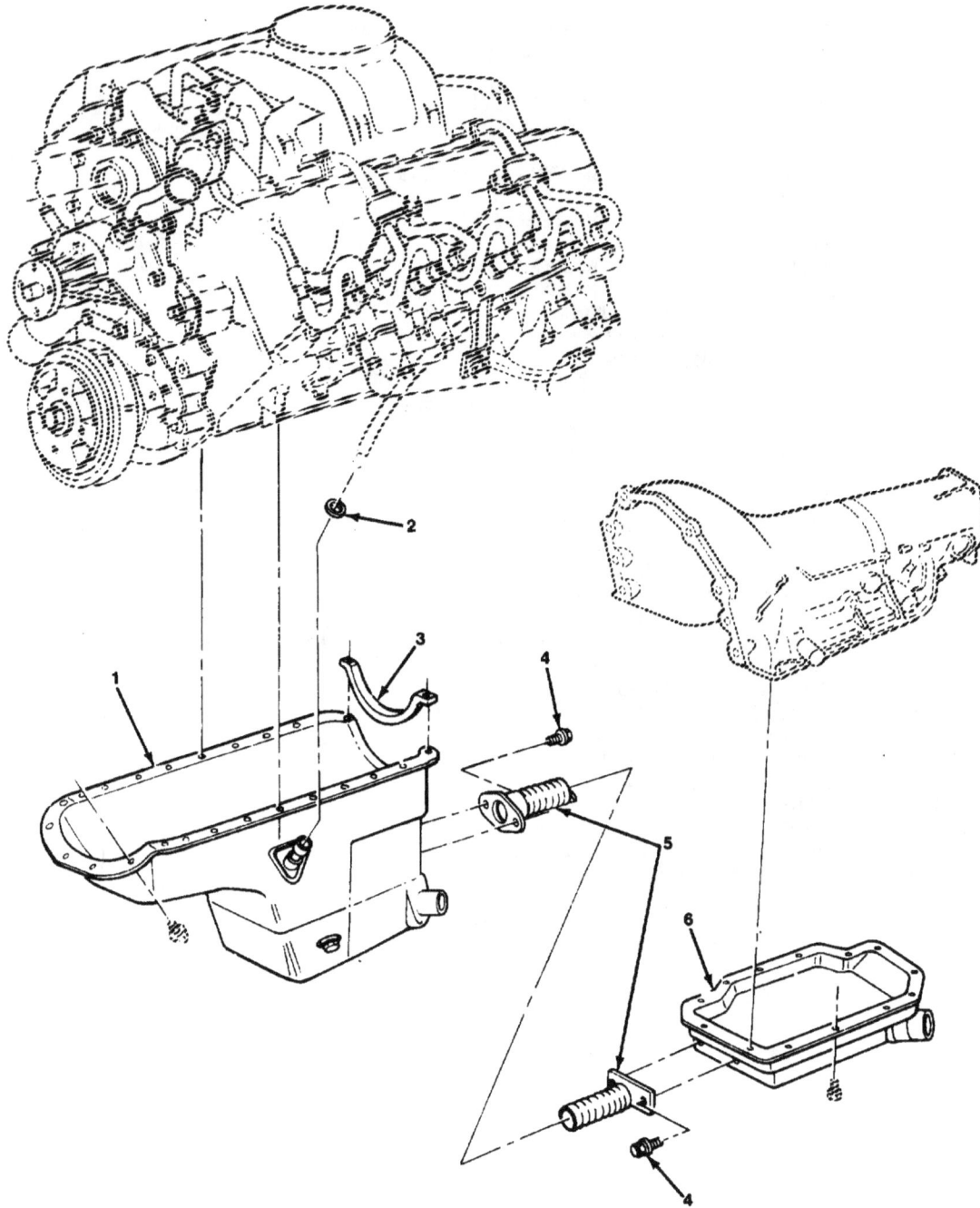

TA511061

FIGURE 254. TRANSMISSION AND OIL PAN.

(1) (2) (3) (4) ITEM SMR NO CODE CAGEC	PART NUMBER	(5) DESCRIPTION AND USABLE ON CODES (UOC)	(6) QTY
		GROUP 3303 WINTERIZATION KITS	
		FIG. 254 TRANSMISSION AND OIL PAN	
1 PAFFF 11862	14067730	OIL PAN	1
2 PAOZZ 24617	274244	PACKING,PREFORMED	1
3 PAOZZ 11862	14022683	SEAL,NONMETALLIC SP	1
4 PAOZZ 11862	9439930	SCREW,CAP,HEXAGON H	4
5 PAOZZ 11862	15599200	PIPE,EXHAUST	1
6 PAOZZ 11862	15599201	OIL PAN AND HEAT EX	1
7 PAOZZ 11862	337185	PLUG,MACHINE THREAD	1
8 PAOZZ 11862	14079550	GASKET	1

END OF FIGURE

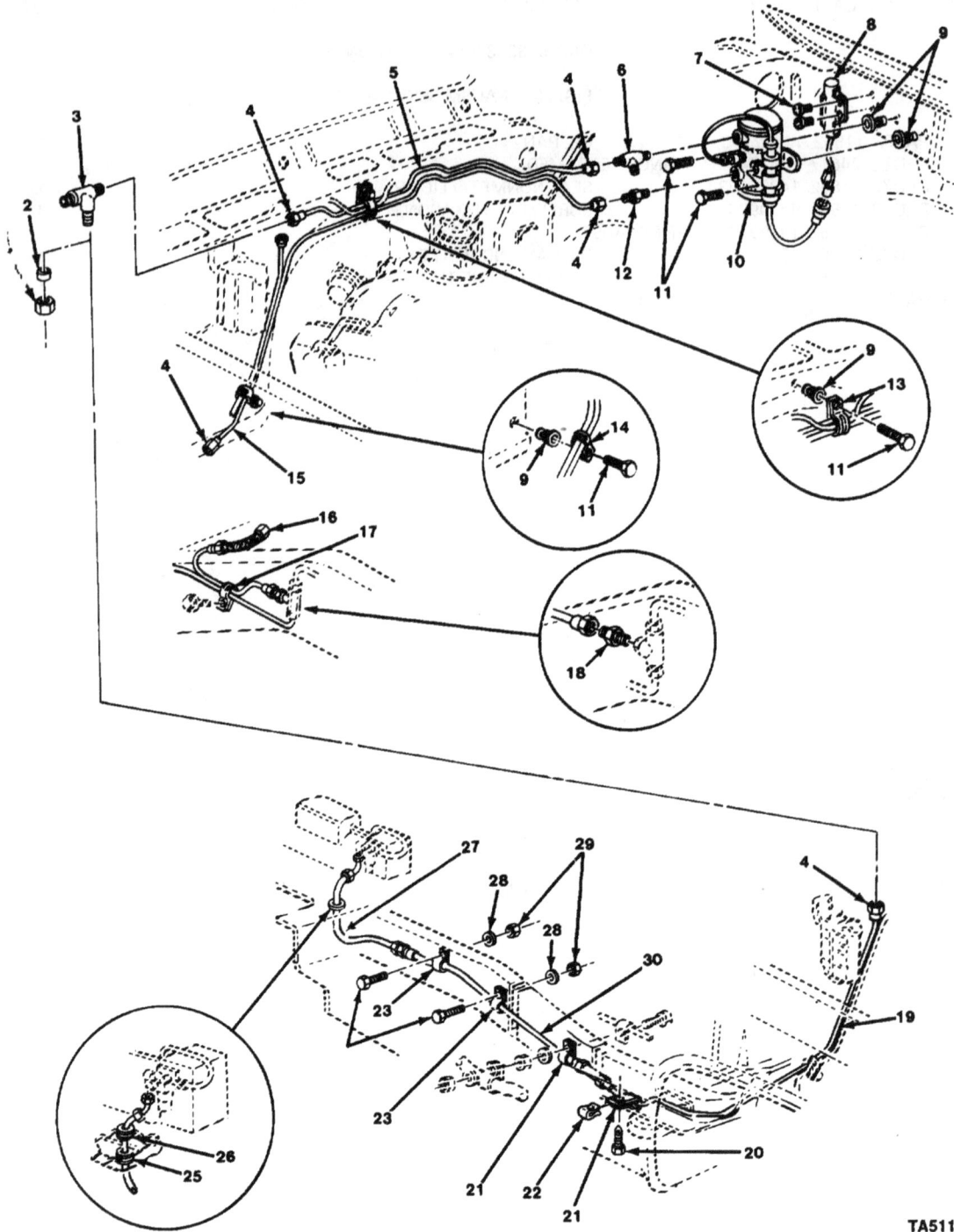

FIGURE 255. AUXILIARY HEATER FUEL PUMP,   LINES, AND RELATED PARTS (M1008).

TA511062

(1) ITEM NO	(2) SMR CODE	(3) CAGEC	(4) PART NUMBER	(5) DESCRIPTION AND USABLE ON CODES (UOC)	(6) QTY
				GROUP 3303 WINTERIZATION KITS	
				FIG. 255 AUXIULIARY HEATER FUEL PUMP, LINES, AND RELATED PARTS (M1008)	
1	PAOZZ	24617	9409761	NUT,PLAIN,CAP UOC:209,230,231	1
2	PAOZZ	11862	9409754	WASHER,FINISHING UOC:209,230,231	1
3	PAOZZ	93061	144F-5	TEE,TUBE UOC:194,208,230,231,252,254,256	1
4	PAOZZ	81343	5 010111B	NUT,TUBE COUPLING UOC:194,208,230,231,252,254,256	6
5	MFOZZ	11862	14072371	PIPE ASM-AUX HTR (44.45" LG) MAKE FROM TUBE, P/N 1324714 UOC:194,208,209,230,231,252,254,256	1
6	PAOZZ	24617	140642	TEE,PIPE TO TUBE UOC:194,208,209,230,231,252,254,256	1
7	PAOZZ	24617	9414713	SCREW,TAPPING,THREA UOC:194,208,230,231,252,254,256	2
8	PAOZZ	11862	14072347	SWITCH,HEATER FUEL UOC:194,208,230,231,252,254,256	1
9	PAOZZ	11862	14005953	SHIELD,EXPANSION UOC:194,208,209,230,231,252,254,256	4
10	PAOZZ	96906	MS51321-1	PUMP,FUEL,ELECTRICA UOC:194,208,209,230,231,252,254,256	1
11	PAOZZ	11862	14007511	BOLT,ASSEMBLED WASH UOC:194,208,209,230,231,252,254,256	2
12	PAOZZ	81240	GM118749	ADAPTER,STRAIGHT,PI UOC:194,208,209,230,231,252,254,256	1
13	PAOZZ	11862	343444	CLAMP,LOOP UOC:194,208,230,231,252,254,256	1
14	PAOZZ	11862	14004512	CLAMP,LOOP UOC:194,208,230,231,252,254,256	1
15	MFOZZ	11862	14072369	PIPE ASM-AUX HTR (65.22" LG) MAKE FROM TUBE, P/N 1324714 UOC:194,208,230,231,252,254,256	1
16	PAOZZ	11862	14072368	HOSE ASSEMBLY,NONME UOC:194,208,230,231,252,254,256	1
17	PAOZZ	11862	25527423	CLAMP,LOOP UOC:194,208,230,231,252,254,256	1
18	PAOZZ	16764	110200	ADAPTER,STRAIGHT,PI UOC:194,208,230,231,252,254,256	1
19	PAOZZ	11862	14063323	TUBE ASSEMBLY,METAL UOC:194,208	1
20	PAOZZ	96906	MS90728-31	BOLT,MACHINE UOC:194,208	1
21	PAOZZ	96906	MS21333-110	CLAMP,LOOP UOC:194,208,230,231,252,254,256	2
22	PAOZZ	24617	1494253	NUT,CLIP-ON UOC:194,208	1
23	PAOZZ	96906	MS21333-46	CLAMP,LOOP	2

(1)	(2)	(3)	(4)	(5)	(6)
ITEM NO	SMR CODE	CAGEC	PART NUMBER	DESCRIPTION AND USABLE ON CODES (UOC)	QTY
				UOC:194,208	
24	PAOZZ	96906	MS90728-59	SCREW,CAP,HEXAGON H	2
				UOC:194,208	
25	PAOZZ	11862	3886908	GROMMET,NONMETALLIC	1
				UOC:194,208	
26	PAOZZ	96906	MS35489-103	GROMMET,NONMETALLIC	1
				UOC:194,208	
27	PAOZZ	11862	14076390	HOSE ASSEMBLY,NONME	1
				UOC:194,208	
28	PAOZZ	11862	11500046	WASHER,LOCK	2
				UOC:194,208	
29	PAOZZ	96906	MS51967-8	NUT,PLAIN,HEXAGON	2
				UOC:194,208	
30	PAOZZ	11862	14063324	HOSE ASSEMBLY,NONME	1
				UOC:194,208	

END OF FIGURE

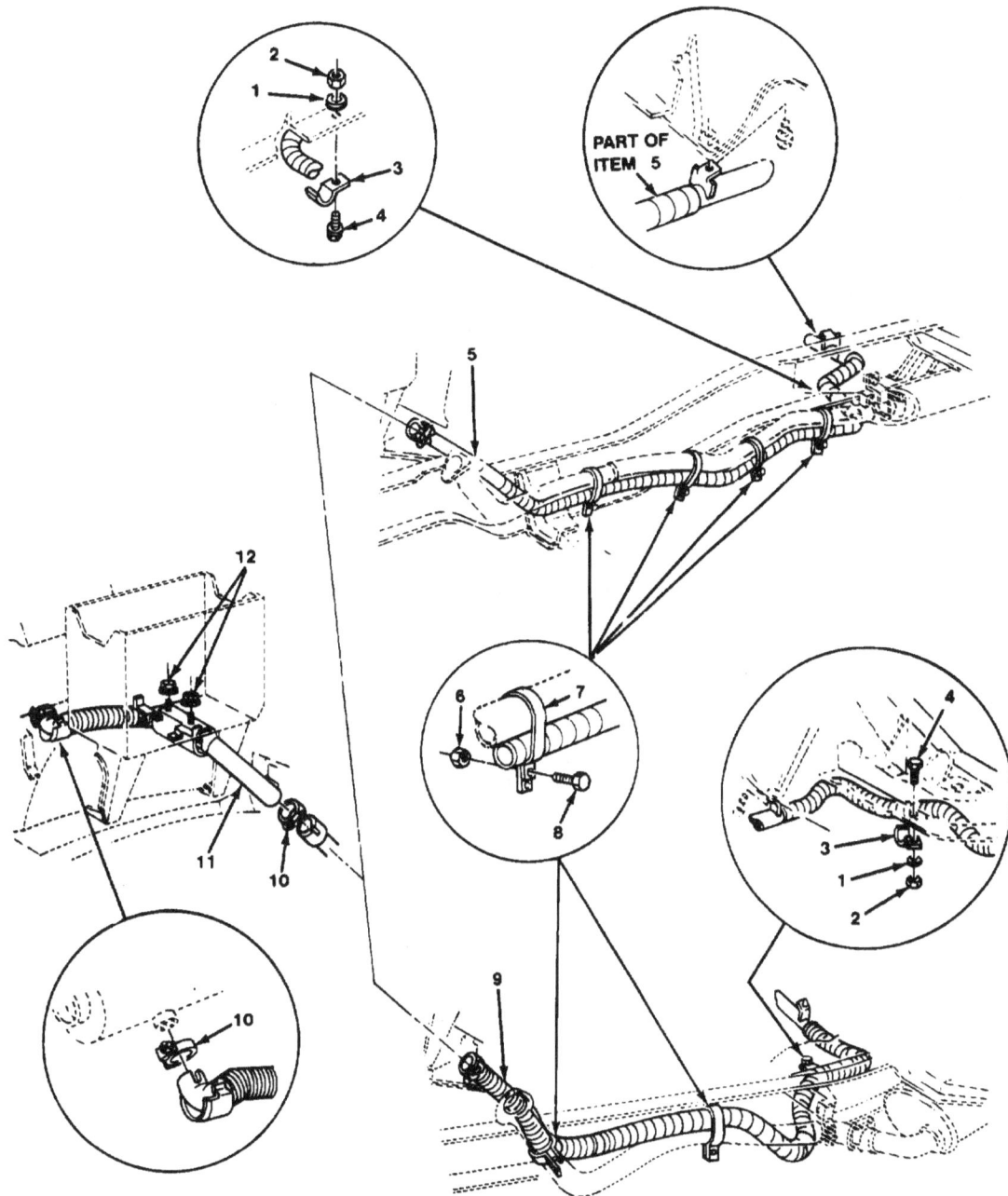

TA511063

FIGURE 256. WARM AIR HEATER EXHAUST (ALL EXCEPT M1010).

(1)	(2)	(3)	(4)	(5)	(6)
ITEM NO	SMR CODE	CAGEC	PART NUMBER	DESCRIPTION AND USABLE ON CODES (UOC)	QTY

GROUP 3303 WINTERIZATION KITS

FIG. 256 WARM AIR HEATER EXHAUST
(ALL EXCEPT M1010)

1	PAOZZ	96906	MS27183-12	WASHER,FLAT	1
				UOC:194,208,209,230,231,252,254,256	
2	PAOZZ	96906	MS51943-33	NUT,SELF-LOCKING,HE	1
				UOC:194,208,209,230,231,252,254,256	
3	PAOZZ	11862	120877	STRAP,RETAINING	1
				UOC:194,208,209,230,231,252,254,256	
4	PAOZZ	24617	9417350	BOLT,ASSEMBLED WASH	1
				UOC:194,208,209,230,231,252,254,256	
5	PAOZZ	11862	14075801	PIPE,EXHAUST	1
				UOC:194,208,230,231,252,254,256	
6	PAOZZ	21450	131245	NUT,SELF-LOCKING,HE	4
				UOC:194,208,230,231,252,254,256	
6	PAOZZ	21450	131245	NUT,SELF-LOCKING,HE	2
				UOC:209	
7	PAOZZ	96906	MS35842-15	CLAMP,HOSE	4
				UOC:194,208,230,231,252,254,256	
7	PAOZZ	96906	MS35842-15	CLAMP,HOSE	2
				UOC:209	
8	PAOZZ	96906	MS90728-8	SCREW,CAP,HEXAGON H	4
				UOC:194,208,230,231,252,254,256	
8	PAOZZ	96906	MS90728-8	SCREW,CAP,HEXAGON H	2
				UOC:209	
9	PAOZZ	11862	15599211	PIPE ASSEMBLY,EXHAU	1
				UOC:209	
10	PAOZZ	63208	650-24	CLAMP,HOSE	2
11	XDOZZ	11862	15599287	PIPE	1
12	PAOZZ	24617	9416918	NUT,PLAIN,EXTENDED	2

END OF FIGURE

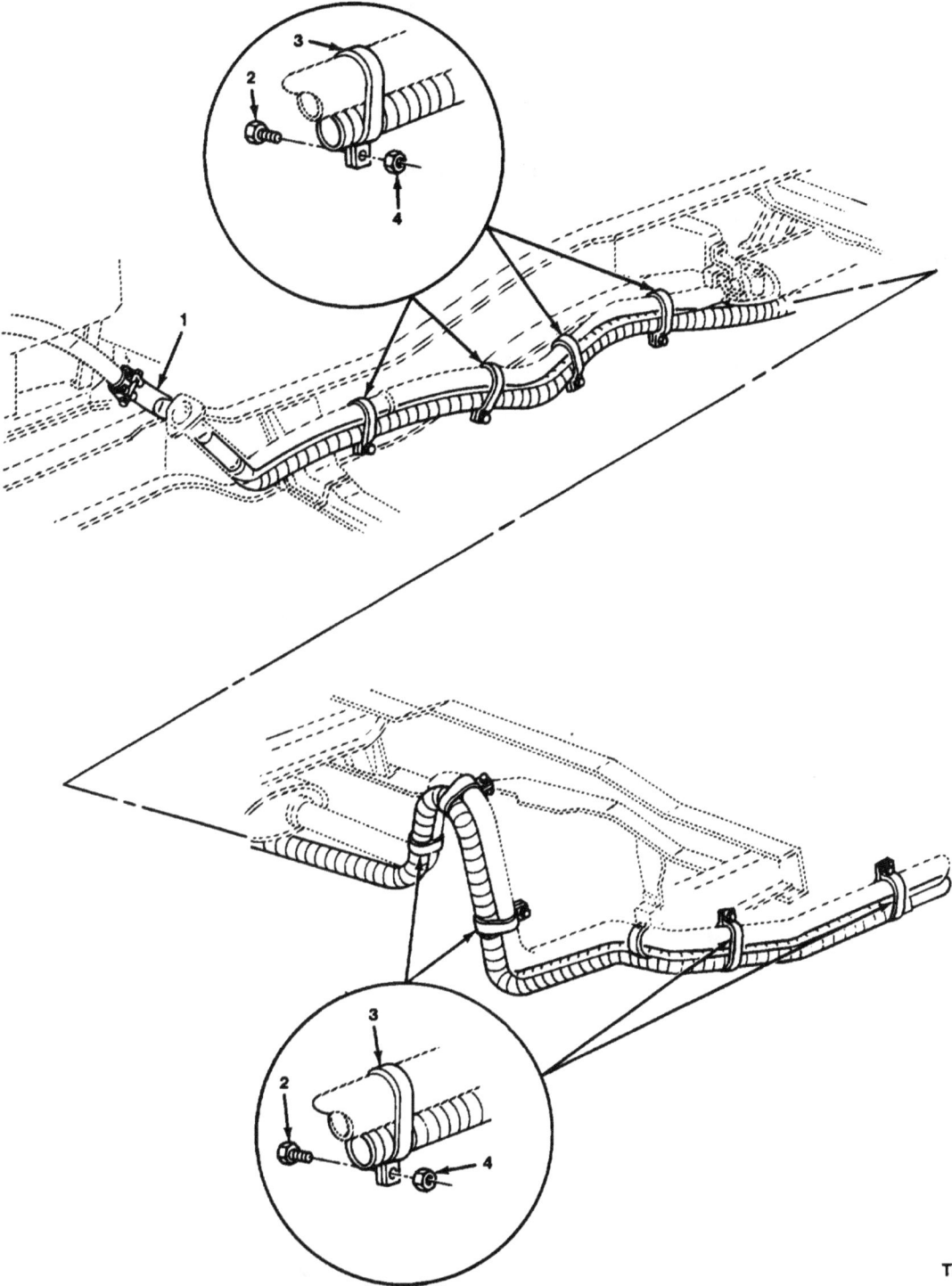

FIGURE 257. WARM AIR HEATER EXHAUST (M1010).

TA511064

(1) ITEM NO	(2) SMR CODE	(3) CAGEC	(4) PART NUMBER	(5) DESCRIPTION AND USABLE ON CODES (UOC)	(6) QTY
				GROUP 3303 WINTERIZATION KITS	
				FIG. 257 WARM AIR HEATER EXHAUST (M1010)	
1	PAOZZ	11862	15599290	PIPE ASSEMBLY,EXHAU UOC:210	1
2	PAOZZ	96906	MS90728-8	SCREW,CAP,HEXAGON H UOC:210	11
3	PAOZZ	96906	MS35842-15	CLAMP,HOSE UOC:210	11
4	PAOZZ	21450	131245	NUT,SELF-LOCKING,HE UOC:210	11

END OF FIGURE

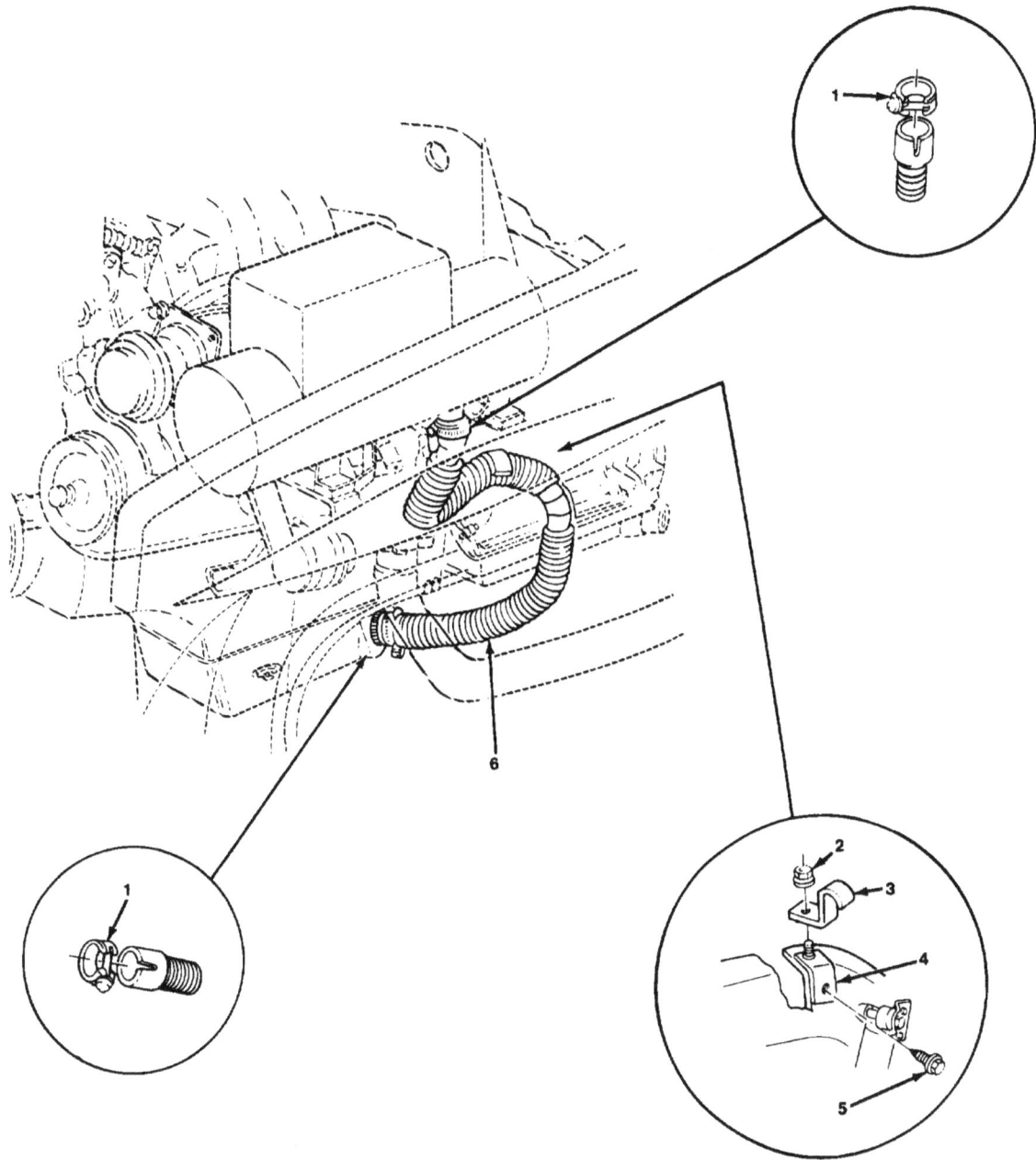

TA511065

FIGURE 258. COOLANT HEATER EXHAUST

(1) (2) (3) (4) ITEM SMR PART NO CODE CAGEC NUMBER	(5) DESCRIPTION AND USABLE ON CODES (UOC)	(6) QTY

GROUP 3303 WINTERIZATION KITS

FIG. 258 COOLANT HEATER EXHAUST

1	PAOZZ 63208 650-24	CLAMP,HOSE		2
2	PAOZZ 24617 9416918	NUT,PLAIN,EXTENDED		1
3	PAOZZ 11862 120877	STRAP,RETAINING		1
		UOC:194,208,230,231,252,254,256		
4	PAOZZ 11862 15599245	BRACKET,ANGLE		1
5	PAOZZ 24617 9416187	SCREW,TAPPING,THREA		1
6	PAOZZ 11862 15599243	PIPE,EXHAUST		1

END OF FIGURE

**ALL EXCEPT
M1009**

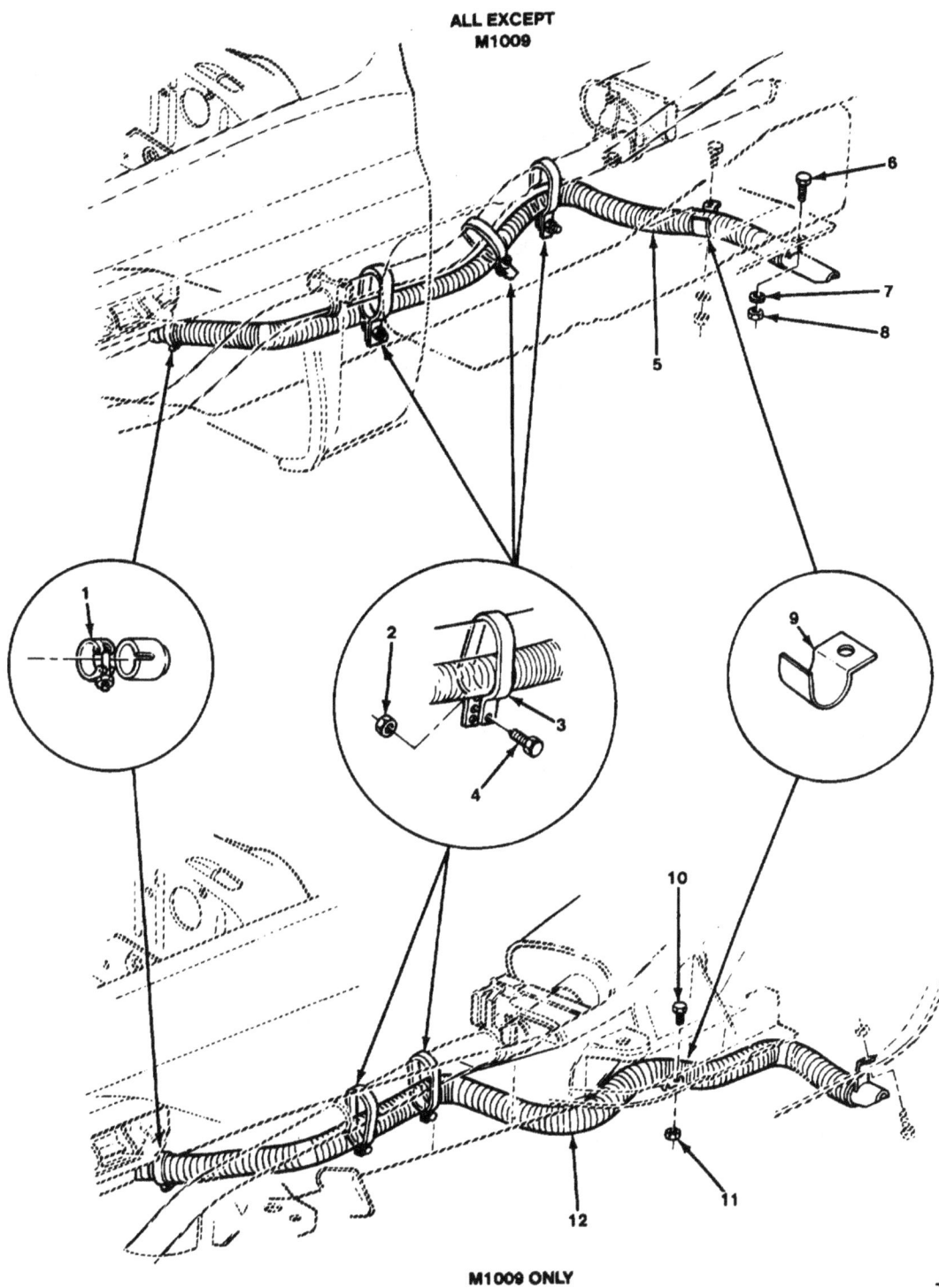

M1009 ONLY

TA511066

FIGURE 259. COOLANT HEATER REAR EXHAUST.

(1) ITEM NO	(2) SMR CODE	(3) CAGEC	(4) PART NUMBER	(5) DESCRIPTION AND USABLE ON CODES (UOC)	(6) QTY
				GROUP 3303 WINTERIZATION KITS	
				FIG. 259 COOLANT HEATER REAR EXHAUST	
1	PAOZZ	63208	650-24	CLAMP,HOSE UOC:194,208,209,210,230,231,252,254	1
2	PAOZZ	21450	131245	NUT,SELF-LOCKING,HE UOC:194,208,210,230,231,252,254,256	3
2	PAOZZ	21450	131245	NUT,SELF-LOCKING,HE UOC:209	2
3	PAOZZ	96906	MS35842-15	CLAMP,HOSE UOC:194,208,210,230,231,252,254,256	3
3	PAOZZ	96906	MS35842-15	CLAMP,HOSE UOC:209	2
4	PAOZZ	96906	MS90728-8	SCREW,CAP,HEXAGON H UOC:194,208,210,230,231,252,254,256	3
4	PAOZZ	96906	MS90728-8	SCREW,CAP,HEXAGON H UOC:209	2
5	PAOZZ	11862	15599215	PIPE,EXHAUST UOC:194,208,230,231,252,254,256	1
5	PAOZZ	11862	14074499	PIPE ASSEMBLY,EXHAU UOC:210	1
6	PAOZZ	24617	9417350	BOLT,ASSEMBLED WASH UOC:194,208,210,230,231,252,254,256	1
7	PAOZZ	96906	MS27183-12	WASHER,FLAT UOC:194,208,230,231,252,254,256	1
7	PAOZZ	96906	MS27183-12	WASHER,FLAT UOC:209	2
8	PAOZZ	96906	MS51943-33	NUT,SELF-LOCKING,HE UOC:194,208,210,230,231,252,254,256	1
9	PAOZZ	11862	120877	STRAP,RETAINING UOC:194,208,210,230,231,252,254,256	1
10	PAOZZ	24617	9417350	BOLT,ASSEMBLED WASH UOC:209	1
11	PAOZZ	96906	MS51943-33	NUT,SELF-LOCKING,HE UOC:209	1
12	PAOZZ	11862	15599212	PIPE,EXHAUST UOC:209	1

END OF FIGURE

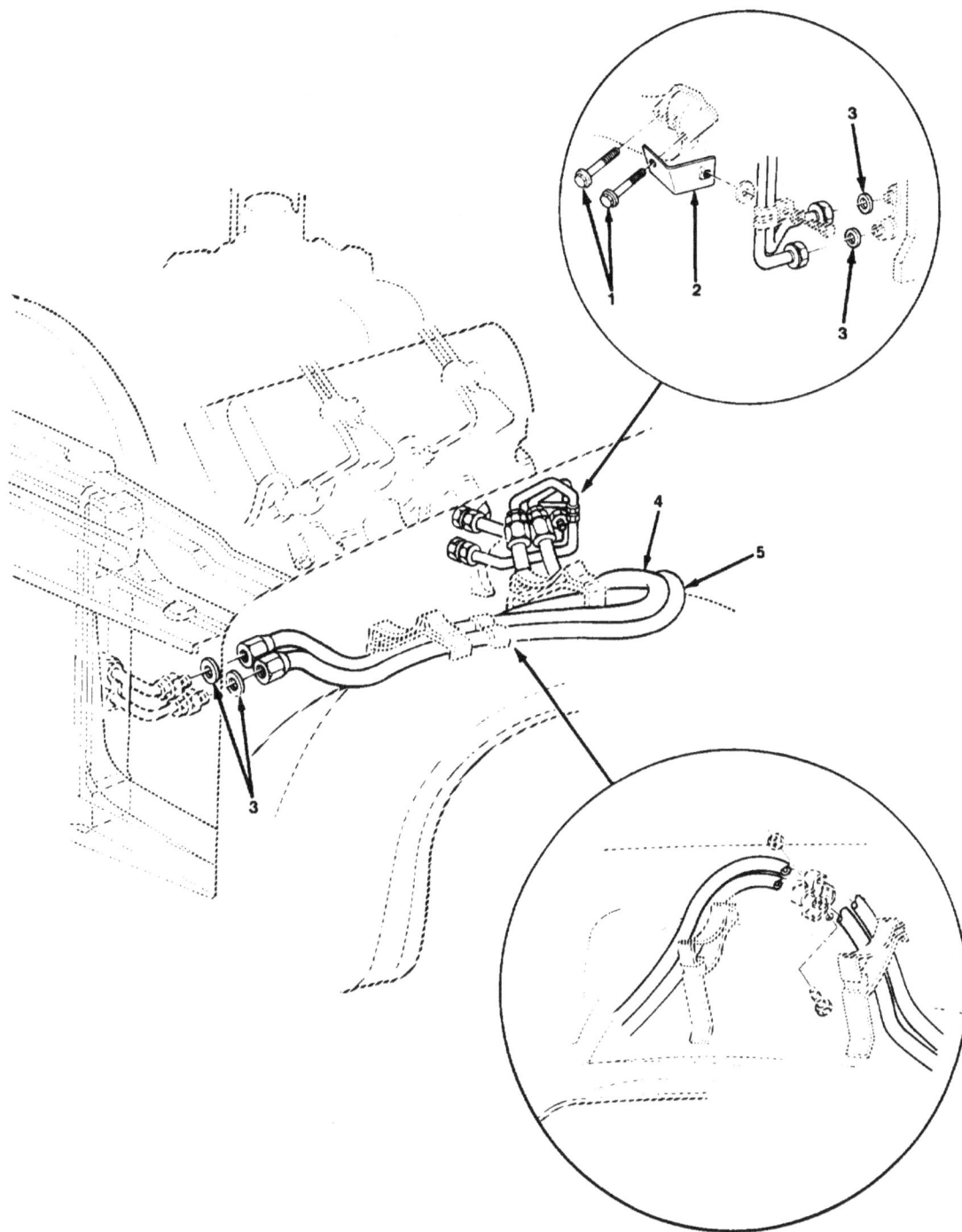

FIGURE 260. ENGINE OIL COOLER LINES.

TA511067

(1)	(2)	(3)	(4)	(5)	(6)
ITEM	SMR		PART		
NO	CODE	CAGEC	NUMBER	DESCRIPTION AND USABLE ON CODES (UOC)	QTY

GROUP 3303 WINTERIZATION KITS

FIG. 260 ENGINE OIL COOLER LINES

1	PAOZZ	11862	14028922	BOLT,MACHINE	2
2	PAOZZ	11862	15599246	BRACKET,ANGLE	1
3	PAOZZ	11862	14055585	SEAL RING,METAL UOC:194,208,209,230,231,252,254	4
4	PAOZZ	11862	14063337	HOSE ASSEMBLY,NONME UOC:194,208,209,230,231,252,254,256	1
5	PAOZZ	11862	14063336	HOSE ASSEMBLY,NONME UOC:194,208,209,230,231,252,254,256	1

END OF FIGURE

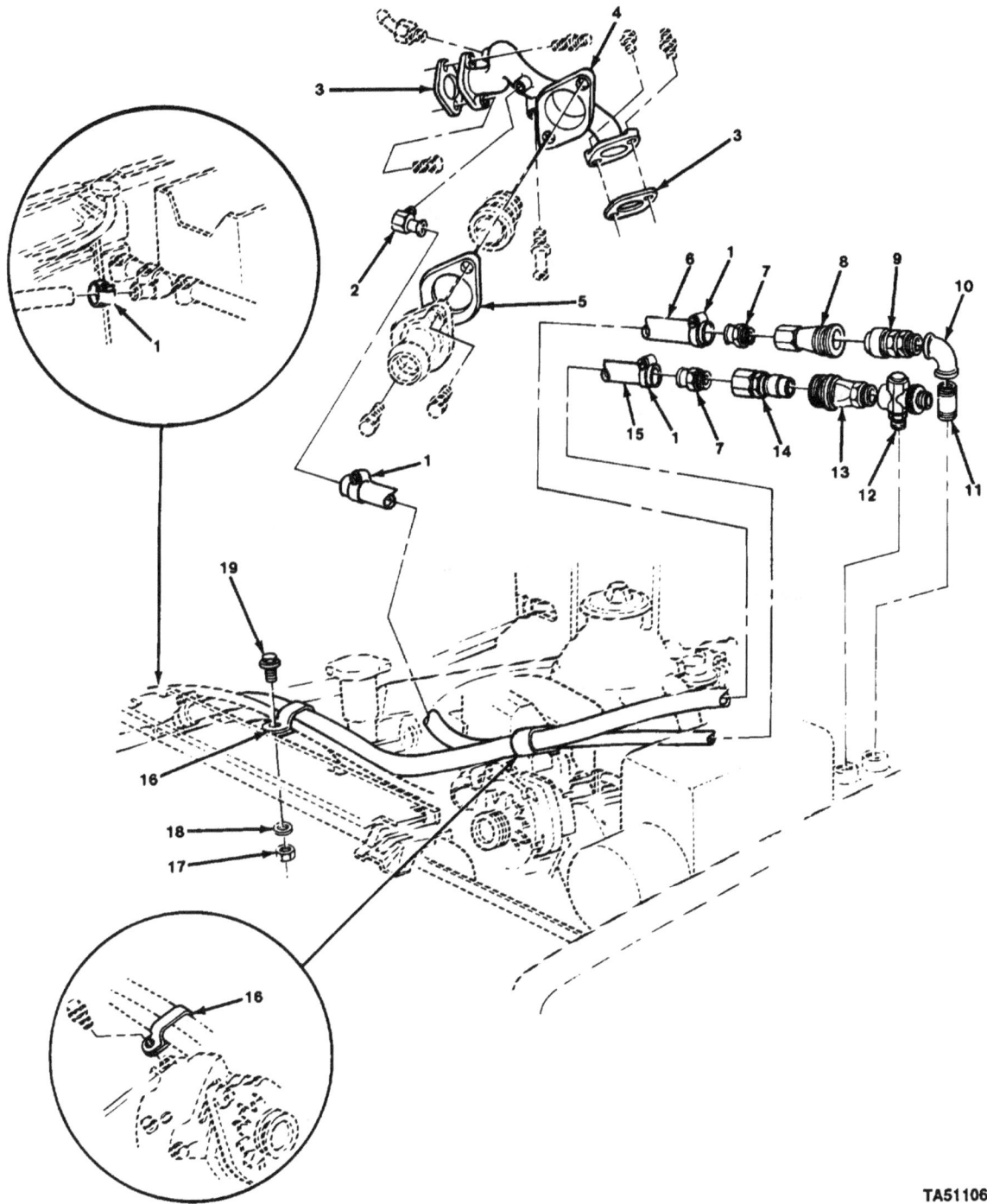

FIGURE 261. ENGINE COOLANT CROSSOVER HOUSING, HOSES, AND FITTINGS.

TA511068

TM9-2320-289-34P

(1) (2) (3) (4) ITEM SMR PART NO CODE CAGEC NUMBER	(5) DESCRIPTION AND USABLE ON CODES (UOC)	(6) QTY
	GROUP 3303 WINTERIZATION KITS	
	FIG. 261 ENGINE COOLANT CROSSOVER HOUSING, HOSES, AND FITTINGS	
1 PAOZZ 96906 MS35842-11 CLAMP,HOSE		4
2 PAOZZ 11862 10005327	ELBOW,PIPE TO HOSE	1
3 PAOZZ 11862 23500846 GASKET		2
4 PAOZZ 11862 14063338	WATER OUTLET,ENGINE	1
5 PAOZZ 11862 14028916 GASKET		1
6 MOOZZ 11862 14063370	HOSE ENG CLT HTR OU (43.00" LG.) MAKE FROM HOSE, P/N MS521304B203R	1
7 PAOZZ 11862 354501	ADAPTER,STRAIGHT,PI	2
8 PAOZZ 73992 8200	COUPLING HALF,QUICK	1
9 PAOZZ 73992 B-84	COUPLING HALF,QUICK UOC:194,208,209,210,230,231,252	1
10 PAOZZ 96906 MS51845-4 ELBOW,PIPE		1
11 PAOZZ 96906 MS51846-64 NIPPLE,PIPE		1
12 PAOZZ 17769 4641	SENSOR ASSEMBLY,ENG	1
13 PAOZZ 73992 8100	COUPLING HALF,QUICK	1
14 PAOZZ 73992 B-85	COUPLING HALF,QUICK	1
15 MOOZZ 11862 14063373	HOSE ENG CLT HTR IN (23.00" LG.) MAKE FROM HOSE, P/N MS521304B203R	1
16 PAOZZ 11862 3825416 CLAMP,LOOP		2
17 PAOZZ 24617 456004	NUT,SELF-LOCKING,HE	1
18 PAOZZ 96906 MS27183-14 WASHER,FLAT	UOC:194,208,209,210,230,231,252	1
19 PAOZZ 11862 9440034 BOLT,MACHINE		1

END OF FIGURE

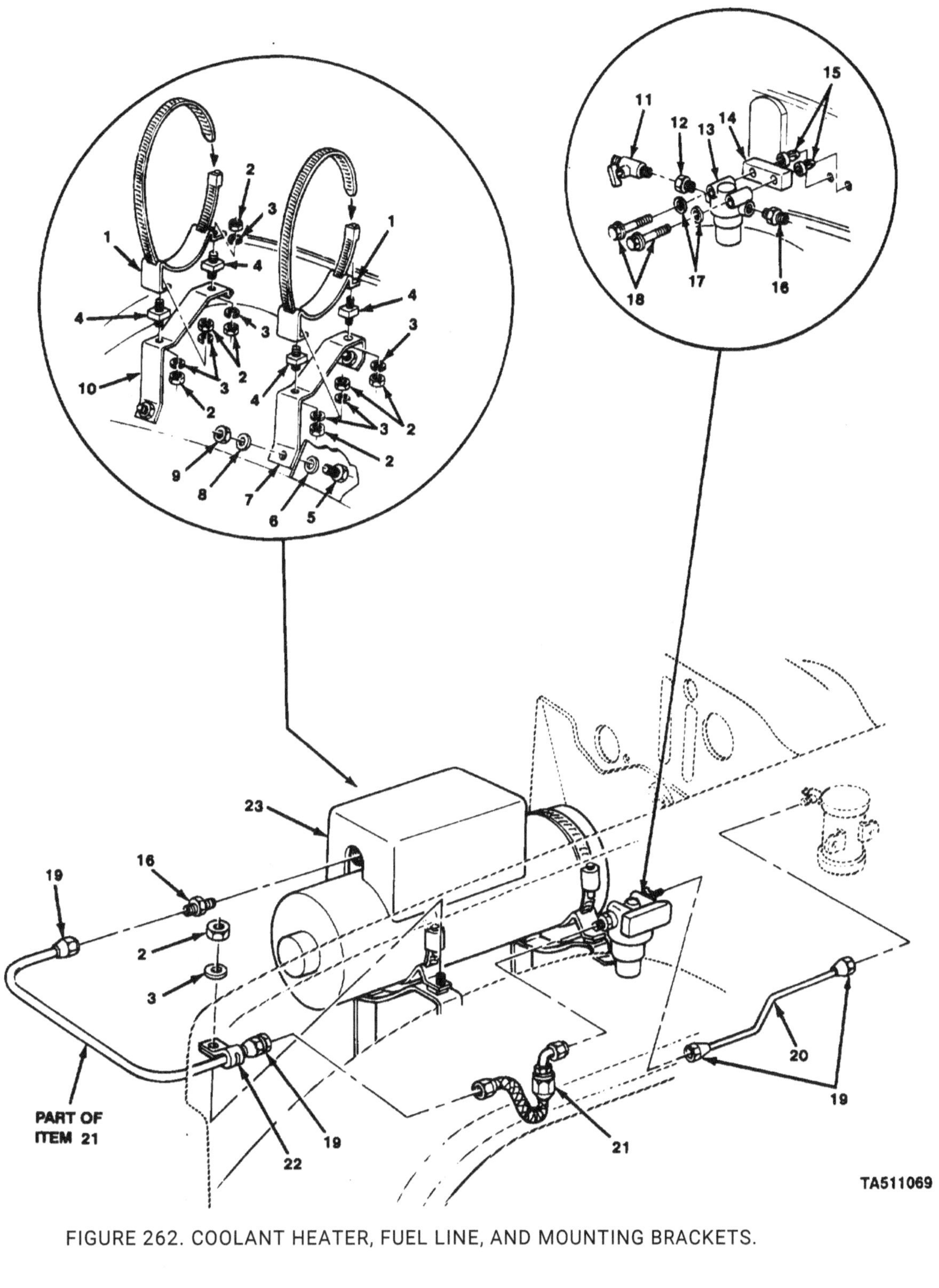

FIGURE 262. COOLANT HEATER, FUEL LINE, AND MOUNTING BRACKETS.

TA511069

(1)	(2)	(3)	(4)	(5)	(6)
ITEM NO	SMR CODE	CAGEC	PART NUMBER	DESCRIPTION AND USABLE ON CODES (UOC)	QTY

GROUP 3303 WINTERIZATION KITS

FIG. 262 COOLANT HEATER, FUEL FILTER, FUEL LINE, AND MOUNTING BRACKETS

1	PAOZZ	11862	14074457	BRACKET	2
2	PAOZZ	96906	MS35691-406	NUT,PLAIN,HEXAGON	8
3	PAOZZ	96906	MS35338-44	WASHER,LOCK	8
4	PAOZZ	19207	10922334	MOUNT,RESILIENT	4
5	PAOZZ	96906	MS90728-32	BOLT,MACHINE	4
6	PAOZZ	24617	9418924	WASHER,FLAT	4
7	PAOZZ	11862	14074458	BRACKET,MOUNTING	1
8	PAOZZ	96906	MS35338-45	WASHER,LOCK	4
9	PAOZZ	96906	MS51967-6	NUT,PLAIN,HEXAGON	4
10	PAOZZ	11862	14074459	BRACKET,MOUNTING	1
11	XDOZZ	11862	9440173	VALVE FUEL SHUTOFF	1
12	PAOZZ	17769	443998	COUPLING,PIPE	1
13	PAOZZ	96906	MS51085-1	FILTER,FLUID	1
14	PAOZZ	11862	14072374	SPACER,HEATER FUEL	1
15	PAOZZ	00613	C-5139-2	RECEPTACLE,TURNLOCK	2
16	PAOZZ	81240	GM118749	ADAPTER,STRAIGHT,PI	2
17	PAOZZ	96906	MS27183-8	WASHER,FLAT	2
18	PAOZZ	24617	456748	SCREW,CAP,HEXAGON H	2
19	PAOZZ	81343	5 010111B	NUT,TUBE COUPLING	4
20	MFOZZ	11862	14072372	PIPE HTR FLTR INL (7.92" LG) MAKE FROM TUBE, P/N 1324714	1
21	PAOZZ	11862	15599234	HOSE ASSEMBLY,NONME	1
22	PAOZZ	96906	MS21333-110	CLAMP,LOOP	1
23	PAOFF	46522	D55395-G1	HEATER,COOLANT,ENGI FCR COMPONENT PARTS SEE FIG 263	1

END OF FIGURE

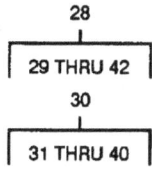

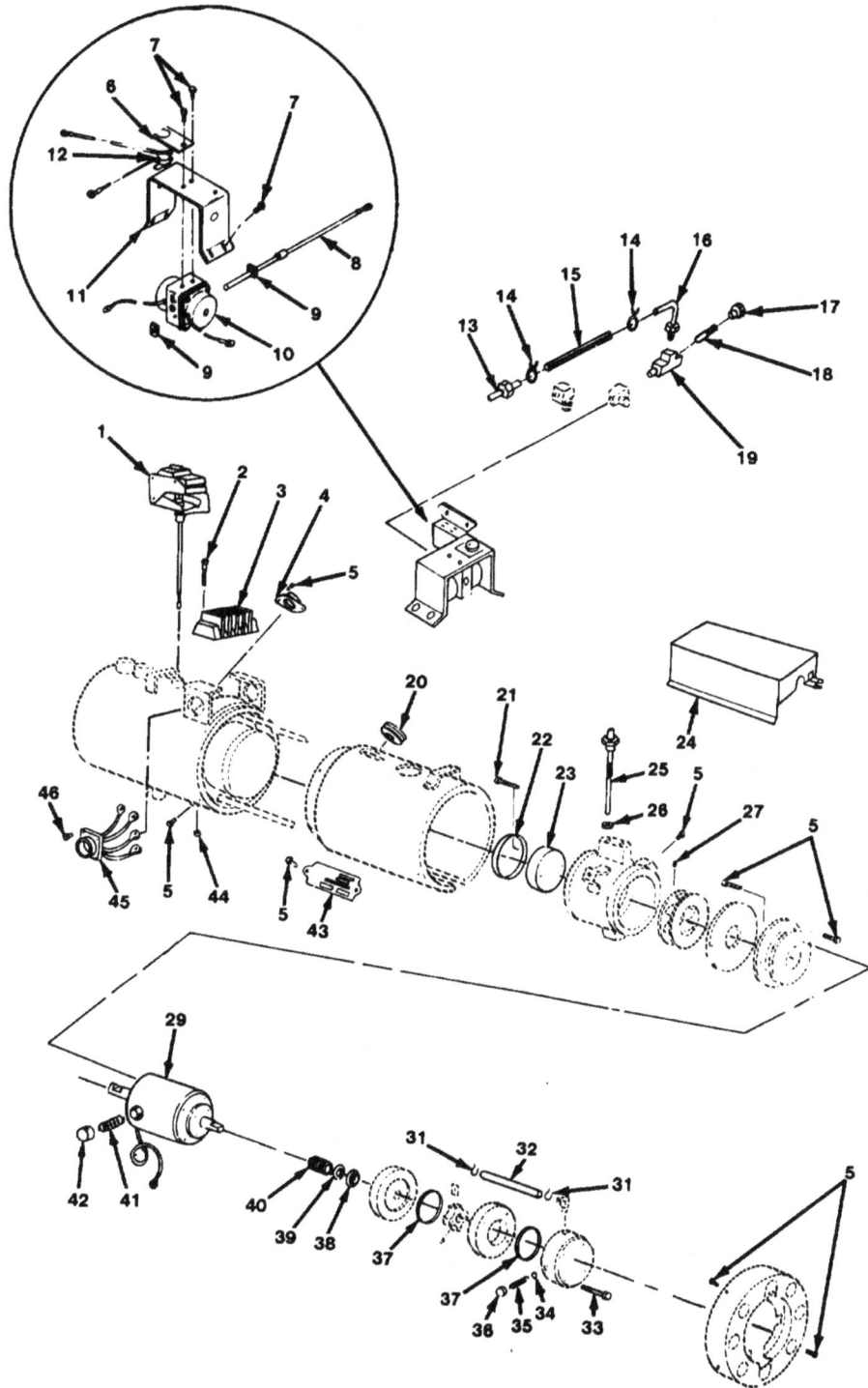

FIGURE 263. ENGINE COOLANT HEATER COMPONENT PARTS.

TA511070

(1)	(2)	(3)	(4)	(5)	(6)
ITEM NO	SMR CODE	CAGEC	PART NUMBER	DESCRIPTION AND USABLE ON CODES (UOC)	QTY

GROUP 3303 WINTERIZATION KITS

FIG. 263 ENGINE COOLANT HEATER
COMPONENT PARTS

ITEM NO	SMR CODE	CAGEC	PART NUMBER	DESCRIPTION AND USABLE ON CODES (UOC)	QTY
1	PAOZZ	19207	8376101	SWITCH,SENSITIVE	1
2	PAFZZ	46522	2251-0832-10-13	SCREW,MACHINE	4
3	PAFZZ	46522	A6005943	TERMINAL BOARD	1
4	PAFZZ	46522	A2322	SWITCH,THERMOSTATIC	1
5	PAFZZ	96906	MS51849-53	SCREW,MACHINE	28
6	PAFZZ	46522	A54201-G1-1	BRACKET,THERMOSTAT	1
7	PAFZZ	46522	2251-0832-06-17	SCREW,LOCKWASHER UOC:208,209,210,230,231,252,254,256	8
8	PAFZZ	46522	A54948-G2	HEATING ELEMENT,ELE	1
9	PAFZZ	46522	30002-83	PUSH ON NUT	2
10	PAFZZ	46522	C54813-G1	VALVE,SOLENOID	1
11	PAFZZ	46522	B54028G1-1	BRACKET-VALVE	1
12	PAFZZ	46522	B54941-G1	THERMOSTAT ASSEMBLY	1
13	PAFZZ	46522	A53181	NIPPLE,TUBE	1
14	PAFZZ	05657	A5	CLAMP,HOSE	2
15	PAFZZ	46522	A52789-G5	HOSE,NONMETALLIC	1
16	PAFZZ	46522	A54907-G1	TUBE ASSEMBLY,METAL	1
17	PAFZZ	46522	A14140	PLUG,PIPE	1
18	PAFZZ	46522	A18167-G2	NOZZLE ASSEMBLY,FIL	1
19	PAFZZ	46522	A14141	TEE,PIPE TO TUBE	1
20	PAFZZ	46522	A54814-G1	GROMMET,NONMETALLIC	1
21	PAFZZ	96906	MS35266-66	SCREW,MACHINE	2
22	PAFZZ	46522	A2612	SPACER,RING	1
23	PAFZZ	46522	A2610	VAPORIZER	1
24	PBFZZ	46522	C54919-G1-1	COVER,ACCESS	1
25	PAOZZ	46522	C2757	IGNITER,SPARK,FUEL	1
26	PAFZZ	46522	A2096	SHIELD,ELECTRICAL C	1
27	PAFZZ	96906	MS51964-49	SETSCREW	1
28	PAFFF	46522	B54827-G1	MOTOR,DIRECT CURREN	1
29	PFFZZ	46522	B2798	.MOTOR,ALTERNATING C	1
30	PAFFF	46522	B2894	.PUMP,ROTARY	1
31	PAFZZ	96906	MS35842-10	..CLAMP,HOSE	2
32	PAFZZ	46522	A2829	..HOSE,PREFORMED	1
33	PAFZZ	46522	A2897	..SCREW,CAP,HEXAGON H	4
34	PAFZZ	96906	MS19060-1014	..BALL,BEARING	1
35	PAFZZ	46522	A2896	..SPRING,HELICAL,COMP	1
36	PAFZZ	46522	A2895	..PLUG,PIPE	1
37	PAFZZ	46522	A2932	..PACKING,PREFORMED	2
38	PAFZZ	46522	A2904	..PACKING WITH RETAIN	1
39	PAFZZ	53335	A2931	..GASKET	1
40	PAFZZ	53335	A930	..SPRING,HELICAL,COMP	1
41	XAFZZ	46522	A2481=XA	.BRUSH,ELECTRICAL CC	2
42	PAFZZ	46522	A2480	.CAP,BRUSH	2
43	PAFZZ	46522	B2714	PLATE,INDENTIFICATIO	1
44	PAFZZ	24617	143935	PLUG,PIPE	1
45	PAFZZ	46522	A54940-G1	RECEPTACLE,HEATER,C	1
46	PAFZZ	46522	2290-0632-05-13	SCREW	4

END OF FIGURE

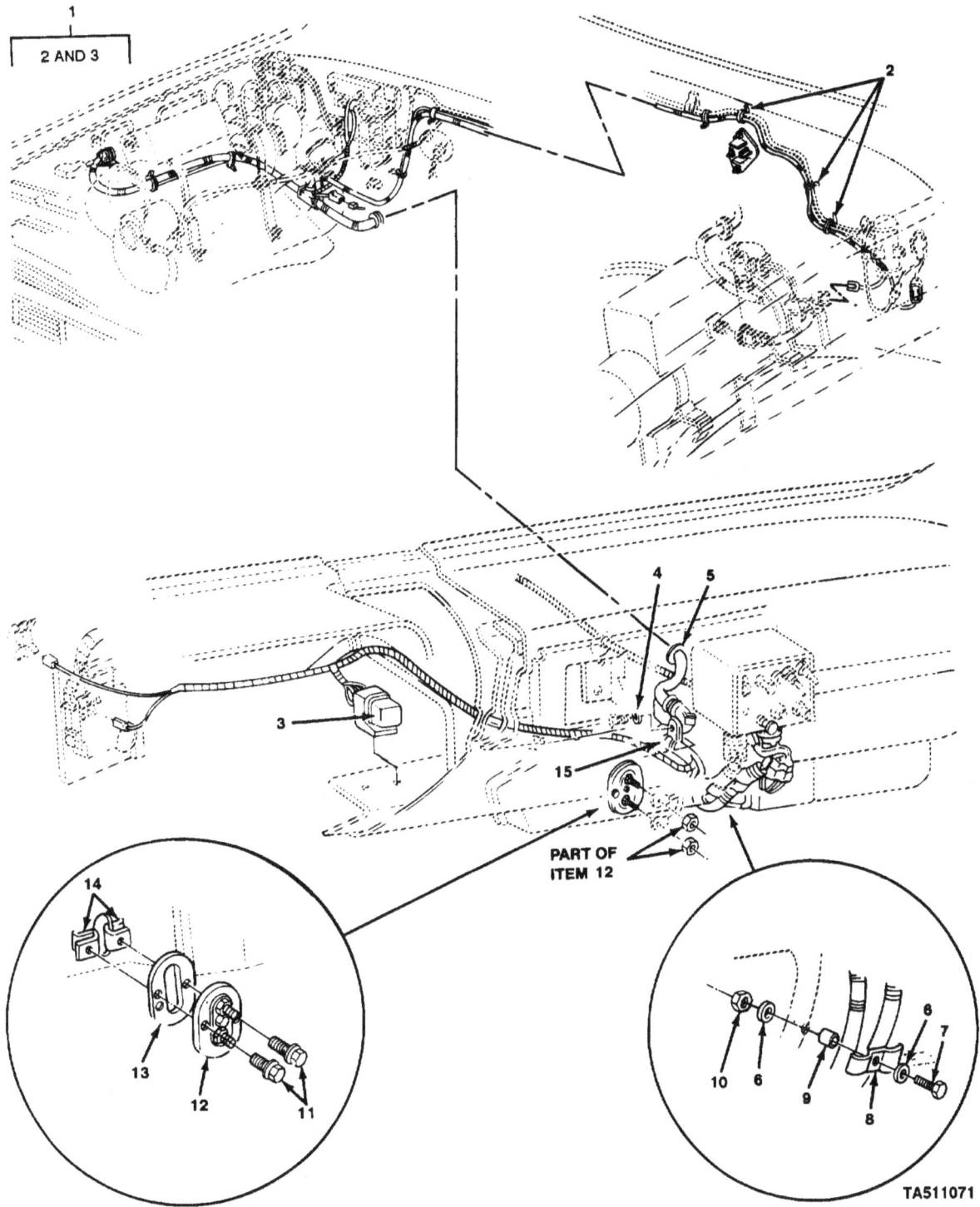

FIGURE 264. WARM AIR HEATER WIRING HARNESS .

TA511071

(1) ITEM NO	(2) SMR CODE	(3) CAGEC	(4) PART NUMBER	(5) DESCRIPTION AND USABLE ON CODES (UOC)	(6) QTY
				GROUP 3303 WINTERIZATION KITS	
				FIG. 264 WARM AIR HEATER WIRING HARNESS	
1	PAOOO	11862	14076236	WIRING HARNESS,BRAN	1
2	PAOZZ	11862	3816659	.STRAP,LINE SUPPORTI	10
3	PAOZZ	11862	2098912	.RELAY	1
4	PAOZZ	96906	MS27183-10	WASHER,FLAT	1
5	PAOZZ	96906	MS35489-121	GROMMET,NONMETALLIC	1
6	PAOZZ	96906	MS9549-10	WASHER,FLAT	2
7	PAOZZ	96906	MS90728-3	SCREW,CAP,HEXAGON H	1
8	PAOZZ	11862	14061352	STRAP,RETAINING	1
9	PAOZZ	11862	15599920	SPACER,SLEEVE	1
10	PAOZZ	96906	MS35691-406	NUT,PLAIN,HEXAGON	1
11	PAOZZ	24617	9415163	SCREW,ASSEMBLED WAS	2
12	PAOZZ	78385	G704410-1	SWITCH,THERMOSTATIC	1
13	PAOZZ	11862	14076270	GASKET	1
14	PAOZZ	11862	9440178	NUT,PLAIN,DODECAGON	2
15	PAOZZ	96906	MS21333-62	CLAMP,LOOP	1

END OF FIGURE

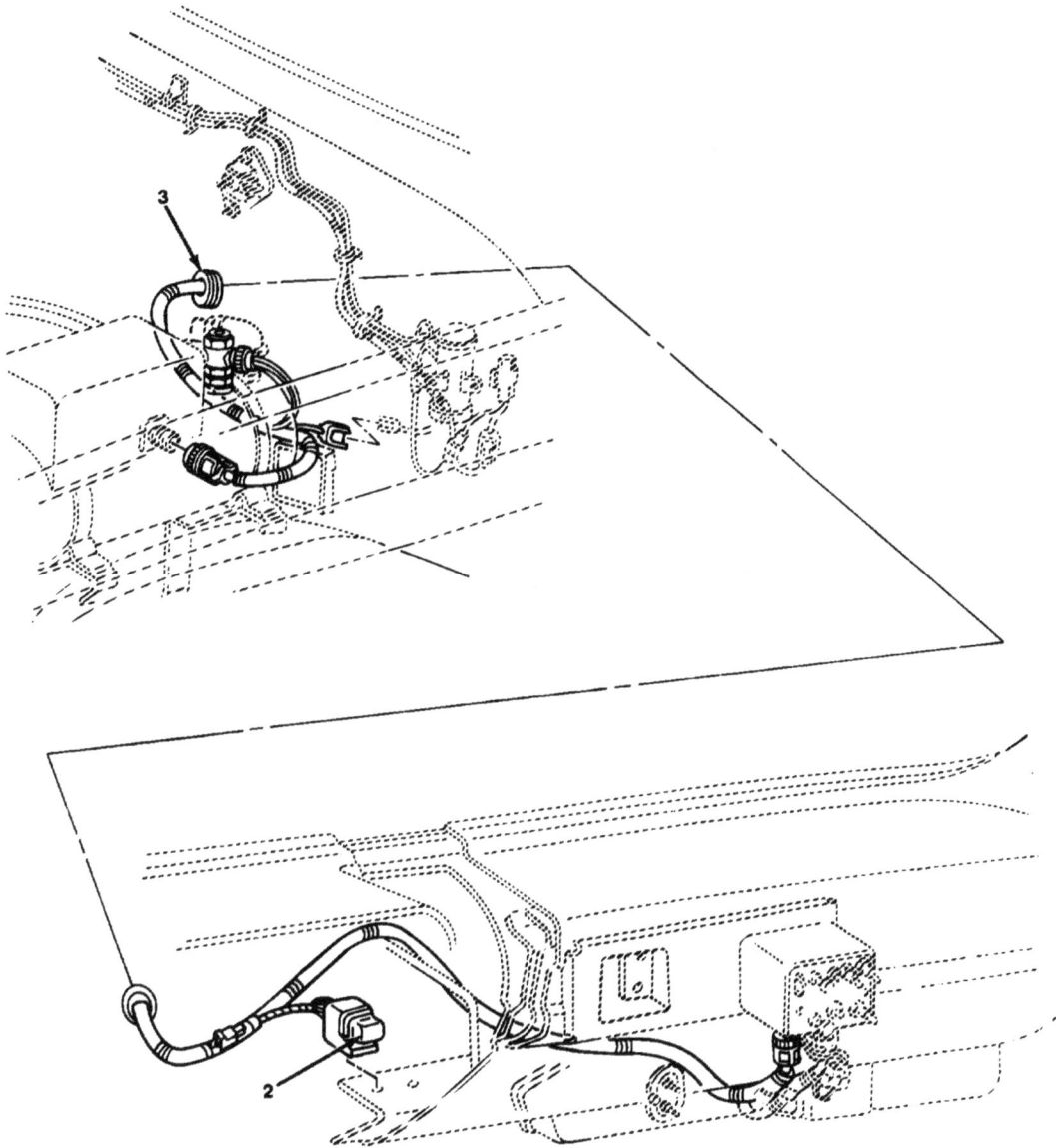

TA511072

FIGURE 265. COOLANT HEATER WIRING HARNESS.

(1) (2) (3) (4)	(5)	(6)
ITEM SMR PART		
NO CODE CAGEC NUMBER	DESCRIPTION AND USABLE ON CODES (UOC)	QTY

GROUP 3303 WINTERIZATION KITS

FIG. 265 COOLANT HEATER WIRING
HARNESS

1	PAOOO	11862	14075888		WIRING HARNESS,BRAN	1
2	PAOZZ	11862	14074461PC6	.RELAY		1
3	PAOZZ	96906	MS35489-121	GROMMET,NONMETALLIC		1

END OF FIGURE

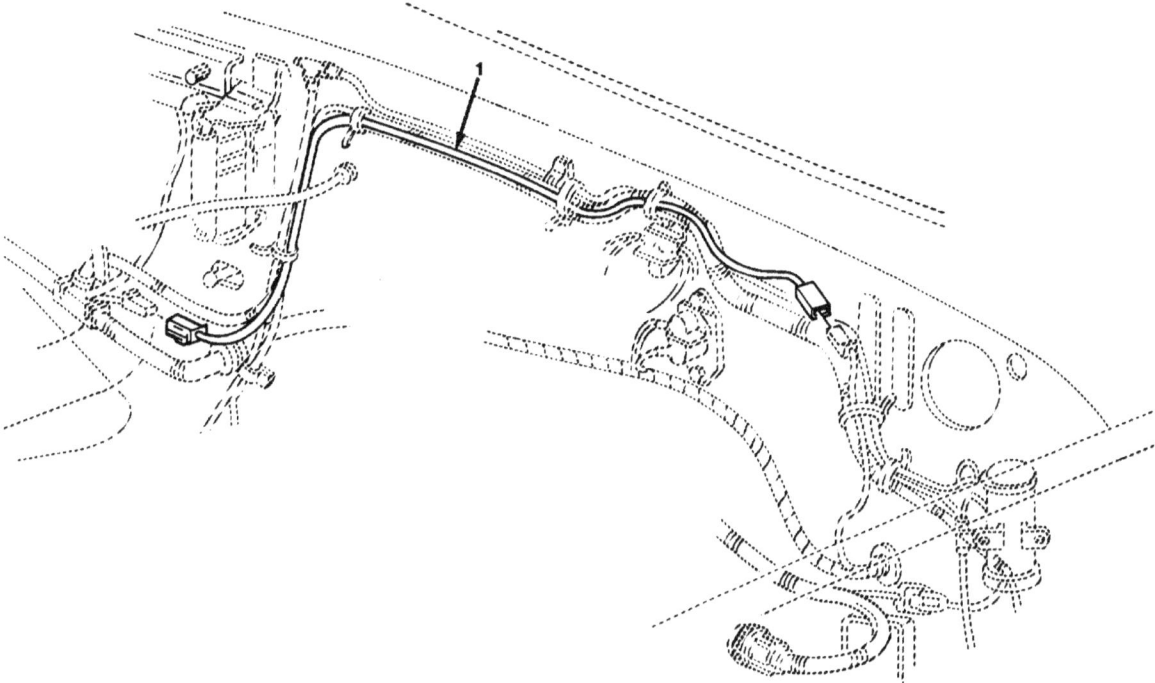

TA511073

FIGURE 266. WARM AIR FUEL PUMP EXTENSION HARNESS (M1010).

(1) (2) (3) (4)			(5)	(6)
ITEM SMR		PART		
NO CODE CAGEC NUMBER			DESCRIPTION AND USABLE ON CODES (UOC)	QTY

GROUP 3303 WINTERIZATION KITS

FIG. 266 WARM AIR FUEL PUMP
EXTENSION HARNESS (M1010)

| 1 | PAOZZ | 11862 | 15599968 | LEAD,ELECTRICAL | 1 |
| | | | | UOC:210 | |

END OF FIGURE

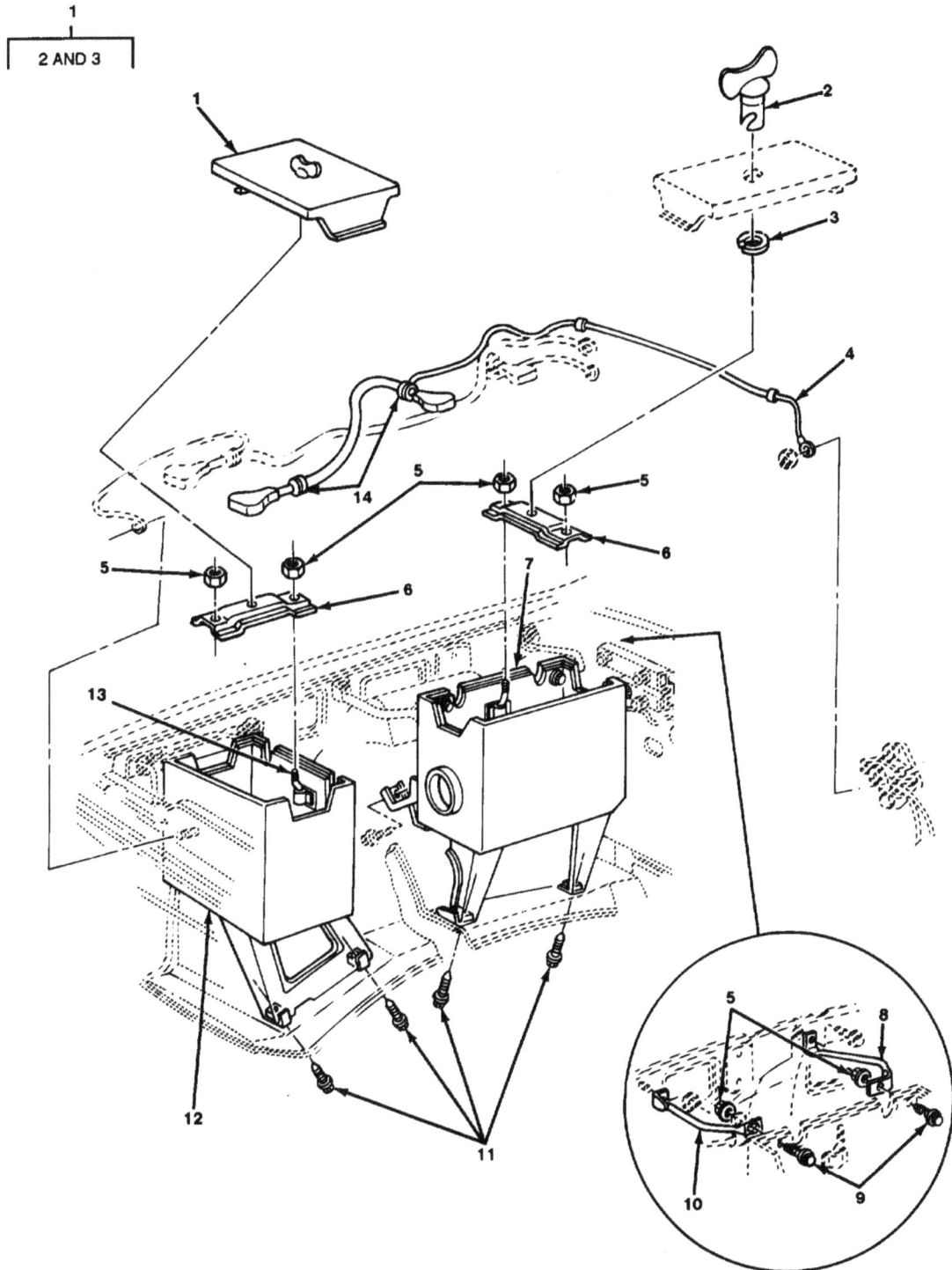

FIGURE 267. BATTERY BOXES AND CABLES.

TA511074

(1) ITEM NO	(2) SMR CODE	(3) CAGEC	(4) PART NUMBER	(5) DESCRIPTION AND USABLE ON CODES (UOC)	(6) QTY
				GROUP 3303 WINTERIZATION KITS	
				FIG. 267 BATTERY BOXES AND CABLES	
1	PAOOO	11862	14076246	COVER,BATTERY BOX	2
2	PAOZZ	72794	AJW7-70	.STUD,TURNLOCK FASTE	2
3	PAOZZ	72794	X-840-SR7C	.RING,RETAINING	2
4	PAOZZ	11862	12039294	LEAD,STORAGE BATTER	1
5	PAOZZ	24617	11506003	NUT,PLAIN,ASSEMLED	6
6	PAOZZ	11862	14076852	RETAINER,BATTERY	2
7	PAOZZ	11862	14076243	BATTERY BOX	1
8	PAOZZ	11862	14076250	SUPPORT,BATTERY BOX	1
9	PAOZZ	11862	9440334	BOLT,ASSEMBLED WASH	2
10	PFOZZ	11862	14076249	SUPPORT,BATTERY BOX	1
11	PAOZZ	11862	2014469	BOLT,ASSEMBLED WASH	4
12	PAOZZ	11862	14076241	BOX AND SUPPORT ASS	1
13	PAOZZ	11862	14076248	BOLT, HOOK	4
14	PAOZZ	96906	MS35489-11	GROMMET,NONMETALLIC	4

END OF FIGURE

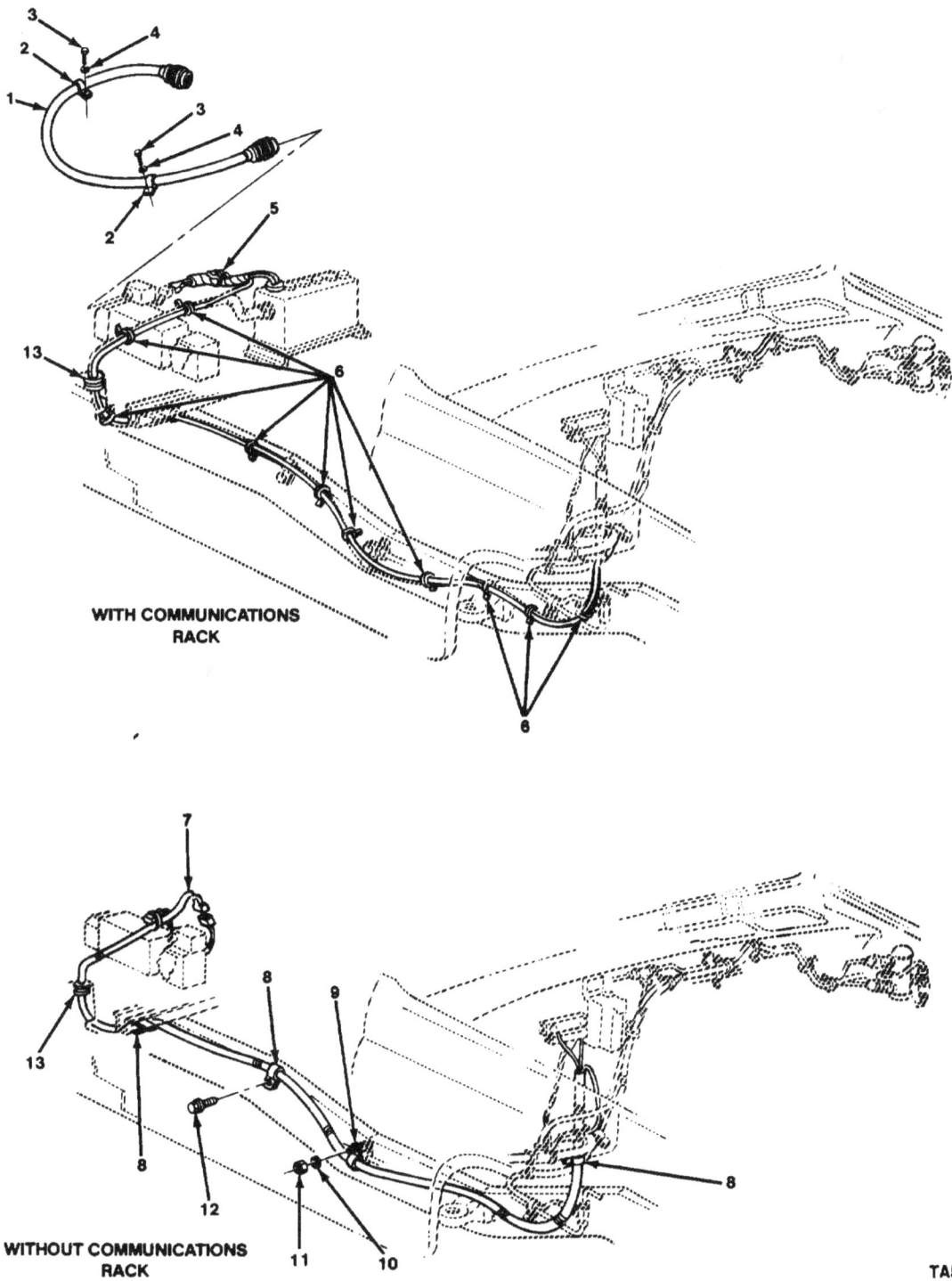

WITH COMMUNICATIONS
RACK

WITHOUT COMMUNICATIONS
RACK

TA511075

FIGURE 268. PERSONNEL HEATER WIRING HARNESS.

(1) (2) (3) (4)	(5)	(6)
ITEM SMR PART		
NO CODE CAGEC NUMBER	DESCRIPTION AND USABLE ON CODES (UOC)	QTY

GROUP 3303 WINTERIZATION KITS

FIG. 268 PERSONNEL HEATER WIRING
HARNESS

1	PAOZZ 11862 14076269	LEAD,ELECTRICAL		1
		UOC:194,208		
2	PAOZZ 96906 MS21333-112	CLAMP,LOOP		2
		UOC:194,208		
3	PAOZZ 24617 9423768	SCREW,CAP,HEXAGON H		2
		UOC:194,208		
4	PAOZZ 24617 9417714	WASHER,FLAT		2
		UOC:194,208		
5	PAOZZ 11862 14076811	WIRING HARNESS,BRAN		1
		UOC:194,208		
6	PAOZZ 11862 3816659	STRAP,LINE SUPPORTI		11
		UOC:194,208		
7	PAOZZ 11862 14076240	WIRING HARNESS,BRAN		1
		UOC:194,208		
8	PAOZZ 96906 MS21333-123	CLAMP,LOOP		3
		UOC:194,208		
9	PAOZZ 96906 MS21333-114	CLAMP,LOOP		1
		UOC:194,208		
10	PAOZZ 96906 MS35338-44	WASHER,LOCK		1
		UOC:194,208		
11	PAOZZ 96906 MS35691-406	NUT,PLAIN,HEXAGON		1
		UOC:194,208		
12	PAOZZ 11862 11504447	SCREW,TAPPING,THREA		2
		UOC:194,208		
13	PAOZZ 11862 3886908	GROMMET,NONMETALLIC		1
		UOC:194,208		

END OF FIGURE

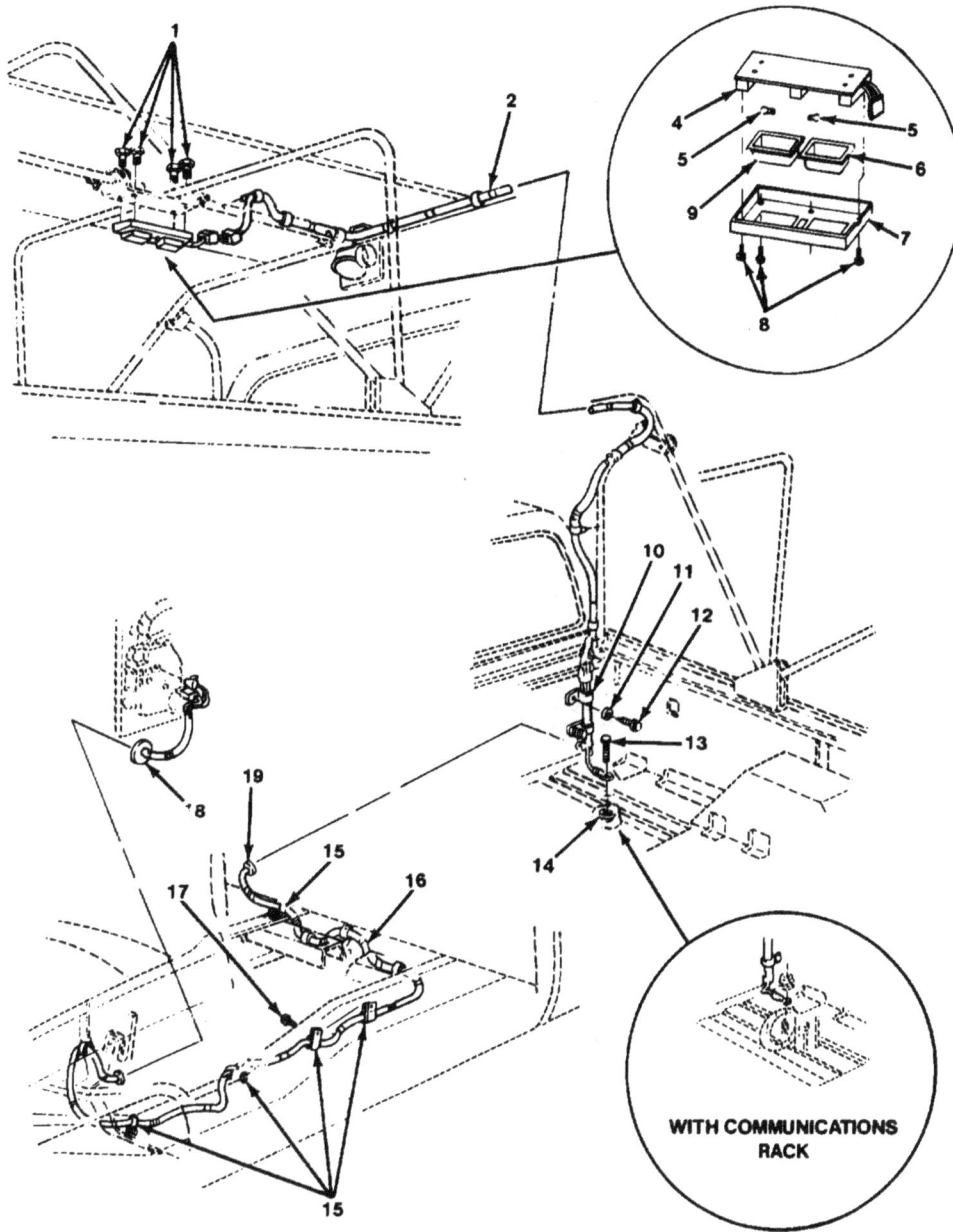

FIGURE 269. DOMELIGHT LAMP AND WIRING HARNESS (M1008 AND M1008A1).

TA511076

(1)	(2)	(3)	(4)	(5)	(6)
ITEM NO	SMR CODE	CAGEC	PART NUMBER	DESCRIPTION AND USABLE ON CODES (UOC)	QTY

GROUP 3303 WINTERIZATION KITS

FIG. 269 DOMELIGHT LAMP AND WIRING
HARNESS (M1008 AND M1008A1)

1	PAOZZ	96906	MS35206-245	SCREW,MACHINE	4
				UOC:194,208	
2	PAOZZ	11862	14076238	LEAD,ELECTRICAL	1
				UOC:194,208	
3	PAOOO	22973	922-900-00	LIGHT,DOME	1
				UOC:194,208	
4	PAOZZ	22973	922-910-00	.MOUNTING BASE,LIGHT	1
				UOC:194,208	
5	PAOZZ	08806	1003	.LAMP,INCANDESCENT	2
				UOC:194,208	
6	PAOZZ	22973	922-901-02	.LENS,LIGHT WHITE	1
				UOC:194,208	
7	PAOZZ	22973	922-904-02	.HOUSING,LIGHT	1
				UOC:194,208	
8	PAOZZ	96906	MS35206-226	.SCREW,MACHINE	6
				UOC:194,208	
9	PAOZZ	22973	922-901-01	.LENS,LIGHT BLUE	1
				UOC:194,208	
10	PAOZZ	96906	MS21333-112	CLAMP,LOOP	2
				UOC:194,208	
11	PAOZZ	24617	9417714	WASHER,FLAT	2
				UOC:194,208	
12	PAOZZ	96906	MS51850-64	SCREW,TAPPING,THREA	2
				UOC:194,208	
13	PAOZZ	96906	MS90728-6	SCREW,CAP,HEXAGON H	1
				UOC:194,208	
14	PAOZZ	24617	271172	NUT,SELF-LOCKING,AS	1
				UOC:194,208	
15	PAOZZ	96906	MS21333-123	CLAMP,LOOP	5
				UOC:194,208	
16	PAOZZ	11862	14076239	WIRING HARNESS,BRAN	1
				UOC:194,208	
17	PAOZZ	11862	11504447	SCREW,TAPPING,THREA	5
				UOC:194,208	
18	PAOZZ	11862	3918889	GROMMET,NONMETALLIC	1
				UOC:194,208	
19	PAOZZ	11862	15591130	GROMMET,NONMETALLIC	1
				UOC:194,208	

END OF FIGURE

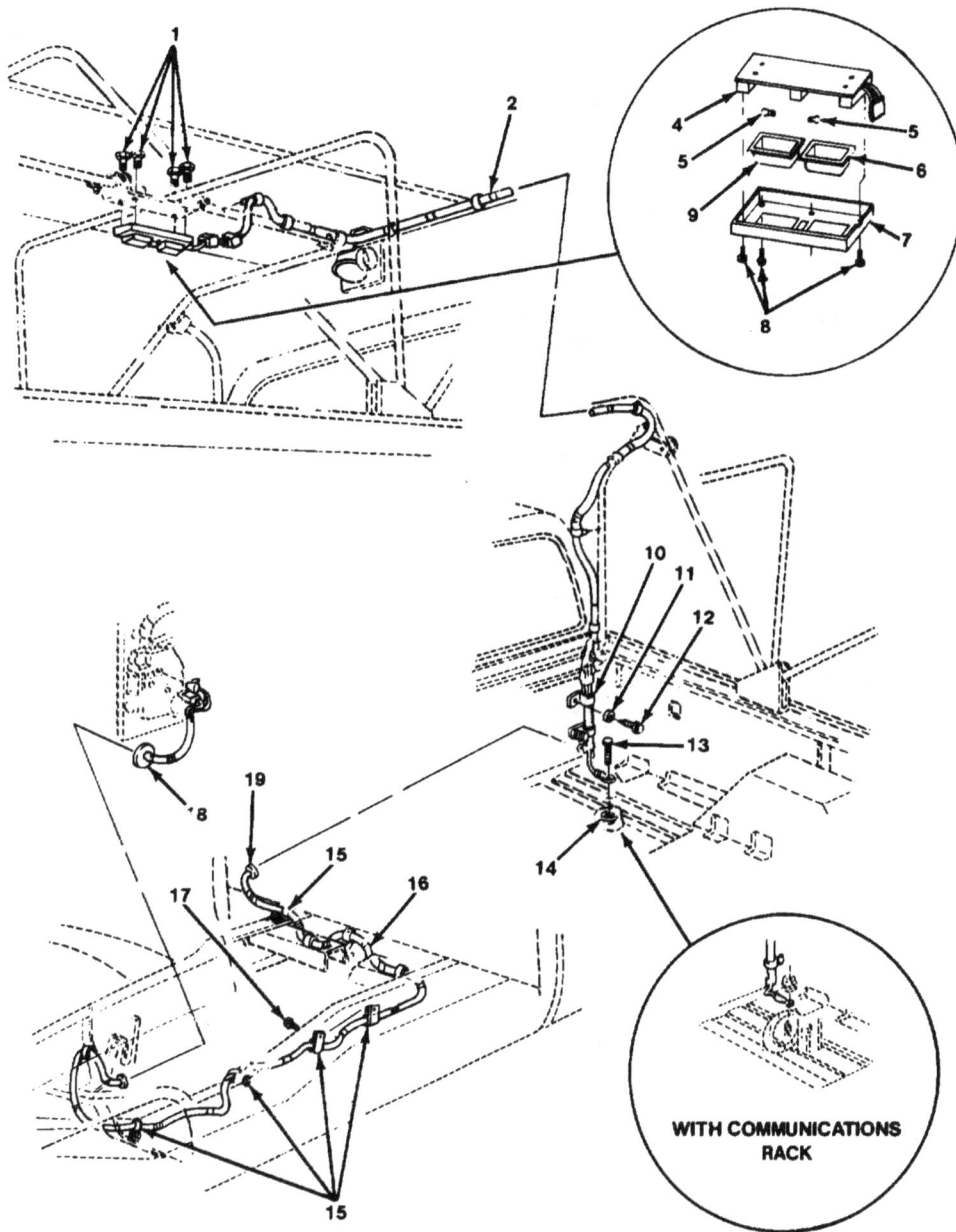

FIGURE 269. DOMELIGHT LAMP AND WIRING HARNESS (M1008 AND M1008A1).

TA511076

(1) (2) (3) (4)			(5)	(6)
ITEM SMR		PART		
NO CODE CAGEC NUMBER			DESCRIPTION AND USABLE ON CODES (UOC)	QTY

GROUP 3303 WINTERIZATION KITS

FIG. 269 DOMELIGHT LAMP AND WIRING HARNESS (M1008 AND M1008A1)

1	PAOZZ 96906 MS35206-245	SCREW,MACHINE		4
		UOC:194,208		
2	PAOZZ 11862 14076238	LEAD,ELECTRICAL		1
		UOC:194,208		
3	PAOOO 22973 922-900-00	LIGHT,DOME		1
		UOC:194,208		
4	PAOZZ 22973 922-910-00	.MOUNTING BASE,LIGHT		1
		UOC:194,208		
5	PAOZZ 08806 1003	.LAMP,INCANDESCENT		2
		UOC:194,208		
6	PAOZZ 22973 922-901-02	.LENS,LIGHT WHITE		1
		UOC:194,208		
7	PAOZZ 22973 922-904-02	.HOUSING,LIGHT		1
		UOC:194,208		
8	PAOZZ 96906 MS35206-226	.SCREW,MACHINE		6
		UOC:194,208		
9	PAOZZ 22973 922-901-01	.LENS,LIGHT BLUE		1
		UOC:194,208		
10	PAOZZ 96906 MS21333-112	CLAMP,LOOP		2
		UOC:194,208		
11	PAOZZ 24617 9417714	WASHER,FLAT		2
		UOC:194,208		
12	PAOZZ 96906 MS51850-64	SCREW,TAPPING,THREA		2
		UOC:194,208		
13	PAOZZ 96906 MS90728-6	SCREW,CAP,HEXAGON H		1
		UOC:194,208		
14	PAOZZ 24617 271172	NUT,SELF-LOCKING,AS		1
		UOC:194,208		
15	PAOZZ 96906 MS21333-123	CLAMP,LOOP		5
		UOC:194,208		
16	PAOZZ 11862 14076239	WIRING HARNESS,BRAN		1
		UOC:194,208		
17	PAOZZ 11862 11504447	SCREW,TAPPING,THREA		5
		UOC:194,208		
18	PAOZZ 11862 3918889	GROMMET,NONMETALLIC		1
		UOC:194,208		
19	PAOZZ 11862 15591130	GROMMET,NONMETALLIC		1
		UOC:194,208		

END OF FIGURE

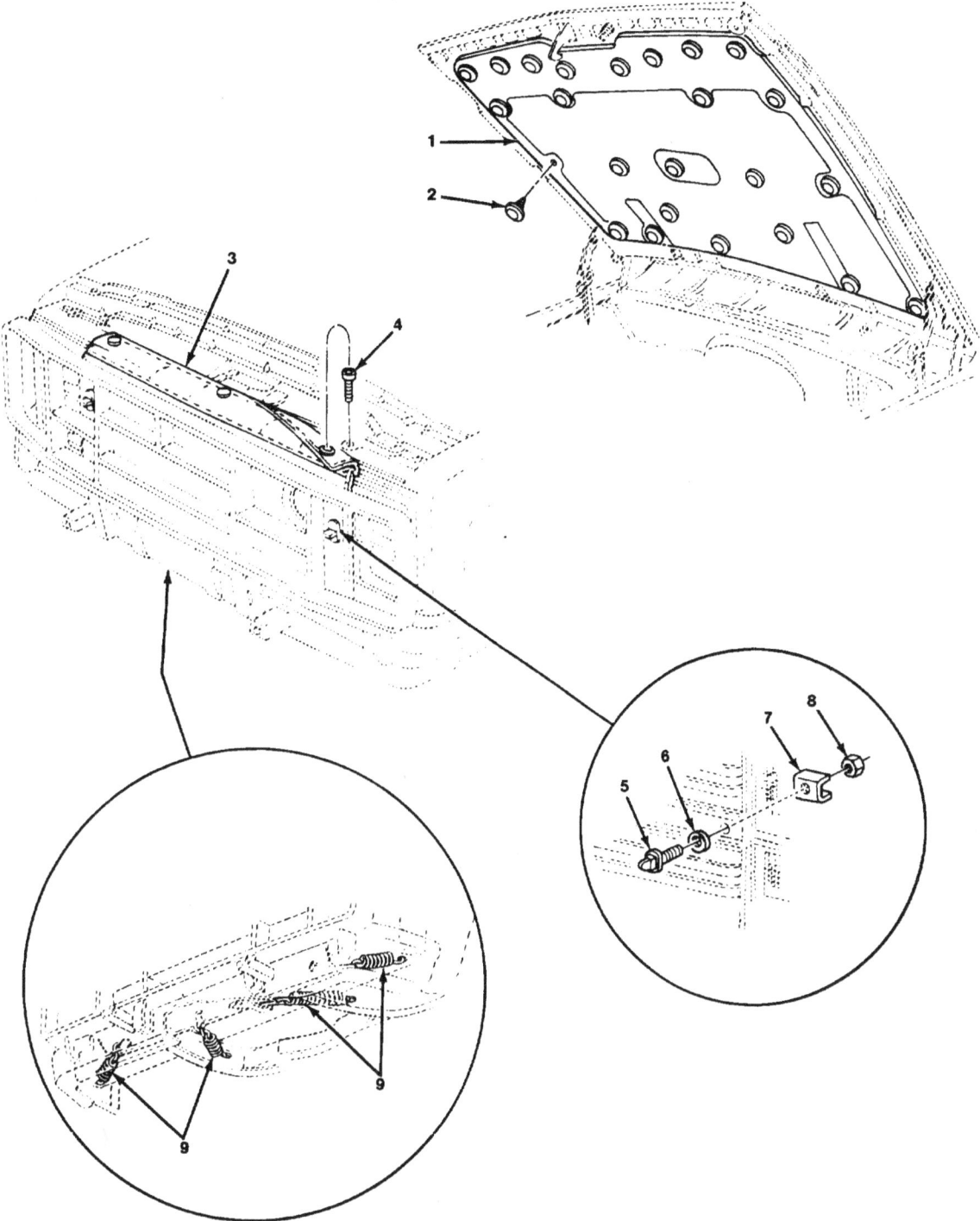

TA511077

FIGURE 270. HOOD AND RADIATOR INSULATORS.

(1) ITEM NO	(2) SMR CODE	(3) CAGEC	(4) PART NUMBER	(5) DESCRIPTION AND USABLE ON CODES (UOC)	(6) QTY
				GROUP 3303 WINTERIZATION KITS	
				FIG. 270 HOOD AND RADIATOR INSULATORS	
1	MOOZZ	11862	15593599	INSULATOR HOOD MAKE FROM INSULATION, P/N 12306178	1
2	PAOZZ	118623	3977775	FASTENER,SPRING TEN UOC:194,208,209,230,231,252,254,256	24
3	PAOZZ	11862	14063365	INSULATOR,WINTERIZA	1
4	PAOZZ	11862	107413	STUD,SELF-LOCKING	3
5	PAOZZ	19207	12255564-2	STUD,TURNBUTTON FAS	2
6	PAOZZ	96906	MS27183-10	WASHER,FLAT	2
7	PFOZZ	11862	14063366	REINFORCEMENT,RADIA	2
8	PAOZZ	21450	131245	NUT,SELF-LOCKING,HE	2
9	PAOZZ	19207	7717066	SPRING,HELICAL,EXTE	4

END OF FIGURE

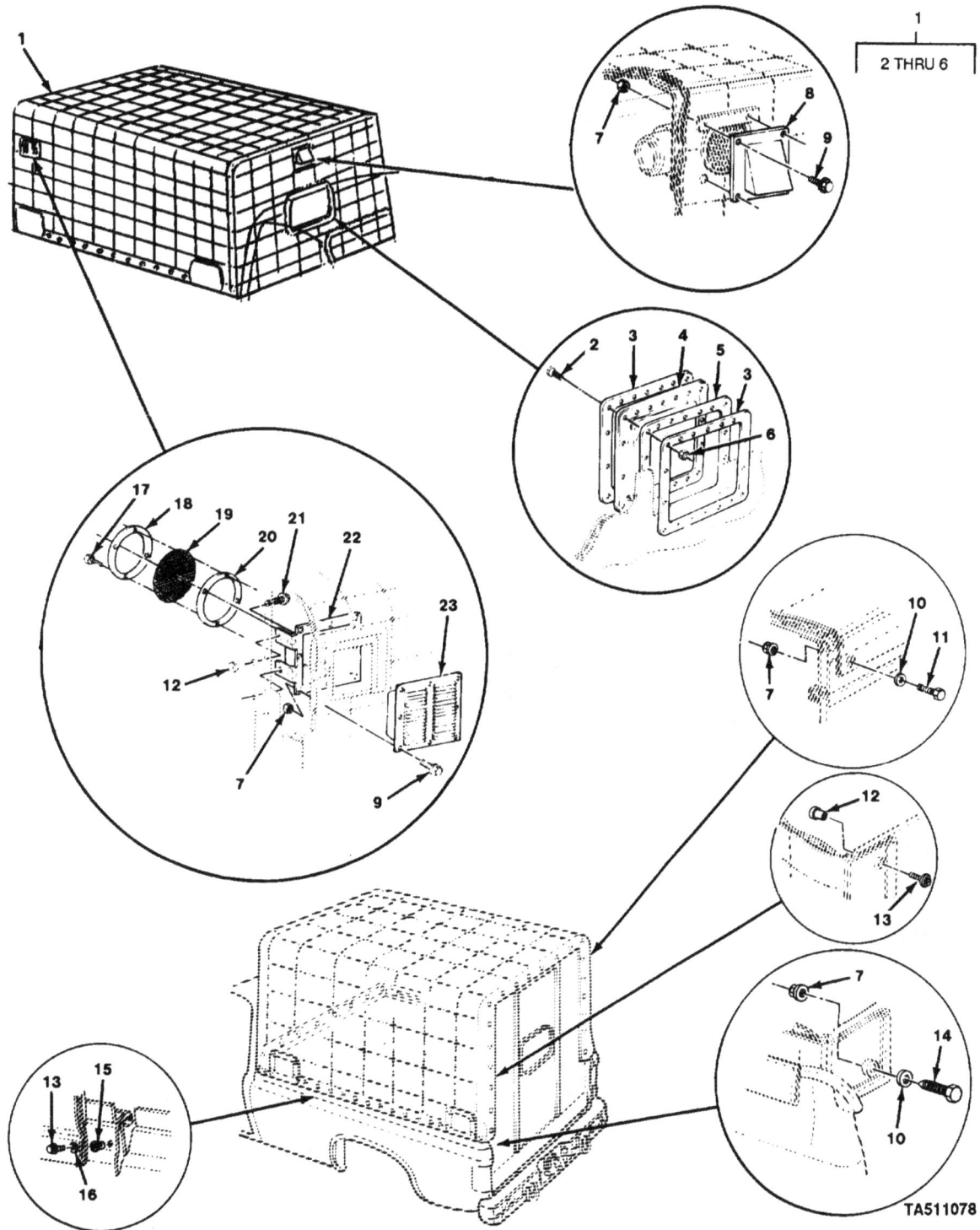

FIGURE 271. ROOF COVER, INTAKE DEFLECTOR, WINDOW, VENT, AND
RELATED PARTS (M1008 AND M1008A1).

TA511078

(1) (2) (3) (4) ITEM SMR PART NO CODE CAGEC NUMBER	(5) DESCRIPTION AND USABLE ON CODES (UOC)	(6) QTY
	GROUP 3303 WINTERIZATION KITS	
	FIG. 271 ROOF COVER, INTAKE DEFLECTOR, WINDOW, VENT, AND RELATED PARTS (M1008 AND M1008A1)	
1 PAOFF 11862 14076228	COVER,FITTED,VEHICU UOC:194,208	1
2 PAOZZ 96906 MS51957-33	.SCREW,MACHINE UOC:194,208	18
3 PAOZZ 11862 14076216	.RETAINER,WINDOW UOC:194,208	2
4 PAOZZ 19207 7353960	.WINDOW,VEHICULAR UOC:194,208	1
5 PAOZZ 11862 14076215	.GASKET UOC:194,208	1
6 PAOZZ 11862 9414031	.NUT,SELF-LOCKING,RO UOC:194,208	18
7 PAOZZ 24617 9416918	NUT,PLAIN,EXTENDED UOC:194,208	20
8 PAOZZ 11862 14075845	VENTILATOR,AIR CIRC UOC:194,208	1
9 PAOZZ 11862 9440033	BOLT,MACHINE UOC:194,208	12
10 PAOZZ 81795 30489	WASHER,FLAT UOC:194,208	8
11 PAOZZ 96906 MS90728-14	SCREW,CAP,HEXAGON H UOC:194,208	6
12 PAOZZ 21450 587227	NUT,PLAIN,PLATE UOC:194,208	12
13 PAOZZ 96906 MS51957-84	SCREW,MACHINE UOC:194,208	44
14 PAOZZ 96906 MS90728-13	SCREW,CAP,HEXAGON H UOC:194,208	2
15 PAOZZ 11862 14005953	SHIELD,EXPENSION UOC:194,208	32
16 PAOZZ 24617 9419265	WASHER,FLAT UOC:194,208	44
17 PAOZZ 11862 20365263	SCREW,TAPPING,THREA UOC:194,208	4
18 PAOZZ 11862 14072367	DOOR,ACCESS UOC:194,208	1
19 PAOZZ 11862 14072366	STRAINER ELEMENT,SE UOC:194,208	1
20 PAOZZ 11862 14075857	GASKET UOC:194,208	1
21 PAOZZ 11862 9440334	BOLT,ASSEMBLED WASH UOC:194,208	2
22 PAOZZ 11862 14072365	COVER,ACCESS UOC:194,208	1
23 PAOZZ 11862 14072364	VENTILATOR,AIR CIRC UOC:194,208	1

END OF FIGURE

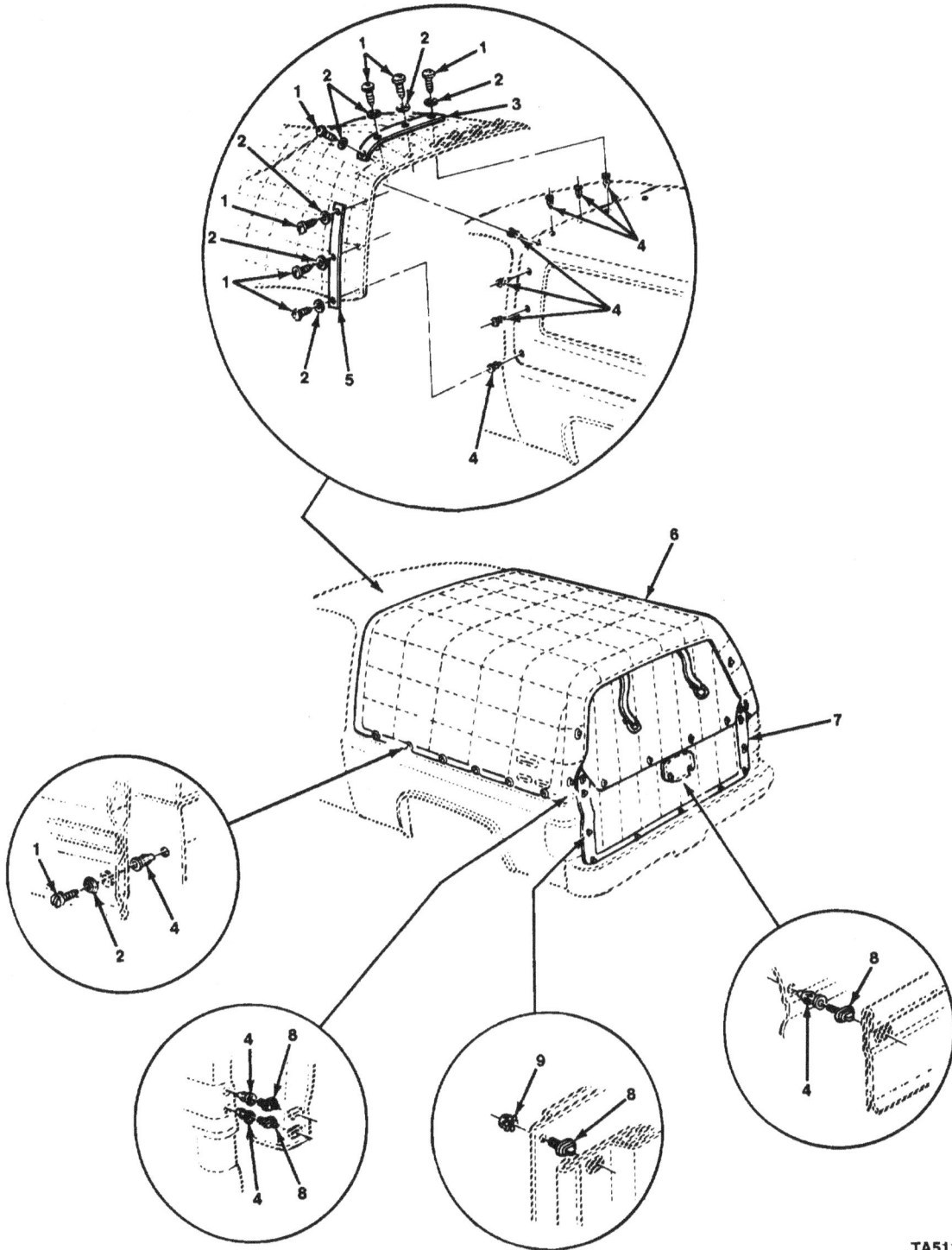

FIGURE 272. ROOF AND ENDGATE COVERS AND MOUNTING PARTS (M1009).

TA511079

(1) (2) (3) (4)	(5)	(6)
ITEM SMR PART NO CODE CAGEC NUMBER	DESCRIPTION AND USABLE ON CODES (UOC)	QTY

GROUP 3303 WINTERIZATION KITS

FIG. 272 ROOF AND ENDGATE COVERS AND
MOUNTING PARTS (M1009)

1	PAOZZ	96906	MS51957-83	SCREW,MACHINE
				UOC:209
2	PAOZZ	24617	9419265	WASHER,FLAT
				UOC:209
3	MOOZ	11862	14075884	RETAINER (1"X 29.6"X 0.54")MAKE FROM STRIP,P/N MS520390079
				UOC:209
4	PAOZZ	11862	14005953	SHIELD,EXPENSION
				UOC:209
5	MOOZZ	11862	14075883	RETAINER (1"X 24.2"X 0.54") MAKE FROM STRIP, P/N MS520390079
				UOC:209
6	PAOFF	11862	14075881	COVER,FITTED,VEHICU
				UOC:209
7	PAOZZ	11862	14075882	CURTAIN,VEHICULAR
				UOC:209
8	PAOZZ	19207	12255564-2	STUD,TURNBUTTON FAS
				UOC:209
9	PAOZZ	24617	271172	NUT,SELF-LOCKING,AS
				UOC:209

QTY column:
- 1: 26
- 2: 26
- 3: 2
- 4: 44
- 5: 2
- 6: 1
- 7: 1
- 8: 20
- 9: 8

END OF FIGURE

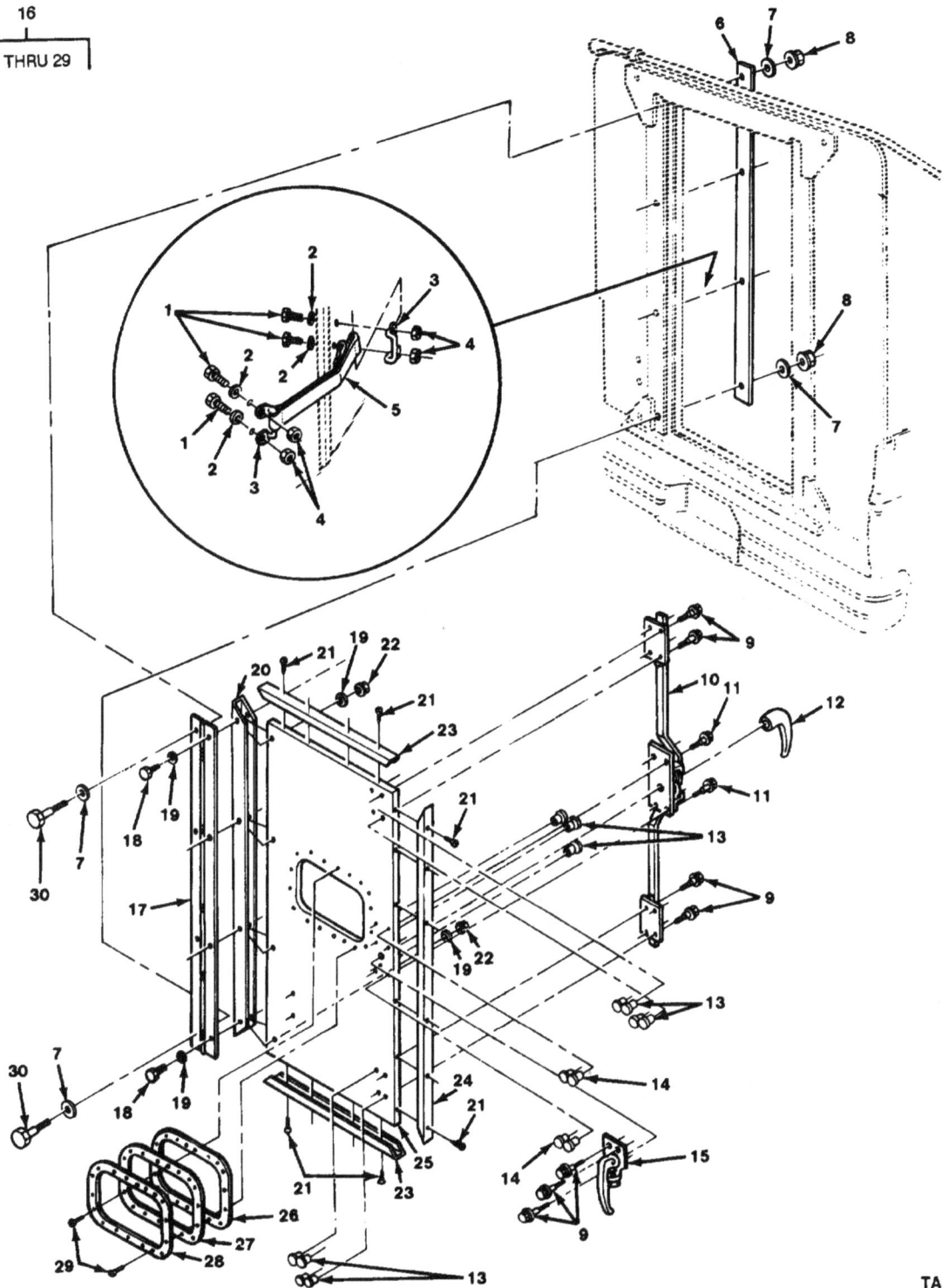

FIGURE 273. REAR DOOR ASSEMBLY (M1008).

TA511080

(1) (2) (3) (4) ITEM SMR PART NO CODE CAGEC NUMBER	(5) DESCRIPTION AND USABLE ON CODES (UOC)	(6) QTY
	GROUP 3303 WINTERIZATION KITS	
	FIG. 273 REAR DOOR ASSEMBLY (M1008)	
1 PAOZZ 96906 MS90728-13	SCREW,CAP,HEXAGON H UOC:194,208	4
2 PAOZZ 81795 30489	WASHER,FLAT UOC:194,208	4
3 PAOZZ 11862 3794767	HOLDER,DOOR UOC:194,208	2
4 PAOZZ 24617 9416918	NUT,PLAIN,EXTENDED UOC:194,208	4
5 PAOZZ 11862 370349	STRAP AND FITTINGS UOC:194,208	1
6 XAOZZ 11862 14076204	MOULDING,REAR DOOR UOC:194,208	1
7 PAOZZ 96906 MS27183-14	WASHER,FLAT UOC:194,208	8
8 PAOZZ 24617 9413534	NUT,SELF-LOCKING,HE UOC:194,208	4
9 PAOZZ 96906 MS51849-78	SCREW,MACHINE UOC:194,208	11
10 PAOZZ 11862 14076882	LATCH ASSEMBLY,DOOR UOC:194,208	1
11 PAOZZ 96906 MS90725-5	SCREW,CAP,HEXAGON H UOC:194,208	4
12 PAOZZ 11862 15599951	HANDLE,DOOR UOC:194,208	1
13 PAOZZ 11862 9411281	STUD,TURNLOCK FASTE UOC:194,208	11
14 PAOZZ 21450 587227	NUT,PLAIN,PLATE UOC:194,208	4
15 PAOZZ 11862 15599950	HANDLE ASSEMBLY,OUT UOC:194,208	1
16 PFOOO 11862 14076875	DOOR,VEHICULAR UOC:194,208	1
17 PAOZZ 11862 14076217	.HINGE,BUTT UOC:194,208	1
18 PAOZZ 96906 MS90728-66	.SCREW,CAP,HEXAGON H UOC:194,208	4
19 PAOZZ 96906 MS27183-14	.WASHER,FLAT UOC:194,208	8
20 XAOZZ 11862 14076210	.FRAME,DOOR UOC:194,208	1
21 PAOZZ 96906 MS35492-54	.SCREW,WOOD UOC:194,208	14
22 PAOZZ 24617 9413534	.NUT,SELF-LOCKING,HE UOC:194,208	4
23 XAOZZ 11862 14076211	.FRAME REAR DOOR UOC:194,208	2
24 XAOZZ 11862 14076884	.FRAME REAR DOOR UOC:194,208	1

(1) (2) (3) (4) ITEM SMR PART NO CODE CAGEC NUMBER	(5) DESCRIPTION AND USABLE ON CODES (UOC)	(6) QTY
25 MOOZZ 11862 14076209	.DOOR,REAR (29.72"X 56.0") MAKE FROM PLYWOOD, P/N NNP530 UOC:194,208	1
26 PAOZZ 19207 7353960	.WINDOW,VEHICULAR UOC:194,208	1
27 PAOZZ 11862 14076215	.GASKET UOC:194,208	1
28 PAOZZ 11862 14076216	.RETAINER,WINDOW UOC:194,208	1
29 PAOZZ 96906 MS35493-37	.SCREW,WOOD UOC:194,208	18
30 PAOZZ 96906 MS90728-66	SCREW,CAP,HEXAG0N H UOC:194,208	4

END OF FIGURE

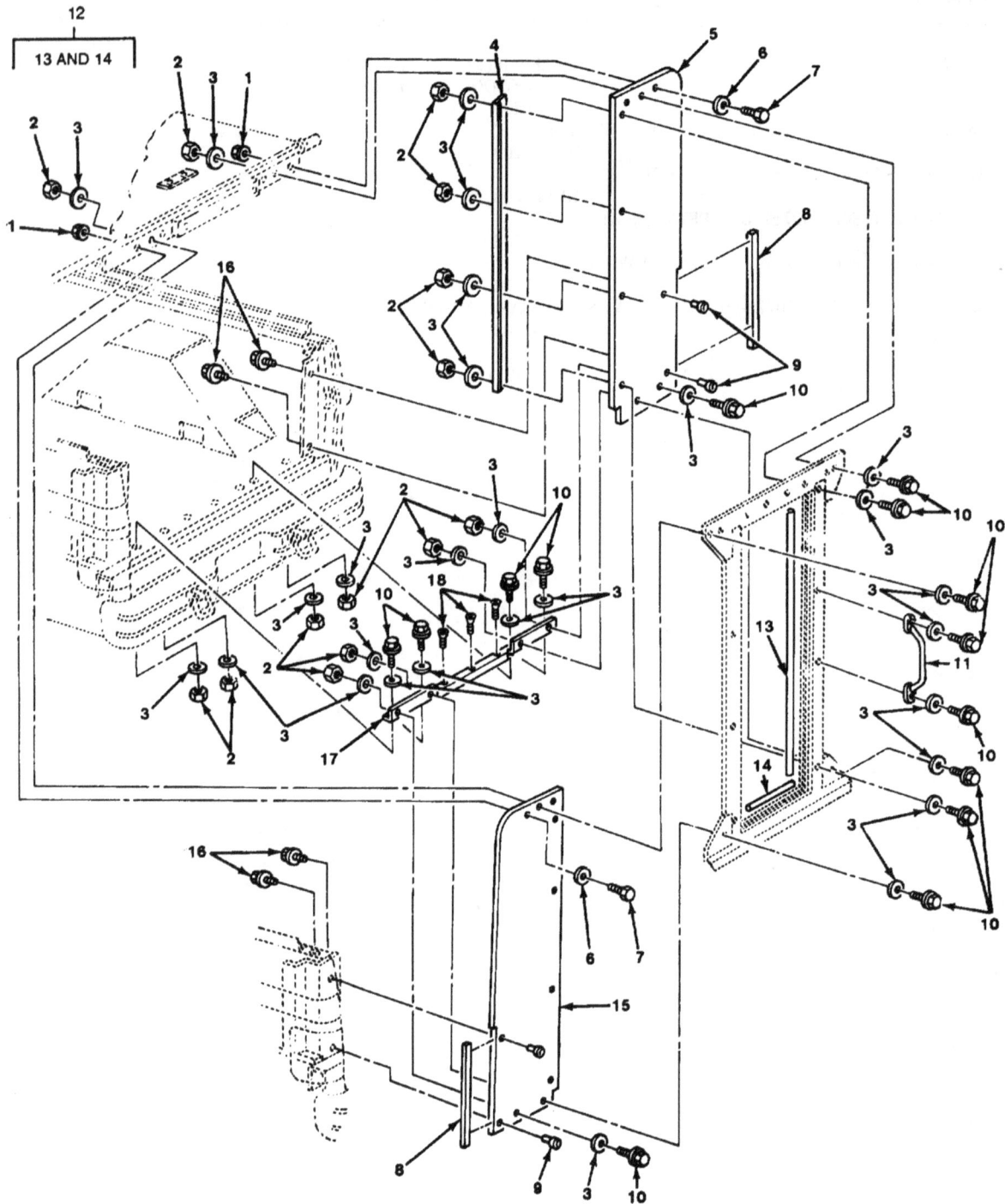

FIGURE 274. REAR PANEL COMPONENTS (M1008).

TA511081

(1) (2) (3) (4) ITEM SMR PART NO CODE CAGEC NUMBER	(5) DESCRIPTION AND USABLE ON CODES (UOC)	(6) QTY
	GROUP 3303 WINTERIZATION KITS	
	FIG. 274 REAR PANEL COMPONENTS (M1008)	
1 PAOZZ 24617 9416918 NUT,PLAIN,EXTENDED	UOC:194,208	2
2 PAOZZ 24617 9413534 NUT,SELF-LOCKING,HE	UOC:194,208	14
3 PAOZZ 96906 MS27183-14 WASHER,FLAT	UOC:194,208	28
4 XAOZZ 11862 14076204 MOULDING,REAR DOOR	UOC:194,208	1
5 MOOZZ 11862 14076872 PANEL ASM RR DR R H (19.32"X 59.09") MAKE FROM PLYWOOD, P/N NNP53C UOC:194,208		1
6 PAOZZ 81795 30489 WASHER,FLAT	UOC:194,208	2
7 PAOZZ 96906 MS90728-13 SCREW,CAP,HEXAGON H	UOC:194,208	2
8 PAOZZ 11862 14076862 RUBBER STRIP	UOC:194,208	2
9 PAOZZ 21450 587227 NUT,PLAIN,PLATE	UOC:194,208	4
10 PAOZZ 96906 MS90728-66 SCREW,CAP,HEXAGON H	UOC:194,208	14
11 PAOZZ 11862 14076221 HANDLE,BOW	UOC:194,208	1
12 PAOOO 11862 15599910 FRAME,DOOR,METAL	UOC:194,208	1
13 XDOZZ 11862 14076212 .SEAL	UOC:194,208	2
14 XDOZZ 11862 14076864 .SEAL	UOC:194,208	2
15 MOOZZ 11862 14076871 PANEL ASM RR CR LH (19.32"X 59.09") MAKE FROM PLYWOOD, P/N NNP53C UOC:194,208		1
16 PAOZZ 11862 9440334 BOLT,ASSEMBLED WASH	UOC:194,208	4
17 PAOZZ 11862 14076208 MOLDING,METAL	UOC:194,208	1
18 PAOZZ 96906 MS35492-54 SCREW,WOOD	UOC:194,208	3

END OF FIGURE

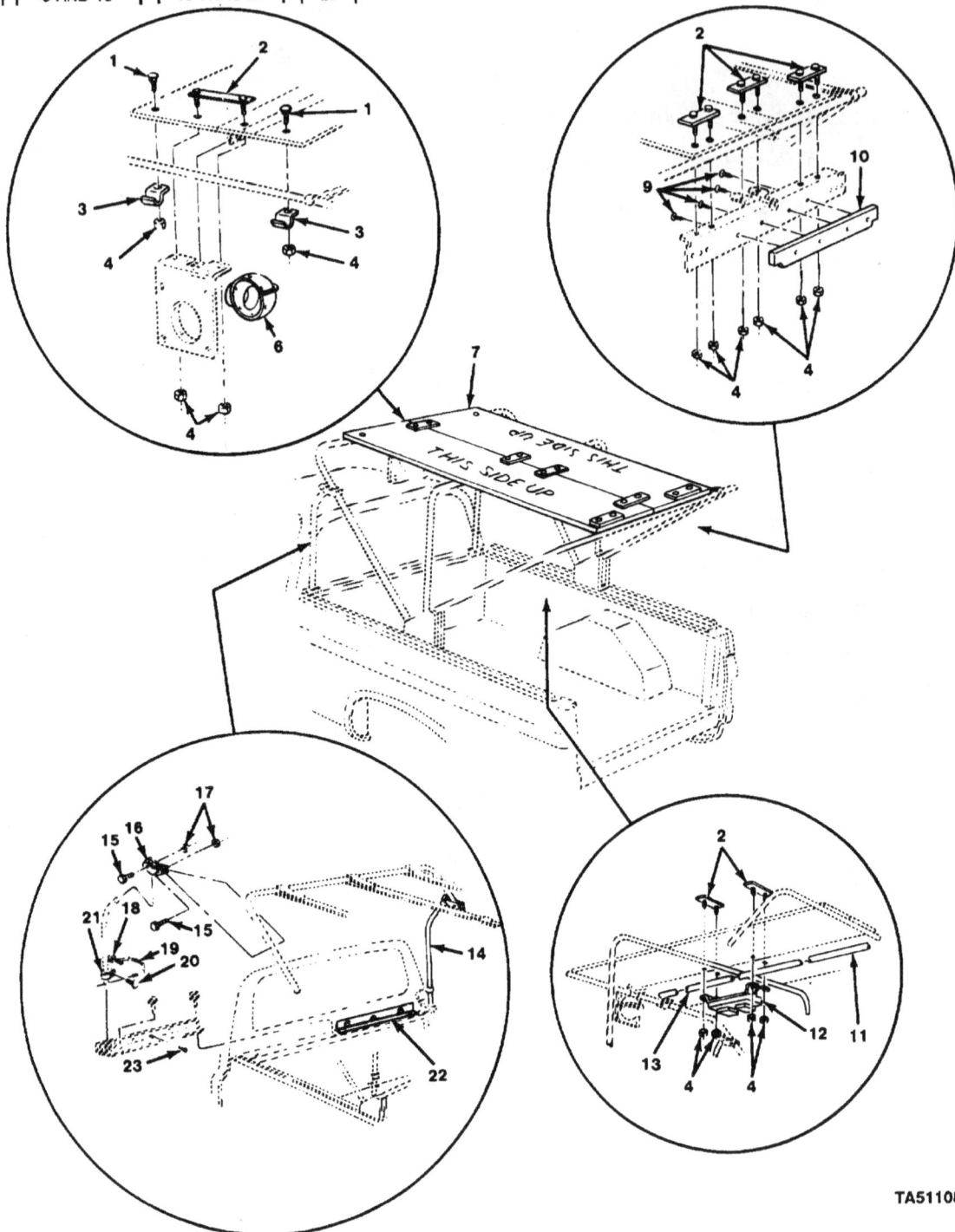

FIGURE 275. ROOF PANEL COMPONENTS (M1008).

TA511082

(1) (2) (3) (4)	(5)	(6)
ITEM SMR PART		
NO CODE CAGEC NUMBER	DESCRIPTION AND USABLE ON CODES (UOC)	QTY

GROUP 3303 WINTERIZATION KITS

FIG. 275 ROOF PANEL COMPONENTS
(M1008)

Item	SMR	CAGEC	Part Number	Description	QTY
1	PAOZZ	96906	MS35751-17	BOLT,SQUARE NECK UOC:194,208	2
2	PAOZZ	11862	14076233	BRACKET,SUPPORT,ROO UOC:194,208	6
3	PAOZZ	11862	14076230	STRAP,RETAINING UOC:194,208	2
4	PAOZZ	24617	9416918	NUT,PLAIN,EXTENDED UOC:194,208	14
5	PAOOO	11862	14075843	ADAPTER ASSEMBLY,AI UOC:194,208	1
6	PAOZZ	19207	7951738	.ADAPTER,AIR INLET UOC:194,208	1
7	MOOZZ	11862	14076222	PANEL ROOF (24.0"X 96.0") MAKE FROM PLYWOOD,P/N NNP530 UOC:194,208	2
8	PFOOO	17769	14076861	PANEL ASSEMBLY,REAR UOC:194,208	1
9	PAOZZ	96906	MS35493-76	.SCREW,WOOD UOC:194,208	2
10	MOOZZ	11862	14076201	.PANEL RAR DOOR (2.28"X 30.88") MAKE FROM PLYWOOD,P/N NNP530 UOC:194,208	1
11	MOOZZ	11862	15599916	SUPPORT (1.00"X 46.50") MAKE FROM TUBE,P/N MILT16343TYPE1 UOC:194,208	1
12	PAOZZ	11862	14076231	BRACKET,SUPPORT,ROO UOC:194,208	1
13	MOOZZ	11862	14076232	SUPPORT (46.50") MAKE FROM TUBE,P/ N MILT16343TYPE1 UOC:194,208	1
14	PAOOO	11862	15599917	SUPPORT,POOF BOW LE UOC:194,208	1
14	PAOOO	11862	15599918	SUPPORT,ROOF BOW RT UOC:194,208	1
15	PAOZZ	96906	MS90728-13	.SCREW,CAP,HEXAGON H UOC:194,208	2
16	PAOZZ	11862	14076226	.STRAP,RETAINING UOC:194,208	1
17	PAOZZ	24617	9416918	.NUT,PLAIN,EXTENDED UOC:194,208	2
18	PAOZZ	24617	9423530	.NUT,PLAIN,ASSEMBLED UOC:194,208	1
19	PAOZZ	19207	11669126-1	.PIN,QUICK RELEASE UOC:194,208	1
20	PAOZZ	96906	MS51957-30	.SCREW,MACHINE UOC:194,208	1
21	PAOZZ	19207	12255561	.CLAMP,LOOP	1

(1) ITEM NO	(2) SMR CODE	(3) CAGEC	(4) PART NUMBER	(5) DESCRIPTION AND USABLE ON CODES (UOC)	(6) QTY
				UOC:194,208	
22	PAOOO	11862	14076299	RAIL,CARGO COVER,VE LEFT	1
				UOC:194,208	
22	PAOOO	11862	14076300	RAIL,CARGO COVER,VE RIGHT	1
				UOC:194,208	
23	PAOZZ	19207	12255564-1	.STUD,TURNBUTTON FAS	4
				UOC:194,208	

END OF FIGURE

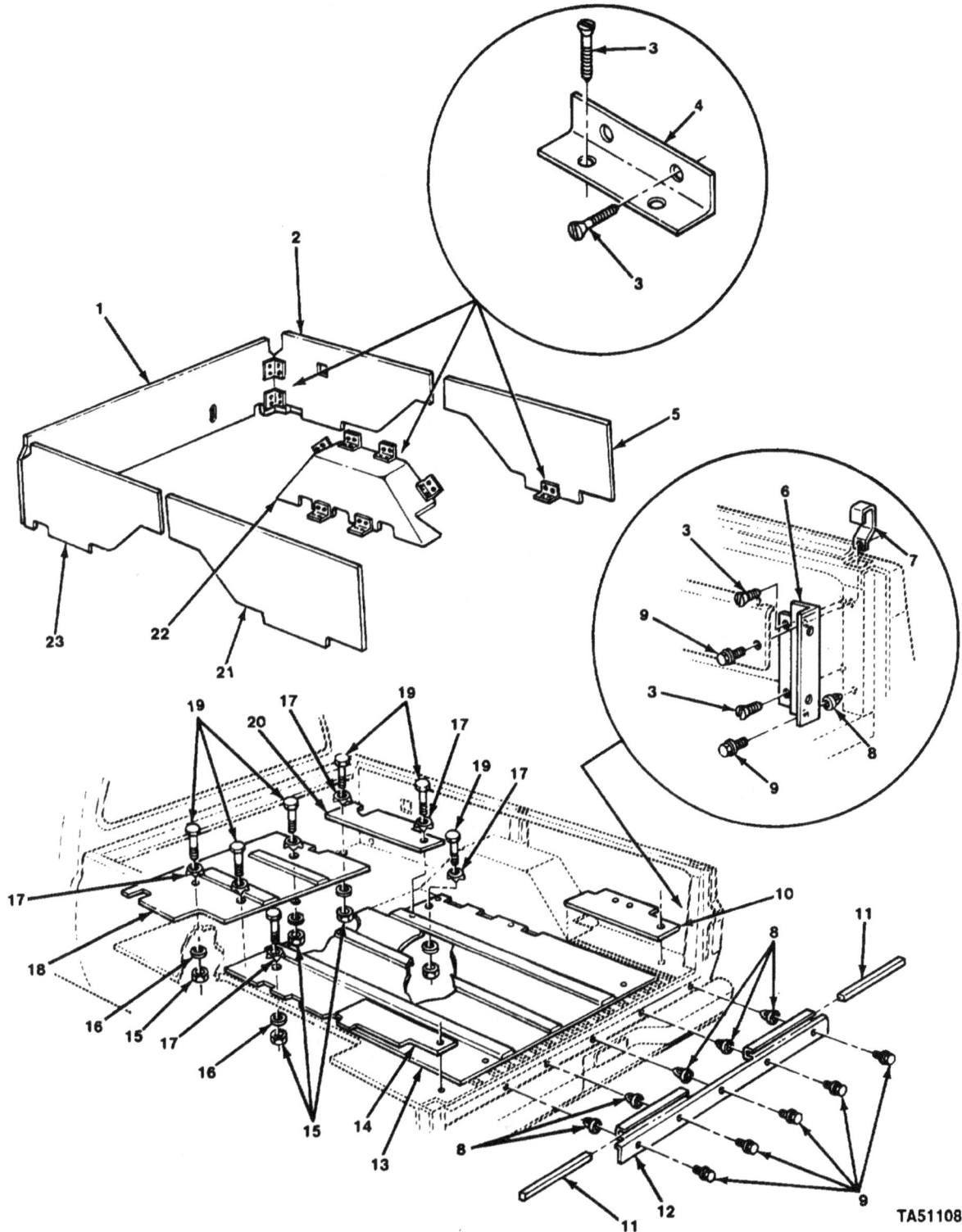

FIGURE 276. FLOOR AND SIDEWALL PANELS (M1008).

TA511083

(1)	(2)	(3)	(4)	(5)	(6)
ITEM NO	SMR CODE	CAGEC	PART NUMBER	DESCRIPTION AND USABLE ON CODES (UOC)	QTY

GROUP 3303 WINTERIZATION KITS

FIG. 276 FLOOR AND SIDEWALL PANELS (M1008)

ITEM NO	SMR CODE	CAGEC	PART NUMBER	DESCRIPTION AND USABLE ON CODES (UOC)	QTY
1	MOOZZ	11862	14076281	INSULATOR (70.00"X 16.74") MAKE FROM PLYWOOD,P/N NNP530 UOC:194,208	1
2	MOOZZ	11862	14076278	INSULATOR (52.26"X 18.28") MAKE FROM PLYWOOD,P/N NNP530 UOC:194,208	1
3	PAOZZ	96906	MS35492-57	SCREW,WOOD UOC:194,208	76
4	PAOZZ	11862	14063349	BRACKET,ANBLE UOC:194,208	18
5	MOOZZ	11862	14076280	INSULATOR (38.12"X 16.80") MAKE FROM PLYWOOD,P/N NNP530 UOC:194,208	1
6	PAOZZ	11862	15599273	BRACKET,MOUNTING LEFT UOC:194,208	1
6	PAOZZ	11862	15599274	BRACKET,MOUNTING RIGHT UOC:194,208	1
7	PAOZZ	11862	14076889	HOOK,SUPPORT UOC:194,208	2
8	PAOZZ	11862	14005953	SHIELD,EXPENSION UOC:194,208	7
9	PAOZZ	11862	9440334	BOLT,ASSEMBLED WASH UOC:194,208	9
10	MOOZZ	11862	14076276	INSULATOR (17.84"X 10.75") MAKE FROM PLYWOOD,P/N NNP530 UOC:194,208	1
11	PAOZZ	11862	14076235	GASKET UOC:194,208	2
12	PFOFF	11862	14076206	MOUNTING,PANEL,INSU UOC:194,208	1
13	MOOZZ	11862	14076272	INUSLATOR (66.62"X 48.00") MAKE FROM PLYWOOD, P/N NNP530 UOC:194,208	1
14	MOOZZ	11862	14076275	INSULATOR (17.84"X 10.75") MAKE FROM PLYWOOD, P/N NNP530 UOC:194,208	1
15	PAOZZ	80205	NAS1408A6	NUT,SELF-LOCKING,HE UOC:194,208	7
16	PAOZZ	96906	MS27183-16	WASHER,FLAT UOC:194,208	7
17	PAOZZ	96906	MS27183-17	WASHER,FLAT UOC:194,208	7
18	MOOZZ	11862	14076271	INSULATOR (46.15"X 28.62") MAKE FROM PLYWOOD, P/N NNP530 UOC:194,208	1
19	PAOZZ	96906	MS35751-46	BOLT,SQUARE NECK UOC:194,208	7

(1) (2) (3) (4)	(5)	(6)
ITEM SMR PART		
NO CODE CAGEC NUMBER	DESCRIPTION AND USABLE ON CODES (UOC)	QTY
20 MOOZZ 11862 14076229	INSULATOR (28.62"X 10.09") MAKE FROM PLYWOOD, P/N NNP530 UOC:194,208	1
21 MOOZZ 11862 14076279	INSULATOR (38.12"X 16.80") MAKE FROM PLYWOOD,P/N NNP530 UOC:194,208	1
22 MOOZZ 11862 14076273	INSULATOR LEFT,MAKE FROM PLYWOOD, P/N NNP530 UOC:194,208	1
22 MOOZZ 11862 14076274	INSULATOR RIGHT, MAKE FROM PLYWOOD, P/N NNP530 UOC:194,208	1
23 MOOZZ 11862 14076277	INSULATOR (52.26"X 18.28") MAKE FROM PLYWOOD,P/N NNP530 UOC:194,208	1

END OF FIGURE

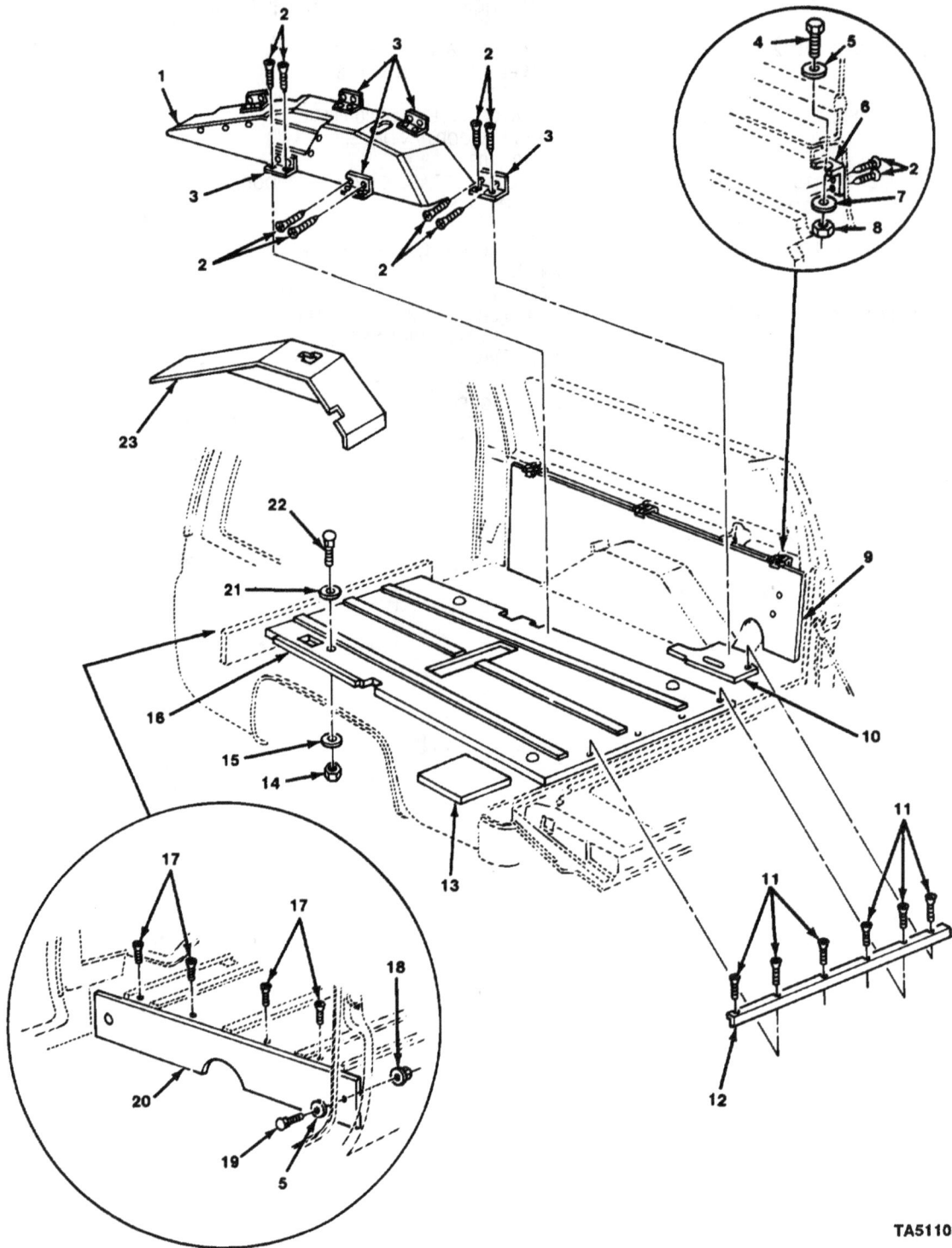

FIGURE 277. FLOOR AND SIDEWALL PANELS (M1009).

TA511084

(1)	(2)	(3)	(4)	(5)	(6)
ITEM NO	SMR CODE	CAGEC	PART NUMBER	DESCRIPTION AND USABLE ON CODES (UOC)	QTY

GROUP 3303 WINTERIZATION KITS

FIG. 277 FLOOR AND SIDEWALL PANELS (M1009)

1	MOOZZ	11862	14063348	INSULATION MAKE FROM PLYWOOD, P/N NNP530 UOC:209	1
2	PAOZZ	96906	MS35492-54	SCREW,WOOD UOC:209	60
3	MOOZZ	11862	14063349	BRACKET,ANGLE MAKE FROM P/N QQS741 UOC:209	12
4	PAOZZ	96906	MS90728-37	BOLT,MACHINE UOC:209	6
5	PAOZZ	96906	MS27183-12	WASHER,FLAT UOC:209	8
6	MOOZZ	11862	14063342	BRACKET MAKE FROM ANGLE, P/N MILS20166 UOC:209	6
7	PAOZZ	11862	9417793	WASHER,FLAT UOC:209	6
8	PAOZZ	24617	9422295	NUT,SELF-LOCKING,CO UOC:209	6
9	MOOZZ	11862	14063343	INSULATION PANEL (LEFT) MAKE FROM PLYWOOD, P/N NNO530 UOC:209	1
9	MOOZZ	11862	14063344	INSULATION PANEL (RIGHT) MAKE FROM PLYWOOD, P/N NNP530 UOC:209	1
10	MOOZZ	11862	14063346	INSULATION FLOOR (RIGHT) MAKE FROM PLYWOOD,P/N NNP530 UOC:209	1
11	PAOZZ	96906	MS35494-83	SCREW,WOOD UOC:209	6
12	MOOZZ	11862	14063350	MOLDING (56.0" LG) MAKE FROM ANGLE, P/N QQS741 UOC:209	1
13	MOOZZ	11862	14063345	INSL-FLR PNL L.R. O (LEFT) MAKE FROM PLYWOOD, P/N NNP530 UOC:209	1
14	PAOZZ	80205	NAS1408A6	NUT,SELF-LOCKING,HE UOC:209	4
15	PAOZZ	96906	MS27183-16	WASHER,FLAT UOC:209	4
16	MOOZZ	11862	14063341	INSULATION ASSEMBLY MAKE FROM PLYWOOD, P/N NNP530 UOC:209	1
17	PAOZZ	96906	MS35492-82	SCREW,WOOD UOC:209	4
18	PAOZZ	24617	11506003	NUT,PLAIN,ASSEMBLED UOC:209	2
19	PAOZZ	96906	MS35751-19	BOLT,SQUARE NECK	2

(1) (2) (3) (4)	(5)	(6)
ITEM SMR PART		
NO CODE CAGEC NUMBER	DESCRIPTION AND USABLE ON CODES (UOC)	QTY

	UOC:209	
20 MOOZZ 11862 14063351	INSULATION FLOOR MAKE FROM PLYWOOD, P/N NNP530	1
	UOC:209	
21 PAOZZ 96906 MS27183-17 WASHER,FLAT		4
	UOC:209	
22 PAOZZ 96906 MS35751-73	BOLT,SQUARE NECK	4
	UOC:209	
23 MOOZZ 11862 14063347	INSULATION (LEFT) MAKE FROM PLYWOOD, P/N NNP530	1
	UOC:209	

END OF FIGURE

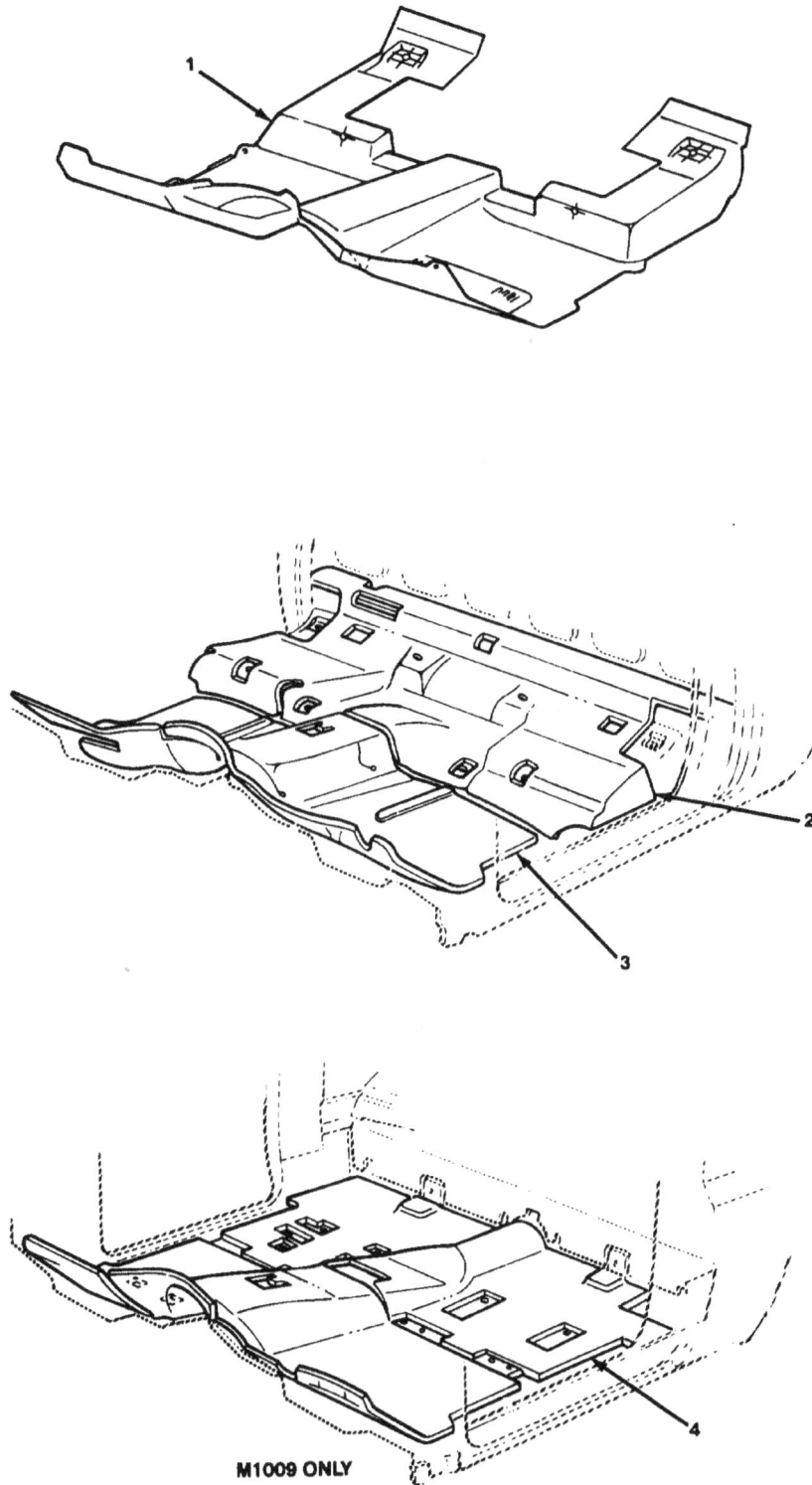

M1009 ONLY

TA511085

FIGURE 278. CAB FLOOR INSULATORS.

(1) (2) (3) (4) ITEM SMR PART NO CODE CAGEC NUMBER	(5) DESCRIPTION AND USABLE ON CODES (UOC)	(6) QTY
	GROUP 3303 WINTERIZATION KITS	
	FIG. 278 CAB FLOOR INSULATORS	
1 PAOZZ 11862 15594983 MAT,FLOOR	UOC:194,208,210,230,231,252,254,256	1
2 MOOZZ 11862 14075862	INSULATOR FL PNL RR (36.0" X 36.0") MAKE FROM INSULATION, P/N 462233 UOC:194,208,230,231,252,254,256	1
2 MOOZZ 11862 14075864	INSULATOR,FLOOR,VEH (36.0" X 36.0") MAKE FROM INSULATION, P/N 462233 UOC:210	1
3 MOOZZ 11862 14076802	INSULATOIR FLR PNL (36.0" X 36.0") MAKE FROM INSULATOR, P/N 462233 UOC:194,208,209,230,231,252,254,256	1
3 MOOZZ 11862 14076803	INSULATION,FLOOR PA (36.0" X 36.0") MAKE FROM INSULATION, P/N 462233 UOC:210	1
4 MOOZZ 11862 14075863	INSUL-FLR PNL RR (36.0" X 36.0") MAKE FROM INSULATOR, P/N 462233 UOC:209	1

END OF FIGURE

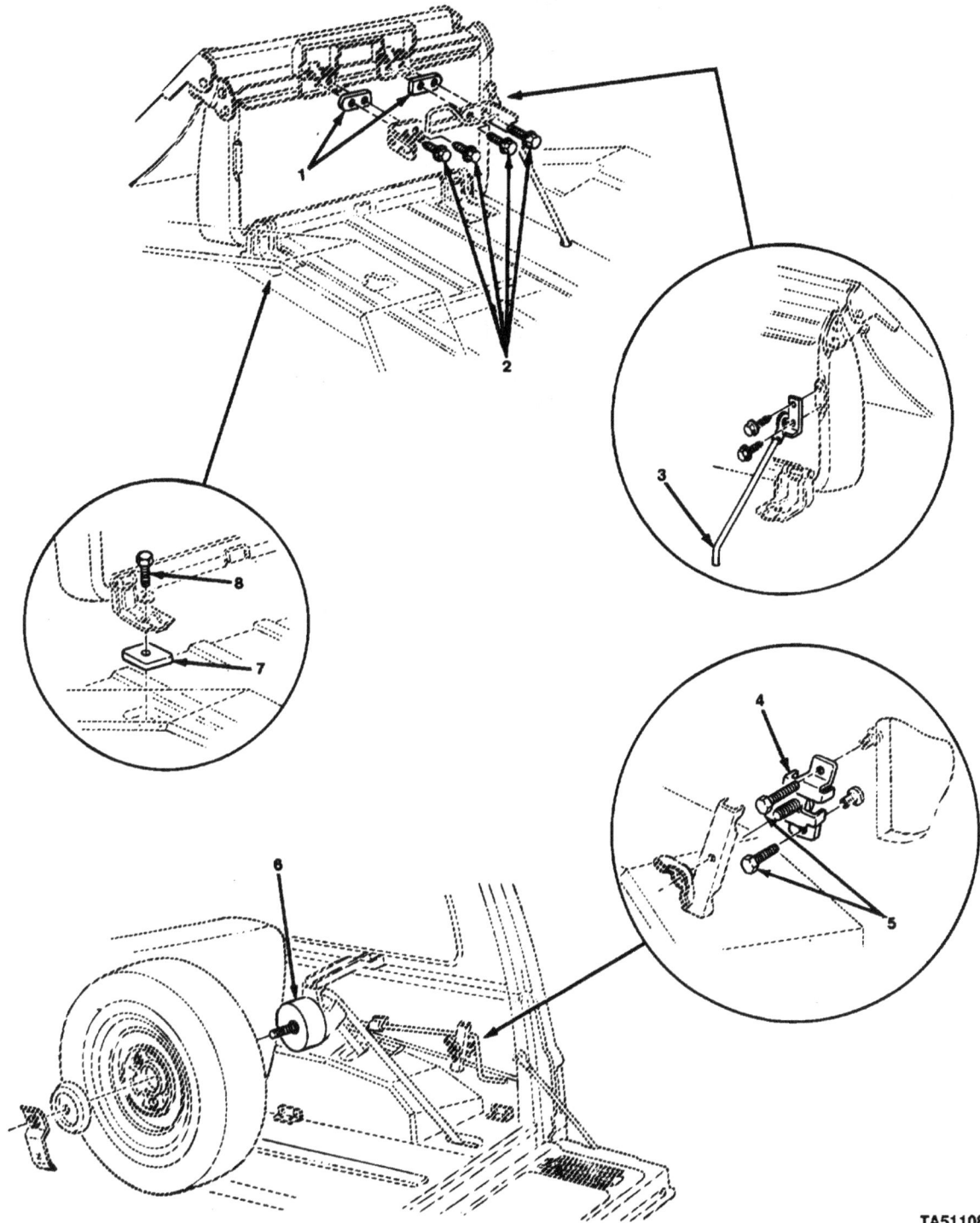

TA511086

FIGURE 279. SEAT AND SPARE TIRE SPACERS (M1009).

(1)	(2)	(3)	(4)	(5)	(6)
ITEM	SMR		PART		
NO	CODE	CAGEC	NUMBER	DESCRIPTION AND USABLE ON CODES (UOC)	QTY

GROUP 3303 WINTERIZATION KITS

FIG. 279 SEAT AND SPARE TIRE SPACERS (M1009)

1	PAOZZ	11862	14063353		PLATE,RESILIENT MOU	2
					UOC:209	
2	XDOZZ	24617	9409103		BOLT,ASSEMBLED WASH	4
					UOC:209	
3	PAOZZ	11862	14063359	LEVER,LOCK-RELEASE		1
					UOC:209	
4	PFOZZ	11862	14074431	BRCKET,MOUNTING		1
					UOC:209	
5	PAOZZ	96906	MS90728-6		SCREW,CAP,HEXAGON H	2
					UOC:209	
6	PAOZZ	11862	14063358		RETAINER,SPARE TIRE	1
					UOC:209	
7	XDOZZ	11862	14063352	SPACER		2
					UOC:209	
8	PAOZZ	96906	MS90728-115		SCREW,CAP,HEXAGON H	2
					UOC:209	

END OF FIGURE

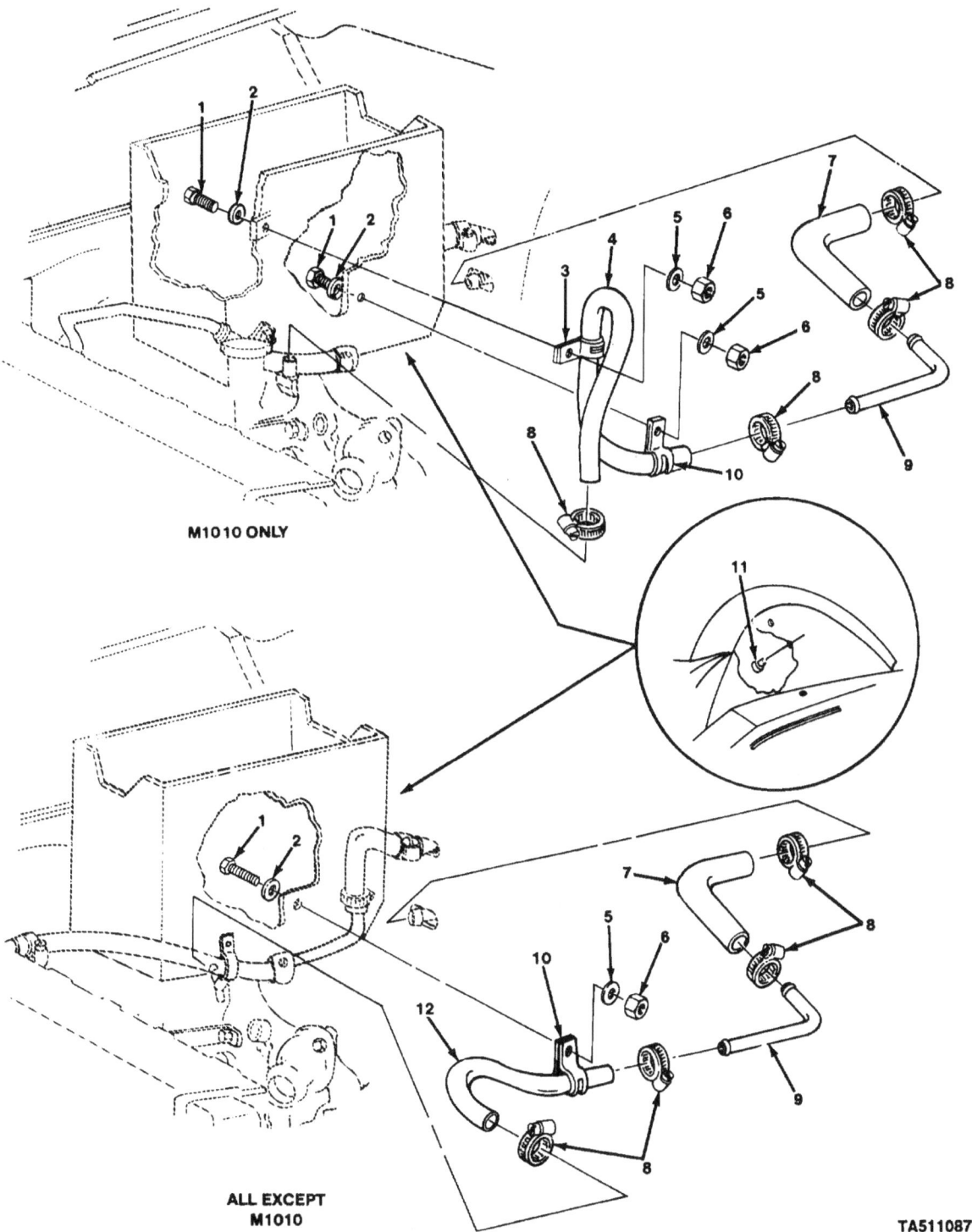

M1010 ONLY

ALL EXCEPT M1010

TA511087

FIGURE 280. BASE HEATER INLET HOSE, PIPE, AND RELATED PARTS.

(1) (2) (3) (4) ITEM SMR PART NO CODE CAGEC NUMBER	(5) DESCRIPTION AND USABLE ON CODES (UOC)	(6) QTY
	GROUP 3303 WINTERIZATION KITS	
	FIG. 280 BASE HEATER INLET HOSE, PIPE, AND RELATED PARTS	
1 PAOZZ 96906 MS90728-34 BOLT,MACHINE	UOC:194,208,209,230,231,252,254,256	1
1 PAOZZ 96906 MS90728-34 BOLT,MACHINE	UOC:210	2
2 PAOZZ 96906 MS35340-45 WASHER,LOCK	UOC:194,208,209,230,231,252,254,256	1
2 PAOZZ 96906 MS35340-45 WASHER, LOCK	UOC:210	2
3 PAOZZ 96906 MS21333-78 CLAMP,LOOP	UOC:210	1
4 MOOZZ 11862 14074444	HOSE,NONMETALLIC (28.75"LG) MAKE FROM HOSE, P/N MS521304B203R UOC:210	1
5 PAOZZ 96906 MS27183-12 WASHER,FLAT	UOC:194,208,209,230,231,252,254,256	1
5 PAOZZ 96906 MS27183-12 WASHER,FLAT	UOC:210	2
6 PAOZZ 96906 MS51943-33 NUT,SELF-LOCKING,HE	UOC:194,208,209,230,231,252,254,256	1
6 PAOZZ 96906 MS51943-33 NUT,SELF-LOCKING,HE	UOC:210	2
7 MOOZZ 11862 14074446	HOSE,PREFORMED (3.20" LG) MAKE FROM HOSE, P/N MS521304B203R	1
8 PAOZZ 96906 MS35842-11 CLAMP,HOSE		4
9 PAOZZ 11862 14074445 TUBE,BENT,METALLIC		1
10 PAOZZ 96906 MS21333-128 CLAMP,LOOP	UOC:194,208,209,230,231,252,254,256	1
11 PAOZZ 11862 3738198	CAP-PLUG,PROTECTIVE	1
12 MOOZZ 11862 14075856	HOSE HEATER INL FRT (24" LG) MAKE FROM HOSE, P/N MS521304B203R UOC:194,208,209,230,231,252,254,256	1

END OF FIGURE

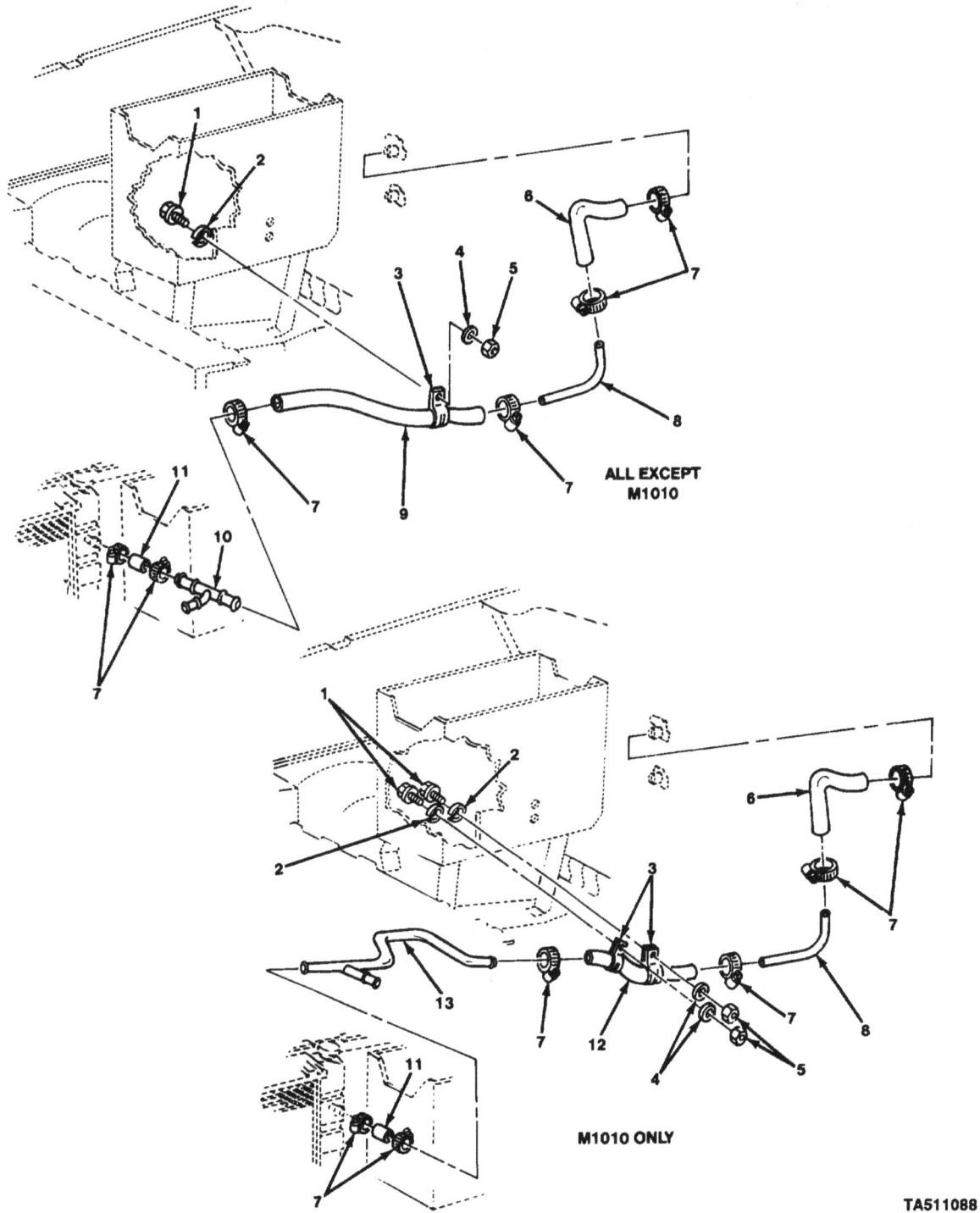

FIGURE 281. BASE HEATER OUTLET HOSE, PIPE, AND RELATED PARTS.

TA511088

(1)	(2)	(3)	(4)	(5)	(6)
ITEM	SMR		PART		
NO	CODE	CAGEC	NUMBER	DESCRIPTION AND USABLE ON CODES (UOC)	QTY

GROUP 3303 WINTERIZATION KITS

FIG. 281 BASE HEATER OUTLET HOSE,
PIPE, AND RELATED PARTS

1	PAOZZ	96906	MS90728-34	BOLT,MACHINE	1
1	PAOZZ	96906	MS90728-34	BOLT,MACHINE	2
				UOC:210	
2	PAOZZ	96906	MS35340-45	WASHER,LOCK	1
2	PAOZZ	96906	MS35340-45	WASHER,LOCK	2
				UOC:210	
3	PAOZZ	96906	MS21333-128	CLAMP,LOOP	2
				UOC:210	
3	PAOZZ	96906	MS21333-128	CLAMP,LOOP	1
4	PAOZZ	96906	MS27183-12	WASHER,FLAT	1
				UOC:194,208,209,210,230,231,252,254	
4	PAOZZ	96906	MS27183-12	WASHER,FLAT	2
				UOC:210	
5	PAOZZ	96906	MS51943-33	NUT,SELF-LOCKING,HE	1
5	PAOZZ	96906	MS51943-33	NUT,SELF-LOCKING,HE	2
				UOC:210	
6	PAOZZ	11862	14074451	HOSE,PREFORMED	1
7	PAOZZ	96906	MS35842-11	CLAMP,HOSE	6
8	PAOZZ	11862	14074450	TUBE,BENT,METALLIC	1
9	MOOZZ	11862	14074449	HOSE HTR OUT INTER (21.5" LG) MAKE FROM HOSE, P/N 48420 UOC:194,208,209,230,231,252,254,256	1
10	PAOZZ	11862	14074448	PIPE ASSEMBLY,HEATE UOC:194,208,209,230,231,252,254,256	1
11	MOOZZ	11862	14074447	HOSE HEATER INLET (2.50" LG) MAKE FROM HOSE, P/N 4842420	1
12	MOOZZ	11862	14074453	HOSE HEATER INLET (11"LG) MAKE FROM HOSE, P/N 482420 UOC:210	1
13	PAOZZ	11862	14074452	TUBE,BENT,METALLIC UOC:210	1

END OF FIGURE

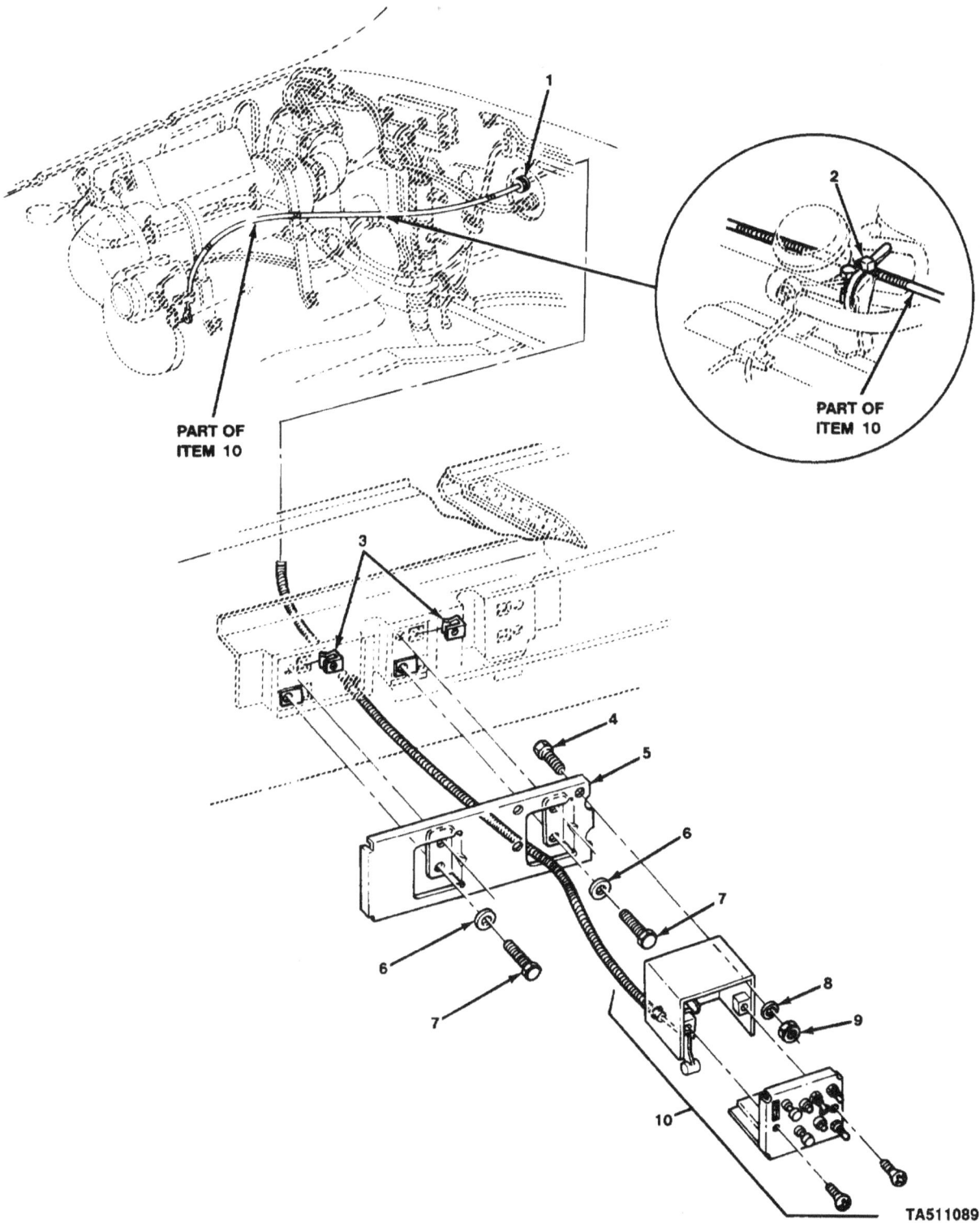

PART OF ITEM 10

PART OF ITEM 10

TA511089

FIGURE 282. HEATER CONTROL MOUNTING AND CABLE.

TM9-2320-289-34P

(1) ITEM NO	(2) SMR CODE	(3) CAGEC	(4) PART NUMBER	(5) DESCRIPTION AND USABLE ON CODES (UOC)	(6) QTY
				GROUP 3303 WINTERIZATION KITS	
				FIG. 282 HEATER CONTROL MOUNTING AND CABLE	
1	PAOZZ	96906	MS35489-5	GROMMET,NONMETALLIC	1
2	PAOZZ	11862	3816659	STRAP,LINE SUPPORT	1
3	PAOZZ	11862	3982098	NUT,PLAIN,HEXAGON	4
4	PAOZZ	96906	MS90728-3	SCREW,CAP,HEXAGON H	3
5	PAOZZ	11862	14072415	BRACKET,MOUNTING	1
6	PAOZZ	96906	MS9549-10	WASHER,FLAT	2
7	PAOZZ	96906	MS90728-6	SCREW,CAP,HEXAGON H	4
8	PAOZZ	96906	MS35338-44	WASHER,LOCK	3
9	PAOZZ	96906	MS35691-406	NUT,PLAIN,HEXAGON	3
10	PAOOO	11862	14076284	CONTROL ASSEMBLY,HE	1

END OF FIGURE

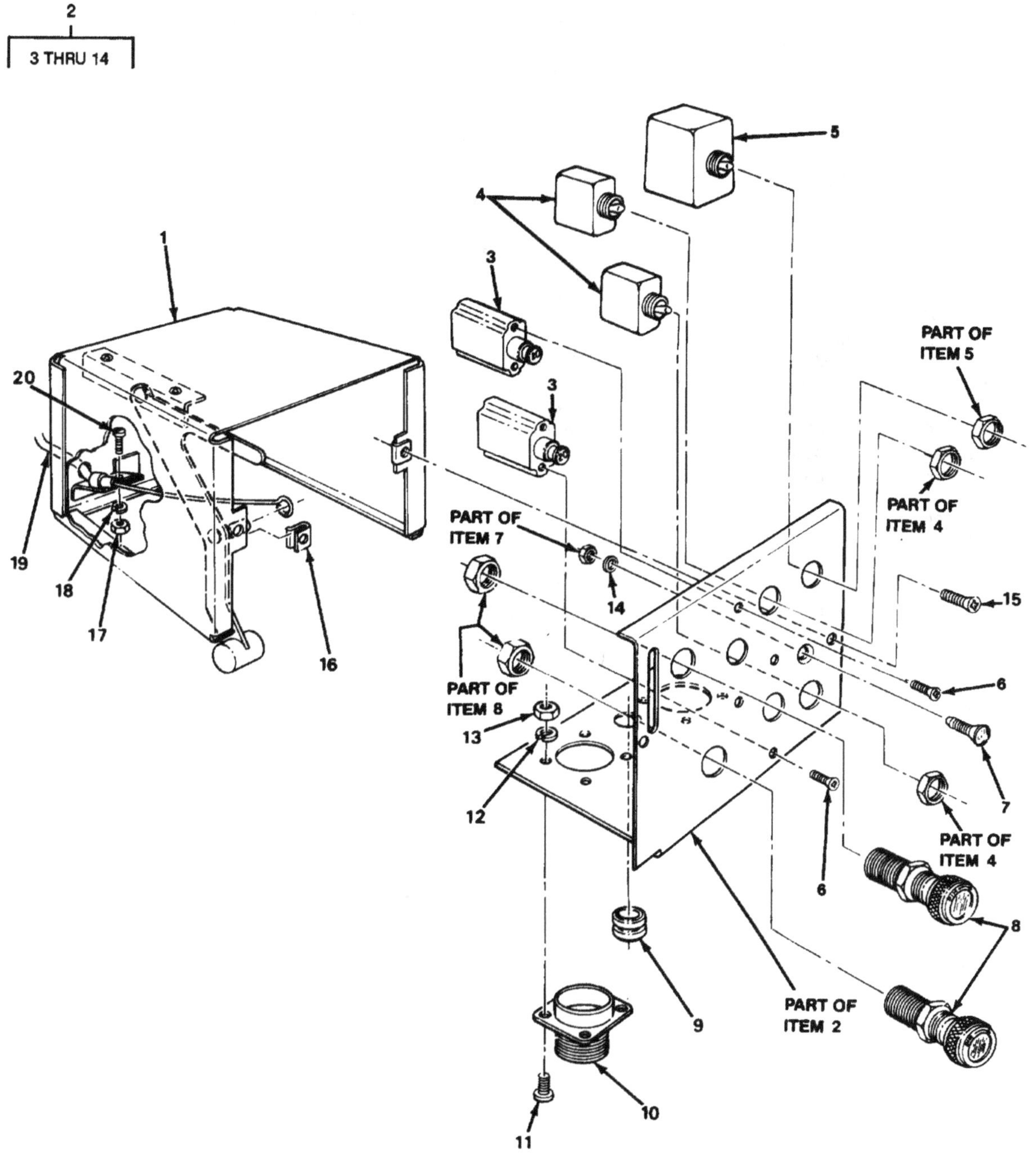

TA511090

FIGURE 283. HEATER CONTROL ASSEMBLY COMPONENT PARTS.

(1) ITEM NO	(2) SMR CODE	(3) CAGEC	(4) PART NUMBER	(5) DESCRIPTION AND USABLE ON CODES (UOC)	(6) QTY
				GROUP 3303 WINTERIZATION KITS	
				FIG. 283 HEATER CONTROL ASSEMBLY COMPONENT PARTS	
1	PAOZZ	11862	14076285	HOUSING,HEATER CONT	1
2	PAOOO	11862	14076287	PANEL ASSEMBLY,HEAT	1
3	PAOZZ	96906	MS25244-P-20	.CIRCUIT BREAKER	2
4	PAOZZ	96906	MS35058-22	.SWITCH,TOGGLE	2
5	PAOZZ	96906	MS25307-312	.SWITCH,TOGGLE	1
6	PAOZZ	96906	MS51957-26	.SCREW,MACHINE	4
7	PAOZZ	81640	L30200R	.LIGHT,INDICATOR	1
8	PAOZZ	96906	MS25331-4-313S	.LIGHT,INDICATOR	2
9	PAOZZ	96906	MS35489-6	.GROMMET,NONMETALLIC	1
10	PAOZZ	96906	MS3452W18-11P	.CONNECTOR,RECEPTACL	2
11	PAOZZ	96906	MS51957-17	.SCREW,MACHINE	8
12	PAOZZ	96906	MS35333-70	.WASHER,LOCK	8
13	PAOZZ	96906	MS35649-244	.NUT,PLAIN,HEXAGON	8
14	PAOZZ	96906	MS27183-42	.WASHER,FLAT	1
15	PAOZZ	96906	MS51957-45	SCREW,MACHINE	2
16	PAOZZ	96906	MS90724-7	NUT,SHEET SPRING	2
17	PAOZZ	96906	MS35649-202	NUT,PLAIN,HEXAGON	1
18	PAOZZ	96906	MS35338-43	WASHER,LOCK	1
19	PAOZZ	96906	14076295	SCREW,MACHINE	1
20	PAOZZ	96906	MS35206-263	SCREW,MACHINE	1

END OF FIGURE

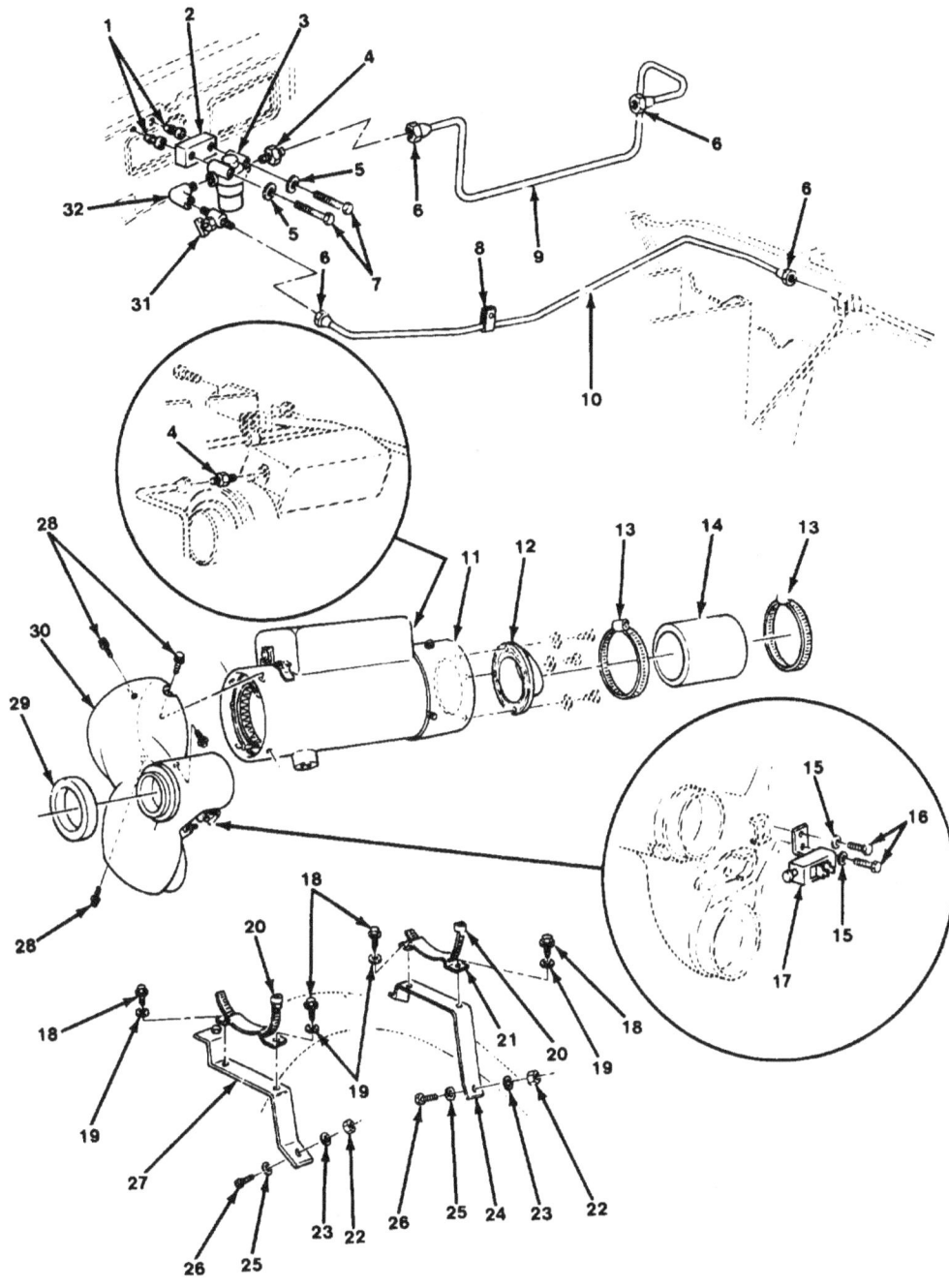

FIGURE 284. HEATER AND MOUNTING PARTS.

TA511091

(1) (2) (3) (4) ITEM SMR PART NO CODE CAGEC NUMBER	(5) DESCRIPTION AND USABLE ON CODES (UOC)	(6) QTY
	GROUP 3303 WINTERIZATION KITS	
	FIG. 284 HEATER AND MOUNTING PARTS	
1 PAOZZ 00613 C-5139-2	RECEPTACLE,TURNLOCK	2
2 PAOZZ 11862 14072374	SPACER,HEATER FUEL	1
3 PAOZZ 96906 MS51085-1 FILTER,FLUID		1
4 PAOZZ 81240 GM118749	ADAPTER,STRAIGHT,PI	2
5 PAOZZ 96906 MS27183-8 WASHER,FLAT		2
6 PAOZZ 81343 5 010111B NUT,TUBE COUPLING		4
7 PAOZZ 24617 456748	SCREW,CAP,HEXAGON H	2
8 PAOZZ 96906 MS21333-110 CLAMP,LOOP		1
9 MFOZZ 11862 14072378	PIPE F/F;TR OULLT (29.99" LG) MAKE FROM TUBE, P/N 1324714	1
10 MFOZZ 11862 15599203	PIPE (43.79" LG) MAKE FROM TUBE, P/N 1324714	1
11 PAOFF 78385 10530B	HEATER,VEHICULAR,CO FOR COMPONENT PARTS SEE FIG'S 246, 248, 249, 250	1
12 PAOZZ 11862 14074466	ADAPTER,HEATER,VEHI	1
13 PAOZZ 96906 MS35842-14 CLAMP,HOSE		2
14 PAOZZ 16632 CHE2015-0001	HOSE,AIR DUCT	1
15 PAOZZ 96906 MS27183-10 WASHER,FLAT		2
16 PAOZZ 24617 9419663	SCREW,TAPPING,THREA	2
17 PAOZZ 11862 15599972	SWITCH,PUSH	1
18 PAOZZ 11862 9439771	BOLT,ASSEMBLED WASH	4
19 PAOZZ 19207 5287638	WASHER,FLAT	4
20 PAOZZ 96906 MS35842-16 CLAMP,HOSE		2
21 PAOZZ 11862 14074469	BRACKET,HEATER,VEHI	2
22 PAOZZ 96906 MS51967-6 NUT,PLAIN,HEXAGON		4
23 PAOZZ 96906 MS35338-45 WASHER,LOCK		4
24 PAOZZ 11862 14074471	BRACKET,MOUNTING	1
25 PAOZZ 24617 9418924	WASHER,FLAT	4
26 PAOZZ 96906 MS90728-32 BOLT,MACHINE		4
27 PAOZZ 11862 14074470	BRACKET,HEATER FRON	1
28 PAOZZ 24617 423532	SCREW,ASSEMBLED WAS	4
29 PAOZZ 16632 CHE2017	GASKET	1
30 PAOZZ 11862 14074473	VALVE ASSEMBLY,HEAT	1
31 XDOZZ 11862 9440173	VALVE FUEL SHUTOFF	1
32 PAOZZ 96906 MS51952-1 ELBOW,PIPE		1

END OF FIGURE

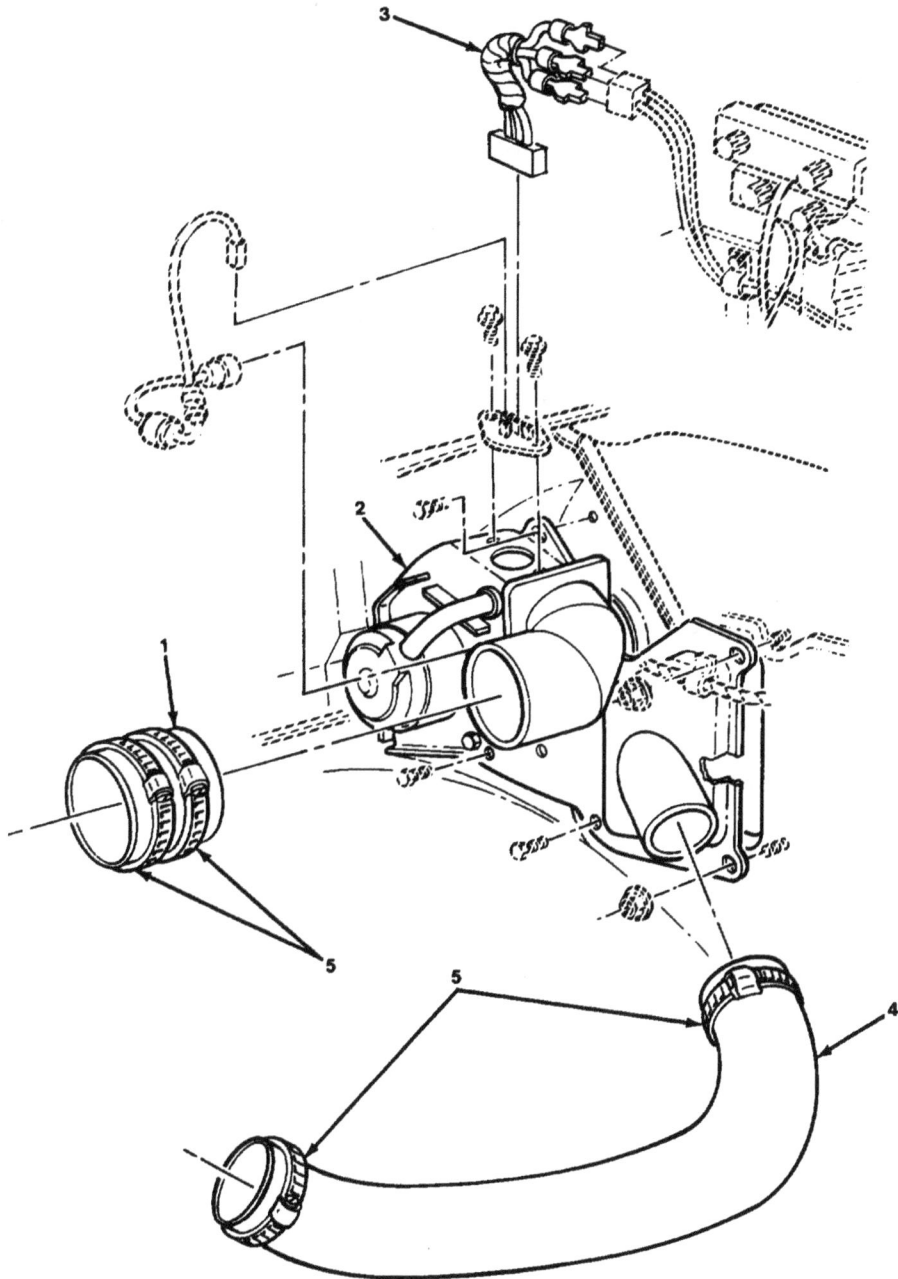

TA511092

FIGURE 285. HEATER BLOWER AND HOSES.

TM9-2320-289-34P

(1) ITEM NO	(2) SMR CODE	(3) CAGEC	(4) PART NUMBER	(5) DESCRIPTION AND USABLE ON CODES (UOC)	(6) QTY
				GROUP 3303 WINTERIZATION KITS	
				FIG. 285 HEATER BLOWER AND HOSES	
1	PAOZZ	16632	CHE2015-0001	HOSE,AIR DUCT	1
2	PAOOO	11862	14074465	BLOWER ASSEMBLY,HEA FOR COMPONENT PARTS SEE FIG 286	1
3	PAOZZ	11862	14076237	LEAD ASSEMBLY,ELECT UOC:194,208,209,210,230,231,252,254	1
4	PAOZZ	16632	CHE2015-0002	HOSE,AIR DUCT	1
5	PAOZZ	96906	MS35842-15	CLAMP,HOSE	4

END OF FIGURE

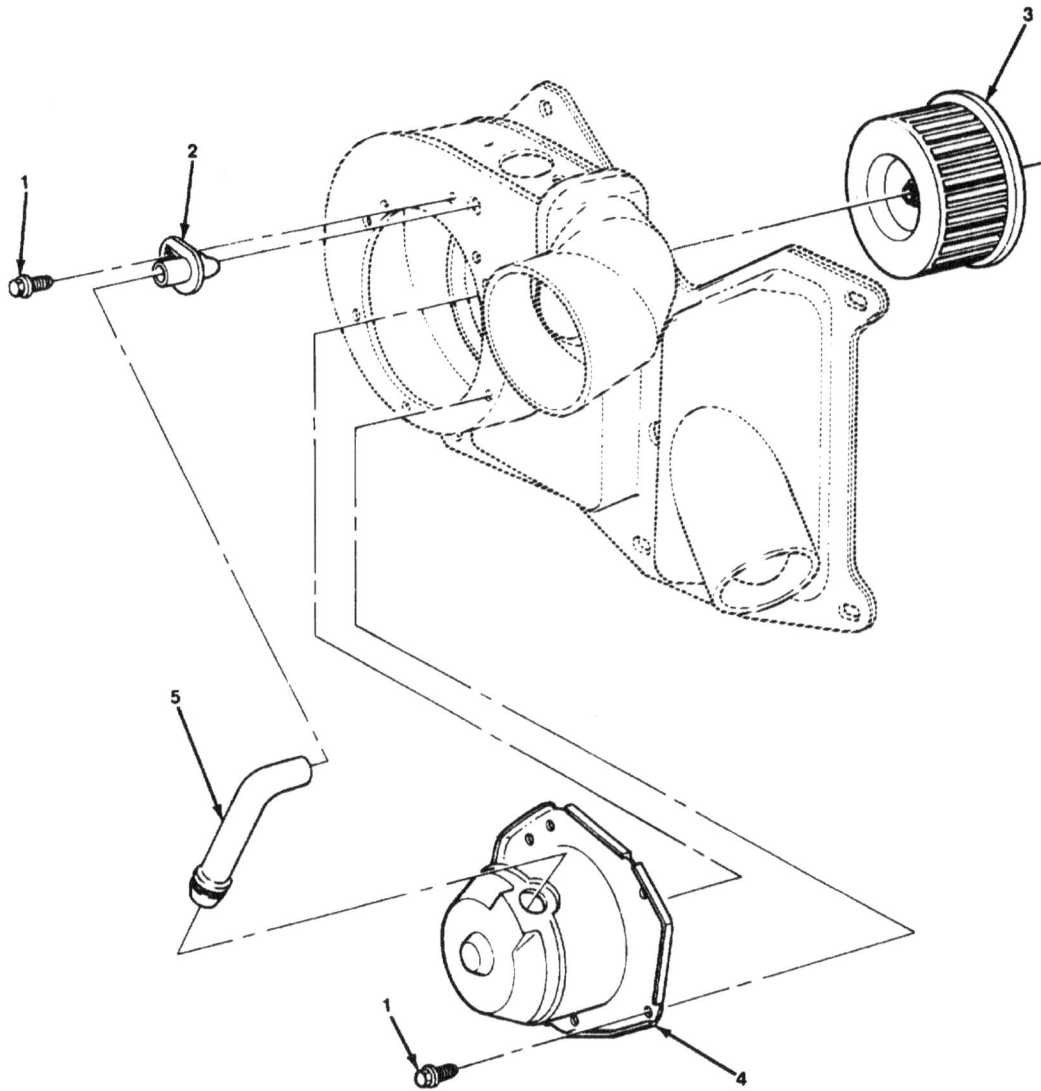

TA511093

FIGURE 286. BLOWER COMPONENT PARTS.

(1) ITEM NO	(2) SMR CODE	(3) CAGEC	(4) PART NUMBER	(5) DESCRIPTION AND USABLE ON CODES (UOC)	(6) QTY
				GROUP 3303 WINTERIZATION KITS	
				FIG. 286 BLOWER COMPONENT PARTS	
1	PAOZZ	24617	9419663	SCREW,TAPPING,THREA	6
2	PAOZZ	11862	3037476	PIPE,AIR CONDITIONI	1
3	PAOZZ	11862	3055734	MOTOR,HEATER FAN	1
4	PAOZZ	11862	22098841	MOTOR,DIRECT CURREN	1
5	PAOZZ	11862	3035192	PIPE,AIR CONDITIONI	1

END OF FIGURE

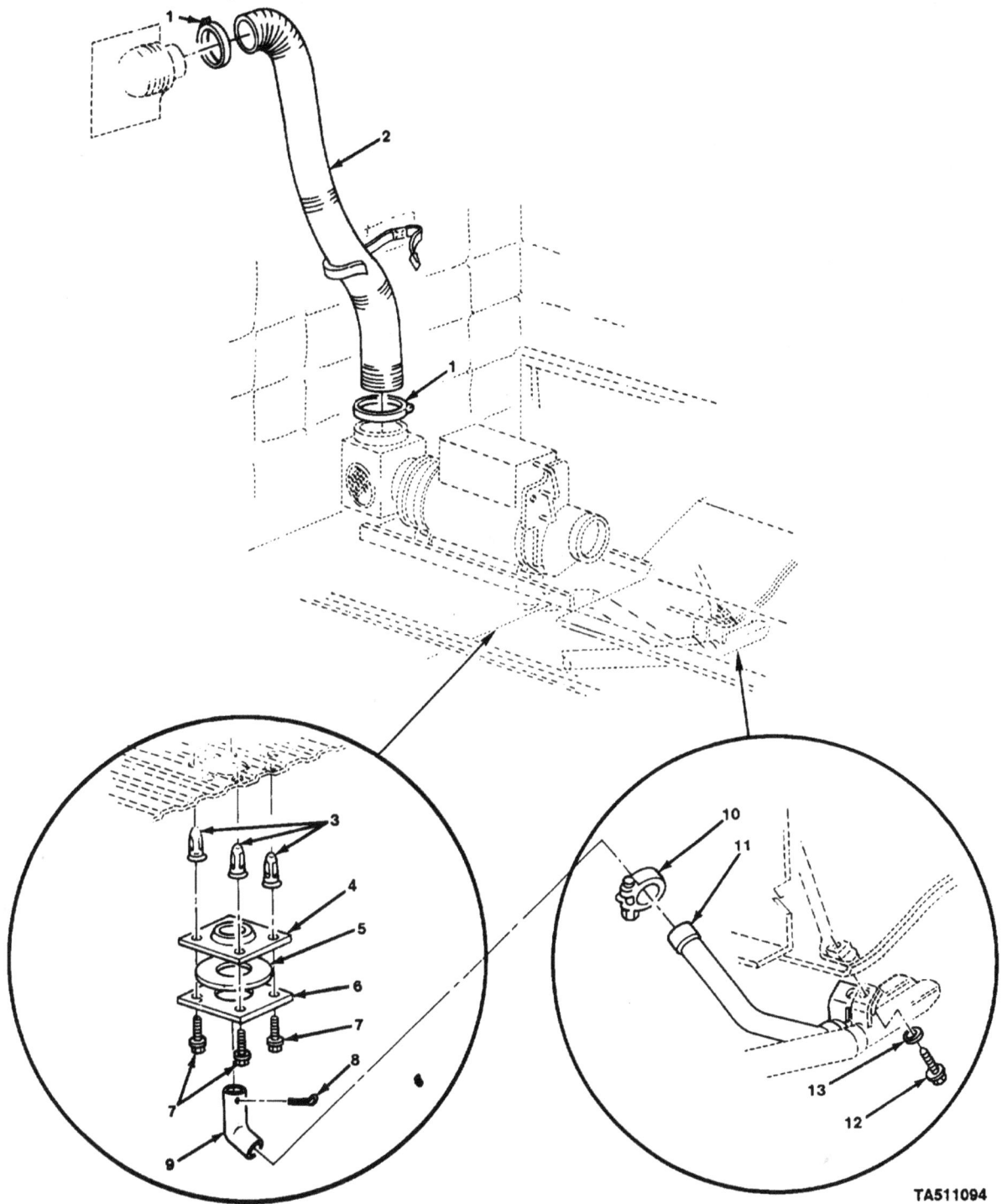

FIGURE 287. INLET HOSE AND EXHAUST PIPE.

TA511094

(1) ITEM NO	(2) SMR CODE	(3) CAGEC	(4) PART NUMBER	(5) DESCRIPTION AND USABLE ON CODES (UOC)	(6) QTY
				GROUP 3303 WINTERIZATION KITS	
				FIG. 287 INLET HOSE AND EXHAUST PIPE	
1	PAOZZ	9696	MS35842-15	CLAMP,HOSE UOC:194,208	2
2	PAOZZ	16632	CHE2013-0001	HOSE,AIR DUCT UOC:194,208	1
3	PAOZZ	11862	14005953	SHIELD,EXPANSION UOC:194,208	4
4	PAOZZ	16632	CHE2016-0002	GASKET UOC:194,208	1
5	PAOZZ	16632	CHE2023	GASKET UOC:194,208	1
6	PAOZZ	11862	15599942	CONNECTOR,EXHAUST P UOC:194,208	1
7	PAOZZ	11862	9440334	BOLT,ASSEMBLED WASH UOC:194,208	4
8	PAOZZ	96906	MS24665-516	PIN,COTTER UOC:194,208	1
9	PAOZZ	11862	15599940	ADAPTER,EXHAUST,HEA UOC:194,208	1
10	PAOZZ	63208	650-24	CLAMP,HOSE UOC:194,208	1
11	PAOZZ	11862	14076805	PIPE,EXHAUST UOC:194,208	1
12	PAOZZ	96906	MS90728-35	BOLT,MACHINE UOC:194,208	1
13	PAOZZ	96906	MS35340-45	WASHER,LOCK UOC:194,208	1

END OF FIGURE

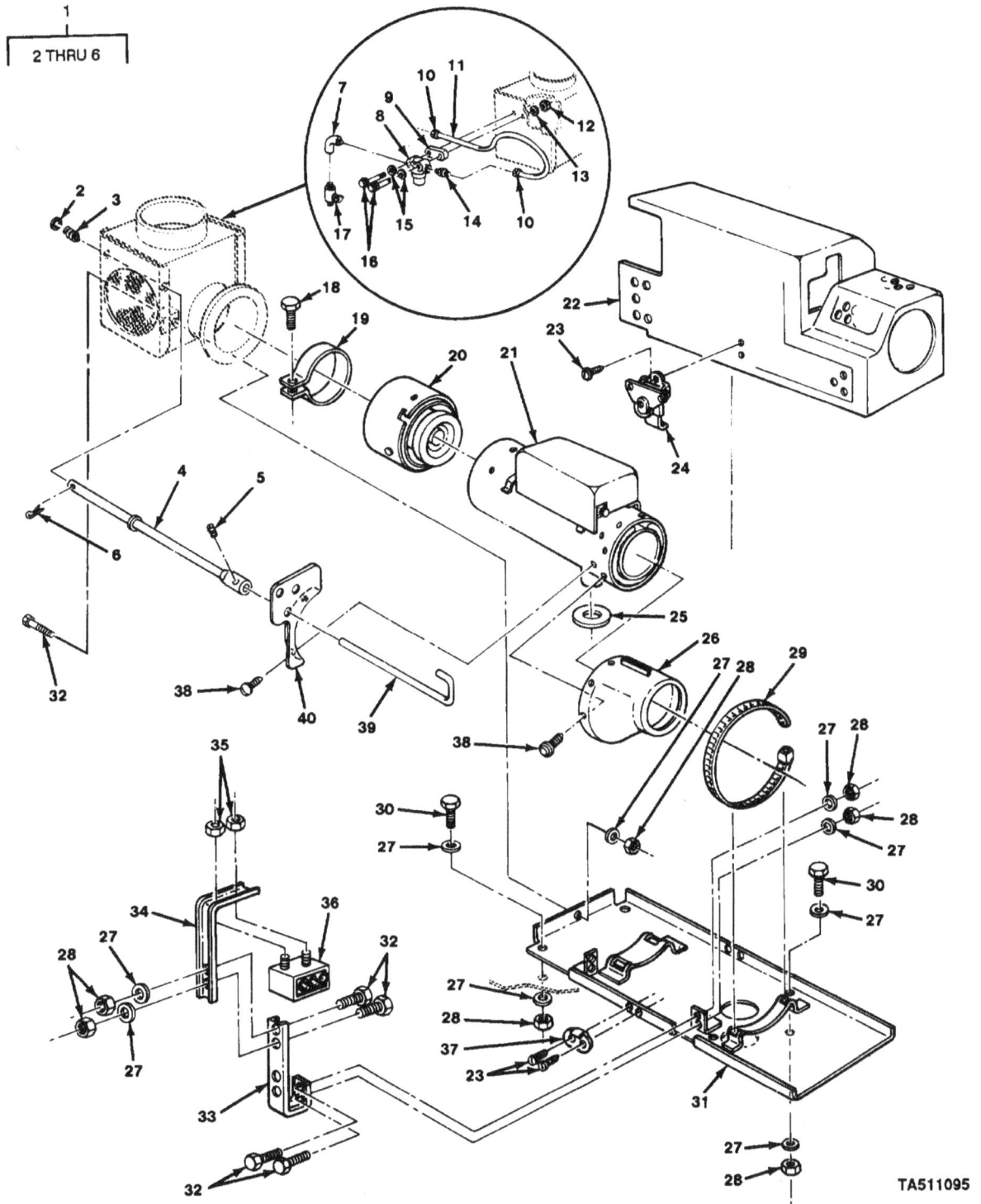

FIGURE 288. PERSONNEL/CARGO HEATER (M1008).

TA511095

(1) (2) (3) (4) ITEM SMR NO CODE CAGEC	PART NUMBER	(5) DESCRIPTION AND USABLE ON CODES (UOC)	(6) QTY
		GROUP 3303 WINTERIZATION KITS	
		FIG. 288 PERSONNEL/CARGO HEATER (M1008)	
1 PAOZZ 11862	14076252	VENTILATOR,AIR CIRC UOC:194,208	1
2 PAOZZ 96906	MS27183-10	.WASHER,FLAT UOC:194,208	1
3 PAOZZ 96906	MS24585-1276	.SPRING,HELICAL,COMP UOC:194,208	1
4 PAOZZ 11862	14076253	.ROD ASSEMBLY,AIR IN UOC:194,208	1
5 PAOZZ 96906	MS51023-49	.SETSCREW UOC:194,208	1
6 PAOZZ 96152	A82-1	.PIN,COTTER UOC:194,208	1
7 PAOZZ 96906	MS51952-1	ELBOW,PIPE UOC:194,208	1
8 PAOZZ 96906	MS51085-1	FILTER,FLUID FOR COMPONENT PARTS SEE FIG 245 UOC:194,208	1
9 PAOZZ 11862	14072374	SPACER,HEATER FUEL UOC:194,208	1
10 PAOZZ 81343	5 010111B	NUT,TUBE COUPLING UOC:194,208	2
11 MFOZZ 11862	14076255	PIPE HTR FUEL (18.44" LG) MAKE FROM TUBE, P/N 1324714 UOC:194,208	1
12 PAOZZ 88044	AN365-1024A	NUT,SELF-LOCKING,HE UOC:194,208	2
13 PAOZZ 24617	9417714	WASHER,FLAT UOC:194,208	2
14 PAOZZ 81240	GM118749	ADAPTER,STRAIGHT,PI UOC:194,208	1
15 PAOZZ 96906	MS27183-8	WASHER,FLAT UOC:194,208	2
16 PAOZZ 24617	456748	SCREW,CAP,HEXAGON H UOC:194,208	2
17 XDOZZ 11862	9440173	VALVE FUEL SHUTOFF UOC:194,208	1
18 PAOZZ 11862	9440334	BOLT,ASSEMBLED WASH UOC:194,208	1
19 PAOZZ 96906	MS35842-16	CLAMP,HOSE UOC:194,208	1
20 PAOZZ 11862	15599952	ADAPTER,AIR DUCT HO UOC:194,208	1
21 PAOFF 78385	10530B	HEATER,VEHICULAR,CO FOR COMPONENT PARTS SEE FIG'S 246, 248, 249, 250 UOC:194,208	1
22 PAOZZ 11862	14076265	SHIELD,HEATER,AUXIL UOC:194,208	1

(1) (2) (3) (4)	(5)	(6)
ITEM SMR PART		
NO CODE CAGEC NUMBER	DESCRIPTION AND USABLE ON CODES (UOC)	QTY

23 PAOZZ 9696 MS20600-B4W4 RIVET,BLIND		8
	UOC:194,208	
24 PAOZZ 82240 SPEC-3-10-L-L CATCH,CLAMPING		2
	UOC:194,208	
25 PAOZZ 16632 CHE2018	GASKET	1
	UOC:194,208	
26 PAOZZ 19207 7524078	ADAPTER ASSEMBLY,CO	1
	UOC:194,208	
27 PAOZZ 96906 MS27183-10	WASHER,FLAT	13
	UOC:194,208	
28 PAOZZ 21450 131245	NUT,SELF-LOCKING,HE	9
	UOC:194,208	
29 PAOZZ 96906 MS35842-16	CLAMP,HOSE	2
	UOC:194,208	
30 PAOZZ 96906 MS90728-8	SCREW,CAP,HEXAGON H	4
	UOC:194,208	
31 PAOZZ 11862 14076259	PLATE,MOUNTING,HEAT	1
	UOC:194,208	
32 PAOZZ 96906 MS90728-5	SCREW,CAP,HEXAGON H	4
	UOC:194,208	
33 PFOZZ 11862 14076261	BRACKET,DOUBLE ANGL	1
	UOC:194,208	
34 PAOZZ 11862 14076262	BRACKET,ANGLE	1
	UOC:194,208	
35 PAOZZ 96906 MS51967-2	NUT,PLAIN,HEXAGON	2
	UOC:194,208	
36 PAOZZ 11862 14076801	CONTROL BOX,ELECTRI	1
	UOC:194,208	
37 PAOZZ 82240 B-1900-334	STRIKE,CATCH	2
	UOC:194,208	
38 PAOZZ 24617 423532	SCREW,ASSEMBLED WAS	6
	UOC:194,208	
39 PAOZZ 11862 14076258	LEVER,MANUAL CONTRO	1
	UOC:194,208	
40 PAOZZ 11862 14076257	BRACKET,HEATER AIR	1
	UOC:194,208	

END OF FIGURE

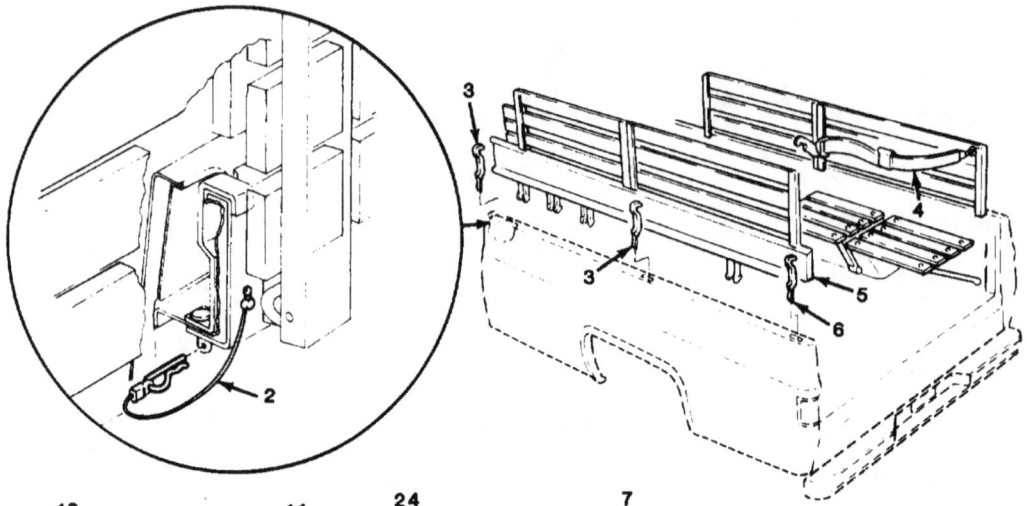

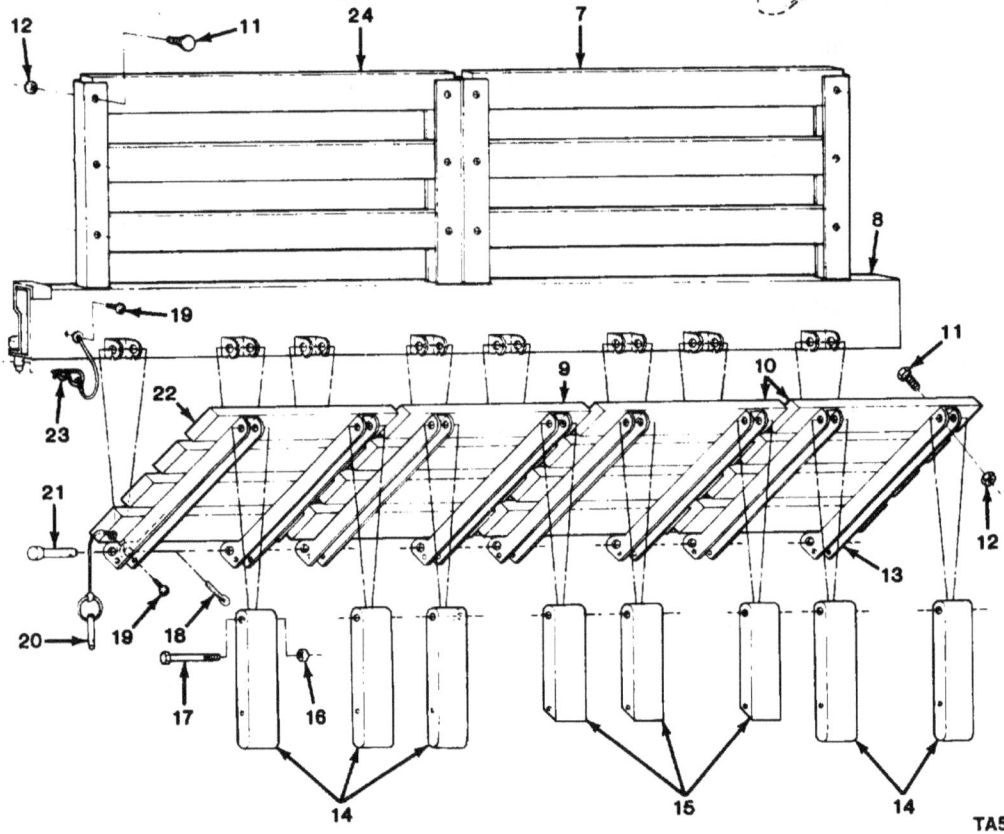

FIGURE 289. TROOP SEAT ASSEMBLY AND COMPONENT PARTS.

TA511096

(1) (2) (3) (4) ITEM SMR PART NO CODE CAGEC NUMBER	(5) DESCRIPTION AND USABLE ON CODES (UOC)	(6) QTY
	GROUP 3307 SPECIAL PURPOSE KITS FIG. 289 TROOP SEAT ASSEMBLY AND COMPONENT PARTS	
1 AOOOO 11862 14072308	SEAT ASM TROOP UOC:194,208	1
2 PAOZZ 11862 14072314	.LANYARD,SEAT UOC:194,208	6
3 PAOZZ 11862 14072325	.STRAP,RETAINING UOC:194,208	4
4 PAOZZ 19207 11682088-1	.STRAP,WEBBING UOC:194,208	1
5 PAOZZ 11862 14075811	.FRAME,SEAT,VEHICULA LEFT UOC:194,208	1
6 PAOZZ 11862 14072324	.CLAMP UOC:194,208	2
7 PAOZZ 19207 12343359-3	.SLAT,TROOP SEAT UOC:194,208	6
8 PAOZZ 11862 14075812	.SUPPORT ASSY,SEAT,R RIGHT UOC:194,208	1
9 PAOZZ 19207 12343359-5	.SLAT,TROOP SEAT FRA UOC:194,208	8
10 PAOZZ 19207 12343359-1	.BOARD,TROOP SEAT UOC:194,208	16
11 PAOZZ 21450 126373	.BOLT,SQUARE NECK UOC:194,208	88
12 PAOZZ 24617 271184	.NUT,PLAIN,ASSEMBLED UOC:194,208	88
13 PAOZZ 11862 14072319	.FRAME,SEAT,VEHICULA UOC:194,208	16
14 PAOZZ 11862 14072323	.LEG,SEAT UOC:194,208	10
15 PAOZZ 11862 14072322	.SEAT,TROOP,LEG UOC:194,208	6
16 PAOZZ 80205 NAS1408A6	.NUT,SELF-LOCKING,HE UOC:194,208	16
17 PAOZZ 96906 MS90728-68	.SCREW,CAP,HEXAGON H UOC:194,208	16
18 PAOZZ 96906 MS24665-283	.PIN,COTTER UOC:194,208	16
19 PAOZZ 96906 MS51850-44	.SCREW,TAPPING,THREA UOC:194,208	14
20 PAOZZ 19207 12255608	.PIN,QUICK RELEASE UOC:194,208	8
21 PAOZZ 19207 7370134	.PIN,STRAIGHT,HEADED UOC:194,208	16
22 PAOZZ 19207 12343359-4	.SLAT,TROOP SEAT FRA UOC:194,208	8
23 PAOZZ 11862 14072380	.CLIP UOC:194,208	6
24 PAOZZ 19207 12343359-2	.SLAT,TROOP SEAT UOC:194,208	6

END OF FIGURE

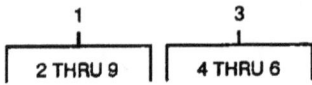

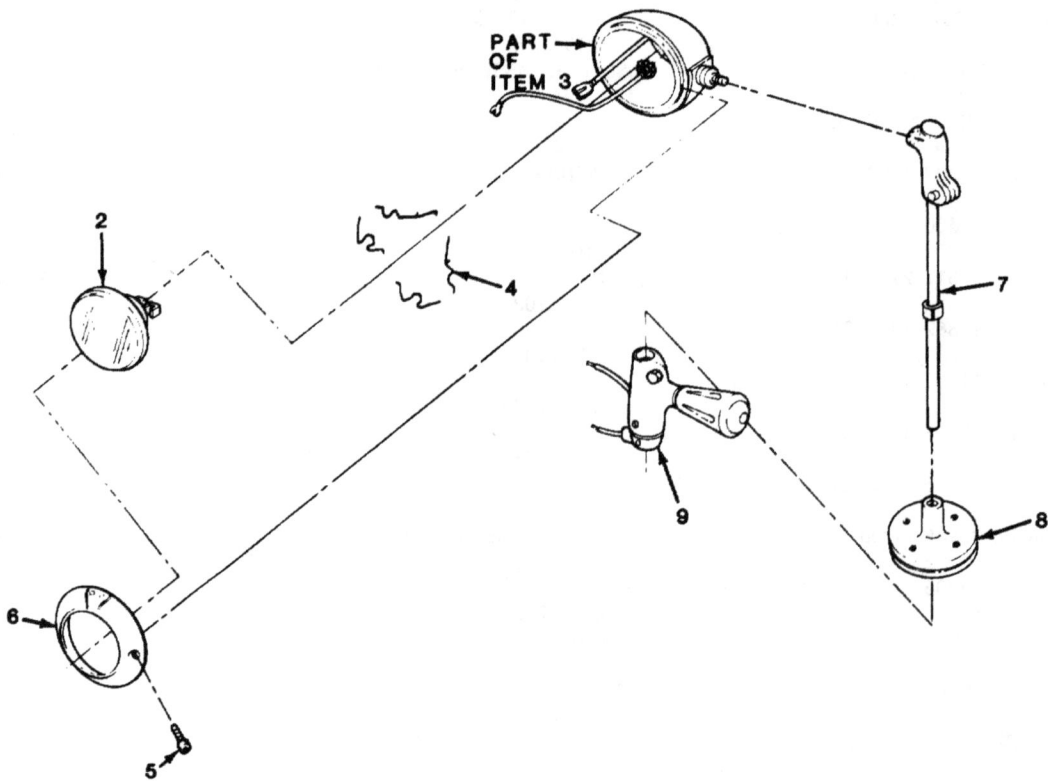

FIGURE 290. SPOTLIGHT ASSEMBLY AND COMPONENT PARTS (M1010).

TA511097

(1) (2) (3) (4) ITEM SMR PART NO CODE CAGEC NUMBER	(5) DESCRIPTION AND USABLE ON CODES (UOC)	(6) QTY
	GROUP 39 SEARCHLIGHT AND ELECTRICAL ILLUMINATING EQUIPMENT	
	GROUP 3901 SEARCHLIGHT OR ILLUMINATING LIGHT ASSEMBLY	
	FIG. 290 SPOTLIGHT ASSEMBLY AND COMPONENT PARTS (M1010)	
1 PAOOO 81349 M4510-1A4435 SPOTLIGHT UOC:210		1
2 PAOZZ 08806 4435	.LAMP,INCANDESCENT UOC:210	1
3 PAOOO 78977 650U-0016 .BRACKET UOC:210		1
4 PAOZZ 78977 6598	..CLIP,RETAINING UOC:210	4
5 PAOZZ 78977 6471	..SCREW,MACHINE UOC:210	1
6 PAOZZ 78977 6566-0004 ..RETAINER,LENS UOC:210		1
7 PAOZZ 78977 6710-BU-2 .TUBE,SPOTLIGHT UOC:210		1
8 PAOZZ 78977 100-7-3	.PARTS KIT,MOUNTING UOC:210	1
9 PAOZZ 78977 6701-0025	.HANDLE,MANUAL CONTR UOC:210	1

END OF FIGURE

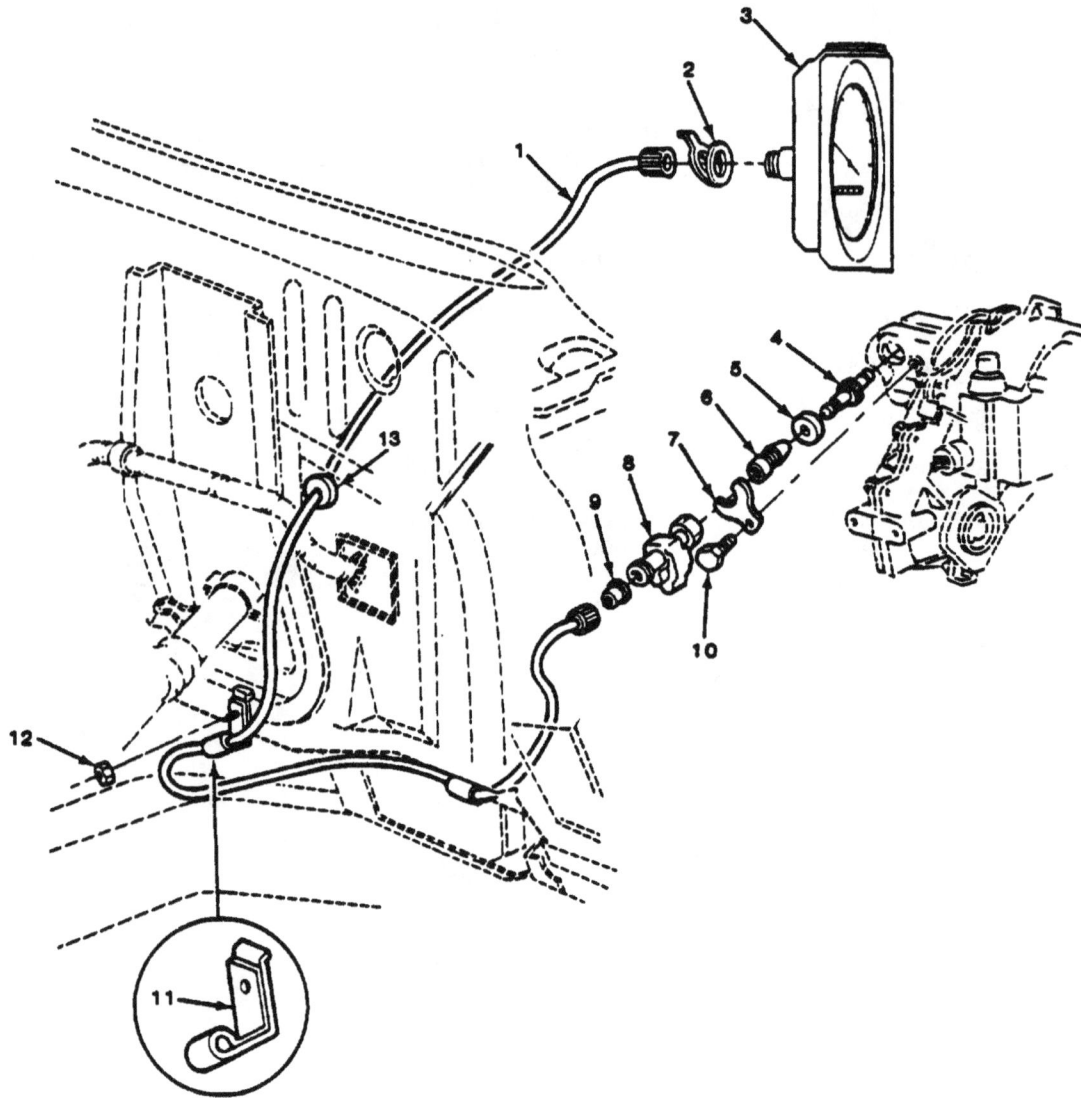

TA511098

FIGURE 291. SPEEDOMETER AND CABLE ASSEMBLY AND RELATED PARTS.

(1) ITEM NO	(2) SMR CODE	(3) CAGEC	(4) PART NUMBER	(5) DESCRIPTION AND USABLE ON CODES (UOC)	(6) QTY
				GROUP 47 GAGES (NONELECTRICAL), WEIGHING AND MEASURING DEVICES	
				GROUP 4701 INSTRUMENTS	
				FIG. 291 SPEEDOMETER AND CABLE ASSEMBLY AND RELATED PARTS	
1	PAOZZ	11862	25033627	SHAFT ASSEMBLY,FLEX	1
2	PAOZZ	11862	25020687	CLIP,SPRING TENSION	1
3	PAOZZ	11862	25052373	SPEEDOMETER	1
4	PAOZZ	11862	1362195	GEAR,SPEEDOMETER	1
				UOC:209	
4	PAOZZ	11862	9780470	GEAR SHAFT,HELICAL	1
				UOC:194,208,210,230	
4	XDOZZ	11862	3866918	GEAR,SPUR	1
				UOC:231,252,254,256	
5	PAOZZ	11862	15562374	SEAL	1
6	PAOZZ	11862	326561	SLEEVE	1
				UOC:213,252,254,256	
6	PAOZZ	11862	1362293	SLEEVE AND SEAL ASS	1
				UOC:194,208,209,210,230	
7	PAOZZ	11862	1254856	RETAINER,SPEEDOMETE	1
8	PAOZZ	11862	368026	ADAPTER,SPEEDOMETER	1
				UOC:209	
9	PAOZZ	11862	14018671	RUBBER ROUND SECTIO	1
10	PAOZZ	11862	8639743	SCREW,CAP,HEXAGON H	1
				UOC:194,208,209,210,230,231,252,254	
11	PAOZZ	11862	378362	CLIP,CABLE,SPEEDOME	1
12	PAOZZ	11862	11506101	NUT	1
13	PAOZZ	11862	474579	GROMMET,NONMENTALLIC	1

END OF FIGURE

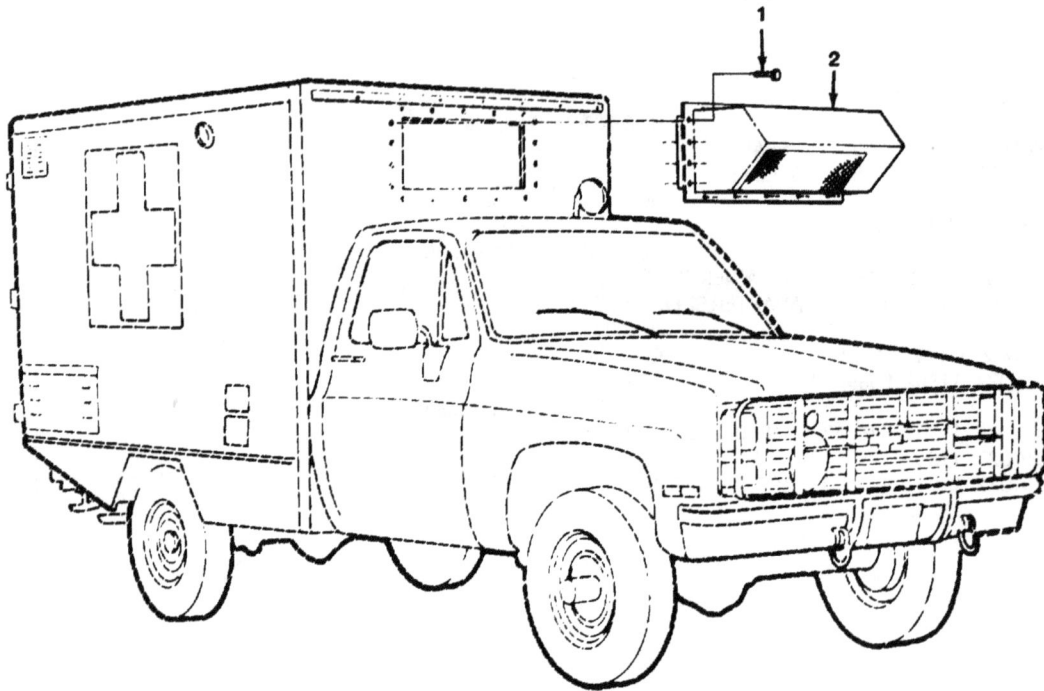

TA511099

FIGURE 292. AIR CONDITIONER ASSEMBLY (M1010).

(1) (2) (3) (4) (5)						(6)
ITEM	SMR		PART			
NO	CODE	CAGEC	NUMBER	DESCRIPTION AND USABLE ON CODES (UOC)		QTY

GROUP 52 REFRIGERATION, AIR
CONDITIONER/HEATER, AND AIR
CONDITIONING COMPONENTS

GROUP 5200 AIR CONDITIONER/HEATER
ASSEMBLY AND GAS COMPRESSOR ASSEMBLY

FIG. 292 AIR CONDITIONER ASSEMBLY
(M1010)

1	PAFZZ	24617	9421073	SCREW,TAPPING,THREA	16
				UOC:210	
2	PFFHH	99688	78401	AIR CONDITIONER FOR COMPONENT	1
				PARTS SEE FIG'S 293, 294, 298, 299	
				UOC:210	

END OF FIGURE

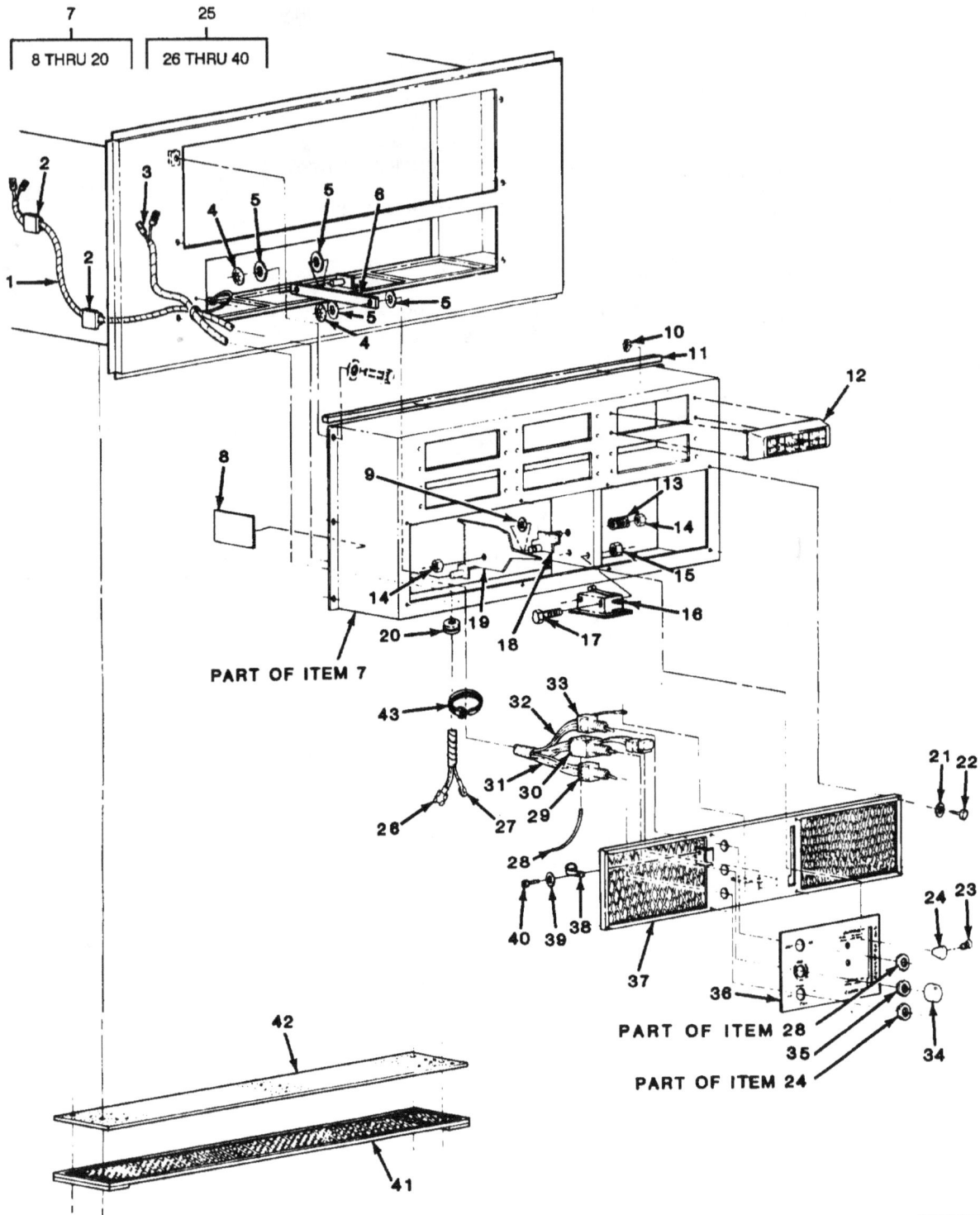

FIGURE 293. AIR OUTLET BOX AND CONTROL PANEL ASSEMBLY
COMPONENT PARTS (M1010).

TA511100

(1) (2) (3) (4) ITEM SMR PART NO CODE CAGEC NUMBER	(5) DESCRIPTION AND USABLE ON CODES (UOC)	(6) QTY
	GROUP 5200 AIR CONDITIONER/HEATER ASSEMBLY AND GAS COMPRESSOR ASSEMBLY	
	FIG. 293 AIR OUTLET BOX AND CONTROL PANEL ASSEMBLY COMPONENT PARTS (M1010)	
1 PAFZZ 99688 65132	WIRING HARNESS,BRAN UOC:210	1
2 PAFZZ 99688 57972	CLIP,SPRING TENSION UOC:210	2
3 PAFZZ 99688 65129	WIRING HARNESS,BRAN UOC:210	1
4 PAFZZ 99688 62229	PUSH ON NUT UOC:210	2
5 PAFZZ 99688 62210	WASHER,FLAT UOC:210	4
6 PAFZZ 99688 46006	CONNECTING LINK,RIG UOC:210	1
7 PAFFF 99688 85925	VENTILATOR,AIR CIRC UOC:210	1
8 PAFZZ 99688 58601	.MARKER,IDENTIFICATI UOC:210	1
9 PAFZZ 99688 62220	.WASHER,SPRING TENSI UOC:210	2
10 PAFZZ 99688 62229	.PUSH ON NUT UOC:210	24
11 PAFZZ 99688 58602	.GASKET UOC:210	3
12 PAFZZ 99688 58599	.VENTILATOR,AIR CIRC UOC:210	6
13 PAFZZ 99688 58600	.SPRING,HELICAL,COMP UOC:210	2
14 PAFZZ 99688 62219	.NUT,SELF-LOCKING,HE UOC:210	3
15 PAFZZ 99688 62222	.NUT,SELF-LOCKING,HE UOC:210	2
16 PAFZZ 99688 65136	.SWITCH,SENSITIVE UOC:210	1
17 PAFZZ 1T998 274898	.SCREW,MACHINE UOC:210	2
18 PAFZZ 99688 85934	.BRACKET,DOUBLE ANGL UOC:210	1
19 PAFZZ 99688 85932	.LEVER,HEATER UOC:210	1
20 PAFZZ 96906 MS35489-135	.GROMMET,NONMETALLIC UOC:210	1
21 PAFZZ 96906 MS35338-43	WASHER,LOCK UOC:210	8
22 PAFZZ 24617 9417222	SCREW,TAPPING,THREA UOC:210	8
23 PAFZZ 24617 9416137	SCREW,TAPPING,THREA	1

(1) ITEM NO	(2) SMR CODE	(3) CAGEC	(4) PART NUMBER	(5) DESCRIPTION AND USABLE ON CODES (UOC)	(6) QTY
				UOC:210	
24	PAFZZ	99688	58668	KNOB	1
				UOC:210	
25	PAFFF	99688	85926	SWITCH ASSEMBLY	1
				UOC:210	
26	PAFZZ	99688	65145	.LEAD,ELECTRICAL (GREEN)	1
				UOC:210	
27	PAFZZ	99688	65133	.LEAD,ELECTRICAL (BLACK)	1
				UOC:210	
28	PAFZZ	99688	58662	.INSULATION SLEEVING	1
				UOC:210	
29	PAFZZ	99688	65138	.SWITCH,TOGGLE	1
				UOC:210	
30	PAFZZ	99688	65139	.SWITCH,THERMOSTATIC	1
				UOC:210	
31	PAFZZ	99688	65146	.WIRING HARNESS,BRAN (YELLOW)	1
				UOC:210	
32	PAFZZ	99688	85935	.WIRING HARNESS,BRAN	1
				UOC:210	
33	PAFZZ	99688	65140	.SWITCH,TOGGLE	1
				UOC:210	
34	PAFZZ	99688	58603	.KNOB	1
				UOC:210	
35	PAFZZ	99688	62028	.NUT,PLAIN,HEXAGON	1
				UOC:210	
36	PAFZZ	99688	58605	.MARKER,IDENTIFICATI	1
				UOC:210	
37	PAFZZ	99688	85954	.PANEL,SKELETAL BLAN	1
				UOC:210	
38	PAFZZ	96906	MS21333-107	.CLAMP,LOOP	1
				UOC:210	
39	PAFZZ	96906	MS35338-44	.WASHER,LOCK	1
				UOC:210	
40	PAFZZ	96906	MS51869-26	.SCREW,TAPPING,THREA	1
				UOC:210	
41	PAFZZ	99688	85929	HOLDER,FILTER,AIR	1
				UOC:210	
42	PAFZZ	99688	58587	FILTER MEDIA,AIR CO	1
				UOC:210	
43	PAFZZ	96906	MS3367-1-9	STRAP,TIEDOWN,ELECT	4
				UOC:210	

END OF FIGURE

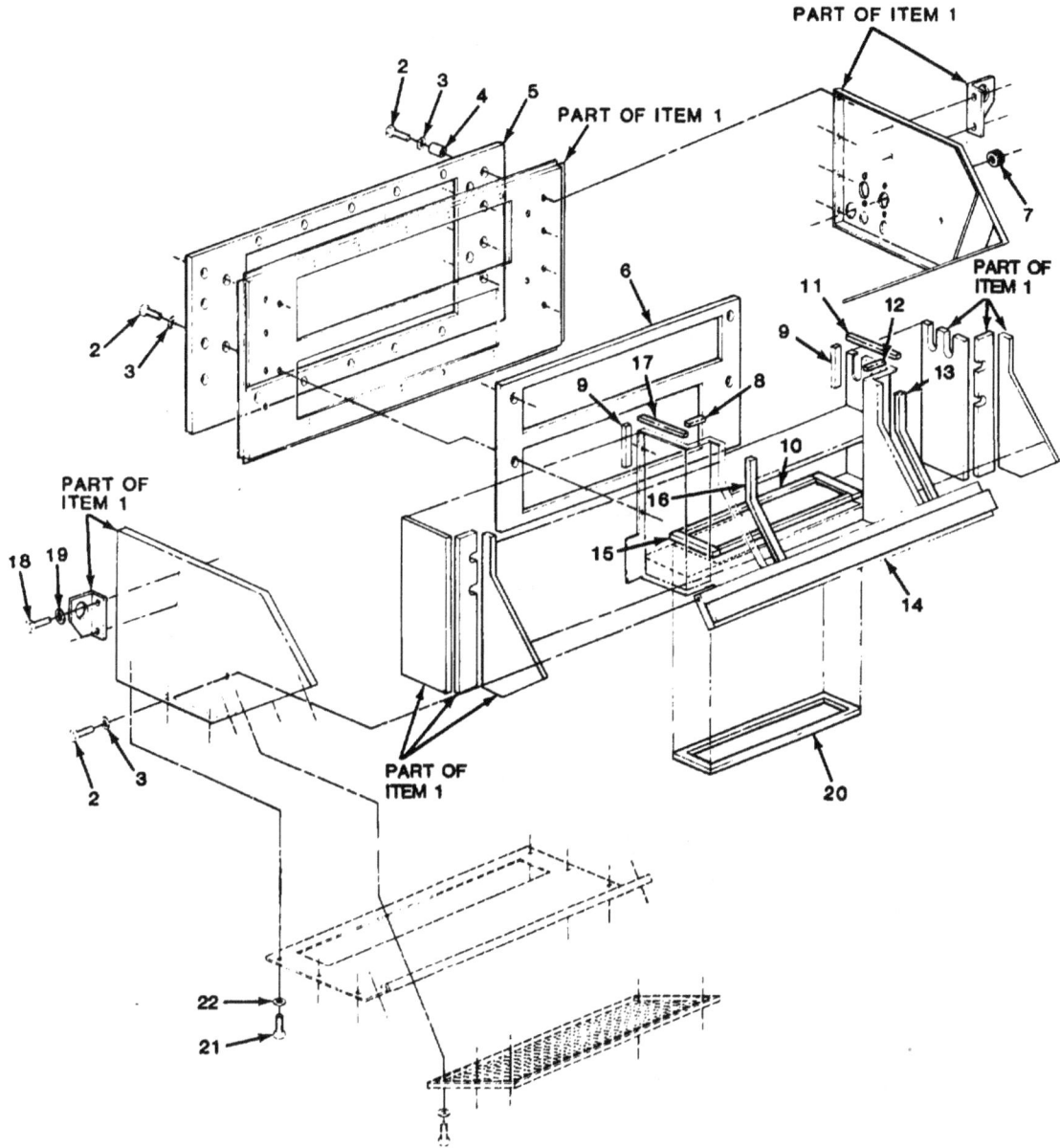

FIGURE 294. MAIN CASE ASSEMBLY AND COMPONENT PARTS (M1010).

TA511101

(1) ITEM NO	(2) SMR CODE	(3) CAGEC	(4) PART NUMBER	(5) DESCRIPTION AND USABLE ON CODES (UOC)	(6) QTY
				GROUP 5200 AIR CONDITIONER/HEATER ASSEMBLY AND GAS COMPRESSOR ASSEMBLY	
				FIG. 294 MAIN CASE ASSEMBLY AND COMPONENT PARTS (M1010)	
1	PAOOO	99688	85927	MAIN CASE,AIR CONDI UOC:210	1
2	PAOZZ	96906	MS51869-26	.SCREW,TAPPING,THREA UOC:210	23
3	PAOZZ	96906	MS35338-44	.WASHER,LOCK UOC:210	23
4	PAOZZ	99688	58606	.SPACER,SLEEVE UOC:210	16
5	PAOZZ	99688	58608	.GASKET UOC:210	1
6	PAOZZ	99688	58618	.GASKET UOC:210	1
7	PAOZZ	96906	MS35489-135	.GROMMET,NONMETALLIC. UOC:210	3
8	PAOZZ	99688	58610	.INSULATION BLANKET, UOC:210	1
9	PAOZZ	99688	58607	.SEAL,NONMETALLIC ST UOC:210	2
10	PAOZZ	99688	58617	.GASKET UOC:210	1
11	PAOZZ	99688	58615	.INSULATION BLANKET, UOC:210	1
12	PAOZZ	99688	58614	.INSULATION BLANKET, UOC:210	1
13	PAOZZ	99688	58613	.INSULATION BLANKET, UOC:210	1
14	PAOZZ	99688	58612	.INSULATION BLANKET, UOC:210	1
15	PAOZZ	99688	58616	.SEAL,RUBBER STRIP UOC:210	1
16	PAOZZ	99688	58611	.INSULATION BLANKET, UOC:210	1
17	PAOZZ	99688	58609	.INSULATION BLANKET, UOC:210	1
18	PAOZZ	96906	MS51871-14	.SCREW,TAPPING,THREA UOC:210	4
19	PAOZZ	96906	MS35338-46	.WASHER,LOCK UOC:210	4
20	PAOZZ	99688	58586	GASKET UOC:210	1
21	PAOZZ	96906	MS51869-26	SCREW,TAPPING,THREA UOC:210	39
22	PAOZZ	96906	MS35338-44	WASHER,LOCK UOC:210	45

END OF FIGURE

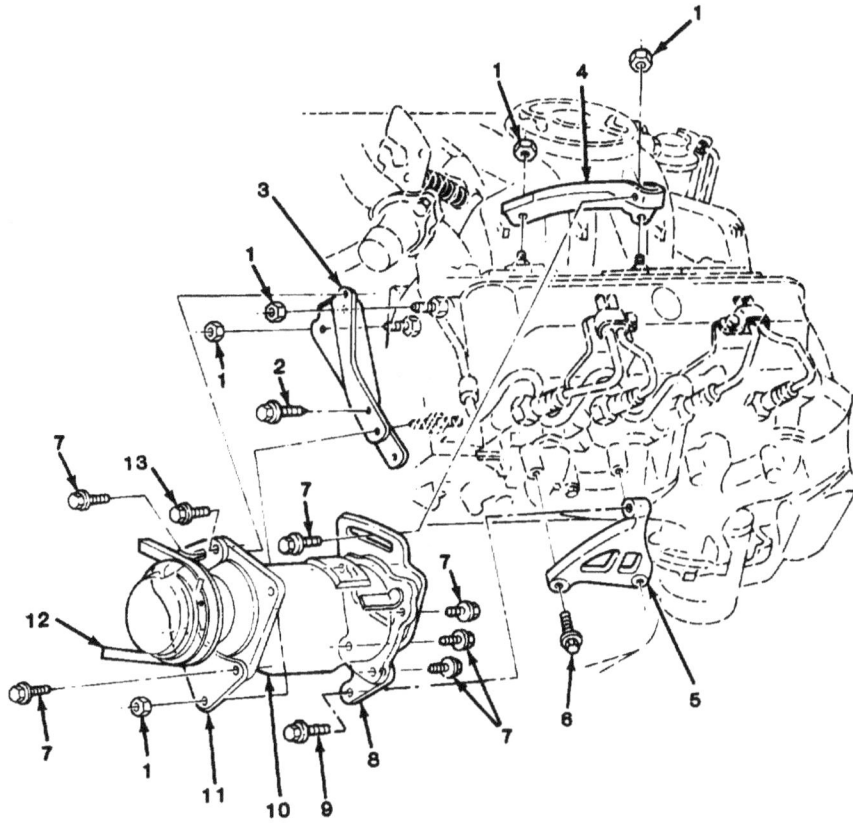

TA511102

FIGURE 295. GAS COMPRESSOR ASSEMBLY AND MOUNTING BRACKETS (M1010).

(1) ITEM NO	(2) SMR CODE	(3) CAGEC	(4) PART NUMBER	(5) DESCRIPTION AND USABLE ON CODES (UOC)	(6) QTY
				GROUP 5200 AIR CONDITIONER/HEATER ASSEMBLY AND GAS COMPRESSOR ASSEMBLY	
				FIG. 295 GAS COMPRESSOR ASSEMBLY AMD MOUNTING BRACKETS (M1010)	
1	PAOZZ	11862	11506101	NUT	5
				UOC:210,230	
2	PAFZZ	11862	11504595	SCREW,CAP,HEXAGON H	1
				UOC:210,230	
3	PAFZZ	11862	14033864	BRACKET,MOUNTING	1
				UOC:210	
4	PAFZZ	11862	14033866	BRACKET,MOUNTING	1
				UOC:210	
5	PAFZZ	11862	14033867	BRACKET,MOUNTING	1
				UOC:210	
6	PAFZZ	11862	14028922	BOLT,MACHINE	2
				UOC:210	
7	PAOZZ	11862	1635490	BOLT,SELF-LOCKING	6
				UOC:210	
8	PAFZZ	11862	14033863	MOUNTING,PLATE	1
				UOC:210	
9	PAOZZ	11862	1635490	BOLT,SELF-LOCKING	1
				UOC:210	
10	PAFFF	11862	12300270	COMPRESSOR,REFRIGER FOR COMPONENT PARTS SEE FIG 296 UOC:210	1
11	PAFZZ	11862	14033861	MOUNTING PLATE	1
				UOC:210	
12	PAOZZ	20796	42-4877	BELT,V	1
				UOC:210,230	
13	PAFZZ	24617	11504986	BOLT,MACHINE	3
				UOC:210	

END OF FIGURE

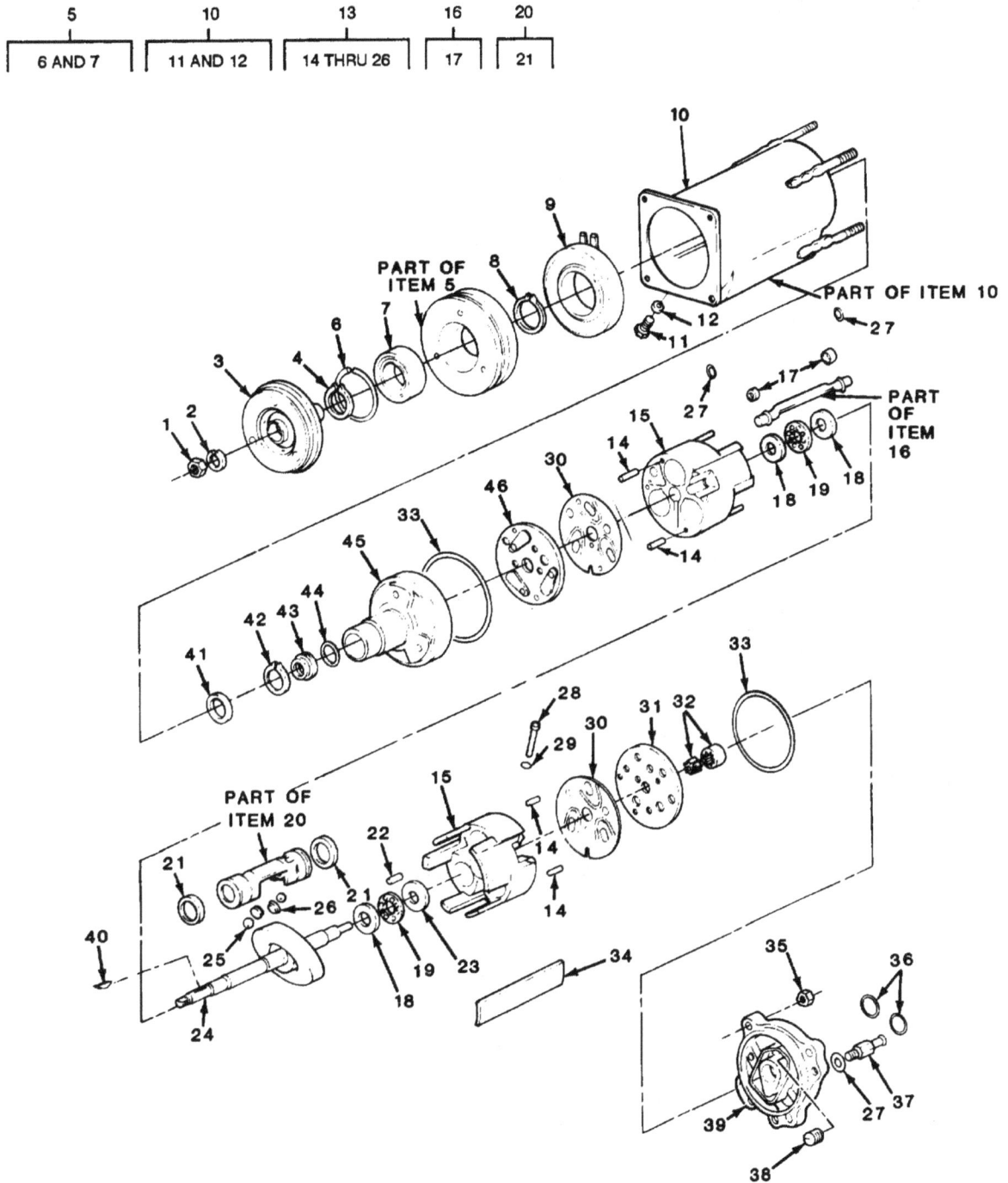

FIGURE 296. GAS COMPRESSOR COMPONENT PARTS (M1010).

TA511103

(1) ITEM NO	(2) SMR CODE	(3) CAGEC	(4) PART NUMBER	(5) DESCRIPTION AND USABLE ON CODES (UOC)	(6) QTY
				GROUP 5200 AIR CONDITIONER/HEATER ASSEMBLY AND GAS COMPRESSOR ASSEMBLY	
				FIG. 296 GAS COMPRESSOR COMPONENT PARTS (M1010)	
1	PAFZZ	11862	6556923	NUT,SELF-LOCKING,EX PART OF KIT P/N 6590589 UOC:210	1
2	PAFZZ	11862	6555328	RING,RETAINING PART OF KIT P/N 6590591 UOC:210	1
3	PAFZZ	11862	5914719	DRIVER ASSEMBLY,CLU UOC:210	1
4	PAFZZ	11862	6555329	RING,RETAINING PART OF KIT P/N 6590591 UOC:210	1
5	PAFFF	11862	6550750	PULLEY,GROOVE UOC:210	1
6	PAFZZ	11862	6555338	.RING,RETAINING UOC:210	1
7	PAFZZ	11862	908287	.BEARING,BALL ANNULA UOC:210	1
8	PAFZZ	11862	6555330	RING,RETAINING UOC:210	1
9	PAFZZ	11862	6550835	COIL,ELECTRICAL UOC:210	1
10	XAFFF	11862	2724176	HOUSING,COMPRESSOR UOC:210	1
11	PAFZZ	11862	5858694	.PLUG,DRAIN,COMPRESS UOC:210	1
12	PAFZZ	11862	5858696	.GASKET UOC:210	1
13	XAFFF	11862	5914809	CYLINDER AND SHAFT UOC:210	1
14	XAFZZ	96906	MS16562-36	.PIN,SPRING UOC:210	4
15	XAFZZ	11862	5914807	.CYLINDER,AIR COMPRE UOC:210	1
16	XAFZZ	11862	6590592=XA	.PARTS KIT,COMPRESSO UOC:210	1
17	XAFZZ	11862	6555440	..SEAL,RUBBER ROUND S UOC:210	2
18	XAFZZ	11862	6556000	.BEARING,WASHER,THRU UOC:210	3
19	XAFZZ	11862	6555259	.RETAINER AND ROLLER UOC:210	2
20	XAFFF	76462	6551150	.RING,PISTON UOC:210	3
21	XAFZZ	11862	6556605=XA	..PACKING,PREFORMED UOC:210	6
22	XAFZZ	11862	6590686	.PIN,STRAIGHT,HEADLE	1

(1) ITEM NO	(2) SMR CODE	(3) CAGEC	(4) PART NUMBER	(5) DESCRIPTION AND USABLE ON CODES (UOC)	(6) QTY
				UOC:210	
23	XAFZZ	11862	6556050	.BEARING,WASHER,THRU (5)	1
				UOC:210	
23	XAFZZ	11862	6556055	.BEARING,WASHER THRU (5.5)	1
				UOC:210	
23	XAFZZ	11862	6556060	.BEARING,WASHER,THRU (6)	1
				UOC:210	
23	XAFZZ	11862	6556065	.BEARING,WASHER,THRU (6.5)	1
				UOC:210	
23	XAFZZ	11862	6556070	.BEARING,WASHER,THRU (7)	1
				UOC:210	
23	XAFZZ	11862	6556075	.BEARING,WASHER,THRU (7.5)	1
				UOC:210	
23	XAFZZ	11862	6556080	.BEARING,WASHER,THRU (8)	1
				UOC:210	
23	XAFZZ	11862	6556085	.BEARING,WASHER,THRU (8.5)	1
				UOC:210	
23	XAFZZ	11862	6556090	.BEARING,WASHER,THRU (9)	1
				UOC:210	
23	XAFZZ	11862	6556095	.BEARING,WASHER,THRU (9.5)	1
				UOC:210	
23	XAFZZ	11862	6556100	.BEARING,WASHER,THRU (10)	1
				UOC:210	
23	XAFZZ	23671	6556105	.BEARING,WASHER,THRU (10.5)	1
				UOC:210	
23	XAFZZ	11862	6556110	.BEARING,WASHER,THRU (11)	1
				UOC:210	
23	XAFZZ	11862	6556115	.BEARING,WASHER,THRU (11.5)	1
				UOC:210	
23	XAFZZ	11862	6556120	.BEARING,WASHER,THRU (12)	1
				UOC:210	
24	XAFZZ	247462	5914705	.SHAFT AND PLATE ASS	1
				UOC:210	
25	XAFZZ	11862	6555258	.BALL,BEARING	6
				UOC:210	
26	XAAFZZ	11862	6557000	.DISK,SHOE,COMPRESSO (0)	3
				UOC:210	
26	XAFZZ	11862	6556175	.DISC,SHOE COMPRESSO (17.5)	1
				UOC:210	
26	XAFZZ	11862	6556180	.DISC,SHOE COMPRESSO (18)	1
				UOC:210	
26	XAFZZ	11862	6556185	.DISC,SHOE COMPRESSO (18.5)	1
				UOC:210	
26	XAFZZ	11862	6556190	.DISC,SHOE COMPRESSO (19)	1
				UOC:210	
26	XAFZZ	11862	6556195	.DISK,SHOE,COMPRESSO (19.5)	1
				UOC:210	
26	XAFZZ	11862	6556200	.DISC,SHOE COMPRESSO (20)	1
				UOC:210	
26	XAFZZ	11862	6556205	.DISC,SHOE COMPRESSOR (20.5)	1
				UOC:210	
26	XAFZZ	11862	6556210	.DISK,SHOE,COMPRESSO (21)	1

(1) (2) (3) (4) (5)					(6)
ITEM SMR			PART		
NO	CODE	CAGEC	NUMBER	DESCRIPTION AND USABLE ON CODES (UOC)	QTY
				UOC:210	
26	XAFZZ	11862	6556215	.DISC,SHOE COMPRESSO (21.5)	1
				UOC:210	
26	XAFZZ	11862	6556220	.DISC,SHOE COMPRESSO (22)	1
				UOC:210	
27	PAFZZ	11862	3093499	PACKING,PREFORMED	3
				UOC:210	
28	PAFZZ	11862	6555298	COUPLING,TUBE	1
				UOC:210	
29	PAFZZ	11862	5929144	PACKING,PREFORMED PART OF KIT P/N 6591956	1
				UOC:210	
30	PAFZZ	11862	6555279	DIAPHRAGM,VALVE,FLA	2
				UOC:210	
31	PAFZZ	27462	5914709	VALVE PLATE ASSEMBL	1
				UOC:210	
32	PAFZZ	27462	6590594	PARTS KIT,PUMP,COMP	1
				UOC:210	
33	KFFZZ	11862	6555293	O-RING COMP SHELL PART OF KIT P/N 6591956	2
				UOC:210	
34	PAFZZ	27462	6555755	COVER,SUCTION,AIR C PART OF KIT P/N 6591956	1
				UOC:210	
35	KFFZZ	11862	6555378	NUT SPECIAL PART OF KIT P/N 6590589	4
				UOC:210	
36	PAFZZ	11862	5887997	PACKING,PREFORMED PART OF KIT P/N 6591956	2
				UOC:210	
37	PAFZZ	27462	5914435	VALVE,VENT	1
				UOC:210	
38	PAFZZ	27462	6555299	STRAINER,ELEMENT,SE	1
				UOC:210	
39	PAFZZ	11862	6557057	CYLINDER HEAD,COMPR	1
				UOC:210	
40	PAFZZ	11862	6555320	KEY,MACHINE	1
				UOC:210	
41	PAFZZ	11862	6556076	RETAINER,PACKING PART OF KIT P/N 6599114	1
				UOC:210	
42	PAFZZ	11862	6556767	RING,RETAINING PART OF KIT P/N 6599114 PART OF KIT P/N 6590591	1
				UOC:210	
43	PAFZZ	11862	6556077	GASKET PART OF KIT P/N 6599114	1
				UOC:210	
44	KFFZZ	11862	6555290	PREFORMED PACKING PART OF KIT P/N 6599114	1
				UOC:210	
45	PAFZZ	27462	6555283	CYLINDER HEAD,COMPR	1
				UOC:210	
46	PAFZZ	11862	5914706	VALVE PLATE,AIR CO	1
				UOC:210	

END OF FIGURE

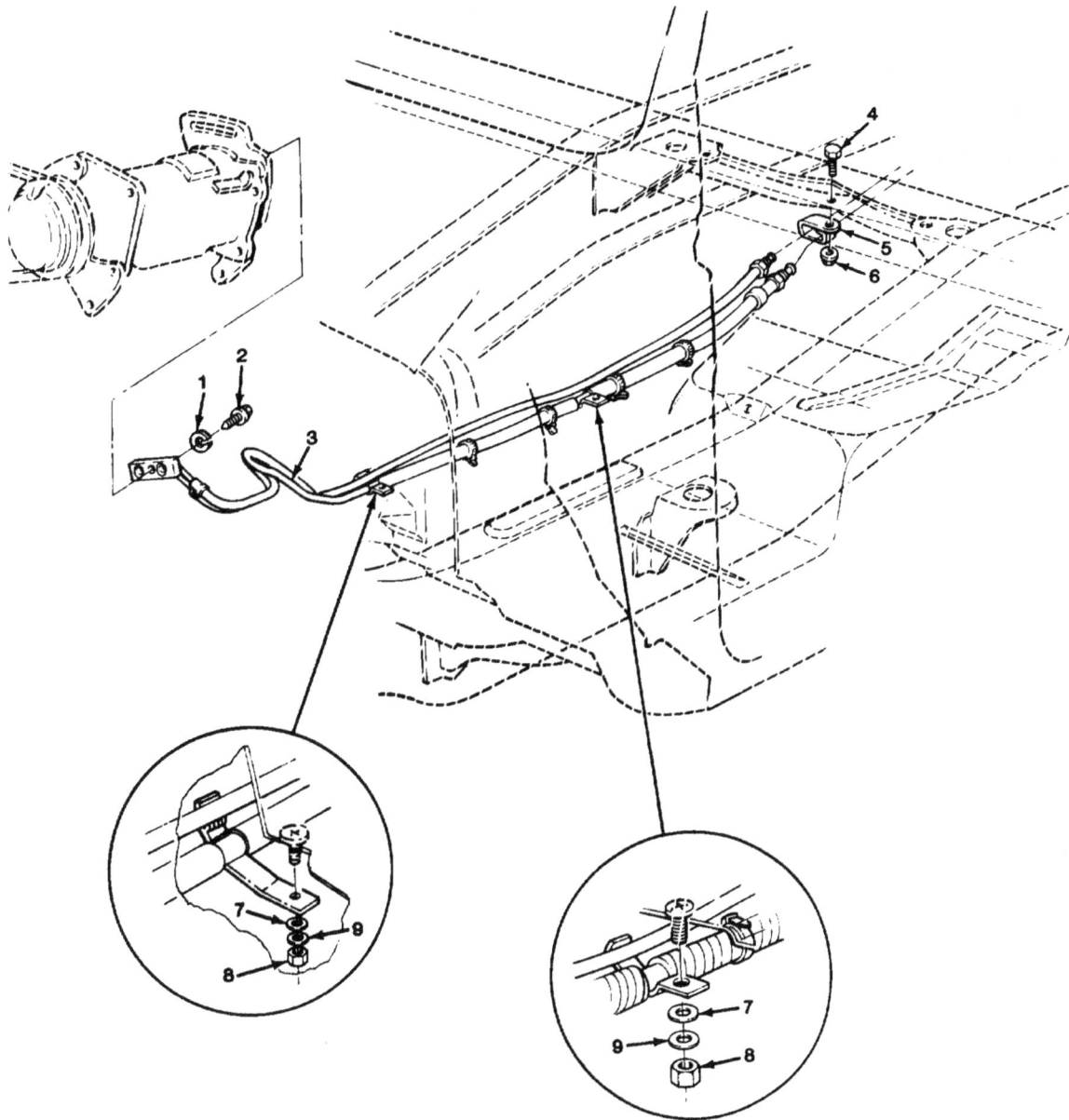

TA511104

FIGURE 297. GAS COMPRESSOR HOSES (M1010).

(1) (2) (3) (4) ITEM SMR PART NO CODE CAGEC NUMBER	(5) DESCRIPTION AND USABLE ON CODES (UOC)	(6) QTY
	GROUP 5217 REFRIGERANT PIPING	
	FIG. 297 GAS COMPRESSOR HOSES (M1010)	
1 PAFZZ 11862 11500046 WASHER,LOCK	UOC:210	1
2 PAFZZ 24617 11500831 BOLT,SELF-LOCKING	UOC:210	1
3 PAFZZ 11862 14072408 TUBE ASSEMBLY,DUPLE	UOC:210	1
4 PAFZZ 96906 MS90728-6 SCREW,CAP,HEXAGON H	UOC:210	1
5 PAFZZ 11862 3825416 CLAMP,LOOP	UOC:210	1
6 PAFZZ 24617 11506003 NUT,PLAIN,ASSEMBLED	UOC:210	1
7 PAFZZ 11862 3903572 WASHER,FLAT	UOC:210	2
8 PAFZZ 21450 131245 NUT,SELF-LOCKING,HE	UOC:210	2
9 PAFZZ 96906 MS27183-10 WASHER,FLAT	UOC:210	2

END OF FIGURE

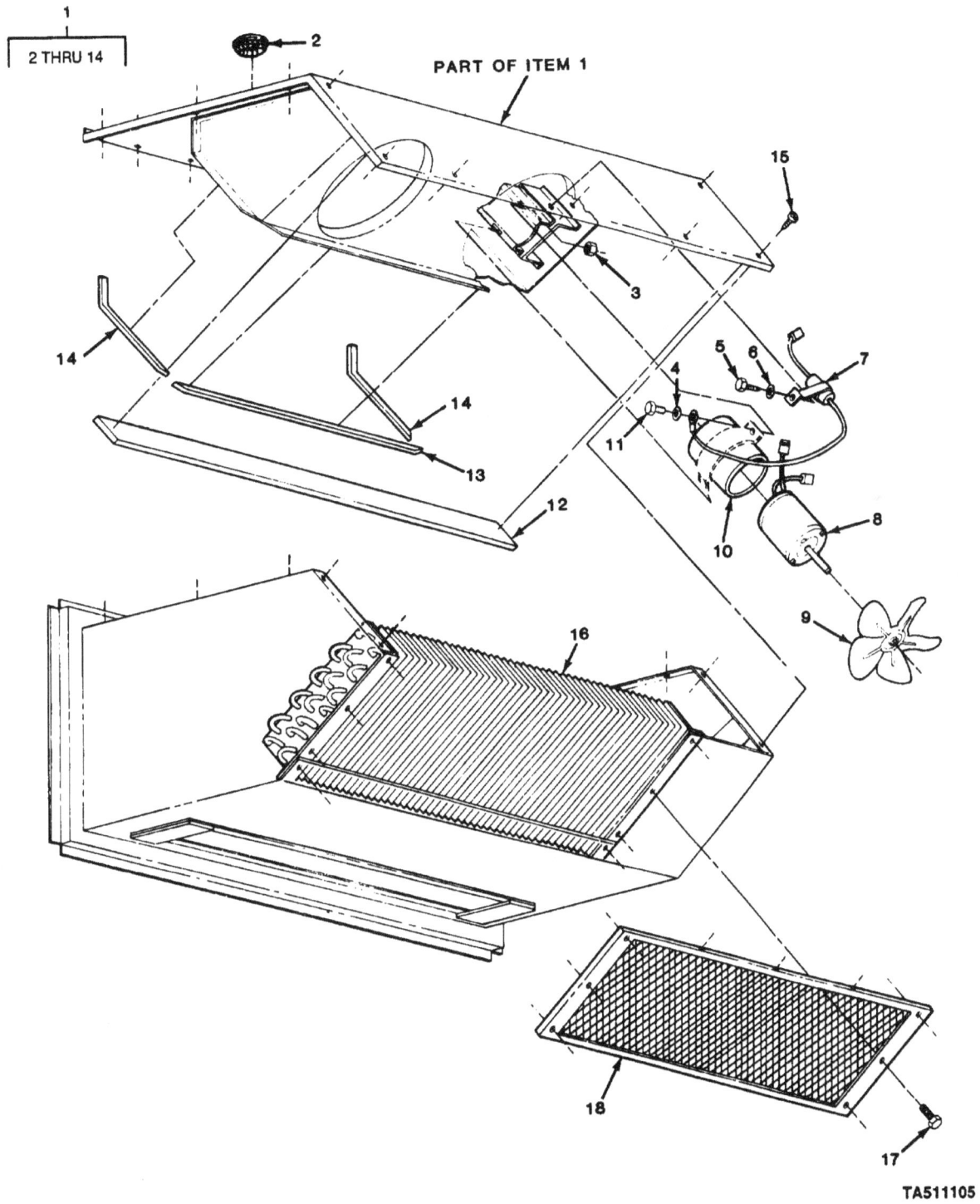

FIGURE 298. CONDENSER, COVER ASSEMBLY AND COMPONENT PARTS (M1010).

TA511105

(1) (2) (3) (4)	(5)	(6)
ITEM SMR PART NO CODE CAGEC NUMBER	DESCRIPTION AND USABLE ON CODES (UOC)	QTY

GROUP 5230 CONDENSER

FIG. 298 CONDENSER, COVER ASSEMBLY,
AND COMPONENT PARTS (M1010)

1	PAOZZ 99688 85923	FAN,CIRCULATING UOC:210	1
2	PAOZZ 03743 S150	.BUTTON,PLUG UOC:210	1
3	PAOZZ 99688 62218	.NUT UOC:210	2
4	PAOZZ 96906 MS35335-33	.WASHER,LOCK UOC:210	2
5	PAOZZ 24617 9421073	.SCREW,TAPPING,THREA UOC:210	2
6	PAOZZ 96906 MS35335-32	.WASHER,LOCK UOC:210	2
7	PAOZZ 99688 65134	.CAPACITOR,FIXED,ELE UOC:210	2
8	PAOZZ 99688 20566	.MOTOR,DIRECT CURREN UOC:210	2
9	PAOZZ 99688 58595	.IMPELLER,FAN,AXIAL UOC:210	2
10	PAOZZ 99688 58701	.RUBBER ROUND SECTIO UOC:210	2
11	PAOZZ 96906 MS90728-10	.SCREW,CAP,HEXAGON H UOC:210	2
12	PAOZZ 99688 58598	.GASKET UOC:210	1
13	PAOZZ 99688 58597	.RUBBER STRIP UOC:210	1
14	PAOZZ 99688 58596	.SEAL,PLAIN UOC:210	2
15	PAOZZ 24617 9440025	SCREW,TAPPING,THREA UOC:210	20
16	PAFZZ 99688 15112	COOLING COIL,AIR,DU UOC:210	1
17	PAOZZ 96906 MS90728-3	SCREW,CAP,HEXAGON H UOC:210	6
18	XDOZZ 99688 85928	GRILLE,METAL UOC:210	1

END OF FIGURE

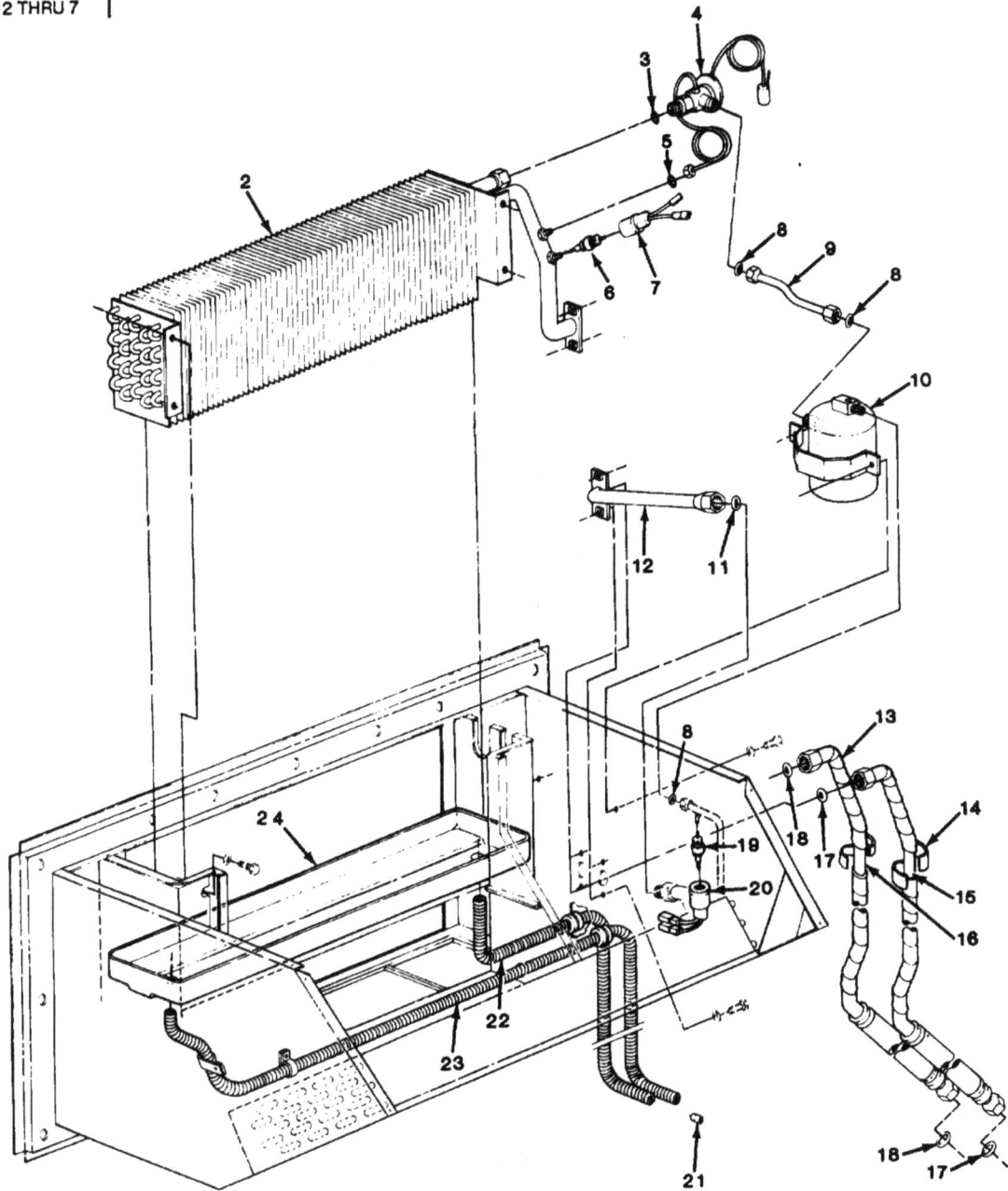

FIGURE 299. EVAPORATOR ASSEMBLY COMPONENT PARTS AND
RELATED PARTS (M1010).

TA511106

(1) ITEM NO	(2) SMR CODE	(3) CAGEC	(4) PART NUMBER	(5) DESCRIPTION AND USABLE ON CODES (UOC)	(6) QTY
				GROUP 5241 EVAPORATOR	
				FIG. 299 EVAPORATOR ASSEMBLY COMPONENT PARTS AND RELATED PARTS (M1010)	
1	PAFFF	99688	85924	EVAPORATOR ASSEMBLY UOC:210	1
2	PAFZZ	99688	15113	.EVAPORATOR COIL,REF UOC:210	1
3	PAFZZ	99688	56470	.PACKING,PREFORMED UOC:210	1
4	PAFZZ	99688	58583	.VALVE,EXPANSION UOC:210	1
5	PAFZZ	99688	56480	.PACKING,PREFORMED UOC:210	1
6	PAFZZ	27783	5137	.CORE,VALVE UOC:210	1
7	PAFZZ	99688	65131	.SWITCH,PRESSURE UOC:210	1
8	PAFZZ	99688	57012	PACKING,PREFORMED UOC:210	3
9	PAFZZ	99688	58582	TUBE ASSEMBLY,METAL UOC:210	1
10	PAFZZ	99688	585810	RECEIVER-DEHYDRATOR UOC:210	1
11	PAFZZ	99688	57277	PACKING,PREFORMED UOC:210	1
12	PAFZZ	99688	58580	TUBE ASSEMBLY,METAL UOC:210	1
13	PAFZZ	99688	58645	TUBING,NONMETALLIC UOC:210	1
14	PAFZZ	99688	58646	TUBING,NONMETALLIC UOC:210	1
15	PAFZZ	99688	58647	TUBE ASSEMBLY,METAL UOC:210	1
16	PAFZZ	99688	58648	TUBE ASSEMBLY,METAL UOC:210	1
17	PAFZZ	99688	57277	PACKING,PREFORMED UOC:210	2
18	PAFZZ	99688	58021	PACKING,PREFORMED UOC:210	2
19	PAFZZ	27783	5137	CORE,VALVE UOC:210	1
20	PAFZZ	99688	65130	SWITCH,PRESSURE UOC:210	1
21	PAFZZ	99688	58579	VALVE,CHECK UOC:210	1
22	PAFZZ	99688	58577	TUBING,NONMETALLIC LEFT UOC:210	1
23	PAFZZ	99688	58578	TUBING,NONMETALLIC UOC:210	1

(1) ITEM NO	(2) SMR CODE	(3) CAGEC	(4) PART NUMBER	(5)	DESCRIPTION AND USABLE ON CODES (UOC)	(6) QTY
24	PAFZZ	99688	58585		PAN,DRIP,AIR CONDIT UOC:210	1

END OF FIGURE

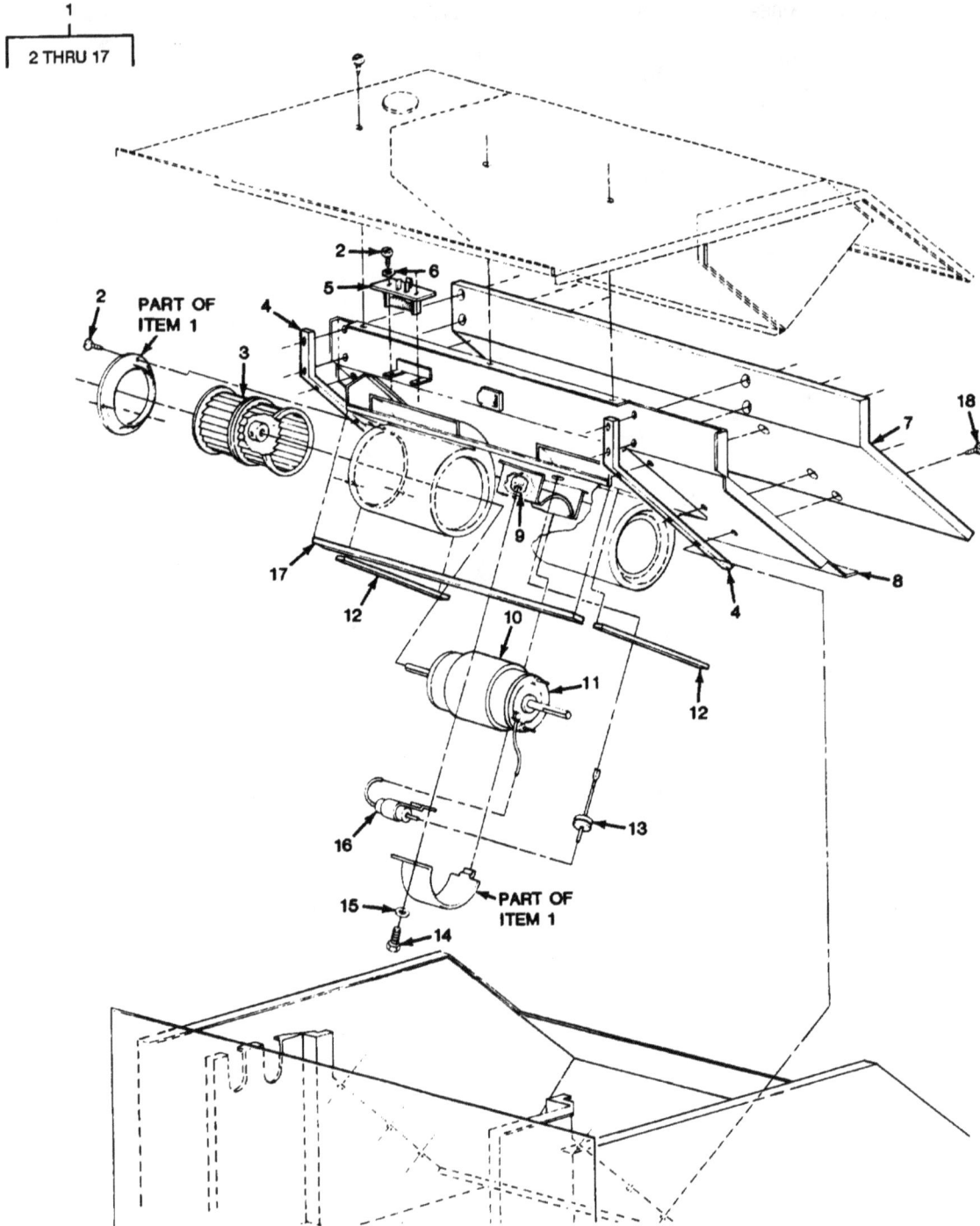

FIGURE 300. BLOWER ASSEMBLY COMPONENT PARTS (M1010).

TA511107

(1) ITEM NO	(2) SMR CODE	(3) CAGEC	(4) PART NUMBER	(5) DESCRIPTION AND USABLE ON CODES (UOC)	(6) QTY
				GROUP 5243 BLOWER ASSEMBLY	
				FIG. 300 BLOWER ASSEMBLY COMPONENT PARTS (M1010)	
1	PAOOO	99688	85922	FAN,CENTRIFUGAL UOC:210	1
2	PAOZZ	24617	9421073	.SCREW,TAPPING,THREA UOC:210	9
3	PAOZZ	99688	58590	.IMPELLER,FAN,CENTRI RIGHT UOC:210	1
3	PAOZZ	99688	58589	.IMPELLER,FAN,CENTRI LEFT UOC:210	1
4	PAOZZ	99688	58592	.SEAL,NONMETALLIC ST UOC:210	2
5	PAOZZ	99688	65158	.RESISTOR ASSEMBLY UOC:210	1
6	PAOZZ	96906	MS35335-32	.WASHER,LOCK UOC:210	2
7	PAOZZ	99688	58594	.INSULATION BLANKET UOC:210	1
8	PAOZZ	99688	85931	.COVER,ACCESS UOC:210	1
9	PAOZZ	99688	62218	.NUT UOC:210	1
10	PAOZZ	99688	58701	.RUBBER ROUND SECTIO UOC:210	1
11	PAOZZ	99688	20565	.MOTOR,DIRECT CURREN UOC:210	1
12	PAOZZ	99688	58591	.RUBBER STRIP UOC:210	2
13	PAOZZ	59875	TD97203	.GROMMET,NONMETALLIC UOC:210	1
14	PAOZZ	96906	MS90728-10	.SCREW,CAP,HEXAGON H UOC:210	1
15	PAOZZ	96906	MS35338-52	.WASHER,LOCK UOC:210	1
16	PAOZZ	99688	65134	.CAPACITOR,FIXED,ELE UOC:210	1
17	PAOZZ	99688	58593	.RUBBER STRIP UOC:210	1
18	PAOZZ	24617	9421073	SCREW,TAPPING,THREA UOC:210	13

END OF FIGURE

(1) (2) (3) (4) ITEM SMR PART NO CODE CAGEC NUMBER	(5) DESCRIPTION AND USABLE ON CODES (UOC)	(6) QTY
	GROUP 94 REPAIR KITS	
	GROUP 9401 REPAIR KITS	
	FIG. KITS	
PAOZZ 27647 11967	PARTS KIT,HUB BODY UOC:209 RING,RETAINING (2) 125-5 RING,RETAINING (1) 125-7 SCREW (6) 125-3	2
PAFZZ 11862 14067590	GASKET AND SEAL SET GASKET (2) 30-9 GASKET (1) 30-14 GASKET (8) 12-12 GASKET (2) 2-10 GASKET SET (2) 11-1 SEAL INLET (16) 7-24	1
PAOZZ 27647 15528	PARTS KIT,FOUR WHEE UOC:194,208,210,230,231,252,254,256 RING,RETAINING (1) 124-7 RING,RETAINING (1) 124-9 SCREW (6) 124-5 SEAL "O" RING (1) 124-8	1
PAHZZ 11862 15537086	SHIM SET UOC:254,256 SHIM (.003) (1) 111-8 SHIM (.005) (1) 111-8 SHIM (.010) (1) 111-8 SHIM (.014) (1) 111-8 SHIM (.016) (1) 111-8 SHIM (.018) (1) 111-8 SHIM (.020) (1) 111-8 SHIM (.022) (1) 111-8 SHIM (.030) (1) 111-8 SHIM (.015) (1) 111-8 SHIM (.021) (1) 111-8 SHIM (.023) (1) 111-8 SHIM (.010) (1) 111-8	1
PAFZZ 11862 15633467	GASKET AND SEAL SET GASKET (1) 31-7 GASKET (1) 13-11 GASKET (1) 13-3 GASKET (1) 13-9 GASKET (1) 7-5 PACKING,PREFORMED (2) 3-7 SEAL,EXHAUST PIPE (2) 27-23 SEAL,PLAIN ENCASED (1) 6-8	1
PAHZZ 84760 26214	PARTS KIT,COIL ARM LEAD,IGNITION,ENGIN(1) 15-31 NUT,PLAIN,HEXAGON (3) 15-29 NUT,SELF-LOCKING,HE(3) 15-37	1

(1) ITEM NO	(2) SMR CODE	(3) CAGEC	(4) PART NUMBER	(5) DESCRIPTION AND USABLE ON CODES (UOC)	(6) QTY
				SOLENOID (1) 15-23	
				SOLENOID (1) 15-23	
				TERMINAL,QUICK DISC (1) 15-39	
				WASHER,FLAT (3) 15-27	
				WASHER,LOCK (6) 15-28	
				WASHER,SHOULDERED (3) 15-26	
	XDFZZ	14892	2770723	PARTS KIT,BRAKE BOO	1
				UOC:194,208,210,230,231,252	
				BUSHING,NONMETALLIC(1) 117-24	
				O RING (1) 117-16	
				PISTON SEAL (1) 117-20	
				SEAL (2) 117-17	
				SEAL HOUSING (1) 117-14	
				SEAL RING SPOOL PC (1) 117-8	
				SEAT INSERT (2) 117-9	
	XDFZZ	14892	2771098	PARTS KIT,BRAKE CHA	1
				UOC:209	
				BOOT,DUST AND MOIST(1) 117-23	
				O RING (1) 117-16	
	PAHZZ	11862	5687182	PARTS KIT,STEERING	1
				PACKING,PREFORMED (1) 131-24	
				SEAL-P/S GR VLV ASM(1) 131-23	
				SEAL-P/S GR VLV ASM(1) 131-19	
				SEAL-P/S SPOOL VLV (1) 131-22	
	PAHZZ	11862	5688C44	GASKET SET	1
				GASKET (1) 133-23	
				MAGNET,PERMANENT (1) 133-9	
				PACKING,PREFORMED (1) 133-26	
				PACKING,PREFORMED (2) 133-19	
				PACKING,PREFORMED (1) 133-8	
				SEAL P/S HSG END (2) 133-25	
				SEAL,PLAIN ENCASED (1) 133-2	
	PAHZZ	19207	5704051	PARTS KIT,HEATER,VE	1
				UOC:210	
				BOLT-HOOK (4) 246-29	
				BURNER ASSEMBLY (1) 246-28	
				CLAMP,RIM CLENCHING(4) 246-30	
				GASKET (1) 246-26	
				NUT,PLAIN,HEXAGON (6) 246-4	
				PACKING,PREFORMED (1) 246-27	
				WASHER,LOCK (14) 246-20	
	PAHZZ	19207	5704052	PARTS KIT,VEHICULAR	1
				UOC:210	
				BOLT-HOOK (4) 246-29	
				CLAMP,RIM CLENCHING(4) 246-30	
				GASKET (1) 246-26	
				NUT,PLAIN,HEXAGON (6) 246-4	
				PACKING,PREFORMED (1) 246-27	
				SCREW,MACHINE (1) 250-5	
				VAPORIZER (1) 250-1	
				WASHER,FLAT (1) 250-3	
				WASHER,FIBER (1) 250-11	

(1) ITEM NO	(2) SMR CODE	(3) CAGEC	(4) PART NUMBER	(5) DESCRIPTION AND USABLE ON CODES (UOC)	(6) QTY
				WASHER,FLAT (1) 250-2	
				WASHER,LOCK (14) 246-20	
				WICK (1) 250-8	
	PAHZZ	19207	5704064	PARTS KIT,VEHICULAR	1
				UOC:210	
				CORE VALVE (1) 248-18	
				DIAPHRAGM (1) 248-29	
				GASKET (2) 248-32	
				PACKING,PREFORMED (1) 248-30	
				PACKING,PREFORMED (1) 248-3	
				PACKING,PREFORMED (3) 248-6	
				SCREEN,INLET (1) 248-40	
				SEAL,PLAIN (2) 248-39	
	PAOZZ	11862	6258545	LINK UNIT,STABILIZE	2
				UOC:254,256	
				BOLT,STAB,LINK (2) 157-10	
				GROMMET STABIL LINK(4) 157-3	
				NUT,SPECIAL (2) 157-1	
				RETAINER,WASHER SPE(4) 157-2	
				SPACER,STAB,LINK (2) 157-11	
	PAFZZ	27462	6590589	PARTS KIT,SHAFT,COM	1
				UOC:210	
				NUT SPECIAL (4) 296-35	
				NUT,SELF-LOCKING,EX(1) 296-1	
	PAFZZ	11862	6590591	PARTS KIT,PISTON AS	1
				UOC:210	
				RING,RETAINING (1) 296-42	
				RING,RETAINING (1) 296-2	
				RING,RETAINING (1) 296-4	
	XDFZZ	11862	6591956	SEAL KIT,OIL	1
				UOC:210	
				COVER,SUCTION,AIR C(1) 296-34	
				O-RING COMP SHELL (2) 296-33	
				PACKING,PREFORMED (2) 296-36	
				PACKING,PREFORMED (1) 196-29	
	PAFZZ	27462	6599114	PARTS KIT,SEAL REPL	1
				UOC:210	
				GASKET (1) 296-43	
				PREFORMED PACKING (1) 296-44	
				RETAINER,PACKING (1) 296-41	
				RING,RETAINING (1) 296-42	
	PAHZZ	11862	7817485	PARTS KIT,STEERING	1
				RING (1) 131-2	
	PAHZZ	11862	7817487	PARTS KIT,STEERING 1	
				GASKET (1) 131-5	
				PACKING,PREFORMED (1) 131-6	
	PAHZZ	11862	7826470	SEAL,PITMAN SHAFT	1
				RING-PIT SFF SNAP (1) 131-44	
				SEAL-PITMAN SHF (1) 131-47	
				SEAL,PITMAN SHAFT (1) 131-46	
				WASHER-SPECIAL (2) 131-45	
	PAHZZ	24617	7828565	SEAL KIT	1

(1) ITEM NO	(2) SMR CODE	(3) CAGEC	(4) PART NUMBER	(5) DESCRIPTION AND USABLE ON CODES (UOC)	(6) QTY
				PACKING,PREFORMED (1) 131-29	
				RING-STUB SHF RET (1) 131-33	
				SEAL-P/S GR STUB (1) 131-31	
				SEAL-P/S GR STUB (1) 131-32	
	PAHZZ	4F733	8623979	SWAP RING KIT	1
				RING,RETAINING (1) 73-31	
				RING,RETAINING (1) 74-4	
				RING,RETAINING (1) 77-25	
				RING,RETAINING (1) 74-19	
				RING,RETAINING (1) 78-10	
				RING,SNAP (1) 73-8	
				RING,SNAP (1) 73-22	
	PAHZZ	11862	8625905	GASKET AND SEAL SET	1
				GASKET (1) 79-5	
				GASKET (1) 77-21	
				GASKET (1) 77-23	
				GASKET (1) 77-6	
				GASKET (1) 77-9	
				GASKET (1) 77-17	
				GASKET (1) 76-13	
				OIL SEAL (1) 79-29	
				SEAL,NONMETALLIC RC(1) 79-7	
				SEAL,SPEEDOMETER (1) 70-9	
				SEAL,SPEEDOMETER (1) 75-3	
				WASHER,FLAT (6) 79-2	
	PAHZZ	11862	8629955	REPAIR PACKAGE	1
				DISK,CLUTCH (5) 73-10	
				DISK,CLUTCH (3) 73-37	
				GASKET (1) 77-21	
				GASKET (1) 77-23	
				GASKET (1) 77-6	
				GASKET (1) 77-11	
				PARTS KIT,CLUTCH,IN (1) 74-9	
				RETAINER,PACKING (4) 74-3	
				RETAINER,PACKING (2) 79-4	
				SEAL RING,METAL (1) 77-34	
				SEAL,KIT (1) 73-4	
				SEAL,KIT (1) 73-26	
	PAFZZ	11862	8629977	PARTS KIT,GASOLINE	1
				PIN,SPRING (1) 77-3	
				PIN,STRAIGHT,HEADED(2) 77-2	

END OF FIGURE

(1) ITEM NO	(2) SMR CODE	(3) CAGEC	(4) PART NUMBER	(5) DESCRIPTION AND USABLE ON CODES (UOC)	(6) QTY
				GROUP 95 GENERAL USE STANDARDIZED PARTS	
				GROUP 9501 BULK MATERIEL	
				FIG. BULK	
1	PAOZZ	81348	QQS741	ANGLE,STRUCTURAL	V
2	PAOZZ	81349	MILS20166	ANGLE,STRUCTURAL	V
3	XDOZZ	74410	RRC271BTY2CLDIA072	CHAIN	V
4	PAOZZ	11862	8919355	CONDUIT,NONMETALLIC	V
5	PAOZZ	11862	8919356	CONDUIT,NONMETALLIC	V
6	PAOZZ	96906	MS18029-13S-8	COVER,TERMINAL BOAR	V
7	PAOZZ	96906	MS18029-4S-8	COVER,TERMINAL BOAR	V
8	PAOZZ	11862	6263877	FELT,MECHANICAL,PRE	V
9	PAOZZ	11862	370389	FILLER BODY,GLASS C	V
10	PBFZZ	11862	6292996	FUSE LINK,ELECTRICA	V
11	PBOZZ	11862	6293923	FUSE LINK,ELECTRICA	V
12	PAOZZ	24234	319029	HOSE,AIR DUCT UOC:210	V
13	PAOZZ	11862	482420	HOSE,NONMETALLIC	V
14	PAOZZ	11862	7828506	HOSE,NONMETALLIC	V
15	PAOZZ	11862	9439104	HOSE,NONMETALLIC	V
16	PAOZZ	11862	9438383	HOSE,NONMETALLIC	V
17	PAOZZ	11862	9438257	HOSE,NONMETALLIC	V
18	PAOZZ	11862	9439162	HOSE,NONMETALLIC	V
19	PAOZZ	11862	9439046	HOSE,NONMETALLIC	V
20	PAOZZ	11862	9438381	HOSE,NONMETALLIC	V
21	PAOZZ	11862	1359744	HOSE,NONMETALLIC	V
22	PAOZZ	96906	MS521304B203R	HOSE,NONMETALLIC	V
23	PAOZZ	11862	9438373	HOSE,NONMETALLIC	V
24	PAOZZ	11862	9439402	HOSE,NONMETALLIC	V
25	PAOZZ	11862	9438315	HOSE,NONMETALLIC	V
26	XDOZZ	11862	9439274	HOSE,NONMETALLIC	V
27	PAOZZ	24617	9438124	HOSE,NONMETALLIC	V
28	PAOZZ	11862	3987364	HOSE,WIPER	V
29	PAOZZ	11862	12306178	INSULATION BLANKET	V
30	PAOZZ	11862	462233	INSLULATOR,FLOOR,VE	V
31	PAFZZ	11862	3696822	PIPE,HYDRAULIC BRAK	V
32	PAOZZ	81348	NNP530	PLYWOOD,CONSTRUCTIO	V
33	PAOZZ	11862	370390	RUBBER STRIP	V
34	PAOZZ	11862	363139	SEAL,FRONT DOOR	V
35	PAOZZ	96906	MS520390079	STRIP,METAL	QTY
36	PAFZZ	11862	15599621	STRUT,COVER BOW	V
37	PAOZZ	96906	MS27212-4-8	TERMINAL BOARD	V
38	PAFZZ	11862	1324714	TUBE,METALLIC	V
39	PAFZZ	11862	3750950	TUBE,METALLIC	V
40	PAFZZ	11862	603827	TUBE,METALLIC	V
41	XDOZZ	81349	MILT16343TYPE1	TUBE,METALLIC	V
42	PAFZZ	11862	465246	TUBE,METALLIC	V
43	PBFZZ	11862	6292997	WIRE,ELECTRICAL	V
44	PBFZZ	11862	6293702	WIRE,ELECTRICAL	V

END OF FIGURE

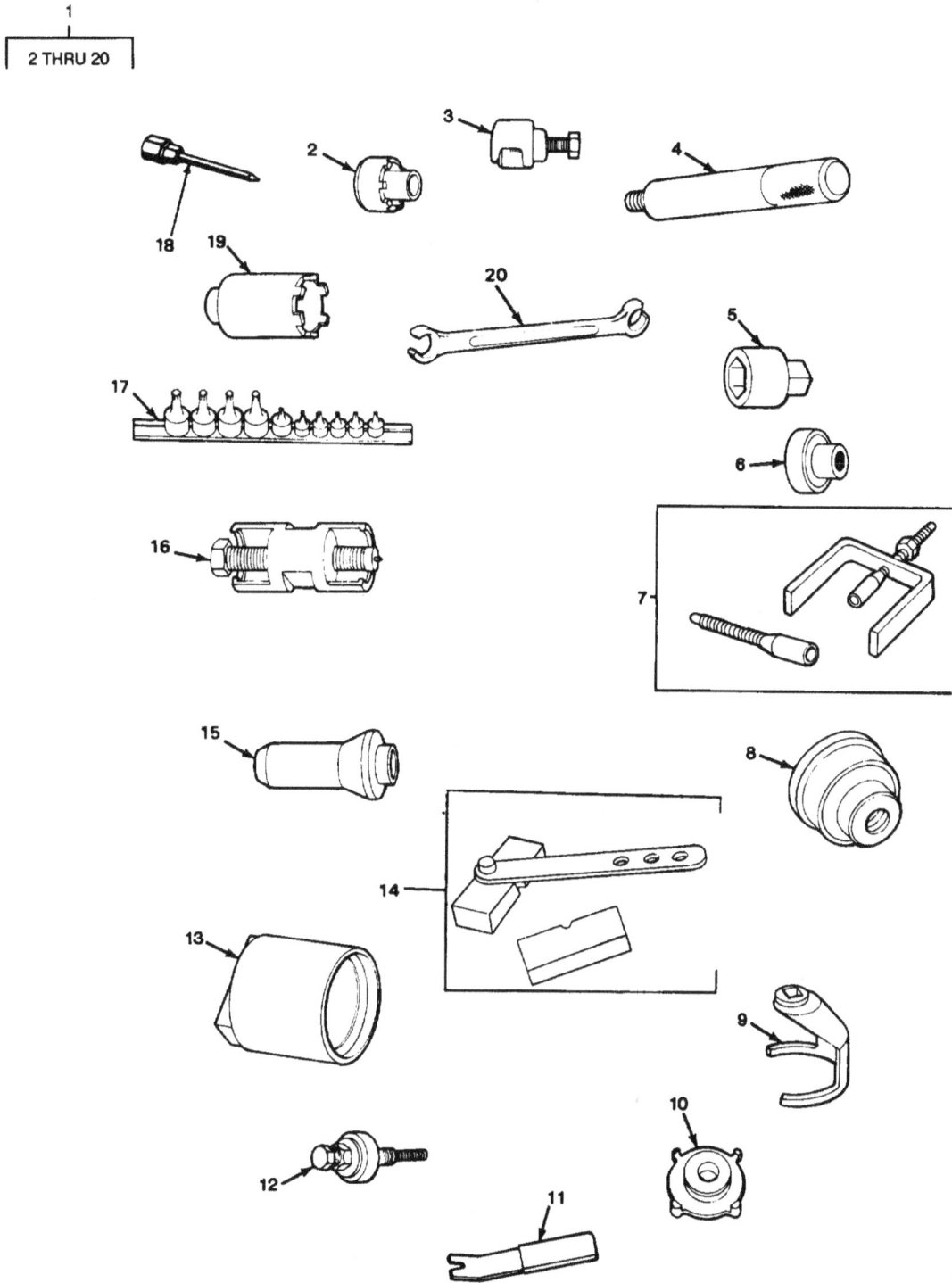

TA511108

FIGURE 301. SPECIAL TOOLS.

(1)	(2)	(3)	(4)	(5)	(6)
ITEM NO	SMR CODE	CAGEC	PART NUMBER	DESCRIPTION AND USABLE ON CODES (UOC)	QTY

GROUP 26 TOOLS AND TEST EQUIPMENT

GROUP 2604 SPECIAL TOOLS

FIG. 301 SPECIAL TOOLS

1	AOOOO	19207	12314542	TOOL KIT,GENERAL ME	
2	PEOZZ	25341	J-2222-C	.WRENCH,WHEEL BEARIN	
				UOC:194,208,210,230,231,252,254,256	
3	PEOZZ	25341	J-6632-01	.PULLER,MECHANICAL	
4	PEOZZ	33287	J-8092	.HANDLE,DRIVE	
5	PEOZZ	25341	J-21757-03	.SOCKET,OIL SWITCH	
6	PEOZZ	25341	J-23445-A	.INSERTER,BEARING AN	
				UOC:194,208,210,230,231,252,254,256	
7	PEOZZ	25341	J-23653-C	.LOCK PLATE,COMPRESS	
8	PEOZZ	25341	J-23690	.INSERTER,BEARING AN	
				UOC:209	
9	PEOZZ	25341	J-24187	.REMOVER,INSTALLER,F	
10	PEOZZ	25341	J-24426	.REMOVER,BEARING AND	
				UOC:194,208,210,230,231,252,254,256	
11	PEOZZ	25341	J-24595-C	.REMOVER,CLIP,RETAIN	
12	PEOZZ	30282	553	.INSTALLER,POWER STE	
13	PEOZZ	25341	J-26878-A	.WRENCH NUT	
				UOC:194,208,210,230,231,252,254,256	
14	PEOZZ	25341	J-33043	.BLOCK,VALVE GAGE	
15	PEOZZ	25341	J-29713	.DRIVE TOOL,SEAL	
				UOC:209	
16	PEOZZ	25341	J-25034-B	.REMOVER,PULLEY,WATE	
17	PEOZZ	33287	J-29843	.SOCKET WRENCH ATTAC	
18	PEOZZ	33627	J25512-2	.COUPLING,GREASE GUN	V
				UOC:194,208,209,210,230,231,252,254	
19	PEOZZ	25341	J-34616	.WRENCH,SPANNER	
19	PEOZZ	25341	J-6893-D	.WRENCH,HUB	
				UOC:194,209	
20	PEOZZ	25341	J-33124	.WRENCH,OPEN END	

END OF FIGURE

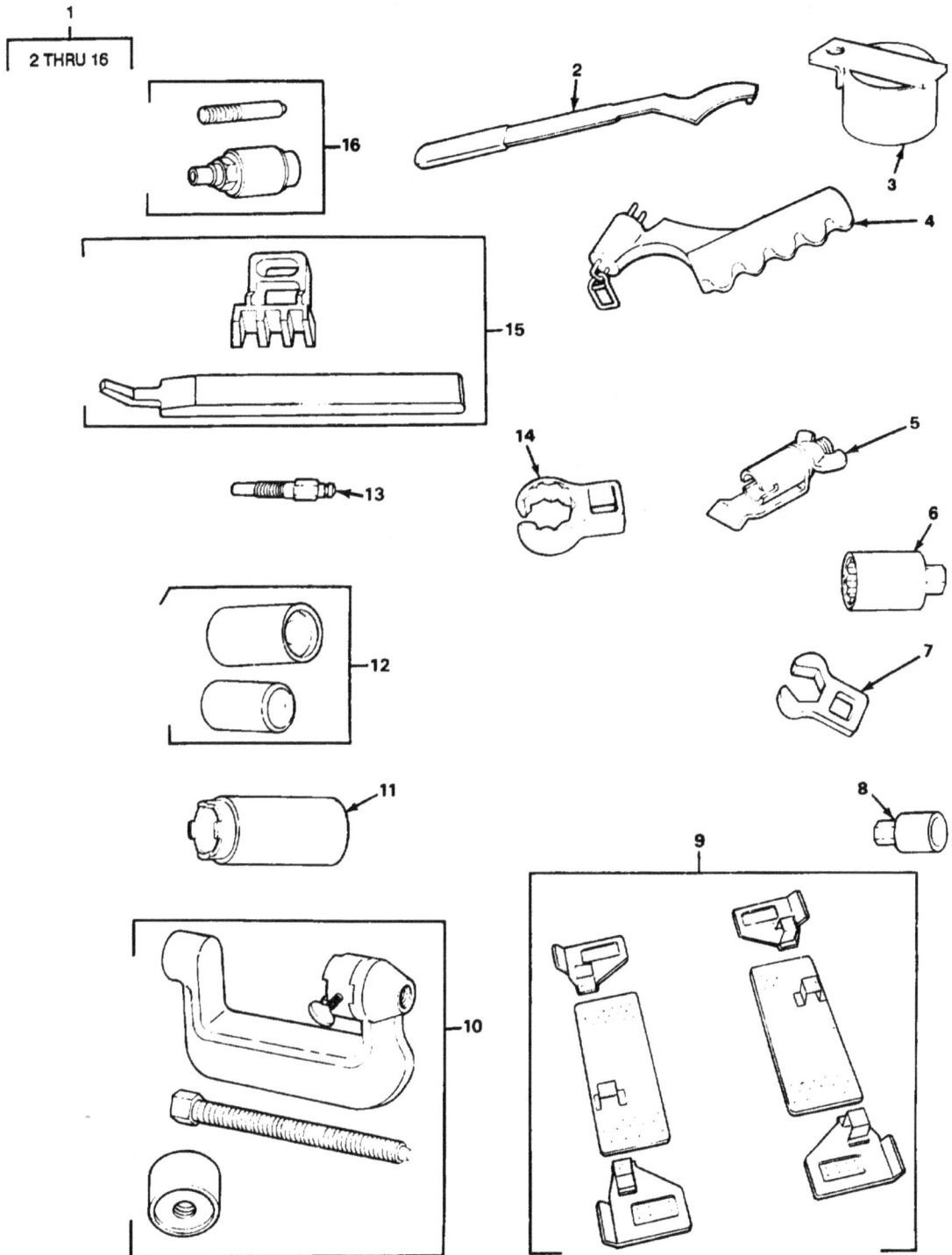

FIGURE 302. SPECIAL TOOLS.

TA511109

(1) ITEM NO	(2) SMR CODE	(3) CAGEC	(4) PART NUMBER	(5) DESCRIPTION AND USABLE ON CODES (UOC)	(6) QTY
				GROUP 2604 SPECIAL TOOLS	
				FIG. 302 SPECIAL TOOLS	
1	ADFFF	19207	12314543	TOOL KIT,GENERAL ME	
2	PEFZZ	25341	J-24429	.WRENCH NUT UOC:194,208,210,230,231,252,254,256	
3	PEFZZ	25341	J-26889	.REMOVER,ACCUMULATOR UOC:194,208,210,230,231,252,254,256	
4	PEFZZ	25341	J-26471	.INSTALLING TOOL,M	
5	PEFZZ	25341	J-29601	.GAGE ASSEMBLY,TIMIN	
6	PEFZZ	25341	J-29873	.ADAPTER,SOCKET WREN	
7	PEFZZ	25341	J-28402	.CROWFOOT ATTACHMENT	
8	PEFZZ	25341	J-26871-A	.ADAPTER,SOCKET WREN UOC:194,208,210,230,231,252,254,256	
9	PEFZZ	25341	J-29664-1	.COVER,MANIFOLD	
10	PEFZZ	25341	J-33122	.REMOVER AND REPLACE UOC:194.208,210,230,231,252,254,256	
11	PEFZZ	25341	J-23447	.SLEEVE,BALL STUD AC UOC:209	
12	PEFZZ	25341	J-23454-D	.ADAPTER SET,BALL JC UOC:194,208,210,230,231,252,254,256	
13	PEFZZ	33287	J-26999-30	.ADAPTER,CYLINDER CC	
14	PEFZZ	25341	J-29698-A	.CROWFOOT ATTACHMENT	
15	PEFZZ	25341	J-33154	.PACKER,SEAL	
16	PEFZZ	25341	J-23996	.INSTALLER AND REMOV UOC:194,208,210,230,231,252,254,256	

END OF FIGURE

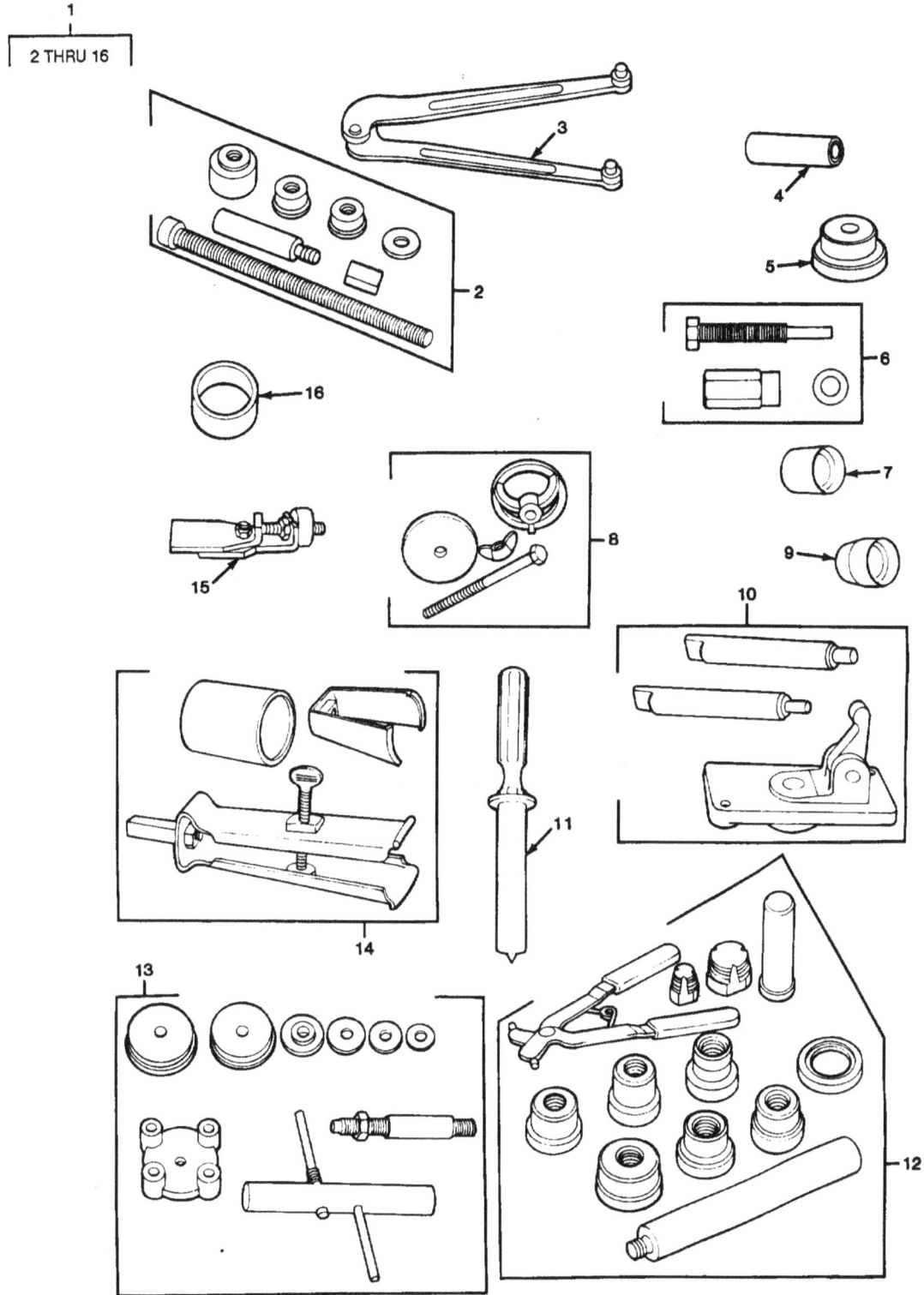

FIGURE 303. SPECIAL TOOLS.

TA511110

(1) ITEM NO	(2) SMR CODE	(3) CAGEC	(4) PART NUMBER	(5) DESCRIPTION AND USABLE ON CODES (UOC)	(6) QTY
				GROUP 2604 SPECIAL TOOLS	
				FIG. 303 SPECIAL TOOLS	
1	ADHHH	19207	12314544	TOOL KIT,GENERAL ME	
2	PEHZZ	33287	J-35178	.PULLER,MECHANICAL	
3	PEFZZ	25341	J-7624	.WRENCH,SPANNER	
4	PEHZZ	25341	J-7728	.INSTALLATION TOOL	
5	PEHZZ	25341	J-8107-2	.WASHER,SHOULDERED	
6	PEHZZ	25341	J-9480-B	.INSTALLER,HUB AND D UOC:210	
7	PEHZZ	25341	J-21362	.PROTECTOR,INNER SEA	
8	PEHZZ	25341	J-23327	.CLUTCH REBUILDER,MC	
9	PEHZZ	25341	J-21363	.PROTECTOR,INNER SEA	
10	PEHZZ	25341	J-21370	.GAGE,PIN SELECTOR	
11	PEFZZ	25341	J-21552	.PUNCH,HAND TOOL	
12	PEHZZ	25341	J-21465-1	.INSERTER,BEARING AN	
13	PEHZZ	25341	J-21777-500	.GAGE,PINION SETTING UOC:209	
14	PEHZZ	25341	J-21795-02	.HOLDING UNIT,GEAR	
15	PEHZZ	25341	J-22779	.GAGE,BEARING PRE-LC UOC:209	
16	PEHZZ	25341	J-21409	.PROTECTOR,SEAL	

END OF FIGURE

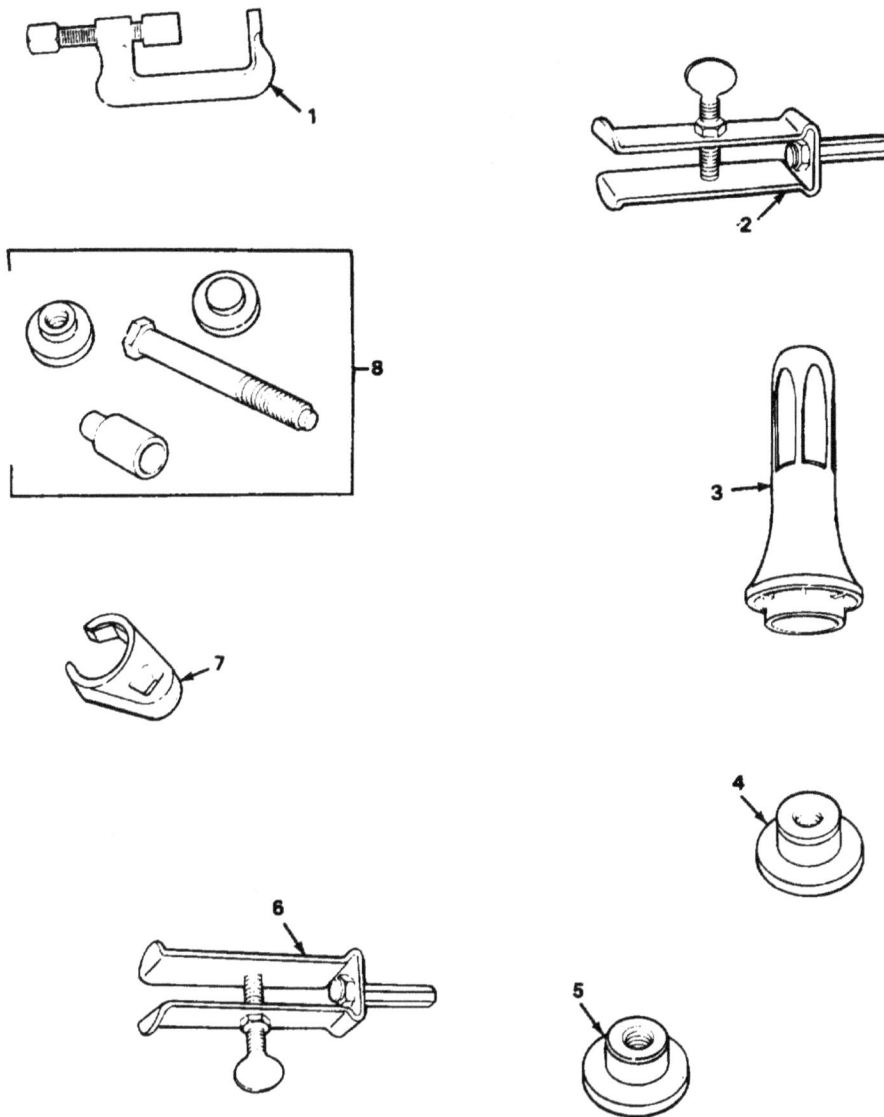

TA511111

FIGURE 304. SPECIAL TOOLS.

　　　　　　　　　　TM9-2320-289-34P

(1) (2) (3) (4) ITEM SMR NO CODE CAGEC	PART NUMBER	(5) DESCRIPTION AND USABLE ON CODES (UOC)	(6) QTY

GROUP 2604 SPECIAL TOOLS

FIG. 304 SPECIAL TOOLS

1	PEHZZ	25341	J-26252	.REMOVER,BUSHING
				UOC:209
2	PEHZZ	25341	J-26941	.PULLER,DRIVE SPROCK
3	PEHZZ	25341	J-29162	.DRIVE TOOL,SEAL
4	PEHZZ	25341	J-29167	.INSERTER,BEARING AN
5	PEHZZ	25341	J-29168	.INSERTER,BEARING AN
6	PEHZZ	25341	J-29369-1	.REMOVER,SPROCKET SU
7	PEHZZ	55719	FC18A	.CROWFOOT ATTACHMENT
8	PEHZZ	25341	J-34502	.COMPRESSOR,CLUTCH
				UOC:194,208,210,230,231,252,254,256

END OF FIGURE

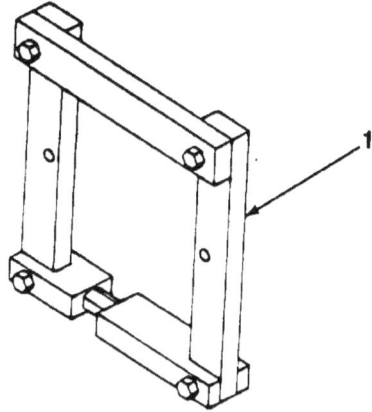

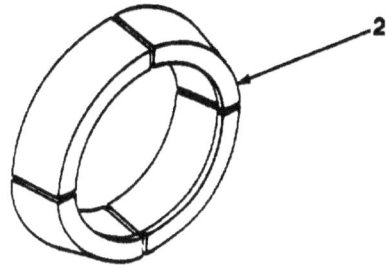

TA511112

FIGURE 305. SPECIAL TOOLS.

(1) (2) (3) (4) ITEM SMR PART NO CODE CAGEC NUMBER	(5) DESCRIPTION AND USABLE ON CODES (UOC)	(6) QTY
	GROUP 2604 SPECIAL TOOLS	
	FIG. 305 SPEICAL TOOLS	
1 PEHZZ 80604 D-167	SPREADER,REAR AXLE UOC:254,256	1
2 PEHZZ 80604 DD-914-95	REMOVER,BEARING AND UOC:254,256	1
	END OF FIGURE	

CROSS-REFERENCE INDEXES

NATIONAL STOCK NUMBER INDEX

STOCK NUMBER	FIG	ITEM	STOCK NUMBER	FIG	ITEM
5330-00-001-1984	77	6	5310-00-044-3342	128	20
5330-00-001-1996	79	29		148	26
5365-00-004-6407	98	19		149	25
5365-00-004-7210	KITS			150	1
3110-00-005-0873	73	7		151	1
	73	23		152	3
5365-00-007-3052	73	16	5315-00-044-3767	83	19
	73	18	4730-00-044-4587	85	4
2520-00-008-9987	73	19	4730-00-044-4655	2	13
5310-00-010-3028	128	18	4730-00-044-4789	31	1
4730-00-011-8538	251	4		108	4
	252	5		109	1
	252	15	5310-00-045-3296	214	10
	255	4			
	262	19		283	18
	284	6	5310-00-045-5001	155	2	293	21
	288	10	4730-00-050-4203	88	1
5310-00-013-1245	39	6		89	10
	87	12		90	11
	256	6		91	10
	256	6		92	10
	257	4		128	13
	259	2		129	4
	259	2		130	5
	270	8		144	3
	288	28		145	12
	297	8	5305-00-050-9237	214	16
5315-00-013-7238	129	5	5310-00-052-6454	115	2
5315-00-013-7258	143	4	5305-00-052-7472	229	6
	144	26	4730-00-053-0266	284	32
	145	32		288	7
4730-00-013-7398	80	1	5305-00-054-5651	283	11
5315-00-014-1195	8	32	5305-00-054-6650	283	6
4730-00-014-2432	16	20	5305-00-054-6654	231	16
	118	7		275	20
4730-00-014-2433	21	4	5305-00-054-6657	271	2
	22	7	5305-00-054-6670	283	15
5310-00-014-5850	245	29	3110-00-055-2100	83	75
	283	14	3110-00-056-9377	83	56
5305-00-014-9926	275	9	5340-00-057-3034	95	9
2910-00-025-3493	245	30		252	2
	262	13		255	21
	284	3		262	22
	288	8		284	8
5306-00-027-0722	277	19	5340-00-057-3037	21	2
5310-00-032-1814	37	61		22	5
5315-00-038-3059	78	3		22	6
	79	15	5340-00-057-3043	95	2
4730-00-043-3750	248	19		104	7
5310-00-044-3342	102	15		105	3

CROSS-REFERENCE INDEXES

NATIONAL STOCK NUMBER INDEX

STOCK NUMBER	FIG	ITEM
5340-00-057-3043	268	2
	269	10
5340-00-057-3052	28	10
	60	5
	64	3
	268	9
5315-00-057-5541	223	22
5305-00-057-9608	228	12
4730-00-058-7558	256	10
	258	1
	259	1
	287	10
5310-00-061-0004	246	3
5365-00-061-2395	98	23
	111	4
5365-00-061-6844	98	23
	111	4
5365-00-061-6845	98	23
	111	5
5365-00-061-6847	98	11
5365-00-061-6849	98	11
5320-00-061-9648	220	2
	221	6
	224	3
5320-00-061-9662	215	2
	216	14
5310-00-062-6828	101	15
5365-00-062-7200	98	11
5365-00-062-7201	98	23
	111	3
5305-00-068-0501	22	19
	77	15
	273	11
5305-00-068-0502	207	3
5305-00-068-0508	9	14
	45	3
	57	6
	61	10
	87	25
	182	4
	207	13
	269	13
	279	5
	282	7
	297	4
5305-00-068-0509	9	
	220	12
	298	11
	300	14
5305-00-068-0510	1	9

STOCK NUMBER	FIG	ITEM
5305-00-068-0510	49	20
	53	3
	140	22
	145	29
	148	2
	149	2
	211	3
	212	3
	215	12
	216	4
5305-00-068-0511	81	4
	81	7
	82	3
	82	5
	82	13
	148	5
	253	30
5310-00-068-5285	128	21
	172	4
5305-00-068-7837	288	32
5305-00-071-1318	272	1
5305-00-071-1784	71	9
5305-00-071-1787	89	13
	90	13
	91	13
	92	13
	110	5
5305-00-071-1788	135	2
	137	6
	145	37
5305-00-071-2055	156	5
5305-00-071-2058	97	12
	109	3
	145	33
5305-00-071-2066	35	7
5305-00-071-2067	135	4
	136	9
	138	2
	145	8
	253	29
5305-00-071-2069	144	17
5305-00-071-2070	187	5
	217	10
5305-00-071-2071	279	8
5305-00-071-2073	96	12
		20
	145	39
5305-00-071-2075	135	8
	136	7
	144	23
	145	9

CROSS-REFERENCE INDEXES

NATIONAL STOCK NUMBER INDEX

STOCK NUMBER	FIG	ITEM	STOCK NUMBER	FIG	ITEM
5305-00-071-2075	153	5	4730-00-087-7273	263	14
5305-00-071-2077	153	3	5310-00-087-7493	1	21
	172	5		151	12
5305-00-071-2081	187	6		215	11
5305-00-071-2083	160	4		216	5
	253	16		223	26
5305-00-071-2087	271	13	4030-00-088-1881	143	6
5305-00-071-2506	223	13	5330-00-089-0978	246	26
	264	7	5330-00-089-0998	246	27
	282	4	3110-00-100-0251	111	16
	298	17	3110-00-100-5805	101	11
5305-00-071-2510	271	14	5310-00-103-2953	247	7
	273	1		293	9
	274	7	5330-00-107-3925	8	18
	275	15		254	8
5305-00-071-2511	271	11	3110-00-108-9247	36	35
5930-00-073-0390	54	4	5330-00-110-8437	6	8
5975-00-074-2072	293	43	5360-00-113-9490	115	17
2540-00-078-6633	144	1		116	19
	144	12	5305-00-115-9406	263	5
	145	10	5305-00-115-9436	248	14
	145	19	5305-00-115-9526	23	17
5310-00-080-6004	81	8		27	4
	82	11	5305-00-115-9934	47	7
	83	67	6105-00-116-5124	249	7
	88	13	5360-00-123-0137	115	15
	145	28		116	17
	148	1	2530-00-125-2769	115	14
	149	1		116	16
	232	1	4730-00-132-4625	21	6
	261	18		22	9
	273	7	3110-00-135-0983	99	4
	273	19		112	9
	274	3	5310-00-137-3018	98	13
5310-00-081-4219	19	32		98	17
	22	35	4720-00-139-4105	263	32
	87	14	4730-00-142-2177	255	12
	231	5		262	16
	253	23		284	4
	256	1		288	14
	259	7	3110-00-142-4387	111	15
	259	7	6240-00-144-4693	42	21
	277	5		43	1
	280	5		43	5
	280	5		51	8
	281	4		52	5
	281	4	5310-00-144-8453	283	16
5365-00-081-5194	84	65	5305-00-145-0828	248	37
5310-00-082-1404	46	2	5305-00-146-2524	19	24
2940-00-082-6034	8	13		41	7

CROSS-REFERENCE INDEXES

NATIONAL STOCK NUMBER INDEX

STOCK NUMBER	FIG	ITEM	STOCK NUMBER	FIG	ITEM	FIG	ITEM
5305-00-146-2524	41	8	5310-00-209-0786	207	12		
	49	12				298	4
	61	1	5310-00-209-0788	246	6		
	69	9	5310-00-209-0965	223	12		
	243	2				252	22
5340-00-150-1658	252	24	5310-00-209-1218	15	36		
5365-00-152-5273	296	2	5310-0-209-2811	130	1		
3110-00-155-6152	83	4	2540-00-216-5722	246	22		
3110-00-155-6735	84	66	4520-00-217-5782	246	38		
2805-00-155-7266	7	22	4320-00-217-5827	263	30		
5306-00-157-8315	96	14	5305-00-217-9183	290	5		
5970-00-160-4001	37	10	4730-00-221-2140	253	14		
5970-00-160-4066	37	14	5120-00-224-7288	302	7		
2520-00-163-0709	73	15	5306-00-225-2864	89	14		
5310-00-166-8567	131	40				90	14
5310-00-167-0680	154	9	5305-00-225-3843	9	3		
	155	8				217	2
5310-00-167-0721	48	6				217	15
	108	6				229	8
2520-00-172-1947	73	21				232	3
5305-00-173-0165	131	9				249	15
4820-00-174-0315	9	6				252	20
5325-00-174-5314	246	5				256	8
	282	1				256	8
5306-00-177-5707	277	22				257	2
5305-00-182-9584	123	6				259	4
5120-00-184-8397	304	7				259	4
5330-00-185-0075	299	5				288	30
3110-00-185-6305	84	13	5306-00-226-4822	108	7		
4730-00-187-4191	263	44				213	16
4730-00-187-4210	97	16	5306-00-226-4824	70	7		
5120-00-189-7898	302	14				76	12
5310-00-190-0752	15	27				77	4
4730-00-196-1991	261	11				119	5
3110-00-198-0492	83	80				252	7
2540-00-198-2478	KITS					255	20
2540-00-200-4249	KITS		5306-00-226-4825	22	36		
5365-00-200-7377	203	22				69	10
2540-00-201-3474	KITS					84	46
2910-00-203-3322	245	32				134	1
5305-00-207-3984	15	91				262	5
5310-00-208-1918	228	6				284	26
	288	12	5306-00-226-4826	19	30		
5310-00-209-0786	45	4				22	38
	47	10				58	3
	52	2				114	4
	53	2	5306-00-226-4827	27	14		
	61	2				79	20
	62	3				217	19
	65	22				237	6

CROSS-REFERENCE INDEXES

NATIONAL STOCK NUMBER INDEX

STOCK NUMBER	FIG	ITEM	STOCK NUMBER	FIG	ITEM	FIG	ITEM
5306-00-226-4827	238	8	5325-00-263-6651	300	13		
	280	1	5310-00-264-1930	125	24		
	280	1	5330-00-265-1089	245	31		
	281	1	4730-00-266-0535	251	6		
	281	1				252	25
5306-00-226-4828	253	15	4730-00-266-0536	251	15		
	287	12				255	18
5306-00-226-4829	214	18	5305-00-269-3219	253	10		
5306-00-226-4830	26	4	5305-00-269-3233	57	25		
	27	2				253	13
	277	4	5305-00-269-3236	116	7		
5306-00-226-4831	79	18	5305-00-272-3533	288	5		
5306-00-226-4832	79	19	5325-00-276-4993	247	1		
	231	1	4130-00-277-3486	296	3		
5306-00-226-4833	118	14	4730-00-277-6339	263	36		
	211	9	4730-00-277-8269	9	5		
	212	9	4730-00-278-4824	252	11		
5305-00-227-1543	44	10	4730-00-278-8886	118	12		
	46	1	5325-00-279-1235	247	2		
3110-00-227-4431	83	45	5325-00-279-1248	255	26		
3110-00-227-4667	111	9	5365-00-282-1633	234	25		
2530-00-228-6992	115	25	5340-00-282-7509	60	4		
5360-00-229-5312	115	24				264	15
4720-00-230-6523	BULK	21	5340-00-282-7539	255	23		
5340-00-231-1964	2				17		
5930-00-234-1390	264	12	5310-00-285-5112	248	35		
5340-00-237-7779	289	23	5340-00-285-8868	167	18		
5315-00-243-1169	144	15	4730-00-288-9390	9	4		
	145	22				119	12
5315-00-243-1170	287	8	4730-00-288-9440	9	8		
5310-00-245-3424	22	32	4730-00-288-9497	248	17		
	114	13	5310-00-291-4619	40	18		
5305-00-245-4144	277	11	5325-00-291-9366	267	14		
4810-00-248-1635	246	8	5315-00-298-1481	102	5		
4730-00-249-3885	261	10				102	20
5305-00-249-5278	263	7				128	10
5305-00-250-5613	15	5	5360-00-310-4493	116	25		
5905-00-251-7145	246	11	5310-00-316-6513	1	14		
6240-00-252-7138	290	2				26	15
5310-00-252-8748	44	3				27	25
	46	8				81	9
5305-00-253-5626	144	16				82	12
	145	23				139	11
3120-00-255-5697	73	33				140	4
5365-00-257-6035	99	2				140	9
	112	10				276	15
5325-00-263-6632	283	9				277	14
5325-00-263-6648	293	20	4730-00-317-4231	261	8	289	16
	294	7	4730-00-319-0454	263	19		

CROSS-REFERENCE INDEXES

NATIONAL STOCK NUMBER INDEX

STOCK NUMBER	FIG	ITEM	STOCK NUMBER	FIG	ITEM
5360-00-327-5879	248	31	5310-00-429-2686	37	60
5340-00-329-4420	298	2	5310-00-429-3110	37	27
2805-00-336-1716	8	27	5310-00-429-3135	37	2
5325-00-337-6636	267	2	5310-00-429-3156	37	66
5930-00-345-5455	246	21	5310-00-432-3959	37	34
5310-00-350-2655	50	7	5305-00-432-4163	245	5
5330-00-352-0327	299	3		247	8
3030-00-357-5506	35	11	5305-00-432-4171	255	7
5330-00-360-7881	181	3	5305-00-432-4201	176	7
2530-00-363-4389	121	3	5305-00-432-7953	51	3
	121	9	5305-00-432-8220	179	16
5310-00-380-1514	118	8	6250-00-433-5946	42	19
4210-00-383-7127	238	3	6210-00-438-4745	283	7
5360-00-392-3453	115	22	5305-00-446-9901	269	12
	116	24	5340-00-449-6408	2	18
3120-00-393-4067	15	76	5305-00-450-0463	37	40
5315-00-393-5865	3	3	5305-00-450-5937	40	42
	7	9	5315-00-450-9163	127	73
4710-00-395-5144	BULK	31	5330-00-451-0118	83	71
5325-00-397-5962	40	38	5310-00-451-0631	37	33
5310-00-400-8585	15	4	5305-00-451-1623	37	17
3110-00-403-1488	131	48	3110-00-451-4601	127	53
5305-00-403-5130	246	10	2540-00-454-0665	263	23
5940-00-405-8976	BULK	7	5340-00-455-5899	208	2
3110-00-406-9608	100	9	4730-00-459-6077	15	110
	110	24	4730-00-460-6725	263	17
	112	7	5305-00-471-0373	36	36
5306-00-407-3737	40	19	4730-00-471-3102	252	12
5310-00-407-9566	19	26	5310-00-472-3214	271	10
	22	21		273	2
	26	3		274	6
	27	8	5330-00-472-6954	263	37
	66	8	5930-00-475-5537	263	1
	69	13	2590-00-476-5459	169	8
	89	15		169	17
	90	15	5305-00-483-0554	50	13
	128	3		193	5
	184	13	5340-00-486-1765	95	4
	211	10	5310-00-490-4639	167	3
	212	10		191	10
	238	7	2910-00-493-2138	13	12
	253	24	2530-00-494-8165	115	14
	262	8		116	16
	284	23	5310-00-498-2381	84	1
5340-00-411-4508	290	4		102	3
1660-00-413-0756	299	6		102	23
	299	19	5365-00-514-1277	79	27
4730-00-415-3172	262	12	5310-00-514-6674	27	15
4710-00-420-4759	BULK	38		44	4
5365-00-423-2811	84	41		46	7

CROSS-REFERENCE INDEXES

NATIONAL STOCK NUMBER INDEX

STOCK NUMBER	FIG	ITEM	STOCK NUMBER	FIG	ITEM	FIG	ITEM
5310-00-514-6674	57	9	5310-00-582-5965	282	8		
	58	1				293	39
	59	2				294	3
	60	10				294	22
	63	3	5310-00-584-5272	128	19		
	233	3				253	19
5310-00-516-2701	37	1	5310-00-584-7889	84	45		
9515-00-516-5756	BULK	35	5310-00-596-5152	36	34		
5310-00-528-7638	284	19	5310-00-596-6897	232	16		
5305-00-533-5542	273	21				261	17
	274	18	5310-00-596-7691	249	1		
	277	2				298	6
4030-00-542-3183	144	25				300	6
	145	31	5310-00-596-7693	246	20		
5305-00-543-2419	1			20		248	23
	140	40				249	3
	142	31	5340-00-604-0566	98	16		
5305-00-543-2866	289	17	5310-00-606-6608	40	14		
5305-00-543-4302	40	2	3110-00-606-9576	111	10		
5930-00-548-5640	214	5	5305-00-614-0248	263	21		
5940-00-549-6581	57	39	5310-00-616-1739	37	67		
5940-00-549-6583	57	39	5320-00-616-4344	229	1		
5310-00-550-3503	57	12	5320-00-616-4346	211	13		
	65	21				212	15
5310-00-550-3715	283	12				213	1
5925-00-553-2274	247	18				215	31
2640-00-555-2829	124	29				216	31
2640-00-555-2840	125	1				218	3
5305-00-557-6612	72	3				219	3
2520-00-557-6619	70	6				220	16
4520-00-567-1886	263	18				221	10
5365-00-571-6816	83	25				222	9
5310-00-579-0079	246	15				227	4
3110-00-580-3708	111	6				245	38
3110-00-580-3709	111	7	5315-00-616-5523	37	35		
3110-00-580-3843	83	44	6240-00-617-0991	50	4		
4730-00-580-6738	24	1	6685-00-618-1822	253	31		
5320-00-582-3521	211	2	5530-00-618-6955	BULK	32		
	212	1	4730-00-619-9362	251	7		
	213	8				255	6
	215	18	5310-00-627-6128	82	6		
	216	26				84	54
5310-00-582-5765	15	37	9330-00-629-8239	299	14		
5310-00-582-5965	9				13		
			5930-00-636-1584	214	6		
	22	18	5310-00-637-4000	293	5		
	45	9	5320-00-637-5014	229	4		
	48	4	5310-00-637-9541	23	18		
	182	6				27	5
	262	3				48	21
	268	10				50	8

CROSS-REFERENCE INDEXES

NATIONAL STOCK NUMBER INDEX

STOCK NUMBER	FIG	ITEM	STOCK NUMBER	FIG	ITEM
5310-00-637-9541	57	26	5340-00-702-2848	280	10
	81	5		281	3
	82	2		281	3
	116	2	5305-00-709-8284	148	12
	128	6		149	11
	140	21	5305-00-709-8517	223	7
	140	37	2910-00-710-6054	251	11
	142	29		255	10
	145	27	5305-00-719-5184	115	4
	148	4	5305-00-719-5219	101	4
	149	4	5305-00-719-5239	110	20
	154	6	5305-00-719-5240	144	11
	155	4		145	25
	157	5	5365-00-720-8064	178	10
	215	14	5305-00-721-5492	84	30
	216	3	5365-00-721-6476	102	6
	223	23		102	22
	253	9	5365-00-721-6876	125	5
	294	19		125	9
3110-00-647-1100	83	11	5365-00-721-7680	78	10
5930-00-655-1512	247	20	5310-00-721-7809	40	33
5930-00-655-1514	247	19	2520-00-722-7074	89	2
	283	4		90	2
5310-00-655-9370	110	6		91	2
	137	5		92	2
5320-00-660-0821	215	9	5305-00-724-6798	263	27
	216	9	3110-00-724-7884	263	34
5120-00-677-2259	301	4	5305-00-725-2317	13	7
5930-00-679-5925	246	13	5305-00-732-0511	253	21
5365-00-682-1762	77	25	5310-00-732-0558	27	6
	115	13		49	22
	116	15		50	9
5310-00-682-5930	40	3		53	5
5306-00-685-2523	98	1		140	19
5340-00-685-5899	288	37		140	38
5310-00-685-6855	249	11		145	26
5306-00-685-7790	276	19		148	3
5305-00-688-2111	71	7		149	3
	119	15		154	5
	142	24		155	3
	149	5		157	4
6210-00-688-5088	247	15		255	29
	283	8	5310-00-732-0559	76	1
3110-00-689-4076	84	18		151	11
3110-00-690-8923	125	22		223	24
5310-00-696-5172	65	4	5365-00-734-2494	84	41
5340-00-697-4703	71	2	5365-00-734-2495	84	41
5340-00-700-1423	114	11	5365-00-734-2496	84	41
	125	18	5340-00-734-2649	84	44
4730-00-701-7737	248	28	5365-00-735-1129	84	10

CROSS-REFERENCE INDEXES

NATIONAL STOCK NUMBER INDEX

STOCK NUMBER	FIG	ITEM	STOCK NUMBER	FIG	ITEM
5315-00-737-0134	289	21	5310-00-775-5139	37	6
5330-00-737-3727	84	11	5310-00-775-5180	37	64
2805-00-752-0158	7	24	5310-00-775-5182	37	63
2540-00-752-4078	288	26	5330-00-777-1454	299	18
5306-00-753-6996	289	11	5310-00-781-4787	37	19
5330-00-753-8036	BULK	33	5305-00-782-9489	273	18
5310-00-754-2005	300	15		273	30
2040-00-754-4176	249	10		274	10
3110-00-756-2022	83	74	5310-00-785-1762	116	1
3110-00-756-4535	107	4		128	5
5330-00-757-1680	15	19	4820-00-785-8153	252	10
5310-00-761-6882	45	8	2910-00-788-0986	15	6
	52	1	5310-00-792-3617	130	8
	53	1	5320-00-801-1548	246	18
	62	2	5365-00-803-7313	40	13
	165	1	5365-00-803-7317	124	7
	288	35	5365-00-804-2027	15	78
5330-00-763-0213	107	3	5365-00-804-9666	40	15
5310-00-763-8894	115	1	5325-00-807-0580	264	5
5310-00-763-8905	148	19		265	3
	146	18	5340-00-809-1490	20	17
	150	11		21	11
	151	23		119	21
	152	2	5340-00-809-1494	59	8
	155	1	5340-00-809-1500	293	38
5310-00-763-8913	154	8	5310-00-809-3078	114	20
	155	7		165	2
5340-00-764-7052	45	10		217	18
4730-00-765-9103	252	13	5310-00-809-3079	135	12
5310-00-768-0318	135	6		144	22
	136	3		148	27
	137	8		149	26
	144	27		150	3
	145	2		151	2
	146	1		152	4
	148	7		179	20
	153	1	5310-00-809-4058	39	7
	154	4		57	17
	160	9		61	11
	172	9		264	4
	174	4		270	6
	181	10		284	15
	186	7		288	2
	217	5		288	27
	253	20		297	9
5935-00-768-7042	263	26	5310-00-809-4061	18	6
4730-00-768-8880	261	13		19	38
5360-00-771-7066	270	9		69	17
5935-00-773-1428	61	9		140	39

CROSS-REFERENCE INDEXES

NATIONAL STOCK NUMBER INDEX

STOCK NUMBER	FIG	ITEM
5310-00-809-4061	142	30
	145	36
	223	10
	253	8
5310-00-809-4085	1	
	26	16
	27	26
	33	7
	34	7
	127	57
	127	79
	139	12
	140	3
	140	45
	142	37
	148	6
	179	18
	276	16
	277	15
5310-00-809-5997	217	9
	276	17
	277	21
5310-00-809-5998	135	5
	136	5
	137	10
	144	18
	145	38
	147	8
	154	2
	160	10
	172	8
	174	3
	253	17
5310-00-809-8546	262	17
	284	5
	288	15
5310-00-809-8836	98	18
	111	1
5310-00-814-0673	231	7
	256	2
	259	8
	259	11
	280	6
	280	6
	281	5
	281	5
5315-00-814-3530	77	3
	244	8
5315-00-814-3531	223	29
5315-00-816-1794	87	17

STOCK NUMBER	FIG	ITEM	FIG	ITEM
5315-00-816-1794	88	12	128	16
5306-00-819-3038	15	21		
5305-00-821-3869	119	16	157	8
		10		
5320-00-822-6257	218	9	219	2
5310-00-823-8803	148	20	149	19
			150	10
			151	24
			152	1
5310-00-823-8804	57	5	217	3
			217	14
			252	21
5305-00-823-9139	289	19		
4730-00-826-4268	20	9	21	9
			22	1
			23	21
5320-00-828-1284	218	13	219	15
5315-00-828-5487	8	29		
5330-00-830-1745	13	11		
5310-00-830-7825	15	26		
5310-00-834-7606	101	3	110	19
			115	3
			136	4
			137	9
			144	28
			145	3
			148	8
			149	7
			153	2
			181	9
			187	4
			203	17
			217	6
5310-00-835-2036	87	18	88	9
			223	11
6240-00-836-2079	211	5	212	5
5315-00-839-2325	288	6		
5315-00-839-5820	84	23	223	4
5315-00-839-5822	7	11		
5310-00-840-6222	148	18		

CROSS-REFERENCE INDEXES

NATIONAL STOCK NUMBER INDEX

STOCK NUMBER	FIG	ITEM	STOCK NUMBER	FIG	ITEM	FIG	ITEM
5310-00-840-6222	149	17	5365-00-884-5640	98	19		
2920-00-841-3254	40	5	2520-00-884-8430	98	30		
5310-00-842-1490	128	17	5360-00-887-1536	15	15		
5315-00-842-3044	87	13	5315-00-887-1539	15	84		
	289	18	5305-00-887-1547	15	95		
5310-00-842-7783	128	9	5305-00-887-1564	15	100		
5315-00-843-7986	36	28	3120-00-888-6630	40	16		
5330-00-843-9235	KITS		6240-00-889-1799	53	11		
4730-00-844-5721	80	7				213	4
4820-00-844-6744	28	16	5310-00-889-2528	207	15		
5320-00-845-9501	213	14	5306-00-889-2943	275	1		
	215	3	5305-00-889-3001	37	47		
	216	15	5940-00-890-2831	44	1		
	220	17				46	10
	227	9	5310-00-891-1709	217	17		
	288	23				238	10
5315-00-846-0126	144	7	5340-00-892-7413	71	4		
	145	16	5310-00-896-0903	18	5		
5330-00-848-0972	131	24				19	37
5330-00-848-4439	133	23				60	8
5310-00-849-6882	144	6				137	4
	145	15				140	42
6240-00-850-4280	41	17				142	38
	226	3				145	4
	296	5	2910-00-897-2465	15	86		
5310-00-851-2674	217	4	5315-00-899-4119	245	35		
	217	12	5365-00-900-0982	178	8		
	220	9	3110-00-900-2560	77	8		
	252	23	5360-00-900-2564	15	105		
3110-00-854-1504	100	3	4730-00-900-3296	245	12		
	112	5	5305-00-901-3110	276	3		
5305-00-855-0960	36	15	5305-00-901-3144	277	17		
5305-00-855-3598	40	30	3110-00-902-1690	84	34		
3110-00-858-0988	110	22	5310-00-903-5966	207	5		
5305-00-866-0934	234	5	5310-00-905-4600	128	4		
5310-00-877-4956	15	44	4730-00-908-3194	10	2		
5310-00-878-7196	250	12				19	40
5310-00-880-7744	142	28				20	27
	211	11				30	17
	212	11				127	50
	253	25				244	1
5310-00-880-7746	48	1				261	1
5340-00-881-5303	12	11				280	8
	22	16				281	7
	118	2	4730-00-908-3195	23	23		
	119	21				28	8
	119	24				86	2
2520-00-884-5635	98	8				95	6
	98	29				104	5
5365-00-884-5639	98	19				105	5

CROSS-REFERENCE INDEXES

NATIONAL STOCK NUMBER INDEX

STOCK NUMBER	FIG	ITEM	STOCK NUMBER	FIG	ITEM	FIG	ITEM
4730-00-908-3195	263	31	5310-00-934-9747	44	4		
4730-00-908-6292	19	10				46	5
	284	13				246	7
4730-00-908-6293	245	16	5310-00-934-9748	283	13		
	256	7	5310-00-934-9751	15	120		
	256	7	5310-00-934-9757	15	29		
	257	3				36	17
	259	3				65	23
	259	3				246	4
	285	5				248	24
	287	1	5310-00-934-9758	40	32		
4730-00-908-6294	245	10				49	3
	284	20				59	9
	288	19				214	9
	288	29				283	17
4730-00-909-8627	19	39	5310-00-934-9764	48	2		
	20	22	5305-00-935-7506	273	29		
	30	1	5330-00-935-9136	8	8		
5340-00-914-1000	262	4				254	2
4030-00-916-2141	144	14	2530-00-937-1275	131	11		
	145	21	5310-00-938-8387	87	26		
5340-00-916-6539	232	9				264	6
	232	14				282	6
	232	19	6240-00-944-1264	42	18		
6240-00-924-7526	51	7				42	20
	53	10				50	17
5940-00-926-8034	BULK	6				63	1
2520-00-928-6150	98	15				68	1
3110-00-929-8365	84	48	5315-00-945-8441	40	34		
4720-00-930-2231	BULK	22	5360-00-947-1184	40	4		
5310-00-931-8167	19	25	5310-00-949-9299	36	33		
	22	20	3110-00-950-4236	234	13		
	26	2	5940-00-950-7783	BULK	37		
	27	9	5305-00-958-5477	249	12		
	32	5	5340-00-958-8457	280	3		
	69	12	5310-00-959-4675	53	4		
	89	16				142	22
	90	16	5310-00-959-4679	58	2		
	114	19				114	5
	119	10				119	6
	134	11				134	12
	173	5				237	5
	175	4				280	2
	184	14				280	2
	262	9				281	2
	284	22				281	2
5310-00-933-4310	289	12				287	13
5310-00-933-8121	207	2	5340-00-960-9340	131	3		
5310-00-933-8123	108	2	5340-00-960-9354	131	8		
	154	3	5330-00-960-9355	131	5		

CROSS-REFERENCE INDEXES

NATIONAL STOCK NUMBER INDEX

STOCK NUMBER	FIG	ITEM	STOCK NUMBER	FIG	ITEM
2530-00-960-9363	KITS		5310-01-012-8962	275	4
6220-00-961-0783	290	1		275	17
3110-00-962-3263	36	25	5930-01-014-0187	49	15
3030-00-967-4898	35	4	5977-01-018-2742	36	7
5310-00-975-2075	102	8	5310-01-018-5634	250	3
	102	25	5305-01-019-1884	124	19
5930-00-978-8805	283	5	5330-01-020-9319	108	10
5320-00-982-3815	215	24	3110-01-020-9786	110	21
	216	19	5310-01-021-9027	124	1
	225	6	5935-01-022-2377	220	18
	228	7	5310-01-023-8927	37	18
5310-00-982-4908	83	68	2520-01-024-0279	89	5
5945-00-983-4374	49	29		90	6
5940-00-983-6082	263	3		91	5
5305-00-984-4983	246	12		92	7
	269	8		93	3
5305-00-984-6189	246	35		93	3
	249	4	5310-01-024-3109	37	62
5305-00-984-6191	38	1	2540-01-025-0433	224	9
	213	18	5315-01-025-0930	214	11
	246	19	5330-01-025-4212	79	12
5305-00-984-6193	68	9	5820-01-026-0983	207	4
	269	1	5975-01-027-0253	207	6
5305-00-984-6195	245	14	5310-01-027-1353	263	9
5305-00-984-6210	40	26	5340-01-027-2628	263	11
	283	20	3110-01-027-4475	124	17
5305-00-984-6214	45	12	2540-01-028-0574	216	17
5340-00-989-1771	268	8	5930-01-028-1949	227	10
	269	15	5340-01-028-9063	218	6
5305-00-990-6444	249	2		219	6
5945-00-992-5415	49	31		220	7
5305-00-993-1848	37	20		221	3
5320-00-994-7076	56	9		224	5
2520-00-997-9818	40	9		225	4
5310-00-998-0608	102	4	3110-01-030-8475	110	10
	102	19		124	15
5930-00-998-9211	49	18	6140-01-031-6882	56	11
2510-00-999-9856	271	4	2530-01-033-1830	131	7
	273	26	2530-01-033-1855	KITS	
2520-01-004-2057	97	10	2530-01-034-1715	KITS	
8145-01-005-2994	253	3	4730-01-034-8228	96	15
2520-01-007-0804	78	13	6130-01-035-6412	213	22
2540-01-008-1501	246	34	2520-01-035-6670	91	11
5330-01-008-6527	248	6	5330-01-036-3861	110	25
6150-01-010-6558	247	4	5340-01-036-7665	231	11
5310-01-012-8962	256	12	5330-01-037-0663	50	5
	258	2	5365-01-037-6813	249	9
	271	7	5340-01-038-3428	231	10
	273	4	2520-01-038-7283	106	1
	274	1	5310-01-038-8500	131	15

CROSS-REFERENCE INDEXES

NATIONAL STOCK NUMBER INDEX

STOCK NUMBER	FIG	ITEM	STOCK NUMBER	FIG	ITEM	FIG	ITEM
4820-01-039-3769	133	6	5310-01-076-6196	50	6		
6220-01-039-9809	213	19	5305-01-076-6308	246	14		
5306-01-040-2041	131	14	5310-01-077-6817	188	3		
2520-01-040-2160	106	2				189	2
3110-01-040-6541	131	38	5930-01-077-7793	248	16		
5330-01-041-4749	296	29	5306-01-078-2719	37	22		
	296	36	5330-01-080-3253	79	7		
2540-01-041-4912	289	10	5306-01-081-0464	40	35		
5340-01-043-5214	231	18	5305-01-081-0465	36	11		
	275	19	5120-01-082-6436	303	3		
5330-01-043-5495	133	2	5120-01-082-6448	303	4		
3030-01-043-6749	295	12	2920-01-082-6458	37	36		
5330-01-044-0703	KITS		3120-01-082-6996	37	21		
5935-01-044-8382	57	24	3120-01-082-9417	37	11		
5340-01-044-8389	231	15	3120-01-082-9418	37	13		
	275	21	3120-01-082-9419	37	45		
5970-01-044-8391	57	27	5330-01-084-2410	109	6		
2990-01-046-1170	26	9	5310-01-084-4491	167	5		
5325-01-050-6192	232	10	5920-01-085-0825	49	28		
	232	12				49	35
	232	18	2510-01-085-0908	73	38		
	270	5	5365-01-085-0910	79	28		
	272	8	5330-01-085-0918	124	20		
2910-01-054-3816	15	10	2930-01-085-0926	28	11		
4820-01-057-5925	248	41	5355-01-085-0995	127	71		
5305-01-057-8031	248	21	5306-01-085-1953	121	1		
5360-01-057-9249	248	20	2590-01-085-1968	36	27		
5305-01-057-9322	248	8	9320-01-085-2889	298	10		
5340-01-059-0114	57	23				300	10
5935-01-059-0117	57	22	2590-01-085-6956	55	1		
5961-01-059-0562	36	22	5330-01-086-3503	97	5		
5330-01-059-4286	57	28	5330-01-086-3504	97	3		
5330-01-061-3000	299	8	5330-01-086-3505	100	4		
4710-01-062-3719	214	7	5330-01-086-3506	125	16		
5925-01-067-2926	283	3	2540-01-086-5433	235	1		
5310-01-069-5243	38	5	5330-01-086-5457	KITS			
	47	5	5305-01-087-1917	245	28		
	50	21	3110-01-087-2653	125	19		
	51	10	5330-01-087-4714	291	5		
	53	14	5910-01-089-1916	36	20		
	68	11	6240-01-089-6149	213	20		
4730-01-069-6408	251	3	5330-01-090-5428	131	29		
	255	3	5310-01-093-2907	101	1		
3040-01-075-1878	98	10	2920-01-093-9216	36	2		
2530-01-075-5080	121	4	2520-01-094-1817	98	12		
	121	10	2520-01-094-1818	98	22		
4730-01-075-7310	21	5	5340-01-094-9025	289	20		
	22	8	2510-01-096-2985	190	6		
5330-01-076-3009	124	27	5365-01-096-6749	148	28		
5330-01-076-6172	50	3				149	27

CROSS-REFERENCE INDEXES

NATIONAL STOCK NUMBER INDEX

STOCK NUMBER	FIG	ITEM	STOCK NUMBER	FIG	ITEM
2530-01-096-6752	115	27	5330-01-106-7938	124	14
3110-01-096-6755	84	51	1430-01-106-8451	273	15
2510-01-096-6758	51	4	5310-01-106-9929	37	26
2530-01-096-6764	122	4	6220-01-107-2613	50	2
2510-01-096-6769	190	11	5360-01-107-3245	263	40
5330-01-096-6775	83	2	5310-01-107-4051	124	1
	84	58	3110-01-109-3058	37	30
3020-01-096-6892	296	5	5330-01-109-7941	263	39
2510-01-096-7680	140	47	2520-01-110-1081	40	6
	142	36	2530-01-110-5304	122	4
5330-01-096-7698	BULK	34	5330-01-111-6531	37	31
5330-01-096-7699	70	9	5330-01-112-1533	13	9
	75	3	5330-01-112-4286	110	7
4720-01-096-7718	BULK	28	5330-01-113-9602	96	11
2530-01-096-7731	124	23	5340-01-114-7712	289	4
5330-01-096-9649	96	5	3110-01-117-2436	84	8
5330-01-096-9650	KITS		2920-01-117-3251	15	31
3020-01-096-9657	110	3	2910-01-117-7252	15	75
2510-01-096-9662	151	13	5310-01-118-2248	15	30
2540-01-096-9664	165	11	5305-01-118-4114	15	35
2530-01-096-9670	124	16	5330-01-118-9504	84	7
2530-01-097-7659	131	16	5315-01-119-3115	40	10
5305-01-097-7894	246	33	5310-01-119-3668	19	31
5310-01-097-8222	273	8		22	37
	273	22		87	8
	274	2		118	23
5310-01-097-9414	142	3		277	8
	156	3	3110-01-119-4441	83	55
2530-01-098-4145	101	14	3120-01-121-2809	40	8
2520-01-098-5200	139	4	2940-01-121-6350	79	10
4730-01-098-5229	101	5	5305-01-121-9352	263	2
2530-01-098-5242	101	8	4810-01-121-9786	263	10
9330-01-098-6554	220	3	2530-01-123-3553	97	6
	221	5	2930-01-123-4941	28	3
	224	2	5920-01-123-5211	49	27
5360-01-099-7921	101	18	5920-01-123-5212	49	26
5310-01-099-7945	101	2	5325-01-123-6798	79	9
5330-01-099-9423	101	17	2530-01-124-3422	118	21
5306-01-100-5149	36	26	4720-01-124-8252	263	15
2920-01-101-2552	KITS		5330-01-125-6132	84	43
5935-01-103-8700	36	9	5330-01-126-0733	85	5
5999-01-104-0445	36	3	5330-01-126-1039	84	36
4540-01-105-5667	263	8	5330-01-126-1040	82	9
2520-01-105-5676	96	9	5330-01-127-2883	84	52
2520-01-105-5679	96	8	2520-01-127-3969	77	5
5305-01-105-6888	37	16	4820-01-130-3387	263	12
5330-01-106-4329	37	42	2990-01-131-2751	263	45
5330-01-106-6101	37	44	4710-01-131-3398	263	16
5930-01-106-6382	263	4	2920-01-131-4932	35	2
5330-01-106-7938	124	14	5305-01-132-2166	50	28

CROSS-REFERENCE INDEXES

NATIONAL STOCK NUMBER INDEX

STOCK NUMBER	FIG	ITEM	STOCK NUMBER	FIG	ITEM
5305-01-132-2166	233	16	5310-01-144-2779	131	41
5310-01-132-8275	262	6	5340-01-144-6245	37	58
	284	25	2920-01-144-6362	37	24
4520-01-132-8908	263	25	5330-01-145-0706	37	57
5310-01-133-7215	231	13	5330-01-145-0724	37	54
3110-01-136-2093	37	43	5330-01-145-0725	37	23
5310-01-136-5532	37	38	2920-01-145-0993	37	7
3110-01-136-5534	37	32	5935-01-145-1974	37	68
5310-01-136-5547	37	56	5977-01-145-4308	263	42
5330-01-136-8334	248	22	5340-01-145-4311	263	6
5305-01-136-8734	250	5	9905-01-145-4320	263	43
5307-01-136-8740	248	26	6105-01-145-4324	263	29
5330-01-138-2106	23	10	4730-01-145-4326	263	13
5330-01-138-5190	73	4	5325-01-145-4349	263	20
	73	26	5330-01-145-5376	37	48
5905-01-139-3620	36	12	5940-01-145-7817	65	7
2530-01-140-6144	115	10	4140-01-145-8099	32	4
5305-01-140-9118	18	1	5340-01-146-1905	263	24
	19	36	6220-01-146-4455	51	1
	84	55	6220-01-146-4469	51	4
	140	11	2520-01-146-5482	71	6
	163	5	2520-01-146-5483	81	1
	176	2	6620-01-146-8006	30	13
	255	24	5340-01-147-2266	76	11
2920-01-141-1488	37	39	2520-01-147-2267	83	64
5330-01-141-7167	84	61	5340-01-147-2268	132	5
2530-01-142-0261	115	12	2590-01-147-2269	146	8
2920-01-142-2631	7	21	2990-01-147-3953	27	13
5305-01-142-3150	263	46	2990-01-147-3954	142	6
5940-01-142-8411	37	65	2520-01-147-4005	93	1
6130-01-142-8507	37	12	2930-01-147-4198	30	7
6130-01-142-8508	37	15	2815-01-147-4200	7	31
5977-01-142-9121	37	53	2815-01-147-4201	7	33
5977-01-142-9122	37	49	2520-01-147-4205	100	6
5310-01-143-0512	127	56	2520-01-147-4207	81	3
	127	77	5330-01-147-4208	72	8
	132	4	2530-01-147-4209	125	21
5310-01-143-0542	25	2	5330-01-147-4212	26	13
5310-01-143-1679	37	52		27	23
5310-01-143-1715	37	25	2530-01-147-4214	114	15
5310-01-143-1719	37	3	5340-01-147-4217	1	18
5305-01-143-6416	37	59	2910-01-147-4218	19	15
5305-01-143-6417	37	9	2910-01-147-4219	18	12
5305-01-143-6466	37	37	2930-01-147-4221	29	4
5305-01-143-6467	37	51	2930-01-147-4222	28	5
5977-01-143-6955	37	46	2920-01-147-4272	33	1
5977-01-143-6996	37	55	2815-01-147-4275	8	21
5305-01-143-7411	37	8	2920-01-147-4278	34	2
5305-01-143-7412	37	5	2590-01-147-4285	8	2
5970-01-144-1291	37	4	2990-01-147-4289	27	22

CROSS-REFERENCE INDEXES

NATIONAL STOCK NUMBER INDEX

STOCK NUMBER	FIG	ITEM	STOCK NUMBER	FIG	ITEM	FIG	ITEM
2990-01-147-4290	27	1	5310-01-148-2682	87	16		
4820-01-147-4294	70	10	5310-01-148-2687	83	63		
6680-01-147-4629	8	4	5340-01-148-2730	108	9		
6680-01-147-5497	28	7	4730-01-148-2755	30	10		
2540-01-147-5537	69	5	4730-01-148-2758	30	15	31	2
2590-01-147-5538	147	9				261	7
2520-01-147-5539	124	25					
2530-01-147-5541	156	9	4720-01-148-2761	134	10		
2510-01-147-5576	191	3	4720-01-148-2762	134	5		
3030-01-147-6410	34	4	4720-01-148-2763	118	18		
2530-01-147-6421	127	70	4720-01-148-2768	BULK	25		
2530-01-147-6423	120	2	3030-01-148-2792	132	1		
2530-01-147-6424	120	10	5340-01-148-2818	27	18		
4730-01-147-6425	105	6	2910-01-148-2910	19	13		
6680-01-147-6583	72	10	2530-01-148-2914	124	22		
3020-01-147-7935	35	3	2520-01-148-2919	93	1		
2930-01-147-8555	31	5	2990-01-148-2928	25	1		
2530-01-147-8556	132	6	2990-01-148-2929	114	9		
2920-01-147-8559	34	8	5995-01-148-2930	114	18		
2920-01-147-8562	35	9	2510-01-148-2937	150	4	151	5
5310-01-147-8743	17	2				152	7
	135	3					
	145	5	2510-01-148-2942	153	4		
	148	17	2540-01-148-2943	154	1		
	149	16	3020-01-148-2948	132	13		
5310-01-147-8744	3	2	3020-01-148-2949	32	2		
5310-01-147-8746	99	10	3020-01-148-2950	32	2		
5310-01-147-8747	100	7	4710-01-148-2969	86	1		
	112	2	4720-01-148-2970	30	4		
5310-01-147-8748	156	2	5306-01-148-3664	1	17		
2990-01-147-9284	10	14	5306-01-148-3667	12	10		
3040-01-147-9321	99	8				132	11
2530-01-147-9329	121	7	5305-01-148-3685	11	10		
2930-01-147-9330	32	6	5305-01-148-3686	3	1		
2815-01-147-9358	7	30	5305-01-148-3687	146	9		
3020-01-147-9359	3	11	5310-01-148-3693	17	3		
5330-01-147-9698	83	57	2815-01-148-3771	4	1		
5310-01-147-9792	35	8	4710-01-148-4989	72	9		
5330-01-147-9808	31	7	5310-01-148-4991	1	8		
2930-01-147-9916	30	11	4720-01-148-5000	9	21		
2510-01-147-9917	230	1	2510-01-148-5049	148	22		
5310-01-148-0245	83	62				149	21
2530-01-148-1463	114	17	5305-01-148-5915	11	5		
2520-01-148-1481	107	2	5310-01-148-5922	255	28		
2510-01-148-1617	173	1				297	1
2610-01-148-1634	126	1	5310-01-148-5956	100	10		
2610-01-148-1635	126	1				112	13
5305-01-148-2646	1	22	4720-01-148-5981	BULK	14		
4710-01-148-2659	12	9	3040-01-148-5982	91	15		
5310-01-148-2676	124	24				92	15

CROSS-REFERENCE INDEXES

NATIONAL STOCK NUMBER INDEX

STOCK NUMBER	FIG	ITEM	STOCK NUMBER	FIG	ITEM	FIG	ITEM
2940-01-148-5992	17	5	5330-01-149-0874	30	9		
5306-01-148-6765	97	15				261	3
	109	4	5365-01-149-0880	8	19		
4720-01-148-6946	118	18				254	7
4720-01-148-6947	118	18	2520-01-149-1221	74	14		
2815-01-148-6966	6	12	5340-01-149-1867	79	16		
2815-01-148-6971	7	4	2520-01-149-1868	73	30		
3020-01-148-6983	3	11	2530-01-149-1886	122	3		
4720-01-148-6984	19	41	4710-01-149-1899	9	25		
4720-01-148-7398	118	18	5305-01-149-1936	128	1		
5306-01-148-7442	8	14				140	48
5306-01-148-7456	2	11				142	35
5306-01-148-7457	147	6	5305-01-149-1938	7	1		
5305-01-148-7460	1	6				31	6
	2	8	5360-01-149-1959	26	14		
	3	12				27	24
	33	3	2530-01-149-3375	120	17		
	132	8	5340-01-149-3376	120	17		
	295	2	2520-01-149-3461	72	6		
5305-01-148-7465	49	4	2510-01-149-3465	191	13		
5310-01-148-7474	8	6	2910-01-149-3786	25	7		
5330-01-148-7492	72	4	2520-01-149-3809	74	7		
5330-01-148-7497	30	14	2530-01-149-3827	97	1		
	261	5				97	1
5330-01-148-7499	117	28	5306-01-149-4398	6	2		
5340-01-148-7528	13	15				66	7
5340-01-148-7529	16	12				127	80
5306-01-148-8198	4	2	5310-01-149-4407	136	2		
5305-01-148-8208	120	15				138	4
4730-01-148-8242	30	10				144	20
5340-01-148-8349	13	6	5310-01-149-4411	7	17		
5340-01-148-8351	114	10	5310-01-149-4413	7	12		
5340-01-148-8352	114	8	5340-01-149-4434	28	9		
4710-01-148-8354	9	2				41	10
3020-01-148-8846	13	2				57	8
3040-01-148-9430	7	15				60	2
3040-01-148-9432	83	76				66	11
2815-01-148-9497	8	22				118	5
2815-01-148-9534	6	6				162	3
2815-01-148-9535	3	8				233	8
2815-01-148-9540	7	13				264	2
3020-01-148-9547	7	38				268	6
3020-01-148-9548	83	26				282	2
2815-01-148-9559	11	12	4720-01-149-4659	119	25		
2815-01-148-9560	11	11	2990-01-149-4966	7	6		
4710-01-148-9580	16	2	2520-01-149-4993	83	65		
4710-01-148-9581	16	17	3020-01-149-5049	73	14		
4710-01-148-9582	16	5	4710-01-149-5075	16	18		
5310-01-149-0870	7	37	4710-01-149-5076	16	3		
5330-01-149-0874	2	5	4710-01-149-5077	16	15		

CROSS-REFERENCE INDEXES

NATIONAL STOCK NUMBER INDEX

STOCK NUMBER	FIG	ITEM	STOCK NUMBER	FIG	ITEM	FIG	ITEM
5315-01-149-5433	3	9	5315-01-149-9712	2	21		
5365-01-149-5434	99	2	5340-01-149-9729	93	5		
	112	10	5340-01-149-9732	68	3		
5365-01-149-5435	99	2	5360-01-150-0196	76	10		
	112	10	5340-01-150-0197	9	24		
3020-01-149-5915	7	35	3040-01-150-0407	74	28		
5306-01-149-6278	11	8	3040-01-150-0409	83	13		
	260	1				83	13
	295	6	2520-01-150-0415	83	31		
5306-01-149-6279	32	3	2815-01-150-0673	7	32		
5306-01-149-6280	6	10	4710-01-150-0842	8	5		
	7	7	4730-01-150-0879	8	12		
	11	2	2930-01-150-0895	30	19		
	30	8	2910-01-150-0950	12	13		
	33	2	4710-01-150-0971	16	4		
	33	4	5306-01-150-1190	31	12		
	34	3	5306-01-150-1197	125	20		
	71	1	5330-01-150-1215	12	12		
	132	7	3120-01-150-1226	98	27		
	295	7	5307-01-150-1228	32	1		
	295	9	5325-01-150-1229	87	15		
5310-01-149-6285	98	21	5340-01-150-1377	9	19		
5325-01-149-6293	17	6				264	8
	114	14	5365-01-150-1386	7	3		
5360-01-149-6308	133	7	5305-01-150-1521	16	9		
5360-01-149-6309	50	36	5355-01-150-1541	49	7		
3040-01-149-6706	73	1	5340-01-150-1545	39	5		
3040-01-149-6759	77	35	5307-01-150-1549	31	3		
5920-01-149-6952	49	33	2920-01-150-1610	49	1		
	222	12	2815-01-150-2181	11	6		
5920-01-149-6953	49	30	2815-01-150-2198	5	4		
	222	14	3020-01-150-2207	7	34		
5305-01-149-7356	49	24	2520-01-150-2279	73	9		
5310-01-149-7793	79	22	2520-01-150-2280	74	5		
5340-01-149-7811	77	20	2910-01-150-3675	25	5		
2805-01-149-7859	3	10	2910-01-150-3676	25	11		
2520-01-149-7861	74	31	2520-01-150-3692	73	28		
2520-01-149-7935	77	14	2530-01-150-3693	76	5		
3020-01-149-7938	74	17	2520-01-150-3823	83	7		
3020-01-149-7941	74	30	2520-01-150-3931	73	11		
2520-01-149-7962	76	2	2520-01-150-3932	73	10		
5340-01-149-7976	2	22				73	20
5340-01-149-7977	2	15	3010-01-150-3933	83	23		
2920-01-149-8606	33	8	2520-01-150-3934	83	40		
5930-01-149-9305	83	51	5310-01-150-4003	1	13		
5930-01-149-9306	49	25				123	4
5306-01-149-9668	2	19				142	2
5305-01-149-9673	31	9				156	4
5365-01-149-9691	7	16	5330-01-150-4022	96	13		
5365-01-149-9710	83	17	5340-01-150-4104	76	9		

CROSS-REFERENCE INDEXES

NATIONAL STOCK NUMBER INDEX

STOCK NUMBER	FIG	ITEM	STOCK NUMBER	FIG	ITEM
5340-01-150-4105	6		11 5360-01-150-6091	73	24
	34	1	5365-01-150-6092	73	31
	38	12	5330-01-150-6239	77	23
	44	11	5340-01-150-6249	9	17
	46	12	5310-01-150-6260	100	10
	47	8		112	13
	50	11	5340-01-150-6275	16	11
	56	10	5935-01-150-6319	75	4
	57	7	4720-01-150-7575	9	10
	57	11	2520-01-150-7609	73	37
	57	33	5330-01-150-7744	8	15
	65	15		254	3
	65	20	5365-01-150-7761	100	2
	69	15		112	6
	209	4	5307-01-150-7764	35	12
	232	2	5340-01-150-7774	16	19
	235	14	5360-01-150-7829	77	28
	251	10	5365-01-150-7830	73	35
	255	9	2520-01-150-8354	113	8
	271	15	5306-01-150-8713	77	19
	272	4	5306-01-150-9493	34	5
	276	8	5306-01-150-9497	79	11
	287	3	5306-01-150-9499	151	7
5340-01-150-4106	16	10	5305-01-150-9500	102	9
5340-01-150-4110	1	16		102	13
5306-01-150-4835	77	7	5360-01-150-9549	79	24
5360-01-150-4963	76	4	2530-01-150-9757	131	1
4820-01-150-4964	78	1	5305-01-150-9781	8	17
5365-01-150-4982	74	25		291	10
5340-01-150-4991	30	2	5315-01-151-1105	78	14
5340-01-150-4992	57	20	5360-01-151-1118	7	14
2815-01-150-5002	1	4	5360-01-151-1120	25	9
5305-01-150-5785	50	32	5360-01-151-1121	25	3
5306-01-150-5893	101	20	2520-01-151-2690	83	1
	101	26	3020-01-151-3444	113	5
5305-01-150-5895	7	36	3020-01-151-3446	99	7
5310-01-150-5919	77	29	2520-01-151-3569	73	27
5310-01-150-5921	79	2	2520-01-151-3570	74	11
5330-01-150-5928	77	21	2520-01-151-3571	74	9
5330-01-150-5944	13	3	2920-01-151-3627	24	3
5340-01-150-5986	1	5	4710-01-151-3662	75	1
5307-01-150-5991	11	4	4710-01-151-3663	77	13
5307-01-150-5992	8	20	2520-01-151-3784	78	9
5307-01-150-5993	6	5	2520-01-151-3857	77	1
5340-01-150-6026	16	7	3020-01-151-3967	83	24
5310-01-150-6034	99	2	2520-01-151-3982	83	33
	112	10	5310-01-151-4137	77	36
5360-01-150-6086	77	39	5315-01-151-4180	13	10
5360-01-150-6087	77	33	3020-01-151-4480	100	1
5360-01-150-6091	73	6		112	8

CROSS-REFERENCE INDEXES

NATIONAL STOCK NUMBER INDEX

STOCK NUMBER	FIG	ITEM	STOCK NUMBER	FIG	ITEM
5306-01-151-4925	13	13	2920-01-152-2414	43	10
5340-01-151-4964	77	38	5306-01-152-2582	119	14
3040-01-151-5663	77	24	3120-01-152-2612	7	2
2530-01-151-5967	117	1	3120-01-152-2613	83	28
2520-01-151-5971	110	13	6680-01-152-2845	133	18
5330-01-151-6106	77	9	5310-01-152-4229	79	23
5365-01-151-6111	128	2	3120-01-152-4239	3	4
	140	49	5306-01-152-4693	25	10
	142	34		295	13
5325-01-151-6117	8	3	5306-01-152-4696	83	5
2510-01-151-6362	148	21		83	9
2510-01-151-6363	148	29	2520-01-152-4742	103	1
3040-01-151-6382	83	66	2910-01-152-5516	19	34
5340-01-151-7409	19	21	2530-01-152-5517	131	36
4730-01-151-7972	118	19	5340-01-152-5641	1	23
3040-01-151-7989	83	70	4710-01-152-5798	79	8
2530-01-151-7995	101	19	5330-01-152-5941	79	5
2530-01-151-7996	101	25	5330-01-152-5942	77	11
2520-01-151-8043	93	4	3120-01-152-5945	7	43
2990-01-151-8115	114	9	3120-01-152-5946	7	40
5305-01-151-8288	50	14	3120-01-152-5948	83	3
5310-01-151-8347	115	19		84	57
	116	21	2530-01-152-7115	115	5
5310-01-151-8353	123	13	5340-01-152-7155	120	14
5330-01-151-8364	77	17	5365-01-152-7439	89	3
5307-01-151-8374	11	9		90	3
5305-01-151-9285	28	2		91	3
5330-01-151-9306	78	8		92	3
5340-01-151-9956	12	3	5365-01-152-7441	99	3
5340-01-151-9957	12	8		112	10
2530-01-152-0180	115	29	2920-01-152-7646	37	29
	116	31	3040-01-152-7786	116	18
2990-01-152-0251	114	12	2530-01-152-7787	116	18
5310-01-152-0598	47	6	2990-01-152-7788	27	17
	57	3	2990-01-152-7828	27	20
	65	5	2510-01-152-7832	149	20
	178	12	2510-01-152-7833	149	28
	207	10	5305-01-152-8193	91	14
	232	7		92	14
	232	13		93	6
	232	21	3120-01-152-8200	7	42
	269	14	3120-01-152-8201	3	4
	272	9	5305-01-152-8945	209	2
5365-01-152-0613	148	16	5315-01-152-9029	76	15
	149	15		77	2
2520-01-152-0941	97	9	5315-01-152-9031	77	37
3120-01-152-1989	3	6	5315-01-152-9032	78	12
3120-01-152-2039	3	5	5315-01-152-9033	78	2
5340-01-152-2382	172	14	5315-01-152-9038	77	26
2530-01-152-2383	127	2	5315-01-152-9039	77	26

CROSS-REFERENCE INDEXES

NATIONAL STOCK NUMBER INDEX

STOCK NUMBER	FIG	ITEM	STOCK NUMBER	FIG	ITEM
5315-01-152-9040	77	26	2530-01-153-9450	116	30
2520-01-152-9171	109	5	2540-01-153-9470	123	11
2530-01-152-9258	116	29	2510-01-153-9473	148	15
2510-01-152-9259	140	35		149	14
	142	26	2510-01-153-9584	153	4
2510-01-152-9260	140	46	3020-01-153-9586	132	13
	142	41	5365-01-153-9591	83	22
2530-01-152-9305	125	17	2520-01-154-1185	77	10
2530-01-152-9306	121	11	2530-01-154-1222	120	4
2530-01-152-9307	121	11		120	13
2530-01-152-9308	116	5	4710-01-154-1230	20	29
2530-01-152-9312	130	4	2510-01-154-1254	148	13
2530-01-152-9313	130	9		149	12
2530-01-152-9314	130	7	2510-01-154-1260	150	12
3120-01-153-0281	96	2		151	21
5315-01-153-0317	124	28	2510-01-154-1261	148	11
5315-01-153-0318	87	20		149	10
5365-01-153-0872	79	6	2530-01-154-1262	128	12
5307-01-153-0873	17	7	2530-01-154-1263	122	5
5360-01-153-0933	74	6	2530-01-154-1292	123	1
5306-01-153-1368	106	3	2540-01-154-1293	123	10
5310-01-153-1381	97	4	2530-01-154-1294	117	1
2530-01-153-1492	116	29	2990-01-154-1323	27	16
2510-01-153-1614	152	8	2990-01-154-1324	26	10
5340-01-153-1631	156	7	2530-01-154-1355	102	14
4710-01-153-1636	16	16	4730-01-154-1366	118	3
2530-01-153-1813	117	25	6150-01-154-1381	57	13
2530-01-153-1814	154	7	5310-01-154-1461	148	23
	155	5		149	22
3040-01-153-1815	99	5		151	9
4820-01-153-1851	8	10	5365-01-154-1484	7	25
6615-01-153-1852	8	9	5340-01-154-2000	97	14
4730-01-153-2718	95	8	3040-01-154-2090	113	1
3120-01-153-7545	3	4	5340-01-154-2107	69	11
5306-01-153-8281	150	6	5310-01-154-2273	42	1
2520-01-153-8390	89	4		50	16
	90	7		51	6
	91	4		52	8
	92	6		158	3
2520-01-153-8430	89	12		164	3
	90	12		165	9
	91	12		167	1
	92	12		169	21
2520-01-153-8431	89	9		194	2
	91	9	5315-01-154-2291	78	4
5310-01-153-9301	130	6	5905-01-154-2354	300	5
5310-01-153-9302	125	14	5945-01-154-3143	25	14
3020-01-153-9435	83	78	2520-01-154-3677	97	18
2520-01-153-9436	100	5	2990-01-154-3743	26	1
2530-01-153-9449	116	30	3040-01-154-3799	76	7

CROSS-REFERENCE INDEXES

NATIONAL STOCK NUMBER INDEX

STOCK NUMBER	FIG	ITEM	STOCK NUMBER	FIG	ITEM
5306-01-154-3971	148	25	5310-01-155-1898	116	3
5310-01-154-3990	125	12	5365-01-155-1932	150	9
5310-01-154-3992	83	15		151	20
5310-01-154-3993	118	20	5365-01-155-1941	89	6
5305-01-154-4318	150	2		90	9
	151	3		91	6
	152	5		92	8
5306-01-154-4320	148	24		93	2
5310-01-154-4341	115	9	5310-01-155-2503	10	4
	116	8	3110-01-155-2598	83	48
5330-01-154-4342	112	4	5340-01-155-2614	118	1
5365-01-154-4365	34	9		118	15
5365-01-154-4366	71	3	5340-01-155-2616	134	13
3120-01-154-4369	74	35	2940-01-155-3190	17	4
5365-01-154-4378	2	14	3120-01-155-3509	83	29
5310-01-154-5263	117	27	3120-01-155-3517	5	6
5340-01-154-5269	119	7	5360-01-155-3652	83	38
5340-01-154-5270	123	2	5340-01-155-3668	119	19
2520-01-154-5463	73	12	5330-01-155-4388	89	8
	73	32		91	8
5935-01-154-6264	283	10	5330-01-155-4393	83	50
5340-01-154-6559	77	27	5330-01-155-4399	97	13
2520-01-154-6683	94	2	3110-01-155-4438	73	29
5977-01-154-6777	234	18	3120-01-155-4462	79	3
2510-01-154-6906	163	1	3120-01-155-4463	79	3
2530-01-154-6952	124	3	3120-01-155-4464	79	3
5330-01-154-7159	8	24	3120-01-155-4465	79	3
5340-01-154-7163	132	9	3120-01-155-4466	79	3
3120-01-154-7174	74	23	3120-01-155-4467	79	3
	74	34	3120-01-155-4468	74	13
2530-01-154-8146	125	23	3120-01-155-4470	83	30
5365-01-154-8514	123	9	5325-01-155-4482	114	3
3120-01-154-8516	74	35	5360-01-155-4555	83	60
3120-01-154-8517	74	35		84	12
5365-01-154-8561	74	32	2910-01-155-5063	19	2
5365-01-154-8562	74	12		20	25
5365-01-154-8577	116	28	2540-01-155-5110	206	5
5307-01-154-9586	6	1	2540-01-155-5111	206	3
4140-01-154-9615	300	3	2510-01-155-5112	186	6
3020-01-155-0183	98	4	4730-01-155-5135	134	7
	98	25	2910-01-155-5136	19	29
3040-01-155-0194	205	8	2910-01-155-5137	19	27
3040-01-155-0195	205	15	2910-01-155-5138	20	20
2510-01-155-0336	152	10		22	30
3040-01-155-0371	170	1	2910-01-155-5139	19	8
3040-01-155-0372	170	9	2590-01-155-5140	19	22
3040-01-155-0373	170	9	2910-01-155-5147	19	9
2540-01-155-0376	235	7	2910-01-155-5148	19	9
5930-01-155-0783	234	26	2990-01-155-5149	26	12
5310-01-155-1897	125	13	2990-01-155-5150	26	11

CROSS-REFERENCE INDEXES

NATIONAL STOCK NUMBER INDEX

STOCK NUMBER	FIG	ITEM	STOCK NUMBER	FIG	ITEM
2990-01-155-5151	26	8	2520-01-155-6828	113	10
4320-01-155-5153	133	21	2520-01-155-6829	113	10
2590-01-155-5177	19	12	2520-01-155-6830	113	10
2510-01-155-5183	150	8	3040-01-155-6912	196	22
	151	16	2520-01-155-6936	93	4
4720-01-155-5194	19	11	2520-01-155-6939	110	18
2510-01-155-5432	190	4	2540-01-155-6947	196	3
2510-01-155-5433	190	2	6140-01-155-6997	56	15
2510-01-155-5434	190	1	6140-01-155-6998	56	12
2510-01-155-5435	190	7	5975-01-155-7084	207	1
2520-01-155-5746	83	69	2540-01-155-7278	272	7
3020-01-155-5779	7	39	2540-01-155-7298	170	6
2540-01-155-5824	206	8	2540-01-155-7299	240	7
2510-01-155-5825	185	3	5995-01-155-7387	68	6
	186	9	2510-01-155-7425	158	1
2520-01-155-5844	83	41	2520-01-155-7453	94	1
2910-01-155-5845	22	34	2520-01-155-7454	94	1
2510-01-155-5846	173	10	2530-01-155-7457	128	8
	175	3	5365-01-155-7465	172	12
2510-01-155-5848	18	9	2540-01-155-7496	155	6
	173	11	2540-01-155-7502	168	3
	191	12	2540-01-155-7503	168	3
2510-01-155-5849	176	20	2540-01-155-7535	137	12
2510-01-155-5850	176	19	2540-01-155-7542	271	1
2510-01-155-5851	176	13	2540-01-155-7543	272	6
2510-01-155-5853	176	10	5306-01-155-7659	238	1
2510-01-155-5854	176	3	5330-01-155-7700	129	3
2510-01-155-5857	176	8		129	8
3040-01-155-5864	196	18		130	3
3020-01-155-5871	98	26	2590-01-155-7711	18	7
2520-01-155-5883	96	7		BULK	8
3020-01-155-5898	83	21	5340-01-155-7744	132	10
5306-01-155-6108	156	1	4720-01-155-7784	BULK	17
5305-01-155-6110	148	30	5945-01-155-7830	15	43
5305-01-155-6113	154	10	2510-01-155-7877	176	20
	155	9	2910-01-155-7878	18	8
3020-01-155-6399	83	12	2990-01-155-7879	18	14
6220-01-155-6515	50	31	2910-01-155-7880	20	7
6220-01-155-6521	50	27	2910-01-155-7881	20	2
6150-01-155-6522	45	15	2590-01-155-7882	18	10
6140-01-155-6530	57	40	2510-01-155-7942	143	1
6140-01-155-6531	57	15	2530-01-155-7943	115	27
6250-01-155-6547	53	15	2910-01-155-7965	20	26
2530-01-155-6709	131	13	2590-01-155-7966	61	7
2540-01-155-6821	196	24	2590-01-155-7967	63	5
2540-01-155-6822	196	4	3020-01-155-8038	83	79
2540-01-155-6823	196	13	4720-01-155-8062	19	11
2540-01-155-6824	196	13	2530-01-155-8460	116	23
2540-01-155-6825	196	12	5306-01-155-8528	240	1
2520-01-155-6827	113	10	5365-01-155-8564	160	7

CROSS-REFERENCE INDEXES

NATIONAL STOCK NUMBER INDEX

STOCK NUMBER	FIG	ITEM	STOCK NUMBER	FIG	ITEM	FIG	ITEM
5365-01-155-8564	172	2	6220-01-156-4475	50	24		
5365-01-155-8576	169	26	6220-01-156-4476	51	11		
3120-01-155-8713	83	73				53	12
2510-01-155-8785	176	6	5975-01-156-4544	127	39		
2540-01-155-8786	196	12	2510-01-156-4854	160	8		
2510-01-155-8787	158	5				172	3
2510-01-155-8799	166	6	2540-01-156-4855	235	6		
2510-01-155-8800	166	5	2540-01-156-4869	170	3		
2510-01-155-8817	191	19	2540-01-156-4870	166	2		
2590-01-155-8871	184	9				166	11
	190	10	2510-01-156-4871	169	24		
5340-01-155-9701	110	15	2510-01-156-4872	167	11		
	111	22	2510-01-156-4873	167	11		
5340-01-155-9861	27	19	2540-01-156-4874	240	8		
5315-01-155-9932	83	59	2530-01-156-4875	115	21		
2910-01-156-0045	23	1	2540-01-156-4885	167	19		
4730-01-156-0055	104	6	2530-01-156-4900	115	30		
2540-01-156-0061	205	14	2530-01-156-4901	115	30		
	206	7	2540-01-156-4903	167	13		
2510-01-156-0062	185	6	2540-01-156-4904	167	13		
5365-01-156-0067	234	11	2540-01-156-4907	169	20		
2540-01-156-0068	234	28	2540-01-156-4908	169	20		
2540-01-156-0069	234	7	5305-01-156-5006	116	27		
2540-01-156-0070	240	10	5340-01-156-5061	215	15		
2540-01-156-0071	240	11				216	2
2540-01-156-0072	240	4	5330-01-156-5141	134	8		
2540-01-156-0073	240	13	5330-01-156-5147	7	5		
2590-01-156-0074	209	1	3120-01-156-5189	74	27		
2590-01-156-0075	209	3				79	14
2590-01-156-0076	208	1	5330-01-156-5236	KITS			
2590-01-156-0077	208	4	6140-01-156-5326	56	5		
2590-01-156-0078	208	4	5306-01-156-5435	108	8		
2590-01-156-0080	208	6	5305-01-156-5438	20	15		
5315-01-156-0081	115	7				21	7
4720-01-156-0085	21	16				22	11
	22	2				45	5
2540-01-156-0088	193	7				57	19
	278	1				58	4
4720-01-156-0547	BULK	15				59	3
4720-01-156-0548	BULK	18				60	3
4720-01-156-0549	BULK	19				63	2
4720-01-156-0550	BULK	20				64	4
2520-01-156-0552	110	4				95	1
2520-01-156-0553	110	11				104	4
2540-01-156-0564	165	10				105	2
2540-01-156-0565	170	3				118	10
2590-01-156-0583	114	12				119	22
2540-01-156-0584	233	4				175	11
6620-01-156-0712	43	6				252	1
3120-01-156-3827	99	11				268	12

CROSS-REFERENCE INDEXES

NATIONAL STOCK NUMBER INDEX

STOCK NUMBER	FIG	ITEM	STOCK NUMBER	FIG	ITEM
5305-01-156-5438	269	17	6240-01-157-0636	51	5
2510-01-156-5881	165	5	5330-01-157-0856	11	1
2540-01-156-5882	233	10	5305-01-157-0880	71	5
2530-01-156-5883	115	16	3120-01-157-0914	7	41
2540-01-156-6107	168	2	2510-01-157-1382	190	8
2540-01-156-6108	168	2	5325-01-157-1698	50	20
2530-01-156-6190	121	2		158	4
2530-01-156-6220	102	12		194	8
6150-01-156-6326	65	19	5360-01-157-1767	205	4
	66	9	5355-01-157-1865	243	6
5315-01-156-6562	69	4	5355-01-157-1866	196	19
	123	14		205	9
5340-01-156-6779	151	4	5360-01-157-1880	110	16
2530-01-156-7016	117	30		111	23
3040-01-156-7182	115	8	5330-01-157-1883	133	26
2540-01-156-7233	127	75	5330-01-157-1884	133	8
2540-01-156-7238	166	9	5330-01-157-1916	117	35
5365-01-156-8061	148	14	5330-01-157-1952	239	1
	149	13	5340-01-157-1955	117	33
2510-01-156-8092	165	29	5306-01-157-1985	83	61
4130-01-156-8096	295	10	5305-01-157-1987	169	23
2510-01-156-8183	160	6		194	3
	172	1	5340-01-157-2101	116	12
6220-01-156-8247	53	7	2540-01-157-2966	205	5
3040-01-156-8307	115	18	5315-01-157-3004	120	16
2530-01-156-8308	115	18	2540-01-157-3032	233	1
2520-01-156-8309	113	10	5310-01-157-3260	131	37
2540-01-156-8315	170	6	5306-01-157-3279	195	5
2540-01-156-8316	234	1		197	5
2530-01-156-8317	118	21		198	4
2910-01-156-8361	23	2		204	4
3020-01-156-8365	110	14	5306-01-157-3330	135	11
5340-01-156-8395	115	16		136	8
6220-01-156-8420	53	7		137	11
5306-01-156-8680	99	1		144	30
	112	11	4140-01-157-3501	300	3
5305-01-156-8692	139	14	5315-01-157-3515	83	36
	140	2	2540-01-157-3525	234	3
3120-01-156-8763	73	13	2540-01-157-3528	205	12
5310-01-156-8852	99	2		206	4
	112	10	5360-01-157-3662	115	20
5365-01-156-8948	151	15	2920-01-157-3765	39	8
5365-01-156-9635	151	8	5930-01-157-4060	54	2
2540-01-156-9675	176	15	2520-01-157-4376	110	9
5305-01-156-9711	151	18	5310-01-157-4855	234	8
3040-01-156-9729	170	4	5330-01-157-4872	234	14
5360-01-156-9730	116	22	2530-01-157-5164	116	20
2540-01-156-9740	235	2	5305-01-157-5625	165	16
3040-01-156-9994	114	6	5310-01-157-5670	140	10
6240-01-157-0635	50	26		196	7

CROSS-REFERENCE INDEXES

NATIONAL STOCK NUMBER INDEX

STOCK NUMBER	FIG	ITEM	STOCK NUMBER	FIG	ITEM	FIG	ITEM
5310-01-157-5672	166	16	2530-01-157-7933	128	11		
	179	2	3040-01-157-7970	115	8		
5330-01-157-5684	166	14	3040-01-157-7997	166	13		
5365-01-157-5752	119	17	3040-01-157-7998	86	4		
5340-01-157-6092	166	7				95	3
2540-01-157-6414	235	5				104	2
5340-01-157-6428	28	18				105	1
5340-01-157-6429	28	17	2540-01-157-8008	145	35		
5330-01-157-6664	120	6	2540-01-157-8009	179	25		
	120	11	3040-01-157-8021	178	14		
5340-01-157-6697	236	2	6220-01-157-9046	51	1		
	237	3	5305-01-157-9720	20	1	59	7
5310-01-157-6698	100	8				180	4
	112	1				203	14
5330-01-157-6757	2	10					
5365-01-157-6764	83	77	5306-01-157-9817	146	2		
5365-01-157-6772	172	16	5340-01-157-9825	20	21		
5310-01-157-6773	7	27	5306-01-157-9882	172	6		
5365-01-157-6778	113	3	5306-01-157-9883	172	13		
5306-01-157-6796	166	19	5306-01-157-9936	166	17		
	179	22	5340-01-158-0297	114	16		
	181	4	5340-01-158-0303	60	6		
	284	18				104	3
5306-01-157-6797	47	2	5340-01-158-0314	20	16		
	167	4				21	1
	168	4				22	4
	169	19	5340-01-158-0321	117	37		
	169	25	5305-01-158-0335	170	2		
	170	7				180	12
5330-01-157-6827	167	9	6210-01-158-0396	43	2		
5330-01-157-6828	167	9	5340-01-158-0503	166	20		
5305-01-157-7388	176	9	2540-01-158-0602	215	28		
5330-01-157-7458	166	8	2540-01-158-1569	233	11		
5330-01-157-7459	166	4	2540-01-158-1570	234	24		
5330-01-157-7464	184	8	2540-01-158-1721	167	22		
	190	12	5360-01-158-1727	234	6		
5340-01-157-7471	166	25	2520-01-158-1760	70	8		
5365-01-157-7476	196	6	2540-01-158-1877	196	25		
5340-01-157-7559	133	9	5365-01-158-2004	185	4		
5310-01-157-7560	146	4				186	8
5310-01-157-7582	102	11	5306-01-158-2023	184	7		
5310-01-157-7589	172	10	5305-01-158-2032	19	3		
5310-01-157-7600	113	4				19	6
5310-01-157-7601	98	7				20	4
	98	28				173	2
5330-01-157-7604	20	3				176	5
5330-01-157-7605	127	24				192	5
5340-01-157-7607	205	6	3120-01-158-2096	123	8		
5340-01-157-7608	205	17	5307-01-158-2110	102	17		
2540-01-157-7907	206	2	5340-01-158-2160	165	3		

CROSS-REFERENCE INDEXES

NATIONAL STOCK NUMBER INDEX

STOCK NUMBER	FIG	ITEM	STOCK NUMBER	FIG	ITEM		STOCK NUMBER	FIG	ITEM
5365-01-158-2182	74	4	2540-01-158-6893	203	15				
5365-01-158-2191	115	26	2540-01-158-6894	203	30				
5365-01-158-2193	125	10	5340-01-158-6895	203	11				
2540-01-158-3576	166	1	2540-01-158-6896	203	2				
6210-01-158-3857	42	11	2510-01-158-6903	175	1				
6150-01-158-3861	41	9	2510-01-158-6904	175	5				
5930-01-158-4428	43	11	2540-01-158-6906	241	5				
5340-01-158-4583	180	3	2540-01-158-6910	127	20				
2540-01-158-4599	145	40	2590-01-158-6912	58	5				
2540-01-158-4600	179	12	2590-01-158-6913	68	4				
2540-01-158-4601	181	8	2540-01-158-6955	203	5				
2540-01-158-4602	178	2	6350-01-158-7035	68	14				
2540-01-158-4603	199	12	4710-01-158-7511	BULK	39				
	200	12	2520-01-158-7530	113	6				
2540-01-158-4604	199	9	2540-01-158-7551	203	13				
	200	10	2510-01-158-7575	160	1				
6210-01-158-4668	42	12	2540-01-158-7583	203	21				
5930-01-158-4808	49	37	5305-01-158-7820	19	7				
5306-01-158-5374	101	7					20	24	
5365-01-158-5381	179	3					158	2	
5306-01-158-6231	149	24					161	5	
5305-01-158-6235	164	6					191	2	
5306-01-158-6243	102	10					194	7	
	102	24	2520-01-158-8481	113	15				
5310-01-158-6257	57	21	2520-01-158-8543	96	17				
5310-01-158-6260	46	15	2540-01-158-8548	144	19				
	50	22	5340-01-158-8549	188	7				
	51	9					189	1	
	56	4	2510-01-158-8550	191	8				
	65	2	2590-01-158-8551	243	4				
	241	1	5340-01-158-8552	127	81				
	267	5	2540-01-158-8553	145	1				
	277	18	2540-01-158-8554	177	2				
	297	6	2540-01-158-8555	182	8				
3120-01-158-6304	74	21	2440-01-158-8556	182	2				
5307-01-158-6312	11	7	2540-01-158-8557	180	9				
4010-01-158-6331	179	5	2540-01-158-8558	199	8				
5340-01-158-6354	196	21	2540-01-158-8559	201	9				
6210-01-158-6575	42	13	2540-01-158-8560	202	11				
5340-00-158-6681	140	33	2540-01-158-8561	200	5				
	142	20	2540-01-158-8562	203	31				
5306-01-158-6682	124	13	2540-01-158-8563	203	8				
5330-01-158-6683	167	8	2530-01-158-8584	173	4				
5330-01-158-6725	96	3	2530-01-158-8585	173	4				
5310-01-158-6780	235	16	2540-01-158-8611	233	15				
5340-01-158-6816	240	2	2540-01-158-8612	233	6				
5365-01-158-6819	263	22	2540-01-158-8613	127	83				
4820-01-158-6836	14	10	5340-01-158-8624	1	2				
2520-01-158-6854	74	8	2540-01-158-8626	42	3				
5340-01-158-6892	203	19	3040-01-158-8703	182	5				

CROSS-REFERENCE INDEXES

NATIONAL STOCK NUMBER INDEX

STOCK NUMBER	FIG	ITEM	STOCK NUMBER	FIG	ITEM	FIG	ITEM
4730-01-158-8717	233	17	5330-01-159-1298	19	19		
2590-01-158-8784	162	5	3120-01-159-1311	179	15		
2540-01-158-8811	203	7				203	12
2540-01-158-8812	201	1	5340-01-159-1321	45	6		
2540-01-158-8813	177	5				46	13
5306-01-158-9018	28	12				57	4
	38	7	5340-01-159-1324	244	4		
	56	14				261	16
	160	2				297	5
	163	3	5360-01-159-1449	115	23		
	165	6	5340-01-159-1460	207	9		
	173	13	3040-01-159-1775	180	5		
	175	8	5340-01-159-1788	79	17		
	176	18	2510-01-159-1790	276	12		
	179	7	2510-01-159-1791	274	12		
	191	16	2510-01-159-1792	271	3		
	196	16				273	28
	203	3	2540-01-159-1793	218	2		
	205	13				219	7
	206	6	6210-01-159-1794	226	1		
	267	11	2510-01-159-1795	227	7		
6130-01-158-9175	222	1	5340-01-159-1796	214	17		
5310-01-158-9205	159	3				227	11
	167	16	2510-01-159-1797	227	1		
5999-01-158-9249	42	4	2510-01-159-1798	218	14		
2520-01-158-9332	99	9				219	16
5970-01-158-9337	193	11	2540-01-159-1799	293	7		
	BULK	30	2540-01-159-1800	283	19		
2540-01-158-9370	242	7	2990-01-159-1801	256	5		
5307-01-158-9932	56	7	2930-01-159-1802	261	4		
5905-01-159-0771	241	2	5970-01-159-1803	270	3		
2910-01-159-0867	13	8	5895-01-159-1804	222	17		
2510-01-159-0868	164	4	2590-01-159-1805	127	10		
2540-01-159-0874	203	9	4720-01-159-1839	30	18		
5930-01-159-0925	49	17	6105-01-159-2223	300	11		
3040-01-159-0930	291	4	5930-01-159-2610	293	16		
3040-01-159-0996	182	11	6105-01-159-6666	298	8		
5365-01-159-1121	99	2	2520-01-159-2732	74	36		
	112	10	5306-01-159-2772	179	21		
5365-01-159-1122	99	2	5305-01-159-2779	194	6		
	112	10	5305-01-159-2780	164	5		
5306-01-159-1130	195	4	5305-01-159-2781	240	9		
	197	2	5305-01-159-2783	182	9		
	198	1	5306-01-159-2784	19	35		
	204	3	5330-01-159-2807	235	4		
5330-01-159-1152	183	1	5330-01-159-2811	83	72		
5330-01-159-1153	167	8	5330-01-159-2816	183	3		
5340-01-159-1185	22	24	2920-01-159-2842	127	8		
3120-01-159-1275	131	30	5325-01-159-2843	57	34		
5330-01-159-1298	18	4	5340-01-159-2901	28	6		

CROSS-REFERENCE INDEXES

NATIONAL STOCK NUMBER INDEX

STOCK NUMBER	FIG	ITEM	STOCK NUMBER	FIG	ITEM
2930-01-159-2902	28	13	5340-01-159-6174	185	2
2540-01-159-2928	144	32	2540-01-159-6198	275	6
2510-01-159-2929	160	11	5305-01-159-6567	203	1
5360-01-159-2931	127	52	5306-01-159-6574	188	1
2540-01-159-2992	136	10		189	5
5340-01-159-2996	187	2	5310-01-159-6586	57	18
2530-01-159-3447	127	1	5310-01-159-6587	127	60
2530-01-159-3448	127	21	5355-01-159-6622	87	29
2590-01-159-3449	127	5	5340-01-159-6626	16	14
5306-01-159-3450	127	33	5307-01-159-6632	35	6
2530-01-159-3451	127	27	6150-01-159-6901	44	6
2530-01-159-3452	127	28		46	6
2530-01-159-3453	127	25	5340-01-159-6905	176	14
2530-01-159-3454	127	45	5640-01-159-6935	9	12
2530-01-159-3455	127	43	2520-01-159-7118	127	31
3010-01-159-3456	127	55	2520-01-159-7119	127	30
2590-01-159-3505	61	8	2510-01-159-7120	179	24
2920-01-159-3589	49	5	2540-01-159-7740	143	2
2530-01-159-3604	127	64	2540-01-159-7741	144	31
5340-01-159-3880	196	5	2540-01-159-7744	200	9
	201	3	2920-01-159-7749	127	11
	202	2	3010-01-159-7750	178	9
	203	29	2530-01-159-7754	120	18
2520-01-159-4173	96	6	2530-01-159-7755	120	18
5340-01-159-4395	140	1	2520-01-159-7757	72	5
	142	1		83	14
5365-01-159-4471	99	2		85	6
	112	10		97	17
5340-01-159-4478	96	16		108	11
5320-01-159-4507	115	11		109	7
5340-01-159-4517	237	13	2510-01-159-7759	141	5
5340-01-159-4518	207	7	2510-01-159-7760	142	15
5340-01-159-4519	237	12	2510-01-159-7761	142	15
5330-01-159-4777	181	6	2520-01-159-7762	98	9
5365-01-159-4833	116	10	5365-01-159-7767	150	7
5365-01-159-4840	107	5	2510-01-159-7768	140	27
2520-01-159-5496	127	40	3040-01-159-7950	170	4
5999-01-159-5603	42	17	2540-01-159-7954	233	14
2540-01-159-5670	127	15	2540-01-159-7963	195	2
5306-01-159-5710	122	1	5305-01-159-8262	127	14
5340-01-159-5762	50	23	5305-01-159-8263	127	12
5340-01-159-5765	51	13	5310-01-159-8264	298	3
3120-01-159-5773	74	35		300	9
4720-01-159-5796	BULK	23	5330-01-159-8336	97	8
3830-01-159-5871	127	23	5330-01-159-8504	182	10
2520-01-159-5887	76	8	5310-01-159-8559	9	11
2510-01-159-5888	142	12		22	17
2510-01-159-5889	142	9		61	12
5360-01-159-5952	49	8		182	7
2530-01-159-5958	128	14		262	2

CROSS-REFERENCE INDEXES

NATIONAL STOCK NUMBER INDEX

STOCK NUMBER	FIG	ITEM	STOCK NUMBER	FIG	ITEM
5310-01-159-8559	264	10	3120-01-159-9386	97	2
	268	11	2510-01-159-9789	191	6
	282	9	3020-01-159-9795	98	5
3120-01-159-8588	83	49	2590-01-159-9809	234	20
5315-01-159-8660	203	20	5340-01-160-0294	127	42
5315-01-159-8666	184	12	5305-01-160-0331	196	14
5310-01-159-8678	99	2	5340-01-160-0367	19	23
	112	10	5315-01-160-0403	83	43
5310-01-159-8679	100	10	3120-01-160-0570	201	11
	112	13		202	15
2540-01-159-8705	240	12	5315-01-160-0575	166	22
2510-01-159-8712	140	28		166	28
2510-01-159-8713	140	28		179	14
2510-01-159-8714	140	25	5306-01-160-0769	191	17
2510-01-159-8715	140	17		199	7
2590-01-159-8716	140	12		200	7
2540-01-159-8721	194	5	4730-01-160-0814	241	11
2540-01-159-8722	240	6	2540-01-160-0849	41	6
4730-01-159-8723	78	5	2590-01-160-1047	233	19
4730-01-159-8724	78	11	2590-01-160-1496	58	9
2530-01-159-8725	128	15	4730-01-160-1505	54	3
2510-01-159-8726	187	7	2530-01-160-1542	102	2
2540-01-159-8727	203	23		102	18
2510-01-159-8728	190	9	2540-01-160-1591	161	2
2510-01-159-8729	190	9	2510-01-160-1592	141	3
2520-01-159-8730	98	2	2930-01-160-1597	28	4
2590-01-159-87573	83	34	3040-01-160-1598	127	41
2540-01-159-8759	176	1	4710-01-160-1599	127	48
2540-01-159-8760	235	9	2590-01-160-1669	65	8
2510-01-159-8761	161	3	2520-01-160-1893	113	10
2510-01-159-8762	159	2	3120-01-160-1894	127	34
2590-01-159-8763	243	5	3120-01-160-1895	127	47
2530-01-159-8764	291	6	5306-01-160-1952	179	4
2510-01-159-8765	50	33	5306-01-160-1968	235	12
2590-01-159-8766	49	23		251	12
2520-01-159-8798	110	12		255	11
2590-01-159-8799	59	1	5305-01-160-1970	191	9
2590-01-159-8800	64	1	5305-01-160-1974	241	10
2530-01-159-8802	114	2	5305-01-160-1975	28	1
3040-01-159-8803	69	14		29	3
2590-01-159-8857	243	1		30	3
2590-01-159-8861	159	5		41	12
5360-01-159-8862	178	13		127	82
2510-01-159-8871	190	3		163	13
2540-01-159-8874	194	1	5305-01-160-2005	108	3
2540-01-159-8875	194	1	5340-01-160-2155	117	29
2540-01-159-8878	114	1	5340-01-160-2171	143	3
2540-01-159-8880	169	5	5340-01-160-2172	181	7
2540-01-159-8881	135	10	5325-01-160-2237	167	20
5365-01-159-8890	102	16	5325-01-160-2238	114	7

CROSS-REFERENCE INDEXES

NATIONAL STOCK NUMBER INDEX

STOCK NUMBER	FIG	ITEM	STOCK NUMBER	FIG	ITEM	FIG	ITEM
5340-01-160-2239	23	16	5310-01-160-4536	186	10		
5340-01-160-2346	161	7	5340-01-160-4691	140	20		
5340-01-160-2367	179	6	5340-01-160-4592	68	7		
5360-01-160-2411	196	20	5340-01-160-4597	68	8		
	196	23	5340-01-160-4600	140	31	142	18
	205	10					
	205	16	5340-01-160-4601	142	19		
5360-01-160-2415	123	7	5325-01-160-4618	16	8		
5340-01-160-2443	165	22	5315-01-160-4639	143	5		
5340-01-160-2445	26	5				144	24
	27	7				145	30
5340-01-160-2470	180	10	5315-01-160-4642	83	58		
5365-01-160-2483	167	23	5365-01-160-4693	74	1		
5340-01-160-2488	176	11	2510-01-160-4969	142	11		
	180	7	2510-01-160-4970	159	1		
2510-01-160-3622	142	9	6220-01-160-5094	211	4		
2510-01-160-3634	180	6				212	4
2540-01-160-3651	198	2	6220-01-160-5187	269	3		
2540-01-160-3652	203	25	6140-01-160-5196	267	4		
2540-01-160-3653	203	28	6680-01-160-5256	59	6		
2540-01-160-3654	203	27	6680-01-160-5276	291	8		
4720-01-160-3664	241	12	5310-01-160-5708	235	10		
6220-01-160-3686	213	17				242	1
6220-01-160-3687	213	3	5310-01-160-5727	127	13		
6220-01-160-3705	213	5	5310-01-160-5728	127	51		
6680-01-160-3870	42	7	4720-01-160-5781	20	23		
5305-01-160-3937	166	24	2510-01-160-5837	173	3		
	196	15				175	5
	199	1	2540-01-160-5838	184	2		
	200	8	2540-01-160-5840	202	7		
	205	11	2590-01-160-5841	49	23		
	206	1	2990-01-160-5873	175	12		
5305-01-160-3938	58	8	2590-01-160-5885	60	1		
	258	5	3040-01-160-5913	182	12		
5306-01-160-3942	152	9	2540-01-160-5918	177	3		
5305-01-160-3945	49	10	5340-01-160-5922	170	8		
5305-01-160-3955	201	13				180	8
	202	17	3020-01-160-6516	83	47		
5320-01-160-3999	49	14	5340-01-160-6874	140	29		
	208	3				142	16
5325-01-160-4028	167	6	5306-01-160-7553	233	20		
5340-01-160-4068	140	14	4140-01-160-7664	300	1		
5340-01-160-4130	203	16	4130-01-160-7695	294	1		
6220-01-160-4247	50	18	4130-01-160-7696	299	1		
6220-01-160-4254	50	30	4130-01-160-7697	299	2		
5340-01-160-4397	7	8	4130-01-160-7705	299	24		
5305-01-160-4494	203	24	4130-01-160-7716	293	42		
5305-01-160-4528	196	10	5340-01-160-7774	151	14		
5310-01-160-4536	28	21	5340-01-160-7778	172	7		
	38	6	2910-01-160-8107	18	13		

CROSS-REFERENCE INDEXES

NATIONAL STOCK NUMBER INDEX

STOCK NUMBER	FIG	ITEM	STOCK NUMBER	FIG	ITEM
5975-01-160-8458	BULK	4	6220-01-161-5016	50	30
4140-01-160-8503	298	1	4820-01-161-5059	299	21
4130-01-160-8506	KITS		5306-01-161-5489	25	13
4130-01-160-8519	296	31	5340-01-161-5522	167	2
4130-01-160-8520	296	34	5310-01-161-6121	15	34
4130-01-160-8521	296	32	5306-01-161-6178	44	12
4130-01-160-8522	KITS			46	11
4130-01-160-8541	296	45		47	11
5360-01-160-8935	127	32		55	5
5310-01-160-9529	28	19		65	14
	191	15	4710-01-161-6406	281	8
	192	4	4820-01-161-6435	299	4
	252	6	6220-01-161-6439	50	18
	255	22	4730-01-161-6618	271	19
5365-01-160-9530	127	69	5310-01-161-7308	245	4
5360-01-160-9838	127	29		247	5
5360-01-160-9839	127	19		293	4
4710-01-161-0138	BULK	40		293	10
2590-01-161-0212	191	11	5360-01-161-7561	166	3
4810-01-161-0817	248	7		166	10
	248	36	5360-01-161-9169	293	13
2590-01-161-1328	41	13	5360-01-161-9171	127	46
2540-01-161-1356	202	1	5340-01-161-9187	172	15
6680-01-161-1439	18	2	5340-01-161-9188	165	7
	19	18	5305-01-162-0015	235	11
5340-01-161-1440	291	11	3120-01-162-0060	196	8
2590-01-161-2119	114	18	4730-01-162-0095	261	2
6210-01-161-2138	42	14	4720-01-162-0119	9	10
5306-01-161-2146	131	35		260	5
5310-01-161-2374	264	14	4720-01-162-0120	9	21
5330-01-161-2516	127	54		260	4
5310-01-161-2531	271	6	4720-01-162-0121	255	27
5305-01-161-2581	171	1	3040-01-162-0255	166	13
	196	11	4720-01-162-0283	281	6
5305-01-161-2583	38	3	5945-01-162-0516	265	2
	65	17	5945-01-162-0517	264	3
5306-01-161-2593	127	58	5930-01-162-0803	49	36
5330-01-161-2608	233	21	5310-01-162-2515	293	35
5340-01-161-2636	140	32	5360-01-162-2849	196	17
5340-01-161-2749	199	5		199	10
	200	3		200	6
5340-01-161-2789	194	4		205	7
6680-01-161-3656	291	7	2510-01-162-3623	163	7
4120-01-161-3706	292	2	5340-01-162-3624	191	5
5305-01-161-3995	57	35	2510-01-162-3625	191	4
5305-01-161-3997	50	19	2530-01-162-3626	164	7
5330-01-161-4019	234	2	5340-01-162-3627	43	3
5340-01-161-4025	119	18	5340-01-162-3628	43	3
3120-01-161-4033	74	35	5930-01-162-3669	49	9
5365-01-161-4055	124	21	2510-01-162-3679	164	8

CROSS-REFERENCE INDEXES

NATIONAL STOCK NUMBER INDEX

STOCK NUMBER	FIG	ITEM	STOCK NUMBER	FIG	ITEM
5330-01-162-3744	51	2	5340-01-162-5883	76	6
5340-01-162-3747	140	13	9340-01-162-5947	228	2
5340-01-162-3748	140	7	9340-01-162-5948	215	13
5340-01-162-3749	140	7		216	6
5305-01-162-3763	36	18	5305-01-162-5995	50	29
5365-01-162-3797	2	9	5305-01-162-5996	201	6
5365-01-162-3849	191	183		202	6
5340-01-162-3850	216	28	5340-01-162-6061	50	12
5340-01-162-3853	211	1	5340-01-162-6062	237	7
	212	2	5340-01-162-6077	53	13
	213	9	2540-01-162-6418	288	22
	215	19	2540-01-162-6493	288	36
	216	25	4710-01-162-7080	281	13
5340-01-162-3854	213	13	4720-01-162-7097	285	4
	214	1	4720-01-162-7098	284	14
5305-01-162-3961	165	25		285	1
	167	14	2590-01-162-7108	191	11
	201	5	2540-01-162-7110	273	10
	202	12	2510-01-162-7111	275	22
2590-01-162-4352	268	1	2510-01-162-7112	181	1
2590-01-162-4353	269	2	2540-01-162-7113	259	5
2510-01-162-4407	275	2	2540-01-162-7114	257	1
2510-01-162-4408	275	12	2540-01-162-7116	228	10
2540-01-162-4411	200	1	2540-01-162-7117	217	7
2540-01-162-4412	271	23	2510-01-162-7119	166	21
2990-01-162-4416	259	12	2510-01-162-7120	166	21
2590-01-162-4417	268	5	2590-01-162-7130	269	16
2590-01-162-4418	268	7	2590-01-162-7139	218	17
9390-01-162-4500	169	27		219	9
5340-01-162-4774	57	30	4710-01-162-7148	299	12
5340-01-162-4775	145	7	4710-01-162-7149	299	9
5340-01-162-4820	238	2	2510-01-162-7224	BULK	29
5340-01-162-4852	162	1	5961-01-162-7228	36	16
4720-01-162-5113	251	16	2590-01-162-7367	165	12
	255	16	5305-01-162-7884	184	11
2510-01-162-5172	163	4		191	14
2510-01-162-5173	163	4	5305-01-162-7885	13	1
2540-01-162-5174	283	1		163	6
2540-01-162-5175	283	2	5305-01-162-7890	147	7
2540-01-162-5176	288	4		197	3
2540-01-162-5177	288	40		198	3
2540-01-162-5178	288	31		203	26
2540-01-162-5189	288	39	5310-01-162-7912	201	14
5340-01-162-5619	215	23		202	16
	216	21	6625-01-162-8124	43	7
	225	8	5305-01-162-8512	245	8
5305-01-162-5703	293	22		247	16
5305-01-162-5707	184	10		292	1
5310-01-162-5732	160	5		298	5
3120-01-162-5787	74	35		300	2

CROSS-REFERENCE INDEXES

NATIONAL STOCK NUMBER INDEX

STOCK NUMBER	FIG	ITEM	STOCK NUMBER	FIG	ITEM	FIG	ITEM
5305-01-162-8512	300	18	2540-01-163-0834	228	9		
5305-01-162-8514	193	8	5360-01-163-0885	163	9		
5306-01-162-8525	50	25	5360-01-163-0886	127	61		
	51	2	2540-01-163-0894	229	5		
5330-01-162-8595	183	2	5340-01-163-0917	163	10		
3120-01-162-8654	150	5	5340-01-163-0919	23	25		
	151	6				59	5
	151	17				60	9
	152	6	6140-01-163-1081	57	1		
5340-01-162-8759	144	33	4720-01-163-1089	252	3		
5340-01-162-8760	144	21				255	30
5340-01-162-8846	83	37	2510-01-163-1139	163	12		
5365-01-162-8868	110	8	2540-01-163-1140	285	2		
5365-01-162-8869	110	8	2540-01-163-1141	251	8		
5365-01-162-8870	110	8				255	8
5365-01-162-8876	179	9	5365-01-163-1142	262	14		
5365-01-162-8884	110	8				284	2
2540-01-162-8983	275	5				288	9
2520-01-162-8985	83	35	2540-01-163-1143	284	21		
	83	42	2510-01-163-1146	137	1		
2530-01-162-8986	115	28	5365-01-163-1147	217	8		
5306-01-162-9678	84	5	2530-01-163-1152	116	4		
	179	11	2815-01-163-1174	254	1		
5305-01-162-9689	162	4	2540-01-163-1175	288	1		
	242	5	2540-01-163-1176	293	12		
5305-01-162-9695	53	8	2990-01-163-1179	258	6		
	176	17	2990-01-163-1180	259	5		
	179	23	2990-01-163-1182	287	11		
5305-01-162-9713	133	22	5995-01-163-1183	265	1		
5340-01-162-9777	19	33	2590-01-163-1184	264	1		
5340-01-162-9781	19	28	2590-01-163-1238	159	4		
5340-01-162-9782	51	13	5340-01-163-1343	38	4		
5340-01-162-9844	166	18	6150-01-163-1384	57	37		
5360-01-162-9935	116	26	6150-01-163-1385	57	36		
5340-01-162-9954	234	15	5340-01-163-1388	27	3		
5330-01-163-0387	98	14	5340-01-163-1389	140	6		
3020-01-163-0498	36	32	5340-01-163-1390	140	15		
4710-01-163-0594	280	9	5340-01-163-1401	120	7		
5340-01-163-0632	238	4				121	6
3120-01-163-0638	150	13				122	2
	151	19	5340-01-163-1411	142	13		
2540-01-163-0765	282	10	5930-01-163-1439	43	13		
2540-01-163-0766	275	22	5340-01-163-1973	70	4		
2990-01-163-0771	256	9	5330-01-163-1992	42	10		
2540-01-163-0772	223	19	5330-01-163-2000	296	41		
3040-01-163-0797	124	4	9390-01-163-2028	169	27		
2530-01-163-0798	124	6	5977-01-163-2032	40	23		
2530-01-163-0799	124	10	5330-01-163-2055	274	8		
2530-01-163-0800	KITS		5305-01-163-2423	294	18		
2540-01-163-0801	289	15	5305-01-163-2438	186	12		

CROSS-REFERENCE INDEXES

NATIONAL STOCK NUMBER INDEX

STOCK NUMBER	FIG	ITEM	STOCK NUMBER	FIG	ITEM
5305-01-163-2438	236	4	5680-01-163-6347	171	9
	237	4	5305-01-163-6466	54	14
5305-01-163-2439	203	18		52	7
5930-01-163-2583	214	4	2510-01-163-7016	163	12
5330-01-163-2614	77	34	2540-01-163-7017	289	8
4820-01-163-2656	79	21	3040-01-163-7018	87	23
2510-01-163-2681	175	7	4730-01-163-7163	80	2
2520-01-163-2708	77	12	2815-01-163-7189	7	29
2510-01-163-2709	169	9	2815-01-163-7190	7	29
2540-01-163-2719	169	14	2815-01-163-7191	7	29
5930-01-163-2779	219	14	4730-01-163-7194	12	6
	227	3		20	10
4710-01-163-2805	246	1		21	12
5970-01-163-2895	36	19		22	13
5977-01-163-2900	40	28		23	19
5977-01-163-2930	40	24		134	2
5977-01-163-2931	40	27	3040-01-163-7208	8	23
5330-01-163-3150	284	29	2920-01-163-7210	36	4
5330-01-163-3151	288	25	2920-01-163-7211	36	1
5340-01-163-3294	273	17	2520-01-163-7213	73	25
2520-01-163-3494	254	6	2540-01-163-7225	287	9
2590-01-163-3529	285	3	2530-01-163-7227	116	6
4140-01-163-3543	36	31	2510-01-163-7228	137	3
2530-01-163-3557	148	10	2510-01-163-7229	175	7
2990-01-163-3575	254	5	2510-01-163-7231	179	8
2590-01-163-3576	293	3	2540-01-163-7232	205	2
2540-01-163-3585	228	13	2540-01-163-7233	205	3
5340-01-163-4337	165	13	5360-01-163-7234	118	17
5355-01-163-4940	49	16		119	3
5305-01-163-5512	40	29	2540-01-163-7281	42	16
5360-01-163-5578	223	17	5930-01-163-7282	49	11
5360-01-163-5579	223	18	2520-01-163-7283	87	21
5360-01-163-5580	223	20	2520-01-163-7284	87	19
5330-01-163-5706	228	3	3040-01-163-7285	125	4
5305-01-163-5761	247	6	3040-01-163-7286	KITS	
	293	40	2590-01-163-7287	213	15
	294	2	2540-01-163-7288	218	15
	294	21		219	10
5330-01-163-5850	276	11	2590-01-163-7290	220	4
5340-01-163-5902	22	31		221	7
5340-01-163-5903	238	5	2540-01-163-7305	169	10
5340-01-163-5908	163	8	2510-01-163-7306	169	1
5325-01-163-5973	236	1	2520-01-163-7314	127	72
5315-01-163-6027	78	6	3040-01-163-7315	87	6
5365-01-163-6137	83	18	2520-01-163-7316	87	2
5365-01-163-6138	296	42	3040-01-163-7347	125	8
5365-01-163-6139	296	6	5340-01-163-7501	291	2
5340-01-163-6145	108	5	2590-01-163-7626	211	16
5365-01-163-6195	145	34		212	16
5930-01-163-6256	293	33	2590-01-163-7669	218	11

CROSS-REFERENCE INDEXES

NATIONAL STOCK NUMBER INDEX

STOCK NUMBER	FIG	ITEM	STOCK NUMBER	FIG	ITEM	FIG	ITEM
2590-01-163-7669	219	17	2590-01-164-0134	159	4		
4720-01-163-7833	BULK	24	2590-01-164-0135	267	1		
2815-01-163-7838	5	1	2815-01-164-0138	5	7		
4730-01-163-7864	2	12	5340-01-164-0746	212	21		
2520-01-163-7866	73	5	5340-01-164-0747	214	12		
2920-01-163-7872	40	21	5340-01-164-0748	290	3		
2540-01-163-7874	215	20	5340-01-164-0958	218	12		
	216	24				219	18
2530-01-163-7878	121	5	5340-01-164-0974	214	2		
	121	8	5310-01-164-1036	249	14		
2510-01-163-7879	215	16	5340-01-164-1048	238	6		
	216	7	5340-01-164-1063	250	9		
2530-01-163-7890	120	5	5340-01-164-1075	238	9		
	120	12	5340-01-164-1076	225	3		
2540-01-163-7948	218	4	5340-01-164-1077	225	5		
2540-01-163-7949	218	5	5340-01-164-1078	220	13		
2510-01-163-8014	216	10				221	11
2540-01-163-8017	204	2	5340-01-164-1100	212	22		
4720-01-163-8039	23	20	5310-01-164-1207	100	10		
5305-01-163-8241	99	6				112	13
	113	2	2510-01-164-1532	215	5		
5310-01-163-8256	293	15	2540-01-164-1562	223	8		
5340-01-163-8520	179	1	5305-01-164-1604	241	14		
2540-01-163-8595	145	6	5310-01-164-1647	69	20		
2590-01-163-8596	224	1	5330-01-164-1653	42	5		
2540-01-163-8597	229	3	5340-01-164-1743	71	8		
2540-01-163-8598	223	2	2540-01-164-1842	145	6		
2540-01-163-8599	223	16	2540-01-164-1843	199	2		
2540-01-163-8600	223	15	2540-01-164-1844	199	4		
2540-01-163-8623	211	12	2520-01-164-1855	69	2		
	212	23	2540-01-164-1886	271	8		
2540-01-163-8624	212	14	2540-01-164-1891	166	26		
2920-01-163-8626	40	17	5360-01-164-1949	200	2		
2540-01-163-8638	223	1	6220-01-164-2271	50	35		
5930-01-163-8851	219	12	5305-01-164-2313	43	12		
	227	6				165	17
5930-01-163-8924	219	13				242	2
	227	5				243	9
5935-01-163-8987	246	16				251	9
5340-01-163-9400	22	22				264	11
2520-01-163-9713	98	6	5306-01-164-2323	107	1		
2530-01-163-9952	117	23	5310-01-164-2336	241	6		
2815-01-163-9999	5	1	5310-01-164-2338	57	32		
2540-01-164-0032	267	12	5325-01-164-2377	62	5		
2530-01-164-0037	101	24				269	19
2530-01-164-0038	101	6	5340-01-164-2397	274	11		
2530-01-164-0039	115	28	5360-01-164-2404	179	19		
2540-01-164-0046	215	8	5360-01-164-2405	179	19		
	216	13	5365-01-164-2411	110	8		
2540-01-164-0122	267	7	6250-01-164-3266	218	8		

CROSS-REFERENCE INDEXES

NATIONAL STOCK NUMBER INDEXES

STOCK NUMBER	FIG	ITEM	STOCK NUMBER	FIG	ITEM
6250-01-164-3266	219	1	5340-01-164-6589	275	16
4810-01-164-3267	296	30	5340-01-164-6591	275	3
4130-01-164-3268	296	46	5340-01-164-6595	279	3
5340-01-164-3269	28	20	5365-01-164-6598	110	8
4730-01-164-3692	296	11	5365-01-164-6616	110	8
4130-01-164-3693	293	41	5365-01-164-6617	110	8
5365-01-164-4525	116	9	5365-01-164-6618	110	8
5307-01-164-4538	40	36	5365-01-164-6619	110	8
5315-01-164-4570	296	40	5365-01-164-6620	110	8
5365-01-164-4601	110	8	5340-01-164-7021	81	10
5365-01-164-4602	110	8	2590-01-164-7024	211	6
5365-01-164-4603	110	8		212	6
5680-01-164-4964	169	6	4730-01-164-7028	10	12
6220-01-164-5227	42	9	2590-01-164-7053	77	32
6220-01-164-5228	42	6	2815-01-164-7054	7	28
5315-01-164-5334	223	30	2920-01-164-7091	36	5
5315-01-164-5336	223	5	2510-01-164-7116	136	1
3110-01-164-5457	98	20	2510-01-164-7118	136	6
5310-01-164-5600	187	3	2510-01-164-7118	135	7
5330-01-164-5602	296	12	2510-01-164-7119	135	1
5330-01-164-5603	271	20	2510-01-164-7120	135	13
5330-01-164-5604	296	43	2540-01-164-7121	281	10
5340-01-164-5797	295	3	2540-01-164-7123	284	12
5340-01-164-5798	140	8	2540-01-164-7124	284	30
5340-01-164-5799	295	5	2530-01-164-7126	116	6
2520-01-164-6228	87	4	2510-01-164-7127	137	7
2540-01-164-6251	217	11	2510-01-164-7128	138	2
5305-01-164-6319	55	2	2510-01-164-7152	169	18
5305-01-164-6320	6	7	2815-01-164-7154	8	30
5305-01-164-6321	214	3	2815-01-164-7155	8	25
	222	16	2520-01-164-7156	74	26
	227	2		74	33
5340-01-164-6410	238	8	2520-01-164-7157	77	30
5340-01-164-6411	237	9	2520-01-164-7158	76	14
5340-01-164-6412	237	10	2590-01-164-7177	22	33
5325-01-164-6431	45	7	2990-01-164-7178	142	6
5340-01-164-6435	271	22	2520-01-164-7234	72	2
5340-01-164-6473	2	6	2540-01-164-7260	235	15
5340-01-164-6473	251	14	5330-01-164-7506	87	27
	255	14	5330-01-164-7509	287	4
5340-01-164-6474	251	13	5325-01-164-7550	270	2
	255	13	5330-01-164-7593	KITS	
5340-01-164-6524	282	5	5340-01-164-7603	295	8
5340-01-164-6526	262	7	2590-01-164-7825	127	62
5340-01-164-6527	262	10	4710-01-164-7850	297	3
5340-01-164-6528	262	1	2520-01-164-7851	87	2
5340-01-164-6529	295	11	2590-01-164-7897	293	1
6105-01-164-6546	241	13	2590-01-164-7898	213	23
5365-01-164-6567	296	4	2815-01-164-7948	2	3
5365-01-164-6568	296	8	5340-01-164-8137	215	25

CROSS-REFERENCE INDEXES

NATIONAL STOCK NUMBER INDEX

STOCK NUMBER	FIG	ITEM	STOCK NUMBER		FIG	ITEM
5340-01-164-8137	216	20	5340-01-165-1494	289 6		
	228	7	2510-01-165-1495	135 9		
5340-01-164-8170	295	4	2510-01-165-1496	191 1		
5340-01-164-8171	44	9	2510-01-165-1497	191 1		
5920-01-164-8260	222	15	5360-01-165-1563	8 28		
5930-01-164-8264	293	25	5360-01-165-1564	69 16		
2540-01-164-8379	234	17	5360-01-165-1574	87 3		
5330-01-164-8385	167	7	5340-01-165-1585	75 7		
5330-01-164-8579	184	15	5340-01-165-1586	75 2		
5365-01-164-8632	110	23	5930-01-165-1657	293 30		
5325-01-164-8651	68	5	5305-01-165-2260	124 5		
5325-01-164-8652	59	4			125	3
5325-01-164-8655	255	25	5306-01-165-2425	5 3		
	268	13	5310-01-165-2475	175 2		
5340-01-164-8757	165	20	5340-01-165-2526	58 7		
	173	6	3110-01-165-2530	36 30		
	175	6	5325-01-165-2551	193 1		
5340-01-164-8761	165	8	5315-01-165-2568	40 11		
2520-01-164-9229	87	24	5306-01-165-3258	69 21		
2540-01-164-9278	289	14	5306-01-165-3310	6 9		
2510-01-164-9280	229	2	5310-01-165-3331	8 7		
2510-01-164-9314	191	8			9	18
2590-01-164-9326	41	2			10	6
6680-01-164-9433	19	20			12	4
6150-01-165-0168	57	16			16	13
4730-01-165-0173	296	28			70	1
3040-01-165-0188	69	19	5310-01-165-3333	5 5		
5310-01-165-0464	50	15	5310-01-165-3346	296 1		
2540-01-165-0466	284	27	5330-01-165-3388	81 6		
5340-01-165-0539	40	25			82	1
9905-01-165-0541	211	17	5330-01-165-3401	184 5		
	212	18	5330-01-165-3402	184 6		
9905-01-165-0542	211	18	5340-01-165-3417	53 6		
	212	17	5365-01-165-3457	16 6		
9905-01-165-0544	213	7	5325-01-165-3475	291 13		
5340-01-165-0564	176	4	5315-01-165-3536	87 5		
5340-01-165-0602	224	7	5340-01-165-3717	240 3		
5340-01-165-0657	177	7	5306-01-165-4283	83 16		
5930-01-165-0732	243	7			297	2
2590-01-165-0747	36	10	5306-01-165-4286	174 1		
2530-01-165-0803	102	1	5330-01-165-4333	74 3		
2540-01-165-0813	289	5			79	4
2540-01-165-0837	203	6	5340-01-165-4351	181 2		
2540-01-165-0895	201	10	5340-01-165-4353	145 7		
	202	5	5325-01-165-4372	273 13		
3010-01-165-1221	112	3	5340-01-165-4379	271 18		
5310-01-165-1327	87	9	5340-01-165-4429	289 3		
5330-01-165-1357	81	2	5365-01-165-4511	110 8		
5330-01-165-1358	117	31	5340-01-165-4543	284 24		
5330-01-165-1378	184	1	5340-01-165-4544	189 4		

CROSS-REFERENCE INDEXES

NATIONAL STOCK NUMBER INDEX

STOCK NUMBER	FIG	ITEM
6105-01-165-4561	241	8
3040-01-165-4562	141	4
5945-01-165-4602	40	1
6160-01-165-4637	56	6
6160-01-165-4638	56	2
5340-01-165-4705	19	1
	20	6
5340-01-165-4791	169	2
5340-01-165-4792	169	11
2815-01-165-4826	5	
		6
2510-01-165-4932	181	5
2520-01-165-4935	69	22
5306-01-165-5582	8	
		31
5306-01-165-5583	31	4
	39	4
	199	11
	200	11
5305-01-165-5591	254	4
5340-01-165-5617	288	34
2540-01-165-5971	220	10
2520-01-165-5974	125	2
2540-01-165-5996	242	6
2530-01-165-6005	129	9
2815-01-165-6143	7	
		23
2540-01-165-6145	235	15
5895-01-165-6792	43	9
2540-01-165-6793	224	11
2920-01-165-6794	40	7
7230-01-165-6795	228	8
5340-01-165-6797	188	4
	189	3
	279	4
5340-01-165-6798	188	6
5340-01-165-6799	188	5
5950-01-165-6803	296	9
9905-01-165-6901	215	17
	216	27
5325-01-165-6975	275	23
5310-01-165-7572	69	18
2520-01-165-7885	87	11
2540-01-165-7931	223	6
2815-01-165-8002	7	
		28
2510-01-165-8101	87	28
2540-01-165-8174	242	3
2540-01-165-8177	214	8
2815-01-165-8216	2	
		1
5305-01-165-8612	233	2
5340-01-165-8986	162	2
2910-01-165-9487	15	80
2520-01-165-9563	76	3

STOCK NUMBER	FIG	ITEM	FIG	ITEM
2930-01-165-9595	160	3		
4820-01-165-9596	15	33		
2530-01-165-9653	129	6		
2530-01-165-9654	129	1		
5935-01-165-9807	36	21		
5340-01-166-0568	288	33		
2815-01-166-0621	8	11		
2590-01-166-1071	40	21		
2510-01-166-1146	165	15		
2540-01-166-1370	203	4	205	1
2540-01-166-1373	201	2	202	4
5340-01-166-1470	70	2		
5305-01-166-1471	213	21		
5305-01-166-1473	222	13		
5320-01-166-1477	228	14		
5340-01-166-1534	12	1		
5305-01-166-1610	215	27		
5306-01-166-1665	228	4		
5330-01-166-1712	299	11	299	17
5355-01-166-1720	293	34		
5315-01-166-1733	40	22		
2540-01-166-2009	267	10		
2540-01-166-2010	267	8		
5340-01-166-2011	211	14	212	20
5945-01-166-2012	222	8		
6150-01-166-2013	220	19		
6150-01-166-2014	220	11		
2510-01-166-2015	214	14		
4130-01-166-2115	298	16		
4520-01-166-2133	286	2		
4520-01-166-2134	286	5		
5340-01-166-2152	256	3	258	3
			259	9
5330-01-166-2154	184	3		
5310-01-166-2615	110	2		
2530-01-166-3033	116	14		
3120-01-166-3677	75	6		
3040-01-166-4497	123	12		
3120-01-166-5638	127	26		
5340-01-166-5639	40	39		
5340-01-166-5640	2	2		
5340-01-166-5652	179	8		
5340-01-166-5654	218	10	219	4
2910-01-166-5655	15	81		

CROSS-REFERENCE INDEXES

NATIONAL STOCK NUMBER INDEX

STOCK NUMBER	FIG	ITEM
3110-01-166-5667	131	28
5340-01-166-5861	197	4
	198	5
	204	5
5330-01-166-5912	184	4
2540-01-166-5913	224	4
4310-01-166-5958	KITS	
4310-01-166-6123	296	39
2510-01-166-6253	140	16
5365-01-166-6324	83	27
3120-01-166-6724	166	23
	166	29
3020-01-166-6802	291	4
5360-01-166-6856	78	7
5306-01-166-8556	174	2
5365-01-166-8901	102	7
	102	21
5340-01-167-0136	166	26
3110-01-167-0195	296	7
5325-01-167-0510	235	8
5680-01-167-1068	169	15
5365-01-167-1122	110	8
6110-01-167-1822	38	8
5360-01-167-1890	36	8
7230-01-167-2075	215	4
	216	16
3110-01-167-2443	74	16
3040-01-167-2836	74	15
2540-01-167-2985	204	1
5355-01-167-4114	247	10
	293	24
3120-01-167-4172	74	29
5306-01-167-4346	83	52
5305-01-167-5498	87	7
3120-01-167-5544	36	24
5315-01-167-5584	15	42
5305-01-167-6246	165	19
5330-01-167-6335	271	5
	273	27
2930-01-167-7250	270	7
5340-01-167-7794	13	18
6695-01-167-8108	42	8
5330-01-167-8123	181	3
5305-01-167-8334	87	22
	271	17
5310-01-167-8344	188	2
	189	6
5325-01-167-8372	234	12
4720-01-167-9137	287	2
5340-01-167-9694	80	3

STOCK NUMBER	FIG	ITEM	FIG	ITEM
5340-01-168-0939	288	24		
4030-01-168-1282	213	12		
5365-01-168-1409	110	8		
2530-01-168-1440	120	3	120	9
5340-01-168-1501	49	19		
5330-01-168-1535	291	9		
5325-01-168-1886	69	3		
2520-01-168-1983	74	20		
2520-01-168-2060	73	39		
5330-01-168-3870	82	8	83	54
5306-01-168-4481	29	2	50	10
			56	3
			57	31
			207	11
			267	9
			271	21
			274	16
			276	9
			287	7
			288	18
5325-01-168-5677	64	2	269	18
5365-01-168-5729	74	22		
2530-01-168-6369	117	32		
5340-01-168-6372	148	9	149	8
5365-01-168-7294	110	8		
3110-01-169-0734	74	18		
5310-01-169-2849	187	1		
2510-01-169-3765	142	33		
2510-01-169-3766	142	8		
2510-01-169-3767	142	14		
2510-01-169-3785	215	7	216	11
5120-01-169-4876	304	5		
5120-01-169-4877	303	12		
5120-01-169-4878	301	6		
3120-01-169-6440	87	10		
2520-01-169-7674	73	3		
4730-01-169-7714	296	38		
5120-01-170-0627	302	2		
5120-01-170-0628	301	13		
5120-01-170-3278	304	4		
5120-01-170-3279	301	8		
5340-01-170-4777	83	39		
5120-01-170-5473	302	20		
5340-01-170-5530	167	2		

CROSS-REFERENCE INDEXES

NATIONAL STOCK NUMBER INDEX

STOCK NUMBER	FIG	ITEM
5330-01-170-6303	163	2
5120-01-170-6664	301	19
5120-01-170-6703	304	8
2530-01-170-7118	101	16
5310-01-170-8765	271	16
	272	2
5310-01-170-9100	29	1
	57	29
	161	4
	282	3
5940-01-171-3195	222	7
5365-01-171-3392	73	8
	73	22
5365-01-171-4041	110	8
5120-01-171-5233	302	6
4130-01-171-5997	299	10
5306-01-171-8076	72	7
5340-01-171-8242	142	7
5360-01-171-8248	200	4
5305-01-171-8252	127	37
5340-01-171-8256	101	12
5340-01-171-8257	140	26
5935-01-171-8273	45	14
6220-01-171-9557	211	8
	212	8
2520-01-172-0394	83	46
2520-01-172-0395	113	10
2520-01-172-0396	113	10
4710-01-172-0471	119	2
2520-01-172-0501	110	17
	111	24
5307-01-172-1526	133	20
5306-01-172-1590	101	23
	101	27
5310-01-172-1591	124	18
5340-01-172-1942	228	15
5340-01-172-1944	172	11
5340-01-172-2012	140	31
	142	18
5340-01-172-2087	19	4
	20	5
2510-01-172-3020	140	43
	142	40
2510-01-172-3022	164	2
6220-01-172-5300	211	7
	212	7
4730-01-172-6683	9	15
2815-01-172-6780	7	15
2510-01-172-6781	139	10
	141	9

STOCK NUMBER	FIG	ITEM	FIG	ITEM
5340-01-173-0045	140	30	142	17
2510-01-173-0046	140	36	142	27
5340-01-173-0047	140	50	141	8
5340-01-173-0048	139	9		
3040-01-173-0049	139	3		
2510-01-173-0050	142	10		
5930-01-173-0825	299	7		
5930-01-173-0826	299	20		
2530-01-173-1248	116	20		
2540-01-173-1249	223	27		
4820-01-173-1250	223	3		
3020-01-173-1332	113	13		
2520-01-173-1362	89	7	91	7
3120-01-173-3728	3	5		
2510-01-173-4189	139	6	141	6
2990-01-173-4197	252	8		
2590-01-173-6991	165	24		
5945-01-173-7760	222	2		
2530-01-173-9671	131	10		
2520-01-174-1418	97	11		
5330-01-174-8090	77	31		
3120-01-174-8153	74	10		
5365-01-174-8626	74	19		
5365-01-174-8657	96	4		
5310-01-174-8766	113	7		
6680-01-175-0565	18	3		
5310-01-175-0624	113	7		
5310-01-175-0625	113	7		
5365-01-175-0683	113	7		
5310-01-175-1035	113	7		
5310-01-175-2539	110	1		
3120-01-175-3813	3	5		
3020-01-175-6477	83	32		
2520-01-175-6492	73	36		
5360-01-175-7269	7	26		
5340-01-175-7656	140	24	142	25
3120-01-176-1041	113	17		
5310-01-176-2466	83	20		
5310-01-176-2467	113	7		
3120-01-176-2493	113	17		
3120-01-176-2493	113	17		
3120-01-176-2496	113	17		
2520-01-176-2839	74	24		
3020-01-177-3055	83	6		

CROSS-REFERENCE INDEXES

NATIONAL STOCK NUMBER INDEX

STOCK NUMBER	FIG	ITEM	STOCK NUMBER	FIG	ITEM
3120-01-177-3064	3	6	5330-01-181-2454	169	12
5310-01-177-9758	113	7	5360-01-181-2482	19	5
4910-01-178-0360	303	16		20	8
4910-01-178-0713	303	9	3020-01-181-3795	84	60
4910-01-178-0722	303	10	2990-01-181-6725	142	32
4910-01-178-0724	303	8	3120-01-181-7938	3	6
5120-01-178-6342	301	17	4910-01-182-2704	302	10
4910-01-178-6551	303	7	4720-01-182-3457	BULK	16
5330-01-178-7351	169	3	5330-01-182-4121	178	11
5307-01-178-7445	31	11	2590-01-182-4455	146	3
4910-01-178-8864	302	3	3120-01-182-8417	79	3
4910-01-178-8865	303	14	5365-01-182-8469	113	16
4910-01-178-8866	303	13	4910-01-183-0044	303	15
2510-01-178-8867	180	11	4730-01-183-2140	15	99
2510-01-178-8877	191	13	5330-01-183-5688	263	38
4910-01-178-9788	302	5	5310-01-183-6930	113	7
3040-01-178-9789	127	49	2520-01-183-8322	84	56
5120-01-179-1032	303	11	2520-01-183-8323	82	7
5120-01-179-1033	302	15	5120-01-183-8576	302	8
5120-01-179-1034	301	10	2530-01-183-8860	115	6
5120-01-179-1318	301	3	4720-01-184-0432	12	14
5330-01-179-2249	101	9	5310-01-184-5418	147	3
4910-01-179-2515	302	12	5310-01-184-5866	169	7
4910-01-179-2516	301	15		169	16
4910-01-179-2517	301	12	5930-01-184-6370	284	17
4910-01-179-2518	301	11	5360-01-185-0341	182	1
5340-01-179-4108	139	9	5310-01-185-3400	101	22
	141	8		101	28
4820-01-179-4869	301	14	5315-01-185-3415	98	3
4910-01-179-4870	302	9		98	24
4910-01-179-5530	304	3	3120-01-185-3509	113	17
5330-01-179-5907	101	10	3120-01-185-3678	113	17
4910-01-179-6338	303	6	5330-01-185-4676	116	11
4910-01-179-6339	302	16	5306-01-185-7048	30	6
4910-01-179-6340	301	9	5306-01-185-7049	132	12
4910-01-179-6341	301	5	5306-01-185-7050	33	9
4910-01-179-6364	304	2	5315-01-186-1352	2	16
5310-01-179-9486	303	5	5305-01-186-5381	72	1
5120-01-180-0558	301	2	4140-01-186-9753	298	9
5340-01-180-2577	140	18	5360-01-187-0301	127	65
2990-01-180-2988	27	21	5310-01-187-7610	142	23
5970-01-180-3731	293	28	5340-01-187-8520	140	32
4910-01-180-6155	301	7		142	19
5120-01-180-7928	304	1	5340-01-187-8673	167	17
5120-01-180-8592	302	4	5305-01-188-0489	15	45
6240-01-180-9022	50	34	5305-01-188-0490	15	45
4910-01-181-0183	304	6	5305-01-188-0491	15	45
3120-01-181-1637	113	17	5305-01-188-0492	15	45
4910-01-181-1958	302	11	5305-01-188-0493	15	45
4910-01-181-1959	301	16	5305-01-188-0494	15	45

CROSS-REFERENCE INDEXES

NATIONAL STOCK NUMBER INDEX

STOCK NUMBER	FIG	ITEM	STOCK NUMBER	FIG	ITEM		FIG	ITEM
5315-01-188-0495	15	106	2910-01-189-0895	15	102			
5305-01-188-0496	15	53	2910-01-189-1747	15	119			
5315-01-188-0765	15	124	2910-01-189-1748	15	86			
5315-01-188-0766	15	124	5315-01-189-2141	15	60			
5365-01-188-0783	15	63	2910-01-189-2142	15	2			
5365-01-188-0784	15	56	2910-01-189-2191	15	51			
5365-01-188-0785	15	62	5910-01-189-3011	298	7			
5360-01-188-0807	15	18					300	16
5305-01-188-0948	15	71	5945-01-189-3307	41	3			
5365-01-188-0958	15	68					41	14
5365-01-188-0959	15	111	5360-01-189-3466	15	9			
5365-01-188-0960	15	116	5970-01-189-4883	15	38			
5365-01-188-0962	15	46	6250-01-189-4981	61	3			
5365-01-188-0993	15	121	5910-01-189-5109	68	2			
7690-01-188-2862	293	8	5910-01-189-5110	65	3			
8040-01-188-2953	41	1	5910-01-189-5152	44	5			
2910-01-188-3176	15	70	5910-01-189-5153	65	16			
3040-01-188-3222	15	123	6250-01-189-6926	61	5			
3040-01-188-3242	15	85	7690-01-189-6932	293	36			
2910-01-188-3243	15	94	5940-01-189-8033	15	40			
2910-01-188-3244	15	97	5940-01-189-9841	15	39			
2910-01-188-3249	15	115	2910-01-190-0069	15	59			
2910-01-188-3250	15	113	5315-01-190-0429	15	52			
2910-01-188-3251	15	114	5925-01-190-1211	222	5			
2910-01-188-3252	15	17	5930-01-190-1212	293	29			
2910-01-188-3254	15	11	5930-01-190-1231	247	22			
2910-01-188-3255	15	8	6140-01-190-2516	56	16,			
2910-01-188-3256	15	47	6140-01-190-2517	56	8			
2910-01-188-3257	15	64	5305-01-190-4070	15	74			
3040-01-188-3261	15	77	5305-01-190-4078	15	45			
2520-01-188-3282	15	1	5305-01-190-4555	15	73			
2910-01-188-3683	15	104	5910-01-190-4600	55	4			
2910-01-188-3775	15	112	5945-01-190-5143	15	23			
5307-01-188-4684	6				4		15	23
5920-01-188-6294	49	32	6110-01-190-5501	293	37			
	49	34	5305-01-190-5745	15	48			
	222	11	5360-01-190-6215	15	117			
5305-01-188-6566	15	45	6150-01-190-6498	220	8			
5305-01-188-6567	15	90	6110-01-191-0127	38	11			
5305-01-188-6568	15	65	5305-01-191-0374	15	98			
5310-01-188-6743	15	13	5325-01-191-3293	191	7			
5360-01-188-6806	15	92	5315-01-191-3393	15	16			
4140-01-188-6977	245	42	5945-01-191-3743	222	3			
6105-01-188-7154	245	40	6230-01-191-3856	226	4			
3110-01-188-7682	15	67	2510-01-191-4259	245	9			
5961-01-188-8315	38	9	2530-01-191-4262	131	43			
5307-01-188-9217	15	50	2590-01-191-4263	127	3			
2520-01-189-0596	89	1	2530-01-191-4264	127	38			
	90	1	2540-01-191-4289	127	17			
4820-01-189-0894	15	108	2540-01-191-4327	245	19			

CROSS-REFERENCE INDEXES

NATIONAL STOCK NUMBER INDEX

STOCK NUMBER	FIG	ITEM	STOCK NUMBER	FIG	ITEM	FIG	ITEM
2540-01-191-4331	245	41	2590-01-191-9275	247	17		
2510-01-191-4334	221	8	2590-01-191-9276	245	39		
2510-01-191-4335	220	15	2530-01-191-9385	116	4		
2510-01-191-4336	220	14	2590-01-191-9426	293	32		
2590-01-191-4357	245	36	2520-01-191-9518	90	1		
5340-01-191-4827	300	8	2540-01-191-9564	231	9		
9330-01-191-4883	299	13	5975-01-191-9851	BULK	5		
2510-01-191-6509	BULK	36	2520-01-192-1257	240	5		
2540-01-191-6510	69	8	4710-01-192-1433	BULK	42		
2590-01-191-6511	208	8	4720-01-192-1631	BULK	12		
2540-01-191-6512	233	12	3040-01-192-1635	84	42		
2540-01-191-6536	217	13	6130-01-192-1643	41	4		
2540-01-191-6537	293	19				41	15
2540-01-191-6538	229	7				65	10
2540-01-191-6539	23	9				66	3
2540-01-191-6540	247	14	2520-01-192-1793	90	10		
2540-01-191-6541	245	34				92	9
2920-01-191-6635	33	11	2540-01-192-1823	25	4		
2510-01-191-6638	221	9	2520-01-192-1919	118	21		
2590-01-191-6649	63	4	2920-01-192-3020	39	1		
2590-01-191-8322	60	11	4720-01-192-3510	245	13		
2540-01-191-8439	147	1	4720-01-192-3520	299	22		
2540-01-191-8440	202	8	4720-01-192-3521	299	23		
2540-01-191-8441	202	9	2540-01-192-3572	146	7		
2920-01-191-8442	35	5	2540-01-192-3573	232	24		
2540-01-191-8443	232	5	2540-01-192-3574	232	23		
2540-01-191-8444	232	6	2540-01-192-3575	232	22		
2540-01-191-8445	232	11	2540-01-192-3576	231	8		
2540-01-191-8446	232	17	2520-01-192-3718	84	25		
2510-01-191-8447	193	6	4710-01-192-3731	299	16		
2610-01-191-8448	15	14	4710-01-192-3732	299	15		
2910-01-191-8453	15	86	2920-01-192-3774	36	14		
2910-01-191-8454	15	86	3040-01-192-4267	15	3		
2910-01-191-8455	15	86	2520-01-192-4314	90	5		
2910-01-191-8456	15	86				92	5
2910-01-191-8457	15	86	5340-01-192-4366	84	29		
2910-01-191-8458	15	86	5995-01-192-4374	222	4		
2910-01-191-8459	15	86	2920-01-192-4375	47	1		
2540-01-191-8462	217	16	2920-01-192-4376	47	9		
4720-01-191-8463	245	43	4730-01-192-4394	252	9		
2910-01-191-8465	15	82	4730-01-192-4434	245	22		
2540-01-191-8467	289	13	3040-01-192-4466	7	15		
2540-01-191-8549	230	2	2910-01-192-4475	KITS			
2540-01-191-8668	197	1	2540-01-192-4479	231	12		
2590-01-191-9269	293	27	2510-01-192-4480	140	44		
2590-01-191-9270	293	26				142	39
2590-01-191-9271	247	24	2990-01-192-4576	245	25		
2590-01-191-9272	247	23	2910-01-192-4585	15	94		
2590-01-191-9273	247	25	2990-01-192-4597	245	21		
2590-01-191-9274	247	21	3020-01-192-4635	84	27		

CROSS-REFERENCE INDEXES

NATIONAL STOCK NUMBER INDEX

STOCK NUMBER	FIG	ITEM	STOCK NUMBER	FIG	ITEM
3040-01-192-4694	293	6	2540-01-193-7892	202	14
6685-01-192-4834	54	1	2540-01-193-7893	201	4
5330-01-192-5779	15	20	2540-01-193-7894	202	3
4820-01-192-5819	252	19	2540-01-193-7895	195	3
2590-01-192-5911	245	37	2540-01-193-7896	195	1
2815-01-192-5962	10	5	5340-01-193-9565	223	31
2510-01-192-5963	1	12	5340-01-193-9654	293	2
4520-01-192-6005	245	23	4730-01-194-0126	14	6
3040-01-192-6024	84	15	2510-01-194-0206	161	1
5340-01-192-6030	15	54	2540-01-194-0261	166	12
2590-01-192-6031	293	31	2520-01-194-0278	88	10
4520-01-192-6073	245	15	4720-01-194-0336	10	3
4710-01-192-7967	10	8	5305-01-194-0614	128	7
2520-01-192-7979	88	11	3120-01-194-0754	231	2
5945-01-192-7985	39	3	5315-01-194-0819	178	7
2520-01-192-8282	82	10	5365-01-194-0879	84	53
4720-01-192-8533	23	22	3010-01-194-0886	84	4
5945-01-192-8653	48	5	3020-01-194-1714	84	38
5330-01-192-8904	245	33	4730-01-194-2002	30	5
5330-01-192-8905	245	27	2520-01-194-2145	84	26
5330-01-192-8906	245	7	5360-01-194-3162	15	7
5330-01-192-9335	245	2	5340-01-194-3188	170	5
2520-01-192-9729	88	16	2540-01-194-3323	245	11
2990-01-192-9730	10	13		284	11
2510-01-192-9752	228	1		288	21
	228	5	5330-01-194-4751	294	5
2540-01-192-9754	178	1	5330-01-194-4752	294	10
2530-01-192-9778	117	34	5330-01-194-4753	294	20
4720-01-192-9823	19	41	5330-01-194-4754	245	24
5330-01-193-0226	300	4	5306-01-194-4977	185	1
5330-01-193-0227	294	9		186	1
5365-01-193-0458	39	11	5365-01-194-5074	174	5
5330-01-193-1840	167	7	5365-01-194-5273	84	33
5305-01-193-2377	15	118	5365-01-194-5274	84	16
5365-01-193-2948	84	49	5365-01-194-5275	84	53
2590-01-193-3443	146	5	5365-01-194-5276	84	53
2540-01-193-3623	215	22	5365-01-194-5277	84	53
	216	22	5340-01-194-5294	8	26
2520-01-193-4053	82	4	5330-01-194-5804	84	6
5680-01-193-5078	215	26	5340-01-194-5805	215	10
	216	18		216	8
5310-01-193-6927	251	2	3120-01-194-6495	84	28
	255	2	4710-01-194-6590	290	7
5360-01-193-7130	69	6	6220-01-194-6591	290	8
5905-01-193-7212	47	4	4710-01-194-6775	86	5
2930-01-193-7802	31	8	3040-01-194-6853	84	22
3020-01-193-7825	84	31	2540-01-194-6875	47	3
2520-01-193-7870	88	7	2520-01-194-6955	84	19
2540-01-193-7884	193	2	3040-01-194-6992	84	67
2540-01-193-7892	201	12	5310-01-194-7069	84	63

CROSS-REFERENCE INDEXES

NATIONAL STOCK NUMBER INDEX

STOCK NUMBER	FIG	ITEM
5310-01-194-7081	88	14
5640-01-194-7193	300	7
2520-01-194-7984	113	11
5310-01-194-9208	251	1
	255	1
5310-01-194-9217	145	41
5310-01-194-9220	88	4
5310-01-194-9221	297	7
5310-01-194-9233	88	3
5310-01-194-9234	178	4
5340-01-194-9279	293	18
5330-01-195-1564	294	15
5640-01-195-4633	294	17
5640-01-195-4634	294	8
5640-01-195-4635	294	16
5640-01-195-4636	294	14
5330-01-195-4880	241	9
5365-01-195-4948	174	5
5310-01-195-5088	84	2
	88	5
5310-01-195-5092	84	62
5320-01-195-5106	43	4
5315-01-195-5234	84	24
5305-01-195-5807	57	10
5365-01-195-5934	294	4
5365-01-195-6203	84	64
5365-01-195-6204	84	64
5306-01-195-6595	1	15
5360-01-195-6699	234	16
4720-01-195-7603	10	7
5306-01-195-7915	1	11
	142	5
5640-01-195-9786	294	11
5306-01-196-0219	111	19
2540-01-196-1622	49	13
5330-01-196-2426	83	8
5340-01-196-3130	127	9
5310-01-196-5587	17	1
5340-01-196-6463	214	13
5315-01-196-6464	228	11
5310-01-196-6465	246	23
5330-01-196-6586	84	9
5640-01-196-7002	294	13
5640-01-196-7003	294	12
6220-01-197-0486	226	2
5315-01-197-0812	84	14
5360-01-197-0870	199	6
5330-01-197-0881	131	6
	133	19
5330-01-197-0897	294	6

STOCK NUMBER	FIG	ITEM	FIG	ITEM
5340-01-197-1199	23	8		
5340-01-197-1259	33	5		
5320-01-197-1394	146	6		
5305-01-197-1475	147	4		
5365-01-197-1481	84	39		
5315-01-197-1482	223	28		
5315-01-197-1483	223	21		
5306-01-197-1492	65	1		
5355-01-197-1501	203	10		
5360-01-197-1505	15	109		
5360-01-197-1506	15	12		
5360-01-197-1507	5	58		
5365-01-197-1524	84	64		
3120-01-197-1535	83	20		
5310-01-197-1538	84	17		
5310-01-197-1539	84	50		
5325-01-197-1547	247	11		
5315-01-197-1548	127	36		
5340-01-197-1550	223	25		
5340-01-197-1585	290	9		
5365-01-197-2286	196	9		
5305-01-197-2319	293	23		
5305-01-197-2320	242	4	298	15
5330-01-197-2454	15	107		
5305-01-197-2535	41	11		
5305-01-197-2536	48	3		
	15		65	18
5306-01-197-3089	38	10	52	3
	11		62	4
			208	5
			271	9
5305-01-197-3112	25	8	165	4
5355-01-197-3172	88	8		
5330-01-197-3229	84	3		
5340-01-197-3244	174	6		
5305-01-197-3287	39	2	243	8
			284	16
			286	1
5305-01-197-3290	186	4		
5340-01-197-3372	85	1		
5340-01-197-3433	10	10		
5340-01-197-3434	9	16		
5340-01-197-3480	84	21		
5365-01-197-3499	1	7		
5325-01-197-3540	88	15		
6220-01-197-3938	290	6		

CROSS-REFERENCE INDEXES

NATIONAL STOCK NUMBER INDEX

STOCK NUMBER	FIG	ITEM	STOCK NUMBER	FIG	ITEM
5310-01-197-4435	247	13	5320-01-200-4017	215	21
	293	14		216	23
5340-01-197-4600	147	2		218	7
5315-01-197-4845	84	37		219	5
3110-01-197-5495	84	35		220	6
3110-01-197-5496	84	59		221	2
5310-01-197-5499	56	13		224	10
	192	1		225	2
3110-01-197-6101	127	44		227	8
5305-01-197-6351	231	14		232	8
5305-01-197-6576	232	4		232	15
5310-01-197-6621	231	17		232	20
	275	18		245	18
3110-01-197-6651	84	40	2910-01-200-4338	14	9
2920-01-197-7229	40	41	5340-01-200-5843	289	2
5365-01-197-8165	125	7	4010-01-200-7581	247	12
5315-01-197-8547	15	10	5305-01-200-7735	215	6
5305-01-197-9418	148	31		216	12
	149	29	2920-01-200-8461	40	20
2910-01-198-0868	15	25	5340-01-200-8473	244	3
4710-01-198-2701	10	9	5305-01-200-9869	248	42
5340-01-198-3434	237	11	5930-01-201-1843	127	35
5305-01-198-4154	88	6	2510-01-201-2420	1	19
5306-01-198-5515	2	20	2520-01-201-2501	106	1
5365-01-198-5516	178	5	6250-01-201-3300	58	6
5325-01-198-8040	157	3	5305-01-201-3334	218	16
5325-01-198-8239	202	10		219	11
5340-01-198-8591	138	1	5305-01-201-3788	222	10
	138	3	5340-01-201-4095	113	4
3120-01-199-1399	84	32	2520-01-201-4096	124	2
3040-01-199-1485	84	47	2520-01-201-4097	69	1
5995-01-199-1579	67	1	2520-01-201-7851	84	20
5340-01-199-2312	26	6	5340-01-201-7954	207	8
	27	10	5320-01-201-9453	224	6
5325-01-199-3461	45	11	5330-01-201-9681	300	12
	55	3	5330-01-201-9682	298	13
	262	15	5340-01-202-2517	90	4
	284	1		92	4
5945-01-199-4431	38	2	3120-01-202-2602	127	6
5340-01-199-4448	209	5	5340-01-202-2622	23	9
5340-01-199-4993	245	3	5310-01-202-2695	185	5
	247	9	3020-01-202-3360	113	9
3020-01-199-7847	234	23	3020-01-202-3361	113	12
2510-01-200-1021	193	6	5930-01-202-3573	85	2
5365-01-200-1290	127	7	5340-01-202-3651	2	4
4210-01-200-2574	165	18	4730-01-202-8523	88	2
2540-01-200-3167	144	29	3120-01-203-0332	127	4
5320-01-200-4017	211	15	5305-01-203-2289	35	1
	212	19	2990-01-203-2426	27	12
	214	15	5360-01-203-6365	245	6

CROSS-REFERENCE INDEXES

NATIONAL STOCK NUMBER INDEX

STOCK NUMBER	FIG	ITEM	STOCK NUMBER	FIG	ITEM
5307-01-203-9081	49	6	5340-01-211-3086	23	6
5306-01-203-9082	235	13	5305-01-211-3103	7	20
5340-01-204-4268	26	7	5310-01-211-3811	277	7
	27	11	2540-01-211-4621	178	3
5330-01-204-4312	300	17	2510-01-211-6608	141	2
5365-01-204-6702	127	22	2510-01-211-6609	139	8
2540-01-205-2509	289	7	2510-01-211-6610	139	2
2540-01-205-2510	289	24	2520-01-211-6755	92	1
2540-01-205-2511	289	22	5305-01-211-7464	14	11
2540-01-205-2512	289	9	3120-01-211-7528	156	6
5310-01-205-2536	124	11	2920-01-212-4771	23	13
5310-01-205-2537	124	11	4720-01-212-4782	252	27
5325-01-205-2545	167	10	2920-01-212-4941	234	21
6685-01-205-3676	24	2	2510-01-212-5819	161	6
5330-01-205-5056	298	12	2520-01-212-6642	73	34
5365-01-205-5966	84	64	5340-01-212-6716	276	6
2920-01-205-6018	36	23	5340-01-212-6717	276	6
4730-01-205-7917	13	14	3040-01-212-7616	7	18
5306-01-205-8882	108	3	2590-01-212-7639	61	6
5360-01-205-8888	163	11	2520-01-213-1680	72	6
2920-01-206-0891	40	12	5340-01-213-5318	127	59
5340-01-206-2995	166	27	5340-01-213-6934	80	4
5340-01-206-2996	166	127	5330-01-213-9811	298	14
5120-01-206-3818	303	2	5330-01-213-9966	23	3
2530-01-206-4860	127	78	5930-01-214-0401	49	2
5360-01-206-6616	291	6	5945-01-214-1490	15	43
5330-01-206-7353	293	11	2540-01-214-2634	275	14
2520-01-207-9169	103	2	2540-01-214-2635	275	14
5330-01-207-9421	180	2	5365-01-214-4927	156	8
5330-01-208-3843	190	5	2590-01-214-7131	247	3
6150-01-208-4507	234	19	5305-01-215-2501	127	16
5930-01-208-6292	23	15	2510-01-215-3931	141	7
4820-01-209-0473	23	14	4730-01-216-0021	288	20
2590-01-209-0934	68	6	2510-01-216-0039	218	1
5365-01-209-6943	124	12		219	8
5315-01-209-7063	223	14	2540-01-216-3188	273	12
5340-01-209-7066	127	76	2530-01-216-4554	124	16
5305-01-209-7069	127	18	6220-01-216-5288	269	9
2910-01-210-1322	23	4	6220-01-216-5289	269	6
2910-01-210-1323	23	7	2520-01-216-5611	94	2
2590-01-210-1357	68	12	6140-01-216-7923	267	6
2520-01-210-1382	93	1	5305-01-217-2046	293	17
6140-01-210-1964	56	11	5340-01-217-2168	273	3
5340-01-210-8824	193	10	5340-01-217-2278	273	5
5305-01-210-9425	167	12	5310-01-217-5205	268	4
5310-01-211-1648	241	7		269	11
2510-01-211-1654	141	1		288	13
2510-01-211-1655	139	1	2940-01-217-8089	8	1
5305-01-211-3031	23	12	5330-01-218-0862	9	1
5305-01-211-3032	23	11		260	3

CROSS-REFERENCE INDEXES

NATIONAL STOCK NUMBER INDEX

STOCK NUMBER	FIG	ITEM	STOCK NUMBER	FIG	ITEM
5360-01-218-1610	199	3	2510-01-225-2237	139	15
5306-01-218-3119	261	19	2510-01-225-2238	165	21
5305-01-218-3139	268	3	2540-01-225-2521	196	25
6145-01-218-3759	BULK	43	5995-01-225-2534	46	18
6145-01-218-3760	BULK	44	6680-01-225-4432	41	5
6145-01-218-3761	BULK	10	6680-01-225-4475	291	3
5340-01-218-5823	276	7	2510-01-225-5865	140	5
4730-01-218-6690	261	14	2520-01-225-5866	139	7
4730-01-218-6691	261	9	2540-01-225-5972	196	1
2540-01-218-6833	178	6	5430-01-225-8971	9	9
2540-01-218-8099	279	6	5365-01-226-2342	267	3
5305-01-219-5399	284	28	5306-01-227-1454	267	13
	288	38	4730-01-227-1929	9	23
5120-01-219-6753	301	19	5306-01-227-9085	39	9
5340-01-219-7272	258	4	5310-01-228-1405	129	2
5340-01-219-7275	276	4		129	7
5365-01-219-7285	264	9	5340-01-228-1659	46	3
2590-01-219-7808	147	5	5320-01-229-8183	46	16
5330-01-220-3117	287	5	2510-01-230-1165	175	10
5330-01-220-6153	264	13	5325-01-230-1844	237	2
5340-01-221-0264	280	11	5306-01-230-3354	8	16
6620-01-221-1942	261	12		46	14
2510-01-221-2094	BULK	9	6680-01-230-5684	291	1
5365-01-221-9717	124	9	2910-01-230-9007	15	104
5340-01-221-9972	260	2	5305-01-230-9846	169	28
2530-01-222-8068	119	9	5340-01-231-0925	93	5
2520-01-223-3754	97	9	5305-01-231-1297	169	22
6140-01-223-9144	57	14	5305-01-231-1298	65	6
5935-01-223-9420	45	13		167	15
2920-01-224-3153	40	37	2510-01-231-2879	175	9
2510-01-224-8839	177	6	2990-01-231-2938	27	22
2510-01-225-0997	149	9	8145-01-231-3747	253	1
2510-01-225-0998	179	17	5320-01-231-3889	46	17
2510-01-225-0999	179	13	2510-01-231-5398	173	7
2510-01-225-1004	140	41	5305-01-231-7384	193	9
2510-01-225-1005	139	5	5365-01-231-7584	13	17
2510-01-225-1006	176	12	2910-01-232-1044	15	103
2540-01-225-1007	165	28	5305-01-232-1436	165	26
2510-01-225-1008	165	27	5330-01-232-2145	15	32
2990-01-225-1021	142	6	3110-01-232-2388	110	21
2540-01-225-1023	176	16	3120-01-232-6781	127	74
2530-01-225-1024	119	8	2510-01-232-7188	139	13
2990-01-225-1028	26	11	2510-01-232-8176	173	9
2990-01-225-1029	26	1	5340-01-232-8179	123	5
2520-01-225-1033	127	72	4710-01-232-8478	255	19
2990-01-225-1052	26	9	5330-01-233-2778	15	57
2910-01-225-1068	25	6	5330-01-233-8597	15	55
2540-01-225-1106	196	2	5365-01-234-0447	5	2
2540-01-225-1108	196	4	5340-01-234-1465	9	7
2530-01-225-2236	119	4	5330-01-234-2615	15	24

CROSS-REFERENCE INDEXES

NATIONAL STOCK NUMBER INDEX

STOCK NUMBER	FIG	ITEM	STOCK NUMBER	FIG	ITEM
6150-01-234-3253	266	1	5310-01-249-4210	144	9
2590-01-234-6468	186	2		145	18
5355-01-235-6616	127	71	2530-01-249-5401	97	1
5995-01-235-6877	66	1	2510-01-249-6434	171	2
5995-01-235-6878	66	10	3120-01-250-0583	157	6
5330-01-236-0476	15	96	5310-01-250-3301	39	10
5330-01-236-1724	23	5	2530-01-250-6472	117	36
5306-01-237-4995	149	23	5310-01-250-7679	13	5
5330-01-237-7512	180	1		34	6
4910-01-238-2551	302	13		35	10
5310-01-238-2983	40	40		132	2
5340-01-238-5923	186	11		291	12
5340-01-238-5924	186	5		295	1
2510-01-238-7051	175	10	2990-01-250-8612	287	6
2510-01-238-7052	175	9	5355-01-251-0633	69	7
2520-01-239-3800	91	1	2815-01-251-1122	7	13
5305-01-239-9265	210	7	5330-01-251-1607	76	13
2540-01-241-4238	186	3	5315-01-251-1701	130	2
2590-01-242-1050	186	3	2540-01-251-1714	201	7
5305-01-242-1148	262	18	2540-01-251-1715	201	8
	284	7	2510-01-251-5487	171	2
	288	16	5340-01-253-2102	208	7
6220-01-242-7557	53	9	5306-01-253-7073	10	1
2590-01-242-8068	274	17	2510-01-254-1075	173	12
5310-01-242-8561	236	3	5305-01-254-2558	31	10
	237	1	6105-01-254-3237	263	28
5945-01-243-1702	55	6	6105-01-254-9496	286	4
2540-01-243-4934	234	9	4730-01-255-2976	119	11
5310-01-244-2259	125	11	2920-01-256-2253	36	13
5330-01-244-2277	234	10	2910-01-256-3698	15	64
5340-01-244-7925	18	11	2990-01-257-1569	56	1
5360-01-245-0405	288	3	3030-01-258-5125	33	10
4730-01-245-6925	252	14	2510-01-259-5587	193	4
5340-01-246-2700	117	4	5307-01-259-7656	270	4
2510-01-246-4236	275	8	5360-01-260-5649	15	88
2510-01-246-4237	273	16	4730-01-263-5361	101	13
2815-01-246-5268	5	7	2930-01-264-3480	28	15
2815-01-246-5269	5	7	5310-01-264-5903	271	12
5305-01-246-5770	233	7		273	14
5306-01-246-7459	68	10		274	9
	177	1	5330-01-264-6537	253	5
	182	3	5340-01-264-6540	253	11
3040-01-247-0893	15	123	5340-01-264-6541	253	27
2590-01-247-3286	193	10	5340-01-264-6543	253	22
4720-01-247-4680	262	21	5340-01-264-6544	253	6
2540-01-248-2477	202	13	5340-01-265-3674	253	28
6220-01-248-6269	50	1	5340-01-265-3676	253	18
5305-01-248-6917	263	33	2815-01-265-7071	2	1
5330-01-249-1629	3	7	5306-01-266-2419	256	4
5360-01-249-4056	263	35		259	6

CROSS-REFERENCE INDEXES

NATIONAL STOCK NUMBER INDEX

STOCK NUMBER	FIG	ITEM	STOCK NUMBER	FIG	ITEM
5306-01-266-2419	259	10	5355-01-280-2975	170	10
5305-01-266-9194	193	3	2510-01-283-1966	142	8
2540-01-267-1360	144	10	2815-01-285-5065	6	3
	145	24	2990-01-287-2158	262	23
4720-01-267-2052	BULK	27	2530-01-287-3980	127	67
2920-01-267-2088	37	41	5365-01-287-5878	117	24
5310-01-267-3043	164	1	5340-01-293-1200	279	1
3040-01-267-4283	144	2	2640-01-302-1388	124	29
	145	11	6220-01-306-4265	52	6
5310-01-267-6293	144	5	5340-01-307-2247	213	10
	145	14	5340-01-307-2248	213	11
5340-01-267-6296	144	4	5340-01-307-2249	211	22
	145	13		212	25
2920-01-267-9423	36	6	5340-01-307-2250	211	21
2520-01-268-1051	253	2		212	24
5307-01-268-2640	101	21	5340-01-307-2251	211	20
5945-01-268-4265	77	16		212	13
2540-01-268-7202	286	3	5340-01-307-2252	211	19
2520-01-268-7413	112	12		212	12
6220-01-268-8795	269	4	5340-01-307-2253	215	29
5310-01-268-8948	165	23		216	29
	243	3	5340-01-307-2254	215	30
5340-01-268-9064	22	10		216	30
	118	4	2815-01-314-6887	7	28
	255	17	4320-01-317-0692	15	83
6220-01-269-0465	269	7	2990-01-320-8915	15	79
5340-01-269-1361	25	12	6350-01-321-7005	68	13
5306-01-269-4319	79	1	4930-01-323-0998	301	18
5305-01-269-4329	41	16	2640-01-323-2632	124	29
	42	2	5306-01-323-5544	93	6
	241	3	5330-01-323-5567	140	24
5307-01-269-4336	1	3	2590-01-323-5857	157	7
	2	7	2540-01-323-6049	235	7
	11	3	2540-01-323-6050	235	7
	30	12	5306-01-323-8967	151	7
5306-01-269-6138	77	18	5340-01-323-9727	114	10
5365-01-269-8614	213	6	2520-01-324-4895	93	1
5365-01-270-1977	127	66	2590-01-324-5042	93	5
5305-01-270-3030	127	68	5340-01-324-6756	140	23
5310-01-270-3712	110	2	5975-01-324-7825	62	6
5310-01-270-5464	110	1	5305-01-324-8862	192	8
5340-01-270-7423	14	2	5340-01-324-9553	140	23
5310-01-271-1793	222	6	2520-01-325-1860	103	2
5340-01-271-3059	253	7	2510-01-325-8216	192	7
6210-01-271-6871	42	15	2510-01-325-9069	157	9
5305-01-273-4486	10	11	2510-01-325-9070	192	2
	123	3	2510-01-325-9071	192	2
	165	14	2510-01-325-9077	192	6
6220-01-276-0635	210	8	2510-01-325-9078	192	6
5330-01-276-3235	296	27	2530-01-325-9112	124	3

CROSS-REFERENCE INDEXES

NATIONAL STOCK NUMBER INDEX

STOCK NUMBER	FIG	ITEM	STOCK NUMBER	FIG	ITEM
2510-01-326-0762	KITS				
5310-01-326-1052	151	24			
5365-01-326-1113	151	8			
5365-01-326-1128	151	9			
2510-01-326-1389	192	3			
2510-01-326-1460	192	3			
2530-01-326-1462	124	23			
2510-01-326-1464	151	21			
2590-01-326-3001	151	22			
5365-01-326-4346	124	26			
2590-01-326-5827	151	22			
2910-01-326-8187	13	4			
6220-01-327-1025	52	6			
6220-01-327-3252	52	4			
6150-01-327-7394	62	1			
5120-01-327-9533	305	2			
5120-01-328-0260	305	1			
2520-01-328-4898	93	1			
3040-01-329-9927	111	17			
2520-01-330-0016	111	25			
2520-01-330-0028	111	18			
2520-01-330-3249	108	9			
3020-01-330-3258	111	21			
2520-01-330-3703	111	20			
5365-01-330-6970	111	12			
5365-01-330-8450	KITS				
5310-01-330-9817	111	2			
5330-01-331-1207	KITS				
5330-01-331-7230	125	6			
2520-01-332-3841	111	14			
3020-01-332-5456	111	13			
2520-01-334-1731	111	11			
5340-01-335-9359	114	8			
5340-01-335-9360	119	19			
3120-01-338-6380	177	2			

PART NUMBER INDEX

CAGEC	PART NUMBER	STOCK NUMBER	FIG	ITEM
79846	ABA64LBA	5320-01-231-3889	46	17
30760	AHC68-31	3110-00-108-9247	36	35
72794	AJW7-70	5325-00-337-6636	267	2
88044	AN365-1024A	5310-01-208-1918	228	6
			288	12
60380	AR-33761	5315-01-197-4845	84	37
02697	ARP583-116	5330-00-451-0118	83	71
46522	A14140	4730-00-460-6725	263	17
46522	A14141	4730-00-319-0454	263	19
46522	A18167-G2	4520-00-567-1886	263	18
46522	A2096	5935-00-768-7042	263	26
46522	A2322	5930-01-106-6382	263	4
46522	A2480	5977-01-145-4308	263	42
46522	A2481=XA		263	41
46522	A2610	2540-00-454-0665	263	23
46522	A2612	5365-01-158-6819	263	22
46522	A2829	4720-00-139-4105	263	32
46522	A2895	4730-00-277-6339	263	36
46522	A2896	5360-01-249-4056	263	35
46522	A2897	5305-01-248-6917	263	33
46522	A2904	5330-01-183-5688	263	38
53335	A2931	5330-01-109-7941	263	39
46522	A2932	5330-00-472-6954	263	37
05657	A5	4730-00-087-7273	263	14
46522	A52789-G5	4720-01-124-8252	263	15
46522	A53181	4730-01-145-4326	263	13
46522	A54201-G1-1	5340-01-145-4311	263	6
46522	A54814-G1	5325-01-145-4349	263	20
46522	A54907-G1	4710-01-131-3398	263	16
46522	A54940-G1	2990-01-131-2751	263	45
46522	A54948-G2	4540-01-105-5667	263	8
46522	A6005943	5940-00-983-6082	263	3
70040	A644C	2940-01-155-3190	17	4
96152	A82-1	5315-00-839-2325	288	6
76760	A85344	5365-00-423-2811	84	41
53335	A930	5360-01-107-3245	263	40
82240	B-1900-334	5340-00-685-5899	288	37
79227	B-6000-1/4IN	4820-01-192-5819	252	19
73992	B-84	4730-01-218-6691	261	9
73992	B-85	4730-01-218-6690	261	14
60380	B2012-0H	3110-00-227-4431	83	45
46522	B2714	9905-01-145-4320	263	43
46522	B2798	6105-01-145-4324	263	29
46522	B2894	4320-00-217-5827	263	30
46522	B54028G1-1	5340-01-027-2628	263	11
46522	B54827-G1	6105-01-254-3237	263	28
46522	B54941-G1	4820-01-130-3387	263	12
00613	C-5139-2	5325-01-199-3461	45	11
			55	3
			262	15
			284	1

PART NUMBER INDEX

CAGEC	PART NUMBER	STOCK NUMBER	FIG	ITEM
16632	CHE2013-0001	4720-01-167-9137	287	2
16632	CHE2015-0001	4720-01-162-7098	284	14
			285	1
16632	CHE2015-0002	4720-01-162-7097	285	4
16632	CHE2016-0002	5330-01-164-7509	287	4
16632	CHE2017	5330-01-163-3150	284	29
16632	CHE2018	5330-01-163-3151	288	25
16632	CHE2023	5330-01-220-3117	287	5
16236	CS-4520-SV-0705	4520-00-217-5782	246	38
60380	C1086Q	3110-00-198-0492	83	80
46522	C2757	4520-01-132-8908	263	25
46522	C54813-G1	4810-01-121-9786	263	10
46522	C54919-G1-1	5340-01-146-1905	263	24
80604	D-167	5120-01-328-0260	305	1
84760	DB2829-4521	2910-01-326-8187	13	4
60380	DC57524	3110-01-232-2388	110	21
5A910	DC8211	5330-01-037-0663	50	5
5A910	DC8218	6220-01-107-2613	50	2
34904	DC8226	5330-01-076-6172	50	3
5A910	DC8228	5310-01-076-6196	50	6
80604	DD-914-95	5120-01-327-9533	305	2
80063	DL-SC-B-691368	5820-01-026-0983	207	4
72210	D23980-J	2530-01-165-9654	129	1
46522	D55395-G1	2990-01-287-2158	262	23
19954	EDS-98477-48	3120-01-185-3678	113	17
70040	FC106	2940-01-217-8089	8	1
55719	FC18A	5120-00-184-8397	304	7
11862	FLW-12		57	2
			65	13
			66	6
11862	FLW-16		65	12
			66	4
11862	FLW-18		65	9
			66	2
11862	FLW-20		65	11
			66	5
60380	FNT5070	3110-01-155-2598	83	48
08805	F48T12/CW/WM	6240-01-089-6149	213	20
60380	F5020	3110-01-119-4441	83	55
78385	G-488934		249	13
78385	G-700038		249	8
78385	G-700637-100	4810-01-161-0817	248	36
78385	G-700637-101	4810-01-161-0817	248	7
78385	G-704183		246	9
78385	G-704195		246	28
78385	G-704196		250	6
78385	G-704213		246	2
78385	G-704232		246	25
78385	G-704234	5935-01-163-8987	246	16
78385	G-704284		250	4
78385	G-704288-1		246	37

PART NUMBER INDEX

CAGEC	PART NUMBER	STOCK NUMBER	FIG	ITEM
78385	G-704293		246	39
78385	G-704373		246	31
78385	G-704398		248	29
78385	G-704553		249	5
78385	G-704554	2540-01-008-1501	246	34
78385	G-720122		249	6
81240	GM118749	4730-00-142-2177	255	12
			262	16
			284	4
			288	14
11862	GM8670757	5306-01-150-4835	77	7
81348	GP2A/LT235/85R16 /E/LTAW	2610-01-148-1635	126	1
78385	G704177		246	32
78385	G704410-1	5930-00-234-1390	264	12
78385	G704676		248	15
78385	G705349		248	27
60038	HM807010	3110-00-227-4667	111	9
60038	HM807040	3110-00-606-9576	111	10
60038	HM88542	3110-00-580-3708	111	6
08806	H6054	6240-01-180-9022	50	34
89749	IF316	5315-00-816-1794	87	17
			88	12
			128	16
25341	J-21362	4910-01-178-6551	303	7
25341	J-21363	4910-01-178-0713	303	9
25341	J-21370	4910-01-178-0722	303	10
25341	J-21409	4910-01-178-0360	303	16
25341	J-21465-1	5120-01-169-4877	303	12
25341	J-21552	5120-01-179-1032	303	11
25341	J-21757-03	4910-01-179-6341	301	5
25341	J-21777-500	4910-01-178-8866	303	13
25341	J-21795-02	4910-01-178-8865	303	14
25341	J-2222-C	5120-01-180-0558	301	2
25341	J-22779	4910-01-183-0044	303	15
25341	J-23327	4910-01-178-0724	303	8
25341	J-23445-A	5120-01-169-4878	301	6
25341	J-23447	4910-01-181-1958	302	11
25341	J-23454-D	4910-01-179-2515	302	12
25341	J-23653-C	4910-01-180-6155	301	7
25341	J-23690	5120-01-170-3279	301	8
25341	J-23996	4910-01-179-6339	302	16
25341	J-24187	4910-01-179-6340	301	9
25341	J-24426	5120-01-179-1034	301	10
25341	J-24429	5120-01-170-0627	302	2
25341	J-24595-C	4910-01-179-2518	301	11
25341	J-25034-B	4910-01-181-1959	301	16
25341	J-26252	5120-01-180-7928	304	1
25341	J-26471	5120-01-180-8592	302	4
25341	J-26871-A	5120-01-183-8576	302	8
25341	J-26878-A	5120-01-170-0628	301	13

PART NUMBER INDEX

CAGEC	PART NUMBER	STOCK NUMBER	FIG	ITEM
25341	J-26889	4910-01-178-8864	302	3
25341	J-26941	4910-01-179-6364	304	2
33287	J-26999-30	4910-01-238-2551	302	13
25341	J-28402	5120-00-224-7288	302	7
25341	J-29162	4910-01-179-5530	304	3
25341	J-29167	5120-01-170-3278	304	4
25341	J-29168	5120-01-169-4876	304	5
25341	J-29369-1	4910-01-181-0183	304	6
25341	J-29601	4910-01-178-9788	302	5
25341	J-29664-1	4910-01-179-4870	302	9
25341	J-29698-A	5120-00-189-7898	302	14
25341	J-29713	4910-01-179-2516	301	15
33287	J-29843	5120-01-178-6342	301	17
25341	J-29873	5120-01-171-5233	302	6
25341	J-33043	4820-01-179-4869	301	14
25341	J-33122	4910-01-182-2704	302	10
25341	J-33124	5120-01-170-5473	302	20
25341	J-33154	5120-01-179-1033	302	15
25341	J-34502	5120-01-170-6703	304	8
25341	J-34616	5120-01-170-6664	301	19
33287	J-35178	5120-01-206-3818	303	2
25341	J-6632-01	5120-01-179-1318	301	3
25341	J-6893-D	5120-01-219-6753	301	19
25341	J-7624	5120-01-082-6436	303	3
25341	J-7728	5120-01-082-6448	303	4
33287	J-8092	5120-00-677-2259	301	4
25341	J-8107-2	5310-01-179-9486	303	5
25341	J-9480-B	4910-01-179-6338	303	6
33627	J25512-2	4930-01-323-0998	301	18
43334	LM104949LM104911	3110-01-027-4475	124	17
43334	LM501349-LM501310	3110-00-690-8923	125	22
43334	LM501349/LM501314	3110-00-135-0983	99	4
			112	9
81640	L30200R	6210-00-438-4745	283	7
9W635	MBH-V	2990-01-173-4197	252	8
81349	MILS20166		BULK	2
81349	MILT16343TYPE1		BULK	41
96906	MS14151-6	5310-01-018-5634	250	3
96906	MS15001-1	4730-00-050-4203	144	3
			145	12
96906	MS16562-147	5315-00-057-5541	223	22
96906	MS16562-33	5315-00-843-7986	36	28
96906	MS16562-35	5315-00-814-3530	77	3
			224	8
96906	MS16562-36		296	14
96906	MS16562-50	5315-00-814-3531	223	29
96906	MS16624-1024	5365-00-720-8064	178	10
96906	MS16624-1031	5365-00-803-7313	40	13
96906	MS16624-1093	5365-00-804-2027	15	78

PART NUMBER INDEX

CAGEC	PART NUMBER	STOCK NUMBER	FIG	ITEM
96906	MS16624-1125	5365-00-721-6876	125	5
			125	9
96906	MS16624-1131	5365-00-803-7317	124	7
96906	MS16624-1175	5365-00-721-6476	102	6
			102	22
96906	MS16627-1087	5365-00-514-1277	79	27
96906	MS16633-1015	5365-00-200-7377	203	22
96906	MS16633-1018	5365-00-282-1633	234	25
96906	MS16633-1021	5365-00-900-0982	178	8
96906	MS16633-1031	5365-00-682-1762	77	25
			115	13
			116	15
96906	MS17131-42	3110-00-647-1100	83	11
96906	MS17131-47	3110-00-056-9377	83	56
96906	MS17829-50	5310-00-245-3424	22	32
			114	13
96906	MS18029-13L-5		44	2
			46	9
96906	MS18029-13S-3		44	15
96906	MS18029-13S-8	5940-00-926-8034	BULK	6
96906	MS18029-24	5940-00-890-2831	44	1
			46	10
96906	MS18029-4S-8	5940-00-405-8976	BULK	7
96906	MS18154-58	5305-00-115-9526	23	17
			27	4
96906	MS18154-96	5305-00-182-9584	123	6
96906	MS19060-1014	3110-00-724-7884	263	34
96906	MS19061-20005	3110-00-950-4236	234	13
96906	MS19061-20007	3110-00-900-2560	77	8
96906	MS2060D-B4W3	5320-00-582-3521	211	2
			212	1
			213	8
			215	18
			216	26
96906	MS20600-B4W4	5320-00-845-9501	213	14
			215	3
			216	15
			220	17
			227	9
			288	23
96906	MS20600-B6W10	5320-00-822-6257	218	9
			219	2
96906	MS20600-B6W12	5320-00-637-5014	229	4
96906	MS20600-B6W5	5320-00-061-9662	215	2
			216	14
96906	MS20600B6W4	5320-00-616-4346	211	13
			212	15
			213	1
			215	31
			216	31
			218	3

PART NUMBER INDEX

CAGEC	PART NUMBER	STOCK NUMBER	FIG	ITEM
96906	MS20600B6W4	5320-00-616-4346	219	3
			220	16
			221	10
			222	9
			227	4
			245	38
96906	MS20600B6W8	5320-00-616-4344	229	1
96906	MS20600MP6W10	5320-00-660-0821	215	9
			216	9
96906	MS20601-B6W6	5320-00-061-9648	220	2
			221	6
			224	3
96906	MS20613-4P4	5320-00-994-7076	56	9
96906	MS20913-6S	4730-00-221-2140	253	14
96906	MS21043-08	5310-00-878-7196	250	12
96906	MS21045-6	5310-00-982-4908	83	68
96906	MS21141-00604	5320-01-200-4017	211	15
			212	19
			214	15
			215	21
			216	23
			218	7
			219	5
			220	6
			221	2
			224	10
			225	2
			227	8
			232	8
			232	15
			232	20
			245	18
96906	MS21141-00607	5320-01-201-9453	224	6
96906	MS21207-10-10	5305-00-057-9608	228	12
96906	MS21318-47	5305-00-253-5626	144	16
			145	23
96906	MS21333-105	5340-00-809-1494	59	8
96906	MS21333-107	5340-00-809-1500	293	38
96906	MS21333-110	5340-00-057-3034	95	9
			252	2
			255	21
			262	22
			284	8
96906	MS21333-111	5340-00-057-3037	21	2
			22	5
			22	6
96906	MS21333-112	5340-00-057-3043	95	2
			104	7
			105	3
			268	2
			169	10

PART NUMBER INDEX

CAGEC	PART NUMBER	STOCK NUMBER	FIG	ITEM
96906	MS21333-114	5340-00-057-3052	28	10
			60	5
			64	3
			268	9
96906	MS21333-116	5340-00-764-7052	45	10
96906	MS21333-123	5340-00-989-1771	268	8
			269	15
96906	MS21333-128	5340-00-702-2848	280	10
			281	3
			281	3
96906	MS21333-45	5340-00-881-5303	12	11
			22	16
			119	21
96906	MS21333-46	5340-00-282-7539	255	23
96906	MS21333-48	5340-00-486-1765	95	4
96906	MS21333-62	5340-00-282-7509	60	4
			264	15
96906	MS21333-78	5340-00-958-8457	280	3
96906	MS21333-98	5340-00-809-1490	20	17
			21	11
			119	21
96906	MS24585-1276	5360-01-245-0405	288	3
96906	MS24627-67	5305-00-052-7472	229	6
96906	MS24662-153	5320-00-982-3815	215	24
			216	19
			225	6
			228	7
96906	MS24662-155	5320-00-828-1284	218	13
			219	15
96906	MS24665-134	5315-00-839-5820	223	4
96906	MS24665-283	5315-00-842-3044	87	13
			289	18
96906	MS24665-353	5315-00-839-5822	7	11
96906	MS24665-357	5315-00-298-1481	102	5
			102	20
			128	10
96906	MS24665-425	5315-00-013-7238	129	5
96906	MS24665-446	5315-00-899-4119	245	35
96906	MS24665-497	5315-00-013-7258	143	4
			144	26
			145	32
96906	MS24665-516	5315-00-243-1170	287	8
96906	MS24665-628	5315-00-846-0126	144	7
			145	16
96906	MS25043-32DA	5935-01-223-9420	45	13
96906	MS25244-P-20	5925-01-067-2926	283	3
96906	MS25307-312	5930-00-978-8805	283	5
96906	MS25331-4-313S	6210-00-688-5088	247	15
			283	8
96906	MS27183-10	5310-00-809-4058	39	7
			57	17

PART NUMBER INDEX

CAGEC	PART NUMBER	STOCK NUMBER	FIG	ITEM
96906	MS27183-10	5310-00-809-4058	61	11
			264	4
			270	6
			284	15
			288	2
			288	27
			297	9
96906	MS27183-11	5310-00-809-3078	114	20
			165	2
			217	18
96906	MS27183-12	5310-00-081-4219	19	32
			22	35
			87	14
			231	5
			253	23
			256	1
			259	7
			259	7
			277	5
			280	5
			280	5
			281	4
			281	4
96906	MS27183-13	5310-00-087-7493	1	21
			151	12
			215	11
			216	5
			223	26
96906	MS27183-14	5310-00-080-6004	81	8
			82	11
			83	67
			88	13
			145	28
			148	1
			149	1
			232	1
			261	18
			273	7
			273	19
			274	3
96906	MS27183-15	5310-00-809-4061	18	6
			19	38
			69	17
			140	39
			142	30
			145	36
			223	10
			253	8
96906	MS27183-16	5310-00-809-4085	1	10
			26	16
			27	26

PART NUMBER INDEX

CAGEC	PART NUMBER	STOCK NUMBER	FIG	ITEM
96906	MS27183-16	5310-00-809-4085	33	7
			34	7
			127	57
			127	79
			139	12
			140	3
			140	45
			142	37
			148	6
			179	18
			276	16
			277	15
96906	MS27183-17	5310-00-809-5997	217	9
			276	17
			277	21
96906	MS27183-18	5310-00-809-5998	135	5
			136	5
			137	10
			144	18
			145	38
			147	8
			154	2
			160	10
			172	8
			174	3
			253	17
96906	MS27183-19	5310-00-809-3079	135	12
			144	22
			148	27
			149	26
			150	3
			151	2
			152	4
			179	20
			253	26
96906	MS27183-20	5310-00-068-5285	128	21
			172	4
96906	MS27183-21	5310-00-823-8803	148	20
			149	19
			150	10
			151	24
			152	1
96906	MS27183-42	5310-00-014-5850	245	29
			283	14
96906	MS27183-6	5310-00-082-1404	46	2
96906	MS27183-8	5310-00-809-8546	262	17
			284	5
			288	15
96906	MS27183-9	5310-00-823-8804	57	5
			217	3
			217	14

PART NUMBER INDEX

CAGEC	PART NUMBER	STOCK NUMBER	FIG	ITEM
96906	MS27183-9	5310-00-823-8804	252	21
96906	MS27212-4-3		44	13
96906	MS27212-4-5		44	8
			46	4
96906	MS27212-4-8	5940-00-950-7783	BULK	37
96906	MS29513-125	5330-00-265-1089	245	31
96906	MS3367-1-9	5975-00-074-2072	293	43
96906	MS3452W18-11P	5935-01-154-6264	283	10
96906	MS35058-22	5930-00-655-1514	247	19
			283	4
96906	MS35190-254	5305-00-958-5477	249	12
96906	MS35206-226	5305-00-984-4983	246	12
			269	8
96906	MS35206-231	5305-00-889-3001	37	47
96906	MS35206-241	5305-00-984-6189	246	35
			249	4
96906	MS35206-243	5305-00-984-6191	38	1
			213	18
			246	19
96906	MS35206-245	5305-00-984-6193	68	9
			269	1
96906	MS35206-247	5305-00-984-6195	245	14
96906	MS35206-263	5305-00-984-6210	40	26
			283	20
96906	MS35206-267	5305-00-984-6214	45	12
96906	MS35207-261	5305-00-990-6444	249	2
96906	MS35207-265	5305-00-993-1848	37	20
96906	MS35265-79	5305-00-543-4302	40	2
96906	MS35266-66	5305-00-614-0248	263	21
96906	MS35333-37	5310-00-579-0079	246	15
96906	MS35333-41	5310-00-167-0721	48	6
			108	6
96906	MS35333-70	5310-00-550-3715	283	12
96906	MS35335-30	5310-00-209-0788	246	6
96906	MS35335-31	5310-00-596-7693	246	20
			248	23
			249	3
96906	MS35335-32	5310-00-596-7691	249	1
			298	6
			300	6
96906	MS35335-33	5310-00-209-0786	45	4
			47	10
			52	2
			53	2
			61	2
			62	3
			65	22
			207	12
			298	4
96906	MS35335-34	5310-00-514-6674	27	15
			44	4

PART NUMBER INDEX

CAGEC	PART NUMBER	STOCK NUMBER	FIG	ITEM
96906	MS35335-34	5310-00-514-6674	46	7
			57	9
			58	1
			59	2
			60	10
			63	3
			233	3
96906	MS35335-35	5310-00-627-6128	82	6
			84	54
96906	MS35335-36	5310-00-550-3503	57	12
			65	21
96906	MS35338-139	5310-00-933-8121	207	2
96906	MS35338-38		15	28
96906	MS35338-43	5310-00-045-3296	214	10
			283	18
			293	21
96906	MS35338-44	5310-00-582-5965	9	13
			22	18
			45	9
			48	4
			182	6
			262	3
			268	10
			282	8
			293	39
			294	3
			294	22
96906	MS35338-45	5310-00-407-9566	19	26
			22	21
			26	3
			27	8
			66	8
			69	13
			89	15
			90	15
			128	3
			184	13
			211	10
			212	10
			238	7
			253	24
			262	8
			284	23
96906	MS35338-46	5310-00-637-9541	23	18
			27	5
			49	21
			50	8
			57	26
			81	5
			82	2
			116	2

PART NUMBER INDEX

CAGEC	PART NUMBER	STOCK NUMBER	FIG	ITEM
96906	MS35338-46	5310-00-637-9541	128	6
			140	21
			140	37
			142	29
			145	27
			148	4
			149	4
			154	6
			155	4
			157	5
			215	14
			216	3
			223	23
			253	9
			294	19
96906	MS35338-47	5310-00-209-0965	223	12
			252	22
96906	MS35338-48	5310-00-584-5272	128	19
			253	19
96906	MS35338-49	5310-00-167-0680	154	9
			155	8
96906	MS35338-52	5310-00-754-2005	300	15
96906	MS35338-53	5310-00-584-7889	84	45
96906	MS35340-43	5310-00-721-7809	40	33
96906	MS35340-44	5310-00-682-5930	40	3
96906	MS35340-45	5310-00-959-4679	58	2
			114	5
			119	6
			134	12
			237	5
			280	2
			280	2
			281	2
			281	2
			287	13
96906	MS35340-46	5310-00-959-4675	53	4
			142	22
96906	MS35340-47	5310-00-655-9370	110	6
			137	5
96906	MS35340-48	5310-00-834-7606	101	3
			110	19
			115	3
			136	4
			137	9
			144	28
			145	3
			148	8
			149	7
			153	2
			181	9
			187	4

PART NUMBER INDEX

CAGEC	PART NUMBER	STOCK NUMBER	FIG	ITEM
96906	MS35340-48	5310-00-834-7606	203	17
			217	6
96906	MS35340-49	5310-00-933-8123	108	2
			154	3
96906	MS35340-50	5310-00-045-5001	155	2
96906	MS35340-51	5310-00-052-6454	115	2
96906	MS35425-72	5310-01-077-6817	188	3
			189	2
96906	MS35478-1073	6240-00-617-0991	50	4
96906	MS35489-103	5325-00-279-1248	255	26
96906	MS35489-11	5325-00-291-9366	267	14
96906	MS35489-121	5325-00-807-0580	264	5
			265	3
96906	MS35489-135	5325-00-263-6648	293	20
			294	7
96906	MS35489-5	5325-00-174-5314	246	5
			282	1
96906	MS35489-6	5325-00-263-6632	283	9
96906	MS35492-54	5305-00-533-5542	273	21
			274	18
			277	2
96906	MS35492-57	5305-00-901-3110	276	3
96906	MS35492-82	5305-00-901-3144	277	17
96906	MS35493-37	5305-00-935-7506	273	29
96906	MS35493-76	5305-00-014-9926	275	9
96906	MS35494-83	5305-00-245-4144	277	11
96906	MS35649-202	5310-00-934-9758	40	32
			49	3
			59	9
			214	9
			283	17
96906	MS35649-205	5310-00-934-9764	48	2
96906	MS35649-244	5310-00-934-9748	283	13
96906	MS35649-262	5310-00-934-9747	44	7
			46	5
			246	7
96906	MS35649-282	5310-00-934-9757	15	29
			36	17
			65	23
			246	4
			248	24
96906	MS35650-302	5310-00-934-9751	15	120
96906	MS35650-3314	5310-00-252-8748	44	3
			46	8
96906	MS35690-824	5310-00-010-3028	128	18
96906	MS35691-1	5310-00-851-2674	217	4
			217	12
			220	9
			252	23
96906	MS35691-21	5310-00-975-2075	102	8
			102	25

PART NUMBER INDEX

CAGEC	PART NUMBER	STOCK NUMBER	FIG	ITEM
96906	MS35691-29	5310-00-835-2036	87	18
			88	9
			223	11
96906	MS35691-406	5310-01-159-8559	9	11
			22	17
			61	12
			182	7
			262	2
			264	10
			268	11
			282	9
96906	MS35691-9	5310-00-891-1709	217	17
			238	10
96906	MS35692-37	5310-00-842-1490	128	17
96906	MS35692-53	5310-00-842-7783	128	9
96906	MS35692-61	5310-00-998-0608	102	4
			102	19
96906	MS35692-94	5310-00-849-6882	144	6
			145	15
96906	MS35751-17	5306-00-889-2943	275	1
96906	MS35751-19	5306-00-027-0722	277	19
96906	MS35751-46	5306-00-685-7790	276	19
96906	MS35751-73	5306-00-177-5707	277	22
96906	MS35756-11	5315-00-616-5523	37	35
96906	MS35790-9	5310-00-685-6855	249	11
96906	MS35842-10	4730-00-908-3195	23	23
			28	8
			86	2
			95	6
			104	5
			105	5
			263	31
96906	MS35842-11	4730-00-908-3194	10	2
			19	40
			20	27
			30	17
			127	50
			244	1
			261	1
			280	8
			281	7
96906	MS35842-13	4730-00-909-8627	19	39
			20	22
			30	1
96906	MS35842-14	4730-00-908-6292	19	10
			284	13
96906	MS35842-15	4730-00-908-6293	245	16
			256	7
			256	7
			257	3
			259	3

PART NUMBER INDEX

CAGEC	PART NUMBER	STOCK NUMBER	FIG	ITEM
96906	MS35842-15	4730-00-908-6293	259	3
			285	5
			287	1
96906	MS35842-16	4730-00-908-6294	245	10
			284	20
			288	19
			288	29
96906	MS39158-23	4730-01-192-4394	252	9
96906	MS39173-2	4730-00-288-9497	248	17
96906	MS45904-57	5310-00-061-0004	246	3
96906	MS45904-68	5310-00-889-2528	204	15
96906	MS51023-49	5305-00-272-3533	288	5
96906	MS51085-1	2910-00-025-3493	245	30
			262	13
			284	3
			288	8
96906	MS51321-1	2910-00-710-6054	251	11
			255	10
96906	MS51335-1	2540-00-078-6633	144	1
			144	12
			145	10
			145	19
96906	MS51845-4	4730-00-249-3885	261	10
96906	MS51846-64	4730-00-196-1991	261	11
96906	MS51849-33	5305-00-227-1543	44	10
			46	1
96906	MS51849-53	5305-00-115-9406	263	5
96906	MS51849-55	5305-00-115-9934	47	7
96906	MS51849-70	5305-01-087-1917	245	28
96906	MS51849-78	5305-00-240-6668	273	9
96906	MS51850-44	5305-00-823-9139	289	19
96906	MS51850-64	5305-00-446-9901	269	12
96906	MS51850-86	5305-00-146-2524	19	24
			41	7
			41	8
			49	12
			61	1
			69	9
			243	2
96906	MS51861-24	5305-00-432-4163	245	5
			247	8
96906	MS51861-38	5305-00-432-7953	51	3
96906	MS51861-45	5305-00-432-4201	176	7
96906	MS51862-26	5305-00-483-0554	50	13
			193	5
96906	MS51862-35	5305-00-432-8220	179	16
96906	MS51863-21	5305-00-115-9436	248	14
96906	MS51863-22	5305-01-076-6308	246	14
96906	MS51863-32	5305-01-097-7894	246	33
96906	MS51869-24	5305-01-158-2032	19	3
			19	6

PART NUMBER INDEX

CAGEC	PART NUMBER	STOCK NUMBER	FIG	ITEM
96906	MS51869-24	5305-01-158-2032	20	4
			173	2
			176	5
			192	5
96906	MS51869-26	5305-01-163-5761	247	6
			293	40
			294	2
			294	21
96906	MS51869-28	5305-01-211-7464	14	11
96906	MS51871-14	5305-01-163-2423	294	18
96906	MS51871-4	5305-01-132-2166	50	28
			233	16
96906	MS51877-4	4730-00-278-8886	118	12
96906	MS51884-9	4730-00-187-4210	97	16
96906	MS51939-2	5340-00-916-6539	232	9
			232	14
			232	19
96906	MS51943-33	5310-00-814-0673	231	7
			256	2
			259	8
			259	11
			280	6
			280	6
			281	5
			281	5
96906	MS51952-1	4730-00-053-0266	284	32
			288	7
96906	MS51957-17	5305-00-054-5651	283	11
96906	MS51957-26	5305-00-054-6650	283	6
96906	MS51957-30	5305-00-054-6654	231	16
			275	20
96906	MS51957-33	5305-00-054-6657	271	2
96906	MS51957-45	5305-00-054-6670	283	15
96906	MS51957-71	5305-00-050-9237	214	16
96906	MS51957-83	5305-00-071-1318	272	1
96906	MS51957-84	5305-00-071-2087	271	13
96906	MS51964-49	5305-00-724-6798	263	27
96906	MS51967-12	5310-00-896-0903	18	5
			19	37
			60	8
			137	4
			140	42
			142	38
			145	4
96906	MS51967-14	5310-00-768-0318	135	6
			136	3
			137	8
			144	27
			145	2
			146	1
			148	7

PART NUMBER INDEX

CAGEC	PART NUMBER	STOCK NUMBER	FIG	ITEM
96906	MS51967-14	5310-00-768-0318	149	6
			153	1
			154	4
			160	9
			172	9
			174	4
			181	10
			186	7
			217	5
			253	20
96906	MS51967-17	5310-00-763-8913	154	8
			155	7
96906	MS51967-2	5310-00-761-6882	45	8
			52	1
			53	1
			62	2
			165	1
			288	35
96906	MS51967-5	5310-00-880-7744	142	28
			211	11
			212	11
			253	25
96906	MS51967-6	5310-00-931-8167	19	25
			22	20
			26	2
			27	9
			32	5
			69	12
			89	16
			90	16
			114	19
			119	10
			134	11
			173	5
			175	4
			184	14
			262	9
			284	22
96906	MS51967-8	5310-00-732-0558	27	6
			49	22
			50	9
			53	5
			140	19
			140	38
			145	26
			148	3
			149	3
			154	5
			155	3
			157	4
			255	29

PART NUMBER INDEX

CAGEC	PART NUMBER	STOCK NUMBER	FIG	ITEM
96906	MS51968-20	5310-00-763-8905	148	19
			149	18
			150	11
			151	23
			152	2
			155	1
96906	MS51968-24	5310-00-763-8894	115	1
96906	MS51968-5	5310-00-880-7746	48	1
96906	MS51968-6	5310-00-905-4600	128	4
96906	MS51968-8	5310-00-732-0559	76	1
			151	11
			223	24
96906	MS51968-9	5310-00-785-1762	116	1
			128	5
96906	MS51971-1	5310-00-903-5966	207	5
96906	MS520390079	9515-00-516-5756	BULK	35
96906	MS521304B203R	4720-00-930-2231	BULK	22
96906	MS52149-1	6140-01-210-1964	56	11
96906	MS52150-30HE	5340-01-204-4268	26	7
			27	11
96906	MS52150-31HE	5340-01-199-2312	26	6
			27	10
96906	MS75004-1	5940-00-549-6581	57	39
96906	MS75004-2	5940-00-549-6583	57	39
96906	MS87006-53	4030-00-916-2141	144	14
			145	21
96906	MS90724-34	5310-01-154-2273	42	1
			50	16
			51	6
			52	8
			158	3
			164	3
			165	9
			167	1
			169	21
			194	2
96906	MS90724-40	5310-00-490-4639	167	3
			191	10
96906	MS90724-7	5310-00-144-8453	283	16
96906	MS90725-5	5305-00-068-0501	22	19
			77	15
			273	11
96906	MS90725-6	5305-00-068-0502	207	3
96906	MS90725-69	5305-00-269-3219	253	10
96906	MS90727-103	5305-00-709-8284	148	12
			149	11
96906	MS90727-109	5305-00-719-5184	115	4
96906	MS90727-111	5305-00-719-5219	101	4
96906	MS90727-116	5305-00-719-5239	110	20
96906	MS90727-117	5305-00-719-5240	144	11
			145	25

PART NUMBER INDEX

CAGEC	PART NUMBER	STOCK NUMBER	FIG	ITEM
96906	MS90727-145	5305-01-194-0614	128	7
96906	MS90727-57	5305-00-269-3233	57	25
			253-13	
96906	MS90727-60	5305-00-269-3236	116	7
96906	MS90727-85	5305-00-709-8517	223	7
96906	MS90728-10	5305-00-068-0509	9	20
			220	12
			298	11
			300	14
96906	MS90728-109	5305-00-071-2066	35	7
96906	MS90728-110	5305-00-732-0511	253	21
96906	MS90728-111	5305-00-071-2067	135	4
			136	9
			138	2
			145	8
96906	MS90728-113	5305-00-071-2069	144	17
			253	29
96906	MS90728-114	5305-00-071-2070	187	5
			217	10
96906	MS90728-115	5305-00-071-2071	279	8
96906	MS90728-117	5305-00-071-2073	96	12
			145	39
96906	MS90728-119	5305-00-071-2075	135	8
			136	7
			144	23
			145	9
			153	5
96906	MS90728-121	5305-00-071-2077	153	3
			172	5
96906	MS90728-125	5305-00-071-2081	187	6
96906	MS90728-127	5305-00-071-2083	160	4
			253	16
96906	MS90728-13	5305-00-071-2510	271	14
			273	1
			274	7
			275	15
96906	MS90728-14	5305-00-071-2511	271	11
96906	MS90728-143	5305-01-160-2005	108	3
96906	MS90728-150	5305-01-154-4318	150	2
			151	3
			152	5
96906	MS90728-152	5305-01-197-9418	148	31
			149	29
96906	MS90728-29	5306-00-226-4822	108	7
			213	16
96906	MS90728-3	5305-00-071-2506	223	13
			264	7
			282	4
			298	17
96906	MS90728-31	5306-00-226-4824	70	7
			76	12

PART NUMBER INDEX

CAGEC	PART NUMBER	STOCK NUMBER	FIG	ITEM
96906	MS90728-31	5306-00-226-4824	77	4
			119	5
			252	7
			255	20
96906	MS90728-32	5306-00-226-4825	22	36
			69	10
			84	46
			134	1
			262	5
			284	26
96906	MS90728-33	5306-00-226-4826	19	30
			22	38
			58	3
			114	4
96906	MS90728-34	5306-00-226-4827	27	14
			79	20
			217	19
			237	6
			283	8
			280	1
			280	1
			281	1
			281	1
96906	MS90728-35	5306-00-226-4828	253	15
			287	12
96906	MS90728-36	5306-00-226-4829	214	18
96906	MS90728-37	5306-00-226-4830	26	4
			27	2
			277	4
96906	MS90728-38	5306-00-226-4831	79	18
96906	MS90728-39	5306-00-226-4832	79	19
			231	1
96906	MS90728-40	5306-00-226-4833	118	14
			211	9
			212	9
96906	MS90728-5	5305-00-068-7837	288	32
96906	MS90728-57	5305-00-721-5492	84	30
96906	MS90728-59	5305-01-140-9118	18	1
			19	36
			84	55
			140	11
			163	5
			176	2
			255	24
96906	MS90728-6	5305-00-068-0508	9	14
			45	3
			57	6
			61	10
			87	25
			182	4
			207	13

PART NUMBER INDEX

CAGEC	PART NUMBER	STOCK NUMBER	FIG	ITEM
96906	MS90728-6	5305-00-068-0508	269	13
			279	5
			282	7
			297	4
96906	MS90728-60	5305-00-068-0510	1	9
			49	20
			53	3
			140	22
			145	29
			148	2
			149	2
			211	3
			212	3
			215	12
			216	4
96906	MS90728-61	5305-00-543-2419	1	20
			140	40
			142	31
96906	MS90728-62	5305-00-068-0511	81	4
			81	7
			82	3
			82	5
			82	13
			148	5
			253	30
96906	MS90728-63	5305-00-688-2111	71	7
			119	15
			142	24
			149	5
96906	MS90728-64	5305-00-725-2317	13	7
96906	MS90728-65	5305-00-821-3869	157	8
96906	MS90728-66	5305-00-782-9489	273	18
			273	30
			274	10
96906	MS90728-68	5305-00-543-2866	289	17
96906	MS90728-8	5305-00-225-3843	9	3
			217	2
			217	15
			229	8
			232	3
			249	15
			252	20
			256	8
			256	8
			257	2
			259	4
			259	4
			288	30
96906	MS90728-83	5305-00-071-1784	71	9
96906	MS90728-86	5305-00-071-1787	89	13
			90	13

PART NUMBER INDEX

CAGEC	PART NUMBER	STOCK NUMBER	FIG	ITEM
96906	MS90728-86	5305-00-071-1787	91	13
			92	13
			110	5
96906	MS90728-87	5305-00-071-1788	135	2
			137	6
			145	37
96906	MS90728-89	5305-00-071-2055	156	5
96906	MS90728-90L	5306-01-196-0219	111	19
96906	MS90728-92	5305-00-071-2058	97	12
			109	3
			145	33
96906	MS9176-13	5340-01-149-7976	2	22
96906	MS9549-10	5310-00-938-8387	87	26
			264	6
			282	6
11862	M127		14	8
11862	M140		14	5
27647	M257	2520-01-165-5974	125	2
60380	M28161X0H	3110-00-902-1690	84	34
81349	M4510-1A4435	6220-00-961-0783	290	1
11862	M495		14	7
11862	M51		14	4
43334	M802048/M802011	3110-00-406-9608	100	9
43334	M88048/M88010	3110-00-854-1504	100	3
			112	5
80205	NAS1408A6	5310-00-316-6513	1	14
			26	15
			27	25
			81	9
			82	12
			139	11
			140	4
			140	9
			276	15
			277	14
			289	16
80205	NAS1409A7	5310-00-840-6222	148	18
			149	17
81348	NNP530	5530-00-618-6955	BULK	32
60380	NTA2435	3110-00-580-3843	83	44
60380	NTA2840	3110-00-756-2022	83	74
93061	NV108P-4	4820-00-785-8153	252	10
82647	PDA-15	5925-00-553-2274	247	18
70040	PF-35	2940-00-082-6034	8	13
OA7R8	PV23135R	6105-01-188-7154	245	40
21335	P207K	3110-00-155-6152	83	4
60380	QAR29523	3110-00-689-4076	84	18
81348	QQS741		BULK	1
35211	RL984A	4730-00-043-3750	248	19
74410	RRC271BTY2CLDIA072		BULK	3

PART NUMBER INDEX

CAGEC	PART NUMBER	STOCK NUMBER	FIG	ITEM
81343	SAEJ513	4730-00-266-0535	251	6
			252	25
80063	SC-B-75180-IV		45	1
			207	14
80063	SC-C-691545	5340-01-159-4518	207	7
80063	SC-D-691375	5975-01-027-0253	207	6
80063	SC-D-691391		45	2
80063	SC-D-866091	5975-01-155-7084	207	1
11862	SH1366S		7	44
82240	SPEC-3-10-L-L	5340-01-168-0939	288	24
70411	SP2489-FM	5430-01-225-8971	9	9
09386	SR104396	2530-01-096-9670	124	16
03743	S150	5340-00-329-4420	298	2
97271	S19587-T	5330-01-155-7700	129	8
72210	S23981-H	2530-01-165-6005	129	9
11862	TC181	3020-01-155-5779	7	39
59875	TD97203	5325-00-263-6651	300	13
74410	TH-0681	5365-01-163-1147	217	8
60380	TRA2840	3110-00-055-2100	83	75
11862	TXD01487		142	21
11862	TX001488		142	4
11862	TX015963		192	9
17875	T14R	2640-00-555-2840	125	1
81348	WC596/9-1	5935-01-022-2377	220	18
81348	WW-P-471ACBBCB	4730-00-044-4655	2	13
72794	X-840-SR7C	5365-01-226-2342	267	3
74410	XA-T-61-SF	2540-01-267-1360	144	10
			145	24
74410	XA-T-61-SR	3040-01-267-4283	144	2
			145	11
74410	XA-T-88	5310-01-267-6293	144	5
			145	14
72800	XAN-225-H	5365-01-163-6137	83	18
72800	XAN-250-H	5365-01-157-6764	83	77
78553	XAN-262-H	5365-01-166-6324	83	27
74410	XB-T-45-1	5310-01-249-4210	144	9
			145	18
74410	XB-766		144	8
			145	17
74410	XB-767-10	5340-01-267-6296	144	4
			145	13
74410	XX123		144	13
			145	20
25022	06-0075	4710-01-062-3719	214	7
25022	07-01-01	5310-01-271-1793	222	6
25022	07-1016	5305-01-239-9265	210	7
25022	08-0475	2540-01-163-3585	228	13
25022	09-0056	2540-01-163-0772	223	19
25022	09-0093	5340-01-166-2011	211	14
			212	20
25022	09-0951	2540-01-163-0834	228	9

PART NUMBER INDEX

CAGEC	PART NUMBER	STOCK NUMBER	FIG	ITEM
25022	09-0952	2540-01-162-7116	228	10
25022	09-0954	5340-01-162-3853	211	1
			212	2
			213	9
			215	19
			216	25
25022	09-0955	2540-01-162-7117	217	7
25022	09-0959	5340-01-162-3854	213	13
			214	1
25022	09-0963	5340-01-196-6463	214	13
25022	09-0965	2540-01-163-7288	218	15
			219	10
6N299	0917425	4730-00-132-4625	21	6
22787	10-9858	5310-00-190-0752	15	27
78977	100-7-3	6220-01-194-6591	290	8
11862	10000462	5340-00-449-6408	2	18
11862	10005327	4730-01-162-0095	261	2
11862	10008936	5306-01-148-6765	97	15
			109	4
11862	10012288-45		244	5
11862	10024396	5307-01-188-4684	6	4
08806	1003	6240-00-850-4280	41	17
			226	3
			269	5
70040	10045847	6685-01-205-3676	24	2
11862	10054241	5330-01-096-7699	70	9
			75	3
14569	1007-2	4730-00-844-5721	80	7
11862	10137454	2815-01-314-6887	7	28
09386	102007	5306-01-158-6682	124	13
89346	103868	4730-00-044-4587	85	4
84760	10394		15	72
09386	104192	2530-01-216-4554	124	16
84760	10453	5330-00-757-1680	15	19
11862	1049600	5330-01-157-6757	2	10
16764	10496538	2920-01-206-0891	40	12
11862	10499310	2920-01-192-3774	36	14
78385	10530B	2540-01-194-3323	245	11
			284	11
			288	21
84760	10541	5360-00-887-1536	15	15
24617	106751	5315-00-393-5865	3	3
			7	9
11862	107413	5307-01-259-7656	270	4
19207	10906697	6685-00-618-1822	253	31
19207	10922334	5340-00-914-1000	262	4
19207	10948233	5930-00-679-5925	246	13
16764	110200	4730-00-266-0536	251	15
			255	18
11862	1105500	2920-01-149-8606	33	8
84760	11056	2910-00-897-2465	15	86

PART NUMBER INDEX

CAGEC	PART NUMBER	STOCK NUMBER	FIG	ITEM
84760	11057	2910-01-189-1748	15	86
84760	11058	2910-01-191-8453	15	86
84760	11059	2910-01-191-8454	15	86
84760	11060	2910-01-191-8455	15	86
84760	11062	2910-01-191-8456	15	86
84760	11063	2910-01-191-8457	15	86
84760	11064	2910-01-191-8458	15	86
84760	11065	2910-01-191-8459	15	86
16764	1113591	2920-01-157-3765	39	8
84760	11141	5315-00-887-1539	15	84
11862	1114373	5945-01-165-4602	40	1
11862	1116423	2920-01-267-9423	36	6
84760	11175	5305-00-207-3984	15	91
17773	11176106-5	5340-00-150-1658	252	24
84760	11331	5306-00-819-3038	15	21
84760	11438	5305-00-887-1547	15	95
21450	114628		246	24
11862	11500046	5310-01-148-5922	255	28
			297	1
11862	11500668	5305-01-266-9194	193	3
11862	11500742	5305-01-160-1974	241	10
24617	11500815	5305-01-149-1938	7	1
			31	6
24617	11500831	5306-01-165-4283	83	16
			297	2
11862	11500921	5306-01-150-1190	31	12
24617	11500997	5305-01-164-1604	241	14
24617	11500999	5305-01-159-2781	240	9
73342	11501033	5310-01-143-0512	127	56
			127	77
			132	4
11862	11501047	5310-01-267-3043	164	1
11862	11501095	5306-01-161-5489	25	13
11862	11501149	5305-01-157-1987	169	23
			194	3
24617	11501151	5305-01-158-6235	164	6
24617	11501153	5305-01-159-2780	164	5
11862	11501812	5305-01-164-6319	55	2
11862	11501937	5310-01-268-8948	165	23
			243	3
24617	11502488	5310-01-155-2503	10	4
11862	11502634	5305-01-162-8514	193	8
11862	11502656	5306-01-149-4398	66	7
11862	11502804	5306-01-165-3310	6	9
11862	11502812	5310-01-154-5263	117	27
24617	11503316	5305-01-150-5895	7	36
24617	11503395	5305-01-158-7820	158	2
			191	2
			194	7
24617	11503396	5305-01-162-9689	162	4
			242	5

PART NUMBER INDEX

CAGEC	PART NUMBER	STOCK NUMBER	FIG	ITEM
24617	11503428	5306-01-152-4696	83	5
			83	9
11862	11503537	5305-01-162-3961	165	25
			167	14
			201	5
			202	12
24617	11503603	5305-01-211-3103	7	20
24617	11503606	5305-01-231-7384	193	9
11862	11503617	5305-01-150-1521	16	9
11862	11503643		33	6
			132	3
11862	11503739	5310-01-158-6257	57	21
24617	11503778	5306-01-162-8525	50	25
			51	12
11862	11504108	5310-01-159-6586	57	18
11862	11504115	5305-01-151-9285	28	2
24617	11504270	5306-01-149-4398	6	2
			127	80
11862	11504447	5305-01-156-5438	20	15
			21	7
			22	11
			45	5
			57	19
			58	4
			59	3
			60	3
			63	2
			64	4
			95	1
			104	4
			105	2
			118	10
			119	22
			175	11
			252	1
			268	12
			269	17
11862	11504512	5306-01-148-3667	12	10
			132	11
11862	11504595	5305-01-148-7460	1	6
			2	8
			3	12
			33	3
			132	8
			295	2
11862	11504655	5305-01-163-6466	51	14
			52	7
11862	11504656	5305-01-162-5995	50	29
11862	11504678	5305-01-209-7069	127	18
11862	11504736	5305-01-162-9695	53	8
11862	11504967	5305-01-149-9673	31	9

PART NUMBER INDEX

CAGEC	PART NUMBER	STOCK NUMBER	FIG	ITEM
24617	11504986	5306-01-152-4693	25	10
			295	13
11862	11505068	5305-01-148-5915	11	5
11862	11505299	5306-01-185-7049	132	12
24617	11506003	5310-01-158-6260	46	15
			50	22
			51	9
			56	4
			65	2
			241	1
			267	5
			277	18
			297	6
11862	11506101	5310-01-250-7679	13	5
			34	6
			35	10
			132	2
			291	12
			295	1
24617	11507029	5305-01-150-9781	8	17
11862	11508017	5305-01-162-7885	13	1
11862	11508164	5305-01-162-7885	163	6
11862	11508353	5306-01-253-7073	10	1
11862	11508446	5310-01-164-2338	57	32
11862	11508534	5306-01-230-3354	8	16
			46	14
11862	11508566	5305-01-160-1975	28	1
			29	3
			30	3
			41	12
			127	82
			163	13
11862	11508581	5305-01-180-5381	72	1
11862	11508600	5306-01-165-5582	8	31
24617	11508687	5306-01-197-1492	65	1
11862	11508858	5305-01-161-3995	57	35
11862	11509088	5340-01-200-8473	244	3
11862	11509121	5305-01-232-1436	165	26
11862	11509135	5305-01-273-4486	10	11
			123	3
			165	14
11862	11509202	5305-01-254-2558	31	10
11862	11509371	5305-01-195-5807	57	10
11862	11509669	5306-01-151-4925	13	13
11862	11513606	5306-01-185-7048	30	6
11862	11513932	5305-01-269-4329	41	16
			42	2
			241	3
11862	11514337	5315-01-251-1701	130	2
11862	1154611	5340-01-196-3130	127	9
11862	1155445	2530-01-166-3033	116	14

PART NUMBER INDEX

CAGEC	PART NUMBER	STOCK NUMBER	FIG	ITEM
08806	1156	6240-00-924-7526	51	7
			53	10
08806	1157	6240-00-889-1799	53	11
			213	4
19207	11588688	5330-00-089-0998	246	27
19207	11608950-4	4730-00-826-4268	20	9
			23	21
11862	11663000		12	2
19207	11663057	5930-00-345-5455	246	21
19207	11663058	4810-00-248-1635	246	8
19207	11663061	5905-00-251-7145	246	11
19207	11669126-1	5340-01-043-5214	231	18
			275	19
19207	11674728	5935-01-059-0117	57	22
19207	11674729	5330-01-059-4286	57	28
19207	11674730	5970-01-044-8391	57	27
19207	11675004	5340-01-059-0114	57	23
19207	11682088-1	5340-01-114-7712	289	4
19207	11682345	5935-01-044-8382	57	24
72582	118754	4730-00-288-9440	9	8
27647	11967	3040-01-163-7286	KITS	
11862	12001184	2590-01-160-1496	58	9
11862	12004005	5920-01-123-5212	49	26
11862	12004007	5920-01-123-5211	49	27
11862	12004008	5920-01-149-6952	49	33
			222	12
11862	12004009	5920-01-085-0825	49	28
			49	35
11862	12004010	5920-01-149-6953	49	30
			222	14
11862	12004011	5920-01-188-6294	49	32
			49	34
			222	11
11862	12006377	6130-01-192-1643	41	4
			41	15
			65	10
			66	3
11862	12013813	6250-01-201-3300	58	6
30780	1203P-4	4730-00-471-3102	252	12
11862	12031346	2590-01-158-6913	68	4
11862	12033862	5945-01-189-3307	41	3
			41	14
11862	12034592		67	2
11862	12039201	2590-01-158-6912	58	5
11862	12039203	2590-01-155-7967	63	5
11862	12039204	2590-01-191-6649	63	4
11862	12039205	2590-01-155-7966	61	7
11862	12039206	2590-01-160-5885	60	1
11862	12039208	2590-01-159-3505	61	8
11862	12039253	6150-01-156-6326	65	19
			66	9

PART NUMBER INDEX

CAGEC	PART NUMBER	STOCK NUMBER	FIG	ITEM
11862	12039254	6150-01-155-6522	45	15
77060	12039255	2590-01-159-8800	64	1
11862	12039257	6150-01-163-1385	57	36
11862	12039267	6150-01-163-1384	57	37
11862	12039271	6150-01-154-1381	57	13
11862	12039272	6150-01-165-0168	57	16
11862	12039293	6140-01-163-1081	57	1
11862	12039294	6140-01-160-5196	267	4
77060	12039297	2920-01-192-4375	47	1
77060	12039298	2920-01-192-4376	47	9
11862	12039304	5995-01-155-7387	68	6
11862	12039305	2590-01-209-0934	68	6
11862	12039308	5995-01-199-1579	67	1
11862	12039309	2590-01-164-9326	41	2
11862	12039310	2590-01-161-1328	41	13
11862	12039311	2590-01-160-1669	65	8
11862	12039434	2590-01-159-8799	59	1
11862	12039464	2590-01-210-1357	68	12
11862	12044586	5995-01-225-2534	46	18
11862	12044637	5995-01-235-6877	66	1
11862	12044638	5995-01-235-6878	66	10
24617	120690	5315-00-839-5820	84	23
11862	120877	5340-01-166-2152	256	3
			258	3
			259	9
11862	12096970	6150-01-327-7394	62	1
84760	12216	5305-00-887-1564	15	100
19207	12255559	5340-01-038-3428	231	10
19207	12255561	5340-01-044-8389	231	15
			275	21
19207	12255564-1	5325-01-165-6975	275	23
19207	12255564-2	5325-01-050-6192	232	10
			232	12
			232	18
			270	5
			272	8
19207	12255567	5340-01-036-7665	231	11
19207	12255608	5340-01-094-9025	289	20
11862	1226174	2590-01-159-8763	243	5
11862	12300197	2510-01-164-7152	169	18
11862	12300270	4130-01-156-8096	295	10
11862	12300852	2520-01-212-6642	73	34
11862	12306178	2510-01-162-7224	BULK	29
19207	12314542		301	1
19207	12314543		302	1
19207	12314544		303	1
11862	12321435		116	13
19207	12338064	8145-01-231-3747	253	1
19207	12338065		253	12
19207	12338066		253	4
19207	12338067	5340-01-271-3059	253	7

PART NUMBER INDEX

CAGEC	PART NUMBER	STOCK NUMBER	FIG	ITEM
19207	12338068	5340-01-264-6544	253	6
19207	12338070	5340-01-264-6543	253	22
19207	12338071	5340-01-265-3676	253	18
19207	12338073	5330-01-264-6537	253	5
19207	12338074	5340-01-264-6540	253	11
19207	12338075	5340-01-264-6541	253	27
19207	12338076	5340-01-265-3674	253	28
19207	12338078	2520-01-268-1051	253	2
19207	12343359-1	2540-01-041-4912	289	10
19207	12343359-2	2540-01-205-2510	289	24
19207	12343359-3	2540-01-205-2509	289	7
19207	12343359-4	2540-01-205-2511	289	22
19207	12343359-5	2540-01-205-2512	289	9
11862	1234418	5325-01-150-1229	87	15
11862	1234726	5365-01-150-7761	100	2
			112	6
84760	12358	2910-00-788-0986	15	6
84760	12360	5305-00-250-5613	15	5
84760	12362	5310-00-400-8585	15	4
11862	1239146	5306-01-152-2582	119	14
11862	1242101	5930-01-163-7282	49	11
11862	1243465	5330-01-154-4342	112	4
11862	1244067	5306-01-246-7459	68	10
			177	1
			182	3
11862	1244707	5315-01-156-6562	69	4
			123	14
84760	12500	5310-00-830-7825	15	26
11862	1252415	2520-01-152-9171	109	5
11862	1252981	3040-01-153-1815	99	5
11862	1253637	6350-01-158-7035	68	14
24617	125384	5310-01-228-1405	129	2
			129	7
11862	1254856	6680-01-161-3656	291	7
11862	1256654	3010-01-165-1221	112	3
14892	125666		117	9
11862	1257087	2520-01-192-1919	118	21
11862	1257203	2530-01-124-3422	118	21
11862	1258709	3020-01-151-4480	100	1
			112	8
11862	1259475	5330-01-147-4208	72	8
11862	1260823	5310-01-157-6698	100	8
			112	1
11862	1260895	5310-01-157-5672	166	16
			179	2
11862	1261219	5930-00-998-9211	49	18
21450	126281	5306-01-166-1665	228	4
21450	126373	5306-00-753-6996	289	11
14892	129472		117	18
14892	129484		117	20
14892	129494		117	12

PART NUMBER INDEX

CAGEC	PART NUMBER	STOCK NUMBER	FIG	ITEM
14892	129495	5365-01-287-5878	117	24
14892	129497		117	6
14892	129596		117	3
14892	129839		117	5
14892	129894		117	22
14892	129959		117	2
25022	13-1104	2510-01-163-8014	216	10
25022	13-1105	2510-01-164-1532	215	5
27647	13109	2530-01-163-0799	124	10
27647	13113	3040-01-163-7347	125	8
11862	1312281	5360-00-392-3453	115	22
			116	24
21450	131245	5310-00-013-1245	39	6
			87	12
			256	6
			256	6
			257	4
			259	2
			259	2
			270	8
			288	28
			297	8
11862	1324714	4710-00-420-4759	BULK	38
27647	13446	5330-01-331-7230	125	6
84760	13521	5310-01-161-6121	15	34
11862	1355003	5310-01-084-4491	167	5
72712	1358938	5306-00-225-2864	89	14
			90	14
11862	1359744	4720-00-230-6523	BULK	21
11862	1359887	5310-01-197-5499	56	13
			192	1
11862	1361699	5340-01-168-1501	49	19
11862	1362195	3020-01-166-6802	291	4
11862	1362293	2530-01-159-8764	291	6
11862	1365065		157	1
11862	137396	4730-00-288-9390	9	4
24617	137397	4730-00-014-2432	16	20
			118	7
72582	137398	4730-00-013-7398	80	1
11862	1377083	5310-01-165-7572	69	18
35510	13771	5305-01-143-6417	37	9
11862	1381477	5306-01-165-3258	69	21
11862	1385607	3040-01-165-0188	69	19
11862	1394293	2520-01-164-1855	69	2
11862	1394892	5310-01-159-8679	100	10
			112	13
11862	1394893	5310-01-148-5956	100	10
			112	13
11862	1394894	5310-01-150-6260	100	10
			112	13
11862	1394895	5310-01-164-1207	100	10

PART NUMBER INDEX

CAGEC	PART NUMBER	STOCK NUMBER	FIG	ITEM
11862	1394895	5310-01-164-1207	112	13
11862	1397647	2520-01-158-9332	99	9
11862	14000077	5340-01-160-2443	165	22
11862	14000091	2510-01-162-7119	166	21
11862	14000092	2510-01-162-7120	166	21
11862	14000093	2540-01-164-1891	166	26
11862	14000094	5340-01-167-0136	166	26
11862	14000172	5310-01-154-3993	118	20
11862	14000209	2510-01-159-7768	140	27
11862	14000211	2510-01-159-8714	140	25
11862	14000212	2510-01-159-8715	140	17
11862	14000213	5340-01-180-2577	140	18
11862	14000214	2510-01-166-6253	140	16
11862	14000217	2930-01-150-0895	30	19
11862	14000395	2590-01-159-8766	49	23
11862	14001067	5340-01-324-9553	140	23
11862	14001068	5340-01-324-6756	140	23
11862	14001197	5310-01-148-3693	17	3
11862	14002543	2530-01-147-6423	120	2
11862	14003417	5330-00-763-0213	107	3
11862	14003948	5305-01-164-6320	6	7
11862	14003968	5310-01-157-6773	7	27
23862	14003974	2920-01-142-2631	7	72
11862	14004512	5340-01-164-6473	251	14
			255	14
11862	14004810	5330-01-148-7499	117	28
11862	14005953	5340-01-150-4105	6	11
			34	1
			38	12
			44	11
			46	12
			47	8
			50	11
			56	10
			57	7
			57	11
			57	33
			65	15
			65	20
			69	15
			209	4
			232	2
			235	14
			251	10
			255	9
			271	15
			272	4
			276	8
			287	3
11862	14006401	3040-01-147-9321	99	8
11862	14007142	2990-01-181-6725	142	32

PART NUMBER INDEX

CAGEC	PART NUMBER	STOCK NUMBER	FIG	ITEM
11862	14007429	2540-01-157-6414	235	5
11862	14007430	2540-01-156-4855	235	6
11862	14007435	5325-01-167-0510	235	8
11862	14007449	5340-01-160-2488	176	11
			180	7
11862	14007511	5306-01-160-1968	235	12
			251	12
			255	11
11862	14007539	5310-01-158-6780	235	16
11862	14007541	2540-01-165-6145	235	15
11862	14007542	2540-01-164-7260	235	15
11862	14007545	2590-01-219-7808	147	5
11862	14007644	2530-01-154-1262	128	12
11862	14008191	5330-01-159-2807	235	4
11862	14008652	5306-01-158-6243	102	10
			102	24
11862	14009313	4730-01-202-8523	88	2
11862	14009321	5340-00-892-7413	71	4
11862	14009322	5365-01-154-4366	71	3
11862	14009324	5305-01-157-0880	71	5
11862	14009626	3120-01-153-0281	96	2
11862	14009982	2530-01-152-9308	116	5
11862	14010707	5340-01-165-4705	19	1
			20	6
11862	14010954	5340-01-187-8673	167	17
11862	14011345	2930-01-085-0926	28	11
11862	14013071	2530-01-156-6220	102	12
11862	14013072	2530-01-165-0803	102	1
11862	14013122	2540-01-165-5996	242	6
11862	14013710	5306-01-157-9883	172	13
11862	14013724	5305-01-159-6567	203	1
11862	14013753	2540-01-159-8874	194	1
11862	14013754	2540-01-159-8875	194	1
11862	14013789	5365-01-155-8576	169	26
11862	14014293	2540-01-159-0874	203	9
11862	14014389	5306-01-157-9882	172	6
11862	14014799	2510-01-164-7116	136	1
11862	14014800	2510-01-164-7117	136	6
11862	14015724	5365-01-214-4927	156	8
11862	14015726	5340-01-153-1631	156	7
11862	14016511	2510-01-249-6434	171	2
11862	14016512	2510-01-251-5487	171	2
11862	14018523	5330-01-170-6303	163	2
11862	14018526	5340-01-163-0917	163	10
11862	14018529	5360-01-205-8888	163	11
11862	14018531	5340-01-163-5908	163	8
11862	14018532	5360-01-163-0885	163	9
75829	14018597	2510-01-159-8871	190	3
11862	14018630	5340-01-157-9825	20	21
11862	14018647		20	18
11862	14018658		20	14

PART NUMBER INDEX

CAGEC	PART NUMBER	STOCK NUMBER	FIG	ITEM
11862	14018671	5330-01-168-1535	291	9
11862	14018700	5305-01-152-8193	91	14
			92	14
			93	6
11862	14020403	2520-01-210-1382	93	1
11862	14020478	2990-01-147-3954	142	6
11862	14020491	2910-01-155-7878	18	8
11862	14020492	2990-01-155-7879	18	14
11862	14020698	5306-01-149-6279	32	3
11862	14020854	5330-01-165-1357	81	2
11862	14020861	2520-01-147-4207	81	3
11862	14021204	2540-01-155-6821	196	24
11862	14021206	2540-01-155-6822	196	4
11862	14021207	2540-01-158-1877	196	25
11862	14021208	2540-01-155-6947	196	3
11862	14021209	2540-01-155-6823	196	13
11862	14021210	2540-01-155-6824	196	13
11862	14021211	5305-01-160-4528	196	10
11862	14021213	5305-01-160-0331	196	14
11862	14021243	2510-01-158-7575	160	1
11862	14021253	2510-01-163-7016	163	12
11862	14021254	2510-01-163-1139	163	12
11862	14021275	2540-01-156-9675	176	15
11862	14021292	5365-01-158-5381	179	3
11862	14021315	5360-01-164-2404	179	19
11862	14021316	5360-01-164-2405	179	19
11862	14021317	5306-01-159-2772	179	21
11862	14021357	2510-01-164-7128	137	2
11862	14021358	2510-01-163-7228	137	3
11862	14021385	2510-01-155-7877	176	20
11862	14021386	2510-01-155-5849	176	20
11862	14021389	2510-01-155-5853	176	10
11862	14022210	5340-01-197-3480	84	21
11862	14022212	5330-01-197-3229	84	3
11862	14022213	2520-01-183-8322	84	56
11862	14022216	2520-01-194-6955	84	19
11862	14022217	5340-01-197-3372	85	1
11862	14022218	2520-01-183-8323	82	7
11862	14022219	5330-01-126-1040	82	9
11862	14022221	3040-01-194-6853	84	22
11862	14022537	5340-01-163-1390	140	15
11862	14022538	5340-01-163-1389	140	6
11862	14022597	2510-01-148-2937	150	4
			151	5
			152	7
11862	14022634	2815-01-251-1122	7	13
11862	14022640	2815-01-147-4201	7	33
11862	14022643	5365-01-150-1386	7	3
11862	14022644	3120-01-152-2612	7	2
11862	14022645	3020-01-149-5915	7	35
11862	14022646	3020-01-150-2207	7	34

PART NUMBER INDEX

CAGEC	PART NUMBER	STOCK NUMBER	FIG	ITEM
11862	14022648	5310-01-149-0870	7	37
11862	14022649	5330-01-156-5147	7	5
11862	14022650	5340-01-160-4397	7	8
11862	14022651	5330-01-150-5944	13	3
11862	14022652	3020-01-148-8846	13	2
11862	14022653	3020-01-148-9547	7	38
11862	14022654	5307-01-151-8374	11	9
11862	14022657	2815-01-150-2181	11	6
11862	14022670	5315-01-149-9712	2	21
11862	14022671	2805-01-149-7859	3	10
11862	14022672	5305-01-148-3686	3	1
11862	14022673	5310-01-147-8744	3	2
11862	14022683	5330-01-150-7744	8	15
			254	3
11862	14022699	3040-01-163-7208	8	23
11862	14022700	4730-01-150-0879	8	12
11862	14022777	3040-01-155-5864	196	18
11862	14022778	3040-01-155-6912	196	22
11862	14022786	5340-01-158-6354	196	21
11862	14022831	2510-01-165-1496	191	1
11862	14022832	2510-01-165-1497	191	1
75829	14022841	2510-01-155-5435	190	7
75829	14022842	2510-01-155-5434	190	1
11862	14022885	5305-01-161-2581	171	1
11862	14023008	2530-01-162-3626	164	7
11862	14023039	2510-01-172-3022	164	2
11862	14023174	3020-01-153-9586	132	13
11862	14023429	2530-01-152-9306	121	11
11862	14023430	2530-01-152-9307	121	11
11862	14023439	5315-01-157-3004	120	16
11862	14023889	3040-01-155-0194	205	8
11862	14023890	3040-01-155-0195	205	15
11862	14024203	5307-01-150-5993	6	5
11862	14024208	2930-01-147-8555	31	5
11862	14024209	5330-01-147-9808	31	7
11862	14024271	2815-01-148-9535	3	8
11862	14024561	5340-01-155-9861	27	19
11862	14024567	5340-01-160-4068	140	14
11862	14024577	2510-01-159-7759	141	5
11862	14024587	2510-01-172-6781	139	10
			141	9
11862	14024588	5340-01-173-0047	140	50
11862	14024593	2510-01-173-0046	140	36
			142	27
11862	14024997	5360-01-151-1120	25	9
11862	14025512	5360-01-175-7269	7	26
11862	14025515	5365-01-154-1484	7	25
11862	14025517	5340-01-166-5640	2	2
11862	14025518	2815-01-164-7948	2	3
11862	14025523	2815-01-150-2198	5	4
11862	14025526	5306-01-165-2425	5	3

PART NUMBER INDEX

CAGEC	PART NUMBER	STOCK NUMBER	FIG	ITEM
11862	14025527	5310-01-165-3333	5	5
11862	14025557	5330-01-150-1215	12	12
11862	14025568	2815-01-148-9560	11	11
11862	14025673	5340-01-157-7607	205	6
11862	14025674	5340-01-157-7608	205	17
11862	14025677	2540-01-156-0061	205	14
			206	7
11862	14025679	2540-01-157-2966	205	5
11862	14025873	2510-01-238-7051	175	10
11862	14025874	2510-01-238-7052	175	9
11862	14026209	2510-01-230-1165	175	10
11862	14026247	5305-01-158-7820	19	7
			20	24
			161	5
11862	14026383	5330-01-157-6827	167	9
11862	14026384	5330-01-157-6828	167	9
11862	14026409	2540-01-156-4903	167	13
11862	14026410	2540-01-156-4904	167	13
11862	14026765	2530-01-147-4209	125	21
11862	14026802	2530-01-152-9313	130	9
11862	14026803	2530-01-152-9312	130	4
11862	14026804	2530-01-152-9314	130	7
11862	14026805	5310-01-153-9301	130	6
11862	14027345	2540-01-155-5824	206	8
11862	14027346	2540-01-157-7907	206	2
11862	14027347	2540-01-155-5110	206	5
11862	14027348	2540-01-155-5111	206	3
11862	14027431	2540-01-155-7502	168	3
11862	14027432	2540-01-155-7503	168	3
11862	14027451	2510-01-160-5837	173	3
			175	5
11862	14027454	2510-01-158-6904	175	5
11862	14027472	2510-01-156-4854	160	8
			172	3
11862	14027542		177	4
11862	14027555	5340-01-160-2346	161	7
11862	14027645	2590-01-161-0212	191	11
11862	14027646	2590-01-162-7108	191	11
11862	14027745	5340-01-158-8552	127	81
11862	14027746	2540-01-158-8613	127	83
11862	14027775	5330-01-164-8385	167	7
11862	14027776	5330-01-193-1840	167	7
11862	14027777	9390-01-162-4500	169	27
11862	14027778	9390-01-163-2028	169	27
11862	14027795	2510-01-158-8550	191	8
11862	14027796	2510-01-164-9314	191	8
11862	14027797	2540-01-158-8562	203	31
11862	14027798	2540-01-158-8563	203	8
11862	14027799	5340-01-160-4130	203	16
11862	14027926	5340-01-197-4600	147	2
11862	14028914	5306-01-148-7442	8	14

PART NUMBER INDEX

CAGEC	PART NUMBER	STOCK NUMBER	FIG	ITEM
11862	14028916	5330-01-148-7497	30	14
			261	5
11862	14028917	2930-01-147-9916	30	11
11862	14028918	2930-01-147-4198	30	7
11862	14028922	5306-01-149-6278	11	8
			260	1
			295	6
11862	14028923	5305-01-148-3685	11	10
11862	14028924	5307-01-158-6312	11	7
11862	14028931	5340-01-150-1545	39	5
11862	14028942	5325-01-151-6117	8	3
11862	14028949	4730-01-163-7864	2	12
11862	14028976	2815-01-164-7155	8	25
11862	14028981	5360-01-151-1118	7	14
11862	14028982	5310-01-149-4413	7	12
11862	14028983	5310-01-149-4411	7	17
11862	14029107	2520-01-164-6228	87	4
11862	14029108	5360-01-165-1574	87	3
11862	14029111	5315-01-165-3536	87	5
11862	14029116	2520-01-164-9229	87	24
11862	14029117	2520-01-163-7283	87	21
11862	14029122	5315-01-153-0318	87	20
11862	14029126	3040-01-151-6382	83	66
11862	14029158	2520-01-192-8282	82	10
11862	14029193	5340-01-159-4395	140	1
			142	1
11862	14029194	5340-01-175-7656	140	34
			142	25
11862	14029200	5340-01-168-6372	148	9
			149	8
11862	14029235		95	10
11862	14029286	2510-01-173-0050	142	10
11862	14029852	2520-01-151-8043	93	4
11862	14029956	2990-01-154-1324	26	10
11862	14030586	2540-01-156-4885	167	19
11862	14031893	5340-01-161-2789	194	4
11862	14032395	2930-01-147-9330	32	6
11862	14032787	5340-01-165-0564	176	4
11862	14032789	2520-01-193-7870	88	7
11862	14032812	5340-01-165-4544	189	4
11862	14032814	5306-01-159-6574	188	1
			189	5
11862	14032995	4710-01-148-2969	86	1
11862	14032996	4710-01-194-6775	86	5
11862	14033573	2590-01-326-5827	151	22
11862	14033574	2590-01-326-3001	151	22
11862	14033818	5307-01-154-9586	6	1
11862	14033820	2815-01-148-6966	6	12
11862	14033822	3040-01-148-9430	7	15
11862	14033823		30	16
11862	14033824	5340-01-159-6626	16	14

PART NUMBER INDEX

CAGEC	PART NUMBER	STOCK NUMBER	FIG	ITEM
11862	14033861	5340-01-164-6529	295	11
11862	14033863	5340-01-164-7603	295	8
11862	14033864	5340-01-164-5797	295	3
11862	14033866	5340-01-164-8170	295	4
11862	14033867	5340-01-164-5799	295	5
11862	14033879	5340-01-147-2268	132	5
11862	14033880	5340-01-154-7163	132	9
11862	14033881	5340-01-155-7744	132	10
11862	14033893	5340-01-151-9956	12	3
11862	14033895	5340-01-151-9957	12	8
11862	14033896	5340-01-148-8349	13	6
11862	14033911	4710-01-149-5075	16	18
11862	14033912	4710-01-148-9580	16	2
11862	14033913	4710-01-148-9581	16	17
11862	14033914	4710-01-149-5076	16	3
11862	14033915	4710-01-153-1636	16	16
11862	14033916	4710-01-150-0971	16	4
11862	14033917	4710-01-149-5077	16	15
11862	14033918	4710-01-148-9582	16	5
11862	14033920	5365-01-165-3457	16	6
11862	14033921	5340-01-150-4106	16	10
11862	14033922	5340-01-150-6275	16	11
11862	14033927	2815-01-163-7189	7	29
11862	1403394-6		1	1
11862	14033945	5340-01-158-8624	1	2
11862	14033947	5340-01-150-5986	1	5
11862	14033948	5307-01-153-0873	17	7
11862	14033953	5340-01-150-7774	16	19
11862	14033955	5340-01-150-6026	16	7
11862	14034410	5310-01-155-1897	125	13
11862	14034413	5310-01-244-2259	125	11
11862	14034501	5340-01-162-3747	140	13
11862	14034503	5340-01-162-3748	140	7
11862	14034504	5340-01-162-3749	140	7
11862	14034513	2990-01-164-7178	142	6
11862	14034543	5340-01-159-1185	22	24
11862	14034546	2910-01-155-5845	22	34
11862	14034547	5340-01-148-2818	27	18
11862	14034558	2510-01-169-3765	142	33
11862	14034571		118	16
11862	14034572		118	22
11862	14034586		118	13
11862	14034599		119	23
11862	14034728	2520-01-163-7314	127	72
11862	14035374	2530-01-154-6952	124	3
11862	14036369	5310-01-148-7474	8	6
11862	14036705		119	20
11862	14036706		119	13
11862	14036711		119	20
11862	14036712		119	13
11862	14036719	5340-01-160-4601	142	19

PART NUMBER INDEX

CAGEC	PART NUMBER	STOCK NUMBER	FIG	ITEM
11862	14036720	5340-01-187-8520	140	32
			142	19
11862	14036723	4720-01-148-7398	118	18
11862	14036736	4720-01-148-2763	118	18
11862	14036744	4720-01-148-2970	30	4
11862	14036751	4720-01-148-6984	19	41
11862	14036773	5340-01-154-5269	119	7
11862	14036775	4730-01-255-2976	119	11
11862	14036779	5340-01-150-4991	30	2
11862	14036784	5340-01-150-0197	9	24
11862	14036789	5340-01-164-5798	140	8
11862	14036792	4710-01-172-0471	119	2
11862	14036797		119	1
11862	14037059	2540-01-155-6825	196	12
11862	14037060	2540-01-155-8786	196	12
11862	14037808	5340-01-163-1388	27	3
11862	14037812	2990-01-147-3953	27	13
11862	14037836	2990-01-203-2426	27	12
11862	14037856	2990-01-154-3743	26	1
11862	14037861	5365-01-197-3499	1	7
11862	14037878	5340-01-152-5641	1	23
11862	14037889	3040-01-163-7315	87	6
11862	14037893	3040-01-163-7018	87	23
11862	14037909		85	3
11862	14037928	5330-01-196-6586	84	9
11862	14037947		83	53
11862	14037948	5315-01-160-4642	83	58
11862	14037949	3020-01-175-6477	83	32
11862	14037956	5315-01-155-9932	83	59
11862	14037957	2520-01-155-5746	83	69
11862	14037958	5330-01-159-2811	83	72
11862	14037959	3040-01-151-7989	83	70
11862	14037961	5315-01-157-3515	83	36
11862	14037962	5340-01-170-4777	83	39
11862	14037963	2520-01-150-3934	83	40
11862	14037964	5315-01-160-0403	83	43
11862	14037965	2520-01-155-5844	83	41
11862	14037966	2520-01-162-8985	83	35
			83	42
11862	14037967	5340-01-162-8846	83	37
11862	14037968	5360-01-155-3652	83	38
11862	14037976	5365-01-153-9591	83	22
11862	14037979	3040-01-148-9432	83	76
11862	14037980	3020-01-155-8038	83	79
11862	14037983		83	10
11862	14037984	5310-01-154-3992	83	15
11862	14037986	5930-01-149-9305	83	51
11862	14037987	5365-01-149-9710	83	17
11862	14037990	2520-01-151-2690	83	1
11862	14037991	5310-01-148-0245	83	62
11862	14037992	2520-01-147-2267	83	64

PART NUMBER INDEX

CAGEC	PART NUMBER	STOCK NUMBER	FIG	ITEM
11862	14037993	2520-01-149-4993	83	65
11862	14037995	3020-01-160-6516	83	47
11862	14037997	3020-01-155-6399	83	12
11862	14038051	5365-01-209-6943	124	12
11862	14038644	2990-01-148-2928	25	1
11862	14038647	2910-01-150-3676	5	11
11862	14039008	2530-01-160-1542	102	2
			102	18
11862	14039547	2520-01-148-1481	107	2
11862	14039710	2510-01-160-3634	180	6
11862	14039712	3040-01-160-5913	182	12
11862	14039716	3040-01-159-1775	180	5
11862	14039763	2540-01-156-0565	170	3
11862	14039764	2540-01-156-4869	170	3
11862	14039765	3040-01-159-7950	170	4
11862	14039766	3040-01-156-9729	170	4
11862	14039767	3040-01-155-0372	170	9
11862	14039768	3040-01-155-0373	170	9
11862	14039924	5306-01-165-4286	174	1
11862	14039948	2930-01-160-1597	28	4
11862	14039949	5340-01-164-3269	28	20
11862	14039950	5340-01-159-2901	28	6
11862	14039963	2590-01-158-8784	162	5
11862	14040525	2590-01-160-5841	49	23
11862	14040692	5340-01-173-0045	140	30
			142	17
11862	14040735		134	4
11862	14040775		104	1
11862	14040813	5340-01-150-4992	57	20
11862	14040817	4730-01-160-1505	54	3
11862	14041258		20	28
11862	14041286	2510-01-173-4189	139	6
			141	6
11862	14041287	2510-01-159-8712	140	28
11862	14041288	2510-01-159-8713	140	28
11862	14042575	2815-01-165-6143	7	23
11862	14043724	5340-01-160-2239	23	16
11862	14043823	2510-01-162-5172	163	4
11862	14043824	2510-01-162-5173	163	4
11862	14043873	6220-01-160-4254	50	30
11862	14043874	6220-01-161-5016	50	30
11862	14043880	2510-01-159-2929	160	11
11862	14043889	2510-01-160-3622	142	9
11862	14043890	2510-01-159-5889	142	9
11862	14044340	5305-01-159-2779	194	6
11862	14044471	2540-01-156-0564	165	10
11862	14044931		233	5
11862	14044937	3120-01-155-3509	83	29
11862	14044971	2815-01-148-9534	6	6
11862	14044995	2990-01-152-7828	27	20
11862	14044996	2990-01-154-1323	27	16

PART NUMBER INDEX

CAGEC	PART NUMBER	STOCK NUMBER	FIG	ITEM
11862	14045233		14	1
11862	14045263	5307-01-150-1549	31	3
11862	14045268	4710-01-150-0842	8	5
11862	14045274	5340-01-164-6436	2	6
11862	14045504	5305-01-167-5498	87	7
11862	14045515	2510-01-164-7118	135	7
11862	14045516	2510-01-164-7119	135	1
11862	14045521	2990-01-155-5149	26	12
11862	14045525	2990-01-155-5150	26	11
11862	14045605		22	23
11862	14045626		80	5
11862	14045628		80	6
11862	14045642	4710-01-148-4989	72	9
11862	14045654	2530-01-163-3557	148	10
11862	14045658	2520-01-163-7316	87	2
11862	14045697	2510-01-165-8101	87	28
11862	14045698	5340-01-160-2155	117	29
11862	14046907	2590-01-324-5042	93	5
11862	14047899	5340-01-197-3434	9	16
11862	14048388	5310-01-157-7600	113	4
11862	14048409	5365-01-157-6778	113	3
11862	14048410	2520-01-194-7984	113	11
11862	14048412	3040-01-154-2090	113	1
11862	14048413	3020-01-173-1332	113	13
11862	14048416	3120-01-176-1041	113	17
11862	14048417	3120-01-185-3509	113	17
11862	14048418	3120-01-176-2493	113	17
11862	14048419	3120-01-176-2494	113	17
11862	14048420		113	17
11862	14048421	3120-01-176-2496	113	17
11862	14048423	3120-01-181-1637	113	17
11862	14048425	2520-01-172-0395	113	10
11862	14048426	2520-01-160-1893	113	10
11862	14048427	2520-01-172-0396	113	10
11862	14048428	2520-01-155-6827	113	10
11862	14048429	2520-01-155-6828	113	10
11862	14048430	2520-01-155-6829	113	10
11862	14048431	2520-01-155-6830	113	10
11862	14048432	2520-01-156-8309	113	10
11862	14048454	3120-01-152-5945	7	43
11862	14048455	3120-01-152-8200	7	42
11862	14048456	3120-01-157-0914	7	41
11862	14048457	3120-01-152-5946	7	40
11862	14049351	5355-01-085-0995	127	71
11862	14049494	4720-01-160-5781	20	23
11862	14049556	5325-01-197-3540	88	15
11862	14049809	5310-01-162-5732	160	5
11862	14049810	5340-01-164-8757	165	20
			173	6
			175	6
11862	14049881	2510-01-254-1075	173	12

PART NUMBER INDEX

CAGEC	PART NUMBER	STOCK NUMBER	FIG	ITEM
11862	14050425	5315-01-151-4180	13	10
11862	14050440	5340-01-163-8520	179	1
11862	14050441	5340-01-197-3433	10	10
11862	14050442	4710-01-192-7967	10	8
11862	14050443	4710-01-198-2701	10	9
11862	14050444	4730-01-164-7028	10	12
11862	14050445	4720-01-195-7603	10	7
11862	14050446	2990-01-192-9730	10	13
11862	14050523	6680-01-147-4629	8	4
11862	14050658	2815-01-163-7190	7	29
11862	14050659	2815-01-163-7191	7	29
11862	14050661	2815-01-164-7054	7	28
11862	14050662	2815-01-165-8002	7	28
11862	14050679	5310-01-205-2536	124	11
11862	14050685	2910-01-160-8107	18	13
11862	14052026	5340-01-172-2087	19	4
			20	5
11862	14052221	2930-01-159-2902	28	13
11862	14053400	3120-01-173-3728	3	5
11862	14053591	3040-01-156-9994	114	6
11862	14053593	2590-01-161-2119	114	18
11862	14053598	5340-01-160-6874	140	29
			142	16
11862	14054120	5340-01-323-9727	114	10
11862	14054122	5340-01-148-8352	114	8
11862	14054173	2990-01-152-0251	114	12
11862	14054174	2990-01-151-8115	114	9
11862	14054220	2520-01-192-9729	88	16
11862	14054257		118	16
11862	14054259	5340-01-161-2636	140	32
11862	14054269	4720-01-148-6947	118	18
11862	14054270	4720-01-148-6946	118	18
11862	14055002	3120-01-152-1989	3	6
11862	14055003	3120-01-177-3064	3	6
11862	14055004	3120-01-181-7938	3	6
11862	14055279	2530-01-163-1152	116	4
11862	14055280	2530-01-191-9385	116	4
11862	14055315	2530-01-157-5164	116	20
11862	14055363	2520-01-152-4742	103	1
11862	14055501	2510-01-192-5963	1	12
11862	14055502	2510-01-201-2420	1	19
11862	14055531	2520-01-194-0278	88	10
11862	14055556	5365-01-157-5752	119	17
11862	14055585	5330-01-218-0862	9	1
			260	3
11862	14055586	4730-01-172-6683	9	15
11862	14055591	5995-01-148-2930	114	18
11862	14055592	5340-01-163-1411	142	13
11862	14056133	5310-01-157-7582	102	11
11862	14056196	5305-01-163-8241	99	6
			113	2

PART NUMBER INDEX

CAGEC	PART NUMBER	STOCK NUMBER	FIG	ITEM
11862	14056297	4730-01-156-0055	104	6
11862	14056299	4730-01-153-2718	95	8
11862	14056723	5305-01-160-3955	201	13
			202	17
11862	14057219	4820-01-158-6836	14	10
11862	14057232	2815-01-150-0673	7	32
11862	14057296	2815-01-172-6780	7	19
11862	14057297	5365-01-149-9691	7	16
11862	14059083	2520-01-158-7530	113	6
11862	14059238	3120-01-160-0570	201	11
			202	15
11862	14060613	5306-01-165-5583	31	4
			39	4
			199	11
			200	11
11862	14061223		22	29
11862	14061227		22	25
11862	14061344	4710-01-148-8354	9	2
11862	14061345	4710-01-149-1899	9	25
11862	14061348	5340-01-150-6249	9	17
11862	14061350	5640-01-159-6935	9	12
11862	14061352	5340-01-150-1377	9	19
			264	8
11862	14061396	2530-01-222-8068	119	9
11862	14061503		13	16
11862	14061505		7	10
11862	14061569	4710-01-148-2659	12	9
11862	14061649	2815-01-147-4275	8	21
11862	14061661	5307-01-150-1228	32	1
11862	14063301	4720-01-192-8533	23	22
11862	14063302	4720-01-163-8039	23	20
11862	14063306	2520-01-154-6683	94	2
11862	14063307	2530-01-154-8146	125	23
11862	14063308	5340-01-154-2000	97	14
11862	14063314		22	15
11862	14063315		21	8
11862	14063317		20	19
11862	14063319	2910-01-155-5138	20	20
			22	30
11862	14063323		252	4
		4710-01-232-8478	255	19
11862	14063324	4720-01-163-1089	252	3
			255	30
11862	14063325	2910-01-155-7965	20	26
11862	14063326	2910-01-155-5063	19	2
			20	25
11862	14063327	2910-01-155-5148	19	9
11862	14063328	4720-01-155-8062	19	11
11862	14063329	2910-01-155-7880	20	7
11862	14063331		87	1
11862	14063332	2520-01-164-7851	87	2

PART NUMBER INDEX

CAGEC	PART NUMBER	STOCK NUMBER	FIG	ITEM
11862	14063333	2910-01-155-5147	19	9
11862	14063334	4720-01-155-5194	19	11
11862	14063335	4720-01-192-9823	19	41
11862	14063336	4720-01-162-0119	9	10
			260	5
11862	14063337	4720-01-162-0120	9	21
			260	4
11862	14063338	2930-01-159-1802	261	4
11862	14063339		16	1
11862	14063340	5340-01-167-7794	13	18
11862	14063341		277	16
11862	14063342		277	6
11862	14063343		277	9
11862	14063344		277	9
11862	14063345		277	13
11862	14063346		277	10
11862	14063347			277
11862	14063348		277	1
11862	14063349	5340-01-219-7275	276	4
			277	3
11862	14063350		277	12
11862	14063351		277	20
11862	14063352		279	7
11862	14063353	5340-01-293-1200	279	1
11862	14063358	2540-01-218-8099	279	6
11862	14063359	5340-01-164-6595	279	3
11862	14063363	2910-01-155-5139	19	8
11862	14063365	5970-01-159-1803	270	3
11862	14063366	2930-01-167-7250	270	7
11862	14063370		261	6
11862	14063373		261	15
11862	14063391	4720-01-156-0085	21	16
			22	2
11862	14063392	2520-01-151-5971	110	13
11862	14063581	5365-01-175-0683	113	7
11862	14063582	5310-01-175-0624	113	7
11862	14063583	5310-01-175-0625	113	7
11862	14063584	5310-01-175-1035	113	7
11862	14063585	5310-01-183-6930	113	7
11862	14063586	5310-01-176-2467	113	7
11862	14063587	5310-01-174-8766	113	7
11862	14063588	5310-01-177-9758	113	7
11862	14063795	2990-01-046-1170	26	9
24617	140642	4730-00-619-9362	251	7
			255	6
11862	14064660	2530-01-155-7457	128	8
11862	14064663	2590-01-156-0583	114	12
11862	14064664	2990-01-148-2929	114	9
11862	14064924	2540-01-165-8174	242	3
11862	14066195	3120-01-162-0060	196	8
11862	14066196	5365-01-197-2286	196	9

PART NUMBER INDEX

CAGEC	PART NUMBER	STOCK NUMBER	FIG	ITEM
11862	14066255	5945-01-154-3143	25	14
11862	14066301	5340-01-166-1534	12	1
11862	14066305	4720-01-184-0432	12	14
11862	14066306		12	5
11862	14066307	5307-01-150-5992	8	20
11862	14066308	2815-01-148-6971	7	4
11862	14066310	2815-01-166-0621	8	11
11862	14066657	2920-01-192-3020	39	1
11862	14066662	6210-01-158-0396	43	2
11862	14066913	5306-01-156-8680	99	1
			112	11
11862	14067429	2990-01-180-2988	27	21
11862	14067430	2990-01-152-7788	27	17
11862	14067457	5360-01-157-1767	205	4
11862	14067590	5330-01-164-7593	KITS	
11862	14067701	3020-01-148-2948	132	13
11862	14067702	3020-01-148-6983	3	11
11862	14067703	3020-01-147-9359	3	11
11862	14067704	3020-01-148-2949	32	2
11862	14067705	3020-01-148-2950	32	2
11862	14067714	2920-01-147-8559	34	8
11862	14067715	2920-01-191-8442	35	5
11862	14067717	5306-01-150-9493	34	5
11862	14067718	5307-01-159-6632	35	6
11862	14067721	2920-01-147-8562	35	9
11862	14067724	2920-01-147-4278	34	2
11862	14067725	5365-01-154-4365	34	9
11862	14067727	4730-01-148-2755	30	10
11862	14067730	2815-01-163-1174	254	1
11862	14067732	4720-01-194-0336	10	3
11862	14067733	2815-01-192-5962	10	5
11862	14067734	2815-01-150-5002	1	4
11862	14067737	4730-01-148-8242	30	10
11862	14067752		151	10
11862	14067753	2510-01-153-1614	152	8
11862	14067756	5340-01-160-7774	151	14
11862	14067759	2990-01-147-4289	27	22
11862	14067762	2520-01-147-4005	93	1
11862	14067763	4720-01-159-1839	30	18
11862	14067764	5340-01-164-7021	81	10
11862	14067765	2520-01-146-5483	81	1
11862	14067770	2540-01-158-8553	145	1
11862	14067771	5340-01-165-4353	145	7
11862	14067772	5340-01-162-4775	145	7
11862	14067773	2540-01-163-8595	145	6
11862	14067774	2540-01-164-1842	145	6
11862	14067775	2540-01-158-4599	145	40
11862	14067776	2540-01-157-8008	145	35
11862	14067779	5340-01-162-8759	144	33
11862	14067780	5340-01-162-8760	144	21
11862	14067782	2510-01-160-1592	141	3

PART NUMBER INDEX

CAGEC	PART NUMBER	STOCK NUMBER	FIG	ITEM
11862	14067783	2510-01-163-1146	137	1
11862	14067784	2510-01-164-7127	137	7
11862	14067785	2540-01-159-2928	144	32
11862	14067786	2540-01-158-8548	144	19
11862	14067787	2540-01-159-7741	144	31
11862	14067789	5365-01-163-6195	145	34
11862	14067790	2510-01-155-7942	143	1
11862	14067791	2510-01-165-1495	135	9
11862	14067792	2510-01-164-7120	135	13
11862	14067793	4030-00-542-3183	144	25
			145	31
11862	14067794	5315-01-160-4639	143	5
			144	24
			145	30
11862	14067795	2540-01-159-7740	143	2
11862	14067796	5340-01-160-2171	143	3
11862	14068905	2530-01-156-4900	115	30
11862	14068906	2530-01-156-4901	115	30
11862	14069597	3040-01-194-6992	84	67
11862	14069636	5305-01-160-4494	203	24
11862	14070348	5365-01-197-8165	125	7
11862	14070352	2530-01-152-9305	125	17
11862	14070396	5365-01-221-9717	124	9
11862	14070448		196	1
11862	14070453		196	2
11862	14070703	2510-01-162-3623	163	7
11862	14071047	5930-01-157-4060	54	2
11862	14071059	2590-01-147-4285	8	2
11862	14071068	2815-01-148-9559	11	12
7X677	14071072	2815-01-265-7071	2	1
11862	14071080	5307-01-178-7445	31	11
11862	14071461	2540-01-193-7892	201	12
			202	14
11862	14071555	2540-01-193-7893	201	4
11862	14071556	2540-01-193-7894	202	3
11862	14071875	3120-01-159-8588	83	49
11862	14071876	2520-01-150-0415	83	31
11862	14071877	2510-01-155-5183	150	8
			151	16
11862	14071882	3120-01-155-8713	83	73
11862	14071884	2520-01-330-3249	108	9
11862	14071950		87	1
11862	14071952	2520-01-165-7885	87	11
11862	14071954	5330-01-164-7506	87	27
11862	14071955	2520-01-192-7979	88	11
11862	14071967	5340-01-147-4217	1	18
11862	14071973	5340-01-160-4591	140	20
11862	14071980	2520-01-328-4898	93	1
11862	14071983	4730-01-194-2002	30	5
11862	14071984	2910-01-148-2910	19	13
11862	14072305	5340-01-159-2996	187	2

PART NUMBER INDEX

CAGEC	PART NUMBER	STOCK NUMBER	FIG	ITEM
11862	14072306	2510-01-159-8726	187	7
11862	14072307	5310-01-169-2849	187	1
11862	14072308		289	1
11862	14072314	5340-01-200-5843	289	2
11862	14072319	2540-01-191-8467	289	13
11862	14072322	2540-01-163-0801	289	15
11862	14072323	2540-01-164-9278	289	14
11862	14072324	5340-01-165-1494	289	6
11862	14072325	5340-01-165-4429	289	3
11862	14072333	6220-01-248-6269	50	1
11862	14072336	6150-01-159-6901	44	6
			46	6
11862	14072337		44	14
11862	14072338	5930-01-158-4428	43	11
11862	14072339	5930-01-158-4808	49	37
11862	14072340	5340-01-160-4592	68	7
11862	14072341	6110-01-167-1822	38	8
11862	14072347	2540-01-163-1141	251	8
			255	8
11862	14072358	5930-01-162-0803	49	36
11862	14072364	2540-01-162-4412	271	23
11862	14072365	5340-01-164-6435	271	22
11862	14072366	4730-01-161-6618	271	19
11862	14072367	5340-01-165-4379	271	18
11862	14072368	4720-01-162-5113	251	16
			255	16
11862	14072369		251	17
			255	15
11862	14072371		251	5
			255	5
11862	14072372		262	20
11862	14072374	5365-01-163-1142	262	14
			284	2
			288	9
11862	14072378		284	9
11862	14072380	5340-00-237-7779	289	23
11862	14072405	2510-01-159-0868	164	4
11862	14072406	5340-01-162-3627	43	3
11862	14072408	4710-01-164-7850	297	3
11862	14072409	5340-01-162-3628	43	3
11862	14072410	5895-01-165-6792	43	9
11862	14072412	5930-01-163-1439	43	13
11862	14072413	5930-01-159-0925	49	17
11862	14072414	2510-01-156-5881	165	5
11862	14072415	5340-01-164-6524	282	5
11862	14072421	5340-01-159-5762	50	23
11862	14072422	2540-01-159-8881	135	10
11862	14072425	5306-01-157-3330	135	11
			136	8
			137	11
			144	30

PART NUMBER INDEX

CAGEC	PART NUMBER	STOCK NUMBER	FIG	ITEM
11862	14072426	5340-01-157-6428	28	18
11862	14072427	5340-01-157-6429	28	17
11862	14072428	5340-01-163-1343	38	4
11862	14072430		28	14
11862	14072431	5340-01-162-6061	50	12
11862	14072432	5340-01-162-4774	57	30
11862	14072433	5340-01-159-5765	51	13
11862	14072434	5340-01-162-9782	51	13
11862	14072435	2540-01-159-2992	136	10
11862	14072436	2540-01-155-7535	137	12
11862	14072439	5340-01-164-6412	237	10
11862	14072440	5340-01-164-6411	237	9
11862	14072441	5340-01-164-6410	237	8
11862	14072442	5340-01-159-4517	237	13
11862	14072445	5340-01-157-6697	236	2
			237	3
11862	14072448	2920-01-152-2414	43	10
11862	14072449	2520-01-155-7454	94	1
11862	14072451	2520-01-155-7453	94	1
11862	14072452	5340-01-159-6174	185	2
11862	14072454	5365-01-158-2004	185	4
			186	8
11862	14072455	2510-01-155-5112	186	6
11862	14072458	5305-01-197-3290	186	4
11862	14072459	2510-01-156-0062	185	6
11862	14072460	2510-01-155-5825	185	3
			186	9
11862	14072461	2540-01-191-8443	232	5
11862	14072462	2540-01-191-8444	232	6
11862	14072463	2540-01-192-3573	232	24
11862	14072464	2540-01-191-8445	232	11
11862	14072465	2540-01-192-3574	232	23
11862	14072466	2540-01-191-8446	232	17
11862	14072467	2540-01-192-3575	232	22
11862	14072468	2540-01-192-3576	231	8
11862	14072469	2540-01-192-4479	231	12
11862	14072470	2540-01-191-9564	231	9
11862	14072471		231	3
11862	14072472		231	4
11862	14072473		231	6
11862	14072475	2540-01-191-8549	230	2
11862	14072479	2510-01-147-9917	230	1
11862	14072481	5340-01-162-6077	53	13
11862	14072485		235	3
11862	14072486		235	3
11862	14072487	2540-01-155-0376	235	7
11862	14072488	2510-01-155-8787	158	5
11862	14072489	2540-01-159-8760	235	9
11862	14072493	2540-01-156-7238	166	9
11862	14072494	2540-01-158-3576	166	1
11862	14072497	2540-01-158-4602	178	2

PART NUMBER INDEX

CAGEC	PART NUMBER	STOCK NUMBER	FIG	ITEM
11862	14072499	2540-01-211-4621	178	3
11862	14072631	2510-01-215-3931	141	7
11862	14072632	2510-01-211-6608	141	2
11862	14072639	2510-01-211-6609	139	8
11862	14072640	2510-01-211-6610	139	2
11862	14072666	5306-01-159-2784	19	35
11862	14072686	5330-01-147-4212	26	13
			27	23
11862	14072690	2510-01-192-4480	140	44
			142	39
11862	14072692	2530-01-147-4214	114	15
11862	14072695	2510-01-172-3020	140	43
			142	40
11862	14072697	5340-01-148-8351	114	10
11862	14072850	2510-01-159-8762	159	2
11862	14072916	2520-01-152-0941	97	9
11862	14072919	2530-01-149-3827	97	1
11862	14072921	5365-01-158-2193	125	10
11862	14072927		125	15
11862	14072930	4730-01-147-6425	105	6
11862	14073962	2510-01-231-2879	175	9
11862	14074319	2540-01-158-9370	242	7
11862	14074431	5340-01-165-6797	188	4
			189	3
			279	4
11862	14074432	2590-01-156-0075	209	3
11862	14074435	2590-01-156-0076	208	1
11862	14074437	5340-00-455-5899	208	2
11862	14074438	2590-01-156-0080	208	6
11862	14074439	2590-01-156-0074	209	1
11862	14074440	5340-01-165-6798	188	6
11862	14074441	2590-01-156-0077	208	4
11862	14074442	2590-01-156-0078	208	4
11862	14074443	5340-01-165-6799	188	5
11862	14074444		280	4
11862	14074445	4710-01-163-0594	280	9
11862	14074446		280	7
11862	14074447		281	11
11862	14074448	2540-01-164-7121	281	10
11862	14074449		281	9
11862	14074450	4710-01-161-6406	281	8
11862	14074451	4720-01-162-0283	281	6
11862	14074452	4710-01-162-7080	281	13
11862	14074453		281	12
11862	14074457	5340-01-164-6528	262	1
11862	14074458	5340-01-164-6526	262	7
11862	14074459	5340-01-164-6527	262	10
11862	14074461PC6	5945-01-162-0516	265	2
11862	14074465	2540-01-163-1140	285	2
11862	14074466	2540-01-164-7123	284	12
11862	14074469	2540-01-163-1143	284	21

PART NUMBER INDEX

CAGEC	PART NUMBER	STOCK NUMBER	FIG	ITEM
11862	14074470	2540-01-165-0466	284	27
11862	14074471	5340-01-165-4543	284	24
11862	14074473	2540-01-164-7124	284	30
11862	14074479	2540-01-159-8727	203	23
11862	14074480	5340-01-160-4597	68	8
11862	14074499	2540-01-162-7113	259	5
11862	14075205	2520-01-150-3823	83	7
11862	14075210	3020-01-151-3967	83	24
11862	14075211	3010-01-150-3933	83	23
11862	14075212	5310-01-176-2466	83	20
11862	14075306	5306-01-167-4346	83	52
11862	14075347	2910-01-156-8361	23	2
11862	14075351	2520-01-155-6939	110	18
11862	14075352	2520-01-172-0501	110	17
			111	24
11862	14075353	5360-01-157-1880	110	16
			111	23
11862	14075354	5340-01-155-9701	110	15
			111	22
11862	14075355	3020-01-156-8365	110	14
11862	14075372	2540-01-163-7232	205	2
11862	14075373	2540-01-163-7233	205	3
11862	14075374	2540-01-160-3654	203	27
11862	14075375	2540-01-160-3652	203	25
11862	14075376	2540-01-160-3653	203	28
11862	14075379	2540-01-163-8017	204	2
11862	14075388	6140-01-155-6997	56	15
11862	14075389	6140-01-155-6998	56	12
11862	14075801	2990-01-159-1801	256	5
11862	14075811	2540-01-165-0813	289	5
11862	14075812	2540-01-163-7017	289	8
11862	14075815	2540-01-166-1370	203	4
			205	1
11862	14075816	2540-01-165-0837	203	6
11862	14075817	2540-01-166-1373	201	2
			202	4
11862	14075818	2540-01-165-0895	201	10
			202	5
11862	14075819	2540-01-161-1356	202	1
11862	14075820	2540-01-158-8812	201	1
11862	14075821	2540-01-158-8558	199	8
11862	14075822	2540-01-159-7744	200	9
11862	14075823	2540-01-164-1843	199	2
11862	14075824	2540-01-164-1844	199	4
11862	14075827	2510-01-162-3625	191	4
11862	14075829	5340-01-162-3624	191	5
11862	14075839	2510-01-159-9789	191	6
11862	14075843	2540-01-162-8983	275	5
11862	14075845	2540-01-164-1886	271	8
11862	14075846	5340-01-162-4820	238	2
11862	14075847	5340-01-163-0632	238	4

PART NUMBER INDEX

CAGEC	PART NUMBER	STOCK NUMBER	FIG	ITEM
11862	14075848	5340-01-163-5903	238	5
11862	14075856		280	12
11862	14075857	5330-01-164-5603	271	20
11862	14075858	6620-01-156-0712	43	6
11862	14075862		278	2
11862	14075863		278	4
11862	14075864		278	2
11862	14075881	2540-01-155-7543	272	6
11862	14075882	2540-01-155-7278	272	7
11862	14075883		272	5
11862	14075884		272	3
11862	14075888	5995-01-163-1183	265	1
11862	14075894	5307-01-158-9932	56	7
11862	14075896	6140-01-156-5326	56	5
11862	14075900	5340-01-164-8171	44	9
11862	14076201		275	10
11862	14076204		273	6
			274	4
11862	14076206	2510-01-159-1790	276	12
11862	14076208	2590-01-242-8068	274	17
11862	14076209		273	25
11862	14076210		273	20
11862	14076211		273	23
11862	14076212		274	13
11862	14076215	5330-01-167-6335	271	5
			273	27
11862	14076216	2510-01-159-1792	271	3
			273	28
11862	14076217	5340-01-163-3294	273	17
11862	14076221	5340-01-164-2397	274	11
11862	14076222		275	7
11862	14076226	5340-01-164-6589	275	16
11862	14076228	2540-01-155-7542	271	1
11862	14076229		276	20
11862	14076230	5340-01-164-6591	275	3
11862	14076231	2510-01-162-4408	275	12
11862	14076232		275	13
11862	14076233	2510-01-162-4407	275	2
11862	14076235	5330-01-163-5850	276	11
11862	14076236	2590-01-163-1184	264	1
11862	14076237	2590-01-163-3529	285	3
11862	14076238	2590-01-162-4353	269	2
11862	14076239	2590-01-162-7130	269	16
11862	14076240	2590-01-162-4418	268	7
11862	14076241	2540-01-164-0032	267	12
11862	14076243	2540-01-164-0122	267	7
11862	14076246	2590-01-164-0135	267	1
11862	14076248	5306-01-227-1454	267	13
11862	14076249	2540-01-166-2009	267	10
11862	14076250	2540-01-166-2010	267	8
11862	14076252	2540-01-163-1175	288	1

PART NUMBER INDEX

CAGEC	PART NUMBER	STOCK NUMBER	FIG	ITEM
11862	14076253	2540-01-162-5176	288	4
11862	14076255		288	11
11862	14076257	2540-01-162-5177	288	40
11862	14076258	2540-01-162-5189	288	39
11862	14076259	2540-01-162-5178	288	31
11862	14076261	5340-01-166-0568	288	33
11862	14076262	5340-01-165-5617	288	34
11862	14076265	2540-01-162-6418	288	22
11862	14076269	2590-01-162-4352	268	1
11862	14076270	5330-01-220-6153	264	13
11862	14076271		276	18
11862	14076272		276	13
11862	14076273		276	22
11862	14076274		276	22
11862	14076275		276	14
11862	14076276		276	10
11862	14076277		276	23
11862	14076278		276	2
11862	14076279		276	21
11862	14076280		276	5
11862	14076281		276	1
11862	14076284	2540-01-163-0765	282	10
11862	14076285	2540-01-162-5174	283	1
11862	14076287	2540-01-162-5175	283	2
11862	14076295	2540-01-159-1800	283	19
11862	14076298	5325-01-163-5973	236	1
11862	14076299	2510-01-162-7111	275	22
11862	14076300	2540-01-163-0766	275	22
11862	14076390	4720-01-162-0121	255	27
11862	14076801	2540-01-162-6493	288	36
11862	14076802		278	3
11862	14076803		278	3
11862	14076805	2990-01-163-1182	287	11
11862	14076811	2590-01-162-4417	268	5
11862	14076838	2540-01-158-8612	233	6
11862	14076847	5905-01-193-7212	47	4
11862	14076848	2540-01-194-6875	47	3
11862	14076852	6140-01-216-7923	267	6
11862	14076856	6160-01-165-4638	56	2
11862	14076857	6160-01-165-4637	56	6
75829	14076859	2510-01-159-8728	190	9
75829	14076860	2510-01-159-8729	190	9
17769	14076861	2510-01-246-4236	275	8
11862	14076862	5330-01-163-2055	274	8
11862	14076864		274	14
11862	14076871		274	15
11862	14076872		274	5
11862	14076875	2510-01-246-4237	273	16
11862	14076882	2540-01-162-7110	273	10
11862	14076884		273	24
11862	14076887	5340-01-197-3244	174	6

PART NUMBER INDEX

CAGEC	PART NUMBER	STOCK NUMBER	FIG	ITEM
11862	14076889	5340-01-218-5823	276	7
11862	14076894	5365-01-195-4948	174	5
75829	14076899	2510-01-157-1382	190	8
11862	14077122	6620-01-146-8006	30	13
11862	14077147	2920-01-191-6635	33	11
11862	14077148	2815-01-147-4200	7	31
11862	14077149	2920-01-147-4272	33	1
11862	14077151	5340-01-197-1259	33	5
11862	14077157	2815-01-148-3771	4	1
11862	14077163	5340-01-202-3651	2	4
11862	14077182	2815-01-164-7154	8	30
11862	14077192	5306-01-149-9668	2	19
11862	14077193	5306-01-148-7456	2	11
11862	14077195	5306-01-198-5515	2	20
11862	14077928	4140-01-145-8099	32	4
11862	14078806	2540-01-200-3167	144	29
11862	14079056	2540-01-159-7963	195	2
11862	14079057	2540-01-193-7895	195	3
11862	14079058	2540-01-193-7896	195	1
11862	14079089	5330-01-163-0387	98	14
11862	14079304	2815-01-165-8216	2	1
11862	14079426	2815-01-148-9497	8	22
11862	14079550	5330-00-107-3925	8	18
			254	8
11862	14081915	2520-01-225-1033	127	72
11862	14089132	2990-01-225-1052	26	9
11862	14090911	5365-01-154-4378	2	14
11862	14094948	4730-01-151-7972	118	19
11862	14095588		97	7
11862	14095597	5306-01-157-1985	83	61
11862	14095623	3020-01-155-5898	83	21
11862	14095676	3020-01-148-9548	83	26
11862	14095677	3120-01-155-4470	83	30
11862	14095678	3120-01-152-2613	83	28
11862	14095679	2520-01-151-3982	83	33
11862	14095680	2590-01-159-8757	83	34
24617	141195	5315-00-014-1195	8	32
24617	142433	4730-00-014-2433	21	4
			22	7
24617	143935	4730-00-187-4191	263	44
93061	144F-5	4730-01-069-6408	251	3
			255	3
84760	14408	5310-00-877-4956	15	44
08806	1445		43	8
11862	1453658	5315-01-186-1352	2	16
11862	1456507	5365-01-155-1941	89	6
			90	9
			91	6
			92	8
			93	2
79470	1461-4	5310-01-196-6465	246	23

PART NUMBER INDEX

CAGEC	PART NUMBER	STOCK NUMBER	FIG	ITEM
80201	14700	5330-01-179-5907	101	10
11862	1488565		134	6
24617	1494253	5310-01-160-9529	28	19
			191	15
			192	4
			252	6
			255	22
99688	15112	4130-01-166-2115	298	16
99688	15113	4130-01-160-7697	299	2
27647	15147	3040-01-163-7285	125	4
27647	15149	2530-01-163-0798	124	6
84760	15228	4730-00-459-6077	15	110
84760	15349	5330-01-138-2106	23	10
11862	15517986	5340-01-213-6934	80	4
11862	15521884	2520-01-223-3754	97	9
11862	15521977	5306-01-156-5435	108	8
11862	15522022	5340-01-164-1743	71	8
11862	15522077	2530-01-096-6752	115	27
11862	15522078	2530-01-155-7943	115	27
11862	15522079	2530-01-162-8986	115	28
11862	15522080	2530-01-164-0039	115	28
11862	15522081	2530-00-228-6992	115	25
11862	15522095	2540-01-154-1293	123	10
11862	15522381	2510-01-225-0997	149	9
11862	15522392	2590-01-164-7177	22	33
11862	15522444		118	13
11862	15522697	2930-01-147-4221	29	4
11862	15522707	2510-01-225-0998	179	17
11862	15522708	2510-01-225-0999	179	13
11862	15522752	2510-01-147-5576	191	3
11862	15522764	5680-01-163-6347	171	3
27647	15528	2530-01-163-0800	KITS	
11862	15530620	5325-01-149-6293	17	6
			114	14
11862	15531547	5305-01-231-1297	169	22
11862	15537018	2815-01-163-7838	5	1
11862	15537020	2815-01-163-9999	5	1
11862	15537086	5365-01-330-8450	KITS	
11862	15537131		108	1
11862	15537219	2530-01-151-7996	101	25
11862	15537684		108	1
11862	15538092	5360-01-165-1564	69	16
11862	15538215	2530-01-225-1024	119	8
11862	15538347	2510-01-225-1005	139	5
11862	15544950	5306-01-227-9085	39	9
11862	15545178	5340-01-160-5922	170	8
			180	8
11862	15548901	4720-01-150-7575	9	10
11862	15548902	4720-01-148-5000	9	21
11862	15549248	2530-01-225-2236	119	4
11862	15552844	5310-01-270-5464	110	1

PART NUMBER INDEX

CAGEC	PART NUMBER	STOCK NUMBER	FIG	ITEM
11862	15554576	5305-01-162-0015	235	11
11862	15554915	2510-01-155-7425	158	1
11862	15557723	5340-01-335-9359	114	8
11862	15559312	6220-01-156-4476	51	11
			53	12
11862	15559316	6220-01-156-4475	50	24
11862	15562374	5330-01-087-4714	291	5
11862	15565342	2540-01-160-5838	184	2
11862	15567924	5340-01-269-1361	25	12
11862	15569009	5340-01-158-6895	203	11
11862	15569010	2540-01-158-6896	203	2
11862	15569071	5330-01-158-6683	167	8
11862	15569072	5330-01-159-1153	167	8
11862	15571581	5365-01-152-0613	148	16
			149	15
11862	15571640	2510-01-154-6906	163	1
11862	15571643	2510-01-155-8799	166	6
11862	15577694	2540-01-167-2985	204	1
11862	15581505	2520-01-330-0016	111	25
11862	15582233	5310-01-205-2537	124	11
11862	15588503	5355-01-159-6622	87	29
11862	15588504	5355-01-197-3172	88	8
11862	15589794	2510-01-211-1654	141	1
11862	15589799	2510-01-211-1655	139	1
11862	15590123	2910-01-225-1068	25	6
11862	15590401	2540-01-156-6107	168	2
11862	15590402	2540-01-156-6108	168	2
11862	15590415		167	21
11862	15590421	5330-01-208-3843	190	5
11862	15590422	2540-01-158-1721	167	22
11862	15590443	5305-01-230-9846	169	28
11862	15590455	5330-01-165-3401	184	5
11862	15590559	2510-01-225-2238	165	21
11862	15591130	5325-01-164-2377	62	5
			269	19
11862	15591138	6680-01-225-4432	41	5
11862	15591247	2540-01-191-8668	197	1
11862	15591248	2540-01-160-3651	198	2
11862	15591700	2520-01-216-5611	94	2
11862	15591702	2540-01-193-7884	193	2
11862	15591703	2540-01-157-3032	233	1
11862	15591704	2540-01-156-0584	233	4
11862	15591705	2590-01-234-6468	186	2
23862	15591706	2540-01-241-4238	186	3
11862	15591707	5340-01-238-5923	186	11
11862	15591709	2590-01-242-1050	186	3
1T998	15591710	5340-01-238-5924	186	5
11862	15591718	5945-01-192-7985	39	3
11862	15593230	2510-01-155-8800	166	5
11862	15593254	2530-01-158-8585	173	4
11862	15593569	2510-01-225-1008	165	27

PART NUMBER INDEX

CAGEC	PART NUMBER	STOCK NUMBER	FIG	ITEM
11862	15593570	5330-01-237-7512	180	1
11862	15593571	5330-01-207-9421	180	2
11862	15593599		270	1
11862	15593831	2540-01-225-1007	165	28
11862	15593849	5340-01-232-8179	123	5
11862	15593866	5360-01-185-0341	182	1
11862	15593907	2990-01-225-1021	142	6
11862	15593908		142	6
11862	15593955	2510-01-225-2237	139	15
11862	15593979	2520-01-225-5866	139	7
11862	15593980	2510-01-232-7188	139	13
11862	15594111		108	1
11862	15594116	5306-01-205-8882	108	3
11862	15594126		109	2
11862	15594158	3040-01-192-6024	84	15
11862	15594176	5930-01-202-3573	85	2
11862	15594177	2530-01-164-7126	116	6
11862	15594178	2530-01-163-7227	116	6
11862	15594195	3040-01-150-0409	83	13
			83	13
11862	15594196	3120-01-197-1535	83	20
11862	15594643	2540-01-225-1023	176	16
11862	15594644	2510-01-225-1006	176	12
11862	15594722	2540-01-225-5972	196	1
11862	15594727	2540-01-225-1106	196	2
11862	15594889	2510-01-178-8877	191	13
11862	15594890	2510-01-149-3465	191	13
11862	15594895	5340-01-210-8824	193	10
11862	15594896	2590-01-247-3286	193	10
11862	15594983	2540-01-156-0088	193	7
			278	1
11862	15595211	2590-01-159-8716	140	12
11862	15595216	2990-01-147-4290	27	1
11862	15595224	2990-01-225-1029	26	1
11862	15595271	2990-01-225-1028	26	11
61928	15596614	5930-01-208-6292	23	15
11862	15596628	2510-01-148-1617	173	1
11862	15596686	2520-01-153-8431	89	9
			91	9
11862	15596975	2510-01-155-5850	176	19
11862	15596976	2510-01-155-5851	176	13
11862	15596994		173	8
11862	15597600	2510-01-156-8183	160	6
			172	1
11862	15597621	2510-01-232-8176	173	9
11862	15597629	5365-01-155-8564	160	7
			172	2
11862	15597653	2540-01-155-7298	170	6
11862	15597654	2540-01-156-8315	170	6
11862	15597667	2510-01-156-4872	167	11
11862	15597668	2510-01-156-4873	167	11

PART NUMBER INDEX

CAGEC	PART NUMBER	STOCK NUMBER	FIG	ITEM
11862	15597787	2520-01-207-9169	103	2
11862	15598706	2510-01-156-8092	165	29
11862	15598708	2510-01-159-8761	161	3
11862	15598709	2510-01-212-5819	161	
11862	15598766	2540-01-225-2521	196	25
11862	15598767	2540-01-225-1108	196	4
11862	15598769	2540-01-160-1591	161	2
11862	15598770	2510-01-194-0206	161	1
11862	15598787	2930-01-165-9595	160	3
11862	15599200	2990-01-163-3575	254	5
11862	15599201	2520-01-163-3494	254	6
11862	15599203		284	10
11862	15599204	3020-01-147-7935	35	3
11862	15599209		21	3
			22	3
11862	15599211	2990-01-163-0771	256	9
11862	15599212	2990-01-162-4416	259	12
11862	15599215	2990-01-163-1180	259	5
11862	15599216		27	1
11862	15599219	5340-01-198-8591	138	1
11862	15599220	5340-01-198-8591	138	3
11862	15599221	5340-01-162-9777	19	33
11862	15599222	5910-01-189-5153	65	16
11862	15599223	5910-01-189-5109	68	2
11862	15599224	5910-01-190-4600	55	4
11862	15599225	5910-01-189-5152	44	5
11862	15599234	4720-01-247-4680	262	21
11862	15599235		70	3
11862	15599243	2990-01-163-1179	258	6
11862	15599245	5340-01-219-7272	258	4
11862	15599246	5340-01-221-9972	260	2
11862	15599248	2520-01-193-4053	82	4
11862	15599250	5365-01-194-5074	174	5
11862	15599251	5961-01-188-8315	38	9
11862	15599259		118	9
11862	15599260		118	6
11862	15599261		118	9
11862	15599262		118	6
11862	15599269	2990-01-231-2938	27	22
11862	15599270	4210-01-200-2574	165	18
11862	15599271	2590-01-191-6511	208	8
11862	15599273	5340-01-212-6716	276	6
11862	15599274	5340-01-212-6717	276	6
11862	15599283	2590-01-163-1238	159	4
11862	15599284	2590-01-164-0134	159	4
11862	15599285	2510-01-160-4970	159	1
11862	15599286	2590-01-159-8861	159	5
11862	15599287		256	11
11862	15599290	2540-01-162-7114	257	1
11862	15599432	5306-01-148-7457	147	6
11862	15599619	3020-01-202-3361	113	12

PART NUMBER INDEX

CAGEC	PART NUMBER	STOCK NUMBER	FIG	ITEM
11862	15599620	3020-01-202-3360	113	9
11862	15599621	2510-01-191-6509	BULK	36
11862	15599623	5340-01-201-4095	113	14
11862	15599677	2540-01-194-0261	166	12
11862	15599678	2540-01-192-9754	178	1
11862	15599687	2520-01-201-2501	106	1
11862	15599900	2990-01-257-1569	56	1
11862	15599901	6140-01-190-2516	56	16
11862	15599902	6140-01-190-2517	56	8
11862	15599910	2510-01-159-1791	274	12
11862	15599915	2590-01-182-4455	146	3
11862	15599916		275	11
11862	15599917	2540-01-214-2634	275	14
11862	15599918	2540-01-214-2635	275	14
11862	15599919	3120-01-194-0754	231	2
11862	15599920	5365-01-219-7285	264	9
11862	15599929	5306-01-194-4977	185	1
			186	1
11862	15599940	2540-01-163-7225	287	9
11862	15599942	2990-01-250-8612	287	6
11862	15599949	6110-01-191-0127	38	11
11862	15599950	1430-01-106-8451	273	15
11862	15599951	2540-01-216-3188	273	12
11862	15599952	4730-01-216-0021	288	20
11862	15599958	5340-01-159-1460	207	9
11862	15599959	5340-01-201-7954	207	8
11862	15599961	5340-01-199-4448	209	5
11862	15599962	5340-01-162-6062	237	7
11862	15599963	5340-01-159-4519	237	12
11862	15599964	5340-01-198-3434	237	11
11862	15599965	5340-01-253-2102	208	7
11862	15599968	6150-01-234-3253	266	1
11862	15599971	5306-01-166-8556	174	2
11862	15599972	5930-01-184-6370	284	17
11862	15599973	5340-00-881-5303	118	2
			119	24
11862	15599975	2540-01-162-4411	200	1
11862	15599976	2540-01-158-8561	200	5
11862	15599986		9	22
11862	15599987	4730-01-227-1929	9	23
11862	15599988	5340-01-234-1465	9	7
11862	15599989	5340-01-228-1659	46	3
11862	15599990	5325-01-230-1844	237	2
11862	15599994	5310-01-242-8561	236	3
			237	1
11862	15599997	2510-01-225-1004	140	41
11862	15599998	2510-01-225-5865	140	5
11862	15599999		22	39
11862	15605040	5305-01-161-3997	50	19
11862	15606404	2510-01-325-8216	192	7
11862	15606405	2510-01-325-9077	192	6

PART NUMBER INDEX

CAGEC	PART NUMBER	STOCK NUMBER	FIG	ITEM
11862	15606406	2510-01-325-9078	192	6
11862	15606407	5305-01-324-8862	192	8
11862	15607227	5310-00-380-1514	118	8
11862	15614462	2510-01-224-8839	177	6
11862	15614467	2510-01-155-8817	191	19
11862	15617126	2510-01-163-2709	169	9
11862	15627452	2540-01-158-8560	202	11
11862	15628608	2510-01-155-5857	176	8
11862	15628618	2510-01-231-5398	173	7
7X677	15633464	5330-01-157-0856	11	1
11862	15633467	5330-01-331-1207	KITS	
11862	15634657	2520-01-325-1860	103	2
11862	15634658		124	13
11862	15634659	2540-01-323-6049	235	7
11862	15634660	2540-01-323-6050	235	7
11862	15634661	5365-01-326-4346	124	26
11862	15634663	2530-01-326-1462	124	23
11862	15635684	2540-01-163-7305	169	10
11862	15635685	2510-01-163-7306	169	1
11862	15637897	3040-01-329-9927	111	17
11862	15637901	3020-01-330-3258	111	21
11862	15637902	2520-01-330-0028	111	18
11862	15637904	2520-01-330-3703	111	20
11862	15641780	2540-01-158-8554	177	2
11862	15646949	2540-01-159-8721	194	5
11862	15668598	2530-01-325-9112	124	3
84760	15699	5360-00-900-2564	15	105
24617	157456	5305-01-197-6351	231	14
16758	16015256	5930-01-165-0732	243	7
11862	16034561	2590-01-158-8551	243	4
11862	1604854	3120-01-232-6781	127	74
30076	160635	5310-00-792-3617	130	8
11862	1610819	5306-01-185-7050	33	9
11862	1623159	5307-01-150-7764	35	12
11862	1635490	5306-01-149-6280	6	10
			7	7
			11	2
			30	8
			33	2
			33	4
			34	3
			71	1
			132	7
			295	7
			295	9
11862	1638274	5340-01-158-0314	20	16
			21	1
			22	4
24617	163881	5320-01-166-1477	228	14
24617	1640810	5305-01-231-1298	65	6
			167	15

PART NUMBER INDEX

CAGEC	PART NUMBER	STOCK NUMBER	FIG	ITEM
11862	1640902	5305-01-160-3945	49	10
11862	16500591	6220-01-155-6515	50	31
11862	16501759	2510-01-159-8765	50	33
24617	165079	5305-01-166-1610	215	27
11862	16604537	2540-01-158-8559	201	9
08806	168	6240-00-144-4693	42	21
			43	1
			43	5
			51	8
			52	5
79410	17-01-014-001	5310-01-154-3990	125	12
25022	17-0177	5340-01-164-8137	215	25
			216	20
			225	7
25022	17-0178	5340-01-162-5619	215	23
			216	21
			225	8
25022	17-0195	2540-01-163-0894	229	5
25022	17-0201	5680-01-193-5078	216	18
25022	17-0202	5680-01-193-5078	215	26
25022	17-0209	2540-01-193-3623	215	22
			216	22
24617	171108	5305-01-201-3334	218	16
			219	11
11862	1727059	5310-01-162-7912	201	14
			202	16
11862	1731168	5340-01-166-5861	197	4
			198	5
			204	5
72582	178917	4730-01-075-7310	21	5
			22	8
11862	17981073	4720-01-149-4659	119	25
11862	17983936	2530-01-287-3980	127	67
11862	17987489	2530-01-147-6421	127	70
11862	18001032	5365-01-159-4833	116	10
11862	18002428	5310-01-151-8347	115	19
			116	21
11862	18003151	2530-01-149-1886	122	3
11862	18004057	5365-01-164-4525	116	9
11862	18004794	2530-01-096-6764	122	4
11862	18004890	2530-01-110-5304	122	4
11862	18007952	2530-01-147-9329	121	7
11862	18009094	3120-01-155-3517	5	6
11862	18009095	2815-01-165-4826	5	6
11862	18009096	3120-01-152-4239	3	4
11862	18009097	3120-01-152-8201	3	4
11862	18009098	3120-01-153-7545	3	4
11862	18013395	2530-01-163-7878	121	5
			121	8
11862	18015381	2530-01-156-6190	121	2
84760	18020	3040-01-192-4267	15	3

PART NUMBER INDEX

CAGEC	PART NUMBER	STOCK NUMBER	FIG	ITEM
84760	18021	2910-01-189-2142	15	2
84760	18493	5310-01-118-2248	15	30
11862	1852519	2920-01-163-7210	36	4
11862	1852890	5977-01-163-2900	40	28
16764	1875645	5961-01-059-0562	36	22
11862	1876358	5340-01-166-5639	40	39
11862	1876359	5977-01-163-2930	40	24
11862	1876458	5977-01-163-2032	40	23
16764	1876681	5305-01-081-0465	36	11
16764	1876806	2920-01-093-9216	36	2
16764	1876873	5977-01-018-2742	36	7
16764	1887021	2920-01-163-7872	40	21
11862	1892163	2590-01-085-6956	55	1
11862	1892941	2920-01-256-2253	36	13
16764	1893445	2920-01-163-8626	40	17
16764	1894023	3120-01-121-2809	40	8
24617	189448	5320-01-197-1394	146	6
84760	18986	2910-01-191-8465	15	82
25022	19-0445		213	2
25022	19-0816	6210-01-159-1794	226	1
25022	19-0817	6220-01-160-5094	211	4
			212	4
25022	19-0818	6220-01-160-3686	213	17
25022	19-0819	2540-01-163-7948	218	4
25022	19-0820	2590-01-164-7898	213	23
25022	19-0831	2540-01-163-7949	218	5
25022	19-0833	6130-01-158-9175	222	1
25022	19-0837	6150-01-190-6498	220	8
25022	19-0843	2540-01-165-5971	220	10
25022	19-0849	5340-01-166-5654	218	10
			219	4
25022	19-0861	6220-01-160-3705	213	5
25022	19-0875	5945-01-173-7760	222	2
25022	19-0876	5945-01-166-2012	222	8
25022	19-0877	5940-01-171-3195	222	7
25022	19-0878	5920-01-164-8260	222	15
25022	19-0880	6150-01-166-2013	220	19
25022	19-0881	6150-01-166-2014	220	11
25022	19-0882	6220-01-172-5300	211	7
			212	7
25022	19-0884	2590-01-164-7024	211	6
			212	6
25022	19-0885	6220-01-171-9557	211	8
			212	8
25022	19-0893	6220-01-160-3687	213	3
25022	19-0909	5925-01-190-1211	222	5
25022	19-0910	5995-01-192-4374	222	4
25022	19-0940	6220-01-197-0486	226	2
25022	19-0941	6230-01-191-3856	226	4
25022	19-0941-1		226	5
25022	19-0942	6250-01-164-3266	218	8

PART NUMBER INDEX

CAGEC	PART NUMBER	STOCK NUMBER	FIG	ITEM
25022	19-0942	6250-01-164-3266	219	1
25022	19-0969	5365-01-269-8614	213	6
24617	1915172	5310-00-596-5152	36	34
16764	1928021	3120-00-888-6630	40	16
24617	1928022	5365-00-804-9666	40	15
24617	1928023	5310-00-606-6608	40	14
08806	194	6240-00-944-1264	42	18
			42	20
			50	17
			63	1
			68	1
11862	1941113	2920-00-841-3254	40	5
16764	1941978	5310-00-949-9299	36	33
11862	1945804	5315-01-165-2568	40	11
11862	1951567	2520-00-997-9818	40	9
16764	1955946	5325-00-397-5962	40	38
24617	1958679	5360-00-947-1184	40	4
16764	1960908	5306-01-081-0464	40	34
16764	1966923	5315-00-945-8441	40	34
16764	1968396	5305-00-450-5937	40	42
11862	1970149	5970-01-163-2895	36	19
16764	1970227	5999-01-104-0445	36	3
16764	1970263	2520-01-110-1081	40	6
11862	1971993	2590-01-085-1968	36	27
11862	1975326	5307-01-164-4538	40	36
11862	1976049		36	29
11862	1976143	3120-01-167-5544	36	24
11862	1976882	2920-01-165-6794	40	7
11862	1976940	5315-01-166-1733	40	22
11862	1977064	5961-01-162-7228	36	16
11862	1977357	5360-01-167-1890	36	8
11862	1978058	4140-01-163-3543	36	31
11862	1978059	3110-01-165-2530	36	30
11862	1978068	3020-01-163-0498	36	32
16764	1978146	5910-01-089-1916	36	20
84760	19832	5330-01-197-2454	15	107
84760	19837	5315-01-188-0495	15	106
16764	1984076	5310-01-238-2983	40	40
11862	1986019	5340-01-165-0539	40	25
11862	1986427	2920-01-205-6018	36	23
11862	1986428	5935-01-165-9807	36	21
11862	1986433	2590-01-165-0747	36	10
11862	1986464	2920-01-224-3153	40	37
11862	1986471	2920-01-197-7229	40	41
11862	1986473	2920-01-200-8461	40	20
11862	1986552	5305-01-162-3763	36	18
11862	1987049	5315-01-119-3115	40	10
11862	1987254	2590-01-166-1071	40	31
11862	1987808	2920-01-163-7211	36	1
11862	1987809	2920-01-164-7091	36	5
11862	1988380	5910-01-189-5110	65	3

PART NUMBER INDEX

CAGEC	PART NUMBER	STOCK NUMBER	FIG	ITEM
84760	19895	4820-01-189-0894	15	108
11862	1990115	2920-01-159-3589	49	5
11862	1995217	5930-01-014-0187	49	15
11862	1997983	2540-01-159-5670	157	15
97271	2-4-3801X	2520-00-928-6150	98	15
11862	20025648	5305-01-162-5996	201	6
			202	6
11862	20030401	5325-01-165-2551	193	1
11862	20056525	5360-01-162-2849	196	17
			199	10
			200	6
			205	7
11862	2014469	5306-01-158-9018	28	12
			38	7
			56	14
			160	2
			163	3
			165	6
			173	13
			175	8
			176	18
			179	7
			191	16
			196	16
			203	3
			205	13
			206	6
			267	11
11862	20171141	2540-01-160-5918	177	3
84760	20222	3120-00-393-4067	15	76
11862	20243999	2540-01-158-4603	199	12
			200	12
11862	20264728	5340-01-165-4792	169	11
11862	20264729	5340-01-165-4791	169	2
11862	20264730	5330-01-181-2454	169	12
11862	20264731	5330-01-178-7351	169	3
11862	20264736	2540-01-163-2719	169	14
11862	20264737	2540-01-159-8880	169	5
11862	20264743	2510-01-155-5432	190	4
11862	20264744	2510-01-155-5433	190	2
11862	20289493	2540-01-251-1715	201	8
11862	20293842	2540-01-160-5840	202	7
11862	20293843	2540-01-251-1714	201	7
11862	20351007	5360-01-197-0870	199	6
11862	20354945	5680-01-167-1068	169	15
11862	20354946	5680-01-164-4964	169	6
11862	20365263	5305-01-167-8334	87	22
			271	7
11862	20369919	2540-01-191-8440	202	8
11862	20369920	2540-01-191-8441	202	9
11862	20410901	2540-01-248-2477	202	13

PART NUMBER INDEX

CAGEC	PART NUMBER	STOCK NUMBER	FIG	ITEM
11862	2043150	5340-01-163-1973	70	4
11862	2043151	5340-01-159-1321	45	6
			46	13
			57	4
11862	2044779	5340-01-166-1470	70	2
11862	20489125	5330-01-161-2608	233	21
84760	20512	2910-01-188-3683	15	104
84760	20523	2910-01-188-3250	15	113
84760	20527	2910-01-188-3775	15	112
84760	20528	5365-01-188-0960	15	116
84760	20529	5365-01-188-0959	15	111
84760	20530	2910-01-189-0895	15	102
99688	20565	6105-01-159-2223	300	11
99688	20566	6105-01-159-2666	298	8
08806	2057	6240-01-157-0636	51	5
08806	2057NA	6240-01-157-0635	50	26
11862	20573776	5325-01-198-8239	202	10
11862	20696927	5325-01-160-4028	167	6
93061	207ACBHS-4	4730-01-245-6925	252	14
84760	20727	4730-01-183-2140	15	99
84760	20803	2910-01-230-9007	15	104
84760	20849	2910-01-054-3816	15	10
84760	20951	2920-01-117-3251	15	31
11862	2098912	5945-01-162-0617	264	3
84760	21194	5305-01-188-0948	15	71
84760	21198	5360-01-197-1505	15	109
22337	212-776	2610-01-148-1634	126	1
84760	21200	2910-01-188-3249	15	115
84760	21201	3040-01-188-3261	15	77
84760	21283	2910-01-188-3244	15	97
84760	21284	5315-01-167-5584	15	42
84760	21287	5305-01-191-0374	15	98
84760	21296	2910-01-188-3251	15	114
84760	21312	2910-01-117-7252	15	75
84760	21323	5945-01-190-5143	15	23
			15	23
84760	21358	5360-01-190-6215	15	117
93061	215PN-4	4730-00-765-9103	252	13
84760	21618	5970-01-189-4883	15	38
84760	21646	5305-01-193-2377	15	118
84760	21661	5305-01-188-0490	15	45
84760	21662	5305-01-188-0491	15	45
84760	21663	5305-01-188-0492	15	45
84760	21664	5305-01-188-0493	15	45
84760	21665	5305-01-188-0494	15	45
84760	21712	5305-01-190-4070	15	74
84760	21763	2910-01-188-3254	15	11
84760	21860	5330-01-192-5779	15	20
84760	21895	2910-01-188-3255	15	8
84760	21917	5360-01-189-3466	15	9
25022	22-0832	2510-01-216-0039	218	1

PART NUMBER INDEX

CAGEC	PART NUMBER	STOCK NUMBER	FIG	ITEM
25022	22-0832	2510-01-216-0039	219	8
25022	22-0858	2540-01-163-8597	229	3
25022	22-0860	2540-01-166-5913	224	4
25022	22-0861	2590-01-163-7669	218	11
			219	17
25022	22-0864	2510-01-192-9752	228	5
25022	22-0865	2590-01-163-8596	224	1
25022	22-0866		221	4
25022	22-0867		220	5
25022	22-0871		245	1
25022	22-0873	7230-01-165-6795	228	8
25022	22-0875	2590-01-163-7287	213	15
25022	22-0876	5340-01-164-1076	225	3
25022	22-0877	5340-01-164-0958	218	12
			219	18
25022	22-0878	5340-01-164-1077	225	5
25022	22-0880	2590-01-163-7290	220	4
			221	7
25022	22-0881	5340-01-164-0746	212	21
25022	22-0884	2540-01-159-1793	218	2
			219	7
03350	22FT832	5310-00-582-5765	15	37
16748	22009290	5977-01-154-6777	234	18
11862	22009291	5360-01-195-6699	234	16
93061	2202P-4-4	4730-00-278-4824	252	11
11862	22020945	6105-01-164-6546	241	13
11862	22021655		233	9
11862	22021679	5325-01-167-8372	234	12
11862	22021690		234	22
11862	22029629	2540-01-156-5882	233	10
11862	22029630	2540-01-191-6512	233	12
11862	22038909	2540-01-164-8379	234	17
11862	22038924	6150-01-208-4507	234	19
11862	22038925	2920-01-212-4941	234	21
11862	22038927	2540-01-243-4934	234	9
11862	22038928	2540-01-156-0068	234	28
11862	22038931	5930-01-155-0783	234	26
11862	22048281	5330-01-157-4872	234	14
11862	22048287	5340-01-162-9954	234	15
11862	22048303	5330-01-161-4019	234	2
72560	22048352	2540-01-159-7954	233	14
11862	22048368	2540-01-156-8316	234	1
11862	22049511	5360-01-158-1727	234	6
11862	22049531	2540-01-156-0069	234	7
11862	22049534	2590-01-159-9809	234	20
11862	22049837	5305-00-866-0934	234	5
11862	22054153	5310-01-157-4855	234	8
11862	22054154		234	4
11862	22054155	5365-01-156-0067	234	11
11862	22054156	5330-01-244-2277	234	10
11862	22054158		234	27

PART NUMBER INDEX

CAGEC	PART NUMBER	STOCK NUMBER	FIG	ITEM
84760	22064	5315-01-191-3393	15	16
11862	22098841	6105-01-254-9496	286	4
84760	22125	5360-01-188-0807	15	18
24234	221968	6150-01-010-6558	247	4
84760	22256	5315-01-197-8547	15	10
14892	2227168	5330-01-157-1916	117	35
14892	2229044	5330-01-165-1358	117	31
14892	2229046	5340-01-157-1955	117	33
14892	2229448	5340-01-158-0321	117	37
14892	2230359	2530-01-163-7890	120	5
			120	12
80201	22306	5330-01-196-2426	83	8
14892	2232072	2530-01-192-9778	117	34
14892	2232073	2530-01-156-7016	117	30
14892	2232076	2530-01-168-6369	117	32
14892	2232077	2530-01-250-6472	117	36
84760	22325	5310-01-188-6743	15	13
84760	22326	2910-01-191-8448	15	14
84760	22327	2910-01-188-3252	15	17
84760	22351	5305-01-118-4114	15	35
84760	22367	2910-01-190-0069	15	59
14892	2238710	2530-01-168-1440	120	3
			120	9
14892	2238739		120	1
14892	2238740		120	8
14892	2238742	2530-01-147-6424	120	10
45152	2239H	5310-00-209-1218	15	36
84760	22397	5315-01-190-0429	15	52
84760	22398	5305-01-190-5745	15	48
94988	224-12V	5945-00-992-5415	49	31
72582	224425	4730-00-277-8269	9	5
11862	22506637	2910-01-200-4338	14	9
11862	22507977	5355-01-150-1541	49	7
46522	2251-0832-06-17	5305-00-249-5278	263	7
46522	2251-0832-10-13	5305-01-121-9352	263	2
11862	22510143	2540-01-156-7233	127	75
11862	22511422	5340-01-148-7528	13	15
11862	22514738	4730-01-155-5135	134	7
11862	22514861	2920-01-150-1610	49	1
11862	22515965	5330-01-159-1298	18	4
			19	19
11862	22516548	6680-01-161-1439	18	2
			19	18
11862	22521054	5310-01-250-3301	39	10
11862	22521550	5310-01-165-3331	8	7
			9	18
			10	6
			12	4
			16	13
			70	1
11862	22527167	5360-01-163-7234	118	17

PART NUMBER INDEX

CAGEC	PART NUMBER	STOCK NUMBER	FIG	ITEM
11862	22527167	5360-01-163-7234	119	3
11862	22529441	6350-01-321-7005	68	13
11862	22535073	5307-01-269-4336	1	3
			2	7
			11	3
			30	12
84760	22591	5330-01-213-9966	23	3
84760	22642	5305-01-188-0496	15	53
84760	22693	5365-01-188-0784	15	56
84760	22721	2910-01-189-2191	15	51
84760	22733	5305-01-190-4078	15	45
84760	22734	5305-01-188-6566	15	45
84760	22813	5360-01-197-1506	15	12
84760	22840	5945-01-155-7830	15	43
84760	22851	2910-01-198-0868	15	25
46522	2290-0632-05-13	5305-01-142-3150	263	46
84760	22900	5365-01-188-0993	15	121
84760	22917	5340-01-192-6030	15	54
11862	2292300	5310-01-270-3712	110	2
84760	22935	2910-01-166-5655	15	81
84760	22937	5365-01-188-0958	15	68
84760	22985	5940-01-189-9841	15	39
84760	22988	2910-01-232-1044	15	103
25022	23-0193	9905-01-165-0542	211	18
			212	17
25022	23-0196	9905-01-165-0541	211	17
			212	18
25022	23-0198	9905-01-165-0544	213	7
25022	23-0200	9905-01-165-6901	215	17
			216	27
73342	23017556	5315-01-152-9029	76	15
			77	2
84760	23056	5365-01-188-0785	15	62
84760	23100	5315-01-188-0765	15	124
84760	23101	3110-01-188-7682	15	67
84760	23107	2910-01-188-3176	15	70
84760	23120	2910-01-192-4585	15	94
84760	23124	5305-01-188-6567	15	90
35510	2313	5310-01-106-9929	37	26
84760	23171	5360-01-197-1507	15	58
84760	23183	4820-01-165-9596	15	33
84760	23238	5360-01-188-6806	15	92
84760	23265	2910-01-165-9487	15	80
84760	23352	5307-01-188-9217	15	50
84760	23426	5365-01-188-0783	15	63
84760	23428	2910-01-188-3256	15	47
84760	23461	2520-01-188-3282	15	1
11862	23500073		7	10
11862	23500074	2815-01-148-9540	7	13
11862	23500075	3040-01-192-4466	7	15
11862	23500076	3040-01-212-7616	7	18

PART NUMBER INDEX

CAGEC	PART NUMBER	STOCK NUMBER	FIG	ITEM
11862	23500111	3120-01-175-3813	3	5
11862	23500133	2930-01-193-7802	31	8
11862	23500139	5330-01-249-1629	3	7
11862	23500298	5365-01-234-0447	5	2
11862	23500391	2815-01-246-5268	5	7
11862	23500392	2815-01-164-0138	5	7
11862	23500393	2815-01-246-5269	5	7
11862	23500396	5365-01-193-0458	39	11
89554	23500832	5307-01-150-5991	11	4
11862	23500846	5330-01-149-0874	2	5
			30	9
			261	3
11862	23501992	2815-01-285-5065	6	3
11862	2353015	5310-01-330-9817	111	2
11862	2353021	5310-00-809-8836	111	1
84760	23566	5365-01-188-0962	15	46
35510	2364	5310-01-024-3109	37	62
84760	23643	5360-01-194-3162	15	7
84760	23685	5315-01-188-0766	15	124
84760	23796	4820-01-209-0473	23	14
35510	2385	5310-00-781-4787	37	19
84760	23861	5945-01-214-1490	15	43
11862	2387523	5310-01-157-7589	172	10
84760	23925	5315-01-189-2141	15	60
72210	24004-J	2530-01-165-9653	129	6
11862	2423517	5310-01-195-5088	84	2
			88	5
84760	24265	5340-01-197-1199	23	8
84760	24267	5340-01-211-3086	23	6
84760	24281	5340-01-202-2622	23	9
84760	24285	2910-01-210-1322	23	4
84760	24322	5305-01-211-3031	23	12
35510	2434	5310-00-775-5139	37	6
84760	24345	2910-01-189-1747	15	119
35510	2435		37	50
84760	24419	5305-01-190-4555	15	73
84760	24433	2910-01-188-3257	15	64
84760	24434	2910-01-256-3698	15	64
84760	24437	5305-01-211-3032	23	11
84760	24566	5305-01-188-6568	15	65
84760	24569	3040-01-188-3242	15	85
84760	24585		15	122
84760	24623	3040-01-188-3222	15	123
84760	24680		15	41
11862	2477054	5320-01-160-3999	49	14
			208	3
72983	248X4	4730-00-900-3296	245	12
84760	24901	5940-01-189-8033	15	40
11862	25004137	6680-01-164-9433	19	20
11862	25004140	6680-01-175-0565	18	3
11862	25011206	4820-01-153-1851	8	10

PART NUMBER INDEX

CAGEC	PART NUMBER	STOCK NUMBER	FIG	ITEM
11862	25011208	6615-01-153-1852	8	9
11862	25015099	5330-01-164-1653	42	5
11862	25017376	6695-01-167-8108	42	8
11862	25020687	5340-01-163-7501	291	2
11862	25022883	6220-01-164-5228	42	6
11862	25022884	6210-01-158-4668	42	12
70040	25023641	5355-01-251-0633	69	7
11862	25033627	6680-01-230-5684	291	1
11862	25037177	6685-01-192-4834	54	1
11862	25041910	2940-01-148-5992	17	5
11862	25042462	2990-01-147-9284	10	14
11862	25052373	6680-01-225-4475	291	3
11862	25052807	2540-01-158-8626	42	3
11862	25053500	6210-01-161-2138	42	14
11862	25053501	6210-01-158-6575	42	13
11862	25053622	2540-01-163-7281	42	16
11862	25053623	5999-01-158-9249	42	4
11862	25076586	6210-01-271-6871	42	15
11862	25078571	2540-01-147-5537	69	5
11862	25078578	2540-01-191-6510	69	8
27647	25113	3040-01-163-0797	124	4
97271	25127-1X	3040-01-075-1878	98	10
35510	2523	5310-00-775-5182	37	63
11862	25515635	2530-01-156-8317	118	21
11862	25516531	5340-01-158-0297	114	16
11862	25518880	4730-01-163-7194	12	6
			20	10
			21	12
			22	13
			23	19
			134	2
11862	25523703	5945-01-243-1702	55	6
11862	25527423	5340-01-268-9064	22	10
			118	4
			255	17
30780	260-6-6		252	26
11862	2600236	5365-00-081-5194	84	65
11862	26004461	2520-01-268-7413	112	12
11862	26005078	2520-01-158-8481	113	15
11862	26013911	2520-01-211-6755	92	1
11862	26013913	2520-01-191-9518	90	1
11862	26016662	5330-01-084-2410	109	6
11862	26019661	5930-01-214-0401	49	2
11862	26020811	3020-01-096-9657	110	3
84760	26070		15	89
84760	26071	5360-01-260-5649	15	88
84760	26214	2920-01-101-2552	KITS	
11862	2622207	2530-01-075-5080	121	4
			121	10
11862	2622667	2530-01-154-1263	122	5
90005	26422-B	2910-00-203-3322	245	32

PART NUMBER INDEX

CAGEC	PART NUMBER	STOCK NUMBER	FIG	ITEM
10001	2642880	8145-01-005-2994	253	3
84760	26833		15	87
84760	26834		15	87
84760	27002		15	22
84760	27015		15	22
72582	271163	5310-01-069-5243	38	5
			47	5
			50	21
			51	10
			53	14
			68	11
24617	271166	5310-00-696-5172	65	4
24617	271172	5310-01-152-0598	47	6
			57	3
			65	5
			178	12
			207	10
			232	7
			232	13
			232	21
			269	14
			272	9
24617	271184	5310-00-933-4310	289	12
84760	27163	5330-01-233-8597	15	55
11862	2724176		296	10
84760	27244	5330-01-234-2615	15	24
84760	27245		15	93
84760	27284	2920-01-212-4771	23	13
84760	27290	2910-01-156-0045	23	1
11862	273487	2640-00-555-2829	124	29
24617	274244	5330-00-935-9136	8	8
			254	2
80201	27467	5330-01-106-7938	124	14
			124	14
1T998	274898	5305-01-217-2046	293	17
84760	27600		15	49
84760	27601	5330-01-236-0476	15	96
84760	27602		15	66
84760	27603		15	69
84760	27607	5330-01-232-2145	15	32
84760	27608		15	101
84760	27609	5330-01-233-2778	15	57
84760	27610		15	61
14892	2770209	2530-01-151-5967	117	1
14892	2770317	2530-01-154-1294	117	1
14892	2770532		117	14
14892	2770614		117	11
14892	2770685		117	16
			117	16
14892	2770698		117	17
14892	2770699		117	15

PART NUMBER INDEX

CAGEC	PART NUMBER	STOCK NUMBER	FIG	ITEM
14892	2770715		117	7
14892	2770720		117	13
14892	2770723		KITS	
14892	2770746		117	10
14892	2770771	2530-01-163-9952	117	23
14892	2770829		117	26
14892	2770972		117	15
35510	2771	5310-00-616-1739	37	67
14892	2771098		KITS	
14892	2771125		117	21
14892	2771165	5340-01-246-2700	117	4
14892	2771247		117	19
84760	27820	5330-01-236-1724	23	5
84760	27833	2910-01-188-3243	15	94
60038	2793-2720	3110-01-197-6651	84	40
84760	27984	2990-01-320-8915	15	79
84760	28396	4320-01-317-0692	15	83
84760	29090	2910-01-210-1323	23	7
77060	2973932	6250-00-433-5946	42	19
97271	29907X	2530-01-149-3827	97	1
07200	29940-0435	5330-01-147-9698	83	57
73680	29940-0445	5330-01-168-3870	82	8
			83	54
6V625	30-600	2640-01-302-1388	124	29
46522	30002-83	5310-01-027-1353	263	9
22593	30103	5330-00-360-7881	181	3
11862	3013475	4730-01-160-0814	241	11
11862	3015545	5310-01-211-1648	241	7
11862	3024673	5340-01-165-3717	240	3
11862	3024867	2540-01-155-7299	240	7
11862	3025501	5340-01-158-6816	240	2
95019	30258	2520-00-884-8430	98	30
97271	30263	2520-00-884-5635	98	8
			98	29
72447	30266	5306-00-685-2523	98	1
72447	30271	5310-00-809-8836	98	18
97271	30272-1	5365-00-061-6847	98	11
97271	30272-2		98	11
97271	30272-3	5365-00-061-6849	98	11
97271	30272-4	5365-00-062-7200	98	11
11862	3027247	2540-01-156-0072	240	4
97271	30273	5310-00-137-3018	98	13
11862	3027308	2540-01-159-8705	240	12
72447	30275	5310-00-137-3018	98	17
97271	30276-1	5365-00-061-6844	98	23
			111	4
97271	30276-2	5365-00-061-2395	98	23
			111	4
97271	30276-3	5365-00-061-6845	98	23
			111	5
97271	30276-4	5365-00-062-7201	98	23

PART NUMBER INDEX

CAGEC	PART NUMBER	STOCK NUMBER	FIG	ITEM
97271	30276-4	5365-00-062-7201	111	3
95019	30291-1	5365-00-884-5640	98	19
95019	30291-2	5365-00-884-5639	98	19
72447	30291-3	5365-00-004-6407	98	19
11862	3029730	2540-01-158-6906	241	5
11862	3030072	2540-01-156-0071	240	11
11862	3030075	2540-01-156-4874	240	8
11862	3035192	4520-01-166-2134	286	5
11862	3036927	4720-01-160-3664	241	12
11862	3037476	4520-01-166-2133	286	2
11862	3037550	6105-01-165-4561	241	8
11862	3039873	5330-01-195-4880	241	9
11862	3042351	5310-01-164-2336	241	6
11862	3048067	2540-01-156-0073	240	13
11862	3048083	2520-01-192-1257	240	5
81795	30489	5310-00-472-3214	271	10
			273	2
			274	6
11862	3054308		239	2
11862	3054315	2540-01-159-8722	240	6
27462	3054316	2540-01-156-0070	240	10
11862	3055734	2540-01-268-7202	286	3
11862	3058097		241	4
61928	3058966	2930-01-264-3480	28	15
97271	30797-1		111	8
97271	30797-2		111	8
97271	30797-3		111	8
97271	30875	5310-00-062-6828	101	15
11862	3093499	5330-01-276-3235	296	27
25022	31-0030	2540-01-163-8623	211	12
			212	23
9W635	31-0040		252	17
9W635	31-0040A		252	18
81343	31-620	6140-01-031-6882	56	11
97271	31026-1	5306-01-172-1590	101	23
			101	27
35510	31256	5310-01-143-1719	37	3
35510	31587	5310-00-429-3110	37	27
11862	3187843	2510-01-153-9584	153	4
11862	3187844	2540-01-155-7496	155	6
11862	3187845	2540-01-148-2943	154	1
11862	3187846	2510-01-148-2942	153	4
24234	319029	4720-01-192-1631	BULK	12
91929	32TS1-5	5930-00-655-1512	247	20
14894	3203465	5340-01-149-3376	120	17
14892	3203466	2530-01-149-3375	120	17
35510	3231	5310-00-032-1814	37	61
11862	326418	2510-01-283-1966	142	8
11862	326439	2510-01-152-9259	140	35
			142	26
11862	326440	2510-01-152-9260	140	46

PART NUMBER INDEX

CAGEC	PART NUMBER	STOCK NUMBER	FIG	ITEM
11862	326440	2510-01-152-9260	142	41
11862	326489	5306-01-160-3942	152	9
11862	326560		70	5
11862	326561	5360-01-206-6616	291	6
11862	326934	5330-01-159-1152	183	1
11862	327005	5330-01-159-2816	183	3
11862	327006	5330-01-162-8595	183	2
11862	327015	5330-01-182-4121	178	11
11862	327062	2540-01-156-4870	166	2
			166	11
11862	327065	5360-01-161-7561	166	3
			166	10
11862	327067	3040-01-155-0371	170	1
11862	327100	5340-01-161-9187	172	15
11862	327225	5365-01-326-1128	151	9
11862	327260	2510-01-169-3767	142	14
24617	327739	5330-01-076-3009	124	27
11862	327959	5340-01-206-2995	166	27
11862	327960	5340-01-206-2996	166	27
11862	328059	5340-01-179-4108	139	9
			141	8
11862	328060	5340-01-173-0048	139	9
			141	8
11862	328107	2510-01-325-9069	157	9
11862	328108		157	11
11862	328111		157	10
11862	328128	3120-01-211-7528	156	6
11862	328130	5306-01-155-6108	156	1
11862	328131	5310-01-147-8748	156	2
11862	328132	2530-01-147-5541	156	9
11862	328169	5340-01-156-6779	151	4
11862	328200	5340-01-171-8257	140	26
11862	329198		233	18
11862	329457	5360-01-160-2411	196	20
			196	23
			205	10
			205	16
11862	329830	5975-01-324-7825	62	6
11862	329842	2540-01-160-0849	41	6
11862	330046	2540-01-191-8439	147	1
11862	330393	2510-01-326-1389	192	3
11862	330394	2510-01-326-1460	192	3
11862	330438	2510-01-155-5846	173	10
			175	3
11862	330485	5325-01-205-2545	167	10
11862	330489	2510-01-325-9070	192	2
11862	330490	2510-01-325-9071	192	2
11862	330492	6220-01-327-3252	52	4
11862	331416	5340-01-155-3668	119	19
11862	331478	5305-01-148-8208	120	15
11862	331638	2540-01-158-8556	182	2

PART NUMBER INDEX

CAGEC	PART NUMBER	STOCK NUMBER	FIG	ITEM
11862	334101	2510-01-162-7112	181	1
11862	334102	2510-01-165-4932	181	5
11862	334104	5330-01-167-8123	181	3
11862	334132	2510-01-159-7120	179	24
11862	334146	5340-01-154-5270	123	2
11862	334150	2590-01-155-8871	184	9
			190	10
11862	334195	5306-01-160-0769	191	17
			199	7
			200	7
11862	334307	3040-01-156-7182	115	8
11862	334308	3040-01-157-7970	115	8
11862	334362	5365-01-165-4511	115	8
11862	334363	5365-01-164-4601	115	8
11862	334364	5365-01-164-6598	115	8
11862	334365	5365-01-164-4602	115	8
11862	334366	5365-01-164-2411	115	8
11862	334367	5365-01-164-4603	115	8
11862	334368	5365-01-162-8868	110	8
11862	334369	5365-01-162-8869	110	8
11862	334370	5365-01-162-8870	110	8
11862	334371	5365-01-171-4041	110	8
11862	334372	5365-01-164-6616	110	8
11862	334373	5365-01-168-7294	110	8
11862	334374	5365-01-168-1409	110	8
11862	334375	5365-01-167-1122	110	8
11862	334376	5365-01-162-8884	110	8
11862	334377	5365-01-164-6617	110	8
11862	334378	5365-01-164-6618	110	8
11862	334379	5365-01-164-6619	110	8
11862	334380	5365-01-164-6620	110	8
23862	334387	5810-01-107-4051	124	1
11862	334521	2590-01-155-5140	19	22
11862	334522	5340-01-160-0367	19	23
11862	334523	2590-01-155-5177	19	12
11862	334532	6680-01-147-6583	72	10
11862	334533	5310-01-164-1647	69	20
11862	334540	2530-01-159-8802	114	2
11862	334541	5325-01-160-2238	114	7
11862	334666	3040-01-173-0049	139	3
11862	334675	5340-01-244-7925	18	11
11862	334963	6680-01-160-5256	59	6
11862	335435	2540-01-157-3528	205	12
			206	4
11862	335452	3040-01-158-8703	182	5
11862	335524	3120-01-159-1311	179	15
			203	12
11862	335551	5305-01-160-1970	191	9
11862	336406	5355-01-157-1865	243	6
14892	3368689	2530-01-183-8860	115	6
11862	336926	3040-01-166-4497	123	12

PART NUMBER INDEX

CAGEC	PART NUMBER	STOCK NUMBER	FIG	ITEM
11862	336989	5360-01-151-1121	25	3
11862	337185	5365-01-149-0880	8	19
			254	7
97271	33734	5310-01-172-1591	124	18
11862	337520	5365-01-159-7767	150	7
11862	337714		233	22
11862	337715	2540-01-156-4907	169	20
11862	337716	2540-01-156-4908	169	20
11862	337751	5340-01-152-2382	172	17
11862	337773	5365-01-157-6772	172	16
11862	337788	5340-01-160-2367	179	6
11862	337899	5340-01-160-2172	181	7
11862	337957	5330-01-323-5567	140	24
11862	338269	2530-01-154-1222	120	4
			120	13
11862	338696	5340-01-155-2616	134	13
11862	338699	5340-01-171-8242	142	7
35510	3395	5310-00-775-5180	37	64
11862	339885	6220-01-306-4265	52	6
11862	339887	6220-01-327-1025	52	6
11862	340034	5315-01-159-8666	184	112
11862	340053	2540-01-158-8813	177	5
11862	341160	5340-01-160-2445	26	5
			27	7
11862	341287	2910-01-155-5137	19	27
11862	341509	5310-01-148-2676	124	24
11862	341510	5315-01-153-0317	124	28
11862	341511	2520-01-147-5539	124	5
11862	341990	5325-01-155-4482	114	3
11862	342221	5306-01-159-1130	195	4
			197	2
			198	1
			204	3
11862	42405	2910-01-149-3786	25	7
11862	342677	5340-01-155-2614	118	15
11862	343124	5325-01-164-6431	45	7
11862	343178	5305-01-149-1936	128	1
			140	48
			142	35
11862	343179	5365-01-151-6111	128	2
			140	49
			142	34
11862	343350	5340-01-270-7423	14	2
11862	343438	5340-01-155-2614	118	1
11862	343444	5340-01-164-6474	251	13
			255	13
97271	34367	3020-01-155-0183	98	4
			98	25
97271	34367	3020-01-155-0183	165	15
11862	343951	2590-01-147-5538	147	9
11862	343978	5365-01-157-7476	196	6

PART NUMBER INDEX

CAGEC	PART NUMBER	STOCK NUMBER	FIG	ITEM
11862	344029	5320-01-159-4507	115	11
11862	344714	5340-01-162-9781	19	28
11862	345683	2530-01-153-1813	117	25
11862	345819	5365-01-162-3849	191	18
97271	34592	5340-00-604-0566	98	16
11862	345943	2530-01-152-9258	116	29
11862	345944	2530-01-153-1492	116	29
11862	346381	3120-01-158-2096	123	8
97271	34729	3120-01-150-1226	98	27
97271	34730	5310-01-157-7601	98	7
			98	28
11862	347347	5325-01-157-1698	50	20
			158	4
			194	8
97271	34801-1		111	8
97271	34801-10		111	8
97271	34801-2		111	8
97271	34801-3		111	8
97271	34801-4		111	8
97271	34801-5		111	8
97271	34801-6		111	8
97271	34801-7		111	8
97271	34801-8		111	8
97271	34801-9		111	8
97271	34822	5306-00-157-8315	96	14
11862	350036	2590-01-147-2269	146	8
11862	350037	2590-01-193-3443	146	5
11862	350371		134	3
11862	350544	3120-01-163-0638	150	13
			151	19
11862	350726	2510-01-096-9662	151	13
11862	351789	5340-01-158-2160	165	3
11862	354501	4730-01-148-2758	30	15
			31	2
			261	7
11862	355561	2540-01-153-9470	123	11
97271	35625	3020-01-155-5871	98	26
11862	357445	5310-01-165-2475	175	2
11862	357490	5330-01-165-1378	184	1
11862	357845	2530-01-156-5883	115	16
11862	357846	5340-01-156-8395	115	16
11862	357889	2530-01-152-7787	116	18
11862	357890	3040-01-152-7786	116	18
97271	35801	2520-01-159-7762	98	9
97271	35830	3020-01-159-9795	98	5
11862	358375	6680-00-147-5497	28	7
11862	358501	5310-00-264-1930	125	24
11862	358543	2510-01-158-6903	175	1
11862	358549	5306-01-158-2023	184	7
11862	358589	5330-01-166-2154	184	3
11862	358590	5330-01-166-5912	184	4

PART NUMBER INDEX

CAGEC	PART NUMBER	STOCK NUMBER	FIG	ITEM
11862	358593	5330-01-165-3402	184	6
11862	359816	5340-01-161-4025	119	18
11862	359847	5340-01-151-7409	19	21
11862	359877	2510-01-153-9473	148	15
			149	14
11862	359878	2510-01-154-1261	148	11
			149	10
11862	359917	2520-01-158-1760	70	8
25022	36-0039	2540-01-163-8638	223	1
25022	36-0043	2540-01-173-1249	223	27
25022	36-0044	4820-01-173-1250	223	3
25022	36-0045	2540-01-163-8598	223	2
25022	36-0046	2540-01-163-8599	223	16
25022	36-0047	2540-01-163-8600	223	15
25022	36-0048	5360-01-163-5578	223	17
25022	36-0049	5360-01-163-5579	223	18
25022	36-0050	5360-01-163-5580	223	20
25022	36-0051	5315-01-164-5334	223	30
25022	36-0054	2540-01-164-1562	223	8
25022	36-0055	2540-01-165-7931	223	6
25022	36-0056	5215-01-164-5336	223	5
25022	36-0066	2540-01-025-0433	224	9
25022	36-0070	2540-01-191-6539	223	9
25022	36-0071	5315-01-209-7063	223	14
25022	36-0072	5340-01-193-9565	223	31
25022	36-0073	5315-01-197-1482	223	28
25022	36-0074	5340-01-197-1550	223	25
25022	36-0075	5315-01-197-1483	223	21
11862	360582	5360-01-165-1563	8	28
11862	361119	5365-01-194-5273	84	33
11862	362275	2510-01-154-1260	150	12
			151	21
11862	362297	2530-01-159-5958	128	14
11862	362298	2530-01-159-8725	128	15
11862	362379	5310-01-165-0464	50	15
11862	362433	5340-01-170-5530	167	2
11862	362434	5340-01-161-5522	167	2
75829	363107	2510-01-096-2985	190	6
11862	363137	5325-01-160-2237	167	20
11862	363139	5330-01-096-7698	BULK	34
80020	36344N24	5315-00-243-1169	144	15
			145	22
97271	36352-1	5330-01-159-8336	97	8
11862	363913	5365-01-156-8061	148	14
			149	13
11862	364372	5365-01-160-2483	167	23
72447	36472	4730-01-034-8228	96	15
11862	365443	5310-01-184-5866	169	7
			169	16
11862	3655180	5325-01-164-8651	68	5
11862	365953		168	1

PART NUMBER INDEX

CAGEC	PART NUMBER	STOCK NUMBER	FIG	ITEM
11862	365953-1		169	4
			169	13
11862	3661804	5325-01-164-8652	59	4
11862	368026	6680-01-160-5276	291	8
11862	368752	2910-01-147-4219	18	12
11862	368786	2530-01-148-1463	114	17
97271	36880	5310-01-093-2907	101	1
35510	36912	5310-01-143-1679	37	52
11862	3694822	5360-00-310-4493	116	25
11862	3696822	4710-00-395-5144	BULK	31
25022	37-0019	5930-01-163-2779	219	14
			227	3
25022	37-0020	5930-01-163-8851	219	12
			227	6
25022	37-0051	5340-01-165-0602	224	7
11862	370053	5306-01-154-4320	148	24
11862	370054	5306-01-154-3971	148	25
11862	370055	2510-01-151-6362	148	21
11862	370056	2510-01-151-6363	148	29
11862	3701679	5315-01-149-5433	3	9
11862	3702366	2805-00-336-1716	8	27
11862	3702807	5310-01-151-8353	123	13
11862	370349	5340-01-217-2278	273	5
11862	370389	2510-01-221-2094	BULK	9
11862	370390	5330-00-753-8036		33
11862	3704871	5340-01-194-5294	8	26
11862	3705044	5330-00-830-1745	13	11
11862	3705444	5340-00-697-4703	71	2
11862	370867	6220-01-156-8420	53	7
11862	370868	6220-01-156-8247	53	7
11862	370873	5340-01-165-3417	53	6
11862	370874	6250-01-155-6547	53	15
11862	3709351	5365-01-194-5275	84	53
11862	3709352	5365-01-194-5277	84	53
11862	3709353	5365-01-194-0879	84	53
11862	3709354	5365-01-194-5276	84	53
11862	3711876	5305-01-150-9500	102	9
			102	13
11862	371603	5305-01-148-3687	146	9
11862	3717658 N.D.	5305-01-156-9711	151	18
11862	3719599	2910-00-493-2138	13	12
11862	3721887	5365-01-152-7439	89	3
			90	3
			91	3
			92	3
11862	372249	2530-01-152-0180	115	29
			116	31
11862	372379	2530-01-140-6144	115	10
55880	372380	3020-01-153-9435	83	78
11862	3725668	5310-01-184-5418	147	3
11862	3727207	5306-01-148-8198	4	2

PART NUMBER INDEX

CAGEC	PART NUMBER	STOCK NUMBER	FIG	ITEM
9K937	37296	5307-01-268-2640	101	21
97271	37299	2530-01-164-0037	101	24
97271	37300	5360-01-099-7921	101	18
97271	37305	5340-01-171-8256	101	12
97271	37307	5330-01-099-9423	101	17
97271	37308	2520-01-159-4173	96	6
97271	37312	5365-01-174-8657	96	4
11862	3738198	5340-01-221-0264	280	11
11862	3750950	4710-01-158-7511	BULK	39
11862	375180	5340-01-194-3188	170	5
11862	3760300	5310-01-155-1898	116	3
11862	376176	3020-01-177-3055	83	6
11862	3762400	2590-00-476-5459	169	8
			169	17
11862	3764438	2530-01-153-1814	154	7
			155	5
11862	3765243	5340-01-161-9188	165	7
11862	3767138	5360-01-159-1449	115	23
11862	376851	5330-01-086-3503	97	5
11862	376852	5310-01-153-1381	97	4
11862	376855	5330-01-086-3504	97	3
11862	376869	5306-01-153-1368	106	3
11862	3773687		134	9
11862	3782730		233	13
11862	3782732		233	3
11862	378362	5340-01-161-1440	291	11
11862	3787240	2520-01-159-7757	72	5
			83	14
			85	6
			97	17
			108	11
			109	7
97271	37879	5306-01-150-5893	101	20
			101	26
11862	3790768	5310-01-147-8743	17	2
			135	3
			145	5
			148	17
			149	16
11862	3792287	5310-01-160-4536	28	21
			38	6
			186	10
11862	3792381	5365-01-231-7584	13	17
11862	3793014	5315-00-450-9163	127	73
11862	379393	2510-01-152-7832	149	20
11862	379394	2510-01-152-7833	149	28
11862	379397	5306-01-237-4995	149	23
11862	379398	5306-01-158-6231	149	24
11862	3794767	5340-01-217-2168	273	3
11862	3798872	2590-01-160-1047	233	19
97271	38000	2530-01-159-7755	120	18

PART NUMBER INDEX

CAGEC	PART NUMBER	STOCK NUMBER	FIG	ITEM
97271	38001	2530-01-159-7754	120	18
97271	38081-1	5310-01-099-7945	101	2
97271	38139	2530-01-170-7118	101	16
11862	3815936	5930-00-073-0390	54	4
11862	3816659	5340-01-149-4434	28	9
			41	10
			57	8
			60	2
			66	11
			118	5
			162	3
			233	8
			264	2
			268	6
			282	2
11862	3820163	5360-01-156-9730	116	22
11862	382105	5310-01-148-2682	87	16
11862	3824124	5306-01-160-7553	233	20
11862	3825416	5340-01-159-1324	244	4
			261	16
			297	5
11862	3827499	5310-01-196-5587	17	1
11862	3833322	5365-01-159-4840	107	5
11862	3835333	2805-00-752-0158	7	24
11862	3838153	5310-01-194-9220	88	4
11862	3850084	5360-01-160-2415	123	7
11862	3853759	5365-00-257-6035	99	2
			112	10
11862	3853760	5365-01-149-5434	99	2
			112	10
11862	3853761	5310-01-150-6034	99	2
			112	10
11862	3853762	5365-01-149-5435	99	2
			112	10
11862	3853912	5365-01-159-4471	99	2
			112	10
11862	3856834	5315-01-156-0081	115	7
11862	3856843	5360-01-157-3662	115	20
11862	3856849	5365-01-158-2191	115	26
11862	3856850	5360-00-229-5312	115	24
11862	3856855	2530-00-494-8165	115	14
			116	16
11862	3856856	2530-00-125-2769	115	14
			116	16
11862	3856857	3040-01-156-8307	115	18
11862	3856858	2530-01-156-8308	115	18
11862	3860095	5330-00-110-8437	6	8
11862	386451	2520-00-722-7074	89	2
			90	2
			91	2
			92	2

PART NUMBER INDEX

CAGEC	PART NUMBER	STOCK NUMBER	FIG	ITEM
11862	3866187	5340-01-158-0303	60	6
			104	3
11862	3866918		291	4
60038	387AS-382A	3110-01-030-8475	110	10
			124	15
11862	3882979	3040-01-148-5982	91	15
			92	15
11862	3886908	5325-01-164-8655	255	25
			268	13
11862	3887347	5360-01-162-9935	116	26
11862	3887751	5365-01-096-6749	148	28
			149	27
11862	3889864	5340-01-159-6905	176	14
11862	3893181	2540-01-159-8878	114	1
11862	3894327	5365-01-162-3797	2	9
11862	3898059	2530-01-156-4875	115	21
11862	3900684	5340-01-158-0503	166	20
11862	3903572	5310-01-194-9221	297	7
11862	3905674	5340-01-160-2470	180	10
11862	3907088	5310-01-159-8678	99	2
			112	10
11862	3907089	5310-01-156-8852	99	2
			112	10
73342	3909063	5310-01-143-0542	25	2
11862	3909351	5340-01-172-1944	172	
11862	3914674	5310-01-157-5670	140	10
			196	7
11862	3918889	5325-01-168-5677	64	2
			269	18
11862	3920486	5340-01-149-9729	93	5
11862	3929059	5315-01-025-0930	214	11
11862	393578	5310-01-147-8746	99	10
11862	3943493	5340-01-172-2012	140	31
			142	18
11862	3943494	5340-01-160-4600	140	31
			142	18
11862	3944769	5340-01-157-7471	166	25
11862	3946246	5340-01-160-7778	172	7
11862	3946247	5365-01-155-7465	172	12
11862	3947770	2805-00-155-7266	7	22
11862	3953987	5310-01-194-7081	88	14
11862	3954730	5305-01-162-7890	147	7
			197	3
			198	3
			203	26
11862	3954735	5310-01-147-9792	35	8
11862	3957093	2590-01-162-7367	165	12
11862	3965121	3120-01-159-9386	97	2
11862	3965137	5307-01-158-2110	102	17
11862	3965138	5365-01-159-8890	102	16
11862	3967845	5365-01-166-8901	102	7.

PART NUMBER INDEX

CAGEC	PART NUMBER	STOCK NUMBER	FIG	ITEM
11862	3967845	5365-01-166-8901	102	21
11862	3967877	3120-01-199-1399	84	32
11862	3967879	2520-01-194-2145	84	6
11862	3967881	5310-01-197-1539	84	50
11862	3967882	3020-01-193-7825	84	31
11862	3967883	5365-01-193-2948	84	49
11862	3967884	3020-01-194-1714	84	38
11862	3967885	5330-00-737-3727	84	11
11862	3967886	5360-01-155-4555	83	60
			84	12
11862	3967887	5310-01-195-5092	84	62
11862	3967888	5310-01-194-7069	84	63
11862	3967895	5310-01-197-1538	84	17
11862	3967898	5365-01-197-1481	84	39
11862	3970076		14	3
11862	3970988	5310-01-154-1461	148	23
			149	22
			151	9
11862	3975202	2510-01-154-1254	148	13
			149	12
11862	3975698	5365-00-734-2496	84	41
11862	3975700	5365-00-734-2494	84	41
11862	3975703	3040-01-192-1635	84	42
11862	3975705	3020-01-192-4635	84	27
11862	3975710	5340-00-734-2649	84	44
11862	3975711	5330-01-125-6132	84	43
11862	3975715	2520-01-201-7851	84	20
11862	3975716	5315-01-197-0812	84	14
11862	3975717	5365-00-735-1129	84	10
11862	3977325	2520-01-157-4376	110	9
11862	3977326	5340-01-163-6145	108	5
11862	3977354	5330-01-112-4286	110	7
11862	3977355	5365-01-164-8632	110	23
11862	3977358	2520-01-156-0552	110	4
11862	3977359	53300-01-036-3861	110	25
11862	3977360	5310-01-166-2615	110	2
11862	3977383	2520-01-040-2160	106	2
11862	3977384	2520-01-038-7283	103	1
11862	3977386	53400-01-148-2730	108	9
11862	3977387	5330-01-020-9319	108	10
11862	3977397	25300-01-096-7731	124	23
11862	3977775	5325-01-164-7550	270	2
11862	3978765	3120-01-152-5948	83	3
			84	57
11862	3978901	5310-01-021-9027	124	1
11862	3979450	3120-01-194-6495	84	28
11862	3979453	5330-01-127-2883	84	52
11862	3979552	2530-01-154-1355	102	14
11862	3979626	5330-01-126-0733	85	5
11862	3979756	5325-01-159-2843	57	34
11862	3982098	5310-01-170-9100	29	1

PART NUMBER INDEX

CAGEC	PART NUMBER	STOCK NUMBER	FIG	ITEM
11862	3982098	5310-01-170-9100	57	29
			161	4
			282	3
11862	3984818	3120-01-156-3827	99	11
11862	3986821	4730-01-158-8717	233	17
11862	3987364	4720-01-096-7718	BULK	28
11862	3988524	2520-01-147-4205	100	6
11862	3988538	5305-01-019-1884	124	19
24617	3990160	5310-01-097-9414	142	3
			156	3
11862	3990892	2540-01-158-8611	233	15
11862	3991022	5306-01-157-9817	146	22
11862	3992925	3120-01-169-6440	87	10
11862	3993087	2910-01-150-3675	25	5
11862	3993729		157	2
11862	3995791	5365-01-159-1121	99	2
			112	10
11862	3995792	5365-01-159-1122	99	2
			112	10
11862	3995843	3020-01-151-3444	113	5
11862	3995881	3110-01-197-5495	84	35
11862	3996270	5940-01-145-7817	65	7
11862	3997718	5340-01-167-9694	80	3
11862	3999200	5340-01-149-7977	2	15
11862	3999572	8040-01-188-2953	41	1
55156	40178	5945-01-199-4431	38	2
11862	404062	3120-01-250-0583	157	6
11862	404234	5360-01-163-0886	127	61
76445	40424	5310-01-153-9302	125	14
86403	4049697	2520-01-035-6670	91	11
11862	406887	2590-01-323-5857	157	7
11862	407217	5305-01-162-7884	184	1
			191	14
86403	4086641	2520-01-098-5200	139	4
11862	409190		127	63
97271	40955	2520-01-155-5883	96	7
93061	411FS-6	4730-01-205-7917	13	14
11862	411700	5325-01-191-3293	191	7
14892	4150514	5340-01-152-7155	120	14
14892	4150515	2530-01-152-7115	115	5
11862	4158246	3120-01-338-6380	177	2
11862	4168122	5340-00-285-8868	167	18
11862	419454	5365-01-160-9530	127	69
20796	42-4877	3030-01-043-6749	295	12
20796	42-5023	3030-01-148-2792	132	1
20796	42-6919	3030-00-967-4898	35	4
20796	42-6921	3030-00-357-5506	35	11
20796	42-6923	3030-01-147-6410	34	4
93061	42F-5		252	16
24617	423532	5305-01-219-5399	284	28
			288	38

PART NUMBER INDEX

CAGEC	PART NUMBER	STOCK NUMBER	FIG	ITEM
20796	43-3226	3030-01-258-5125	33	10
11862	4303911	5360-01-159-8862	178	13
11862	431615	5305-01-163-5512	40	29
43334	43307-P-S1568A	3110-01-117-2436	84	8
35510	4340	5310-00-429-3135	37	2
11862	4410574	5340-01-157-6092	166	7
08806	4411	6240-00-836-2079	211	
			212	5
24617	443342	5310-00-044-3342	149	25
			150	1
			151	1
			152	3
08806	4435	6240-00-252-7138	290	2
24617	443767	5315-00-044-3767	83	19
24617	443945	5305-01-197-1475	147	4
17769	443998	4730-00-415-3172	262	12
24617	444034	4730-00-580-6738	24	1
24617	444620	4730-00-132-4625	22	9
30379	444789	4730-00-044-4789	31	1
			108	4
			109	1
24617	447143	5305-01-200-7735	215	6
			216	12
11862	447164	5305-00-855-0960	36	15
11862	4495180	2540-01-158-8555	182	8
11862	4497001	5310-01-194-9233	88	3
24617	451786	5340-01-162-5883	76	6
60038	453X	3110-00-142-4387	111	15
24617	453610	3110-00-185-6305	24	13
24617	455956	5305-00-145-0828	248	37
24617	456004	5310-00-596-6897	232	16
			261	17
24617	456697	5306-01-159-5710	122	1
24617	456748	5305-01-242-1148	262	18
			284	7
			288	16
11862	458025	2510-01-155-5854	176	3
11862	458026	2510-01-155-8785	176	6
79260	45823	2990-01-155-5151	26	8
11862	458300	5306-01-323-5544	93	6
11862	458418	2520-01-173-1362	89	7
			91	7
11862	4587931	5330-01-157-5684	166	14
11862	458859	5330-01-086-3505	100	4
11862	458860	5330-01-155-4399	97	3
11862	458877	2520-01-174-1418	97	11
11862	458878	2520-01-154-3677	97	18
11862	458985	5305-01-162-5707	184	10
11862	459021	5340-01-150-4110	1	16
11862	459454	6150-01-158-3861	41	9
11862	459461	5360-01-149-6309	50	36

PART NUMBER INDEX

CAGEC	PART NUMBER	STOCK NUMBER	FIG	ITEM
99688	46006	3040-01-192-4694	293	6
99688	46023	2540-01-191-4331	245	41
11862	460308	5306-01-148-3664	1	17
11862	460340	5305-01-155-6110	148	30
11862	460354	2510-01-148-5049	148	22
			149	21
11862	460373	2510-01-159-7760	142	15
11862	460374	2510-01-159-7761	142	15
11862	460397	5340-01-158-6681	140	33
			142	20
11862	460835	3010-01-194-0886	84	4
99688	46090	4520-01-192-6005	245	23
11862	461610	2510-01-156-4871	169	24
11862	462233	5970-01-158-9337	193	11
			BULK	30
11862	462787	2520-01-150-8354	113	8
11862	462794	5306-01-158-5374	101	7
11862	462798	4730-01-263-5361	101	13
23862	462809	2520-01-105-5679	96	8
11862	462811	5330-01-096-9649	96	5
11862	462853	2530-01-151-7995	101	19
23862	462857	5330-01-113-9602	96	11
11862	462863	5310-01-149-6285	98	21
35510	4629JA	2920-01-131-4932	35	2
11862	464039	2530-01-249-5401	97	1
17769	4641	6620-01-221-1942	261	12
11862	464718	2510-01-326-1464	151	21
11862	465246	4710-01-192-1433	BULK	42
11862	465482	5330-01-194-5804	84	6
11862	465483	5330-01-118-9504	84	7
11862	465536	5355-01-157-1866	196	19
			205	9
11862	466157	5330-01-157-7464	184	8
			190	12
11862	466578	5310-01-165-1327	87	9
11862	466717	2590-01-173-6991	165	24
11862	467117	5310-00-209-2811	130	1
11862	467247	2510-01-191-8447	193	6
11862	467248	2510-01-200-1021	193	6
11862	467282	2530-01-154-1292	123	1
11862	467283	5340-01-209-7066	127	76
11862	467299	5340-01-158-8549	188	7
			189	1
11862	467509	5340-01-163-5902	22	31
11862	467524	5340-01-163-9400	22	22
11862	467525	2910-01-155-5136	19	29
11862	467911	2510-01-163-2681	175	7
11862	467912	2510-01-163-7229	175	7
11862	467913	2530-01-158-8584	173	4
11862	467983	2540-01-158-4604	199	9
			200	10

PART NUMBER INDEX

CAGEC	PART NUMBER	STOCK NUMBER	FIG	ITEM
11862	468234	2540-01-192-1823	25	4
11862	468481	3120-01-162-8654	150	5
			151	6
			151	17
			152	6
11862	468484	4710-01-154-1230	20	29
11862	468661	2530-01-155-8460	116	23
11862	468673	2530-01-153-9449	116	30
11862	468674	2530-01-153-9450	116	30
11862	468675	5305-01-156-5006	116	27
60038	469	3110-00-100-0251	111	16
11862	469302	5355-01-163-4940	49	16
11862	469339	5340-01-165-2526	58	7
11862	469694	5330-01-085-0918	124	20
11862	470205	5355-01-235-6616	127	71
11862	470949	2540-01-157-8009	179	25
11862	470974	5315-01-159-8660	203	20
11862	470984	2540-01-158-7551	203	13
11862	470993	5330-01-159-8504	182	10
11862	471009	2510-01-096-6769	190	11
11862	471010	2540-01-158-7583	203	21
11862	471018	5340-01-158-6892	203	19
11862	471020	2540-01-158-6955	203	5
11862	471021	2540-01-158-8811	203	7
11862	471079	2540-01-158-6894	203	30
11862	471083	5306-01-157-3279	195	5
			197	5
			198	4
			204	4
11862	471089	2990-01-160-5873	175	12
11862	471657	5365-01-156-9635	151	8
11862	471658	5365-01-326-1113	151	8
11862	471663	5306-01-323-8967	151	7
11862	471665	5306-01-150-9499	151	7
11862	471667	5306-01-153-8281	150	6
11862	471727	2520-01-192-3718	84	25
11862	471777		108	1
11862	471873	2520-01-156-0553	110	11
11862	471896	5365-01-194-5274	84	16
11862	472450	5340-01-162-4852	162	1
11862	472536	2520-01-201-4096	124	2
11862	473187	5355-01-197-1501	203	10
11862	473197	2540-01-158-6893	203	15
11862	473894	2540-01-158-4601	181	8
11862	473917	2540-01-158-8557	180	9
11862	473995	2510-01-178-8867	180	11
11862	474022	2510-01-259-5587	193	4
11862	474071	3110-01-197-5496	84	59
11862	474102	2590-01-164-7825	127	62
11862	474309	5365-01-161-4055	124	21
11862	474579	5325-01-165-3475	291	13

PART NUMBER INDEX

CAGEC	PART NUMBER	STOCK NUMBER	FIG	ITEM
78385	474669		248	32
11862	474935		104	8
			105	4
11862	474955	2910-01-152-5516	19	34
11862	474957		21	14
11862	474978	5310-01-326-1052	151	24
11862	475922	6220-01-242-7557	53	9
11862	475968	5330-01-164-8579	184	15
78385	476220		248	33
78385	476229	5310-00-285-5112	248	35
78385	476339	5360-00-327-5879	248	31
11862	476781	5365-00-734-2495	84	41
11862	476916	2910-01-155-7881	20	2
11862	476927	5330-01-157-7604	20	3
11862	477249	5330-01-154-7159	8	24
11862	477361	5930-01-149-9306	49	25
11862	477402	4730-00-826-4268	21	9
			22	1
11862	480534	2510-01-155-5848	18	9
			173	11
			191	12
11862	480567	5310-01-194-9217	145	41
11862	4813235	5360-01-181-2482	19	5
			20	8
11862	482420		BULK	13
11862	482995		244	2
78385	484044		248	34
11862	487425		244	2
11862	4876466	5340-01-159-3880	196	5
			201	3
			202	2
			203	29
11862	488369	5365-01-155-1932	150	9
			151	20
78385	488755	5320-00-801-1548	246	18
78385	488773		246	36
78385	488935	2040-00-754-4176	249	10
11862	4914356	2540-01-157-3525	234	3
11862	4918446	2540-01-158-1570	234	24
11862	4918562	2540-01-158-1569	233	11
11862	4939314	3020-01-199-7847	234	23
11862	4993563	2530-01-157-7933	128	11
81343	5 010111B	4730-00-011-8538	251	4
			252	5
			252	15
			255	4
			262	19
			284	6
			288	10
97271	5-297X	2520-01-004-2057	97	10
9W635	500-4	4720-01-212-4782	252	27

PART NUMBER INDEX

CAGEC	PART NUMBER	STOCK NUMBER	FIG	ITEM
97271	500381-3	5310-01-185-3400	101	22
			101	28
97271	500598-21	5315-01-185-3415	98	3
			98	24
11862	500890	5905-01-159-0771	241	2
99688	50090	5325-00-279-1235	247	2
25022	51-0979	5340-01-028-9063	218	6
			219	6
			220	7
			221	3
			224	5
			225	
25022	51-0985	2540-01-165-6793	224	11
25022	51-0988	2540-01-158-0602	215	28
25022	51-0989	2540-01-028-0574	216	17
25022	51-0991	5340-01-164-1078	220	13
			221	11
25022	51-0998	5340-01-156-5061	215	15
			216	2
25022	51-0999	2540-01-164-0046	215	8
			216	13
25022	51-1304	5340-01-159-1796	17	
			227	11
25022	51-1305	2510-01-166-2015	214	14
25022	51-1603	9340-01-162-5948	215	13
			216	6
25022	51-1604	9340-01-162-5947	228	2
25022	51-1721	2510-01-163-7879	215	16
			216	7
25022	51-1724	5330-01-163-5706	228	3
25022	51-1901	5930-00-636-1584	214	6
25022	51-1902	5930-00-548-5640	14	5
25022	51-1926	5930-01-028-1949	227	10
25022	51-1942	6220-01-039-9809	213	19
25022	51-1959	5930-01-163-8924	219	13
			227	5
25022	51-1991	6130-01-035-6412	213	2
25022	51-2204	2510-01-169-3785	215	7
			216	11
25022	51-2294	7230-01-167-2075	215	4
			216	16
25022	51-2297	9330-01-098-6554	220	3
			221	5
			224	2
27783	5137	1660-00-413-0756	299	6
			299	19
11862	517900	5310-01-147-8747	100	7
			112	2
25022	52-0914	2540-01-163-7874	215	20
			216	24
35510	52066	5310-01-023-8927	37	18

PART NUMBER INDEX

CAGEC	PART NUMBER	STOCK NUMBER	FIG	ITEM
11862	5234530	2815-01-147-9358	7	30
11862	52351724	5310-01-160-5708	235	10
			242	1
19207	5287638	5310-00-528-7638	24	19
19207	5294507	5310-00-350-2655	50	7
35510	5413	5310-00-516-2701	37	1
48018	543319	2520-01-153-9436	100	5
11862	5454797	5310-01-154-4341	115	9
			116	8
11862	5461145	5360-00-123-0137	115	15
			116	17
11862	5461156	2530-01-173-1248	116	20
11862	5461984	5360-00-113-9490	115	17
			116	19
11862	5462496	5365-01-154-8577	116	28
11862	5468226	5306-01-085-1953	121	1
11862	5468767	2530-00-363-4889	121	3
			121	9
11862	5469497	5340-01-157-2101	116	12
11862	5469581	5340-01-163-1401	120	7
			121	6
			122	2
11862	5470497	5330-01-185-4676	116	11
94988	552-12V	5945-00-983-4374	49	29
30282	553	4910-01-179-2517	301	12
11862	556742	5930-01-162-3669	49	9
11862	556743	5360-01-159-5952	49	8
11862	560613	5340-01-148-7529	16	12
11862	560614	5325-01-160-4618	16	8
11862	560625	4730-01-194-0126	14	6
11862	5613939	2920-01-151-3627	24	3
99688	56470	5330-00-352-0327	299	3
99688	56480	5330-00-185-0075	299	5
11862	5667628	5310-00-166-8567	131	40
11862	5671921	3110-00-451-4601	127	53
11862	5672489		131	19
11862	5682898		131	44
11862	5683373		131	47
11862	5686527	5330-01-090-5428	131	29
11862	5686539	5340-00-960-9354	131	8
11862	5686550		131	23
11862	5686611	5330-00-848-0972	131	24
11862	5686814		131	2
11862	5686815	5340-00-960-9340	131	3
11862	5687182	2530-00-960-9363	KITS	
11862	5687209		131	34
11862	5687958	5330-00-960-9355	131	5
11862	5687973	5310-01-038-8500	131	15
11862	5688014		133	17
11862	5688015	5330-01-157-1883	133	26
11862	5688035	5330-01-157-1884	133	8

PART NUMBER INDEX

CAGEC	PART NUMBER	STOCK NUMBER	FIG	ITEM
11862	5688037	5360-01-149-6308	133	7
11862	5688044	5330-01-044-0703	KITS	
48018	5688049	5330-00-848-4439	133	23
11862	5688109		133	25
11862	5689357		133	4
11862	5689358		133	16
11862	5690517		131	45
11862	5692682	5340-01-157-7559	133	9
52788	5694191	5365-01-204-6702	127	22
11862	5695513	2530-00-937-1275	131	11
11862	5695774	2530-01-173-9671	131	10
11862	5695777		131	12
11862	5696151	5305-00-173-0165	131	9
11862	5697702	5310-01-144-2779	131	41
11862	5697804	3110-00-403-1488	131	48
99688	57012	5330-01-061-3000	299	8
19207	5704051	2540-00-201-3474	KITS	
19207	5704052	2540-00-200-4249	KITS	
19207	5704064	2540-00-198-2478	KITS	
11862	5713268	3040-01-157-8021	178	14
11862	5713274	5310-01-194-9234	178	4
11862	5713276	5365-01-198-5516	178	5
11862	5717887	5315-01-194-0819	178	7
99688	57277	5330-01-166-1712	299	11
			299	17
34623	5740201	3120-01-152-2039	3	5
34623	5740572	5305-01-188-0489	15	45
99688	57972	5340-01-193-9654	293	2
99688	58021	5330-00-777-1454	299	18
99688	58577	4720-01-192-3520	299	22
99688	58578	4720-01-192-3521	299	23
99688	58579	4820-01-161-5059	299	21
99688	58580	4710-01-162-7148	299	12
99688	585810	4130-01-171-5997	299	10
99688	58582	4710-01-162-7149	299	9
99688	58583	4820-01-161-6435	299	4
99688	58585	4130-01-160-7705	299	24
99688	58586	5330-01-194-4753	294	20
11862	5858694	4730-01-164-3692	296	11
11862	5858696	5330-01-164-5602	296	12
99688	58587	4130-01-160-7716	293	42
99688	58589	4140-01-157-3501	300	3
99688	58590	4140-01-154-9615	300	3
99688	58591	5330-01-201-9681	300	12
99688	58592	5330-01-193-0226	300	4
99688	58593	5330-01-204-4312	300	17
99688	58594	5640-01-194-7193	300	7
99688	58595	4140-01-186-9753	298	9
99688	58596	5330-01-213-9811	298	14
99688	58597	5330-01-201-9682	298	13
99688	58598	5330-01-205-5056	298	12

PART NUMBER INDEX

CAGEC	PART NUMBER	STOCK NUMBER	FIG	ITEM
99688	58599	2540-01-163-1176	293	12
99688	58600	5360-01-161-9169	293	13
99688	58601	7690-01-188-2862	293	8
99688	58602	5330-01-206-7353	293	11
99688	58603	5355-01-166-1720	293	34
99688	58605	7690-01-189-6932	293	36
99688	58606	5365-01-195-5934	294	4
99688	58607	5330-01-193-0227	294	9
99688	58608	5330-01-194-4751	294	5
99688	58609	5640-01-195-4633	294	17
99688	58610	5640-01-195-4634	294	8
99688	58611	5640-01-195-4635	294	16
99688	58612	5640-01-195-4636	294	14
99688	58613	5640-01-196-7002	294	13
99688	58614	5640-01-196-7003	294	12
99688	58615	5640-01-195-9786	294	11
99688	58616	5330-01-195-1564	294	15
99688	58617	5330-01-194-4752	294	10
99688	58618	5330-01-197-0897	294	6
99688	58620	4730-01-192-4434	245	22
99688	58623	452001-192-6073	245	15
99688	58624	5330-01-192-8904	245	33
99688	58625	4720-01-192-3510	245	13
99688	58627	5360-01-203-6365	245	6
99688	58628		245	17
99688	58629		245	26
99688	58632	5330-01-192-8905	245	27
99688	58634	4140-01-188-6977	245	42
99688	58635	5330-01-194-4754	245	24
99688	58636	5330-01-192-9335	245	2
99688	58637	5330-01-192-8906	245	7
99688	58643	4010-01-200-7581	247	12
99688	58645	9330-01-191-4883	299	13
99688	58646	9330-00-629-8239	299	14
99688	58647	4710-01-192-3732	299	15
99688	58648	4710-01-192-3731	299	16
99688	58649	2990-01-192-4597	245	21
99688	58651	2990-01-192-4573	245	25
99688	58659	5340-01-199-4993	245	3
			247	9
99688	58660	5325-00-276-4993	247	1
99688	58661	5325-01-197-1547	247	11
99688	58662	5970-01-180-3731	293	28
99688	58668	5355-01-167-4114	247	10
			293	24
99688	58701	9320-01-085-2889	298	10
			300	10
21450	587227	5310-01-264-5903	271	12
			273	14
			274	9
35510	58754	5305-00-450-0463	37	40

PART NUMBER INDEX

CAGEC	PART NUMBER	STOCK NUMBER	FIG	ITEM
11862	587575	5360-01-149-1959	26	14
			27	24
11862	5887997	5330-01-041-4749	296	36
27462	5914435		296	37
27462	5914705		296	24
11862	5914706	4130-01-164-3268	296	46
27462	5914709	4130-01-160-8519	296	31
11862	5914719	4130-00-277-3486	296	3
11862	5914807		296	15
11862	5914809		296	13
35510	59225	5310-00-451-0631	37	33
11862	5929144	5330-01-041-4749	296	29
11862	5965748	5330-01-162-3744	51	2
11862	5965771	6220-01-146-4469	51	4
11862	5965772	2510-01-096-6758	51	4
11862	5965775	6220-01-146-4455	51	1
11862	5965776	6220-01-157-9046	51	1
11862	5966249	5305-01-150-5785	50	32
11862	5968095	6220-01-164-2271	50	35
35510	59982	5310-00-429-3156	37	66
79470	60X3	4730-00-701-7737	248	28
11862	603827	4710-01-161-0138	BULK	40
97271	620058	5330-01-179-2249	101	9
97271	620062-B	5330-01-158-6725	96	3
97271	620132	2530-01-164-0038	101	6
97271	620180	2530-01-098-4145	101	14
99688	62028	5310-01-162-2515	293	35
99688	62210	5310-00-637-4000	293	5
99688	62218	5310-01-159-8264	298	3
			300	9
99688	62219	5310-01-197-4435	247	13
			293	14
99688	62220	5310-00-103-2953	247	7
			293	9
99688	62222	5310-01-163-8256	293	15
99688	62229	5310-01-161-7308	245	4
			247	5
			293	4
			293	10
84760	62538	3040-01-247-0893	15	123
58499	6258	6140-01-155-6531	57	15
11862	6258213	2540-01-196-1622	49	13
11862	6258340	2520-01-159-8798	110	12
11862	6258364	2590-01-159-8857	243	1
11862	6258545	2510-01-326-0762	KITS	
11862	6258561	5330-01-157-7458	166	8
11862	6258562	5330-01-157-7459	166	4
11862	6259033	5330-01-157-6664	120	6
			120	11
11862	6259071	5340-01-335-9360	119	19
11862	6259074	5330-01-155-7700	129	3

PART NUMBER INDEX

CAGEC	PART NUMBER	STOCK NUMBER	FIG	ITEM
11862	6259074	5330-01-155-7700	130	3
11862	6259083	5330-01-126-1039	84	36
97271	6259149	5310-01-148-2687	83	63
11862	6259153	3040-01-199-1485	84	47
11862	6259192	5365-01-195-6203	84	64
11862	6259193	5365-01-195-6204	84	64
11862	6259194	5365-01-205-5966	84	64
11862	6259195	5365-01-197-1524	84	64
11862	6259423	2940-01-121-6850	79	10
11862	6260421	2540-01-096-9664	165	11
11862	6260631	2590-01-155-7882	18	10
11862	6260830	2530-01-148-2914	124	2
23862	6260885	2530-01-142-0261	115	12
58499	6262	6140-01-155-6530	57	40
11862	6262029	2540-01-159-8759	176	1
11862	6262054	5340-01-165-8986	162	2
11862	6262211	5305-01-148-2646	1	22
11862	6262212	5310-01-148-4991	1	8
19207	6262328	5340-00-700-1423	125	18
11862	6262755	2910-01-147-4218	19	15
11862	6263870		19	14
11862	6263871		19	17
11862	6263877	2590-01-155-7711	18	7
			BULK	8
11862	6263945	2510-01-155-0336	152	10
11862	6264100	2930-01-147-4222	28	5
11862	6264131	5340-01-163-4337	165	13
11862	6264951	5365-01-154-8514	123	9
11862	6270213	2510-01-159-5888	142	12
11862	6270214	2510-01-160-4969	142	11
11862	6270410	5340-01-149-9732	68	3
11862	6270704	5325-01-198-8040	157	3
11862	6270977	3020-01-151-3446	99	7
11862	6271322	5365-01-156-8948	151	15
11862	6271341	3040-01-165-4562	141	4
11862	6271344	2510-01-096-7680	140	47
			142	36
11862	6271386	5340-01-154-2107	69	11
11862	6271387	2520-01-201-4097	69	1
11862	6271391	5325-01-168-1886	69	3
11862	6271394	2520-01-165-4935	69	22
11862	6271989	5315-01-160-0575	166	22
			166	28
			179	14
11862	6272627		166	15
11862	6273213	5340-01-192-4366	84	29
23862	6273214	5330-01-141-7167	84	61
11862	6273325	5330-01-157-1952	239	1
11862	6273934	2530-01-123-3553	97	6
11862	6273948	5330-01-086-3506	125	16
11862	6273951	5330-01-150-4022	96	13

PART NUMBER INDEX

CAGEC	PART NUMBER	STOCK NUMBER	FIG	ITEM
11862	6273992	3020-01-181-3795	84	60
11862	6274031	5310-01-157-7560	146	4
11862	6274036	2540-01-192-3572	146	7
11862	6274179	2510-01-169-3766	142	8
11862	6274550	5340-01-164-8761	165	8
11862	6274836	2540-01-158-4600	179	12
11862	6274847	5340-01-166-5652	179	8
11862	6274848	2510-01-163-7231	179	8
11862	6274849	5365-01-162-8876	179	9
11862	6274850	4010-01-158-6331	179	5
11862	6274853	5330-01-159-4777	181	6
11862	6274854	5340-01-165-4351	181	2
11862	6274890	5340-01-158-4583	180	3
11862	6274970	2510-01-162-3679	164	8
11862	6287160	6140-01-223-9144	57	14
11862	6292996	6145-01-218-3761	BULK	10
11862	6292997	6145-01-218-3759	BULK	43
11862	6293702	6145-01-218-3760	BULK	44
11862	6293923		BULK	11
11862	6298886	6250-01-189-4981	61	3
11862	6410785	2930-01-123-4941	28	3
70040	6433429	6680-01-160-3870	42	7
11862	6437746	5325-01-123-6798	79	9
11862	644536	5310-01-175-2539	110	1
78977	6471	5305-00-217-9183	290	5
11862	6471831	2910-01-159-0867	13	8
70040	6474942A	6625-01-162-8124	43	7
11862	6497475	6210-01-158-3857	42	11
11862	6497476	6220-01-164-5227	42	9
11862	6497483	5330-01-163-1992	42	10
63208	650-24	4730-00-058-7558	256	10
			258	1
			259	1
			287	10
78977	650U-0016	5340-01-164-0748	290	3
99688	65129	2590-01-163-3576	293	3
99688	65130	5930-01-173-0826	299	20
99688	65131	5930-01-173-0825	299	7
99688	65132	2590-01-164-7897	293	1
99688	65133	2590-01-191-9269	293	27
99688	65134	5910-01-189-3011	298	7
			300	16
99688	65136	5930-01-159-2610	293	16
99688	65138	5930-01-190-1212	293	29
99688	65139	5930-01-165-1657	293	30
99688	65140	5930-01-163-6256	293	33
99688	65145	2590-01-191-9270	293	26
99688	65146	2590-01-192-6031	293	31
99688	65148	2590-01-191-4357	245	36
99688	65149	2590-01-191-9276	245	39
99688	65151	2590-01-191-9275	247	17

PART NUMBER INDEX

CAGEC	PART NUMBER	STOCK NUMBER	FIG	ITEM
99688	65152	2590-01-191-9271	247	24
99688	65153	2590-01-191-9274	247	21
99688	65154	2590-01-191-9273	247	25
99688	65155	2590-01-191-9272	247	23
99688	65158	5905-01-154-2354	300	5
99688	65171	5930-01-190-1231	247	22
11862	6550750	3020-01-096-6892	296	5
11862	6550835	5950-01-165-6803	296	9
76462	6551150		296	20
11862	6555258		296	25
11862	6555259		296	19
11862	6555279	4810-01-164-3267	296	30
27462	6555283	4130-01-160-8541	296	45
11862	6555290		296	44
11862	6555293		296	33
11862	6555298	4730-01-165-0173	296	28
27462	6555299	4730-01-169-7714	296	38
11862	6555320	5315-01-164-4570	296	40
11862	6555328	5365-00-152-5273	296	2
11862	6555329	5365-01-164-6567	296	4
11862	6555330	5365-01-164-6568	296	8
11862	6555338	5365-01-163-6139	296	6
11862	6555378		296	35
11862	6555440		296	17
27462	6555755	4130-01-160-8520	296	34
11862	6556000		296	18
11862	6556050		296	23
11862	6556055		296	23
11862	6556060		296	23
11862	6556065		296	23
11862	6556070		296	23
11862	6556075		296	23
11862	6556076	5330-01-163-2000	296	41
11862	6556077	5330-01-164-5604	296	43
11862	6556080		296	23
11862	6556085		296	23
11862	6556090		296	23
11862	6556095		296	23
11862	6556100		296	23
23671	6556105		296	23
11862	6556110		296	23
11862	6556115		296	23
11862	6556120		296	23
11862	6556175		296	26
11862	6556180		296	26
11862	6556185		296	26
11862	6556190		296	26
11862	6556195		296	26
11862	6556200		296	26
11862	6556205		296	26
11862	6556210		296	26

PART NUMBER INDEX

CAGEC	PART NUMBER	STOCK NUMBER	FIG	ITEM
11862	6556215		296	26
11862	6556220		296	26
11862	6556605=XA		296	21
11862	6556767	5365-01-163-6138	296	42
11862	6556923	5310-01-165-3346	296	1
11862	6557000		296	26
11862	6557057	4310-01-166-6123	296	39
78977	6566-0004	6220-01-197-3938	290	6
27462	6590589	4130-01-160-8522	KITS	
11862	6590591	4310-01-166-5958	KITS	
11862	6590592=XA		296	16
27462	6590594	4130-01-160-8521	296	32
11862	6590686		296	22
11862	6591956		KITS	
78977	6598	5340-00-411-4508	290	4
27462	6599114	4130-01-160-8506	KITS	
97271	660182-5	2520-01-158-8543	96	17
97271	660182-6	2520-01-105-5676	96	9
78977	6701-0025	5340-01-197-1585	290	9
63632	6704001	2910-01-150-0950	12	13
78977	6710-BU-2	4710-01-194-6590	290	7
73342	6771005	5330-01-080-3253	79	7
79470	6820	4820-00-174-0315	9	6
97271	700013L		96	1
19207	7001423	5340-00-700-1423	114	11
97271	701056 1X	2520-01-332-3841	111	14
97271	701064 X	5365-01-330-6970	111	12
78385	701328		248	38
78385	702129	5365-01-037-6813	249	9
78385	702903	5330-00-089-0978	246	26
78385	702942		248	18
78385	703546		246	30
78385	703547		246	29
11862	7040173	3040-01-162-0255	166	13
11862	7040174	3040-01-157-7997	166	13
78385	704181		250	7
78385	704189		250	10
78385	704191		250	11
78385	704192		250	1
78385	704225	2540-00-216-5722	246	22
78385	704283		250	13
78385	704285	5340-01-164-1063	250	9
78385	704363	4710-01-163-2805	246	1
78385	704371-1		250	8
78385	704390		248	5
78385	704391-1		248	10
78385	704395		248	2
78385	704396		248	1
78385	704397	5305-01-057-8031	248	21
78385	704401	5930-01-077-7793	248	16
78385	704402		248	4

PART NUMBER INDEX

CAGEC	PART NUMBER	STOCK NUMBER	FIG	ITEM
78385	704406	5330-01-008-6527	248	6
78385	704408		249	16
78385	704432		248	3
78385	704434	5360-01-057-9249	248	20
78385	704447		248	13
78385	704501		246	17
78385	704579	4820-01-057-5925	248	41
78385	704580		248	30
78385	704678		250	2
78385	704789		248	11
78385	704898		248	40
78385	705117	5305-01-136-8734	250	5
78385	705311		248	9
78385	705344	5330-01-136-8334	248	22
78385	705345	5307-01-136-8740	248	26
78385	705347		248	25
97271	706045X	2520-01-094-1817	98	12
97271	706046X	3110-01-164-5457	98	20
97271	706047X	2520-01-094-1818	98	22
97271	706395X	2530-01-098-5242	101	8
97271	70696-4X	2520-01-163-9713	98	6
97271	706968X	5340-01-159-4478	96	16
35510	71237	6130-01-142-8507	37	12
35510	71238	6130-01-142-8508	37	15
97271	71701-X		96	10
11862	718368	2520-01-163-7284	87	19
78385	719675	5305-00-403-5130	246	10
97271	72166 X	3020-01-332-5456	111	13
97271	72166 5X	2520-01-334-1731	111	11
35510	73009	5310-00-429-2686	37	60
97271	73493-4X	2520-01-159-8730	98	2
19207	7353960	2510-00-999-9856	271	4
			273	26
78385	735411		248	39
35510	73543	5305-00-451-1623	37	17
78385	735447-5		248	12
35510	73545	5970-00-160-4066	37	14
35510	73546	5970-00-160-4001	37	10
35510	73547	3120-01-082-6996	37	21
19207	7357009	4210-00-383-7127	238	3
19207	7358030	4030-00-088-1881	143	6
78385	736861	5310-01-164-1036	249	14
19207	7370134	5315-00-737-0134	289	21
24617	7451155	3110-00-406-9608	110	24
			112	7
11862	7451809	3110-00-756-4535	107	4
11862	7451870	3110-01-020-9786	110	21
11862	7451888	3110-00-858-0988	110	22
11862	7455617	3110-01-087-2653	125	19
35510	74732	5310-01-136-5547	37	56
19207	7524078	2540-00-752-4078	288	26

PART NUMBER INDEX

CAGEC	PART NUMBER	STOCK NUMBER	FIG	ITEM
35510	75348	5305-01-143-6416	37	59
11862	7591126	5355-01-280-2975	170	10
35510	76985	5310-00-432-3959	37	34
19207	7717066	5360-00-771-7066	270	9
19207	7731428	5935-00-773-1428	61	9
11862	7740374	5306-01-160-1952	179	4
35510	77916	3120-01-082-9419	37	45
11862	7800580	5310-01-160-5727	127	13
11862	7804410	5360-01-160-8935	127	32
11862	7804414	2920-01-159-7749	127	11
11862	7804439	5340-01-160-0294	127	42
11862	7804440	2530-01-159-3455	127	43
11862	7805700	3110-01-197-6101	127	44
11862	7805822	2530-01-159-3454	127	45
11862	7805950	3120-01-160-1894	127	34
11862	7806140	2520-01-024-0279	89	5
			90	6
			91	5
			92	7
			93	3
			93	3
11862	7806185	5306-01-159-3450	127	33
11862	7806433	5305-01-148-7465	49	4
11862	7806867	5305-01-159-8262	127	14
11862	7807271	5306-01-161-2146	131	35
11862	7807845		131	21
11862	7808195	5330-01-043-5495	133	2
11862	7809057	2520-01-153-8390	89	4
			90	7
			91	4
			92	6
11862	7809128	5305-01-215-2501	127	16
11862	7809232	4820-01-039-3769	133	6
11862	7809408	5330-01-161-2516	127	54
11862	7809409	5360-01-159-2931	127	52
11862	7809925	2530-01-206-4860	127	78
11862	7810567	5360-01-161-9171	127	46
11862	7812136	2540-01-158-6910	127	20
11862	7812526	2590-01-159-3449	127	5
11862	7814387	5330-01-157-7605	127	24
11862	7814503	5330-01-197-0881	131	6
			133	19
11862	7815848		89	11
			90	8
			92	11
11862	7815849	2520-01-153-8430	89	12
			90	12
			91	12
			92	12
11862	7816509		131	39
11862	7816516	5306-01-040-2041	131	14

PART NUMBER INDEX

CAGEC	PART NUMBER	STOCK NUMBER	FIG	ITEM
11862	7817355	2530-01-033-1830	131	7
11862	7817454	5310-01-160-5728	127	51
11862	7817484	2530-01-155-6709	131	13
11862	7817485	2530-01-034-1715	KITS	
11862	7817486	2530-01-097-7659	131	16
11862	7817487	2530-01-033-1855	KITS	
11862	7817529		131	4
52788	7817725		131	25
11862	7818809		131	17
11862	7819517	3120-01-203-0332	127	4
11862	7819738	3830-01-159-5871	127	23
11862	7819849		131	27
11862	7819898	5360-01-160-9839	127	19
11862	7819986		131	22
11862	7825561		133	13
11862	7826012	5310-01-157-3260	131	37
11862	7826470		131	46
		5330-01-096-9650	KITS	
11862	7826542	2530-01-152-5517	131	36
11862	7826850	3110-01-040-6541	131	38
11862	7827077	2530-01-159-3452	127	28
11862	7827111	3120-01-166-5638	127	26
11862	7827942	5330-01-155-4388	89	8
			91	8
11862	7828012	3120-01-159-1275	131	30
11862	7828271	5340-01-213-5318	127	59
11862	7828506	4720-01-148-5981	BULK	14
24617	7828565	5330-00-843-9235	KITS	
11862	7829923	5330-01-156-5141	134	8
11862	7830209	2530-01-191-4264	127	38
11862	7830236		133	3
11862	7830239	5305-01-162-9713	133	22
11862	7830314	5305-01-171-8252	127	37
11862	7830335	5975-01-156-4544	127	39
11862	7830375	2590-01-159-1805	127	10
11862	7830377	5305-01-159-8263	127	12
11862	7830384	5315-01-197-1548	127	36
11862	7830913	4320-01-155-5153	133	21
11862	7830927	2520-01-239-3800	91	1
11862	7831388	5307-01-172-1526	133	20
11862	7831538	3040-01-160-1598	127	41
11862	7831570	3040-01-178-9789	127	49
11862	7831571	3010-01-159-3456	127	55
11862	7832112	3120-01-160-1895	127	47
11862	7832311	5365-01-200-1290	127	7
11862	7832422		131	20
11862	7832426		131	31
11862	7832427		131	32
11862	7832428		131	33
11862	7832429		131	26
11862	7832729	3110-01-166-5667	131	28

PART NUMBER INDEX

CAGEC	PART NUMBER	STOCK NUMBER	FIG	ITEM
11862	7832808	3120-01-202-2602	127	6
11862	7832907	5306-01-161-2593	127	58
11862	7832984	2590-01-191-4263	127	3
11862	7834183	6680-01-152-2845	133	18
11862	7834284	2530-01-191-4262	131	43
11862	7834333		131	42
11862	7836172	2530-01-159-3451	127	27
11862	7836369		133	5
11862	7837032	2530-01-152-2383	127	2
11862	7837185	2530-01-159-3448	127	21
11862	7837284		133	12
11862	7837321		133	1
11862	7837322		133	10
11862	7837378	5360-01-160-9838	127	29
11862	7838665	2520-01-155-6936	93	4
52788	7838936	2530-01-147-8556	132	6
11862	7838941	4720-01-148-2762	134	5
11862	7838942	4720-01-148-2761	134	10
11862	7839341		133	11
11862	7839499		133	24
11862	7839667		133	15
11862	7839669		133	14
99688	78401	4120-01-161-3706	292	2
11862	7840235	5340-01-202-2517	90	4
			92	4
11862	7840274	5930-01-201-1843	127	35
99688	784030	2540-01-191-4327	245	19
99688	78404	2540-01-163-8624	212	14
99688	78405		245	20
99688	78406	4720-01-191-8463	245	43
11862	7842684	2530-01-159-3453	127	25
11862	7842688	4710-01-160-1599	127	48
11862	7843294	2920-01-159-2842	127	8
11862	7843368	2520-01-159-7118	127	31
11862	7843369	2530-01-159-3447	127	1
11862	7843400	2540-01-191-4289	127	17
11862	7843467	2520-01-159-5496	127	40
11862	7844074	2520-01-148-2919	93	1
11862	7845102	2520-01-189-0596	89	1
			90	1
11862	7845119	2520-01-192-1793	90	10
			92	9
11862	7845127	2520-01-192-4314	90	5
			92	5
11862	7845130	2520-01-159-7119	127	30
11862	7846418	5307-01-203-9081	49	6
11862	7846626		131	18
11862	7846740	5340-01-231-0925	93	5
11862	7846959	2530-01-150-9757	131	1
11862	7846970	5310-01-159-6587	127	60
11862	7849302	2990-01-149-4966	7	6

PART NUMBER INDEX

CAGEC	PART NUMBER	STOCK NUMBER	FIG	ITEM
35510	78895	3110-01-136-5534	37	32
35510	78983	2920-01-141-1488	37	39
35510	78997	5306-01-078-2719	37	22
35510	79035	2920-01-082-6458	37	36
35510	79070	3120-01-082-9418	37	13
35510	79071	3120-01-082-9417	37	11
35510	79313	5305-01-105-6888	37	16
35510	79403	5330-01-106-6101	37	44
35510	79404	5330-01-106-4329	37	42
35510	79405	5330-01-111-6531	37	31
35510	79406	3110-01-136-2093	37	43
35510	79469	5310-01-143-1715	37	25
35510	79486	3110-01-109-3058	37	30
35510	79517	2920-01-267-2088	37	41
19207	7951738	2540-01-159-6198	275	6
35510	79627	5310-01-136-5532	37	38
35510	7983	5970-01-144-1291	37	4
11862	800091	5977-01-163-2931	40	27
16764	800549	5305-00-471-0373	36	36
16764	801810	5306-01-100-5149	36	26
16764	801815	5935-01-103-8700	36	9
73992	8100	4730-00-768-8880	261	13
73992	8200	4730-00-317-4231	261	8
24617	821453	5310-00-291-4619	40	18
11862	830478	5905-01-139-3620	36	12
19207	8338687	5365-00-571-6816	83	25
19207	8376101	5930-00-475-5537	263	1
11862	838839	5315-00-828-5487	8	29
99688	85922	4140-01-160-7664	300	1
99688	85923	4140-01-160-8503	298	1
99688	85924	4130-01-160-7696	299	1
99688	85925	2540-01-159-1799	293	7
99688	85926	5930-01-164-8264	293	25
99688	85927	4130-01-160-7695	294	1
99688	85928		298	18
99688	85929	4130-01-164-3693	293	41
99688	85931	5340-01-191-4827	300	8
99688	85932	2540-01-191-6537	293	19
99688	85934	5340-01-194-9279	293	18
99688	85935	2590-01-191-9426	293	32
99688	85941	2510-01-191-4259	245	9
99688	85942	2540-01-191-6541	245	34
99688	859430	2590-01-192-5911	245	37
99688	85944	2540-01-191-6540	247	14
99688	859450	2590-01-214-7131	247	3
99688	85954	6110-01-190-5501	293	37
11862	8611710	5340-01-165-1585	75	7
11862	8620318	5340-01-150-4104	76	9
11862	8622045	5315-01-163-6027	78	6
11862	8622361	5305-00-557-6612	72	3
11862	8623039	5310-01-152-4229	79	23

PART NUMBER INDEX

CAGEC	PART NUMBER	STOCK NUMBER	FIG	ITEM
11862	8623041	5360-01-150-0196	76	10
73342	8623078	5315-00-038-3059	78	3
			79	15
73342	8623104	3110-00-005-0873	73	7
			73	23
73342	8623105	5365-01-171-3392	73	8
			73	22
73342	8623112	5365-00-007-3052	73	16
			73	18
11862	8623120	5365-01-150-6092	73	31
73342	8623122	2520-00-008-9987	73	19
11862	8623131	5330-01-163-2614	77	34
11862	8623145	5360-01-153-0933	74	6
11862	8623146	2520-01-150-2280	74	5
11862	8623149	5365-01-150-7830	73	35
11862	8623151	2520-01-150-7609	73	37
11862	8623152	2520-01-175-6492	73	36
11862	8623153	5365-01-158-2182	74	4
11862	8623157	4710-01-151-3663	77	13
11862	8623174	5330-01-150-5928	77	21
11862	8623202	3020-01-149-7938	74	17
11862	8623216	5330-01-165-3388	81	6
			82	1
73342	8623262	2520-01-127-3969	77	5
11862	8623263	5330-00-001-1984	77	6
11862	8623292	5340-01-159-1788	79	17
73342	8623295	2520-01-007-0804	78	13
11862	8623298	4730-01-159-8723	78	5
11862	8623300	3120-01-155-4462	79	3
11862	8623301	3120-01-155-4463	79	3
11862	8623302	3120-01-182-8417	79	3
11862	8623303	3120-01-155-4464	79	3
11862	8623304	3120-01-155-4465	79	3
11862	8623305	3120-01-155-4466	79	3
11862	8623306	3120-01-155-4467	79	3
11862	8623368	5310-01-149-7793	79	22
11862	8623422	4820-01-163-2656	79	21
11862	8623430	5330-01-150-6239	77	23
11862	8623437	5365-01-160-4693	74	1
11862	8623463	5340-01-165-1586	75	2
11862	8623489	5360-01-150-6086	77	39
11862	8623507	5315-01-151-1105	78	14
11862	8623561	5330-01-152-5942	77	11
11862	8623592	4820-01-150-4964	78	1
11862	8623653	5330-01-151-9306	78	8
11862	8623654	5360-01-166-6856	78	7
11862	8623664	5340-01-154-6559	77	27
11862	8623666	5360-01-150-7829	77	28
11862	8623671	5330-01-174-8090	77	31
11862	8623741	5330-01-151-8364	77	17
11862	8623744	5310-01-150-5919	77	29

TM9-2320-289-34P
CROSS-REFERENCE INDEXES

PART NUMBER INDEX

CAGEC	PART NUMBER	STOCK NUMBER	FIG	ITEM
11862	8623842	2520-01-149-1221	74	14
11862	8623849	2520-01-150-3931	73	11
11862	8623851	2520-01-150-2279	73	9
23862	8623917	5330-01-138-5190	73	4
			73	26
11862	8623918	2520-01-151-3571	74	9
11862	8623920	3120-01-167-4172	74	29
11862	8623921	3110-01-167-2443	74	16
11862	8623922	3110-01-169-0734	74	18
11862	8623941	3120-01-166-3677	75	6
11862	8623942	3120-01-174-8153	74	10
11862	8623944	3120-01-156-5189	74	27
			79	14
11862	8623978	5330-01-152-5941	79	5
4F733	8623979	5365-00-004-7210	KITS	
11862	8623983	4820-01-147-4294	70	10
11862	8624091	2530-01-150-3693	76	5
11862	8624101	5360-01-150-6091	73	6
			73	24
11862	8624112	5340-01-147-2266	76	11
11862	8624136	5340-01-149-7811	77	20
11862	8624138	3040-01-151-5663	77	24
11862	8624139	5315-01-152-9038	77	26
11862	8624140	5315-01-152-9039	77	26
11862	8624141	5315-01-152-9040	77	26
11862	8624145		77	22
11862	8624196	3120-01-156-8763	73	13
11862	8624199	2510-01-085-0908	73	38
11862	8624256	5360-01-150-6087	77	33
11862	8624336	2520-01-149-7962	76	2
11862	8624482	5315-01-152-9033	78	2
11862	8624772	3040-01-149-6706	73	1
11862	8624781	3120-00-255-5697	73	33
11862	8624783	3120-01-155-4468	74	13
11862	8624908	2520-01-169-7674	73	3
11862	8624976	2520-01-159-5887	76	8
11862	8625197	2520-00-172-1947	73	21
11862	8625221	5365-01-174-8626	74	19
11862	8625324	5315-01-154-2291	78	4
11862	8625401	3120-01-154-8516	74	35
11862	8625402	3120-01-159-5773	74	35
11862	8625403	3120-01-154-4369	74	35
11862	8625404	3120-01-161-4033	74	35
11862	8625405	3120-01-162-5787	74	35
11862	8625406	3120-01-154-8517	74	35
11862	8625544	5315-01-152-9032	78	12
11862	8625546	4730-01-159-8724	78	11
11862	8625648	5360-01-150-9549	79	24
11862	8625650		79	25
11862	8625717	3020-01-149-5049	73	14
73342	8625718	2520-00-163-0709	73	15

TM9-2320-289-34P
CROSS-REFERENCE INDEXES

PART NUMBER INDEX

CAGEC	PART NUMBER	STOCK NUMBER	FIG	ITEM
11862	8625736	3040-01-167-2836	74	15
11862	8625773	2520-01-165-9563	76	3
11862	8625814	2520-01-150-3692	73	28
11862	8625905	5330-01-156-5236	KITS	
11862	8625913	2520-01-164-7156	74	26
			74	33
11862	8625955	5340-01-149-1867	79	16
11862	8625990	2520-01-151-3570	74	11
11862	8625997	2520-01-154-5463	73	12
			73	32
11862	8626112	5365-01-168-5729	74	22
11862	8626173	5365-01-150-4982	74	25
11862	8626281	5310-01-150-5921	79	2
73342	8626309	5330-00-001-1996	79	29
11862	8626356	5330-01-165-4333	74	3
			79	4
11862	8626372	3120-01-154-7174	74	23
			74	34
11862	8626807	3020-01-149-7941	74	30
11862	8626809	3040-01-150-0407	74	28
11862	8626816	5365-01-154-8562	74	12
11862	8626848		73	17
11862	8626878	3040-01-149-6759	77	35
11862	8626879	5315-01-152-9031	77	37
11862	8626881	5340-01-151-4964	77	38
11862	8626884	5310-01-151-4137	77	36
11862	8626885		79	26
11862	8626902	2520-01-164-7157	77	30
11862	8626916	5330-01-025-4212	79	12
11862	8626938	2520-01-158-6854	74	8
11862	8626979	2520-01-164-7158	76	14
11862	8627153	2590-01-164-7053	77	32
11862	8627334	2520-01-149-1868	73	30
11862	8627385	2520-01-149-3809	74	7
11862	8627387	2520-01-151-3569	73	27
11862	8627509	2520-01-149-7935	77	14
11862	8627592		73	2
11862	8627644		79	13
73342	8627650	2520-00-557-6619	70	6
11862	8627657	5365-01-154-8561	74	32
11862	8627989	2520-01-163-7213	73	25
11862	8627990	2520-01-163-7866	73	5
11862	8627996		75	5
11862	8629080	2520-01-151-3784	78	9
11862	8629227	2520-01-151-3857	77	1
11862	8629487	3110-01-155-4438	73	29
11862	8629503	5935-01-150-6319	75	4
11862	8629523	5365-01-153-0872	79	6
11862	8629526	4710-01-152-5798	79	8
11862	8629796	2520-01-154-1185	77	10
11862	8629935	5945-01-268-4265	77	16

PART NUMBER INDEX

CAGEC	PART NUMBER	STOCK NUMBER	FIG	ITEM
11862	8629955	5330-01-086-5457	KITS	
11862	8629961	5365-01-085-0910	79	28
11862	8629977	2910-01-192-4475	KITS	
72590	8633059	2520-01-168-1983	74	20
11862	8633060	3040-01-154-3799	76	7
11862	8633075		74	2
11862	8633096	5360-01-150-4963	76	4
11862	8633203	2520-01-149-3461	72	6
11862	8633208	5306-01-150-9497	79	11
11862	8633257	2520-01-159-2732	74	36
11862	8633296	2520-01-163-2708	77	12
11862	8633408	4710-01-151-3662	75	1
11862	8637742	4730-01-163-7163	80	2
11862	8639743	5305-01-150-9781	291	10
11862	8640496	3040-01-157-7998	86	4
			95	3
			104	2
			105	1
11862	8653985	3120-01-158-6304	74	21
11862	8655020	2520-01-213-1680	72	6
11862	8655059	2520-01-176-2839	74	24
11862	8655111	2520-01-146-5482	71	6
11862	8655280	5306-01-269-4319	79	1
11862	8655619	2520-01-150-3932	73	10
			73	20
72590	8655621	2520-01-168-2060	73	39
11862	8655625	5330-01-148-7492	72	4
11862	8656942	2520-01-164-7234	72	2
11862	8657163	5330-01-251-1607	76	13
11862	8670393	5330-01-151-6106	77	9
11862	8675627	2520-01-149-7861	74	31
19207	8701347	5935-01-171-8273	45	14
19207	8720780-1	6105-00-116-5124	249	7
11862	8742340	3040-01-159-0996	182	11
11862	8782501	5305-01-159-2783	182	9
11862	8785295	5340-01-165-0657	177	7
60038	88510	3110-00-580-3709	111	7
11862	8906127	2590-01-191-8322	60	11
11862	8906150	2590-01-212-7639	61	6
11862	8909518	6250-01-189-6926	61	5
11862	8914822		61	4
11862	8917605		60	7
11862	8919163		57	38
11862	8919355	5975-01-160-8458	BULK	4
11862	8919356	5975-01-191-9851	BULK	5
11862	8985418	5360-01-193-7130	69	6
11862	8986000	5999-01-159-5603	42	17
11862	908287	3110-01-167-0195	296	7
97271	911105-5130	2520-01-324-4895	93	1
11862	915449	6220-01-160-4247	50	18
11862	915450	6220-01-161-6439	50	18

PART NUMBER INDEX

CAGEC	PART NUMBER	STOCK NUMBER	FIG	ITEM
11862	915908	6220-01-155-6521	50	27
11862	918656	2540-01-156-9740	235	2
22973	922-900-00	6220-01-160-5187	269	3
22973	922-901-01	6220-01-216-5288	269	9
22973	922-901-02	6220-01-216-5289	269	6
22973	922-904-02	6220-01-269-0465	269	7
22973	922-910-00	6220-01-268-8795	269	4
11862	94009398	5310-01-164-5600	187	3
24617	9409103		279	2
24617	9409613	5310-01-167-8344	188	2
			189	6
11862	9409754	5310-01-193-6927	251	2
			255	2
24617	9409761	5310-01-194-9208	251	1
			255	1
24617	9411031	4730-01-098-5229	101	5
11862	9411281	5325-01-165-4372	273	13
24617	9412978	5365-00-721-7680	78	10
24617	9413534	5310-01-097-8222	273	8
			273	22
			274	2
11862	9414031	5310-01-161-2531	271	6
24617	9414224	5305-00-855-3598	40	30
24617	9414238	5305-01-166-1471	213	21
24617	9414411		19	16
24617	9414712	5305-01-162-9695	176	17
			179	23
24617	9414713	5305-00-432-4171	255	7
24617	9414714	5305-01-164-6321	214	3
			222	16
			227	2
24617	9414724	5305-01-210-9425	167	12
24617	9414849	5365-01-182-8469	113	16
24617	9415153	5305-01-197-6576	232	4
24617	9415163	5305-01-164-2313	43	12
			165	17
			242	2
			243	9
			251	9
			264	11
24617	9415319	5305-01-057-9322	248	8
24617	9416137	5305-01-197-2319	293	23
24617	9416187	5305-01-160-3938	58	8
			258	5
24617	9416918	5310-01-012-8962	256	12
			258	2
			271	7
			273	4
			274	1
			275	4
			275	17

PART NUMBER INDEX

CAGEC	PART NUMBER	STOCK NUMBER	FIG	ITEM
24617	9417222	5305-01-162-5703	293	22
24617	9417325	5305-01-161-2581	196	11
24617	9417350	5306-01-266-2419	256	4
			259	6
			259	10
24617	9417714	5310-01-217-5205	268	4
			269	11
			288	13
11862	9417793	5310-01-211-3811	277	7
11862	9417901	4730-00-050-4203	88	1
			89	10
			90	11
			91	10
			92	10
			128	13
			129	4
			130	5
11862	9418355	3110-00-100-5805	101	11
24617	9418719	5305-01-166-1473	222	13
24617	9418924	5310-01-132-8275	262	6
			284	25
11862	9418931	5310-01-187-7610	142	23
11862	9418944	5305-01-149-7356	49	24
11862	9419138	5305-01-155-6113	154	10
			155	9
24617	9419163	5305-01-246-5770	233	7
24617	9419265	5310-01-170-8765	271	16
			272	2
23862	9419284	3110-00-929-8365	84	48
24617	9419320	5305-01-161-2583	38	3
			65	17
24617	9419327	5305-01-157-9720	20	1
			59	7
			180	4
			203	14
72582	9419454	5310-01-133-7215	231	13
24617	9419663	5305-01-197-3287	39	2
			243	8
			284	16
			286	1
24617	9419699	5306-01-155-8528	240	1
24617	9419950	5305-01-200-9869	248	42
24617	9420408	5305-01-157-5625	165	16
24617	9420621	5310-01-158-9205	159	3
			167	16
11862	9420818	5305-01-167-6246	165	19
24617	9421073	5305-01-162-8512	245	8
			247	16
			292	1
			298	5
			300	2

PART NUMBER INDEX

CAGEC	PART NUMBER	STOCK NUMBER	FIG	ITEM
24617	9421073	5305-01-162-8512	300	18
24617	9421432	5305-01-152-8945	209	2
24617	9421745	5340-00-231-1964	2	17
24617	9421985	5305-01-165-8612	233	2
24617	9422295	5310-01-119-3668	19	31
			22	37
			87	8
			118	23
			277	8
11862	9422299	5310-01-150-4003	1	13
			123	4
			142	2
			156	4
24617	9422301	5310-01-149-4407	136	2
			138	4
			144	20
11862	9422303	5310-00-044-3342	102	15
			128	20
			148	26
11862	9422308	5310-00-498-2381	84	1
			102	3
			102	23
23862	9422715	3110-00-155-6735	84	66
24617	9422956	5305-01-197-3112	25	8
			165	4
24617	9423530	5310-01-197-6621	231	17
			275	18
24617	9423768	5305-01-218-3139	268	3
11862	9424320	5305-01-156-8692	139	14
			140	2
11862	9424955	4730-01-154-1366	118	3
24617	9425117	5306-01-155-7659	238	1
24617	9426277	5306-01-203-9082	235	13
24617	9426623	5305-01-201-3788	222	10
24617	9427468	2520-01-172-0394	83	46
16764	9427815	5306-00-407-3737	40	19
11862	9428004	3110-01-096-6755	84	51
11862	9428613	5305-01-197-2535	41	11
11862	9430895	5306-01-269-6138	77	18
11862	9431663	5305-01-157-7388	176	9
24617	9431995	5305-01-198-4154	88	6
24617	9432075	4730-00-288-9390	118	11
			119	12
24617	9432194	5305-01-163-2439	203	18
11862	9436175	5320-01-229-8183	46	16
16764	9436831	3110-00-962-3263	36	25
11862	9437207	4820-00-844-6744	28	16
11862	9437242	5306-01-162-9678	84	5
			179	11
11862	9437415	5315-01-195-5234	84	24
11862	9437702	5306-01-161-6178	44	12

PART NUMBER INDEX

CAGEC	PART NUMBER	STOCK NUMBER	FIG	ITEM
11862	9437702	5306-01-161-6178	46	11
			47	11
			55	5
			65	14
11862	9438039		179	10
24617	9438124	4720-01-267-2052	BULK	27
11862	9438150	5305-01-151-8288	50	14
11862	9438227		21	15
			22	40
11862	9438257	4720-01-155-7784	BULK	17
11862	9438315	4720-01-148-2768	BULK	25
11862	9438373	4720-01-159-5796	BULK	23
11862	9438381	4720-01-156-0550	BULK	20
11862	9438383	4720-01-182-3457	BULK	16
11862	9438916	5305-01-160-3937	166	24
			196	15
			199	1
			200	8
			205	11
			206	1
11862	9439001		21	13
11862	9439004		22	26
11862	9439010		20	13
			22	14
11862	9439046	4720-01-156-0549	BULK	19
11862	9439048		86	3
11862	9439059		22	27
11862	9439068		20	11
11862	9439088		95	5
11862	9439091		95	7
11862	9439092		23	24
11862	9439104	4720-01-156-0547	BULK	15
11862	9439117		21	10
			22	12
11862	9439120		22	28
11862	9439128		20	12
11862	9439162	4720-01-156-0548	BULK	18
11862	9439274		BULK	26
11862	9439363		12	7
11862	9439402	4720-01-163-7833	BULK	24
11862	9439637	5305-00-821-3869	119	16
11862	9439757	5310-01-202-2695	185	5
11862	9439770	5306-01-157-6797	47	2
			167	4
			168	4
			169	19
			169	25
			170	7
11862	9439771	5306-01-157-6796	166	17
			179	22
			181	4

PART NUMBER INDEX

CAGEC	PART NUMBER	STOCK NUMBER	FIG	ITEM
11862	9439771	5306-01-157-6796	284	18
11862	9439772	5305-01-158-0335	170	2
			180	12
11862	9439930	5305-01-165-5591	254	4
24617	9440025	5305-01-197-2320	242	4
			298	15
11862	9440033	5306-01-197-3089	38	10
			52	3
			62	4
			208	5
			271	9
11862	9440034	5306-01-218-3119	261	19
11862	9440166	5305-01-197-2536	48	3
			65	18
11862	9440173		262	11
			284	31
			288	17
11862	9440178	5310-01-161-2374	264	14
11862	9440224	5306-01-171-8076	72	7
11862	9440280	5305-01-203-2289	35	1
11862	9440300	5305-01-163-2438	186	12
			236	4
			237	4
11862	9440334	5306-01-168-4481	29	2
			50	10
			56	3
			57	31
			207	11
			267	9
			271	21
			274	16
			276	9
			287	7
			288	18
11862	9440338	5306-01-150-8713	77	19
11862	9440344	5306-01-195-7915	1	11
			142	5
11862	9440356	5306-01-195-6595	1	15
11862	9441669	5320-01-195-5106	43	4
76680	9449-K	5330-01-096-6775	83	2
			84	58
27647	9477	5305-01-165-2260	124	5
			125	3
35510	95152	5305-01-143-6466	37	37
14892	951965		117	8
35510	95279		37	28
35510	95300	2920-01-145-0993	37	7
11862	9590271	5306-01-164-2323	107	1
11862	9591270	2640-01-323-2632	124	29
11862	9601750	5306-01-157-9936	166	17
11862	9645073	5360-01-164-1949	200	2

PART NUMBER INDEX

CAGEC	PART NUMBER	STOCK NUMBER	FIG	ITEM
09386	96735	5306-01-150-1197	125	20
11862	9702916	3010-01-159-7750	178	9
11862	9703344	2540-01-218-6833	178	6
11862	9711038	5360-01-218-1610	199	3
11862	9721917	3120-01-166-6724	166	23
			166	29
11862	9728247	5340-01-162-9844	166	18
24617	9749363	5360-01-187-0301	127	65
11862	9754764	5365-01-270-1977	127	66
11862	9762199	2530-01-159-3604	127	64
11862	9767270	5305-01-270-3030	127	68
11862	9775809	5365-01-152-7441	99	3
			112	10
11862	9776705	5330-01-112-1533	13	9
11862	9780470	3040-01-159-0930	291	4
11862	9785074	5340-01-163-0919	23	25
			59	5
			60	9
11862	9795470	3040-01-159-8803	69	14
72055	98226	5945-01-192-8653	48	5
11862	9826897	5360-01-171-8248	200	4
11862	9831062	2540-01-086-5433	235	1
11862	9834636	5340-01-161-2749	199	5
			200	3
25022	99-1088-0	4030-01-168-1282	213	12
25022	99-1265-01		215	1
25022	99-1265-02		216	1
25022	99-4256-0	5930-01-163-2583	214	4
25022	99-4256-2	2540-01-165-8177	214	8
25022	99-4257-1	5340-01-172-1942	228	15
25022	99-4276-0		210	3
25022	99-4282-1	2510-01-159-1797	227	1
25022	99-4290-0		210	4
25022	99-4299-0		210	5
25022	99-4308-1	2590-01-163-7626	211	16
			212	16
25022	99-4315-1	5340-01-164-0747	214	12
25022	99-4317-0		210	2
25022	99-4324-0	2510-01-192-9752	288	1
25022	99-4325-0		221	1
25022	99-4340-0	2540-01-164-6251	217	11
25022	99-4341-1	5340-01-164-1075	238	9
25022	99-4342-0		210	6
25022	99-4356-1	5340-01-164-0974	214	2
25022	99-4358-0		217	1
25022	99-4360-1	5895-01-159-1804	222	17
25022	99-4361-0		210	1
25022	99-4408-1	2510-01-159-1795	227	7
25022	99-4420-1	2590-01-162-7139	218	17
			219	9
25022	99-4420-2	2510-01-159-1798	218	14

PART NUMBER INDEX

CAGEC	PART NUMBER	STOCK NUMBER	FIG	ITEM
25022	99-4420-2	2510-01-159-1798	219	16
25022	99-4447-0		220	1
25022	99-4456-0	2540-01-191-8462	217	16
25022	99-4469-0	2540-01-191-6536	217	13
25022	99-4519-1	5945-01-191-3743	222	3
25022	99-4530-1	2510-01-164-9280	229	2
25022	99-4577-1	5340-01-307-2247	213	10
25022	99-4578-1	5340-01-307-2248	213	11
25022	99-4579-1	5340-01-307-2249	211	22
			212	25
25022	99-4580-1	5340-01-307-2250	211	21
			212	24
25022	99-4581-1	5340-01-307-2251	211	20
			212	13
25022	99-4582-1	5340-01-307-2252	211	19
			212	12
25022	99-4583-1	5340-01-307-2253	215	29
			216	29
25022	99-4584-1	5340-01-307-2254	215	30
			216	30
25022	99-4593-0		225	1
25022	99-4602-1	5340-01-164-1048	238	6
25022	99-4675-1	5340-01-194-5805	215	10
			216	8
25022	99-4680-1	5340-01-162-3850	216	28
25022	99-4683-1	5315-01-196-6464	228	11
25022	99-4888-0	5340-01-164-1100	212	22
25022	99-4892-0	2540-01-191-6538	229	7
25022	99-4893-01	2510-01-191-4334	221	8
25022	99-4893-02	2510-01-191-6638	221	9
25022	99-4894-01	2510-01-191-4335	220	15
25022	99-4894-02	2510-01-191-4336	220	14
13548	99012R	6220-01-276-0635	210	8
35510	99459	5305-01-143-7411	37	8
35510	99478	2920-01-144-6362	37	24
35510	99479	5977-01-143-6996	37	55
35510	99481	5935-01-145-1974	37	68
35510	99508	5977-01-143-6955	37	46
35510	99511	5330-01-145-0706	37	57
35510	99512	5340-01-144-6245	37	58
35510	99513	5940-01-142-8411	37	65
35510	99514	5330-01-145-5376	37	48
35510	99515	5330-01-145-0725	37	23
35510	99516	5977-01-142-9122	37	49
35510	99517	5977-01-142-9121	37	53
35510	99518	5330-01-145-0724	37	54
27647	9952		124	8
35510	99524	5305-01-143-6467	37	51
35510	99525	5305-01-143-7412	37	5
76760	99780	5330-01-155-4393	83	50
35510	99789	2920-01-152-7646	37	29

FIGURE AND ITEM NUMBER INDEX

FIG ITEM	STOCK NUMBER	CAGEC	PART NUMBER
BULK 1		81348	QQS741
BULK 2		81349	MILS20166
BULK 3		74410	RRC271BTY2CLDIAO 72
BULK 4	5975-01-160-8458	11862	8919355
BULK 5	5975-01-191-9851	11862	8919356
BULK 6	5940-00-926-8034	96906	MS18029-13S-8
BULK 7	5940-00-405-8976	96906	MS18029-4S-8
BULK 8	2590-01-155-7711	11862	6263877
BULK 9	2510-01-221-2094	11862	370389
BULK 10	6145-01-218-3761	11862	6292996
BULK 11		11862	6293923
BULK 12	4720-01-192-1631	24234	319029
BULK 13		11862	482420
BULK 14	4720-01-148-5981	11862	7828506
BULK 15	4720-01-156-0547	11862	9439104
BULK 16	4720-01-182-3457	11862	9438383
BULK 17	4720-01-155-7784	11862	9438257
BULK 18	4720-01-156-0548	11862	9439162
BULK 19	4720-01-156-0549	11862	9439046
BULK 20	4720-01-156-0550	11862	9438381
BULK 21	4720-00-230-6523	11862	1359744
BULK 22	4720-00-930-2231	96906	MS5213C4B203R
BULK 23	4720-01-159-5796	11862	9438373
BULK 24	4720-01-163-7833	11862	9439402
BULK 25	4720-01-148-2768	11862	9438315
BULK 26		11862	9439274
BULK 27	4720-01-267-2052	24617	9438124
BULK 28	4720-01-096-7718	11862	3987364
BULK 29	2510-01-162-7224	11862	12306178
BULK 30	5970-01-158-9337	11862	462233
BULK 31	4710-00-395-5144	11862	3696822
BULK 32	5530-00-618-6955	81348	NNP530
BULK 33	5330-00-753-8036	11862	370390
BULK 34	5330-01-096-7698	11862	363139
BULK 35	9515-00-516-5756	96906	MS52039C079
BULK 36	2510-01-191-6509	11862	15599621
BULK 37	5940-00-950-7783	96906	MS27212-4-8
BULK 38	4710-00-420-4759	11862	1324714
BULK 39	4710-01-158-7511	11862	3750950
BULK 40	4710-01-161-0138	11862	603827
BULK 41		81349	MILT16343TYPE1
BULK 42	4710-01-192-1433	11862	465246
BULK 43	6145-01-218-3759	11862	6292997
BULK 44	6145-01-218-3760	11862	6293702
KITS		11862	6591956
KITS		14892	2770723
KITS		14892	2771098
KITS	2510-01-326-0762	11862	6258545
KITS	2530-00-960-9363	11862	5687182
KITS	2530-01-033-1855	11862	7817487

FIGURE AND ITEM NUMBER INDEX

FIG	ITEM	STOCK NUMBER		CAGEC	PART NUMBER
KITS		2530-01-034-1715	11862	7817485	
KITS		2530-01-163-0800	27647	15528	
KITS		2540-00-198-2478	19207	5704064	
KITS		2540-00-200-4249	19207	5704052	
KITS		2540-00-201-3474	19207	5704051	
KITS		2910-01-192-4475	11862	8629977	
KITS		2920-01-101-2552	84760	26214	
KITS		3040-01-163-7286	27647	11967	
KITS		4130-01-160-8506	27462	659914	
KITS		4130-01-160-8522	27462	6590589	
KITS		4310-01-166-5958	11862	6590591	
KITS		5330-00-843-9235	24617	7828565	
KITS		5330-01-044-0703	11862	5688044	
KITS		5330-01-086-5457	11862	8629955	
KITS		5330-01-096-9650	11862	7826470	
KITS		5330-01-156-5236	11862	8625905	
KITS		5330-01-164-7593	11862	14067590	
KITS		5330-01-331-1207	118692	15633467	
KITS		5365-00-004-7210	4F733	8623979	
KITS		5365-01-330-8450	11862	15537086	
1	1			11862	1403394-6
1	2	5340-01-158-8624	11862	14033945	
1	3	5307-01-269-4336	11862	22535073	
1	4	2815-01-150-5002	11862	14067734	
1	5	5340-01-150-5986	11862	14033947	
1	6	5305-01-148-7460	11862	11504595	
1	7	5365-01-197-3499	11862	14037861	
1	8	5310-01-148-4991	11862	6262212	
1	9	5305-00-068-0510	96906	MS90728-60	
1	10	5310-00-809-4085	96906	MS27183-16	
1	11	5306-01-195-7915	11862	9440344	
1	12	2510-01-192-5963	11862	14055501	
1	13	5310-01-150-4003	11862	9422299	
1	14	5310-00-316-6513	80205	NAS1408A6	
1	15	5306-01-195-6595	11862	9440356	
1	16	5340-01-150-4110	11862	459021	
1	17	5306-01-148-3664	11862	460308	
1	18	5340-01-147-4217	11862	14071967	
1	19	2510-01-201-2420	11862	14055502	
1	20	5305-00-543-2419	96906	MS90728-61	
1	21	5310-00-087-7493	96906	MS27183-13	
1	22	5305-01-148-2646	11862	6262211	
1	23	5340-01-152-5641	11862	14037878	
2	1	2815-01-165-8216	11862	14079304	
2	1	2815-01-265-7071	7X677	14071072	
2	2	5340-01-166-5640	11862	14025517	
2	3	2815-01-164-7948	11862	14025518	
2	4	5340-01-202-3651	11862	14077163	
2	5	5330-01-149-0874	11862	23500846	
2	6	5340-01-164-6436	11862	14045274	
2	7	5307-01-269-4336	11862	22535073	

FIGURE AND ITEM NUMBER INDEX

FIG	ITEM	STOCK NUMBER	CAGEC	PART NUMBER
2	8	5305-01-148-7460	11862	11504595
2	9	5365-01-162-3797	11862	3894327
2	10	5330-01-157-6757	11862	1049600
2	11	5306-01-148-7456	11862	14077193
2	12	4730-01-163-7864	11862	14028949
2	13	4730-00-044-4655	81348	WW-P-471ACBBCB
2	14	5365-01-154-4378	11862	14090911
2	15	5340-01-149-7977	11862	3999200
2	16	5315-01-186-1352	11862	1453658
2	17	5340-00-231-1964	24617	9421745
2	18	5340-00-449-6408	11862	10000462
2	19	5306-01-149-9668	11862	14077192
2	20	5306-01-198-5515	11862	14077195
2	21	5315-01-149-9712	11862	14022670
2	22	5340-01-149-7976	96906	MS9176-13
3	1	5305-01-148-3686	11862	14022672
3	2	5310-01-147-8744	11862	14022673
3	3	5315-00-393-5865	24617	106751
3	4	3120-01-152-4239	11862	18009096
3	4	3120-01-152-8201	11862	18009097
3	4	3120-01-153-7545	11862	18009098
3	5	3120-01-152-2039	34623	5740201
3	5	3120-01-173-3728	11862	14053400
3	5	3120-01-175-3813	11862	23500111
3	6	3120-01-152-1989	11862	14055002
3	6	3120-01-177-3064	11862	14055003
3	6	3120-01-181-7938	11862	14055004
3	7	5330-01-249-1629	11862	23500139
3	8	2815-01-148-9535	11862	14024271
3	9	5315-01-149-5433	11862	3701679
3	10	2805-01-149-7859	11862	14022671
3	11	3020-01-147-9359	11862	14067703
3	11	3020-01-148-6983	11862	14067702
3	12	5305-01-148-7460	11862	11504595
4	1	2815-01-148-3771	11862	14077157
4	2	5306-01-148-8198	11862	3727207
5	1	2815-01-163-7838	11862	15537018
5	1	2815-01-163-9999	11862	15537020
5	2	5365-01-234-0447	11862	23500298
5	3	5306-01-165-2425	11862	14025526
5	4	2815-01-150-2198	11862	14025523
5	5	5310-01-165-3333	11862	14025527
5	6	2815-01-165-4826	11862	18009095
5	6	3120-01-155-3517	11862	18009094
5	7	2815-01-164-0138	11862	23500392
5	7	2815-01-246-5268	11862	23500391
5	7	2815-01-246-5269	11862	23500393
6	1	5307-01-154-9586	11862	14033818
6	2	5306-01-149-4398	24617	11504270
6	3	2815-01-285-5065	11862	23501992
6	4	5307-01-188-4684	11862	10024396

FIGURE AND ITEM NUMBER INDEX

FIG	ITEM	STOCK NUMBER	CAGEC	PART NUMBER
6	5	5307-01-150-5993	11862	14024203
6	6	2815-01-148-9534	11862	14044971
6	7	5305-01-164-6320	11862	14003948
6	8	5330-00-110-8437	11862	3860095
6	9	5306-01-165-3310	11862	11502804
6	10	5306-01-149-6280	11862	1635490
6	11	5340-01-150-4105	11862	14005953
6	12	2815-01-148-6966	11862	14033820
7	1	5305-01-149-1938	24617	11500815
7	2	3120-01-152-2612	11862	14022644
7	3	5365-01-150-1386	11862	14022643
7	4	2815-01-148-6971	11868	14066308
7	5	5330-01-156-5147	11862	14022649
7	6	2990-01-149-4966	11862	7849302
7	7	5306-01-149-6280	11862	1635490
7	8	5340-01-160-4397	11862	14022650
7	9	5315-00-393-5865	24617	106751
7	10		11862	14061505
7	10		11862	23500073
7	11	5315-00-839-5822	96906	MS24665-353
7	12	5310-01-149-4413	11862	14028982
7	13	2815-01-148-9540	11862	2350074
7	13	2815-01-251-1122	11862	14022634
7	14	5360-01-151-1118	11862	14028981
7	15	3040-01-148-9430	11862	14033822
7	15	3040-01-192-4466	11862	23500075
7	16	5365-01-149-9691	11862	14057297
7	17	5310-01-149-4411	11862	14028983
7	18	3040-01-212-7616	11862	23500076
7	19	2815-01-172-6780	11862	14057296
7	20	5305-01-211-3103	24617	11503603
7	21	2920-01-142-2631	23862	14003974
7	22	2805-00-155-7266	11862	3947770
7	23	2815-01-165-6143	11862	14042575
7	24	2805-00-752-0158	11862	3835333
7	25	5365-01-154-1484	11862	14025515
7	26	5360-01-175-7269	11862	14025512
7	27	5310-01-157-6773	11862	14003968
7	28	2815-01-164-7054	11862	14050661
7	28	2815-01-165-8002	11862	14050662
7	28	2815-01-314-6887	11862	10137454
7	29	2815-01-163-7189	11862	14033927
7	29	2815-01-163-7190	11862	14050658
7	29	2815-01-163-7191	11862	14050659
7	30	2815-01-147-9358	11862	5234530
7	31	2815-01-147-4200	11862	14077148
7	32	2815-01-150-0673	11862	14057232
7	33	2815-01-147-4201	11862	14022640
7	34	3020-01-150-2207	11862	14022646
7	35	3020-01-149-5915	11862	14022645
7	36	5305-01-150-5895	24617	11503316

FIGURE AND ITEM NUMBER INDEX

FIG	ITEM	STOCK NUMBER	CAGEC	PART NUMBER
7	37	5310-01-149-0870	11862	14022648
7	38	3020-01-148-9547	11862	14022653
7	39	3020-01-155-5779	11862	TO181
7	40	3120-01-152-5946	11862	14048457
7	41	3120-01-157-0914	11862	14048456
7	42	3120-01-152-8200	11862	14048455
7	43	3120-01-152-5945	11862	14048454
7	44		11862	SH1366S
8	1	2940-01-217-8089	70040	F0106
8	2	2590-01-147-4285	11862	14071059
8	3	5325-01-151-6117	11862	14028942
8	4	6680-01-147-4629	11862	14050523
8	5	4710-01-150-0842	11862	14045268
8	6	5310-01-148-7474	11862	14036369
8	7	5310-01-165-3331	11862	22521550
8	8	5330-00-935-9136	246147	274244
8	9	6615-01-153-1852	11862	25011208
8	10	4820-01-153-1851	11862	25011206
8	11	2815-01-166-0621	11862	14066310
8	12	4730-01-150-0879	11862	14022700
8	13	2940-00-082-6034	70040	PF-35
8	14	5306-01-148-7442	11862	14028914
8	15	5330-01-150-7744	11862	14022683
8	16	5306-01-230-3354	11862	11508534
8	17	5305-01-150-9781	24617	11507029
8	18	5330-00-107-3925	11862	14079550
8	19	5365-01-149-0880	11862	337185
8	20	5307-01-150-5992	11862	14066307
8	21	2815-01-147-4275	11862	14061649
8	22	2815-01-148-9497	11862	14079426
8	23	3040-01-163-7208	11862	14022699
8	24	5330-01-154-7159	11862	477249
8	25	2815-01-164-7155	11862	14028976
8	26	5340-01-194-5294	11862	3704871
8	27	2805-00-336-1716	11862	3702366
8	28	5360-01-165-1563	118662	360582
8	29	5315-00-828-5487	11862	838839
8	30	2815-01-164-7154	11862	14077182
8	31	5306-01-165-5582	11862	11508600
8	32	5315-00-014-1195	24617	141195
9	1	5330-01-218-0862	11862	14055585
9	2	4710-01-148-8354	11862	14061344
9	3	5305-00-225-3843	96906	MS90728-8
9	4	4730-00-288-9390	11862	137396
9	5	4730-00-277-8269	72582	224425
9	6	4820-00-174-0315	79470	6820
9	7	5340-01-234-1465	11862	15599988
9	8	4730-00-288-9440	72582	118754
9	9	5430-01-225-8971	70411	SP2489-FM
9	10	4720-01-150-7575	11862	15548901
9	10	4720-01-162-0119	11862	14063336

FIGURE AND ITEM NUMBER INDEX

FIG	ITEM	STOCK NUMBER	CAGEC	PART NUMBER
9	11	5310-01-159-8559	96906	MS35691-406
9	12	5640-01-159-6935	11862	14061350
9	13	5310-00-582-5965	96906	MS35338-44
9	14	5305-00-068-0508	96906	MS90728-6
9	15	4730-01-172-6683	11862	14055586
9	16	5340-01-197-3434	11862	14047899
9	17	5340-01-150-6249	11862	14061348
9	18	5310-01-165-3331	11862	22521550
9	19	5340-01-150-1377	11862	14061352
9	20	5305-00-068-0509	96906	MS90728-10
9	21	4720-01-148-5000	11862	15548902
9	21	4720-01-162-0120	11862	14063337
9	22		11862	15599986
9	23	4730-01-227-1929	11862	15599987
9	24	5340-01-150-0197	11862	14036784
9	25	4710-01-149-1899	11862	14061345
10	1	5306-01-253-7073	11862	11508353
10	2	4730-00-908-3194	96906	MS35842-11
10	3	4720-01-194-0336	11862	14067732
10	4	5310-01-155-2503	24617	11502488
10	5	2815-01-192-5962	11862	14067733
10	6	5310-01-165-3331	11862	22521550
10	7	4720-01-195-7603	11862	14050445
10	8	4710-01-192-7967	11862	14050442
10	9	4710-01-198-2701	11862	14050443
10	10	5340-01-197-3433	11862	14050441
10	11	5305-01-273-4486	11862	11509135
10	12	4730-01-164-7028	11862	14050444
10	13	2990-01-192-9730	11862	14050446
10	14	2990-01-147-9284	11862	25042462
11	1	5330-01-157-0856	7X677	15633464
11	2	5306-01-149-6280	11862	1635490
11	3	5307-01-269-4336	11862	22535073
11	4	5307-01-150-5991	89554	23500832
11	5	5305-01-148-5915	11862	11505068
11	6	2815-01-150-2181	11862	14022657
11	7	5307-01-158-6312	11862	14028924
11	8	5306-01-149-6278	11862	14028922
11	9	5307-01-151-8374	11862	14022654
11	10	5305-01-148-3685	11862	14028923
11	11	2815-01-148-9560	11862	14025568
11	12	2815-01-148-9559	11862	14071068
12	1	5340-01-166-1534	11862	14066301
12	2		11862	11663000
12	3	5340-01-151-9956	11862	14033893
12	4	5310-01-165-3331	11862	22521550
12	5		11862	14066306
12	6	4730-01-163-7194	11862	25518880
12	7		11862	9439363
12	8	5340-01-151-9957	11862	14033895
12	9	4710-01-148-2659	11862	14061569

FIGURE AND ITEM NUMBER INDEX

FIG	ITEM	STOCK NUMBER		CAGEC	PART NUMBER
12	10	5306-01-148-3667	11862 11504512		
12	11	5340-00-881-5303	96906 MS21333-45		
12	12	5330-01-150-1215	11862 14025557		
12	13	2910-01-150-0950	63632 6704001		
12	14	4720-01-184-0432	11862 14066305		
13	1	5305-01-162-7885	11862 11508017		
13	2	3020-01-148-8846	11862 14022652		
13	3	5330-01-150-5944	11862 14022651		
13	4	2910-01-326-8187	84760 DB2829-4521		
13	5	5310-01-250-7679	11862 11506101		
13	6	5340-01-148-8349	11862 14033896		
13	7	5305-00-725-2317	96906 MS90728-64		
13	8	2910-01-159-0867	11862 6471831		
13	9	5330-01-112-1533	11862 9776705		
13	10	5315-01-151-4180	11862 14050425		
13	11	5330-00-830-1745	11862 3705044		
13	12	2910-00-493-2138	11862 3719599		
13	13	5306-01-151-4925	11862 11509669		
13	14	4730-01-205-7917	93061 411FS-6		
13	15	5340-01-148-7528	11862 22511422		
13	16			11862	14061503
13	17	5365-01-231-7584	11862 3792381		
13	18	5340-01-167-7794	11862 14063340		
14	1			11862	14045233
14	2	5340-01-270-7423	11862 343350		
14	3			11862	3970076
14	4			11862	M51
14	5			11862	M140
14	6	4730-01-194-0126	11862 560625		
14	7			11862	M495
14	8			11862	M127
14	9	2910-01-200-4338	11862 22506637		
14	10	4820-01-158-6836	11862 14057219		
14	11	5305-01-211-7464	96906 MS51869-28		
15	1	2520-01-188-3282	84760 23461		
15	2	2910-01-189-2142	84760 18021		
15	3	3040-01-192-4267	84760 18020		
15	4	5310-00-400-8585	84760 12362		
15	5	5305-00-250-5613	84760 12360		
15	6	2910-00-788-0986	84760 12358		
15	7	5360-01-194-3162	84760 23643		
15	8	2910-01-188-3255	84760 21895		
15	9	5360-01-189-3466	84760 21917		
15	10	2910-01-054-3816	84760 20849		
15	10	5315-01-197-8547	84760 22256		
15	11	2910-01-188-3254	84760 21763		
15	12	5360-01-197-1506	84760 22813		
15	13	5310-01-188-6743	84760 22325		
15	14	2910-01-191-8448	84760 22326		
15	15	5360-00-887-1536	84760 10541		
15	16	5315-01-191-3393	84760 22064		

FIGURE AND ITEM NUMBER INDEX

FIG	ITEM	STOCK NUMBER	CAGEC	PART NUMBER
15	17	2910-01-188-3252	84760	22327
15	18	5360-01-188-0807	84760	22125
15	19	5330-00-757-1680	84760	10453
15	20	5330-01-192-5779	84760	21860
15	21	5306-00-819-3038	84760	11331
15	22		84760	27002
15	22		84760	27015
15	23	5945-01-190-5143	84760	21323
15	23	5945-01-190-5143	84760	21323
15	24	5330-01-234-2615	84760	27244
15	25	2910-01-198-0868	84760	22851
15	26	5310-00-830-7825	84760	12500
15	27	5310-00-190-0752	22787	10-9858
15	28		96906	MS35338-38
15	29	5310-00-934-9757	96906	MS35649-282
15	30	5310-01-118-2248	84760	18493
15	31	2920-01-117-3251	84760	20951
15	32	5330-01-232-2145	84760	27607
15	33	4820-01-165-9596	84760	23183
15	34	5310-01-161-6121	84760	13521
15	35	5305-01-118-4114	84760	22351
15	36	5310-00-209-1218	45152	2239H
15	37	5310-00-582-5765	03350	22FT832
15	18	5970-01-189-4883	84760	21618
15	39	5940-01-189-9841	84760	22985
15	40	5940-01-189-8033	84760	24901
15	41		84760	24680
15	42	5315-01-167-5584	84760	21284
15	43	5945-01-155-7830	84760	22840
15	43	5945-01-214-1490	84760	23861
15	44	5310-00-877-4956	84760	14408
15	45	5305-01-188-0489	34623	5740572
15	45	5305-01-188-0490	84760	21661
15	45	5305-01-188-0491	84760	21662
15	45	5305-01-188-0492	84760	21663
15	45	5305-01-188-0493	84760	21664
15	45	5305-01-188-0494	84760	21665
15	45	5305-01-188-6566	84860	22734
15	45	5305-01-190-4078	84760	22733
15	46	5365-01-188-0962	84760	23566
15	47	2910-01-188-3256	84760	23428
15	48	5305-01-190-5745	84760	22398
15	49		84760	27600
15	50	5307-01-188-9217	84760	23352
15	51	2910-01-189-2191	84760	22721
15	52	5315-01-190-0429	84760	22397
15	53	5305-01-188-0496	84760	22642
15	54	5340-01-192-6030	84760	22917
15	55	5330-01-233-8597	84760	27163
15	56	5365-01-188-0784	84760	22693
15	57	5330-01-233-2778	84760	27609

FIGURE AND ITEM NUMBER INDEX

FIG	ITEM	STOCK NUMBER	CAGEC	PART NUMBER
15	58	5360-01-197-1507	84760	23171
15	59	2910-01-190-0069	84760	22367
15	60	5315-01-189-2141	84760	23925
15	61		84760	27610
15	62	5365-01-188-0785	84760	23056
15	63	5365-01-188-0783	84760	23426
15	64	2910-01-188-3257	84760	24433
15	64	2910-01-256-3698	84760	24434
15	65	5305-01-188-6568	84760	24566
15	66		84760	27602
15	67	3110-01-188-7682	84760	23101
15	68	5365-01-188-0958	84760	22937
15	69		84760	27603
15	70	2910-01-188-3176	84760	23107
15	71	5305-01-188-0948	84760	21194
15	72		84760	10394
15	73	5305-01-190-4555	84760	24419
15	74	5305-01-190-4070	84760	21712
15	75	2910-01-117-7252	84760	21312
15	76	3120-00-393-4067	84760	20222
15	77	3040-01-188-3261	84760	21201
15	78	5365-00-804-2027	96906	MS16624-1093
15	79	2990-01-320-8915	84760	27984
15	80	2910-01-165-9487	84760	23265
15	81	2910-01-166-5655	84760	22935
15	82	2910-01-191-8465	84760	18986
15	83	4320-01-317-0692	84760	28396
15	84	5315-00-887-1539	84760	11141
15	85	3040-01-188-3242	84760	24569
15	86	2910-00-897-2465	84760	11056
15	86	2910-01-189-1748	84760	11057
15	86	2910-01-191-8453	84760	11058
15	86	2910-01-191-8454	84760	11059
15	83	2910-01-191-8455	84760	11060
15	86	2910-01-191-8456	84760	11062
15	86	2910-01-191-8457	84760	11063
15	86	2910-01-191-8458	84760	11064
15	86	2910-01-191-8459	84760	11065
15	87		84760	26833
15	87		84760	26834
15	88	5360-01-260-5649	84760	26071
15	89		84760	26070
15	90	5305-01-188-6567	84760	23124
15	91	5305-00-207-3984	84760	11175
15	92	5360-01-188-6806	84760	23238
15	93		84760	27245
15	94	2910-01-188-3243	84760	27833
15	94	2910-01-192-4585	84760	23120
15	95	5305-00-887-1547	84760	11438
15	96	5330-01-236-0476	84760	27601
15	97	2910-01-188-3244	84760	21283

FIGURE AND ITEM NUMBER INDEX

FIG	ITEM	STOCK NUMBER	CAGEC	PART NUMBER
15	98	5305-01-191-0374	84760	21287
15	99	4730-01-183-2140	84760	20727
15	100	5305-00-887-1564	84760	12216
15	101		84760	27608
15	102	2910-01-189-0895	84760	20530
15	103	2910-01-232-1044	84760	22988
15	104	2910-01-188-3683	84760	20512
15	104	2910-01-230-9007	84760	20803
15	105	5360-00-900-2564	84760	15699
15	106	5315-01-188-0495	84760	19837
15	107	5330-01-197-2454	84760	19832
15	108	4820-01-189-0894	84760	19895
15	109	5360-01-197-1505	84760	21198
15	110	4730-00-459-6077	84760	15228
15	111	5365-01-188-0959	84760	20529
15	112	2910-01-188-3775	84760	20527
15	113	2910-01-188-3250	84760	20523
15	114	2910-01-188-3251	84760	21296
15	115	2910-01-188-3249	84760	21200
15	116	5365-01-188-0960	84760	20528
15	117	5360-01-190-6215	84760	21358
15	118	5305-01-193-2377	84760	21646
15	119	2910-01-189-1747	84760	24345
15	120	5310-00-934-9751	96906	MS35650-302
15	121	5365-01-188-0993	84760	22900
15	122		84760	24585
15	123	3040-01-188-3222	84760	24623
15	123	3040-01-247-0893	84760	62538
15	124	5315-01-188-0765	84760	23100
15	124	5315-01-188-0766	84760	23685
16	1		11862	14063339
16	2	4710-01-148-9580	11862	14033912
16	3	4710-01-149-5076	11862	14033914
16	4	4710-01-150-0971	11862	14033916
16	5	4710-01-148-9582	11862	14033918
16	6	5365-01-165-3457	11862	14033920
16	7	5340-01-150-6026	11862	14033955
16	8	5325-01-160-4618	11862	560614
16	9	5305-01-150-1521	11862	11503617
16	10	5340-01-150-4106	11862	14033921
16	11	5340-01-150-6275	11862	14033922
16	12	5340-01-148-7529	11862	560613
16	13	5310-01-165-3331	11862	22521550
16	14	5340-01-159-6626	11862	14033824
16	15	4710-01-149-5077	11862	14033917
16	16	4710-01-153-1636	11862	14033915
16	17	4710-01-148-9581	11862	14033913
16	18	4710-01-149-5075	11862	14033911
16	19	5340-01-150-7774	11862	14033953
16	20	4730-00-014-2432	24617	137397
17	1	5310-01-196-5587	11862	3827499

FIGURE AND ITEM NUMBER INDEX

FIG	ITEM	STOCK NUMBER	CAGEC	PART NUMBER
17	2	5310-01-147-8743	11862	3790768
17	3	5310-01-148-3693	11862	14001197
17	4	2940-01-155-3190	70040	A6440
17	5	2940-01-148-5992	11862	25041910
17	6	5325-01-149-6293	11862	15530620
17	7	5307-01-153-0873	11862	14033948
18	1	5305-01-140-9118	96906	MS90728-59
18	2	6680-01-161-1439	11862	22516548
18	3	6680-01-175-0565	11862	25004140
18	4	5330-01-159-1298	11862	22515965
18	5	5310-00-896-0903	96906	MS51967-12
18	6	5310-00-809-4061	96906	MS27183-15
18	7	2590-01-155-7711	11862	6263877
18	8	2910-01-155-7878	11862	14020491
18	9	2510-01-155-5848	11862	480534
18	10	2590-01-155-7882	11862	6260631
18	11	5340-01-244-7925	11862	334675
18	12	2910-01-147-4219	11862	368752
18	13	2910-01-160-8107	11862	14050685
18	14	2990-01-155-7879	11862	14020492
19	1	5340-01-165-4705	11862	14010707
19	2	2910-01-155-5063	11862	14063326
19	3	5305-01-158-2032	96906	MS51869-24
19	4	5340-01-172-2087	11862	14052026
19	5	5360-01-181-2482	11862	4813235
19	6	5305-01-158-2032	96906	MS51869-24
19	7	5305-01-158-7820	11862	14026247
19	8	2910-01-155-5139	11862	14063363
19	9	2910-01-155-5147	11862	14063333
19	9	2910-01-155-5148	11862	14063327
19	10	4730-00-908-6292	96906	MS35842-14
19	11	4720-01-155-5194	11862	14063334
19	11	4720-01-155-8062	11862	14063328
19	12	2590-01-155-5177	11862	334523
19	13	2910-01-148-2910	11862	14071984
19	14		11862	6263870
19	15	2910-01-147-4218	11862	6262755
19	16		24617	9414411
19	17		11862	6263871
19	18	6680-01-161-1439	11862	22516548
19	19	5330-01-159-1298	11862	22515965
19	20	6680-01-164-9433	11862	25004137
19	21	5340-01-151-7409	11862	359847
19	22	2590-01-155-5140	11862	334521
19	23	5340-01-160-0367	11862	334522
19	24	5305-00-146-2524	96906	MS51850-86
19	25	5310-00-931-8167	96906	MS51967-6
19	26	5310-00-407-9566	96906	MS35338-45
19	27	2910-01-155-5137	11862	341287
19	28	5340-01-162-9781	11862	344714
19	29	2910-01-155-5136	11862	467525

FIGURE AND ITEM NUMBER INDEX

FIG	ITEM	STOCK NUMBER	CAGEC	PART NUMBER		
19	30	5306-00-226-4826	96906	MS90728-33		
19	31	5310-01-119-3668	24617	9422295		
19	32	5310-00-081-4219	96906	MS27183-12		
19	33	5340-01-162-9777	11862	15599221		
19	34	2910-01-152-5516	11862	474955		
19	35	5306-01-159-2784	11862	14072666		
19	36	5305-01-140-9118	96906	MS90728-59		
19	37	5310-00-896-0903	96906	MS51967-12		
19	38	5310-00-809-4061	96906	MS27183-15		
19	39	4730-00-909-8627	96906	MS35842-13		
19	40	4730-00-908-3194	96906	MS35842-11		
19	41	4720-01-148-6984	11862	14036751		
19	41	4720-01-192-9823	11862	14063335		
20	1	5305-01-157-9720	24617	9419327		
20	2	2910-01-155-7881	11862	476916		
20	3	5330-01-157-7604	11862	476927		
20	4	5305-01-158-2032	96906	MS51869-24		
20	5	5340-01-172-2087	11862	14052026		
20	6	5340-01-165-4705	11862	14010707		
20	7	2910-01-155-7880	11862	14063329		
20	8	5360-01-181-2482	11862	4813235		
20	9	4730-00-826-4268	19207	11608950-4		
20	10	4730-01-163-7194	11862	25518880		
20	11				11862	9439068
20	12				11862	9439128
20	13				11862	9439010
20	14				11862	14018658
20	15	5305-01-156-5438	11862	11504447		
20	16	5340-01-158-0314	11862	1638274		
20	17	5340-00-809-1490	96906	MS21333-98		
20	18				1862	14018647
20	19				11862	14063317
20	20	2910-01-155-5138	11862	14063319		
20	21	5340-01-157-9825	11862	14018630		
20	22	4730-00-909-8627	96906	MS35842-13		
20	23	4720-01-160-5781	11862	14049494		
20	24	5305-01-158-7820	11862	14026247		
20	25	2910-01-155-5063	11862	14063326		
20	26	2910-01-155-7965	11862	14063325		
20	27	4730-00-908-3194	96906	MS35842-11		
20	28				11862	14041258
20	29	4710-01-154-1230	11862	468484		
21	1	5340-01-158-0314	11862	1638274		
21	2	5340-00-057-3037	96906	MS21333-111		
21	3				11862	15599209
21	4	4730-00-014-2433	24617	142433		
21	5	4730-01-075-7310	72582	178917		
21	6	4730-00-132-4625	6N299	0917425		
21	7	5305-01-156-5438	11862	11504447		
21	8				11862	14063315
21	9	4730-00-826-4268	11862	477402		

FIGURE AND ITEM NUMBER INDEX

FIG	ITEM	STOCK NUMBER	CAGEC	PART NUMBER
21	10		11862	9439117
21	11	5340-00-809-1490	96906	MS21333-98
21	12	4730-01-163-7194	11862	25518880
21	13		11862	9439001
21	14		11862	474957
21	15		11862	9438227
21	16	4720-01-156-0085	11862	14063391
22	1	4730-00-826-4268	11862	477402
22	2	4720-01-156-0085	11862	14063391
22	3		11862	15599209
22	4	5340-01-158-0314	11862	1638274
22	5	5340-00-057-3037	96906	MS21333-111
22	6	5340-00-057-3037	96906	MS21333-111
22	7	4730-00-014-2433	24617	142433
22	8	4730-01-075-7310	72582	178917
22	9	4730-00-132-4625	24617	444620
22	10	5340-01-268-9064	11862	25527423
22	11	5305-01-156-5438	11862	11504447
22	12		11862	9439117
22	13	4730-01-163-7194	11862	25518880
22	14		11862	9439010
22	15		11862	14063314
22	16	5340-00-881-5303	96906	MS21333-45
22	17	5310-01-159-8559	96906	MS35691-406
22	18	5310-00-582-5965	96906	MS35338-44
22	19	5305-00-068-0501	96906	MS90725-5
22	20	5310-00-931-8167	96906	MS51967-6
22	21	5310-00-407-9566	96906	MS35338-45
22	22	5340-01-163-9400	11862	467524
22	23		11862	14045605
22	24	5340-01-159-1185	11862	14034543
22	25		11862	14061227
22	26		11862	9439004
22	27		11862	9439059
22	28		11862	9439120
22	29		11862	14061223
22	30	2910-01-155-5138	11862	14063319
22	31	5340-01-163-5902	11862	467509
22	32	5310-00-245-3424	96906	MS17829-50
22	33	2590-01-164-7177	11862	15522392
22	34	2910-01-155-5845	11862	14034546
22	35	5310-00-081-4219	96906	MS27183-12
22	36	5306-00-226-4825	96906	MS90728-32
22	37	5310-01-119-3668	24617	9422295
22	38	5306-00-226-4826	96906	MS90728-33
22	39		11862	15599999
22	40		11862	9438227
23	1	2910-01-156-0045	84760	27290
23	2	2910-01-156-8361	11862	14075347
23	3	5330-01-213-9966	84760	22591
23	4	2910-01-210-1322	84760	24285

FIGURE AND ITEM NUMBER INDEX

FIG	ITEM	STOCK NUMBER	CAGEC	PART NUMBER
23	5	5330-01-236-1724	84760	27820
23	6	5340-01-211-3086	84760	24267
23	7	2910-01-210-1323	84760	29090
23	8	5340-01-197-1199	84760	24265
23	9	5340-01-202-2622	84760	24281
23	10	5330-01-138-2106	84760	15349
23	11	5305-01-211-3032	84760	24437
23	12	5305-01-211-3031	84760	24322
23	13	2920-01-212-4771	84760	27284
23	14	4820-01-209-0473	84760	23796
23	15	5930-01-208-6292	61928	15596614
23	16	5340-01-160-2239	11862	14043724
23	17	5305-00-115-9526	96906	MS18154-58
23	18	5310-00-637-9541	96906	MS35338-46
23	19	4730-01-163-7194	11862	25518880
23	20	4720-01-163-8039	11862	14063302
23	21	4730-00-826-4268	19207	11608950-4
23	22	4720-01-192-8533	11862	14063301
23	23	4730-00-908-3195	96906	MS35842-10
23	24		11862	9439092
23	25	5340-01-163-0919	11862	9785074
24	1	4730-00-580-6738	24617	444034
24	2	6685-01-205-3676	70040	10045847
24	3	2920-01-151-3627	11862	5613939
25	1	2990-01-148-2928	11862	14038644
25	2	5310-01-143-0542	73342	3909063
25	3	5360-01-151-1121	11862	336989
25	4	2540-01-192-1823	11862	468234
25	5	2910-01-150-3675	11862	3993087
25	6	2910-01-225-1068	11862	15590123
25	7	2910-01-149-3786	11862	342405
25	8	5305-01-197-3112	24617	9422956
25	9	5360-01-151-1120	11862	14024997
25	10	5306-01-152-4693	24617	11504986
25	11	2910-01-150-3676	11862	14038647
25	12	5340-01-269-1361	11862	15567924
25	13	5306-01-161-5489	11862	11501095
25	14	5945-01-154-3143	11862	14066255
26	1	2990-01-154-3743	11862	14037856
26	1	2990-01-225-1029	11862	15595224
26	2	5310-00-931-8167	96906	MS51967-6
26	3	5310-00-407-9566	96906	MS35338-45
26	4	5306-00-226-4830	96906	MS90728-37
26	5	5340-01-160-2445	11862	341160
26	6	5340-01-199-2312	96906	MS52150-31HE
26	7	5340-01-204-4268	96906	MS52150-30HE
26	8	2990-01-155-5151	79260	45823
26	9	2990-01-046-1170	11862	14063795
26	9	2990-01-225-1052	11862	14089132
26	10	2990-01-154-1324	11862	14029956
26	11	2990-01-155-5150	11862	14045525

FIGURE AND ITEM NUMBER INDEX

FIG	ITEM	STOCK NUMBER	CAGEC	PART NUMBER
23	11	2990-01-225-1028	11862	15595271
26	12	2990-01-155-5149	11862	14045521
26	13	5330-01-147-4212	11862	14072686
26	14	5360-01-149-1959	11862	587575
26	15	5310-00-316-6513	80205	NAS1408A6
26	16	5310-00-809-4085	96906	MS27183-16
27	1		11862	15599216
27	1	2990-01-147-4290	11862	15595216
27	2	5306-00-226-4830	96906	MS90728-37
27	3	5340-01-163-1388	11862	14037808
27	4	5305-00-115-9526	96906	MS18154-58
27	5	5310-00-637-9541	96906	MS35338-46
27	6	5310-00-732-0558	96906	MS51967-8
27	7	5340-01-160-2445	11862	341160
27	8	5310-00-407-9566	96906	MS35338-45
27	9	5310-00-931-8167	96906	MS51967-6
27	10	5340-01-199-2312	96906	MS52150-31HE
27	11	5340-01-204-4268	96906	MS52150-30HE
27	12	2990-01-203-2426	11862	14037836
27	13	2990-01-147-3953	11862	14037812
27	14	5306-00-226-4827	96906	MS90728-34
27	15	5310-00-514-6674	96906	MS35335-34
27	16	2990-01-154-1323	11862	14044996
27	17	2990-01-152-7788	11862	14067430
27	18	5340-01-148-2818	11862	14034547
27	19	5340-01-155-9861	11862	14024561
27	20	2990-01-152-7828	11862	14044995
27	21	2990-01-180-2988	11862	14067429
27	22	2990-01-147-4289	11862	14067759
27	22	2990-01-231-2938	11862	15599269
27	23	5330-01-147-4212	11862	14072686
27	24	5360-01-149-1959	11862	587575
27	25	5310-00-316-6513	80205	NAS1408A6
27	26	5310-00-809-4085	96906	MS27183-16
28	1	5305-01-160-1975	11862	11508566
28	2	5305-01-151-9285	11862	11504115
28	3	2930-01-123-4941	11862	6410785
28	4	2930-01-160-1597	11862	14039948
28	5	2930-01-147-4222	11862	6264100
28	6	5340-01-159-2901	11862	14039950
28	7	6680-01-147-5497	11862	358375
28	8	4730-00-908-3195	96906	MS35842-10
28	9	5340-01-149-4434	11862	3816659
28	10	5340-00-057-3052	96906	MS21333-114
28	11	2930-01-085-0926	11862	14011345
28	12	5306-01-158-9018	11862	2014469
28	13	2930-01-159-2902	11862	14052221
28	14		11862	14072430
28	15	2930-01-264-3480	61928	3058966
28	16	4820-00-844-6744	11862	9437207
28	17	5340-01-157-6429	11862	14072427

FIGURE AND ITEM NUMBER INDEX

FIG	ITEM	STOCK NUMBER	CAGEC	PART NUMBER
28	18	5340-01-157-6428	11862	14072426
28	19	5310-01-160-9529	24617	1494253
28	20	5340-01-164-3269	11862	14039949
28	21	5310-01-160-4536	11862	3792287
29	1	5310-01-170-9100	11862	3982098
29	2	5306-01-168-4481	11862	9440334
29	3	5305-01-160-1975	11862	11508566
29	4	2930-01-147-4221	11862	15522697
30	1	4730-00-909-8627	96906	MS35842-13
30	2	5340-01-150-4991	11862	14036779
30	3	5305-01-160-1975	11862	11508566
30	4	4720-01-148-2970	11862	14036744
30	5	4730-01-194-2002	11862	14071983
30	6	5306-01-185-7048	11862	11513606
30	7	2930-01-147-4198	11862	14028918
30	8	5306-01-149-6280	11862	1635490
30	9	5330-01-149-0874	11862	23500846
30	10	4730-01-148-2755	11862	14067727
30	10	4730-01-148-8242	11862	14067737
30	11	2930-01-147-9916	11862	14028917
30	12	5307-01-269-4336	11862	22535073
30	13	6620-01-146-8006	11862	14077122
30	14	5330-01-148-7497	11862	14028916
30	15	4730-01-148-2758	11862	354501
30	16		11862	14033823
30	17	4730-00-908-3194	96906	MS35842-11
30	18	4720-01-159-1839	11862	14067763
30	19	2930-01-150-0895	11862	14000217
31	1	4730-00-044-4789	30379	444789
31	2	4730-01-148-2758	11862	354501
31	3	5307-01-150-1549	11862	14045263
31	4	5306-01-165-5583	11862	14060613
31	5	2930-01-147-8555	11862	14024208
31	6	5305-01-149-1938	24617	11500815
31	7	5330-01-147-9808	11862	14024209
31	8	2930-01-193-7802	11862	23500133
31	9	5305-01-149-9673	11862	11504967
31	10	5305-01-254-2558	11862	11509202
31	11	5307-01-178-7445	11862	14071080
31	12	5306-01-150-1190	11862	11500921
32	1	5307-01-150-1228	11862	14061661
32	2	3020-01-148-2949	11862	14067704
32	2	3020-01-148-2950	11862	14067705
32	3	5306-01-149-6279	11862	14020698
32	4	4140-01-145-8099	11862	14077928
32	5	5310-00-931-8167	96906	MS51967-6
32	6	2930-01-147-9330	11862	14032395
33	1	2920-01-147-4272	11862	14077149
33	2	5306-01-149-6280	11862	1635490
33	3	5305-01-148-7460	11862	11504595
33	4	5306-01-149-6280	11862	1635490

FIGURE AND ITEM NUMBER INDEX

FIG	ITEM	STOCK NUMBER	CAGEC	PART NUMBER
33	5	5340-01-197-1259	11862	14077151
33	6		11862	11503643
33	7	5310-00-809-4085	96906	MS27183-16
33	8	2920-01-149-8606	11862	1105500
33	9	5306-01-185-7050	11862	1610819
33	10	3030-01-258-5125	20796	43-3226
33	11	2920-01-191-6635	11862	14077147
34	1	5340-01-150-4105	11862	14005953
34	2	2920-01-147-4278	11862	14067724
34	3	5306-01-149-6280	11862	1635490
34	4	3030-01-147-6410	20796	42-6923
34	5	5306-01-150-9493	11862	14067717
34	6	5310-01-250-7679	11862	11506101
34	7	5310-00-809-4085	96906	MS27183-16
34	8	2920-01-147-8559	11862	14067714
34	9	5365-01-154-4365	11862	14067725
35	1	5305-01-203-2289	11862	9440280
35	2	2920-01-131-4932	35510	46290A
35	3	3020-01-147-7935	11862	15599204
35	4	3030-00-967-4898	20796	42-6919
35	5	2920-01-191-8442	11862	14067715
35	6	5307-01-159-6632	11862	14067718
35	7	5305-00-071-2066	96906	MS90728-109
35	8	5310-01-147-9792	11862	3954735
35	9	2920-01-147-8562	11862	14067721
35	10	5310-01-250-7679	11863	11506101
35	11	3030-00-357-5506	20796	42-6921
35	12	5307-01-150-7764	11862	1623159
36	1	2920-01-163-7211	11862	1987808
36	2	2920-01-093-9216	16764	1876806
36	3	5999-01-104-0445	16764	1970227
36	4	2920-01-163-7210	11862	1852519
36	5	2920-01-164-7091	11862	1987809
36	6	2920-01-267-9423	11862	1116423
36	7	5977-01-018-2742	16764	1876873
36	8	5360-01-167-1890	11862	1977357
36	9	5935-01-103-8700	16764	801815
36	10	2590-01-165-0747	11862	1986433
36	11	5305-01-081-0465	16764	1876681
36	12	5905-01-139-3620	11862	830478
36	13	2920-01-256-2253	11862	1892941
36	14	2920-01-192-3774	11862	10499310
36	15	5305-00-855-0960	11862	447164
36	16	5961-01-162-7228	11862	1977064
36	17	5310-00-934-9757	96906	MS35649-282
36	18	5305-01-162-3763	11862	1986552
36	19	5970-01-163-2895	11862	1970149
36	20	5910-01-089-1916	16764	1978146
36	21	5935-01-165-9807	11862	1986428
36	22	5961-01-059-0562	16764	1875645
36	23	2920-01-205-6018	11862	1986427

FIGURE AND ITEM NUMBER INDEX

FIG	ITEM	STOCK NUMBER	CAGEC	PART NUMBER
36	24	3120-01-167-5544	11862	1976143
36	25	3110-00-962-3263	16764	9436831
36	26	5306-01-100-5149	16764	801810
36	27	2590-01-085-1968	11862	1971993
36	28	5315-00-843-7986	96906	MS16562-33
36	29		11862	1976049
36	30	3110-01-165-2530	11862	1978059
36	31	4140-01-163-3543	11862	1978058
36	32	3020-01-163-0498	11862	1978068
36	33	5310-00-949-9299	16764	1941978
36	34	5310-00-596-5152	24617	1915172
36	35	3110-00-108-9247	30760	AH068-31
36	36	5305-00-471-0373	16764	800549
37	1	5310-00-516-2701	35510	5413
37	2	5310-00-429-3135	35510	4340
37	3	5310-01-143-1719	35510	31256
37	7	5970-01-144-1291	35510	7983
37	5	5305-01-143-7412	35510	99525
37	6	5310-00-775-5139	35510	2434
37	7	2920-01-145-0993	35510	95300
37	8	5305-01-143-7411	35510	99459
37	9	5305-01-143-6417	35510	13771
37	10	5970-00-160-4001	35510	73546
37	11	3120-01-082-9417	35510	79071
37	12	6130-01-142-8507	35510	71237
37	13	3120-01-082-9418	35510	79070
37	14	5970-00-160-4066	35510	73545
37	15	6130-01-142-8508	35510	71238
37	16	5305-01-105-6888	35510	79313
37	17	5305-00-451-1623	35510	73543
37	18	5310-01-023-8927	35510	52066
37	19	5310-00-781-4787	35510	2385
37	20	5305-00-993-1848	96906	MS35207-265
37	21	3120-01-082-6996	35510	73547
37	22	5306-01-078-2719	35510	78997
37	23	5330-01-145-0725	35510	99515
37	24	2920-01-144-6362	35510	99478
37	25	5310-01-143-1715	35510	79469
37	26	5310-01-106-9929	35510	2313
37	27	5310-00-429-3110	35510	31587
37	28		35510	95279
37	29	2920-01-152-7646	35510	99789
37	30	3110-01-109-3058	35510	79486
37	31	5330-01-111-6531	35510	79405
37	32	3110-01-136-5534	35510	78895
37	33	5310-00-451-0631	35510	59225
37	34	5310-00-432-3959	35510	76985
37	35	5315-00-616-5523	96906	MS35756-11
37	36	2920-01-082-6458	35510	79035
37	37	5305-01-143-6466	35510	95152
37	38	5310-01-136-5532	35510	79627

FIGURE AND ITEM NUMBER INDEX

FIG	ITEM	STOCK NUMBER	CAGEC	PART NUMBER
37	39	2920-01-141-1488	35510	78983
37	40	5305-00-450-0463	35510	58754
37	41	2920-01-267-2088	35510	79517
37	42	5330-01-106-4329	35510	79404
37	43	3110-01-136-2093	35510	79406
37	44	5330-01-106-6101	35510	79403
37	45	3120-01-082-9419	35510	77916
37	46	5977-01-143-6955	35510	99508
37	47	5305-00-889-3001	96906	MS35206-231
37	48	5330-01-145-5376	35510	99514
37	49	5977-01-142-9122	35510	99516
37	50		35510	2435
37	51	5305-01-143-6467	35510	99524
37	52	5310-01-143-1679	35510	36912
37	53	5977-01-142-9121	35510	99517
37	54	5330-01-145-0724	35510	99518
37	55	5977-01-143-6996	35510	99479
37	56	5310-01-136-5547	35510	74732
37	57	5330-01-145-0706	35510	99511
37	58	5340-01-144-6245	35510	99512
37	59	5305-01-143-6416	35510	75348
37	60	5310-00-429-2686	35510	73009
37	61	5310-00-032-1814	35510	3231
37	62	5310-01-024-3109	35510	2364
37	63	5310-00-775-5182	35510	2523
37	64	5310-00-775-5180	35510	3395
37	65	5940-01-142-8411	35510	99513
37	66	5310-00-429-3156	35510	59982
37	67	5310-00-616-1739	35510	2771
37	68	5935-01-145-1974	35510	99481
38	1	5305-00-984-6191	96906	MS35206-243
38	2	5945-01-199-4431	55156	40178
38	3	5305-01-161-2583	24617	9419320
38	4	5340-01-163-1343	11862	14072428
38	5	5310-01-069-5243	72582	271163
38	6	5310-01-160-4536	11862	3792287
38	7	5306-01-158-9018	11862	2014469
38	8	6110-01-167-1822	11862	14072341
38	9	5961-01-188-8315	11862	15599251
38	10	5306-01-197-3089	11862	9440033
38	11	6110-01-191-0127	11862	15599949
38	12	5340-01-150-4105	11862	14005953
39	1	2920-01-192-3020	11862	14066657
39	2	5305-01-197-3287	24617	9419663
39	3	5945-01-192-7985	11862	15591718
39	4	5306-01-165-5583	11862	14060613
39	5	5340-01-150-1545	11862	14028931
39	6	5310-00-013-1245	21450	131245
39	7	5310-00-809-4058	96906	MS27183-10
39	8	2920-01-157-3765	16764	1113591
39	9	5306-01-227-9085	11862	15544950

FIGURE AND ITEM NUMBER INDEX

FIG	ITEM	STOCK NUMBER	CAGEC	PART NUMBER
39	10	5310-01-250-3301	11862	22521054
39	11	5365-01-193-0458	11862	23500396
40	1	5945-01-165-4602	11862	1114373
40	2	5305-00-543-4302	96906	MS35265-79
40	3	5310-00-682-5930	96906	MS35340-44
40	4	5360-00-947-1184	24617	1958679
40	5	2920-00-841-3254	11862	1941113
40	6	2520-01-110-1081	16764	1970263
40	7	2920-01-165-6794	11862	1976882
40	8	3120-01-121-2809	16764	1894023
40	9	2520-00-997-9818	11862	1951567
40	10	5315-01-119-3115	11862	1987049
40	11	5315-01-165-2568	11862	1945804
40	12	2920-01-206-0891	16764	10496538
40	13	5365-00-803-7313	96906	MS16624-1031
40	14	5310-00-606-6608	24617	1928023
40	15	5365-00-804-9666	24617	1928022
40	16	3120-00-888-6630	16764	1928021
40	17	2920-01-163-8626	16734	1893445
40	18	5310-00-291-4619	24617	821453
40	19	5306-00-407-3737	16764	9427815
40	20	2920-01-200-8461	11862	1986473
40	21	2920-01-163-7872	16764	1887021
40	22	5315-01-166-1733	11862	1976940
40	23	5977-01-163-2032	11862	1876458
40	24	5977-01-163-2930	11862	1876359
40	25	5340-01-165-0539	11862	1986019
40	26	5305-00-984-6210	96906	MS35206-263
40	27	5977-01-163-2931	11862	800091
40	28	5977-01-163-2900	11862	1852890
40	29	5305-01-163-5512	11862	431615
40	30	5305-00-855-3598	24617	9414224
40	31	2590-01-166-1071	11862	1987254
40	32	5310-00-934-9758	96906	MS35649-202
40	33	5310-00-721-7809	96906	MS35340-43
40	34	5315-00-945-8441	16764	1966923
40	35	5306-01-081-0464	16764	1960908
40	36	5307-01-164-4538	11862	1975326
40	37	2920-01-224-3153	11862	1986464
40	38	5325-00-397-5962	16764	1955946
40	39	5340-01-166-5639	11862	1876358
40	40	5310-01-238-2983	16764	1984076
40	41	2920-01-197-7229	11862	1986471
40	42	5305-00-450-5937	16764	1968396
41	1	8040-01-188-2953	11862	3999572
41	2	2590-01-164-9326	11862	12039309
41	3	5945-01-189-3307	11862	12033862
41	4	6130-01-192-1643	11862	12006377
41	5	6680-01-225-4432	11862	15591138
41	6	2540-01-160-0849	11862	329842
41	7	5305-00-146-2524	96906	MS51850-86

FIGURE AND ITEM NUMBER INDEX

FIG	ITEM	STOCK NUMBER	CAGEC	PART NUMBER
41	8	5305-00-146-2524	96906	MS51850-86
41	9	6150-01-158-3861	11862	459454
41	10	5340-01-149-4434	11862	3816659
41	11	5305-01-197-2535	11862	9428613
41	12	5305-01-160-1975	11862	11508566
41	13	2590-01-161-1328	11862	12039310
41	14	5945-01-189-3307	11862	12033862
41	15	6130-01-192-1643	11862	12006377
41	16	5305-01-269-4329	11862	11513932
41	17	6240-00-850-4280	08806	1003
42	1	5310-01-154-2273	96906	MS90724-34
42	2	5305-01-269-4329	11862	11513932
42	3	2540-01-158-8626	11862	25052807
42	4	5999-01-158-9249	11862	25053623
42	5	5330-01-164-1653	11862	25015099
42	6	6220-01-164-5228	11862	25022883
42	7	6680-01-160-3870	70040	6433429
42	8	6695-01-167-8108	11862	25017376
42	9	6220-01-164-5227	11862	6497476
42	10	5330-01-163-1992	11862	6497483
42	11	6210-01-158-3857	11862	6497475
42	12	6210-01-158-4668	11862	25022884
42	13	6210-01-158-6575	11862	25053501
42	14	6210-01-161-2138	11862	25053500
42	15	6210-01-271-6871	11862	25076586
42	16	2540-01-163-7281	11862	25053622
42	17	5999-01-159-5603	11862	8986000
42	18	6240-00-944-1264	08806	194
42	19	6250-00-433-5946	77060	2973932
42	20	6240-00-944-1264	08806	194
42	21	6240-00-144-4693	08806	168
43	1	6240-00-144-4693	08806	168
43	2	6210-01-158-0396	11862	14066662
43	3	5340-01-162-3627	11862	14072406
43	3	5340-01-162-3628	11862	14072409
43	4	5320-01-195-5106	11862	9441669
43	5	6240-00-144-4693	08806	168
43	6	6620-01-156-0712	11862	14075858
43	7	6625-01-162-8124	70040	6474942A
43	8		08806	1445
43	9	5895-01-165-6792	11862	14072410
43	10	2920-01-152-2414	11862	14072448
43	11	5930-01-158-4428	11862	14072338
43	12	5305-01-164-2313	24617	9415163
43	13	5930-01-163-1439	11862	14072412
44	1	5940-00-890-2831	96906	MS18029-24
44	2		96906	MS18029-13L-5
44	3	5310-00-252-8748	96906	MS35650-3314
44	4	5310-00-514-6674	96906	MS35335-34
44	5	5910-01-189-5152	11862	15599225
44	6	6150-01-159-6901	11862	14072336

FIGURE AND ITEM NUMBER INDEX

FIG	ITEM	STOCK NUMBER	CAGEC	PART NUMBER
44	7	5310-00-934-9747	96906	MS35649-262
44	8		96906	MS27212-4-5
44	9	5340-01-164-8171	11862	14075900
44	10	5305-00-227-1543	96906	MS51849-33
44	11	5340-01-150-4105	11862	14005953
44	12	5306-01-161-6178	11862	9437702
44	13		96906	MS27212-4-3
44	14		11862	14072337
44	15		96906	MS18029-13S-3
45	1		80063	SC-B-75180-IV
45	2		80063	SC-D-691391
45	3	5305-00-068-0508	96906	MS90728-6
45	4	5310-00-209-0786	96906	MS35335-33
45	5	5305-01-156-5438	11862	11504447
45	6	5340-01-159-1321	11862	2043151
45	7	5325-01-164-6431		343124
45	8	5310-00-761-6882	96906	MS51967-2
45	9	5310-00-582-5965	96906	MS35338-44
45	10	5340-00-764-7052	96906	MS21333-116
45	11	5325-01-199-3461	00613	C-5139-2
45	12	5305-00-984-6214	96906	MS35206-267
45	13	5935-01-223-9420	96906	MS25043-32DA
45	14	5935-01-171-8273	19207	8701347
45	15	6150-01-155-6522	11862	12039254
46	1	5305-00-227-1543	96906	MS51849-33
46	2	5310-00-082-1404	96906	MS27183-6
46	3	5340-01-228-1659	11862	15599989
46	4		96906	MS27212-4-5
46	5	5310-00-934-9747	96906	MS35649-262
46	6	6150-01-159-6901	11862	14072336
46	7	5310-00-514-6674	96906	MS35335-34
46	8	5310-00-252-8748	96906	MS35650-3314
46	9		96906	MS18029-13L-5
46	10	5940-00-890-2831	96906	MS18029-24
46	11	5306-01-161-6178	11862	9437702
46	12	5340-01-150-4105	11862	14005953
46	13	5340-01-159-1321	11862	2043151
46	14	5306-01-230-3354	11862	11508534
46	15	5310-01-158-6260	24617	11506003
46	16	5320-01-229-8183	11862	9436175
46	17	5320-01-231-3889	79846	ABA64LBA
46	18	5995-01-225-2534	11862	12044586
47	1	2920-01-192-4375	77060	12039297
47	2	5306-01-157-6797	11862	9439770
47	3	2540-01-194-6875	11862	14076848
47	4	5905-01-193-7212	11862	14076847
47	5	5310-01-069-5243	11862	271163
47	6	5310-01-152-0598	24617	271172
47	7	5305-00-115-9934	96906	MS51849-55
47	8	5340-01-150-4105	11862	14005953
47	9	2920-01-192-4376	77060	12039298

FIGURE AND ITEM NUMBER INDEX

FIG	ITEM	STOCK NUMBER	CAGEC	PART NUMBER
47	10	5310-00-209-0786	96906	MS35335-33
47	11	5306-01-161-6178	11862	9437702
48	1	5310-00-880-7746	96906	MS51968-5
48	2	5310-00-934-9764	96906	MS35649-205
48	3	5305-01-197-2536	11862	9440166
48	4	5310-00-582-5965	96906	MS35338-44
48	5	5945-01-192-8653	72055	98226
48	6	5310-00-167-0721	96906	MS35333-41
49	1	2920-01-150-1610	11862	22514861
49	2	5930-01-214-0401	11862	26019661
49	3	5310-00-934-9758	96906	MS35649-202
49	4	5305-01-148-7465	11862	7806433
49	5	2920-01-159-3589	11862	1990115
49	6	5307-01-203-9081	11862	7846418
49	7	5355-01-150-1541	11862	22507977
49	8	5360-01-159-5952	11862	556743
49	9	5930-01-162-3669	11862	556742
49	10	5305-01-160-3945	11862	1640902
49	11	5930-01-163-7282	11862	1242101
49	12	5305-00-146-2524	96906	MS51850-86
49	13	2540-01-196-1622	11862	6258213
49	14	5320-01-160-3999	11862	2477054
49	15	5930-01-014-0187	11862	1995217
49	16	5355-01-163-4940	11862	469302
49	17	5930-01-159-0925	11862	14072413
49	18	5930-00-998-9211	11862	1261219
49	19	5340-01-168-1501	11862	1361699
49	20	5305-00-068-0510	96906	MS90728-60
49	21	5310-00-637-9541	96906	MS35338-46
49	22	5310-00-732-0558	96906	MS51967-8
49	23	2590-01-159-8766	11862	14000395
49	23	2590-01-160-5841	11862	14040525
49	24	5305-01-149-7356	11862	9418944
49	25	5930-01-149-9306	11862	477361
49	26	5920-01-123-5212	11862	12004005
49	27	5920-01-123-5211	11862	12004007
49	28	5920-01-085-0825	11862	12004009
49	29	5945-00-983-4374	94988	552-12V
49	30	5920-01-149-6953	94988	12004010
49	31	5945-00-992-5415	94988	224-12V
49	32	5920-01-188-6294	11862	12004011
49	33	5920-01-149-6952	11862	12004008
49	34	5920-01-188-6294	11862	12004011
49	35	5920-01-085-0825	11862	12004009
49	36	5930-01-162-0803	11862	14072358
49	37	5930-01-158-4808	11862	14072339
50	1	6220-01-248-6269	11862	14072333
50	2	6220-01-107-2613	5A910	D08218
50	3	5330-01-076-6172	34904	DC8226
50	4	6240-00-617-0991	96906	MS35478-1073
50	5	5330-01-037-0663	5A910	DC8211

FIGURE AND ITEM NUMBER INDEX

FIG	ITEM	STOCK NUMBER	CAGEC	PART NUMBER
50	6	5310-01-076-6196	5A910	DC8228
50	7	5310-00-350-2655	19207	5294507
50	8	5310-00-637-9541	96906	MS35338-46
50	9	5310-00-732-0558	96906	MS51967-8
50	10	5306-01-168-4481	11862	9440334
50	11	5340-01-150-4105	11862	14005953
50	12	5340-01-162-6061	11862	14072431
50	13	5305-00-483-0554	96906	MS51862-26
50	14	5305-01-151-8288	11862	9438150
50	15	5310-01-165-0464	11862	362379
50	16	5310-01-154-2273	96906	MS90724-34
50	17	6240-00-944-1264	08806	194
50	18	6220-01-160-4247	11862	915449
50	18	6220-01-161-6439	11862	915450
50	19	5305-01-161-3997	11862	15605040
50	20	5325-01-157-1698	11862	347347
50	21	5310-01-069-5243	72582	271163
50	22	5310-01-158-6260	24617	11506003
50	23	5340-01-159-5762	11862	14072421
50	24	6220-01-156-4475	11862	15559316
50	25	5306-01-162-8525	24617	11503778
50	26	6240-01-157-0635	08806	2057NA
50	27	6220-01-155-6521	11862	915908
50	28	5305-01-132-2166	96906	MS51871-4
50	29	5305-01-162-5995	11862	11504656
50	30	6220-01-160-4254	11862	14043873
50	30	6220-01-161-5016	11862	14043874
50	31	6220-01-155-6515	11862	16500591
50	32	5305-01-150-5785	11862	5966249
50	33	2510-01-159-8765	11862	16501759
50	34	6240-01-180-9022	08806	H6054
50	35	6220-01-164-2271	11862	5968095
50	36	5360-01-149-6309	11862	459461
51	1	6220-01-146-4455	11862	5965775
51	1	6220-01-157-9046	11862	5965776
51	2	5330-01-162-3744	11862	5965748
51	3	5305-00-432-7953	96906	MS51861-38
51	4	2510-01-096-6758	11862	5965772
51	4	6220-01-146-4469	11862	5965771
51	5	6240-01-157-0636	08806	2057
51	6	5310-01-154-2273	96906	MS90724-34
51	7	6240-00-924-7526	08806	1156
51	8	6240-00-144-4693	08806	168
51	9	5310-01-158-6260	24617	11506003
51	10	5310-01-069-5243	72582	271163
51	11	6220-01-156-4476	11862	15559312
51		5306-01-162-8525	24617	11503778
51	13	5340-01-159-5765	11862	14072433
51	13	5340-01-162-9782	11862	14072434
51	14	5305-01-163-6466	11862	11504655
52	1	5310-00-761-6882	96906	MS51967-2

FIGURE AND ITEM NUMBER INDEX

FIG	ITEM	STOCK NUMBER	CAGEC	PART NUMBER
52	2	5310-00-209-0786	96906	MS35335-33
52	3	5306-01-197-3089	11862	9440033
52	4	6220-01-327-3252	11862	330492
52	5	6240-00-144-4693	08806	168
52	6	6220-01-306-4265	11862	339885
52	6	6220-01-327-1025	11862	339887
52	7	5305-01-163-6466	11862	11504655
52	8	5310-01-154-2273	96906	MS90724-34
53	1	5310-00-761-6882	96906	MS51967-2
53	2	5310-00-209-0786	96906	MS35335-33
53	3	5305-00-068-0510	96906	MS90728-60
53	4	5310-00-959-4675	96906	MS35340-46
53	5	5310-00-732-0558	96906	MS51967-8
53	6	5340-01-165-3417	11862	370873
53	7	6220-01-156-8247	11862	370868
53	7	6220-01-156-8420	11862	370867
53	8	5305-01-162-9695	11862	11504736
53	9	6220-01-242-7557	11862	475922
53	10	6240-00-924-7526	08806	1156
53	11	6240-00-889-1799	08806	1157
53	12	6220-01-156-4476	11862	15559312
53	13	5340-01-162-6077	11862	14072481
53	14	5310-01-069-5243	72582	271163
53	15	6250-01-155-6547	11862	370874
54	1	6685-01-192-4834	11862	25037177
54	2	5930-01-157-4060	11862	14071047
54	3	4730-01-160-1505	11862	14040817
54	4	5930-00-073-0390	11862	3815936
55	1	2590-01-085-6956	11862	1892163
55	2	5305-01-164-6319	11862	11501812
55	3	5325-01-199-3461	00613	C-5139-2
55	4	5910-01-190-4600	11862	15599224
55	5	5306-01-161-6178	11862	9437702
55	6	5945-01-243-1702	11862	25523703
56	1	2990-01-257-1569	11862	15599900
56	2	6160-01-165-4638	11862	14076856
56	3	5306-01-168-4481	11862	9440334
56	4	5310-01-158-6260	24617	11506003
56	5	6140-01-156-5326	11862	14075896
56	6	6160-01-165-4637	11862	14076857
56	7	5307-01-158-9932	11862	14075894
56	8	6140-01-190-2517	11862	15599902
56	9	5320-00-994-7076	96906	MS20613-4P4
56	10	5340-01-150-4105	11862	14005953
56	11	6140-01-031-6882	81343	31-620
56	11	6140-01-210-1964	96906	MS52149-1
56	12	6140-01-155-6998	11862	14075389
56	13	5310-01-197-5499	11862	1359887
56	14	5306-01-158-9018	11862	2014469
56	15	6140-01-155-6997	11862	14075388
56	16	6140-01-190-2516	11862	15599901

FIGURE AND ITEM NUMBER INDEX

FIG	ITEM	STOCK NUMBER	CAGEC	PART NUMBER
57	1	6140-01-163-1081	11862	12039293
57	2		11862	FLW-12
57	3	5310-01-152-0598	24617	271172
57	4	5340-01-159-1321	11862	2043151
57	5	5310-00-823-8804	96906	MS27183-9
57	6	5305-00-068-0508	96906	MS90728-6
57	7	5340-01-150-4105	11862	14005953
57	8	5340-01-149-4434	11862	3816659
57	9	5310-00-514-6674	96906	MS35335-34
57	10	5305-01-195-5807	11862	11509371
57	11	5340-01-150-4105	11862	14005953
57	12	5310-00-550-3503	96906	MS35335-36
57	13	6150-01-154-1381	11862	12039271
57	14	6140-01-223-9144	11862	6287160
57	15	6140-01-155-6531	58499	6258
57	16	6150-01-165-0168	11862	12039272
57	17	5310-00-809-4058	96906	MS27183-10
57	18	5310-01-159-6586	11862	11504108
57	19	5305-01-156-5438	11862	11504447
57	20	5340-01-150-4992	11862	14040813
57	21	5310-01-158-6257	11862	11503739
57	22	5935-01-059-0117	19207	11674728
57	23	5340-01-059-0114	19207	11675004
57	24	5935-01-044-8382	19207	11682345
57	25	5305-00-269-3233	96906	MS90727-57
57	26	5310-00-637-9541	96906	MS35338-46
57	27	5970-01-044-8391	19207	11674730
57	28	5330-01-059-4286	19207	11674729
57	29	5310-01-170-9100	11862	3982098
57	30	5340-01-162-4774	11862	14072432
57	31	5306-01-168-4481	11862	9440334
57	32	5310-01-164-2338	11862	11508446
57	33	5340-01-150-4105	11862	14005953
57	34	5325-01-159-2843	11862	3979756
57	35	5305-01-161-3995	11862	11508858
57	36	6150-01-163-1385	11862	12039257
57	37	6150-01-163-1384	11862	12039267
57	38		11862	8919163
57	39	5940-00-549-6581	96906	MS75004-1
57	39	5940-00-549-6583	96906	MS75C04-2
57	40	6140-01-155-6530	58433	6262
58	1	5310-00-514-6674	96906	MS35335-34
58	2	5310-00-959-4679	96906	MS35340-45
58	3	5306-00-226-4826	96906	MS90728-33
58	4	5305-01-156-5438	11862	11504447
58	5	2590-01-158-6912	11862	12039201
58	6	6250-01-201-3300	11862	12013813
58	7	5340-01-165-2526	11862	469339
58	8	5305-01-160-3938	24617	9416187
58	9	2590-01-160-1496	11862	12001184
59	1	2590-01-159-8799	11862	12039434

FIGURE AND ITEM NUMBER INDEX

FIG	ITEM	STOCK NUMBER	CAGEC	PART NUMBER
59	2	5310-00-514-6674	96906	MS35335-34
59	3	5305-01-156-5438	11862	11504447
59	4	5325-01-164-8652	11862	3661804
59	5	5340-01-163-0919	11862	9785074
59	6	6680-01-160-5256	11862	334963
59	7	5305-01-157-9720	24617	9419327
59	8	5340-00-809-1494	96906	MS21333-105
59	9	5310-00-934-9758	96906	MS35649-202
60	1	2590-01-160-5885	11862	12039206
60	2	5340-01-149-4434	11862	3816659
60	3	5305-01-156-5438	11862	11504447
60	4	5340-00-282-7509	96906	MS21333-62
60	5	5340-00-057-3052	96906	MS21333-114
60	6	5340-01-158-0303	11862	3866187
60	7		11862	8917605
60	8	5310-00-896-0903	96906	MS51967-12
60	9	5340-01-163-0919	11862	9785074
60	10	5310-00-514-6674	96906	MS35335-34
60	11	2590-01-191-8322	11862	8906127
61	1	5305-00-146-2524	96906	MS51850-86
61	2	5310-00-209-0786	96906	MS35335-33
61	3	6250-01-189-4981	11862	6298886
61			11862	8914822
61	5	6250-01-189-6926	11862	8909518
61	6	2590-01-212-7639	11862	8906150
61	7	2590-01-155-7966	11862	12039205
61	8	2590-01-159-3505	11862	12039208
61	9	5935-00-773-1428	19207	7731428
61	10	5305-00-068-0508	96906	MS90728-6
61	11	5310-00-809-4058	96906	MS27183-10
61	12	5310-01-159-8559	96906	MS35691-406
62	1	6150-01-327-7394	11862	12096970
62	2	5310-00-761-6882	96906	MS51967-2
62	3	5310-00-209-0786	96906	MS35335-33
62	4	5306-01-197-3089	11862	9440033
62	5	5325-01-164-2377	11862	15591130
62	6	5975-01-324-7825	11862	329830
63	1	6240-00-944-1264	08806	194
63	2	5305-01-156-5438	11862	11504447
63	3	5310-00-514-6674	96906	MS35335-34
63	4	2590-01-191-6649	11862	12039204
63	5	2590-01-155-7967	11862	12039203
64	1	2590-01-159-8800	77060	12039255
64	2	5325-01-168-5677	11862	3918889
64	3	5340-00-057-3052	96906	MS21333-114
64	4	5305-01-156-5438	11862	11504447
65	1	5306-01-197-1492	24617	11508687
65	2	5310-01-158-6260	24617	11506003
65	3	5910-01-189-5110	11862	1988380
65	4	5310-00-696-5172	24617	271166
65	5	5310-01-152-0598	24617	271172

FIGURE AND ITEM NUMBER INDEX

FIG	ITEM	STOCK NUMBER	CAGEC		PART NUMBER
65	6	5305-01-231-1298	24617	1640810	
65	7	5940-01-145-7817	11862	3996270	
65	8	2590-01-160-1669	11862	12039311	
65	9		11862	FLW-18	
65	10	6130-01-192-1643	11862	12006377	
65	11		11862	FLW-20	
65	12		11862	FLW-16	
65	13		11862	FLW-12	
65	14	5306-01-161-6178	11862	9437702	
65	15	5340-01-150-4105	11862	14005953	
65	16	5910-01-189-5153	11862	15599222	
65	17	5305-01-161-2583	24617	9419320	
65	18	5305-01-197-2536	11862	9440166	
65	19	6150-01-156-6326	11862	12039253	
65	20	5340-01-150-4105	11862	14005953	
65	21	5310-00-550-3503	96906	MS35335-36	
65	22	5310-00-209-0786	96906	MS35335-33	
65	23	5310-00-934-9757	96906	MS35649-282	
66	1	5995-01-235-6877	11862	12044637	
66	2		11862	FLW-18	
66	3	6130-01-192-1643	11862	12006377	
66	4		11862	FLW-16	
66	5		11862	FLW-20	
66	6		11862	FLW-12	
66	7	5306-01-149-4398	11862	11502656	
66	8	5310-00-407-9566	96906	MS35338-45	
66	9	6150-01-156-6326	11862	12039253	
66	10	5995-01-235-6878	11862	12044638	
66	11	5340-01-149-4434	11862	3816659	
67	1	5995-01-199-1579	11862	12039308	
67	2			12034592	12034592
68	1	6240-00-944-1264	08806	194	
68	2	5910-01-189-5109	08806	15599223	
68	3	5340-01-149-9732	11862	6270410	
68	4	2590-01-158-6913	11862	12031346	
68	5	5325-01-164-8651	11862	3655180	
68	6	2590-01-209-0934	11862	12039305	
68	6	5995-01-155-7387	11862	12039304	
68	7	5340-01-160-4592	11862	14072340	
68	8	5340-01-160-4597	11862	14074480	
68	9	5305-00-984-6193	96906	MS35206-245	
68	10	5306-01-246-7459	11862	1244067	
68	11	5310-01-069-5243	72582	271163	
68	12	2590-01-210-1357	11862	12039464	
68	13	6350-01-321-7005	11862	22529441	
68	14	6350-01-158-7035	11862	1253637	
69	1	2520-01-201-4097	11862	6271387	
69	2	2520-01-164-1855	11862	1394293	
69	3	5325-01-168-1886	11862	6271391	
69	4	5315-01-156-6562	11862	1244707	
69	5	2540-01-147-5537	11862	25078571	

FIGURE AND ITEM NUMBER INDEX

FIG	ITEM	STOCK NUMBER	CAGEC	PART NUMBER
69	6	5360-01-193-7130	11862	8985418
69	7	5355-01-251-0633	70040	25023641
69	8	2540-01-191-6510	11862	25078578
69	9	5305-00-146-2524	96906	MS51850-86
69	10	5306-00-226-4825	96906	MS90728-32
69	11	5340-01-154-2107	11862	6271386
69	12	5310-00-931-8167	96906	MS51967-6
69	13	5310-00-407-9566	96906	MS35338-45
69	14	3040-01-159-8803	11862	9795470
69	15	5340-01-150-4105	11862	14005953
69	16	5360-01-165-1564	11862	15538092
69	17	5310-00-809-4061	96906	MS27183-15
69	18	5310-01-165-7572	11862	1377083
69	19	3040-01-165-0188	11862	1385607
69	20	5310-01-164-1647	11862	334533
69	21	5306-01-165-3258	11862	1381477
69	22	2520-01-165-4935	11862	6271394
70	1	5310-01-165-3331	11862	22521550
70	2	5340-01-166-1470	11862	2044779
70	3		11862	15599235
70	4	5340-01-163-1973	11862	2043150
70	5		11862	326560
70	6	2520-00-557-6619	73342	8627650
70	7	5306-00-226-4824	96906	MS90728-31
70	8	2520-01-158-1760	11862	359917
70	9	5330-01-096-7699	11862	10054241
70	10	4820-01-147-4294	11862	8623983
71	1	5306-01-149-6280	11862	1635490
71	2	5340-00-697-4703	11862	3705444
71	3	5365-01-154-4366	11862	14009322
71	4	5340-00-892-7413	11862	14009321
71	5	5305-01-157-0880	11862	14009324
71	6	2520-01-146-5482	11862	8655111
71	7	5305-00-688-2111	96906	MS90728-63
71	8	5340-01-164-1743	11862	15522022
71	9	5305-00-071-1784	96906	MS90728-83
72	1	5305-01-186-5381	11862	11508581
72	2	2520-01-164-7234	11862	8656942
72	3	5305-00-557-6612	11862	8622361
72	4	5330-01-148-7492	11862	8655625
72	5	2520-01-159-7757	11862	3787240
72	6	2520-01-149-3461	11862	8633203
72	6	2520-01-213-1680	11862	8655020
72	7	5306-01-171-8076	11862	9440224
72	8	5330-01-147-4208	11862	1259475
72	9	4710-01-148-4989	11862	14045642
72	10	6680-01-147-6583	11862	334532
73	1	3040-01-149-6706	11862	8624772
73	2		11862	8627592
73	3	2520-01-169-7674	11862	8624908
73	4	5330-01-138-5190	23862	8623917

FIGURE AND ITEM NUMBER INDEX

FIG	ITEM	STOCK NUMBER	CAGEC	PART NUMBER
73	5	2520-01-163-7866	11862	8627990
73	6	5360-01-150-6091	11862	8624101
73	7	3110-00-005-0873	73342	8623104
73	8	5365-01-171-3392	73342	8623105
73	9	2520-01-150-2279	11862	8623851
73	10	2520-01-150-3932	11862	8655619
73	11	2520-01-150-3931	11862	8623849
73	12	2520-01-154-5463	11862	8625997
73	13	3120-01-156-8763	11862	8624196
73	14	3020-01-149-5049	11862	8625717
73	15	2520-00-163-0709	73342	8625718
73	16	5365-00-007-3052	73342	8623112
73	17		11862	8626848
73	18	5365-00-007-3052	73342	8623112
73	19	2520-00-008-9987	73342	8623122
73	20	2520-01-150-3932	11862	8655619
73	21	2520-00-172-1947	11862	8625197
73	22	5365-01-171-3392	73342	8623105
73	23	3110-00-005-0873	73342	8623104
73	24	5360-01-150-6091	11862	8624101
73	25	2520-01-163-7213	11862	8627989
73	26	5330-01-138-5190	23862	8623917
73	27	2520-01-151-3569	11862	8627387
73	28	2520-01-150-3692	11862	8625814
73	29	3110-01-155-4438	11862	8629487
73	30	2520-01-149-1868	11862	8627334
73	31	5365-01-150-6092	11862	8623120
73	32	2520-01-154-5463	11862	8625997
73	33	3120-00-255-5697	11862	8624781
73	34	2520-01-212-6642	11862	12300852
73	35	5365-01-150-7830	11862	8623149
73	36	2520-01-175-6492	11862	8623152
73	37	2520-01-150-7609	11862	8623151
73	38	2510-01-085-0908	11862	8624199
73	39	2520-01-168-2060	72590	8655621
74	1	5365-01-160-4693	11862	8623437
74	2		11862	8633075
74	3	5330-01-165-4333	11862	8626356
74	4	5365-01-158-2182	11862	8623153
74	5	2520-01-150-2280	11862	8623146
74	6	5360-01-153-0933	11862	8623145
74	7	2520-01-149-3809	11862	8627385
74	8	2520-01-158-6854	11862	8626938
74	9	2520-01-151-3571	11862	8623918
74	10	3120-01-174-8153	11862	8623942
74	11	2520-01-151-3570	11862	8625990
74	12	5365-01-154-8562	11862	8626816
74	13	3120-01-155-4468	11862	8624783
74	14	2520-01-149-1221	11862	8623842
74	15	3040-01-167-2836	11862	8625736
74	16	3110-01-167-2443	11862	8623921

FIGURE AND ITEM NUMBER INDEX

FIG	ITEM	STOCK NUMBER	CAGEC	PART NUMBER
74	17	3020-01-149-7938	11862	8623202
74	18	3110-01-169-0734	11862	8623922
74	19	5365-01-174-8626	11862	8625221
74	20	2520-01-168-1983	72590	8633059
74	21	3120-01-158-6304	11862	8653985
74	22	5365-01-168-5729	11862	8626112
74	23	3120-01-154-7174	11862	8626372
74	24	2520-01-176-2839	11862	8655059
74	25	5365-01-150-4982	11862	8626173
74	26	2520-01-164-7156	11862	8625913
74	27	3120-01-156-5189	11862	8623944
74	28	3040-01-150-0407	11862	8626809
74	29	3120-01-167-4172	11862	8623920
74	30	3020-01-149-7941	11862	8626807
74	31	2520-01-149-7861	11862	8675627
74	32	5365-01-154-8561	11862	8627657
74	33	2520-01-164-7156	11862	8625913
74	34	3120-01-154-7174	11862	8626372
74	35	3120-01-154-4369	11862	8625403
74	35	3120-01-154-8516	11862	8625401
74	35	3120-01-154-8517	11862	8625406
74	35	3120-01-159-5773	11862	8625402
74	35	3120-01-161-4033	11862	8625404
74	35	3120-01-162-5787	11862	8625405
74	36	2520-01-159-2732	11862	8633257
75	1	4710-01-151-3662	11862	8633408
75	2	5340-01-165-1586	11862	8623463
75	3	5330-01-096-7699	11862	10054241
75	4	5935-01-150-6319	11862	8629503
75	5		11862	8627996
75	6	3120-01-166-3677	11862	8623941
75	7	5340-01-165-1585	11862	8611710
76	1	5310-00-732-0559	96906	MS51968-8
76	2	2520-01-149-7962	11862	8624336
76	3	2520-01-165-9563	11862	8625773
76	4	5360-01-150-4963	11862	8633096
76	5	2530-01-150-3693	11862	8624091
76	6	5340-01-162-5883	24617	451786
76	7	3040-01-154-3799	11862	8633060
76	8	2520-01-159-5887	11862	8624976
76	9	5340-01-150-4104	11862	8620318
76	10	5360-01-150-0196	11862	8623041
76	11	5340-01-147-2266	11862	8624112
76	12	5306-00-226-4824	96906	MS90728-31
76	13	5330-01-251-1607	11862	8657163
76	14	2520-01-164-7158	11862	8626979
76	15	5315-01-152-9029	73342	23017556
77	1	2520-01-151-3857	11862	8629227
77	2	5315-01-152-9029	73342	23017556
77	3	5315-00-814-3530	96906	MS16562-35
77	4	5306-00-226-4824	96906	MS90728-31

FIGURE AND ITEM NUMBER INDEX

FIG	ITEM	STOCK NUMBER	CAGEC	PART NUMBER
77	5	2520-01-127-3969	73342	8623262
77	6	5330-00-001-1984	11862	8623263
77	7	5306-01-150-4835	11862	GN8670757
77	8	3110-00-900-2560	96906	MS19061-20007
77	9	5330-01-151-6106	11862	8670393
77	10	2520-01-154-1185	11862	8629796
77	11	5330-01-152-5942	11862	8623561
77	12	2520-01-163-2708	11862	8633296
77	13	4710-01-151-3663	11862	8623157
77	14	2520-01-149-7935	11862	8627509
77	15	5305-00-068-0501	96906	MS90725-5
77	16	5945-01-268-4265	11862	8629935
77	17	5330-01-151-8364	11862	8623741
77	18	5306-01-269-6138	11862	9430895
77	19	5306-01-150-8713	11862	9440338
77	20	5340-01-149-7811	11862	8624136
77	21	5330-01-150-5928	11862	8623174
77	22		11862	8624145
77	23	5330-01-150-6239	11862	8623430
77	24	3040-01-151-5663	11862	8624138
77	25	5365-00-682-1762	96906	MS16633-1031
77	26	5315-01-152-9038	11862	8624139
77	26	5315-01-152-9039	11862	8624140
77	26	5315-01-152-9040	11862	8624141
77	27	5340-01-154-6559	11862	8623664
77	28	5360-01-150-7829	11862	8623666
77	29	5310-01-150-5919	11862	8623744
77	30	2520-01-164-7157	11862	8626902
77	31	5330-01-174-8090	11862	8623671
77	32	2590-01-164-7053	11862	8627153
77	33	5360-01-150-6087	11862	8624256
77	34	5330-01-163-2614	11862	8623131
77	35	3040-01-149-6759	11862	8626878
77	36	5310-01-151-4137	11862	8626884
77	37	5315-01-152-9031	11862	8626879
77	38	5340-01-151-4964	11862	8626881
77	39	5360-01-150-6086	11862	8623489
78	1	4820-01-150-4964	11862	8623592
78	2	5315-01-152-9033	11862	8624482
78	3	5315-00-038-3059	73342	8623078
78	4	5315-01-154-2291	11862	8625324
78	5	4730-01-159-8723	11862	8623298
78	6	5315-01-163-6027	11862	8622045
78	7	5360-01-166-6856	11862	8623654
78	8	5330-01-151-9306	11862	8623653
78	9	2520-01-151-3784	11862	8629080
78	10	5365-00-721-7680	24617	9412978
78	11	4730-01-159-8724	11862	8625546
78	12	5315-01-152-9032	11862	8625544
78	13	2520-01-007-0804	73342	8623295
78	14	5315-01-151-1105	11862	8623507

FIGURE AND ITEM NUMBER INDEX

FIG	ITEM	STOCK NUMBER	CAGEC	PART NUMBER
79	1	5306-01-269-4319	11862	8655280
79	2	5310-01-150-5921	11862	8626281
79	3	3120-01-155-4462	11862	8623300
79	3	3120-01-155-4463	11862	8623301
79	3	3120-01-155-4464	11862	8623303
79	3	3120-01-155-4465	11862	8623304
79	3	3120-01-155-4466	11862	8623305
79	3	3120-01-155-4467	11862	8623306
79	3	3120-01-182-8417	11862	8623302
79	4	5330-01-165-4333	11862	8626356
79	5	5330-01-152-5941	11862	8623978
79	6	5365-01-153-0872	11862	8629523
79	7	5330-01-080-3253	73342	6771005
79	8	4710-01-152-5798	11862	8629526
79	9	5325-01-123-6798	11862	6437746
79	10	2940-01-121-6350	11862	6259423
79	11	5306-01-150-9497	11862	8633208
79	12	5330-01-025-4212	11862	8626916
79	13		11862	8627644
79	14	3120-01-156-5189	11862	8623944
79	15	5315-00-038-3059	73342	8623078
79	16	5340-01-149-1867	11862	8625955
79	17	5340-01-159-1788	11862	8623292
79	18	5306-00-226-4831	96906	MS90728-38
79	19	5306-00-226-4832	96906	MS90728-39
79	20	5306-00-226-4827	96906	MS90728-34
79	21	4820-01-163-2656	11862	8623422
79	22	5310-01-149-7793	11862	8623368
79	23	5310-01-152-4229	11862	8623039
79	24	5360-01-150-9549	11862	8625648
79	25		11862	8625650
79	26		11862	8626885
79	27	5365-00-514-1277	96906	MS16627-1087
79	28	5365-01-085-0910	11862	8629961
79	29	5330-00-001-1996	73342	8626309
80	1	4730-00-013-7398	72582	137398
80	2	4730-01-163-7163	11862	8637742
80	3	5340-01-167-9694	11862	3997718
80	4	5340-01-213-6934	11862	15517986
80	5		11862	14045626
80	6		11862	14045628
80	7	4730-00-844-5721	14569	1007-2
81	1	2520-01-146-5483	11862	14067765
81	2	5330-01-165-1357	11862	14020854
81	3	2520-01-147-4207	11862	14020861
81	4	5305-00-068-0511	96906	MS90728-62
81	5	5310-00-637-9541	96906	MS35338-46
81	6	5330-01-165-3388	11862	8623216
81	7	5305-00-068-0511	96906	MS90728-62
81	8	5310-00-080-6004	96906	MS27183-14
81	9	5310-00-316-6513	80205	NAS1408A6

FIGURE AND ITEM NUMBER INDEX

FIG	ITEM	STOCK NUMBER	CAGEC	PART NUMBER
81	10	5340-01-164-7021	11862	14067764
82	1	5330-01-165-3388	11862	8623216
82	2	5310-00-637-9541	96906	MS35338-46
82	3	5305-00-068-0511	96906	MS90728-62
82	4	2520-01-193-4053	11862	15599248
82	5	5305-00-068-0511	96906	MS90728-62
82	6	5310-00-627-6128	96906	MS35335-35
82	7	2520-01-183-8323	11862	14022218
82	8	5330-01-168-3870	73680	29940-0445
82	9	5330-01-126-1040	11862	14022219
82	10	2520-01-192-8282	11862	14029158
82	11	5310-00-080-6004	96906	MS27183-14
82	12	5310-00-316-6513	80205	NAS1408A6
82	13	5305-00-068-0511	96906	MS90728-62
83	1	2520-01-151-2690	11862	14037990
83	2	5330-01-096-6775	76680	9449-K
83	3	3120-01-152-5948	11862	3978765
83	4	3110-00-155-6152	21335	P207K
83	5	5306-01-152-4696	24617	11503428
83	6	3020-01-177-3055	11862	376176
83	7	2520-01-150-3823	11862	14075205
83	8	5330-01-196-2426	80204	22306
83	9	5306-01-152-4696	24617	11503428
83	10		11862	14037983
83	11	3110-00-647-1100	96906	MS17131-42
83	12	3020-01-155-6399	11862	14037997
83	13	3040-01-150-0409	11862	15594195
83	13	3040-01-150-0409	11862	15594195
83	14	2520-01-159-7757	11862	3787240
83	15	5310-01-154-3992	11862	14037984
83	16	5306-01-165-4283	24617	11500831
83	17	5365-01-149-9710	11862	14037987
83	18	5365-01-163-6137	72800	XAN-225-H
83	19	5315-00-044-3767	24617	443767
83	20	3120-01-197-1535	11862	15594196
83	20	5310-01-176-2466	11862	14075212
83	21	3020-01-155-5898	11862	14095623
83	22	5365-01-153-9591	11862	14037976
83	23	3010-01-150-3933	11862	14075211
83	24	3020-01-151-3967	11862	14075210
83	25	5365-00-571-6816	19207	8338687
83	26	3020-01-148-9548	11862	14095676
83	27	5365-01-166-6324	78553	XAN-262-H
83	28	3120-01-152-2613	11862	14095678
83	29	3120-01-155-3509	11862	14044937
83	30	3120-01-155-4470	11862	14095677
83	31	2520-01-150-0415	11862	14071876
83	32	3020-01-175-6477	11862	14037949
83	33	2520-01-151-3982	11862	14095679
83	34	2590-01-159-8757	11862	14095680
83	35	2520-01-162-8985	11862	14037966

FIGURE AND ITEM NUMBER INDEX

FIG	ITEM	STOCK NUMBER	CAGEC		PART NUMBER
83	36	5315-01-157-3515	11862		14037961
83	37	5340-01-162-8846	11862		14037967
83	38	5360-01-155-3652	11862		14037968
83	39	5340-01-170-4777	11862		14037962
83	40	2520-01-150-3934	11862		14037963
83	41	2520-01-155-5844	11862		14037965
83	42	2520-01-162-8985	11862		14037966
83	43	5315-01-160-0403	11862		14037964
83	44	3110-00-580-3843	60380		NTA2435
83	45	3110-00-227-4431	60380		B2012-OH
83	46	2520-01-172-0394	24617		9427468
83	47	3020-01-160-6516	1862		14037995
83	48	3110-01-155-2598	60380		FNT5070
83	49	3120-01-159-8588	11862		14071875
83	50	5330-01-155-4393	76760		99780
83	51	5930-01-149-9305	11862		14037986
83	52	5306-01-167-4346	11862		14075306
83	53		11862		14037947
83	54	5330-01-168-3870	73680		29940-0445
83	55	3110-01-119-4441	60380		F5020
83	56	3110-00-056-9377	96906		MS17131-47
83	57	5330-01-147-9698	0720		29940-0435
83	58	5315-01-160-4642	11862		14037948
83	59	5315-01-155-9932	11862		14037956
83	60	5360-01-155-4555	11862		3967886
83	61	5306-01-157-1985	11862		14095597
83	62	5310-01-148-0245	11862		14037991
83	63	5310-01-148-2687	97271		6259149
83	64	2520-01-147-2267	11862		14037992
83	65	2520-01-149-4993	11862		14037993
83	66	3040-01-151-6382	11862		14029126
83	67	5310-00-080-6004	96906		MS27183-14
83	68	5310-00-982-4908	96906		MS21045-6
83	69	2520-01-155-5746	11862		14037957
83	70	3040-01-151-7989	11862		14037959
83	71	5330-00-451-0118	02697		AFF583-116
83	72	5330-01-159-2811	11862		14037958
83	73	3120-01-155-8713	11862		14071882
83	74	3110-00-756-2022	60380		NTA2840
83	75	3110-00-055-2100	60380		TRA2840
83	76	3040-01-148-9432	11862		14037979
83	77	5365-01-157-6764	72800		XAN-250-H
83	78	3020-01-153-9435	55880		372380
83	79	3020-01-155-8038	11862		14037980
83	80	3110-00-198-0492	60380		01086Q
84	1	5310-00-498-2381	5310-00-498-2381	11862	
84	2	5310-01-195-5088	11862		2423517
84	3	5330-01-197-3229	11862		14022212
84	4	3010-01-194-0886	11862		460835
84	5	5306-01-162-9678	11862		9437242
84	6	5330-01-194-5804	11862		465482

FIGURE AND ITEM NUMBER INDEX

FIG	ITEM	STOCK NUMBER	CAGEC	PART NUMBER
84	7	5330-01-118-9504	11862	465483
84	8	3110-01-117-2436	43334	43307-P-S1568A
84	9	5330-01-196-6586	11862	14037928
84	10	5365-00-735-1129	11862	3975717
84	11	5330-00-737-3727	11862	3967885
84	12	5360-01-155-4555	11862	3967886
84	13	3110-00-185-6305	24617	453610
84	14	5315-01-197-0812	11862	3975716
84	15	3040-01-192-6024	11862	15594158
84	16	5365-01-194-5274	11862	471896
84	17	5310-01-197-1538	11862	3967895
84	18	3110-00-689-4076	60380	QAR29523
84	19	2520-01-194-6955	11862	14022216
84	20	2520-01-201-7851	11862	3975715
84	21	5340-01-197-3480	11862	14022210
84	22	3040-01-194-6853	11862	14022221
84	23	5315-00-839-5820	24617	120690
84	24	5315-01-195-5234	11862	9437415
84	25	2520-01-192-3718	11862	471727
84	26	2520-01-194-2145	11862	3967879
84	27	3020-01-192-4635	11862	3975705
84	28	3120-01-194-6495	11862	3979450
84	29	5340-01-192-4366	11862	6273213
84	30	5305-00-721-5492	96906	MS90728-57
84	31	3020-01-193-7825	11862	3967882
84	32	3120-01-199-1399	11862	3967877
84	33	5365-01-194-5273	11862	361119
84	34	3110-00-902-1690	60380	M28161X0H
84	35	3110-01-197-5495	11862	3995881
84	36	5330-01-126-1039	11862	6259083
84	37	5315-01-197-4845	60380	AR-33761
84	38	3020-01-194-1714	11862	3967884
84	39	5365-01-197-1481	11862	3967898
84	40	3110-01-197-6651	60038	2793-2720
84	41	5365-00-423-2811	76760	A85344
84	41	5365-00-734-2494	11862	3975700
84	41	5365-00-734-2495	11862	476781
84	41	5365-00-734-2496	11862	3975698
84	42	3040-01-192-1635	11862	3975703
84	43	5330-01-125-6132	11862	3975711
84	44	5340-00-734-2649	11862	3975710
84	45	5310-00-584-7889	96906	MS35338-53
84	46	5306-00-226-4825	96906	MS90728-32
84	47	3040-01-199-1485	11862	6259153
84	48	3110-00-929-8365	23862	9419284
84	49	5365-01-193-2948	23862	3967883
84	50	5310-01-197-1539	23862	3967881
84	51	3110-01-096-6755	11862	9428004
84	52	5330-01-127-2883	11862	3979453
84	53	5365-01-194-0879	11862	3709353
84	53	5365-01-194-5275	11862	3709351

FIGURE AND ITEM NUMBER INDEX

FIG	ITEM	STOCK NUMBER	CAGEC	PART NUMBER
84	53	5365-01-194-5276	11862	3709354
84	53	5365-01-194-5277	11862	3709352
84	54	5310-00-627-6128	96906	MS35335-35
84	55	5305-01-140-9118	96906	MS90728-59
84	56	2520-01-183-8322	11862	14022213
84	57	3120-01-152-5948	11862	3978765
84	58	5330-01-096-6775	76680	9449-K
84	59	3110-01-197-5496	11862	474071
84	60	3020-01-181-3795	11862	6273992
84	61	5330-01-141-7167	23862	6273214
84	62	5310-01-195-5092	11862	3967887
84	63	5310-01-194-7069	11862	3967888
84	64	5365-01-195-6203	11862	6259192
84	64	5365-01-195-6204	11862	6259193
84	64	5365-01-197-1524	11862	6259195
84	64	5365-01-205-5966	11862	6259194
84	65	5365-00-081-5194	11862	2600236
84	66	3110-00-155-6735	23862	9422715
84	67	3040-01-194-6992	11862	14069597
85	1	5340-01-197-3372	11862	14022217
85	2	5930-01-202-3573	11862	15594176
85	3		11862	14037909
85	4	4730-00-044-4587	89346	103868
85	5	5330-01-126-0733	11862	3979626
85	6	2520-01-159-7757	11862	3787240
86	1	4710-01-148-2969	11862	14032995
86	2	4730-00-908-3195	96906	MS35842-10
86	3		11862	9439048
86	4	3040-01-157-7998	11862	8640496
86	5	4710-01-194-6775	11862	14032996
87	1		11862	14063331
87	1		11862	14071950
87	2	2520-01-163-7316	11862	14045658
87	2	2520-01-164-7851	11862	14063332
87	3	5360-01-165-1574	11862	14029108
87	4	2520-01-164-6228	11862	14029107
87	5	5315-01-165-3536	11862	14029111
87	6	3040-01-163-7315	11862	14037889
87	7	5305-01-167-5498	11862	14045504
87	8	5310-01-119-3668	24617	9422295
87	9	5310-01-165-1327	11862	466578
87	10	3120-01-169-6440	11862	3992925
87	11	2520-01-165-7885	11862	14071952
87	12	5310-00-013-1245	21450	131245
87	13	5315-00-842-3044	96906	MS24665-283
87	14	5310-00-081-4219	96906	MS27183-12
87	15	5325-01-150-1229	11862	1234418
87	16	5310-01-148-2682	11862	382105
87	17	5315-00-816-1794	89749	IF316
87	18	5310-00-835-2036	96906	MS35691-29
87	19	2520-01-163-7284	11862	718368

FIGURE AND ITEM NUMBER INDEX

FIG	ITEM	STOCK NUMBER	CAGEC	PART NUMBER
87	20	5315-01-153-0318	11862	14029122
87	21	2520-01-163-7283	11862	14029117
87	22	5305-01-167-8334		20365263
87	23	3040-01-163-7018	11862	14037893
87	24	2520-01-164-9229	11862	14029116
87	25	5305-00-068-0508	96906	MS90728-6
87	26	5310-00-938-8387	96906	MS9549-10
87	27	5330-01-164-7506	11862	14071954
87	28	2510-01-165-8101	11862	14045697
87	29	5355-01-159-6622	11862	15588503
88	1	4730-00-050-4203	11862	9417901
88	2	4730-01-202-8523	11862	14009313
88	3	5310-01-194-9233	11862	4497001
88	4	5310-01-194-9220	11862	3838153
88	5	5310-01-195-5088	11862	2423517
88	6	5305-01-198-4154	24617	9431995
88	7	2520-01-193-7870	11862	14032789
88	8	5355-01-197-3172	11862	15588504
88	9	5310-00-835-2036	96906	MS35691-29
88	10	2520-01-194-0278	11862	14055531
88	11	2520-01-192-7979	11862	14071955
88	12	5315-00-816-1794	89749	IF316
88	13	5310-00-080-6004	96906	MS27183-14
88	14	5310-01-194-7081	11862	3953987
88	15	5325-01-197-3540	11862	14049556
88	16	2520-01-192-9729	11862	14054220
89	1	2520-01-189-0596	11862	7845102
89	2	2520-00-722-7074	11862	386451
89	3	5365-01-152-7439	11862	3721887
89	4	2520-01-153-8390	11862	7809057
89	5	2520-01-024-0279	11862	7806140
89	6	5365-01-155-1941	11862	1456507
89	7	2520-01-173-1362	11862	458418
89	8	5330-01-155-4388	11862	7827942
89	9	2520-01-153-8431	11862	15596686
89	10	4730-00-050-4203	11862	9417901
89	11		11862	7815848
89	12	2520-01-153-8430	11862	7815849
89	13	5305-00-071-1787	96906	MS90728-86
89	14	5306-00-225-2864	72712	1358938
89	15	5310-00-407-9566	96906	MS35338-45
89	16	5310-00-931-8167	96906	MS51967-6
90	1	2520-01-189-0596	11862	7845102
90	1	2520-01-191-9518	11862	26013913
90	2	2520-00-722-7074	11862	386451
90	3	5365-01-152-7439	11862	3721887
90	4	5340-01-202-2517	11862	7840235
90	5	2520-01-192-4314	11862	7845127
90	6	2520-01-024-0279	11862	7806140
90	7	2520-01-153-8390	11862	7809057
90	8		11862	7815848

FIGURE AND ITEM NUMBER INDEX

FIG	ITEM	STOCK NUMBER	CAGEC	PART NUMBER
90	9	5365-01-155-1941	11862	1456507
90	10	2520-01-192-1793	11862	7845119
90	11	4730-00-050-4203	11862	9417901
90	12	2520-01-153-8430	11862	7815849
90	13	5305-00-071-1787	96906	MS90728-86
90	14	5306-00-225-2864	11862	1358938
90	15	5310-00-407-9566	96906	MS35338-45
90	16	5310-00-931-8167	96906	MS51967-6
91	1	2520-01-239-3800	11862	7830927
91	2	2520-00-722-7074	11862	386451
91	3	5365-01-152-7439	11862	3721887
91	4	2520-01-153-8390	11862	7809057
91	5	2520-01-024-0279	11862	7806140
91	6	5365-01-155-1941	11862	1456507
91	7	2520-01-173-1362	11862	458418
91	8	5330-01-155-4388	11862	7827942
91	9	2520-01-153-8431	11862	15596686
91	10	4730-00-050-4203	11862	9417901
91	11	2520-01-035-6670	86403	4049697
91	12	2520-01-153-8430	11862	7815849
91	13	5305-00-071-1787	96906	MS90728-86
91	14	5305-01-152-8193	11862	14018700
91	15	3040-01-148-5982	11862	3882979
92	1	2520-01-211-6755	11862	26013911
92	2	2520-00-722-7074	11862	386451
92	3	5365-01-152-7439	11862	3721887
92	4	5340-01-202-2517	11862	7840235
92	5	2520-01-192-4314	11862	7845127
92	6	2520-01-153-8390	11862	7809057
92	7	2520-01-024-0279	11862	7806140
92	8	5365-01-155-1941	11862	1456507
92	9	2520-01-192-1793	11862	7845119
92	10	4730-00-050-4203	11862	9417901
92	11		11862	7815848
92	12	2520-01-153-8430	11862	7815849
92	13	5305-00-071-1787	96906	MS90728-86
92	14	5305-01-152-8193	11862	14018700
92	15	3040-01-148-5982	11862	3882979
93	1	2520-01-147-4005	11862	14067762
93	1	2520-01-148-2919	11862	7844074
93	1	2520-01-210-1382	11862	14020403
93	1	2520-01-324-4895	97271	911105-5130
93	1	2520-01-328-4898	11862	14071980
93	2	5365-01-155-1941	11862	1456507
93	3	2520-01-024-0279	11862	7806140
93	3	2520-01-024-0279	11862	7806140
93	4	2520-01-151-8043	11862	14029852
93	4	2520-01-155-6936	11862	7838665
93	5	2590-01-324-5042	11862	14046907
93	5	5340-01-149-9729	11862	3920486
93	5	5340-01-231-0925	11862	7846740

FIGURE AND ITEM NUMBER INDEX

FIG	ITEM	STOCK NUMBER	CAGEC	PART NUMBER
93	6	5305-01-152-8193	11862	14018700
93	6	5306-01-323-5544	11862	458300
94	1	2520-01-155-7453	11862	14072451
94	1	2520-01-155-7454	11862	14072449
94	2	2520-01-154-6683	11862	14063306
94	2	2520-01-216-5611	11862	15591700
95	1	5305-01-156-5438	11862	11504447
95	2	5340-00-057-3043	96906	MS21333-112
95	3	3040-01-157-7998	11862	8640496
95	4	5340-00-486-1765	96906	MS21333-48
95	5		11862	9439088
95	6	4730-00-908-3195	96906	MS35842-10
95	7		11862	9439091
95	8	4730-01-153-2718	11862	14056299
95	9	5340-00-057-3034	96906	MS21333-110
95	10		11862	14029235
96	1		97271	700013L
96	2	3120-01-153-0281	11862	14009626
96	3	5330-01-158-6725	97271	620062-B
96	4	5365-01-174-8657	97271	37312
96	5	5330-01-096-9649	11862	462811
96	6	2520-01-159-4173	97271	37308
96	7	2520-01-155-5883	97271	40955
96	8	2520-01-105-5679	23862	462809
96	9	2520-01-105-5676	97271	660182-6
96	10		97271	71701-X
96	11	5330-01-113-9602	23862	462857
96	12	5305-00-071-2073	96906	MS90728-117
96	13	5330-01-150-4022	11862	6273951
96	14	5306-00-157-8315	97271	34822
96	15	4730-01-034-8228	72447	36472
96	16	5340-01-159-4478	97271	706968X
96	17	2520-01-158-8543	97271	660182-5
97	1	2530-01-149-3827	11862	14072919
97	1	2530-01-149-3827	97271	29907X
97	1	2530-01-249-5401	11862	464039
97	2	3120-01-159-9386	11862	3965121
97	3	5330-01-086-3504	11862	376855
97	4	5310-01-153-1381	11862	376852
97	5	5330-01-086-3503	11862	376851
97	6	2530-01-123-3553	11862	6273934
97	7		11862	14095588
97	8	5330-01-159-8336	97271	36352-1
97	9	2520-01-152-0941	11862	14072916
97	9	2520-01-223-3754	11862	15521884
97	10	2520-01-004-2057	97271	5-297X
97	11	2520-01-174-1418	11862	458877
97	12	5305-00-071-2058	96906	MS90728-92
97	13	5330-01-155-4399	11862	458860
97	14	5340-01-154-2000	11862	14063308
97	15	5306-01-148-6765	11862	10008936

FIGURE AND ITEM NUMBER INDEX

FIG	ITEM	STOCK NUMBER	CAGEC	PART NUMBER
97	16	4730-00-187-4210	96906	MS51884-9
97	17	2520-01-159-7757	11862	3787240
97	18	2520-01-154-3677	11862	458878
98	1	5306-00-685-2523	72447	30266
98	2	2520-01-159-8730	97271	73493-4X
98	3	5315-01-185-3415	97271	500598-21
98	4	3020-01-155-0183	34367	34367
98	5	3020-01-159-9795	97271	35830
98	6	2520-01-163-9713	97271	70696-4X
98	7	5310-01-157-7601	97271	34730
98	8	2520-00-884-5635	97271	30263
98	9	2520-01-159-7762	97271	35801
98	10	3040-01-075-1878	97271	25127-1X
98	11		97271	30272-2
98	11	5365-00-061-6847	97271	30272-1
98	11	5365-00-061-6849	97271	30272-3
98	11	5365-00-062-7200	97271	30272-4
98	12	2520-01-094-1817	97271	706045X
98	13	5310-00-137-3018	97271	30273
98	14	5330-01-163-0387	11862	14079089
98	15	2520-00-928-6150	97271	2-4-3801X
98	16	5340-00-604-0566	97271	34592
98	17	5310-00-137-3018	72447	30275
98	18	5310-00-809-8836	72447	30271
98	19	5365-00-004-6407	72447	30291-3
98	19	5365-00-884-5639	95019	30291-2
98	19	5365-00-884-5640	95019	30291-1
98	20	3110-01-164-5457	97271	706046X
98	21	5310-01-149-6285	11862	462863
98	22	2520-01-094-1818	97271	706047X
98	23	5365-00-061-2395	97271	30276-2
98	23	5365-00-061-6844	97271	30276-1
98	23	5365-00-061-6845	97271	30276-3
98	23	5365-00-062-7201	97271	30276-4
98	24	5315-01-185-3415	97271	500598-21
98	25	3020-01-155-0183	97271	34367
98	26	3020-01-155-5871	97271	35625
98	27	3120-01-150-1226	97271	34729
98	28	5310-01-157-7601	97271	34730
98	29	2520-00-884-5635	97271	30263
98	30	2520-00-884-8430	95019	30258
99	1	5306-01-156-8680	11862	14066913
99	2	5310-01-150-6034	11862	3853761
99	2	5310-01-156-8852	11862	3907089
99	2	5310-01-159-8678	11862	3907088
99	2	5365-00-257-6035	11862	3853759
99	2	5365-01-149-5434	11862	3853760
99	2	5365-01-149-5435	11862	3853762
99	2	5365-01-159-1121	11862	3995791
99	2	5365-01-159-1122	11862	3995792
99	2	5365-01-159-4471	11862	3853912

FIGURE AND ITEM NUMBER INDEX

FIG	ITEM	STOCK NUMBER	CAGEC	PART NUMBER
99	3	5365-01-152-7441	11862	9775809
99	4	3110-00-135-0983	43334	LM501349/LM50131 4
99	5	3040-01-153-1815	11862	1252981
99	6	5305-01-163-8241	11862	14056196
99	7	3020-01-151-3446	11862	6270977
99	8	3040-01-147-9321	11862	14006401
99	9	2520-01-158-9332	11862	1397647
99	10	5310-01-147-8746	11862	393578
99	11	3120-01-156-3827	11862	3984818
100	1	3020-01-151-4480	11862	1258709
100	2	5365-01-150-7761	11862	1234726
100	3	3110-00-854-1504	43334	M88048/M88010
100	4	5330-01-086-3505	11862	458859
100	5	2520-01-153-9436	48018	543319
100	6	2520-01-147-4205	11862	3988524
100	7	5310-01-147-8747	11862	517900
100	8	5310-01-157-6698	11862	1260823
100	9	3110-00-406-9608	43334	M802048/M802011
100	10	5310-01-148-5956	11862	1394893
100	10	5310-01-150-6260	11862	1394894
100	10	5310-01-159-8679	11862	1394892
100	10	5310-01-164-1207	11862	1394895
101	1	5310-01-093-2907	97271	36880
101	2	5310-01-099-7945	97271	38081-1
101	3	5310-00-834-7606	96906	MS35340-48
101	4	5305-00-719-5219	96906	MS90727-111
101	5	4730-01-098-5229	24617	9411031
101	6	2530-01-164-0038	97271	620132
101	7	5306-01-158-5374	11862	462794
101	8	2530-01-098-5242	97271	706395X
101	9	5330-01-179-2249	97271	620058
101	10	5330-01-179-5907	80201	14700
101	11	3110-00-100-5805	11862	9418355
101	12	5340-01-171-8256	97271	37305
101	13	4730-01-263-5361	11862	462798
101	14	2530-01-098-4145	97271	620180
101	15	5310-00-062-6828	97271	30875
101	16	2530-01-170-7118	97271	38139
101	17	5330-01-099-9423	97271	37307
101	18	5360-01-099-7921	97271	37300
101	19	2530-01-151-7995	11862	462853
101	20	5306-01-150-5893	97271	37879
101	21	5307-01-268-2640	9K937	37296
101	22	5310-01-185-3400	97271	500381-3
101	23	5306-01-172-1590	97271	31026-1
101	24	2530-01-164-0037	97271	37299
101	25	2530-01-151-7996	11862	15537219
101	26	5306-01-150-5893	97271	37879
101	27	5306-01-172-1590	97271	31026-1
101	28	5310-01-185-3400	97271	500381-3

FIGURE AND ITEM NUMBER INDEX

FIG	ITEM	STOCK NUMBER	CAGEC	PART NUMBER
102	1	2530-01-165-0803	11862	14013072
102	2	2530-01-160-1542	11862	14039008
102	3	5310-00-498-2381	24617	9422308
102	4	5310-00-998-0608	96906	MS35692-61
102	5	5315-00-298-1481	96906	MS24665-357
102	6	5365-00-721-6476	96906	MS16624-1175
102	7	5365-01-166-8901	11862	3967845
102	8	5310-00-975-2075	96906	MS35691-21
102	9	5305-01-150-9500	11862	3711876
102	10	5306-01-158-6243	11862	14008652
102	11	5310-01-157-7582	11862	14056133
102	12	2530-01-156-6220	11862	14013071
102	13	5305-01-150-9500	11862	3711876
102	14	2530-01-154-1355	11862	3979552
102	15	5310-00-044-3342	11862	9422303
102	16	5365-01-159-8890	11862	3965138
102	17	5307-01-158-2110	11862	3965137
102	18	2530-01-160-1542	11862	14039008
102	19	5310-00-998-0608	96906	MS35692-61
102	20	5315-00-298-1481	96906	MS24665-357
102	21	5365-01-166-8901	11862	3967845
102	22	5365-00-721-6476	96906	MS16624-1175
102	23	5310-00-498-2381	24617	9422308
102	24	5306-01-158-6243	11862	14008652
102	25	5310-00-975-2075	96906	MS35691-21
103	1	2520-01-152-4742	11862	14055363
103	2	2520-01-207-9169	11862	15597787
103	2	2520-01-325-1860	11862	15634657
104	1		11862	14040775
104	2	3040-01-157-7998	11862	8640496
104	3	5340-01-158-0303	11862	3866187
104	4	5305-01-156-5438	11862	11504447
104	5	4730-00-908-3195	96906	MS35842-10
104	6	4730-01-156-0055	11862	14056297
104	7	5340-00-057-3043	96906	MS21333-112
104	8		11862	474935
105	1	3040-01-157-7998	11862	8640496
105	2	5305-01-156-5438	11862	11504447
105	3	5340-00-057-3043	96906	MS21333-112
105	4		11862	474935
105	5	4730-00-908-3195	96906	MS35842-10
105	6	4730-01-147-6425	11862	14072930
106	1	2520-01-038-7283	11862	3977384
106	1	2520-01-201-2501	11862	15599687
106	2	2520-01-040-2160	11862	3977383
106	3	5306-01-153-1368	11862	376869
107	1	5306-01-164-2323	11862	9590271
107	2	2520-01-148-1481	11862	14039547
107	3	5330-00-763-0213	11862	14003417
107	4	3110-00-756-4535	11862	7451809
107	5	5365-01-159-4840	11862	3833322

FIGURE AND ITEM NUMBER INDEX

FIG	ITEM	STOCK NUMBER	CAGEC	PART NUMBER
108	1		11862	15537131
108	1		11862	15537684
108	1		11862	15594111
108	1		11862	471777
108	2	5310-00-933-8123	96906	MS35340-49
108	3	5305-01-160-2005	96906	MS90728-143
108	3	5306-01-205-8882	11862	15594116
108	4	4730-00-044-4789	30379	444789
108	5	5340-01-163-6145	11862	3977326
108	6	5310-00-167-0721	96906	MS35333-41
108	7	5306-00-226-4822	96906	MS90728-29
108	8	5306-01-156-5435	11862	15521977
108	9	2520-01-330-3249	11862	14071884
108	9	5340-01-148-2730	11862	3977386
108	10	5330-01-020-9319	11862	3977387
108	11	2520-01-159-7757	11862	3787240
109	1	4730-00-044-4789	30379	444789
109	2		11862	15594126
109	3	5305-00-071-2058	96906	MS90728-92
109	4	5306-01-148-6765	11862	10008936
109	5	2520-01-152-9171	11862	1252415
109	6	5330-01-084-2410	11862	26016662
109	7	2520-01-159-7757	11862	3787240
110	1	5310-01-175-2539	11862	644536
110	1	5310-01-270-5464	11862	15552844
110	2	5310-01-166-2615	11862	3977360
110	2	5310-01-270-3712	11862	2292300
110	3	3020-01-096-9657	11862	26020811
110	4	2520-01-156-0552	11862	3977358
110	5	5305-00-071-1787	96906	MS90728-86
110	6	5310-00-655-9370	96906	MS35340-47
110	7	5330-01-112-4286	11862	3977354
110	8	5365-01-162-8868	11862	334368
110	8	5365-01-162-8869	11862	334369
110	8	5365-01-162-8870	11862	334370
110	8	5365-01-162-8884	11862	334376
110	8	5365-01-164-2411	11862	334366
110	8	5365-01-164-4601	11862	334363
110	8	5365-01-164-4602	11862	334365
110	8	5365-01-164-4603	11862	334367
110	8	5365-01-164-6598	11862	334364
110	8	5365-01-164-6616	11862	334372
110	8	5365-01-164-6617	11862	334377
110	8	5365-01-164-6618	11862	334378
110	8	5365-01-164-6619	11862	334379
110	8	5365-01-164-6620	11862	334380
110	8	5365-01-165-4511	11862	334362
110	8	5365-01-167-1122	11862	334375
110	8	5365-01-168-1409	11862	334374
110	8	5365-01-168-7294	11862	334373
110	8	5365-01-171-4041	11862	334371

FIGURE AND ITEM NUMBER INDEX

FIG	ITEM	STOCK NUMBER	CAGEC	PART NUMBER
110	9	2520-01-157-4376	11862	3977325
110	10	3110-01-030-8475	60038	387AS-382A
110	11	2520-01-156-0553	11862	471873
110	12	2520-01-159-8798	11862	6258340
110	13	2520-01-151-5971	11862	14063392
110	14	3020-01-156-8365	11862	14075355
110	15	5340-01-155-9701	11862	14075354
110	16	5360-01-157-1880	11862	14075353
110	17	2520-01-172-0501	11862	14075352
110	18	2520-01-155-6939	11862	14075351
110	19	5310-00-834-7606	96906	MS35340-48
110	20	5305-00-719-5239	96906	MS90727-116
110	21	3110-01-020-9786	11862	7451870
110	21	3110-01-232-2388	60380	DC57524
110	22	3110-00-858-0988	11862	7451888
110	23	5365-01-164-8632	11862	3977355
110	24	3110-00-406-9608	24617	7451155
110	25	5330-01-036-3861	11862	3977359
111	1	5310-00-809-8836	11862	2353021
111	2	5310-01-330-9817	11862	2353015
111	3	5365-00-062-7201	97271	30276-4
111	4	5365-00-061-2395	97271	30276-2
111	4	5365-00-061-6844	97271	30276-1
111	5	5365-00-061-6845	97271	30276-3
111	6	3110-00-580-3708	60038	HM88542
111	7	3110-00-580-3709	60038	88510
111	8		97271	30797-1
111	8		97271	30797-2
111	8		97271	30797-3
111	8		97271	34801-1
111	8		97271	34801-10
111	8		97271	34801-2
111	8		97271	34801-3
111	8		97271	34801-4
111	8		97271	34801-5
111	8		97271	34801-6
111	8		972171	34801-7
111	8		97271	34801-8
111	8		97271	34801-9
111	9	3110-00-227-4667	60038	HM807010
111	10	3110-00-606-9576	60038	HM807040
111	11	2520-01-334-1731	97271	72166 5X
111	12	5365-01-330-6970	97271	701064 X
111	13	3020-01-332-5456	97271	72166 X
111	14	2520-01-332-3841	97271	701056 1X
111	15	3110-00-142-4387	60038	453X
111	16	3110-00-100-0251	60038	469
111	17	3040-01-329-9927	11862	15637897
111	18	2520-01-330-0028	11862	15637902
111	19	5306-01-196-0219	96906	MS90728-90L
111	20	2520-01-330-3703	11862	15637904

FIGURE AND ITEM NUMBER INDEX

FIG	ITEM	STOCK NUMBER	CAGEC	PART NUMBER
111	21	3020-01-330-3258	11862	15637901
111	22	5340-01-155-9701	11862	14075354
111	23	5360-01-157-1880	11862	14075353
111	24	2520-01-172-0501	11862	14075352
111	25	2520-01-330-0016	11862	15581505
112	1	5310-01-157-6698	11862	1260823
112	2	5310-01-147-8747	11862	517900
112	3	3010-01-165-1221	11862	1256654
112	4	5330-01-154-4342	11862	1243465
112	5	3110-00-854-1504	43334	M88048/M88010
112	6	5365-01-150-7761	11862	1234726
112	7	3110-00-406-9608	24617	7451155
112	8	3020-01-151-4480	11862	1258709
112	9	3110-00-135-0983	43334	LM501349/LM50131 4
112	10	5310-01-150-6034	11862	3853761
112	10	5310-01-156-8852	11862	3907089
112	10	5310-01-159-8678	11862	3907088
112	10	5365-00-257-6035	11862	3853759
112	10	5365-01-149-5434	11862	3853760
112	10	5365-01-149-5435	11862	3853762
112	10	5365-01-152-7441	11862	9775809
112	10	5365-01-159-1121	11862	3995791
112	10	5365-01-159-1122	11862	3995792
112	10	5365-01-159-4471	11862	3853912
112	11	5306-01-156-8680	11862	14066913
112	12	2520-01-268-7413	11862	26004461
112	13	5310-01-148-5956	11862	1394893
112	13	5310-01-150-6260	11862	1394894
112	13	5310-01-159-8679	11862	1394892
112	13	5310-01-164-1207	11862	1394895
113	1	3040-01-154-2090	11862	14048412
113	2	5305-01-163-8241	11862	14056196
113	3	5365-01-157-6778	11862	14048409
113	4	5310-01-157-7600	11862	14048388
113	5	3020-01-151-3444	11862	3995843
113	6	2520-01-158-7530	11862	14059083
113	7	5310-01-174-8766	11862	14063587
113	7	5310-01-175-0624	11862	14063582
113	7	5310-01-175-0625	11862	14063583
113	7	5310-01-175-1035	11862	14063584
113	7	5310-01-176-2467	11862	14063586
113	7	5310-01-177-9758	11862	14063588
113	7	5310-01-183-6930	11862	14063585
113	7	5365-01-175-0683	11862	14063581
113	8	2520-01-150-8354	11862	462787
113	9	3020-01-202-3360	11862	15599620
113	10	2520-01-155-6827	11862	14048428
113	10	2520-01-155-6828	11862	14048429
113	10	2520-01-155-6829	11862	14048430
113	10	2520-01-155-6830	11862	14048431

FIGURE AND ITEM NUMBER INDEX

FIG	ITEM	STOCK NUMBER	CAGEC	PART NUMBER
113	10	2520-01-156-8309	11862	14048432
113	10	2520-01-160-1893	11862	14048426
113	10	2520-01-172-0395	11862	14048425
113	10	2520-01-172-0396	11862	14048427
113	11	2520-01-194-7984	11862	14048410
113	12	3020-01-202-3361	11862	15599619
113	13	3020-01-173-1332	11862	14048413
113	14	5340-01-201-4095	11862	15599623
113	15	2520-01-158-8481	11862	26005078
113	16	5365-01-182-8469	24617	9414849
113	17		11862	14048420
113	17	3120-01-176-1041	11862	14048416
113	17	3120-01-176-2493	11862	14048418
113	17	3120-01-176-2494	11862	14048419
113	17	3120-01-176-2496	11862	14048421
113	17	3120-01-181-1637	11862	14048423
113	17	3120-01-185-3509	11862	14048417
113	17	3120-01-185-3678	19954	EDS-98477-48
114	1	2540-01-159-8878	11862	3893181
114	2	2530-01-159-8802	11862	334540
114	3	5325-01-155-4482	11862	341990
114	4	5306-00-226-4826	96906	MS90728-33
114	5	5310-00-959-4679	96906	MS35340-45
114	6	3040-01-156-9994	11862	14053591
114	7	5325-01-160-2238	11862	334541
114	8	5340-01-148-8352	11862	14054122
114	8	5340-01-335-9359	11862	15557723
114	9	2990-01-148-2929	11862	14064664
114	9	2990-01-151-8115	11862	14054174
114	10	5340-01-148-8351	11862	14072697
114	10	5340-01-323-9727	11862	14054120
114	11	5340-00-700-1423	19207	7001423
114	12	2590-01-156-0583	11862	14064663
114	12	2990-01-152-0251	11862	14054173
114	13	5310-00-245-3424	96906	MS17829-5C
114	14	5325-01-149-6293	11862	15530620
114	15	2530-01-147-4214	11862	14072692
114	16	5340-01-158-0297	11862	25516531
114	17	2530-01-148-1463	11862	368786
114	18	2590-01-161-2119	11862	14053593
114	18	5995-01-148-2930	11862	14055591
114	19	5310-00-931-8167	96906	MS51967-6
114	20	5310-00-809-3078	96906	MS27183-11
115	1	5310-00-763-8894	96906	MS51968-24
115	2	5310-00-052-6454	96906	MS35340-51
115	3	5310-00-834-7606	96906	MS35340-48
115	4	5305-00-719-5184	96906	MS90727-109
115	5	2530-01-152-7115	14892	4150515
115	6	2530-01-183-8860	14892	3368689
115	7	5315-01-156-0081	11862	3856834
115	8	3040-01-156-7182	11862	334307

FIGURE AND ITEM NUMBER INDEX

FIG	ITEM	STOCK NUMBER	CAGEC	PART NUMBER
115	8	3040-01-157-7970	11862	334308
115	9	5310-01-154-4341	11862	5454797
115	10	2530-01-140-6144	11862	372379
115	11	5320-01-159-4507	11862	344029
115	12	2530-01-142-0261	23862	6260885
115	13	5365-00-682-1762	96906	MS16633-1031
115	14	2530-00-125-2769	11862	3856856
115	14	2530-00-494-8165	11862	3856855
115	15	5360-00-123-0137	11862	5461145
115	16	2530-01-156-5883	11862	357845
115	16	5340-01-156-8395	11862	357846
115	17	5360-00-113-9490	11862	5461984
115	18	2530-01-156-8308	11862	3856858
115	18	3040-01-156-8307	11862	3856857
115	19	5310-01-151-8347	11862	18002428
115	20	5360-01-157-3662	11862	3856843
115	21	2530-01-156-4875	11862	3898059
115	22	5360-00-392-3453	11862	1312281
115	23	5360-01-159-1449	11862	3767138
115	24	5360-00-229-5312	11862	3856850
115	25	2530-00-228-6992	11862	15522081
115	26	5365-01-158-2191	11862	3856849
115	27	2530-01-096-6752	11862	15522077
115	27	2530-01-155-7943	11862	15522078
115	28	2530-01-162-8986	11862	15522079
115	28	2530-01-164-0039	11862	15522080
115	29	2530-01-152-0180	11862	372249
115	30	2530-01-156-4900	11862	14068905
115	30	2530-01-156-4901	11862	14068906
116	1	5310-00-785-1762	96906	MS51968-9
116	2	5310-00-637-9541	96906	MS35338-46
116	3	5310-01-155-1898	11862	3760300
116	4	2530-01-163-1152	11862	14055279
116	4	2530-01-191-9385	11862	14055280
116	5	2530-01-152-9308	11862	14009982
116	6	2530-01-163-7227	11862	15594178
116	6	2530-01-164-7126	11862	15594177
116	7	5305-00-269-3236	96906	MS90727-60
116	8	5310-01-154-4341	11862	5454797
116	9	5365-01-164-4525	11862	18004057
116	10	5365-01-159-4833	11862	18001032
116	11	5330-01-185-4676	11862	5470497
116	12	5340-01-157-2101	11862	5469497
116	13		11862	12321435
116	14	2530-01-166-3033	11862	1155445
116	15	5365-00-682-1762	96906	MS16633-1031
116	16	2530-00-125-2769	11862	3856856
116	16	2530-00-494-8165	11862	3856855
116	17	5360-00-123-0137	11862	5461145
116	18	2530-01-152-7787	11862	357889
116	18	3040-01-152-7786	11862	357890

FIGURE AND ITEM NUMBER INDEX

FIG	ITEM	STOCK NUMBER	CAGEC	PART NUMBER
116	19	5360-00-113-9490	11862	5461984
116	20	2530-01-157-5164	11862	14055315
116	20	2530-01-173-1248	11862	5461156
116	21	5310-01-151-8347	11862	18002428
116	22	5360-01-156-9730	11862	3820163
116	23	2530-01-155-8460	11862	468661
116	24	5360-00-392-3453	11862	1312281
116	25	5360-00-310-4493	11862	3694822
116	26	5360-01-162-9935	11862	3887347
116	27	5305-01-156-5006	11862	468675
116	28	5365-01-154-8577	11862	5462496
116	29	2530-01-152-9258	11862	345943
116	29	2530-01-153-1492	11862	345944
116	30	2530-01-153-9449	11862	468673
116	30	2530-01-153-9450	11862	468674
116	31	2530-01-152-0180	11862	372249
117	1	2530-01-151-5967	14892	2770209
117	1	2530-01-154-1294	14892	2770317
117	2		14892	129959
117	3		14892	129596
117	4	5340-01-246-2700	14892	2771165
117	5		14892	129839
117	6		14892	129497
117	7		14892	2770715
117	8		14892	951965
117	9		14892	125666
117	10		14892	2770746
117	11		14892	2770614
117	12		14892	129494
117	13		14892	2770720
117	14		14892	2770532
117	15		14892	2770699
117	15		14892	2770972
117	16		14892	2770685
117	16		14892	2770685
117	17		14892	2770698
117	18		14892	129472
117	19		14892	2771247
117	20		14892	129484
117	21		14892	2771125
117	22		14892	129894
117	23	2530-01-163-9952	14892	2770771
117	24	5365-01-287-5878	14892	129495
117	25	2530-01-153-1813	11862	345683
117	26		14892	2770829
117	27	5310-01-154-5263	11862	11502812
117	28	5330-01-148-7499	11862	14004810
117	29	5340-01-160-2155	11862	14045698
117	30	2530-01-156-7016	14892	2232073
117	31	5330-01-165-1358	14892	2229044
117	32	2530-01-168-6369	14892	2232076

FIGURE AND ITEM NUMBER INDEX

FIG	ITEM	STOCK NUMBER	CAGEC	PART NUMBER
117	33	5340-01-157-1955	14892 2229046	
117	34	2530-01-192-9778	14892 2232072	
117	35	5330-01-157-1916	14892 2227168	
117	36	2530-01-250-6472	14892 2232077	
117	37	5340-01-158-0321	14892 2229448	
118	1	5340-01-155-2614	11862 343438	
118	2	5340-00-881-5303	11862 15599973	
118	3	4730-01-154-1366	11862 9424955	
118	4	5340-01-268-9064	11862 25527423	
118	5	5340-01-149-4434	11862 3816659	
118	6		11862	15599260
118	6		11862	15599262
118	7	4730-00-014-2432	24617 137397	
118	8	5310-00-380-1514	11862 15607227	
118	9		11862	15599259
118	9		11862	15599261
118	10	5305-01-156-5438	11862 11504447	
118	11	4730-00-288-9390	24617 9432075	
118	12	4730-00-278-8886	96906 MS51877-4	
118	13		11862	14034586
118	13		11862	15522444
118	14	5306-00-226-4833	96906 MS90728-40	
118	15	5340-01-155-2614	11862 342677	
118	16		11862	14034571
118	16		11862	14054257
118	17	5360-01-163-7234	11862 22527167	
118	18	4720-01-148-2763	11862 14036736	
118	18	4720-01-148-6946	11862 14054270	
118	18	4720-01-148-6947	11862 14054269	
118	18	4720-01-148-7398	11862 14036723	
118	19	4730-01-151-7972	11862 14094948	
118	20	5310-01-154-3993	11862 14000172	
118	21	2520-01-192-1919	11862 1257087	
118	21	2530-01-124-3422	11862 1257203	
118	21	2530-01-156-8317	11862 25515635	
118	22		11862	14034572
118	23	5310-01-119-3668	24617 9422295	
119	1		11862	14036797
119	2	4710-01-172-0471	11862 14036792	
119	3	5360-01-163-7234	11862 22527167	
119	4	2530-01-225-2236	11862 15549248	
119	5	5306-00-226-4824	96906 MS90728-31	
119	6	5310-00-959-4679	96906 MS35340-45	
119	7	5340-01-154-5269	11862 14036773	
119	8	2530-01-225-1024	11862 15538215	
119	9	2530-01-222-8068	11862 14061396	
119	10	5310-00-931-8167	96906 MS51967-6	
119	11	4730-01-255-2976	11862 14036775	
119	12	4730-00-288-9390	24617 9432075	
119	13		11862	14036706
119	13		11862	14036712

FIGURE AND ITEM NUMBER INDEX

FIG	ITEM	STOCK NUMBER	CAGEC	PART NUMBER		
119	14	5306-01-152-2582	11862	1239146		
119	15	5305-00-688-2111	96906	MS90728-63		
119	16	5305-00-821-3869	11862	9439637		
119	17	5365-01-157-5752	11862	14055556		
119	18	5340-01-161-4025	11862	359816		
119	19	5340-01-155-3668	11862	331416		
119	19	5340-01-335-9360	11862	6259071		
119	20				11862	14036705
119	20				11862	14036711
119	21	5340-00-809-1490	96906	MS21333-98		
119	21	5340-00-881-5303	96906	MS21333-45		
119	22	5305-01-156-5438	11862	11504447		
119	23				11862	14034599
119	24	5340-00-881-5303	11862	15599973		
119	25	4720-01-149-4659	11862	17981073		
120	1				14892	2238739
120	2	2530-01-147-6423	11862	14002543		
120	3	2530-01-168-1440	14892	2238710		
120	4	2530-01-154-1222	11862	338269		
120	5	2530-01-163-7890	14892	2230359		
120	6	5330-01-157-6664	11862	6259033		
120	7	5340-01-163-1401	11862	5469581		
120	8				14892	2238740
120	9	2530-01-168-1440	14892	2238710		
120	10	2530-01-147-6424	14892	2238742		
120	11	5330-01-157-6664	11862	6259033		
120	12	2530-01-163-7890	14892	2230359		
120	13	2530-01-154-1222	11862	338269		
120	14	5340-01-152-7155	14892	4150514		
120	15	5305-01-148-8208	11862	331478		
120	16	5315-01-157-3004	11862	14023439		
120	17	2530-01-149-3375	14892	3203466		
120	17	5340-01-149-3376	14894	3203465		
120	18	2530-01-159-7754	97271	38001		
120	18	2530-01-159-7755	97271	38000		
121	1	5306-01-085-1953	11862	5468226		
121	2	2530-01-156-6190	11862	18015381		
121	3	2530-00-363-4389	1862	5468767		
121	4	2530-01-075-5080	11862	2622207		
121	5	2530-01-163-7878	11862	18013395		
121	6	5340-01-163-1401	11862	5469581		
121	7	2530-01-147-9329	11862	18007952		
121	8	2530-01-163-7878	11862	18013395		
121	9	2530-00-363-4389	11862	5468767		
121	10	2530-01-075-5080	11862	2622207		
121	11	2530-01-152-9306	11862	14023429		
121	11	2530-01-152-9307	11862	14023430		
122	1	5306-01-159-5710	24617	456697		
122	2	5340-01-163-1401	11862	5469581		
122	3	2530-01-149-1886	11862	18003151		
122	4	2530-01-096-6764	11862	18004794		

FIGURE AND ITEM NUMBER INDEX

FIG	ITEM	STOCK NUMBER	CAGEC	PART NUMBER
122	4	2530-01-110-5304	11862	18004890
122	5	2530-01-154-1263	11862	2622667
123	1	2530-01-154-1292	11862	467282
123	2	5340-01-154-5270	11862	334146
123	3	5305-01-273-4486	11862	11509135
123	4	5310-01-150-4003	11862	9422299
123	5	5340-01-232-8179	11862	15593849
123	6	5305-00-182-9584	96906	MS18154-96
123	7	5360-01-160-2415	11862	3850084
123	8	3120-01-158-2096	11862	346381
123	9	5365-01-154-8514	11862	6264951
123	10	2540-01-154-1293	11862	15522095
123	11	2540-01-153-9470	11862	355561
123	12	3040-01-166-4497	11862	336926
123	13	5310-01-151-8353	11862	3702807
123	14	5315-01-156-6562	11862	1244707
124	1	5310-01-021-9027	11862	3978901
124	1	5310-01-107-4051	23862	334387
124	2	2520-01-201-4096	11862	472536
124	3	2530-01-154-6952	11862	14035374
124	3	2530-01-325-9112	11862	15668598
124	4	3040-01-163-0797	27647	25113
124	5	5305-01-165-2260	27647	9477
124	6	2530-01-163-0798	27647	15149
124	7	5365-00-803-7317	96906	MS16624-1131
124	8		27647	9952
124	9	5365-01-221-9717	11862	14070396
124	10	2530-01-163-0799	27647	13109
124	11	5310-01-205-2536	11862	14050679
124	11	5310-01-205-2537	11862	15582233
124	12	5365-01-209-6943	11862	14038051
124	13		11862	15634658
124	13	5306-01-158-6682	09386	102007
124	14	5330-01-106-7938	80201	27467
124	14	5330-01-106-7938	80201	27467
124	15	3110-01-030-8475	60038	387AS-382A
124	16	2530-01-096-9670	09386	SR104396
124	16	2530-01-216-4554	09386	104192
124	17	3110-01-027-4475	43334	LM104949LM104911
124	18	5310-01-172-1591	97271	33734
124	19	5305-01-019-1884	11862	3988538
124	20	5330-01-085-0918	11862	469694
124	21	5365-01-161-4055	11862	474309
124	22	2530-01-148-2914	11862	6260830
124	23	2530-01-096-7731	11862	3977397
124	23	2530-01-326-1462	11862	15634663
124	24	5310-01-148-2676	11862	341509
124	25	2520-01-147-5539	11862	341511
124	26	5365-01-326-4346	11862	15634661
124	27	5330-01-076-3009	24617	327739
124	28	5315-01-153-0317	11862	341510

FIGURE AND ITEM NUMBER INDEX

FIG	ITEM	STOCK NUMBER	CAGEC	PART NUMBER
124	29	2640-00-555-2829	11862	273487
124	29	2640-01-302-1388	6V625	30-600
124	29	2640-01-323-2632	11862	9591270
125	1	2640-00-555-2840	17875	T148
125	2	2520-01-165-5974	27647	M257
125	3	5305-01-165-2260	27647	9477
125	4	3040-01-163-7285	27647	15147
125	5	5365-00-721-6876	96906	MS16624-1125
125	6	5330-01-331-7230	27647	13446
125	7	5365-01-197-8165	11862	14070348
125	8	3040-01-163-7347	27647	13113
125	9	5365-00-721-6876	96906	MS16624-1125
125	10	5365-01-158-2193	11862	14072921
125	11	5310-01-244-2259	11862	14034413
125	12	5310-01-154-3990	79410	17-01-014-001
125	13	5310-01-155-1897	11862	14034410
125	14	5310-01-153-9302	76445	40424
125	15		11862	14072927
125	16	5330-01-086-3506	11862	6273948
125	17	2530-01-152-9305	11862	14070352
125	18	5340-00-700-1423	19207	6262328
125	19	3110-01-087-2653	11862	7455617
125	20	5306-01-150-1197	09386	96735
125	21	2530-01-147-4209	11862	14026765
125	22	3110-00-690-8923	43334	LM501349-LM501310
125	23	2530-01-154-8146	11862	14063307
125	24	5310-00-264-1930	11862	358501
126	1	2610-01-148-1634	22337	212-776
126	1	2610-01-148-1635	81348	GF2A/LT235/85R16/E/LTAW
127	1	2530-01-159-3447	11862	7843369
127	2	2530-01-152-2383	11862	7837032
127	3	2590-01-191-4263	11862	7832984
127	4	3120-01-203-0332	11862	7819517
127	5	2590-01-159-3449	11862	7812526
127	6	3120-01-202-2602	11862	7832808
127	7	5365-01-200-1290	11862	7832311
127	8	2920-01-159-2842	11862	7843294
127	9	5340-01-196-3130	11862	1154611
127	10	2590-01-159-1805	11862	7830375
127	11	2920-01-159-7749	11862	7804414
127	12	5305-01-159-8263	11862	7830377
127	13	5310-01-160-5727	11862	7800580
127	14	5305-01-159-8262	11862	7806867
127	15	2540-01-159-5670	11862	1997983
127	16	5305-01-215-2501	11862	7809128
127	17	2540-01-191-4289	11862	7843400
127	18	5305-01-209-7069	11862	11504678
127	19	5360-01-160-9839	11862	7819898
127	20	2540-01-158-6910	11862	7812136

FIGURE AND ITEM NUMBER INDEX

FIG	ITEM	STOCK NUMBER	CAGEC	PART NUMBER
127	21	2530-01-159-3448	11862	7837185
127	22	5365-01-204-6702	52788	5694191
127	23	3830-01-159-5871	11862	7819738
127	24	5330-01-157-7605	11862	7814387
127	25	2530-01-159-3453	11862	7842684
127	26	3120-01-166-5638	11862	7827111
127	27	2530-01-159-3451	11862	7836172
127	28	2530-01-159-3452	11862	7827077
127	29	5360-01-160-9838	11862	7837378
127	30	2520-01-159-7119	11862	7845130
127	31	2520-01-159-7118	11862	7843368
127	32	5360-01-160-8935	11862	7804410
127	33	5306-01-159-3450	11862	7806185
127	34	3120-01-160-1894	11862	7805950
127	35	5930-01-201-1843	11862	7840274
127	36	5315-01-197-1548	11862	7830384
127	37	5305-01-171-8252	11862	7830314
127	38	2530-01-191-4264	11862	7830209
127	39	5975-01-156-4544	11862	7830335
127	40	2520-01-159-5496	11862	7843467
127	41	3040-01-160-1598	11862	7831538
127	42	5340-01-160-0294	11862	7804439
127	43	2530-01-159-3455	11862	7804440
127	44	3110-01-197-6101	11862	7805700
127	45	2530-01-159-3454	11862	7805822
127	46	5360-01-161-9171	11862	7810567
127	47	3120-01-160-1895	11862	7832112
127	48	4710-01-160-1599	11862	7842688
127	49	3040-01-178-9789	11862	7831570
127	50	4730-00-908-3194	96906	MS35842-11
127	51	5310-01-160-5728	11862	7817454
127	52	5360-01-159-2931	11862	7809409
127	53	3110-00-451-4601	11862	5671921
127	54	5330-01-161-2516	11862	7809408
127	55	3010-01-159-3456	11862	7831571
127	56	5310-01-143-0512	73342	11501033
127	57	5310-00-809-4085	96906	MS27183-16
127	58	5306-01-161-2593	11862	7832907
127	59	5340-01-213-5318	11862	7828271
127	60	5310-01-159-6587	11862	7846970
127	61	5360-01-163-0886	11862	404234
127	62	2590-01-164-7825	11862	474102
127	63		11862	409190
127	64	2530-01-159-3604	11862	9762199
127	65	5360-01-187-0301	24617	9749363
127	66	5365-01-270-1977	11862	9754764
127	67	2530-01-287-3980	11862	17983936
127	68	5305-01-270-3030	11862	9767270
127	69	5365-01-160-9530	11862	419454
127	70	2530-01-147-6421	11862	17987489
127	71	5355-01-085-0995	11862	14049351

FIGURE AND ITEM NUMBER INDEX

FIG	ITEM	STOCK NUMBER	CAGEC	PART NUMBER
127	71	5355-01-235-6616	11862	470205
127	72	2520-01-163-7314	11862	14034728
127	72	2520-01-225-1033	11862	14081915
127	73	5315-00-450-9163	11862	3793014
127	74	3120-01-232-6781	11862	1604854
127	75	2540-01-156-7233	11862	22510143
127	76	5340-01-209-7066	11862	467283
127	77	5310-01-143-0512	73342	11501033
127	78	2530-01-206-4860	11862	7809925
127	79	5310-00-809-4085	96906	MS27183-16
127	80	5306-01-149-4398	24617	11504270
127	81	5340-01-158-8552	11862	14027745
127	82	5305-01-160-1975	11862	11508566
127	83	2540-01-158-8613	11862	14027746
128	1	5305-01-149-1936	11862	343178
128	2	5365-01-151-6111	11862	343179
128	3	5310-00-407-9566	96906	MS35338-45
128	4	5310-00-905-4600	96906	MS51968-6
128	5	5310-00-785-1762	96906	MS51968-9
128	6	5310-00-637-9541	96906	MS35338-46
128	7	5305-01-194-0614	96906	MS90727-145
128	8	2530-01-155-7457	11862	14064660
128	9	5310-00-842-7783	96906	MS35692-53
128	10	5315-00-298-1481	96906	MS24665-357
128	11	2530-01-157-7933	11862	4993563
128	12	2530-01-154-1262	11862	14007644
128	13	4730-00-050-4203	11862	9417901
128	14	2530-01-159-5958	11862	362297
128	15	2530-01-159-8725	11862	362298
128	16	5315-00-816-1794	89749	IF316
128	17	5310-00-842-1490	96906	MS35692-37
128	18	5310-00-010-3028	96906	MS35690-824
128	19	5310-00-584-5272	96906	MS35338-48
128	20	5310-00-044-3342	11862	9422303
128	21	5310-00-068-5285	96906	MS27183-20
129	1	2530-01-165-9654	72210	D23980-J
129	2	5310-01-228-1405	24617	125384
129	3	5330-01-155-7700	11862	6259074
129	4	4730-00-050-4203	11862	9417901
129	5	5315-00-013-7238	96906	MS24665-425
129	6	2530-01-165-9653	72210	24004-J
129	7	5310-01-228-1405	24617	125384
129	8	5330-01-155-7700	97271	S19587-T
129	9	2530-01-165-6005	72210	S23981-H
130	1	5310-00-209-2811	11862	467117
130	2	5315-01-251-1701	11862	11514337
130	3	5330-01-155-7700	11862	6259074
130	4	2530-01-152-9312	11862	14026803
130	5	4730-00-050-4203	11862	9417901
130	6	5310-01-153-9301	11862	14026805
130	7	2530-01-152-9314	11862	14026804

FIGURE AND ITEM NUMBER INDEX

FIG	ITEM	STOCK NUMBER	CAGEC	PART NUMBER
130	8	5310-00-792-3617	30076	160635
130	9	2530-01-152-9313	11862	14026802
131	1	2530-01-150-9757	11862	7846959
131	2		11862	5686814
131	3	5340-00-960-9340	11862	5686815
131	4		11862	7817529
131	5	5330-00-960-9355	11862	5687958
131	6	5330-01-197-0881	11862	7814503
131	7	2530-01-033-1830	11862	7817355
131	8	5340-00-960-9354	11862	5686539
131	9	5305-00-173-0165	11862	5696151
131	10	2530-01-173-9671	11862	5695774
131	11	2530-00-937-1275	11862	5695513
131	12		11862	5695777
131	13	2530-01-155-6709	11862	7817484
131	14	5306-01-040-2041	11862	7816516
131	15	5310-01-038-8500	11862	5687973
131	16	2530-01-097-7659	11862	7817486
131	17		11862	7818809
131	18		11862	7846626
131	19		11862	5672489
131	20		11862	7832422
131	21		11862	7807845
131	22		11862	7819986
131	23		11862	5686550
131	24	5330-00-848-0972	11862	5686611
131	25		52788	7817725
131	26		11862	7832429
131	27		11862	7819849
131	28	3110-01-166-5667	11862	7832729
131	29	5330-01-090-5428	11862	5686527
131	30	3120-01-159-1275	11862	7828012
131	31		11862	7832426
131	32		11862	7832427
131	33		11862	7832428
131	34		11862	5687209
131	35	5306-01-161-2146	11862	7807271
131	36	2530-01-152-5517	11862	7826542
131	37	5310-01-157-3260	11862	7826012
131	38	3110-01-040-6541	11862	7826850
131	39		11862	7816509
131	40	5310-00-166-8567	11862	5667628
131	41	5310-01-144-2779	11862	5697702
131	42		11862	7834333
131	43	2530-01-191-4262	11862	7834284
131	44		11862	5682898
131	45		11862	5690517
131	46		11862	7826470
131	47		11862	5683373
131	48	3110-00-403-1488	11862	5697804
132	1	3030-01-148-2792	20796	42-5023

FIGURE AND ITEM NUMBER INDEX

FIG	ITEM	STOCK NUMBER	CAGEC	PART NUMBER		
132	2	5310-01-250-7679	11862	11506101		
132	3				11862	11503643
132	4	5310-01-143-0512	73342	11501033		
132	5	5340-01-147-2268	11862	14033879		
132	6	2530-01-147-8556	52788	7838936		
132	7	5306-01-149-6280	11862	1635490		
132	8	5305-01-148-7460	11862	11504595		
132	9	5340-01-154-7163	11862	14033880		
132	10	5340-01-155-7744	11862	14033881		
132	11	5306-01-148-3667	11862	11504512		
132	12	5306-01-185-7049	11862	11505299		
132	13	3020-01-148-2948	11862	14067701		
132	13	3020-01-153-9586	11862	14023174		
133	1				11862	7837321
133	2	5330-01-043-5495	11862	7808195		
133	3				11862	7830236
133	4				11862	5689357
133	5				11862	7836369
133	6	4820-01-039-3769	11862	7809232		
133	7	5360-01-149-6308	11862	5688037		
133	8	5330-01-157-1884	11862	5688035		
133	9	5340-01-157-7559	11862	5692682		
133	10				11862	7837322
133	11				11862	7839341
133	12				11862	7837284
133	13				11862	7825561
133	14				11862	7839669
133	15				11862	7839667
133	16				11862	5689358
133	17				11862	5688014
133	18	6680-01-152-2845	11862	7834183		
133	19	5330-01-197-0881	11862	7814503		
133	20	5307-01-172-1526	11862	7831388		
133	21	4320-01-155-5153	11862	7830913		
133	22	5305-01-162-9713	11862	7830239		
133	23	5330-00-848-4439	48018	5688049		
133	24				11862	7839499
133	25				11862	5688109
133	26	5330-01-157-1883	11862	5688015		
134	1	5306-00-226-4825	96906	MS90728-32		
134	2	4730-01-163-7194	11862	25518880		
134	3				11862	350371
134	4				11862	14040735
134	5	4720-01-148-2762	11862	7838941		
134	6				11862	1488565
134	7	4730-01-155-5135	11862	22514738		
134	8	5330-01-156-5141	11862	7829923		
134	9				11862	3773687
134	10	4720-01-148-2761	11862	7838942		
134	11	5310-00-931-8167	96906	MS51967-6		
134	12	5310-00-959-4679	96906	MS35340-45		

FIGURE AND ITEM NUMBER INDEX

FIG	ITEM	STOCK NUMBER	CAGEC	PART NUMBER
134	13	5340-01-155-2616	11862	338696
135	1	2510-01-164-7119	11862	14045516
135	2	5305-00-071-1788	96906	MS90728-87
135	3	5310-01-147-8743	11862	3790768
135	4	5305-00-071-2067	96906	MS90728-111
135	5	5310-00-809-5998	96906	MS27183-18
135	6	5310-00-768-0318	96906	MS51967-14
135	7	2510-01-164-7118	11862	14045515
135	8	5305-00-071-2075	96906	MS90728-119
135	9	2510-01-165-1495	11862	14067791
135	10	2540-01-159-8881	11862	14072422
135	11	5306-01-157-3330	11862	14072425
135	12	5310-00-809-3079	96906	MS27183-19
135	13	2510-01-164-7120	11862	14067792
136	1	2510-01-164-7116	11862	14014799
136	2	5310-01-149-4407	24617	9422301
136	3	5310-00-768-0318	96906	MS51967-14
136	4	5310-00-834-7606	96906	MS35340-48
136	5	5310-00-809-5998	96906	MS27183-18
136	6	2510-01-164-7117	11862	14014800
136	7	5305-00-071-2075	96906	MS90728-119
136	8	5306-01-157-3330	11862	14072425
136	9	5305-00-071-2067	96906	MS90728-111
136	10	2540-01-159-2992	11862	14072435
137	1	2510-01-163-1146	11862	14067783
137	2	2510-01-164-7128	11862	14021357
137	3	2510-01-163-7228	11862	14021358
137	4	5310-00-896-0903	96906	MS51967-12
137	5	5310-00-655-9370	96906	MS35340-47
137	6	5305-00-071-1788	96906	MS90728-87
137	7	2510-01-164-7127	11862	14067784
137	8	5310-00-768-0318	96906	MS51967-14
137	9	5310-00-834-7606	96906	MS35340-48
137	10	5310-00-809-5998	96906	MS27183-18
137	11	5306-01-157-3330	11862	14072425
137	12	2540-01-155-7535	11862	14072436
138	1	5340-01-198-8591	11862	15599219
138	2	5305-00-071-2067	96906	MS90728-111
138	3	5340-01-198-8591	11862	15599220
138	4	5310-01-149-4407	24617	9422301
139	1	2510-01-211-1655	11862	15589799
139	2	2510-01-211-6610	11862	14072640
139	3	3040-01-173-0049	11862	334666
139	4	2520-01-098-5200	86403	4086641
139	5	2510-01-225-1005	11862	15538347
139	6	2510-01-173-4189	11862	14041286
139	7	2520-01-225-5866	11862	15593979
139	8	2510-01-211-6609	11862	14072639
139	9	5340-01-173-0048	11862	328060
139	9	5340-01-179-4108	11862	328059
139	10	2510-01-172-6781	11862	14024587

FIGURE AND ITEM NUMBER INDEX

FIG	ITEM	STOCK NUMBER	CAGEC	PART NUMBER
139	11	5310-00-316-6513	80205	NAS1408A6
139	12	5310-00-809-4085	96906	MS27183-16
139	13	2510-01-232-7188	11862	15593980
139	14	5305-01-156-8692	11862	9424320
139	15	2510-01-225-2237	11862	15593955
140	1	5340-01-159-4395	11862	14029193
140	2	5305-01-156-8692	11862	9424320
140	3	5310-00-809-4085	96906	MS27183-16
140	4	5310-00-316-6513	80205	NAS1408A6
140	5	2510-01-225-5865	11862	15599998
140	6	5340-01-163-1389	11862	14022538
140	7	5340-01-162-3748	11862	14034503
140	7	5340-01-162-3749	11862	14034504
140	8	5340-01-164-5798	11862	14036789
140	9	5310-00-316-6513	80205	NAS1408A6
140	10	5310-01-157-5670	11862	3914674
140	11	5305-01-140-9118	96906	MS90728-59
140	12	2590-01-159-8716	11862	15595211
140	13	5340-01-162-3747	11862	14034501
140	14	5340-01-160-4068	11862	14024567
140	15	5340-01-163-1390	11862	14022537
140	16	2510-01-166-6253	11862	14000214
140	17	2510-01-159-8715	11862	14000212
140	18	5340-01-180-2577	11862	14000213
140	19	5310-00-732-0558	96906	MS51967-8
140	20	5340-01-160-4591	11862	14071973
140	21	5310-00-637-9541	96906	MS35338-46
140	22	5305-00-068-0510	96906	MS90728-60
140	23	5340-01-324-6756	11862	14001068
140	23	5340-01-324-9553	11862	14001067
140	24	5330-01-323-5567	11862	337957
140	25	2510-01-159-8714	11862	14000211
140	26	5340-01-171-8257	11862	328200
140	27	2510-01-159-7768	11862	14000209
140	28	2510-01-159-8712	11862	14041287
140	28	2510-01-159-8713	11862	14041288
140	29	5340-01-160-6874	11862	14053598
140	30	5340-01-173-0045	11862	14040692
140	31	5340-01-160-4600	11862	3943494
140	31	5340-01-172-2012	11862	3943493
140	32	5340-01-161-2636	11862	14054259
140	32	5340-01-187-8520	11862	14036720
140	33	5340-01-158-6681	11862	460397
140	34	5340-01-175-7656	11862	14029194
140	35	2510-01-152-9259	11862	326439
140	36	2510-01-173-0046	11862	14024593
140	37	5310-00-637-9541	96906	MS35338-46
140	38	5310-00-732-0558	96906	MS51967-8
140	39	5310-00-809-4061	96906	MS27183-15
140	40	5305-00-543-2419	96906	MS90728-61
140	41	2510-01-225-1004	11862	15599997

FIGURE AND ITEM NUMBER INDEX

FIG	ITEM	STOCK NUMBER	CAGEC	PART NUMBER
140	42	5310-00-896-0903	96906	MS51967-12
140	43	2510-01-172-3020	11862	14072695
140	44	2510-01-192-4480	11862	14072690
140	45	5310-00-809-4085	96906	MS27183-16
140	46	2510-01-152-9260	11862	326440
140	47	2510-01-096-7680	11862	6271344
140	48	5305-01-149-1936	11862	343178
140	49	5365-01-151-6111	11862	343179
140	50	5340-01-173-0047	11862	14024588
141	1	2510-01-211-1654	11862	15589794
141	2	2510-01-211-6608	11862	14072632
141	3	2510-01-160-1592	11862	14067782
141	4	3040-01-165-4562	11862	6271341
141	5	2510-01-159-7759	11862	14024577
141	6	2510-01-173-4189	11862	14041286
141	7	2510-01-215-3931	11862	14072631
141	8	5340-01-173-0048	11862	328060
141	8	5340-01-179-4108	11862	328059
141	9	2510-01-172-6781	11862	14024587
142	1	5340-01-159-4395	11862	14029193
142	2	5310-01-150-4003	11862	9422299
142	3	5310-01-097-9414	24617	3990160
142	4		11862	TX001488
142	5	5306-01-195-7915	11862	9440344
142	6		11862	15593908
142	6	2990-01-147-3954	11862	14020478
142	6	2990-01-164-7178	11862	14034513
142	6	2990-01-225-1021	11862	15593907
142	7	5340-01-171-8242	11862	338699
142	8	2510-01-169-3766	11862	6274179
142	8	2510-01-283-1966	11862	326418
142	9	2510-01-159-5889	11862	14043890
142	9	2510-01-160-3622	11862	14043889
142	10	2510-01-173-0050	11862	14029286
142	11	2510-01-160-4969	11862	6270214
142	12	2510-01-159-5888	11862	6270213
142	13	5340-01-163-1411	11862	14055592
142	14	2510-01-169-3767	11862	327260
142	15	2510-01-159-7760	11862	460373
142	15	2510-01-159-7761	11862	460374
142	16	5340-01-160-6874	11862	14053598
142	17	5340-01-173-0045	11862	14040692
142	18	5340-01-160-4600	11862	3943494
142	18	5340-01-172-2012	11862	3943493
142	19	5340-01-160-4601	11862	14036719
142	19	5340-01-187-8520	11862	14036720
142	20	5340-01-158-6681	11862	460397
142	21		11862	TX001487
142	22	5310-00-959-4675	96906	MS35340-46
142	23	5310-01-187-7610	11862	9418931
142	24	5305-00-688-2111	96906	MS90728-63

FIGURE AND ITEM NUMBER INDEX

FIG	ITEM	STOCK NUMBER	CAGEC	PART NUMBER
142	25	5340-01-175-7656	11862	14029194
142	26	2510-01-152-9259	11862	326439
142	27	2510-01-173-0046	11862	14024593
142	28	5310-00-880-7744	96906	MS51967-5
142	29	5310-00-637-9541	96906	MS35338-46
142	30	5310-00-809-4061	96906	MS27183-15
142	31	5305-00-543-2419	96906	MS90728-61
142	32	2990-01-181-6725	11862	14007142
142	33	2510-01-169-3765	11862	14034558
142	34	5365-01-151-6111	11862	343179
142	35	5305-01-149-1936	11862	343178
142	36	2510-01-096-7680	11862	6271344
142	37	5310-00-809-4085	96906	MS27183-16
142	38	5310-00-896-0903	96906	MS51967-12
142	39	2510-01-192-4480	11862	14072690
142	40	2510-01-172-3020	11862	14072695
142	41	2510-01-152-9260	11862	326440
143	1	2510-01-155-7942	11862	14067790
143	2	2540-01-159-7740	11862	14067795
143	3	5340-01-160-2171	11862	14067796
143	4	5315-00-013-7258	96906	MS24665-497
143	5	5315-01-160-4639	11862	14067794
143	6	4030-00-088-1881	19207	7358030
144	1	2540-00-078-6633	96906	MS51335-1
144	2	3040-01-267-4283	74410	XA-T-61-SR
144	3	4730-00-050-4203	96906	MS15001-1
144	4	5340-01-267-6296	74410	XB-767-10
144	5	5310-01-267-6293	74410	XA-T-88
144	6	5310-00-849-6882	96906	MS35692-94
144	7	5315-00-846-0126	96906	MS24665-628
144	8		74410	XB-766
144	9	5310-01-249-4210	74410	XB-T-45-1
144	10	2540-01-267-1360	74410	XA-T-61-SF
144	11	5305-00-719-5240	96906	MS90727-117
144	12	2540-00-078-6633	96906	MS51335-1
144	13		74410	XX123
144	14	4030-00-916-2141	96906	MS87006-53
144	15	5315-00-243-1169	80020	36344N24
144	16	5305-00-253-5626	96906	MS21318-47
144	17	5305-00-071-2069	96906	MS90728-113
144	18	5310-00-809-5998	96906	MS27183-18
144	19	2540-01-158-8548	11862	14067786
144	20	5310-01-149-4407	24617	9422301
144	21	5340-01-162-8760	11862	14067780
144	22	5310-00-809-3079	96906	MS27183-19
144	23	5305-00-071-2075	96906	MS90728-119
144	24	5315-01-160-4639	11862	14067794
144	25	4030-00-542-3183	11862	14067793
144	26	5315-00-013-7258	96906	MS24665-497
144	27	5310-00-768-0318	96906	MS51967-14
144	28	5310-00-834-7606	96906	MS35340-48

FIGURE AND ITEM NUMBER INDEX

FIG	ITEM	STOCK NUMBER	CAGEC	PART NUMBER
144	29	2540-01-200-3167	11862	14078806
144	30	5306-01-157-3330	11862	14072425
144	31	2540-01-159-7741	11862	14067787
144	32	2540-01-159-2928	11862	14067785
144	33	5340-01-162-8759	11862	14067779
145	1	2540-01-158-8553	11862	14067770
145	2	5310-00-768-0318	96906	MS51967-14
145	3	5310-00-834-7606	96906	MS35340-48
145	4	5310-00-896-0903	96906	MS51967-12
145	5	5310-01-147-8743	11862	3790768
145	6	2540-01-163-8595	11862	14067773
145	6	2540-01-164-1842	11862	14067774
145	7	5340-01-162-4775	11862	14067772
145	7	5340-01-165-4353	11862	14067771
145	8	5305-00-071-2067	96906	MS90728-111
145	9	5305-00-071-2075	96906	MS90728-119
145	10	2540-00-078-6633	96906	MS51335-1
145	11	3040-01-267-4283	74410	XA-T-61-SR
145	12	4730-00-050-4203	96906	MS15001-1
145	13	5340-01-267-6296	74410	XB-767-10
145	14	5310-01-267-6293	74410	XA-T-88
145	15	5310-00-849-6882	96906	MS35692-94
145	16	5315-00-846-0126	96906	MS24665-628
145	17		74410	XB-766
145	18	5310-01-249-4210	74410	XB-T-45-1
145	19	2540-00-078-6633	96906	MS51335-1
145	20		74410	XX123
145	21	4030-00-916-2141	96906	MS87006-53
145	22	5315-00-243-1169	80020	36344N24
145	23	5305-00-253-5626	96906	MS21318-47
145	24	2540-01-267-1360	74410	XA-T-61-SF
145	25	5305-00-719-5240	96906	MS90727-117
145	26	5310-00-732-0558	96906	MS51967-8
145	27	5310-00-637-9541	96906	MS35338-46
145	28	5310-00-080-6004	96906	MS27183-14
145	29	5305-00-068-0510	96906	MS90728-60
145	30	5315-01-160-4639	11862	14067794
145	31	4030-00-542-3183	11862	14067793
145	32	5315-00-013-7258	96906	MS24665-497
145	33	5305-00-071-2058	96906	MS90728-92
145	34	5365-01-163-6195	11862	14067789
145	35	2540-01-157-8008	11862	14067776
145	36	5310-00-809-4061	96906	MS27183-15
145	37	5305-00-071-1788	96906	MS90728-87
145	38	5310-00-809-5998	96906	MS27183-18
145	39	5305-00-071-2073	96906	MS90728-117
145	40	2540-01-158-4599	11862	14067775
145	41	5310-01-194-9217	11862	480567
146	1	5310-00-768-0318	96906	MS51967-14
146	2	5306-01-157-9817	11862	3991022
146	3	2590-01-182-4455	11862	15599915

FIGURE AND ITEM NUMBER INDEX

FIG	ITEM	STOCK NUMBER	CAGEC	PART NUMBER
146	4	5310-01-157-7560	11862	6274031
146	5	2590-01-193-3443	11862	350037
146	6	5320-01-197-1394	24617	189448
146	7	2540-01-192-3572	11862	6274036
146	8	2590-01-147-2269	11862	350036
146	9	5305-01-148-3687	11862	371603
147	1	2540-01-191-8439	11862	330046
147	2	5340-01-197-4600	11862	14027926
147	3	5310-01-184-5418	11862	3725668
147	4	5305-01-197-1475	24617	443945
147	5	2590-01-219-7808	11862	14007545
147	6	5306-01-148-7457	11862	15599432
147	7	5305-01-162-7890	11862	3954730
147	8	5310-00-809-5998	96906	MS27183-18
147	9	2590-01-147-5538	11862	343951
148	1	5310-00-080-6004	96906	MS27183-14
148	2	5305-00-068-0510	96906	MS90728-60
148	3	5310-00-732-0558	96906	MS51967-8
148	4	5310-00-637-9541	96906	MS35338-46
148	5	5305-00-068-0511	96906	MS90728-62
148	6	5310-00-809-4085	96906	MS27183-16
148	7	5310-00-768-0318	96906	MS51967-14
148	8	5310-00-834-7606	96906	MS35340-48
148	9	5340-01-168-6372	11862	14029200
148	10	2530-01-163-3557	11862	14045654
148	11	2510-01-154-1261	11862	359878
148	12	5305-00-709-8284	96906	MS90727-103
148	13	2510-01-154-1254	11862	3975202
148	14	5365-01-156-8061	11862	363913
148	15	2510-01-153-9473	11862	359877
148	16	5365-01-152-0613	11862	15571581
148	17	5310-01-147-8743	11862	3790768
148	18	5310-00-840-6222	80205	NAS1409A7
148	19	5310-00-763-8905	96906	MS51968-20
148	20	5310-00-823-8803	96906	MS27183-21
148	21	2510-01-151-6362	11862	370055
148	22	2510-01-148-5049	11862	460354
148	23	5310-01-154-1461	11862	3970988
148	24	5306-01-154-4320	11862	370053
148	25	5306-01-154-3971	11862	370054
148	26	5310-00-044-3342	11862	9422303
148	27	5310-00-809-3079	96906	MS27183-19
148	28	5365-01-096-6749	11862	3887751
148	29	2510-01-151-6363	11862	370056
148	30	5305-01-155-6110	11862	460340
148	31	5305-01-197-9418	96906	MS90728-152
149	1	5310-00-080-6004	96906	MS27183-14
149	2	5305-00-068-0510	96906	MS90728-60
149	3	5310-00-732-0558	96906	MS51967-8
149	4	5310-00-637-9541	96906	MS35338-46
149	5	5305-00-688-2111	96906	MS90728-63

FIGURE AND ITEM NUMBER INDEX

FIG	ITEM	STOCK NUMBER	CAGEC	PART NUMBER
149	6	5310-00-768-0318	96906	MS51967-14
149	7	5310-00-834-7606	96906	MS35340-48
149	8	5340-01-168-6372	11862	14029200
149	9	2510-01-225-0997	11862	15522381
149	10	2510-01-154-1261	11862	359878
149	11	5305-00-709-8284	96906	MS90727-103
149	12	2510-01-154-1254	11862	3975202
149	13	5365-01-156-8061	11862	363913
149	14	2510-01-153-9473	11862	359877
149	15	5365-01-152-0613	11862	15571581
149	16	5310-01-147-8743	11862	3790768
149	17	5310-00-840-6222	80205	NAS1409A7
149	18	5310-00-763-8905	96906	MS51968-20
149	19	5310-00-823-8803	96906	MS27183-21
149	20	2510-01-152-7832	11862	379393
149	21	2510-01-148-5049	11862	460354
149	22	5310-01-154-1461	11862	3970988
149	23	5306-01-237-4995	11862	379397
149	24	5306-01-158-6231	11862	379398
149	25	5310-00-044-3342	24617	443342
149	26	5310-00-809-3079	96906	MS27183-19
149	27	5365-01-096-6749	11862	3887751
149	28	2510-01-152-7833	11862	379394
149	29	5305-01-197-9418	96906	MS90728-152
150	1	5310-00-044-3342	24617	443342
150	2	5305-01-154-4318	96906	MS90728-150
150	3	5310-00-809-3079	96906	MS27183-19
150	4	2510-01-148-2937	11862	14022597
150	5	3120-01-162-8654	11862	468481
150	6	5306-01-153-8281	11862	471667
150	7	5365-01-159-7767	11862	337520
150	8	2510-01-155-5183	11862	14071877
150	9	5365-01-155-1932	11862	488369
150	10	5310-00-823-8803	96906	MS27183-21
150	11	5310-00-763-8905	96906	MS51968-20
150	12	2510-01-154-1260	11862	362275
150	13	3120-01-163-0638	11862	350544
151	1	5310-00-044-3342	24617	443342
151	2	5310-00-809-3079	96906	MS27183-19
151	3	5305-01-154-4318	96906	MS90728-150
151	4	5340-01-156-6779	11862	328169
151	5	2510-01-148-2937	11862	14022597
151	6	3120-01-162-8654	11862	468481
151	7	5306-01-150-9499	11862	471665
151	7	5306-01-323-8967	11862	471663
151	8	5365-01-156-9635	11862	471657
151	8	5365-01-326-1113	11862	471658
151	9	5310-01-154-1461	11862	3970988
151	9	5365-01-326-1128	11862	327225
151	10		11862	14067752
151	11	5310-00-732-0559	96906	MS51968-8

FIGURE AND ITEM NUMBER INDEX

FIG	ITEM	STOCK NUMBER	CAGEC	PART NUMBER	
151	12	5310-00-087-7493	96906	MS27183-13	
151	13	2510-01-096-9662	11862	350726	
151	14	5340-01-160-7774	11862	14067756	
151	15	5365-01-156-8948	11862	6271322	
151	16	2510-01-155-5183	11862	14071877	
151	17	3120-01-162-8654	11862	468481	
151	18	5305-01-156-9711	11862		3717658 N.D.
151	19	3120-01-163-0638	11862	350544	
151	20	5365-01-155-1932	11862	488369	
151	21	2510-01-154-1260	11862	362275	
151	21	2510-01-326-1464	11862	464718	
151	22	2590-01-326-3001	11862	14033574	
151	22	2590-01-326-5827	11862	14033573	
151	23	5310-00-763-8905	96906	MS51968-20	
151	24	5310-00-823-8803	96906	MS27183-21	
151	24	5310-01-326-1052	11862	474978	
152	1	5310-00-823-8803	96906	MS27183-21	
152	2	5310-00-763-8905	96906	MS51968-20	
152	3	5310-00-044-3342	24617	443342	
152	4	5310-00-809-3079	96906	MS27183-19	
152	5	5305-01-154-4318	96906	MS90728-150	
152	6	3120-01-162-8654	11862	468481	
152	7	2510-01-148-2937	11862	14022597	
152	8	2510-01-153-1614	11862	14067753	
152	9	5306-01-160-3942	11862	326489	
152	10	2510-01-155-0336	11862	6263945	
153	1	5310-00-768-0318	96906	MS51967-14	
153	2	5310-00-834-7606	96906	MS35340-48	
153	3	5305-00-071-2077	96906	MS90728-121	
153	4	2510-01-148-2942	11862	3187846	
153	4	2510-01-153-9584	11862	3187843	
153	5	5305-00-071-2075	96906	MS90728-119	
154	1	2540-01-148-2943	11862	3187845	
154	2	5310-00-809-5998	96906	MS27183-18	
154	3	5310-00-933-8123	96906	MS35340-49	
154	4	5310-00-768-0318	96906	MS51967-14	
154	5	5310-00-732-0558	96906	MS51967-8	
154	6	5310-00-637-9541	96906	MS35338-46	
154	7	2530-01-153-1814	11862	3764438	
154	8	5310-00-763-8913	96906	MS51967-17	
154	9	5310-00-167-0680	96906	MS35338-49	
154	10	5305-01-155-6113	11862	9419138	
155	1	5310-00-763-8905	96906	MS51968-20	
155	2	5310-00-045-5001	96906	MS35340-50	
155	3	5310-00-732-0558	96906	MS51967-8	
155	4	5310-00-637-9541	96906	MS35338-46	
155	5	2530-01-153-1814	11862	3764438	
155	6	2540-01-155-7496	11862	3187844	
155	7	5310-00-763-8913	96906	MS51967-17	
155	8	5310-00-167-0680	96906	MS35338-49	
155	9	5305-01-155-6113	11862	9419138	

FIGURE AND ITEM NUMBER INDEX

FIG	ITEM	STOCK NUMBER	CAGEC	PART NUMBER
156	1	5306-01-155-6108	11862	328130
156	2	5310-01-147-8748	11862	328131
156	3	5310-01-097-9414	24617	3990160
156	4	5310-01-150-4003	11862	9422299
156	5	5305-00-071-2055	96906	MS90728-89
156	6	3120-01-211-7528	11862	328128
156	7	5340-01-153-1631	11862	14015726
156	8	5365-01-214-4927	11862	14015724
156	9	2530-01-147-5541	11862	328132
157	1		11862	1365065
157	2		11862	3993729
157	3	5325-01-198-8040	11862	6270704
157	4	5310-00-732-0558	96906	MS51967-8
157	5	5310-00-637-9541	96906	MS35338-46
157	6	3120-01-250-0583	11862	404062
157	7	2590-01-323-5857	11862	406887
157	8	5305-00-821-3869	96906	MS90728-65
157	9	2510-01-325-9069	11862	328107
157	10		11862	328111
157	11		11862	328108
158	1	2510-01-155-7425	11862	15554915
158	2	5305-01-158-7820	24617	11503395
158	3	5310-01-154-2273	96906	MS90724-34
158	4	5325-01-157-1698	11862	347347
158	5	2510-01-155-8787	11862	14072488
159	1	2510-01-160-4970	11862	15599285
159	2	2510-01-159-8762	11862	14072850
159	3	5310-01-158-9205	24617	9420621
159	4	2590-01-163-1238	11862	15599283
159	4	2590-01-164-0134	11862	15599284
159	5	2590-01-159-8861	11862	15599286
160	1	2510-01-158-7575	11862	14021243
160	2	5306-01-158-9018	11862	2014469
160	3	2930-01-165-9595	11862	15598787
160	4	5305-00-071-2083	96906	MS90728-127
160	5	5310-01-162-5732	11862	14049809
160	6	2510-01-156-8183	11862	15597600
160	7	5365-01-155-8564	11862	15597629
160	8	2510-01-156-4854	11862	14027472
160	9	5310-00-768-0318	96906	MS51967-14
160	10	5310-00-809-5998	96906	MS27183-18
160	11	2510-01-159-2929	11862	14043880
161	1	2510-01-194-0206	11862	15598770
161	2	2540-01-160-1591	11862	15598769
161	3	2510-01-159-8761	11862	15598708
161	4	5310-01-170-9100	11862	3982098
161	5	5305-01-158-7820	11862	14026247
161	6	2510-01-212-5819	11862	15598709
161	7	5340-01-160-2346	11862	14027555
162	1	5340-01-162-4852	11862	472450
162	2	5340-01-165-8986	11862	6262054

FIGURE AND ITEM NUMBER INDEX

FIG	ITEM	STOCK NUMBER	CAGEC	PART NUMBER
162	3	5340-01-149-4434	11862	3816659
162	4	5305-01-162-9689	24617	11503396
162	5	2590-01-158-8784	11862	14039963
163	1	2510-01-154-6906	11862	15571640
163	2	5330-01-170-6303	11862	14018523
163	3	5306-01-158-9018	11862	2014469
163	4	2510-01-162-5172	11862	14043823
163	4	2510-01-162-5173	11862	14043824
163	5	5305-01-140-9118	96906	MS90728-59
163	6	5305-01-162-7885	11862	11508164
163	7	2510-01-162-3623	11862	14070703
163	8	5340-01-163-5908	11862	14018531
163	9	5360-01-163-0885	11862	14018532
163	10	5340-01-163-0917	11862	14018526
163	11	5360-01-205-8888	11862	14018529
163	12	2510-01-163-1139	11862	14021254
163	12	2510-01-163-7016	11862	14021253
163	13	5305-01-160-1975	11862	11508566
164	1	5310-01-267-3043	11862	11501047
164	2	2510-01-172-3022	11862	14023039
164	3	5310-01-154-2273	96906	MS90724-34
164	4	2510-01-159-0868	11862	14072405
164	5	5305-01-159-2780	24617	11501153
164	6	5305-01-158-6235	24617	11501151
164	7	2530-01-162-3626	11862	14023008
164	8	2510-01-162-3679	11862	6274970
165	1	5310-00-761-6882	96906	MS51967-2
165	2	5310-00-809-3078	96906	MS27183-11
165	3	5340-01-158-2160	11862	351789
165	4	5305-01-197-3112	24617	9422956
165	5	2510-01-156-5881	11862	14072414
165	6	5306-01-158-9018	11862	2014469
165	7	5340-01-161-9188	11862	3165243
165	8	5340-01-164-8761	11862	6274550
165	9	5310-01-154-2273	96906	MS90724-34
165	10	2540-01-156-0564	11862	14044471
165	11	2540-01-096-9664	11862	6260421
165	12	2590-01-162-7367	11862	3957093
165	13	5340-01-163-4337	11862	6264131
165	14	5305-01-273-4486	11862	11509135
165	15	2510-01-166-1146	11862	343915
165	16	5305-01-157-5625	24617	9420408
165	17	5305-01-164-2313	24617	9415163
165	18	4210-01-200-2574	11862	15599270
165	19	5305-01-167-6246	11862	9420818
165	20	5340-01-164-8757	11862	14049810
165	21	2510-01-225-2238	11862	15590559
165	22	5340-01-160-2443	11862	14000077
165	23	5310-01-268-8948	11862	11501937
165	24	2590-01-173-6991	11862	466717
165	25	5305-01-162-3961	11862	11503537

FIGURE AND ITEM NUMBER INDEX

FIG	ITEM	STOCK NUMBER	CAGEC	PART NUMBER
165	26	5305-01-232-1436	11862	11509121
165	27	2510-01-225-1008	11862	15593569
165	28	2540-01-225-1007	11862	15593831
165	29	2510-01-156-8092	11862	15598706
166	1	2540-01-158-3576	11862	14072494
166	2	2540-01-156-4870	11862	327062
166	3	5360-01-161-7561	11862	327065
166	4	5330-01-157-7459	11862	6258562
166	5	2510-01-155-8800	11862	15593230
166	6	2510-01-155-8799	11862	15571643
166	7	5340-01-157-6092	11862	4410574
166	8	5330-01-157-7458	11862	6258561
166	9	2540-01-156-7238	11862	14072493
166	10	5360-01-161-7561	11862	327065
166	11	2540-01-156-4870	11862	327062
166	12	2540-01-194-0261	11862	15599677
166	13	3040-01-157-7997	11862	7040174
166	13	3040-01-162-0255	11862	7040173
166	14	5330-01-157-5684	11862	4587931
166	15		11862	6272627
166	16	5310-01-157-5672	11862	1260895
166	17	5306-01-157-9936	11862	9601750
166	18	5340-01-162-9844	11862	9728247
166	19	5306-01-157-6796	11862	9439771
166	20	5340-01-158-0503	11862	3900684
166	21	2510-01-162-7119	11862	14000091
166	21	2510-01-162-7120	11862	14000092
166	22	5315-01-160-0575	11862	6271989
166	23	3120-01-166-6724	11862	9721917
166	24	5305-01-160-3937	11862	9438916
166	25	5340-01-157-7471	11862	3944769
166	26	2540-01-164-1891	11862	14000093
166	26	5340-01-167-0136	11862	14000094
166	27	5340-01-206-2995	11862	327959
166	27	5340-01-206-2996	11862	327960
166	28	5315-01-160-0575	11862	6271989
166	29	3120-01-166-6724	11862	9721917
167	1	5310-01-154-2273	96906	MS90724-34
167	2	5340-01-161-5522	11862	362434
167	2	5340-01-170-5530	11862	362433
167	3	5310-00-490-4639	96906	MS90724-40
167	4	5306-01-157-6797	11862	9439770
167	5	5310-01-084-4491	11862	1355003
167	6	5325-01-160-4028	11862	20696927
167	7	5330-01-164-8385	11862	14027775
167	7	5330-01-193-1840	11862	14027776
167	8	5330-01-158-6683	11862	15569071
167	8	5330-01-159-1153	11862	15569072
167	9	5330-01-157-6827	11862	14026383
167	9	5330-01-157-6828	11862	14026384
167	10	5325-01-205-2545	11862	330485

FIGURE AND ITEM NUMBER INDEX

FIG	ITEM	STOCK NUMBER	CAGEC	PART NUMBER
167	11	2510-01-156-4872	11862	15597667
167	11	2510-01-156-4873	11862	15597668
167	12	5305-01-210-9425	24617	9414724
167	13	2540-01-156-4903	11862	14026409
167	13	2540-01-156-4904	11862	14026410
167	14	5305-01-162-3961	11862	11503537
167	15	5305-01-231-1298	24617	1640810
167	16	5310-01-158-9205	24617	9420621
167	17	5340-01-187-8673	11862	14010954
167	18	5340-00-285-8868	11862	4168122
167	19	2540-01-156-4885	11862	14030586
167	20	5325-01-160-2237	11862	363137
167	21		11862	15590415
167	22	2540-01-158-1721	11862	15590422
167	23	5365-01-160-2483	11862	364372
168	1		11862	365953
168	2	2540-01-156-6107	11862	15590401
168	2	2540-01-156-6108	11862	15590402
168	3	2540-01-155-7502	11862	14027431
168	3	2540-01-155-7503	11862	14027432
168	4	5306-01-157-6797	11862	9439770
169	1	2510-01-163-7306	11862	15635685
169	2	5340-01-165-4791	11862	20264729
169	3	5330-01-178-7351	11862	20264731
169	4		11862	365953-1
169	5	2540-01-159-8880	11862	20264737
169	6	5680-01-164-4964	11862	20354946
169	7	5310-01-184-5866	11862	365443
169	8	2590-00-476-5459	11862	3762400
169	9	2510-01-163-2709	11862	15617126
169	10	2540-01-163-7305	11862	15635684
169	11	5340-01-165-4792	11862	20264728
169	12	5330-01-181-2454	11862	20264730
169	13		11862	365953-1
169	14	2540-01-163-2719	11862	20264736
169	15	5680-01-167-1068	11862	20354945
169	16	5310-01-184-5866	11862	365443
169	17	2590-00-476-5459	11862	3762400
169	18	2510-01-164-7152	11862	12300197
169	19	5306-01-157-6797	11862	9439770
169	20	2540-01-156-4907	11862	337715
169	20	2540-01-156-4908	11862	337716
169	21	5310-01-154-2273	96906	MS90724-34
169	22	5305-01-231-1297	11862	15531547
169	23	5305-01-157-1987	11862	11501149
169	24	2510-01-156-4871	11862	461610
169	25	5306-01-157-6797	11862	9439770
169	26	5365-01-155-8576	11862	14013789
169	27	9390-01-162-4500	11862	14027777
169	27	9390-01-163-2028	11862	14027778
169	28	5305-01-230-9846	11862	15590443

FIGURE AND ITEM NUMBER INDEX

FIG	ITEM	STOCK NUMBER	CAGEC	PART NUMBER
170	1	3040-01-155-0371	11862	327067
170	2	5305-01-158-0335	11862	9439772
170	3	2540-01-156-0565	11862	14039763
170	3	2540-01-156-4869	11862	14039764
170	4	3040-01-156-9729	11862	14039766
170	4	3040-01-159-7950	11862	14039765
170	5	5340-01-194-3188	11862	375180
170	6	2540-01-155-7298	11862	15597653
170	6	2540-01-156-8315	11862	15597654
170	7	5306-01-157-6797	11862	9439770
170	8	5340-01-160-5922	11862	15545178
170	9	3040-01-155-0372	11862	14039767
170	9	3040-01-155-0373	11862	14039768
170	10	5355-01-280-2975	11862	7591126
171	1	5305-01-161-2581	11862	14022885
171	2	2510-01-249-6434	11862	14016511
171	2	2510-01-251-5487	11862	14016512
171	3	5680-01-163-6347	11862	15522764
172	1	2510-01-156-8183	11862	15597600
172	2	5365-01-155-8564	11862	15597629
172	3	2510-01-156-4854	11862	14027472
172	4	5310-00-068-5285	96906	MS27183-20
172	5	5305-00-071-2077	96906	MS90728-121
172	6	5306-01-157-9882	11862	14014389
172	7	5340-01-160-7778	11862	3946246
172	8	5310-00-809-5998	96906	MS27183-18
172	9	5310-00-768-0318	96906	MS51967-14
172	10	5310-01-157-7589	11862	2387523
172	11	5340-01-172-1944	11862	3909351
172	12	5365-01-155-7465	11963	3946247
172	13	5306-01-157-9883	11862	14013710
172	14	5340-01-152-2382	11862	337751
172	15	5340-01-161-9187	11862	327100
172	16	5365-01-157-6772	11862	337773
173	1	2510-01-148-1617	18862	15596628
173	2	5305-01-158-2032	96906	MS51869-24
173	3	2510-01-160-5837	11862	14027451
173	4	2530-01-158-8584	11862	467913
173	4	2530-01-158-8585	11862	15593254
173	5	5310-00-931-8167	96906	MS51967-6
173	6	5340-01-164-8757	11862	14049810
173	7	2510-01-231-5398	11862	15628618
173	8		11862	15596994
173	9	2510-01-232-8176	11862	15597621
173	10	2510-01-155-5846	11862	330438
173	11	2510-01-155-5848	11862	480534
173	12	2510-01-254-1075	11862	14049881
173	13	5306-01-158-9018	11862	2014469
174	1	5306-01-165-4286	11862	14039924
174	2	5306-01-166-8556	11862	15599971
174	3	5310-00-809-5998	96906	MS27183-18

FIGURE AND ITEM NUMBER INDEX

FIG	ITEM	STOCK NUMBER	CAGEC	PART NUMBER
174	4	5310-00-768-0318	96906	MS51967-14
174	5	5365-01-194-5074	11862	15599250
174	5	5365-01-195-4948	11862	14076894
174	6	5340-01-197-3244	11862	14076887
175	1	2510-01-158-6903	11862	358543
175	2	5310-01-165-2475	11862	357445
175	3	2510-01-155-5846	11862	330438
175	4	5310-00-931-8167	96906	MS51967-6
INS	5	2510-01-158-6904	11862	14027454
175	5	2510-01-160-5837	11862	14027451
175	6	5340-01-164-8757	11862	14049810
175	7	2510-01-163-2681	11862	467911
175	7	2510-01-163-7229	11862	467912
175	8	5306-01-158-9018	11862	2014469
175	9	2510-01-231-2879	11862	14073962
175	9	2510-01-238-7052	11862	14025874
175	10	2510-01-230-1165	11862	14026209
175	10	2510-01-238-7051	11862	14025873
175	11	5305-01-156-5438	11862	11504447
175	12	2990-01-160-5873	11862	471089
176	1	2540-01-159-8759	11862	6262029
176	2	5305-01-140-9118	96906	MS90728-59
176	3	2510-01-155-5854	11862	458025
176	4	5340-01-165-0564	11862	14032787
176	5	5305-01-158-2032	96906	MS51869-24
176	6	2510-01-155-8785	11862	458026
176	7	5305-00-432-4201	96906	MS51861-45
176	8	2510-01-155-5857	11862	15628608
176	9	5305-01-157-7388	11862	9431663
176	10	2510-01-155-5853	11862	14021389
176	11	5340-01-160-2488	11862	14007449
176	12	2510-01-225-1006	11862	15594644
176	13	2510-01-155-5851	11862	15596976
176	14	5340-01-159-6905	11862	3889864
176	15	2540-01-156-9675	11862	14021275
176	16	2540-01-225-1023	11862	15594643
176	17	5305-01-162-9695	24617	9414712
176	18	5306-01-158-9018	11862	2014469
176	19	2510-01-155-5850	11862	15596975
176	20	2510-01-155-5849	11862	14021386
176	20	2510-01-155-7877	11862	14021385
177	1	5306-01-246-7459	11862	1244067
177	2	2540-01-158-8554	11862	15641780
177	2	3120-01-338-6380	11862	4158246
177	3	2540-01-160-5918	11862	20171141
177	4		11862	14027542
177	5	2540-01-158-8813	11862	340053
177	6	2510-01-224-8839	11862	15614462
177	7	5340-01-165-0657	11862	8785295
178	1	2540-01-192-9754	11862	15599678
178	2	2540-01-158-4602	11862	14072497

FIGURE AND ITEM NUMBER INDEX

FIG	ITEM	STOCK NUMBER	CAGEC	PART NUMBER
178	3	2540-01-211-4621	11862	14072499
178	4	5310-01-194-9234	11862	5713274
178	5	5365-01-198-5516	11862	5713276
178	6	2540-01-218-6833	11862	9703344
178	7	5315-01-194-0819	11862	5717887
178	8	5365-00-900-0982	96906	MS16633-1021
178	9	3010-01-159-7750	11862	9702916
178	10	5365-00-720-8064	96906	MS16624-1024
178	11	5330-01-182-4121	11862	327015
178	12	5310-01-152-0598	24617	271172
178	13	5360-01-159-8862	11862	4303911
178	14	3040-01-157-8021	11862	5713268
179	1	5340-01-163-8520	11862	14050440
179	2	5310-01-157-5672	11862	1260895
179	3	5365-01-158-5381	11862	14021292
179	4	5306-01-160-1952	11862	1740374
179	5	4010-01-158-6331	11862	6274850
179	6	5340-01-160-2367	11862	337788
179	7	5306-01-158-9018	11862	2014469
179	8	2510-01-163-7231	11862	6274848
179	8	5340-01-166-5652	11862	6274847
179	9	5365-01-162-8876	11862	6274849
179	10		11862	9438039
179	11	5306-01-162-9678	11862	9437242
179	12	2540-01-158-4600	11862	6274836
179	13	2510-01-225-0999	11862	15522708
179	14	5315-01-160-0575	11862	6271989
179	15	3120-01-159-1311	11862	335524
179	16	5305-00-432-8220	96906	MS51862-35
179	17	2510-01-225-0998	11862	15522707
179	18	5310-00-809-4085	96906	MS27183-16
179	19	5360-01-164-2404	11862	14021315
179	19	5360-01-164-2405	11862	14021316
179	20	5310-00-809-3079	96906	MS27183-19
179	21	5306-01-159-2772	11862	14021317
179	22	5306-01-157-6796	11862	9439771
179	23	5305-01-162-9695	24617	9414712
179	24	2510-01-159-7120	11862	334132
179	25	2540-01-157-8009	11862	470949
180	1	5330-01-237-7512	11862	15593570
180	2	5330-01-207-9421	11862	15593571
180	3	5340-01-158-4583	11862	6274890
180	4	5305-01-157-9720	24617	9419327
180	5	3040-01-159-1775	11862	14039716
180	6	2510-01-160-3634	11862	14039710
180	7	5340-01-160-2488	11862	14007449
180	8	5340-01-160-5922	11862	15545178
180	9	2540-01-158-8557	11862	473917
180	10	5340-01-160-2470	11862	3905674
180	11	2510-01-178-8867	11862	473995
180	12	5305-01-158-0335	11862	9439772

FIGURE AND ITEM NUMBER INDEX

FIG	ITEM	STOCK NUMBER	CAGEC	PART NUMBER
181	1	2510-01-162-7112	11862	334101
181	2	5340-01-165-4351	11862	6274854
181	3	5330-00-360-7881	22593	30103
181	3	5330-01-167-8123	11862	334104
181	4	5306-01-157-6796	11862	9439771
181	5	2510-01-165-4932	11862	334102
181	6	5330-01-159-4777	11862	6274853
181	7	5340-01-160-2172	11862	337899
181	8	2540-01-158-4601	11862	473894
181	9	5310-00-834-7606	96906	MS35340-48
181	10	5310-00-768-0318	96906	MS51967-14
182	1	5360-01-185-0341	11862	15593866
182	2	2540-01-158-8556	11862	331638
182	3	5306-01-246-7459	11862	1244067
182	4	5305-00-068-0508	96906	MS90728-6
182	5	3040-01-158-8703	11862	335452
182	6	5310-00-582-5965	96906	MS35338-44
182	7	5310-01-159-8559	96906	MS35691-406
182	8	2540-01-158-8555	11862	4495180
182	9	5305-01-159-2783	11862	8782501
182	10	5330-01-159-8504	11862	470993
182	11	3040-01-159-0996	11862	8742340
182	12	3040-01-160-5913	11862	14039712
183	1	5330-01-159-1152	11862	326934
183	2	5330-01-162-8595	11862	327006
183	3	5330-01-159-2816	11862	327005
184	1	5330-01-165-1378	11862	357490
184	2	2540-01-160-5838	11862	15565342
184	3	5330-01-166-2154	11862	358589
184	4	5330-01-166-5912	11862	358590
184	5	5330-01-165-3401	11862	15590455
184	6	5330-01-165-3402	11862	358593
184	7	5306-01-158-2023	11862	358549
184	8	5330-01-157-7464	11862	466157
184	9	2590-01-155-8871	11862	334150
184	10	5305-01-162-5707	11862	458985
184	11	5305-01-162-7884	11862	407217
184	12	5315-01-159-8666	11862	340034
184	13	5310-00-407-9566	96906	MS35338-45
184	14	5310-00-931-8167	96906	MS51967-6
184	15	5330-01-164-8579	11862	475968
185	1	5306-01-194-4977	11862	15599929
185	2	5340-01-159-6174	11862	14072452
185	3	2510-01-155-5825	11862	14072460
185	4	5365-01-158-2004	11862	14072454
185	5	5310-01-202-2695	11862	9439757
185	6	2510-01-156-0062	11862	14072459
186	1	5306-01-194-4977	11862	15599929
186	2	2590-01-234-6468	11862	15591705
186	3	2540-01-241-4238	23862	15591706
186	3	2590-01-242-1050	11862	15591709

FIGURE AND ITEM NUMBER INDEX

FIG	ITEM	STOCK NUMBER	CAGEC	PART NUMBER
186	4	5305-01-197-3290	11862	14072458
186	5	5340-01-238-5924	1T998	15591710
186	6	2510-01-155-5112	11862	14072455
186	7	5310-00-768-0318	96906	MS51967-14
186	8	5365-01-158-2004	11862	14072454
186	9	2510-01-155-5825	11862	14072460
186	10	5310-01-160-4536	11862	3792287
186	11	5340-01-238-5923	11862	15591707
186	12	5305-01-163-2438	11862	9440300
187	1	5310-01-169-2849	11862	14072307
187	2	5340-01-159-2996	11862	14072305
187	3	5310-01-164-5600	11862	94009398
187	4	5310-00-834-7606	96906	MS35340-48
187	5	5305-00-071-2070	96906	MS90728-114
187	6	5305-00-071-2081	96906	MS90728-125
187	7	2510-01-159-8726	11862	14072306
188	1	5306-01-159-6574	11862	14032814
188	2	5310-01-167-8344	24617	9409613
188	3	5310-01-077-6817	96906	MS35425-72
188	4	5340-01-165-6797	11862	14074431
188	5	5340-01-165-6799	11862	14074443
188	6	5340-01-165-6798	11862	14074440
188	7	5340-01-158-8549	11862	467299
189	1	5340-01-158-8549	11862	467299
189	2	5310-01-077-6817	96906	MS35425-72
189	3	5340-01-165-6797	11862	14074431
189	4	5340-01-165-4544	11862	14032812
189	5	5306-01-159-6574	11862	14032814
189	6	5310-01-167-8344	24617	9409613
190	1	2510-01-155-5434	75829	14022842
190	2	2510-01-155-5433	11862	20264744
190	3	2510-01-159-8871	75829	14018597
190	4	2510-01-155-5432	11862	20264743
190	5	5330-01-208-3843	11862	15590421
190	6	2510-01-096-2985	75829	363107
190	7	2510-01-155-5435	75829	14022841
190	8	2510-01-157-1382	75829	14076899
190	9	2510-01-159-8728	75829	14076859
190	9	2510-01-159-8729	75829	14076860
190	10	2590-01-155-8871	11862	334150
190	11	2510-01-096-6769	11862	471009
190	12	5330-01-157-7464	11862	466157
191	1	2510-01-165-1496	11862	14022831
191	1	2510-01-165-1497	11862	14022832
191	2	5305-01-158-7820	24617	11503395
191	3	2510-01-147-5576	11862	15522752
191	4	2510-01-162-3625	11862	14075827
191	5	5340-01-162-3624	11862	14075829
191	6	2510-01-159-9789	11862	14075839
191	7	5325-01-191-3293	11862	411700
191	8	2510-01-158-8550	11862	14027795

FIGURE AND ITEM NUMBER INDEX

FIG	ITEM	STOCK NUMBER	CAGEC	PART NUMBER
191	8	2510-01-164-9314	11862	14027796
191	9	5305-01-160-1970	11862	335551
191	10	5310-00-490-4639	96906	MS90724-40
191	11	2590-01-161-0212	11862	14027645
191	11	2590-01-162-7108	11862	14027646
191	12	2510-01-155-5848	11862	480534
191	13	2510-01-149-3465	11862	15594890
191	13	2510-01-178-8877	11862	15594889
191	14	5305-01-162-7884	11862	407217
191	15	5310-01-160-9529	24617	1494253
191	16	5306-01-158-9018	11862	2014469
191	17	5306-01-160-0769	11862	334195
191	18	5365-01-162-3849	11862	345819
191	19	2510-01-155-8817	11862	15614467
192	1	5310-01-197-5499	11862	1359887
192	2	2510-01-325-9070	11862	330489
192	2	2510-01-325-9071	11862	330490
192	3	2510-01-326-1389	11862	330393
192	3	2510-01-326-1460	11862	330394
192	4	5310-01-160-9529	24617	1494253
192	5	5305-01-158-2032	96906	MS51869-24
192	6	2510-01-325-9077	11862	15606405
192	6	2510-01-325-9078	11862	15606406
192	7	2510-01-325-8216	11862	15606404
192	8	5305-01-324-8862	11862	15606407
192	9		11862	TX015963
193	1	5325-01-165-2551	11862	20030401
193	2	2540-01-193-7884	11862	15591702
193	3	5305-01-266-9194	11862	11500668
193	4	2510-01-259-5587	11862	474022
193	5	5305-00-483-0554	96906	MS51862-26
193	6	2510-01-191-8447	11862	467247
193	6	2510-01-200-1021	11862	467248
193	7	2540-01-156-0088	11862	15594983
193	8	5305-01-162-8514	11862	11502634
193	9	5305-01-231-7384	24617	11503606
193	10	2590-01-247-3286	11862	15594896
193	10	5340-01-210-8824	11862	15594895
193	11	5970-01-158-9337	11862	462233
194	1	2540-01-159-8874	11862	14013753
194	1	2540-01-159-8875	11862	14013754
194	2	5310-01-154-2273	96906	MS90724-34
194	3	5305-01-157-1987	11862	11501149
194	4	5340-01-161-2789	11862	14031893
194	5	2540-01-159-8721	11862	15646949
194	6	5305-01-159-2779	11862	14044340
194	7	5305-01-158-7820	24617	11503395
194	8	5325-01-157-1698	11862	347347
195	1	2540-01-193-7896	11862	14079058
195	2	2540-01-159-7963	11862	14079056
195	3	2540-01-193-7895	11862	14079057

FIGURE AND ITEM NUMBER INDEX

FIG	ITEM	STOCK NUMBER	CAGEC	PART NUMBER
195	4	5306-01-159-1130	11862	342221
195	5	5306-01-157-3279	11862	471083
196	1		11862	14070448
196	1	2540-01-225-5972	11862	15594722
196	2		11862	14070453
196	2	2540-01-225-1106	11862	15594727
196	3	2540-01-155-6947	11862	14021208
196	4	2540-01-155-6822	11862	14021206
196	4	2540-01-225-1108	11862	15598767
196	5	5340-01-159-3880	11862	4876466
196	6	5365-01-157-7476	11862	343978
196	7	5310-01-157-5670	11862	3914674
196	8	3120-01-162-0060	11862	14066195
196	9	5365-01-197-2286	11862	14066196
196	10	5305-01-160-4528	11862	14021211
196	11	5305-01-161-2581	24617	9417325
196	12	2540-01-155-6825	11862	14037059
196	12	2540-01-155-8786	11862	14037060
196	13	2540-01-155-6823	11862	14021209
196	13	2540-01-155-6824	11862	14021210
196	14	5305-01-160-0331	11862	14021213
196	15	5305-01-160-3937	11862	9438916
196	16	5306-01-158-9018	11862	2014469
196	17	5360-01-162-2849	11862	20056525
196	18	3040-01-155-5864	11862	14022777
196	19	5355-01-157-1866	11862	465536
196	20	5360-01-160-2411	11862	329457
196	21	5340-01-158-6354	11862	14022786
196	22	3040-01-155-6912	11862	14022778
196	23	5360-01-160-2411	11862	329457
196	24	2540-01-155-6821	11862	14021204
196	25	2540-01-158-1877	11862	14021207
196	25	2540-01-225-2521	11862	15598766
197	1	2540-01-191-8668	11862	15591247
197	2	5306-01-159-1130	11862	342221
197	3	5305-01-162-7890	11862	3954730
197	4	5340-01-166-5861	11862	1731168
197	5	5306-01-157-3279	11862	471083
198	1	5306-01-159-1130	11862	342221
198	2	2540-01-160-3651	11862	15591248
198	3	5305-01-162-7890	11862	3954730
198	4	5306-01-157-3279	11862	471083
198	5	5340-01-166-5861	11862	1731168
199	1	5305-01-160-3937	11862	9438916
199	2	2540-01-164-1843	11862	14075823
199	3	5360-01-218-1610	11862	9711038
199	4	2540-01-164-1844	11862	14075824
199	5	5340-01-161-2749	11862	9834636
199	6	5360-01-197-0870	11862	20351007
199	7	5306-01-160-0769	11862	334195
199	8	2540-01-158-8558	11862	14075821

FIGURE AND ITEM NUMBER INDEX

FIG	ITEM	STOCK NUMBER	CAGEC	PART NUMBER
199	9	2540-01-158-4604	11862	467983
199	10	5360-01-162-2849	11862	20056525
199	11	5306-01-165-5583	11862	14060613
199	12	2540-01-158-4603	11862	20243999
200	1	2540-01-162-4411	11862	15599975
200	2	5360-01-164-1949	11862	9645073
200	3	5340-01-161-2749	11862	9834636
200	4	5360-01-171-8248	11862	9826897
200	5	2540-01-158-8561	11862	15599976
200	6	5360-01-162-2849	11862	20056525
200	7	5306-01-160-0769	11862	334195
200	8	5305-01-160-3937	11862	9438916
200	9	2540-01-159-7744	11862	14075822
200	10	2540-01-158-4604	11862	467983
200	11	5306-01-165-5583	11862	14060613
200	12	2540-01-158-4603	11862	20243999
201	1	2540-01-158-8812	11862	14075820
201	2	2540-01-166-1373	11862	14075817
201	3	5340-01-159-3880	11862	4876466
201	4	2540-01-193-7893	11862	14071555
201	5	5305-01-162-3961	11862	11503537
201	6	5305-01-162-5996	11862	20025648
201	7	2540-01-251-1714	11862	20293843
201	8	2540-01-251-1715	11862	20289493
201	9	2540-01-158-8559	11862	16604537
201	10	2540-01-165-0895	11862	14075818
201	11	3120-01-160-0570	11862	14059238
201	12	2540-01-193-7892	11862	14071461
201	13	5305-01-160-3955	11862	14056723
201	14	5310-01-162-7912	11862	1727059
202	1	2540-01-161-1356	11862	14075819
202	2	5340-01-159-3880	11862	4876466
202	3	2540-01-193-7894	11862	14071556
202	4	2540-01-166-1373	11862	14075817
202	5	2540-01-165-0895	11862	14075818
202	6	5305-01-162-5996	11862	20025648
202	7	2540-01-160-5840	11862	20293842
202	8	2540-01-191-8440	11862	20369919
202	9	2540-01-191-8441	11862	20369920
202	10	5325-01-198-8239	11862	20573776
202	11	2540-01-158-8560	11862	15627452
202	12	5305-01-162-3961	11862	11503537
202	13	2540-01-248-2477	11862	20410901
202	14	2540-01-193-7892	11862	14071461
202	15	3120-01-160-0570	11862	14059238
202	16	5310-01-162-7912	11862	1727059
202	17	5305-01-160-3955	11862	14056723
203	1	5305-01-159-6567	11862	14013724
203	2	2540-01-158-6896	11862	15569010
203	3	5306-01-158-9018	11862	2014469
203	4	2540-01-166-1370	11862	14075815

FIGURE AND ITEM NUMBER INDEX

FIG	ITEM	STOCK NUMBER	CAGEC	PART NUMBER
203	5	2540-01-158-6955	11862	471020
203	6	2540-01-165-0837	11862	14075816
203	7	2540-01-158-8811	11862	471021
203	8	2540-01-158-8563	11862	14027798
203	9	2540-01-159-0874	11862	14014293
203	10	5355-01-197-1501	11862	473187
203	11	5340-01-158-6895	11862	15569009
203	12	3120-01-159-1311	11862	335524
203	13	2540-01-158-7551	11862	470984
203	14	5305-01-157-9720	24617	9419327
203	15	2540-01-158-6893	11862	473197
203	16	5340-01-160-4130	11862	14027799
203	17	5310-00-834-7606	96906	MS35340-48
203	18	5305-01-163-2439	24617	9432194
203	19	5340-01-158-6892	11862	471018
203	20	5315-01-159-8660	11862	470974
203	21	2540-01-158-7583	11862	471010
203	22	5365-00-200-7377	96906	MS16633-1015
203	23	2540-01-159-8727	11862	14074479
203	24	5305-01-160-4494	11862	14069636
203	25	2540-01-160-3652	11862	14075375
203	26	5305-01-162-7890	11862	3954730
203	27	2540-01-160-3654	11862	14075374
203	28	2540-01-160-3653	11862	14075376
203	29	5340-01-159-3880	11862	4876466
203	30	2540-01-158-6894	11862	471079
203	31	2540-01-158-8562	11862	14027797
204	1	2540-01-167-2985	11862	15577694
204	2	2540-01-163-8017	11862	14075379
204	3	5306-01-159-1130	11862	342221
204	4	5306-01-157-3279	11862	471083
204	5	5340-01-166-5861	11862	1731168
205	1	2540-01-166-1370	11862	14075815
205	2	2540-01-163-7232	11862	14075372
205	3	2540-01-163-7233	11862	14075373
205	4	5360-01-157-1767	11862	14067457
205	5	2540-01-157-2966	11862	14025679
205	6	5340-01-157-7607	11862	14025673
205	7	5360-01-162-2849	11862	20056525
205	8	3040-01-155-0194	11862	14023889
205	9	5355-01-157-1866	11862	465536
205	10	5360-01-160-2411	11862	329457
205	11	5305-01-160-3937	11862	9438916
205	12	2540-01-157-3528	11862	335435
205	13	5306-01-158-9018	11862	2014469
205	14	2540-01-156-0061	11862	14025677
205	15	3040-01-155-0195	11862	14023890
205	16	5360-01-160-2411	11862	329457
205	17	5340-01-157-7608	11862	14025674
206	1	5305-01-160-3937	11862	9438916
206	2	2540-01-157-7907	11862	14027346

FIGURE AND ITEM NUMBER INDEX

FIG	ITEM	STOCK NUMBER	CAGEC	PART NUMBER
206	3	2540-01-155-5111	11862	14027348
206	4	2540-01-157-3528	11862	335435
206	5	2540-01-155-5110	11862	14027347
206	6	5306-01-158-9018	11862	2014469
206	7	2540-01-156-0061	11862	14025677
206	8	2540-01-155-5824	11862	14027345
207	1	5975-01-155-7084	80063	SC-D-866091
207	2	5310-00-933-8121	96906	MS35338-139
207	3	5305-00-068-0502	96906	MS90725-6
207	4	5820-01-026-0983	80063	DL-SC-B-691368
207	5	5310-00-903-5966	96906	MS51971-1
207	6	5975-01-027-0253	80063	SC-D-691375
207	7	5340-01-159-4518	80063	SC-C-691545
207	8	5340-01-201-7954	11862	15599959
207	9	5340-01-159-1460	11862	15599958
207	10	5310-01-152-0598	24617	271172
207	11	5306-01-168-4481	11862	9440334
207	12	5310-00-209-0786	96906	MS35335-33
207	13	5305-00-068-0508	96906	MS90728-6
207	14		80063	SC-B-75180-IV
207	15	5310-00-889-2528	96906	MS45904-68
208	1	2590-01-156-0076	11862	14074435
208	2	5340-00-455-5899	11862	14074437
208	3	5320-01-160-3999	11862	2477054
208	4	2590-01-156-0077	11862	14074441
208	4	2590-01-156-0078	11862	14074442
208	5	5306-01-197-3089	11862	9440033
208	6	2590-01-156-0080	11862	14074438
208	7	5340-01-253-2102	11862	15599965
208	8	2590-01-191-6511	11862	15599271
209	1	2590-01-156-0074	11862	14074439
209	2	5305-01-152-8945	24617	9421432
209	3	2590-01-156-0075	11862	14074432
209	4	5340-01-150-4105	11862	14005953
209	5	5340-01-199-4448	11862	15599961
210	1		25022	99-4361-0
210	2		25022	99-4317-0
210	3		25022	99-4276-0
210	4		25022	99-4290-0
210	5		25022	99-4299-0
210	6		25022	99-4342-0
210	7	5305-01-239-9265	25022	07-1016
210	8	6220-01-276-0635	13548	99012R
211	1	5340-01-162-3853	25022	09-0954
211	2	5320-00-582-3521	96906	MS20600-B4W3
211	3	5305-00-068-0510	96906	MS90728-60
211	4	6220-01-160-5094	25022	19-0817
211	5	6240-00-836-2079	08806	4411
211	6	2590-01-164-7024	25022	19-0884
211	7	6220-01-172-5300	25022	19-0882
211	8	6220-01-171-9557	25022	19-0885

FIGURE AND ITEM NUMBER INDEX

FIG	ITEM	STOCK NUMBER	CAGEC	PART NUMBER
211	9	5306-00-226-4833	96906	MS90728-40
211	10	5310-00-407-9566	96906	MS35338-45
211	11	5310-00-880-7744	96906	MS51967-5
211	12	2540-01-163-8623	25022	31-0030
211	13	5320-00-616-4346	96906	MS20600B6W4
211	14	5340-01-166-2011	25022	09-0093
211	15	5320-01-200-4017	96906	MS21141-U0604
211	16	2590-01-163-7626	25022	99-4308-1
211	17	9905-01-165-0541	25022	23-0196
211	18	9905-01-165-0542	25022	23-0193
211	19	5340-01-307-2252	25022	99-4582-1
211	20	5340-01-307-2251	25022	99-4581-1
211	21	5340-01-307-2250	25022	99-4580-1
211	22	5340-01-307-2249	25022	99-4579-1
212	1	5320-00-582-3521	96906	MS20600-B4W3
212	2	5340-01-162-3853	25022	09-0954
212	3	5305-00-068-0510	96906	MS90728-60
212	4	6220-01-160-5094	25022	19-0817
212	5	6240-00-836-2079	08806	4411
212	6	2590-01-164-7024	25022	19-0884
212	7	6220-01-172-5300	25022	19-0882
212	8	6220-01-171-9557	25022	19-0885
212	9	5306-00-226-4833	96906	MS90728-40
212	10	5310-00-407-9566	96906	MS35338-45
212	11	5310-00-880-7744	96906	MS51967-5
212	12	5340-01-307-2252	25022	99-4582-1
212	13	5340-01-307-2251	25022	99-4581-1
212	14	2540-01-163-8624	99688	78404
212	15	5320-00-616-4346	96906	MS20600B6W4
212	16	2590-01-163-7626	25022	99-4308-1
212	17	9905-01-165-0542	25022	23-0193
212	18	9905-01-165-0541	25022	23-0196
212	19	5320-01-200-4017	96906	MS21141-U0604
212	20	5340-01-166-2011	25022	09-0093
212	21	5340-01-164-0746	25022	22-0881
212	22	5340-01-164-1100	25022	99-4888-0
212	23	2540-01-163-8623	25022	31-0030
212	24	5340-01-307-2250	25022	99-4580-1
212	25	5340-01-307-2249	25022	99-4579-1
213	1	5320-00-616-4346	96906	MS20600B6W4
213	2		25022	19-0445
213	3	6220-01-160-3687	25022	19-0893
213	4	6240-00-889-1799	08806	1157
213	5	6220-01-160-3705	25022	19-0861
213	6	5365-01-269-8614	25022	19-0969
213	7	9905-01-165-0544	25022	23-0198
213	8	5320-00-582-3521	96906	MS20600-B4W3
213	9	5340-01-162-3853	25022	09-0954
213	10	5340-01-307-2247	25022	99-4577-1
213	11	5340-01-307-2248	25022	99-4578-1
213	12	4030-01-168-1282	25022	99-1088-0

FIGURE AND ITEM NUMBER INDEX

FIG	ITEM	STOCK NUMBER	CAGEC	PART NUMBER
213	13	5340-01-162-3854	25022	09-0959
213	14	5320-00-845-9501	96906	MS20600-B4W4
213	15	2590-01-163-7287	25022	22-0875
213	16	5306-00-226-4822	96906	MS90728-29
213	17	6220-01-160-3686	25022	19-0818
213	18	5305-00-984-6191	96906	MS35206-243
213	19	6220-01-039-9809	25022	51-1942
213	20	6240-01-089-6149	08805	F48T12/CW/WM
213	21	5305-01-166-1471	24617	9414238
213	22	6130-01-035-6412	25022	51-1991
213	23	2590-01-164-7898	25022	19-0820
214	1	5340-01-162-3854	25022	09-0959
214	2	5340-01-164-0974	25022	99-4356-1
214	3	5305-01-164-6321	24617	9414714
214	4	5930-01-163-2583	25022	99-4256-0
214	5	5930-00-548-5640	25022	51-1902
214	6	5930-00-636-1584	25022	51-1901
214	7	4710-01-062-3719	25022	06-0075
214	8	2540-01-165-8177	25022	99-4256-2
214	9	5310-00-934-9758	96906	MS35649-202
214	10	5310-00-045-3296	96906	MS35338-43
214	11	5315-01-025-0930	11862	3929059
214	12	5340-01-164-0747	25022	99-4315-1
214	13	5340-01-196-6463	25022	09-0963
214	14	2510-01-166-2015	25022	51-1305
214	15	5320-01-200-4017	96906	MS21141-U0604
214	16	5305-00-050-9237	96906	MS51957-71
214	17	5340-01-159-1796	25022	51-1304
214	18	5306-00-226-4829	96906	MS90728-36
215	1		25022	99-1265-01
215	2	5320-00-061-9662	96906	MS20600-B6W5
215	3	5320-00-845-9501	96906	MS20600-B4W4
215	4	7230-01-167-2075	25022	51-2294
215	5	2510-01-164-1532	25022	13-1105
215	6	5305-01-200-7735	24617	447143
215	7	2510-01-169-3785	25022	51-2204
215	8	2540-01-164-0046	25022	51-0999
215	9	5320-00-660-0821	96906	MS20600MP6W10
215	10	5340-01-194-5805	25022	99-4675-1
215	11	5310-00-087-7493	96906	MS27183-13
215	12	5305-00-068-0510	96906	MS90728-60
215	13	9340-01-162-5948	25022	51-1603
215	14	5310-00-637-9541	96906	MS35338-46
215	15	5340-01-156-5061	25022	51-0998
215	16	2510-01-163-7879	25022	51-1721
215	17	9905-01-165-6901	25022	23-0200
215	18	5320-00-582-3521	96906	MS20600-B4W3
215	19	5340-01-162-3853	25022	09-0954
215	20	2540-01-163-7874	25022	52-0914
215	21	5320-01-200-4017	96906	MS21141-U0604
215	22	2540-01-193-3623	25022	17-0209

FIGURE AND ITEM NUMBER INDEX

FIG	ITEM	STOCK NUMBER	CAGEC	PART NUMBER
215	23	5340-01-162-5619	25022	17-0178
215	24	5320-00-982-3815	96906	MS24662-153
215	25	5340-01-164-8137	25022	17-0177
215	26	5680-01-193-5078	25022	17-0202
215	27	5305-01-166-1610	24617	165079
215	28	2540-01-158-0602	25022	51-0988
215	29	5340-01-307-2253	25022	99-4583-1
215	30	5340-01-307-2254	25022	99-4584-1
215	31	5320-00-616-4346	96906	MS20600B6W4
216	1		25022	99-1265-02
216	2	5340-01-156-5061	25022	51-0998
216	3	5310-00-637-9541	96906	MS35338-46
216	4	5305-00-068-0510	96906	MS90728-60
216	5	5310-00-087-7493	96906	MS27183-13
216	6	9340-01-162-5948	25022	51-1603
216	7	2510-01-163-7879	25200	51-1721
216	8	5340-01-194-5805	25022	99-4675-1
216	9	5320-00-660-0821	96906	MS20600MP6W10
216	10	2510-01-163-8014	25022	13-1104
216	11	2510-01-169-3785	25022	51-2204
216	12	5305-01-200-7735	24617	447143
216	13	2540-01-164-0046	25022	51-0999
216	14	5320-00-081-9662	96906	MS20600-B6W5
216	15	5320-00-845-9501	96906	MS20600-B4W4
216	16	7230-01-167-2075	25022	51-2294
216	17	2540-01-028-0574	25200	51-0989
216	18	5680-01-193-5078	25022	17-0201
216	19	5320-00-982-3815	96906	MS24662-153
216	20	5340-01-164-8137	25022	17-0177
216	21	5340-01-162-5619	25022	17-0178
216	22	2540-01-193-3623	25022	17-0209
216	23	5320-01-200-4017	96906	MS21141-U0604
216	24	2540-01-163-7874	25022	52-0914
216	25	5340-01-162-3853	25022	09-0954
216	26	5320-00-582-3521	96906	MS20600-B4W3
216	27	9905-01-165-6901	25022	23-0200
216	28	5340-01-162-3850	25022	99-4680-1
216	29	5340-01-307-2253	25022	99-4583-1
216	30	5340-01-307-2254	25022	99-4584-1
216	31	5320-00-616-4346	96906	MS20600B6W4
217	1		25022	99-4358-0
217	2	5305-00-225-3843	96906	MS90728-8
217	3	5310-00-823-8804	96906	MS27183-9
217	4	5310-00-851-2674	96906	MS35691-1
217	5	5310-00-768-0318	96906	MS51967-14
217	6	5310-00-834-7606	96906	MS35340-48
217	7	2540-01-162-7117	25022	09-0955
217	8	5365-01-163-1147	74410	TH-0681
217	9	5310-00-809-5997	96906	MS27183-17
217	10	5305-00-071-2070	96906	MS90728-114
217	11	2540-01-164-6251	25022	99-4340-0

FIGURE AND ITEM NUMBER INDEX

FIG	ITEM	STOCK NUMBER	CAGEC	PART NUMBER
217	12	5310-00-851-2674	96906	MS35691-1
217	13	2540-01-191-6536	25022	99-4469-0
217	14	5310-00-823-8804	96906	MS27183-9
217	15	5305-00-225-3843	96906	MS90728-8
217	16	2540-01-191-8462	25022	99-4456-0
217	17	5310-00-891-1709	96906	MS35691-9
217	18	5310-00-809-3078	96906	MS27183-11
217	19	5306-00-226-4827	96906	MS90728-34
218	1	2510-01-216-0039	25022	22-0832
218	2	2540-01-159-1793	25022	22-0884
218	3	5320-00-616-4346	96906	MS20600B6W4
218	4	2540-01-163-7948	25022	19-0819
218	5	2540-01-163-7949	25022	19-0831
218	6	5340-01-028-9063	25022	51-0979
218	7	5320-01-200-4017	96906	MS21141-U0604
218	8	6250-01-164-3266	25022	19-0942
218	9	5320-00-822-6257	96906	MS20600-B6W10
218	10	5340-01-166-5654	25022	19-0849
218	11	2590-01-163-7669	25022	22-0861
218	12	5340-01-164-0958	25022	22-0877
218	13	5320-00-828-1284	96906	MS24662-155
218	14	2510-01-159-1798	25022	99-4420-2
218	15	2540-01-163-7288	25022	09-0965
218	16	5305-01-201-3334	24617	171108
218	17	2590-01-162-7139	25022	99-4420-1
219	1	6250-01-164-3266	25022	19-0942
219	2	5320-00-822-6257	96906	MS20600-B6W10
219	3	5320-00-616-4346	96906	MS20600B6W4
219	4	5340-01-166-5654	25022	19-0849
219	5	5320-01-200-4017	96906	MS21141-U0604
219	6	5340-01-028-9063	25022	51-0979
219	7	2540-01-159-1793	25022	22-0884
219	8	2510-01-216-0039	25022	22-0832
219	9	2590-01-162-7139	25022	99-4420-1
219	10	2540-01-163-7288	25022	09-0965
219	11	5305-01-201-3334	24617	171108
219	12	5930-01-163-8851	25022	37-0020
219	13	5930-01-163-8924	25022	51-1959
219	14	5930-01-163-2779	25022	37-0019
219	15	5320-00-828-1284	96906	MS24662-155
219	16	2510-01-159-1798	25022	99-4420-2
219	17	2590-01-163-7669	25022	22-0861
219	18	5340-01-164-0958	25022	22-0877
220	1		25022	99-4447-0
220	2	5320-00-061-9648	96906	MS20601-B6W6
220	3	9330-01-098-6554	25022	51-2297
220	4	2590-01-163-7290	25022	22-0880
220	5		25022	22-0867
220	6	5320-01-200-4017	96906	MS21141-U0604
220	7	5340-01-028-9063	25022	51-0979
220	8	6150-01-190-6498	25022	19-0837

FIGURE AND ITEM NUMBER INDEX

FIG	ITEM	STOCK NUMBER	CAGEC	PART NUMBER
220	9	5310-00-851-2674	96906	MS35691-1
220	10	2540-01-165-5971	25022	19-0843
220	11	6150-01-166-2014	25022	19-0881
220	12	5305-00-068-0509	96906	MS90728-10
220	13	5340-01-164-1078	25022	51-0991
220	14	2510-01-191-4336	25022	99-4894-02
220	15	2510-01-191-4335	25022	99-4894-01
220	16	5320-00-616-4346	96906	MS20600B6W4
220	17	5320-00-845-9501	96906	MS20600-B4W4
220	18	5935-01-022-2377	81348	WC596/9-1
220	19	6150-01-166-2013	25022	19-0880
221	1		25022	99-4325-0
221	2	5320-01-200-4017	96906	MS21141-U0604
221	3	5340-01-028-9063	25022	51-0979
221	4		25022	22-0866
221	5	9330-01-098-6554	25022	51-2297
221	6	5320-00-061-9648	96906	MS20601-B6W6
221	7	2590-01-163-7290	25022	22-0880
221	8	2510-01-191-4334	25022	99-4893-01
221	9	2510-01-191-6638	25022	99-4893-02
221	10	5320-00-616-4346	96906	MS20600B6W4
221	11	5340-01-164-1078	25022	51-0991
222	1	6130-01-158-9175	25022	19-0833
222	2	5945-01-173-7760	25022	19-0875
222	3	5945-01-191-3743	25022	99-4519-1
222	4	5995-01-192-4374	25022	19-0910
222	5	5925-01-190-1211	25022	19-0909
222	6	5310-01-271-1793	25022	07-01-01
222	7	5940-01-171-3195	25022	19-0877
222	8	5945-01-166-2012	25022	19-0876
222	9	5320-00-616-4346	96906	MS20600B6W4
222	10	5305-01-201-3788	24617	9426623
222	11	5920-01-188-6294	11862	12004011
222	12	5920-01-149-6952	11862	12004008
222	13	5305-01-166-1473	24617	9418719
222	14	5920-01-149-6953	11862	12004010
222	15	5920-01-164-8260	25022	19-0878
222	16	5305-01-164-6321	24617	9414714
222	17	5895-01-159-1804	25022	99-4360-1
223	1	2540-01-163-8638	25022	36-0039
223	2	2540-01-163-8598	25022	36-0045
223	3	4820-01-173-1250	25022	36-0044
223	4	5315-00-839-5820	96906	MS24665-134
223	5	5315-01-164-5336	25022	36-0056
223	6	2540-01-165-7931	25022	36-0055
223	7	5305-00-709-8517	96906	MS90727-85
223	8	2540-01-164-1562	25022	36-0054
223	9	2540-01-191-6539	25022	36-0070
223	10	5310-00-809-4061	96906	MS27183-15
223	11	5310-00-835-2036	96906	MS35691-29
223	12	5310-00-209-0965	96906	MS35338-47

FIGURE AND ITEM NUMBER INDEX

FIG	ITEM	STOCK NUMBER	CAGEC	PART NUMBER
223	13	5305-00-071-2506	96906	MS90728-3
223	14	5315-01-209-7063	25022	36-0071
223	15	2540-01-163-8600	25022	36-0047
223	16	2540-01-163-8599	25022	36-0046
223	17	5360-01-163-5578	25022	36-0048
223	18	5360-01-163-5579	25022	36-0049
223	19	2540-01-163-0772	25022	09-0056
223	20	5360-01-163-5580	25022	36-0050
223	21	5315-01-197-1483	25022	36-0075
223	22	5315-00-057-5541	96906	MS16562-147
223	23	5310-00-637-9541	96906	MS35338-46
223	24	5310-00-732-0559	96906	MS51968-8
223	25	5340-01-197-1550	25022	36-0074
223	26	5310-00-087-7493	96906	MS27183-13
223	27	2540-01-173-1249	25022	36-0043
223	28	5315-01-197-1482	25022	36-0073
223	29	5315-00-814-3531	96906	MS16562-50
223	30	5315-01-164-5334	25022	36-0051
223	31	5340-01-193-9565	25022	36-0072
224	1	2590-01-163-8596	25022	22-0865
224	2	9330-01-098-6554	25022	51-2297
224	3	5320-00-061-9648	96906	MS20601-B6W6
224	4	2540-01-166-5913	25022	22-0860
224	5	5340-01-028-9063	25022	51-0979
224	6	5320-01-201-9453	96906	MS21141-U0607
224	7	5340-01-165-0602	25022	37-0051
224	8	5315-00-814-3530	96906	MS16562-35
224	9	2540-01-025-0433	25022	36-0066
224	10	5320-01-200-4017	96906	MS21141-U0604
224	11	2540-01-165-6793	25022	51-0985
225	1		25022	99-4593-0
225	2	5320-01-200-4017	96906	MS21141-U0604
225	3	5340-01-164-1076	25022	22-0876
225	4	5340-01-028-9063	25022	51-0979
225	5	5340-01-164-1077	25022	22-0878
225	6	5320-00-982-3815	96906	MS24662-153
225	7	5340-01-164-8137	25022	17-0177
225	8	5340-01-162-5619	25022	17-0178
226	1	6210-01-159-1794	25022	19-0816
226	2	6220-01-197-0486	25022	19-0940
226	3	6240-00-850-4280	08806	1003
226	4	6230-01-191-3856	25022	19-0941
226	5		25022	19-0941-1
227	1	2510-01-159-1797	25022	99-4282-1
227	2	5305-01-164-6321	24617	9414714
227	3	5930-01-163-2779	25022	37-0019
227	4	5320-00-616-4346	96906	MS20600B6W4
227	5	5930-01-163-8924	25022	51-1959
227	6	5930-01-163-8851	25022	37-0020
227	7	2510-01-159-1795	25022	99-4408-1
227	8	5320-01-200-4017	96906	MS21141-U0604

FIGURE AND ITEM NUMBER INDEX

FIG	ITEM	STOCK NUMBER	CAGEC	PART NUMBER		
227	9	5320-00-845-9501	96906	MS20600-B4W4		
227	10	5930-01-028-1949	25022	51-1926		
227	11	5340-01-159-1796	25022	51-1304		
228	1	2510-01-192-9752	25022	99-4324-0		
228	2	9340-01-162-5947	25022	51-1604		
228	3	5330-01-163-5706	25022	51-1724		
228	4	5306-01-166-1665	21450	126281		
228	5	2510-01-192-9752	25022	22-0864		
228	6	5310-00-208-1918	88044	AN365-1024A		
228	7	5320-00-982-3815	96906	MS24662-153		
228	8	7230-01-165-6795	25022	22-0873		
228	9	2540-01-163-0834	25022	09-0951		
228	10	2540-01-162-7116	25022	09-0952		
228	11	5315-01-196-6464	25022	99-4683-1		
228	12	5305-00-057-9608	96906	MS21207-10-10		
228	13	2540-01-163-3585	25022	08-0475		
228	14	5320-01-166-1477	24617	163881		
228	15	5340-01-172-1942	25022	99-4257-1		
229	1	5320-00-616-4344	96906	MS20600B6W8		
229	2	2510-01-164-9280	25022	99-4530-1		
229	3	2540-01-163-8597	25022	22-0858		
229	4	5320-00-637-5014	96906	MS20600-B6W12		
229	5	2540-01-163-0894	25022	17-0195		
229	6	5305-00-052-7472	96906	MS24627-67		
229	7	2540-01-191-6538	25022	99-4892-0		
229	8	5305-00-225-3843	96906	MS90728-8		
230	1	2510-01-147-9917	11862	14072479		
230	2	2540-01-191-8549	11862	14072475		
231	1	5306-00-226-4832	96906	MS90728-39		
231	2	3120-01-194-0754	11862	15599919		
231	3				11862	14072471
231	4				11862	14072472
231	5	5310-00-081-4219	96906	MS27183-12		
231	6				11862	14072473
231	7	5310-00-814-0673	96906	MS51943-33		
231	8	2540-01-192-3576	11862	14072468		
231	9	2540-01-191-9564	11862	14072470		
231	10	5340-01-038-3428	19204	12255559		
231	11	5340-01-036-7665	19207	12255567		
231	12	2540-01-192-4479	11862	14072469		
231	13	5310-01-133-7215	72582	9419454		
231	14	5305-01-197-6351	24617	157456		
231	15	5340-01-044-8389	19207	12255561		
231	16	5305-00-054-6654	96906	MS51957-30		
231	17	5310-01-197-6621	24617	9423530		
231	18	5340-01-043-5214	19207	11669126-1		
232	1	5310-00-080-6004	96906	MS27183-14		
232	2	5340-01-150-4105	11862	14005953		
232	3	5305-00-225-3843	96906	MS90728-8		
232	4	5305-01-197-6576	24617	9415153		
232	5	2540-01-191-8443	11862	14072461		

FIGURE AND ITEM NUMBER INDEX

FIG	ITEM	STOCK NUMBER	CAGEC	PART NUMBER
232	6	2540-01-191-8444	11862	14072462
232	7	5310-01-152-0598	24617	271172
232	8	5320-01-200-4017	96906	MS21141-U0604
232	9	5340-00-916-6539	96906	MS51939-2
232	10	5325-01-050-6192	19207	12255564-2
232	11	2540-01-191-8445	11862	14072464
232	12	5325-01-050-6192	19207	12255564-2
232	13	5310-01-152-0598	24617	271172
232	14	5340-00-916-6539	96906	MS51939-2
232	15	5320-01-200-4017	96906	MS21141-U0604
232	16	5310-00-596-6897	24617	456004
232	17	2540-01-191-8446	11862	14072466
232	18	5325-01-050-6192	19207	12255564-2
232	19	5340-00-916-6539	96906	MS51939-2
232	20	5320-01-200-4017	96906	MS21141-U0604
232	21	5310-01-152-0598	24617	271172
232	22	2540-01-192-3575	11862	14072467
232	23	2540-01-192-3574	11862	14072465
232	24	2540-01-192-3573	11862	14072463
233	1	2540-01-157-3032	11862	15591703
233	2	5305-01-165-8612	24617	9421985
233	3	5310-00-514-6674	96906	MS35335-34
233	4	2540-01-156-0584	11862	15591704
233	5		11862	14044931
233	6	2540-01-158-8612	11862	14076838
233	7	5305-01-246-5770	24617	9419163
233	8	5340-01-149-4434	11862	3816659
233	9		11862	22021655
233	10	2540-01-156-5882	11862	22029629
233	11	2540-01-158-1569	11862	4918562
233	12	2540-01-191-6512	11862	22029630
233	13		11862	3782730
233	14	2540-01-159-7954	72560	22048352
233	15	2540-01-158-8611	11862	3990892
233	16	5305-01-132-2166	96906	MS51871-4
233	17	4730-01-158-8717	11862	3986821
233	18		11862	329198
233	19	2590-01-160-1047	11862	3798372
233	20	5306-01-160-7553	11862	3824124
233	21	5330-01-161-2608	11862	20489125
233	22		11862	337714
233	23		11862	3782732
234	1	2540-01-156-8316	11862	22048368
234	2	5330-01-161-4019	11862	22048303
234	3	2540-01-157-3525	11862	4914356
234	4		11862	22054154
234	5	5305-00-866-0934	11862	22049837
234	6	5360-01-158-1727	11862	22049511
234	7	2540-01-156-0069	11862	22049531
234	8	5310-01-157-4855	11862	22054153
234	9	2540-01-243-4934	11862	22038927

FIGURE AND ITEM NUMBER INDEX

FIG	ITEM	STOCK NUMBER	CAGEC	PART NUMBER
234	10	5330-01-244-2277	11862	22054156
234	11	5365-01-156-0067	11862	22054155
234	12	5325-01-167-8372	11862	22021679
234	13	3110-00-950-4236	96906	MS19061-20005
234	14	5330-01-157-4872	11862	22048281
234	15	5340-01-162-9954	11862	22048287
234	16	5360-01-195-6699	11862	22009291
234	17	2540-01-164-8379	11862	22038909
234	18	5977-01-154-6777	16748	22009290
234	19	6150-01-208-4507	11862	22038924
234	20	2590-01-159-9809	11862	22049534
234	21	2920-01-212-4941	11862	22038925
234	22		11862	22021690
234	23	3020-01-199-7847	11862	4939314
234	24	2540-01-158-1570	11862	4918446
234	25	5365-00-282-1633	96906	MS16633-1018
234	26	5930-01-155-0783	11862	22038931
234	27		11862	22054158
234	28	2540-01-156-0068	11862	22038928
235	1	2540-01-086-5433	11862	9831062
235	2	2540-01-156-9740	11862	918656
235	3		11862	14072485
235	3		11862	14072486
235	4	5330-01-159-2807	11862	14008191
235	5	2540-01-157-6414	11862	14007429
235	6	2540-01-156-4855	11862	14007430
235	7	2540-01-155-0376	11862	14072487
235	7	2540-01-323-6049	11862	15634659
235	7	2540-01-323-6050	11862	15634660
235	8	5325-01-167-0510	11862	14007435
235	9	2540-01-159-8760	11862	14072489
235	10	5310-01-160-5708	11862	52351724
235	11	5305-01-162-0015	11862	15554576
235	12	5306-01-160-1968	11862	14007511
235	13	5306-01-203-9082	24617	9426277
235	14	5340-01-150-4105	11862	14005953
235	15	2540-01-164-7260	11862	14007542
235	15	2540-01-165-6145	11862	14007541
235	16	5310-01-158-6780	11862	14007539
236	1	5325-01-163-5973	11862	14076298
236	2	5340-01-157-6697	11862	14072445
236	3	5310-01-242-8561	11862	15599994
236	4	5305-01-163-2438	11862	9440300
237	1	5310-01-242-8561	11862	15599994
237	2	5325-01-230-1844	11862	15599990
237	3	5340-01-157-6697	11862	14072445
237	4	5305-01-163-2438	11862	9440300
237	5	5310-00-959-4679	96906	MS35340-45
237	6	5306-00-226-4827	96906	MS90728-34
237	7	5340-01-162-6062	11862	15599962
237	8	5340-01-164-6410	11862	14072441

FIGURE AND ITEM NUMBER INDEX

FIG	ITEM	STOCK NUMBER	CAGEC	PART NUMBER
237	9	5340-01-164-6411	11862	14072440
237	10	5340-01-164-6412	11862	14072439
237	11	5340-01-198-3434	11862	15599964
237	12	5340-01-159-4519	11862	15599963
237	13	5340-01-159-4517	118623	14072442
238	1	5306-01-155-7659	24617	9425117
238	2	5340-01-162-4820	11862	14075846
238	3	4210-00-383-7127	19207	7357009
238	4	5340-01-163-0632	11862	14075847
238	5	5340-01-163-5903	11862	14075848
238	6	5340-01-164-1048	25022	99-4602-1
238	7	5310-00-407-9566	96906	MS35338-45
238	8	5306-00-226-4827	96906	MS90728-34
238	9	5340-01-164-1075	25022	99-4341-1
238	10	5310-00-891-1709	96906	MS35691-9
239	1	5330-01-157-1952	11862	6273325
239	2		11862	3054308
240	1	5306-01-155-8528	24617	9419699
240	2	5340-01-158-6816	11862	3025501
240	3	5340-01-165-3717	11862	3024673
240	4	2540-01-156-0072	11862	3027247
240	5	2520-01-192-1257	11862	3048083
240	6	2540-01-159-8722	11862	3054315
240	7	2540-01-155-7299	11862	3024867
240	8	2540-01-156-4874	11862	3030075
240	9	5305-01-159-2781	24617	11500999
240	10	2540-01-156-0070	27462	3054316
240	11	2540-01-156-0071	11862	3030072
240	12	2540-01-159-8705	11862	3027308
240	13	2540-01-156-0073	11862	3048067
241	1	5310-01-158-6260	24617	11506003
241	2	5905-01-159-0771	11862	500890
241	3	5305-01-269-4329	11862	11513932
241	4		11862	3058097
241	5	2540-01-158-6906	11862	3029730
241	6	5310-01-164-2336	11862	3042351
241	7	5310-01-211-1648	11862	3015545
241	8	6105-01-165-4561	11862	3037550
241	9	5330-01-195-4880	11862	3039873
241	10	5305-01-160-1974	11862	11500742
241	11	4730-01-160-0814	11862	3013475
241	12	4720-01-160-3664	11862	3036927
241	13	6105-01-164-6546	11862	22020945
241	14	5305-01-164-1604	24617	11500997
242	1	5310-01-160-5708	11862	52351724
242	2	5305-01-164-2313	24617	9415163
242	3	2540-01-165-8174	11862	14064924
242	4	5305-01-197-2320	24617	9440025
242	5	5305-01-162-9689	24617	11503396
242	6	2540-01-165-5996	11862	14013122
242	7	2540-01-158-9370	11862	14074319

FIGURE AND ITEM NUMBER INDEX

FIG	ITEM	STOCK NUMBER	CAGEC	PART NUMBER
243	1	2590-01-159-8857	11862	6258364
243	2	5305-00-146-2524	96906	MS51850-86
243	3	5310-01-268-8948	11862	11501937
243	4	2590-01-158-8551	11862	16034561
243	5	2590-01-159-8763	11862	1226174
243	6	5355-01-157-1865	11862	336406
243	7	5930-01-165-0732	16758	16015256
243	8	5305-01-197-3287	24617	9419663
243	9	5305-01-164-2313	24617	9415163
244	1	4730-00-908-3194	96906	MS35842-11
244	2		11862	482995
244	2		11862	487425
244	3	5340-01-200-8473	11862	11509088
244	4	5340-01-159-1324	11862	3825416
244	5		11862	10012288-45
245	1		25022	22-0871
245	2	5330-01-192-9335	99688	58636
245	3	5340-01-199-4993	99688	58659
245	4	5310-01-161-7308	99688	62229
245	5	5305-00-432-4163	96906	MS51861-24
245	6	5360-01-203-6365	99688	58627
245	7	5330-01-192-8906	99688	58637
245	8	5305-01-162-8512	24617	9421073
245	9	2510-01-191-4259	99688	85941
245	10	4730-00-908-6294	96906	MS35842-16
245	11	2540-01-194-3323	78385	10530B
245	12	4730-00-900-3296	72983	248X4
245	13	4720-01-192-3510	99688	58625
245	14	5305-00-984-6195	96906	MS35206-247
245	15	4520-01-192-6073	99688	58623
245	16	4730-00-908-6293	96906	MS35842-15
245	17		99688	58628
245	18	5320-01-200-4017	96906	MS21141-U0604
245	19	2540-01-191-4327	99688	784030
245	20		99688	78405
245	21	2990-01-192-4597	99688	58649
245	22	4730-01-192-4434	99688	58620
245	23	4520-01-192-6005	99688	46090
245	24	5330-01-194-4754	99688	58635
245	25	2990-01-192-4576	99688	58651
245	26		99688	58629
245	27	5330-01-192-8905	99688	58632
245	28	5305-01-087-1917	96906	MS51849-70
245	29	5310-00-014-5850	96906	MS27183-42
245	30	2910-00-025-3493	96906	MS51085-1
245	31	5330-00-265-1089	96906	MS29513-125
245	32	2910-00-203-3322	90005	26422-B
245	33	5330-01-192-8904	99688	58624
245	34	2540-01-191-6541	99688	85942
245	35	5315-00-899-4119	96906	MS24665-446
245	36	2590-01-191-4357	99688	65148

FIGURE AND ITEM NUMBER INDEX

FIG	ITEM	STOCK NUMBER	CAGEC	PART NUMBER
245	37	2590-01-192-5911	99688	859430
245	38	5320-00-616-4346	96906	MS20600B6W4
245	39	2590-01-191-9276	99688	65149
245	40	6105-01-188-7154	0A7R8	PV23135R
245	41	2540-01-191-4331	99688	46023
245	42	4140-01-188-6977	99688	58634
245	43	4720-01-191-8463	99688	78406
246	1	4710-01-163-2805	78385	704363
246	2		78385	G-704213
246	3	5310-00-061-0004	96906	MS45904-57
246	4	5310-00-934-9757	96906	MS35649-282
246	5	5325-00-174-5314	96906	MS35489-5
246	6	5310-00-209-0788	96906	MS35335-30
246	7	5310-00-934-9747	96906	MS35649-262
246	8	4810-00-248-1635	19207	11663058
246	9		78385	G-704183
246	10	5305-00-403-5130	78385	719675
246	11	5905-00-251-7145	19207	11663061
246	12	5305-00-984-4983	96906	MS35206-226
246	13	5930-00-679-5925	19207	10948233
246	14	5305-01-076-6308	96906	MS51863-22
246	15	5310-00-579-0079	96906	MS35333-37
246	16	5935-01-163-8987	78385	G-704234
246	17		78385	704501
246	18	5320-00-801-1548	78385	488755
246	19	5305-00-984-6191	96906	MS35206-243
246	20	5310-00-596-7693	96906	MS35335-31
246	21	5930-00-345-5455	19207	11663057
246	22	2540-00-216-5722	78385	704225
246	23	5310-01-196-6465	79470	1461-4
246	24		21450	114628
246	25		78385	G-704232
246	26	5330-00-089-0978	78385	702903
246	27	5330-00-089-0998	19207	11588688
246	28		78385	G-704195
246	29		78385	703547
246	30		78385	703546
246	31		78385	G-704373
246	32		78385	G704177
246	33	5305-01-097-7894	96906	MS51863-32
246	34	2540-01-008-1501	78385	G-704554
246	35	5305-00-984-6189	96906	MS35206-241
246	36		78385	488773
246	37		78385	G-704288-1
246	38	4520-00-217-5782	16236	CS-4520-SV-0705
246	39		78385	G-704293
247	1	5325-00-276-4993	99688	58660
247	2	5325-00-279-1235	996888	50090
247	3	2590-01-214-7131	99688	859450
247	4	6150-01-010-6558	24234	221968
247	5	5310-01-161-7308	99688	62229

FIGURE AND ITEM NUMBER INDEX

FIG	ITEM	STOCK NUMBER	CAGEC	PART NUMBER
247	6	5305-01-163-5761	96906	MS51869-26
247	7	5310-00-103-2953	99688	62220
247	8	5305-00-432-4163	96906	MS51861-24
247	9	5340-01-199-4993	99688	58659
247	10	5355-01-167-4114	99688	58668
247	11	5325-01-197-1547	99688	58661
247	12	4010-01-200-7581	99688	58643
247	13	5310-01-197-4435	99688	62219
247	14	2540-01-191-6540	99688	85944
247	15	6210-00-688-5088	96906	MS25331-4-313S
247	16	5305-01-162-8512	24617	9421073
247	17	2590-01-191-9275	99688	65151
247	18	5925-00-553-2274	82647	PDA-15
247	19	5930-00-655-1514	96906	MS35058-22
247	20	5930-00-655-1512	91929	32TS1-5
247	21	2590-01-191-9274	99688	65153
247	22	5930-01-190-1231	99688	65171
247	23	2590-01-191-9272	99688	65155
247	24	2590-01-191-9271	99688	65152
247	25	2590-01-191-9273	99688	65154
248	1		78385	704396
248	2		78385	704395
248	3		78385	704432
248	4		78385	704402
248	5		78385	704390
248	6	5330-01-008-6527	78385	704406
248	7	4810-01-161-0817	78385	G-700637-101
248	8	5305-01-057-9322	24617	9415319
248	9		78385	705311
248	10		78385	704391-1
248	11		78385	704789
248	12		78385	735447-5
248	13		78385	704447
248	14	5305-00-115-9436	96906	MS51863-21
248	15		78385	G704676
248	16	5930-01-077-7793	78385	704401
248	17	4730-00-288-9497	96906	MS39173-2
248	18		78385	702942
248	19	4730-00-043-3750	35211	RL984A
248	20	5360-01-057-9249	78385	704434
248	21	5305-01-057-8031	78385	704397
248	22	5330-01-136-8334	78385	705344
248	23	5310-00-596-7693	96906	MS35335-31
248	24	5310-00-934-9757	96906	MS35649-282
248	25		78385	705347
248	26	5307-01-136-8740	78385	705345
248	27		78385	G705349
248	28	4730-00-701-7737	79470	60X3
248	29		78385	G-704398
248	30		78385	704580
248	31	5360-00-327-5879	78385	476339

FIGURE AND ITEM NUMBER INDEX

FIG	ITEM	STOCK NUMBER	CAGEC	PART NUMBER
248	13		78385	474669
248	33		78385	476220
248	34		78385	484044
248	35	5310-00-285-5112	78385	476229
248	36	4810-01-161-0817	78385	G-700637-100
248	37	5305-00-145-0828	24617	455956
248	38		78385	701328
248	39		78385	735411
248	40		78385	704898
248	41	4820-01-057-5925	78385	704579
248	42	5305-01-200-9869	24617	9419950
249	1	5310-00-596-7691	96906	MS35335-32
249	2	5305-00-990-6444	96906	MS35207-261
249	3	5310-00-596-7693	96906	MS35335-31
249	4	5305-00-984-6189	96906	MS35206-241
249	5		78385	G-704553
249	6		78385	G-720122
249	7	6105-00-116-5124	19207	8720780-1
249	8		78385	G-700038
249	9	5365-01-037-6813	78385	702129
249	10	2040-00-754-4176	78385	488935
249	11	5310-00-685-6855	96906	MS35790-9
249	12	5305-00-958-5477	96906	MS35190-254
249	13		78385	G-488934
249	14	5310-01-164-1036	78385	736861
249	15	5305-00-225-3843	96906	MS90728-8
249	16		78385	704408
250	1		78385	704192
250	2		78385	704678
250	3	5310-01-018-5634	96906	MS14151-6
250	4		78385	G-704284
250	5	5305-01-136-8734	78385	705117
250	6		78385	G-704196
250	7		78385	704181
250	8		78385	704371-1
250	9	5340-01-164-1063	78385	704285
250	10		78385	704189
250	11		78385	704191
250	123	5310-00-878-7196	96906	MS21043-08
250	13		78385	704283
251	1	5310-01-194-9208	24617	9409761
251	2	5310-01-193-6927	11862	9409754
251	3	4730-01-069-6408	93061	144F-5
251	4	4730-00-011-8538	81343	5 010111B
251	5		11862	14072371
251	6	4730-00-266-0535	81343	SAEJ513
251	7	4730-00-619-9362	24617	140642
251	8	2540-01-163-1141	11862	14072347
251	9	5305-01-164-2313	24617	9415163
251	10	5340-01-150-4105	11862	14005953
251	11	2910-00-710-6054	96906	MS51321-1

FIGURE AND ITEM NUMBER INDEX

FIG	ITEM	STOCK NUMBER	CAGEC	PART NUMBER
251	12	5306-01-160-1968	11862	14007511
251	13	5340-01-164-6474	11862	343444
251	14	5340-01-164-6473	11862	14004512
251	15	4730-00-266-0536	16764	110200
251	16	4720-01-162-5113	11862	14072368
251	17		11862	14072369
252	1	5305-01-156-5438	11862	11504447
252	2	5340-00-057-3034	96906	MS21333-110
252	3	4720-01-163-1089	11862	14063324
252	4		11862	14063323
252	5	4730-00-011-8538	81343	5 010111B
252	6	5310-01-160-9529	24617	1494253
252	7	5306-00-226-4824	96906	MS90728-31
252	8	2990-01-173-4197	9W635	MBH-V
252	9	4730-01-192-4394	96906	MS39158-23
252	10	4820-00-785-8153	93061	NV108P-4
252	11	4730-00-278-4824	93061	2202P-4-4
252	12	4730-00-471-3102	30780	1203P-4
252	13	4730-00-765-9103	93061	215PN-4
252	14	4730-01-245-6925	93061	207ACBHS-4
252	15	4730-00-011-8538	81343	5 010111B
252	16		93061	42F-5
252	17		9W635	31-0040
252	18		9W635	31-0040A
252	19	4820-01-192-5819	79227	B-6000-1/4IN
252	20	5305-00-225-3843	96906	MS90728-8
252	21	5310-00-823-8804	96906	MS27183-9
252	22	5310-00-209-0965	96906	MS35338-47
252	23	5310-00-851-2674	96906	MS35691-1
252	24	5340-00-150-1658	17773	11176106-5
252	25	4730-00-266-0535	81343	SAEJ513
252	26		30780	260-6-6
252	27	4720-01-212-4782	9W635	500-4
253	1	8145-01-231-3747	19207	12338064
253	2	2520-01-268-1051	19207	12338078
253	3	8145-01-005-2994	10001	2642880
253	4		19207	12338066
253	5	5330-01-264-6537	19207	12338073
253	6	5340-01-264-6544	19207	12338068
253	7	5340-01-271-3059	19207	12338067
253	8	5310-00-809-4061	96906	MS27183-15
253	9	5310-00-637-9541	96906	MS35338-46
253	10	5305-00-269-3219	96906	MS90725-69
253	11	5340-01-264-6540	19207	12338074
253	12		19207	12338065
253	13	5305-00-269-3233	96906	MS90727-57
253	14	4730-00-221-2140	96906	MS20913-6S
253	15	5306-00-226-4828	96906	MS90728-35
253	16	5305-00-071-2083	96906	MS90728-127
253	17	5310-00-809-5998	96906	MS27183-18
253	18	5340-01-265-3676	19207	12338071

FIGURE AND ITEM NUMBER INDEX

FIG	ITEM	STOCK NUMBER	CAGEC	PART NUMBER
253	19	5310-00-584-5272	96906	MS35338-48
253	20	5310-00-768-0318	96906	MS51967-14
253	21	5305-00-732-0511	96906	MS90728-110
253	22	5340-01-264-6543	19207	12338070
253	23	5310-00-081-4219	96906	MS27183-12
253	24	5310-00-407-9566	96906	MS35338-45
253	25	5310-00-880-7744	96906	MS51967-5
253	26	5310-00-809-3079	96906	MS27183-19
253	27	5340-01-264-6541	19207	12338075
253	28	5340-01-265-3674	19207	12338076
253	29	5305-00-071-2069	96906	MS90728-113
253	30	5305-00-068-0511	96906	MS90728-62
253	31	6685-00-618-1822	19207	10906697
254	1	2815-01-163-1174	118623	14067730
254	2	5330-00-935-9136	24617	274244
254	3	5330-01-150-7744	11862	14022683
254	4	5305-01-165-5591	11862	9439930
254	5	2990-01-163-3575	11862	15599200
254	6	2520-01-163-3494	11862	15599201
254	7	5365-01-149-0880	11862	337185
254	8	5330-00-107-3925	11862	14079550
255	1	5310-01-194-9208	24617	9409761
255	2	5310-01-193-6927	11862	9409754
255	3	4730-01-069-6408	93061	144F-5
255	4	4730-00-011-8538	81343	5 010111B
255	5		11862	14072371
255	6	4730-00-619-9362	24617	140642
255	7	5305-00-432-4171	24617	9414713
255	8	2540-01-163-1141	11862	14072347
255	9	5340-01-150-4105	11862	14005953
255	10	2910-00-710-6054	96906	MS51321-1
255	11	5306-01-160-1968	11862	14007511
255	12	4730-00-142-2177	81240	GM118749
255	13	5340-01-164-6474	11862	343444
255	14	5340-01-164-6473	11862	14004512
255	15		11862	14072369
255	16	4720-01-162-5113	11862	14072368
255	17	5340-01-268-9064	11862	25527423
255	18	4730-00-266-0536	16764	110200
255	19	4710-01-232-8478	11862	14063323
255	20	5306-00-226-4824	96906	MS90728-31
255	21	5340-00-057-3034	96906	MS21333-110
255	22	5310-01-160-9529	24617	1494253
255	23	5340-00-282-7539	96906	MS21333-46
255	24	5305-01-140-9118	96906	MS90728-59
255	25	5325-01-164-8655	11862	3886908
255	26	5325-00-279-1248	96906	MS35489-103
255	27	4720-01-162-0121	11862	14076390
255	28	5310-01-148-5922	11862	11500046
255	29	5310-00-732-0558	96906	MS51967-8
255	30	4720-01-163-1089	11862	14063324

FIGURE AND ITEM NUMBER INDEX

FIG	ITEM	STOCK NUMBER	CAGEC	PART NUMBER
256	1	5310-00-081-4219	96906	MS27183-12
256	2	5310-00-814-0673	96906	MS51943-33
256	3	5340-01-166-2152	11862	120877
256	4	5306-01-266-2419	24617	9417350
256	5	2990-01-159-1801	11862	14075801
256	6	5310-00-013-1245	21450	131245
256	6	5310-00-013-1245	21450	131245
256	7	4730-00-908-6293	96906	MS35842-15
256	7	4730-00-908-6293	96906	MS35842-15
256	8	5305-00-225-3843	96906	MS90728-8
256	8	5305-00-225-3843	96906	MS90728-8
256	9	2990-01-163-0771	11862	15599211
256	10	4730-00-058-7558	63208	650-24
256	11		11862	15599287
256	12	5310-01-012-8962	24617	9416918
257	1	2540-01-162-7114	11862	15599290
257	2	5305-00-225-3843	96906	MS90728-8
257	3	4730-00-908-6293	96906	MS35842-15
257	4	5310-00-013-1245	21450	131245
258	1	4730-00-058-7558	63208	650-24
258	2	5310-01-012-8962	24617	9416918
258	3	5340-01-166-2152	11862	120877
258	4	5340-01-219-7272	11862	15599245
258	5	5305-01-160-3938	24617	9416187
258	6	2990-01-163-1179	11862	15599243
259	1	4730-00-058-7558	63208	650-24
259	2	5310-00-013-1245	21450	131245
259	2	5310-00-013-1245	21450	131245
259	3	4730-00-908-6293	96906	MS35842-15
259	3	4730-00-908-6293	96906	MS35842-15
259	4	5305-00-225-3843	96906	MS90728-8
259	4	5305-00-225-3843	96906	MS90728-8
259	5	2540-01-162-7113	11862	14074499
259	5	2990-01-163-1180	11862	15599215
259	6	5306-01-266-2419	24617	9417350
259	7	5310-00-081-4219	96906	MS27183-12
259	7	5310-00-081-4219	96906	MS27183-12
259	8	5310-00-814-0673	96906	MS51943-33
259	9	5340-01-166-2152	11862	120877
259	10	5306-01-266-2419	24617	9417350
259	11	5310-00-814-0673	96906	MS51943-33
259	12	2990-01-162-4416	11862	15599212
260	1	5306-01-149-6278	11862	14028922
260	2	5340-01-221-9972	11862	15599246
260	3	5330-01-218-0862	11862	14055585
260	4	4720-01-162-0120	11862	14063337
260	5	4720-01-162-0119	11862	14063336
261	1	4730-00-908-3194	96906	MS35842-11
261	2	4730-01-162-0095	11862	10005327
261	3	5330-01-149-0874	11862	23500846
261	4	2930-01-159-1802	11862	14063338

FIGURE AND ITEM NUMBER INDEX

FIG	ITEM	STOCK NUMBER	CAGEC	PART NUMBER
261	5	5330-01-148-7497	11862	14028916
261	6		11862	14063370
261	7	4730-01-148-2758	11862	354501
261	8	4730-00-317-4231	73992	8200
261	9	4730-01-218-6691	73992	B-84
261	10	4730-00-249-3885	96906	MS51845-4
261	11	4730-00-196-1991	96906	MS51846-64
261	12	6620-01-221-1942	17769	4641
261	13	4730-00-768-8880	73992	8100
261	14	4730-01-218-6690	73992	B-85
261	15		11862	14063373
261	16	5340-01-159-1324	11862	3825416
261	17	5310-00-596-6897	24617	456004
261	18	5310-00-080-6004	96906	MS27183-14
261	19	5306-01-218-3119	11862	9440034
262	1	5340-01-164-6528	11862	14074457
262	2	5310-01-159-8559	96906	MS35691-406
262	3	5310-00-582-5965	96906	MS35338-44
262	4	5340-00-914-1000	19207	10922334
262	5	5306-00-226-4825	96906	MS90728-32
262	6	5310-01-132-8275	24617	9418924
262	7	5340-01-164-6526	11862	14074458
262	8	5310-00-407-9566	96906	MS35338-45
262	9	5310-00-931-8167	96906	MS51967-6
262	10	5340-01-164-6527	11862	14074459
262	11		11862	9440173
262	12	4730-00-415-3172	17769	443998
262	13	2910-00-025-3493	96906	MS51085-1
262	14	5365-01-163-1142	11862	14072374
262	15	5325-01-199-3461	00613	C-5139-2
262	16	4730-00-142-2177	81240	GM118749
262	17	5310-00-809-8546	96906	MS27183-8
262	18	5305-01-242-1148	24617	456748
262	19	4730-00-011-8538	81343	5 010111B
262	20		11862	14072372
262	21	4720-01-247-4680	11862	15599234
262	22	5340-00-057-3034	96906	MS21333-110
262	23	2990-01-287-2158	46522	D55395-G1
263	1	5930-00-475-5537	19207	8376101
263	2	5305-01-121-9352	46522	2251-0832-10-13
263	3	5940-00-983-6082	46522	A6005943
263	4	5930-01-106-6382	46522	A2322
263	5	5305-00-115-9406	96906	MS51849-53
263	6	5340-01-145-4311	46522	A54201-G1-1
263	7	5305-00-249-5278	46522	2251-0832-06-17
263	8	4540-01-105-5667	46522	A54948-G2
263	9	5310-01-027-1353	46522	30002-83
263	10	4810-01-121-9786	46522	C54813-G1
263	11	5340-01-027-2628	46522	B54028G1-1
263	12	4820-01-130-3387	46522	B54941-G1
263	13	4730-01-145-4326	46522	A53181

FIGURE AND ITEM NUMBER INDEX

FIG	ITEM	STOCK NUMBER	CAGEC	PART NUMBER
263	14	4730-00-087-7273	05657	A5
263	15	4720-01-124-8252	46522	A52789-G5
263	16	4710-01-131-3398	46522	A54907-G1
263	17	4730-00-460-6725	46522	A14140
263	18	4520-00-567-1886	46522	A18167-G2
263	19	4730-00-319-0454	46522	A14141
263	20	5325-01-145-4349	46522	A54814-G1
263	21	5305-00-614-0248	96906	MS35266-66
263	22	5365-01-158-6819	46522	A2612
263	23	2540-00-454-0665	46522	A2610
263	24	5340-01-146-1905	46522	C54919-G1-1
263	25	4520-01-132-8908	46522	C2757
263	26	5935-00-768-7042	46522	A2096
263	27	5305-00-724-6798	96906	MS51964-49
263	28	6105-01-254-3237	46522	B54827-G1
263	29	6105-01-145-4324	46522	B2798
263	30	4320-00-217-5827	46522	B2894
263	31	4730-00-908-3195	96906	MS35842-10
263	32	4720-00-139-4105	46522	A2829
263	33	5305-01-248-6917	46522	A2897
263	34	3110-00-724-7884	96906	MS19060-1014
263	35	5360-01-249-4056	46522	A2896
263	36	4730-00-277-6339	46522	A2895
263	37	5330-00-472-6954	46522	A2932
263	38	5330-01-183-5688	46522	A2904
263	39	5330-01-109-7941	53335	A2931
263	40	5360-01-107-3245	53335	A930
263	41		46522	A2481=XA
263	42	5977-01-145-4308	46522	A2480
263	43	9905-01-145-4320	46522	B2714
263	44	4730-00-187-4191	24617	143935
263	45	2990-01-131-2751	46522	A54940-G1
263	46	5305-01-142-3150	46522	2290-0632-05-13
264	1	2590-01-163-1184	11862	14076236
264	2	5340-01-149-4434	11862	3816659
264	3	5945-01-162-0517	11862	2098912
264	4	5310-00-809-4058	96906	MS27183-10
264	5	5325-00-807-0580	96906	MS35489-121
264	6	5310-00-938-8387	96906	MS9549-10
264	7	5305-00-071-2506	96906	MS90728-3
264	8	5340-01-150-1377	11862	14061352
264	9	5365-01-219-7285	11862	15599920
264	10	5310-01-159-8559	96906	MS35691-406
264	11	5305-01-164-2313	24617	9415163
264	12	5930-00-234-1390	78385	G704410-1
264	13	5330-01-220-6153	11862	14076270
264	14	5310-01-161-2374	11862	9440178
264	15	5340-00-282-7509	96906	MS21333-62
265	1	5995-01-163-1183	11862	14075888
265	2	5945-01-162-0516	11862	14074461PC6
265	3	5325-00-807-0580	96906	MS35489-121

FIGURE AND ITEM NUMBER INDEX

FIG	ITEM	STOCK NUMBER	CAGEC	PART NUMBER
266	1	6150-01-234-3253	11862	15599968
267	1	2590-01-164-0135	11862	14076246
267	2	5325-00-337-6636	72794	AJW7-70
267	3	5365-01-226-2342	72794	X-840-SR7C
267	4	6140-01-160-5196	11862	12039294
267	5	5310-01-158-6260	24617	11506003
267	6	6140-01-216-7923	11862	14076852
267	7	2540-01-164-0122	11862	14076243
267	8	2540-01-166-2010	11862	14076250
267	9	5306-01-168-4481	11862	9440334
267	10	2540-01-166-2009	11862	14076249
267	11	5306-01-158-9018	11862	2014469
267	12	2540-01-164-0032	11862	14076241
267	13	5306-01-227-1454	11862	14076248
267	14	5325-00-291-9366	96906	MS35489-11
268	1	2590-01-162-4352	11862	14076269
268	2	5340-00-057-3043	96906	MS21333-112
268	3	5305-01-218-3139	24617	9423768
268	4	5310-01-217-5205	24617	9417714
268	5	2590-01-162-4417	11862	14076811
268	6	5340-01-149-4434	11862	3816659
268	7	2590-01-162-4418	11862	14076240
268	8	5340-00-989-1771	96906	MS21333-123
268	9	5340-00-057-3052	96906	MS21333-114
268	10	5310-00-582-5965	96906	MS35338-44
268	11	5310-01-159-8559	96906	MS35691-406
268	12	5305-01-156-5438	11862	11504447
268	13	5325-01-164-8655	11862	3886908
269	1	5305-00-984-6193	96906	MS35206-245
269	2	2590-01-162-4353	11862	14076238
269	3	6220-01-160-5187	22973	922-900-00
269	4	6220-01-268-8795	22973	922-910-00
269	5	6240-00-850-4280	08806	1003
269	6	6220-01-216-5289	22973	922-901-02
269	7	6220-01-269-0465	22973	922-904-02
269	8	5305-00-984-4983	96906	MS35206-226
269	9	6220-01-216-5288	22973	922-901-01
269	10	5340-00-057-3043	96906	MS21333-112
269	11	5310-01-217-5205	24617	9417714
269	12	5305-00-446-9901	96906	MS51850-64
269	13	5305-00-068-0508	96906	MS90728-6
269	14	5310-01-152-0598	24617	271172
269	15	5340-00-989-1771	96906	MS21333-123
269	16	2590-01-162-7130	11862	14076239
269	17	5305-01-156-5438	11862	11504447
269	18	5325-01-168-5677	11862	3918889
269	19	5325-01-164-2377	11862	15591130
270	1		11862	15593599
270	2	5325-01-164-7550	11862	3977775
270	3	5970-01-159-1803	11862	14063365
270	4	5307-01-259-7656	11862	107413

FIGURE AND ITEM NUMBER INDEX

FIG	ITEM	STOCK NUMBER	CAGEC	PART NUMBER
270	5	5325-01-050-6192	19207	12255564-2
270	6	5310-00-809-4058	96906	MS27183-10
270	7	2930-01-167-7250	11862	14063366
270	8	5310-00-013-1245	21450	131245
270	9	5360-00-771-7066	19207	7717066
271	1	2540-01-155-7542	11862	14076228
271	2	5305-00-054-6657	96906	MS51957-33
271	3	2510-01-159-1792	11862	14076216
271	4	2510-00-999-9856	19207	7353960
271	5	5330-01-167-6335	11862	14076215
271	6	5310-01-161-2531	11862	9414031
271	7	5310-01-012-8962	24617	9416918
271	8	2540-01-164-1886	11862	14075845
271	9	5306-01-197-3089	11862	9440033
271	10	5310-00-472-3214	81795	30489
271	11	5305-00-071-2511	96906	MS90728-14
271	12	5310-01-264-5903	21450	587227
271	13	5305-00-071-2087	96906	MS51957-84
271	14	5305-00-071-2510	96906	MS90728-13
271	15	5340-01-150-4105	11862	14005953
271	16	5310-01-170-8765	24617	9419265
271	17	5305-01-167-8334	11862	20365263
271	18	5340-01-165-4379	11862	14072367
271	19	4730-01-161-6618	11862	14072366
271	20	5330-01-164-5603	11862	14075857
271	23	5306-01-168-4481	11862	9440334
271	22	5340-01-164-6435	11862	14072365
271	23	2540-01-162-4412	11862	14072364
272	1	5305-00-071-1318	96906	MS51957-83
272	2	5310-01-170-8765	24617	9419265
272	3		11862	14075884
272	4	5340-01-150-4105	11862	14005953
272	5		11862	14075883
272	6	2540-01-155-7543	11862	14075881
272	7	2540-01-155-7278	11862	14075882
272	8	5325-01-050-6192	19207	12255564-2
272	9	5310-01-152-0598	24617	271172
273	1	5305-00-071-2510	96906	MS90728-13
273	2	5310-00-472-3214	81795	30489
273	3	5340-01-217-2168	11862	3794767
273	4	5310-01-012-8962	24617	9416918
273	5	5340-01-217-2278	11862	370349
273	6		11862	14076204
273	7	5310-00-080-6004	96906	MS27183-14
273	8	5310-01-097-8222	24617	9413534
273	9	5305-00-240-6668	96906	MS51849-78
273	10	2540-01-162-7110	11862	14076882
273	11	5305-00-068-0501	96906	MS90725-5
273	12	2540-01-216-3188	11862	15599951
273	13	5325-01-165-4372	11862	9411281
273	14	5310-01-264-5903	21450	587227

FIGURE AND ITEM NUMBER INDEX

FIG	ITEM	STOCK NUMBER	CAGEC	PART NUMBER
273	15	1430-01-106-8451	11862	15599950
273	16	2510-01-246-4237	11862	14076875
273	17	5340-01-163-3294	11862	14076217
273	18	5305-00-782-9489	96906	MS90728-66
273	19	5310-00-080-6004	96906	MS27183-14
273	20		11862	14076210
273	21	5305-00-533-5542	96906	MS35492-54
273	22	5310-01-097-8222	24617	9413534
273	23		11862	14076211
273	24		11862	14076884
273	25		11862	14076209
273	26	2510-00-999-9856	19207	7353960
273	27	5330-01-167-6335	11862	14076215
273	28	2510-01-159-1792	11862	14076216
273	29	5305-00-935-7506	96906	MS35493-37
273	30	5305-00-782-9489	96906	MS90728-66
274	1	5310-01-012-8962	24617	9416918
274	2	5310-01-097-8222	24617	9413534
274	3	5310-00-080-6004	96906	MS27183-14
274	4		11862	14076204
274	5		11862	14076872
274	6	5310-00-472-3214	81795	30489
274	7	5305-00-071-2510	96906	MS90728-13
274	8	5330-01-163-2055	11862	14076862
274	9	5310-01-264-5903	21450	587227
274	10	5305-00-782-9489	96906	MS90728-66
274	11	5340-01-164-2397	11862	14076221
274	12	2510-01-159-1791	11862	15599910
274	13		11862	14076212
274	14		11862	14076864
274	15		11862	14076871
274	16	5306-01-168-4481	11862	9440334
274	17	2590-01-242-8068	11862	14076208
274	18	5305-00-533-5542	96906	MS35492-54
275	1	5306-00-889-2943	96906	MS35751-17
275	2	2510-01-162-4407	11862	14076233
275	3	5340-01-164-6591	11862	14076230
275	4	5310-01-012-8962	24617	9416918
275	5	2540-01-162-8983	11862	14075843
275	6	2540-01-159-6198	19207	7951738
275	7		11862	14076222
275	8	2510-01-246-4236	17769	14076861
275	9	5305-00-014-9926	96906	MS35493-76
275	10		11862	14076201
275	11		11862	15599916
275	12	2510-01-162-4408	11862	14076231
275	13		11862	14076232
275	14	2540-01-214-2634	11862	15599917
275	14	2540-01-214-2635	11862	15599918
275	15	5305-00-0781-2510	96906	MS90728-13
275	16	5340-01-164-6589	11862	14076226

FIGURE AND ITEM NUMBER INDEX

FIG	ITEM	STOCK NUMBER	CAGEC	PART NUMBER
275	17	5310-01-012-8962	24617	9416918
275	18	5310-01-197-6621	24617	9423530
275	19	5340-01-043-5214	19207	11669126-1
275	20	5305-00-054-6654	96906	MS51957-30
275	21	5340-01-044-8389	19207	12255561
275	22	2510-01-162-7111	11862	14076299
275	22	2540-01-163-0766	11862	14076300
275	23	5325-01-165-6975	19207	12255564-1
276	1		11862	14076281
276	2		11862	14076278
276	3	5305-00-901-3110	96906	MS35492-57
276	4	5340-01-219-7275	11862	14063349
276	5		11862	14076280
276	6	5340-01-212-6716	11862	15599273
276	6	5340-01-212-6717	11862	15599274
276	7	5340-01-218-5823	11862	14076889
276	8	5340-01-150-4105	11862	14005953
276	9	5306-01-168-4481	11862	9440334
276	10		11862	14076276
276	11	5330-01-163-5850	11862	14076235
276	12	2510-01-159-1790	11862	14076206
276	13		11862	14076272
276	14		11862	14076275
276	15	5310-00-316-6513	80205	NAS1408A6
276	16	5310-00-809-4085	96906	MS27183-16
276	17	5310-00-809-5997	96906	MS27183-17
276	18		11862	14076271
276	19	5306-00-685-7790	96906	MS35751-46
276	20		11862	14076229
276	21		11862	14076279
276	22		11862	14076273
276	22		11862	14076274
276	23		11862	14076277
277	1		11862	14063348
277	2	5305-00-533-5542	96906	MS35492-54
277	3		11862	14063349
277	4	5306-00-226-4830	96906	MS90728-37
277	5	5310-00-081-4219	96906	MS27183-12
277	6		11862	14063342
277	7	5310-01-211-3811	11862	9417793
277	8	5310-01-119-3668	24617	9422295
277	9		11862	14063343
277	9		11862	14063344
277	10		11862	14063346
277	11	5305-00-245-4144	96906	MS35494-83
277	12		11862	14063350
277	13		11862	14063345
277	14	5310-00-316-6513	80205	NAS1408A6
277	15	5310-00-809-4085	96906	MS27183-16
277	16		11862	14063341
277	17	5305-00-901-3144	96906	MS35492-82

FIGURE AND ITEM NUMBER INDEX

FIG	ITEM	STOCK NUMBER	CAGEC	PART NUMBER
277	18	5310-01-158-6260	24617	11506003
277	19	5306-00-027-0722	96906	MS35751-19
277	20		11862	14063351
277	21	5310-00-809-5997	96906	MS27183-17
277	22	5306-00-177-5707	96906	MS35751-73
277	23		11862	14063347
278	1	2540-01-156-0088	11862	15594983
278	2		11862	14075862
278	2		11862	14075864
278	3		11862	14076802
278	3		11862	14076803
278	4		11862	14075863
279	1	5340-01-293-1200	11862	14063353
279	2		24617	9409103
279	3	5340-01-164-6595	11862	14063359
279	4	5340-01-165-6797	11862	14074431
279	5	5305-00-068-0508	96906	MS90728-6
279	6	2540-01-218-8099	11862	14063358
279	7		11862	14063352
279	8	5305-00-071-2071	96906	MS90728-115
280	1	5306-00-226-4827	96906	MS90728-34
280	1	5306-00-226-4827	96906	MS90728-34
280	2	5310-00-959-4679	96906	MS35340-45
280	2	5310-00-959-4679	96906	MS35340-45
280	3	5340-00-958-8457	96906	MS21333-78
280	4		11862	14074444
280	5	5310-00-081-4219	96906	MS27183-12
280	5	5310-00-081-4219	96906	MS27183-12
280	6	5310-00-814-0673	96906	MS51943-33
280	6	5310-00-814-0673	96906	MS51943-33
280	7		11862	14074446
280	8	4730-00-908-3194	96906	MS35842-11
280	9	4710-01-163-0594	11862	14074445
280	10	5340-00-702-2848	96906	MS21333-128
280	11	5340-01-221-0264	11862	3738198
280	12		11862	14075856
281	1	5306-00-226-4827	96906	MS90728-34
281	1	5306-00-226-4827	96906	MS90728-34
281	2	5310-00-959-4679	96906	MS35340-45
281	2	5310-00-959-4679	96906	MS35340-45
281	3	5340-00-702-2848	96906	MS21333-128
281	3	5340-00-702-2848	96906	MS21333-128
281	4	5310-00-081-4219	96906	MS27183-12
281	4	5310-00-081-4219	96906	MS27183-12
281	5	5310-00-814-0673	96906	MS51943-33
281	5	5310-00-814-0673	96906	MS51943-33
281	6	4720-01-162-0283	11862	14074451
281	7	4730-00-908-3194	96906	MS35842-11
281	8	4710-01-161-6406	11862	14074450
281	9		11862	14074449
281	10	2540-01-164-7121	11862	14074448

FIGURE AND ITEM NUMBER INDEX

FIG	ITEM	STOCK NUMBER	CAGEC	PART NUMBER
281	11		11862	14074447
281	12		11862	14074453
281	13	4710-01-162-7080	11862	14074452
282	1	5325-00-174-5314	96906	MS35489-5
282	2	5340-01-149-4434	11862	3816659
282	3	5310-01-170-9100	11862	3982098
282	4	5305-00-071-2506	96906	MS90728-3
282	5	5340-01-164-6524	11862	14072415
282	6	5310-00-938-8387	96906	MS9549-10
282	7	5305-00-068-0508	96906	MS90728-6
282	8	5310-00-582-5965	96906	MS35338-44
282	9	5310-01-159-8559	96906	MS35691-406
282	10	2540-01-163-0765	11862	14076284
283	1	2540-01-162-5174	11862	14076285
283	2	2540-01-162-5175	11862	14076287
283	3	5925-01-067-2926	96906	MS25244-P-20
283	4	5930-00-655-1514	96906	MS35058-22
283	5	5930-00-978-8805	96906	MS25307-312
283	6	5305-00-054-6650	96906	MS51957-26
283	7	6210-00-438-4745	81640	L30200R
283	8	6210-00-688-5088	96906	MS25331-4-313S
283	9	5325-00-263-6632	96906	MS35489-6
283	10	5935-01-154-6264	96906	MS3452W18-11P
283	11	5305-00-054-5651	96906	MS51957-17
283	12	5310-00-550-3715	96906	MS35333-70
283	13	5310-00-934-9748	96906	MS35649-244
283	14	5310-00-014-5850	96906	MS27183-42
283	15	5305-00-054-6670	96906	MS51957-45
283	16	5310-00-144-8453	96906	MS90724-7
283	17	5310-00-934-9758	96906	MS35649-202
283	18	5310-00-045-3296	96906	MS35338-43
283	19	2540-01-159-1800	11862	14076295
283	20	5305-00-984-6210	96906	MS35206-263
284	1	5325-01-199-3461	00613	C-5139-2
284	2	5365-01-163-1142	11862	14072374
284	3	2910-00-025-3493	96906	MS51085-1
284	4	4730-00-142-2177	81240	GM118749
284	5	5310-00-809-8546	96906	MS27183-8
284	6	4730-00-011-8538	81343	5 010111B
284	7	5305-01-242-1148	24617	456748
284	8	5340-00-057-3034	96906	MS21333-110
284	9		11862	14072378
284	10		11862	15599203
284	11	2540-01-194-3323	78385	10530B
284	12	2540-01-164-7123	11862	14074466
284	13	4730-00-908-6292	96906	MS35842-14
284	14	4720-01-162-7098	16632	CHB2015-0001
284	15	5310-00-809-4058	96906	MS27183-10
284	16	5305-01-197-3287	24617	9419663
284	17	5930-01-184-6370	11862	15599972
284	18	5306-01-157-6796	11862	9439771

FIGURE AND ITEM NUMBER INDEX

FIG	ITEM	STOCK NUMBER	CAGEC	PART NUMBER
284	19	5310-00-528-7638	19207	5287638
284	20	4730-00-908-6294	96906	MS35842-16
284	21	2540-01-163-1143	11862	14074469
284	22	5310-00-931-8167	96906	MS51967-6
284	23	5310-00-407-9566	96906	MS35338-45
284	24	5340-01-165-4543	11862	14074471
284	25	5310-01-132-8275	24617	9418924
284	26	5306-00-226-4825	96906	MS90728-32
284	27	2540-01-165-0466	11862	14074470
284	28	5305-01-219-5399	24617	423532
284	29	5330-01-163-3150	16632	CHB2017
284	30	2540-01-164-7124	11862	14074473
284	31		11862	9440173
284	32	4730-00-053-0266	96906	MS51952-1
285	1	4720-01-162-7098	16632	CHB2015-0001
285	2	2540-01-163-1140	11862	14074465
285	3	2590-01-163-3529	11862	14076237
285	4	4720-01-162-7097	16632	CHB2015-0002
285	5	4730-00-908-6293	96906	MS35842-15
286	1	5305-01-197-3287	24617	9419663
286	2	4520-01-166-2133	11862	3037476
286	3	2540-01-268-7202	11862	3055734
286	4	6105-01-254-9496	11862	22098841
286	5	4520-01-166-2134	11862	3035192
287	1	4730-00-908-6293	96906	MS35842-15
287	2	4720-01-167-9137	16632	CHB2013-0001
287	3	5340-01-150-4105	11862	14005953
287	4	5330-01-164-7509	16632	CHB2016-0002
287	5	5330-01-220-3117	16632	CHB2023
287	6	2990-01-250-8612	11862	15599942
287	7	5306-01-168-4481	11862	9440334
287	8	5315-00-243-1170	96906	MS24665-516
287	9	2540-01-163-7225	11862	15599940
287	10	4730-00-058-7558	63208	650-24
287	11	2990-01-163-1182	11862	14076805
287	12	5306-00-226-4828	96906	MS90728-35
287	13	5310-00-959-4679	96906	MS35340-45
288	1	2540-01-163-1175	11862	14076252
288	2	5310-00-809-4058	96906	MS27183-10
288	3	5360-01-245-0405	96906	MS24585-1276
288	4	2540-01-162-5176	11862	14076253
288	5	5305-00-272-3533	96906	MS51023-49
288	6	5315-00-839-2325	96152	A82-1
288	7	4730-00-053-0266	96906	MS51952-1
288	8	2910-00-025-3493	96906	MS51085-1
288	9	5365-01-163-1142	11862	14072374
288	10	4730-00-011-8538	81343	5 010111B
288	11		11862	14076255
288	12	5310-00-208-1918	88044	AN365-1024A
288	13	5310-01-217-5205	24617	9417714
288	14	4730-00-142-2177	81240	GM118749

FIGURE AND ITEM NUMBER INDEX

FIG	ITEM	STOCK NUMBER	CAGEC	PART NUMBER
288	15	5310-00-809-8546	96906	MS27183-8
288	16	5305-01-242-1148	24617	456748
288	17		11862	9440173
288	18	5306-01-168-4481	11862	9440334
288	19	4730-00-908-6294	96906	MS35842-16
288	20	4730-01-216-0021	11862	15599952
288	21	2540-01-194-3323	78385	10530B
288	22	2540-01-162-6418	11862	14076265
288	23	5320-00-845-9501	96906	MS20600-B4W4
288	24	5340-01-168-0939	82240	SPEC-3-10-L-L
288	25	5330-01-163-3151	16632	CHB2018
288	26	2540-00-752-4078	19207	7524078
288	27	5310-00-809-4058	96906	MS27183-10
288	28	5310-00-013-1245	21450	131245
288	29	4730-00-908-6294	96906	MS35842-16
288	30	5305-00-225-3843	96906	MS90728-8
288	31	2540-01-162-5178	11862	14076259
288	32	5305-00-068-7837	96906	MS90728-5
288	33	5340-01-166-0568	11862	14076261
288	34	5340-01-165-5617	11862	14076262
288	35	5310-00-761-6882	96906	MS51967-2
288	36	2540-01-162-6493	11862	14076801
288	37	5340-00-685-5899	82240	B-1900-334
288	38	5305-01-219-5399	24617	423532
288	39	2540-01-162-5189	11862	14076258
288	40	2540-01-162-5177	11862	14076257
289	1		11862	14072308
289	2	5340-01-200-5843	11862	14072314
289	3	5340-01-165-4429	11862	14072325
289	4	5340-01-114-7712	19207	11682088-1
289	5	2540-01-165-0813	11862	14075811
289	6	5340-01-165-1494	11862	14072324
289	7	2540-01-205-2509	19207	12343359-3
289	8	2540-01-163-7017	11862	14075812
289	9	2540-01-205-2512	19207	12343359-5
289	10	2540-01-041-4912	19207	12343359-1
289	11	5306-00-753-6996	21450	126373
289	12	5310-00-933-4310	24617	271184
289	13	2540-01-191-8467	11862	14072319
289	14	2540-01-164-9278	11862	14072323
289	15	2540-01-163-0801	11862	14072322
289	16	5310-00-316-6513	80205	NAS1408A6
289	17	5305-00-543-2866	96906	MS90728-68
289	18	5315-00-842-3044	96906	MS24665-283
289	19	5305-00-823-9139	96906	MS51850-44
289	20	5340-01-094-9025	19207	12255608
289	21	5315-00-737-0134	19207	7370134
289	22	2540-01-205-2511	19207	12343359-4
289	23	5340-00-237-7779	11862	14072380
289	24	2540-01-205-2510	19207	12343359-2
290	1	6220-00-961-0783	81349	M4510-1A4435

FIGURE AND ITEM NUMBER INDEX

FIG	ITEM	STOCK NUMBER	CAGEC	PART NUMBER
290	2	6240-00-252-7138	08806	4435
290	3	5340-01-164-0748	78977	6500-0016
290	4	5340-00-411-4508	78977	6598
290	5	5305-00-217-9183	78977	6471
290	6	6220-01-197-3938	78977	6566-0004
290	7	4710-01-194-6590	78977	6710-BU-2
290	8	6220-01-194-6591	78977	100-7-3
290	9	5340-01-197-1585	78977	6701-0025
291	1	6680-01-230-5684	11862	25033627
291	2	5340-01-163-7501	11862	25020687
291	3	6680-01-225-4475	11862	25052373
291	4		11862	3866918
291	4	3020-01-166-6802	11862	1362195
291	4	3040-01-159-0930	11862	9780470
291	5	5330-01-087-4714	11862	15562374
291	6	2530-01-159-8764	11862	1362293
291	6	5360-01-206-6616	11862	326561
291	7	6680-01-161-3656	11862	1254856
291	8	6680-01-160-5276	11862	368026
291	9	5330-01-168-1535	11862	14018671
291	10	5305-01-150-9781	11862	8639743
291	11	5340-01-161-1440	11862	378362
291	12	5310-01-250-7679	11862	11506101
291	13	5325-01-165-3475	11862	474579
292	1	5305-01-162-8512	24617	9421073
292	2	4120-01-161-3706	99688	78401
293	1	2590-01-164-7897	99688	65132
293	2	5340-01-193-9654	99688	57972
293	3	2590-01-163-3576	99688	65129
293	4	5310-01-161-7308	99688	62229
293	5	5310-00-637-4000	99688	62210
293	6	3040-01-192-4694	99688	46006
293	7	2540-01-159-1799	99688	85925
293	8	7690-01-188-2862	99688	58601
293	9	5310-00-103-2953	99688	62220
293	10	5310-01-161-7308	99688	62229
293	11	5330-01-206-7353	99688	58602
293	12	2540-01-163-1176	99688	58599
293	13	5360-01-161-9169	99688	58600
293	14	5310-01-197-4435	99688	62219
293	15	5310-01-163-8256	99688	62222
293	16	5930-01-159-2610	99688	65136
293	17	5305-01-217-2046	1T998	274898
293	18	5340-01-194-9279	99688	85934
293	19	2540-01-191-6537	99688	85932
293	20	5325-00-263-6648	96906	MS35489-135
293	21	5310-00-045-3296	96906	MS35338-43
293	22	5305-01-162-5703	24617	9417222
293	23	5305-01-197-2319	24617	9416137
293	24	5355-01-167-4114	99688	58668
293	25	5930-01-164-8264	99688	85926

FIGURE AND ITEM NUMBER INDEX

FIG	ITEM	STOCK NUMBER	CAGEC	PART NUMBER
293	26	2590-01-191-9270	99688	65145
293	27	2590-01-191-9269	99688	65133
293	28	5970-01-180-3731	99688	58662
293	29	5930-01-190-1212	99688	65138
293	30	5930-01-165-1657	99688	65139
293	31	2590-01-192-6031	99688	65146
293	32	2590-01-191-9426	99688	85935
293	33	5930-01-163-6256	99688	65140
293	34	5355-01-166-1720	99688	58603
293	35	5310-01-162-2515	99688	62028
293	36	7690-01-189-6932	99688	58605
293	37	6110-01-190-5501	99688	85954
293	38	5340-00-809-1500	96906	MS21333-107
293	39	5310-00-582-5965	96906	MS35338-44
293	40	5305-01-163-5761	96906	MS51869-26
293	41	4130-01-164-3693	99688	85929
293	42	4130-01-160-7716	99688	58587
293	43	5975-00-074-2072	96906	MS3367-1-9
294	1	4130-01-160-7695	99688	85927
294	2	5305-01-163-5761	96906	MS51869-26
294	3	5310-00-582-5965	96906	MS35338-44
294	4	5365-01-195-5934	99688	58606
294	5	5330-01-194-4751	99688	58608
294	6	5330-01-197-0897	99688	58618
294	7	5325-00-263-6648	96906	MS35489-135
294	8	5640-01-195-4634	99688	58610
294	9	5330-01-193-0227	99688	58607
294	10	5330-01-194-4752	99688	58617
294	11	5640-01-195-9786	99688	58615
294	12	5640-01-196-7003	99688	58614
294	13	5640-01-196-7002	99688	58613
294	14	5640-01-195-4636	99688	58612
294	15	5330-01-195-1564	99688	58616
294	16	5640-01-195-4635	99688	58611
294	17	5640-01-195-4633	99688	58609
294	18	5305-01-163-2423	96906	MS51871-14
294	19	5310-00-637-9541	96906	MS35338-46
294	20	5330-01-194-4753	99688	58586
294	21	5305-01-163-5761	96906	MS51869-26
294	22	5310-00-582-5965	96906	MS35338-44
295	1	5310-01-250-7679	11862	11506101
295	2	5305-01-148-7460	11862	11504595
295	3	5340-01-164-5797	11862	14033864
295	4	5340-01-164-8170	11862	14033866
295	5	5340-01-164-5799	11862	14033867
295	6	5306-01-149-6278	11862	14028922
295	7	5306-01-149-6280	11862	1635490
295	8	5340-01-164-7603	11862	14033863
295	9	5306-01-149-6280	11862	1635490
295	10	4130-01-156-8096	11862	12300270
295	11	5340-01-164-6529	11862	14033861

FIGURE AND ITEM NUMBER INDEX

FIG	ITEM	STOCK NUMBER	CAGEC	PART NUMBER
295	12	3030-01-043-6749	20796	42-4877
295	13	5306-01-152-4693	24617	11504986
296	1	5310-01-165-3346	11862	6556923
296	2	5365-00-152-5273	11862	6555328
296	3	4130-00-277-3486	11862	5914719
296	4	5365-01-164-6567	11862	6555329
296	5	3020-01-096-6892	11862	6550750
296	6	5365-01-163-6139	11862	6555338
296	7	3110-01-167-0195	11862	908287
296	8	5365-01-164-6568	11862	6555330
296	9	5950-01-165-6803	11862	6550835
296	10		11862	2724176
296	11	4730-01-164-3692	11862	5858694
296	12	5330-01-164-5602	11862	5858696
296	13		11862	5914809
296	14		96906	MS16562-36
296	15		11862	5914807
296	16		11862	6590592=XA
296	17		11862	6555440
296	18		11862	6556000
296	19		11862	6555259
296	20		76462	6551150
296	21		11862	6556605=XA
296	22		11862	6590686
296	23		11862	6556050
296	23		11862	6556055
296	23		11862	6556060
296	23		11862	6556065
296	23		11862	6556070
296	23		11862	6556075
296	23		11862	6556080
296	23		11862	6556085
296	23		11862	6556090
296	23		11862	6556095
296	23		11862	6556100
296	23		11862	6556110
296	23		11862	6556115
296	23		11862	6556120
296	23		23671	6556105
296	24		27462	5914705
296	25		11862	6555258
296	26		11862	6556175
296	26		11862	6556180
296	26		11862	6556185
296	26		11862	6556190
296	26		11862	6556195
296	26		11862	6556200
296	26		11862	6556205
296	26		11862	6556210
296	26		11862	6556215
296	26		11862	6556220

FIGURE AND ITEM NUMBER INDEX

FIG	ITEM	STOCK NUMBER	CAGEC	PART NUMBER
296	16		11862	6557000
296	27	5330-01-276-3235	11862	3093499
296	28	4730-01-165-0173	11862	6555298
296	29	5330-01-041-4749	11862	5929144
296	30	4810-01-164-3267	11862	6555279
296	31	4130-01-160-8519	27462	5914709
296	32	4130-01-160-8521	27462	6590594
296	33		11862	6555293
296	34	4130-01-160-8520	27462	6555755
296	35		11862	6555378
296	36	5330-01-041-4749	11862	5887997
296	37		27462	5914435
296	38	4730-01-169-7714	27462	6555299
296	39	4310-01-166-6123	11862	6557057
296	40	5315-01-164-4570	11862	6555320
296	41	5330-01-163-2000	11862	6556076
296	42	5365-01-163-6138	11862	6556767
296	43	5330-01-164-5604	11862	6556077
296	44		11862	6555290
296	45	4130-01-160-8541	27462	6555283
296	46	4130-01-164-3268	11862	5914706
297	1	5310-01-148-5922	11862	11500046
297	2	5306-01-165-4283	24617	11500831
297	3	4710-01-164-7850	11862	14072408
297	4	5305-00-068-0508	96906	MS90728-6
297	5	5340-01-159-1324	11862	3825416
297	6	5310-01-158-6260	24617	11506003
297	7	5310-01-194-9221	11862	3903572
297	8	5310-00-013-1245	21450	131245
297	9	5310-00-809-4058	96906	MS27183-10
298	1	4140-01-160-8503	99688	85923
298	2	5340-00-329-4420	J3743	S150
298	3	5310-01-159-8264	99688	62218
298	4	5310-00-209-0786	96906	MS35335-33
298	5	5305-01-162-8512	24617	9421073
298	6	5310-00-596-7691	96906	MS35335-32
298	7	5910-01-189-3011	99688	65134
298	8	6105-01-159-2666	99688	20566
298	9	4140-01-186-9753	99688	58595
298	10	9320-01-085-2889	99688	58701
298	11	5305-00-068-0509	96906	MS90728-10
298	12	5330-01-205-5056	99688	58598
298	13	5330-01-201-9682	99688	58597
298	14	5330-01-213-9811	99688	58596
298	15	5305-01-197-2320	24617	9440025
298	16	4130-01-166-2115	99688	15112
298	17	5305-00-071-2506	96906	MS90728-3
298	18		99688	85928
299	1	4130-01-160-7696	99688	85924
299	2	4130-01-160-7697	99688	15113
299	3	5330-00-352-0327	99688	56470

FIGURE AND ITEM NUMBER INDEX

FIG	ITEM	STOCK NUMBER	CAGEC	PART NUMBER
299	4	4820-01-161-6435	99688	58583
299	5	5330-00-185-0075	99688	56480
299	6	1660-00-413-0756	27783	5137
299	7	5930-01-173-0825	99688	65131
299	8	5330-01-061-3000	99688	57012
299	9	4710-01-162-7149	99688	58582
299	10	4130-01-171-5997	99688	585810
299	11	5330-01-166-1712	99688	57277
299	12	4710-01-162-7148	99688	58580
299	13	9330-01-191-4883	99688	58645
299	14	9330-00-629-8239	99688	58646
299	15	4710-01-192-3732	99688	58647
299	16	4710-01-192-3731	99688	58648
299	17	5330-01-166-1712	99688	57277
299	18	5330-00-777-1454	99688	58021
299	19	1660-00-413-0756	27783	5137
299	20	5930-01-173-0826	99688	65130
299	21	4820-01-161-5059	99688	58579
299	22	4720-01-192-3520	99688	58577
299	23	4720-01-192-3521	99688	58578
299	24	4130-01-160-7705	99688	58585
300	1	4140-01-160-7664	99688	85922
300	2	5305-01-162-8512	24617	9421073
300	3	4140-01-154-9615	99688	58590
300	3	4140-01-157-3501	99688	58589
300	4	5330-01-193-0226	99688	58592
300	5	5905-01-154-2354	99688	65158
300	6	5310-00-596-7691	96906	MS35335-32
300	7	5640-01-194-7193	99688	58594
300	8	5340-01-191-4827	99688	85931
300	9	5310-01-159-8264	99688	62218
300	10	9320-01-085-2889	99688	58701
300	11	6105-01-159-2223	99688	20565
300	12	5330-01-201-9681	99688	58591
300	13	5325-00-263-6651	59875	TD97203
300	14	5305-00-068-0509	96906	MS90728-10
300	15	5310-00-754-2005	96906	MS35338-52
300	16	5910-01-189-3011	99688	65134
300	17	5330-01-204-4312	99688	58593
300	18	5305-01-162-8512	24617	9421073
301	1		19207	12314542
301	2	5120-01-180-0558	25341	J-2222-C
301	3	5120-01-179-1318	25341	J-6632-01
301	4	5120-00-677-2259	33287	J-8092
301	5	4910-01-179-6341	25341	J-21757-03
301	6	5120-01-169-4878	25341	J-23445-A
301	7	4910-01-180-6155	25341	J-23653-C
301	8	5120-01-170-3279	25341	J-23690
301	9	4910-01-179-6340	25341	J-24187
301	10	5120-01-179-1034	25341	J-24426
301	11	4910-01-179-2518	25341	J-24595-C

FIGURE AND ITEM NUMBER INDEX

FIG	ITEM	STOCK NUMBER	CAGEC	PART NUMBER
301	12	4910-01-179-2517	30282	553
301	13	5120-01-170-0628	25341	J-26878-A
301	14	4820-01-179-4869	25341	J-33043
301	15	4910-01-179-2516	25341	J-29713
301	16	4910-01-181-1959	25341	J-25034-B
301	17	5120-01-178-6342	33287	J-29843
301	18	4930-01-323-0998	33627	J25512-2
301	19	5120-01-170-6664	25341	J-34616
301	19	5120-01-219-6753	25341	J-6893-D
302	1		19207	12314543
302	2	5120-01-170-0627	25341	J-24429
302	3	4910-01-178-8864	25341	J-26889
302	4	5120-01-180-8592	25341	J-26471
302	5	4910-01-178-9788	25341	J-29601
302	6	5120-01-171-5233	25341	J-29873
302	7	5120-00-224-7288	25341	J-28402
302	8	5120-01-183-8576	25341	J-26871-A
302	9	4910-01-179-4870	25341	J-29664-1
302	10	4910-01-182-2704	25341	J-33122
302	11	4910-01-181-1958	25341	J-23447
302	12	4910-01-179-2515	25341	J-23454-D
302	13	4910-01-238-2551	33287	J-26999-30
302	14	5120-00-189-7898	25341	J-29698-A
302	15	5120-01-179-1033	25341	J-33154
302	16	4910-01-179-6339	25341	J-23996
302	20	5120-01-170-5473	25341	J-33124
303	1		19207	12314544
303	2	5120-01-206-3818	33287	J-35178
303	3	5120-01-082-6436	25341	J-7624
303	4	5120-01-082-6448	25341	J-7728
303	5	5310-01-179-9486	25341	J-8107-2
303	6	4910-01-179-6338	25341	J-9480-B
303	7	4910-01-178-6551	25341	J-21362
303	8	4910-01-178-0724	25341	J-23327
303	9	4910-01-178-0713	25341	J-21363
303	10	4910-01-178-0722	25341	J-21370
303	11	5120-01-179-1032	25341	J-21552
303	12	5120-01-169-4877	25341	J-21465-1
303	13	4910-01-178-8866	25341	J-21777-500
303	14	4910-01-178-8865	25341	J-21795-02
303	15	4910-01-183-0044	25341	J-22779
303	16	4910-01-178-0360	25341	J-21409
304	1	5120-01-180-7928	25341	J-26252
304	2	4910-01-179-6364	25341	J-26941
304	3	4910-01-179-5530	25341	J-29162
304	4	5120-01-170-3278	25341	J-29167
304	5	5120-01-169-4876	25341	J-29168
304	6	4910-01-181-0183	25341	J-29369-1
304	7	5120-00-184-8397	55719	FC18A
304	8	5120-01-170-6703	25341	J-34502
305	1	5120-01-328-0260	80604	D-167

TM9-2320-289-34P
CROSS-REFERENCE INDEXES

FIGURE AND ITEM NUMBER INDEX

FIG	ITEM	STOCK NUMBER	CAGEC	PART NUMBER
305	2	5120-01-327-9533	80604	DD-914-95

www.ingramcontent.com/pod-product-compliance
Lightning Source LLC
Chambersburg PA
CBHW080412030426
42335CB00020B/2428